MW00838151

Conservation Biology

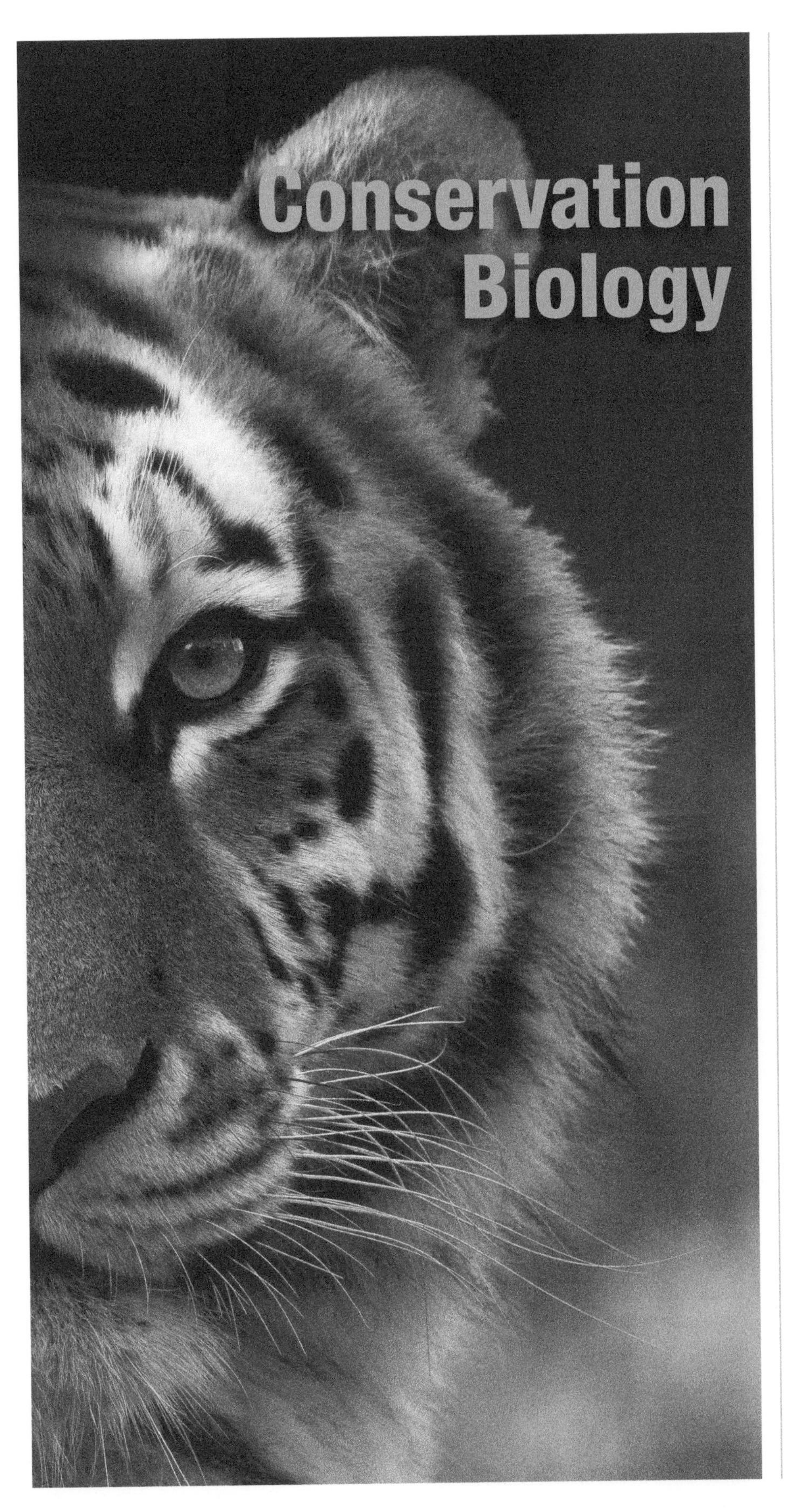

Bradley J. Cardinale
University of Michigan

Richard B. Primack
Boston University

James D. Murdoch
University of Vermont

SINAUER ASSOCIATES

NEW YORK OXFORD
OXFORD UNIVERSITY PRESS

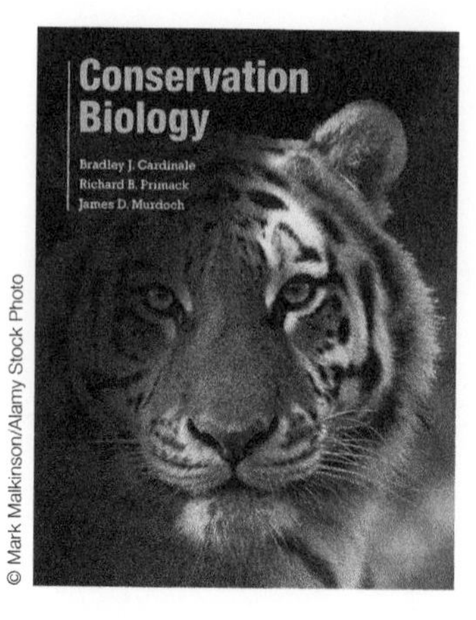

© Mark Malkinson/Alamy Stock Photo

About the Cover

The tiger (*Panthera tigris*) is the largest species among the Felidae. As an apex predator that requires a large geographic range, it is one of the most endangered species in the world. But it is also one of the most beautiful species in existence.

Conservation Biology

Oxford University Press is a department of the University of Oxford. It furthers the University's objective of excellence in research, scholarship, and education by publishing worldwide. Oxford is a registered trade mark of Oxford University Press in the UK and certain other countries.

Published in the United States of America by Oxford University Press
198 Madison Avenue, New York, NY 10016, United States of America

Sinauer Associates is an imprint of Oxford University Press.

For titles covered by Section 112 of the US Higher Education Opportunity Act, please visit www.oup.com/us/he for the latest information about pricing and alternate formats.

Address editorial correspondence to:

Sinauer Associates
23 Plumtree Road
Sunderland, MA 01375 U.S.A.

Address orders, sales, license, permissions, and translation inquiries to:

Oxford University Press U.S.A.
2001 Evans Road
Cary, NC 27513 U.S.A.
Orders: 1-800-445-9714

Library of Congress Cataloging-in-Publication Data

CIP data is on file at the Library of Congress ISBN number 9781605357140

Printed in the United States of America

This book is dedicated to our children, to their children, and to the children of all who will use this book as a part of their training to become conservation biologists.

If we collectively succeed in saving some portion of the world's biodiversity, then our brief time on Earth will have been meaningful and well spent.

If we fail, then at least we will be remembered by our children as moral ancestors who dedicated ourselves to a higher purpose.

Brief Contents

Contents

Preface

Earth is home to a spectacular variety of life. Humans share the planet with more than a million described species, and there are probably another 8 million multicellular companions that are unknown to science, and as many as 1 trillion types of unicellular prokaryotic organisms we have yet to discover. This amazing biodiversity is the most striking feature of our planet, and the only thing that makes Earth unique.

But the variety of life on our planet is in trouble. Humans now dominate every ecosystem on Earth, and we have co-opted the vast majority of limited resources for ourselves. As this book goes to press, an inferno is destroying the Amazon rainforest at an alarming rate as illegal clearing and burning makes way for cattle ranching. So it is no surprise that biodiversity is being lost at rates that are unprecedented in human history.

We are at a crossroads. We have just a few generations to learn how to be good stewards of our natural resources, and to live sustainably alongside the other nonhuman species that share our planet. If we do not succeed, much of the biodiversity on Earth, perhaps the majority of it, will be lost for good.

Why did we write this book?

We wrote this book to inspire the next generation of conservation biologists to help humans become better stewards of the world's biodiversity. We also wrote this book to fill a gap in the education and training of conservation biologists. While most universities have entry level classes in conservation that are taught as part of their traditional programs of biology, ecology, and environmental science, the number of universities that offer more advanced courses is comparatively small. It is even more rare for universities to offer interdisciplinary courses in conservation that extend beyond the discipline's historical roots in natural sciences to include the disciplines of social sciences, engineering, and humanities that have become central to the modern mission of conservation. The lack of advanced, interdisciplinary training has resulted in a paucity of graduates who have the full set of skills needed to be successful in jobs. In fact, the skillset of graduates in some countries is so underdeveloped that many conservation organizations have had to design their own training programs to prepare their entering workforce.

Who is this book for?

This book is for aspiring conservation biologists who seek advanced training. The book was developed to support classes designed for upper-division undergraduates who have already had some introduction to environmental science, ecology, wildlife biology, forestry, or other field related to conservation. In addition, the book was developed to support classes designed for beginning graduate students, such as those in the growing number of professional master's programs that provide advanced degrees in environmental science, policy, management, or sustainability.

Our text is also relevant for those who want a more interdisciplinary perspective than is typical of introductory conservation textbooks. Conservation is no longer a bunch of ecologists, wildlife biologists, or other natural scientists trying to save their favorite species in a dwindling habitat. The modern practice of conservation relies on numerous disciplines from the social sciences that account for human behaviors, values, needs, and decision making. Modern conservation relies on disciplines from engineering and architecture to help plan, design, and construct practical solutions to problems. Modern conservation also relies on disciplines from the humanities that compose law and policy, and that communicate effectively through literature, art, and photography. Interdisciplinary approaches to conservation are integrated throughout the text, and form the foundation for several individual chapters.

What pedagogical approaches does the book use?

To meet the conservation challenges of the future, the next generation of practitioners will need to be better equipped than our predecessors. They will need a broader tool kit that includes an ever-increasing body of quantitative methods and predictive models. In addition, they will need a greater variety of qualitative tools such as stories, narratives, and art that serve as effective forms of communication. This text uses a variety of pedagogical approaches to develop the practitioner's tool kit:

- **Success Stories**, included in every chapter, detail conservation biology's most important achievements to date. These stories provide inspiration, and remind us that we can succeed.
- **Conservation in Practice** boxes detail some of the most common methods, models, and techniques that form the quantitative foundations of the field of conservation.
- **Case in Point** boxes describe case studies that reinforce key topics covered in the chapters. They are designed to help translate concepts into real-world applications.
- **Challenges & Opportunities** boxes summarize contemporary challenges, controversies, and opportunities that future practitioners will face and need to resolve over the course of their careers.
- **Suggested Exercises** provide opportunities to practice the methods, models, and techniques described in the text, and to explore datasets and sources of information that practitioners commonly use.
- **Supplemental Videos** are used to reinforce topics in each chapter with compelling real-world examples and applications. These can be assigned with readings, or used to augment lectures and discussions.
- **Chapter Summaries** succinctly recap the take-home points of each chapter.
- **For Discussion** questions can be assigned, or used to facilitate in-class discussions that get students thinking more deeply about the key points and issues from each chapter.
- **Suggested Readings** provide a list of additional papers students can read, or that instructors can assign, if there is a desire to dig deeper.

In addition to the pedagogical approaches, students are provided with a companion website that organizes all supplemental online material and keeps the weblinks current. Instructors resources are also available online to aid instructors in creating presentations and to assess student understanding.

What is our hope for this book?

It took us a long two years to develop the first edition of this textbook, and there were many ups and downs along the way. But we were determined to complete the project because of our optimism that the current trajectory of biodiversity loss can be altered for the better. We need a new generation of individuals who have both a passion for nature's living things, and the skills needed to help conservation succeed more often than it has in the past. For those who have both the passion and the skills, we believe you will have a rewarding career filled with purpose. Our sincere hope is that this book will help you become the next great champion for biodiversity.

Acknowledgments

We want to thank the large number of individuals who provided us with data, figures, and tables that are used throughout this textbook. The next generation of practitioners will stand atop your work. We extend our sincere gratitude to the many individuals who graciously reviewed various chapters to help us improve this book. This includes several anonymous reviewers, as well as the following outstanding individuals:

Dr. Matthew S. Becker, *Zambian Carnivore Programme*
Andrew F. Bennett, *La Trobe University*
Nate Bickford, *University of Nebraska, Kearney*
Robert Boyd, *Auburn University*
Joseph W. Bull, *University of Kent*
Abigail E. Cahill, *Albion College*
J. Baird Callicott, *University of North Texas*
Scott Connelly, *University of Georgia*
Brian L. Cypher, *California State University, Stanislaus*
Laura E. DeWald, *Western Carolina University*
Dr. Nicole Duplaix, *Oregon State University*
Prof. Aaron M. Ellison, *Harvard University*
Joel T. Heinen, *Florida International University*
Ray Hilborn, *University of Washington*
Joshua D. Holbrook, *Montreat College*
Vanessa Hull, *University of Florida*
Daniel Karp, *University of California, Davis*
Christopher Kellner, PhD, *Arkansas Tech University*
Richard K. Kessler, *Campbellsville University*
Jason A. Koontz, *Augustana College*
Jay T. Lennon, *Indiana University*
Mark Manteuffel, PhD, *Washington University*
Heidi Marcum, PhD, *Baylor University*
Stephen G. Mech, *Albright College*
Curt Meine, *University of Wisconsin, Madison*
Ben A. Minteer, *Arizona State University*
Vincent Nijman, *Oxford Brookes University*
Liba Pejchar, *Colorado State University*
Charles Perrings, *Arizona State University*
Andrew Rassweiler, *Florida State University*
Michael Reed, *Tufts University*
Dov F. Sax, *Brown University*
D. Alexander Wait, *Missouri State University*

We are indebted to the team at Sinauer Associates/Oxford University Press that included Dean Scudder (President, Sinauer Associates), Martha Lorantos (Senior Production Editor), Peter Lacey (Senior Production Editor, Media), Meg Britton Clark (Production Specialist), Joan Gemme (Production Manager), Mark Siddall (Photo Researcher), Michele Beckta (Permissions Supervisor), Tracy Marton (Senior Production Editor), artist Jan Troutt, and copy editors Lou Doucette and Carol Wigg. You took our imperfect text and partially developed ideas and turned them into a polished and beautifully illustrated book that we all can feel proud of. Thank you!

Lastly, we thank our families for putting up with us for two years as we completed this project. Your love and patience were inspirational, and we were touched.

Bradley J. Cardinale
Richard B. Primack
James D. Murdoch

August 2019

Media & Supplements

to accompany

Conservation Biology

 oup.com/he/cardinale1e

For the Student

The Companion Website for *Conservation Biology*, First Edition, provides students with study and review tools to help them master the material presented in the textbook, all free of charge and requiring no access code.

The site includes the following resources:

Videos and Audio – Engaging media clips provide information and inspiration on a range of conservation biology topics.

Suggested Exercises – Students can learn essential skills and practice analyzing real data with these thoughtful exercises.

Web Links – Connect with the many organizations and institutions discussed in the text with helpful web links.

Chapter Outlines – Useful study aids summarize each chapter's key concepts.

Flashcards – Students can gain mastery over each chapter's key vocabulary and concepts with this helpful flashcard set.

For the Instructor

(*Instructor resources are available to adopting instructors online. Registration is required. Please contact your Oxford University Press representative to request access.*)

The Instructor Resources for *Conservation Biology*, First Edition, include a variety of tools to help instructors incorporate visual resources into their lectures and course materials and enhance student learning. The site includes:

Figure JPGs – Includes all of the figures and tables from the textbook, formatted for optimal legibility when projected. Complex images are provided in both whole and split versions.

Figure PowerPoint Slides – Features all the figures and tables from each chapter, with titles on each slide, and complete captions in the "notes" field.

Suggested Exercise Resources – For many of the Suggested Exercises, instructors have access to helpful materials, such as preformatted spreadsheets for data-analysis exercises and answers to exercise assessments.

eBook (ISBN 978-1-60535-882-6)

Conservation Biology, First Edition, is available as an eBook via several different eBook providers, including RedShelf and VitalSource. Please visit the Oxford University Press website at **oup.com/ushe** for more information.

Conservation Biology

Part I

Foundations of Conservation Biology

© Katesalin Pagkaihang/Shutterstock.com

1

The State of Our Planet

By many if not most measures, humans are the most remarkably successful species to ever inhabit planet Earth. We have evolved one of the largest brains per body size that exists in the animal kingdom, which has given us an unparalleled intellectual capacity and ability to reason. Our species is one of the more social organisms on the planet (particularly among mammals), which has led to cooperation, advanced learning, and a division of labor that has fostered our success. Our mastery of tools and our technological advances are unparalleled among other life-forms. These, and other qualities, have allowed our species population to grow rapidly and to conquer and dominate more of the planet than almost any previous organism.

But there is one measure that humans do not yet excel in, compared with other species—longevity. *Homo sapiens* have existed on this planet for a mere 200,000 years, which is but a fraction of the time that most species survive on Earth. Will humans be able to sustain themselves for another million years and achieve the longevity that is typical for one of Earth's species? Many of the very qualities that have allowed us to be so remarkably successful thus far are eroding the **natural infrastructure** of the planet—the collection of genes, species, and biological communities that compose the natural and managed ecosystems on which people depend.

Orangutans are presently found only in the rain forests of Borneo and Sumatra, where they are threatened with extinction by deforestation.

If we are to ultimately become one of longest-lived species in Earth's history, humans are going to have to use our intellect to learn how to live more sustainably on finite resources. We are going to have to use our tools and technology to conserve the biophysical processes that regulate life on the planet. And we will have to do so while using our social cooperation to generate a decent quality of life for our fellow humans. If we succeed, then *Homo sapiens* will indeed become the most remarkable species by all measures. If we fail, then our era will be recorded in the geological record as one of the greatest periods of biological loss in the history of life on Earth.

This chapter gives an overview of the current state of biology on our planet. It is the precursor to all other chapters in this book, as it establishes some basic facts about the forms of environmental change that are occurring across the globe as a result of human domination of the planet and the subsequent impact that these changes are having on biodiversity. We begin with the state of the human species, emphasizing gains that have allowed our species to be successful, while also emphasizing the social inequalities that are preventing many from having a good quality of life. We then review the state of our global environment, emphasizing how human gains for some have come at a cost to the environment. This, in turn, leads us to overview the state of biodiversity, including biodiversity losses that are occurring as a result of environmental change. We end the chapter with some reasons to be optimistic that we can, in fact, become the most remarkable species ever to inhabit planet Earth—a species that can provide a high quality of life to our own kind while, at the same time, protecting the myriad of amazing life-forms that share our planet.

State of the Human Species

Modern humans first appeared in the fossil record about 200,000 years ago, which is but a tiny fraction of the 3.5-billion-year history of life on Earth (**Figure 1.1**). For most of our brief history, our ancestors lived as hunters and gatherers—a dangerous way of life that kept population sizes small, probably less than 10 million individuals across the entire globe.

As humans transformed from hunter-gatherer to agrarian societies, communities evolved that could support more people. By roughly 1 CE, the world population had expanded to 300 million, and it continued to increase at a constant and steady rate for hundreds of years. Then the **Industrial Revolution** hit. Living standards in the eighteenth century rose dramatically in the developing countries of Europe and in the United States, and improvements in technology and medicine reduced the prevalence of famines and epidemics in many parts of the world. Population growth accelerated quickly, and humans reached their first 1 billion individuals about 1800.

While it took nearly 200,000 years for *Homo sapiens* to reach the first 1 billion, it took just 130 years to reach the second billion, 30 years to reach the third billion, and 15 years to reach the fourth billion. In a span of less than 200 years, our species increased more than sevenfold to the current mark of 7.6 billion in 2017 (see Figure 1.1). This staggering rate of growth is almost impossible to convey with graphs or words (**Video 1.1**). No large vertebrate animal in the history of the planet has ever grown as quickly, and with as much success, as *Homo sapiens*, and that is a testament to our unique intellect, innovation, and technological advances as a species.

▶ **Video 1.1** See the staggering growth of the human population over time.
oup-arc.com/e/cardinale-v1.1

The human population recently passed an inflection point called the **demographic transition**, where global growth has now shifted from unconstrained,

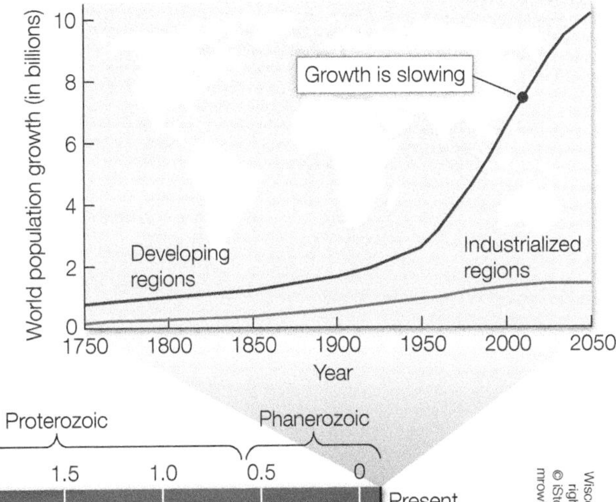

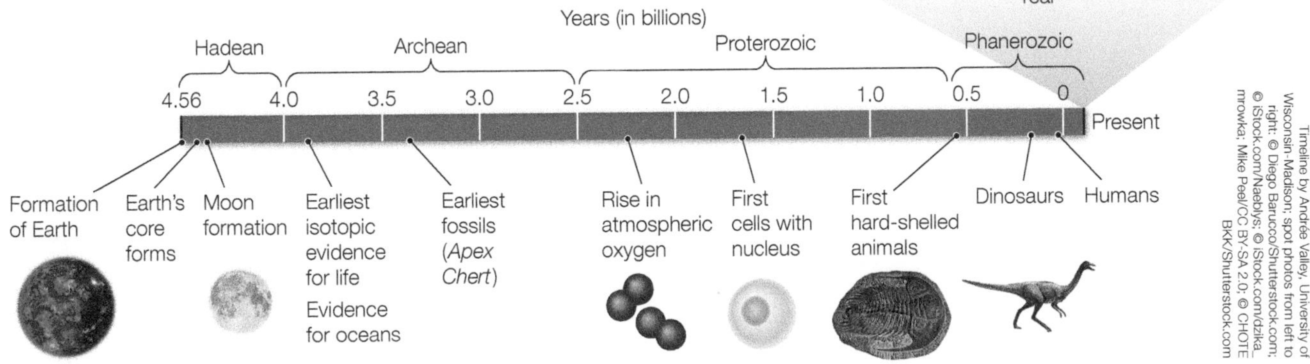

Figure 1.1 Human dominance is but a small fraction of Earth's 4.6-billion-year history. Although *Homo sapiens* evolved 200,000 years ago, nearly all of human population growth has occurred in the last 200 years. Now at 7.6 billion, the population has passed an inflection point (red dot) where growth has slowed and is heading toward a steady state. (After T. P. Soubbotina et al. 2000. *Beyond Economic Growth: Meeting the Challenges of Global Development.* Washington, DC: World Bank.)

exponential increases to reduced logistic growth that is headed toward a steady state (see Figure 1.1). This inflection point represents the transition from high birth and death rates to lower birth and death rates that have occurred as countries have transitioned from preindustrial to industrialized economic systems. There are many potential triggers of the demographic transition. Social triggers include reduced fertility rates caused by improved access by the population to contraception, a reduction in subsistence agriculture, an increase in the status and education of women, and a reduction in the value of children's work.[1] Ecological triggers include limitation by essential resources like energy, food, or water that become increasingly scarce as the population approaches Earth's carrying capacity.[2,3]

Whatever the triggers—whether social, ecological, or both—it is now clear that growth of the global human population has slowed and will soon reach some type of steady state, perhaps as early as the middle of this century. There have been at least 65 attempts to estimate Earth's **carrying capacity** for humans (**Figure 1.2**). While these range from extreme pessimism (1 billion) to unconstrained optimism (1000 billion), the credible interval of the estimates

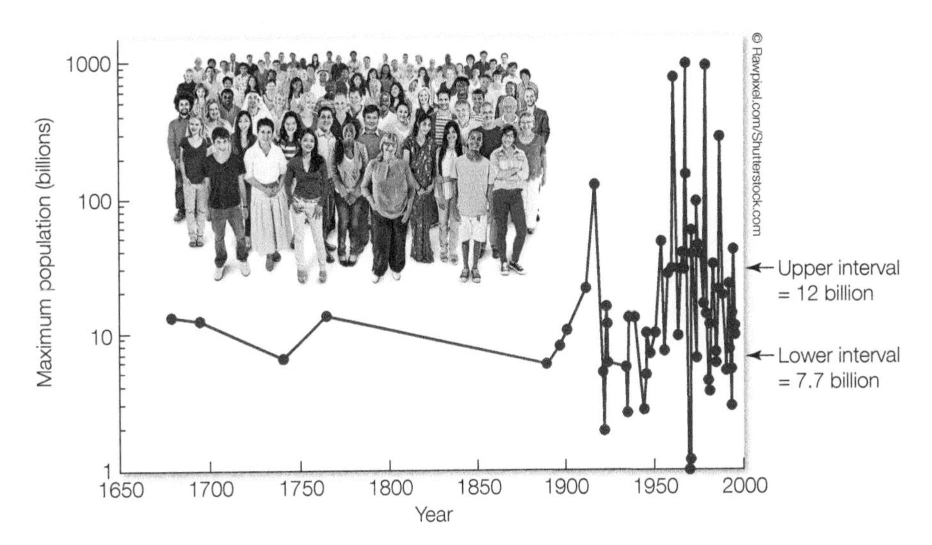

Figure 1.2 This graph shows 65 estimates of Earth's carrying capacity for humans (*y*-axis) that have been published through time (*x*-axis). The credible interval for these estimates ranges from 7.7 to 12 billion, closely matching projections from the United Nations of 7.8 to 12.5 billion. (After J. E. Cohen. 1995. *Science* 269: 341–346.)

ranges from 7.7 to 12 billion,[4] which closely matches projections from the United Nations. Given that the human population is already at the lower bound of these estimates, and given that the world's current growth rate of 1.14% has a potential doubling time of 61 years, those reading this book will likely be the first people in history to know what Earth's carrying capacity actually is for humans.

Given the staggering growth of the human population, it is perhaps surprising that the "average" person born in the modern era will live a longer, healthier, and more prosperous life than the average person born in any other period of human history. Mortality rates in premodern countries for children under the age of five were routinely 300 to 500 deaths per 1000 live births.[5] But during the twentieth century, modern medicine, central heating, improved hygiene, cleaner water, and access to more food have all helped reduce infant mortality to historically low rates (**Figure 1.3A**).[6] Those born in the premodern world had life expectancies of a mere 30 years.[5] Since 1900 the global average life expectancy has more than doubled, exceeding 76 years at the start of the twenty-first century. The average person in the modern era will also have a better standard of living compared with his or her ancestors. Since the 1800s, the global gross domestic product (GDP) per capita has increased more than tenfold (**Figure 1.3B**).[7] Modern technology and the green revolution have led to nearly a doubling of per capita caloric intake in modernized countries,[8] and the per capita calorie supply has been increasing steadily in nearly all regions of the globe since the 1960s.[9]

While conditions have improved for the "average" person on the planet, the benefits of economic development have been unequally distributed across the globe, leading to vast inequalities among countries and among those living in individual countries (**Video 1.2**). For example, although Americans constitute just 5% of the world's population, they consume 24% of the world's energy.[10,11] If everyone on the planet consumed as much energy, food, water, minerals, and space as the average U.S. citizen, we would need four Earths to sustain them. There are also huge disparities among individuals. The wealthiest 10% of the world's population accounts for 59% of all consumption on the planet,

▶ **Video 1.2** Consider the argument for a more equitable future for humanity.
oup-arc.com/e/cardinale-v1.2

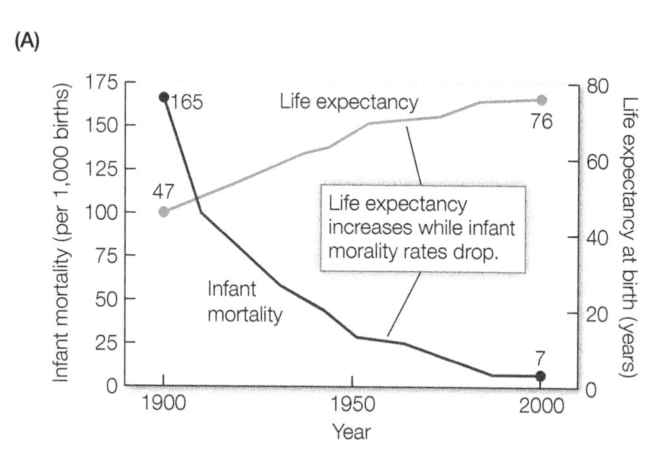

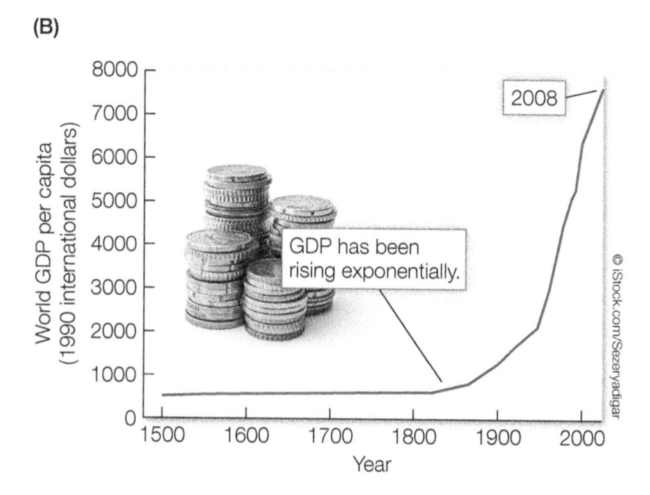

Figure 1.3 The average child born in the modern era will have (A) a greater chance of survival and a higher life expectancy and (B) greater per capita wealth than any generation in human history. The resulting population growth and consumption, particularly in fully developed countries, will be a major challenge for conservation efforts. (A, data from U.S. Census Bureau, *Statistical Abstract of the United States*. 1999. Compiled from 1900–1970, U.S. Public Health Service, *Vital Statistics of the United States*, Vols. 1 and 2; 1971–1997, U.S. National Center for Health Statistics, *Vital Statistics of the United States*; *National Vital Statistics Report* [NVSR] [formerly Monthly Vital Statistics Report]; and unpublished data; B, data from A. Maddison. 2009. *Statistics on World Population, GDP and Per Capita GDP, 1–2008 AD.* © Angus Maddison/University of Groningen.)

Figure 1.4 While the "average" person born in the modern era will live a longer, healthier, and more prosperous life than his or her ancestors, more than 2 billion people still live in extreme poverty and are malnourished. People whose basic needs are not met do not have the luxury of caring about biodiversity loss. Therefore, bringing communities out of poverty, particularly in developing nations, will be a major challenge for conservation efforts.

whereas 1.3 billion people continue to live in absolute poverty, surviving on less than US$1.25 per day (**Figure 1.4**).[12] Still another billion people live in sufficient poverty that they are malnourished, with food shortages disproportionately represented in less-developed countries of the world. While the number of poor and malnourished people is lower than recorded a few decades ago (suggesting that some progress has been made), the need to alleviate the suffering of those living in extreme poverty is still considered to be the greatest challenge to achieving a sustainable human population on the planet.

Suggested Exercise 1.1 Use the Global Footprint Network's website to estimate your ecological footprint, which is the number of planets we'd need if everyone lived like you. **oup-arc.com/e/cardinale-ex1.1**

State of the Global Environment

With our vastly expanded population size, increased affluence, and improved technologies, human activities now control almost all physical and biological processes that regulate life on the planet. While we do not wish to present a "doom-and-gloom" narrative in this book, it is necessary to establish some basic facts at the beginning of this book about what is happening across the globe as a result of human domination of the planet. These facts form the template for actions to save biodiversity.

Land

More than 50% of the inhabitable land surface of the planet has now been transformed in some way to support the growing human population.[13] To date, 5 billion hectares (50 million km^2) of natural habitat have been converted to croplands and pastures,[14] and an additional 350 million hectares (3.5 million km^2) of the world's land surface have been converted into urban environments.[15]

The vast majority of this habitat loss has occurred in the past 100 to 200 years (**Figure 1.5**). For 2 of the world's 14 biomes (Mediterranean forests and temperate forests), more than two-thirds of the habitat had been converted to other land uses—mostly agriculture—by 1950.[16] For 4 other biomes (temperate broadleaf forests, tropical and subtropical dry forests, flooded grasslands and savannas, tropical and subtropical grasslands), more than half of the land area had been lost by 1990. If business continues as usual, projections are that 70%–80% of many types of natural biomes will be lost by 2050 as agricultural and urban environments continue to expand.[14] The one notable exception is temperate forests, which are expected to be in recovery in many developed countries after decades of logging.[16] This provides a tremendous opportunity to restore biodiversity that was previously lost in these types of forest habitats.

Of those terrestrial habitats that have not been converted to other uses, much of what remains is degraded. Land pollution—the introduction into

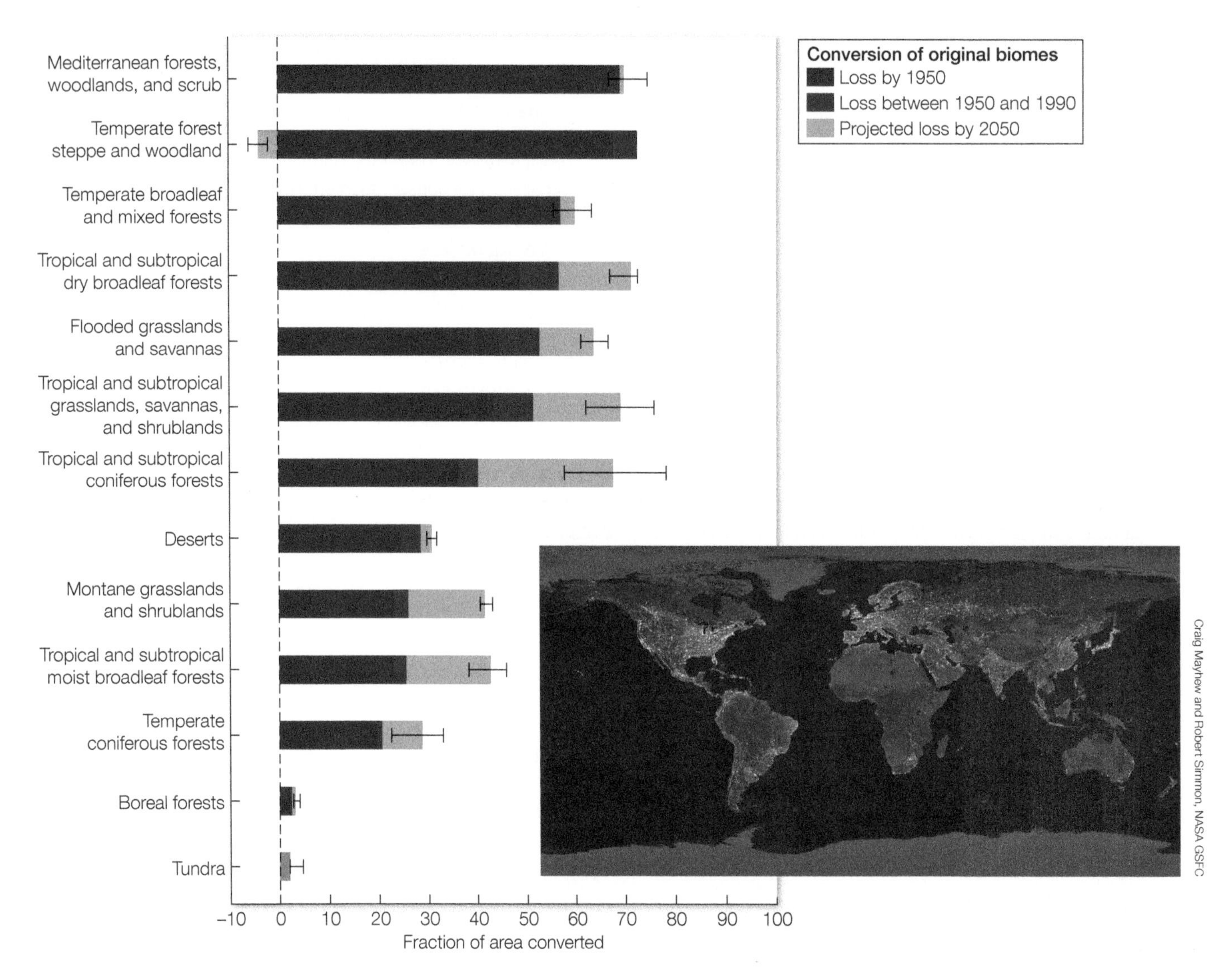

Figure 1.5 This figure shows the fraction of various terrestrial biomes that have been destroyed and converted to other land uses (cultivation, urbanization, etc.). Human transformation of the planet is readily apparent from NASA's global view of Earth at night, which shows the lights of cities and villages. (After Millennium Ecosystem Assessment. 2005. *Ecosystems and Human Well-being: Synthesis*. Washington, DC: Island Press.)

the environment of biologically harmful materials (e.g., trash, pesticides, fertilizers, heavy metals, pharmaceuticals, and health care products)—is widespread. Estimates by the U.S. Environmental Protection Agency indicate that expenditures for soil and groundwater cleanup needed to mitigate human health hazards at more than 300,000 sites in the United States will exceed $200 billion.[17] Many of the contaminants common to these sites, such as metals and volatile organic compounds, are known or suspected to cause cancer or adverse neurological, reproductive, or developmental conditions. A report by China's Ministry of Environmental Protection and the Ministry of Land Resources[18] showed that more than 16% of the country's soil was degraded, including 20% of farmlands polluted by factory waste, irrigation with polluted water, or overuse of fertilizers and pesticides. These areas represent habitats that not only are dangerous to human health, but also are degraded habitat for fish and wildlife.

Desertification is rampant, occurring when soil degradation by human activity causes relatively dry areas of land to become increasingly arid and lose their water bodies, vegetation, and wildlife.[19] Drylands account for about 40% of Earth's land area and are home to more than 2 billion people. Between 6 and 12 million km^2 of drylands (10%–20% of the total) are already heavily degraded, and more than a billion people are under threat from further desertification.[20]

In addition to pollution and desertification, most of the remaining land-based habitats on Earth are highly fragmented. A recent analysis of global forests showed that nearly 20% of the world's remaining forest cover is within 100 m of an agricultural, urban, or other modified environment where human activities, altered microclimate, and nonforest species may influence and degrade forest ecosystems.[21] More than 70% of the world's forests are within 1 km of a forest edge, subject to the degrading effects of habitat fragmentation. Thus, most terrestrial life-forms on the planet are being forced to live in rapidly shrinking habitats that are fragmented and degraded.

Climate

Despite the denial of climate change by select demographic groups and by some government officials,[22] there is consensus among scientists that Earth's climate is changing rapidly and that these changes are almost entirely due to human activities.[23,24] Anthropogenic emission of greenhouse gases such as carbon dioxide, methane, and nitrous oxide have increased dramatically in the postindustrial era, reaching concentrations that are unprecedented in at least the last 800,000 years (**Figure 1.6A**). The primary contributor to greenhouse gas emission is the burning of fossil fuels—coal, oil, and natural gas—for energy, but agriculture, forest clearing, and certain industrial activities also make significant contributions.

Detailed observations of Earth's surface temperature assembled and analyzed by several independent research groups showed that the planet's average surface temperature has warmed by 1.4°F (0.8°C) (**Figure 1.6B**). Evidence of this warming trend has been corroborated by a variety of other observations that indicate warming in other parts of the Earth system, including the cryosphere (snow- and ice-covered regions), the lower atmosphere, and the oceans.[25]

Warming of Earth's climate is associated with a suite of other types of environmental change, including rising sea levels, ocean acidification, decreased snow and ice cover, and an increased frequency and severity of extreme events like fires and hurricanes (**Figure 1.6C–F**). This suite of changes is not only having widespread impacts on human systems, but also substantially altering Earth's natural systems. Parmesan and Yohe[26] showed that 87% of species are shifting their ranges toward Earth's poles at an average pace of 6.1 km per decade.

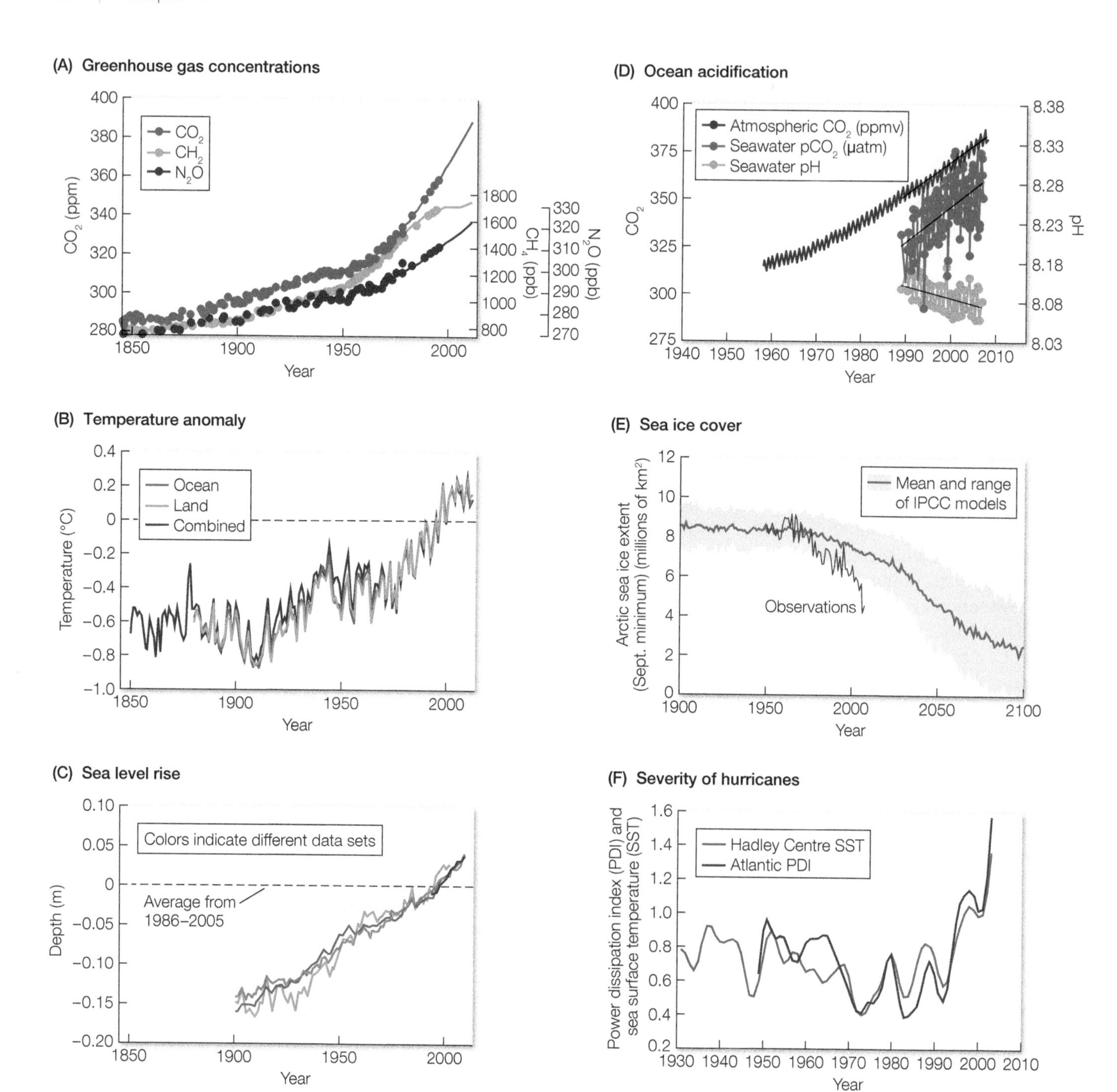

Figure 1.6 The environmental changes associated with (A) increased concentration of greenhouse gases and (B) global warming include (C) sea level rise, (D) ocean acidification, (E) reduced ice cover, and possibly (F) an increased frequency and severity of storm events. (A–C, after IPCC. 2014: Climate Change 2014: *Synthesis Report. Contrib. of WG I, II and III to the 5th Assess. Rpt. of the Intergov. Panel on Climate Change.* R. K. Pachauri and L. A. Meyer [eds.]. Geneva, Switzerland: IPCC; D, after PMEL/NOAA [www.pmel.noaa.gov/co2/story/OA+Observations+and+Data], based on J. E. Dore et al. 2009. PNAS 106: 12235–12240; C. D. Keeling et al. 1976. Tellus 28: 538–551; K. W. Thoning et al. 1989. *J Geophys Res* 94: 8549–8565; E, after J. Stroeve et al. 2007. *Geophys Res Lett* 34, L09501: 1–5.; F, after K. Emanuel. 2005. *Nature* 436: 686–688.)

Climate-driven species redistribution is influencing food security and patterns of disease transmission that directly affect humans.[27] It is also altering the phenology (seasonal life-cycles) of species, leading to what are termed *biological asynchronies* that threaten their survival. For example, Visser and colleagues[28] showed that shifting seasonality, such as the earlier arrival of spring, is causing species that once depended on each other for survival to be out of sync. Examples include flowers blooming before their pollinators have hatched, and chicks of birds starving because their insect food source hatched earlier than normal. Some have argued that entire biomes, such as coral reefs that are home to one-quarter of marine species, may cease to exist on a warmer planet and be replaced by other types of habitat.

Evidence for widespread geographic range shifts, altered phenologies, examples of biological asynchrony, and the projected loss of entire biomes all suggest that climate change could become one of the greatest drivers of biological change on the planet.[29] However, most of those biological changes are still playing out across the globe, and we do not yet know what the ultimate impacts will be.

Oceans

Oceans and seas cover 70% of the Earth's surface and represent 99% of the living space on the planet by volume. They also generate roughly half of all the oxygen we breathe, produce a third of the oil and gas we burn, and absorb nearly 30% of the carbon dioxide we generate. But the growing human population is pushing the ability of oceans to sustain human activities.

This was the conclusion of the first **World Ocean Assessment (WOA)**, part of the global reporting and assessment protocols that were established after the 2002 World Summit on Sustainable Development.[30] The first WOA, written by more than 600 scientists and released in 2015, emphasized that by all measures, oceans are changing rapidly. Ocean waters are warming and becoming more acidic, making it harder for calcifying forms of marine life like coral, shellfish, sea urchins, and certain forms of plankton to form the calcium carbonate structures they need to survive. Coastal waters have experienced large increases in pollution from both land-based activities and marine industries such as aquaculture. Excess nutrient flow from streams and rivers is now a leading cause of degraded water quality worldwide, promoting blooms of harmful microorganisms that generate excessive biochemical oxygen demand and 245,000 km^2 of dead zones in 400+ coastal habitats around the world.[31] High heavy metal concentrations and levels of other toxic substances found in marine mammals and fish have made them unfit for human consumption and illustrate the contamination of once-unspoiled ocean waters.

In a comprehensive study analyzing changes in Earth's oceans over a five-year period,[32] researchers found that nearly 66% of the ocean and 77% of exclusive economic zones (areas of the sea where nations have exclusive rights for exploration and use of marine resources) have shown increased human impact, driven mostly by effects on marine ecosystems imposed by fishing, climate change, and ocean- and land-based stressors (Figure 1.7). Only 10% of the ocean experiences low human impact and decreasing pressures, which means the oceans are improving in only a few places on Earth. The increasing number of marine protected areas (MPAs) has potential to offset some of the human impacts on the world's oceans, but the effectiveness of MPAs has been limited by design, management, and compliance challenges.[33]

Freshwater

Freshwater is essential to all life. But very little of the water on Earth is accessible by plants or animals, as most is either saline or frozen. Globally, humans now appropriate more than half of all accessible freshwater runoff, with most of that (70%) going for use in agriculture.[34] To meet increasing demands for the limited supply of freshwater, humans constructed more than 45,000 (>15 m high) dams worldwide during the twentieth century,[35] and by the turn of this

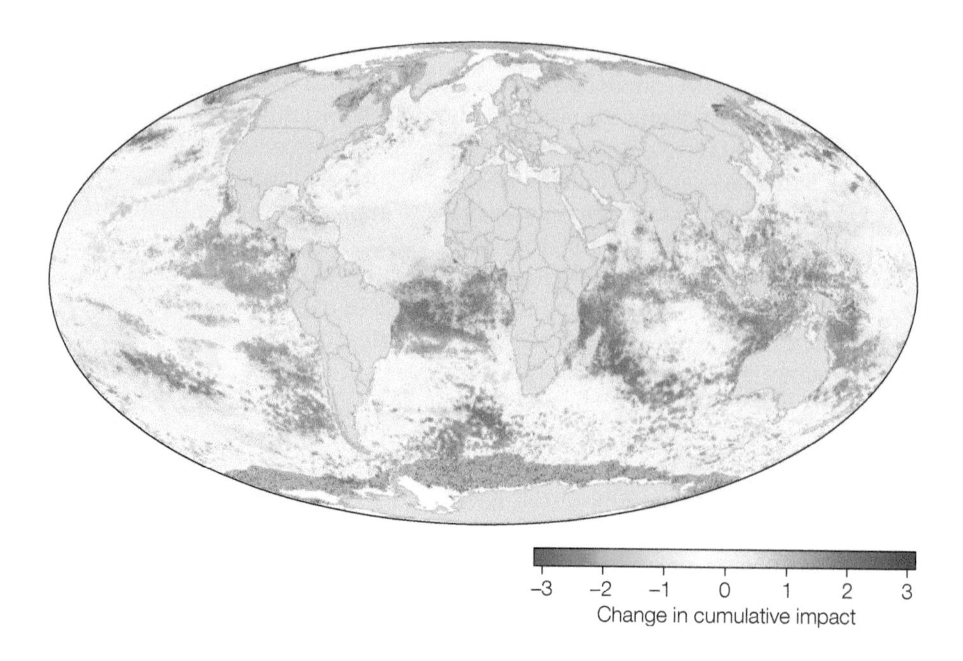

Figure 1.7 In a comprehensive study analyzing changes in Earth's oceans over a five-year period, researchers found that nearly two-thirds of ocean areas has experienced increased impact. This graphic shows the change from 2008 to 2013, with shades of red indicating an increased impact, while blue shows a decrease. (After B. S. Halpern et al. 2015. *Nat Comm* 6: 7615. CC BY 4.0)

century the flows of about two-thirds of all of Earth's rivers were regulated. Damming has altered the global flux of water and sediment from continents to oceans through the world's river basins, with major rivers such as the Colorado, Nile, and Ganges now used so extensively that little water reaches the sea. A recent analysis of 6374 existing large dams and 3377 planned or proposed dams found that on a global basis, 48% of river volume is moderately to severely affected by either flow regulation or fragmentation, or both (**Figure 1.8**).[36] Assuming completion of all dams planned and under construction, this percentage of affected river volume would nearly double to 93%, largely because of major dam construction in the Amazon Basin.

Some of the greatest rates of freshwater habitat loss have occurred for wetlands, which is concerning, given their contributions to biodiversity and ecosystem services. A recent review of 189 reports of change in wetland area[37] found that long-term loss of natural wetlands averaged 54%–57% but may have been as high as 87% since 1700 CE. The rate of wetland loss during the twentieth and early twenty-first centuries has accelerated 3.7-fold, with a 64%–71% loss of wetlands since 1900. Loss of inland wetlands has been larger and occurred faster than loss of coastal wetlands. Although the rate of wetland loss in Europe has slowed, and losses in North America have remained low since the 1980s, the rate continues to be especially high in Asia, where large-scale conversion of coastal and inland wetlands continues. Record keeping on wetland habitat change is especially poor for Africa and the Neotropics; however, the few records that exist suggest these areas are losing wetlands at a rate of 1% to 2% of the total area per year.

Human alteration to regional hydrologic regimes has led to rather dramatic landscape transformations. Withdrawal of groundwater and diversion of water for agriculture have caused once massive inland water bodies, such as the Aral Sea and Lake Chad, to be greatly reduced in volume, resulted in the demise of fisheries, the production of a drier and more continental local climate, a decrease in regional water quality, and an increase in human diseases.[38,39] Global conflicts over water are likely to be increasingly common in the coming century as the growing human population imposes even more stress on water quality and availability. Modification of the global water cycle has already posed significant

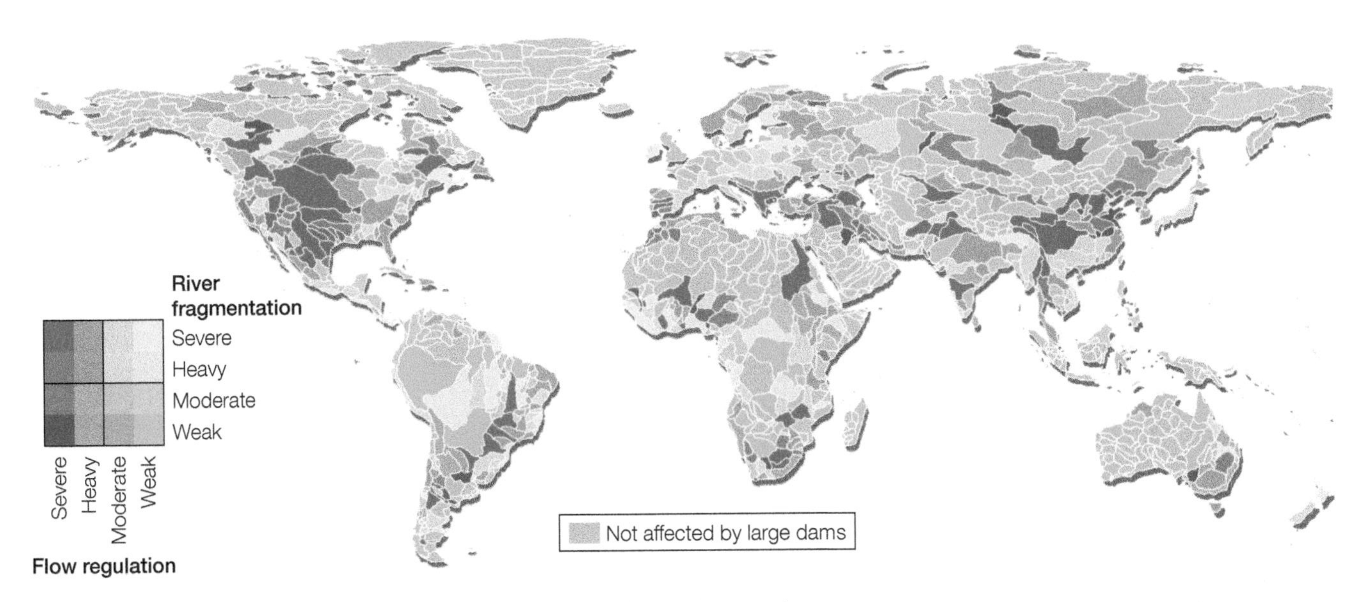

Figure 1.8 This map shows the combined effect of current and future dams on river ecosystems across the globe. (From G. Grill et al. 2015. *Environ Res Lett* 10: 1–15. CC BY 3.0)

threats to freshwater biodiversity on a global scale, and these threats are projected to get substantially worse.

State of the World's Biodiversity

What has been a great era for *Homo sapiens* has not been so great for most other organisms on the planet, and in fact, many are now struggling to survive in the face of human domination. Biodiversity—the variety of life on Earth, including all of its genes, populations, species, communities, and biomes—has changed dramatically across the Earth as the result of human activities. In Chapter 8 we will describe in more detail the state of the world's biodiversity. Here, we give a brief overview.

Global extinctions

Perhaps the most treacherous form of biotic change is global extinction. Once a species goes globally extinct, its evolutionary history, genetic information, and ecological functions are lost. While zoos, aquaria, and botanical gardens are working to "bring back" species that are extinct in the wild, and although modern technology is now exploring the potential to resurrect extinct species by using DNA from laboratory or museum specimens (Chapter 16), these programs are only likely to work for a small number of species, often at great cost. Therefore, global extinction is the final stage of biodiversity loss.

The Red List of Threatened Species kept by the International Union for Conservation of Nature (IUCN) currently includes 866 species that are considered extinct and an additional 69 species that are considered extinct in the wild (Chapter 8). Bull and Maron[40] tallied 1359 known extinctions that have occurred during the Holocene (11,000 years ago to the present) and 786 extinctions documented since 1500. This includes 79 species of mammals, 129 species of birds, 21 species of reptiles, 34 amphibian species, 81 fish species, 359 species of invertebrates, and 86 plant species. Collectively, the sum total of all documented extinctions represents less than 1% of the roughly 1.74 million species that have already been described[41] and an even smaller fraction of the 8.7 million types of multicellular

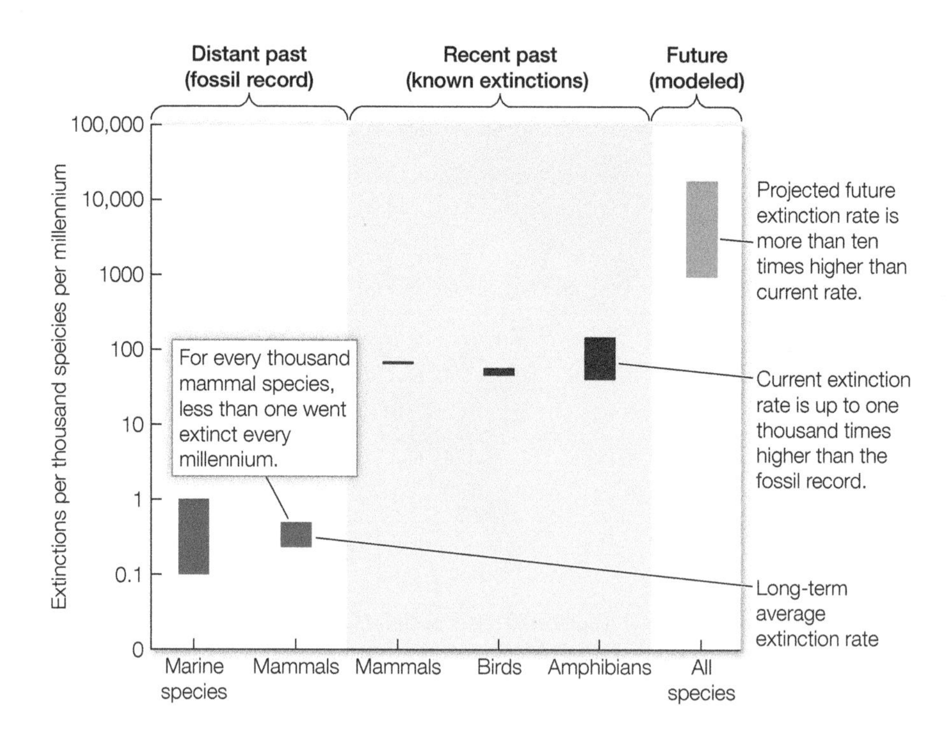

Figure 1.9 Accelerating rates of extinction. Background rates of extinction recorded in the fossil record for marine species and mammals average 0.1 to 1 extinction per 1000 species per millennium (left). Rates of extinction documented over the past few centuries are 100 to 1000 times higher than background rates, and they are projected to be more than 10,000 times background rates in the near future. (From Millennium Ecosystem Assessment. 2005. *Ecosystems and Human Well-being: Synthesis*. Washington, DC: Island Press.)

organisms that are estimated to exist on Earth.[42] However, while global extinctions represent only a small fraction of the world's biodiversity, the rate of extinction documented in the modern era is 100 to 1000 times faster than the historical rate that has been recorded in the fossil record[41,43] (**Figure 1.9**). Moreover, about 36% of known species are currently considered to be threatened, vulnerable, or in danger of global extinction,[44] a number that has doubled since 2000.

The narrative for global extinction is a "good news, bad news" story. The good news is that the vast majority of contemporary species that have shared the planet with *Homo sapiens* still exist somewhere on the planet; thus, we still have the opportunity to conserve them. The bad news is that a large fraction of contemporary species are not doing well and many face impending extinction; thus, their conservation is pressing.

Suggested Exercise 1.2 Go to the data portal on the Living Planet Index's website and perform a search using the name of your favorite species. After the search, use the map or list view to find a record closest to where you live. Is the population of your favorite species increasing or decreasing? **oup-arc.com/e/cardinale-ex1.2**

Local extirpations

While the total fraction of species that have gone globally extinct is low overall, the fraction of species that have experienced local **extirpation** from some portion of their geographic range is much higher. The PREDICTS project (Projecting Responses of Ecological Diversity in Changing Terrestrial Systems, oup-arc.com/e/cardinale-w1.1) is a major collaboration by scientists around the world to quantify how humans are influencing biodiversity. To date, the project has collated 2.5 million records of biodiversity at more than 21,000 sites that differ in the nature or intensity of human impacts relating to land use.[45,46] Newbold and colleagues[47] summarized results of the PREDICTS project to date, quantifying changes in biodiversity in 380 data sets that allowed a human-affected site to be explicitly compared with a spatial reference that served as a control. They found that, on average, 76% of species have been locally extirpated in the most human-affected terrestrial habitats on Earth, and an average 14% have been lost across all habitats for which data are available.

In the subset of Earth's habitats that have not yet been entirely converted from natural systems to other uses by humans (e.g., nature reserves) or that are now recovering from historic destruction (e.g., natural forests), analyses show mixed results. In some sites the numbers of species are declining because of human impacts such as fragmentation or climate change. In others, they are increasing as natural migrations and introduced species outpace local extirpations.[48,49] Therefore, our perspective on local extirpations depends on whether we are talking about habitats that have been directly altered by humans, or about habitats that are still intact but are being indirectly influenced by human activities.

Population decay

Focusing on global extinctions and local extirpations fails to give a complete picture of what is happening to species across the planet, as it only represents whether a species exists or not. What we really want to know is whether species populations are healthy and sustainable. The answer is that many, if not most, species across the planet are unhealthy and experiencing **population decay**. The **Living Planet Index** (**LPI**), which measures the health of 14,152 populations of 3706 vertebrate species in terrestrial, freshwater, and marine habitats, recently showed a 58% global decline between 1970 and 2012.[50,51] This means that, on average, animal populations are roughly half the size that they were just 42 years ago.

A 2017 analysis[52] of 27,600 vertebrate species showed that 32% are decreasing in population size and geographic range. Among the 177 mammals for which there was detailed information, all have lost at least 30% of their geographic ranges, and nearly half of the species have experienced more than 80% range shrinkage. Many species that were thought to be healthy just a few decades ago are now threatened or endangered. Since the 1980s, the giraffe (*Giraffa camelopardalis*) population has fallen by 40%, from at least 152,000 animals to just 98,000 in 2015. In the last decade, savanna elephant (*Loxodonta africana*) numbers have fallen by 30%, cheetahs (*Acinonyx jubatus*) are down to their last 7000 individuals, and orangutans (*Pongo* spp.) to their last 5000. This led the authors of the study to state:

> As much as 50 percent of the number of animal individuals that once shared Earth with us are already gone, as are billions of populations... [and the biosphere is] being ravaged by a serious and rapid wave of population declines and extinctions.[52]

One of the most striking examples of population decay comes from a long-term study of insects in Germany.[53] Researchers there monitored the biomass of insects, using malaise traps that were deployed in 63 nature protection areas over a 27-year period. Their analysis revealed that the biomass of flying insects has declined by 76%–82% over the period of record, consistent across a wide variety of taxonomic groups and independent of habitat type. This trend is deeply concerning given that insects are dominant pollinators of agricultural crops, are the primary source of food for many vertebrate species, and are major sources of biological control of pests and disease.

Mass mortality events

Mass mortality events (**MMEs**) are rapid, catastrophic die-offs of organisms that kill more than 90% of a population in a short time. They have been known to result in the death of more than a billion individuals for some species and to produce

700 million tons of dead biomass in a single event. A 2015 analysis of 727 MMEs affecting 2407 animal populations across the globe showed that since 1940, reports of MMEs have been increasing for all types of animals, including mammals, birds, amphibians, reptiles, fishes, and invertebrates.[54] A sharp increase in the occurrence of amphibian and reptile MMEs began in the 1970s, coincident with the growing awareness of global declines in amphibian populations. Analyses reveal that the magnitudes of MMEs—meaning, the number of animals that died during each event—are growing larger for birds, fishes, and marine invertebrates but decreasing for reptiles and amphibians. The increase in MMEs appears to be associated with a rise in the emergence of disease, biotoxicity, starvation, and events produced by the interaction of multiple stressors.

Genetic diversity

Species extinctions and population decay probably mask the real extent of biodiversity loss, because ultimately the origin of diversity is the genetic material that resides within individual organisms. This material is not easily seen or tracked in most biological monitoring programs. Nevertheless, Frankham[55] reviewed the literature and found that 22 of 23 studies of plants and animals showed a positive correlation between genetic variation and species population sizes; that 16 of 19 studies involving mammals, birds, reptiles, and insects found positive correlations between genetic variation and habitat area; and that genetic diversity tends to be higher for species with wider geographic ranges. The fact that so many species are currently threatened or endangered because their population sizes are in decay, while their range distributions are being reduced by habitat loss and fragmentation, suggests that genetic diversity is generally in decline. Genetic diversity is ultimately the biological library from which all diversity is made, and loss of this library is going to severely limit species' abilities to adapt and rebound from their current predicament. It also severely limits our options for saving them.

Willoughby and colleagues[56] more recently conducted a large, systematic review of 5165 published articles where researchers had measured **heterozygosity** (genetic variability) in wild populations of vertebrates. When the authors compared heterozygosity between threatened and non-threatened species, they found that heterozygosity was significantly reduced in threatened species. This suggests that healthy species populations are more genetically diverse than wildlife populations that are now typical of the Anthropocene (see below).

Biotic homogenization

Biotic homogenization is the process by which exotic species replace native flora and fauna, in turn causing ecosystems to lose their biological uniqueness. Homogenization can occur naturally, such as when species migrate on their own from one habitat to another. Or it can be promoted by human activities, such as when people accidentally or intentionally introduce nonnative species to a habitat, or when human-induced climate change leads to unnatural migrations of taxa toward polar regions of the globe.[57,58]

Biotic homogenization tends to reduce biodiversity by diminishing the distinctiveness of flora and fauna in different locations. Homogenization often leads to a relatively small number of "winners"—usually generalist species that are highly flexible in their habitat requirements, so they can replace a large number of "losers," which often are unique and specialized species that are locally adapted to a habitat and have a limited geographic range.[59]

Some of the best examples of biotic homogenization come from studies of freshwater habitats. Rahel[60] compared the present-day freshwater fish faunas in the United States with those that were here before European settlement, to assess the degree of fish homogenization among U.S. states. He showed that, on average, any two randomly chosen states have 15 more species in common now than before Europeans settled North America. The 89 pairs of states that formerly had no species in common now share an average of 25 species. This pattern is largely the result of species introductions associated with fish stocking for recreational purposes (e.g., brown trout, *Salmo trutta;* rainbow trout, *Oncorhynchus mykiss;* and smallmouth bass, *Micropterus dolomieu*) or aquaculture (e.g., common carp, *Cyprinus carpio*).

Dornelas and colleagues recently analyzed 100 time-series data sets where researchers had monitored 35,613 species of mammals, birds, fishes, invertebrates, and plants measured at 430,324 locations around the planet (mostly in oceans). These authors found that in 79 out of 100 time-series data sets, ecological communities were being homogenized at a rate of roughly 10% of their species per decade.[49] Collectively, these analyses show that biotic homogenization of the world's flora and fauna is widespread and happening quickly.

Seeds of a Good Anthropocene

Human domination over Earth's ecosystems as well as their biophysical processes is now so pervasive that the current geological epoch is being called the **Anthropocene**.[61] This epoch will, in fact, be the first time in the history of all life on Earth that any single species will co-opt more than half of all the available space, food, and water on the planet.[62] This means that all of the remaining unicellular and multicellular creatures that inhabit Earth will somehow need to survive on less than half of the most biologically limiting resources. This is ultimately the dilemma that gave rise to the field of conservation biology (Chapter 2).

Conserving biodiversity in the coming decades is going to become an even greater challenge as humans become increasingly disconnected from nature. The millennials will be the first generation in the history of humanity in which the majority of people will grow up and spend most of their lives within megacities, with little to no connection to nature beyond a city park or zoo. Millennials will be the first generation in the history of humanity to live in a world where the most abundant habitat type on the land surface of the planet will be agriculture. Millennials may be the first generation in the history of humanity to live in a world where we chemically control the formation of clouds and rainstorms and use **geoengineering** to attempt global manipulations of our atmosphere and oceans to maintain a habitable climate for humans. How will we protect natural habitats, or wild species, when people have never seen or experienced them and when the vast majority of habitats will be controlled and engineered for human benefit?

The path we are on now is creating a planet that is becoming so covered with plantations, farms, cities, and human-dominated habitats that wild creatures and wild places, if they still exist in the future, will do little more than linger at the margins of engineered land and seascapes. Is this the kind of planet you want to live on? Or would you rather live on a planet that sustains both people and nature, leaving room for wild creatures to live and thrive in habitats that are free of human domination and interference? The decisions you make now will shape the face of all life on Earth for millions of years to come. And the future of biological diversity—the library of all genes, populations, species, communities, and ecosystems—will depend on what the current generation decides to do with their Earth.

Six reasons to be optimistic

While the enormity of our challenge is daunting, there are many reasons to feel optimistic. First, while biodiversity is being severely degraded by human activities, the vast majority of modern species are still represented somewhere on the planet. This means that most biodiversity still exists, and it's not too late to save it! Second, world population growth has begun to slow, indicating that some challenges may be behind us. Of course, certain parts of the world have yet to reach the demographic transition, and the pending population explosions in Asia and Africa have the potential to wipe out habitats and organisms that inhabit those parts of the globe. It is possible these regions of the world will come to look much like countries in Europe and the United States—heavily managed landscapes with only a fraction of their biological past. On the other hand, we can learn from our past successes and failures in parts of the world that have already passed the inflection point and use that knowledge to improve the lives of people while still protecting nature in rapidly developing parts of the globe.

Third, we have momentum. All around the world there is a growing understanding that nature is not inexhaustible as once thought, and stakeholders as diverse as religious organizations, financial institutions, and government organizations are acknowledging that stewardship of our natural resources is a top priority for the sustainability of humanity. The World Bank—one of the great vendors of international development—has recognized environmental stewardship as one of the three pillars of sustainable development and has partnered with the United Nations to accomplish its Sustainable Development Goals (Chapter 17). Leaders of many of the world's major religions have urged their followers to be better stewards of the environment (Chapter 5). In the *Laudato Si': On Care for Our Common Home,* Pope Francis spoke to Catholics:

> It is not enough, however, to think of different species merely as potential "resources" to be exploited, while overlooking the fact that they have value in themselves. Each year sees the disappearance of thousands of plant and animal species which we will never know, which our children will never see, because they have been lost forever. The great majority become extinct for reasons related to human activity. Because of us, thousands of species will no longer give glory to God by their very existence, nor convey their message to us. We have no such right.[63]

The *Islamic Declaration on Climate Change,* signed by 60 world leaders of the Muslim faith, agreed:

> God—whom we know as Allah—has created the universe in all its diversity, richness, and vitality: the stars, the sun and moon, the earth and all its communities of living beings.... What will future generations say of us, who leave them a degraded planet as our legacy? How will we face our Lord and Creator?[64]

Indeed, there is a will to achieve sustainability in our use of natural resources, and a desire to protect biodiversity, that has never existed before.

Fourth, we have increasingly realized that conservation of nature and human prosperity often go hand in hand and that what is good for the environment can be good for people too. Indeed, the past several decades have shown that ecosystems and their resident organisms produce valuable goods and services (Chapter 6) that drive the health and economic prosperity of human societies (Chapter 7). This realization in no way diminishes the intrinsic values of biodiversity. Many people still choose to conserve because of moral convictions about stewardship and the belief that species have a right to exist. Many still conserve because of the beauty of biodiversity and the wonder and inspiration it evokes (Chapter 5). But contemporary conservation has expanded beyond its historical reliance on intrinsic valuation to consider utilitarian arguments as well.

Fifth, the science that underlies conservation biology is maturing rapidly. As is true for most disciplines, the tools and techniques used during the early years of conservation biology were crude. Models were often qualitative, built on narratives, individual case studies, and trial and error. But as the field has progressed and the amount of information has amassed, conservation biology has begun to transition toward an evidence-based, predictive science (Chapter 2). That's not to say that we have all the information we need nor that our scientific models are yet accurate enough. But it does mean that we now have a solid understanding of the modern threats to biodiversity (Chapters 8–12), as well as the scientific tools needed to conserve biodiversity at all levels, from genes to ecosystems, in both natural and engineered habitats (Chapters 13–17).

Sixth, and last, humans are remarkably innovative. We have a strong history of solving environmental problems that seemed insurmountable at the time (**Figure 1.10**). It wasn't long ago that widespread use of chlorofluorocarbons (CFCs) in aerosol sprays and refrigerants threatened to deplete the ozone layer of the upper atmosphere. It wasn't long ago that emissions of sulfur and

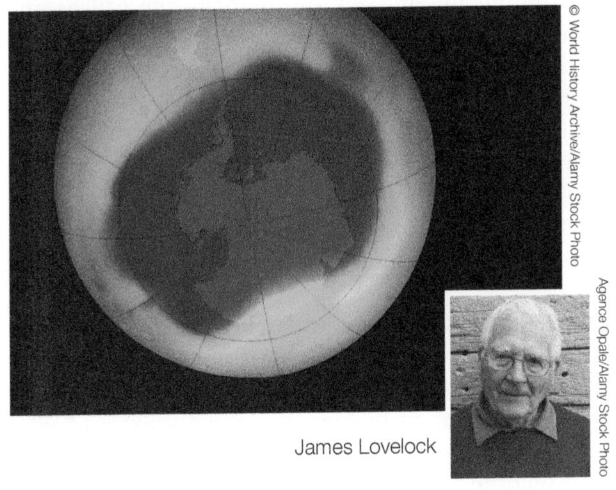

Figure 1.10 Examples of successful efforts to solve global environmental problems and the individuals who made the difference. These successes should provide optimism that we can solve the current biodiversity crisis.

Gretchen Daily: science.in2pic.com/CC BY-SA 3.0; E.O. Wilson: Michael Dwyer/Alamy Stock Photo; Nalini Nadkarni: courtesy of University of Utah Communications Office; Aldo Leopold: USDA/CC BY 2.0; Jane Goodall and Daniel Pauly: public domain; Sandra Diaz: courtesy of Sandra Diaz; Rhino: © Cathy Withers-Clarke/Shutterstock.com nitrous oxides threatened to kill the forests of many industrialized countries with acid rain. It wasn't long ago that widespread use of pesticides like DDT threatened to eliminate birds. But these seemingly impossible challenges are now the success stories told by environmental historians.[65] They were solved, in part, because people rose up to become champions of the cause, to generate the momentum needed by society for change, and to organize the science and engineering needed for a solution.

The next generation of champions

Just as we have solved past environmental problems, we can solve the current biodiversity crisis as well. Not only do we have momentum in society and scientific tools at our disposal, we have success stories that show us that species can be saved (**Box 1.1**). But unlike some of our past environmental problems, biodiversity loss is a chronic form of change driven by a complex weave of environmental and social problems that will take decades, if not centuries, to resolve. Over this period, we will need a new generation of people—people from a wide variety of disciplines—who can push forward the legacies begun by our past champions of conservation. We will need more scientists like Gretchen Daily, who championed the value of nature's benefits and services to people, and E. O. Wilson, whose Half-Earth Project is trying to set aside large portions of the planet to save biodiversity (see Figure 1.10). We need more researchers like Nalini Nadkarni, who pioneered the study of Costa Rican rain forest canopies; Sandra Diaz, who helped establish international programs to protect biodiversity and ecosystem services; and Daniel Pauly, whose work has helped us manage ocean fisheries more sustainably. We need more esteemed biologists like Jane Goodall to use their accomplishments to become the voices of threatened species. We need a new generation of writers like Aldo Leopold and of artists like Ansel Adams who can convey the beauty of nature to people who have never experienced it. And we need a new generation of explorers like Jacques Cousteau who can instill in us the wonder of life on the planet and the thrill of finding new living organisms.

This book is all about training the next generation who will become protectors of the world's biodiversity. Though they will stand on the shoulders of giants, the next generation will need to be better equipped than their predecessors. The next generation of practitioners will need to have a broader tool kit at their disposal to better understand and manage the causes and consequences of biodiversity change. That tool kit will need to include both qualitative tools such as stories, narratives, and art, as well as an ever-increasing body of quantitative methods and models from genetics, population modeling, reserve design, and others. This text will provide compelling examples, stories, visuals, and other qualitative tools that practitioners can use to communicate their points. In addition, it will provide in-depth training in modern quantitative tools that conservation biologists need to be successful. Those quantitative tools will then be reinforced with practice exercises and suggested activities.

The next generation of practitioners will also need to be more interdisciplinary and understand that conservation is not just a bunch of ecologists, wildlife biologists, or other natural scientists trying to save their favorite species in a dwindling habitat. Rather, conservation biology is a social science as well. People can't possibly care about conservation when their fundamental needs are not being met or when wildlife are eating their crops, endangering their children, or destroying their property. The conservation biologist of the future is going to have to think more about how to meet the needs of people, including

BOX 1.1 Success Story

The Indian Rhino

The greater one-horned rhinoceros (*Rhinoceros unicornis*, also called the Indian rhino) is the largest of all rhino species. It once ranged across the entire northern part of the Indian subcontinent, grazing throughout the rich alluvial grasslands of the Terai and Brahmaputra basins. During the nineteenth century, its habitat range shrunk to just 11 sites in northern India and southern Nepal because of expansion of agriculture and excessive hunting.

By the end of the twentieth century, fewer than 200 individuals of the one-horned rhinoceros remained, and the species was at dire risk of extinction. The prominent horn for which these rhinos are known was the cause of their downfall. The hard, hairlike growth is revered for medicinal use in China, Taiwan, Hong Kong, and Singapore and is valued in North Africa and the Middle East as an ornamental dagger handle. Indian rhinos were killed and poached by the thousands to support international demand for their horns.

In 1975, *R. unicornis* was listed in the Convention on International Trade in Endangered Species of Wild Fauna and Flora (CITES)—an international treaty signed by 183 parties that protect endangered plants and animals by prohibiting international trade that threatens their survival. With the help of the World Wildlife Fund (WWF), the Indian and Nepalese governments took major steps toward rhinoceros conservation. National parks and captive breeding programs were established. Grassroots antipoaching programs were formed, and these were complemented by government formation of armed patrols, such as the Gaida Gasti—a patrol group of 130 armed men who established a network of guard posts throughout the Chitwan National Park.

Today, populations of the Indian rhino have increased to about 3500 individuals in northeastern India and the Terai grasslands of Nepal. Thanks to dedicated conservation efforts and improved protection programs, the species has been pulled back from the brink of extinction. Though it is still considered threatened, recovery of the greater one-horned rhino is among the greatest conservation success stories in Asia.

Indian rhinos, which had dwindled to only 200 individuals a few decades ago because of poaching, have now bounced back from the edge of extinction thanks to grassroots antipoaching programs.

how to resolve human-wildlife conflict, and how to protect livelihoods, all while calculating and communicating the value of nature to society. Therefore, many examples in this text are specifically dedicated to topics such as conflict resolution, effective communication, and methods for evaluating trade-offs in human needs and desires.

We sincerely hope you enjoy this book, and that you will benefit from the variety of approaches, tools, and techniques that are presented throughout. Now that you have the background needed to understand the current state of our planet, Chapter 2 will explain how the discipline of conservation biology has emerged as an interdisciplinary field that aims to improve the modern biodiversity crisis by protecting species, their genetic diversity, populations, habitats, and ecosystems while at the same time maintaining human well being.

Summary

1. No other large vertebrate in the history of the planet has ever grown as quickly, and with as much success, as has *Homo sapiens*. Our intellectual capacity, social cooperation, tools, and technology have allowed the human population to increase more than sevenfold in less than 200 years. The human population now co-opts more than half of all available space, food, and water on the planet, and human activities are the dominant control over all biophysical processes that support life.

2. By most standards, the average person born today will be healthier, live longer, and be wealthier than his or her ancestors. However, health and prosperity are unequally distributed around the world, with nearly a quarter of all people malnourished or living in poverty. One of the greatest challenges for conservation in the Anthropocene is to protect the magnificent variety of life that is the most striking feature of our planet, while also meeting the needs of people. This will be particularly difficult in areas where the human population is projected to grow in the coming decades (such as in Africa and Asia).

3. But there is hope and reason for optimism. Just as we have solved many prior environmental problems, we can solve the biodiversity crisis as well. We have momentum in society. We have the scientific and engineering tools at our disposal. And we have a long history of individuals who have stepped up to become champions of the cause. But we need a new generation who can expand conservation's reach and carry the torch of past generations.

4. The next generation of conservation biologists will need to have better scientific training and a broader tool kit available to them. The next generation will also need to be more interdisciplinary and understand that conservation problems are social problems. Because of this, the conservation biologist of the future will need to have better skills in decision-making, conflict resolution, communication, and methods for valuating human needs and desires than did prior generations.

For Discussion

1. Has the remarkable success of humans led to a better quality of life? Are you generally happier with your life than were your parents or grandparents?

2. What kind of planet do you want to live on? Would "nature" on your ideal planet be wild, or would it be managed and controlled? How much of the Earth would humans dominate, and how much would be set aside to allow other species to exist?

3. Given that we know most species that exist on Earth have not yet been described, should we invest limited resources in trying to find and identify new species, or in protecting those we already know about?

4. Do people need to somehow "connect" with nature in order to care about conservation? If yes, how can the next generation of conservation biologists help the average person connect?

Suggested Readings

Barnosky, A. D. et al. 2011. Has the Earth's sixth mass extinction already arrived? *Nature* 471: 51–57. An outstanding summary of the current status of global biodiversity change.

Dirzo, R. et al. 2014. Defaunation in the Anthropocene. *Science* 345: 401–406. A second outstanding summary of the current status of global biodiversity change.

Kennedy, D. (Ed.). 2008. *Science Magazine's State of the Planet 2008–2009*. Washington, DC: Island Press. Excellent overview of the major environmental problems we face, and what progress has been made.

Millennium Ecosystem Assessment. 2005. *Ecosystems and Human Well-being: Synthesis*. Washington, DC: Island Press. From the World Resources Institute, this is the most comprehensive survey of the status of Earth's ecosystems.

Ogden, L. E. 2016. Conservation biology: A new hope? *BioScience* 66: 1088–1088. A reminder that hope-filled stories are better motivators for conservation than "doom-and-gloom" messages.

Waldron, A. et al. 2017. Reductions in global biodiversity loss predicted from conservation spending. *Nature* 551: 364. Evidence that conservation biologists do make a difference!

Visit the
Conservation Biology
Companion Website
oup.com/he/cardinale1e
for videos, exercises, links, and other study resources.

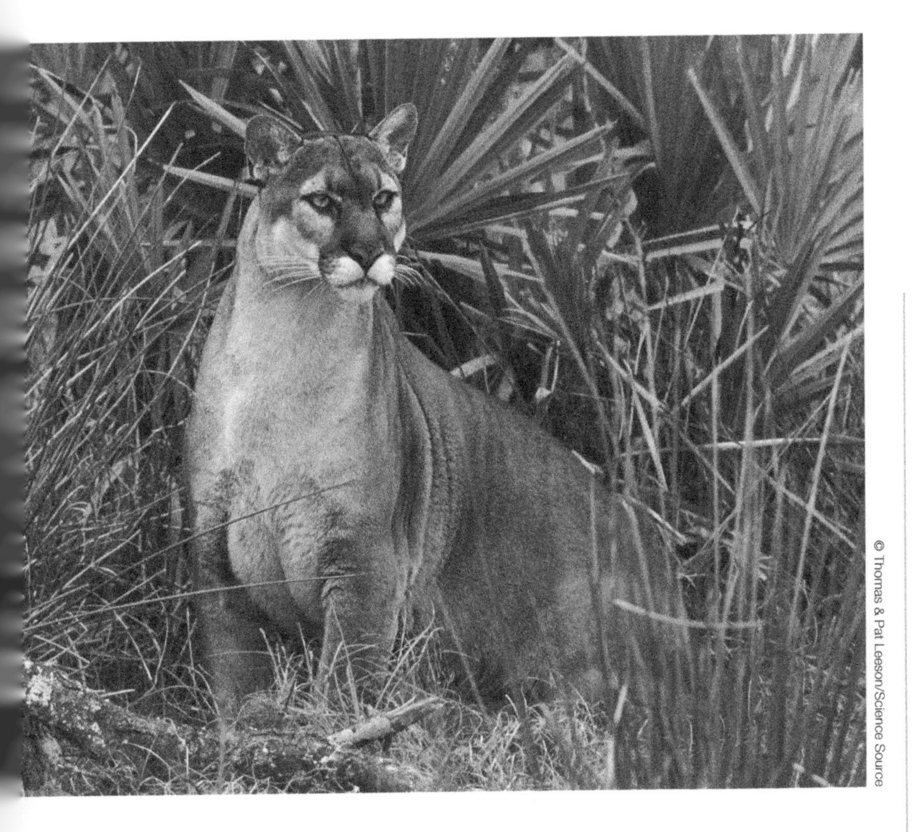

© Thomas & Pat Leeson/Science Source

2

The Rise of Conservation Biology

The Florida panther is a subspecies of *Puma concolor* (also known as mountain lion, cougar, or puma) and represents the only known breeding population of puma in the eastern United States.

In Chapter 1 we established that humans have been one of the most remarkably successful species to ever inhabit the Earth. But our very success and domination of the planet have led to a myriad of environmental changes that are making it difficult for many other organisms to survive. Indeed, the Anthropocene is already marked as an era of unprecedented biological change, characterized by large declines of wildlife populations, local extirpations of many species, rapidly increasing rates of global extinction, and homogenization of biological communities across the planet's oceans and land surface.

The current biodiversity crisis—and the need to protect the biological library of genes, species, communities, and ecosystems that has taken 3.5 billion years to evolve—has become increasingly evident to scientists, the public, and policy makers. Conservation biology is an interdisciplinary field that brings together experts from the natural sciences, social sciences, humanities, and economics as well as practices from natural resource management for a common purpose: to protect Earth's biodiversity. The field has three primary goals:

1. To investigate and describe the full variety of life on the planet
2. To evaluate and predict the effects of human activities on biodiversity
3. To develop practical solutions to protect and manage biodiversity sustainably

Since emerging as a distinct field in the mid-1980s, conservation biology has grown dramatically. It now includes tens of thousands of practitioners around the world, hundreds of millions of supporters, multiple professional societies, numerous governmental and nongovernmental organizations, a wealth of companies and corporations, and a large number of colleges and universities that offer degree and vocational programs worldwide. While conscious efforts to conserve and protect global biodiversity are a recent phenomenon, the ideals behind conservation have deep historic roots that date back to the beginning of human history. To understand how we got to the point of a global, well-coordinated discipline, it is useful to look back at the origins of conservation.

Origins of Conservation

Many of the ideals that are espoused in modern conservation biology originated in the belief systems of premodern societies. But true conservation efforts began with formation of royal game and nature reserves, sustainable forestry practices by British colonies in East India, and the formation of national parks in Mongolia and the United States. In this section, we detail these precursors of contemporary conservation.

Premodern societies

Although we may think of environmental degradation as a product of recent times, significant environmental degradation has always accompanied humankind. Throughout the world, early human societies in Europe, Asia, Africa, Australia, and the Americas radically altered the environments in which they lived and affected both the species living in those environments and the ecological processes they performed. Prehistoric humans were a dominant driver of extinction[1,2] as hunter-gatherers overexploited populations of woolly mammoths, large flightless birds, saber-toothed tigers, and other large megafauna (**Figure 2.1**). For tens of thousands of years, early peoples used fire to change landscapes, get rid of undesirable species, and create cultivatable land.

In the classical Greek period, Aristotle commented on the widespread destruction of forests in the Baltic region caused by land clearing for agriculture, while Plato lamented the condition of Greek land, comparing the once fertile

Figure 2.1 Human impacts on biodiversity are not just a modern phenomenon. Expansion of hunter-gatherer societies was a dominant driver of extinction of megafaunal species, including the woolly mammoth from Arctic and northern lands; many of the flightless birds, like the moa, from New Zealand; and the saber-toothed cats from Europe and Asia.

soils to "the skeleton of a body wasted by disease." At the same time in southern Asia, large tracts of forests were felled to meet the growing need for timber to build trading ships to serve expanding mercantile centers such as Constantinople (now Istanbul). The barren landscapes that we associate with much of Turkey, Syria, Iraq, and Iran, which in earlier times were known as the "land of perpetual shade," are now unnatural deserts resulting from massive overexploitation of woodlands.

While human beings have always degraded their environment and endangered other species, there has been much debate about whether prehistoric societies also practiced select forms of conservation. Conservation requires acts that preserve, protect, or restore natural resources in order to prevent their depletion and to ensure their long-term benefits to future generations. It requires restraint and incurs costs, and forms of resource management that require neither restraint nor cost are not conservation.[3]

The above definition of conservation is important because, if we are careless about definitions, we risk generating false narratives. For example, a common narrative about prehistoric societies is that they were **noble savages**—an idea that romanticizes primitive cultures as symbolizing the innate goodness of humanity, free from the corrupting influence of civilization.[4] A common corollary of this narrative is that modern humans, with their advanced civilization and technology, are greedy destroyers of nature, whereas most primitive peoples lived in harmony with nature, managing their natural resources wisely to live within their means. But what is the evidence that premodern cultures knowingly, and willingly, used resources in a way that showed restraint and incurred costs—all to preserve them for future generations?

Suggested Exercise 2.1 Visit the Companion Website for a link to "Pristine Myths, Noble Savages and Conservation." How do inaccurate stereotypes of historical societies influence our potential to succeed with modern conservation efforts? **oup-arc.com/e/cardinale-ex2.1**

There is certainly evidence that the belief systems, religions, and myths of many early human hunter-gatherer societies and stable agricultural societies tended to emphasize conservation themes and the wise use of natural resources. For example, many religions of the world that formed in the Axial and Middle Ages had belief systems in which people were considered to be both physically and spiritually connected to the plants and animals in their surrounding environment.[5] Confucianism and Taoism in China, and Buddhism and Hinduism in India, all emphasized the value of sacred wilderness areas for their ability to provide spiritual experiences and encouraged the protection of these habitats. The teachings of Islam, Judaism, and Christianity throughout the Middle East, Europe, central Asia, and Africa also emphasized the idea that people are given the sacred responsibility to be guardians of nature. Indeed, many early religious philosophies cultivated a direct connection between the natural and spiritual worlds and emphasized that the connection breaks when the natural world is altered or destroyed by human activity.

Some traditional societies also had philosophies that seemed to espouse certain conservationist ideals (Figure 2.2). Hunting and gathering societies like the Penan of Borneo gave thousands of names to individual trees, animals, and places in their surroundings to create a cultural landscape they considered to be vital to the well-being of the tribe. Such a close relationship between humans and nature was eloquently described at the Fourth World Wilderness Congress in 1987 by a delegate from the Kuna people of Panama:

Figure 2.2 Many traditional societies espoused conservationist ideals in their belief systems. Depicted here is Pachamama—the earth/time mother of nature revered by Indigenous people of the Andes. In Inca mythology, she is a fertility goddess who presides over planting and harvesting and has the power to sustain life on Earth.

> For the Kuna culture, the land is our mother and all living things that we live on are her brothers in such a manner that we must take care of her and live in a harmonious manner with her, because the extinction of one thing is also the end of another.[6]

Other hunter-gatherer societies are known to have developed rather sophisticated forms of management that showed restraint and sacrifice to protect natural resources. One example is found in the native Tlingit and Haida tribes that originated 10,000 to 12,000 years ago near the mouths of the Skeena and Nass Rivers in British Columbia, Canada. The Tlingit and Haida had an elaborate set of rules to govern access to sockeye salmon (*Oncorhynchus nerka*) that limited access to particular streams by an individual clan or house group. Streams were assigned by the *yisati*, the clan's eldest male, who also set limits on how many fish could be taken and at what times.[3]

The idea of **intergenerational sustainability** appears to have originated in China during the Song dynasty (960–1279 CE). During this period, the practice of setting aside "sacred" or *fengshui* forests around villages developed as part of a religious philosophy that attempted to promote harmony between humans and their surrounding environment (Figure 2.3).[7] People were instructed to maintain these sacred forests for many generations. Indeed, a village history from China's Anhui province instructs, "Every family must take care of the mountains and waters around. Plant trees and bamboo as shelters...Keep an eye on the environment and protect it from damage. This is a chore for people of one hundred generations to take."[8] The idea of intergenerational sustainability evolved independently in the Iroquois nation, who developed the **seventh generation principle**—a philosophy that the decisions we make today should result in a sustainable world seven generations into the future.[9] The first recorded concepts of the seventh generation principle date back to somewhere between 1142 and 1500 CE, when the Great Law was developed for formation of the Iroquois Confederacy.

Clearly, the belief systems of many early human societies embodied certain ideals of conservation. And there is evidence that some societies willingly forewent the exploitation of resources in ways that showed restraint and incurred costs in order to preserve them for future generations. However, there is

Figure 2.3 The Qingyin Pavilion is located in Mount Emei, which is one of China's Four Sacred Buddhist Mountains that are regarded as places of enlightenment. The pavilion is built on an outcrop in the middle of a fast-flowing stream so one can appreciate nature's music.

reasonable evidence to suggest that conservation, though apparent, was not the norm in premodern societies. A number of scholars[10–13] have argued that the majority of primitive societies were not, in fact, people who showed great care for their natural resources, nor who lived sustainably within their means. In his book *Collapse: How Societies Choose to Fail or Succeed*, Jared Diamond[10] argued that virtually wherever humans have expanded to or settled, environmental destruction has been the rule, and overexploitation of natural resources has generally caused human societies to move or collapse. While the low population densities of early human societies and their propensity to move after exhausting resources kept the scales of environmental destruction and impacts on biodiversity relatively limited, resource management that showed both restraint and cost—that is, conservation—was not common in premodern societies. Rather, conservation seems to have first appeared as a common practice in medieval times with the advent of game and nature reserves.

First game and nature reserves

In many cultures, the practice of conservation began not with the intent to protect species or their habitats for intergenerational sustainability. Rather, in some cultures it began with more selfish motives: certain types of people were prohibited from using resources in particular areas so that those resources were reserved for extraction by a privileged few. The earliest examples of such a practice occurred throughout Europe and Asia where royalty and other wealthy individuals set aside land as hunting preserves, forbidding common people to kill game animals and thereby protecting the animals so they could be hunted by the elite.[14] In fact, the word *forest* is a term of European origin that originally referred to areas of land that had been legally set aside for specific purposes like hunting by royalty.

Some of the earliest royal forests were established in England, Wales, and Scotland during the medieval period by kings who were avid huntsmen. William the Conqueror, the first Norman king of England, had a great love of hunting and established a system of forest law that served to protect game animals and their forest habitat from destruction. Offences in forest law were divided into two categories: trespass against the vert (the vegetation of the forest), and trespass against the venison (the game). Trespass against the vert precluded the construction of any buildings, the clearing of forest land for agriculture, and the felling of trees or clearing of shrubs. Trespass against the venison prohibited the hunting of deer, boar, rabbits, and wolf as well as "beasts of chase" (fox, marten, and roe deer) and "fowls of warren" (pheasant and partridge).[14]

Inhabitants of the forests were generally forbidden to have hunting weapons. However, after kings realized that royal forests could be useful sources of income, local nobles were often granted a hunting license to take a certain amount of game. The common inhabitants of the forest were sometimes granted a variety of rights of their own, such as the right to take firewood, the right to pasture swine in the forest, the right to cut turf for fuel, and various other rights to harvest useful products.[15]

The forests were overseen by wardens, who supervised foresters and rangers who were responsible for preserving the forest and game. Management of royal forests sometimes included the translocation of game from one location to another, or stocking to increase population sizes. The foresters and rangers were also responsible for apprehending offenders of the law, who were turned over to forest courts with justices who oversaw offenses. Violators were fined, imprisoned, or even killed.

After the concept of royal forests was introduced by the Normans to England in the eleventh century, it spread rapidly. By the height of this practice in the late twelfth and early thirteenth centuries, one-third of the land area of southern England was set aside as royal forest; at one point in the twelfth century, all of Essex was afforested, and on his accession Henry II declared all of Huntingdonshire to be protected forest.[16]

Chinese royalty similarly established hunting regulations and protected game preserves with walls and guards. Genghis Khan codified regulations that forbade the hunting of animals between March and October during the breeding season.[17] By protecting animals in the summer, Khan also provided a safety net for the winter, and hunters had to limit their kill to what they needed for food and no more. During the Qing dynasty (1644–1911), a government agency called the Bureau of Imperial Gardens and Hunting Parks was established to manage royal reserves that were divided into (1) hunting enclosures, (2) enclosures for provision of the royal household, and (3) enclosures for military training.[8] Although regulations were mostly expressions of royal privilege, they demonstrate the early recognition that natural resources like forests and animal populations could be overexploited, and needed to be regulated to manage them sustainably.

It is debatable whether the system of royal forests set up around the world actually led to conservation of species, and some have argued that this view has been overly romanticized.[18] However, it is clear that the royal forests were a form of management in which use of resources was restrained (forcibly) and where some (common people) incurred costs to preserve the resource for use by others (royalty or wealthy individuals). Although the intent was rarely one of intergenerational sustainability of the forest resources, many reserves in the royal forests persisted for generations. In fact, the reserves continue to play a major role in European conservation, as many of those forests—some dating back to the twelfth century—are still intact today.

Foundations of forest conservation

While the royal forest system was primarily set up to conserve game populations for hunting by the elite, many conservation ideas that were established in the royal forests period began to show up in the management of areas being harvested for lumber. But the initial motivation for conservation in forests extended well beyond the effort to ensure a stable supply of wood. Rather, conservation was intended to mitigate against the detrimental impacts of logging on the environment. Early principles of conservation in forestry were applied in the British colonies of East India where scientific officers began to argue that protection of forests was necessary to prevent soil erosion, provide water for irrigation and drinking, maintain wood supplies, and prevent famine. Some colonial administrators also argued that certain intact forests should remain uncut because of their necessary role in ensuring a steady supply of rainfall in adjacent agricultural areas.[19]

In the forests of British India, a conservation ethic began to evolve that was based on three principles: (1) human activity damaged the environment; (2) there was a civic duty to maintain the environment for future generations; and (3) scientific, empirically based methods should be applied to ensure this duty was carried out. Such arguments led directly to the establishment of conservation ordinances. For example, on the Indian Ocean island of Mauritius, the French colonial administration in 1769 stipulated that 25% of landholdings should remain forested to prevent erosion, degraded areas should be planted with trees, and forests growing within 200 meters of water should be protected. In order to prevent water pollution and the destruction of fish populations, subsequent colonial governments passed laws in the late eighteenth century regulating the pollutants being discharged by sugar mills and other factories.[20]

One of the strongest proponents of early forest conservation was Sir James Ranald Martin (1796–1874), a surgeon in India who worked for the British East India Company. Martin was a prolific writer and publicized the environmental impacts of deforestation, particularly the impacts on human and public health.

He promoted this view in numerous reports that demonstrated the scale of damage caused by large-scale deforestation and desiccation, and he lobbied extensively for the institutionalization of forest conservation activities in British India through establishment of forest departments.[20]

The Madras Board of Revenue initiated local conservation efforts in 1842, headed by Alexander Gibson, a professional botanist who developed the first state-run forest conservation program based on scientific principles.[19] Following this, British scientists working in India issued a report in 1852 urging the establishment of forest reserves throughout the vast subcontinent—reserves that would be managed by professional foresters who would work to avert environmental calamities and economic losses. The British East India Company, which saw that conservation made economic sense, embraced the report. In 1855, Governor-General Lord Dalhousie introduced the first permanent, large-scale forest conservation program in the world. The British colonial system of forest reserves in India was soon replicated in other parts of the colonial world, including Southeast Asia, Australia, and Africa. It also had a strong influence on forestry practices that developed later in the United States and Canada.

Foundation of the "national park"

The idea of establishing nature reserves—land set aside and managed not just for religious purposes, or for hunting or extraction, but rather to protect a threatened species—has been around for only about four centuries. The formation of nature reserves became common in the sixteenth and seventeenth centuries as royalty and wealthy individuals became increasingly aware of the problem of species extinction. For example, the loss of wild cattle (*Bos primigenius*, also known as aurochs) from Europe in 1627 and the extinction of the dodo bird (*Raphus cucullatus*) in Mauritius in the 1680s made the realities of extinction apparent to everyone (**Figure 2.4**).

One of the earliest known nature reserves, the Białowieża Forest, was established by the Polish king in 1561 to address the problem of the decline and

(A)

(B)

Figure 2.4 (A) Roelandt Savery's figure of the dodo in his picture of the fall of Adam, in the Royal Gallery at Berlin. This illustration was painted using a live dodo that was brought to Europe in the early seventeenth century, before the species went extinct. (B) One of Europe's first nature reserves was established to protect the European bison (*Bison bonasus*) in Poland. (A, painting by Roelandt Savery, 1626.)

▶ **Video 2.1** Watch this video about Białowieża Forest, one of the earliest known nature reserves.
oup-arc.com/e/cardinale-v2.1

possible extinction of the European bison (*Bison bonasus*).[21] The Białowieża Forest represented one of the earliest deliberate European efforts to conserve a threatened species, and it was soon replicated with the Wood of The Hague—a reserve established by a Dutch prince in 1576—and by Karpfstock, a reserve established in Switzerland in 1576[3] (**Video 2.1**).

But in the case of early nature reserves, the areas established were small and had already been altered extensively by human habitation and development. The idea that large, untouched expanses of wilderness might be set aside for the purpose of conservation was not something that had generally been considered. That changed dramatically in the 1700s and 1800s when an entirely new paradigm in conservation was established: the idea of the national park.

Credit for the first national park goes to the Mongolians, who in 1783 declared a 673 km^2 area of Bogd Khan Mountain a nature reserve.[22] Bogd Khan Mountain had been declared a holy site and protected from logging and hunting as far back as the thirteenth century. But in the late 1700s it was set aside under the Qing dynasty to be preserved simply for its natural beauty, and it was later declared a biosphere reserve by the United Nations Educational, Scientific and Cultural Organization (UNESCO) in 1996.

Credit for setting aside large tracts of land to establish a system of national parks goes to the United States. The establishment of the U.S. national park system began in 1806 when John Colter, a member of the Lewis and Clark Expedition, left the expedition to become an independent fur trapper and explorer. As he was exploring an area of the western United States that is now Wyoming, he came across a region that he described as an area of incredible scenic beauty that included geothermal features, geysers, boiling pots of mud, petrified forests, and hot springs. Colter's stories about the area were not believed by most people, who named the "imaginary" place Colter's Hell. As a result, the area remained largely unknown for several more decades.[23]

In 1830, a mountain man and military scout named Jim Bridger explored the area again. Bridger reported observing boiling springs, spouting water, and a mountain of glass and yellow rock. But his reports were also largely ignored, apparently because he was known as one who told tall tales.[3]

Despite public skepticism, persistent rumors of the region's incredible natural wonders eventually led to the first detailed expedition to the Yellowstone area in 1870. The expedition was led by the surveyor general of Montana, Henry Washburn, and included Nathaniel Langford (who later became known as National Park Langford) and a U.S. Army detachment commanded by Lt. Gustavus Doane. The expedition spent about a month exploring the region, collecting specimens, and naming sites of interest[23] (**Video 2.2**).

▶ **Video 2.2** Watch this introduction to Yellowstone National Park.
oup-arc.com/e/cardinale-v2.2

During the expedition, team members were astounded by the natural beauty of the area, its remarkable wonders, and the abundant fish and wildlife. As they discussed what should be done with the area, Cornelius Hedges, a Montana writer and lawyer who was a member of the Washburn expedition, suggested they should not abandon the area to commercial development. Rather, Yellowstone should be set aside as a "national park" for the enjoyment of U.S. citizens as well as visitors from around the world. This proposition was likely inspired by Henry David Thoreau and George Catlin, whose writings about outdoor museums and wilderness parks had set the stage for a national discussion about parks.

Today, it might not sound outlandish to propose the formation of a national park. But in the 1870s, this suggestion by Hedges was considered truly crazy. Never before had any government in the world considered setting aside a large, economically valuable, untamed and unexplored landscape just so people could come visit it, view it, and enjoy it. That a government might preserve the best of its natural heritage simply to make it accessible to everyone was a truly revolutionary idea.[3]

Even so, members of the expedition were convinced by Hedges to support his proposition to form a national park, and they returned to the East to initiate a campaign for Yellowstone's preservation. Hedges wrote detailed articles about his observations that were published in newspapers between 1870 and 1871. His articles reinforced sentiments of then Montana Territorial Governor Thomas Francis Meagher, who had also commented that the region should be protected.[23]

The public education and lobbying proved successful. In 1871, American geologist Ferdinand Hayden was charged with leading America's first federally funded geologic survey into the Yellowstone region of northwestern Wyoming. During his exploration, Hayden compiled a comprehensive report detailing the remarkable features of the landscape. Importantly, his report was augmented by large-format photographs taken by the famous Civil War photographer William Henry Jackson and by dazzling paintings of Yellowstone by Thomas Moran. Hayden's beautifully illustrated report helped convince the U.S. Congress to withdraw this region from public auction and protect it. On March 1, 1872, Congress approved a bill creating Yellowstone National Park as "a public park or pleasuring-ground for the benefit and enjoyment of the people" (**Figure 2.5**).

The formation of Yellowstone National Park was a monumental turning point in the conservation movement. Never before had a government made a decision to forgo the utilitarian benefits of development and extraction solely to preserve the scenic beauty of a splendid landscape for the purpose of public enjoyment. The formation of Yellowstone National Park had a rippling effect, with the U.S. concept of the national park soon imitated throughout the world: Australia established its first park in 1879, Canada in 1885, numerous countries in Africa starting in 1890, and Sweden in 1903. The national park was also the inspiration for marine protected areas (MPAs), which are a modern conservation success story (**Box 2.1**).

Figure 2.5 The astounding beauty and unique geothermal features of Yellowstone contributed to its becoming the first national park in the U.S. The unique thermophilic (heat-loving) bacteria later discovered in these hot springs produced the enzymes that led to a modern revolution in genetic engineering and biotechnology.

BOX 2.1 Success Story

Marine Protected Areas

Marine protected areas (MPAs) are protected areas of seas, oceans, estuaries, or large lakes. They include marine reserves, fully protected marine areas, no-take zones, marine sanctuaries, ocean sanctuaries, marine parks, and locally managed marine areas. Inspired by national parks that protect wildlife and their habitats on land, MPAs have become an important tool for protecting ocean wildlife and their aquatic habitats.

The concept of protecting marine habitats from fishing and other human activities is not new. In certain island nations in Oceania (Polynesia, Melanesia, and Micronesia), measures to regulate and manage fisheries have been in use for centuries. However, during the 1950s and early 1960s, as marine ecosystems became more heavily exploited by fishing and affected by other human activities, the worldwide need to devise methods to manage and protect marine environments and resources became more apparent.[54]

The first conference on marine protected areas was sponsored by the International Union for the Conservation of Nature and Natural Resources (IUCN, also known as the World Conservation Union) in Tokyo in 1975.[55] The report of that conference called attention to the increasing pressures imposed by humans on marine environments and pleaded for the establishment of a system of MPAs that were representative of the world's marine ecosystems. In 1980, the IUCN, with the World Wildlife Fund and the United Nations Environment Programme, published the World Conservation Strategy, which emphasized the importance of marine environments and ecosystems in the overall goal of adopting conservation measures to ensure sustainable development. In 1982, the IUCN Commission on National Parks and Protected Areas (CNPPA, now the World Commission on Protected Areas) organized a series of workshops at the Third World Congress on National Parks, in Indonesia, to promote the creation and management of marine and coastal protected areas. This workshop resulted in the publication of *Marine and Coastal Protected Areas: A Guide for Planners and Managers*.[56]

Since the 1980s, the number of marine protected areas has increased rapidly. Nearly 7000 MPAs have now been established worldwide, which collectively protect about 4% of the world's oceanic habitat. Over the last decade, there has been a global trend toward establishment of very large MPAs that are 250,000 km^2 in size or larger.[57] This process began in 2000, when the Northwestern Hawaiian Islands Coral Reef Ecosystem Reserve, now known as the Papahānaumokuākea Marine National Monument, a World Heritage site, was formed. Since 2004, 10 large MPAs have been established that represent more than 80% of the worldwide ocean protection, most of them in the Pacific.[58]

While the marine protected area network is still in its infancy, the rising use of MPAs to protect the world's oceanic species and their habitats is a conservation success story that rivals the national parks' protection of the world's terrestrial species and habitats.

© NatPar Collection/Alamy Stock Photo

The Papahānaumokuākea Marine National Monument is a World Heritage site that was formed in 2000. It is one of ten very large marine protected areas in the world.

Intellectual foundations in the United States

Aside from the formation of the national park, several philosophies that underlie modern conservation, and events that influenced the formalization of conservation as a discipline, developed in the United States during the nineteenth and twentieth centuries. It's useful for future practitioners to understand, in some detail, the U.S. history of conservation that led to these philosophies and events.

Europeans first colonizing North America found a landscape that, by comparison with a highly exploited Europe, seemed pristine. Though Native Americans had made extensive use of fire to manage lands for both agriculture and game, their low population densities, limited industrial technologies, and propensity to move their settlements precluded widespread environmental destruction.

But during the early American colonial period (1500–1600s), more permanent settlements began to clear-cut New England forests to meet the domestic and European demand for lumber and to make way for agricultural land and farms to feed a population that reached 2.4 million by the late 1700s.[24] During this period, the North American fur trade became a lucrative enterprise that resulted in development of elaborate trade networks and companies set up to meet overseas demand for pelts and fur (**Figure 2.6A**).

By the end of the eighteenth century, habitat loss in North America, coupled with overexploitation of wildlife from trapping and hunting, had led to dramatic population declines of species like beaver (*Castor canadensis*), otter (*Lontra canadensis*), mink (*Neovison vison*), pine marten (*Martes americana*), fisher (*Pekania pennanti*), and wolverine (*Gulo gulo*).[25]

Early in the eighteenth century, settlers began to move west in search of new land and resources. Westward expansion was an explicit priority of early U.S. presidents, who felt the nation's future depended on settlement of new land and exploitation of new resources. This expansion was facilitated in the early nineteenth century by the Louisiana Purchase (1803), which doubled the size of the country. The War of 1812 subsequently secured existing U.S. boundaries and defeated native tribes of the Old Northwest, the region of the Ohio and Upper Mississippi Valleys. The Indian Removal Act of 1830 forcibly relocated Native American populations from the Southeast to the present states of Arkansas and Oklahoma. Then, in 1862, President Lincoln financially incentivized western migration by signing the Homestead Act, which gave American settlers ownership of 160 acres of public land in exchange for a five-year commitment to

(A)

Library and Archives Canada/C-001229

(B)

(C)

Wisc. Land and Lumber Co. Collection, IXL Museum, Hermansville, MI and Bently Hist. Lib., U of Mich.

Figure 2.6 Examples of gross overexploitation of nature: (A) Fur trader in Fort Chipewyan, Northwest Territories in the 1890s. (B) Man stands atop an enormous pile of buffalo skulls. (C) Logging thrived around the Great Lakes from the 1850s through the 1880s, clear-cutting whole forests from the landscape.

permanent residence. Collectively, these events gave rise to the term **manifest destiny**—the widely held belief that American settlers were destined to move west and occupy the land from coast to coast across North America and to become prosperous as they did so.

Western expansion across the United States coincided with the start of the American Industrial Revolution, and the two combined to greatly increase demand for timber. As Americans settled the Great Plains, they needed materials from the lumber-rich parts of the nation to build their cities. The burgeoning railroad industry needed lumber to build railcars and stations, fashion ties, and power trains. Even the coal-mining industry needed lumber to support its mining structures and create its own railbeds. To meet vastly increasing demand, the timber industry moved westward from the overexploited forests of New England into the Great Lakes and their tributary waterways, areas that were densely covered with virgin timber.

The predominant view of those participating in the westward expansion was that nature was limitless. The forests, prairies, and waterways were teaming with fish and wildlife, and they seemed inexhaustible. This false perception of unlimited natural resources led to numerous examples of grotesque overexploitation of fauna like the American bison (*Bison bison*), once the most abundant large mammal recorded in modern times, which was nearly hunted to extinction for meat, pelts, and other products (Figure 2.6B).

But by the mid-1800s there was widespread recognition that many wildlife populations in North America were in sharp decline and that rampant deforestation (Figure 2.6C) and agricultural development had led to unsustainable levels of erosion as well as widespread pollution of waterways by lumber mills. In 1871, the slash-and-burn forestry practices of Midwestern logging companies triggered public outcry after the Peshtigo forest fire in Wisconsin, the deadliest wildfire in recorded history, burned 6216 km^2 and killed somewhere between 1500 and 2500 people.

The unconstrained and sometimes shocking impacts of early Americans on the nation's forests, waterways, and wildlife led to the rise of **romantic transcendentalism**. Romantic transcendentalism began with the writings of Ralph Waldo Emerson, who published dozens of essays detailing his views about the relationship that should exist between humans and nature. In his 1836 *Nature*,[26] his first and most influential essay, Emerson suggested that the divine, or God, permeates all of nature. Therefore, nature not only provides material needs, but is the temple in which people commune with the spiritual world and achieve enlightenment. Emerson also believed that nature fulfilled one's aesthetic sentiment:

> To the attentive eye, each moment of the year has its own beauty, and in the same field, it beholds, every hour, a picture which was never seen before, and which shall never be seen again.[26]

R. W. Emerson

Another leading figure of romantic transcendentalism was Henry David Thoreau—an essayist, poet, and naturalist who was best known for his 1854 book *Walden*.[27] In *Walden*, Thoreau detailed his experiences over the course of two years and two months, living in a cabin he built near Walden Pond in Massachusetts, amidst woodland owned by his friend and mentor Ralph Waldo Emerson. Thoreau compressed his time at Walden Pond into a single calendar year and used the passage of four seasons to outline his philosophy of life, politics, and nature while endorsing the values of austerity, simplicity, and solitude. Thoreau repeatedly emphasized the minimalism of his lifestyle and the contentment he derived from it, and contrasted his own freedom with the imprisonment of others who devote their lives to material prosperity:

> I went to the woods because I wished to live deliberately, to front only the essential facts of life, and see if I could not learn what it had to teach, and not, when I came to die, discover that I had not lived.[27]

Thoreau's words articulated the concerns of many of his contemporaries that industrialization and development were permanently destroying the world around them and that the protection of nature was necessary to guarantee there would be places of peace and solitude where the simplicities of life could be enjoyed. Thoreau's experiment at Walden Pond inspired generations of others to follow his example and retire to an isolated spot—if only for a short time—to ponder the world and their place in it. His work also inspired a movement of minimalist lifestyles, driven by the recognition that great happiness and contentment in life can be achieved by those who live simply, with few material belongings (**Video 2.3**).

H. D. Thoreau

Video 2.3 Thoreau's simple life at Walden.
oup-arc.com/e/cardinale-v2.3

George Perkins Marsh's 1864 publication of *Man and Nature; or, Physical Geography as Modified by Human Action*[28] is considered by many to be the first great landmark in modern conservation literature.[29] Marsh wrote about observations he made as a youth growing up in Vermont. As he witnessed the destruction of New England's forests, he argued that humans were the agents of change, or "disturbing agents," and that human actions were causing widespread disruption of natural "harmonies":

> All nature is linked together by invisible bonds, and every organic creature, however low, however feeble, however dependent, is necessary to the well-being of some other among the myriad forms of life with which the Creator has peopled the Earth.[28]

Marsh's view contrasted with the conventional wisdom held by geographers at the time, who argued that changes to Earth were entirely the result of natural phenomena. Publication of Marsh's book provided direction and definition to the nascent conservation movement through the remainder of the 1800s and into the early 1900s.

Wilderness advocate, naturalist, and author John Muir later used transcendental themes in his campaigns to preserve natural areas. According to Muir, natural areas such as forest groves, mountaintops, and waterfalls have spiritual values that are superior to the material benefits obtained by exploitation.[30] Muir argued that all human beings benefit from the beauty of nature, saying:

> Everybody needs beauty as well as bread, places to play in and pray in, where nature may heal and give strength to body and soul.[30]

Muir was one of the first American conservationists to explicitly state that nature has **intrinsic value**—value that stems from its very existence and is entirely separate from any utilitarian value to humanity. Muir argued that nonhuman species have an equal place in God's creation of nature and to destroy them was to undo God's handiwork.[31]

John Muir spent much of his life lobbying for protection of land in the Sierra Nevada of California, and his activism helped preserve the Yosemite Valley, Sequoia National Park, and many other wilderness areas. In 1892, Muir founded and become the first president of the Sierra Club, an environmental organization that, to this day, continues to use Muir's ideas in their mission "to explore, enjoy, and protect the wild places of the earth."

Romantic transcendentalism, augmented by John Muir's views on the intrinsic value of nature, ultimately gave rise to the **preservationist ethic**, which represented the first of three environmental philosophies to shape conservation policies in the United States. The preservationist ethic is the view that the primary

John Muir

role of conservation is to preserve wilderness—large areas that remain essentially unoccupied, unmanaged, and unmodified by human beings. The preservationist ethic played a major role in establishment of the U.S. system of national parks, as well as the formation of the National Park Service that was established in 1872 to manage them. For this, Muir is often called the Father of the National Parks.

The preservationist philosophy was also instrumental in the passage in 1891 of the Forest Reserve Act, which gave U.S. presidents the authority to create forest reserves. More than 809,000 km^2 have been established by various presidents, and these forest reserves ultimately became the U.S. national forest system. Romantic transcendentalism and the preservationist ethic, though without the overt religious rationale of their early proponents, are still the basis for management by the U.S. National Park Service as well as for activism by many private conservation organizations throughout the world, whose goals are to save natural areas in a pristine state for their inherent value.

In the late 1800s, the United States saw the rise of a second philosophy of conservation—the **resource conservation ethic**—that stood in direct contrast with ideals of romantic transcendentalism and the preservationist ethic. Ironically, the catalyst for this transition was Theodore Roosevelt, the twenty-sixth president of the United States and perhaps the most important and influential conservationist in American history. Roosevelt was an avid outdoorsman who was friends with John Muir and heavily influenced by his preservationist ideals. Roosevelt had a love for nature and ascended to the presidency at a time when many private citizens were becoming increasingly vocal in opposition to the consumptive use of natural resources on public lands. Roosevelt himself viewed large mining and lumber corporations as unfair monopolies that profited from public resources. He was known as a "trust buster," and one of the goals of his administration became to use the power of the federal government to better manage public land for the good of the people.

Roosevelt signed into law bills that created five national parks; established the first 51 bird reserves, four game preserves, and 150 national forests; and established the U.S. Forest Service. The area of the country he placed under protection totaled more than 930,000 km^2. The irony of his legacy lies in his appointment of Gifford Pinchot as the first chief of the U.S. Forest Service. Pinchot was a U.S.-born forest scientist who began his career with the U.S. Department of Agriculture, but he quickly rose in prominence for his development of forest management policies. Pinchot had been trained in the German traditions of forest management by his mentor Dietrich Brandis, who worked with the British Imperial Forestry Service in colonial India. By the 1800s, German scientists like Brandis had developed advanced techniques for silviculture (tree growing) and were using accurate and predictive models to maximize the commercial yield of wood. They had already developed and begun to use the concept of **maximum sustainable yield**, which sought to harvest forest products as quickly as they regenerate. And they had begun to treat forestry much like agriculture, growing single-species stands of fast-growing trees to maximize wood production.

Teddy Roosevelt

Despite his appointment by Theodore Roosevelt, and despite the influence of John Muir early in his career, Pinchot saw timber in national forests as a resource to be utilized with careful application of science-based management.[29] He is known for developing the resource conservation ethic[32]—an ethic that argued that the proper use of natural resources is whatever will further "the greatest good of the greatest number [of people] for the longest time." Pinchot's resource conservation ethic was based on three principles: First, resources should be fairly distributed among individuals in the present generation as well as between present and future generations. This was the beginning of **sustainable forest management** in the United States. Second, resources should be used efficiently and not wasted. Third, resources should be managed scientifically using the best available data. Near the end of his career with the U.S. Forest Service, Pinchot established the Yale School of Forestry at Yale University—the first school of its kind in North America, and one that was deeply embedded in German techniques of forest management.[3]

Gifford Pinchot

After a period of global instability that included World War I (1914–1918), the Great Depression (1929–1941), and World War II (1939–1945), a third philosophy—the **evolutionary-ecological land ethic** of conservation—began to emerge in the United States.[33] The evolutionary-ecological land ethic was developed by Aldo Leopold, who was born and raised in Iowa. Leopold was trained as a forester at the Yale School of Forestry and then joined the U.S. Forest Service, where he quickly rose in rank to supervisor of Carson National Forest in New Mexico at the age of 24.

During his early years as a government forester, Leopold embraced Pinchot's resource conservation ethic. But he eventually became dissatisfied with the contemporary model of forest management, coming to believe it was inadequate to protect and sustain natural resources, because it viewed the land merely as a collection of individual goods that can be used in different ways. In fact, after visiting Germany later in life to learn more about their forest management practices, and being dismayed by their obsession with spruce monocultures, Leopold wrote, "Never before or since have the forests of a whole nation been converted into a new species within a single generation." The Germans had "taught the world to plant trees like cabbages."[34]

Leopold soon abandoned the resource conservation ethic and instead began to consider nature as a landscape that was organized as a system of interrelated processes.[35] He came to the conclusion that the most important goal of conservation is not to maximize a sustainable yield of a particular product. Rather, it is to maintain the health of natural ecosystems and the ecological processes they perform.[36] Leopold did consider humans to be part of the ecological community, as opposed to standing apart from and exploiting nature as the resource conservation ethic encouraged. In line with that thinking, he was committed to the idea that humans should be intimately involved in land management, and he sought middle ground between exploitation and total control over nature, on the one hand, and complete preservation of land with no human presence or activity, on the other (**Video 2.4**).

Video 2.4 Learn about Aldo Leopold and his land ethic.
oup-arc.com/e/cardinale-v2.4

Aldo Leopold's ideas were regularly published during his lifetime, but the posthumous publication of his now famous book *A Sand County Almanac* (1949) made them broadly available to the general public. His ideas had a major impact on American conservation, including passage of the U.S. Wilderness Act that was signed into law in 1964, creating the National Wilderness Preservation System that recognized wilderness as "an area where the earth and its community of life are untrammeled by man, [but] where man himself is a visitor who does not remain." The evolutionary-ecological land ethic has been influential with many conservation organizations worldwide, often taking the form of what is now called **ecosystem-based management** (Chapter 15), which has the goal to preserve the health, integrity, and functioning of entire ecosystems.

Contemporary Conservation Biology

Contemporary conservation biology began to emerge in the 1970s following a sustained period of worldwide economic growth that followed World War II. The so-called Golden Age (1950s–1970s) was characterized by unprecedented growth of manufacturing and industry; vast development of public works in energy, sanitation, and transportation systems; unrestrained urban sprawl; and widespread introduction of fertilizers, pesticides, and heavy machinery in agriculture. Economic growth in the United States, western Europe, and east Asia

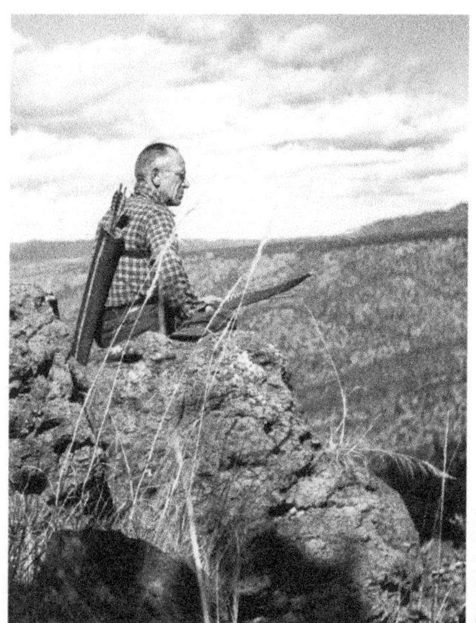

Aldo Leopold

as well as many countries that had been devastated by the war (Japan, West Germany, France, Italy, and Greece) was often promoted by deregulation of businesses and industry to encourage postwar prosperity.

The prosperity brought about by economic expansion was accompanied by a litany of environmental disasters in the 1950s, 60s, and 70s. These environmental disasters made clear to the general public the magnitude of environmental damage and costs that were being imposed on society by unrestrained growth and unscrupulous business practices. Table 2.1 gives a partial list of events that collectively helped trigger the modern environmental movement and formation of conservation biology as a global discipline.

TABLE 2.1 Examples of key environmental disasters that led to the formation of conservation biology as a global discipline

Year	Disaster	Description
1952	London's Great Smog	Thousands died and 100,000 fell ill after a blanket of smog covered London for five days in 1952. Cold weather combined with windless conditions allowed airborne nitrogen oxides, sulfur dioxide, and soot from coal burning to form a thick layer of smog over the city. Recent research has shown that 12,000 premature deaths can be partially attributed to this smog.
1953	Love Canal	In the 1940s, the Hooker Chemical Company began dumping 21,000 tons of industrial waste into the Love Canal near Niagara Falls, NY. After the site was later sold and developed, Love Canal residents reported strange odors and blue goo bubbling into their basements. High rates of asthma, miscarriages, mental disabilities, and other health problems brought Love Canal into the spotlight in 1978, and subsequent surveys found that 56% of children born there between 1974 and 1978 had birth defects. New York State Health Commissioner David Axelrod called the incident a "national symbol of a failure to exercise a sense of concern for future generations." The U.S. government soon after passed the Superfund law.
1954	Bikini Atoll	Between 1946 and 1958, the United States detonated 23 nuclear devices at Bikini Atoll. In 1954, one of those detonations exposed 23 crew members of the Japanese fishing vessel Lucky Dragon 5 to radioactive fallout, leading to acute radiation sickness. Years of nuclear testing so contaminated the soil and water at Bikini Atoll that the site was ultimately rendered a nuclear wasteland, with radiation levels in 2016 continuing to be above the established safety levels for human habitation. The United States ultimately paid the islanders and their descendants $125 million in compensation for damage caused by the nuclear testing program and displacement from their home island.
1956	Minamata disease	In 1956, Chisso Corporation's industrial wastewater containing highly toxic methylmercury was released into Minamata Bay and the Shiranui Sea. As the mercury bioaccumulated in shellfish and fish that were eaten by the local populace, individuals began to show signs of neurological disorder, including numbness and difficulty seeing, hearing, and swallowing. Initial symptoms were followed by severe convulsions, coma, and eventual death, and 2265 fatalities are recognized as a direct consequence of this pollution event.
1962	DDT	After World War II, use of the pesticide dichlorodiphenyltrichloroethane (DDT) was rampant. DDT was promoted as a wonder chemical: the solution to pest problems large and small. But in her groundbreaking 1962 book *Silent Spring*, Rachel Carson highlighted the dangers of DDT and the disastrous consequences of the indiscriminate use of insecticides on the environment and its wildlife. Carson's book, and testimony before the U.S. Congress, helped trigger the establishment of the U.S. Environmental Protection Agency, which subsequently banned the use of DDT.
1969	Santa Barbara oil spill	The third-largest oil spill in U.S. history fouled the Santa Barbara Channel off the coast of California. Prominent media coverage, with pictures of seabirds, dolphins, sea lions, and elephant seals covered in oil, sparked public outrage.
1969	Cuyahoga River	After being polluted for decades with industrial waste, an oil slick on Cleveland's Cuyahoga River caught fire. A subsequent picture in *Time* magazine of flames leaping from the water to engulf a ship helped catalyze the environmental movement in the United States.
1976	Seveso disaster	In July of 1976, an explosion at a chemical manufacturing plant north of Milan, Italy, released tetrachlorodibenzo-p-dioxin (TCDD) into the atmosphere, adversely affecting the nearby town of Seveso. Shortly thereafter 3300 animals died and many more were euthanized to prevent the spread of contamination into the food chain. Children were hospitalized with skin inflammation, and nearly 500 people were found to have skin lesions.
1978	Portsall oil spill	A huge crude oil tanker, the *Amoco Cadiz*, ran aground on Portsall Rocks off the coast of Brittany, France, split into three parts, and leaked 1,604,500 barrels of crude oil. The spill contributed to the largest loss of marine life ever registered from an oil spill. Millions of dead sea urchins and mollusks washed ashore after the event. More than 20,000 dead birds were recovered, and dead oysters were approximated at 9000 tons.

At the same time that these environmental catastrophes were occurring, researchers began drawing attention to further potential threats to the environment and human population. Paul Ehrlich's 1968 book *The Population Bomb*[37] revived Malthusian concerns about the impact of unconstrained population growth on natural resources. In 1972, a group of scientists and political leaders known as the Club of Rome published their report *The Limits to Growth*,[38] which drew attention to the growing pressure of human activities on natural resources. That same year, the United Nations Conference on the Human Environment was held in Stockholm and drew together representatives of governments from around the world to discuss the state of the global environment. This conference led to the creation of national and international environmental agencies and the United Nations Environment Programme (UNEP).

Formation of conservation biology as a discipline

By the early 1970s, scientists, the public, and policy makers around the world were well aware of the magnitude of the habitat loss and environmental degradation that were threatening natural ecosystems and their biodiversity. Even so, there was no central forum or organization to address the issue. The growing number of people thinking about conservation issues and conducting research needed to be able to communicate with each other to develop new ideas and approaches. Ecologist Michael Soulé organized the First International Conference on Conservation Biology in 1978, which met at the San Diego Wild Animal Park so that wildlife conservationists, zoo managers, and academics could discuss their common interests.[39] At that meeting, Soulé proposed a new interdisciplinary approach that could help save plants and animals from the threat of human-caused extinctions. Subsequently, Soulé, along with colleagues Paul Ehrlich of Stanford University and Jared Diamond of the University of California at Los Angeles, began to develop conservation biology as a discipline that would combine the practical experience of ecology, wildlife, forestry, fisheries, and national park management with the theories of population biology and biogeography. In 1985, this core group of scientists met at the Second International Conference on Conservation Biology, after which they founded the Society for Conservation Biology, representing the first global community of conservation professionals.

> For the record, the Society for Conservation Biology originated about 5 PM on May 8, 1985, in Ann Arbor, Michigan at the conclusion of the Second Conference on Conservation Biology...[40]

The conservation biology explosion

Immediately after its formation as a discipline in the 1980s, conservation biology began to grow rapidly. The new field seemed to fill a gap that policy makers, funding agencies, academic institutions, and the general public had been seeking. Within only a few years, the number of professional journals focused on conservation biology grew to more than two dozen, and membership in the newly formed Society of Conservation Biology increased by thousands, ultimately growing to more than 5000 scientists, resource managers, educators, government and private conservation workers, and students worldwide. In 2000, the society approved the creation of seven regional sections: Africa, Asia, Europe, Latin America, North America, Marine, and Oceania. Each of these regional sections created multiple chapters that were charged with achieving local conservation success.

New funding initiatives in conservation biology were established by international organizations like the World Bank, numerous national government agencies, and charitable organizations. For example, the Global Environment Facility, a special program established by the United Nations and the World

Bank on the eve of the 1992 Rio Earth Summit, has provided $12.5 billion in grants and leveraged $58 billion in cofinancing for almost 4000 conservation and environmental protection projects in over 165 countries. Major charitable organizations, such as the MacArthur Foundation, Ford Foundation, and Pew Charitable Trusts, have made conservation a high priority in their funding portfolios, allowing expansion of conservation programs in developing countries.

Formation of conservation biology as a discipline, and the rise of funding opportunities, fostered a major overhaul of the global educational system. Within a short time, more than 150 of the world's top universities had established degree-granting programs in some area of conservation, at both undergraduate and graduate levels. As the educational system expanded, conservation biology quickly moved beyond its roots in the biological sciences (population biology, taxonomy, ecology, and genetics) and fields of resource management (forestry, fisheries, and wildlife management) to include a much richer variety of disciplines (**Figure 2.7**). The number of practitioners from social sciences (human psychology, sociology, environmental economics, geography, and education) and the humanities (anthropology, history, law and politics, philosophy, literature, and the arts) has grown dramatically. These disciplines have proven crucial to the goals of conservation because the knowledge generated by the natural sciences is ultimately ineffective unless it becomes integrated into human social systems and decision making.

A global agenda for biodiversity conservation also emerged. In 1988, a group of experts working with UNEP proposed an international convention to focus on biodiversity sustainability. A few years later, this idea led to the Convention on Biological Diversity (CBD)—a multilateral treaty that was opened for signatures at the Earth Summit held in Rio de Janeiro in 1992. The CBD was adopted by nearly all countries a year later, all of whom agreed that conservation of biodiversity is "a common concern of humankind" and who pledged to common goals for the international conservation of biodiversity, the sustainable use of its components, and the fair and equitable sharing of benefits arising from genetic resources. The CBD was subsequently amended to include the Cartagena Protocol on Biosafety (adopted 2000), which seeks to protect biodiversity from the potential risks of organisms modified through modern biotechnology; the Gran Canaria Declaration (adopted 2002), which adopted a 16-point plan to slow the rate of plant extinctions around the world; and the Nagoya Protocol (adopted 2010), which established rules for access to genetic resources and for fair and equitable benefits of their development and use.

In 2010, the Conference of the Parties—the governing body of an international convention that included the CBD—met in Nagoya, Aichi Prefecture, Japan, and adopted a revised strategic plan for biodiversity. This plan included the Aichi Biodiversity Targets, which represented 20 measurable targets for

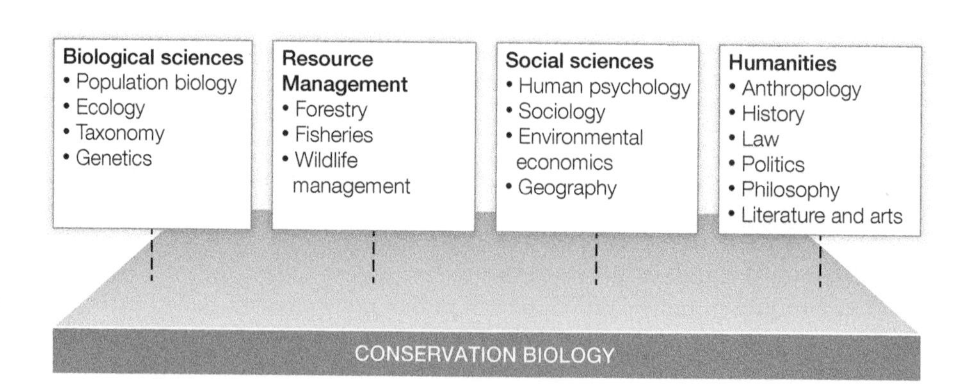

Figure 2.7 The growing interdisciplinary nature of conservation biology. (After G. P. Dietl. 2016. *Front Ecol Evol* 4: 21.)

curbing biodiversity loss, to be accomplished by 2020. In support of the plan, the United Nations decreed 2010 the Year of Biodiversity, and the period 2011–2020 the Decade on Biodiversity. During this period, several major international initiatives have formed. The Intergovernmental Science-Policy Platform on Biodiversity and Ecosystem Services (IPBES) was adopted in 2012 by 94 world governments. IPBES provides a mechanism to synthesize, review, assess, and evaluate information generated by governments, academia, scientific organizations, and nongovernmental organizations as well as Indigenous peoples and local communities on biodiversity and ecosystem services for the conservation and sustainable use of biodiversity.

In 2015, 194 countries of the United Nations General Assembly adopted the development agenda titled *Transforming Our World: The 2030 Agenda for Sustainable Development.* This document laid out 17 global goals with 169 targets for achieving global sustainability. Among them was the conservation and sustainable use of the oceans, seas, and marine resources (Goal 14) and the protection and sustainable management of terrestrial ecosystems to reverse land degradation and halt biodiversity loss (Goal 15).

Agreements in the Convention on Biological Diversity and sustainability goals of the United Nations have also been adopted by many conservation organizations, including large, established groups such as The Nature Conservancy, World Wildlife Fund, and BirdLife International. These organizations have become international in scope, with networks of projects and personnel throughout the world. The key point is that biodiversity conservation has become a prominent global theme, and most governments and major conservation organizations have now committed to common goals for curbing biodiversity loss.

Core principles and values of conservation

Now that conservation biology is well established globally as a field, what are its core principles and values? A few decades ago, Michael Soulé published a foundational essay titled "What Is Conservation Biology"[41] in which he laid out key characteristics of the discipline, as well as the core principles and values upon which the discipline should be founded. Soulé argued that conservation biology was more than just environmentalism, in which decision making was based largely on philosophical, religious, and ethical arguments (Figure 2.8). Rather, Soulé argued that conservation biology was a new scientific discipline—but one that differs from most other biological sciences in that it is a **crisis discipline**. In crisis disciplines—such as those related to human health (e.g., cancer biology or heart surgery) or national security (e.g., foreign policy or war game theory)—practitioners must take action even in the absence of complete information, because waiting to collect the necessary data may result in irreversible loss. Conservation biologists are often asked for advice and input by government and private agencies regarding issues such as the design of nature reserves, potential effects of introduced species, propagation of rare and endangered species, or ecological impacts of development. These decisions are usually politically and economically charged and cannot wait for detailed studies that could take months or years. Therefore, the "expert" is expected to provide quick, unambiguous answers even in the absence of complete knowledge. This is a major challenge for conservation biologists, who must walk a fine line between scientific credibility (and thus conservatism and possibly inaction) and taking action and providing advice based on general and perhaps incomplete knowledge, thereby risking their scientific reputations.

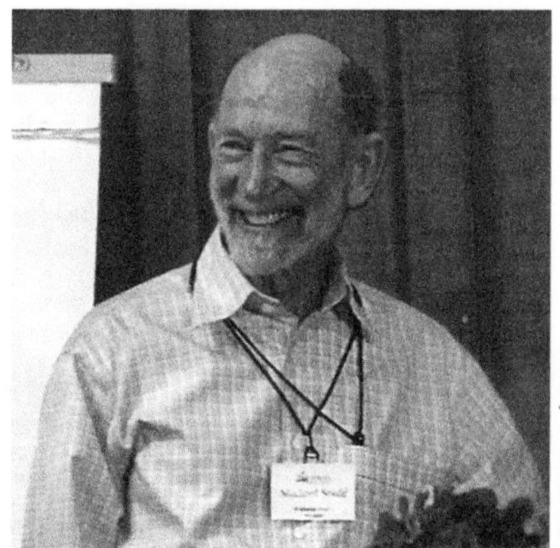
Courtesy of Michael Soulé

Michael Soulé

After describing some of the key characteristics of conservation biology, Soulé went on to outline what he called the functional and normative postulates

Conservation as...

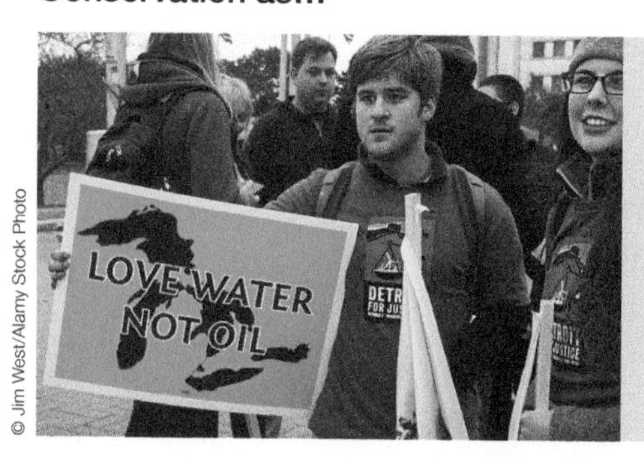

© Jim West/Alamy Stock Photo

Environmentalism

Motivations

- Humans are destroying nature.
- Destruction of nature is bad.
- We must save us from ourselves.

Inferences

- Philosophical and ethical arguments
- Focus on intrinsic values of nature

© Ann and Steve Toon/Alamy Stock Photo

A normative discipline (subjective, value based)

Motivation

- Humans activities are changing the planet.
- Environmental change is endangering biodiversity.
- Loss of biodiversity is bad.

Inferences

- Focus on both intrinsic and utilitarian values of nature
- Science used to justify conservation and guide management

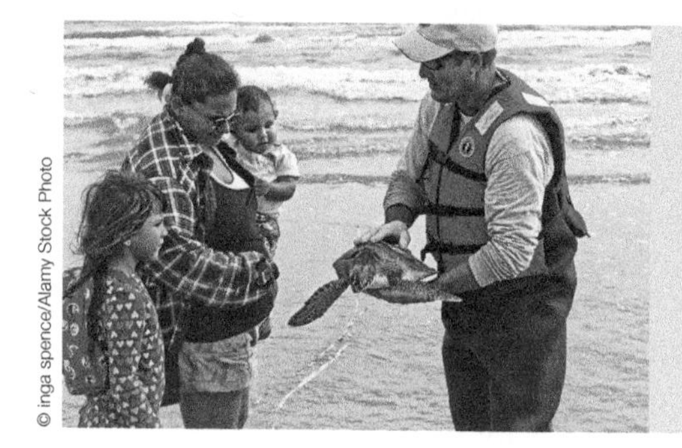

© inga spence/Alamy Stock Photo

A positive discipline (objective, evidence based)

Motivation

- Humans activities are changing the planet.
- Change is neither good nor bad, it just leads to societal trade-offs.
- Trade-offs can be quantified with data.

Inferences

- Science used to quantify trade-offs and help guide management
- Structured decision making integrates science with human values

Figure 2.8 Three distinct views of conservation. (Top) Conservation as environmentalism. (Middle) Conservation as a normative discipline. (Bottom) Conservation as a positive discipline.

that, in his view, represent the core principles and values of the field (**Box 2.2**). Soulé's normative postulates (values) included the following:

1. **Diversity of organisms is good.** Just as doctors and medical researchers use their knowledge to prevent illness because they value human health and want to prevent disease, Soulé argued that conservation biologists use their knowledge to prevent the loss of biodiversity because they believe the preservation of species and ecosystems to be *good* and untimely extinction to be *bad*.
2. **Ecological complexity is good.** Simplification of ecosystems by humans is *bad*, and nature and wilderness are preferable to artificial habitats. While biodiversity may be partially preserved in zoos and gardens, the ecological complexity that exists in natural communities is better at conserving biodiversity, but it will be lost without the preservation of natural areas.

BOX 2.2 Challenges & Opportunities

Two Views on the Functional Postulates (Core Principles) and Normative Postulates (Core Values) of Conservation Biology

While these two views about the functional and normative postulates of conservation are sometimes presented in the literature as if they are diametrically opposed, both viewpoints have certain strengths and weaknesses. We believe that the two views are likely to resonate with different groups of people. As such, they represent an opportunity for practitioners to find complementary motivations for conservation that can maximize success and impact.

Soulé. What Is Conservation Biology?
***BioScience* 1985[41]**

Functional postulates

- Many species that constitute natural communities are products of coevolution.
- Ecological processes often have thresholds, after which they become discontinuous, chaotic, or suspended.
- Genetic and demographic processes have thresholds below which nonadaptive, random forces prevail over adaptive, deterministic forces.
- Nature reserves are inherently disequilibrial for large, rare organisms.

Normative postulates

- Diversity of organisms is good.
- Ecological complexity is good.
- Evolution is good.
- Biotic diversity has intrinsic value, irrespective of its instrumental or utilitarian value.

Kareiva and Marvier. What Is Conservation Science?
***BioScience* 2012[50]**

Functional postulates

- Pristine nature, untouched by humans, no longer exists.
- The fate of nature and that of people are deeply intertwined.
- Nature can be surprisingly resilient.
- Human communities can avoid the tragedy of the commons.
- The "flat world" means that local conservation efforts are deeply connected to global forces.

Normative postulates

- Conservation must occur within human-altered landscapes.
- Conservation will be a durable success only if people support conservation goals.
- Conservationists must work with corporations.
- Only by seeking to jointly maximize conservation and economic objectives is conservation likely to succeed.
- Conservation must not infringe on human rights and must embrace the principles of fairness and gender equity.

3. **Evolution is good.** Evolutionary processes that allow species to adapt, and which are ultimately the foundation of all genetic diversification and speciation, are *good*. Therefore, continued evolution of populations in nature should be supported, in part by preserving genetic diversity and allowing dispersal and exchange of genetic material among populations.
4. **Biodiversity has intrinsic value.** There is inherent value in nonhuman life, and species and the ecosystems in which they live possess value by their very existence (intrinsic value) regardless of their economic, scientific, or aesthetic value to human society.

Many of Soulé's functional and normative postulates have been explicitly or implicitly accepted by practitioners in conservation biology, and they remain highly influential even today. But more than 30 years have passed since Soulé wrote his classic essay, and many of the ideas presented in his paper have been debated and revised as the field has matured. Some of Soulé's functional postulates have been questioned on scientific grounds, such as the idea that most species are highly specialized and irreplaceable,[42–44] or that nature is widely characterized by irreversible tipping points.[45–48] Soulé's normative postulates have also been the subject of much debate, as they represent little more than philosophical positions that can be neither proven nor disproven. As such, Soulé's early attempts to define conservation biology as a science did little to clarify how the field differs from environmentalism (see Figure 2.8).

In 2012, Peter Kareiva (then chief scientist for The Nature Conservancy) and Michelle Marvier (a professor at Santa Clara University) published a paper in which they revisited Soulé's classic paper and claimed that conservation biology's set of core principles needed to be updated and revised (see Box 2.2). Their proposed revisions were guided by two general arguments: First, they argued that while conservation biology remains a crisis discipline, it is now more objective and evidence based than in was in the 1980s (a *positive discipline*, see Figure 2.8). The authors argued that conservation biology has shifted from a reactive analysis of individual crises where decisions were once made by expert opinion and trial and error, to a proactive science in which we try to anticipate crises in advance and be prepared with scientifically grounded contingency plans.[49] As such, the field no longer needs to rely so heavily on normative postulates.

Second, Kareiva and Marvier[50] argued that conservation biology's core values are now more human-centric than Soulé's original postulates. In their view, "the major shortcoming of Soulé's framing of conservation is its inattention to human well-being." Instead, Kareiva and Marvier argued that the fate of people and nature are now deeply intertwined, and we need to consider a more personal role for people in conservation of human-dominated landscapes. This led them to a set of revised normative posulates (see Box 2.2):

1. **Conservation must occur within human-altered landscapes.** Human domination of the planet is now sufficiently pervasive that pristine habitats free of human influence no longer exist. The idea of saving biodiversity by setting aside large swaths of land or ocean may still be the best form of conservation, but it is an ever-decreasing option in many parts of the world.

2. **Conservation will be a durable success only if people support conservation goals.** Conservation is, at its core, an expression of human values. For better or worse, people's attitudes and actions ultimately determine what will be consumed, what will be converted, and what will be left behind for future generations. Therefore, the ethical, psychological, and physical reasons for people's actions and views of nature must be considered in conservation planning and decision making.

3. **Conservationists must work with corporations.** This one is hard to swallow for many practitioners, since corporate practices are often the most damaging to the environment and destructive to biodiversity. But Kareiva and Marvier argued there is a reality we must face: Through the resources they use and the wastes they produce, corporations drive most of what happens to our lands and waters. Furthermore, corporations represent livelihoods. One cannot realistically expect corporations to go away, nor do we want them to if we care about people having jobs. Therefore, conservationists need to stop vilifying large corporations and start working with them to improve their practices, reduce their impacts, and instill better environmental ethics.

4. **Only by seeking to jointly maximize conservation and economic objectives is conservation likely to succeed.** Conservation actions and policies can negatively affect people's lives and livelihoods. There are many examples of people losing jobs and of communities being unjustly displaced and disrupted for the creation of protected areas (Chapter 14). The fact that conservation sometimes disadvantages people highlights the need for conservation practitioners to learn how to better find win-win scenarios, for people and for nature.

5. **Conservation must not infringe on human rights and must embrace the principles of fairness and gender equity.** More often than not, the people who suffer in the name of conservation are poor and politically marginalized. Conservation biologists need to have a much better sense of social and environmental justice. When there are costs to conservation, conservationists must find ways to ensure that those costs are borne by people who can afford them and ensure there is equality for marginalized peoples.

Kareiva and Marvier's paper stimulated considerable discussion and debate among practitioners. Some felt the paper misrepresented the history of conservation biology, and they argued that the field has always been evidence based and focused on human well-being.[51] Others argued that Kareiva and Marvier's proposition to focus less on preserving wilderness, and more on conserving diversity in human-dominated habitats, was misguided and that it ignored a long history of successful conservation efforts.[52] Still others felt that Kareiva and Marvier's revised postulates shifted the motivation for conservation too far toward utilitarian values, particularly economic, and ignored the intrinsic values of nature that so often motivate people to conserve.[52,53] The debate that followed seemed to pit Soulé's[41] paper against Kareiva and Marvier's paper, often presenting the two as if they were diametrically opposing views that were vying for the heart and soul of conservation.

But as often happens with debates after the arguments settle, many practitioners came to the conclusion that both points of view have certain strengths and weaknesses, and they don't necessarily need to represent a dichotomy. People are motivated to conserve biodiversity for a whole variety of reasons, some of which are based on normative (subjective, value based) viewpoints, and others of which are based on positive (objective, evidence-based) viewpoints (see Figure 2.8). While one type of motivation is sometimes emphasized over the other, rarely do we treat those motivations as if they were mutually exclusive in any real act of decision making. In fact, more often than not, they are complementary to each other, because different people respond to different value systems. But to properly take account of these differing value systems, practitioners need to understand how to use decision making tools that integrate objective evidence-based views with subjective value-based views. One such tool is called *structured decision making.*

Structured Decision Making

Structured decision making (SDM) is an approach for the careful and organized analysis of problems relating to natural resource management decisions that integrate objective, evidence-based views with subjective, value-based views. SDM has roots in economics that date back to the 1940s, but the methods have expanded for use in many different fields, with contributions from cybernetics, business, mathematics, human psychology, risk analysis, and statistics. In the field of conservation biology, the goals of SDM are to:

1. integrate data-driven scientific evidence with human values,
2. make conservation and management decisions transparent, and
3. help ensure that decisions are repeatable so that different managers who are faced with the same information can come to the same decision.

The primary role of science in SDM is to help generate alternative actions, predict the consequences of those actions, and then coherently integrate value judgments and technical judgments through reasoned use of decision analysis tools.

Rather than being a single method, SDM is used to describe a suite of decision-making tools that share a set of common features. These include: (1) engaging stakeholders, experts, and policy makers in a joint decision-making process, (2) allowing all parties to clearly articulate and evaluate their differing objectives and desired outcomes by (3) working through a formal process to identify and evaluate alternative management options (4) using data and models to quantify the consequences of different management actions while also (5) optimizing trade-offs among alternatives that occur due to different human values. One of the most common forms of SDM used in conservation biology, called the PrOACT model, is described in **Box 2.3**.

Suggested Exercise 2.2 After reading Box 2.3, you are encouraged to visit the Companion Website for a practice exercise in the use of PrOACT. **oup-arc.com/e/cardinale-ex2.2**

The ultimate goal of SDM is to bring together stakeholders that have differing views, goals, and value systems to work toward a common conservation objective. This is a theme we will emphasize in multiple parts of this book. For example, in Part II we will discuss three human value systems that are commonly used to justify the conservation of nature and its biodiversity (intrinsic, relational, and instrumental). Chapter 5 will give an overview of these value systems and detail how they are influenced by individuals' ethical worldviews (anthropocentrism, biocentrism, and ecocentrism) that control the way they see their relationship to nature and its biota. In Chapters 6 and 7, we'll turn attention toward more utilitarian arguments for conservation to learn about how biodiversity affects the functioning of ecosystems, the goods and services they provide to society, and the quantifiable economic values these have to society. Finally, we'll describe additional decision-making tools such as cost-effectiveness analysis (CEA), cost-benefit analysis (CBA), SMART analysis, and multicriteria decision analysis (MCDA). When students master tools that take into account different value systems and ethical worldviews, then it becomes easier to generate win-win scenarios for both humans and nature, and the chances of achieving successful conservation efforts improve greatly.

Our final take-home message of Chapter 2 is this: it is a great time to be a conservation biologist! The field is more rigorous, more committed to finding win-win scenarios for people and nature, more inclusive of a variety of disciplines and expertise, and growing more quickly in membership, job prospects, and global impact than ever before. So let's now put history aside and focus on getting you ready to join this exciting field!

BOX 2.3 Conservation in Practice

Structured Decision Making (SDM)

Structured decision making (**SDM**) is a general term used to describe a suite of methods and models that help frame and organize the analysis of problems in natural resource management. The goal of SDM is to bring together stakeholders that have differing views, goals, and value systems, and provide a framework that allows them to work toward a common conservation objective while optimizing trade-offs among their differing objectives in a transparent and repeatable way.

One framework for structured decision making is called the PrOACT model, which is an acronym that helps one remember the five-step process for decision making:

- ***Pr**oblem definition*. What specific decision has to be made? What are the spatial and temporal scopes of the decision? Will the decision be iterated over time?
- ***O**bjectives*. What are the management objectives? Ideally, the objectives are stated in quantitative terms that relate to metrics that can be measured. Setting objectives should be informed by legal and regulatory mandates as well as stakeholder viewpoints. Because we rarely know precisely how management actions will affect natural systems, objectives are frequently made in the face of uncertainty, which makes choosing among alternatives more difficult. A good decision-making process will confront *uncertainty* explicitly and evaluate the likelihood of different outcomes and their possible consequences.
- ***A**lternatives*. What are the different management actions to choose from? This element requires explicit articulation of the alternatives available to the decision maker. The range of permissible options is often constrained by legal or political considerations, but structured assessment may lead to creative new alternatives.
- ***C**onsequences*. What are the consequences of different management actions? How much of the objectives would each alternative achieve? In structured decision making, we predict the consequences of the alternative actions with some type of model. Depending on the information available or the quantification desired for a structured decision process, consequences may be modeled with highly scientific computer applications or with personal judgment elicited carefully and transparently. Ideally, models are quantitative, but they need not be; the important thing is that they link actions to consequences.
- ***T**rade-offs*. If there are multiple objectives, how do they trade off with each other? In most complex decisions, the best we can do is choose intelligently between less-than-perfect alternatives. Numerous tools are available to help determine the relative importance or weights among conflicting objectives and to then compare alternatives across multiple attributes to find the best compromise solutions. Trade-offs always entail some measure of *risk tolerance*. Understanding the level of risk a decision maker is willing to accept, or the risk response determined by law or policy, will make the decision-making process more objectives driven, transparent, and defensible.

The ultimate goal of a structured decision-making framework is to (1) integrate data-driven science with human values, (2) make conservation and management decisions transparent, and (3) help ensure that decisions are repeatable so that different managers who are faced with the same information can come to the same decision. The primary role of science in this process is to help generate alternative actions, predict the consequences of those actions, and coherently integrate value judgments and technical judgments through reasoned use of decision analysis tools.

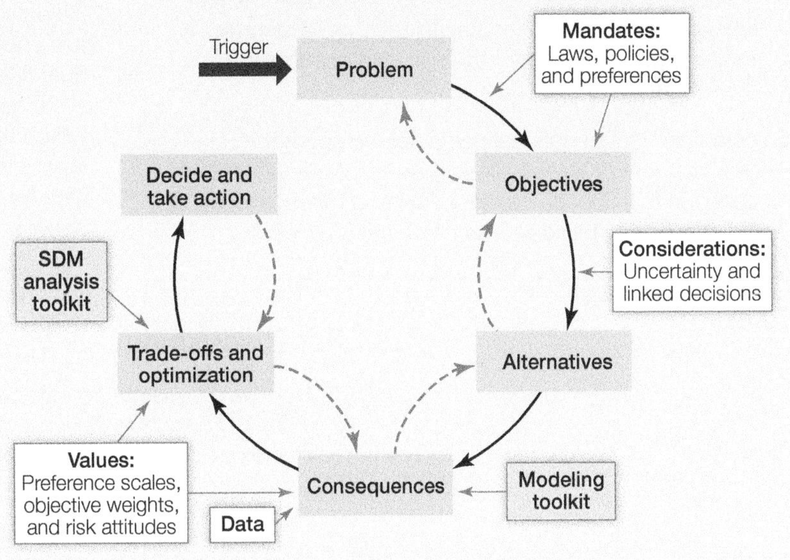

The steps of structured decision making (SDM): The ProACT sequence. (After J. F. Cochrane. 2008. *Structured Decision-Making Fact Sheet*. USGS.)

Summary

1. Conservation biology is an interdisciplinary field that aims to prevent the modern biodiversity crisis by protecting species, their genetic diversity, populations, habitats, and ecosystems while, at the same time, maintaining human well-being. The field seeks to investigate and describe the full variety of life on the planet, to evaluate and predict the effects of human activities on biodiversity, and to develop practical approaches to protect and manage biodiversity sustainably.

2. Environmental destruction and biodiversity loss have accompanied human societies as far back as prehistoric times. While the belief systems, religions, and myths of many early human hunter-gatherer societies and stable agricultural societies tended to emphasize conservation themes and the wise use of natural resources, few early human societies actually practiced conservation.

3. Some of the earliest examples of conservation can be seen in the formation of sacred sites, such as the *fengshui* forests in China, which were part of a religious philosophy to promote harmony between humans and their surrounding environment. Early examples of conservation for the purpose of protecting species included establishment of game and nature reserves set up as royal forests in Europe and Asia to protect the hunting rights of royalty and other wealthy individuals.

4. Conservation changed dramatically in the 1700s after Mongolia established the first national park. This was followed by the establishment of an entire system of national parks in the United States, which was the first time any government had set aside large, economically valuable tracts of land for the purpose of public enjoyment.

5. Establishment of national parks helped spark the development of three philosophies of conservation that continue to influence conservation and management approaches to this day. The preservationist ethic argues that the primary role of conservation is to preserve wilderness—large areas that remain essentially unoccupied, unmanaged, and unmodified by human beings. The resource conservation ethic argues that natural resources are to be extracted and used efficiently and justly for the good of people. The evolutionary-ecological land ethic argues that the goal of conservation is to manage nature in a way that maintains the health of natural ecosystems and the ecological processes they perform.

6. Contemporary conservation biology began in the late 1970s when Michael Soulé organized the First International Conference on Conservation Biology and proposed a new interdisciplinary approach that could help save plants and animals from the threat of human-caused extinctions. The Society for Conservation Biology was founded in 1985 after the Second International Conference on Conservation Biology, and it soon became wildly successful. The new professional society grew rapidly, large funding initiatives were formed, universities around the world established new degree programs, the field expanded to include the social sciences and humanities, and an international agenda for biodiversity conservation was signed onto by nearly every country and every major conservation organization.

7. The core principles and values that guide contemporary conservation biology are a mix of normative (subjective, value-based) postulates that assume biodiversity is inherently good and must be saved, and positive (objective, evidence-based) postulates that do not assume inherent value in nature, but instead focus on quantifying the trade-offs among different human value systems. The most influential practitioners of the future will be those who understand the variety of philosophical viewpoints that are used to motivate conservation and who have mastered tools from both the normative and positive views of the discipline.

For Discussion

1. How is conservation biology fundamentally different from other branches of biology, such as physiology, genetics, or cell biology? How is conservation biology different from environmentalism?

2. What do you think are the major conservation and environmental problems facing the world today? What are the major problems facing your local community? What ideas for solving these problems can you suggest?

3. Consider the public land management and private conservation organizations with which you are familiar. Would you consider their guiding philosophies to be closest to the resource conservation ethic, the preservation ethic, or the

evolutionary-ecological land ethic? What factors allow them to be successful or limit their effectiveness? Take some time to learn more about these organizations through their publications and websites.

4. How would you characterize your own viewpoint about the conservation of biodiversity and the environment? Which of the religious, philosophical, and scientific viewpoints of conservation biology described in this chapter do you agree or disagree with? How do you, or could you, put your viewpoint into practice?

Suggested Readings

Beissinger, S. R. et al. (Eds.). 2016. *Science, Conservation, and National Parks.* Chicago, IL: University of Chicago Press. Set of chapters describes the fascinating history of national parks and their impacts on conservation.

Bennett, N. J. et al. 2017. Conservation social science: Understanding and integrating human dimensions to improve conservation. *Biological Conservation* 205: 93–108. Excellent paper outlining how integration of social and natural science makes conservation efforts more successful.

Caro, T. et al. 2012. Conservation in the Anthropocene. *Conservation Biology* 26: 185–188. Emphasizes that although humanity now dominates most areas on Earth, conservation still requires planning for the places where human influence is minimal.

Leopold, A. 1949. *A Sand County Almanac.* New York, NY: Oxford University Press. Want to be inspired? This is an inspirational must-read for every budding practitioner of conservation.

Meine, C. 2013. Conservation movement, historical. In S. Levin (Ed.), *Encyclopedia of Biodiversity* (2nd ed.). Oxford, UK: Elsevier. A well-written and thorough description of the history of conservation.

Van Dyke, F. 2008. The history and distinctions of conservation biology. Chapter 1, pages 1–28, in *Conservation Biology* (2nd ed.). Dordrecht, The Netherlands: Springer. Another well-written and thorough description of the history of conservation.

Key Journals in the Field

Animal Conservation
Aquatic Conservation
Biodiversity and Conservation
Biological Conservation
BioScience
Bird Conservation International
Conservation Biology
Conservation Evidence
Conservation Genetics
Conservation Letters
Diversity and Distributions
Ecological Applications
Ecology
Ecology Letters
Environmental Conservation
Global Change Biology
Insect Conservation and Diversity
Journal for Nature Conservation
Journal of Applied Ecology
Nature
Oryx
PLOS Biology
PLOS ONE
Proceedings of the National Academy of Sciences
Proceedings of the Royal Society B
Science
Trends in Ecology & Evolution
Tropical Conservation Science

Visit the
Conservation Biology
Companion Website
oup.com/he/cardinale1e
for videos, exercises, links, and other study resources.

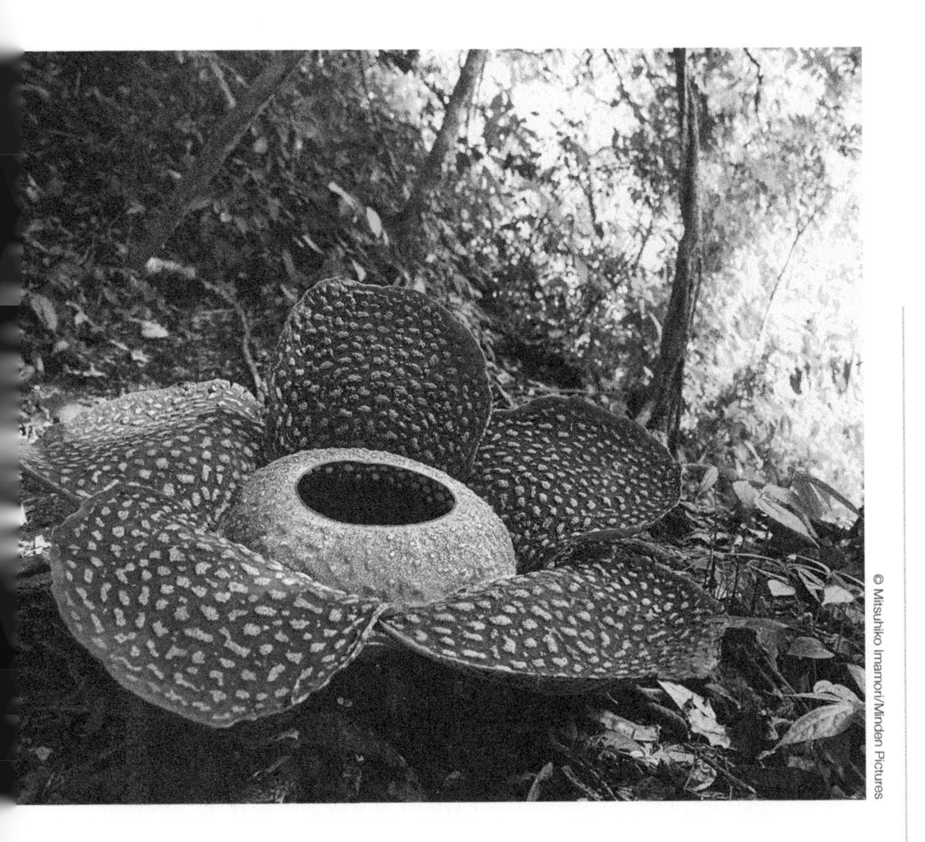

© Mitsuhiko Imamori/Minden Pictures

3

Biodiversity Concepts and Measurement

Rafflesia arnoldii is thought to be the largest flower on the planet, growing up to 3 feet in diameter and weighing up to 24 pounds. The flower emits pungent odors of decomposing flesh to attract pollinators, giving it the nickname corpse flower. Several of the 28 species of *Rafflesia* are endangered in Southeast Asia because of habitat loss.

What exactly is meant by the word *biodiversity*? This seemingly simple question is harder to answer than you might think. If you were to look in a variety of journal articles or textbooks dealing with biodiversity, you would be likely to see dozens of different definitions of the term. The surplus of options would grow considerably larger if you expanded your search to public domains such as websites or popular magazines. Most definitions would have common words and themes, yet many would differ in substantial and important ways. To understand why there is so much diversity in the use of the word *biodiversity*, it is useful to understand the history of the term.

The phrase **biological diversity** was first used in the early 1980s by tropical conservation biologist Thomas Lovejoy to describe the millions of different species that comprise life on Earth. Soon after, scientists decided that *biological diversity* was too long and cumbersome to effectively communicate to a wide audience. So in 1985, the two words were shortened to **biodiversity**, which was first used by Walter Rosen while planning the 1986 National Forum on Biological Diversity, organized by the U.S. National Research Council. Soon after that, the term appeared in a 1988 publication when the renowned biologist Edward O. Wilson used it as the title of his book based on the forum proceedings, *Biodiversity*.[1]

Perhaps because of its simplicity, the term *biodiversity* quickly became popular among biologists, environmentalists, political leaders, and the public. But as it grew in popularity, people started to use it to refer to many different aspects of life on Earth. For example, it was sometimes used to refer to biological variation per se—the number of different types of organisms living in a system of interest. Other times, it was used to refer to biological composition—the particular types of organisms living in a system. The term was further used to describe the variation and composition of life at all levels of biological hierarchy, from genes to landscapes, and across all spatial and temporal scales, from microbes presently living on a sand grain to the sum total of all organisms to ever inhabit the planet. In some instances, the term *biodiversity* was used as a synonym for *life* or *wilderness*.

The problem with using a popularized term like *biodiversity* in such broad and varied ways is that it becomes hard to exclude anything from a concept that refers to everything. Sarkar[2] argued that when we interpret *biodiversity* loosely to refer to so many different aspects of life spanning all biological levels and scales, the term becomes synonymous with "all of biology." At that point, the term is scientifically useless because it becomes impossible to formulate an objective definition that allows us to actually measure biodiversity and protect it.

Rather than dwell on the meaning of a word, we begin this chapter by suggesting that students embrace a broad definition of *biodiversity* that allows for effective communication of ideas about the variety of life on Earth to a broad audience. At the same time, students should understand that as practitioners in the future, you will routinely use far more precise definitions that allow one to quantify specific aspects of biological variation for groups of organisms that co-occur at a specific place and time. The remainder of this chapter is dedicated to laying out some of the key components of biodiversity that conservation biologists routinely attempt to manage, describing how those components are quantified at appropriate scales.

Hierarchical Levels of Biodiversity

Biological variation is a continuum, and it is notoriously difficult to divide life into discrete categories. Nevertheless, we tend to do so anyway, if for no other reason than to help organize our thinking, facilitate research, and define boundaries for conservation and management (**Figure 3.1**). Three common components of biodiversity are the focus of conservation efforts:

1. **Genetic diversity** refers to the genetic variation within species, both among individuals within single populations and among geographically distinct populations.

2. **Species diversity** refers to the variety of species that comprise a biological community—the collection of species that occupy and interact in a particular location.

3. **Community and ecosystem diversity** refers to the different biological communities and their associated ecosystems that comprise whole landscapes.

All three levels of biodiversity are necessary for the continued survival of life as we know it, and all are important to people[3,4] (Chapters 5–7). Genetic diversity is ultimately the source of all variation in life-forms on Earth, and it is required for organisms to maintain reproductive vitality, resistance to disease, and the ability to adapt to changing conditions.[5] Species diversity reflects the entire range

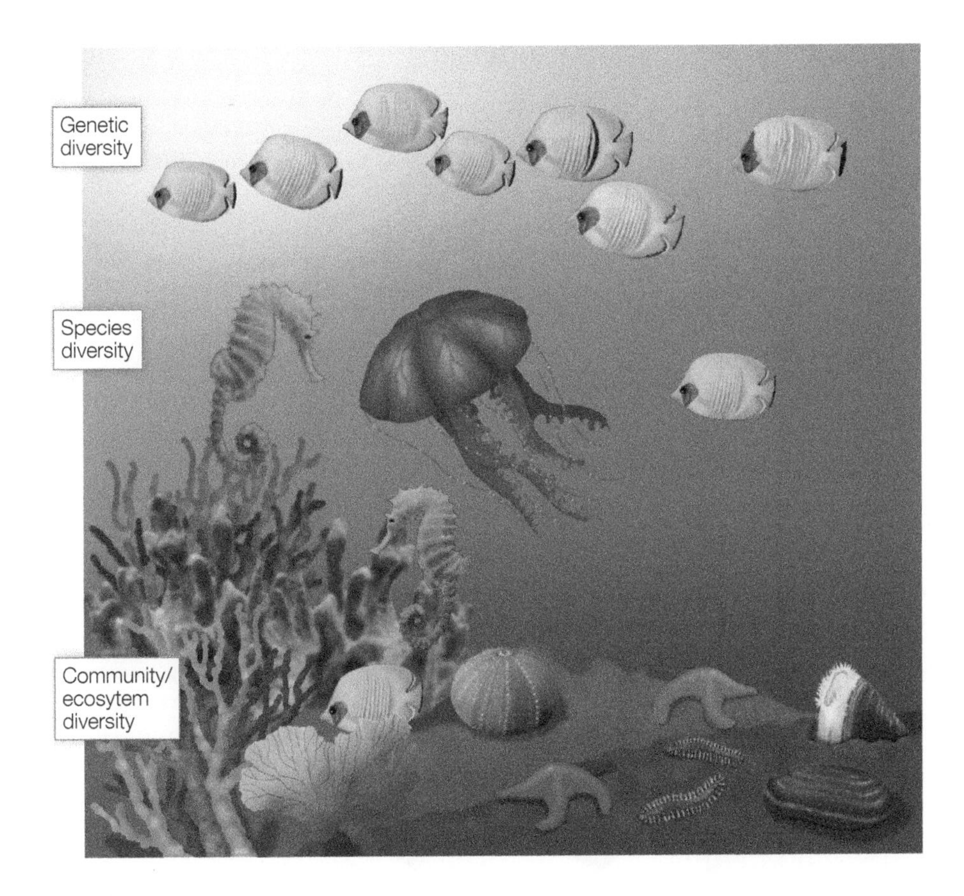

Figure 3.1 Hierarchical levels of biodiversity. Biodiversity includes genetic diversity (the genetic variation that creates phenotypic variation within species), species diversity (the variety of species that comprise a biological community), and community/ecosystem diversity (the different biological communities and their associated ecosystems that comprise whole landscapes).

of evolutionary and ecological adaptations of organisms to their environment, and it provides people with resources and resource alternatives. Community diversity underlies the proper functioning of whole ecosystems, which provide crucial services to people, such as water for drinking and agriculture, flood control, protection from soil erosion, and filtering of air and water. The sections that follow describe these components of biodiversity in greater detail and show how they are routinely measured.

Genetic Diversity

Genetic diversity is the total number of genetic characteristics in the genetic makeup of a species. It represents variation in genetic characteristics among individuals in a single population, among subpopulations of the same species, or among species.

Sources of genetic variation

A **population** is a group of individual organisms of the same species living within a given (or particular) geographical area at the same time. The geographical area is usually defined to be of a size in which individuals are likely to find mates and reproduce. Species that are geographically widespread are often subdivided into distinct breeding groups that live in different geographical areas. These geographically separated groups are sometimes referred to as different populations of the same species or as subpopulations of the species.

Biological variation among individuals within a population, and among geographically separate populations, is regulated, in large part, by their genetic

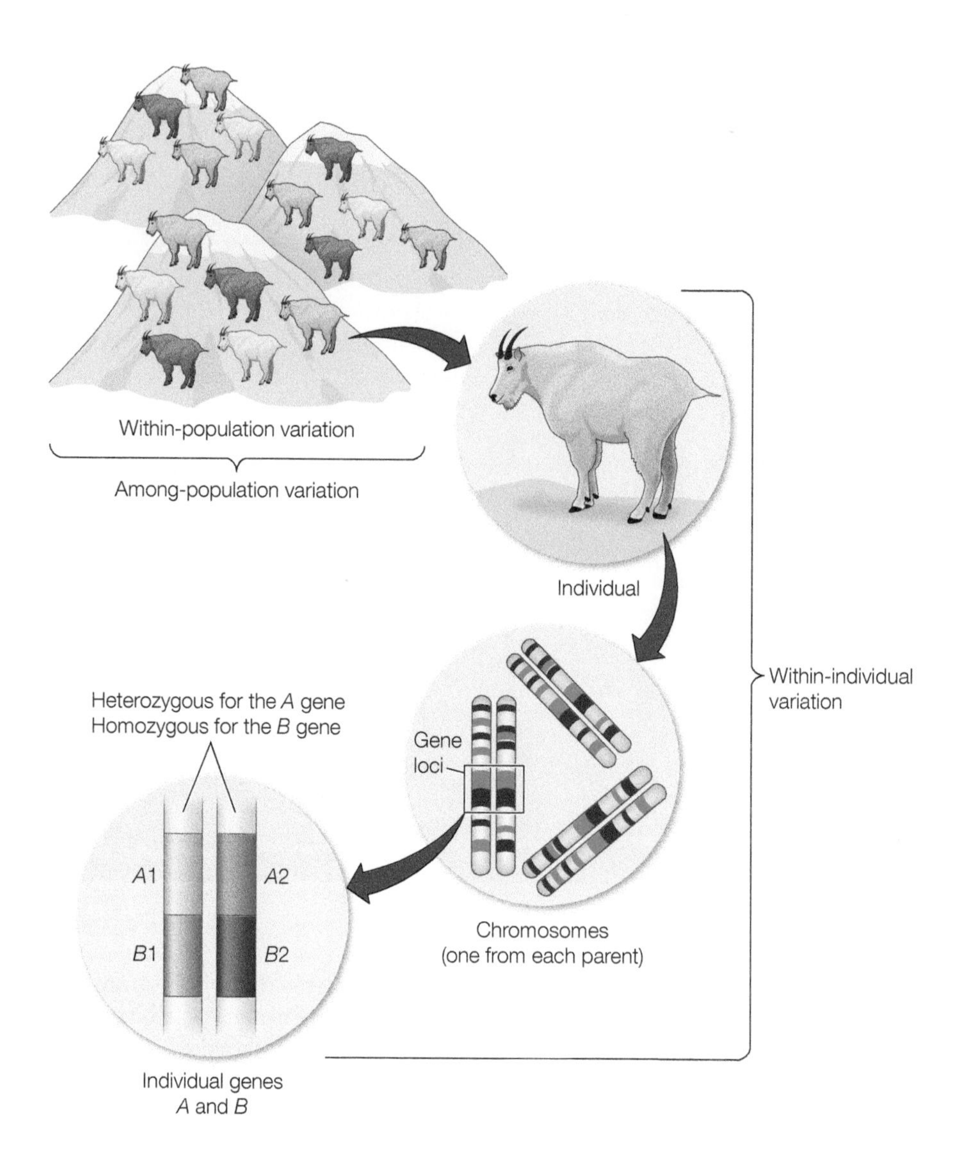

Figure 3.2 Genetic variation occurs between individuals that comprise a population of a single species. It is due to variation in the alleles found at particular gene loci, and to variation between chromosomes. Genetic variation also occurs between spatially separated populations of the species.

Video 3.1 View this refresher video on chromosomes, DNA, genes, and alleles.
oup-arc.com/e/cardinale-v3.1

diversity (Figure 3.2). While individuals in a population of a sexually reproducing species all have the same set of homologous chromosomes, genetic diversity among individuals in a population arises because the **genes** on those chromosomes are represented by different **alleles** (**Video 3.1**). Alleles are different forms of the same gene—sequences of DNA that are found at the same gene **locus** (physical, fixed location where a gene sits on a chromosome) but that code for the production of different proteins. Different proteins can produce different **phenotypes** that cause variation in the morphological, physiological, and biochemical characteristics of an individual. For diploid organisms, a group that includes nearly all mammals and many other vertebrates, an individual carries two copies of each chromosome, one inherited from the mother and one inherited from the father. When alleles, let us say of the *A* gene, are the same at the *A* gene locus (e.g., *A1A1*), the individual organism is said to be **homozygous** with respect to that gene. When the alleles are different (e.g., *A1A2*), the individual is said to be **heterozygous** with respect to that gene. At the locus level, a heterozygous individual offers more genetic diversity to future generations. The collection of alleles across multiple gene loci is referred to as an individual's **genotype**.

While the collection of alleles that make up an individual is referred to as the genotype, the total array of genes and alleles that are found across individuals in an entire population is called the **gene pool**. In practice, conservation biologists usually focus on measuring alleles found at a select subset of gene loci that are of practical interest for management—for example, those loci that influence the reproductive success, growth, or development of the species (see Chapter 13). To measure the alleles found at gene loci in a population, a biologist will typically collect a sample of individuals from the population and determine the genotype of each individual, using any of several common methods (e.g., protein electrophoresis or DNA sequencing). Based on the genotypes of this sample of individuals, the biologist will estimate the genetic diversity of the population, which provides the raw material that allows species to adapt to changing conditions and an uncertain future.

Measures of genetic diversity

Figure 3.3 illustrates how we might quantify genetic diversity for a population of frogs living in a pond. Assume that we randomly sampled just five frogs from the pond. Of course, this is a small sample of the population, and if we were doing a real study, we would gather a larger, more representative sample. But for sake of illustration, assume we performed DNA sequencing on tissue samples from just five frogs to determine their genotypes at two gene loci. Locus *A* represents a gene that codes for a protein that influences body size, and our sequencing of the gene detects two alleles—*A1* and *A2*. Locus *B* represents a gene that codes for a protein that influences skin color, and our sequencing of the gene detects three alleles—*B1*, *B2*, and *B3*.

One of the first things we might do after genotyping the frogs is calculate the **allele frequencies**, that is, the proportional representation of all the alleles in the population. For example, in Figure 3.3, allele *A1* represents 6 out of the 10 *A* alleles measured, which gives a frequency of 0.60 (or 60% of all alleles). Allele frequencies tell us which alleles are numerically dominant in the population, and which are rare. Conservation biologists often compare the observed (measured) allele frequencies with those we might expect for a population under a

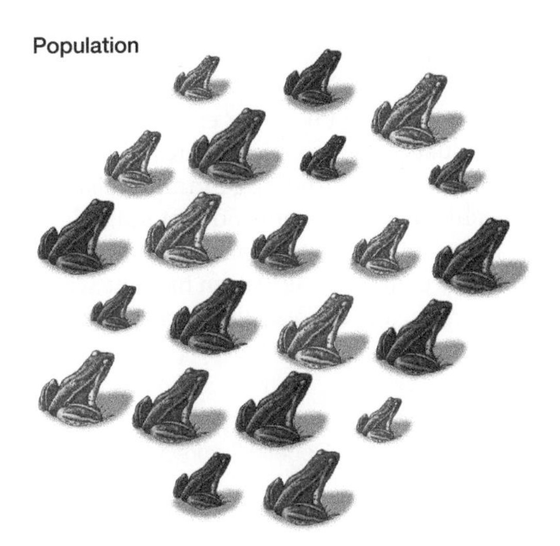

Sample

Individual *i*	Locus *A* genotype	Locus *B* genotype	h_i
1	*A1A1*	*B1B1*	0/2 = 0.0
2	*A1A2*	*B1B2*	2/2 = 1.0
3	*A1A2*	*B1B3*	2/2 = 1.0
4	*A1A1*	*B1B1*	0/2 = 0.0
5	*A2A2*	*B1B2*	1/2 = 0.5

Allele	Frequency
A1	6/10 = 0.60
A2	4/10 = 0.40
B1	7/10 = 0.70
B2	2/10 = 0.20
B3	1/10 = 0.10

Polymorphism $(P) = \frac{n_p}{k} = 1$ *Heterozygosity* $(H) = \frac{1}{N}\sum_{i=1}^{N} h_i = 0.50$

k = number of loci considered (2)
n_p = number of polymorphic loci (2)
N = number of individuals (5)
h_i = proportion of loci k at which individual i is heterozygous

Figure 3.3 Measures of genetic diversity. This example shows how to calculate polymorphism and heterozygosity of a hypothetical population of frogs (left). Assume that five frogs are sampled randomly such that they are representative of the larger population of frogs. Each of the five frogs is then genotyped to determine which alleles it has at each of two loci: at locus *A* it has either allele *A1* or *A2*, while at locus *B* it has either allele *B1*, *B2*, or *B3*. After genotyping each frog, one can calculate the proportion of loci at which each individual is heterozygous, h_i, which is then used to calculate the average heterozygosity, *H*, of frogs that were sampled. One can also calculate the degree of polymorphism, *P*, of the sampled frogs, as well as the frequency of each allele in the sampled population.

particular set of assumptions, such as the assumption of random mating among males and females in a large population (e.g., Hardy-Weinberg equilibrium, Chapter 16). If the observed allele frequencies are lower than what we might expect based on a set of assumptions, that could signal a population whose genetic diversity is dangerously low, leading us to take management action to try and correct the difference between observed and expected values.

We can also use the genotypes of the sampled frogs to estimate genetic diversity for the population. **Polymorphism** (P) is a common measure of genetic diversity in a population; it quantifies the fraction of gene loci in which alternative alleles of a gene occur (i.e., that are polymorphic). Polymorphism is easy to calculate as:

$$P = \frac{n_p}{k} \tag{3.1}$$

where n_p is the number of polymorphic loci, and k is the total number of loci being evaluated in the population. In our example of frogs from Figure 3.3, gene sequencing revealed two alleles for locus A ($A1$ and $A2$) and three alleles for locus B ($B1$, $B2$, and $B3$). Both loci are polymorphic—that is, each gene has multiple forms, so $n_p = 2$ and $P = 2 / 2 = 1$. To make measurements of P more meaningful, a locus is usually considered to be polymorphic in a population only if the frequency of the most common allele is less than some arbitrary threshold, usually 0.95.[6]

Another useful measure of genetic diversity in a population is **heterozygosity** (H), which quantifies the proportion of gene loci at which the average individual in the population is heterozygous. To calculate H, we start by calculating the heterozygosity of each individual sampled frog, h_i, which is the proportion of all loci considered where the individual is heterozygous (has >1 allele at a locus). In Figure 3.3, individual 1 is homozygous for both loci A and B, which means it has no heterozygosity; thus, $h_i = 0$. In contrast, individual 2 is heterozygous at both loci, which means it has an $h_i = 1$ (all gene loci are heterozygous). Heterozygosity of the population (H) is then the average heterozygosity of all of the individuals in the population:

$$H = \frac{1}{N} \sum_{i=1}^{N} h_i \tag{3.2}$$

where N is the total number of individuals sampled (5). In our example of frogs, $H = 2.5 / 5 = 0.5$, which means that individuals in the population are, on average, heterozygous at 50% of the gene loci measured. For the frog example, 100% of the loci in the population are polymorphic ($P = 1$), but only 50% of individuals in the population are heterozygous ($H = 0.50$). Maintaining healthy levels of polymorphism and heterozygosity is a key goal of conservation biology, which helps ensure the reproductive success, growth, and adaptability of rare or threatened species (Chapters 4, 13, and 16).

Suggested Exercise 3.1 Practice calculating polymorphism and heterozygosity using the online exercise Measures of Genetic Diversity. **oup-arc.com/e/cardinale-ex3.1**

Species Diversity

Species diversity is perhaps the most widely used measure of biodiversity in conservation and natural resource management. Even so, it has been one of the hardest forms of biodiversity to define and has been quantified by a wide variety of metrics. Here we present some of the more common definitions and measurements used in conservation biology.

The species concept

When most people think of biodiversity, they think about some aspect of species diversity. **Species** are fundamental units of evolution and primary targets of some of the most powerful examples of conservation legislation. But biologists have admittedly had a difficult time nailing down what exactly a species is. At least three concepts of a species are used routinely:

1. **Morphological species concept**: The concept that a group of individuals that is morphologically, physiologically, or biochemically distinct from other groups in some important characteristic comprises a species.
2. **Biological species concept**: The concept that a group of individuals that can potentially breed among themselves in the wild and that do not breed with individuals of other groups comprises a species.
3. **Evolutionary species concept**: The concept that a group of individuals that share unique similarities of their DNA and hence their evolutionary past comprises a species.

The morphological concept of species is one commonly used by **taxonomists**—biologists who specialize in the identification of unknown specimens and the classification of species (Figure 3.4). The biological concept of species is used less often because it is difficult to establish relationships among individuals, and information about which individuals actually have the potential to breed with one another is rarely available. Historically, the evolutionary concept of species has required access to expensive DNA-sequencing equipment, which has precluded its use in the field; however, as the technology

(A)

Richard B. Primack

(B)

Courtesy of Jeremiah Trimble, Museum of Comparative Zoology, Harvard University © President and Fellows of Harvard College

Figure 3.4 (A) A plant ecologist prepares a museum specimen using a plant press. The flattened and dried plant will later be mounted on heavy paper with a label giving detailed collection information. (B) An ornithologist at the Museum of Comparative Zoology, Harvard University, classifies collections of black-cowled orioles (*Icterus prosthemelas*) from Mexico, and Baltimore orioles (*Icterus galbula*) that occur throughout eastern North America.

BOX 3.1 Conservation in Practice

Naming and Classifying Species

One of the goals of conservation biology is to describe and classify the full variety of life on Earth. **Taxonomy** is the science of describing and classifying living things. The goal of modern taxonomy is to create a system of classification that reflects the evolution of groups of species from their ancestors. By identifying the relationships between species, taxonomists help conservation biologists identify species or groups that may be evolutionarily unique and/or particularly worthy of conservation efforts. Information about the taxonomy, ecology, morphology, distribution, and status of species is being organized into central databases accessible via the internet, such as the Tree of Life (oup-arc.com/e/cardinale-w3.1). In modern classification, the following groupings apply:

- Similar species are grouped into a **genus** (plural, *genera*): the Blackburnian warbler (*Setophaga fusca*) and many similar warbler species belong to the genus *Setophaga*.
- Similar genera are grouped into a **family**: all wood warbler genera belong to the family Parulidae.
- Similar families are grouped into an **order**: all songbird families belong to the order Passeriformes.
- Similar orders are grouped into a **class**: all bird orders belong to the class Aves.
- Similar classes are grouped into a **phylum** (plural, *phyla*): all vertebrate classes belong to the phylum Chordata.
- Similar phyla are grouped into a **kingdom**: all animal classes belong to the kingdom Animalia.*

Biologists throughout the world have agreed to use a standard set of scientific, or Latin, names when discussing species. The use of scientific names avoids the confusion that can occur when common names are used; the Latin names are standard across countries and languages. Scientific species names consist of two words. This naming system, known as **binomial nomenclature**, was developed in the eighteenth century by the Swedish biologist Carolus Linnaeus. In the scientific name for the Blackburnian warbler, *Setophaga fusca*, the genus name is *Setophaga* and *fusca* is the species name. The genus name is somewhat similar to a person's family name in that many people can have the same family name (Sullivan), while the species name is similar to a person's given name (Margaret).

Scientific names are written in a standard way to avoid confusion. The first letter of the genus name is always capitalized, whereas the species name is almost always lowercased. Scientific names are italicized in print or underlined when handwritten. Sometimes scientific names are followed by a person's name, as in *Homo sapiens* Linnaeus, indicating that Linnaeus was the person who first proposed the scientific name given to the human species. When many species in a single genus are being discussed, or if the identity of a species within a genus is uncertain, the abbreviations spp. or sp., respectively, are sometimes used (e.g., "*Setophaga* spp." indicates several species of *Setophaga*). If a species has no close relatives, it may be the only species in its genus. Similarly, a genus that is unrelated to any other genera may form its own family.

for sequencing has advanced and costs have dropped, it is becoming increasingly common to define species according to how much they differ in genetic material. Even so, many field biologists still learn to recognize individuals that look different from other individuals and that might represent a different species, and these are sometimes referred to as **morphospecies** or something similar until taxonomists can give them official scientific names (Box 3.1).[7]

Problems in distinguishing and identifying species are more common than people realize.[8,9] **Phenotypic variation**—variability in phenotypes that exist in a population—can cause a single biological species to be represented by individuals that have observable morphological differences yet still have the ability to reproduce. Alternatively, closely related "sibling" species may appear very similar in morphology and physiology yet be genetically and biologically distinct and unable to interbreed (Figure 3.5). Such a situation can lead to **cryptic biodiversity**—the widespread existence of undescribed species that have been incorrectly classified and grouped together because they have similar appearances.[10,11] To complicate matters further, individuals of related

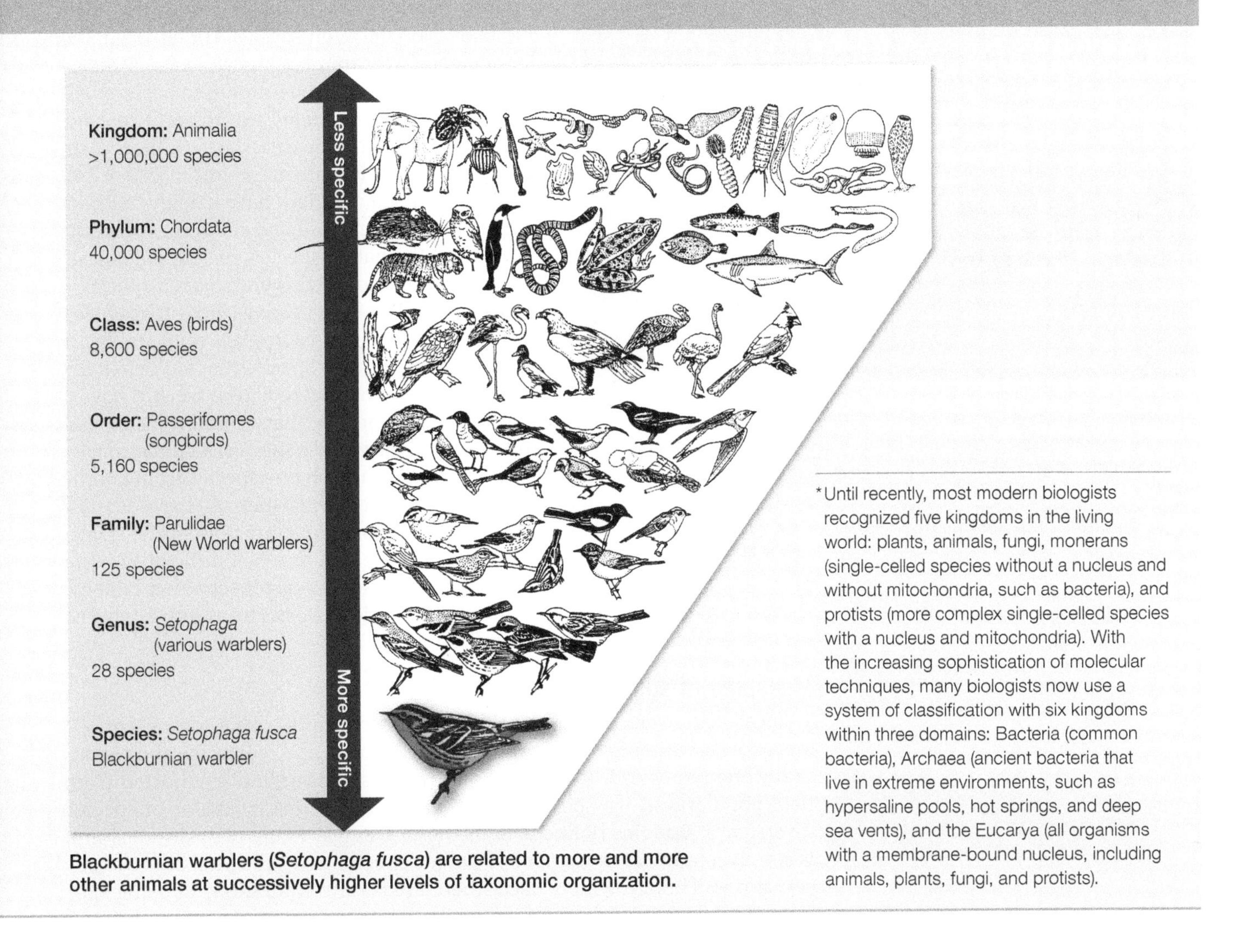

Blackburnian warblers (*Setophaga fusca*) are related to more and more other animals at successively higher levels of taxonomic organization.

*Until recently, most modern biologists recognized five kingdoms in the living world: plants, animals, fungi, monerans (single-celled species without a nucleus and without mitochondria, such as bacteria), and protists (more complex single-celled species with a nucleus and mitochondria). With the increasing sophistication of molecular techniques, many biologists now use a system of classification with six kingdoms within three domains: Bacteria (common bacteria), Archaea (ancient bacteria that live in extreme environments, such as hypersaline pools, hot springs, and deep sea vents), and the Eucarya (all organisms with a membrane-bound nucleus, including animals, plants, fungi, and protists).

Figure 3.5 The western meadowlark (*Sturnella neglecta*; left) and the eastern meadowlark (*Sturnella magna*; right) look almost identical and sometimes even occur in the same place. However, they are distinct species because they have different songs and do not interbreed.

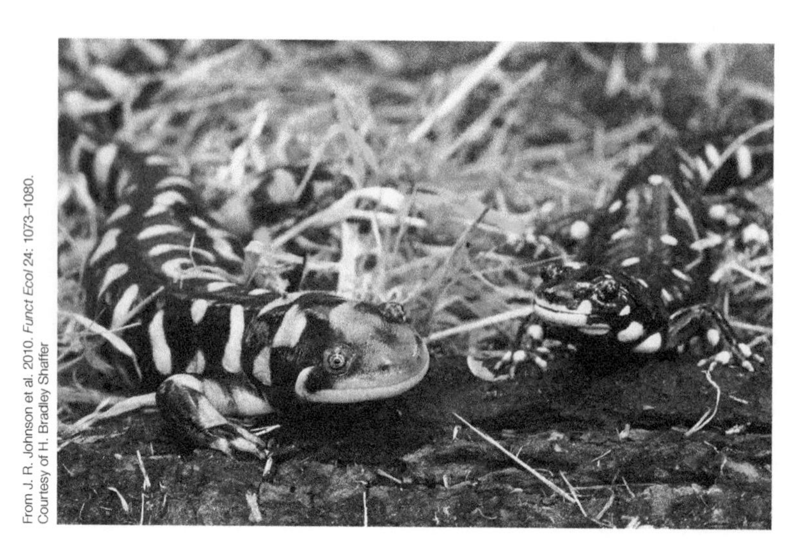

From J. R. Johnson et al. 2010. *Funct Ecol* 24: 1073–1080. Courtesy of H. Bradley Shaffer

Figure 3.6 The hybrid tiger salamander (left) is larger than its parent species, the California tiger salamander (right), and is increasing in abundance. Note the much larger head of the hybrid salamander.

but distinct species may occasionally mate and produce **hybrids**—intermediate forms that blur the distinction between species. For example, the endangered California tiger salamander (*Ambystoma californiense*) and the introduced barred tiger salamander (*A. mavortium*) are thought to have evolved from a common ancestor 5 million years ago, yet they readily mate in California (**Figure 3.6**). These hybrid salamanders have a higher fitness and are better able to tolerate environmental pollution than the native species, further complicating the conservation of this endangered species.[12]

DNA-based methods such as molecular phylogenetics and DNA bar coding are now making it possible to distinguish species that look virtually identical or have hybridized.[13] But we still have a long way to go before we have correctly cataloged and classified the world's species. And our inability to clearly distinguish one species from another, whether due to similarities of characteristics or to confusion over the correct scientific name, often slows down efforts at species protection and makes it difficult to write precise, effective laws. To be effective practitioners of conservation biology, students must be able to use all three definitions of *species* and to apply the measures described in the next section.

Measures of species diversity

To measure species diversity, field biologists typically collect samples or observations of organisms from a particular location and record the abundance of different species (**Figure 3.7**). From records of abundance collected from field populations, it is possible to calculate numerous measures of species diversity.[14,15] **Species richness (*S*)** is the easiest way to quantify diversity and is simply a tally of the number of unique species in a collection or set of observations. Though widely used, there are certain limitations of using species richness as a measure of diversity. First, estimates of species richness are strongly influenced by sampling effort because the probability of finding new species increases as a researcher collects or observes more individuals. To deal with this problem, one can construct species accumulation curves (**Figure 3.8A**) that tally the cumulative number of species found in a system as a function of the total sampling effort.[16] Sampling effort can be quantified in any number of ways and may include the number of samples collected, the number of individuals observed, the amount of time spent collecting, or the cumulative area of land or volume of water or soil sampled. The accumulation curve tells us whether the sampling effort is sufficient to find all of the species in a system, thus producing a reliable estimate of species richness.

Suggested Exercise 3.2 Practice making a species accumulation curve with the online exercise Sampling Species Richness. **oup-arc.com/e/cardinale-ex3.2**

In many instances, biologists are not able to afford the sampling effort needed to find all of the species in a system of interest. For example, the diversity of insects living in the canopy of a tropical forest may be so great that

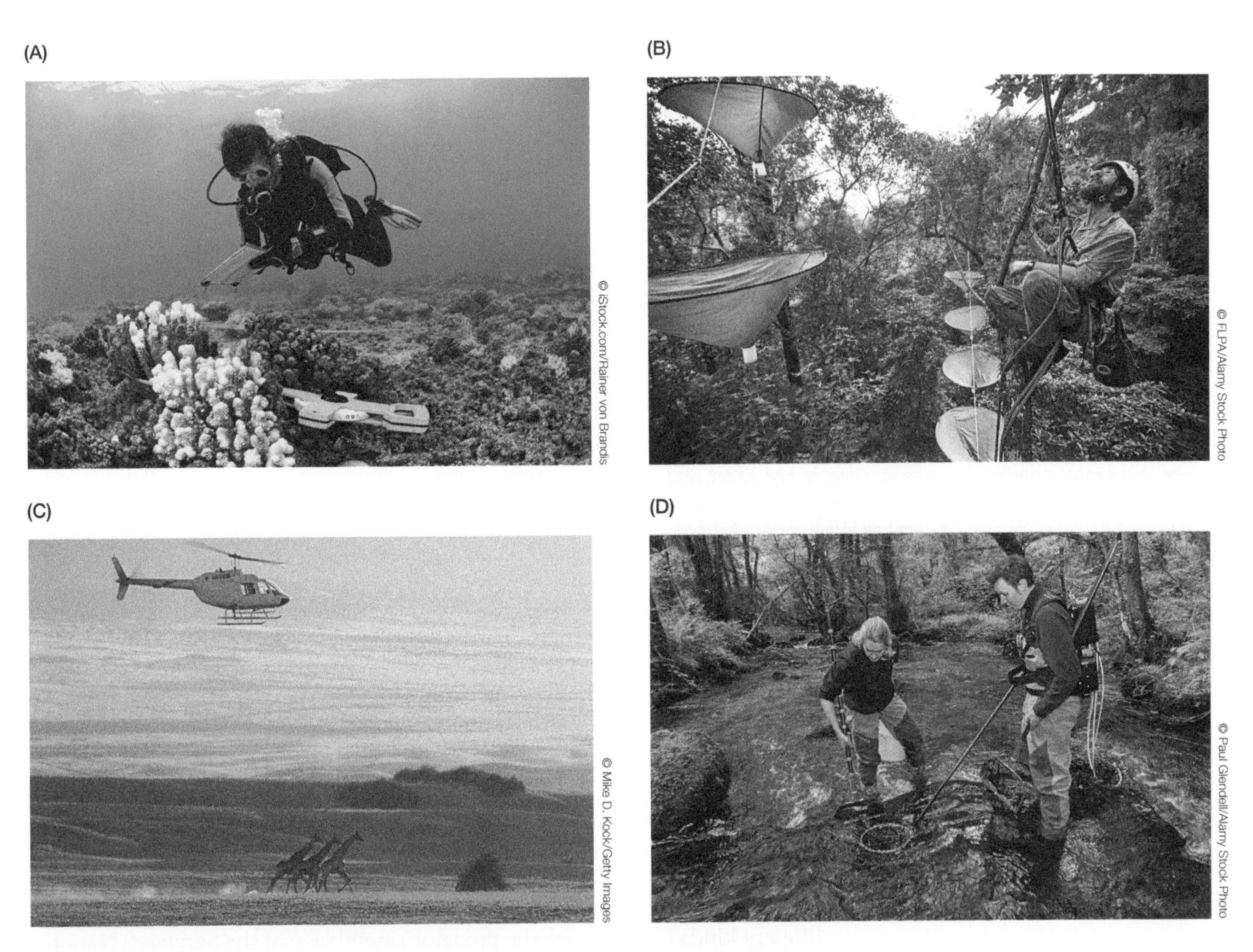

Figure 3.7 Biologists measuring biodiversity in different ecosystems. (A) A marine biologist counting coral species along a transect. (B) An entomologist tallying insect species in a tropical forest canopy. (C) Wildlife biologists performing surveys of animal abundance in African grasslands. (D) Ichthyologists electroshocking to count fish species in a stream.

no reasonable amount of time, money, or energy could possibly produce an accumulation curve that identifies the millions of potential species. In these instances, the only choice is to build a statistical model that extrapolates beyond the accumulation curve to estimate what would be found if sampling were exhaustive (**Figure 3.8B**). This is called the extrapolation curve,[17] and several advanced techniques have been developed to estimate the asymptotic richness of a collection that has been undersampled.[18] Undersampling also makes it hard to compare species richness between two collections, because it becomes impossible to know whether the collections truly differ in species richness or only seem to differ in richness because one collection was sampled more than the other (**Figure 3.8C**). To deal with this problem, biologists calculate rarefaction curves that allow them to compare species richness among two or more collections after standardizing the amount of sampling effort.[16,18]

The second limitation of using species richness as a sole measure of biodiversity is that it can give an inaccurate picture of biodiversity when there are large differences in the relative abundance of species. Consider a hypothetical example where the abundance of top predators was quantified from visual surveys in two national parks in Africa—Kruger National Park and the Serengeti

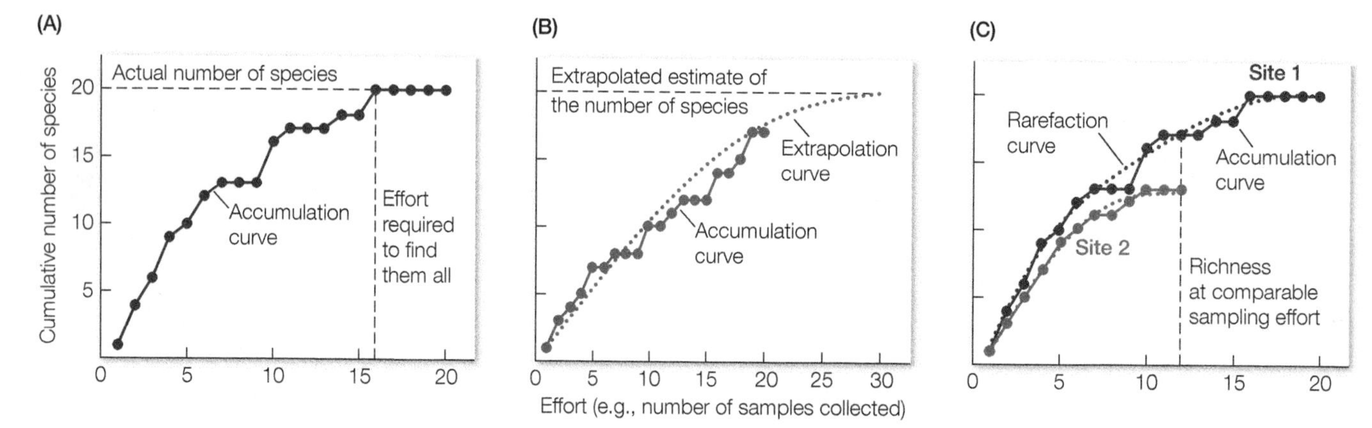

Figure 3.8 Types of sampling curves that conservation biologists use to estimate species richness at a site (or collection) and to compare biodiversity among sites (collections). Panel A shows a species accumulation curve. It quantifies the number of species found per sampling effort, which can be expressed in several ways, such as the number of samples collected and analyzed, the amount of time spent sampling, or the number of individuals identified. The accumulation curve is used to show when sampling is complete and all species have been identified. When sampling is incomplete, an extrapolation curve, shown in panel B, can be fit to data from an accumulation curve to estimate how many species there might be. Several advanced techniques have been developed to estimate the asymptotic richness of a collection that has been undersampled.[18] Panel C shows how to compare species richness at two sites that have received different sampling efforts. Site 1 would be expected to have more species based simply on the fact that it has been sampled more heavily. However, to determine whether site 1 truly has more diversity, we can "rarify" the accumulation curve by randomly selecting the same number of samples used to measure richness in the less sampled site (site 2). The resulting rarefaction curve can be used to compare species richness at the same level of sampling effort.

National Park (**Figure 3.9**). Sampling at both parks revealed the same total abundance of predators ($N = 9$ individuals) divided among four predator species (hyena, lion, leopard, and cheetah). But compared with the predator assemblage at Kruger National Park, where abundance is more evenly distributed among the different kinds of predators, the predator assemblage at the Serengeti National Park is dominated by hyenas. While both parks have the same species richness ($S = 4$), most biologists would agree that Kruger National Park is more diverse. But how do we capture a measure of biodiversity that accounts for differences in the abundance of species?

Simpson[19] developed a measure of species diversity that simultaneously considers species richness and evenness. Simpson's diversity index (D_S) measures the probability that any two individuals, drawn at random from a collection, will produce two different species:

$$D_S = \sum_{i=1}^{S} p_i^2 \tag{3.3}$$

where p_i is the proportional abundance of species i in the collection, and S is the total number of species. The value of Simpson's diversity index ranges between 0 and 1, with values of D_S close to 1 representing low diversity (any two randomly chosen individuals are more likely to be the same species), and values close to 0 indicating high diversity (any two randomly chosen individuals are less likely to be the same species). The lower value of D_S for Kruger National Park means the park has a greater diversity of predator species (see Figure 3.9).

Another popular diversity index that accounts for both species richness and the relative abundance of the species present in a collection is called the Shannon diversity index (H'), also known as the Shannon-Wiener index, calculated as:

$$H' = -\sum_{i=1}^{S} p_i \times \ln(p_i) \tag{3.4}$$

n_i is the number of individuals belonging to species *i*.

p_i is the proportion of individuals belonging to species *i*.

Species *i*	n_i	p_i	$\ln(p_i)$	$p_i \times \ln(p_i)$	p_i^2
Hyena	3	3/9 = 0.33	–1.10	–0.37	0.11
Lion	2	2/9 = 0.22	–1.50	–0.33	0.05
Leopard	3	3/9 = 0.33	–1.10	–0.37	0.11
Cheetah	1	1/9 = 0.11	–2.20	–0.24	0.01

Species richness (*S*) = 4

Simpson's diversity index (D_S) = $\Sigma_{i=1}^{S} p_i^2 = 0.28$

Shannon diversity index (*H′*) = $-\sum_{i=1}^{S} p_i \times \ln(p_i) = 1.31$

Simpson's evenness $(E_{\frac{1}{D}}) = \frac{\frac{1}{D_S}}{S} = 0.88$

Species *i*	n_i	p_i	$\ln(p_i)$	$p_i \times \ln(p_i)$	p_i^2
Hyena	6	6/9 = 0.67	–0.41	–0.27	0.44
Lion	1	1/9 = 0.11	–2.20	–0.24	0.01
Leopard	1	1/9 = 0.11	–2.20	–0.24	0.01
Cheetah	1	1/9 = 0.11	–2.20	–0.24	0.01

Species richness (*S*) = 4

Simpson's diversity index (D_S) = 0.48

Shannon diversity index (*H′*) = 1.00

Simpson's evenness $(E_{\frac{1}{D}}) = 0.52$

Kruger National Park

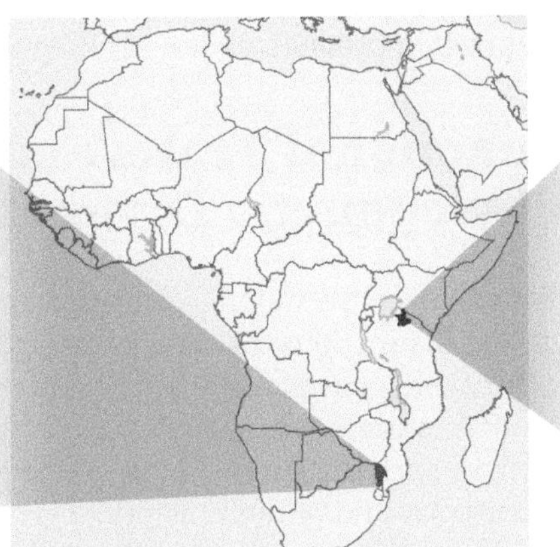

Serengeti National Park

Figure 3.9 Measures of species diversity. This example compares species diversity for a hypothetical group of predators in Kruger and Serengeti National Parks in Africa. After the predators are surveyed visually (by helicopter or on the ground), both parks are found to have four predator species, thus equal species richness. However, the Serengeti National Park (right) is dominated mostly by hyenas, whereas abundances among predator species at Kruger National Park (left) are more evenly distributed. Diversity metrics that consider species evenness in addition to richness can be used to represent the higher diversity of species at Kruger National Park.

where p_i is the proportion of individuals in a collection belonging to species *i*. The maximum value of *H′* is ln(*S*), which occurs when every species in the collection has the same relative abundance (i.e., there is maximum evenness). Note that *H′* is higher in Kruger National Park, representing its higher level of biodiversity (see Figure 3.9).

Suggested Exercise 3.3 Visit the Companion Website for a video tutorial on how to set up a spreadsheet to calculate the Shannon Diversity index. Then use data from Figure 3.9 to reproduce H′ for Kruger and Serengeti National Park. **oup-arc.com/e/cardinale-ex3.3**

The fact that metrics like the Simpson's and Shannon diversity indices incorporate both species richness and evenness into a single measure is both a strength and a weakness. It is a strength because these metrics provide simple, synthetic summaries of species diversity that are easy to compare between two collections of species. But it is also a weakness because combining different measures of diversity into a single metric can make it hard to know whether diversity differs among collections because of differences in species richness, differences in species evenness, or both. Because of the confounding of richness and evenness in metrics like D_S and H', some conservation biologists prefer to stick to two separate numbers: one for species richness, and a second for species evenness.

If one wants to quantify the evenness of a community independent of its richness, then Simpson's evenness (or equitability) index can be used. Simpson's evenness index ($E_{1/D}$) is calculated by taking Simpson's diversity index (D_S) and expressing it as a proportion of the maximum value that D_S could assume if individuals in the community were completely evenly distributed (D_{max}, which equals S—as in a case where there was one individual per species):

$$E_{1/D} = \frac{1}{D_S} \times \frac{1}{S} \tag{3.5}$$

Values of $E_{1/D}$ range from 0 to 1, with 1 being complete evenness of species in a community. Note that $E_{1/D}$ confirms that Kruger National Park has a greater evenness of predator species and, in turn, higher diversity (see Figure 3.9).

Functional Traits and Phylogenetic Diversity

Sometimes conservation biologists are more interested in preserving the variety of ecological functions that species perform in ecosystems, or the evolutionary history that has generated diversity, which may maximize the potential for adaptation into the future. In these instances, we are less interested in measures of species diversity per se, and more interested in measures that quantify the ecological roles species play and/or their history of genetic and evolutionary divergence.

Functional traits are those that define species in terms of their ecological roles in a community, including how they interact with the environment and with other species. A functional trait can be any measurable aspect of the biology of a species that is known to influence the ecological role that species plays in a community. For example, functional traits could be the different shapes and sizes of flowers that insects pollinate, the heights of trees in a canopy that influence their competitiveness for light, or the gape sizes of predatory fishes that influence which types of prey species they eat. Once a researcher decides on which functional traits are important and collects measures of those traits on the species of interest, the next step is to quantify how similar or different species are in "trait space." **Figure 3.10** illustrates how to measure trait distances separating species for measures of just two functional traits. When many functional traits are measured, one usually calculates the Euclidean distance separating each pair of species in an n-dimensional matrix, where n is the number of traits measured. The trait distances that separate species can then be used to make a **phenogram**, which is a diagram depicting relationships among species based on their overall similarity or dissimilarity in functional traits, without regard to the evolutionary history of the species. Once a phenogram is constructed, it is relatively simple to sum up the branch lengths among any species of interest to determine which sets of species have the largest or smallest functional diversity (FD).

A similar approach can be taken when characterizing the evolutionary diversity of species. In this case, a phylogenetic tree is constructed using genetic differences among species (e.g., the number of base pairs that differ). A **phylogenetic tree** is a branching diagram or "tree" showing the evolutionary relationships among species based on similarities and differences in their

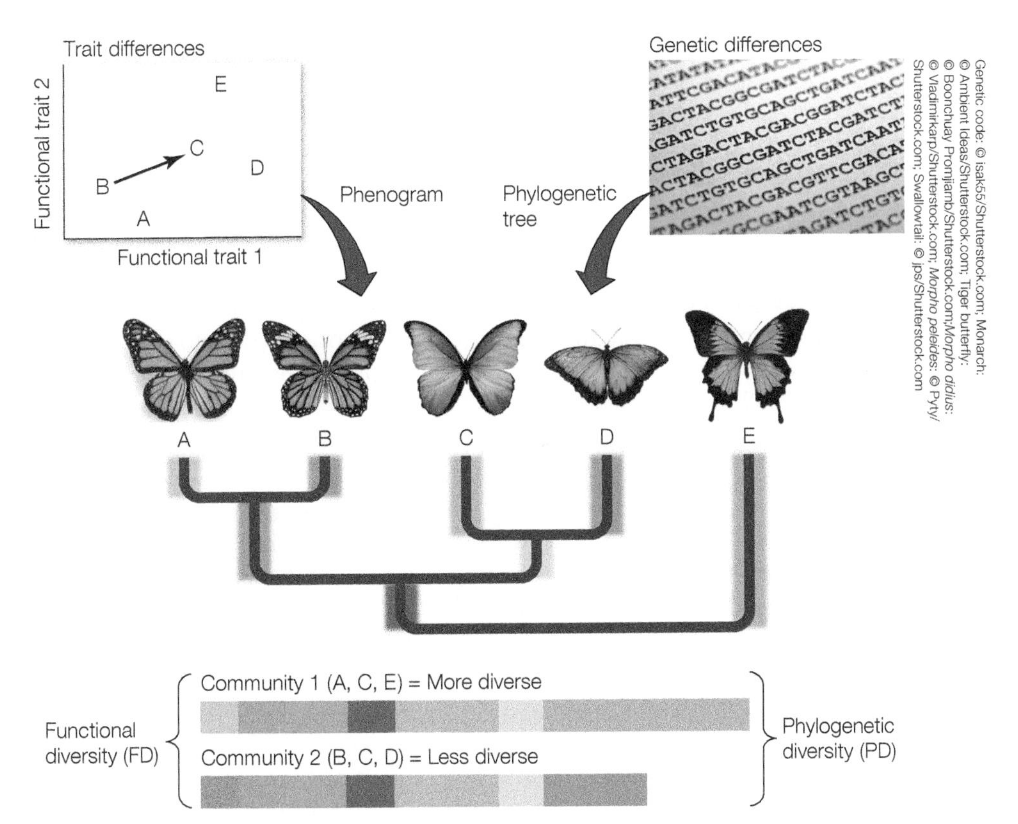

Figure 3.10 Measures of functional and phylogenetic diversity. Functional diversity among species is measured with the aid of a phenogram, which is constructed from measures of species similarity/dissimilarity in functional traits. After measuring multiple traits of interest on a set of species, one calculates the distance separating species in an *n*-dimensional matrix. This is shown in the upper left panel as the Euclidian distance separating butterfly species B and C for measures of just two functional traits. The distances that separate species are then used to make a tree (i.e., phenogram) that depicts relationships among species based on their overall similarity in traits. One can sum up the branch lengths of the phenogram to get a measure of functional diversity among species (greater branch lengths = higher FD). A similar approach can be taken when characterizing the phylogenetic diversity of species; however, the tree is called a phylogenetic tree, and we use genetic differences among species (e.g., the number of base pairs that differ) as the measure of similarity/dissimilarity. Summing branch lengths that separate species gives a measure of phylogenetic diversity, which is the amount of genetic differentiation between any set of species that has resulted since their divergence from a common ancestor.

genetic characteristics. The branching shows relationships among the species: species joined together in the tree have a common ancestor, and species that are closer together have a more recent common ancestor than those that are farther apart on the tree. Summing the branch lengths that separate species gives a measure of **phylogenetic diversity**—the amount of genetic differentiation between any set of species that has resulted since their divergence from a common ancestor (**Video 3.2**).

▶ **Video 3.2** How to make a phylogenetic tree.
oup-arc.com/e/cardinale-v3.2

Community and Ecosystem Diversity

A **biological community** is a collection of species that occupy and interact in a particular location. A biological community, together with its associated abiotic environment, is called an **ecosystem**. Many properties of an ecosystem and how it functions are controlled by the physical and chemical characteristics of the environment, such as the topography, hydrology, temperature, and geology. Other properties of an ecosystem and how it functions are controlled by the organization and complexity of the biological community that resides in it. The organization and complexity of a community is collectively referred to as **community structure**. Here we describe three aspects of community structure that are common foci of efforts to conserve whole communities or ecosystems: trophic structure, food web complexity, and species composition.

Trophic structure

Biological communities within ecosystems can be organized into trophic levels that represent different ways in which organisms acquire nutrients and energy from their environment (Figure 3.11):

- **Primary (1°) producers** obtain energy directly from the sun through the process of photosynthesis. In terrestrial environments, flowering plants, gymnosperms, and ferns are responsible for most photosynthesis, while in aquatic environments, seaweeds, single-celled algae, and cyanobacteria (blue-green algae) are the most important. All of these species use solar energy to build the organic molecules they need to live and grow. Without the primary producers, species at the higher trophic levels could not exist.
- **Primary (1°) consumers**, also known as **herbivores**, eat photosynthetic species. For example, in terrestrial environments, gazelles and grasshoppers eat grass, while in aquatic environments, crustaceans and certain kinds of fish eat algae. The intensity of grazing by herbivores often determines the relative abundance of plant species and even the mass of plant material present.
- **Higher consumers**—also known as secondary (2°) and tertiary (3°) consumers, carnivores, or predators—kill and eat other animals. Secondary consumers (e.g., foxes) eat herbivores (e.g., rabbits), whereas 3° consumers (e.g., bass) eat other carnivores (e.g., frogs). Carnivores are usually

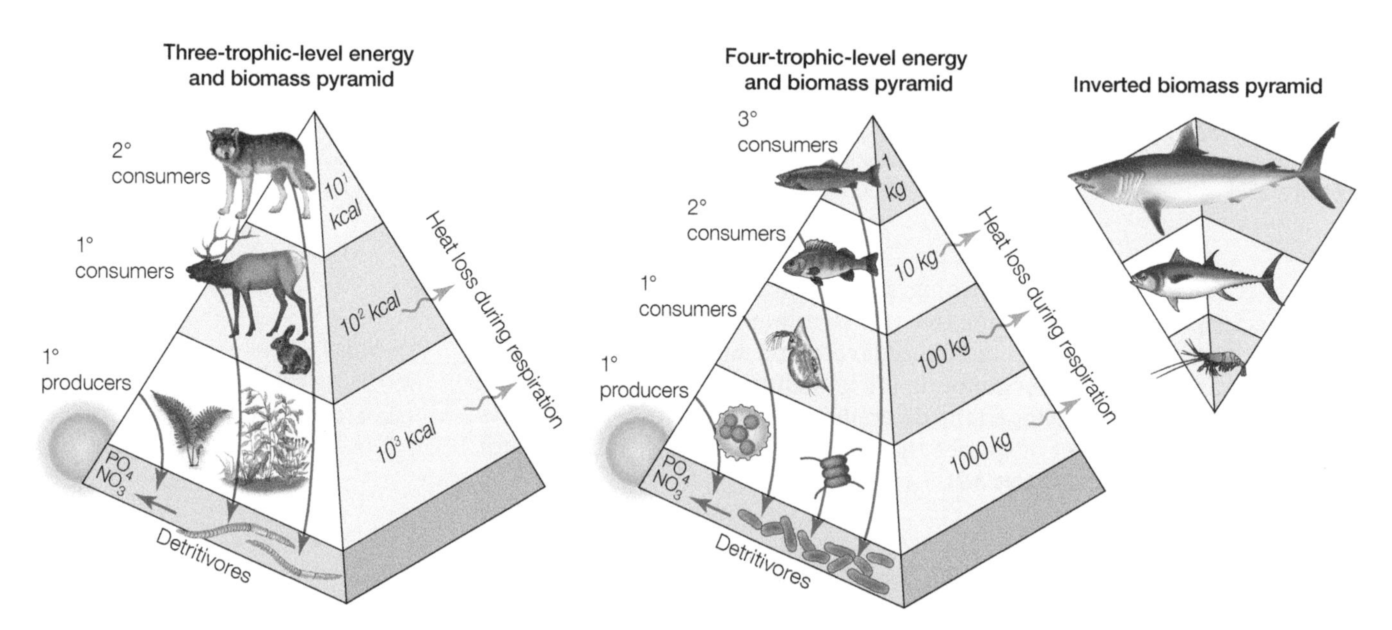

Figure 3.11 Examples of trophic pyramids that describe the amount of energy and biomass at each trophic level of a community. While energy always decreases at higher trophic levels because of heat loss during respiration, and while biomass usually decreases at higher trophic levels, it is possible to find *inverted* biomass pyramids in some ecosystems where the basal trophic levels that support the food web are short-lived and turn over quickly.

predators, though some species combine direct predation with scavenging behavior, and other species, known as **omnivores**, include a portion of plant foods in their diets.

- **Detritivores,** also known as **decomposers**, feed on dead plant and animal tissues and wastes (detritus), breaking down complex tissues and organic molecules. In the process, decomposers release biologically essential minerals such as nitrates and phosphates back into the soil and water, where they can be taken up again by plants and algae. The most important decomposers are fungi and bacteria, but many other kinds of species play roles in breaking down organic materials. For example, vultures and other scavengers tear apart and feed on dead animals, dung beetles feed on and bury animal dung, and worms break down fallen leaves and other organic matter. If decomposers were not present, organic material would accumulate, and plant growth would slowly stop.
- **Parasites** live on or within a host organism for all or part of their lives and may cause harm to the host or might just weaken it, while **pathogens** definitely cause harm to the cell. Pathogens and parasites include a wide variety of protists, bacteria, fungi, viruses, and animals (e.g., flatworms and flukes) that infect plants and animals at all trophic levels. The effects of disease-causing organisms are especially important when the host species is at high density and under stress, which might occur in a zoo or in a habitat that has been fragmented or degraded by human activities. The spread of disease from captive or domestic species, such as domestic dogs, to wild species, such as lions, can be a threat to rare species.

A **trophic pyramid** depicts the amount of biomass in, or energy available to, each trophic level in a community. Energy—the number of kilocalories available for reproduction and growth—always declines at each successive trophic level. Loss of energy occurs for two reasons. First, there are feeding inefficiencies such that all of the biomass produced at a lower trophic level is rarely fully consumed at a higher trophic level. For example, much of the biomass produced by plants is cellulose that is not digestible by herbivores. As a result, less energy is available to the herbivores than was stored by the plants. More important, much of the biomass consumed by a higher trophic level must be used for cellular respiration to maintain an organism's metabolism. These calories are ultimately lost as heat to the atmosphere, which means there is less energy to fuel the production of new tissue.

As a rule of thumb, **ecological efficiency**—the percentage of energy transferred from one trophic level to the next—averages 10%. That is, 10% of the energy produced by a lower trophic level becomes available in the next trophic level. Ecological efficiency varies among ecosystems and types of organisms. Organisms that eat higher-quality, more digestible food tend to have higher ecological efficiencies. In addition, ectothermic organisms that do not have to maintain body temperature through respiration tend to have higher efficiencies.

Because less and less energy is transferred to each successive trophic level in a community, the greatest biomass (weight of living tissue) is usually that of the primary producers, followed by herbivores, and then by higher consumers. Thus, energy and biomass often show the same pattern, each decreasing at higher trophic levels. There are, however, exceptions in some ecosystems where the biomass pyramid becomes inverted, leading to top predators having the greatest total biomass. Inverted pyramids, such as those that support an unusually high biomass of sharks in some marine ecosystems,[20] can occur when species at the basal trophic levels are short-lived and turn over rapidly such that they support high levels of consumption despite having low total biomass.

The number of trophic levels in a system is an important aspect of biodiversity that can influence the structure of a community through trophic cascades.

A **trophic cascade** occurs when higher consumers in a community suppress the biomass of their prey, in turn releasing the next lower trophic level from predation. In other words, a predator can have an effect that *cascades* across multiple trophic levels. A predator may kill a prey species—a direct effect—but the cascade is the *indirect* effect of that predation (i.e., from lower numbers of the prey species or changes in its behavior) on the plant communities at the next trophic level. A classic example of a trophic cascade occurs in marine kelp forests that develop in the high-latitude coastal waters of the world's oceans. Kelp forests consist of a number of species of marine brown algae, such as giant kelp (*Macrocystis pyrifera*). These habitats rank among the most productive ecosystems on Earth,[21] and enormous numbers of ocean fish, shellfish, and invertebrates depend on these forests for food and shelter during different stages of their lives.[22,23] Kelp forests also reduce the impact of waves and currents upon coastal shorelines, preventing erosion of coastal land.

Despite their recognized value, kelp forests have disappeared over the last century across much of northwestern North America. A contributing cause for kelp forest declines began over a century ago, with the harvesting of sea otters (*Enhydra lutris*). Sea otters, once widespread throughout the Pacific, were all but exterminated by fur traders. Sea otters eat large quantities of shellfish—as much as 25% of their body weight each day. In the absence of otters, populations of mussels, abalone, other shellfish, and sea urchins exploded, providing a greater harvest for the shellfish industry. However, sea urchins feed voraciously on kelp—in particular, they eat the holdfasts of the kelp stipes that anchor the algae to hard substrates on the bottom of the subtidal habitat. Left unchecked by sea otters, urchins have created large denuded areas, called urchin barrens, where kelp forests once flourished (**Figure 3.12**).

The sea otter is now protected in the United States and Canada and has begun to recolonize parts of its former range. The return and reintroduction of sea otters has once again initiated a trophic cascade involving reduced herbivory by sea urchins and leading to regrowth of giant kelp beds.[24] Enhanced production of kelp has increased populations of fishes and many other species that inhabit kelp beds, as well as some of the benefits kelp forests provide to society, such as protection of coastlines. The disappearance of kelp beds with the loss of sea otters is just one

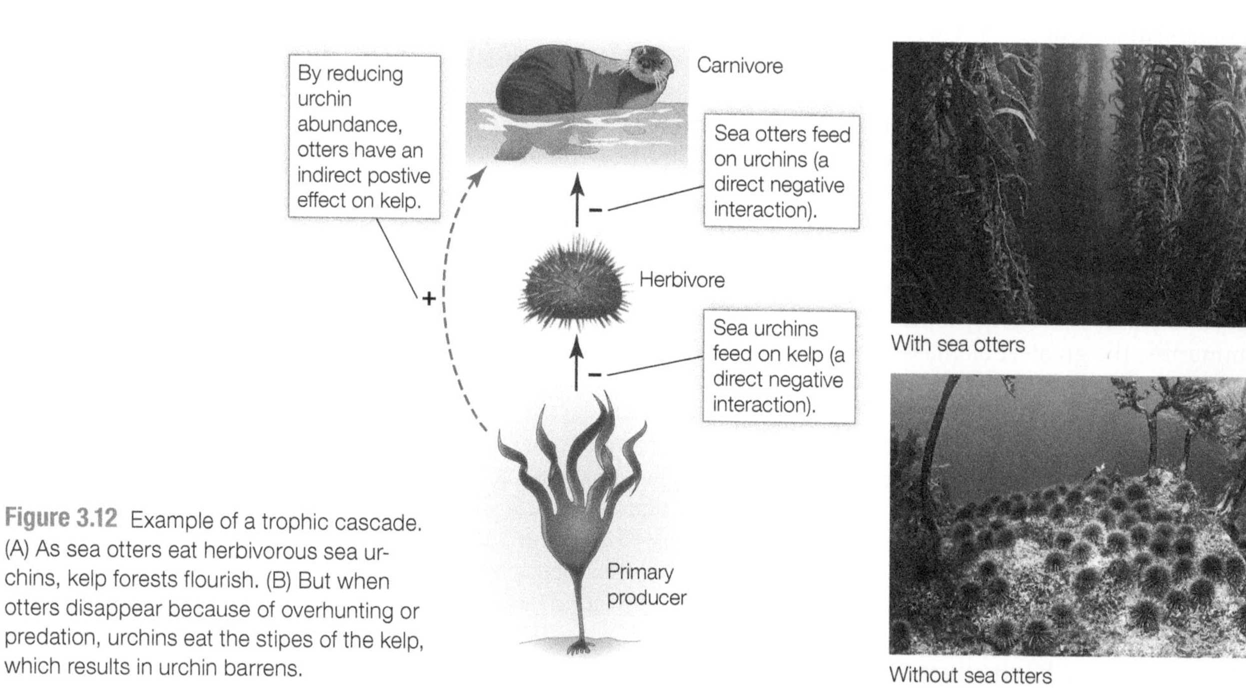

Figure 3.12 Example of a trophic cascade. (A) As sea otters eat herbivorous sea urchins, kelp forests flourish. (B) But when otters disappear because of overhunting or predation, urchins eat the stipes of the kelp, which results in urchin barrens.

of many examples of how extirpation of species from top trophic levels can lead to trophic cascades that profoundly affect nearly every aspect of an ecosystem.[25]

Food web complexity

Many conservation efforts have begun to shift from a narrow focus on individual species to broader approaches that attempt to conserve entire biological communities and functional ecosystems in whole landscapes (Chapters 14 and 15). These broader efforts rely on the fact that much of the functioning and stability of communities and ecosystems stems from interactions among species that form complex interaction networks. The complexity of species interactions in networks is an important attribute of healthy communities that often protects them from **extinction cascades**—a series of secondary extinctions that are triggered by the primary extinction of a key species in an ecosystem. Complex interaction networks can also enhance the capture and utilization of biological resources that keep whole ecosystems productive and sustainable. Therefore, an increasing number of conservation efforts are being directed at conserving highly connected communities, and the metrics and tools that summarize holistic attributes of ecological networks are being used as indicators to guide and assess conservation objectives.[26]

Although species in a community can generally be organized into trophic levels that provide information on biomass and energy flows, a more thorough description of a biological community is a **food web**, in which species are linked by a complex of species interactions. Food webs depict feeding relationships between species that occupy different trophic levels in a community (**Figure 3.13A**); however, food webs can be used to describe any type of connection among species, including competition, mutualism, herbivory, predation, parasitism, and disease. Food web diagrams are typically generated from a species interaction matrix (**Figure 3.13B**), which is an $S \times S$ matrix of all species in a community, where each entry represents whether an interaction occurs between any two species.

(A) Food web diagram

3° consumers (piscivorous fish)

2° consumers (planktivourous fish)

1° consumers (zooplanton)

1° producers (algae)

Number of species (S) = 92
Number of links (L) = 997

Linkage density (D) = $\frac{L}{S}$ = 11   Connectance (C) = $\frac{L}{S^2}$ = 0.12

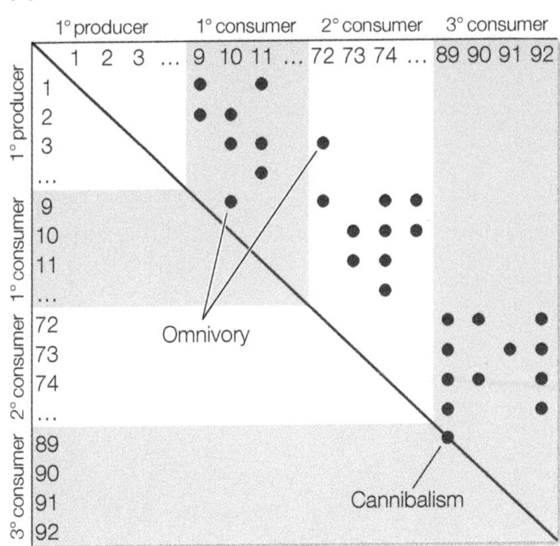

Figure 3.13 Measures of food web complexity. This example is for a hypothetical freshwater lake food web that contains 92 species spanning four trophic levels (1° producers to 3° consumers). (A) The food web can be expressed as a diagram that shows feeding links between species and among trophic levels. (B) The food web can also be described by an interaction matrix showing the presence of a link among species in the community (a dot in the matrix). The food web diagram and matrix can be used to identify species with a disproportionately large number of links (such as omnivores and others that feed at multiple trophic levels). They can also be used to calculate the linkage density and connectance of species in the food web. (Image produced with FoodWeb3D, written by R. J. Williams and provided by the Pacific Ecoinformatics and Computational Ecology Lab: www.foodwebs.org; I. Yoon et al. 2004. *Proc SPIE* 5295: 124–132.)

These entries can be qualitative (e.g., presence vs. absence of an interaction) or quantitative (e.g., measure of feeding rate by one species on another). Entries in the interaction matrix are obtained through a variety of means, including observational studies (watching which bird eats which insect, for instance, or which bee species pollinates which flower), sampling and analyses of individuals (e.g., investigating gut contents of a fish to see what species it eats, or dissecting snails to identify their parasites), and experiments that quantify interactions (perhaps removing a snake to quantify its impact on mice, or adding seeds of a plant to quantify the impacts of competition on other plant species).

Once an interaction matrix and its food web are completed, several metrics can be used to describe the complexity of the web. **Linkage density** (D), or food web complexity, represents the average number of interactions between species:

$$D = \frac{L}{S} \tag{3.6}$$

where L is the total number of links in the species interaction matrix, and S is the total number of species. **Connectance** (C) is the proportion of all possible links between species that are actually realized in the community:

$$C = \frac{L}{S^2} \tag{3.7}$$

Linkage density and connectance are two of the most commonly used metrics to describe the complexity of ecological networks, and they have been widely used to assess the risk of extinction in food webs.[27] For example, Kouhei and Morimasa[28] analyzed the structure of the food web that supports the endangered lycaenid butterfly *Shijimiaeoides divinus asonis*. Of the 27 species that comprise the web, 15 species had 4 or more trophic links whose loss might severely reduce connectance, which would in turn jeopardize the persistence of *S. asonis*. Therefore, the authors suggested that to conserve the endangered butterfly, management plans had to also protect the 15 most highly connected species in the food web. Nearly identical conclusions have been reached with analyses of endangered species in plant-pollinator networks.[29]

In another example, de Visser and colleagues[30] studied the food web properties of the Serengeti ecosystem (in Tanzania) in order to estimate extinction risk for 95 species of vertebrates and invertebrates exposed to increased land-use pressure and overexploitation by human activities. They found that the most poorly connected species are expected to be lost first, followed by the more highly connected species, as human impacts progress. They also identified a select subset of clusters of species, which they called nodes, that are especially prone to **secondary extinctions** should a currently threatened species die out and initiate a trophic cascade. These clusters were deemed high priority for management efforts.

Suggested Exercise 3.4 Visit the Companion Website and select the food web simulator. When you remove feeding links, what happens to the top predator? Is it more or less likely to go extinct? **oup-arc.com/e/cardinale-ex3.4**

Recently, researchers have moved beyond the use of topological descriptors of a food web (e.g., linkage density and connectance) to explore how artificial intelligence (AI) algorithms might help identify species and species clusters that are disproportionately important to the persistence of a community or ecosystem. These algorithms have been adapted from those used by Google, which count the number and quality of links to a webpage to determine how important that page is to the reliability and stability of the entire network of websites. McDonald-Madden and colleagues[31] applied those same AI algorithms to analysis of both real and hypothetical food webs and found the algorithms do a better job than topological metrics at identifying highly connected species that are disproportionately important for persistence of the food web. The authors subsequently showed how consideration of the network-wide impacts of individual species could significantly improve conservation outcomes.

Network theory, along with its associated measures and tools, is becoming an increasingly important contributor to conservation efforts, particularly in applications involving the management of whole communities and ecosystems. Network theory is increasingly being used to quantify extinction risk and establish conservation priorities in food webs, is being used to maximize the design of multispecies habitat networks,[32,33] and is being used to rewild nature with ecological communities that are typical of historical food webs.[34]

Species composition

Within most biological communities, there are certain types of species that have impacts on the community or ecosystem that are disproportionate to their abundance. The importance of these species is not always captured by typical measures of food web complexity; therefore, conservation biologists need to pay particular attention to the composition of species in a community and work to identify species that are disproportionately important for conservation.

While there are numerous characteristics of species that might make them disproportionately important, two special cases are worth mention. First, **ecosystem engineers** are species that create new physical habitat or that extensively modify existing environments through their biological activities.[35] A classic example of an ecosystem engineer is the beaver (*Castor* spp.),[36] which builds dams across streams that, in turn, create entirely new wetland ecosystems that are home to unique biological communities and that have unique ecological functions. Beavers are not particularly abundant and, in fact, may only be represented by a single breeding pair at a dammed site. As beavers were hunted to extinction in many locations, the wetland habitats they created, their suite of resident species, and the ecological functions and processes they supported ceased to exist. But restoration and recovery of beaver populations is quickly becoming a conservation success story (Box 3.2).

There are many other examples of organisms that create physical habitats that wouldn't otherwise exist in their absence (coral reefs, kelp forests, prairie dog burrows), yet it is nearly impossible to capture the importance of these ecosystem engineers with any of the measures of biodiversity previously mentioned in this chapter. Some of these examples are obvious; others are species that are small and seemingly insignificant. For example, net-spinning caddisflies (Hydropschyidae) are insects whose small larvae (<1 cm) spend their portion of the life cycle living on the bottom of streams where they construct silk catch nets to filter food particles out of the water. The silk nets bind rocks together on the streambed, causing the streambed to become more stable and resistant to erosion during floods, which serves to protect entire communities of benthic species (**Video 3.3**). Conservation biologists must use their expert knowledge of the natural history of organisms to identify species of special importance and concern for conservation.

Video 3.3 Watch one species of ecosystem engineers modify its environment.
oup-arc.com/e/cardinale-v3.3

Another type of species that requires special attention in conservation is the **keystone species**. Like ecosystem engineers, keystones are species on which many other species in an ecosystem largely depend, such that if the keystone species were removed, the ecosystem would change drastically. But unlike ecosystem engineers, whose impacts are driven by their formation of habitat,

BOX 3.2 Success Story

Reintroduction of Beavers

Through their construction of elaborate damming systems, beavers create wetland habitats that support thousands of other species. The beavers' building of dams also stores huge quantities of water, which can slow peak flows during flooding events, potentially lessening the impact of damaging floods.

In the seventeenth century, beavers were hunted to extinction throughout the United Kingdom. Today, beavers are making a comeback, boosted in part by studies showing how they can resurrect damaged landscapes into biodiverse ecosystems that also benefit people.

For example, in a 2017 study published in the international journal *Science of the Total Environment*,[40] researchers tracked the effects that a small group of reintroduced beavers had on a wetland close to Blairgowrie in Tayside, Scotland. That wetland had been drained for farming and was degraded and devoid of wildlife after decades of heavy use.

Twelve years after reintroduction of beavers, researchers wrote that they had found an "almost unrecognizable" landscape, with plant species diversity up nearly 50% and increased wildlife across the entire 30-acre site. They further concluded that reintroducing beavers offers a passive but innovative solution to the problem of wetland habitat loss that complements the value of beavers for water storage and flow attenuation during floods.

The recognized benefits of beavers on the landscape have initiated a broader effort to rewild the Scottish countryside. Beavers are now being reintroduced to multiple locations, including the River Otter in Devon and the Forest of Dean, where it is hoped they will help prevent flooding around the village of Lydbrook, Gloucestershire.

Secretary of State for Environment, Food and Rural Affairs Michael Gove of the United Kingdom said, "The beaver has a special place in English heritage and the Forest of Dean proposal is a fantastic opportunity to help bring this iconic species back to the countryside 400 years after it was driven to extinction."[41]

Reintroduction of beavers throughout Scotland is a conservation success story that demonstrates how ecological systems can heal themselves when missing ingredients are brought back into the mix.

Dam building by beavers creates diverse pond ecosystems that store water, reduce the probability of flooding, and increase ground water recharge.

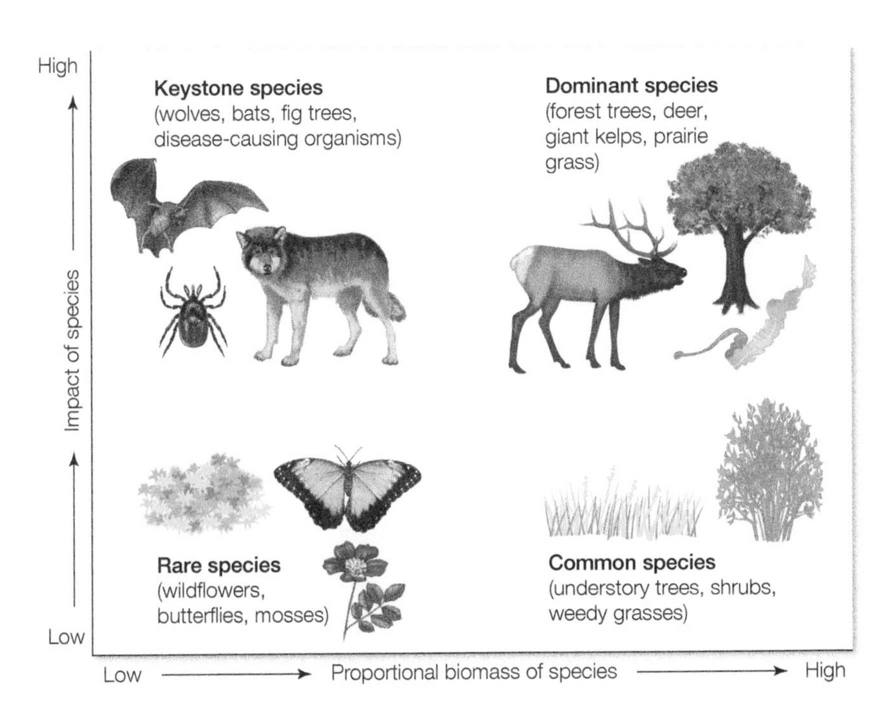

Figure 3.14 Keystone species determine the ability of large numbers of other species to persist within a biological community. Although keystone species make up only a small percentage of the total biomass, a community's composition would change radically if one of them were to disappear. Rare species have minimal biomass and seldom have significant impact on the community. Dominant species constitute a large percentage of the biomass and affect many other species in proportion to this large biomass. Some species, however, have a relatively low impact on the community organization despite being both common and heavy in biomass. (After M. E. Power et al. 1996. *Bioscience* 46: 609–620.)

keystone species influence a community through their interactions with other species—interactions that are stronger and more important than one might predict from their biomass[37] (**Figure 3.14**). For example, the colorful seastar *Pisaster* is an intertidal predator of mollusks along the coasts of North American. *Pisaster* is not especially abundant; however, its importance for controlling species diversity in the intertidal habitat is far greater than one might predict from its relatively small biomass. Selective predation by *Pisaster* on the competitively dominant mollusk species allows a wide variety of intertidal species to persist rather than be outcompeted. If *Pisaster* is removed, diversity of the intertidal community collapses (**Figure 3.15**).[38]

Protecting keystone species is a high priority for conservation efforts because loss of keystones can create extinction cascades that result in a degraded ecosystem with much lower biodiversity at all trophic levels. This may

Figure 3.15 When ochre sea stars (*Pisaster ochraceus*) are present, competition among mollusk species is reduced, and many species can persist in the same habitat. When *P. ochraceus* is experimentally removed, mollusks increase in abundance, and competition eventually eliminates most of the species. (After R. T. Paine. 1966. *Am Nat* 100: 65–75; R. T. Paine. 1974. *Oecologia* 15: 93–120.)

BOX 3.3 Conservation in Practice

Quantifying Biodiversity at Multiple Spatial Scales

In this chapter, we have described common metrics that conservation biologists use to quantify biodiversity at three levels of biological organization (genetic diversity, species diversity, and community/ecosystem diversity). All of these metrics can be further calculated at multiple spatial scales that correspond to the goals of conservation and management actions.

Many conservation and management actions take place at the spatial scale of a biological community, where species interact in a geographically bounded location defined by their ecosystem. Famous ecologist Robert Whittaker[42,43] referred to the community scale of biodiversity as "local" and referred to the amount of biological variation at this scale as **alpha diversity** (α). Whittaker distinguished the local scale of biodiversity from the larger "regional" scale, where collections of many different communities and ecosystems compose entire landscapes. He called the regional scale of biodiversity **gamma diversity** (γ).

Whittaker argued that the total amount of biodiversity across any given region such as a landscape (γ) is determined by two different things: the mean biodiversity within individual communities (α), and the differentiation among communities across a landscape. He referred to the differentiation among communities as **beta diversity** (β), which represents the amount of turnover or uniqueness of communities from one location to the next.

The two scales of biodiversity (alpha and gamma) can be related to one another by the following expression:

$$\gamma = \alpha \times \beta$$

If one has estimates of biodiversity for any two of the three terms in this equation, then it is straightforward to calculate the third. For example, imagine we have estimates of local species richness for communities of herbaceous alpine plants that interact as a community on individual mountaintops. Now imagine that the species that comprise these alpine plant communities change among different mountaintops across a broader region, such as an entire mountain range. If the average number of plant species in a plant community on a single mountain is 4, but the region as a whole has 10 species, then we can calculate the turnover of species:

$$\beta = \frac{\gamma}{\alpha} = \frac{10}{4} = 2.5$$

Beta diversity (β) quantifies the amount of change in species composition from one location to another. When β is 1, all communities have the same species. As β approaches the number of communities being compared (N = 3 alpine communities being compared), the species compositions of communities become more different (see Figure).

Beta diversity (β) is one of the measures that conservation biologists routinely use to quantify the diversity of communities and ecosystems across an entire landscape. While β is often used with estimates of species richness, it can be used with any measure of genetic, species, or community diversity described in this chapter. The equations for alpha, beta, and gamma diversity can also be used to estimate biodiversity at different temporal scales where beta represents the turnover of species (or some other measure of biodiversity) between short time scales (alpha) and longer time scales (gamma). This approach has been used to quantify the homogenization of biological communities over time as rare local species go extinct and nonnative species become more widespread because of human activity.[44]

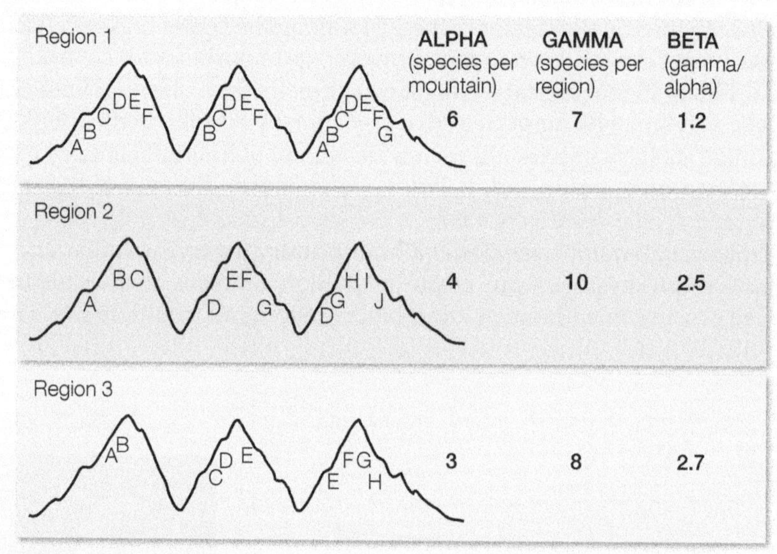

Biodiversity indexes for three regions, each consisting of three separate mountains. Each letter represents a population of a species; some species are found on only one mountain, while other species are found on two or three mountains. Alpha, gamma, and beta diversity values are shown for each region. If funds were available to protect only one mountain range, region 2 should be selected because it has the greatest gamma (total) diversity. However, if only one mountain could be protected, a mountain in region 1 should be selected because these mountains have the highest alpha (local) diversity, that is, the greatest average number of species per mountain. Each mountain in region 3 has a more distinct assemblage of species than the mountains in the other two regions, as shown by the higher beta diversity. If region 3 were selected for protection, the relative priority of the individual mountains should then be judged based on the relative rarity of the assemblages.

already be happening in certain tropical forests where overharvesting of primates has drastically reduced populations that foster biodiversity of trees by acting as seed dispersers.[39]

The Multiple Scales of Biodiversity

Many conservation efforts focus on managing some aspect of biodiversity at a relatively small spatial scale (single community or ecosystem) for short periods of time (years to decades). For example, we might set up a management plan to maximize the genetic diversity of trees in a national park, allowing them to adapt to climate change over the next 20 years. Or we might set up a 10-year plan to improve water quality in a particular lake so that we can conserve the diversity of fish species.

In contrast to shorter-term conservation goals that focus on individual communities in specific locations (a park or lake), other conservation efforts focus on managing biodiversity at the scale of whole landscapes over much broader time frames (centuries to millennia). For example, we might wish to manage a network of grasslands to maximize biodiversity and carbon storage for the next 1000 years. At these larger spatial scales, landscapes that we are attempting to manage may be composed of many different communities and ecosystems, each having a distinct gene pool, species composition, and food web structure. At the longer time scales, the genes, species, and community structures that contribute to biodiversity may also change from one time point to another. This can happen naturally as species evolve, adapt, and migrate. Or turnover can be facilitated by human influence, such as through the introduction and spread of nonnative species.

The change in biodiversity among different communities or ecosystems in a landscape is called **spatial turnover**. Similarly, the change in biodiversity between any two points in time is called **temporal turnover**. Spatial and temporal turnover are important phenomena because they both increase biodiversity by contributing to the uniqueness of genes, species, and communities in space and time. Because quantifying biological turnover is key to managing biodiversity, we end this chapter by introducing students to metrics that quantify biodiversity at multiple scales and that can be used to measure biological turnover (Box 3.3). Once you master the definitions and measures of *genetic, species,* and *community* that are taught in this chapter, and you understand how to calculate and compare those measures at different spatial and temporal scales, you will be ready to quantify biodiversity in nearly every conservation and management scenario imaginable.

Summary

1. Biodiversity is the variety of living organisms on Earth. It includes variation in the types of genes and species that comprise the communities and ecosystems that make up whole landscapes.
2. Genetic diversity is the source of all variation in lifeforms on Earth and is required for organisms to maintain reproductive vitality, resistance to disease, and the ability to adapt to changing conditions. Genetic diversity is often quantified with measures of polymorphism and genetic heterozygosity.
3. Species are generally defined by their differences in morphology, reproductive isolation, or genetic divergence from a relative. Species diversity is quantified using measures of richness, dominance, or combined diversity metrics.
4. Functional trait diversity measures the variety of ecological roles played by species in an ecosystem, while phylogenetic diversity measures the amount of evolutionary variation. These aspects of diversity are measured with phenograms and phylogenetic trees.

Summary (*continued*)

5. The organization and complexity of a community is collectively referred to as community structure. The three key aspects of community structure that are common foci for conservation are trophic structure, food web complexity, and species composition.

6. Keystone species and ecosystem engineers are important in maintaining biodiversity, despite their low relative numbers and/or biomass in a community; if these species are lost from a community, other species might also be lost in a process known as an extinction cascade.

7. Conservation efforts tend to focus on any of three scales of biodiversity. Alpha diversity is the "local" scale that represents biological variation within a community. Gamma diversity is the larger "regional" scale where collections of many different communities and ecosystems compose entire landscapes. Beta diversity is the amount of turnover or uniqueness of communities from one location to the next.

For Discussion

1. Do you think *biodiversity* is a useful term? What are the pros and cons of conservation biologists using a generic term to describe the variety of life on Earth while, at the same time, having a plethora of definitions and metrics (Equations 3.1–3.7) to quantify variation at various hierarchical levels of biology (genes, species, communities)?

2. Conservation efforts usually target genetic variation (polymorphism, heterozygosity), species diversity (species richness, evenness, functional trait diversity, phylogenetic diversity), or biological communities (trophic structure, food web complexity, species composition). Which component of biodiversity do you think is most important, and why? Which is least important?

3. Certain types of species have a disproportionately large impact on their biological communities and ecosystems, and they require special consideration in conservation efforts. Two in particular—keystone species and ecosystem engineers—were mentioned in the text. Can you identify a keystone species and an ecosystem engineer for all trophic levels and all kingdoms of the living world (e.g., animals, plants, protists, fungi, bacteria)?

4. Do you think it is possible to manage a property, such as a degraded rangeland, a forest plantation, or a polluted lake, in a way that simultaneously restores all aspects of biodiversity—genes, species, and community? What kinds of trade-offs might exist?

5. Beta diversity—which is the turnover of genes, species, or entire communities across locations in space or among different points in time—can increase biodiversity by making sites biologically unique. However, beta diversity is known to be declining across the world. What do you think is the reason? What biological processes are likely to be eroding turnover in space and/or time?

Suggested Readings

Estes, J. A. et al. 2011. Trophic downgrading of planet Earth. *Science* 333: 301–306. Shows how species interactions can change over decades, affecting multiple trophic levels.

Gotelli, N. J. and Colwell, R. K. 2001. Quantifying biodiversity: Procedures and pitfalls in the measurement and comparison of species richness. *Ecology Letters* 4: 379–391. A must read for anyone who needs to compare biodiversity among sites, such as when choosing among areas to protect.

Janzen, D. H. et al. 2009. Integration of DNA barcoding into an ongoing inventory of complex tropical biodiversity. *Molecular Ecology Resources* 9: 1–26. Shows how DNA techniques can reveal large numbers of cryptic species.

Jones, C. G. et al. 1994. Organisms as ecosystem engineers. *Oikos* 69: 373–386. Defines and provides examples of different types of ecosystem engineers of which conservation biologists should be aware.

Joppa, L. N. et al. 2011. The population ecology and social behavior of taxonomists. *Trends in Ecology & Evolution* 26: 551–553. Shows how the number of taxonomists and the number of species described per year are increasing through time.

Laikre, L. et al. 2010. Neglect of genetic diversity in implementation of the Convention on Biological Diversity. *Conservation Biology* 24: 86–88. A compelling case that more emphasis on genetic diversity needs to be part of conservation efforts.

Power, M. E. et al. 1996. Challenges in the quest for keystones. *BioScience* 46: 609–620. A classic paper that describes how extinction of keystone species impacts whole ecosystems.

Ripple, W. J. et al. 2014. Status and ecological effects of the world's largest carnivores. *Science* 343: 1241484. A good summary showing how the population size and range of many large predators have been severely reduced, with extensive ecosystem consequences.

Valentini, A. et al. 2009. DNA barcoding for ecologists. *Trends in Ecology & Evolution* 24: 110–117. Describes how DNA technology has been used to improve species identification.

Visit the
Conservation Biology
Companion Website
oup.com/he/cardinale1e
for videos, exercises, links, and other study resources.

4

Global Patterns and Drivers of Biodiversity

The northern white rhinoceros (*Ceratotherium simum cottoni*), formerly found in several countries in eastern and central Africa, is considered critically endangered and may, in fact, be extinct in the wild.

Imagine you are traveling through space and come across Earth for the first time. What would you be most struck by? Would it be the water that gives our planet the nickname the Blue Marble? Probably not. We've now found water on the moon,[1] and on several other planets in our own solar system.[2,3] And a single survey of the Milky Way found more than 270 planets in the so-called habitable zone, warm enough for liquid water.[4] If you were fortunate enough to be a space traveler, you'd probably have seen hundreds of blue marble planets by now.

Would you instead be struck by the mountain ranges, rugged canyons, sand dunes, or other geological features of Earth that are visible from space? Again, probably not. Geologists tell us there are few, if any, landforms on the planet that are wholly unique to Earth.[5,6] You probably would have seen them all before on numerous other planets.

Based on our current understanding of the universe, the only thing a space traveler is likely to be struck by, and the one thing that is still wholly unique to planet Earth, is its remarkable variety of life. Ever since the first prokaryotic organisms evolved almost 4 billion years ago, the variety of carbon-based life-forms on this planet has steadily increased, punctuated by only a handful of mass extinction events.

To conserve the spectacular variety of life that is the sole defining feature of our planet requires that practitioners understand why biodiversity exists, and understand what controls it. In this chapter, you will learn about the geologic history of biodiversity on Earth, about global patterns of biodiversity in the modern era, and about the primary ecological and evolutionary processes that control biodiversity. With a knowledge of Earth's history and modern patterns in hand, and an understanding of the primary controls, you will be in a good position to protect the future of biodiversity.

Biodiversity through Geologic Time

Biodiversity on our planet has varied greatly throughout Earth's history, fluctuating between dramatic lows after mass extinctions, to intervening peaks of diversification. It is worth knowing the geologic history of life on Earth if for no other reason than to understand the causes and consequences of past extinctions, and to be aware of the origins of all modern life-forms that have followed since.

Formation of early life

Earth is roughly 4.6 billion years old.[7] The earliest undisputed evidence of life on Earth comes from filament-like fossils that resemble the remains of microbial mats on rocks from Australia that date back to 3.5 billion years ago[8] (**Figure 4.1**). In addition, stromatolites, which are evidence of growth layers of cyanobacteria, have been found on exposed rock outcrops in Greenland and potentially date back to 3.7 billion years ago.[9] These cyanobacteria represent some of the earliest photosynthetic organisms on the planet and were largely responsible for the mass transfer of oxygen into Earth's early atmosphere that allowed aerobic life to thrive. Recent evidence suggests that prokaryotic life (bacteria and archaea—unicellular organisms lacking a membrane-bound nucleus, mitochondria, or other organelles) may have begun even earlier. In 2017, fossilized microorganisms were discovered in hydrothermal vent precipitates in Quebec, Canada, that may be as much as 4.3 billion years old.[10] If these findings hold up to scientific scrutiny, they will suggest there was a near-instantaneous emergence of life after formation of the world's oceans 4.4 billion years ago, very soon after the formation of Earth itself.

While life may have evolved quickly after formation of the Earth, the first 1.7 billion years of life were rather mundane. During the early history of Earth, evolution produced little beyond prokaryotic bacteria and simple eukaryotic algae that had the most basic cellular nuclei and organelles. But beginning around 635 million years ago (mya), the fossil record shows a great rise and fall of mysterious creatures of the Ediacaran fauna (see Figure 4.1). The Ediacaran fauna are an important landmark in the evolution of life on Earth, representing the first known metazoans—animals that were made up of more than one cell type. These early animals were primitive, soft-bodied organisms with strange frond- and disc-shaped body forms that would be recognized today only in select types of echinoderms (animals with radial symmetry like sea stars and urchins). But the primitive animals of the Ediacaran had the variety of cell types needed to serve as the template on which future evolution and diversification of body forms would proceed.

Radiation events

The first true explosion of biodiversity on the planet occurred during the Cambrian period roughly 542 mya (see Figure 4.1). The Cambrian explosion is perhaps the most remarkable biological event in the history of our planet. During this period, all major animal body plans ever known appeared in a relatively short time span of about 30 million years: bilateral symmetry led to the development of fast-moving

Era	Period	Onset	Major physical changes on Earth	Major biological changes on Earth
Cenozoic	Quaternary	2.6 mya	Cold/dry climate; repeated glaciations	Humans evolve; many large mammals become extinct 
	Tertiary	65.5 mya	Continents near current positions; climate cools	Diversification of birds, mammals, flowering plants, and insects 
Mesozoic	Cretaceous	145.5 mya	Laurasian continents attached to one another; Gondwana begins to drift apart; meteorite strikes near current Yucatán Peninsula at end of period	Dinosaurs continue to diversify; mass extinction at end of period (76% of species lost) 
	Jurassic	201.6 mya	Two large continents form: Laurasia (north) and Gondwana (south); climate warm	Dinosaurs dominate; radiation of ray-finned fishes; first fossils of flowering plants 
	Triassic	251.0 mya	Pangaea begins to drift apart; hot/humid climate	Dinosaurs evolve and radiate; origin of mammals; cone-bearing plants (gymnosperms) dominate landscape; mass extinction at end of period (80% of species lost) 
Paleozoic	Permian	299 mya	O_2 levels 50% higher than present; by end of period continents aggregate to form Pangaea, and O_2 levels drop rapidly	Radiation of reptiles; origin of most modern groups of insects; mass extinction at end of period (~96% of species lost) 
	Carboniferous	359 mya	Climate cools; marked latitudinal climate gradients	First seed plants appear; vast forests of vascular plants form; reptiles appear; amphibians dominate 
	Devonian	416 mya	Continents collide at end of period; giant meteorite probably strikes Earth	Jawed fishes diversify; first insects and amphibians; mass extinction at end of period (75% of species lost) 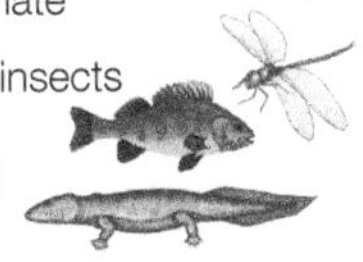
	Silurian	444 mya	Sea levels rise; two large land masses emerge; hot/humid climate	Jawless fishes diversify; first ray-finned fishes; vascular plants form and invade land 
	Ordovician	488 mya	Massive glaciation; sea level drops 50 meters	Marine algae abundant; colonization of land by diverse fungi, plants, and animals; mass extinction at end of period (86% of species lost) 
	Cambrian	542 mya	Atmospheric O_2 levels approach current levels	Rapid diversification of multicellular animals (Cambrian explosion) 
Precambrian	Ediacaran	635 mya	Climactic shift from icehouse to greenhouse conditions; changes in ocean chemistry	Diverse algae and soft-bodied metazoans (animals with multiple cell types) appear 
	Proterozoic	2.5 bya	Atmospheric O_2 levels increase from negligible to about 18%; "snowball Earth" from about 750 to 580 mya	Origin of photosynthesis, multicellular organisms, and eukaryotes 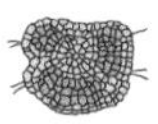
	Archean	3.8 bya	Earth accumulates more atmosphere (still almost no O_2); meteorite impacts greatly reduced	Origin of life; prokaryotic cells appear 
	Hadean	4.6 bya	Formation of Earth; cooling of Earth's surface; atmosphere contains almost no free O_2; oceans form; Earth under almost continuous bombardment from meteorites	Life not yet present

Note: mya, million years ago; bya, billion years ago.

Figure 4.1 A geological and ecological history of life on Earth, starting with the formation of the planet some 4600 million years ago. (After D. E. Sadava et al. 2017. *Life: The Science of Biology*, 11e. Sunderland, MA: Oxford University Press/Sinauer.)

species and radical new ecological interactions such as predation; the proliferation of mineralized skeletons led to a varied assortment of spines, tubes, and shells that illustrate how animals evolved to live and defend themselves in a myriad of different ways; and elaborate body parts developed, including compound eyes, complex appendages such as jointed legs and claws, and specialized structures for particular functions, such as feathery gills for breathing.

Biologists have argued for decades over what ignited the evolutionary burst during the Cambrian. Some say that oxygen levels in the atmosphere and oceans may have suddenly crossed a threshold that allowed for increased rates of metabolism and, in turn, the emergence of larger, more complex animals. Without the "Great Oxidation Event," it is unlikely that animals would have ever ventured onto land, for lack of an ozone layer. Others say that the biodiversity explosion sprang from the development of a few key evolutionary innovations, such as the evolution of eyes and vision. Still others say it sprang from **adaptive radiation**, which is the diversification of organisms into forms filling different ecological niches, such as predation and carnivory that may have set off a burst of new body types and behaviors. While most hypotheses have been posited as stand-alone ideas, recent evidence suggests the Cambrian explosion probably emerged out of a complex interplay of multiple factors, including genetic mutations and small environmental changes that triggered major evolutionary developments that, in turn, gave rise to an exponential rise in ecological opportunities that allowed a variety of new body forms to succeed[11] (**Video 4.1**).

Video 4.1 What caused the Cambrian explosion?
oup-arc.com/e/cardinale-v4.1

After the Cambrian explosion, there were two other noteworthy radiation events in the history of life on Earth, both more modest in their innovations. One gave rise to vascular plants during the early Silurian period (440 mya). The rise of vascular plants and, in particular, the expansion of seed-bearing plants onto land led to a green transformation of the planet that ultimately allowed for the proliferation of land animals (see Figure 4.1). The third major radiation event occurred during the Triassic period, which gave rise to many modern forms of biota, including corals, amphibians, reptiles, birds, and mammals. While virtually all the major body types and higher groups of living organisms to ever inhabit the planet appeared in the Cambrian, the Silurian and Triassic radiation events were important in the sense that they greatly increased the number of families, genera, and species of multicellular organisms.

Historical extinctions

More than 90% of all species that have ever lived on Earth are now extinct. Indeed, extinction has been the norm, and as old species have gone extinct, new species have evolved to replace them or fill ever-changing ecological niches. Biologists have used the fossil record to assess the *life spans* of species, that is, how long a typical species tends to survive on Earth. For example, paleobiologist David Raup analyzed the records of 8500 fossil genera and subgenera of invertebrates, most of which were marine, and found that the average species persists in the fossil record for about 11 million years. Similar analyses have been performed for bivalves, echinoderms, various kinds of oceanic plankton (e.g., diatoms, dinoflagellates), and mammals. While life spans differ among these groups of organisms, the general conclusion of analyses is that the average species persists for roughly 1 to 10 million years before it goes extinct or evolves into a new species.[12] Pimm and colleagues[13] used the shorter end of life spans to estimate a benchmark that represents the background rate of extinction for organisms in the fossil record. They argued that if a species lasts for 1 million years, it follows that 1 species extinction (E) tends to occur per million species every year (called a million species year, or MSY). Thus, they estimated that the background rate of extinction in the fossil record is 1 E/MSY. Several researchers have since worked to improve this estimate,[14–16] and more recent analyses suggest that typical rates of background extinction in the fossil record may be closer to 0.1 E/MSY.[17]

But extinction rates have not been constant throughout Earth's history. In fact, there have been several times in the last 500 million years where abnormally large numbers of species have died out simultaneously, or within a relatively short time period with respect to the geologic record (Figure 4.2). Paleobiologists have identified five unusually large mass extinction events that wiped out half or more of all life-forms in the geological blink of an eye. Records for the Big Five, as they are called, largely stem from the fossil record of marine organisms that, because of their hard body parts and shells made of calcium and silicon, tend to preserve better than many other types of organisms. While many smaller-scale extinction events have occurred, the Big Five represent periods in geologic history that have shaped modern life more than other historical events.

ORDOVICIAN–SILURIAN MASS EXTINCTION As the first of the Big Five mass extinction events, the Ordovician-Silurian actually represented two peak dying

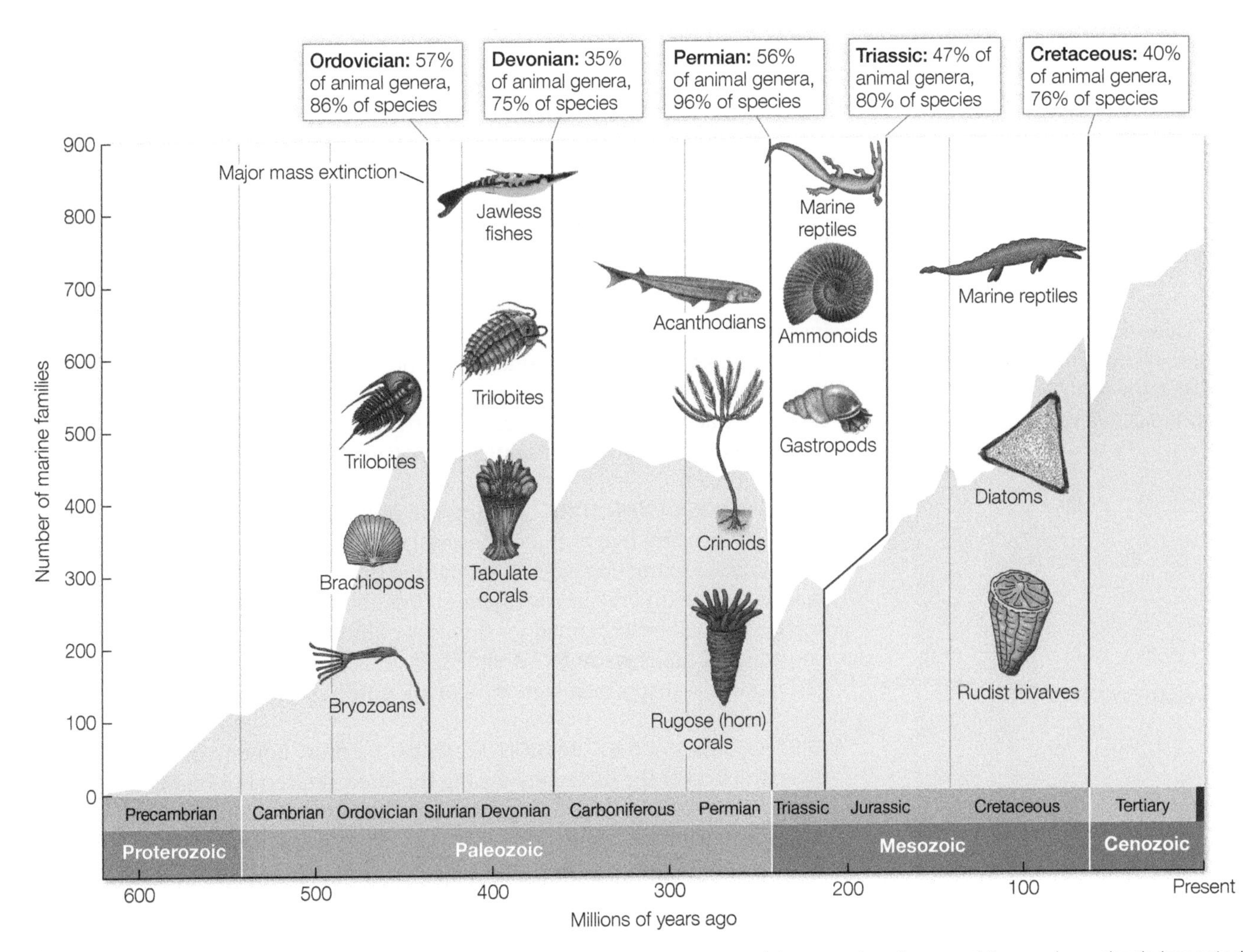

Figure 4.2 The diversity of marine animal families through geologic time, as measured in the fossil record. The Big Five mass extinction events are noted at top, showing the percentage of animal genera and of all species that went extinct during that event. Images provide examples of some of the marine animals impacted by these mass extinction events. (Curve after D. M. Raup and J. J. Sepkoski. 1982. *Science* 215: 1501–1503; data on genera and species loss from A. D. Barnosky et al. 2011. *Nature* 471: 51–57.)

periods that were separated by a few hundred thousand years. One period was associated with glacially driven cooling, and the second followed a drop in sea levels. These were the first events known to significantly affect animal-based communities, and together they represented the second-largest mass extinction ever recorded. Extinction was global during this period, eliminating 57% of marine genera and nearly 86% of marine species.[18] Almost all major taxonomic groups were affected during this extinction event, though trilobites, brachiopods, bryozoans, and graptolites were particularly hard hit.[19]

DEVONIAN MASS EXTINCTION The Devonian extinction was a sequence of two large, and perhaps several more small, events that began about 375 mya and spread over several million years. The Kellwasser event primarily affected marine organisms, with greatest effects on conodonts (jawless chordate fishes resembling eels), trilobites, brachiopods, and reef-building organisms like rugose and tabulate corals. The collapse of reef systems was so severe, it took more than 100 million years for new scleractinian, or "stony," corals to reappear in the fossil record. The Kellwasser event was followed by a second extinction called the Hangenberg event, which led to extinction of all placoderms (early jawed fishes) and most sarcopterygians (lobe-finned fishes) and a complete turnover of vertebrate biota.[20] By the end of the Devonian, one-third of all animal families and three-quarters of all species had died out, including 97% of all vertebrate species.[21] The exact causes of these extinctions are unclear, but leading hypotheses include changes in sea level and ocean hypoxia, possibly triggered by global cooling or oceanic volcanism.[20]

PERMIAN-TRIASSIC (P-T) MASS EXTINCTION The P-T mass extinction was the worst of the Big Five and has been dubbed the Great Dying. During this period a staggering 56% of all genera and 96% of all species went extinct, including 70% of terrestrial vertebrate species.[23] It is the only known mass extinction of insects,[24] with 57% of all insect families and 83% of all genera going extinct. All animal life today is descended from the 4% of species that survived the P-T extinction event. Suggested causes include a large meteor impact, massive volcanism, runaway greenhouse gases, and climate change triggered by sudden release of methane from the seafloor[25] (**Video 4.2**).

Video 4.2 Watch this video about events of the Permian-Triassic mass extinction.
oup-arc.com/e/cardinale-v4.2

TRIASSIC MASS EXTINCTION During the final 18 million years of the Triassic period, there were two or three phases of extinction that comprise the Triassic-Jurassic mass extinction event. During this time, 47% of all genera and 80% of species went extinct. Climate change, flood basalt eruptions, and an asteroid impact have all been blamed for this loss of life. At the end of the Triassic period, several major body forms vanished, including nearly all ammonites and gastropods, and approximately 80% of all reptile species.

CRETACEOUS MASS EXTINCTION Perhaps the most famous of the Big Five is the extinction of the dinosaurs during the late Cretaceous 65 mya, after which mammals achieved dominance in terrestrial communities. While the Cretaceous-Tertiary mass extinction—also known as the K-T extinction—is famed for the death of the dinosaurs, about 76% of all species (40% of genera) perished at the end of the Cretaceous, including a large fraction of marine reptiles, ammonoids, diatoms and certain types of bivalves. The event was likely triggered by a 180-kilometer-wide meteor striking present-day Yucatán, Mexico.

Mass extinctions have been tremendously deadly events, and it has taken 10 million years or more for life on Earth to rebound after these events.[26] But if there is a silver lining to mass extinctions, it is that they have repeatedly opened up opportunities for new life-forms to emerge and proliferate. Dinosaurs appeared after the Permian-Triassic extinction, which allowed many forms of reptiles to become remarkably successful. And when the dinosaurs were wiped out by the Cretaceous extinction, mammals rapidly diversified and came to dominance on the planet. Some people suggest that humans are presently causing the sixth mass extinction in the history of life on Earth.[27] We discuss modern changes to biodiversity in Chapter 8, Extinction, and review the evidence for claims of a sixth mass extinction.

Global Patterns of Biodiversity

While there are more species on Earth now than any previous time in geologic history (see Figure 4.2), modern biodiversity is unequally distributed across the planet. In this section, we detail estimates of how much biodiversity presently exists, and describe where it is found.

How many species are there?

To date, roughly 1.24 million eukaryotic species (organisms that have cells with membrane-bound organelles, unlike prokaryotic bacteria and archaea) have been discovered, described, and scientifically cataloged.[28,29] This includes 1.05 million species on land and 0.19 million from the world's oceans.[29] Of the species that have been cataloged, the vast majority have been animals and plants (**Figure 4.3**). Insects alone represent more than 750,000 described species, and of these, more than 350,000 are beetles.

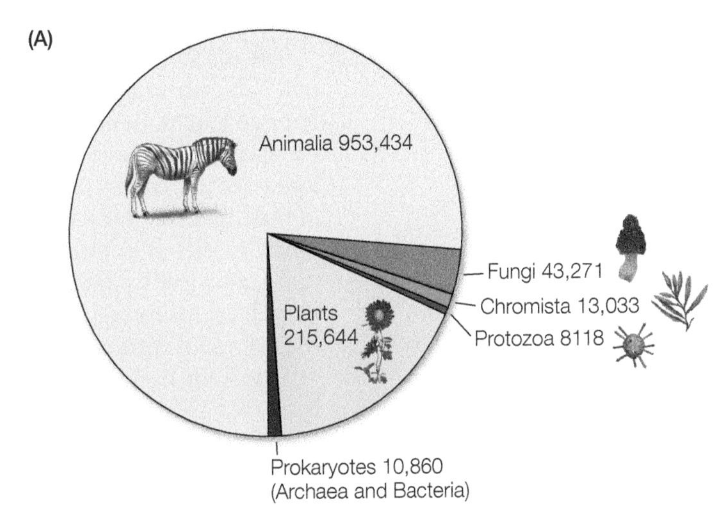

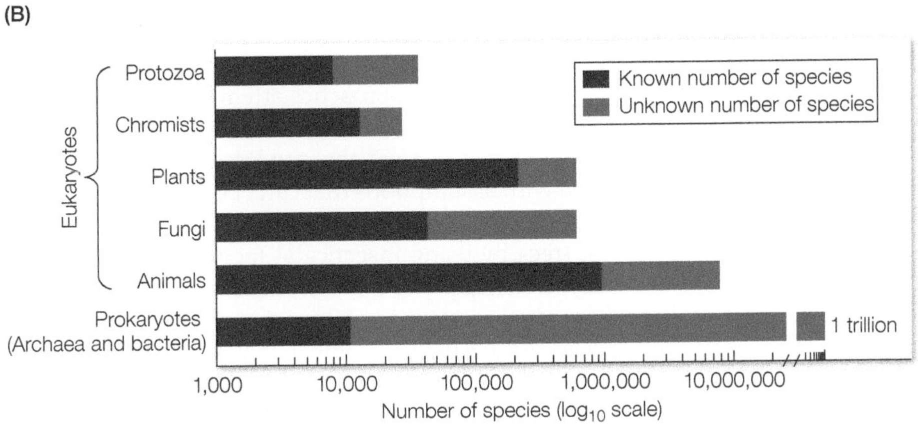

Figure 4.3 (A) About 1.24 million eukaryotic species have now been identified and described by scientists. The majority of these are eukaryotic animals and plants. (B) Recent estimates suggest that described species (red bars) represent only a fraction of those that actually exist on Earth. A large fraction of eukaryotic species (animals, plants, fungi, chromists, and protozoa) have yet to be discovered (blue bars), and somewhere between a billion and a trillion types of prokaryotes (archaea and bacteria) are still unknown to science. (After C. Mora et al. 2011. *PLOS Biol* 9: e1001127. Data in B for prokaryotic bacteria and archae from K. J. Locey and J. T. Lennon. 2016. *Proc Nat Acad Sci* 113: 5970–5975.)

Certain groups of organisms are woefully under-represented in our catalog of described species. The number of types of prokaryotes (unicellular organisms that lack membrane-bound nuclei or organelles) lies at about 11,000 (see Figure 4.3A), which almost certainly represents a vast underestimate. Most prokaryotes defy traditional taxonomic and biological concepts of a species (see Chapter 3), and the genomic techniques needed to properly describe and catalog them have only recently become economically viable. In addition, biologists have shown clear biases for the discovery of larger organisms (vertebrates, insects, vascular plants), leaving most of the smaller species on the planet undetected, undescribed, and unnamed.

About 16,000 new species are still being described every year,[28] which is remarkable given that humans have now explored and come to dominate most locations on the planet. In groups such as insects, spiders, mites, nematodes, and fungi, the number of described species is still increasing at the rate of 1%–2% per year. Even for groups of large organisms that are relatively well known, such as birds, mammals, and flowering plants, steady numbers of new species are being discovered each year.[30] In the past few decades, dozens of new monkey species have been found in Brazil, dozens of new species of lemurs have been discovered in Madagascar, and hundreds of new species of amphibians have been discovered in tropical forests.

The discovery of new species, and sometimes of entire communities that no one had ever imagined could exist, is one of the most exciting things that can happen to a modern biologist. For example, discovery of a previously unknown group of deep-sea tubeworms led to a scientific revolution that completely altered our understanding of life on Earth and the potential for life in other parts of the universe (**Box 4.1**). The discovery of the coelacanth (*Latimeria chalumnae*), a fish species that was thought to have gone extinct 65 mya, was a conservation success story (**Box 4.2**). The coelacanth was a "missing link" in the fossil record, representing the transition from fish that were constrained to swim in the ocean, to vertebrates that were able to invade and walk on land.

Given the high rate of new discovery, and the fact that many types of organisms are vastly understudied, most biologists agree that the 1.24 million species described to date represent just a fraction of life-forms that are likely to exist on Earth (see Figure 4.3B). But estimating the true number of life-forms on Earth is a daunting challenge. Polling taxonomic experts and asking them to give their best guess has produced widely varying estimates that range from 3 to 100 million species globally.[31] Biologists have taken a variety of approaches to try and narrow this range, with most approaches attempting to extrapolate biodiversity from well-known groups of organisms to lesser-known groups, or from well-studied areas of the planet to areas that are lesser known. But these approaches are often riddled with assumptions that can't be verified. For example, entomologists have attempted to sample entire insect communities in the canopies of tropical forest trees by using insecticidal fogging.[32] Using results of these intensive collection efforts performed on a small number of trees and tree species, they have discovered an average of 9 species of specialized beetles feeding on each plant species. Assuming the number of specialist beetles holds across all 55,000 species of known tropical trees and lianas (woody vines), one can estimate there may be 495,000 species of beetles living in the tropical canopy. Of beetles that have been described, canopy beetles represent an estimated 44% of all species; therefore, there may be as many as 1.1 million species of beetles in tropical forests. And given that beetles represent about 20% of all known insects, there may be as many as 5 million species of insect in tropical forests alone. Calculations such as these, performed for many different groups of organisms (insects, mammals, birds, vascular plants, etc.), have produced estimates of 5 to 10 million species of plants and animals for the entire Earth.[33]

BOX 4.1 Case in Point

Discovery of a New Way to Make Life

Throughout the history of biology, the dominant view was that all life ultimately derives from the sun. Photosynthesis—the process by which plants convert inorganic carbon (CO_2) into organic carbon, that is, carbohydrates ($[CH_2O]_n$)—was the only known way to generate new biological tissue:

$$6CO_2 + 6H_2O + \text{sunlight (energy)} \rightarrow [CH_2O]_n + 6O_2$$

If photosynthesis is the only process for creating organic carbon, then plants only exist because of sunlight, and animals only exist because of plants. It follows that places with no sunlight must be biological deserts.

That view of biology was radically altered by the discovery of deep-sea tubeworms living on hydrothermal vents on the ocean floor. The story behind this discovery began in 1900 when a strange tube-dwelling worm was dredged from deep waters around Indonesia. Though the worm resembled an annelid (segmented worm), it lacked any obvious segmentation. More perplexing, it lacked a mouth, gut, and anus. These characteristics were later confirmed by collection of more intact specimens in 1964, and biologists then assigned the strange new animal, which had no known way of feeding and digestion, to its own phylum, Pogonophora (see Figure).

Then, in 1977 the world's first deep-sea submersible—the *Alvin*—discovered hydrothermal vents on the ocean floor in the Galápagos rift of the Pacific Ocean. These vents were underwater openings in the Earth's crust that were gushing excessively hot water (in excess of 150°C [302°F]) and were flush with dissolved minerals such as sulfides. To everyone's surprise, the hydrothermal vents were teeming with life. They were colonized with rich biological communities composed of large, complex animals, including 2-meter-long pogonophoran tubeworms, clams, crabs, and even fishes. How could animals like these possibly live on the bottom of the ocean, where there was no sunlight and there were no plants?

It was soon discovered that pogonophoran tubeworms, and other hydrothermal vent animals, had evolved a specialized organ called a trophosome that houses symbiotic bacteria that provide food to their host. Those symbiotic bacteria were found to be chemosynthetic, able to use a novel process to convert inorganic carbon (CO_2) from seawater into organic carbon ($[CH_2O]_n$) by using energy produced not by sunlight, but by the oxidation of hydrogen sulfide (H_2S) collected from the vents:

$$6CO_2 + 12H_2S \text{ (energy)} \rightarrow [CH_2O]_n + 6H_2O + 12S$$

Sergei Winogradsky had proposed that bacterial chemosynthesis might be possible as far back as the 1890s, but no one could have anticipated that bacteria and animals would evolve a partnership allowing rich communities of higher animal life to thrive in areas with no sunlight and no plants. Chemosynthesis may have been the first type of metabolism to evolve on Earth, leading the way for cellular respiration and photosynthesis to develop later. It has subsequently been proposed that if life exists on other planets, the primary means of energy production is likely to be through similar forms of chemosynthesis.

A 2004 television series hosted by Bill Nye "the Science Guy" named chemosynthesis as one of the 100 greatest scientific discoveries of all time.

NOAA Okeanos Explorer Program, Galápagos Rift Expedition 2011

Deep-sea tubeworms (*Pogonophora*) living on hydrothermal vents on the ocean floor led to the discovery of bacterial chemosynthesis, which is a way to fix carbon and make biological tissue without sunlight.

Recently, Mora and colleagues[29] developed a new approach to estimating global biodiversity—one that may provide more realistic estimates of the true numbers of many groups of organisms. The authors first generated accumulation curves showing the rate at which new taxa have been discovered through time at various levels of the taxonomic hierarchy (phyla, classes, orders, families, genera, and species). Higher taxonomic groupings (e.g., phyla) tend to be

BOX 4.2 **Success Story**

Rediscovery of an Extinct Species

In 1938, fish biologists throughout the world were stunned by the report of a strange fish caught in the Indian Ocean. This fish, subsequently named *Latimeria chalumnae*, belonged to a group of marine fish known as coelacanths that were common in ancient seas but were thought to have gone extinct 65 million years ago.

Coelacanths are of particular interest to evolutionary biologists because they show certain features of muscles and bones in their fins that are comparable to the limbs of the first vertebrates that are known to have crawled onto land (a so-called missing link). Biologists searched the Indian Ocean for 14 years before another coelacanth was found, off the island of Grande Comore between Madagascar and the African coast. Subsequent investigations showed there was one population of about 300 individuals living in underwater caves just offshore Grande Comore and perhaps a few other small populations on the east coast of Africa. In recent years, the Union of the Comoros has implemented a conservation plan to protect the coelacanths, including a ban on catching and selling the fish. In a remarkable addition to this story, in 1997 a marine biologist working in Indonesia was astonished to see a dead coelacanth for sale in a local fish market. Subsequent investigations demonstrated that this was a completely different species of coelacanth (*Latimeria menadoensis*)[69] previously unknown to science but well known to the local fishermen. That population is now estimated to be about 10,000, illustrating how much is still waiting to be discovered in the world's oceans.

The coelacanth (*Latimeria chalumnae*), which was thought to have gone extinct 65 mya, was rediscovered in the Indian Ocean in 1938.

well enough studied that the rate of discovery of new taxa has slowed in recent decades, and in many cases, we've reached an asymptotic estimate of taxonomic diversity at these higher levels (**Figure 4.4A**). In contrast, accumulation curves at lower taxonomic levels (e.g., genera) tend to show that discovery rates are still increasing and that we don't have good estimates of the numbers of genera and species for most groups of organisms (**Figure 4.4B**).

The clever part of Mora and colleagues' approach is they went on to show that the number of taxa in one level of the biological hierarchy can be predicted by the number of taxa in another part of the hierarchy (**Figure 4.4C**). In other words, if you have good estimates of the number of phyla (or classes or families) that have been described for a group of organisms (e.g., animals), then you can use those estimates to predict the number of taxa that are likely to occur for that group at lower levels of taxonomy, such as genera or species. After using well-studied groups of organisms to validate their approach and show that it works, the authors applied their method to major groups of organisms for which estimates of species are largely unknown.

The analyses by Mora and colleagues[29] predicted there are likely 8.7 million species of eukaryotic organisms globally, of which 2.2 million are marine. Of these, there are 7.7 million species of animals predicted to live on Earth, as well as 611,000 types of fungi, 298,000 types of plants, and nearly 64,000 types

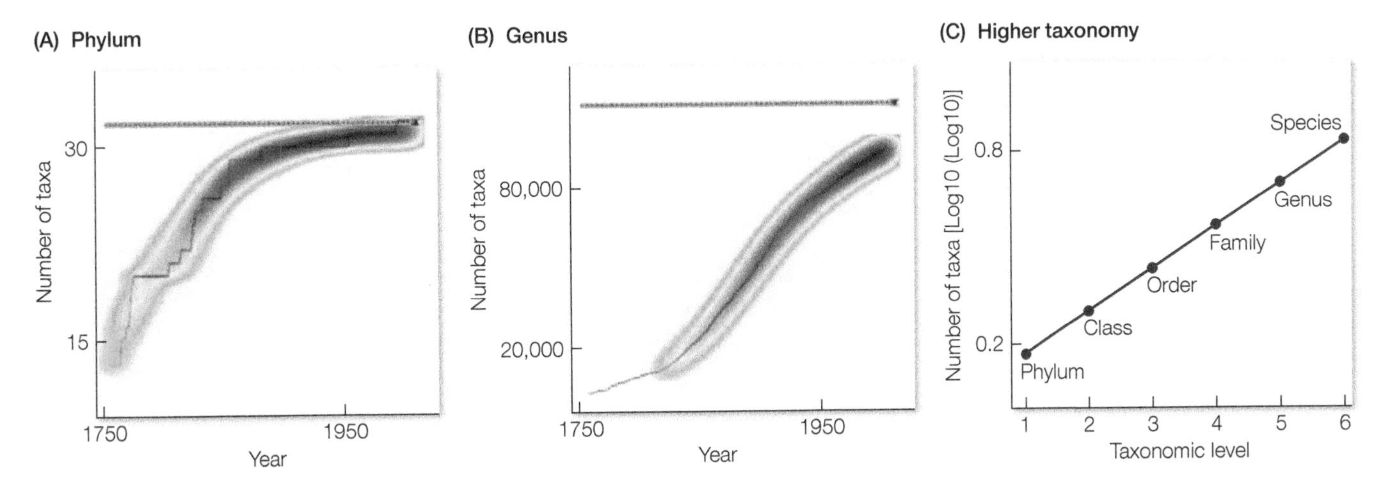

Figure 4.4 Predicting the global number of species of animals from their higher taxonomy. (A) The number of new phyla discovered reached a plateau about 1950. However, (B) the number of new genera being discovered is still going up. This means we don't know for sure how many genera there are. However, (C) for well-studied groups of organisms that have been fully described, there is a predictable relationship between diversity at the level of phyla and diversity at the level of genera. Using this relationship, scientists can estimate how many genera are likely to exist for less-studied groups. (From C. Mora et al. 2011. *PLOS Biol* 9: e1001127/CC BY 4.0.)

of protozoans and chromista (e.g., brown algae, diatoms, and certain kinds of molds that have chlorophyll c and accessory photopigments not found in plants). These results suggest than 86% of all eukaryotic species on the planet, and 91% of those that live in the world's oceans, have yet to be discovered (see Figure 4.3B). Thus, even after 250 years of taxonomic classification of the world's biology dating back to Linnaeus, we have only succeeded in describing and indexing a small fraction of species on the entire Earth (14%), and even fewer in the ocean (9%). This result left the authors with a sobering thought:

> Considering current rates of description of eukaryote species in the last 20 years (i.e., 6,200 species per year...), the average number of new species described per taxonomist's career (i.e., 24.8 species...) and the estimated average cost to describe animal species (i.e., US$48,500 per species) ... describing Earth's remaining species may take as long as 1,200 years and would require 303,000 taxonomists at an approximated cost of US$364 billion.[29]

The method used by Mora and colleagues to estimate eukarotic diversity didn't work so well for prokaryotes (bacteria and archaea), because the discovery of higher taxonomic groupings is still accelerating; thus, we don't have reliable estimates of diversity for prokaryotes even at the highest taxonomic levels. Therefore, microbiologists have taken other approaches to predict the variety of microbial life on Earth. In 2016, Locey and Lennon[34] analyzed data sources that had sampled over 20,000 geographic sites for bacteria, archaea, and microscopic fungi. Using these samples, the researchers worked out a scaling relationship that linked the number of individual organisms to the total number of species. From this scaling relationship, and their best estimates of microbial abundance, the authors estimated there are between 100 billion and 1 trillion different types of microbial species on Earth.

The discovery of new microbes has much potential to influence human well-being. For example, in the field of human medicine, we have just recently learned that the bacterial cells hosted on the human body outnumber human cells three to one. Each person has a unique microbial fauna, and many microbes have evolved as symbionts that help keep the human body healthy (Box 4.3).

BOX 4.3 **Case in Point**

Discovery of the Human Biome

Over the past several decades, scientists have discovered that the microbial cells on a human body outnumber human cells almost three to one. The human body is composed of roughly 37 trillion cells; yet, it is host to an estimated 100 trillion bacterial cells. The vast majority of these microbes have still not been successfully cultured, identified, or studied.

In 2008, the U.S. National Institutes of Health (NIH) launched a 5-year, $115 million initiative called the Human Microbiome Project. The goal of this initiative was to identify and characterize the community of microorganisms that are found in association with the human body (the human microbiome) and to determine how changes in the human microbiome influence human health and disease. For the project, more than 5000 samples from men and women were collected, including tissues and body sites such as mouth, nose, skin, lower intestine (stool), and vagina. All DNA, human and microbial, was analyzed with DNA-sequencing machines. The microbial genome data were extracted by identifying the bacteria-specific ribosomal RNA, 16S rRNA.

The Human Microbiome Project discovered that the human body is populated by vast numbers of viruses, bacteria, fungi, and mites. Researchers estimated that more than 10,000 microbial species make up the human ecosystem. Each person is host to more than 500 different microbial species, and individuals have their own unique combination of microbes (see Figure).

It was found that more of the genes responsible for human survival are contributed by microbes than by humans themselves. For example, bacterial protein-coding genes in the human body are 360 times more abundant than human genes. It was also discovered that microbial communities are often responsible for keeping people healthy. Gut microbes that inhabit healthy people's large intestines help with digestion and various immune processes. Some of these bacteria may play a beneficial role in secreting antimicrobial compounds that control harmful bacteria. Changes to the human microbiome can adversely affect health, increasing the risk of infection and contributing to chronic health problems like Crohn's disease and irritable bowel syndrome. A mother's microbiome may even influence her children's health.

In an age where people seem obsessed with antimicrobial medicines, soaps, disinfectants, and antiseptics, and where parents go to great lengths to create sterile environments for their children, the Human Microbiome Project showed that human bodies are a biome, and the unique community of organisms that has developed on and in each individual can have profound impacts on one's health.

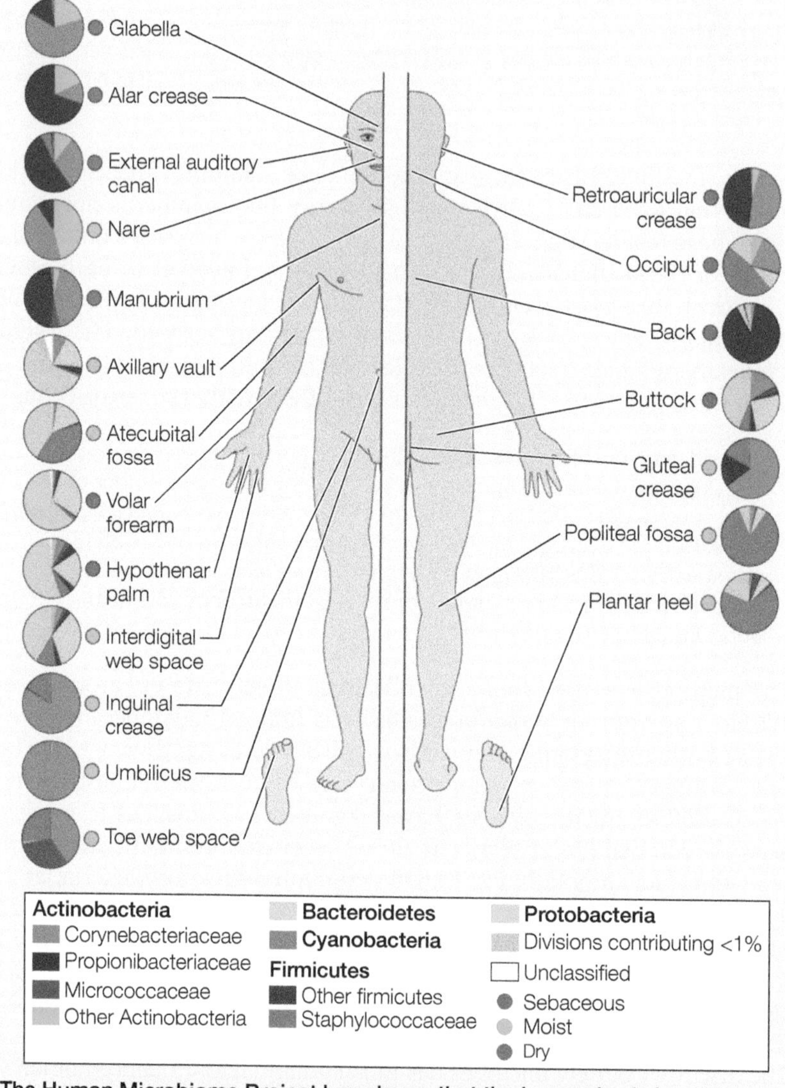

The Human Microbiome Project has shown that the human body is an ecosystem that hosts more than 10,000 unique types of bacteria. Bacterial cells on the human body outnumber human cells almost three to one. (After E. A. Grice and J. A. Segre. 2011. *Nat Rev Microbiol* 9: 244–253; E. A. Grice et al. 2009. *Science* 324: 1190–1192.)

Realms, biomes, and ecoregions

Biologists often organize the diversity of Earth's life-forms into biogeographic realms that represent large regions across the globe where organisms share common geologic and evolutionary histories. Eight biogeographic realms are recognized, corresponding roughly to continents: the Palearctic, Afrotropical, Indo-Malay, Australasian, Oceanic, Nearctic, Neotropic, and Antarctic (**Figure 4.5A**). Each biogeographic realm is composed of biomes—areas that can be identified by their dominant vegetation type on land (see Figure 4.5A), or by their dominant physical processes in aquatic habitats (**Figure 4.5B**). While biomes in different biogeographic realms often share similar physical processes and major types of organisms (e.g., tropical forest biomes are found in several biogeographic realms), the species can differ dramatically among biomes in different biogeographic realms because of their unique evolutionary histories.

Terrestrial biomes tend to change predictably along major environmental gradients (**Figure 4.6**). Holdridge[35] classified terrestrial biomes into "life zones" using two abiotic factors (precipitation and temperature) and one biological variable (potential evapotranspiration) that allowed him to predict the types of biomes that exist in areas of differing latitude, altitude, and what he called humidity provinces. Whittaker[36,37] simplified this classification and showed that terrestrial biomes can be arranged by temperature and precipitation, which have large impacts on the physiology, and thus geographic distribution, of different types of species. For example, in tropical latitudes of the world where temperatures are high, biomes change predictably along precipitation gradients from tropical rain forests, to tropical seasonal forests, to thorn scrub and woodlands, to deserts—each with a unique type of vegetation and species composition. Knowing how biomes vary along major environmental gradients is useful because it can help predict changes in species and community types that might result from human impacts such as global climate change (see Chapter 12).

In 2001, a group of scientists at the World Wildlife Fund reclassified Earth's land-based biomes into 867 terrestrial ecoregions.[38] *Ecoregion* was defined as a unit of land containing a distinct assemblage of natural communities and species, with boundaries that approximate the original extent of natural communities prior to major land-use change. The authors argued these ecoregions were representative of habitats and species assemblages that existed prior to human domination of the planet and, thus, could be used for conservation planning at regional to global scales. The World Wildlife Fund map of ecoregions is now one of the most comprehensive descriptions of landscape-level diversity, and it is a widely accepted base map for conservation planning.

The most diverse ecoregions

The biodiversity described by biologists to date is unequally distributed across the planet, and certain types of biomes and ecoregions are more diverse than others. The most species-rich environments are tropical forests, coral reefs, large tropical lakes and river systems, and regions with Mediterranean climates.

TROPICAL FORESTS While the world's tropical forests occupy only 7% of the land area, they contain more than half the world's species.[39] Insects found in tropical forests likely constitute the majority of the world's species.[33] About 40% of the total 216,000 species of flowering plants, gymnosperms, and ferns occur in tropical forests in the Americas, Africa, Madagascar, Southeast Asia, New Guinea, and Australia and on various tropical islands. About 30% of the world's bird species—1300 species in the American tropics, 400 species in tropical Africa, and 900 in tropical Asia—depend on tropical forests. This figure is probably

(A)

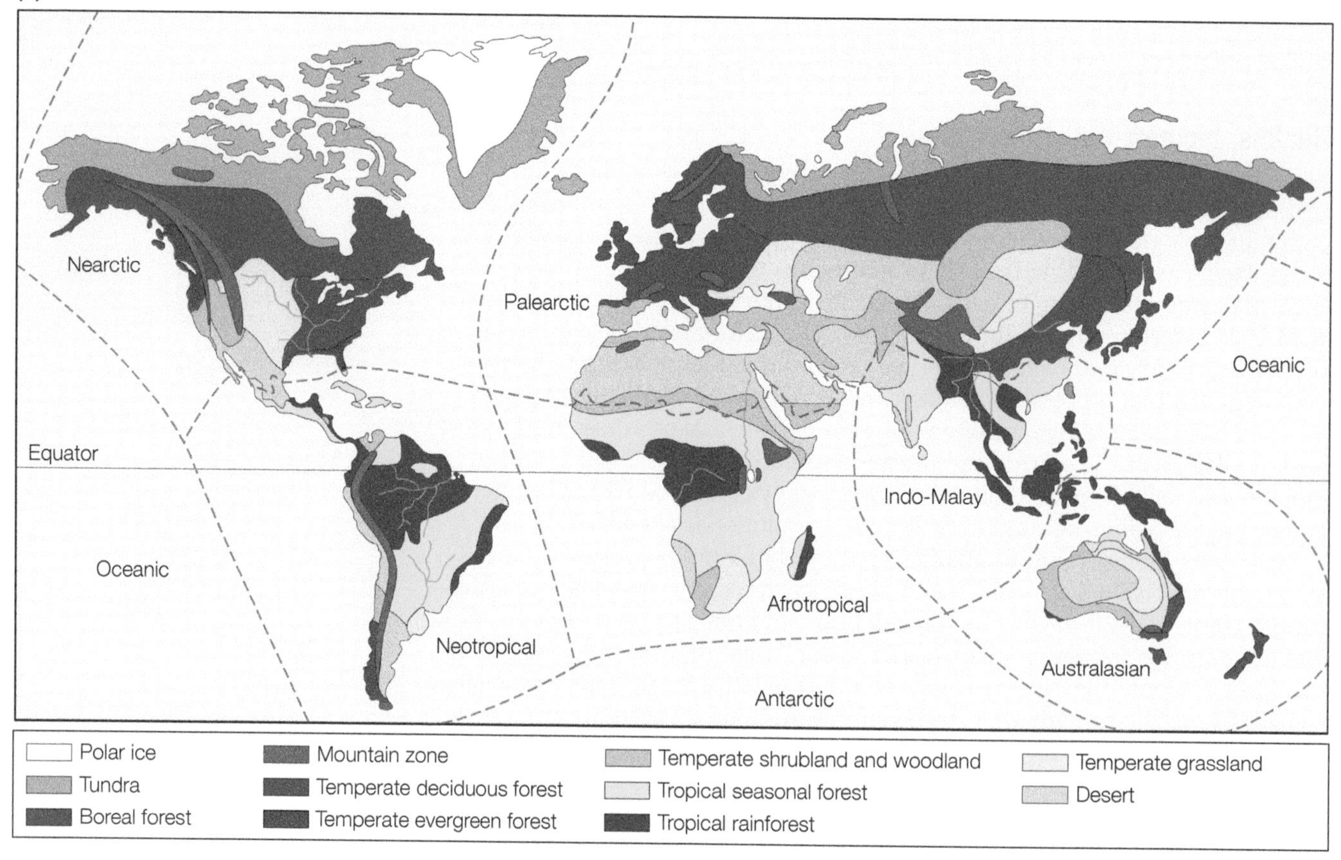

(B)

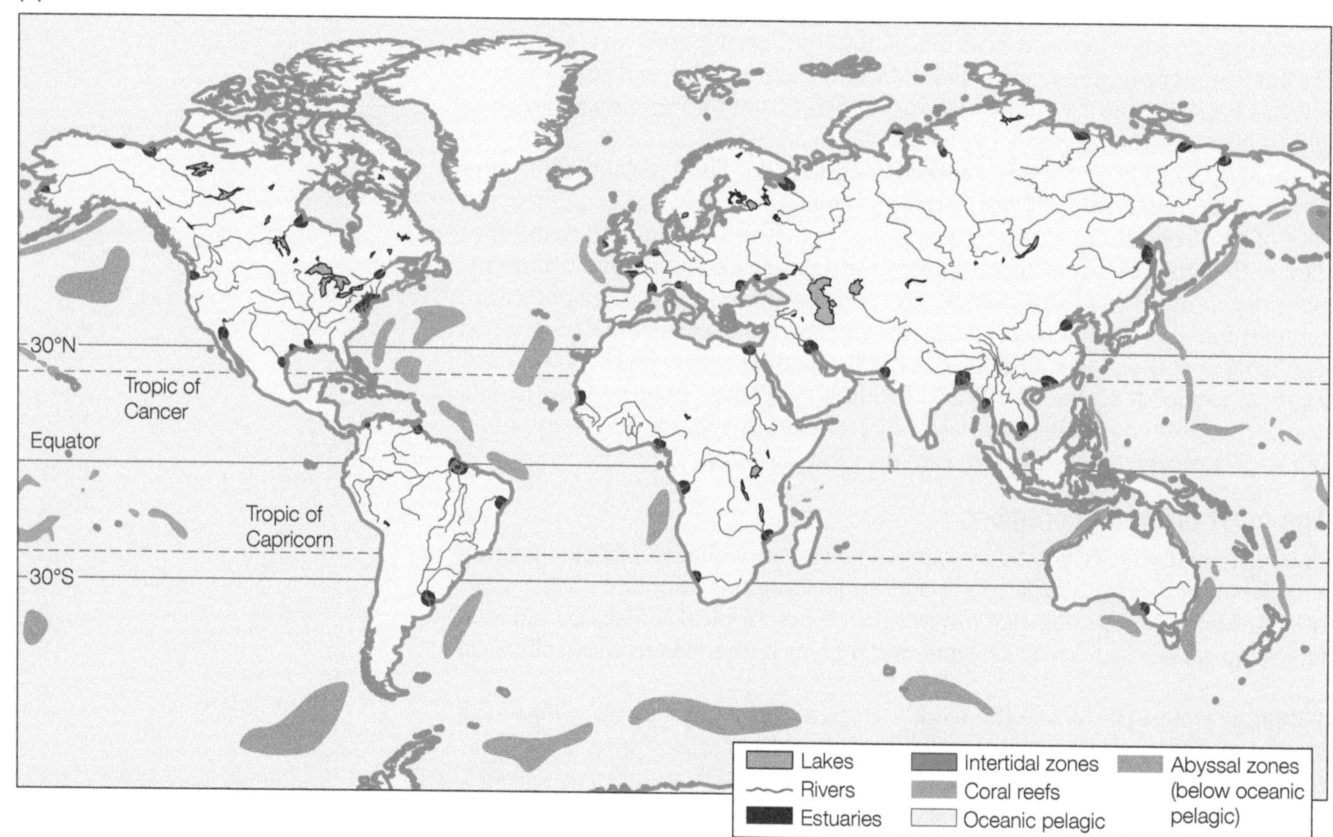

Figure 4.5 Realms and biomes. (A) The globe divided into Earth's major biogeographic realms and terrestrial biomes. (B) Common marine and freshwater biomes. (A, from Millennium Ecosystem Assessment, 2005. *Ecosystems and Human Well-being: Biodiversity Synthesis*. Washington, DC: World Resources Institute.)

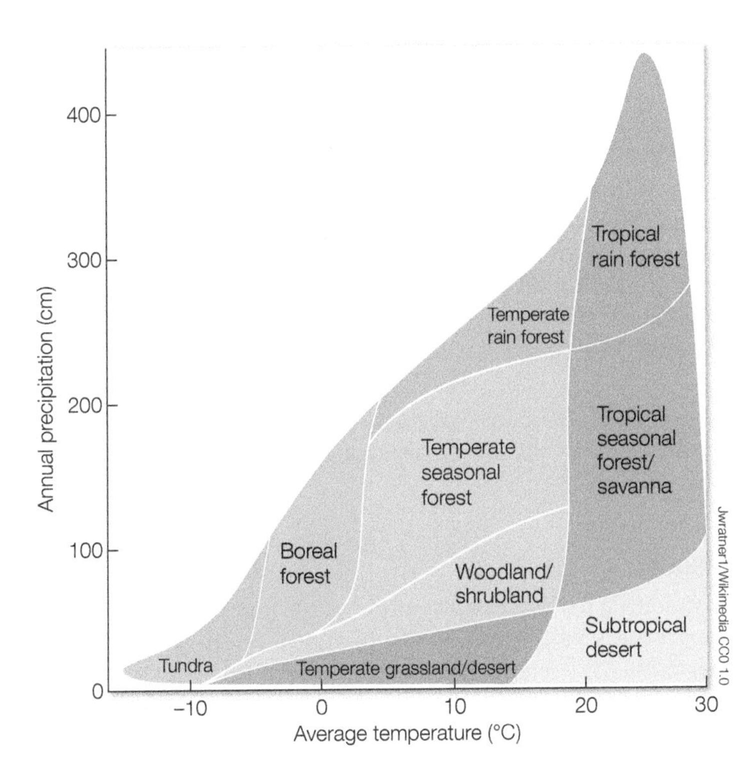

Figure 4.6 Terrestrial biomes, which are defined by dominant types of plant communities, change predictably along environmental gradients such as temperature (*x*-axis) and precipitation (*y*-axis). Note that if you know the temperature and precipitation for any location on Earth, or you can predict what temperature and precipitation may be in the future, then you can likely predict the type of biome and vegetation that will be there. This is useful, for example, when predicting how dominant types of vegetation, and communities of species, will respond to human stressors such as climate change.

an underestimate, since it does not include species that are only partially dependent on tropical forests (such as migratory birds).

CORAL REEFS The marine equivalent of tropical forests is coral reefs, which are believed to have the highest biodiversity of any marine ecosystem on the planet. Marine ecosystems are home to 28 of the 35 animal phyla that exist today; one-third of these phyla exist only in the marine environment.[40] Much of this is due to the diversity of coral reefs. Coral reefs occupy less than 0.1% of ocean surface area, yet they are home to at least 25% of all marine species,[41] including fish, mollusks, worms, crustaceans, echinoderms, sponges, tunicates, and cnidarians. The world's largest coral reef, Australia's Great Barrier Reef, contains over 400 species of corals, 1500 species of fish, 4000 species of mollusks, and 6 species of turtles and provides breeding sites for some 252 species of birds.

LARGE TROPICAL LAKES AND RIVER SYSTEMS Freshwater ecosystems make up just 0.01% of the world's water and approximately 0.8% of the Earth's surface. Yet, this tiny fraction of global water supports an estimated 125,000 species,[42] representing 10% of all known organisms and roughly one-third of all known vertebrates. Large lakes and river systems are disproportionately important for certain groups of organisms. Freshwater fishes comprise almost 45% of all fishes and 25% of all mollusks on Earth, which is truly remarkable given how small freshwater ecosystems are in area and volume compared with the world's oceans. Other major groups that depend on freshwater include reptiles, amphibians, insects, plants, and mammals.

MEDITERRANEAN COMMUNITIES Great diversity is found among plant species in southwestern Australia, the cape region of South Africa, California, central Chile, and the Mediterranean basin, all of which are characterized by a Mediterranean climate of moist winters and hot, dry summers.[43] Of these regions, the Mediterranean basin is the largest in area (2.1 million km^2) and has the most plant species (22,500). Though the Mediterranean basin covers just 0.7% of the world's ocean area, it is one of the major reservoirs of marine and coastal biodiversity, with 28% of endemic species and 7.5% of the world's marine fauna and 18% of its marine flora. The Cape Floristic Region of South Africa has an extraordinary concentration of unique plant species (9000) in a relatively small area (78,555 km^2). The shrub and herb communities in these areas are rich in species apparently because of their combination of geological age, complex site characteristics (e.g., topography and soils), and environmental fluctuations. The frequency of fire in these areas also may favor rapid speciation and prevent the domination of just a few species.

Suggested Exercise 4.1 Go to the Companion Website to investigate *Terrestrial Ecoregions of the World.* **oup-arc.com/e/cardinale-ex4.1**

Biodiversity hotspots

A species that is found in a particular ecoregion but nowhere else is said to be **endemic** to that region. Certain areas on the planet have far more endemic species than others (Table 4.1). For example, the tropical Andes have particularly high numbers of endemic plants, birds, and amphibians. In contrast, a large fraction of endemic mammals is concentrated in Mesoamerica, and a great many

TABLE 4.1 Numbers of endemic species of plants, birds, mammals, reptiles and amphibians in regions that have exceptional degrees of endemism[a]

Region	Size (%)[b]	Plants	Birds	Mammals	Reptiles	Amphibians
Tropical Andes	314,500 (25)	20,000	677	68	218	604
Mesoamerica	231,999 (20)	5,000	251	210	391	307
Caribbean	29,840 (11)	7,000	148	49	418	164
Atlantic Forest of Brazil	91,930 (7.5)	8,000	181	73	60	253
Choco/Darien/ Western Ecuador	16,471 (24)	2,250	85	60	63	210
Cerrado of Brazil	356,630 (20)	4,400	29	19	24	45
Madagascar	59,038 (10)	9,704	199	84	301	187
West African forests	126,500 (10)	2,250	90	45	46	89
Cape Floristic Province	18,000 (24)	5,682	6	9	19	19
Mediterranean Basin	110,000 (5)	13,000	47	46	110	32
Sundaland	125,000 (8)	15,000	139	115	268	179
Wallacea	52,020 (15)	1,500	249	123	122	35
Philippines	9,023 (3)	5,832	183	111	159	65
Indo-Burma	100,000 (5)	7,000	140	73	201	114
South-central China	64,000 (8)	3,500	36	75	16	51
Western Ghats/ Sri Lanka	12,450 (7)	2,180	40	38	161	116
Southwest Australia	33,336 (11)	4,331	19	7	50	24
Polynesia/Miconesia	10,024 (22)	3,334	174	9	37	3

Source: Data from N. Myers et al. 2000. *Nature* 403: 853–858.
[a]1% or greater of global total for endemic plants or vertebrates.
[b]Size (%) refers to remaining primary vegetation (in km^2) and to percent remaining of original extent.

endemic reptiles are concentrated in the Caribbean. Owing to their isolation, islands often have high proportions of endemic species. However, because islands sometimes have relatively impoverished biotas compared with mainland areas, islands with high endemism are not always those that have the highest levels of species richness.

Areas with abnormally high levels of species richness are called biodiversity hotspots. Norman Myers and his colleagues identified 25 biodiversity hotspots that contain as much as 44% of all species of vascular plants, and 35% of all vertebrate species, in areas that together comprise just 1.4% of the land surface of the Earth (Figure 4.7).[44] Others have expanded or refined the number of hotspots to encompass the greatest amount of diversity in the smallest area of land and oceans. Because these hotspots often overlap with areas where significant levels of biodiversity are threatened from activities like habitat loss, the hotspot concept represents an alternative to conservation strategies that focus on saving ecoregions or areas of high endemism. We discuss biodiversity hotspots more in Chapter 14.

Latitudinal gradients

One of the most prominent spatial patterns of biodiversity across the planet is the latitudinal gradient in species richness.[45] Many, if not most, groups of

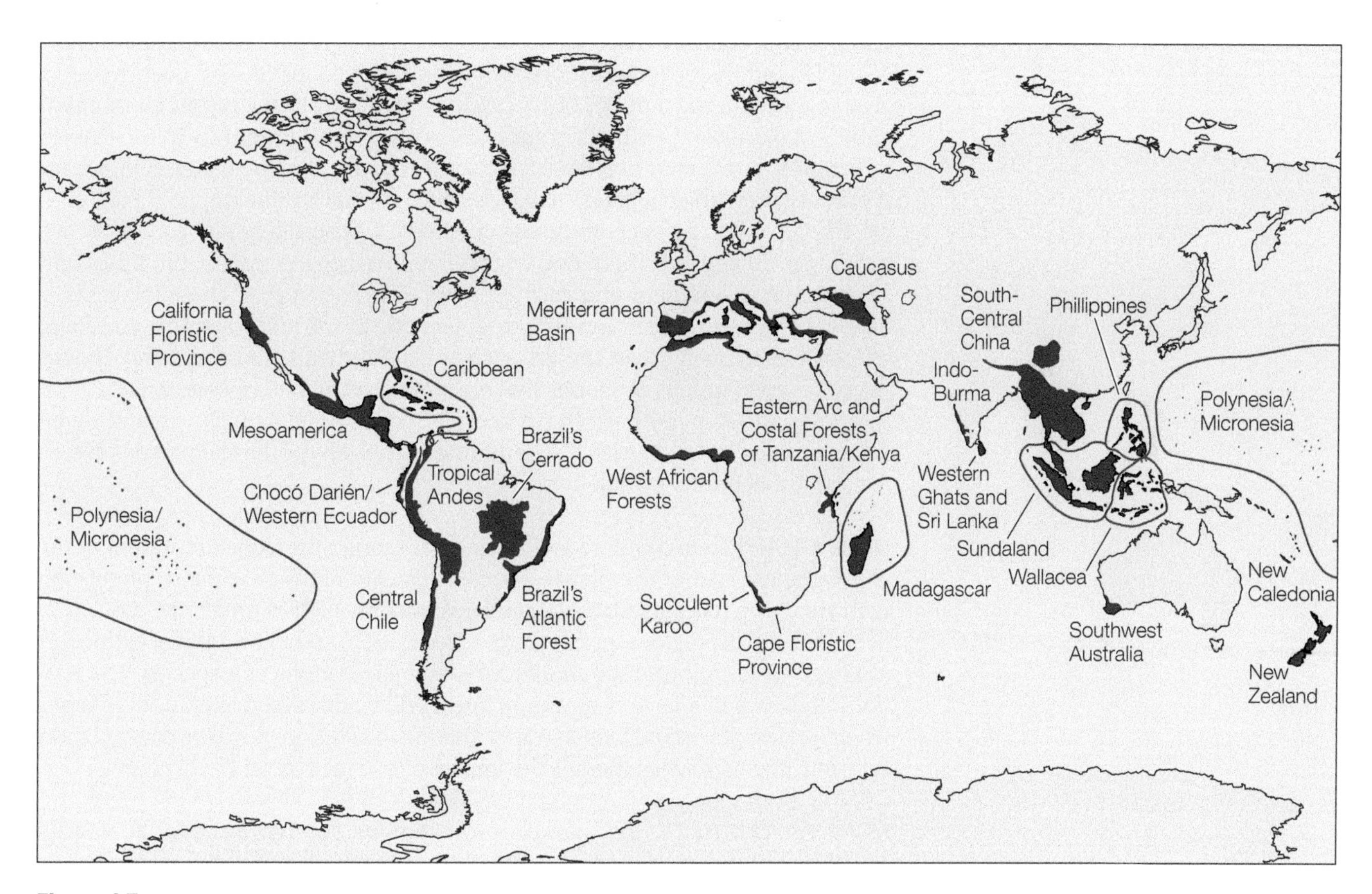

Figure 4.7 Biodiversity is unequally distributed across the planet. This map shows Myers et al.'s original 25 hotspots of biodiversity (in red) that contain 44% of all species of vascular plants, and 35% of all vertebrate species, in areas that comprise just 1.4% of the land surface of the Earth. More refined maps of hotspots are discussed in Chapter 14. (After N. Myers et al. 2000. *Nature* 403: 853–858.)

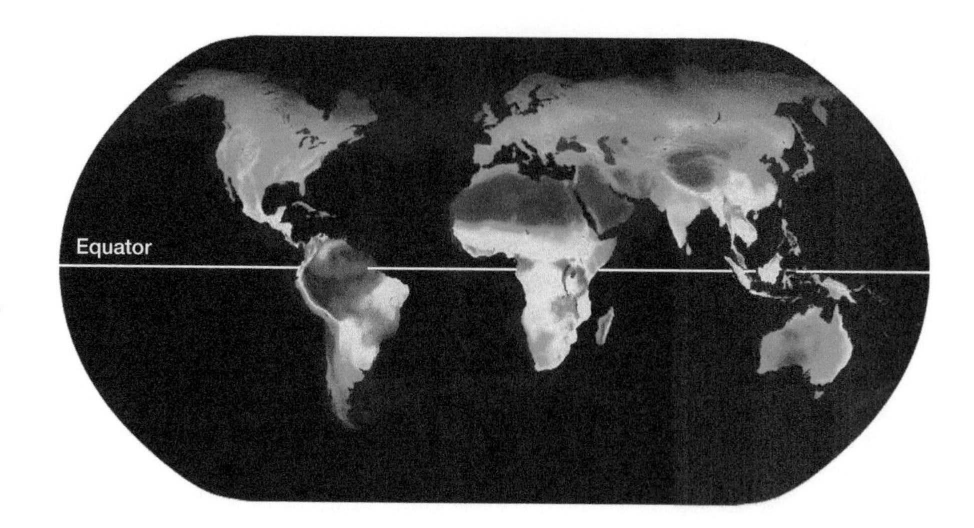

Figure 4.8 As is true for most groups of organisms, vertebrate animals follow a latitudinal gradient; species richness increases from the poles to the equator. In this figure, warmer colors (orange and red) represent more diversity, while cooler colors (yellow, green, and blue) represent less diversity. (Courtesy of C. Jenkins. © SavingSpecies/Globaïa, 2012. www.savingspecies.org/projects/)

organisms tend to increase in species diversity from the poles toward the tropics (**Figure 4.8**). For example, Thailand has 265 species of mammals, while France has only 93, despite the fact that both countries have roughly the same land area (**Table 4.2**). The contrast is particularly striking for trees and other flowering plants: 10 hectares of forest in Amazonian Peru or Brazil might have 300 or more tree species, whereas an equivalent forest area in temperate Europe or the United States would probably contain 30 species or less. Even within a given continent, the number of species increases toward the equator, and many groups of plants and animals occur exclusively or predominantly in the tropics.

Patterns of diversity in terrestrial species are paralleled by patterns in marine species, again with an increase in species diversity toward the tropics.[46] For example, the Great Barrier Reef off the eastern coast of Australia has 50 genera of reef-building coral at its northern end where it approaches the tropics, but it has only 10 genera at its southern end, farthest away from the tropics. These increases in richness of coastal species toward the tropics and in warmer waters are paralleled by increases in open ocean species, such as plankton and predatory fish,[47] though there are select groups of species that are most diverse in temperate waters.

More than 30 hypotheses have been proposed to explain the greater diversity of species in the tropics.[45] A sampling of these hypotheses includes the effects of greater area, productivity, and time.

MORE AREA When compared with temperate zones, the tropics represent a larger geographic area of the planet. Species diversity tends to increase predictably with area, which occurs for any number of reasons, such as greater heterogeneity of biomes and ecoregions, greater rates of speciation, and lower rates of extinction, as larger areas support more individuals and populations of a species. The relationship between area and species richness will be discussed later in this chapter when we describe island biogeography theory. In addition, it will be covered more fully in Chapter 9, which details the impact of habitat loss on biodiversity.

MORE PRODUCTIVITY Because tropical regions receive more solar energy, are warmer, and receive more rainfall than temperate regions, tropical plant communities have higher rates of biomass production than temperate communities (in kilograms of living material produced each year per hectare). High rates of productivity support not only larger populations of plants, but also a greater variety of herbivores and carnivores. Large population sizes also reduce probabilities of extinction.

TABLE 4.2 Number of native mammal species in selected tropical and temperate countries paired for comparable size

Tropical country	Area (1000 km²)	Number of mammal species	Temperate country	Area (1000 km²)	Number of mammal species
Brazil	8512	394	Canada	9220	193
DRC[a]	2345	450	Argentina	2777	320
Mexico	1972	491	Algeria	2382	92
Indonesia	1919	515	Iran	1648	140
Colombia	1139	359	South Africa	1185	247
Venezuela	912	323	Chile	751	91
Thailand	514	265	France	544	93
Philippines	300	153	United Kingdom	245	50
Rwanda	26	151	Belgium	30	58

Source: B. Groombridge and M. D. Jenkins 2010. *World Atlas of Biodiversity: Earth's Living Resources in the 21st Century*, pp. 296–305, Berkeley, CA: University of California Press.
[a]DRC = Democratic Republic of the Congo.

MORE TIME Tropical regions have had a more stable climate that has allowed the processes of evolution and speciation to occur uninterrupted for long periods of time, allowing a greater degree of evolutionary specialization and local adaptation to occur. In contrast, temperate regions have been repeatedly interrupted by glaciation events that have destroyed local species or forced them to disperse, thus limiting the amount of time available for species accumulation.

Drivers of Biodiversity

The geologic trends and global patterns of biodiversity described in this chapter have all been driven by a common suite of evolutionary and ecological processes. In order to conserve the great variety of life on Earth, practitioners must understand how to protect the evolutionary processes that generate diversity and appropriately manage the ecological processes that maintain diversity once it is formed.

Evolutionary drivers

The famous biologist Theodosius Dobzhansky once said, "Nothing in biology makes sense except in the light of evolution." His quote is just as true for conservation biology as it is for any other biological discipline. Practitioners must understand how to maximize the evolutionary processes that generate biological variation, and minimize the evolutionary processes that destroy biological variation at both genetic and species levels.

GENETIC VARIATION Evolution is a process that causes organisms to change through time. In order for evolution to occur, there must be variation in the phenotypes of a population of organisms. The phenotype is the set of observable characteristics or traits in the population, such as the organism's morphology, biochemical or physiological properties, or behavior. These observable characteristics or traits must ultimately have a genetic basis such that they can be passed from parents to progeny.

Ultimately, all variation in inheritable phenotypes originates from changes in DNA. If there were no variation in the DNA of organisms in a population, then all of them would be identical with respect to their inheritable phenotypes, and there would be no evolution. But errors that occur during DNA replication, or damage to DNA that occurs from events like exposure to UV radiation, can lead to genetic mutations. In addition to mutations, sexual reproduction and the swapping of genetic material among chromosomes during recombination can lead to novel sets of genetic information.

Together, mutation, sexual reproduction, and recombination are constantly producing novel sequences and combinations of nucleotides and variation in DNA. Without these processes, there can be no genetic variation, no variation in phenotypes, and ultimately, no biodiversity. Given that mutation and recombination are essential for the creation of diversity, conservation efforts and management plans frequently work to protect, even maximize, these processes to ensure there is genetic variation in a population that will allow it to change and adapt through time. We will return to this topic in Chapters 13 and 16 when we discuss conservation tools used to manage genetic diversity.

NATURAL SELECTION Once there is genetic and phenotypic variation in a population of organisms, two evolutionary processes can occur—natural selection and genetic drift. Natural selection is a process that acts on the phenotype of an individual organism and causes organisms with maladaptive phenotypes to have lower fitness (survival or reproduction) than organisms with traits that are better adapted to their environment. If individuals with maladaptive traits die or fail to reproduce, then the better-adapted organisms are the ones that are most likely to pass on their genes, and thus their phenotypic traits, to their offspring. Over successive generations, maladapted organisms die out, and the most well-adapted individuals come to prominence in a population. Darwin called this change through generations the "survival of the fittest."

Natural selection can alter the phenotype of a population in a multitude of ways that can increase or decrease biodiversity (**Figure 4.9**). When the median phenotype of a population is maladapted to the environment, disruptive selection can favor the two extremes of a phenotype in a population. When the extreme phenotypes in a population are maladapted, then stabilizing selection can favor the median phenotype. When only one tail of the trait distribution is maladapted, selection can lead to directional shifts in the phenotype of a population.

Biologists have repeatedly documented the process of natural selection occurring in their labs, in nature, and even in the homes of ordinary people. In our own backyards, domesticated dogs are a prime example of evolution by natural selection.[48] All domesticated dog species came from a single ancestor—the gray wolf[49] (**Figure 4.10A**). But the variety of sizes, shapes, and types of dogs we have today is the product of individuals being selectively bred by humans over millennia for desired behaviors, sensory capabilities, and physical attributes. Essentially, people have acted as

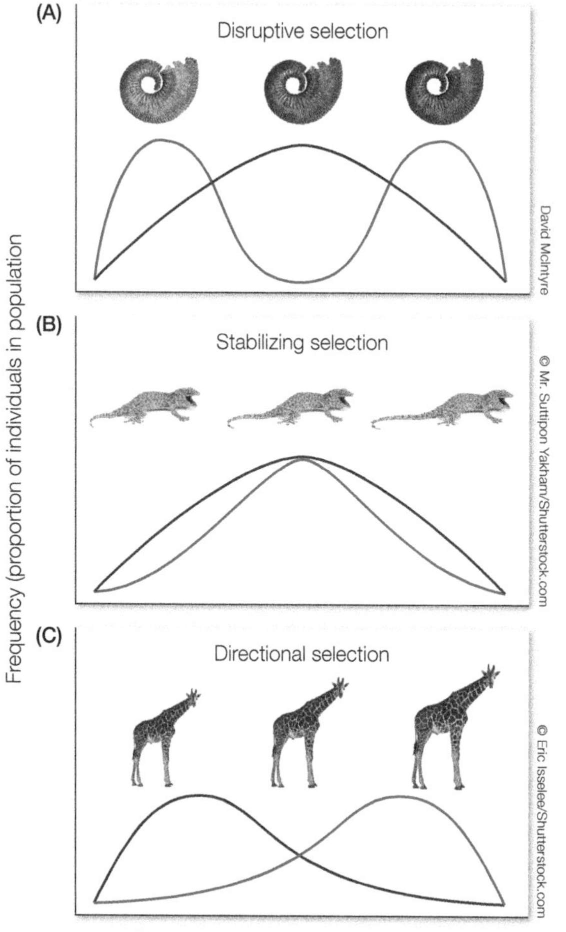

Figure 4.9 Natural selection can cause organisms to change in several possible ways. In disruptive selection (A), the selective force reduces fitness of the average phenotype (red line), causing the population to evolve into different phenotypes (blue line). In stabilizing selection (B), the selective force reduces the fitness of the extreme phenotypes, forcing the population to evolve toward the average (average tail length). In directional selection (C), the selective force reduces the fitness of one phenotype in the population, forcing the distribution in the direction of an alternative phenotype (larger neck size). (After Andrew Z. Colvin/Wikimedia CC BY-SA 3.0.)

Figure 4.10 Examples of evolution by natural selection. (A) In our own backyards, humans have selectively bred dogs for a variety of desired traits, all of which originally came from a wolf ancestor. (B) In the lab, biologists have tracked the accrual of genetic mutations and documented natural selection in bacteria. (C) In nature, pests like the Colorado potato beetle have developed resistance to pesticides.

the agents of natural selection by giving higher fitness (selective breeding) to varieties that have certain phenotypes (desired traits).

In 1988, biologist Richard Lenski began a laboratory experiment in which he inoculated several laboratory cultures with cells of the bacterium *Escherichia coli* that were genetically identical (**Figure 4.10B**). As the cultures began to grow and cells to divide, mutations started to accumulate, and Lenski began to see both genetic and phenotypic diversity in the population of cells. Some of these mutations and phenotypes proved to be detrimental to the cells, and those lineages went extinct. Other mutations and phenotypes allowed certain types of cells to have higher growth and reproduction in the laboratory conditions, and these lineages were more successful.[50] Over thousands of generations, Lenski and his colleagues showed that bacterial populations that began with a single cell evolved in ways that made them better adapted to the laboratory environment in which they were grown.

Instances of natural selection operating in nature are pervasive, and some of the best examples come from the development of pesticide resistance in agriculture. A well-studied example is that of the Colorado potato beetle, *Leptinotarsa decemlineata* (Say), which is one of the most harmful insect pests of potatoes (**Figure 4.10C**). Its current range covers about 16 million km^2 in North America, Europe, and Asia, and it can cause tens of millions of dollars of crop damage in a given year. Many efforts to control the pest have failed, in part because some individuals have been able to survive each type of new pesticide application because of their unique genetic mutations. Various populations have now evolved resistance to 52 different kinds of insecticide.[51]

GENETIC DRIFT Along with natural selection, genetic drift is one of the basic mechanisms that result in a population changing its genetic content through time. But unlike natural selection, which acts to weed from a population the individuals that are least fit, genetic drift is a stochastic process that leads to changes in the genotypes and phenotypes of a population due to chance events.[52] Although genetic drift can occur in all populations, its effects are greatest in populations that have few individuals, because small populations are more prone to chance events.

To give an example of genetic drift, imagine there is a species of lizard that has a small population of 10 individuals living in a forest (**Figure 4.11**). This population has some genetic variation that has produced three color morphs—one dark green, the second light green, and the third brown. Now imagine that by chance alone, some event (e.g., windfall or storm) wipes out all individuals that are brown. The genes that code for brown coloration, and the brown phenotype, would be forever lost.

Suggested Exercise 4.2 Visit the Companion Website and set up a spreadsheet to model genetic drift. **oup-arc.com/e/cardinale-ex4.2**

The consequences of genetic drift can be exacerbated by genetic bottlenecks and founder effects. A **genetic bottleneck** is an extreme change in the genetic and phenotypic makeup of a species that occurs when the size of a population is severely reduced.[53] Events like natural disasters (e.g., earthquakes, floods, fires) can decimate a population, killing most individuals and leaving behind a small, nonrandom assortment of survivors that have highly reduced levels of genetic and phenotypic diversity. Genetic bottlenecks make it difficult for species to recover from disasters. In addition, bottlenecks can lead to **mutational meltdown**,

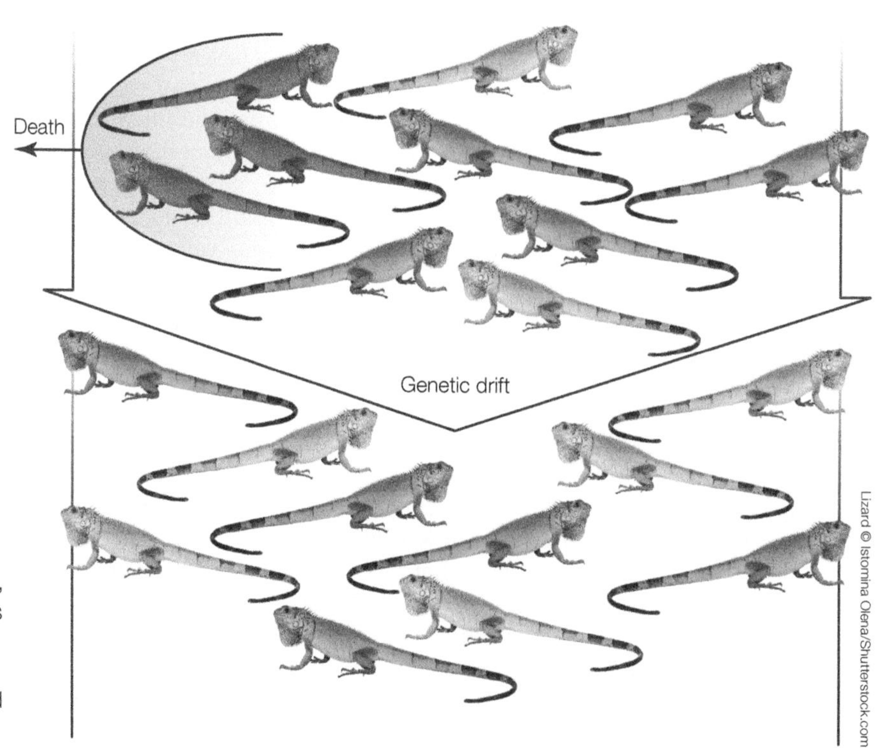

Figure 4.11 Evolution by genetic drift. In a small population of lizards that have different skin colors, there is random chance that all brown lizards will be killed by some event (e.g., windfall or storm) even though they are no less fit than the other lizards. When this happens, the genetic diversity of the population is reduced, and the brown phenotype is eliminated from the next generation.

which occurs when detrimental mutations accumulate in a small population. As mutations accumulate, they can reduce the fitness (survival and reproduction) of individuals, causing the population to get smaller, accumulate more mutations, and eventually spiral to extinction.

Founder effects occur when a small group of individuals breaks off from a larger population to establish a colony.[54] The new colony is isolated from the original population, and the founding individuals may not represent the full genetic or phenotypic diversity of the original population. Something like this might occur if, for example, one pregnant dragonfly were blown to sea, arrived on an island with no dragonflies, and established a new population. The founder effect is similar to a bottleneck in the sense that it can greatly reduce diversity. But it differs in that it occurs via a different mechanism (colonization rather than catastrophe).

Genetic drift can be a major contributor to the loss of biodiversity, particularly when small populations go through genetic bottlenecks or exhibit founder effects. Genetic drift has almost certainly accelerated losses of biodiversity throughout Earth's history, particularly during periods of mass extinction. Minimizing these events is one of the greatest goals of modern conservation (Chapters 13 and 16).

SPECIATION Mutation and recombination, natural selection, and genetic drift are the primary evolutionary processes that ultimately control the formation of new species (speciation). There are four basic modes of speciation in nature, which differ from each other primarily by the extent to which populations undergoing speciation are isolated from one another (**Figure 4.12**):

- *Allopatric speciation.* During allopatric speciation, a single population is divided into two separate populations by the formation of a geographic barrier (e.g., a mountain or river or separation of continental plates). Over time, the isolated populations begin to diverge genetically and phenotypically as a result of natural selection and/or drift. At some point, divergence is sufficiently large that even if the populations were brought back into contact, they could not successfully interbreed to form viable offspring (**Figure 4.13**). Allopatric speciation through geographic isolation is thought by some to be the most prominent mechanism producing biodiversity on Earth.[55]
- *Peripatric speciation.* Peripatric speciation is a subtype of allopatric speciation. In this form of speciation, small peripheral populations become isolated and are somehow prevented from exchanging genes with the main population. The isolation may be due to a physical or geographic barrier, but it may be due to other factors, such as behavior. The key difference between allopatric and peripatric speciation is that the latter is due to a founder effect whereby the small, isolated population becomes different genetically and phenotypically based on genetic drift alone.

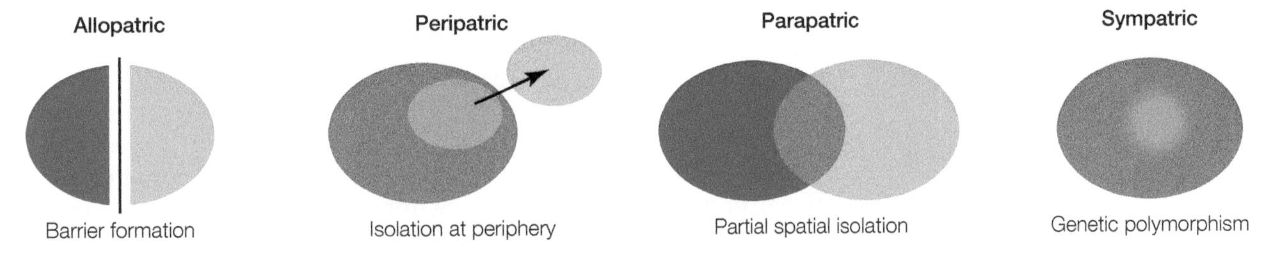

Figure 4.12 Four modes of speciation. Colors represent distinct populations. (After Ilmari Karonen/Wikimedia/CC BY-SA 3.0, adapted from D. Krempels. 2006. *Lecture Notes for Evolution and Biodiversity* [BIL 160 Section HJ], Univ. of Miami.).

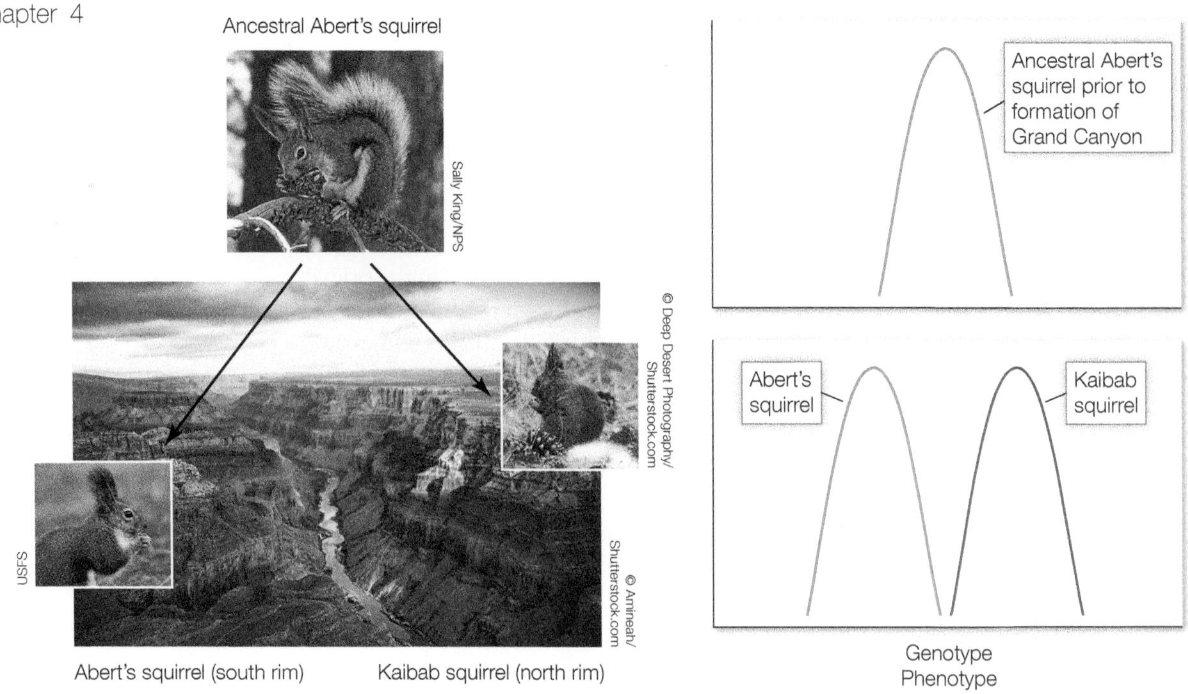

Figure 4.13 An example of allopatric speciation (geographic isolation). Before the Colorado River cut open the Grand Canyon, a single population of Abert's squirrels existed, with individuals having similar genotypes and phenotypes. After the Colorado River formed the Grand Canyon, the single population of Abert's squirrels was split into two groups. On the north rim, the Kaibab squirrel evolved. On the south rim, the Abert's squirrel evolved. These two groups are now biologically distinct species with different genotypes and phenotypes.

- *Parapatric speciation.* In parapatric speciation, diverging populations are only partially separated, with individuals of each population crossing boundaries and coming into contact from time to time. However, despite limited reproduction and gene flow among the populations, divergence is sufficient to cause the hybrid offspring to have reduced fitness. This leads to selection of behaviors or mechanisms that prevent interbreeding among populations and ultimately leads to speciation despite the fact that individuals occasionally interbreed.
- *Sympatric speciation.* Sympatric speciation is the process by which new species evolve from a single ancestral species even while inhabiting the same geographic region. This mode of speciation is common in plants, which are prone to acquiring multiple homologous sets of chromosomes, resulting in polyploidy. The polyploid offspring occupy the same environment as the parent plants (hence sympatry) but are reproductively isolated.

Aside from the four modes described above, speciation can also be induced artificially through animal husbandry, agriculture, or laboratory experiments (see Figure 4.10). Such human-engineered species can contribute significantly to biodiversity in some landscapes (e.g., agricultural or urban habitats) and have sometimes been used to offset or restore species losses.

While a common view is that speciation is a slow process that occurs over time scales of centuries to millennia, recent work has shown that this view is not correct. Evolution by natural selection or drift can produce new species over the course of just a few generations.[56–58] Such examples of **rapid evolution** allow speciation to occur on time scales ranging from days to years, depending on the generation time of the organism.[59] This finding puts the process of speciation squarely within the purview of conservation biology, and in fact, an emerging goal of conservation is to manage the evolutionary processes of natural selection and drift so as to maximize the potential for speciation to occur naturally.

Ecological drivers

One of the fundamental questions that ecologists ask is, Why do so many species coexist? Why is it that, after a variety of species evolve, the single best one doesn't simply take over an entire ecoregion or biome? What is it that prevents one species from becoming the competitively dominant species, and what allows species to coexist after they evolve? The main ecological drivers that maintain biodiversity are competition and niche differentiation, predation and disturbance, and dispersal and colonization.

COMPETITION AND NICHE DIFFERENTIATION One of the earliest attempts to answer the question about why species coexist was a theoretical exploration of the topic by Alfred Lotka and Vito Volterra who, together, developed a widely used model of species competition. Their model was a relatively simple modification of the logistic model of population growth, extended to include the effects that each species can have on each other's growth rates. The equations for their model were:

$$\frac{dN_1}{dt} = r_1 N_1 \left(\frac{K_1 - N_1 - \alpha_{12} \times N_2}{K_1} \right) \quad (4.1)$$

$$\frac{dN_2}{dt} = r_2 N_2 \left(\frac{K_2 - N_2 - \alpha_{21} \times N_1}{K_2} \right) \quad (4.2)$$

where N_i is the population size of species i, r_i is the per capita population growth rate, K_i is the carrying capacity, and α_{ij} is called the competition coefficient. The competition coefficient is a ratio that represents the per capita effect of species j on the growth of species i relative to the effect that species i would have on itself. In other words, it measures the strength of interspecific (between species) competition relative to the strength of intraspecific (within species) competition.

To determine what is required for the coexistence of competing species, Lotka and Volterra solved the equations to show the conditions that allow both species to have positive values of N_i (positive population sizes) when those populations have reached an equilibrium (i.e., populations are no longer changing, or $dN_i/dt = 0$). The take-home message of this analysis was that coexistence of competing species requires that the value for each α_{ij} be less than 1. In words, this means that species can coexist so long as intraspecific competition is greater than interspecific competition.[60] The only way this can happen is when species exhibit **niche differences**, which means they either use different resources or use the same resource differently in space or time (**Figure 4.14**). Niche differentiation tends to reduce competition for resources among different species.

Predictions from the Lotka-Volterra competition equations were confirmed with real organisms by Russian ecologist Georgy Gause, who performed a set of laboratory competition experiments using two species of *Paramecium*.[61] Gause showed that when the two species were forced to compete for the same limiting resource in a homogeneous environment, they were unable to coexist, and one species would competitively exclude the other. However, when the experimental environment was varied in such a way as to allow the species to divide their resources, they were able to coexist. His work led to the **competitive exclusion principle**, which states that no two species can coexist when using the same resource, at the same time, in the same location.

Suggested Exercise 4.3 Go to the Companion Website and run a simulation of the Lotka-Volterra competition equations to determine how competition influences species coexistence. **oup-arc.com/e/cardinale-ex4.3**

Figure 4.14 An example of niche partitioning among warblers that search for insects and nest in different parts of the same tree species. Niche partitioning reduces competition for food and reproductive sites, allowing the species to coexist. (After R. H. MacArthur. 1958. *Ecology* 39: 599–619.)

While the competitive exclusion principle stood unchallenged for several decades, in 1961 G. Evelyn Hutchinson wrote a famous essay called "The Paradox of the Plankton," which questioned the role that competition and niche differences play in allowing species to coexist. Hutchinson was essentially perplexed by how lakes can harbor hundreds of species of phytoplankton—photosynthetic algae that live in the top few meters of the water—in spite of the fact that all of those phytoplankton require, and potentially compete for, the same set of inorganic resources (nutrients and light). Lotka-Volterra competition equations and Gause's competitive exclusion principle said there could be no more species than there are resources that can be divvied up. Yet, here was an example where many more species coexisted than there were resources, and they did so in the same location, at the same time.

Hutchinson proposed that forces other than competition must also regulate species coexistence. Specifically, he proposed that there must be periodic events that impose mortality on species and that this keeps them from ever reaching an equilibrium where competitive exclusion might occur. In lakes, he argued, the fall ice cover and spring turnover of water essentially "reset" the systems such that no one species ever has the time to win.

PREDATION AND DISTURBANCE After Hutchinson posed the paradox of the plankton, ecologists began to study how mortality-inducing events, such as those caused by predation or abiotic disturbances, might also regulate species coexistence. Initially, it was thought that if predation or disturbance could reduce the abundance of species populations and prevent an ecosystem from reaching an equilibrium, then competition would not have the time to play out, and competitive exclusion would not occur.

Many popular hypotheses were developed in the 1960s–80s that assumed mortality events could prevent competitive exclusion by reducing population densities and, in turn, weaken competition. Perhaps the most famous of these hypotheses—one that is still widely taught in ecology textbooks—is the intermediate disturbance hypothesis (IDH).[62] The IDH was developed by Joseph Connell, who was attempting to come up with an explanation for the remarkable variety of species that occur in tropical forests and coral reefs. His hypothesis argued that the diversity of competing species should peak at intermediate frequencies or intensities of disturbance because very frequent or intense disturbances would eliminate disturbance-intolerant species, while rare or weak disturbances would allow sufficient time for competitively dominant species to take over the ecosystem and exclude competitively inferior species. Though intuitively appealing, the IDH (and similar hypotheses) are based on the incorrect assumption that disturbances weaken competition by reducing species' densities. In fact, disturbance can cause competition to become more intense, making competitively inferior species even more susceptible to exclusion in harsh or fluctuating environments.[63] This is, perhaps, why empirical support for the IDH has been limited[64] and why some have called for ecologists to stop teaching the concept to students and practitioners.[65]

It is now generally accepted that mortality-inducing events such as predation and abiotic disturbance enhance species coexistence and biological diversity only when they create new niche opportunities for species in space or time. The primary way they create new niche opportunities is by giving species the chance to express life history trade-offs, which are biological trade-offs in the investments they must make in traits that affect their growth, reproduction, and survivorship. No single species can be the best at everything—if it could, it would take over the Earth. Instead, species must allocate their limiting resources in ways that maximize the biological trade-offs they face.

Figure 4.15 shows two examples of how predation or abiotic disturbance can enhance biodiversity by allowing species to express life history trade-offs. The example for predation comes from the classic experiments of Robert Paine—a marine biologist whose painstaking research in the intertidal zone of the Pacific Northwest of the United States showed why predators can be critical to the maintenance of species diversity. In the intertidal system he studied, the California mussel (*Mytilus californianus*) is the dominant competitor for space. The mussel attaches itself to rocks with remarkably strong byssal threads that allow it to hold tight to space and outcompete and eliminate other sessile species, such as barnacles, limpets, anemones, and algae. But Paine showed that the predacious sea star (*Pisaster ochraceus*) has a particular affinity for mussels and selectively feeds on mussels over other species.[66] The result is that inferior competitors that are less susceptible to predation are able to coexist in the system (see Figure 4.15A). Paine went on to coin the term *keystone predator* (see Chapter 3), arguing that top-down trophic interactions have large impacts on the biodiversity of a system by allowing species to express trade-offs between their competitive ability and their susceptibility to higher consumers.

Abiotic disturbance can have impacts similar to those of predators, but it does so by allowing the expression of a different set of biological trade-offs. For example, a common trade-off that has been documented for plants is their ability to compete for resources locally with other species versus their ability to disperse

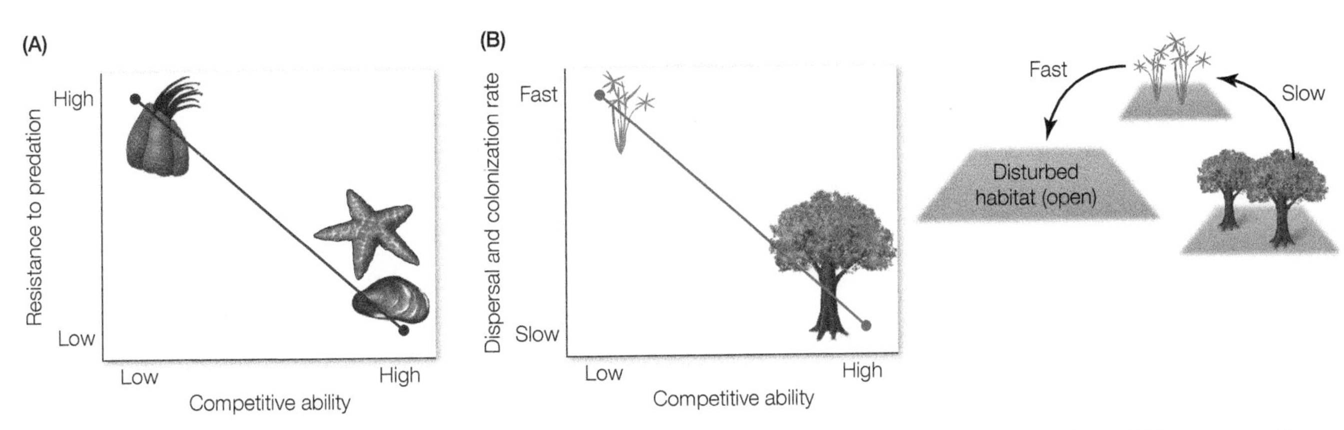

Figure 4.15 Examples of how life history trade-offs allow coexistence and enhance biodiversity. (A) The barnacle and the mussel coexist because of a competition-predation trade-off. The more competitive species for space (the mussel) is less resistant to predation by the sea star. As sea stars eats mussels, this opens up space for the less competitive barnacles. Therefore, the less competitive barnacles can coexist so long as they are more resistant to predation. (B) The tree and the weed coexist by a competition-dispersal trade-off. The tree is more competitive than the weed and will take over space if given enough time. But the weed can coexist with the tree so long as it has a higher rate of dispersal that allows it to colonize open habitats before the tree arrives and takes over through competition.

and colonize open habitats[67] (see Figure 4.15B). Species that allocate a large fraction of their resources to roots or leaves, which they need in order to acquire inorganic resources (soil nutrients and light), allocate less to seeds. In contrast, species that are inferior competitors invest more in dispersal and colonization. Abiotic disturbances create open habitat that allows inferior competitors, but fast colonizers, to persist in an ecosystem and coexist with superior competitors.

DISPERSAL AND COLONIZATION When species fail to exhibit niche partitioning, then coexistence is not possible in a "closed" system, which does not allow declining populations to be maintained by new individuals that disperse from other locations. However, in many real-life situations, dispersal and recolonization can help prevent the local extinction of species and maintain their populations, even in areas where they are being outcompeted by other taxa. One of the best-known hypotheses to explain how dispersal and colonization affect species diversity and coexistence is Robert MacArthur and E. O. Wilson's **theory of island biogeography**.[68]

MacArthur and Wilson were originally trying to explain the richness of species on islands (**Figure 4.16**), but the theory has since been used more broadly in conservation biology. Their model argues that the number of species on an

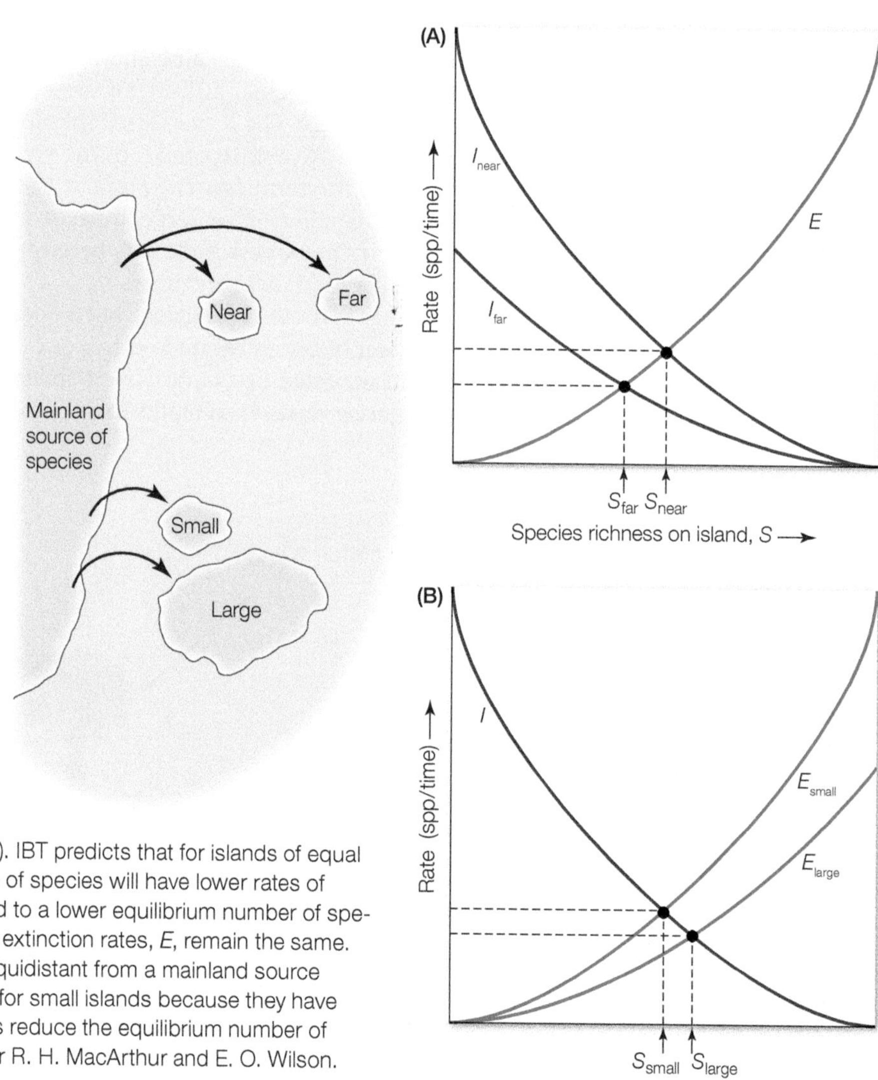

Figure 4.16 Island biogeography theory (IBT). IBT predicts that for islands of equal size (A), those farther from a mainland source of species will have lower rates of immigration, *I*. Lower rates of immigration lead to a lower equilibrium number of species on the island (dot labeled S_{far}), assuming extinction rates, *E*, remain the same. IBT further predicts that for islands that are equidistant from a mainland source of species (B), extinction rates will be greater for small islands because they have smaller population sizes. High extinction rates reduce the equilibrium number of species on the island (dot labeled S_{small}). (After R. H. MacArthur and E. O. Wilson. 1963. *Evolution* 17: 373–387.)

island represents an equilibrium between two opposing processes: immigration and extinction. In their model, species immigrate to an island randomly from a mainland pool. The rate at which new species arrive at the island is determined by three factors:

1. The distance of the island from the mainland
2. The number of species in the mainland pool that have not already established themselves on the island
3. The probability that a given species will disperse from the mainland to the island

The rate at which species on the island go extinct is determined by three different factors:

1. The area of the island
2. The number of species present on the island
3. The probability that a given species on the island will go extinct

In the simplest version of the model, all species have an equal probability of reaching the island and of going extinct once they have colonized there. Immigration is considered inversely related to distance—that is, the rate of immigration goes down as an island gets further away from the mainland because it becomes harder for species to disperse long distances. In addition, immigration rates decline as more species establish themselves on the island because the total number of species on the mainland is finite, and there are fewer species that haven't already colonized. Based on these considerations, we can write a simple equation for the rate of immigration to an island:

$$I = \frac{c(P-S)}{fD} \tag{4.3}$$

where I = the immigration rate (species per time), P = total number of species in the mainland pool, S = species richness of the island, D = distance of the island from the mainland, c = colonization probability, or the probability that a given species will make it to the island (assumed to be equal for all species), and f = a scaling factor for distance. Values for c and f must be determined from actual data, and they depend on the types of species and islands under study. Based on the work of MacArthur and Wilson, reasonable values of $c = 0.10$ and $f = 0.01$.

Note from Equation 4.3 that as the distance between an island and the mainland grows (the denominator gets larger), the rate of immigration goes down. What this means is that, for a given extinction rate, islands that are far away from a mainland will have a lower equilibrium abundance of species than islands closer to the mainland (see Figure 4.16). Equation 4.3 also shows that the rate of immigration is greatest when few mainland species have reached the island (i.e., the numerator $P - S$ is large), and it slows as more species reach the island ($P - S$ is small).

Because equilibrium species richness requires a balance between immigration and extinction, we need to write an equation to describe extinction rates:

$$E = \frac{qS}{A^m} \tag{4.4}$$

where E = extinction rate (species per time), S = species richness of the island, A = area of the island, q = extinction probability for a given species (assumed to be equal for all species), and m = a power scaling factor for area. Values of q and m must be determined from actual data, and work by MacArthur and Wilson suggested values of $q = 0.20$ and $m = 0.25$.

Note that in Equation 4.4, as the area of an island gets larger (the denominator gets larger), the rate of extinction goes down. What this means is that, for a given immigration rate and distance from the mainland, larger islands have a higher equilibrium abundance of species than smaller islands. Also note that as S increases (the numerator gets larger as the number of species on the island grows), the extinction rate, E, will increase.

For Equations 4.3 and 4.4, there is always some equilibrium value of S at which immigration and extinction are equal (i.e., $I = E$). This equilibrium value of species richness occurs because each new species that immigrates to the island is counter-balanced by a species that goes extinct from the island, and each species that goes extinct is counter-balanced by a new arrival. This leads to an important point of the model: The equilibrium number of species is determined by the balance between immigration rates and extinction rates.

Suggested Exercise 4.4 Visit the Companion Website to complete an exercise that will sharpen your understanding of the island biogeography theory. **oup-arc.com/e/cardinale-ex4.4**

A second important point from the model is that the particular species inhabiting an island continue to change even after species richness has reached equilibrium. That is, established species on the island continue to go extinct and be replaced by an equal number of new species that immigrate from the mainland. This means a biologist revisiting the same island at different times would find different sets of species even though s/he would see comparable total numbers of species. The ratio of immigration to extinction at equilibrium is called the *turnover rate,* and the prediction of continual turnover of species is an important feature of MacArthur and Wilson's model. Species richness can be held constant by the balance of immigration and extinction even as the composition of species changes.

Island biogeography theory is important because it shows that biodiversity can be maintained in locations in spite of extinctions, so long as individuals immigrate from areas that have healthy populations of a larger species pool. This theory forms the foundation for more-advanced spatial models that consider how movement of individuals among habitats can influence species coexistence and biodiversity, such as metapopulation and metacommunity models that are widely used tools in conservation planning and management (Chapters 9, and 13–16).

Synopsis

In this final section of Chapter 4, we have described a suite of ecological and evolutionary processes that control biodiversity at levels ranging from genes, to species, to communities. Some of these processes systematically increase diversity; others systematically decrease diversity; still others can increase or decrease biodiversity depending on how they influence the expression of biological trade-offs (Table 4.3). Later, in Part IV of this book, we will cover numerous management tools, techniques, and models that are available to practitioners to maximize those ecological and

TABLE 4.3 Evolutionary and ecological processes the control changes in biodiversity

Driver of change	Increase diversity	Decrease diversity
EVOLUTIONARY		
Mutation	✓	
Recombination	✓	
Speciation	✓	
Natural selection		✓
Genetic drift		✓
ECOLOGICAL		
Adaptive radiation	✓	
Competition		✓
Predation and disturbance	✓	✓
Dispersal and colonization	✓	✓

evolutionary processes that increase biodiversity while, at the same time, minimizing those processes that decrease biodiversity. For example, in Chapter 13 we will show how to calculate the effective population sizes needed to maximize recombination and minimize genetic drift. In Chapter 13 we will also introduce population viability analyses that consider how competition, predation, and stochastic disturbances influence the persistence time of threatened and endangered species. In Chapters 14 and 15 we will show how to design protected areas, landscapes, and networks that maximize the benefits of disturbance, dispersal, and colonization while minimizing negative impacts of competition or predation. Then in Chapters 13 and 16 we will show how to reduce genetic drift to maintain healthy populations. Practitioners who understand why biodiversity exists, and which tools and techniques can be used to control it, will be in the best position to conserve the spectacular variety of life that is still the sole defining feature of our planet.

Summary

1. Ever since the first prokaryotic organisms evolved almost 4 billion years ago, the variety of life-forms on this planet has been shaped by a handful of radiation and extinction events. The most prominent radiation event, the Cambrian explosion, produced nearly every body type, lifestyle, and ecological niche that exists today. The Big Five extinction events have each wiped out 75% or more of known species but have allowed for new life-forms to emerge and proliferate.
2. To date, 1.24 million eukaryotic species and about 10,000 prokaryotic species have been discovered, described, and scientifically cataloged. The majority of these have been eukaryotic animals and plants. Recent estimates suggest there may be 8.7 million species of eukaryotic organisms globally, suggesting we have yet to discover 86% of these species on the planet. Microbiologists estimate there may still be 100 billion to 1 trillion species of microbes on Earth that have yet to be discovered.
3. The variety of life on Earth is organized into eight biogeographic realms with species that share common geologic and evolutionary histories. Biogeographic realms can be subdivided into 14 or more terrestrial biomes, composed of organisms that share similar abiotic requirements, and 7 aquatic biomes that are defined by their physical processes (e.g., ocean currents) or biota.
4. Biodiversity is unequally distributed across the planet, with certain types of biomes and ecoregions being more diverse than others. The most species-rich environments on the planet are tropical forests, coral reefs, large tropical lakes and river systems, and regions with Mediterranean climates. In addition to these highly diverse ecoregions, scientists have mapped out areas with high rates of endemism (existence of species that are found nowhere else), as well as biodiversity hotspots (small areas that contain disproportionate numbers of species). Protection and management of ecoregions, areas of high endemism, and biodiversity hotspots represent complementary ways to conserve biodiversity.
5. Geologic trends and global patterns of biodiversity have all been driven by a common set of evolutionary and ecological processes. Evolutionary processes like mutation, recombination, and speciation systematically generate new diversity, whereas natural selection and genetic drift both act to decrease diversity. Ecological processes such as competition, predation, abiotic disturbance, and dispersal can either maintain or eliminate diversity, depending on whether they generate niche opportunities that allow species to coexist.
6. Many models, tools, and techniques used by conservation biologists seek to maximize evolutionary and ecological processes that increase biodiversity and to minimize processes that decrease biodiversity. For example, island biogeography theory—a conceptual model widely used to design nature reserves of differing sizes and spacing—seeks to maximize dispersal and colonization while minimizing competitive exclusion.

For Discussion

1. Some scientists have argued that life may exist on other planets, pointing to the potential for bacteria to flourish under the surface of Mars. Based on your reading of this chapter and your understanding of how life evolved on planet Earth, what characteristics would you look for while trying to discover life on another planet?
2. The geologic history of life preserved in the fossil record seems to suggest that biodiversity on Earth has been generated not by a smooth and continual diversification of life-forms, but by a relatively small number of radiation and extinction events. Does this suggest that major geologic events that are beyond human control influence biodiversity more than human activities do? Why or why not?
3. Are you surprised at recent estimates that there may be 8.7 million species of eukaryotes (e.g., plants and animals whose cells have organelles)? How about the estimate of 100 billion to 1 trillion types of microbes? Do these estimates seem too high or too low to you?
4. Is it hard to believe that, in an era when humans dominate nearly every ecosystem on Earth, we have yet to discover 86% of the eukaryotic species on the planet, more than 90% of species that live in the oceans, and nearly all microbes? If these estimates are correct, then why isn't taxonomy (the discipline in biology that finds, identifies, and catalogs new species) a greater scientific priority? Do we even need to classify all of the world's species? If yes, do we need to do it with expert taxonomists, or is taxonomy a skill set that every conservation biologist should possess?
5. Conservation biologists have had many arguments over whether the goals of conservation are best achieved through protection of different biomes (areas defined by species assemblages that have a common history), protection of the most diverse ecosystems (tropical forests, coral reefs, or similar systems that have large amounts of diversity), or protection of hotspots (geographic areas of high diversity and endemism). Do you see these as complementary or mutually exclusive? Which do you think gives the most value (e.g., highest return per dollar spent or habitat conserved)?
6. Do you think it is important for practitioners of conservation biology to understand processes that drive evolution, or are the time scales of evolution too long to be relevant to the discipline? How does your answer change when you consider how climate change will affect your management decisions of the next 100 years?
7. As a practitioner, what levers (i.e., management actions) do you have in your control to influence ecological processes such as competition, predation, disturbance, and dispersal? Are these levers sufficiently sensitive that you might be able to minimize the negative effects of ecological processes on diversity while simultaneously maximizing positive effects?

Suggested Readings

Benton, M. J. and Twitchett, R. J. 2003. How to kill (almost) all life: The end-Permian extinction event. *Trends in Ecology & Evolution* 18: 358–365. A great review of the greatest mass extinction of all time.

De Vos, J. M. et al. 2015. Estimating the normal background rate of species extinction. *Conservation Biology* 29: 452–462. Excellent description of how "background" rates of extinction are calculated from the fossil record.

Dodd, M. S. et al. 2017. Evidence for early life in Earth's oldest hydrothermal vent precipitates. *Nature* 543: 60–64. May be evidence for the earliest known life-forms on Earth.

Fox, D. 2016. What sparked the Cambrian explosion? *Nature* 530: 268–270. A solid review of the most important biological event to ever occur on the planet.

Jenkins, C. N. et al. 2013. Global patterns of terrestrial vertebrate diversity and conservation. *Proceedings of the National Academy of Sciences* 110: 2602–2610. Astonishingly detailed patterns of diversity are presented.

Lowman, M. D. et al. 2006. *It's a Jungle Up There: More Tales from the Treetops*. New Haven, CT: Yale University Press. Anecdotes of adventures while exploring the diversity of the tropical forest canopy.

Mora, C. et al. 2011. How many species are there on Earth and in the ocean? *PLoS Biology* 9: e1001127. Represents the best attempt to date to estimate the number of species that exist globally.

Pimm, S. L. and Brown, J. H. 2004. Domains of diversity. *Science* 304: 831–833. Various theories of global diversity are critically examined.

Wilson, E. O. 2010. Within one cubic foot: Miniature surveys of biodiversity. *National Geographic* 217: 62–83. A study of the remarkable amount of life that can be found in a mere cubic foot on land or in the sea.

Visit the
Conservation Biology
Companion Website
oup.com/he/cardinale1e
for videos, exercises, links, and other study resources.

Part II

Importance of Biodiversity

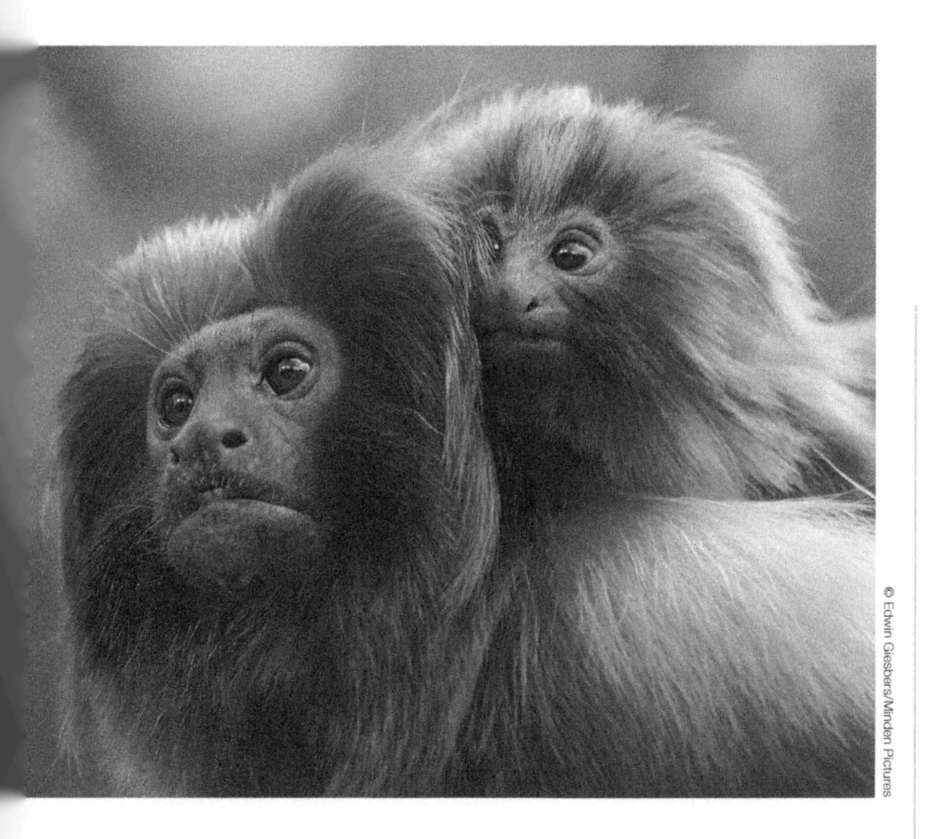

© Edwin Giesbers/Minden Pictures

5

The Many Values of Biodiversity

One of the most colorful of all monkeys, the golden lion tamarin (*Leontopithecus rosalia*) has suffered from the extensive loss of its forest habitat and is now considered endangered.

In Part I of this book, we introduced you to the foundations of conservation biology. We began by covering the current state of our planet (Chapter 1) and the conditions that gave rise to conservation biology as an interdisciplinary field of study (Chapter 2). We then defined what biodiversity is and how it is quantified (Chapter 3). We ended Part I by describing the historical trends and global patterns in biodiversity, as well as the evolutionary and ecological drivers that control how biodiversity is generated and maintained (Chapter 4).

With these fundamentals of conservation now covered, it is time to turn attention in Part II of this book to detailing why biodiversity is important and why the success of humankind is ultimately linked to the great variety of life that exists on this planet. Part II is perhaps the most important part of this book. Unless people feel that biodiversity is important and value it, our efforts to save species, their genetic diversity, and the natural habitats and wild areas they depend on will undoubtedly fail. Therefore, practitioners must work hard to understand people's value systems, which are a function of their culture, religion, upbringing, and other personal and social factors that collectively influence their ethical worldviews on the environment.

In this chapter, we begin Part II of the book by discussing how human values and ethics influence information and decision making in conservation. Next we cover three major value systems—intrinsic, relational, and instrumental—that are commonly used to justify the conservation of nature and its biodiversity. We then discuss three common ethical worldviews—anthropocentrism, biocentrism, and ecocentrism—that influence how people see their relationship with nature and its biota, which in turn influences how and why they value biodiversity in their lives. If you can understand people's value systems and ethical worldviews, then your chances of achieving successful conservation improves greatly. So let's get started.

Values and Ethics

Values are based on an individual's judgment about the relative worth, merit, usefulness, or importance of a thing or action. In contrast, ethics are a systematic way of organizing individual values into a set of principles that guide human conduct and behavior. It is important to recognize that all decisions about biological conservation are, at their core, based on individual value systems, which are themselves heavily influenced by personal ethical views about the natural world.

The idea that all conservation decisions are based on individual values and ethics will feel perfectly natural to some. To others—particularly those who were trained (or are being trained) as scientists—it may feel uncomfortable to allow values and ethics to be infused into decision making. In part, this uncomfortableness for scientists, or those who want to view conservation as a science, stems from the fact that we have been taught that science is free of any value systems or judgments that might taint a researcher's objectivity or compromise the reliability of studies or data. The idea that science should be "value neutral" further implies that scientists should not advocate particular applications of their science to specific problems, because the role of a scientist is to simply provide value-neutral information and expertise. Advocacy is often equated with a loss of objectivity and a reduction in the reliability of scientific information.

While it is true that objectivity is one goal of science, we need to acknowledge that science itself is entirely a human enterprise and that true objectivity can never be attained—even by scientists. In the field of conservation biology, studies are undoubtedly influenced by the value systems and ethical views of individual researchers. This should not be viewed as a negative characteristic of science, nor should anyone think this problem is unique to research in conservation biology. Doctors who study the human body have deeply held values and ethical views that lead them to believe cancer, AIDS, and other human ailments are inherently "bad" and should be eradicated. Meteorologists who study tornadoes, hurricanes, and floods have deeply held beliefs that natural disasters are "bad" for society and that their damaging effects should be minimized. Yet, individual values and ethical beliefs do not necessarily keep us from producing reliable scientific information about human disease, nor do they prevent us from developing predictive models that tell us when and where the next natural disaster will strike. This is because scientists are well aware that values and beliefs exist, and many aspects of the scientific method are specifically designed to safeguard against the potential biases they create or to detect and correct for them when they occur. Transparent, repeatable methods for generating and interpreting data, rather than true "objectivity," are what give science credibility as a way to generate knowledge.

The key point is that there is no such thing as a purely "objective" method for making conservation and management decisions. All stakeholders involved in the decision-making process, including the scientists who bring data to bear on decisions, are products of acquired or enculturated value systems and ethical beliefs. And it is important to acknowledge that, very often, human values and ethical beliefs will supersede the importance of scientific data when people make decisions about nature and its biodiversity. Therefore, it is important to understand the types of value systems that influence different people's decision making.

Types of Value Systems

Environmental philosophers have historically divided human value systems into two types: intrinsic and instrumental.[1] Something is said to have intrinsic value when we value it simply "in itself" or "for its own sake." For example, most people would agree that a human life has intrinsic value—value in itself that is completely independent of how much money that human might make, or how much s/he might give or bring to others through working or personal relationships. Other things commonly believed to have intrinsic value include love, wisdom, pleasure, beauty, and as we will argue, biodiversity (**Figure 5.1**).

In contrast to intrinsic value, instrumental value is the value that something has in helping us get something else we want. For example, paper money has no value in and of itself (aside from its value to collectors). Rather, it is a means to an end in the sense that it allows us to purchase other things that we need (e.g., food, clothes, shelter) or want (e.g., communication, entertainment, transportation). In conservation, nature and its biodiversity have instrumental value

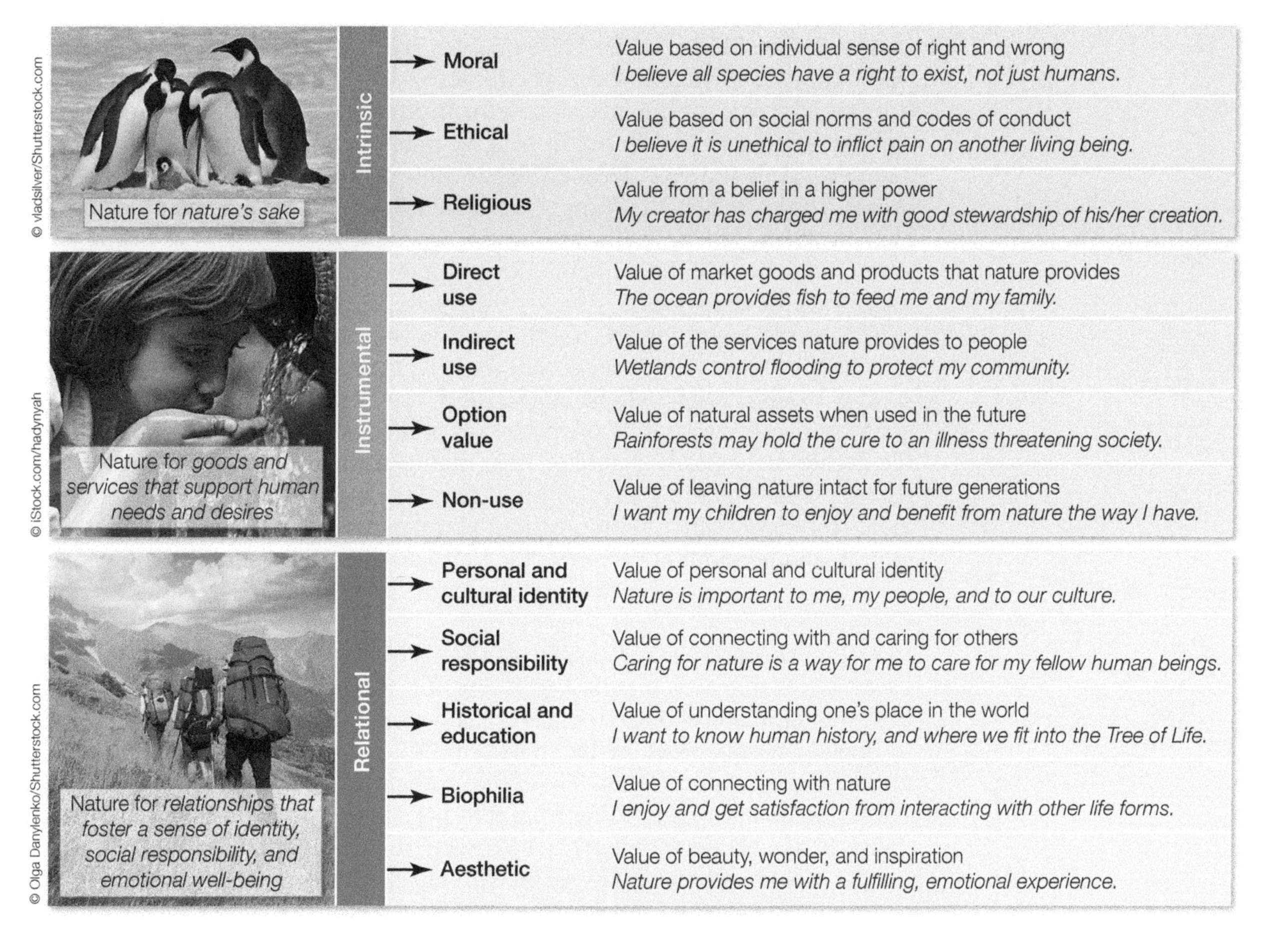

Figure 5.1 A classification of three types of value systems that are used to justify conservation of nature and its biodiversity.

▶ **Video 5.1** Watch this explanation of the difference between intrinsic and instrumental values.
oup-arc.com/e/cardinale-v5.1

because they are the sources of many goods and services that support human needs (e.g., food, lumber, fuel) and wants (e.g., recreation, insurance, and security) (Chapter 6).

Instrumental and intrinsic values represent two extremes of a continuum in the way humans value things (**Video 5.1**). Recently, some have argued that these two views do not accurately represent some of the reasons why people value nature and its biodiversity, or what people believe to be the right way to act toward the environment. It has been proposed that we need to augment our framework of value systems with a third type of value—relational values.[2,3] Relational values are desirable (sought-after) relationships among people, between people and nonhuman organisms, or between people and land that bring happiness to an individual by enhancing personal or cultural identity, social responsibility, and emotional well-being.[2] Collectively, intrinsic, instrumental, and relational values are the three value systems most commonly used to justify the conservation of nature and its biodiversity. We discuss each of these more fully below.

Intrinsic values

While the intrinsic value of human life, love, or wisdom is seldom debated in society, the intrinsic value of nature and its biodiversity has had a mixed history in terms of its recognition and impact. In the 1970s and 80s there were several major international conferences and agreements (e.g., the 1972 Stockholm Declaration of the United Nations[4] and the 1980 International Union for Conservation of Nature's *World Conservation Strategy*[5]) that focused almost exclusively on the instrumental value of nature for humanity and largely ignored intrinsic values. Then, in the 1980s and 90s, international agreements like the UN *Tokyo Declaration* and visionary initiatives like the *Earth Charter*[6] began to strongly advocate for the inclusion of intrinsic values in conservation decision making. The *Earth Charter* boldly said we must "recognize that all beings are interdependent and every form of life has value regardless of its worth to human beings." But somewhat bafflingly, intrinsic values were subsequently ignored by the United Nations; both the outcome document of Rio+20, *The Future We Want*,[7] and the 2015 Sustainable Development Goals[8] failed to mention the intrinsic value of nature.

This mixed history in the use of intrinsic values to justify conservation efforts parallels a much broader debate about whether conservation and management of biodiversity should be motivated by the intrinsic or instrumental value of nature (see Chapter 2, "Core principles and values of conservation"). While the debate continues, it should be clear that intrinsic values of nature resonate strongly with some people and often rank among the most powerful arguments for conservation within certain populations. Here we cover the moral, ethical, and religious arguments that are routinely used to justify the intrinsic value of nature and its biodiversity (see Figure 5.1).

MORAL ARGUMENTS FOR CONSERVATION While morals and ethics are often used interchangeably for distinguishing the difference between "good and bad" or "right and wrong," there is a distinction between the two. Ethics refer to rules or norms provided by an external source like one's nationality, peer group, religion, or profession. For example, the professional ethics of a medical doctor (do no harm) may differ from those of a lawyer (protect client confidentiality) or a public servant (take no bribes) because the disciplines have different norms. Morals, however, have to do with our personal conduct, and they refer to an individual's own innate sense of what is good versus bad. Morals are a personal compass of what is right and wrong, and someone who lacks this compass is said to be immoral.

The most common moral argument for conservation is that *all species have a right to exist*—a value judgment that stems from an innate belief that we have a moral responsibility to actively protect species from population decay or extinction as the result of human activities, regardless of their value or importance to humans. The corollary to this belief is that any individual who knowingly or willfully contributes to the destruction of a species is immoral and lacks a personal compass of right versus wrong.

Moral justifications for the protection of biodiversity are sufficiently strong that they are sometimes incorporated into the political and legal systems of countries. For example, in the United States the intrinsic value of species was established by the Endangered Species Act in 1973, which not only gives species a de facto right to exist, but initially mandated that profits and economic values must be set aside when they threaten to extinguish a species. Ecuador enshrined the Rights for Nature as a part of its new constitution in 2008, which established the intrinsic value of nature and the right of all of life-forms to exist. Similarly, in 2010, Bolivia's Plurinational Legislative Assembly passed the Law of the Rights of Mother Earth, which declared both Mother Earth and all of her life systems as titleholders of inherent rights specified in the law[9] (**Figure 5.2**).

ETHICAL ARGUMENTS FOR CONSERVATION Unlike moral values, which stem from an individual's innate sense of right versus wrong, ethical values are established by social norms and codes of conduct. The ethics that a person adheres to are influenced by external factors like the nation, society, peer group, religion, and profession to which the individual belongs. These social norms and codes of conduct can change through time as external factors change.

Ethical arguments for conservation can focus on the humane treatment of single organisms, such as the treatment of individual animals that have the potential to feel pain and suffering (individualistic ethics). Or they can focus on whole populations, species, communities, or ecosystems that, as collections of individual organisms, do not experience pain or suffering (holistic or transorganismic ethics). Individualistic and holistic ethical arguments are sometimes in conflict with one another. For example, while individualistic arguments for conservation might seek to minimize the pain and suffering of single animals, holistic arguments sometimes require pain and suffering by individuals, such as when natural selection is needed to weed out individuals in a population that are

(A)

El Deber

(B)

© imageBROKER/Alamy Stock Photo

(C)

© iStock.com/photomaru

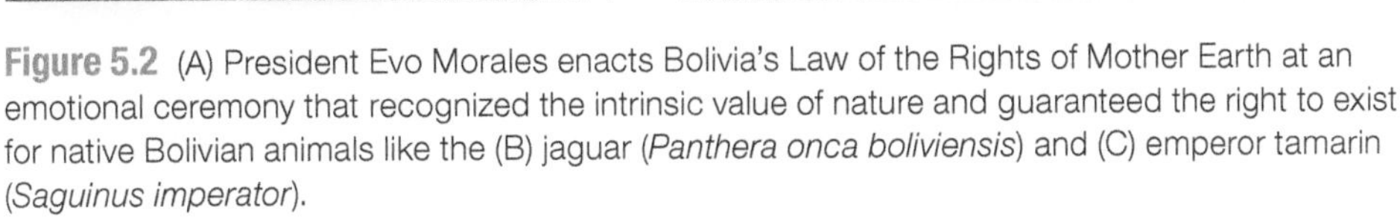

Figure 5.2 (A) President Evo Morales enacts Bolivia's Law of the Rights of Mother Earth at an emotional ceremony that recognized the intrinsic value of nature and guaranteed the right to exist for native Bolivian animals like the (B) jaguar (*Panthera onca boliviensis*) and (C) emperor tamarin (*Saguinus imperator*).

less fit, or when conservation requires the lethal eradication of individual invasive animals (e.g., the North American beaver in Patagonia and Tierra del Fuego).

While some conservation biologists have argued that individualistic ethics are beyond the realm of conservation biology,[10] there are numerous instances where individualistic ethical arguments influence public perception in ways that alter social norms and codes of conduct that, in turn, encourage the conservation of whole populations or species. Fox hunting is but one example. The English custom of tracking, chasing, and killing a fox using trained hounds dates back to the sixteenth century when foxes were considered vermin and hunted by farmers to curb attacks on farm animals.[11] But after the seventeenth-century decline of the United Kingdom's deer population, foxes became a prey of choice for recreational hunting (**Figure 5.3A**). Hunting grew in popularity throughout the eighteenth and nineteenth centuries, but a public debate emerged in the late 1900s over the ethics of fox hunting. Critics of the practice argued that fox hunts are an unusually cruel form of hunting, where animals are killed in ways that inflict inhumane

Figure 5.3 While most human societies consider it unethical for a person to knowingly inflict pain and suffering on another human being, some believe it is also unethical for humans to knowingly inflict pain and suffering on other nonhuman life-forms. This includes not only the pain and suffering inflicted directly by human activities like hunting (A) and trapping (B), but also that inflicted indirectly through human impacts like climate change (C) and pollution (D).

levels of pain and prolonged suffering. They further argued that the sport was being continued simply for pleasure rather than for population control of a pest species. In contrast, proponents argued that hunts were still needed for population control to protect farm animals and that the hunts had become important to rural economies and cultural traditions.

In 1999, a government inquiry (the Burns inquiry)[12] concluded that of the nearly 217,000 foxes in England and Wales, the roughly 25,000 killed each year by hunting hounds represented only a small fraction of the animals shot and killed by landowners, farmers, and game wardens each year. The inquiry acknowledged that hunts played a role in the cultural life of rural communities but also noted that role was less important than those performed by a village pub or church and that some residents in rural communities regarded the hunts as "divisive, intrusive, and disruptive." In response to the report and changing public perceptions about fox hunting, the British House of Commons passed the Hunting Act of 2004, which made "hunting wild mammals with a pack of dogs (3 or more)" unlawful in England and Wales.

The point here is that an individualistic ethical argument about the pain and suffering experienced by individual foxes altered social norms and codes of conduct about an activity (fox hunting) that had been considered socially acceptable for centuries. The change in social norms led to a ban on hunting of wild mammals, which subsequently led to protections for not only the entire fox population, but other hunted mammals as well.

Many ethical arguments for or against conservation are similarly steeped in social norms about what is or is not a humane way to treat individual organisms that have the potential to feel pain, show emotion, and experience suffering and loss (**Figure 5.3B–D**). When people see pictures of a lion maimed by a snare, a starving polar bear searching for food, or a seal being strangled by pollution, their ethical views about the pain and suffering of individual animals can influence social norms about activities like hunting and trapping, climate change, and pollution that affect entire species or communities of organisms. Conservation efforts that use individualistic ethical views to influence social norms have led to major conservation success stories (**Box 5.1**).

RELIGIOUS ARGUMENTS FOR CONSERVATION Many religions teach that humans have a responsibility to act as good stewards of the Earth and, thus, human activities that knowingly contribute to extinction and loss of biodiversity are wrong.[13,14] For example, Hinduism assigns divinity to certain species and recognizes a basic kinship between these species and humans (including the transmigration of souls from an individual of one species to an individual of another). A primary ethical concept in Hinduism and other Indian religions, such as Jainism and Buddhism, is the concept of **ahimsa**—the nonviolence and kindness to individual living beings. To live by this ideal, many religious people become vegetarians and live materially simple lives.

The prophet Muhammad, founder of Islam, also taught human responsibility to steward nature. He is quoted in a translation from the Sahih Muslim: "The world is green and beautiful and God has appointed you as His stewards over it. He sees how you acquit yourselves." Belief in the value of God's creation implies that one must be a good steward who preserves biodiversity: human beings have been given responsibility for God's creation and must preserve, not destroy, what they have been given.

In Christian religions, human responsibility for nature is established in the first chapter of Genesis where, after creating all living things, God assigns intrinsic value by stating that his creation is good. He then commands mankind to have dominion over every living thing on Earth. The phrase "have dominion over"

BOX 5.1 Success Story

The Public Makeover of Sharks

Sharks have become one of the most threatened groups of marine organisms, with populations of many species decimated over the last few decades by overexploitation. In 2000, the global catch and mortality of sharks from reported and unreported landings, discards, and shark finning had risen to 1.44 million metric tons, or 100 million sharks harvested from the oceans annually.[62] Much of the overharvesting was driven by demand for shark fin soup. It was a delicacy in Chinese and Vietnamese cuisine traditionally reserved for the wealthy and served on special occasions such as weddings and banquets. However, the delicacy became increasingly sought after as income levels of Chinese rose in the late 1900s, causing the shark fin trade to double between 1985 and 2001.[63]

Former NBA professional basketball star Yao Ming joined forces with WildAid in a public service campaign to educate the public about the plight of sharks and to reduce demand for shark fin soup in China.

Because sharks are long-lived and have low birth rates, their populations have been unable to rebound from overharvesting, and their long-term survival depends on measures to conserve them. But conservation efforts have often been hampered by public perception of sharks as merciless, indiscriminate killers—a perception promoted by the many horror movies that have personified sharks as man-eaters and instilled a fear of sharks into the public (e.g., *Jaws*, 1975; *Deep Blue Sea*, 1999; *Open Water*, 2003; *The Reef*, 2010; *The Shallows*, 2016; *The Meg*, 2018).

In 2006, the California-based environmental organization WildAid enlisted Yao Ming—a Chinese professional basketball player selected by the National Basketball Association's Houston Rockets as the first overall pick in the 2002 draft—as spokesperson for a public relations campaign designed to give sharks a public makeover (see the Figure). As a highly successful athlete with major product endorsements, Yao had been named by Forbes as the top celebrity in China for six years in a row (2004–2009).

When Yao and WildAid began their 2006 campaign, news reports[64] suggested that 75% of Chinese who were surveyed were unaware that shark fin soup actually involved sharks because the Mandarin translation is "fish wing soup." Additionally, 19% believed the fins grew back. But with a sequence of newspaper and magazine ads, billboards, and TV commercials (**Video 5.2**), Yao and WildAid helped increase public awareness of how fins are removed from sharks while they are still alive and the sharks are discarded into the ocean where they suffocate and drown because they can no longer swim (see above).

---

suggests that God placed mankind in a role that is superior to all other species and gave mankind the authority to exercise control over the Earth and its flora and fauna. This phrasing has been interpreted by some to imply that the use of natural resources is solely for the benefit of humans. But some scholars have interpreted this phrase as a mandate by God to be good stewards of his creation.[15,16] Thus, God gave mankind the authority to rule, but also commanded them to rule well by caring for creation as he himself would and by using the plants, animals, and other components of the natural world wisely and in a just manner.

Over 80% of people in the world follow a specific faith; there are at least 2 billion Christians, 1.34 billion Muslims, 950 million Hindus, and 200 million Buddhists worldwide. Because so many people's ethical values are influenced by a religious faith, religious arguments are often one of the most powerful motivating

"People said it was impossible to change China, but the evidence we are now getting says consumption of shark fin soup in China is down by 50 to 70 percent in the last two years," Peter Knights, executive director of WildAid, told the *Washington Post* in 2013.[65] "It is a myth that people in Asia don't care about wildlife. Consumption is based on ignorance rather than malice."

As consumption continued to decrease through 2016, Yao Ming argued that ethical norms in China about shark fin soup had begun to change: "Now it's something almost shameful for young middle-class people to eat."[64]

Yao Ming's efforts in China paralleled many similar efforts to change public perception about sharks in other parts of the world. For example, in 2006, the award-winning Canadian documentary *Sharkwater* helped debunk historical stereotypes of sharks as inherently violent, using stunning photography to show the important ecological role that sharks play in the seas.

In 2011, British chef, restaurateur, and television personality Gordon Ramsay released the documentary *Shark Bait*, which showed the extreme cruelty of shark finning and the impact it has on shark populations around the world (**Video 5.3**). Large numbers of public service announcements were put out by media outlets (e.g., National Geographic, **Video 5.4**), animal protection organizations (e.g., Humane Society of the United States, **Video 5.5**), and other celebrities (e.g., watch Malaysian celebrities, **Video 5.6**). In addition, new conservation organizations formed to focus on the policy and management of sharks (Shark Angels, oup-arc.com/e/cardinale-w5.1; Shark Savers, oup-arc.com/e/cardinale-w5.2; Shark Trust, oup-arc.com/e/cardinale-w5.3; Predators in Peril, oup-arc.com/e/cardinale-w5.4).

Collectively, these and other efforts have made huge progress toward giving sharks a public makeover, and they have begun to shift cultural norms against the purchase of animal products like shark fins.

But Peter Knights, chief executive of WildAid, cautioned that despite success, we cannot become complacent: "While consumers in mainland China have changed their behavior in response to awareness campaigns and a government banquet ban, shark fin soup remains on the menu in Hong Kong and Taiwan, and consumption is growing in places like Thailand, Vietnam, Indonesia and Macau."[65]

Video 5.2 WildAid PSA: *Shark Fin Soup*
oup-arc.com/e/cardinale-v5.2

Video 5.3 Gordon Ramsey: *Shark Bait*
oup-arc.com/e/cardinale-v5.3

Video 5.4 National Geographic: *The Misunderstood Shark*
oup-arc.com/e/cardinale-v5.4

Video 5.5 Humane Society of the United States: *Shark Finning Cruelty*
oup-arc.com/e/cardinale-v5.5

Video 5.6 Malaysian celebrities say "No" to shark fin soup.
oup-arc.com/e/cardinale-v5.6

forces for conservation efforts.[13] Leaders of many of the world's major religions have stated that their faiths are not only consistent with, but mandate, the conservation of nature (Box 5.2). Leading conservation biologists have begun to embrace religion as a partner in efforts to preserve the natural world (**Video 5.7**, **Video 5.8**). Major conservation organizations have started working with religious leaders and faith communities to articulate the sacred value of Earth and its biodiversity and work together to protect it. Examples include WWF's Sacred Earth Program (oup-arc.com/e/cardinale-w5.5) and the Alliance of Religions and Conservation (oup-arc.com/e/cardinale-w5.6). And professional societies have established sections on the intersection between conservation and religion; for example, the Society for Conservation Biology's Religion and Conservation Biology Working Group (oup-arc.com/e/cardinale-w5.7).

Video 5.7 E.O. Wilson: *Can Religion Help Conservation?*
oup-arc.com/e/cardinale-v5.7

Video 5.8 Jane Goodall: *On Faith and Conservation*
oup-arc.com/e/cardinale-v5.8

BOX 5.2 Case in Point

Religion and Conservation

In September 1986, an interfaith ceremony was held in the Basilica of St. Francis, in Assisi, Italy. It included "Declarations on Nature" by representatives of the five participating religions—Buddhism, Christianity, Hinduism, Islam, and Judaism. For the first time in history, leaders of these faiths jointly declared that their religions mandate the conservation of nature. Excerpts from the five declarations follow.

Leaders of the world's religions increasingly prioritize the conservation of natural resources and respect for all life.

The Buddhist Declaration on Nature
Venerable Lungrig Namgyal Rinpoche, Abbot, Gyuto Tantric University

The simple underlying reason why beings other than humans need to be taken into account is that, like human beings, they too are sensitive to happiness and suffering.... Many have held up usefulness to human beings as the sole criterion for the evaluation of an animal's life. Upon closer examination, one discovers that this mode of evaluation of another's life and right to existence has also been largely responsible for human indifference as well as cruelty to animals.

We regard our survival as an undeniable right. As co-inhabitants of this planet, other species too have this right for survival. And since human beings as well as other non-human sentient beings depend upon the environment as the ultimate source of life and wellbeing, let us share the conviction that the conservation of the environment, the restoration of the imbalance caused by our negligence in the past, be implemented with courage and determination.

The Christian Declaration on Nature
Father Lanfranco Serrini, Minister General, Order of Friars Minor (Franciscans)

To praise the Lord for his creation is to confess that God the Father made all things visible and invisible; it is to thank him for the many gifts he bestows on all his children.... By reason of its created origin, each creature according to its species and all together in the harmonious unity of the universe manifest God's infinite truth and beauty, love and goodness, wisdom and majesty, glory and power.

Man's dominion cannot be understood as license to abuse, spoil, squander or destroy what God has made to manifest his glory. That dominion cannot be anything other than a stewardship in symbiosis with all creatures.... Every human act of irresponsibility towards creatures is an abomination. According to its gravity, it is an offence against that divine wisdom which sustains and gives purpose to the interdependent harmony of the universe.

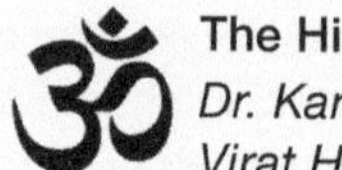

The Hindu Declaration on Nature
Dr. Karan Singh, President, Virat Hindu Samaj

The Hindu viewpoint on nature is permeated by a reverence for life, and an awareness that the great forces of nature—the earth, the sky, the air, the water and fire—as well as various orders of life including plants and trees, forests and animals, are all bound to each other within the great rhythms of nature. The divine is not exterior to creation, but expresses itself through natural phenomena. The *Mahabharata* says that "even if there is only one tree full of flowers and fruits in a village, that place becomes worthy of worship and respect."

Source: Text excerpts from Worldwide Fund for Nature, ed. 1986. *The Assisi Declarations: Messages on Man and Nature from Buddhism, Christianity, Hinduism, Islam and Judaism*. Gland, Switzerland.

Instrumental values

To many people, intrinsic arguments are by far the most convincing reasons for conservation. They are rooted in most deeply held moral and ethical beliefs that appeal to a general respect for other living beings, they have foundations in the values expressed by most world religions and environmental philosophies, and they are readily understood by the general public.[13,17] Even so, over the past few decades, **instrumental value** systems, also called **utilitarian value** systems, have begun to play an increasingly prominent role in decisions about the conservation

The Muslim Declaration on Nature
Dr. Abdullah Omar Nasseef, Secretary General, Muslim World League

The essence of Islamic teaching is that the entire universe is God's creation. Allah makes the waters flow upon the earth, upholds the heavens, makes the rain fall and keeps the boundaries between day and night.... It is God who created the plants and the animals in their pairs and gave them the means to multiply.

For the Muslim, mankind's role on Earth is that of a *khalifa*, viceregent or trustee of God. We are God's stewards and agents on Earth. We are not masters of this Earth; it does not belong to us to do what we wish. It belongs to God and He has entrusted us with its safekeeping.... The *khalifa* is answerable for his/her actions, for the way in which he/she uses or abuses the trust of God.

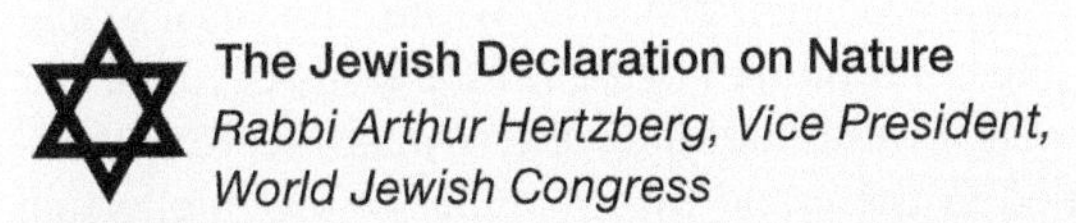

The Jewish Declaration on Nature
Rabbi Arthur Hertzberg, Vice President, World Jewish Congress

The encounter of God and man in nature is conceived in Judaism as a seamless web with man as the leader and custodian of the natural world.... Now, when the whole world is in peril, when the environment is in danger of being poisoned and various species, both plant and animal, are becoming extinct, it is our Jewish responsibility to put the defense of the whole of nature at the very centre of our concern.

We have a responsibility to life, to defend it everywhere, not only against our own sins but also against those of others. We are all passengers together in this same fragile and glorious world. Let us safeguard our rowboat—and let us row together.

of nature and its biodiversity. Instrumental value is the value something has as a means to an end—the ability to get something else that we want and value. Nature and its biodiversity have instrumental value in the form of the goods and services they provide to humanity, which meet people's material needs and desires (see Figure 5.1).

The increasing use of instrumental values in conservation has been driven by several factors. First, some have argued that intrinsic value systems have not been sufficient to motivate widespread conservation across the planet, because intrinsic values are not always shared among people and cultures. Second, many now recognize that—for better or worse—modern lifestyles (especially those in the developed world) depend on ecologically destructive economic systems that run counter to living environmentally responsible lives based on intrinsic values. Conservation based on intrinsic values is often seen as inhibiting progress or limiting to the progress of people's financial well-being. This is why economic arguments for development, resource extraction, and destruction of nature sometimes prevail over moral, ethical, and religious arguments for conservation.

While instrumental values represent an increasingly prominent argument for conservation, they are not without controversy. Many conservation practitioners feel that putting an instrumental value on nature or its biodiversity is unethical or counterproductive to the goals of protecting species.[18] For example, conventional economics—a common tool used to quantify instrumental value (see Chapter 7)—often suggests that species with low population sizes, unattractive appearances, no immediate use to people, and no relationship to other species of economic importance have little value to society. Such qualities may characterize a substantial proportion of the world's species, particularly microbes, insects and other invertebrates, fungi, and nonflowering plants. Halting profitable developments or making costly attempts to preserve these species may not have any obvious economic justification, and in fact, traditional cost-benefit analyses (see Chapter 7) may argue for destroying endangered species that stand in the way of "progress."

Gretchen Daily—a prominent Stanford biologist, once wrote, "Just as it would be absurd to calculate the full value of a human being on the basis of his or her wage-earning power ... there exists no absolute value of ecosystem services waiting to be discovered and revealed to the world by a member of the intellectual community."[19] Her warning is sage advice for all practitioners in conservation; nature and its biodiversity are valuable to people for a multitude of reasons, and instrumental values are but one method for valuation in a tool kit that practitioners can use to accomplish their work. In the subsequent chapters of Part II of the book, we develop this tool kit by introducing you to the variety of **ecosystem services** (see Chapter 6) that nature provides to humanity and by introducing you to the field of **ecological economics** (see Chapter 7), which quantifies the instrumental values of nature. Here, we simply overview the major categories of instrumental value, which include direct use, indirect use, option, and non-use values.

DIRECT USE VALUES **Direct use values** are the value of goods and services that ecosystems provide to people and which are used directly by individuals. Examples include raw materials (timber, fuelwood, skins, feathers, etc.), many types of natural food items (e.g., fruits, vegetables, fish, meat), and medicines (e.g., drugs synthesized from plants or animals). Direct use values can be further divided into **consumptive use value**, for goods that are consumed locally, and **productive use value**, for products that are sold in markets. The goods and services that contribute to direct use values are often traded or sold in existing economic markets, so their values are relatively straightforward to calculate and monitor using well-established methods.[20–22]

INDIRECT USE VALUES Some of the goods and services ecosystems provide to society are not used directly by humans. Rather, **indirect use values** are those benefits that people receive from nature without the need to harvest, destroy, or consume the resource.[20] Examples of indirect use values include flood protection by wetlands, water purification by forests, pollination of food crops by insects, and pest and disease control by microbes. These "ecosystem services" are generally regulated by basic ecological processes (production of biological tissue, decomposition, and nutrient recycling) that allow ecosystems to function properly and sustain life. These ecological processes are called *ecosystem functions*. Thus, indirect use values are typically controlled by ecosystem functions that produce ecosystem services.

Ecosystem functions and services are generally public goods in the sense that they often belong to the society in general, without private ownership. However, these benefits may be crucial to the continued availability of the natural products on which entire economies depend. If natural ecosystems are not available to provide these benefits, substitutes must be found—often at great expense—or local and even regional economies may face a decline in prosperity or even collapse. Because many ecosystem functions and services are not goods that are traded or sold in the traditional economic sense, they do not have value in traditional economic markets and do not typically appear in the statistics of national economies, such as the GDP.[23] However, methods for valuation of ecosystem functions and services—both economic values and other measures of social value—are rapidly developing, and we cover these more fully in Chapters 6 and 7.

OPTION VALUE In cost-benefit analysis and social welfare economics, the term **option value** refers to the value that is placed on a person's or organization's willingness to pay for maintaining or preserving a public asset or service even if there is no immediate plan to use the asset or service. Sometimes referred to as **insurance value**, option value represents a willingness by an individual or organization to leave some resource, such as a forest or a lake, alone and intact so that it will be available for use by someone else or their descendants at some point in the future. In conservation, option value represents one's willingness to pay for preserving threatened species, wildlife habitats, wilderness areas, or other natural resources so that they might be available for use in the future. Future benefits include things like discovery of new medicines, unearthing of new technologies, finding of new species that help control emerging pests or infectious disease, or availability of genetic resources needed to ensure the future success of food crops in a changing climate.

NON-USE VALUE **Non-use value** is the value that people and governments assign to goods and services even if they never have used and never will use them. Two types of non-use value are typically recognized: existence and bequest values.[21] **Existence value** is the benefit people receive from simply knowing that goods and services exist. This benefit, often reflected as a sense of well-being, is generally measured by someone's willingness to pay simply to protect species (e.g., bald eagles, whales), wildlife habitats (tropical rain forests, etc.), wilderness areas (like Yellowstone National Park), or other natural resources, even if those things are never utilized or experienced. **Bequest value** is a special case of existence value and represents the value that the current generation places on ensuring the availability of species, wildlife habitats, wilderness areas, or other natural resources for the benefit of future generations. Bequest and existence values are often calculated by how much money people and their governments are willing to pay to protect some aspect of biodiversity; when quantified properly, they often outweigh people's perception of the use values of nature.[24] For this reason, these non-use values are potentially powerful tools in conservation (more in Chapter 7).

Relational values

When making personal decisions about nature and biodiversity, people consider more than just the inherent worth of life-forms (intrinsic value) or ability to satisfy material wants and needs (instrumental value). Indeed, people consider additional factors that contribute to their happiness, contentment, and potential to have meaningful and satisfying lives.[25] These factors often fall under the category of **relational values**—desirable (sought-after) relationships among people, between people and nonhuman organisms, or between people and land that enhance one's personal or cultural identity, social responsibility, and emotional well-being (see Figure 5.1).[2]

Relational values are rooted in the concept of *eudaimonia*—an old idea in philosophy and psychology that human fulfillment not only requires hedonic happiness (the ability to achieve pleasure and avoid pain), but also eudaimonic happiness (emotional well-being, self-realization, and a sense of purpose).[26] The concept of eudaimonia originated with Aristotle, who stressed that feeling good is not enough to have a good life; rather, full human potential and true happiness are only realized when one achieves a sense of purpose, experiences personal growth, and attains self-acceptance.

Relational values contribute to one's eudaimonic happiness. They are prominent in many traditional societies (e.g., Native Americans and First Nations), Eastern philosophies (e.g., Confucianism, Buddhism), and social movements (e.g., *buen vivir*—the "good living" in Latin American countries).[2] Catholic Pope Francis recently featured relational values prominently in his encyclical *Laudato Si': On Care for Our Common Home*,[27] and they have increasingly been incorporated into the mandates of international treaties and biodiversity initiatives, such as the Intergovernmental Science-Policy Platform on Biodiversity and Ecosystem Services (IPBES).[3] Increased use of relational values by policy makers, NGOs, researchers, and the private sector has potential to ensure that environmental decisions account for people's relationships with nature and conserve the potential for present and future generations to achieve fulfilling lives. Some of the more common types of relational values are:

PERSONAL AND CULTURAL IDENTITY The personal and cultural identities of many people around the world are intimately linked to nature and its biodiversity. Personal identity is the concept you develop about yourself over the course of your life. This may include aspects of your life that you have no control over, such as where you grew up or the color of your skin, as well as choices you make in life, or your beliefs and how you spend your time. In contrast, cultural identity is the feeling of belonging to a group. It is part of a person's self-identity, but it is more broadly related to one's nationality, ethnicity, religion, social class, generation, local community, or any kind of social group that has its own distinct culture. While personal identity tells us who we are, cultural identity tells us how we relate to others.

The personal identities of many people are linked to their livelihoods (what they do), and a growing number of livelihoods are dependent on nature and biodiversity. For example, international tourism accounted for US$7.6 trillion (9.9%) of world GDP in 2014 and more than 277 million jobs worldwide.[28] In this industry, nature travel and ecotourism is the fastest-growing segment, with tourism revenue and corresponding jobs expected to rise by 45% and 29%, respectively. This rapid growth parallels growth projected for industries such as sustainable fisheries, sustainable forestry, renewable energy, and organic farming—all of which represent shifts of developed countries toward "green" economies.[29] Thus, the personal livelihoods of an ever-increasing number of people are being associated with jobs that produce goods or services that benefit the environment, conserve natural resources, or make business practices more environmentally friendly.

Furthermore, the personal identities of vast numbers of people are linked to nature through their hobbies and recreational activities. Many of us call ourselves hikers, anglers, hunters, birders, or nature lovers, and studies suggest that large fractions of people in developed countries associate themselves with

(A)

© Lyakhova Evgeniya/Shutterstock.com

(B)

© Universal Images Group North America LLC/ DeAgostini/Alamy Stock Photo

Figure 5.4 Examples of nature fostering relational values, desirable relationships that give one a sense of personal and cultural identity. (A) A father enjoys time with his daughter as he teaches her the art of fly-fishing, and (B) Yanomami tribesmen and women take part in communal fishing in the Venezuelan Amazon region.

these types of outdoor recreational activities.[30,31] For example, nearly 11% of the world's population in industrialized countries identify themselves as recreational anglers who fish for leisure and the benefits of personal interactions with friends and family (**Figure 5.4A**).[32]

Many people's cultural identities are also closely tied to natural locations. For example, sacred natural sites include areas recognized by Indigenous and traditional peoples, as well as areas recognized by religions or faiths as places for worship and remembrance. They represent some of the world's oldest conservation areas and often contain high levels of biological and cultural diversity.[33] For this reason, sacred natural sites are recognized by the United Nations Convention on Biological Diversity as areas that require protection, and international conservation organizations like the World Wildlife Fund and The Nature Conservancy have begun to integrate sacred sites into their networks of protected areas.

Other locations considered important for both natural beauty and cultural significance have been declared World Heritage sites by the United Nations Educational, Scientific and Cultural Organization (UNESCO). For example, Australia's Great Barrier Reef and Uluru/Ayers Rock are sites of remarkable biological variety and beauty; both are national parks that are central to the identity of modern Australians and have great cultural significance to Australia's Aboriginal people, whose connections to these sites go back more than 60,000 years.

Suggested Exercise 5.1 Go to the UNESCO World Heritage website. How many sites in your country are protected by UNESCO as a natural site? You can also read about how World Heritage sites are nominated and chosen. **oup-arc.com/e/cardinale-ex5.1**

Cultural identities can also be linked to social activities that enhance cultural ties, guard traditional customs, and promote mutual respect between people and nature. One example is transhuman pastoralism on the Iberian Peninsula, where shepherds migrate their sheep seasonally, not just to prevent overgrazing of land, but to reinforce their cultural identity through active ritual care. Another example is tribal salmon fishing on the west coast of North America, or communal fishing by Indigenous people of the Amazon—both of which not only provide sustenance, but are social events used to teach traditional customs and enhance cultural ties (**Figure 5.4B**).

Caring for culturally important places and perpetuating cultural practices to protect core values are by no means limited to Indigenous people. Most of us have likely had some type of social experience in nature (e.g., a camping trip, stays at the family cabin, or enjoyment of a favorite hiking, fishing, or hunting spot) that has become part of our personal and cultural identity. Indeed, when asked about benefits of nature, people of diverse backgrounds say a key benefit is nature's ability to foster family and cultural relationships.[34,35]

SOCIAL RESPONSIBILITY Nature and its biodiversity can also contribute to one's emotional well-being and sense of purpose by providing individuals with a sense of social responsibility. Social responsibility extends beyond the relationships that form one's personal and cultural identity. It instead seeks to improve the human collective by caring for nature and all of its inhabitants, including fellow human beings with whom one has no direct relationship. Social responsibility is premised on the idea that one's lifestyle and actions affect others, even if we don't know those people or have any direct relationship with them. For example, the carbon dioxide that you emit when heating your home or driving your car doesn't just affect you, your relatives, or those in your cultural group. Rather, the CO_2 you emit disperses into the atmosphere, and as it does so, it increases the average temperature of the planet in ways that affect all of humankind. Identical arguments can be made for the rubbish and wastewater we each produce, the overfishing and soil erosion we each contribute to with our dietary choices, the deforestation we each cause by drinking coffee and orange juice grown with nonsustainable practices, and so on. Many of our individual activities contribute to, or alternatively prevent, the destruction of nature and the loss of its biodiversity whether we see the consequences of our actions or not.

One of the most important and exciting developments in conservation biology involves supporting the **sustainable development** of disadvantaged people in ways that are linked to the protection of biodiversity (Chapter 17). Turning poachers into park guards who participate in the ecotourism industry, or helping poor farmers establish sustainable plots of cash crops to achieve economic independence, sometimes reduces the need to overharvest wild species and damage ecosystems. Working with Indigenous people to establish legal title to their land gives them the means to protect the biological communities in which they live (Chapter 15). In developed countries, the **environmental justice** movement seeks to empower poor and politically marginalized people, who are often members of minority groups, to protect their own environments; in the process their well-being and the protection of biodiversity are enhanced.[36] Working for the social and political benefit of poor and powerless people is not only compatible with efforts to preserve the natural environment, but also often a necessary component of an effective conservation program.[37]

HISTORY AND EDUCATION Some people value nature and its biodiversity because it is a biological library and history of our origin and trajectory as a species. Indeed, three of the most fundamental questions in science are:

1. How did life originate?
2. How have humans evolved through time?
3. How are we related to other organisms with whom we share the tree of life?

Thousands of biologists have worked to answer these questions, and we have already learned a great deal, but there are important details and complexities that remain to be discovered. New techniques of molecular biology are allowing for greater detail and insight into the relationships among living species as well as our relationship to extinct species known to us only from fossils. Our growing knowledge about biology is one of humanity's greatest achievements, and this knowledge has been facilitated by the conservation of species and their habitats.

However, when species go extinct, important clues are lost from the biological library, and mysteries about our place in the world, and our relationship to other organisms, become harder to solve. For example, if *Homo sapiens'* closest living relatives—chimpanzees, bonobos, gorillas, and orangutans—go extinct, we will lose important clues about the origin of humanity, as well as insights into topics such as the evolution of intelligence, social cooperation, speech, religion, brain size, violence, toolmaking, use of fire, and other things that have helped make humans unique as a species.

BIOPHILIA The term **biophilia** refers to the "love of life or living systems." It was first used by Erich Fromm to describe a psychological orientation of being attracted to all that is alive and vital.[38] Biologist Edward O. Wilson popularized the term in his 1984 book *Biophilia*,[39] in which he argued that humans have an innate tendency to seek connections with nature and other forms of life. Wilson proposed that attractions and positive feelings toward other organisms are deeply rooted in our biology and may have evolved through natural selection as survival was enhanced by innate responses to beneficial organisms (e.g., species that provide food or contribute to positive emotions) and innate fears of harmful organisms (e.g., venomous spiders and snakes). Biophilia may help explain why ordinary people will nurse sick and wounded animals back to health; enjoy watching birds and other animals in the wild, at zoos, and on television; or keep plants and flowers in their homes.

It is quite possible that biophilia, and a feeling of connectedness to other organisms, are behaviors that we learn from our family and surroundings, rather than an innate, evolved trait. If true, that might help explain why the increased dependence of the human species on technology has led to a weakening of the connections with nature in modern society.[40] Such changes in modern, urban society, characterized by electronic activities and food from supermarkets, could translate into a loss of human respect for the natural world, a decreased appreciation for the diversity of life-forms, and more environmentally destructive behaviors. Therefore, reestablishing the human connection with nature and increased emphasis on environmental education has become an important theme in conservation.[41,42]

AESTHETIC Aesthetic value is the value that an object possesses by virtue of its capacity to elicit pleasure based on appearance. Nature is filled with astounding beauty (**Figure 5.5**). Nearly 4 billion years of evolution have produced an unbelievable variety of plants, animals, fungi, and microorganisms that display an incredible variety of colors, body forms, and behaviors. This great variety of life inspires wonder and awe in people, and observing wildlife and natural landscapes brings enjoyment to their life. This enjoyment of nature can even have psychological benefits by making people feel happier, more satisfied, and more fulfilled.[43,44] This is another powerful reason why outdoor recreational activities such as hiking, canoeing, and mountain climbing are intellectually and emotionally satisfying.

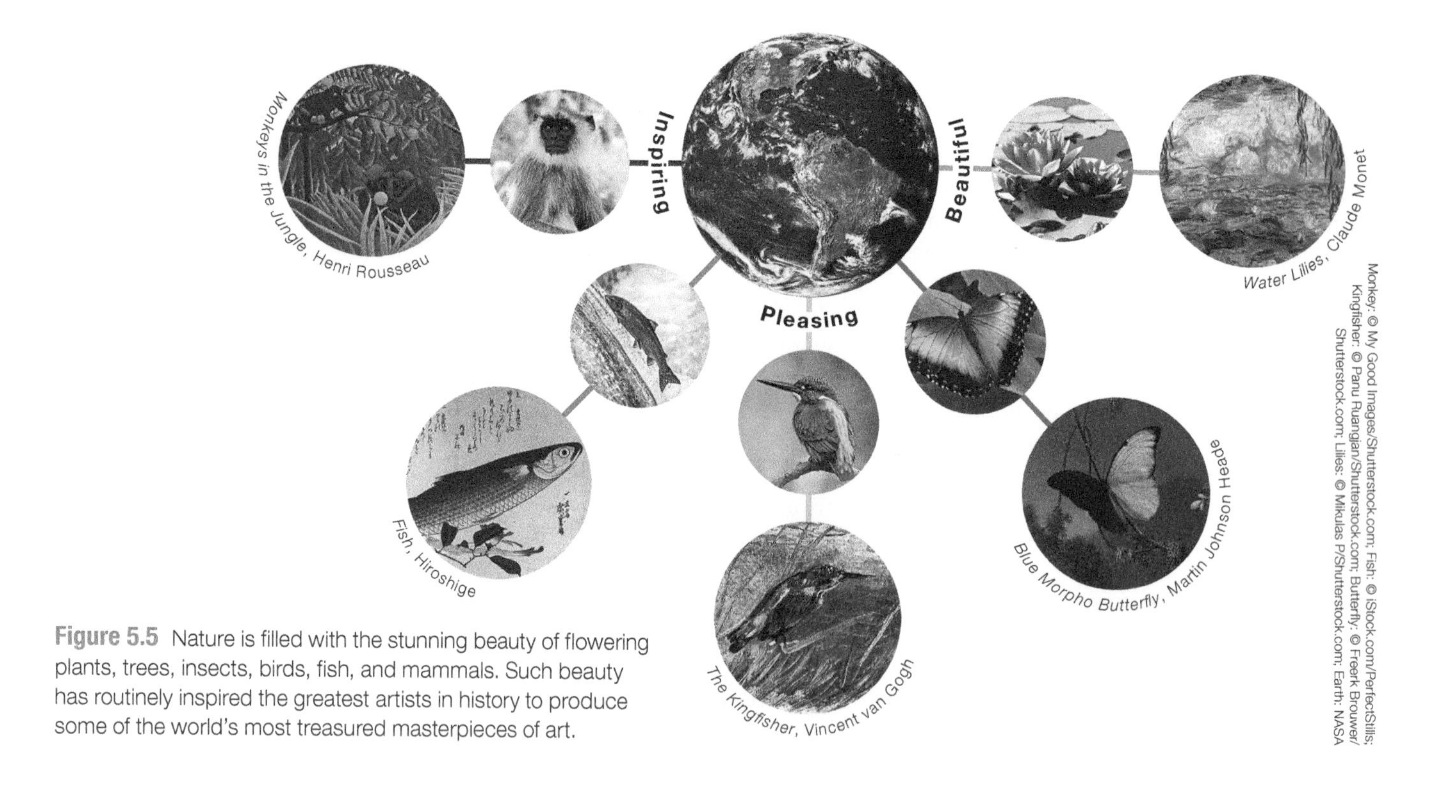

Figure 5.5 Nature is filled with the stunning beauty of flowering plants, trees, insects, birds, fish, and mammals. Such beauty has routinely inspired the greatest artists in history to produce some of the world's most treasured masterpieces of art.

Throughout history, poets, writers, painters, and musicians of all cultures have drawn inspiration from nature and its biodiversity. Many of the world's architectural wonders have been inspired by nature (**Figure 5.6**). Philosophers too go to nature to find insights into human existence and our place in the wider universe. Preserving biodiversity preserves sources of inspiration for all artists, designers, and philosophers and for everyone who appreciates and benefits from their works.

Suggested Exercise 5.2 Visit the Companion Website for an activity on biomimicry (the process of imitating nature to find a solution for a human challenge). Can you think of additional examples of nature-inspired engineering and design? **oup-arc.com/e/cardinale-ex5.2**

(A) Lotus temple, New Delhi

(B) Sagrada Familia, Barcelona

(C) Beijing National Stadium

Figure 5.6 Many of the world's architectural wonders have been inspired by nature. (A) The Lotus Temple in New Delhi was designed by architect Fariborz Sahba, who was inspired by the shape of the lotus flower. (B) The central nave of Barcelona's spectacular La Sagrada Familia was modeled after tree trunks so that those who enter feel as though they're walking into forest—just as famed Spanish architect Antoni Gaudi intended. (C) Beijing National Stadium in China, nicknamed the Bird's Nest, was designed by Swiss architecture firm Herzog & de Meuron for the 2008 Olympic Games.

Ethical Worldviews

The value systems that people use when making decisions on conservation are very much influenced by their ethical views of the natural world. Ethical worldviews control how people view themselves with respect to nature, and what the value of nature is for themselves and the rest of humankind. The field of **environmental ethics** is a branch of philosophy that studies the foundation of environmental values as well as societal attitudes, actions, and policies to protect and sustain biodiversity and ecological systems.[45,46] Environmental ethicists often describe three ethical worldviews that influence how people view themselves with respect to nature, biodiversity, and nonhuman organisms: anthropocentrism, biocentrism, and ecocentrism (**Figure 5.7**).

Anthropocentrism

Anthropocentrism is an ethical worldview that considers human beings to be the primary holders of moral standing in the world. Those who embrace this worldview see nature primarily in terms of its value and benefit to humans and their experiences (see Figure 5.7). The roots of anthropocentrism lie primarily in Western culture, and some have attributed them to Western religious beliefs that promote the ideas that only human beings are worthy of ethical consideration and that nature is primarily a means for human gain. In his classic essay "The Historical Roots of our Ecological Crisis,"[47] Lynn White argued that Judeo-Christian religions, especially Christianity, have been most responsible for promoting environmentally destructive behaviors, because doctrines that establish anthropocentrism occur in the first chapter of Genesis near the end of the biblical creation story:

> Then God said "Let us make man in our image, after our likeness. Let them have dominion over the fish of the sea, the birds of the air, and the cattle, and over all the wild animals and all the creatures that crawl on the ground." (Gen. 1:26)
>
> God blessed them, saying: "Be fertile and multiply; fill the earth and subdue it. Have dominion over the fish of the sea, the birds of the air, and all the living things that move on the earth." (Gen. 1:28)

Because Jews and Christians believed for centuries that it was not only their God-given right but their religious duty to dominate all other forms of life, White argued, anthropocentric attitudes and behaviors developed uniquely in Western cultures that emphasized these religious views, and many views held today remain artifacts of this historical perspective. Certain populations in the Christian faith promote rather extreme forms of anthropocentrism, such as the ideals of **human exceptionalism**, or **human supremacism**, espoused by creation "science" organizations like the Discovery Institute, which teach that humans are biologically and morally superior to all other life-forms. Such teachings are generally presented as a resistance to a growing movement by Asian religions, scientific teachings of evolution, and forms of environmental activism that challenge traditional Christian views that humans are biologically unique and superior and have sole moral and ethical rights.

White's essay was important because it stimulated a deeper examination of the relationship between religion and human attitudes toward nature[48] and played a role in helping shift Christian religions toward more environmentally conscientious views.[49,50] However, it is too simplistic to think that human-centric views of nature are solely the product of religious doctrines. Human-centric views of nature have been promoted by patriarchal social structures that encourage

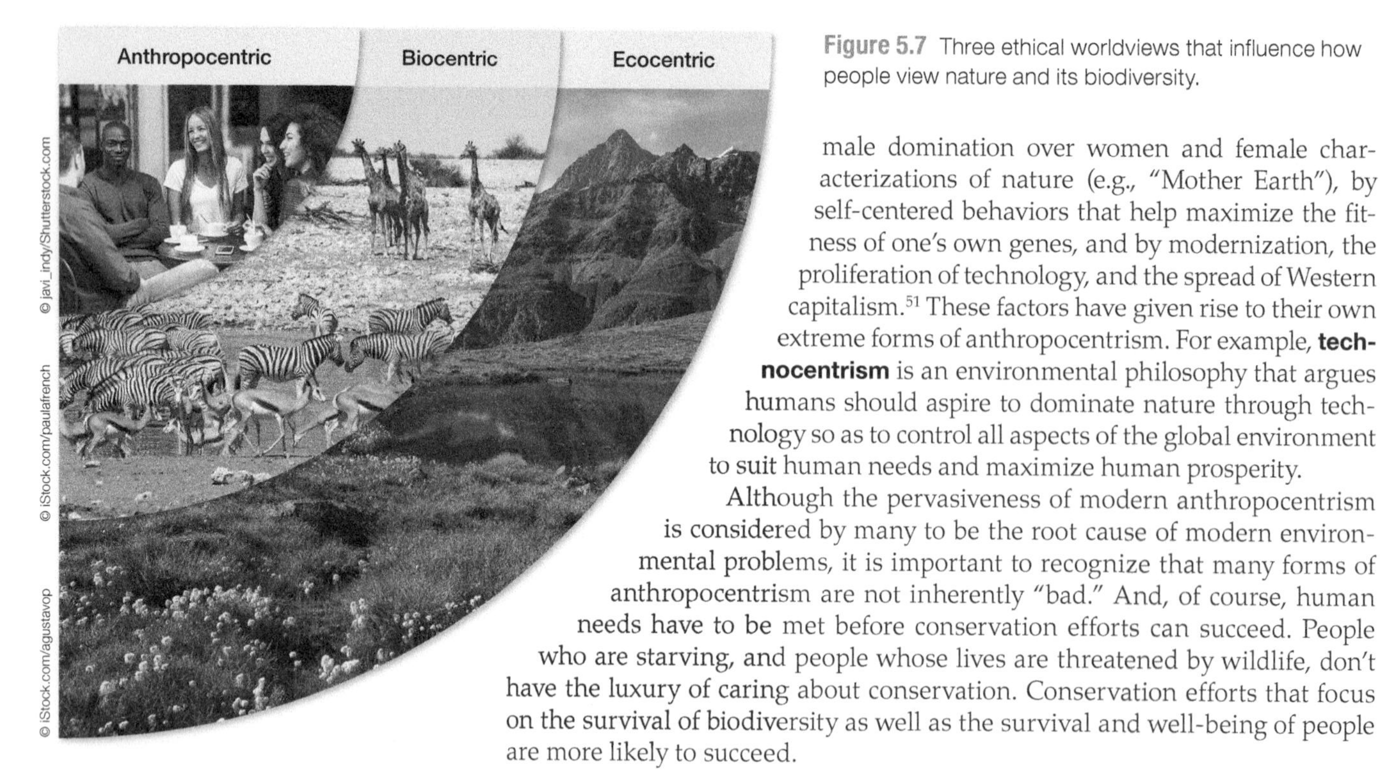

Figure 5.7 Three ethical worldviews that influence how people view nature and its biodiversity.

male domination over women and female characterizations of nature (e.g., "Mother Earth"), by self-centered behaviors that help maximize the fitness of one's own genes, and by modernization, the proliferation of technology, and the spread of Western capitalism.[51] These factors have given rise to their own extreme forms of anthropocentrism. For example, **technocentrism** is an environmental philosophy that argues humans should aspire to dominate nature through technology so as to control all aspects of the global environment to suit human needs and maximize human prosperity.

Although the pervasiveness of modern anthropocentrism is considered by many to be the root cause of modern environmental problems, it is important to recognize that many forms of anthropocentrism are not inherently "bad." And, of course, human needs have to be met before conservation efforts can succeed. People who are starving, and people whose lives are threatened by wildlife, don't have the luxury of caring about conservation. Conservation efforts that focus on the survival of biodiversity as well as the survival and well-being of people are more likely to succeed.

Biocentrism

Biocentrism is an ethical worldview that extends inherent value and moral consideration to all living beings, not just humans.[51] Biocentrists believe that each individual organism has intrinsic value and a right to exist (see Figure 5.7), and they often express the view that humans are not inherently "superior" to other organisms in a moral or ethical sense. Those who espouse the biocentrist philosophy often emphasize the interconnectedness between humans and other living beings and that actions that negatively affect other living beings can adversely affect humans as well.

Like anthropocentrism, biocentrism has religious roots. In Islam, biocentric ethics are part of the belief that all of creation belongs to Allah (God) and exists to praise him. Human destruction of living things prevents the Earth from praising God to its fullest. In Hinduism, nonhuman life is respected because all living things are believed to have souls (*ātman*) and to cyclically go through births and rebirths, including reincarnation into nonhuman forms (*saṃsāra*).[52] Jainism, which exists in tandem with Hinduism, encourages people to practice ahimsa (nonviolence), which includes not causing physical, mental, or spiritual harm to any organism.[53] Harming another living being that has spiritual energy is akin to harming oneself. In Buddhism, teachings emphasize that all living things are important and that humans are not above other creatures, but part of existence.[54]

Nonreligious precursors for modern biocentrism arose in Albert Schweitzer's *Reverence for Life*[55] and Peter Singer's ethics of *Animal Liberation*.[56] In contrast with conventional ethics, which concerns itself exclusively with human beings, Schweitzer argued that any distinctions between "high and low" or "valuable and less valuable" individuals belonging to other-than-human life-forms were arbitrary and subjective. Singer agreed that simply being a member of the species *Homo sapiens* is an arbitrary requisite for moral importance and proposed that sentience (the capacity to feel, perceive, or experience) was a better criterion to apply to any organism. Singer argued that if we accept sentience as a criterion for moral importance, it follows that we must extend the same basic moral consideration (i.e., principles of equality) to all other sentient beings as we do to human beings.

Principles of modern biocentrism were most clearly laid out in Paul Taylor's 1986 book *Respect for Nature: A Theory of Environmental Ethics.*[57] Taylor argued that biocentrism is an "attitude of respect for nature" whereby one attempts to live one's life in a way that respects the welfare and inherent worth of all living creatures. Taylor espoused his belief that humans are members of a community of life in which all members of all species are morally and ethically equal. This community consists of a system of interdependent members where each individual organism has a purpose and a reason for being, and each organism is inherently "good" and "valuable."

Biocentrism has been criticized by some as an antihuman philosophy that is founded in logical contradictions. In his essay "A Critique of Anti-Anthropocentric Biocentrism,"[58] Richard Watson argued that if humans are put on a par with other organisms, truly equal with all other living beings, then the evolution of human actions and culture should be considered every bit as natural as the ways in which all other animals thrive or behave. In turn, it would be a contradiction to also argue that humans should change their behavior to refrain from causing harm to other organisms, because the very act of doing so would only serve to set humans apart from other organisms by assigning more power to them than other organisms. This would be akin to reverting back to the basic tenets of anthropocentrism that assign a special status of morality and responsibility to humans.

Ecocentrism

Ecocentrism is an ethical worldview that extends moral consideration to transorganismic entities like species, biotic communities, ecosystems, and biodiversity (see Figure 5.7). While the distinction between biocentrism and ecocentrism is sometimes poorly defined, ecocentrism is distinct in that it recognizes Earth's living and nonliving systems as a whole, rather than recognizing just the value of individual organisms (biocentrism). Thus, ecocentrism is the more holistic of the environmental philosophies, considering Earth as a biophysical system that includes humans, nonhuman life, and the sum of physical and biological processes that are required to sustain life.

Numerous world religions, and many Indigenous cultures around the world, espouse beliefs and teachings that emphasize the need for people to *live in harmony with nature,* to *care for Mother Earth,* and to *tread lightly on the land and water.* Although the extent to which people and cultures have achieved sustainable use of their natural resources is debatable, teachings that convey an ecocentric philosophy and need to care for the health of entire ecosystems are particularly common in cultures where people's livelihoods are closely tied to the land and the natural resources it provides (see Chapter 2).

The foundation of modern ecocentrism as an environmental ethic is usually credited to Aldo Leopold,[59] who believed that principles of ecology and evolution broadened the boundaries of a biological community to include the variety of plants and animals that share evolutionary kinship and that live together on the soils and waters where they are found (see Chapter 2). Given this belief, Leopold wrote that a proper land ethic "changes the role of *Homo sapiens* from conqueror of the land community to plain member and citizen of it. It implies respect for his fellow-members and also respect for the community as such." Leopold's land ethic considers *Homo sapiens* to be a part of nature, not separate from it. Hence, human-caused changes to nature are no less natural than changes imposed by any other species. However, Leopold argued, because *Homo sapiens* is a moral species, capable of ethical deliberation, conscientious choice, evolutionary kinship, and good citizenry, any changes we impose on nature should be ethically evaluated with respect to their impacts on other species, and with respect to how those changes will affect the ecological processes that are required to sustain life—not just human life, but all life.

Leopold's land ethic has served as the motivation for numerous environmental movements. These include the **deep ecology** movement (not to be confused with the scientific discipline of ecology) coined by the environmental philosopher Arne Naess in 1973,[60] which espouses the view that the natural world is a balance of complex relationships in which the existence of humans is closely linked to the existence of other species in coupled human-natural systems. Therefore, human interference with or destruction of the natural world poses a threat not only to other species, but to humans. Deep ecology's core principle is the belief that the living environment as a whole should be respected and regarded as having certain inalienable legal rights to live and flourish, independent of its instrumental values for human use. When summarizing the deep ecology philosophy, Austrian-born physicist Fritjof Capra wrote:

> Deep ecology does not separate humans—or anything else—from the natural environment. It does see the world not as a collection of isolated objects but as a network of phenomena that are fundamentally interconnected and interdependent. Deep ecology recognizes the intrinsic value of all human beings and views humans as just one particular strand in the web of life.[61]

Summary

1. Three types of value systems are commonly used to justify conservation. Intrinsic values stem from moral, ethical, or religious beliefs that biodiversity has value "in itself" or "for its own sake." Instrumental values stem from the goods and services that nature provides to meet human needs and desires. Relational values are desirable relationships, among people, between people and nonhuman organisms, or between people and land, that bring happiness to individuals by enhancing their personal or cultural identity, social responsibility, and emotional well-being.
2. For many people, the intrinsic and relational values of nature are by far the most convincing reasons to conserve biodiversity. These values have foundations in most world religions and environmental philosophies, appeal to a general respect for life, and are readily understood by the general public.
3. Instrumental values are now playing an increasingly important role in conservation, including direct use values, indirect use values, option value, and existence value. When used effectively and in combination with other value systems, instrumental values have potential to further the goals of conservation biology.
4. The value systems that people use when making decisions about conservation are very much influenced by their ethical views of the natural world. Environmental ethicists often describe three ethical worldviews that influence how people view themselves with respect to nature: anthropocentrism, biocentrism, and ecocentrism.
5. Anthropocentrism assumes that humans are "above" nature and "superior" to all other species. Therefore, humans alone deserve moral consideration, and the purpose of nature is primarily to meet human needs and desires.
6. Biocentrism assumes that individual organisms have both instrumental and intrinsic value and that all living creatures deserve moral consideration. Therefore, people have a responsibility to respect and care for other living things.
7. Ecocentrism assumes that humans are part of nature—just one of many species that interact in a coupled human-natural system. Therefore, entire ecosystems have value, and both species and their environments deserve moral consideration. As members of an ethical species and good community members, humans have a responsibility to manage ecosystems in ways that sustain biodiversity.

For Discussion

1. Which value system (intrinsic, instrumental, or relational) and which ethical worldview (anthropocentrism, biocentrism, or ecocentrism) do your views on nature most closely align with? How have your views been influenced by your upbringing, culture, education, and life experiences?
2. Of all the values listed in Figure 5.1, which do you think resonate most strongly with your friends, family members, or others that you associate with at work or school? Are the values complementary—that is, can they be used simultaneously to accomplish the same goal? Or are they mutually exclusive?
3. Do you think that all living beings—big and small, ugly and beautiful—have a "right to exist"? What about mosquitoes, ticks, or other species that represent pests and diseases of humans? How might your views on these types of organisms change as you gain a better understanding of the role they play in ecosystems?
4. Do human beings have a responsibility to treat all organisms equally? What about animals that seemingly lack self-awareness (e.g., worms). What about plants, which lack a nervous system? How about microbes like bacteria?
5. Should physical entities, such as rivers, lakes, and mountains, have rights? Where should we draw the line of moral responsibility?

Suggested Readings

Carson, R. L. 1965. *The Sense of Wonder.* New York, NY: Harper & Row. Rachel Carson's final book urges adults to teach children about the natural world.

Chan, K. M. A. et al. 2016. Opinion: Why protect nature? Rethinking values and the environment. *Proceedings of the National Academy of Sciences* 113: 1462–1465. Develops the concept of relational values and argues why they are needed to complement intrinsic and instrumental value systems.

Curry, P. 2011. *Ecological Ethics: An Introduction* (2nd ed.). Boston, MA: Polity Press. Excellent introduction to the ethical arguments used in conservation.

Hitzhusen, G. E. and Tucker, M. E. 2013. The potential of religion for Earth Stewardship. *Frontiers in Ecology and the Environment* 11: 368–376. Describes how conservation biologists and religious groups can work together for common goals.

McKibben, B. 2007. *Deep Economy: The Wealth of Communities and the Durable Future.* New York, NY: Henry Holt. Eloquent argument that economic growth is neither sustainable nor improving our lives, with an exploration of alternatives.

Pascual, U. et al. 2017. Valuing nature's contributions to people: The IPBES approach. *Current Opinion in Environmental Sustainability* 26–27: 7–16. Describes the value framework being used by the Intergovernmental Science-Policy Platform on Biodiversity and Ecosystem Services.

Teel, T. T. and Manfredo, M. J. 2010. Understanding the diversity of public interests in wildlife conservation. *Conservation Biology* 24: 128–139. Describes how the public has diverse attitudes toward wildlife that can affect government conservation efforts.

Vonk, M. 2013. Sustainability and values: Lessons from religious communities. *Social Sciences Directory* 2: 120–130. Examines the values that underlie the sustainability of several long-standing religious communities.

Wilson, E. O. 1994. *Biophilia.* Cambridge, MA: Harvard University Press. Stimulating book that argues mankind has evolved to have a deep love for nature and its living creatures.

Visit the
Conservation Biology
Companion Website
oup.com/he/cardinale1e
for videos, exercises, links, and other study resources.

USGS Bee Inventory and Monitoring Lab

6

Biodiversity and Ecosystem Services

Bees provide an important ecosystem service in the form of pollination. The western bumblebee (*Bombus occidentalis*), shown here, is one of several North American bumblebee species that has suffered sharp declines in recent years.

Do you recall the biosphere experiments from the 1990s? The most famous of these was Biosphere 2, which was originally constructed by Space Biosphere Ventures and funded by philanthropist Ed Bass, who provided more than $200 million for construction costs (**Figure 6.1**). Biosphere 2 was to be the second self-sufficient biosphere after Earth itself (Biosphere 1) and was originally designed to demonstrate the viability of closed ecological systems that could sustain human life. Biologists and engineers spent four years (1987–1991) designing and landscaping the 0.013 km^2 Biosphere 2 to have all of the different types of life and ecosystems needed to sustain humans: agricultural systems stocked with a variety of species to provide food, diverse rain forests to regulate the artificial atmosphere, even a miniocean to control climate and temperature. On September 26, 1991, eight researchers were locked inside the enclosed, airtight environment. Almost immediately after, levels of CO_2 inside Biosphere 2 began to increase and fluctuate wildly due to unplanned release of gas from concrete.[1] Levels of oxygen inside the biosphere began a precipitous decline from 21% to a dangerously low 14% as plants failed to photosynthesize as much as had been hoped. Most of the vertebrate species and all of the pollinating insects died within a year, and the researchers themselves experienced sustained weight loss of 17%. The "mission" was terminated after just 24 months when it became clear that health and welfare of the biospherians could no longer be maintained.

(A)

© ClassicStock/Alamy Stock Photo

(B)

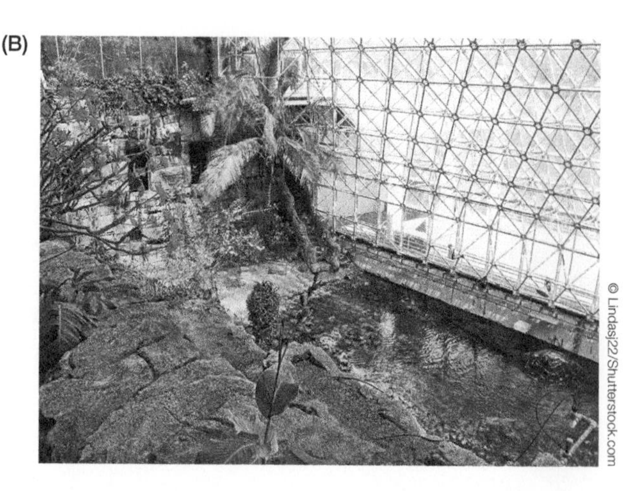

© Lindasj22/Shutterstock.com

Figure 6.1 (A) Biosphere 2, located near Tucson, Arizona, was constructed to be the largest materially enclosed ecosystem on Earth. It was originally intended to demonstrate the potential of using closed systems to sustain human life on another planet. (B) Biologists and engineers spent four years designing and landscaping Biosphere 2 to have all of the different ecosystems needed to sustain life, including this mini-ocean to help control the climate. But after eight researchers were locked inside the airtight environment on September 26, 1991, levels of CO_2 increased, O_2 declined, and most vertebrate species and pollinating insects began to die. As researchers began to experience significant weight loss, headaches, and exhaustion, it became clear that human health and welfare could no longer be maintained inside the Biosphere, and the experiment was terminated. Biosphere 2 was a sobering reminder that we don't yet have the basic understanding needed to design ecological systems that can maintain human life.

Some have viewed Biosphere 2 as a monumental failure, whereas others have viewed it as a magnificent learning experience. Regardless of one's view, it was a sobering reminder that we still don't have even a basic understanding of how to design a self-contained ecological system that can mimic the biological processes, and deliver the basic goods needed to sustain human life. Obviously, this means we're not yet in a position to put a human colony on another planet such as Mars. More importantly, it means we don't yet understand how to live sustainably on our own planet.

Earth, like Biosphere 2, is a materially closed ecosystem. No materials are lost, and no materials are gained, and pretty much everything that is required to sustain human life is made for us by other living creatures that share the planet with us. Without photosynthetic bacteria, algae and plants, there would be no breathable atmosphere. Without microbes, fungi, and animals, there would be no organic soils to grow crops, and nothing to pollinate those crops if they did exist. Without essential players in our planet's global ecosystem, the oceans would produce no fish, and forests would produce no wood. There would be no fossil fuel, no renewable biofuel, and even if we had fuel to burn, there would be nothing to clean the pollutants from combustion out of the water we drink or the air we breathe.

Nature has provided the goods and services needed to sustain human life for so long that most of us take them for granted. But growing evidence suggests that Earth's ecosystems and the biological diversity that provides our natural goods and services are being eroded.[2]

In this chapter, we describe the variety of benefits ecosystems provide to humanity, which are collectively called **ecosystem services**. Ecosystem services represent a subset of the values of biodiversity, discussed in Chapter 5, that are anthropocentric (instrumental and relational). In Chapter 7, we will focus even more narrowly on the subset of ecosystem services from this chapter that can be valued using tools from the field of ecological economics.*

*Parts of this opener were originally published in the February 20, 2013 edition of *The Scientist*.

What Are Ecosystem Services?

Ecosystem services can be defined as "nature's contributions to people." They represent the plethora of benefits that humans receive from both natural and managed ecosystems, including things like the production of consumable goods (e.g., production of fish from oceans, or wood from forests), nonconsumable services (e.g., protection from natural disasters like floods, or biological control of pests and disease), and cultural services (e.g., recreation, or ecotourism).[3] Many ecosystem services are essential to human survival (e.g., production of O_2, food, and regulation of a livable climate), and without them, the planet would not be able to sustain human life. In fact, one could just as well define ecosystem services as the collection of biological processes that are required to sustain human life and provide a good standard of living on Earth.

Despite their indisputable importance, many people don't know about ecosystem services; nor do they understand or appreciate their value (**Video 6.1**). In part, this is because nature has been providing these services for free since the moment humans first walked on the planet, and it is human nature to undervalue things that are both abundant and free. Just as we sometimes tend to undervalue our health or relationships until we lose them, we fail to recognize the importance of the quality of air and water until they become so polluted that we get sick, we fail to think about flood protection until we lose our homes to a hurricane or typhoon, and we pay little attention to natural control of disease until the outbreak of a deadly virus threatens a global pandemic.

Video 6.1 Watch this video about the services ecosystems provide to humanity.
oup-arc.com/e/cardinale-v6.1

In addition to taking ecosystem services for granted, people tend to underappreciate them because their influence on our lives is often indirect, and we don't see them in our daily lives. This is particularly true in developed countries where the goods and services people need to support their daily lives are quite separated from where those goods and services are first produced. For example, when people buy food at the grocery store, they don't consider which ocean their fish came from, or how many other species it took to produce the meat they are about to cook and eat. People don't tend to think about how many types of bacteria and fungi it took to make the organic soils needed to grow their crops, or how many different types of pollinators it took to generate the fruits and vegetables they put into their shopping cart. When people take out their trash or flush their toilets, they don't consider how many different types of organisms it will take to break down and recycle their wastes so that they do not build to toxic levels. The numerous species and myriads of biological processes required to put food on the table, clean up wastes, produce air, and generate raw building materials are simply not things we contemplate in modern society.

But while ecosystem services are abundant and free, and even though we may not see them directly, it is important to recognize that pretty much everything we eat, drink, and breathe is a product of living organisms and the biological processes they perform in ecosystems. As we lose ecosystems and their resident life-forms that control these biological processes, we also risk losing nature's contributions that directly affect the potential for Biosphere 1 (Earth) to sustain human life.

History of Ecosystem Services

The importance of nature to human well-being has long been recognized, perhaps as early as the Stone Age when men and women hunted and gathered food on the plains of Africa. The belief systems, religions, and myths of many early hunter-gatherer, and subsequent agrarian, societies tended to emphasize themes about conservation and the wise use of natural resources because human well-being was entirely dependent on the biosphere (see Chapter 2). Some of the earliest written records about ecosystem services (though they were not called this at the time) stem from the classical Greek period; Plato commented on how land clearing and deforestation for agriculture throughout the Baltic region had caused widespread destruction of once fertile soils, led to large-scale erosion, and reduced the availability of spring water.[4]

While the concept of nature's services is old, the term *ecosystem services* was popularized in the 1990s by the convergence of three areas of study (**Figure 6.2**):

1. Multiple disciplines began to estimate the monetary and nonmonetary value of ecosystems and their biodiversity for human society.
2. An area of research now known as the field of biodiversity and ecosystem functioning began to provide mechanistic evidence linking biodiversity to ecosystem processes and, ultimately, services.
3. The Millennium Ecosystem Assessment critiqued the state of Earth's ecosystems and showed that the services they provide are rapidly deteriorating in the modern era.

Below we summarize each of these three areas of study.

Ecosystem valuation

Precursors for the modern valuation of ecosystem services began with the rise of two fields of study in the mid-twentieth century. First, the field of ecosystem ecology began soon after Arthur Tansley coined the term *ecosystem*,[5] using it to refer to a community of living organisms (plants, animals, microbes) that interact with the nonliving components of their environment (air, water, and

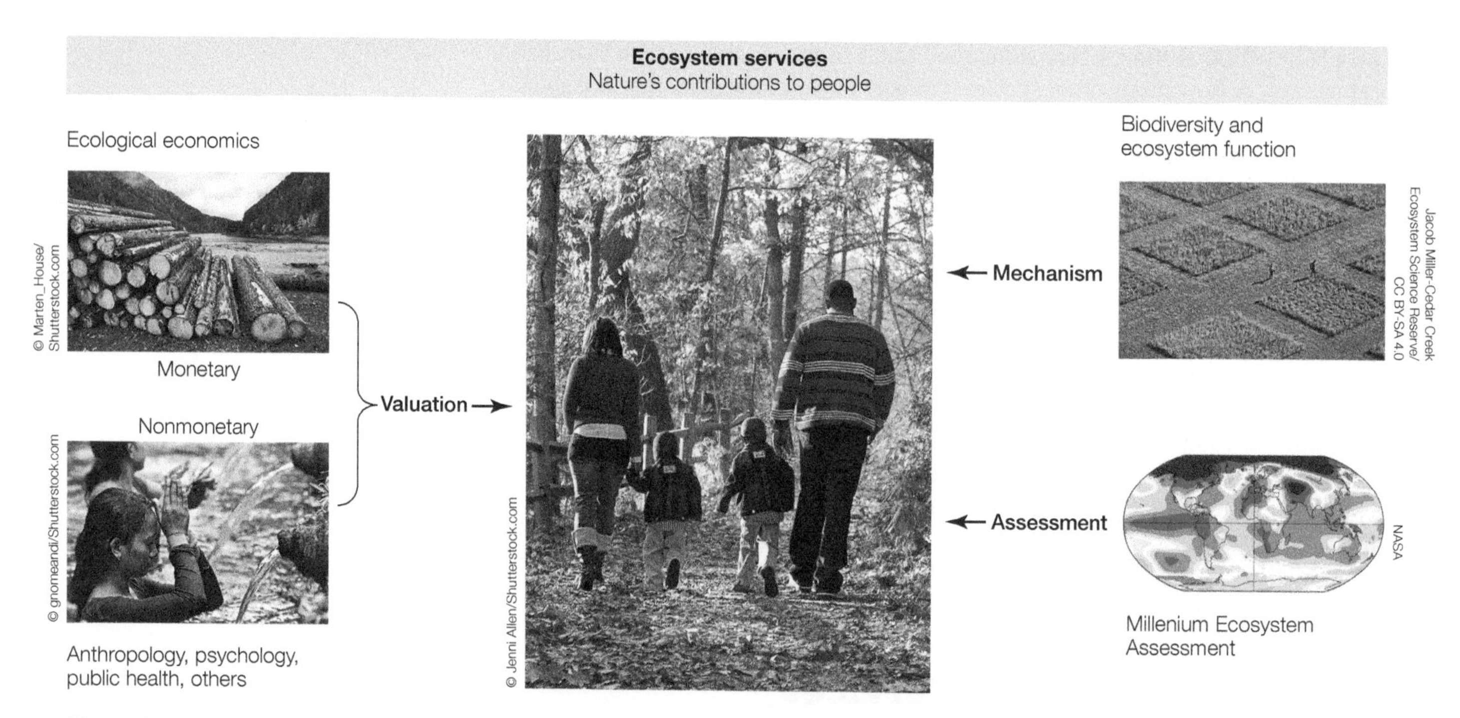

Figure 6.2 While the concept of nature's services is old, the term *ecosystem services* was popularized in the 1990s by the convergence of three areas of study: (1) multiple disciplines began to estimate the monetary and nonmonetary values of ecosystems and their biodiversity, (2) the field of biodiversity and ecosystem functioning began to identify the mechanistic evidence linking biodiversity to ecosystem processes and services, and (3) the Millennium Ecosystem Assessment critiqued the state of Earth's ecosystems and showed that the services they provide are rapidly deteriorating in the modern era.

mineral soil). This concept proved to be a useful one for a group of ecologists who wanted to study nature quantitatively as a system.[6] Subsequent studies from ecosystem ecology, such as the whole-forest clear-cuts performed at Hubbard Brook in New Hampshire by Bormann and Likens,[7] showed that the dominant organisms that comprise ecosystems ultimately control the fluxes of water, nutrients, and energy that are essential to all life-forms living in those ecosystems.

As ecosystem ecology was developing as a science, so too was the field of environmental economics. In 1952, the nonprofit organization Resources for the Future was established in Washington, DC. The early focus at Resources for the Future was on natural resource scarcity, which was prompted by the U.S. President's Materials Policy Commission (the Paley Commission) that was asked to review the future supply of minerals, energy, and agricultural resources in light of the formidable demands made on these resources by World War II. Resources for the Future's early work culminated in the influential book *Scarcity and Growth* by Barnett and Morse,[8] which focused attention on the long-term implications of resource scarcity for economic growth and human well-being.

Soon after the publication of *Scarcity and Growth*, a sequence of environmental catastrophes (see Table 2.1 in Chapter 2) brought the concept of **externalities** to public attention. Externalities are the detrimental effects of a business practice that are not paid for by the business but, rather, are born by the public. For example, when thousands of people died in 1952 as a result of London's Great Smog, and an additional 12,000 people later died prematurely from diseases caused by smog, the coal-burning power plants and industries that originally polluted the city's air with toxic levels of soot, nitrogen oxides, and sulfur oxides were not required to pay for the costs of health care or taking care of families who suffered losses. Instead, people paid the costs of pollution, which leads to distorted incentives whereby businesses reap the profits of their activities, but society pays some of their operational costs.[9]

In the wake of numerous environmental catastrophes in the 1950s through 1970s that included massive oil spills and chemical pollution, it became abundantly clear that free market economic systems do not properly account for the cost of externalities and, as a result, the public is forced to pay for environmental damage caused by unethical business practices. Because markets do not account for externalities, it also became clear that markets do not properly account for the value of ecosystems being destroyed, and thus, traditional markets have no way to determine changes in the supply or demand of resources provided by ecosystems based on the benefits they provide to people.

The recognition that traditional markets fail to account for externalities and the supply and demand of nature's benefits is what stimulated efforts to bridge the disciplines of ecosystem ecology and environmental economics. In 1982, Ann-Mari Jansson organized a symposium in Sweden funded by the Wallenberg foundation on Integrating Ecology and Economics. Of the 48 participants at that meeting, many went on to form the International Society for Ecological Economics, as well as its founding peer-reviewed journal *Ecological Economics*.[10] The primary goal of the new professional society was to promote an interdisciplinary research program focused on figuring out how nature and its goods and services can be properly valued by society (see Figure 6.2).

Even as the nascent field of ecological economics was being formed, some individuals began framing the benefits of nature as utilitarian benefits in order to increase public awareness.[4,11] The term *environmental services* was introduced in a 1970 report of the Study of Critical Environmental Problems, which described the importance of services like insect pollination, fisheries, climate regulation, and flood control.[12] In 1973, nature's benefits to people were described as natural capital by British economist E. F. Schumacher in his collection of essays published in *Small Is Beautiful: A Study of Economics As If People Mattered*.[13] In subsequent years, multiple variants of the term were used, but by the mid-1980s, the term *ecosystem services* had become well established in the scientific literature and was being used to describe the benefits provided to people by nature.[14]

Throughout the 1990s, the concept and use of ecosystem services became increasingly mainstream in both the scientific and popular literature. Gretchen Daily's edited book *Nature's Services: Societal Dependence on Natural Ecosystems*[15] played a key role in popularizing the concept. That book was a collection of chapters written by world-renowned scientists who reviewed the value of ecosystem services, the damage that had been done to them, and the implications for society. The authors concluded that the entire human economy depends on goods and services provided by Earth's natural systems and that human activities were disrupting the functioning of natural systems and imperiling the delivery of these services.

In 1997, Robert Costanza and colleagues published a lightning rod paper that was as surprising as it was controversial.[16] These authors undertook a massive effort to estimate the value of all the world's ecosystems in terms of the goods and services they provide to people across the planet. For the entire biosphere, Costanza and his colleagues concluded that the 1994 value of nature's services was "estimated to be in the range of US$16–54 trillion ($10^{12}$) per year, with an average of US$33 trillion per year." This number was huge, and it suggested that the total value of nature—nearly all of which falls outside of traditional economic markets—equaled or exceeded the entire global gross national product at the time of US$18 trillion per year. The controversy that followed this paper was substantial, with many criticizing the assumptions that were made by the authors in their calculations, as well as some of the methods used for analyses. But during this controversy, most people recognized that even if Costanza and colleagues were off by a few orders of magnitude, the value of natural ecosystems was far greater than anyone had expected them to be.

But even as the field of ecological economics was rising to prominence as a way to value nature, there was growing awareness that monetary valuation provides, at best, a partial representation of the things people value about the natural world. Many people view the intrinsic values of nature and its biodiversity as their primary motivation for conservation; yet, intrinsic values are usually difficult if not impossible to quantify, and they certainly cannot be valued in monetary terms (see Chapter 5). Many relational values, such as cultural experiences, aesthetic beauty, and one's sense of purpose in life, contribute substantially to human well-being and happiness. Well-being and happiness can be studied and quantified (e.g., one can survey people to ask how happy or satisfied they are), but again, not necessarily in monetary terms.[17] Furthermore, many would argue that monetary valuation is not applicable to many aspects of human health, since the value of a life, or the physical and mental well-being of one's self or one's family members are often seen as priceless. One of the most important advances that has occurred over the past two decades is that economic valuation of ecosystem services has increasingly been complemented by nonmonetary valuation of ecosystem services (see Figure 6.2). Disciplines from the health sciences (e.g., human medicine, public health) and the noneconomic social sciences (e.g., anthropology, human psychology, sociology) have increasingly helped quantify how ecosystems and their biodiversity influence people's physical, mental, and emotional well-being.

Biodiversity and ecosystem function

Another factor that contributed to development of the ecosystem services concept was a discipline called biodiversity and ecosystem function, which began to study the mechanistic basis for links between biodiversity and nature's contributions to people. The field of biodiversity and ecosystem function represented a shift in the focus of community ecology research that began in the 1980s. Prior to that time, most research in community ecology had focused on explaining why biodiversity exists, asking how the variety of genes, species, and functional traits that comprise an ecosystem are shaped by other forces like species interactions, disturbances, abiotic gradients, and so on (see Chapter 4). The predominance of this historical focus is clear from most any introductory text in ecology, evolutionary biology, or biogeography, where biodiversity is almost always plotted on the *y*-axis of a graph showing how it responds to, or is controlled by, something else.

But during the 1980s, the focus of biodiversity research began to change. Prompted by accelerating rates of extinction, new areas of study began to ask what species "do" in ecosystems, and what will change once they are lost. For example, the field of ecological engineering put species on the *x*-axis (as an independent variable, or causal factor) to explore how they influence the physical formation of habitats.[18] The field of ecological stoichiometry put species on the *x*-axis as a causal factor to explore how they influence the fluxes of biologically important elements in biogeochemical cycles.[19] The study of trophic cascades put the number of predators on the *x*-axis as a causal factor to explore how they regulate the productivity of entire ecosystems.[20] Research from these fields, and others, began to show that the loss of certain life-forms was likely to substantially alter the functioning of entire ecosystems in ways that might have consequences for humanity.

By the 1990s, several international initiatives were increasingly focused on the more specific question of how the diversity of life can affect ecosystems. The Scientific Committee on Problems of the Environment (SCOPE), established by the International Council for Science (ICSU), produced an influential book reviewing the state of knowledge on how biodiversity influences the functioning of whole ecosystems.[21] The United Nations Environmental Programme commissioned the Global Biodiversity Assessment to evaluate, among other things, biodiversity's role in regulating ecosystem and landscape processes.[22] An international research program called DIVERSITAS formed and produced a global research agenda focused on integrating biodiversity science with human well-being.[23] Collectively, these international initiatives helped steer major funding and research programs such that, by the mid-1990s, hundreds of experiments were being performed in nearly every ecosystem on Earth to examine how the diversity of genes, species, and biological traits of organisms influences ecological processes like biomass production and nutrient cycling.[24,25]

The Millennium Ecosystem Assessment

The factor that ultimately made ecosystem services a global agenda was the Millennium Ecosystem Assessment, MEA (see Figure 6.2). The MEA was commissioned in 2000 by United Nations Secretary-General Kofi Annan to assess the health of Earth's ecosystems, and it involved more than 1300 participants representing 95 countries. Together, this group summarized the state of human impacts on the world's ecosystems, the consequences of ecosystem change for human well-being, and the scientific basis for action needed to enhance the conservation and sustainable use of ecosystems and their contribution to human well-being.

The final report of the MEA,[2] which was published in 2005, highlighted four main conclusions:

1. Between 1950 and 2000, humans changed ecosystems more rapidly and extensively than in any comparable period of time in human history. These changes, which largely stemmed from rapidly growing demands for food, freshwater, timber, fiber, and fuel, led to substantial and largely irreversible loss in the diversity of life on Earth.

2. Changes to ecosystems have contributed to substantial gains in human well-being and economic development; however, these gains have been achieved at great cost—namely, the degradation of many ecosystem services, and the exacerbation of poverty for some groups of people. Approximately 60% (15 out of 24) of ecosystem services evaluated by the MEA have been degraded and are being used unsustainably. These problems substantially diminish the benefits that future generations will obtain from ecosystems.
3. The degradation of ecosystem services could grow significantly worse during the first half of this century and will be a barrier to achieving most sustainable development goals (see Chapter 17), including the eradication of poverty and hunger, reductions of child mortality, improvements in maternal health, and control over widespread diseases like malaria.
4. The challenge of reversing the degradation of Earth's ecosystems while still meeting increased demands for services will require significant changes to national and international policies, as well as institutions and practices that are only in their early stages.

The MEA put the concept of ecosystem services firmly on the international policy agenda, and its release changed the dialogue about ecosystem services in at least four ways. First, the concept of ecosystem services quickly became the basis for a new, rapidly expanding field of interdisciplinary study.[26] Since 2005 there has been an exponential increase in the scholarly literature that seeks to measure, assess, and value society's dependence on nature. There are now graduate degree programs in ecosystem services offered at major universities around the world, professional societies that have formed chapters (groups of practitioners) who study ecosystem services, and entire journals dedicated to the topic of ecosystem services.

Second, the MEA stimulated a number of new initiatives that were charged with better quantifying ecosystem services and infusing those values into policy making. One example is the Economics of Ecosystems and Biodiversity (TEEB) initiative that was launched in 2007 by Germany and the European Commission to develop a global study on the economics of biodiversity loss. One of the products of TEEB was a suite of case studies that estimated the economic value of natural ecosystems that could be shared with decision makers to help them recognize the value of biodiversity and the growing cost of ecosystem degradation.

Suggested Exercise 6.1 Go to the TEEB website and locate the case studies for your part of the world. What story could best explain nature's value to one of your regional policy makers? **oup-arc.com/e/cardinale-ex6.1**

Third, the MEA stimulated the formation of new mechanisms to incorporate ecosystem services into markets and financial planning by individuals, private businesses, and government agencies. For example, numerous ecosystem markets[27] and payments for ecosystem services schemes[4] were developed, creating financial incentives for landowners to conserve watersheds in order to provide clean water to downstream users or to conserve forests to provide carbon sequestration and storage to help offset CO_2 emissions into the atmosphere. While markets and payment schemes are still in early stages of development, these represent a vast change of perspective from purely market-oriented economic systems because they explicitly account for externalities and account for the value of ecosystems measured by their benefits to people.

Last, the MEA stimulated new international treaties such as the Intergovernmental Science-Policy Platform on Biodiversity and Ecosystem Services (IPBES). IPBES, which is an agreement signed in 2012 by 100 governments, was charged with assessing the state of the world's biodiversity and the ecosystem services it provides to society and with responding to requests from policy and decision makers.[28] IPBES takes a more inclusive approach to valuation of ecosystem services than initiatives like TEEB by considering both the monetary and nonmonetary value of ecosystems and their biodiversity (**Video 6.2**).

Video 6.2 Watch this video about the Intergovernmental Panel on Biodiversity and Ecosystem Services.
oup-arc.com/e/cardinale-v6.2

Types of Ecosystem Services

The Millennium Ecosystem Assessment proposed a framework for organizing ecosystem services into four general categories (**Figure 6.3**):

1. **Supporting services** are the ecological processes that control the functioning of ecosystems and production of all other services.
2. **Provisioning services** are the products people obtain from ecosystems.
3. **Regulating services** are benefits obtained from the regulation of ecosystem processes, which help reduce harmful variation and provide insurance.
4. **Cultural services** are the nonmaterial benefits people obtain from ecosystems.

Despite the MEA being more than a decade old, the simplicity and intuitive organization of its framework has allowed it to remain one of the most widely used frameworks in science, policy, and management. In the sections that follow, we expand on each category of ecosystem services and provide examples for each.

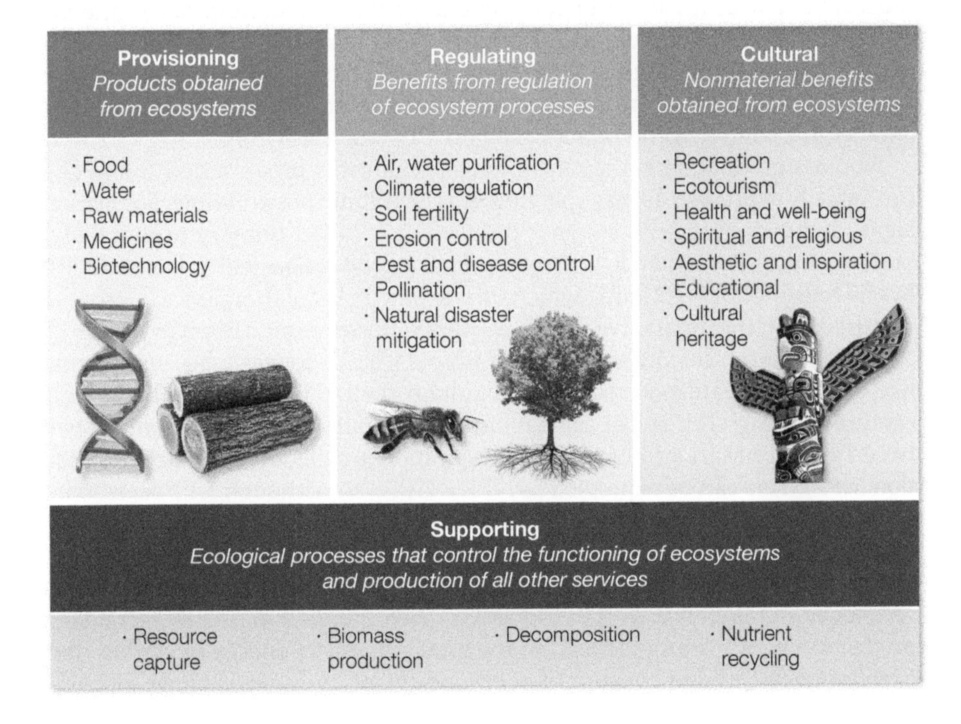

Figure 6.3 The Millennium Ecosystem Assessment's framework organizes ecosystem services into four categories: supporting, provisioning, regulating, and cultural.

Supporting services

Every ecosystem service is ultimately controlled by the stocks and fluxes of energy (e.g., sunlight) and matter (e.g., carbon, nitrogen, phosphorus) that flow through the living and nonliving components of an ecosystem. Stocks represent the "pools" or mass of matter stored per area or volume (e.g., grams per square meter or cubic meter). Stocks of matter can be in living tissue, such as the amount of carbon in the biomass of trees (grams C per cubic meter of wood), or in the nonliving parts of the ecosystem, such as the amount of phosphorus stored in rocks (grams P per cubic meter of sedimentary deposits). Fluxes represent the rate at which matter flows between different stocks, often expressed as a mass per area or volume per unit time (e.g., grams per square meter per year). Fluxes can occur between a nonliving and living component of an ecosystem, such as when trees assimilate CO_2 from the atmosphere by photosynthesis to create organic carbon (e.g., grams CO_2 per square meter of forest per year). Fluxes can also occur between two living components of an ecosystem, such as when an invasive herbivore pest eats these leaves off a tree (e.g., grams C per square meter of forest per year).

Supporting services, which are sometimes referred to as ecosystem functions, are the ecological processes that control the stocks and fluxes of energy and matter in ecosystems. The sizes of all stocks and rates of all fluxes are influenced by living organisms that capture food, grow, reproduce, and then become involved in fluxes themselves as they are eaten, decomposed, and recycled by other organisms. The key ecological processes that control the fluxes of energy, nutrients, and organic matter through various pools in an ecosystem include resource capture and biomass production, and decomposition and nutrient recycling.

RESOURCE CAPTURE AND BIOMASS PRODUCTION Resource capture is the ability of organisms to sequester (e.g., capture and assimilate) biologically essential resources from their surrounding environment. For primary producers like plants, algae, and cyanobacteria, resource capture is the assimilation of inorganic resources they use for growth, such as sunlight, nutrients, and water. For heterotrophic organisms like animals, resource capture is the consumption of food in the form of plants (by herbivores) or animal prey (by predators). The efficiency by which organisms capture their resources has a major impact on the fluxes of energy and matter through an ecosystem. Indeed, biological control agents that maximize the flux of energy, carbon, and nutrients from one pool (the pest) to another (the control agent) provide the best ecosystem service.

Biomass production is the rate at which organisms produce new living tissue, usually expressed as mass per unit area or volume per unit time. For plants, algae, and cyanobacteria, biomass production is called **primary productivity**, or 1° productivity (Figure 6.4). **Gross primary production** (**GPP**) is controlled by the rate of photosynthesis, which dictates the mass of inorganic CO_2 that is captured from the air and water and converted into organic carbon in tissues of the primary producer (cells, leaves, stems, roots, etc.). However, because plants, like all other organisms, respire to maintain their cells, some of the carbon captured during GPP is lost as heat during respiration (R). The difference between gross primary production and respiration is called **net primary production** (**NPP**). This can be expressed as NPP = GPP – R. Although NPP represents a smaller amount of carbon than is captured for GPP, NPP is the actual amount of new plant biomass that is produced per unit area or volume per time.

For heterotrophic organisms like animals, biomass production is called secondary productivity, or 2° productivity (see Figure 6.4). Gross secondary production (GSP) is proportional to the total amount of energy or carbon that is assimilated through consumption of food (plant, prey, or detritus). However, because an animal loses some energy and carbon through heat loss and

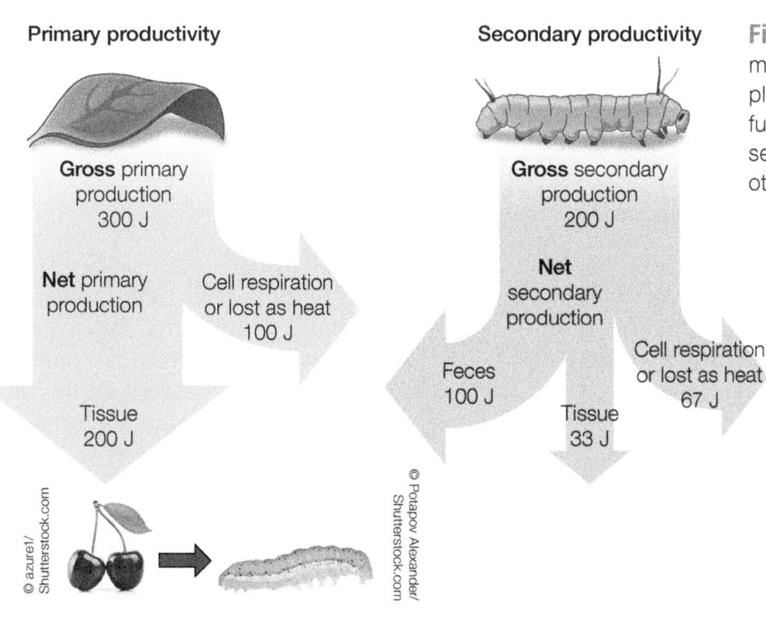

Figure 6.4 Fluxes of energy that control rates of primary and secondary production in autotrophs (e.g., plants and algae) and heterotrophs (e.g., bacteria, fungi, animals). Biomass production is a supporting service that influences the production of nearly all other ecosystem services.

respiration, net secondary production (NSP) is less than GSP (NSP = GSP − R). Animals also lose carbon and energy through waste products like feces or urine. Therefore, the amount of biomass produced by animals per unit area or volume per time is the difference between NSP and the loss of waste products.

Understanding the distinction between gross production and net production is important because they both influence the delivery of ecosystem services. For example, while plants sequester carbon in proportion to GPP, it is NPP that ultimately determines how much of this carbon gets stored in their tissues. To get accurate estimates of carbon storage in plants, one must be careful to measure NPP (not GPP). Furthermore, different types of plants (C3, C4, CAM) have different rates of photosynthesis that influence their sequestration of carbon during GPP. As a result, ecosystems that are dominated by different plant groups or species differ widely in the amount of carbon they sequester and remove from the atmosphere, which affects the regulation of climate. Similarly, animal species differ widely not only in their rates of gross secondary production, but also in their rates of respiration. For example, endothermic animals that maintain internal body temperatures through metabolism tend to have high rates of respiration, which also means they convert much less of their food to new biomass (1%–3%).[29] In contrast, ectothermic animals that do not metabolically regulate body temperatures can convert more than 50% of their food to biomass. This is important because many ecosystem services like the production of food, pest and disease control, or erosion control are directly related to the ability of species to produce new biomass. Species that grow and reproduce slowly will have a very different ability to deliver these services than those that grow and reproduce quickly.

DECOMPOSITION AND NUTRIENT RECYCLING Decomposition is the process by which dead organic matter (detritus) is broken down and decayed. Nutrient recycling is the process by which biologically essential elements (e.g., nitrogen, phosphorus, and potassium) are broken down from their solid forms in organic tissue into soluble inorganic forms (a process called **mineralization**). Once those elements are in their soluble inorganic forms, they can be taken up and reused again by plants and microbes (a process called **immobilization**). Decomposition and nutrient recycling are essential to the proper functioning of ecosystems. Without these two ecological processes, the nutrients required for the production of new biomass would get trapped and would not be available for growth of living organisms. Fertile organic soils wouldn't exist, and the production of new biomass in ecosystems would grind to a halt. Therefore, every ecosystem service that depends on the biomass of a living organism is also controlled—directly or indirectly—by decomposition and nutrient recycling.

Decomposition begins at the moment of death, and the process is often described by an exponential decay curve of the form:

$$y = e^{-kt} \tag{6.1}$$

where y is the amount of mass remaining of a dead item (a dead animal, dead leaf, dead wood, etc.), t is the amount of time that has passed since death, and k is called the decay constant, which represents the percent mass loss per unit time. Different types of detritus have different decay constants. For example, the value of k for dead wood, which contains a lot of complex lignin that is hard for microbes to break down and decompose, is very low. In contrast, the value of k for a dead animal is high because animals have tissues that are not only rich in nutrients, but dominated by proteins that are comparably easy to decompose.

The rate of decomposition is regulated by two factors: (1) autolysis, which is the breaking down of tissues by the body's own internal chemicals and enzymes, and (2) putrefaction, which is the breakdown of tissues by other organisms. The primary organisms that control the process of decomposition are bacteria and fungi, though larger scavengers (e.g., mites, insects, birds, mammals) play an important role when tissues are accessible. Some larger scavengers have evolved niches that allow them to specialize on consumption of dead tissue (e.g., carrion beetles and the family Sarcophagidae, called flesh flies, which lay eggs and hatch maggots on dead tissue). Other scavengers are more opportunistic (e.g., coyotes, hyenas, vultures).

Decomposing tissue can be sufficiently important to the functioning of an ecosystem that entire food webs depend on it. One U.S. example comes from Pacific salmon, which return from the oceans to swim upstream, spawn, and lay their eggs.[30] After spawning, the adult salmon die, and their carcasses are fed on by a variety of other species. Using analyses of stable isotopes of carbon (^{13}C) and nitrogen (^{15}N), researchers have found that salmon are not just consumed by microbes, insects, birds, and mammals that catch then from the streams; they also support insects, reptiles, and even the growth of trees far away from the spawning sites as scavengers either defecate or drag the carrion upland, where it decomposes and releases vital nutrients (**Figure 6.5**).

THE VALUE OF SUPPORTING SERVICES It is important to note that, in and of themselves, the supporting services of ecosystems do not have human value. In order for their value to be determined, they must be placed in the context of one or more of the provisioning, regulating, or cultural services that are described next. To understand why, consider the supporting service of net primary production. In the context of wood production in a forest (a provisioning service), NPP has positive value for people because it leads to higher yields and potential harvest of more lumber. But in the context of recreation and tourism on a coral reef (a cultural service), primary production of algae may have negative value and actually represent an **ecosystem disservice** because algae can overgrow and kill corals, in turn reducing revenue from diving and tourism. Although NPP is the same ecological process that produces trees and algae, the value of this ecological process can only be determined in the context of how it influences a specific provisioning, regulating, or cultural service that we value as people.

Figure 6.5 Supporting services like decomposition and nutrient recycling are crucial to the proper functioning of ecosystems, and they support entire food webs. Here, Pacific salmon that are captured in a stream are transported into terrestrial habitats where they are deposited by predators like birds or bears. As they decompose, their nutrient recycling fuels the growth of everything from spiders to lizards to mice to trees. (From T. Levi et al. 2012. *PLOS BIOL* 10: e1001303.)

Provisioning services

Provisioning services represent the material outputs from ecosystems that are used by people. Many provisioning services are bought and sold in existing markets, and thus they can be quantified economically based on principles of supply and demand (discussed in Chapter 7). Some examples of provisioning services follow.

FOOD Despite there being thousands of edible plants, fungi, and animals in the world, most of the global human population lives on a diet composed of a small number of staples derived from vegetables or animal products. These staples include rice, wheat, maize (corn), millet, sorghum, roots and tubers (potatoes, cassava, yams, and taro), and animal products such as meat, milk, eggs, cheese, and fish.[31] Just 15 crop plants provide 90% of the world's food energy intake (exclusive of meat), with rice, maize, and wheat representing staples for about 80% of the world population and comprising almost two-thirds of human food consumption.

Most staple food is produced using modern and conventional farming practices, which in developed countries involve cultivation of plant or animal monocultures that are heavily managed with fertilizers, biocides, and antibiotics needed to keep the cultivations viable and healthy.[32] A key role of biodiversity in modern farming practice is to serve as a genetic library from which food scientists can search for and develop new strains of plants and animals that have desirable traits, such as increased productivity, resistance to pests and disease, and tolerance to environmental change.

While conventional farming practices have allowed food production to be one of the few ecosystem services that has shown a consistent upward trend

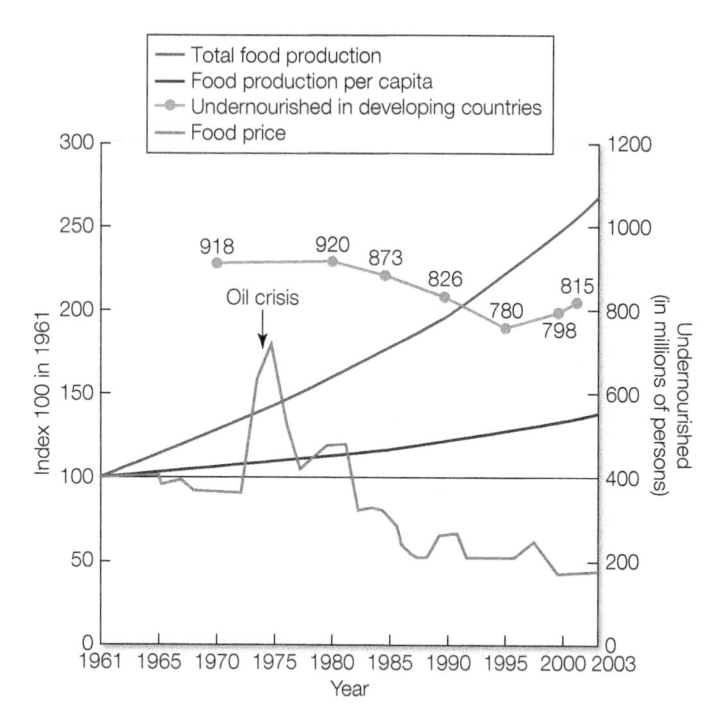

Figure 6.6 Food production is one of the few provisioning services of ecosystems that has shown a consistent upward trend. Total and per capita food production in agroecosystems has consistently increased, while food prices have decreased since 1960. (After Millennium Ecosystem Assessment, 2005. *Ecosystems and Human Well-being: Biodiversity Synthesis*. Washington, DC: World Resources Institute.)

(**Figure 6.6**), there are two limitations of conventional farming. First, the externalities of conventional farming are often large. It is widely acknowledged that modern agriculture has been a primary cause of habitat loss, biodiversity loss, chemical and nutrient pollution, eutrophication in rivers and lakes, and coastal hypoxia (depletion of oxygen) in the world's oceans.[2,33] Because of these externalities, there is a growing demand for the adoption of less intensive, more sustainable farming practices that are less damaging to the environment, such as the adoption of mixed polyculture farming that is common in developing countries (see Chapter 15). Second, many developing nations have not yet adopted large-scale conventional farming, and their populations continue to rely heavily on natural ecosystems to grow, collect, hunt, or harvest food. Wild plants are essential for rural subsistence households that represent at least a billion people worldwide.[31] In Ghana, for example, the leaves and fruits of over 300 species of wild plants are used for food. In rural Swaziland, more than 220 species of wild plants complement the diet of domesticated cultivars. In India, Malaysia, and Thailand, cultures traditionally ate thousands of wild plant species. Forest foods in many tropical and subtropical areas of the world provide a variable, year-round supply of essential minerals and vitamins through wild leaves, seeds and nuts, fruits, roots and tubers, mushrooms, and honey.

Wild animals, including insects, amphibians, reptiles, birds, fish, rodents, and larger mammals, are often the only source of animal protein for rural people, with some 62 developing countries relying on wildlife for at least one-fifth of their animal protein.[31] In parts of the Peruvian Amazon, for example, more than 85% of dietary animal protein is from the wild. Hunting and collecting wild animals for meat constitutes about 40% of the protein consumption in Botswana and about 80% in the Democratic Republic of the Congo (formerly Zaire). Throughout the world, 130 million tons of fish, crustaceans, and mollusks, mainly wild species, are harvested each year, with 100 million tons constituting marine catch and 30 million tons constituting freshwater catch.[34] Fish alone provide more than 3 billion people with almost 20% of their average intake of animal protein.[35] Wild capture fisheries continue to dominate world output of fish protein, but aquaculture accounts for a growing percentage of total fish supply.[36]

WATER The water cycle, which controls the availability and distribution of freshwater used for drinking and irrigation, is heavily controlled by biological processes. Biological control over the water cycle stems primarily from the influence of plants on **evapotranspiration**—which is the sum of evaporation and plant transpiration. Plant transpiration is the movement of water from the soil water table, through a plant, followed by subsequent loss of water as vapor through stomata in the leaves. During a growing season, the leaf of a plant can transpire many times more water than its own weight. A hectare of corn gives off 4613 to 6111 liters of water each day, while a large oak tree can transpire 151,000 liters per year.

Global water budgets show that of the 110,000 km^3 of precipitation that falls on the terrestrial surface of the planet each year, about 65,200 km^3 (60%) comes from evapotranspiration. Of this, transpiration by plants accounts for 61% of evapotranspiration and is responsible for the return of 39% of precipitation back to the atmosphere.[37] This amount is considerably higher in areas of the world that are densely vegetated, such as in tropical rain forests, where 70% of atmospheric moisture and subsequent rainfall is controlled by plant transpiration.[37]

Because of the role that plants play in transpiration, land-use practices that destroy vegetation (e.g., deforestation) can reduce evapotranspiration and groundwater recharge into the aquifers people use for drinking and irrigation,[38] with dramatic consequences for surface runoff for watersheds.[39] For example, the southeast Tibetan Plateau (SETP) is a geographic region that encompasses the southeast Tibet Autonomous Region, northern Yunnan, western Sichuan, southwest Gansu, and southeast Qinghai. SETP is a huge mountainous area, with elevations ranging from 3000 m to more than 7000 m above sea level, that was once covered by one of China's most expansive coniferous forests. However, between the 1950s and 1990s, most of the forest was cleared for agriculture and urbanization.[40] Deforestation reduced rates of evapotranspiration and groundwater recharge and led to surface runoff in 1998 that was responsible for some of the most severe flooding in Chinese history in the Yangtze valley, affecting 223 million people and causing more than US$36 billion in economic losses.[41] Damaging alteration of the hydrologic cycle in places like the SETP led the Chinese government to embark on a US$83 billion reforestation program of 7 million hectares.

RAW MATERIALS Ecosystems provide a variety of raw materials for construction and biofuel. Wood is one of the most significant products obtained from natural ecosystems, with a 2016 export value estimated at about $385 billion per year by the FAO (oup-arc.com/e/cardinale-w6.1). The total value of timber and other wood products (wood pellets, pulp, and paper) is far greater because most wood is used locally and is not exported. In tropical countries such as Indonesia, Brazil, and Malaysia, timber products earn billions of dollars per year.[42] According to the United Nations Food and Agriculture Organization (FAO), global production of nearly all major wood products has been growing rapidly since the global economic downturn of 2008–2009.[43,44] Wood panel and sawn wood production has grown in all regions of the world, with global production reaching record highs of 399 million m^3 for panels, and 452 million m^3 for sawn wood. Production of wood pellets, which are used as a biofuel in many countries, has increased in recent years, mainly owing to demand generated by bioenergy targets set by the European Commission. In 2015, global production of wood pellets reached 28 million metric tons, of which more than half was traded internationally. China has grown in importance as both a producer and a consumer of forest products, recently overtaking Canada in sawn wood production. Production of wood fiber, pulp, and paper products has continued to grow globally. South America, in particular, has expanded wood pulp production with an increasing number of new pulp mills being built in Brazil, Chile, and Uruguay. These three countries currently account for 14% of global wood pulp production and 74% of exports.[44]

The range of raw materials that are harvested from natural ecosystems and then sold in the marketplace is much broader than just timber and wood products. Ecosystems also provide animal products such as hides, skins, feathers, and furs (from alligator, fox, bear, rabbit, etc.) that are widely used to make clothing and fashion accessories. Natural products are used to make dyes and perfumes that are used in cosmetics, as well as gums and resins that are common in food and personal health care products.[34] Several biogenic minerals (silicates, carbonates, collagen, and chitin) are mined for applications in medicine and nanotechnology,[45] as well as for bioremediation of environmental pollutants like uranium and other dangerous radionuclides.[46] It is important to note that many of the raw materials mined from nature do not appear in the gross domestic products of countries, because they are neither bought nor sold beyond the village or local region and do not appear in the national or international marketplace. However, if rural people are unable to obtain these products (as might occur following environmental degradation, overexploitation of natural resources, or even creation of a protected reserve), their standard of living will decline, possibly to the point that they are forced to relocate.

MEDICINAL RESOURCES Natural ecosystems are the source of many medicines, with about 80% of the world's population relying principally on traditional medicines derived from plants and animals as their primary source of treatment.[47] The chemical properties of many native species have proven to be of great value in developed nations that have modernized medical care. For example, the rose periwinkle (*Catharanthus roseus*) found in Madagascar is responsible for two drugs that are effective in treating Hodgkin's disease, leukemia, and other blood cancers. Treatment using these drugs has increased the survival rate for childhood leukemia from 10% to 90%. The antimalarial drug quinine, which was first isolated from the bark of the flowering plant *Cinchona succirubra*, was used for centuries in the treatment of malaria.[48]

Cancer is one of the most devastating diseases worldwide, responsible for 8.2 million deaths in 2012. Yet, there has been a 23% drop in cancer death rates since 1991, owing to the development of improved drugs. Of the small molecules that were developed for chemotherapy between 1940 and 2014, 49% were natural products.[49] Three out of every four anticancer drugs introduced over the past six decades have been derived from, or inspired by, natural products.[49] Paclitaxel (also called Taxol), which is the most widely used breast cancer drug, was first isolated from the Pacific yew plant (*Taxus brevifolia*). The drug was discovered in 1962 when the U.S. Department of Agriculture collected and analyzed bark of the plant as part of their chemical screening program at the National Cancer Institute. Because a large volume of bark from mature trees is required to produce just 1 gram of paclitaxel, natural supplies cannot meet current demand (about 50,000 treatments per year each require 2 grams of the drug). Therefore, the drug is now produced synthetically.[48] (**Video 6.3**).

▶ **VIDEO 6.3** Watch this informative talk about the importance of biodiversity for human medicine.
oup-arc.com/e/cardinale-v6.3

Many other medicines that are now produced synthetically were first discovered in wild species that were used for traditional medicine.[34] One of the best-known examples is aspirin. Extracts of the bark of willow tree (*Salix* sp.) were used by the ancient Greeks and by tribes of Native Americans to treat pain, which led to the discovery of acetylsalicylic acid—the painkilling ingredient in modern aspirin, one of our most important and widely used medicines. Similarly, the use of coca (*Erythroxylum coca*) by natives of the Andean highlands eventually led to the development of synthetic derivatives such as Novocain (procaine) and lidocaine, commonly used as local anesthetics in dentistry and surgery. Many other important medicines were first identified in animals, with venomous animals such as rattlesnakes, spiders, bees, lizards, and cone snails having been especially rich sources of chemicals with medical applications. Venom from the golden orb spider (*Nephila* sp.) is used in drugs to treat epilepsy; exenatide, a drug used to treat type 2 diabetes, comes from an enzyme in the saliva of the Gila monster (*Heloderma suspectum*); and venom from the deadly cone snail (*Conus* sp.) led to identification of Prialt (ziconotide), a painkiller 1000 times more powerful than morphine.

All of the 20 most frequently used pharmaceuticals in the United States are based on chemicals first identified in natural products. These drugs not only save millions of lives, but are economically important and have a combined sales revenue of $6 billion per year. About 25% of the prescriptions filled in the United

States contain active ingredients derived from plants, and many of the most important antibiotics, including penicillin and tetracycline, are derived from fungi and other microorganisms.[34] More recently, the fungus-derived drug cyclosporine has proved to be a crucial element in the success of heart and kidney transplants. Exploration of the natural world is constantly leading to discovery of new chemicals that are the foundation for the next generation of medicines.

BIOTECHNOLOGY Nature and its biodiversity provide the genetic library on which many biotechnologies are discovered and built. Perhaps the most famous example is the polymerase chain reaction (PCR). PCR is a technique used in molecular biology to amplify a segment of DNA, generating thousands to millions of copies of a particular DNA sequence. Most PCR methods rely on thermal cycling, which involves exposing the reactants to cycles of repeated heating and cooling, which permits different temperature-dependent reactions to proceed in a predictable sequence. In order for these temperature-dependent reactions to work, a heat-stable DNA polymerase is required to stabilize the reactions. That heat-stable polymerase was originally discovered in a species of thermophilic (heat-loving) bacteria—*Thermus aquaticus*—that was found to be living in the hot springs of Yellowstone National Park. *T. aquaticus* thrives at temperatures of 70°C (158°F) because it has evolved a heat-stable enzyme, called Taq polymerase, that is essential to its metabolism. Taq polymerase was the biological molecule that made PCR,[50] and all subsequent genetic technologies, possible.

Development of PCR was a eureka moment in biology that revolutionized our ability to study anything related to genetics. PCR is now a common and indispensable tool used for DNA sequencing (e.g., sequencing of the human genome); diagnosis and monitoring of hereditary diseases like certain forms of cancer; analysis of genetic fingerprints for DNA profiling (e.g., in forensic science and parentage testing); and detection of pathogens in nucleic acid tests for the diagnosis of infectious diseases. PCR was deemed such an important advance in biotechnology that Kary Mullis, who invented the process, was awarded the 1993 Nobel Prize in Chemistry.

Regulating services

Regulating services are the benefits that ecosystems provide to people by regulating the stability of ecosystem processes through time (see Figure 6.3). These services tend to reduce harmful variation in ways that provide a form of insurance to human well-being. Regulating services affect water, soil, plants, climate, and more.

WATER PURIFICATION Aquatic ecosystems like swamps, lakes, rivers, floodplains, tidal marshes, mangroves, estuaries, the coastal shelf, and the open ocean are important for filtering out waterborne diseases and for breaking down and immobilizing toxic pollutants, such as heavy metals and pesticides that have been released into the environment by human activities. Aquatic ecosystems also play an important role in processing, storing, and recycling the large amount of nutrients that enter the ecosystem as sewage or agricultural runoff. Some have estimated that the waste treatment services performed by natural ecosystems could be worth as much as $2 trillion per year, worldwide.[16] These estimates have been calculated based on the cost to replace the water purification functions of damaged or degraded natural systems, such as with construction of water treatment facilities that are often orders of magnitude more expensive than the cost of preserving natural ecosystems that perform water purification for free.

An excellent example of the value of nature for water purification is New York City's water supply system—one of the largest, largely unfiltered, municipal water supply systems in the world. This complex system relies on water being purified by natural infiltration in the forested ecosystems of the Catskill Mountains. After filtration, the water flows by gravity downhill through a sequence of reservoirs, aqueducts, and tunnels to meet the daily needs of New York City's more than 9 million residents and visitors (Box 6.1). New York did spend hundreds of millions of dollars to create and maintain this system, but the city would have had to spend billions of dollars to build an artificial system to perform the same functions.

CLIMATE REGULATION Plant communities are important for moderating global, regional, and local climate conditions.[51] At a global scale, photosynthetic organisms on land and in the oceans are the primary control over the carbon cycle that maintains Earth's atmosphere and climate. Land plants store about 560 pentagrams (Pg) of carbon in their tissues (wood, leaves, roots), which represents about 28% of all carbon found on land, in the atmosphere, and at the surface of the ocean (1 Pg C = 10^{15} g carbon = 1 billion metric tons carbon). Land plants also scrub huge amounts of carbon dioxide from the atmosphere, sequestering as much as 120 pentagrams of carbon per year through photosynthesis. The oceans store an additional 38,000 pentagrams of carbon (85% of all carbon that is not bound up in Earth's crust), either as dissolved carbon or as organic carbon that is bound in the tissues of plankton. Oceans absorb about 30% of the carbon dioxide generated by people each year, and photosynthesis by oceanic plankton generates roughly half of all oxygen in Earth's atmosphere.[35] Because of these large influences on the carbon cycle, any reduction in plant life on land, or in the ocean, results in slower uptake of CO_2 and less storage of carbon, which contributes to rising levels of CO_2 in the atmosphere, which promotes global warming.[52,53]

At the regional level, plants take up soil water that has fallen as rain and then transpire it back into the atmosphere, from which it can fall again as rain. The loss of vegetation from large forested regions such as the Amazon Basin and western Africa has resulted in reductions of annual rainfall and altered weather patterns over large areas. For example, in 2014–2017 a lack of rain led to a severe drought that affected much of southeastern Brazil, including cities like São Paulo—one of the largest metropolitan areas in the Western Hemisphere, with 20 million residents. With major reservoirs operating at their lowest capacity in nearly 100 years, millions of residents in São Paulo were subjected to water rationing and daily water shutoffs that extended for 12 hours at a time. The lack of water led to a drinking water crisis that most acutely affected people living in poorer districts. These districts, which were located at higher altitudes on the outskirts of the city, were affected by reduced water pressures imposed by São Paulo's water utility to conserve water, and residents there lacked the resources needed to compensate by purchasing bottled water or water storage tanks. The result was social unrest that sometimes turned violent. The water crisis in São Paulo was partially caused by an El Niño–Southern Oscillation (ENSO) event in the Pacific Ocean. But research suggests the water crisis was made far worse by rampant deforestation throughout the Amazon Basin that reduced evapotranspiration by plants, leading to significant reductions in local precipitation.[54–56] Most climate models suggest droughts like these will become more frequent in southeastern Brazil.

Plants also provide climate regulatory services at more local levels. For example, trees provide shade to humans and livestock, and they reduce local temperatures as they evaporate water from their leaf surfaces during photosynthesis. This cooling effect reduces the need for fans and air conditioners and increases people's comfort and work efficiency. For example, the simulation results of Simpson and McPherson[57] indicated that two trees shading the west-facing exposure of a house and one tree shading the east-facing exposure reduced annual energy use for cooling by 10% to 50% and peak electrical use by

BOX 6.1 Success Story

How New York City Keeps Its Drinking Water Clean

Beginning in the 1830s, the city of New York began to create one of the most extensive municipal water systems in the world. Generations of city leaders went beyond the city boundaries into rural areas of the Catskill Mountains to build a series of reservoirs, tunnels, and aqueducts that would supply the city with pure, pristine water from undeveloped watersheds.

(A)

(B)

To protect its drinking water, New York City decided it would be a smart and profitable investment to conserve the Catskill Mountains that purify water naturally (A), as opposed to building expensive water treatment facilities (B).

But in the 1980s, industrialized agriculture began to undermine the economic vitality of small family farms in the Catskill Mountains, and large-scale farm operations began to spread across the watersheds. Nutrient use from fertilizers increased, erosion accelerated, and pathogen contamination began to grow. Farmers also began selling off the forested portions of their land to take advantage of exurban development.

By the end of the 1980s, the quality of water reaching New York City was rapidly deteriorating, and public health officials were warning both elected officials and the general public that the city was going to have to substantially increase the treatment of its water for it to remain drinkable. The costs for advanced treatment water facilities were estimated to be $4 billion for initial construction and $200 million annually to operate. Not only were the construction costs prohibitive at the time, the operating costs would have doubled the price of water in New York City, potentially with major adverse impacts on low-income families.

City officials pondered their alternatives. It seemed that allowing Catskill drinking water purity to deteriorate and then spending massive sums to clean it up was not an ideal option. But perhaps buying land in the Catskill Mountains and conserving its ability to produce good water would be a smart and profitable investment for New York City.

After 18 months of negotiations and mutual work between the city and the Catskill farming community, an innovative and far-reaching agreement was crafted. The city began to add to their portfolio of publicly held lands in the watershed, focused on buying and protecting lands that were threatened by development and on restoring those located along crucial stream corridors (see Figure).

To control nonpoint source pollution from privately held farmlands and other rural landscapes, the city organized an unprecedented program of regulatory enforcement. The Catskill farmers created a program they called Whole Farm Planning. Its title was designed to capture the fact that it incorporated environmental planning into the business strategy of the farm—a pollution control plan was developed for each farm, by the farmer and local farm and agricultural experts.

To ensure pollution control efforts would reach critical mass, the program set a goal of obtaining a participation rate of 85% of Catskill farmers within five years. Thus, while the program was voluntary for any individual farmer, the Catskill farm community as a whole committed to reaching a goal that would ensure that the city met its pollution reduction objectives. After five years, 93% of all Catskill farmers were full program participants.

The results of these efforts were staggering:

- There was a 75% to 80% reduction in farm pollution loading.
- The quality of the city's drinking water improved, and New York City was able to avoid spending billions of dollars on advanced treatment of drinking water.
- The program eventually paid for itself many times over through its cost savings, which played a critical role in helping to stabilize water and sewer tariffs, providing major benefits to low-income households.
- The program was wildly popular with the public and helped build strong urban support for future watershed protection efforts by New York City.

On a broader scale, the Catskill program spurred watershed protection and environmentally friendly farm programs throughout the United States and catalyzed interest in nontraditional facility construction approaches among the U.S. water industry. Ecosystem service programs like the one used in New York are a way of capturing the environmental profits from the services that rural ecosystems provide to urban areas and then funneling those profits back into the rural landscapes and the rural communities that provide them. This creates a cycle of mutually supportive economic and ecological investments between urban and rural areas, leading to a more sustainable future for both.

up to 23% for houses in California. Huang and colleagues[58] conducted a simulation study that suggested a 25% increase in tree cover would reduce annual cooling energy use by 40%, 25%, and 25% for an average house in Sacramento, Phoenix, and Lake Charles, respectively. A recent analysis of houses in Canada suggested shading by trees can reduce the annual cooling costs of a typical single-family home by as much as 30%.[59]

SOIL FERTILITY The production of organic matter and recycling of vital nutrients is essential for soil fertility and plant growth. The productivity of organic matter in soil, and the mineralization of nutrients like nitrogen and phosphorus, is controlled by fungi, bacteria, and small animals (microarthropods, worms, etc.) that break down dead plant and animal matter, which they use as energy. In the process, the fungi, bacteria, and small animals create fertile soils that allow plants to grow. In certain instances, such as when farmers add fertilizers to fields, the biological processes that generate soil fertility can be replaced. But this is rarely possible at the scale of whole forests, grasslands, and other major ecosystems where biological processes are the dominant controls over soil fertility.

Certain types of plants have evolved specialized relationships with soil microorganisms that help fertilize soil. Legumes, for example, contain symbiotic bacteria, called rhizobia, that live in root nodules (swellings on the roots). These bacteria have the special ability to fix nitrogen from atmospheric, molecular nitrogen (N_2) into ammonia (NH_3) that can be used for plant growth. Other plants have evolved mutualistic (mutually beneficial) relationships with mycorrhizal fungi. Mycorrhizal fungi grow hyphae (branching filaments that make up the mycelium of a fungus) that extend from the soil into plant roots, and these hyphae increase the ability of roots to absorb water and minerals. In return, the plants provide the mutualists with photosynthetic products that help them grow.

EROSION Terrestrial plants have long been known to stabilize soils and prevent erosion, mostly through development of extensive rooting systems that help hold onto soil and keep it clumped together.[60] Aquatic macrophytes, like the brown algae that form kelp forests along marine coastal zones, are known to dampen waves, attenuating energy that erodes beaches and shoreline habitat.[61] More recently, we have learned that biofilms generated by microbes, as well as the biogenic structures created by animals, can reduce erosion.[62] For example, the silk filtration nets that are spun across rocks by larval caddisflies (Hydropsychidae) to collect food in streams can bind rocks together and prevent them from eroding downstream during floods[63] (see Video 3.3 in Chapter 3). When human activities disturb plants' rooting systems, microbial biofilms, or the biogenic structures of ecosystem engineers, rates of soil erosion, and even the chance of landslides, can increase considerably. For example, the Fatal Landslide Event Inventory of China contains records of 1911 nonseismically triggered landslides resulting in 28,139 deaths in China between 1950 and 2016. While excessive exploitation of natural resources and loss of hillside vegetation had increased the frequency of large fatal landslide events (fatalities > 30) since the 1950s, a decreasing trend since 2000 has been attributed to landslide mitigation and prevention projects that focus on revegetation of hillsides and improving soil stability with extensive rooting systems.[64]

BIOLOGICAL CONTROL OF PESTS AND DISEASE A wide variety of predators, parasites, and pathogens provide biological control of undesirable pests or disease organisms. For example, in agricultural systems, wild insects like lady beetles or praying mantises,[65] and many bird and bat species,[66,67] feed on herbivorous pests (aphids, caterpillars, beetle larvae, etc.) that attack and ruin

(A)

(B)

(C)

Figure 6.7 Biological control of insect pests by predators, parasitoids, and pathogens is an important regulating service of ecosystems worth billions of dollars in crop protection. (A) A predacious lady beetle eats aphid herbivores on a crop. (B) A parasitoid wasp oviposits an egg into an aphid. The egg will hatch into a larva, which will develop inside the aphid and consume the aphid from the inside out. (C) A fungal pathogen kills aphids and absorbs their tissues for nutrition.

crops (**Figure 6.7**). Such reduction of insect pests not only increases crop yields, but reduces the need to spray pesticides, which can harm public health. Losey and Vaughn[65] estimated that native predatory insects provide biological control services worth US$400 billion per year globally in terms of avoided crop damage. Boyles and colleagues[68] estimated that loss of insect-eating bats in North America due to white-nose syndrome could lead to agricultural losses of more than $3.7 billion per year. Thus, the ecosystem services provide by biocontrol agents can be financially significant.

Biocontrol agents can also control the prevalence and spread of disease-causing organisms, particularly zoonotic diseases (those transferred from animals to humans), which represent more than 60% of the 1415 pathogens that are known to infect humans. Zoonotic diseases have been responsible for some of the largest pandemics in history (e.g., bubonic plague, HIV) and are responsible for many emerging infectious diseases (e.g., Zika, Ebola, SARS, bird flu, West Nile fever). Many zoonotic diseases are vector borne, meaning they are transferred to humans through other animals (ticks, fleas, mosquitoes); these cause more than 1 million deaths annually.

A good example of a vector-borne disease that is controlled biologically is malaria. Malaria is caused by a protozoan parasite (genus *Plasmodium*) that is transmitted to humans by 100 different species of *Anopheles* mosquitoes.[69] In 2016, an estimated 216 million cases of malaria occurred worldwide and caused 445,000 deaths, mostly in children (oup-arc.com/e/cardinale-w6.2). Chemical strategies to control the *Plasmodium* parasite or the *Anopheles* vector were historically effective; however, the emergence of drug-resistant parasites and insecticide-resistant mosquito strains, along with numerous health and environmental side effects of chemical agents, led the World Health Assembly to call for adoption and development of alternative approaches to controlling vector-borne diseases. Integrated vector management efforts are now oriented toward controlling *Anopheles* mosquitoes at either the larval stages or the adult stages, or both, through means of biological control. Dozens of species of bacteria and fungi, as well as numerous types of viruses, are known to attack and kill the *Plasmodium* parasite or the larval and adult stages of the *Anopheles* vector. Microsporidian parasites and nematode worms are known to parasitize and castrate the mosquito. And numerous kinds of fish are known to prey on the aquatic larvae of *Anopheles*. The use of naturally occurring biological agents has few adverse environmental impacts, and the side effects on humans, domestic animals, and wildlife are minimal.

For these reasons, the biological control of the malaria vector is now considered a fundamental part of the global malaria eradication program recently launched by the World Health Organization.[69]

NATURAL DISASTER MITIGATION Natural and managed ecosystems play an important role in protecting societies from the worst of natural disasters like hurricanes and floods. Flooding is currently the most common natural disaster in the world, killing thousands of people per year. The incidence of major floods has increased manyfold during the last few decades, which can be attributed to the concentration of people living in coastal areas and to the destruction of wetlands and upland ecosystems.[70,71]

Unprecedented catastrophic floods and landslides in Bangladesh, India, the Philippines, Thailand, and Central America have been associated with recent extensive logging in upstream watersheds. Flood damage to India's agricultural areas has led to massive government and private tree-planting programs in the Himalayas. Removal and filling in of coastal mangrove swamps is part of the explanation for the severe devastation associated with the 2004 Indian Ocean tsunami, in which about 300,000 people lost their lives. In the industrial nations of the world, wetlands protection has become a priority in order to prevent flooding of developed areas. In certain locations, wetlands act as sponges for floodwaters and are estimated to have a value of $6000 per hectare per year in reducing flood damage and in other ecosystem services, which is three times the value of farmland created on the same site.[2]

The conversion of wetland and floodplain habitat to farmland along the Mississippi, Missouri, and Red Rivers, and along the Rhine River in Europe, is considered a major factor in the massive, damaging floods in past years. Perhaps the most dramatic example occurred during the devastating flooding of New Orleans, Louisiana, in 2005 after Hurricane Katrina struck the Mississippi Delta, which had undergone heavy conversion of wetlands for urban, industrial, and agricultural development. Loss of these wetlands greatly exacerbated the floods that caused 1836 fatalities and $125 billion in damage in the wake of Hurricane Katrina, which still ranks as one of the worst natural disasters in U.S. history (**Box 6.2**). In contrast to what happened in New Orleans after Hurricane Katrina, coastal wetlands along the East Coast of the United States helped thwart $625 million worth of property damage during Hurricane Sandy in 2012.[72] In more than half the zip codes along the East Coast, wetlands helped reduce the cost of damages by 22%. Even in heavily urbanized New York, where wetlands covered just 2% of the land, an estimated $138 million in property was saved by the existence of these wetlands. Arkema and colleagues estimated that wetlands and other natural habitats (coral reefs, dunes, kelp forests, etc.) protect about 67% of the U.S. coastline, protecting millions of people and more than $4 billion in property within 1 km of the coast that is vulnerable to storms.[73]

POLLINATION Pollination is perhaps the best-known ecosystem service performed by animals. Usually performed by insects, birds, or bats, animal pollination influences the reproductive success of 87% of the world's flowering plants.[74] Worldwide, 1500 different crops require insect pollination,[75] with 3% to 8% of global crop production (in tonnage) depending on insect pollination.[76] The fruit, vegetable, and seed production from 87 of the leading global food crops relies on animal pollination.[75]

In temperate regions of the world, the majority of animal pollination is provided by honeybees (*Apis mellifera*), bumblebees (*Bombus* spp.), solitary bees, wasps, butterflies, moths, and hover flies. In tropical regions, birds and bats are also important.[75] Managed bumblebees are most commonly used in enclosed production systems (e.g., greenhouses), but other managed species, especially

BOX 6.2 Case in Point

Wetland Loss and Hurricane Katrina

As the whirling maelstrom approached the coast, more than a million people evacuated to higher ground. Some 200,000 remained, however—the carless, the homeless, the aged and infirm.... The storm hit Breton Sound with the fury of a nuclear warhead, pushing a deadly storm surge into Lake Pontchartrain.... As it reached 25 feet (eight meters) over parts of the city, people climbed onto roofs to escape it. Thousands drowned in the murky brew that was soon contaminated by sewage and industrial waste.... It took two months to pump the city dry, and by then the Big Easy was buried under a blanket of putrid sediment, a million people were homeless.... It was the worst natural disaster in the history of the United States.[158]

This *National Geographic* excerpt might be any of the narratives of the devastation wreaked on New Orleans by Hurricane Katrina but for one fact: it was published almost a year before the hurricane happened. Other major publications, including *Scientific American, Popular Mechanics*, and the *Houston Chronicle*, had written similar articles warning of the pending catastrophe, echoing an earlier 1998 Louisiana task force report by government officials, engineers, and scientists.

How did New Orleans reach the point at which impending disaster could be so clearly predicted? And why was action to prevent it not taken? The answer to the first question lies partly in the area's geology and ecology and partly in human behavior. For millennia, sediment carried by the muddy Mississippi River was deposited in the delta's marshes and swamps, nourishing the plants that grow there. Until recently, these wetlands absorbed the fury of storms before they reached New Orleans. The construction of levees, or raised walls, along the river margin to prevent flooding, the filling of coastal wetlands for development, and the creation of canals for ship traffic and oil drilling, however, changed the equation. Levees prevented sediment from reaching the coastal wetlands, and the wetland landscape subsided below sea level. Over 12,800 km (8000 miles) of canals were cut into the marshes, increasing soil erosion and introducing salt water into the freshwater wetlands. A total of 4877 km^2 of wetlands were lost over the past 80 years, an area three-quarters the size of Delaware. By the 1990s, New Orleans had become a shallow depression almost 5 m (14 feet) below sea level, ringed by levees to protect it from the Mississippi River to its south and Lake Pontchartrain to the north and stripped of its barrier of protective wetlands (see Figure).

Divisions among stakeholders, many of whom benefited from the status quo, led to slow and underfunded efforts to restore wetlands prior to Katrina. Even after the hurricane, disagreements among those with competing interests—shipping, oil, fishing, tourism, conservation, cities, and others—have slowed restoration of wetlands.

However, wetland restoration appears to be gaining momentum. In 2012, more than 6 years after Hurricane Katrina, the state's Coastal Protection and Restoration Authority released a master plan for the coast that, if carried out completely, would cost $50 billion over 50 years. The plan relies on a mix of wetland restoration, levee construction, bank stabilization, and other measures, but the biggest piece of the budget, $17.9 billion, is dedicated to wetland restoration, in recognition of the cost-effective ecosystem services they provide.

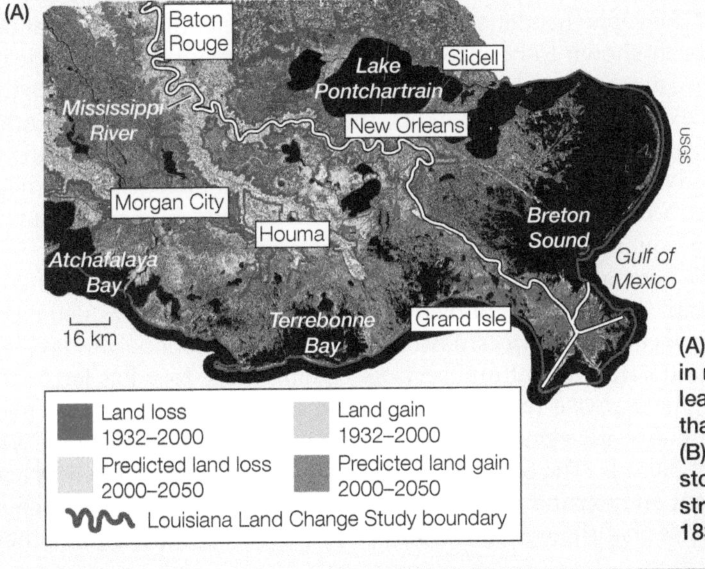

(B)

© FEMA/Alamy Stock Photo

(A) Loss of coastal wetlands from 1932 to 2000 (shown in red) was severe in the southern Louisiana area, leaving New Orleans with no natural infrastructure that could provide flood protection from hurricanes. (B) When hurricane Katrina hit New Orleans in 2005, storm surge caused 53 breaches to flood-protection structures, submerging 80% of the city and causing 1836 fatalities and $125 billion in damage.

honeybees, are predominantly used for field and orchard crops (e.g., apples and almonds).[77] Globally, evidence is emerging that wild bees and other insects are more important to crop pollination than previously believed.[78]

Pollination is an often-cited example of nature's multifaceted services to people; it has both monetary and nonmonetary values. A recent summary of data suggests that crop pollination by insects is worth US$17 billion of crop production in the United States, US$40 billion in China, US$2.5 billion in New Zealand, US$1.8 billion in Egypt, and US$167 billion worldwide.[77] Recent work further suggests that natural pollinators are critically important for avoiding malnutrition, particularly for overcoming micronutrient deficiencies in vitamin A, zinc, iron, folate, and calcium.[79] Each year, vitamin A deficiency causes an estimated 800,000 deaths in women and children, including 20%–24% of child mortality from measles, diarrhea, and malaria and 20% of all causes of maternal mortality. As much as 50% of the production of plant-derived sources of vitamin A requires pollination throughout much of Southeast Asia, whereas other essential micronutrients such as iron and folate have lower dependencies, scattered throughout Africa, Asia, and Central America. Micronutrient deficiencies are three times as likely to occur in areas of highest pollination dependence for vitamin A, suggesting that disruptions in pollination could have serious implications for the accessibility of micronutrients for public health.[80] Given its importance, several major international initiatives, including the IPBES, have made the assessment of pollinating services one of their highest priorities.

Cultural services

Cultural ecosystem services were defined by the Millennium Ecosystem Assessment[2] as "the nonmaterial benefits people obtain from ecosystems through spiritual enrichment, cognitive development, reflection, recreation, and aesthetic experiences." Since publication of the MEA in 2005, we have learned three things about cultural services. First, we've learned that cultural services often yield disproportionate power as a motivation for conservation. For example, when researchers have assessed the demand for ecosystem services, or the motivation behind conservation projects, they have generally found that cultural services—especially the conservation of species and habitats of cultural significance—are among the most common motivations for protecting ecosystem services.[81,82]

Second, while many types of cultural services were once thought to be "soft" or "fuzzy" concepts that defied quantification and scientific study, that perception has been shown to be incorrect.[83] It has become increasingly clear that, not only are most cultural services just as easily quantified as services that are valued in existing economic markets, the methods needed to do so are readily available in fields like public health, human psychology, and anthropology. As the study of ecosystem services has matured and biologists have begun to interact with researchers from the social sciences and humanities, improved methods for quantification have rapidly been applied.

Last, we have learned that the cultural services of ecosystems are far more variable than initially thought by those who wrote the MEA, which listed just four types of cultural services: aesthetic, spiritual, educational, and recreational. The proliferation of cultural services beyond these four has led to numerous attempts to revise the list of described cultural services.[84–86] It has even prompted entire revisions of human value systems (see "Relational values," Chapter 5). Because the concept of relational values was proposed as a way to complement and expand on many of the cultural services of ecosystems,[17,87] it is no surprise that there is considerable overlap between the two. Rather than revisit the numerous relational values that people use to justify the conservation of nature (see Chapter 5), here we focus on a select few cultural services that have been given the most attention in the literature on ecosystem services.

RECREATION Ecosystems provide many recreational services for humans, including activities like hiking, biking, boating, fishing, hunting, photography, and bird watching.[88] The monetary value of these activities is considerable. For example, it has been estimated that more than 250 million people spend 7 billion hours per year enjoying nature at national parks, state parks, wildlife refuges, and other protected public lands in the United States.[89] A 2011 study published for the U.S. National Fish and Wildlife Foundation[90] quantified the economic value of these activities and made these conservative estimates:

- The sum total of outdoor recreational activities generated a total economic activity of about $1.06 trillion per year, producing more than 9 million jobs in the United States.
- The total contribution from hunting, fishing, wildlife viewing, and the "human-powered" recreations (hiking, camping, skiing, and bicycling) in the United States was over $730 billion a year, generating $88 billion in federal and state tax revenues.[90]
- The economic total of outdoor recreational sales (gear and trips combined) was $325 billion per year, which exceeded the annual returns from pharmaceutical and medicine manufacturing ($162 billion), legal services ($253 billion), and power generation and supply ($283 billion).

Across international sites known for their conservation value or exceptional beauty, the nonconsumptive recreational values of ecosystems are often very large (Table 6.1) and can dwarf the economic value of other major industrial enterprises like farming, logging, or mining.[91] For example, a 2014 study of the Moreton Bay Marine Park near Queensland, Australia, showed that recreational fishing generates almost US$23 million annually, which led the authors of the study to write, "The economic value of the recreational fishery in Moreton Bay may far outweigh the value of the commercial fishery."[92]

TABLE 6.1 Select examples that illustrate the value of ecosystem services associated with outdoor recreational activities

Activity	Location	Year	Users	Annual value[a]	Reference
Wildlife viewing	National Wildlife Refuge System, U.S.A.	2011	46.5 million visitors	$2.6 billion	E. Carver and J. Caudill, *U.S. Fish & Wildlife Service Report* (2013)[155]
Recreational hunting	Public land in Michigan, U.S.A.	2002	7.7–12.1 million hunting trips	$109.6 million	S. Knoche and F. Lupi, *J Wildl Manage* (2012)[156]
Recreational fishing	Moreton Bay Marine Park, Australia	2008	337,000 fishing trips	$22.9 million	S. Pascoe et al., *Tourism Manage* (2014)[92]
Bird watching	Kuscenneti National Park, Turkey	2001	About 10,000 visitors who come to the wetlands to view migratory birds	$143 million	S. Gurluk and E. Rehber, *J Environ Man* (2008)[93]
Boating	Lake Ontario & St. Lawrence River, U.S.A. and Canada	2002	32,702 boat owners; 1.3 million recreation days	$123.3 million	N. A. Connelly et al., *J Am Water Res Assoc* (2007)[157]
Mountain biking	Moab, Utah, U.S.A.	1994	900 mountain bikers who used biking trails in Moab	$2.2 million	K. Chakraborty and J. E. Keith, *J Environ Plan Man* (2000)[94]

[a]All values are in 2018 U.S. dollars.

The Kuscenneti National Park—a wetland area in Turkey that is considered a paradise for migratory birds—is estimated to generate US$143 million annually in revenue from bird watchers alone.[93] Needless to say, if this wetland were damaged and the birds departed, the economic value of the park would plummet, with consequences for the local economy. Mountain bikers who recreate in Moab—a small city in Utah that is known for its network of mountain trails—contribute more than $2 million annually to the local economy of the town.[94] In the United States, more than 46 million people visit the National Wildlife Refuge System for wildlife viewing, generating $2.6 billion in economic activity each year. Importantly, all of these values are underestimates of the ecosystem services associated with these systems, because each one focuses on just one set of activities (e.g., bird watching, hunting, fishing) that is associated with just one particular user group. When future studies are able to tally the sum total of all recreational values associated with many different user interests, the dollar values will surely rise substantially. For now, suffice it to say that the value of outdoor recreational activities is not only quantifiable, but surprisingly large.

ECOTOURISM **Ecotourism** is a special category of recreation that involves people visiting places and spending money wholly or in part to experience and enjoy spectacular biological communities (such as rain forests, African savannas, coral reefs, deserts, the Galápagos Islands, and the Everglades) and to view particular "flagship" species (such as elephants, on safari trips).[95] Tourism is among the world's largest industries (on the scale of the petroleum and motor vehicle industries). The International Ecotourism Society (oup-arc.com/e/cardinale-w6.3) reports that ecotourism currently represents roughly 20% of the $600 billion dollar per year tourist industry. Nature travel and ecotourism is the fastest-growing segment of the international tourism industry.[96]

Ecotourism has traditionally been a key industry in eastern African countries such as Kenya and Tanzania and has become increasingly important in Latin America. Select countries like Belize and Costa Rica have structured their economies to focus on ecotourism as a key form of economic development. Costa Rica, for example, has seen ecotourism skyrocket over the past two decades, with nearly 2.7 million foreign visitors spending US$2.9 billion in international tourist receipts in 2015, up 8% from the previous year.[97] The vibrant ecotourism industry has helped Costa Rica achieve some of the highest standards of living, per capita incomes, and literacy rates in Latin America.

The revenue provided by ecotourism has the potential to provide one of the most immediate justifications for protecting biodiversity, particularly when ecotourism activities are integrated into overall management plans. In integrated conservation-development projects, local communities develop accommodations, expertise in nature guiding, local handicraft outlets, and other sources of income so they can benefit from the tourist economy. However, in order for ecotourism to produce the incentives to work for conservation, it must provide a significant and secure income for the protected area and surrounding community. Unfortunately, in the typical ecotourist package, only 20%–40% of the retail price of the trip remains in the destination country, and only 0.01%–1% is paid in entrance fees to native parks.[95] Another danger of ecotourism is the **love it to death syndrome**, which occurs when overly abundant and assertive tourists themselves become the source of environmental destruction. As an example, when Adélie penguins (*Pygoscelis adeliae*) in the Antarctic were exposed to the activities of tourists, hatching success was reduced by 47%.[98] Similar problems occur in other ecosystems where people trample vegetation, break and damage coral, or impose undue stress on wildlife.

HEALTH AND WELL-BEING One of the most striking advances in the study of ecosystem services has been the discovery that ecosystems affect human health. An increasing body of evidence suggests that people who spend time in natural ecosystems are healthier than people who do not, and exposure to nature can increase one's psychological well-being, cognitive function, and physical health. Reviews by Keniger and colleagues,[99] and more recently by Frumkin and colleagues,[100] both concluded that people who spend time in nature have (1) reduced levels of stress, anxiety, and depression, which leads to increased happiness, better sleep, and more satisfaction with life; (2) better social interactions that lead to more connectedness with other people; (3) reduced mental fatigue and longer attention spans that promote improved cognitive function and productivity; and (4) lower blood pressures and improved immune functions that are associated with reduced obesity, diabetes, and risk of heart failure (**Figure 6.8**).

In many instances, the health benefits of nature are attributable to more active lifestyles; those who sit on their couches at home and play video games are substantially less healthy that those who choose to spend significant portions of their time engaged in outdoor recreational activities. Recognizing the value of outdoor recreation and the opportunities it provides to improve individual fitness, an increasing number of physical and mental health professions are promoting the use of ecotherapy—spending time in nature-based recreation, sometimes with a therapist—as part of a broader approach to health and fitness.

While nature provides recreational opportunities to improve fitness, evidence suggests that the benefits of nature for human health extend beyond a more active lifestyle. One example comes from a 2013 study published in the *American Journal of Preventive Medicine*.[101] In that study, researchers looked at how the loss of 100 million trees to the invasive emerald ash borer (*Agrilus planipennis*) influenced human mortality related to cardiovascular and lower-respiratory-tract illness. The authors showed that, even after accounting for

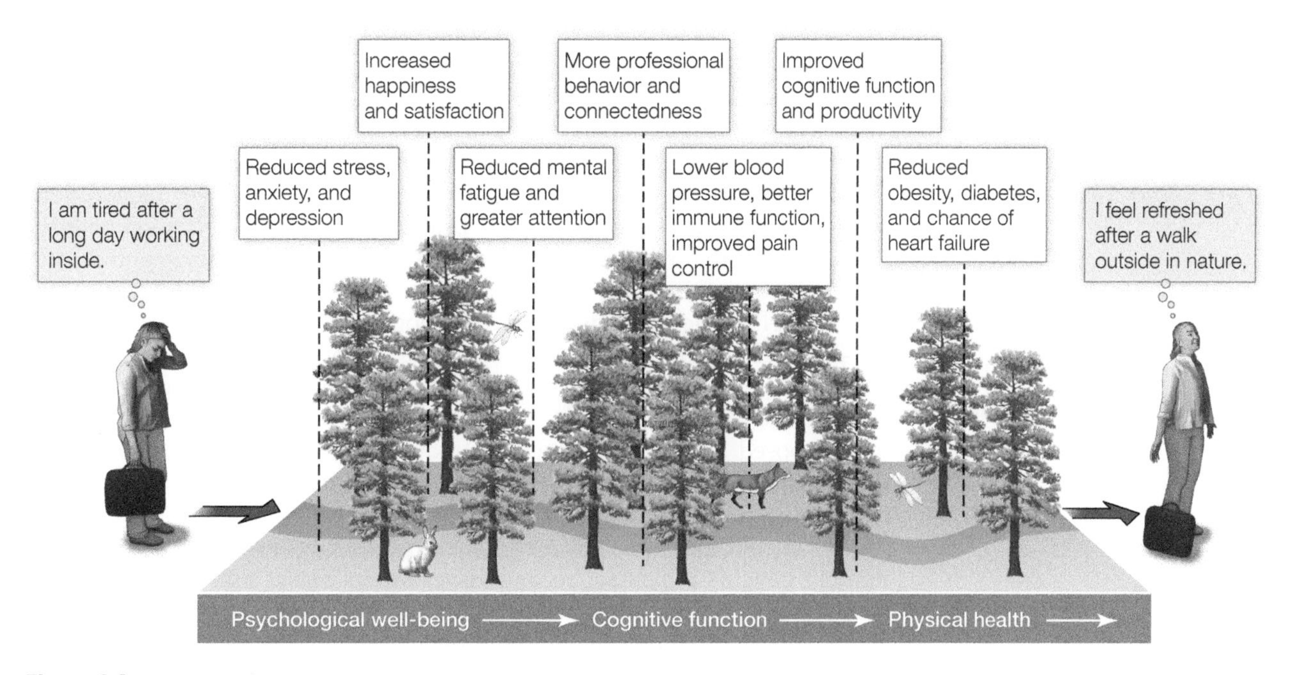

Figure 6.8 Two recent reviews (H. Frumkin et al., 2017, and L. Keniger et al., 2013) summarized literature linking outdoor activities and human health. Both concluded there is strong evidence that spending time in nature improves psychological well-being, cognitive function, and physical health.

differences in people's lifestyles and socioeconomic status, cardiovascular and respiratory illnesses leading to death were more common in counties that had experienced extensive tree die-offs due to infestation by the emerald ash borer. The borer was associated with an additional 6113 human deaths related to illness of the lower respiratory system, and 15,080 cardiovascular-related deaths. The authors speculated that tree die-offs may have influenced human health by degrading air quality, increasing stress, and modifying localized temperatures. Their finding adds to the growing evidence that natural environments provide public health benefits, and it is particularly important as emerald ash borer continues to spread throughout the United States.

Other studies suggest that biodiversity within natural areas can have mental health benefits that extend beyond active lifestyles. Fuller and colleagues[102] surveyed 312 people as they exited green spaces in the United Kingdom. Interestingly, the survey revealed that people "sense" biodiversity and are able to accurately identify nature areas that are more biologically diverse in plant, insect, and bird diversity. And people who recreated in areas that were more biologically diverse conveyed a greater sense of self-reflection and identity. Not surprisingly, there is a budding field of scientific inquiry that examines how exposure to nature influences brain function and happiness.[103]

Alternative frameworks

The four-category framework of ecosystem services established by the Millennium Ecosystem Assessment in 2005 (see Figure 6.3) remains one of the, if not the, most widely used frameworks by agencies and organizations that are involved in management and policy making. We have just reviewed examples of ecosystem services that fall under each category of the MES framework; however, it is important for students to be aware that many other frameworks have been proposed as ways of classifying and organizing nature's benefits to people. Some frameworks are based on the methods used to quantify ecosystem services.[104] Other classification schemes are based on types of human values[105] or the biological processes that generate services.[106] Still others focus on classification schemes that are specific to a field, such as urban planning.[107]

One recently proposed framework that may someday supersede that of the MEA was developed and approved for use by the Intergovernmental Science-Policy Platform on Biodiversity and Ecosystem Services, IPBES (**Figure 6.9**). IPBES was established in 2012 as an independent intergovernmental body open to all member countries of the United Nations. It was formed with the goal of "strengthening the science-policy interface for biodiversity and ecosystem services for the conservation and sustainable use of biodiversity, long-term human well-being and sustainable development" (oup-arc.com/e/cardinale-w6.4). The IPBES framework was constructed through multidisciplinary workshops and open review by a broad range of countries and stakeholders over a period of two years.[108,109] The final framework was intended to build on the one proposed by the MEA, but also to overcome some perceived limitations of the MEA framework. Namely, development of the MEA was heavily dominated by natural scientists, and some felt its framework failed to fully embrace the variety of disciplines (natural, social, health, engineering, etc.) that study nature's benefits to people. Second, development of the MEA framework was mostly led by researchers in Western cultures that emphasized Western science and value systems, often from an economic perspective. Yet, it has increasingly been recognized that goods and services are only part of the high quality of life emphasized by many non-Western countries and Indigenous cultures.[110]

Therefore, the IPBES framework was developed to specifically embrace a greater variety of scientific disciplines, more diverse stakeholder perspectives (the scientific community, governments, international organizations, and civil

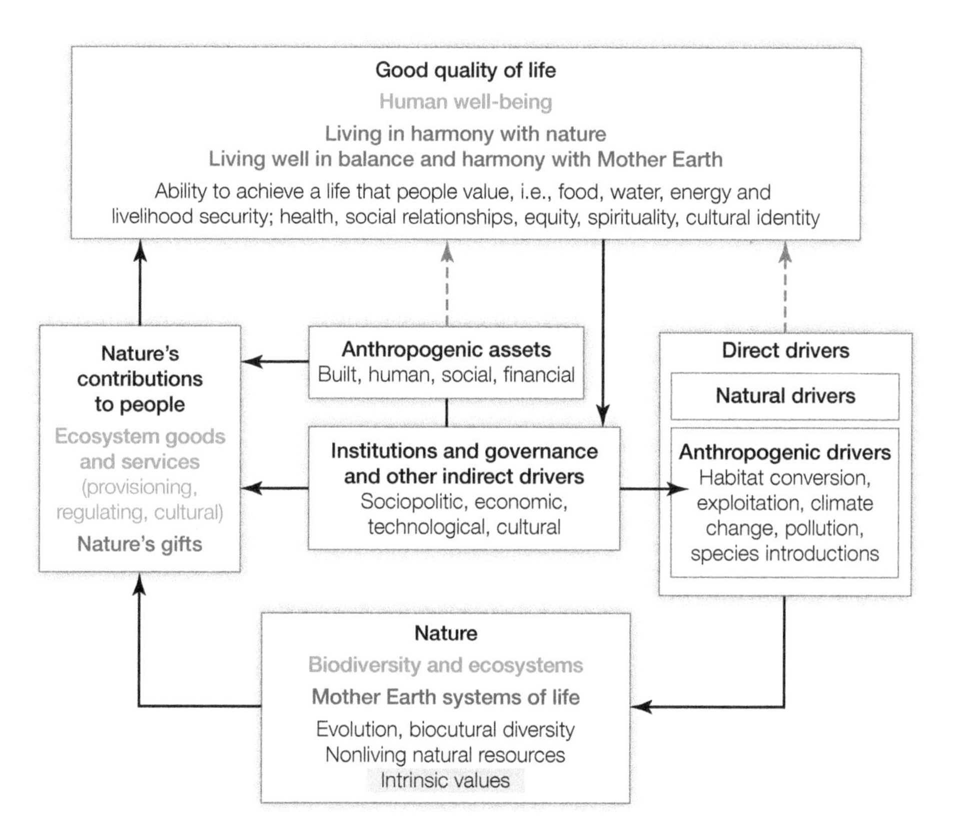

Figure 6.9 The conceptual framework adopted by the Intergovernmental Science-Policy Platform on Biodiversity and Ecosystem Services (IPBES). Boxes denote the elements of nature and society that are at the main focus of IPBES, while arrows show how those elements influence each other. In each of the boxes, the headlines in black are categories that should be intelligible and relevant to all stakeholders. However, concepts in green are specific to Western cultures, whereas concepts common in other knowledge systems (e.g., Eastern cultures, Indigenous peoples) are shown in blue. (After UNEP. 2014. IPBES-2/17.)

society at different levels), and different knowledge systems (Western science and indigenous, local, and practitioners' knowledge). The framework proposes that the ecosystem goods and services emphasized by Western cultures in the MEA are just one form of nature's benefits to people (see Figure 6.9, green text) and that other cultures and knowledge systems tend to focus more on relational values that represent the nonmaterial gifts of Mother Nature (e.g., social relationships, equity, spirituality, cultural identity). Therefore, a good quality of life is influenced not only by human well-being, as defined by having human physical and emotional needs met (per the MEA framework), but also by a higher purpose that extends beyond oneself, which is to live in harmony and balance with others and with nature (an aspect of well-being emphasized in many Eastern as well as Indigenous cultures, see Figure 6.9, blue text). Thus, the IPBES framework incorporates elements of both hedonic and eudaimonic well-being (see Chapter 5).

The pros and cons of various classification frameworks have been discussed at length,[111,112] and even the IPBES framework has proven controversial.[113] It seems unlikely that a consensus view will emerge in the near future, so it is good to be aware that multiple frameworks exist and to understand some of the pros and cons of common frameworks (e.g., MEA and IPBES) that one is likely to see being used.

Biotic Control of Ecosystem Services

Given the importance of ecosystem services for the well-being of humanity, a large and increasing body of research has focused on trying to determine the biological characteristics of ecosystems that are most important in regulating ecosystem services. Figuring out which species, as well as genetic and functional traits, are the best ecosystem service providers has proven to be a daunting challenge. Nevertheless, we've made significant progress in the past few decades and are getting must closer to understanding what must be conserved to maximize the delivery and reliability of ecosystem services to people.

Functional traits

Morphological, physiological, and behavioral traits often control the ecological roles played by species in their environment. In turn, those functional traits also control the impact species have on ecosystem services.

Several fields of study have highlighted the importance of functional traits for regulating ecosystem services. For example, in the 1980s, the field of ecological stoichiometry showed how the biochemical composition of plant and animal tissue (a physiological trait) influences sequestration of biologically essential elements like carbon, nitrogen, and phosphorus from the environment.[19] A well-known example is that of *Daphnia*—a genus of small crustacean that often dominates the zooplankton of lakes. *Daphnia* have a lifestyle that is characterized by fast reproduction and high rates of growth, both of which require large quantities of RNA and ATP (energy) to make new tissue. Because RNA and ATP are molecules rich in phosphorus, *Daphnia* tend to sequester a lot of phosphorus, which gets stored in their tissues.[114] As *Daphnia* sequester phosphorus from the lake to meet their own biological requirements, they can cause other biological processes in a lake to become limited by the reduced availability of phosphorus. Two such processes are photosynthesis, which produces oxygen and sequesters carbon, and the production of biomass by higher consumers in a lake.[115] A few decades ago, it would have been hard to imagine that the physiological traits of a millimeter-sized "water flea" could control all of a lake's supporting services, as well as its provisioning services, such as the production of fish.

Oftentimes, ecosystem services are the products of the functional traits of a relatively small number of species. As a general rule, most species are rare, and ecosystems tend to be dominated at any particular time by a small number of species (which is why we give names like "hemlock forest" or "giant kelp bed"). In instances where a small number of species dominate an ecosystem, the provision of ecosystem services is a product of the functional traits of those dominant species, and changes in ecosystem services can be predicted based on the density or biomass of the dominant species.

However, there are species—such as keystone species or ecosystem engineers (see Chapter 3)—whose impacts on ecosystem services are disproportionate to their density or biomass. For example, after humans exterminated the top predator, wolves (*Canis lupus*), from Yellowstone National Park in the mid-1920s, populations of large herbivores like elk (*Cervus canadensis*) ballooned, resulting in nearly 70 years of overgrazing of vast tracks of land in the park.[116] Losses of woody vegetation led to the decline of other species, such as songbirds, which use the vegetation for habitat, and beavers, which use it as food. But within a few years of the reintroduction of wolves to Yellowstone in 1995, some of the impacts began to be reversed. Elk began to exhibit a behavioral trait related to their fear of predators, avoiding valleys and open meadows where the risk of predation by wolves was high. Their nonlethal, predator-avoidance strategy reduced grazing pressure, leading to increased biomass production of woody plants such as willow, alder, and aspen.[117,118] Regrowth of woody vegetation along stream banks facilitated a resurgence of beaver populations that began to engineer new pond habitats that altered channel morphology and may have reduced sediment erosion from stream banks.[119–121] While some of these proposed effects are controversial and still being studied,[122,123] the point is that the impacts of rare species (keystones like wolves, and engineers like beavers) can be magnified by the behavioral traits of common species (fear and predator avoidance by elk) to not only influence supporting services of ecosystems (biomass production of woody vegetation), but potentially influence regulating services such as stream erosion (**Video 6.4**).

Video 6.4 This video shows how keystone species influence ecosystem services.
oup-arc.com/e/cardinale-v6.4

To help practitioners identify functional traits that are important for ecosystem services, and to predict how those functional traits will be affected by environmental change, scientists have begun building databases to summarize the functional traits of a wide variety of species. For example, the Plant Trait Database (TRY, oup-arc.com/e/cardinale-w6.5) represents a network of vegetation scientists around the world who have collected information on more than 3 million traits in nearly 70,000 plant species. Other similar databases are available, including the U.S. Department of Agriculture Plants Database (oup-arc.com/e/cardinale-w6.6), the World Agroforestry Tree species database (oup-arc.com/e/cardinale-w6.7), TraitNet (oup-arc.com/e/cardinale-w6.8), and the Global Biotraits Database (oup-arc.com/e/cardinale-w6.9). These trait databases have been used to understand why some species are more sensitive to human activities than others,[124] to predict what traits make some species more invasive[125] or prone to extinction than others,[126] and to predict species-specific responses to environmental change.[127]

Biodiversity

There is now strong evidence that the genetic diversity of species, as well as the number and evenness of species that comprise ecological communities, regulates the supporting services of ecosystems. One of the earliest studies to demonstrate this was led by David Tilman and colleagues, who planted between 1 and 16 native grassland plant species in 168 large (81m^2) outdoor plots at the University of Minnesota's Cedar Creek Ecosystem Science Reserve. The researchers found that as the number of species increased, the efficiency with which plants captured soil nitrogen increased, leading to higher rates of net primary production (the amount of new plant biomass accrued per year in the plots).[128] The effect of plant species richness on biomass production grew stronger through time (Figure 6.10A). And as the experiment was continued for more than a decade, the researchers further found that greater plant species richness led to lower year-to-year variation in total plant biomass, which resulted in greater ecosystem stability (Figure 6.10B).[129]

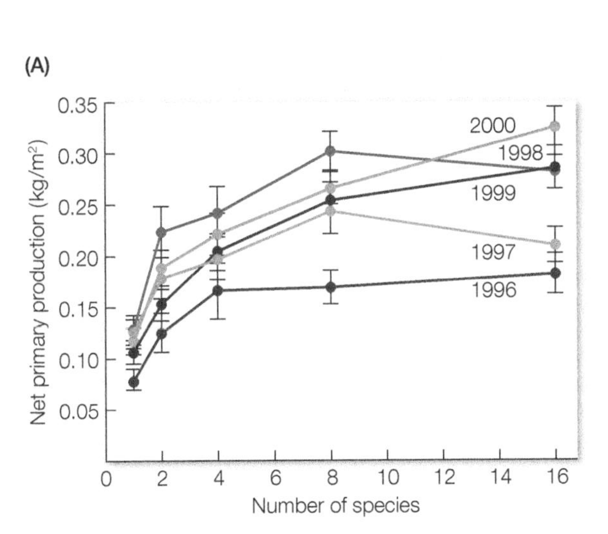

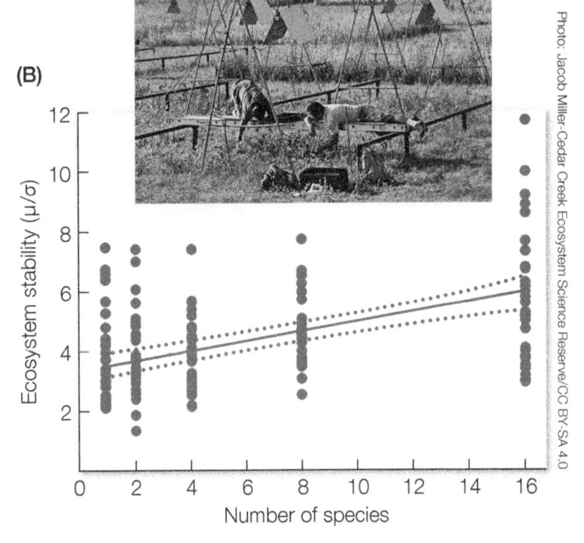

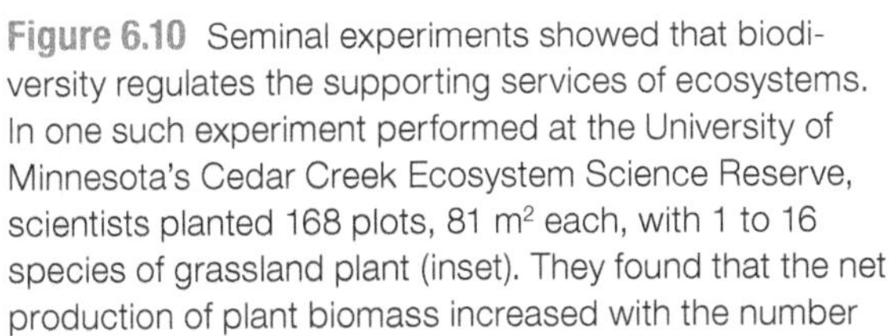

Figure 6.10 Seminal experiments showed that biodiversity regulates the supporting services of ecosystems. In one such experiment performed at the University of Minnesota's Cedar Creek Ecosystem Science Reserve, scientists planted 168 plots, 81 m^2 each, with 1 to 16 species of grassland plant (inset). They found that the net production of plant biomass increased with the number of species in a plot, with effects growing stronger through time (A). They also found that plant richness reduced annual variation in plant biomass for more than a decade, thus helping to make the ecosystem more stable (B). (A, after D. Tilman et al. 2001. *Science* 294: 843–845; B, after D. Tilman et al. 2006. *Nature* 441: 629–632.)

Seminal studies like that of Tilman and colleagues prompted hundreds of other researchers to experimentally manipulate the variety of plants, algae, insects, bacteria, fungi, fish, and other types of species in greenhouse pots, lab aquaria, outdoor plots, or artificial ponds designed to mimic a wide variety of ecosystems around the world. By 2009, several hundred scientific papers had been published reporting the results of more than 600 experiments that had manipulated more than 500 types of organisms in nearly every major freshwater, marine, and terrestrial ecosystem on Earth.[130] In addition to this body of work, researchers branched out and began studying how other aspects of biodiversity, such as species evenness[131] or genetic diversity,[132] influence the supporting services of ecosystems.

Recent papers summarizing results to date have generally concluded that, as a general rule, ecosystems with greater amounts of genetic and species diversity (both richness and evenness) tend to be more efficient at capturing biologically essential resources and converting those into new biomass.[132–134] Those ecosystems also tend to be more stable in their production of biomass through time.[135,136] We have also learned that biodiversity has a stronger effect on the supporting services of ecosystems than do many other prominent forms of global change (e.g., climate change and nutrient pollution),[137–139] and the impacts of biodiversity loss on the supporting services of large natural ecosystems are even greater than what had been predicted from small-scale experiments.[139]

More recently, we've begun to get a better picture of how biodiversity per se impacts the provisioning and regulating services of ecosystems. Cardinale and colleagues reviewed more than 1700 papers to summarize the evidence needed to test claims about how biodiversity influences 32 types of provisioning and regulating services.[130] They found that about 50% of all claims about biodiversity's impact on provisioning and regulating services cannot yet be tested, because either there is insufficient data or studies published to date have been inconclusive. However, for about 40% of the services, evidence is not only sufficient but strong that biodiversity at the genetic or species level controls the service (**Table 6.2**). For example, the synthesis confirmed the following:

- Crops composed of more genetic diversity tend to have greater yields in agricultural ecosystems.
- Oceans with a greater variety of fish species have more stable fisheries yields through time.
- Forests composed of a greater variety of tree species produce more total wood.
- Ecosystems with more plant species are more resistant to invasive pests and disease.
- Ecosystems with more plant species sequester more carbon from the atmosphere and produce more fertile soils.

Why do these effects of biodiversity occur? At least four biological mechanisms that allow communities with a greater variety of genes, species, or functional traits to provide higher levels of an ecosystem service have been identified. First, **niche partitioning** (sometimes called niche complementarity) occurs when species divide limited resources in space or time, leading to greater efficiency and productivity. Cardinale gave an example of how niche partitioning among freshwater algae can influence the regulating service of water purification in streams (**Figure 6.11A**).[140] He showed experimentally that as different species of algae partition the habitats they occupy in streams—some growing in areas of slow flow while others grow in areas of high flow, and some occupying disturbed habitats while others occupy undisturbed habitats—this type of habitat partitioning allows the community as a whole to remove more nutrient

TABLE 6.2 A summary of evidence linking biodiversity to the provisioning and regulating services of ecosystems

Category	Ecosystem service	Service-providing organism(s)	Diversity level	Study type	*N*	Relationship
PROVISIONING						
Food	Crop yield	Plants	Genetic	Exp	575	+
Fisheries	Stability of fisheries yield	Fish	Species	Obs	8	+
Wood	Wood production	Trees	Species	Exp	53	+
Fodder	Fodder yield	Plants	Species	Exp	271	+
REGULATING						
Biocontrol	Control of herbivorous pests	Plants	Species	Obs	40	–
				Exp	100	–
	Control of herbivorous pests	Natural enemies (*predators, parasitoids, pathogens*)	Species	Exp/Obs	266	–
			Species/trait	Obs	18	–
	Resistance to plant invasion	Plants	Species	Exp	120	+
	Disease prevalence (*on plants*)	Plants	Species	Exp	107	–
Climate	Carbon sequestration	Plants	Species	Exp	479	+
Soil	Soil nutrient mineralization	Plants	Species	Exp	103	+
	Soil organic matter	Plants	Species	Exp	85	+

The columns on the left give the categories and types of ecosystem service. The organisms providing the services and the levels of diversity are noted. Study types include experimental (Exp) and observational (Obs), with the number of studies (*N*) noted. The final column gives the relationship between biodiversity and the service that has been supported by the studies (+ is a positive relationship where the service increases as biodiversity increases, and – is a negative relationship where the service declines as biodiversity increases).

pollution from the water than any single species of algae could accomplish on its own (**Figure 6.11B**).

Second, **positive interactions** such as facilitation can allow diverse communities to be more efficient and productive, in turn, enhancing services. A good example comes from the work of Losey and Denno,[141,142] who performed multiple studies looking at how two different insect predators—a foliar-foraging lady beetle (*Coccinella septempunctata*) and a ground-foraging carabid beetle (*Harpalus pennsylvanicus*)—interact to control populations of an invasive pest of forage crops (the pea aphid, *Acyrthosiphon pisum*). When the lady beetle was the only predator present, pea aphids would simply drop off their host plant onto the ground to avoid predation, only to crawl up later to feed on and kill the plant. When the carabid beetle was the only predator present, the aphids would simply avoid predation by remaining on the plant. However, when both the lady beetle and carabid beetle were present, aphids had no escape and were eaten whether they stayed on the plant or dropped to the ground. As a result, facilitation among the two species of beetle predators led to improved biological control of the aphid pest.

The **selection effect** (sometimes called the sampling effect, or the lottery effect in various literatures) is another biological mechanism that can improve ecosystem services. It occurs when communities with a greater variety of species are more likely to have one or a select few that are especially efficient or productive, and these come to dominate the ecological processes that generate ecosystem services. For example, Denoth and colleagues[143] have shown that the odds of successful biological control of weeds or insect pests

(A)

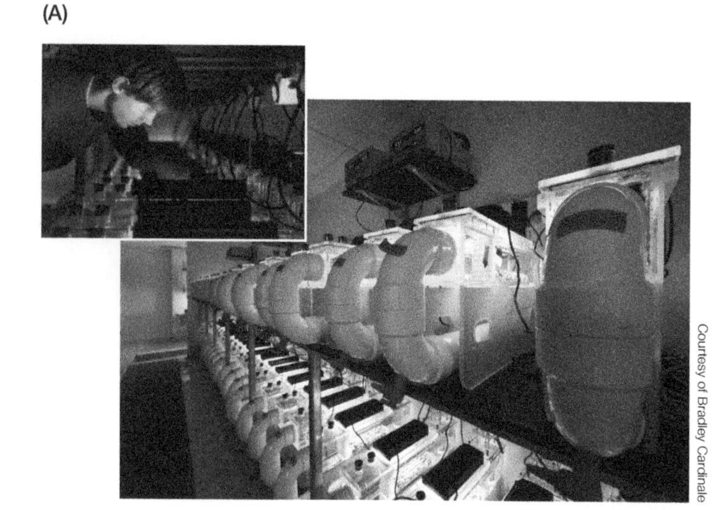

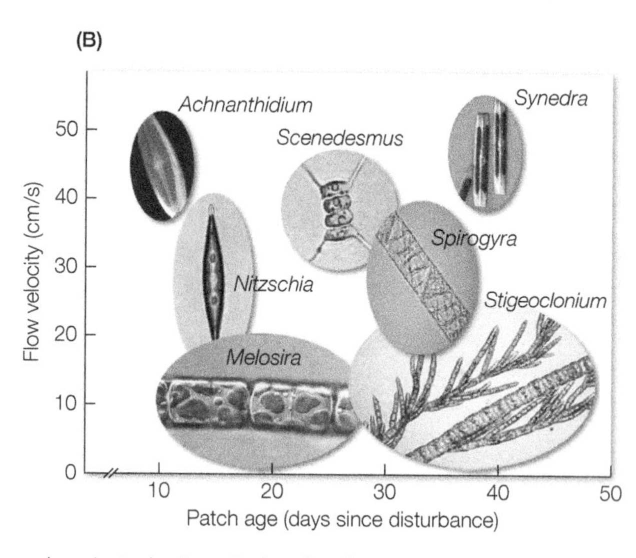

Figure 6.11 Niche partitioning among species of algae improves water quality. (A) Researchers manipulated the number of species of green algae and diatoms in artificial streams, called flumes, to determine how biodiversity influences removal of nitrogen pollution from the water. The inset shows a technician taking a sample of algae to analyze them for ^{15}N-labeled nitrate, NO^{3-}, which was added to flumes to simulate pollution. (B) Streams with more species were able to take up more nitrate because of niche partitioning; some species were adapted for environments with high flow in the streams (*y*-axis, top), others for low flow (*y*-axis, bottom). Some species were adapted for living in frequently disturbed habitats (*x*-axis, left), whereas others were adapted for living in more-stable habitats (*x*-axis, right). Because each species dominated a different type of habitat, several species were complementary in their ability to use nitrate from different parts of the streams. (Inset from B. Cardinale et al. 2012. *Nature* 486: 59–67; B, from B. Cardinale. 2011. *Nature* 472: 86-U113)

increase with the number of biocontrol agents that are released, and this occurs because the probability that at least one agent will exert strong control increases as more agents are released.

Last, the **dilution effect** is a biological mechanism that reduces the risk of infection by zoonotic pathogens (i.e., pathogens transmitted to humans through animal reservoirs) by reducing the prevalence of the pathogen among hosts that carry and transmit the disease. The dilution effect was first described by Richard Ostfeld and his colleagues from the Cary Institute of Ecosystem Studies while studying the influence of biodiversity on the transmission of Lyme disease from ticks to humans. More recently, the dilution effect has been considered as a potential explanation for wildlife diseases like bird malaria[144,145] and disease severity of fungal and viral pathogens among plants.[146] Despite its growing popularity in the literature, some researchers have questioned how often the dilution effect occurs, how important it is relative to other controls of disease, and whether or not the mechanism is sufficiently well established for conservation biologists to claim it as a general benefit of biodiversity (**Box 6.3**).

In addition to providing higher levels of ecosystem services at any given point in time (via niche partitioning, positive interactions, selection, and dilution effects), biodiversity can help ensure the consistent delivery of ecosystem services by reducing their volatility through time. Volatility is an important measure for people whose livelihoods rely on services being sustained through time, such as foresters who must sustain lumber production for decades, or fishermen who must harvest fish or shellfish year after year. Volatility is often measured as the coefficient of variation, CV, which in this context quantifies the volatility of an ecosystem service through time (how much a service varies through time relative to the mean level of the service being delivered). It is quantified as the ratio of the standard deviation, δ, to the mean, μ, where the standard deviation can be broken down into two subcomponents that generate variation: the sum of

BOX 6.3 Challenges & Opportunities

The Dilution Effect

The dilution effect is a mechanism by which biodiversity can change the prevalence of pathogens and diseases that affect plants, wildlife, and even humans. The effect was first described by Richard Ostfeld, a disease ecologist at the Cary Institute of Ecosystem Studies, who studied the prevalence of Lyme disease in the eastern United States.[164] Lyme disease is caused by the bacterium *Borrelia burgdorferi*, which is transmitted to humans through the bite of infected ticks. Early signs of Lyme disease are a characteristic "bulls-eye" rash found at roughly 80% of bite sites, followed by flu-like symptoms of fever, chills, headache, fatigue, and joint pain. If left untreated, Lyme disease can become debilitating, even deadly (see Figure).

Ticks that transmit the *Borrelia* bacterium will bite and take a blood meal from many different mammal species, including humans. But some mammal species are more "competent" hosts than others, meaning the *Borrelia* bacterium is more likely to survive in their bloodstream and later be transmitted to additional ticks. For example, the white-footed mouse (*Peromyscus leucopus*) is a highly competent host that can infect hundreds of ticks per hectare with the *Borrelia* bacterium. In contrast, the Virginia opossum (*Didelphis virginiana*) is a far less competent host for the bacterium, tending to infect just five ticks per hectare[165] (see Figure).

The dilution effect argues that the prevalence of a disease like Lyme can become diluted (e.g., made less frequent) by the presence and abundance of less-competent hosts. For example, if the mammal community of a forest is filled mostly with highly competent white-footed mice, then tick infection rates can be quite high, and the risk of a human hiker being bitten by a tick infected with Lyme is high. But if the forest mammal community is composed of less-competent hosts (opossums, skunks, raccoons), then the prevalence of Lyme in ticks gets diluted, and the likelihood of a person contracting the disease goes down.

The dilution effect has been touted by some as an important ecosystem service by which biodiversity protects humans from zoonotic diseases.[166,167] However, this claim has been challenged by some who think the current scientific literature has been biased toward studies that were predisposed to find evidence of dilution effects, and who believe the conditions required for the effect to occur may be uncommon or limited to specific kinds of disease.[168,169] Skeptics have also pointed out that ecosystems are often the source of human disease and that there are well-known mechanisms by which biodiversity can promote disease.

Determining the extent to which biodiversity protects people from disease, and whether or not mechanisms like the dilution effect are common and widespread, represents an important opportunity for the next generation of conservation biologists.

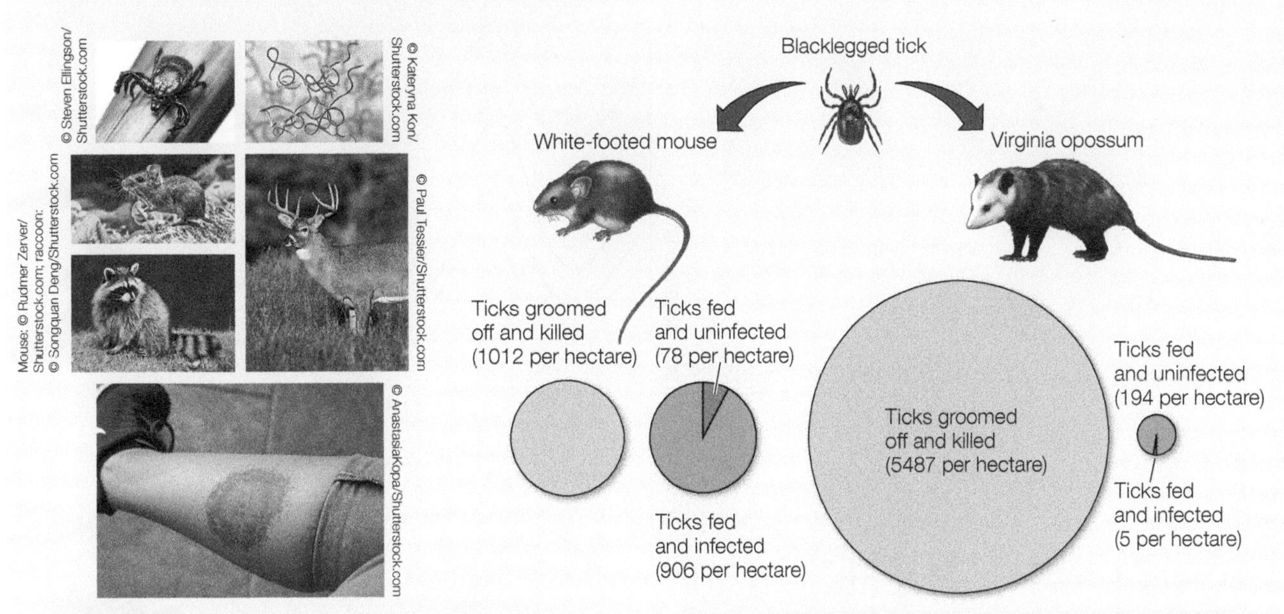

The dilution effect for Lyme disease. Competent hosts like the white-footed mouse carry the disease-causing bacterium, *Borrelia burgdorferi*, which they transmit in high frequency to ticks. The infected ticks can then transmit the disease to people. However, transmission rates can be reduced if competent hosts are replaced by noncompetent hosts (e.g., opossums, skunks, raccoons) that reduce the prevalence of *Borrelia* in ticks. This subsequently reduces risk for humans. (After F. Keesing et al., 2010. *Nature* 468: 647–652.)

the variances among the species (or genes, or functional traits) that provide the service, and the sum of the covariances among the species providing the service:

$$CV = \frac{\sigma}{\mu} = \frac{\sqrt{\sum var + 2cov}}{\mu} \quad (6.2)$$

Equation 6.2 makes clear what is needed to stabilize the delivery of an ecosystem service. First, anything that increases the mean delivery of a service (described in the previous paragraph) will cause the denominator to increase and the coefficient of variation to decrease. Second, anything that decreases the numerator in Equation 6.2 will reduce variation in the service through time (e.g., decrease volatility). This can occur if the species that provide the service are less variable through time. For example, a diverse community may be more likely to be dominated by a particularly stable species. Alternatively, the numerator in Equation 6.2 can be reduced if the various species that provide a service covary less through time. This can occur when the populations are in **population asynchrony**, that is, they fluctuate out of sync. For example, different species may respond differently to changing environmental conditions (e.g., some species populations increase in response to warming temperatures while others decline), or species interactions like competition or predation can cause some species to increase as others decrease. The effect of population asynchrony on volatility has often been compared to a well-diversified stock portfolio where each stock responds differently to market conditions. As the individual stocks fluctuate out of sync through time, they cause the total value of the portfolio to be more stable through time. Similarly, the **portfolio effect**[147] in a diversified ecological community occurs when each species responds differently to changing conditions through time, allowing the total amount of an ecosystem service delivered by the community to remain stable through time (Figure 6.12).

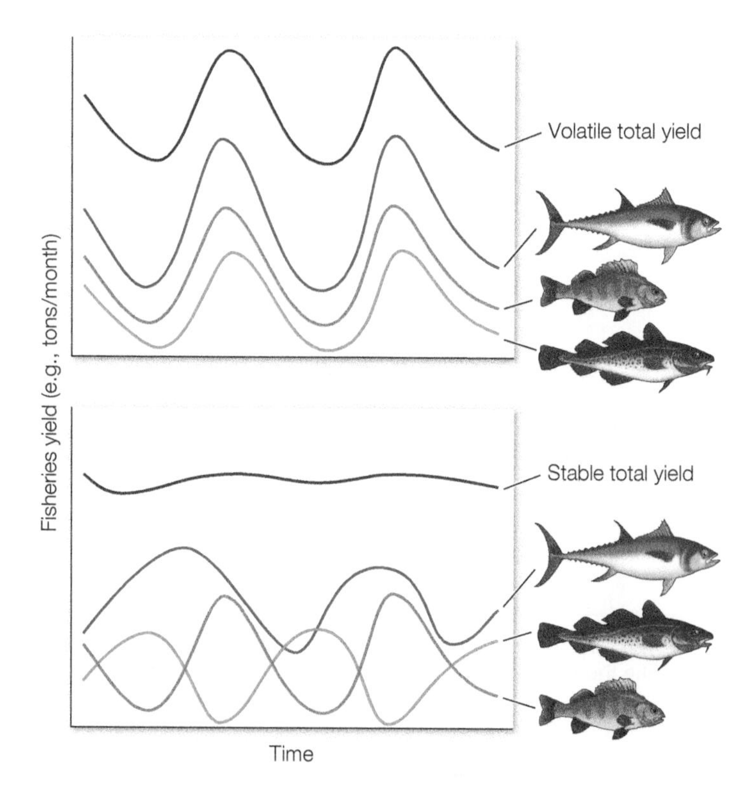

Figure 6.12 Example of how asynchronous dynamics generate the portfolio effect of biodiversity. In the top graph, the populations of three commercial fish species fluctuate in sync, causing their total production to be volatile through time. In the bottom graph, the three species fluctuate out of sync, which causes their total production to be less volatile through time.

Ecosystem Markets and Payments for Ecosystem Services

Because of the growing recognition of the value of ecosystem services to humanity, and the need to develop new ways for society to value ecosystems, it has become increasingly common to establish markets and payment schemes for ecosystem services. Although the terms *ecosystem markets* and *payments for ecosystem services* are used interchangeably to refer to the suite of economic tools used to reward the conservation of ecosystem services, the concepts differ. A market is a regular gathering of people established for the purpose of buying and selling goods and/or services. Several ecosystem markets have been established to allow regular trading of permits or credits related to ecosystem services. An example is the set of carbon markets for emissions trading that were established to buy/sell the ecosystem service of carbon sequestration, which allows countries to meet their obligations to the Kyoto Protocol to mitigate climate change.[148] In contrast to markets, payments for ecosystem services are the suite of economic arrangements that are used in ecosystem markets to reward the conservation of ecosystem services. There are currently six economic arrangements used for these types of payments:

1. *Direct public payments.* Payments that governments make to providers of ecosystem services are called direct public payments. This form of payment for ecosystem services is perhaps the most common, with governments around the world paying landowners to steward their land in ways that will generate ecosystem services for others. For example, the Conservation Reserve Program in the United States pays over $1.5 billion to farmers each year in exchange for their protection of endangered wildlife habitat, open space, and/or wetlands.[149] China and Vietnam have similar programs to fund erosion control,[150] while numerous countries in Latin America have established water funds as conservation financing mechanisms that gather investments from water users and direct the funding toward the protection and restoration of key lands upstream that filter and regulate water supply (**Figure 6.13**).[151]
2. *Direct private payments.* These function much like direct public payments, except that nonprofit organizations or for-profit companies take the place of the government as the buyer of the ecosystem service. For example, the soft drink giant Coca-Cola has numerous manufacturing plants in Latin America that require large quantities of clean water to make their products. Because of this, they

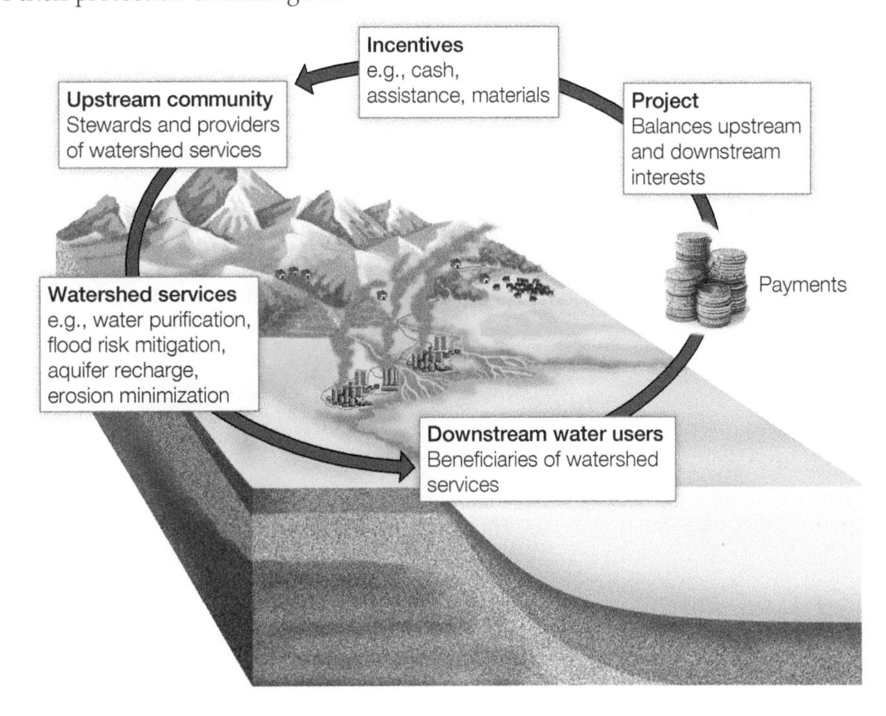

Figure 6.13 Payments for ecosystem services are a set of economic tools used to reward the conservation of those ecosystem services. Here, the downstream users of a river make payments to upstream users to incentivize them to conserve and be good stewards of the watershed. In return for these payments, downstream users get better water quality and flood control. (After G. Bennett, G. et al. 2013. *Charting New Waters: State of Watershed Payments 2012.* Washington, DC: Forest Trends.)

have invested in 50 water funds across 11 countries in Latin American and the Caribbean, setting up a system to pay landowners who live upstream of their processing plants to manage their watersheds in ways that reduce soil erosion and to maintain water quality through actions like replanting trees, keeping living trees standing, and using sustainable agricultural techniques.[152]

3. *Tax incentives.* In exchange for protecting resources that will provide ecosystem services, individuals receive indirect government compensation through tax breaks or incentives. In the United States, for instance, individual landowners receive tax incentives to put their land under conservation easements that provide long-term legal protection of that land.[153]

4. *Cap-and-trade markets.* In these markets a government or regulatory body first sets a limit or "cap" on the amount of environmental degradation or pollution that will be permitted in a given area and then allows firms or individuals to buy and sell credits within that cap. Examples of cap-and-trade markets include carbon emissions trading markets established by many countries to meet obligations for the Kyoto Protocol[148] and wetland mitigation banking systems that were set up by countries to meet obligations to the Ramsar Convention for international cooperation on the conservation of wetlands.

5. *Voluntary markets.* Buyers and sellers engage in transactions for ecosystem services on a voluntary basis (i.e., not because they are forced to trade by regulation or to meet a mandatory cap). Generally speaking, businesses and/or individual consumers engage in voluntary markets for reasons of philanthropy, risk management, and/or preparation for participation in a regulatory market. For example, there are voluntary carbon markets that allow individuals to purchase carbon credits to offset their lifestyle choices (e.g., travel, energy use, or consumption) or allow businesses that purchase carbon credits to take responsibility for offsetting their own emissions.

6. *Certification programs.* These programs reward consumers for buying certain kinds of products that were produced and brought to market in a particular manner. When a consumer pays the extra cost of products that have been labeled as being "ecologically friendly," "organic," or "sustainable," they are essentially choosing to pay for the protection of ecosystem services that are associated with those labels. Certification programs like these, which reward producers who protect ecosystem services, have been developed for a variety of products, including wood, paper, coffee, and food.

Ecosystem markets and payments for ecosystem services are intended to overcome market failures to address externalities and, in turn, to provide incentives to protect valuable goods and services that are not included in traditional markets. However, several factors presently limit these programs from being more widely utilized, including insecure land tenure of many poor people, complex and often bureaucratic project procedures, high project transaction costs (expenses incurred when buying or selling a good or service), and a lack of stable, permanent markets. If these problems can be overcome, ecosystem markets and payments for ecosystem services have the potential to be major social and economic tools that enable people, businesses, and governments to have more sustainable relationships with nature.[154]

Summary

1. Ecosystem services are the variety of benefits that ecosystems provide to human society. They include: (1) supporting services, which are ecological processes like biomass production, decomposition, and nutrient cycling that control the functioning of ecosystems and underlie all other types of services; (2) provisioning services that represent the products people obtain from ecosystems, such as food, water, and wood; (3) regulating services, which are the benefits that stem from regulation of ecosystem processes, including air and water purification, pest control, pollination, and natural disaster mitigation; and (4) cultural services, which are the nonmaterial benefits of ecosystems, such as recreation, tourism, human health, and well-being.
2. The concept of ecosystem services dates back to the ancient Greeks. However, it was popularized in the 1990s by three events: (1) a merger of the fields of ecosystem ecology and environmental economics into the new discipline of ecological economics, (2) the rise of the new field of biodiversity and ecosystem functioning that became a global research agenda, and (3) the Millennium Ecosystem Assessment, which evaluated the state of Earth's ecosystems and the services they provide to humanity.
3. Ecosystem services are controlled by the collection of genes, species, and functional traits that comprise the ecological communities that live in ecosystems. Functional traits include a variety of morphological, physiological, and behavioral characteristics of organisms that control the ecological roles played by species and, in turn, ecosystem services.
4. The diversity of genes, species, and functional traits in ecological communities influences the delivery of ecosystem services through biological mechanisms such as niche partitioning, facilitation, and selection effects. Diversity controls the volatility of ecosystem services through population asynchrony and the portfolio effect.
5. Ecosystem markets and payments for ecosystem services represent a suite of economic tools used to reward the conservation of ecosystem services. Ecosystem markets, such as carbon markets, establish a regular gathering of people for the purpose of buying and selling ecosystem goods and/or services. Payments for ecosystem services are the suite of economic arrangements that are used in ecosystem markets to reward the conservation of ecosystem services.

For Discussion

1. How would you describe the concept of ecosystem services to your family, a business partner, or a politician? Would your description change, or would you emphasize different types of services, depending on who you were talking with?
2. Which types of ecosystem services—supporting, provisioning, regulating, or cultural—resonate most strongly with you as arguments for conservation? Which do you think resonate most strongly with others?
3. What economic arrangements used in payments for ecosystem services do you think are most effective for conservation? Why?
4. Assume that a green space near your home is being considered for development by your local government. How would you use the concept of ecosystem services to argue for conservation of this space?

Suggested Readings

Cardinale, B. J. et al. 2012. Biodiversity loss and its impact on humanity. *Nature* 486: 59–67. A comprehensive summary of how biodiversity loss affects ecosystem services.

Chan, K. M. A. et al. 2017. Payments for ecosystem services: Rife with problems and potential—for transformation towards sustainability. *Ecology Economics* 140: 110–122. Discusses some problems with payments for ecosystem services and provides ideas on how to correct them.

Daily, G. C. 1997. *Nature's Services: Societal Dependence on Natural Ecosystems*. Washington, DC: Island Press. This influential book popularized the idea that nature has value to humans.

Duffy, E. et al. 2017. Biodiversity effects in the wild are common and as strong as key drivers of productivity. *Nature* 549: 261–265. Compares the role of biodiversity change to other forms of environmental change on supporting services in natural ecosystems.

Frumkin, H. et al. 2017. Nature contact and human health: A research agenda. *Environmental Health Perspectives* 125: 18. Summarizes the known health benefits of interacting with nature.

Hanley, N. et al. 2015. Measuring the economic value of pollination services: Principles, evidence and knowledge gaps. *Ecosystem Services* 14: 124–132. Details the value of insects for pollination.

Larigauderie, A. 2015. The Intergovernmental Platform on Biodiversity and Ecosystem Services (IPBES): A call to action. *Gaia* 24: 73. Good introduction to one of the premier international agreements on ecosystem services.

Lele, S. et al. 2013. Ecosystem services: Origins, contributions, pitfalls, and alternatives. *Conservation and Society* 11: 343–358. Tells about the history of ecosystem services and discusses the pros and cons of using ecosystem services to value nature.

Losey, J. E. and Vaughan, M. 2006. The economic value of ecological services provided by insects. *BioScience* 56: 311–323. Tallies the value of insects for biological control.

Mace, G. M. et al. 2012. Biodiversity and ecosystem services: A multilayered relationship. *Trends in Ecology & Evolution* 27: 19–26. A good summary of ecosystem services.

Visit the
Conservation Biology
Companion Website
oup.com/he/cardinale1e
for videos, exercises, links, and other study resources.

7

Ecological Economics

The Philippine eagle (*Pithecophaga jefferyi*) is one of the most endangered bird species in the world. It is believed that fewer than 500 pairs survive in the wild.

Part II of this book has detailed the many reasons why people want to conserve biodiversity, as well as how people assess what to conserve, how much of it to conserve, and when and where to conserve it (**Figure 7.1**). Chapter 5 described the broad array of value systems that people use to justify conservation of biodiversity, including intrinsic, relational, and instrumental values. These value systems are a function of people's ethical worldviews (biocentrism, ecocentrism, and anthropocentrism) that, in turn, influence people's perspectives on biodiversity and the decisions they make to protect it (or not). Chapter 6 looked more closely at a subset of value systems (instrumental and relational values) that represent the goods and services ecosystems provide to people. These goods and services benefit people in terms of human health, prosperity, and safety. Here in Chapter 7, we focus more narrowly on the instrumental value of ecosystems and their biodiversity, describing the methods of valuation and decision making that can be used by practitioners to properly evaluate people's preferences and to assess trade-offs among the many things people value.

This chapter draws heavily on principles from economics, particularly from the subfield of ecological economics. Many decisions on whether or not to protect ecosystems and their resident genes, species, and biological communities come

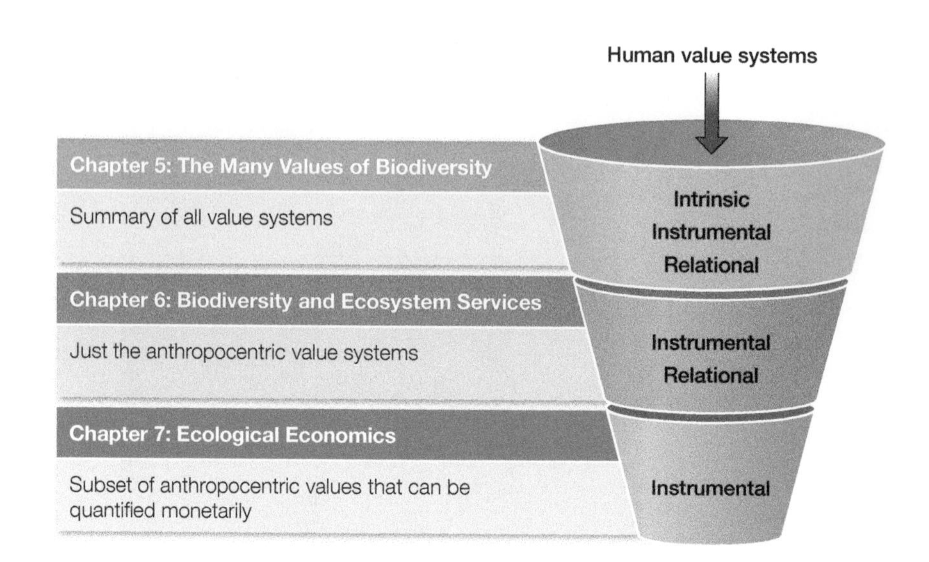

Figure 7.1 A summary of the organization of Part II of this book, which covered the importance of biodiversity. Chapter 5 was the broadest chapter, covering three major human value systems (intrinsic, instrumental, and relational) that influence conservation decisions. Chapter 6 focused more narrowly on the subset of anthropocentric values where nature and its biodiversity benefit people. Here in Chapter 7 we focus on the subset of anthropogenic values that can be quantified economically.

down to an estimate of how much it will cost to do so and whether conservation is worth the money, compared with other things people value or other options that are available. Ecological economics is a decision-making science that helps us predict how people value and allocate the resources nature provides to society.

Some would argue that any attempt to place a monetary value on biodiversity is morally wrong, since many aspects of the natural world are unique and priceless.[1] Supporters of this position believe there is no way to assign monetary value to natural wonders, beauty, or the value of living things. Others will argue that placing a monetary value on biodiversity is counterproductive to conservation because nature will sometimes have little economic value compared with other options such as development. But not everyone shares these viewpoints. Many stakeholders and decision makers, including governments and corporations who wield large power and influence over environmental policies, base their decisions on economic valuation.

Therefore, successful practitioners of conservation must understand methods of economic valuation and decision making in order to make good conservation decisions. Conservation biologists also need the skills to work with those who hold diverse perspectives, ranging from those who believe that biodiversity holds intrinsic value for its own sake, to those who believe decision making about biodiversity boils down to economic arguments. This chapter will give you the tools to deal with both.

Principles of Ecological Economics

Economics is the social science discipline that seeks to understand the decisions people make about the use of scarce resources given their preferences and values. Different subfields in economics address different resource allocation problems involving the supply, demand, and allocation of goods and services. The subfield of **ecological economics** addresses the supply, demand, and allocation of goods and services that are supplied to society by Earth's natural ecosystems (**Video 7.1**). Ecological economics assumes that the human economy, which is based on man-made capital, is part of a larger economic system that also includes the natural capital of Earth's ecosystems (**Figure 7.2**). The entire economic system, composed of both man-made and natural capital,

▶ **VIDEO 7.1** This video explains key concepts of ecological economics.
oup-arc.com/e/cardinale-v7.1

is assumed to be subject to the same physical laws that constrain all living systems. For example, living organisms must obey the law of mass balance, whereby matter can be neither created nor destroyed. Whatever an organism takes in (e.g., what it eats) can be transformed into a useful product (e.g., growth or reproduction) or eliminated as a waste (e.g., urine or feces). But the inputs must equal the outputs. Likewise, ecological economics argues that economic transactions can alter the form of natural resources in space or time, such as when raw materials are harvested and transformed into a product. But those economic transactions must obey the laws of mass balance to account not only for the useful product (e.g., economic growth), but also for any environmental costs and damages that result from the waste products of mass balance (e.g., pollution).[2]

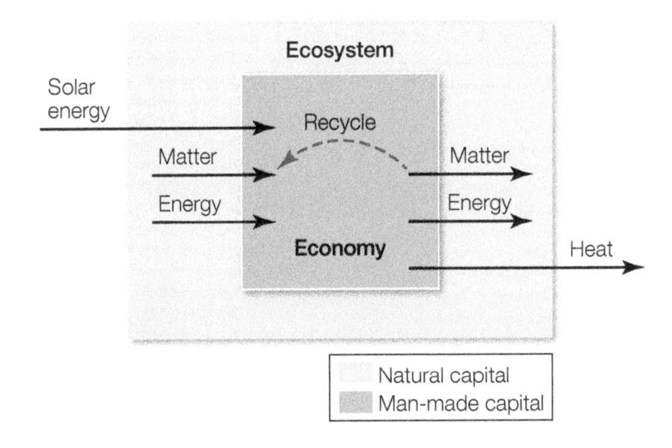

Figure 7.2 Ecological economics addresses the interrelationship between ecological and economic systems. Ecological economics assumes that the human economy, which is based on man-made capital, is but one part of a larger economic system that also includes the natural capital of Earth's ecosystems. (After R. Goodland and H. Daly. 1990. In *Planet Under Stress: The Challenge of Global Change*. C. Mungall and D. J. McLaren [Eds.], pp. 269–281. Royal Society of Canada. Toronto, Canada: Oxford University Press.)

Unfortunately, the environmental costs and damages associated with economic transactions have historically been ignored in traditional methods for economic accounting.[3] In part, this is due to fallacies associated with certain principles on which classical economics was founded. Much modern economic thought is founded on the principle of the **voluntary transaction**—the idea that a monetary transaction takes place only when it is beneficial to both parties involved. For example, a baker who sells loaves of bread for US$40 will find few customers. Likewise, a customer who is willing to pay only 4 cents (US$0.04) for a loaf will soon go hungry. A transaction between seller and buyer will only occur when a mutually agreeable price is set that benefits both parties: perhaps US$4 for that loaf of bread. Adam Smith, the eighteenth-century philosopher whose ideas are the foundation of much modern economic thought, wrote: "It is not upon the benevolence of the butcher, the baker, or the brewer that we eat our daily bread, but upon his own self-interest."[4] All parties involved in an exchange expect to improve their own condition. Smith argued that the sum of each individual acting in his or her self-interest results in a more prosperous society. He likened this effect to an "invisible hand" guiding the market—turning selfish, uncoordinated actions into increased prosperity and efficient use of scarce resources.

The problem with Smith's principle of the voluntary transaction is that it assumes all costs and benefits of free exchange are accepted and borne by the participants involved in the transaction. We know that is not true. As discussed in Chapter 6, many exchanges violate this assumption by producing **externalities**[5]—detrimental effects of a transaction that are not paid for by the individuals or businesses involved, but rather are born by the public at large. Externalities lead to **market failures** that cause resources to be misallocated, allowing a few individuals or businesses to benefit at the expense of the larger society. Market failures are particularly common with **open-access resources** (e.g., water and air) that are collectively owned by society at large and available for everyone to use. Because of the lack of property rights, these resources are often subject to few regulations, and the people, industries, and governments that use and damage these resources pay little more than a minimal cost or sometimes pay nothing at all. For example, when the Swiss company Nestlé—the largest bottled-water company in the world—takes advantage of areas that have lax water laws, it makes billions selling water that it pays practically nothing for. Yet, it doesn't pay the consequences of depleting aquifers that leave local residents with no clean water supply.[6]

Figure 7.3 Clean air is an example of an open-access resource that can be compromised when externalities (e.g., pollution) generate market failures that cause the public to bear the costs (e.g., health care, welfare losses) of business transactions.

© Hung Chung Chih/Shutterstock.com

One of the most striking examples of market failure with an open-access resource is air and water pollution (**Figure 7.3**). More than 9 million people die prematurely from pollution-related diseases each year. This is 16% of all deaths worldwide and represents more deaths than caused by wars, obesity, smoking, and malnutrition combined.[7] Pollution-related diseases cause productivity losses that reduce gross domestic product (GDP) in low- to middle-income countries by 2% per year, and they increase health care spending by 1.7% in high-income countries and up to 7% in countries that are heavily polluted and rapidly developing. Welfare losses due to pollution are estimated to be US$4.6 trillion per year, which represents 6.2% of global economic output. Yet, the industries that generate the pollution that causes deaths do not pay the economic costs. Rather, the costs are paid by individuals, families, and governments. The end result of market failures like these is that society as a whole becomes *less* prosperous from certain types of economic activities, rather than more prosperous.

Market failures also occur in **common-pool resources**—resources owned by national, regional, or local governments as public goods, or by communal groups as common property. The phrase most often used to describe market failure in a common-pool resource is the **tragedy of the commons**.[8] This phrase refers to a social dilemma where individuals who use a common-pool resource act independently and according to their own self-interests, yet the collective action of many individuals pursuing their own self-interests causes them to deplete or spoil the resource, to the detriment of all. The concept originated in an 1833 essay by British economist William Lloyd, who used a hypothetical example of overgrazing on unregulated common land (also known as a *common*) in the British Isles.[9] The concept became widely known over a century later when American ecologist Garrett Hardin used the tragedy of the commons as a metaphor to explain such wide-ranging problems as overexploitation of national parks, pollution, and human population growth.[8]

Since the publication of Hardin's paper, the tragedy of the commons has been used to explain a variety of market failures that result from overuse and degradation of natural resources. Fisheries are often cited as a classic example of the tragedy of the commons.[10] The migratory nature of most fish species makes it difficult to establish and protect rights to fish in the sea. While numerous treaties and conventions have attempted to impose restrictions on harvest (e.g., the United Nations Law of the Sea), agreements are often not honored and cannot be enforced. This allows fisheries in some locations to proliferate to an unrestricted number of users, free of any limits on their access. Individual fishers tend to have little incentive to practice conservation, for they know that if they do not catch the available fish, someone else probably will. The result is overexploitation of the fishery (covered in Chapter 10).

Although common-pool resource systems have been known to collapse as a result of overuse (such as in overfishing), numerous examples exist where members of a community have evolved cooperation or institutional arrangements that allow them to self-regulate the exploitation of those resources prudently without collapse. Elinor Ostrom—an American political economist and one of just two women to ever win the Nobel Prize in Economics—has shown that it is possible to manage natural resources in ways that are both economically and ecologically sustainable and thus avoid the tragedy of the commons (**Video 7.2**). Under certain conditions, people will self-organize social structures that help prevent market failures from occurring and erect good management strategies that facilitate the sustainable use of a common-pool resource.[11,12] In addition, it is clear that when individuals and organizations are required to bear the cost of negative externalities that result from their business transactions, they are much more likely to stop damaging, or at least minimize their damage to, the environment.[13] Therefore, effective practitioners of conservation biology should have sufficient training in ecological economics so that they can help stakeholders (individuals, businesses, and policy makers) be aware of the full costs and benefits that economic transactions have on biodiversity and the services that ecosystems provide to society.[14] The first step toward this training is to understand how costs and benefits influence decision-making.

VIDEO 7.2 Watch this video about the tragedy of the commons and how it can be avoided.
oup-arc.com/e/cardinale-v7.2

Decision-Making Analyses

In this section we describe three of the most common forms of decision analysis that are used in ecological economics to estimate the value of ecosystems and to compare their services to alternative options. All three of the decision-making methods described in this section are designed to deal with small-scale resource allocation problems. They cover the options for meeting particular targets of conservation, or for implementing single projects, and are helpful in allowing practitioners to think about optimizing conservation methods discussed in Parts III and IV of this book.

Cost-effectiveness analysis (CEA) is a decision-making tool that combines an ecological measure of program output (effectiveness) with an economic measure of program input (costs) to find the least-cost means to a conservation objective. The outputs of CEA are often nonmonetary units, such as the probability of a species survival, standardized to a unit cost of action. In contrast, cost-benefit analysis (CBA) is the primary decision-making tool when one wants to use a single currency, like money, to guide the decision-making process. The goal of CBA is to maximize the economic benefits of a decision or project while minimizing cost. Multicriteria decision analysis (MCDA) is a type of structured decision making (introduced in Chapter 2) that provides a formal process for weighting and evaluating trade-offs among the variety of things that people value (e.g., cost, time, environmental justice, ethics). It is often used as an alternative to CBA when making economic and policy decisions involving many different stakeholders who hold conflicting goals, such as differing views between the monetary and nonmonetary values of biodiversity.

Cost-effectiveness analysis

Cost-effectiveness analysis (CEA) allows one to determine the most efficient allocation of resources by combining an ecological measure of a conservation program or action (effectiveness) with an economic measure of program input (costs).[15] These measures of cost, *C*, and effectiveness, *E*, are usually expressed as a cost-effectiveness ratio (CER):

$$CER = \frac{C_i}{E_i} \tag{7.1}$$

where C_i and E_i are the cost and effectiveness of conservation action or program *i*. The goal of a CEA is to minimize the value of Equation 7.1.

Figure 7.4A gives an example of a cost-effectiveness analysis. Assume you have a current conservation plan that costs $100,000 to maintain parcels of land that collectively protect 10 threatened plant species. The CER for this baseline scenario, as calculated from Equation 7.1, is $10,000 per species. Now assume you are considering four alternative conservation plans with land purchase options that cost different amounts of money and that would protect different numbers of threatened plant species. The CER shows that plan 3 has the lowest cost per species conserved ($6,000 per species) and costs $10,000 less than the current baseline to conserve an additional 5 species. Each alternative conservation plan can be plotted on a cost-effectiveness plane (**Figure 7.4B**) to show how alternatives compare with the current baseline (center of coordinate system). In practice, most alternatives fall into quadrants 1 and 3 of the cost-effectiveness plane because cost and effectiveness are usually related to one another. However, occasionally one can find alternatives in quadrant 4 where effectiveness can be increased at a lower cost (e.g., plan 3).

Cost-effectiveness is sometimes expressed as an incremental, or marginal, change in benefits that stem from an incremental (marginal) change in

(A)

Scenario	Cost, *C* ($ per acre of land)	Effectiveness, *E* (number of species protected)	Cost-Effectiveness Ratio (CER)
Current conservation plan (baseline)	$100,000	10 species	$10,000 per species
Alternative conservation **Plan 1**	$400,000	32 species	$12,500 per species
Alternative conservation **Plan 2**	$200,000	8 species	$25,000 per species
Alternative conservation **Plan 3**	$90,000	15 species	$6,000 per species
Alternative conservation **Plan 4**	$50,000	6 species	$8,333 per species

(B)

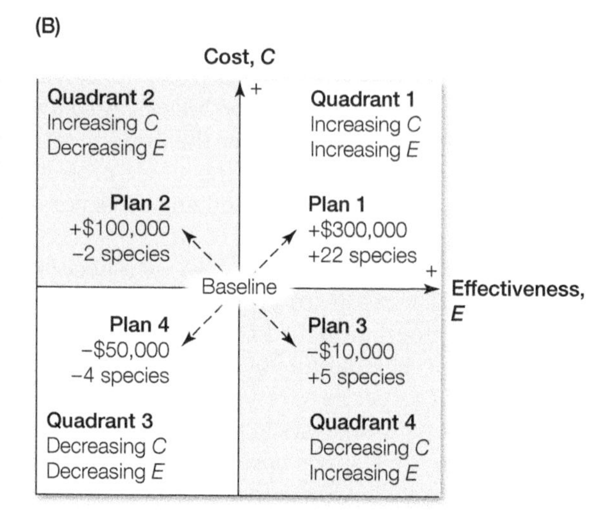

Figure 7.4 An example of a cost-effectiveness analysis. (A) Assume a current conservation plan costs $100,000 per year to protect 10 species. Now assume four alternative plans are being considered (plans 1–4). Each alternative plan has a cost associated with purchasing land that can serve as a protected area, and they differ in the number of threatened plant species that would be protected. The cost-effective ratio can be used to compare alternatives. (B) Alternatives can be plotted on a cost-effectiveness plane to show how they compare with the current conservation baseline (center). In practice, most alternatives fall into quadrants 1 and 3 because cost and effectiveness are correlated. However, occasionally one can find alternatives in quadrant 4 where effectiveness can be increased at a lower cost (e.g., plan 3).

a conservation action or program. This incremental change is usually measured relative to a baseline or current state as the incremental cost-effectiveness ratio (ICER):

$$ICER = \frac{C_i - C_B}{E_i - E_B} \tag{7.2}$$

where C_i and E_i refer to the cost and effectiveness of conservation action/plan *i*, while C_B and E_B refer to the cost and effectiveness of the baseline, or existing status. Equation 7.2 gives the incremental benefit above and beyond what is provided by the existing program.

CEA has a long history of use in health economics, where it has been used to compare the cost and effectiveness of alternative health care options, in turn aiding decisions about which medical care should be offered.[16] The tool has increasingly been used for applications in conservation biology. For example, Montgomery and colleagues[17] used CEA to compare different management strategies for the northern spotted owl (*Strix occidentalis caurina*). The authors evaluated three conservation management proposals for the northern spotted owl, each of which was estimated to have a different effect on the probability of the owl's survival, and each of which had a marginal cost of protection estimated as the welfare loss and reduction in timber harvest. Laycock and colleagues[18] used CEA to perform an analysis of the United Kingdom's Biodiversity Action Plan, which consisted of a suite of action plans for protecting threatened species and their habitats.

A strength of CEA is that it is flexible with respect to which measures of success or effectiveness a practitioner wishes to use for comparison. However, it provides only a relative measure of cost that can be compared among alternative plans, which is why it is often used as a precursor to a more formal cost-benefit analysis.

Cost-benefit analysis

Another important tool in public policy decision making on resource allocation is cost-benefit analysis (CBA) (**Video 7.3**). The goal of a CBA is to determine whether a decision, policy, or project is sound by determining whether the total expected benefits outweigh the total expected costs. CBAs are standard practice in most developed countries, and they are increasingly carried out in developing countries as well. Many international donor agencies that fund development and conservation projects, such as the World Bank, require such evaluations before projects are funded. Most CBAs are based on a common set of assumptions and involve the same set of generalized steps (**Figure 7.5**)[19]:

VIDEO 7.3 Watch this tutorial on CBA.
oup-arc.com/e/cardinale-v7.3

STEP 1: *Define the scope of the project.* The first step of a CBA is to detail the scope of the proposed policy or project so that boundaries can be defined for the analyses. Specific, measurable goals and objectives are written into a CBA to ensure that the project is focused and to ensure that outcomes are measurable and can be monitored. During this stage, it is essential to list all of the stakeholders that will be involved or affected so that one can decide whose costs and whose benefits need to be accounted for. In addition, it is important to determine the geographic scope of the analysis in order to bound the groups that will be affected by the policy or project and to estimate the time frame of the analyses and determine how far into the future one will need to estimate the costs and benefits.

STEP 2: *Specify alternative options.* All CBAs specify alternative options for a policy or project, at least one of which is the "do nothing" option. This

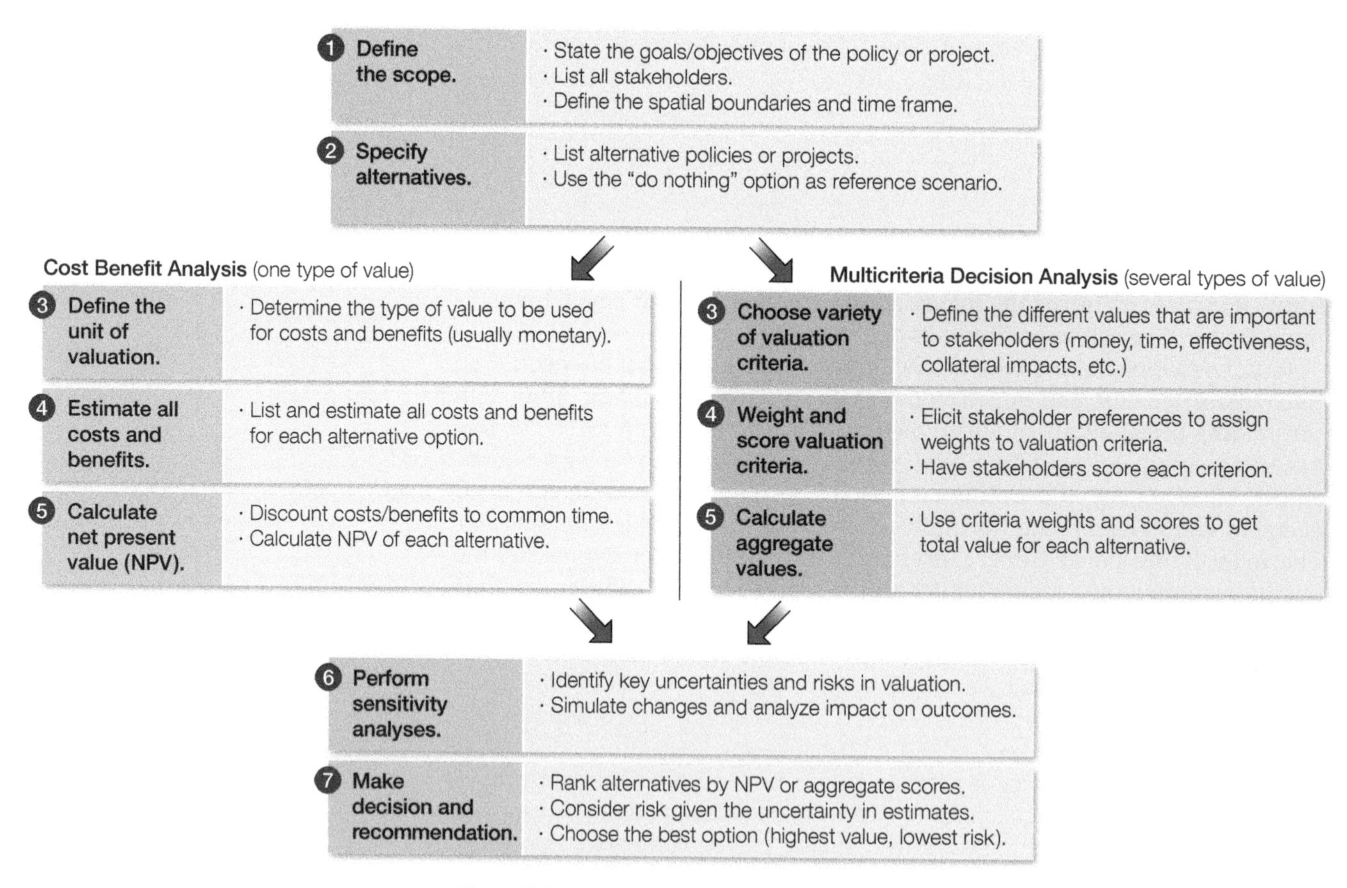

Figure 7.5 Two contrasting methods of decision analysis. Cost-benefit analyses (CBA) use a single currency, usually money, to make a decision on a policy or project. Multicriteria decision analysis (MCDA) optimizes trade-offs among multiple criteria, where money is but one thing that is valued by stakeholders.

option is typically used as the baseline or reference scenario against which the impact of a policy or project is judged. For example, assume a CBA is considering the following three options, each of which is associated with a net annual economic benefit:

Options	Annual benefit
Do nothing = reference scenario	\$10,000
Policy A	\$30,000
Policy B	\$35,000

When compared with the do-nothing reference scenario, both policy options are preferable because they generate net economic benefit. However, policy B is more preferable than policy A because it generates more benefit.

STEP 3: *Define the unit of valuation.* A key assumption of a CBA is there is a single, common unit of measurement that allows one to determine whether a certain project or venture is worth the cost it would take to enact. In principle, CBAs can be based on any aspect of value (e.g., time, money, energy). In practice, the most widely used measurement is money. In order to reach a conclusion about the desirability of a policy or project, one should express all aspects of each alternative policy or project—both positive and negative—in terms of the chosen unit of valuation.

STEP 4: *Estimate all costs and benefits.* Once a common unit of measurement is decided on, the fourth step of a CBA is to make a thorough, itemized list of all costs and benefits of the proposed policy or project. When generating such a list, it is important to keep track not only of the direct costs (e.g., operation and staffing costs) and benefits (e.g., income produced), but also of the indirect costs (e.g., time spent on a project) and benefits (e.g., referrals and customer satisfaction). CBAs should also account for both tangible and intangible costs/benefits. Monetary benefits of ecosystem services considered in a CBA can be generated by any of the valuation methods that are described later in this chapter (see the section Methods of Valuing Ecosystem Services).

When listing the costs and benefits of a policy or project, it is important to avoid double counting. For example, when quantifying the ecosystem service provided by bats that eat and control agricultural pests, it is possible to estimate the value of this service using different methods. One might, for example, use the market value of a crop to estimate the value of pest control based on crop loss caused by a pest. Alternatively, one might estimate the value of pest control via the cost of pesticides a farmer needs to use if natural control agents (e.g., insects, bats, birds) are not available to provide the service. While both of these methods are valid ways to estimate the value of pest control, it would be inappropriate for a CBA to "double count" by including both the value of crop production and the value of pesticide use avoided.

STEP 5: *Calculate net present value (NPV).* The costs and benefits of any CBA must be expressed for a comparable point in time, which is usually taken to be the present. Converting costs and benefits to present values is necessary for two reasons: First, inflation makes it difficult to compare value at different points in time. The purchasing power that US$1 had 10 years ago is not the same as the purchasing power of that same dollar now, nor will it be the same as the purchasing power 10 years from now. Therefore, the purchasing power of a U.S. dollar at different points in time must be standardized so that the values are comparable. Second, earning potential makes money more valuable now. A U.S. dollar available to be spent now can be invested and used to earn interest for 10 years. As a result, it is worth more than the same dollar will be worth 10 years from now.

To estimate the present value of costs or benefits that have been incurred in the past, one can use any number of published indices that compare the purchasing power of money at different points in time (**Video 7.4**). For example, the Consumer Price Index (CPI) published by the U.S. Bureau of Labor Statistics can be used to convert the value of dollars spent in the past into their present value. In order to estimate the present value of costs or benefits that might be incurred in the future, economists use a tool called **discounting**. Assume that a U.S. dollar invested for t years will grow to a value of $(1 + r)^t$, where r is the interest rate. The amount of money that you would need to deposit now so that it would grow to be one dollar t years in the future is $(1 + r)^{-t}$. This value is called the discounted value, or present value, of the dollar. There is no common interest rate that is used for discounting; it usually ranges between 2% and 7% depending on the assumptions of the CBA.

Once the present values of all costs and benefits have been calculated, it is straightforward to sum up these costs and benefits to obtain the NPV for each alternative policy or project being considered. This is done by subtracting the sum total of all costs from the sum total of all benefits.

VIDEO 7.4 Watch this tutorial on discounting.
oup-arc.com/e/cardinale-v7.4

STEP 6: *Perform sensitivity analyses.* It is important to check how sensitive the results are to the accuracy of estimates being used in the CBA, as well as the assumptions being made. For example, if there is a monetary estimate of an ecosystem service that is questionable, or known to have a high uncertainty, then it might make sense to re-estimate the NPV of the policy or project after simulating a 10%, 20%, or even greater change in the estimate. If the overall viability of the project, and its ranking relative to other alternatives, remains unchanged, then you know the CBA is insensitive to this particular estimate. Similarly, one could vary the interest rate assumed in discounting, to determine whether assumptions about the growth of costs and benefits in the future influence one's conclusions about the viability of a project. If they do, then one should consider these assumptions to be associated with some degree of risk. When these risks have important consequences, a formal risk assessment could be undertaken using Monte Carlo simulations to assess the probabilities that a project will be viable under a varying set of assumptions.[20]

STEP 7: *Make a decision and recommendation.* The final decision in a CBA is based on the NPV and its associated risk for each alternative policy or project. The goal is to choose the alternative that has the highest net present value and lowest possible risk.

Although CBA is a widely practiced decision-making tool, this form of decision making has been criticized from perspectives outside of economics, particularly from those in the environmental sciences and noneconomic social sciences. Some of the more common criticisms include the following: (1) Estimating the costs and benefits in a CBA is more of an art form than a science, and the large degree of imprecision in valuation estimates often limits the utility of the method. (2) Standard CBAs do not distinguish between subpopulations in a society—those who received benefits and those who endure the costs of a project—so CBAs can be insensitive to issues of environmental injustice that result when market failures generate externalities that disproportionately impact one part of society (e.g., the poor) more than another (e.g., the rich). (3) Standard CBAs do not take any particular note of irreversible actions in development decisions. For example, while a policy or project may drive a species to extinction, this irreversible action does not carry any additional weight in decision making within a standard CBA, which simply calculates whether the benefits of the project outweigh the costs of causing that extinction. (4) CBA is a rather coarse tool that can be easily manipulated by those performing the analyses to give a desired outcome. For example, it is relatively easy to manipulate the final project decision by changing the project boundary or altering the discount rate. For this reason, transparency and scrutiny of the assumptions by all stakeholders influenced by the final recommendation are keys to success.

Multicriteria decision analysis

The shortcomings of cost-benefit analysis have led many to call for improved methods of decision making. One such method—multicriteria decision analysis (MCDA)—has become common (**Video 7.5**). MCDA is another type of structured decision making, like that introduced in Chapter 2, which provides a formal process for evaluating multiple, often conflicting criteria when making decisions about a project or policy option. Economic value is usually one of the criteria considered in an MCDA; however, other criteria that sometimes lie in direct conflict with economic values (e.g., time, environmental justice, quality of project) can also be considered when evaluating options.

VIDEO 7.5 Watch this tutorial on MCDA.
oup-arc.com/e/cardinale-v7.5

Conflicting criteria can complicate decision making. For example, when we look to purchase a car, we might consider criteria such as the cost, comfort, safety, and fuel economy. Yet, because the least expensive car is rarely the most comfortable or safest one, some of these criteria are in conflict with one another.

Another example is an investment portfolio. The goal of creating an investment portfolio is to either maximize returns subject to an acceptable level of risk, or to minimize risk subject to an acceptable rate of return. But because stocks that have the greatest potential of returns tend to be the most volatile, these two goals (maximize return, minimize risk) are in direct conflict with one another.

While we often weigh multiple criteria implicitly when making decisions, and we may be comfortable basing our decisions only on intuition, intuition is not enough when stakes are high or when there are multiple stakeholders who will be affected by a decision. In these instances, it is necessary to have a more formal process for decision making that can properly structure the problem and explicitly evaluate multiple criteria. MCDA is such a process. It includes multiple criteria that are important to stakeholders, has stakeholders score the importance of those criteria with respect to each other, and then calculates the overall values of several alternative policies or projects. Unlike a cost-benefit analysis, MCDA can show how different alternatives affect different criteria.

The initial steps (1 and 2) and final steps (6 and 7) of an MCDA and are similar to those of a typical cost-benefit analysis (see Figure 7.5).[21] What makes MCDA different from a CBA are the middle steps (3–5):

STEP 3: *Choose the variety of valuation criteria.* In contrast to a CBA, which uses a single unit of valuation, an MCDA considers a variety of valuation criteria in the analyses. Step 3 of the MCDA process is for all stakeholders involved in making a decision about a policy or project to work together to define the different values that are important to them. One of these valuation criteria is usually the net economic value of the project, but other criteria deemed to be important might include time to complete the project, ability to measure the effectiveness or quality of the project, collateral impacts of externalities, and effects on social justice. In principle, any number of criteria may be used in the analysis, and the important thing is that all stakeholders agree that these criteria are the appropriate ones to consider. In practice, considering a large number of valuation criteria (more than 6–8) can become cumbersome to agree on, and it often leads to redundancy among criteria.

STEP 4: *Weight and score the valuation criteria.* After defining a set of values, the next step is for stakeholders to assign weights to the different valuation criteria. Different stakeholders will almost certainly wish to assign different weights to different criteria. For example, some stakeholders may place a higher value on economic criteria than aspects of environmental justice; thus, they would want to assign a higher weight to the former in the analyses. Numerous weighting options exist, one of which is to simply assign to each criterion a value that represents the mean (or median) weight of all stakeholders. Once the weighting of valuation criteria has been completed, the next step is for stakeholders to score the criteria. To score criteria, one chooses a scale (e.g., a 10-point scale, or a 100-point scale) and then assigns ratings as 10 (or 100) = most important, 1 = least important. The simplest way to aggregate the scores of different stakeholders is to take a mean (or median) score for each criterion.

STEP 5: *Calculate aggregate scores.* The fifth step of an MCDA is to multiply the criteria weights by their scores to get the aggregate value for each criterion being considered. For example, say we are considering two alternative projects, and stakeholders have decided on the weighting values for four criteria that will be considered: (1) the net economic value of the project, (2) the disproportionately negative impact the project might have on a poor Indigenous population

Criteria, i	Mean weight, W_i	Alternative 1: Mean score, S_i	Alternative 1: Aggregate score, $W_i \times S_i$	Alternative 2: Mean score, S_i	Alternative 2: Aggregate score, $W_i \times S_i$
Economic value	30	81	2430	40	1200
Environmental justice	30	75	2250	95	2850
Probability of success	20	80	1600	90	1800
Time to completion	20	68	1360	30	600

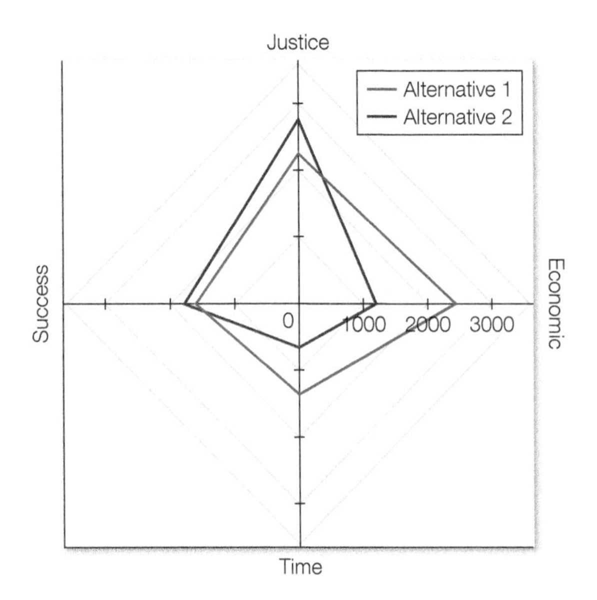

Figure 7.6 An example of output of a multicriteria decision analysis (MCDA) that attempts to optimize trade-offs among four different criteria being considered for two alternative projects. The web diagram (lower left) clearly shows differences and trade-offs among the two alternatives.

(i.e., environmental justice), (3) the probability that the project will prove successful, and (4) the time it takes to complete the project (**Figure 7.6**). For alternative project 1, stakeholders score the net economic value of the project highest, and the other three criteria lower, because the project is believed to have a high probability of success and low probability of negative externalities that might affect the Indigenous population. For project 2, however, economic value scores lower, and other factors score higher, because the project is expected to have a lower chance of success and a high probability of creating externalities that negatively affect the Indigenous population. While the MCDA does not produce a final number that can be used to make a decision (as does CBA), it does make the trade-offs among criteria immediately apparent and open for discussion to help stakeholders understand the choices they are making when choosing among alternative projects.

While MCDA is a transparent method that accommodates complex situations and intentions, it has limitations. Namely, it relies on personal and sometimes arbitrary valuations of criteria. Even so, it may still be the best form of decision analysis when those involved have conflicting objectives, and is preferred over cost-benefit analysis when arbitration and cooperation are essential among stakeholders.[22]

Total Economic Value (TEV) Framework

Now that we have introduced common forms of decision making, it's time to turn attention to the methods of valuation that are used to get the numbers that go into those decision-making analyses. Any proper method of valuation of biodiversity must account for the full value of all goods and services that ecosystems provide to society. The value of certain kinds of services can be captured in existing markets where they are traded and sold. The value of other kinds of services are not captured in existing markets and must instead be quantified using tools from nonmarket analyses. The value of yet another kind of ecosystem service cannot be quantified in the present but must instead be valued as an option that is saved for use in the future. All of these values must be summed up to get a proper accounting of the value of ecosystems and the services provided by their resident organisms.

The total economic value (TEV) framework attempts to organize the types of value provided by different ecosystem services, in turn providing a "checklist" that can be used whenever one attempts to comprehensively evaluate the

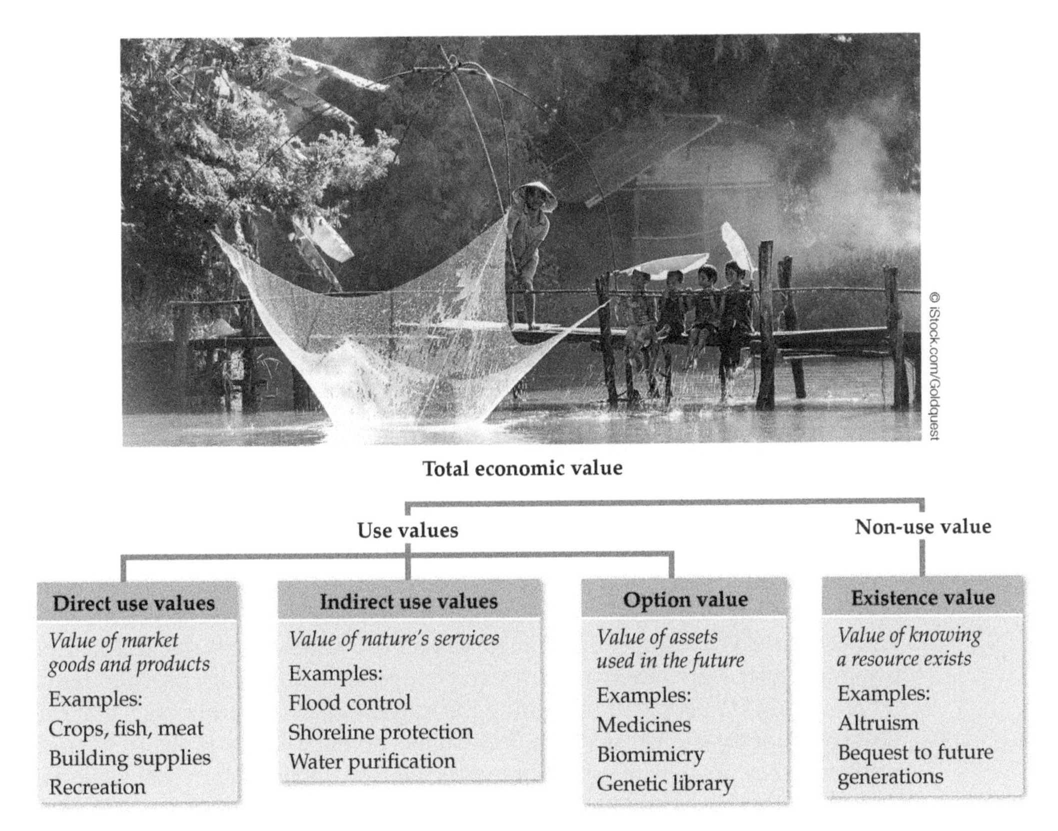

Figure 7.7 The total economic value (TEV) framework is used to quantify the many different values of an ecosystem and its goods and services to people. (From M. J. Groom et al. 2006. *Principles of Conservation Biology*, 3rd ed. Sunderland, MA: Sinauer Associates; based on data in L. Emerton. 1999. *Evaluating Eden Series Discussion Paper No. 5*. London, UK: International Institute for Environment and Development.)

economic benefits that ecosystems provide to society (**Figure 7.7**). A number of different TEV frameworks have been proposed in recent decades,[23–25] and these continue to evolve as new values are discovered and added to older frameworks.[26] Although frameworks vary in detail and application, nearly all begin with the recognition that economic value comes from both the use of an environmental resource (*use value*) and from the value of its existence even in the absence of use (*non-use value*).[27] Use values are generally grouped according to whether they are *direct* or *indirect* and whether they are being used now or represent an *option* for future use. In contrast, nonuse values arise from the continued *existence* of the resource that are unrelated to use. These categories are described further below.

Direct use value

Direct use values are the economic benefits of an ecosystem that are obtained through direct use of the good or service by an individual. Direct use values include both **consumptive** and **nonconsumptive use** of ecosystem services. Consumptive uses involve the extraction of a natural resource for a human purpose, such as the harvesting of fish from a bay for food or the cutting of trees from a forest to produce timber. In contrast, nonconsumptive uses are services that are not extracted but are still used and valued by an individual. Examples include scenic beauty that drives recreation and tourism, or the use of water routes for transportation. Although nonconsumptive uses do not diminish the amount of the resource that is available to others (no extraction is involved), nonconsumptive use can still diminish the quality of a resource through adverse effects like pollution or overuse.

Many of the provisioning services of ecosystems (described in Chapter 6)—such as the production of fisheries, harvesting of timber and other raw materials from a forest, and development and sales of medicines—fall under the category of consumptive direct use values. These services are relatively easy to quantify with traditional economic techniques because they represent goods that are bought and sold in existing markets. Other provisioning services, such as people's extraction of resources to supply themselves with consumable goods like fuelwood, bushmeat, vegetables, fruit, and building materials, are neither bought nor sold in the marketplace.[28,29] Similarly, certain types of cultural ecosystem services (e.g., outdoor recreation) represent nonconsumptive uses of ecosystems that are neither bought nor sold in markets. Ecosystem goods and services that are not traded in markets must be valued with nonmarket techniques that are detailed later in this chapter.

Indirect use value

Indirect use values are the benefits that individuals derive from ecosystems indirectly, rather than via direct use or extraction of a resource.[30] The indirect benefits of ecosystem services usually arise through their support or protection of goods or services that do have direct measurable value. For example, coastal and estuarine wetlands, such as tropical mangroves and temperate marshlands, provide "natural protection" to people by mitigating the impacts of storms and floods that can destroy property and cause death. Individuals do not extract "protection" from wetlands; they do, however, benefit indirectly from the protection that wetlands provide to their lives and property. Similarly, people do not directly extract the ecological processes in ecosystems that generate clear air or water or that produce organic soils. Even so, people indirectly benefit from the ability of ecosystems to provide individuals with clean drinking water, breathable air, and fertile soils that can grow plants. Indirect use values are notoriously difficult to quantify and value; yet, they represent some of the most important services ecosystems provide to humanity, and they often dwarf the direct use value of ecosystems.[3]

Option value

Option value refers to an individual's willingness to pay for maintaining or preserving an ecosystem service so that it can be used for benefit at some time in the future.[31] The concept of option value is perhaps best understood by thinking about risk and considering the value that individuals place on managing risk. If an individual understands that an ecosystem service is important but also perceives there to be uncertainty about the availability or reliability of that service in the future, then he or she might be willing to pay to protect and conserve that service now so that it is more likely to be available for use at some time in the future. Uncertainty about the future availability or reliability of a service could arise for any number of reasons. For example, the species that are essential for providing the service might go extinct, or changing environmental conditions might make it harder to retain the service. Option value is essentially the value of an insurance policy that helps mitigate risk in the face of uncertainty (see Box 7.1 for examples).

Existence value

Existence value is the non-use value that people place on simply knowing that something exists, even if they will never use it or see it. For example, many people throughout the world care about wildlife, plants, or entire ecosystems and are concerned about their protection even though they may not expect, need, or even desire to visit and see these things personally. Existence value is the amount that people are willing to pay to prevent species from going extinct, habitats from being destroyed, and genetic variation from being lost.[32] Existence

BOX 7.1 Case in Point

The Option Value of Biodiversity

A lady beetle preys on herbivorous aphids that are attacking a plant.

Option values of biodiversity call for maintaining biological variation in order to protect unknown future benefits. There are numerous examples of the option value of biodiversity. Here are just a few:

GENETIC IMPROVEMENT OF CROPS Genetic improvement of cultivated plants is necessary to guard against pesticide-resistant insects and increasingly virulent strains of fungi, viruses, and bacteria.[48] Many catastrophic crop failures have been linked to a loss of genetic variability. For example, the 1846 potato blight in Ireland, the 1922 wheat failure in the Soviet Union, and the 1984 outbreak of citrus canker in Florida were all exacerbated by low genetic variability among crop plants. To overcome such problems, scientists constantly breed new, resistant varieties of agricultural species to substitute for susceptible varieties. The source of resistance often comes from genes obtained from wild relatives of crop plants and from local varieties of the domestic species grown by traditional farmers. Development of new crop varieties has a huge economic impact, and the option value of future improvements is similarly great.[49] Genetic improvements in U.S. crops are responsible for increasing harvest values by an average of $8–$15 billion per year.[50] In developing countries, genetic improvements of rice, wheat, and other crops have increased harvests by an estimated US$6–$11 billion per year. Crop breeding has taken on added urgency because of climate change; new varieties of crop plants will need to be developed to tolerate heat stress, summer drought, and higher concentrations of CO_2.

BIOLOGICAL CONTROL Wild species often have option value as biological control agents. Biologists can sometimes control an exotic species by searching the pest species' native habitat for a control agent (predator, parasite, or pathogen). This control agent is then brought to the new locality where, after proper study to make sure it does not have effects on nontarget species, it can be released to control populations of the pest. A classic example is the case of the prickly pear cactus (*Opuntia inermis*), a South American species introduced into Australia for use as a hedgerow plant. The cactus spread out of control and took over millions of hectares of rangeland. In the prickly pear's native habitat, the larvae of a particular moth species (*Cactoblastis cactorum*) feed on the cactus. The moth was successfully introduced into Australia, where it has reduced the cactus to comparative rarity and allowed native species to recover. By conserving biodiversity, options for finding pest control agents in the future are protected (see Figure).

USEFUL PRODUCTS A small proportion of species have enormous potential to be sources of future medicines that treat or cure human disease, to generate important chemicals or industrial products, or to inspire an engineering solution to a key problem. But if these potentially important species were to go extinct before being discovered, it could be a tremendous loss to the global economy, even if the majority of the world's species were preserved. Aldo Leopold once commented:

> *If the biota, in the course of aeons, has built something we like but do not understand, then who but a fool would discard seemingly useless parts? To keep every cog and wheel is the first precaution of intelligent tinkering.*[51]

The diversity of the world's species can be compared to a manual on how to keep the Earth running effectively. The loss of a species is like tearing a page out of the manual. If we ever need the information from that page in the manual to save ourselves and the Earth's other species, the information will have been irretrievably lost.

value is often motivated by a concern that other people (e.g., future generations) or other species should continue to be able to benefit from the conserved resource. An example of this is **bequest value,** which is how much people are willing to pay for the knowledge that something will exist for future generations, even if that something will not be used.

Large animals like whales, pandas, elephants, manatees, lions, and many birds provide good examples of the existence value of nature. Often referred to as **charismatic megafauna**, species like these can elicit strong emotional responses in people who express a willingness to pay for the preservation of these species even though they never expect to see them personally, nor directly benefit from their existence in anyway (Figure 7.8). For example, people who have been asked about the value of threatened species have suggested they would be willing to pay between US$60 and US$92 per year to ensure that species like the grizzly bear (*Ursus arctos*), steelhead trout (*Oncorhynchus mykiss*), and northern spotted owl (*Strix occidentalis caurina*) continue to exist. Another survey of U.S. citizens indicated that they would be willing to spend $31 per person per year to protect their national symbol, the bald eagle (*Haliaeetus leucocephalus*)—a bird whose populations have suffered significant declines in the past but are now rebounding, and which were removed from protection under the U.S. Endangered Species Act in 2007.[33] This amounts to more than $9 billion per year in total if multiplied by the number of people in the United States. Existence value can also be attached to ecosystems, such as tropical rain forests or coral reefs, and to areas of scenic beauty. Growing numbers of people contribute large sums of money annually to organizations like The Nature Conservancy ($620 million in 2016), the World Wildlife Fund ($155 million in 2017), and the Sierra Club ($72 million 2017) to ensure the continued existence of habitats—many of which they may never visit or experience themselves.

Estimates of TEV

In one of the grandest attempts to estimate the total economic value of ecosystems, Robert Costanza and colleagues tallied the value of 17 ecosystem services provided by 16 biomes across the planet.[34] The services considered in their analyses included direct use values, such as food production and recreation, and indirect use values, such as pollination, biological control of pests, and climate regulation, as well as select types of option and existence values. In their 1997 paper, the authors concluded, "For the entire biosphere, the value (most of which is outside the market) is estimated to be in the range of US$16–54 trillion per year, with an average of US$33 trillion per year." In 2014, select authors of the

Figure 7.8 Charismatic megafauna often have high existence value, where people are willing to pay for their continued existence even though they will never experience them personally.

TABLE 7.1 Examples of studies that have attempted to quantify the total economic value (TEV) of ecosystems by calculating the various types of values

Ecosystem	TEV	Direct use value	Indirect use value	Options value	Existence value
Forests in Mexico[36]	$42,685,004,179 nationally per year	$34,267,588,011 (80.3%)	$7,628,704,777 (17.9%)	$667,556,950 (1.6%)	$121,154,441 (<1%)
Forests in Turkey[75]	$1,831,809,157 nationally per year	$1,494,762,715 (81.6%)	$196,078,142 (10.7%)	$139,260,044 (7.6%)	$1,708,256 (<1%)
Tropical forests in Brazil[76]	$2,012 per ha per year	$948 (47.1%)	$715 (35.5%)	$31 (1.5%)	$318 (15.8%)
Mangrove forests in Indonesia[37]	$379,616 per ha per year	$13,973 (3.7%)	$362,835 (95.6%)	$1,240 (<1%)	$1,568 (<1%)
UNESCO World Heritage Site—Viñales Valley, Cuba[77]	$281,803 per ha	$14,075 (5.0%)	$34,586 (12.3%)	$64,534 (22.9%)	$168,608 (59.8%)

Note: Monetary estimates from the original studies have been converted to 2018 U.S. dollars. Values in parentheses represents percentages of TEV.

original paper updated their estimate of the global value of ecosystem services to $125–$145 trillion per year,[35] which is larger than the global gross domestic product in 2011. Another way of saying this is that people are obtaining, mostly for free, more ecosystem services than the global economy could pay for.

While Costanza and colleagues' numbers have been controversial, their general conclusion that the total economic value of ecosystems is exceptionally large has generally been born out in other studies. Indeed, studies performed at national or regional scales on individual ecosystems, where data are more complete and accounting methods are more reliable, have corroborated this general conclusion (Table 7.1). For example, the estimated value of forests in Mexico, not including the direct use value of timber, is US$42 billion,[36] representing more than 4% of the country's gross domestic product.

Another general conclusion of studies that have attempted to quantify TEV is that a large fraction of the value of ecosystems tends to lie outside of their direct use value. For example, of the estimated US$380,000 value per hectare of mangrove forests in Indonesia, the largest fraction stems from indirect use value associated with the protection that mangroves provide against floods and property damage.[37] In most countries for which data are available, the market values of ecosystems associated with timber and fuelwood production represent less than one-third of the total economic value of those ecosystems, including nonmarketed values such as carbon sequestration, watershed protection, and recreation (Figure 7.9). Estimates of option value have proven to be a substantial fraction of TEV for ecosystems anticipated to produce future pharmaceuticals, and existence values have proven to be a large part of TEV for systems that hold high cultural value (see Table 7.1).[3,38]

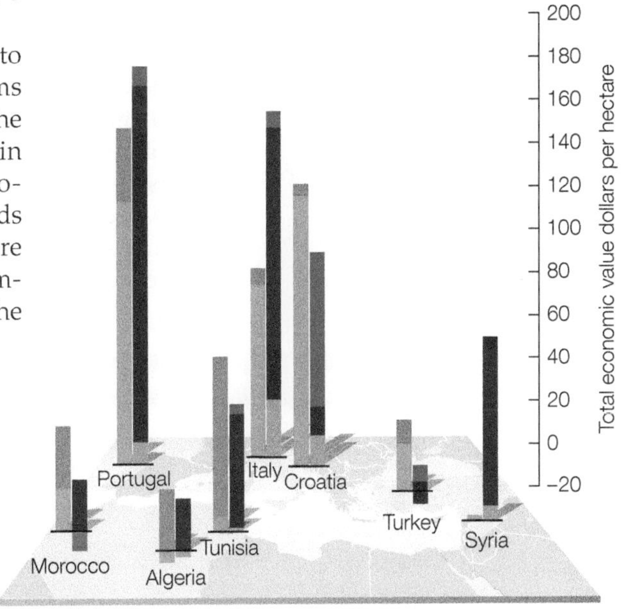

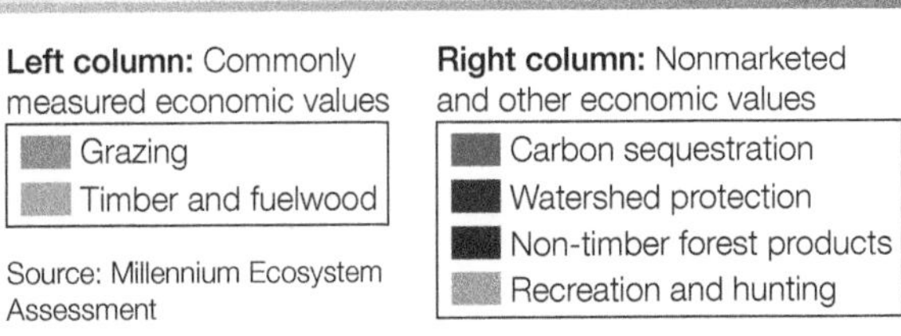

Figure 7.9 The value of forest ecosystem services in select countries. Note that in most countries considered, the market value of goods is less than the nonmarketed value of important ecosystem services. (After Millennium Ecosystem Assessment. 2005. *Ecosystems and Human Well-being: Synthesis*. Washington, DC: Island Press.)

A good example of TEV being used to justify biodiversity conservation comes from a government-funded study of coral reefs in the Maldives.[39] The Maldives is a tropical nation in the Indian Ocean composed of 26 ring-shaped atolls. These atolls have some of the highest biodiversity in the world, and the nation boasts the seventh-largest area of coral reefs in the world. The Maldives Ministry of Housing, Transport and Environment initiated the Atoll Ecosystem Conservation project to summarize the economic value of coastal and marine diversity and compare the value of economic sectors like fisheries and tourism that depend on coastal and marine biodiversity. The study concluded that tourism generated about US$764 million per year (67% of GDP), while fisheries were valuated at US$67 million (8.5% of GDP). The indirect use value of coral reefs, mainly through shoreline protection, was estimated at US$2 billion to US$3 billion, and their existence value was estimated to be US$232 million. These values were sufficiently large that the president of the Maldives at the time used the report to emphasize the immense economic value that biodiversity has for the national economy, instructing the tourism and fisheries sectors to make conservation a chief priority.

Methods of Valuing Ecosystem Services

Now that we have described the components of total economic value (TEV), let's turn attention toward the methods that allow us to value ecosystem services using conventional markets as well as nonmarket techniques. The tools and techniques that are available to quantify ecosystem goods and services that contribute to TEV generally fall into one of three categories: methods that rely on the revealed preferences of consumers, methods that rely on the stated preferences of consumers, and methods that rely on benefits transfer (Figure 7.10). We now describe each of these in turn.

Revealed preference methods

Revealed preference methods estimate the value of ecosystem goods or services using direct observations of the amount people are willing to pay for them. In other words, the value of an ecosystem service is revealed through some form of consumer behavior. Revealed preference methods can be based on real (existing) markets where goods and services are bought and sold, or they can be based on surrogate (implicit) markets where the values of goods that are bought and sold in markets are used as a proxy to value ecosystem services that are related to those goods. Revealed preference methods can also rely on cost-based

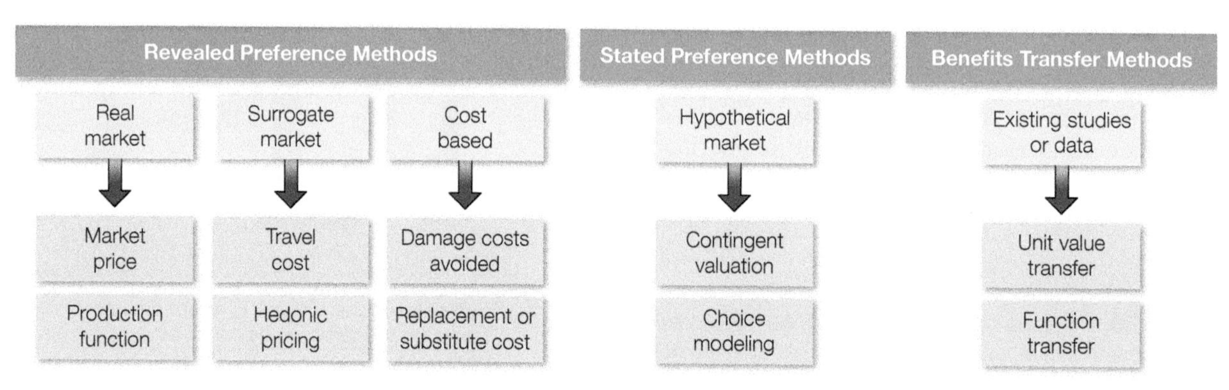

Figure 7.10 A classification of methods that can be used to assign monetary value to various ecosystem services.

techniques that estimate the values of ecosystem services based either on the costs of avoiding damages due to lost services or on the costs of replacing or providing a substitute for the ecosystem service.

Real market techniques

Some ecosystem goods, such as harvested fish or wood, are bought and sold in real (i.e., existing) markets. The value of goods that are traded in real markets can be estimated by standard economic techniques, such as market pricing or production function analyses.

MARKET PRICING The **market pricing method** gives the economic value of either the quantity or quality of an ecosystem good or service that is bought and sold in a commercial market (Table 7.2). The standard tool for measuring the value of resources traded in the marketplace is a supply-and-demand curve, which can be used to estimate **consumer surplus** and **producer surplus** based on market price and quantity data (Figure 7.11). Supply and demand is the most fundamental concept of economics. Demand refers to the quantity of a good or service (x-axis) that people are willing to buy at a given price (y-axis). The relationship between the price and quantity demanded is known as the demand relationship (red line labeled D). The **law of demand** states that if all other factors remain equal, people's demand for a good or service will decrease as the price of the good or service increases. This is why the demand curve slopes downward.

Supply represents the quantity (or quality) of the good or service that nature can offer to the market (e.g., tons of fish that can be harvested and sold). The quantity supplied refers to the amount of a good or service that producers are willing to supply when receiving a certain price. The correlation between price and how much of a good or service is supplied to the market is known as the supply curve (blue line labeled S). Because producers would like to sell more of a good or service at a higher price, the supply curve slopes upward.

At an equilibrium where there are no disruptions to a market, the intersection of the supply and demand curves (point B) sets a market price (P) and quantity (Q) of production for a good or service. At this equilibrium, the net economic benefit, or **economic surplus** of the good or service, can be calculated as the sum of consumer and producer surpluses. Consumer surplus is defined as the difference between the total amount that consumers are willing to pay for a good or service (indicated by the demand curve) and the total amount they actually do pay (i.e., the market price, P). Consumer surplus, which is given by the area under the demand curve and above the price (in triangle ABP), is a measure of the economic benefit that people gain from using the ecosystem good or service. In contrast, the economic benefit to producers is measured as producer surplus, which is the area above the supply curve and below the market price (triangle BEP). This area defines the difference between the total amount earned from a good (price times quantity sold) and the costs of production. The sum of the economic benefits to both consumers and producers represents the total market value of an ecosystem good or service.

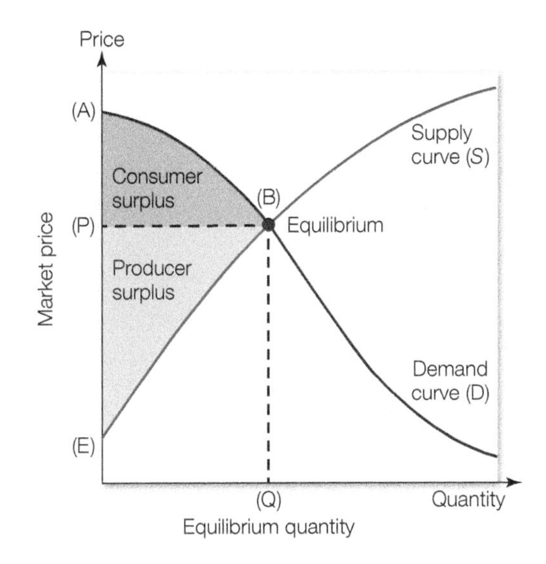

Figure 7.11 The market pricing method quantifies the value of an ecosystem service by using the supply curve (blue line S) and demand curve (red line D) to estimate the total net economic benefit (economic surplus) of the service. The intersection of the supply and demand curves (point B) sets a price P and quantity Q for the service. The net economic benefit of the service can be calculated as the sum of consumer surplus (triangle ABP) and producer surplus (triangle BEP). (SilverStar/Wikimedia/CC BY 2.5.)

TABLE 7.2 Summary of valuation methods used to assign monetary value to ecosystem services

Method	Overview	Types of value	Example applications
Market price	Estimates value of ecosystem services that are bought and sold in commercial markets. ▶ VIDEO 7.6 *Valuation of Ecosystem Services: Market Based Valuation Method* oup-arc.com/e/cardinale-v7.6	Direct use value	Calculating the economic value of wild fisheries[56] or forest timber harvests[57]
Production function	Estimates value of ecosystem services that contribute to the production of goods that are bought and sold in commercial markets. ▶ VIDEO 7.7 *Economic Valuation Theory* oup-arc.com/e/cardinale-v7.7	Direct and indirect use values	Estimating value of natural pollinators for crop yield,[58] or wetland habitat for fish production[30]
Travel cost	Estimates people's willingness to pay to visit and recreate at an ecosystem. ▶ VIDEO 7.8 *Valuation of Ecosystem Services: Travel Cost Method* oup-arc.com/e/cardinale-v7.8	Indirect use value	Estimating recreational value of coral reefs,[59] fishing,[60] or a natural wonder[61]
Hedonic pricing	Evaluates the impact of ecosystem services on properties bought and sold in commercial markets. ▶ VIDEO 7.9 *Valuation of Ecosystem Services: Hedonic Pricing Method* oup-arc.com/e/cardinale-v7.9	Indirect use value	Use of housing sales prices to estimate the value of lake water quality,[62] urban green spaces,[63] or views of nature[64]
Damage cost avoided	Estimates the cost of avoiding damages that would occur because of loss of an ecosystem service. ▶ VIDEO 7.10 *Valuation of Ecosystem Services: Avoided Cost Method* oup-arc.com/e/cardinale-v7.10	Indirect use and option values	Estimating cost of destroying mangrove forests for storm protection[65] or cost of an invasive species[66]
Replacement or substitute cost	Estimates the cost of replacing an ecosystem service or of providing a substitute service. ▶ VIDEO 7.11 *Valuation of Ecosystem Services: Replacement Cost Method* oup-arc.com/e/cardinale-v7.11	Indirect use and option values	Estimating cost to replace pollution removal from a river[67] or to substitute infrastructure for natural flood control[68]
Contingent valuation	Estimates people's willingness to pay for an ecosystem service on a hypothetical market. ▶ VIDEO 7.12 *Valuation of Ecosystem Services: Contingent Valuation* oup-arc.com/e/cardinale-v7.12	Existence value	Estimating the value of a threatened species[69] or impact of an oil spill on a pristine ecosystem[70]
Choice modeling	Quantifies people's preferences between alternative levels of ecosystem services at a given price or cost to the individual. ▶ VIDEO 7.13 *Valuation of Ecosystem Services: Choice Experiments* oup-arc.com/e/cardinale-v7.13	Indirect use, option, and existence values	Estimating people's willingness to pay for bird watching,[71] or acceptability of alternative management scenarios[72]
Benefits transfer	Estimates values of an ecosystem service by transferring information from studies already completed in another location and/or context. ▶ VIDEO 7.14 *Valuation of Ecosystem Services: Benefits Transfer Study* oup-arc.com/e/cardinale-v7.14	All	Estimating value of threatened species in developing countries using studies performed in developed countries,[73] or willingness to pay for water quality in one location based on studies performed at another[74]

Box 7.2 gives an example of how the market pricing method can be used to measure the value of an ecosystem service. In this example, we estimate the economic benefit that would result from increased harvest of yellow perch (*Perca flavescens*) if pollution were cleaned up in Lake Erie. The total economic benefit of perch harvest is the sum of consumer and producer surpluses for the supply-and-demand curves representing the quantity of fish harvested from the lake and sold in markets. To estimate the economic benefits of cleaning the lake, one must estimate the difference between economic surplus before the cleanup and economic surplus after the cleanup.

PRODUCTION FUNCTION The **production function method** is used to estimate the economic value of an ecosystem good or service that contributes, along with other inputs, to the production of a marketed good (see Table 7.2). For example, water quality affects the productivity of irrigated agricultural crops, or the costs of purifying municipal drinking water. Thus, the economic benefits of improved water quality can be measured via the increased revenues that result from greater agricultural productivity, or the decreased costs of providing people with clean drinking water. If a natural resource is a factor that contributes to production of a good, then changes in the quantity or quality of the resource will result in changes in production costs and/or the productivity of other inputs. This, in turn, may affect the price and/or quantity supplied of the final good.

The production function method consists of the following two-step procedure:[30] First, the physical effects of changes in an ecological service on an economic activity is determined. Second, the impact of these changes is valued in terms of the corresponding change in a marketed output of the relevant

BOX 7.2 Conservation in Practice

An Example of the Market Pricing Method: Fisheries and Pollution[52]

(This example is adapted from an example of market pricing that can be found at oup-arc.com/e/cardinale-w7.1)

Say hypothetically you want to measure the total economic benefit that would result from increased harvest of yellow perch if pollution were cleaned up in Lake Erie. The total economic benefit of perch harvest for any one condition in Lake Erie (polluted or unpolluted) would be the sum of consumer and producer surplus for the supply and demand curves representing the quantify of fish harvested from the lake and sold in markets. To estimate the economic benefits of cleaning up the lake, one must estimate the difference between economic surplus before the cleanup and economic surplus after the cleanup.

STEP 1: The first step is to use market data to estimate the demand function and consumer surplus for the fish before the closure of the lake for cleanup. To simplify the example, assume a linear demand function, where the initial market price is $5 per pound and the maximum willingness to pay is $10 per pound. The Figure shows the area the researcher wants to estimate—the consumer surplus, or economic benefit to consumers, before Lake Erie is cleaned up.

At $5 per pound, consumers purchased 10,000 pounds of fish per year. Thus, consumers spent a total of $50,000 on fish per year. However, some consumers were willing to pay more than $5.00 per pound and thus received a net economic benefit from purchasing the fish. This is shown by the shaded area on the graph, the consumer surplus. This area is calculated as the area of a right triangle, or ($10 – $5) × 10,000 / 2 = $25,000. This is the total consumer surplus received from the fish before the closure.

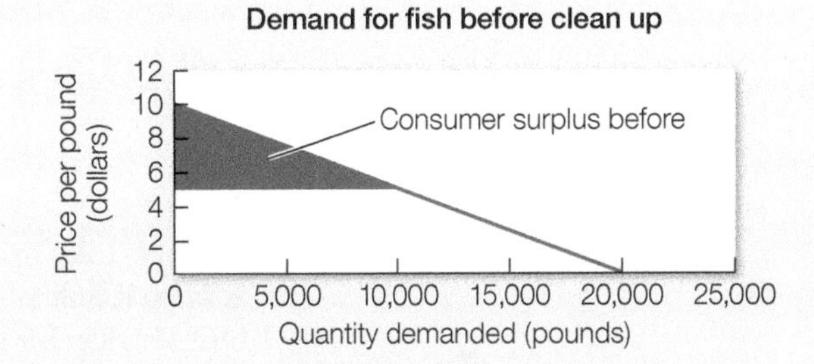

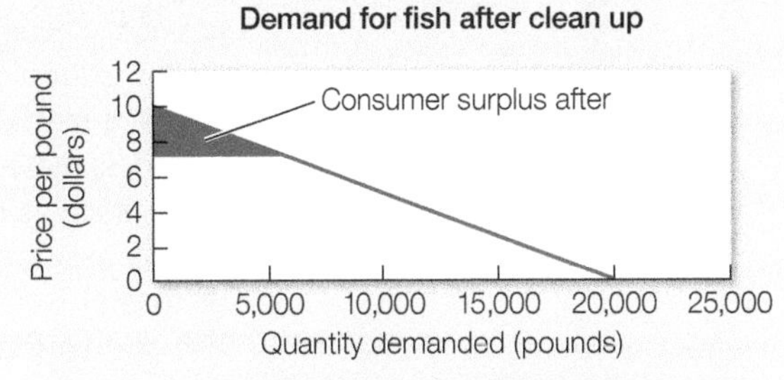

The demand curves relating the price of fish (*y*-axis) to the quantity demanded (*x*-axis) before cleaning up the lake (top panel) and after cleaning up the lake (bottom panel). The difference between consumer surplus before and after represents the value of cleaning up the lake for fish sales. (Courtesy of D. M. King and M. Mazzotta.)

(Continued on next page)

BOX 7.2 Conservation in Practice *(continued)*

STEP 2: The second step is to estimate the market demand function and consumer surplus for the fish after the closure. After the closure, the market price of fish rose from \$5 to \$7 per pound, and the total quantity demanded decreased to 6000 pounds per year. Thus, the economic benefit decreased, as shown in the Figure. The new consumer surplus is calculated as (\$10 – \$7) × 6000 / 2 = \$9,000.

STEP 3: The third step is to estimate the loss in economic benefits to consumers, by subtracting benefits after the closure, \$9,000, from benefits before the closure, \$25,000. Thus, the loss in benefits to consumers is \$16,000.

STEP 4: Because this is a marketed good, the researcher must also consider the losses to producers, in this case the commercial fishermen. This is measured by the loss in producer surplus. As with consumer surplus, the researcher must measure the producer surplus before and after the closure and calculate the difference. Thus, the next step is to estimate the producer surplus before the closure.

Producer surplus is measured by the difference between the total revenues earned from a good and the total variable costs of producing it. Before the closure, 10,000 pounds of fish were caught per year. Fishermen were paid \$1 per pound, so their total revenues were \$10,000 per year. The variable cost to harvest the fish was \$0.50 per pound, so total variable cost was \$5,000 per year. Thus, the producer surplus before the closure was \$10,000 – \$5,000 = \$5,000.

STEP 5: Next, the researcher would measure the producer surplus after the closure. After the closure, 6000 pounds were harvested per year. If the wholesale price remained at \$1, the total revenues after the closure would be \$6,000 per year. If the variable cost increased to \$0.60 because boats had to travel farther to fish, the total variable cost after the closure was \$3,600. Thus, the producer surplus after the closure was \$6,000 – \$3,600 = \$2,400.

STEP 6: The next step is to calculate the loss in producer surplus due to the closure. This is equal to \$5,000 – \$2,400 = \$2,600. Note that this example is based on assumptions that greatly simplify the analysis, for the sake of clarity. Certain factors might make the analysis more complicated. For example, some fishermen might switch to another fishery after the closure, and thus losses would be lower.

STEP 7: The final step is to calculate the total economic losses due to the closure—the sum of lost consumer surplus and lost producer surplus. The total loss is \$16,000 + \$2,600 = \$18,600. Thus, the benefits of cleaning up pollution in order to reopen the area are equal to \$18,600.

activity. Like any other input, its value derives from the value of the good or service it helps to produce. The value of a change in the provision of an ecosystem service, for example, is the impact it has on the value of the good/service it helps to produce. Formally, it is the called value of the marginal physical product of the ecosystem service.

For example, say we want to quantify the factors that influence the production of a marketed good, such as the amount of fish catch off a coastal zone. For fisheries, there are a number of standard inputs that might be used to predict fish catch, such as the number of vessels used or amount of time spent fishing. Now further assume that *F*—the amount of fish caught—is also influenced by the area of coastal wetlands, such as marshlands or mangroves, that augment offshore fish populations by serving as both a spawning ground and a nursery for fry. If we assume simple linear relationships between the inputs and fish catch, then the influence of each factor contributing to fish catch can be predicted as:

$$F = b_0 + b_1 \times \mathit{vessels} + b_2 \times \mathit{time\ fishing} + b_3 \times \mathit{area\ of\ wetlands} \ldots + b_i \times \mathit{other\ inputs} \quad (7.3)$$

where each coefficient b_i quantifies the influence a factor on the right side of the equation has on fish catch (on the left side of the equation). In this way, the ecosystem services provided by coastal wetlands can be quantified as a productive input to the function. Note that relationships between the market variable (left side) and an input on the right side need not be linear, and in fact, production functions are often more complex and nonlinear.

Surrogate market techniques

Some ecosystem goods and services are not directly bought and sold in markets; thus, one cannot use real market techniques. Even so, the prices people are willing to pay can sometimes be estimated from the sale of related goods that are themselves bought and sold in markets. For example, while there may be no direct market for living coral as a good, people may spend more money on travel, accommodations, food, and other marketed items in order to visit a healthy coral reef with lots of biodiversity. While there is no direct market for a beautiful aesthetic view of nature, people may nevertheless pay a higher price for a home that has a spectacular view of the ocean. In the absence of a market, people's willingness to pay for a nonmarket good or service can sometimes be estimated from surrogate markets that people do pay for (travel costs, housing prices). The travel cost method and hedonic pricing method are two techniques commonly used for **surrogate market techniques**.

TRAVEL COST The **travel cost method** is most often used to estimate the economic values of ecosystem services that are associated with tourism and recreation (see Table 7.2). The method attempts to quantify the ecosystem services of visited sites, which are not traded in existing markets, based on the travel and time costs that individuals incur to visit the sites.[27] The basic premise is that the time and travel expenses people incur to visit a site represent the "price" of access to the site, which is then equated to the value of the site. People's willingness to pay to visit can be estimated based on the number of visits they make at different travel costs.

The travel cost method generally uses one of two approaches to collect data. Individual travel cost methods use surveys to collect data on the costs incurred by each individual traveling to the recreational site or amenity. Zonal travel cost methods define zones surrounding the site that represent different travel costs (concentric circles, zip codes, etc.) and then collect information on the number of visitors from each zone. The number of visits made to a site is then plotted against the "price" paid for a visit (Figure 7.12). This plot is essentially a demand curve that quantifies people's willingness to pay for the site and its related services based on the quantity of visits demanded at differing prices.[24] The law of demand states that people's demand for a good or service will decrease as the

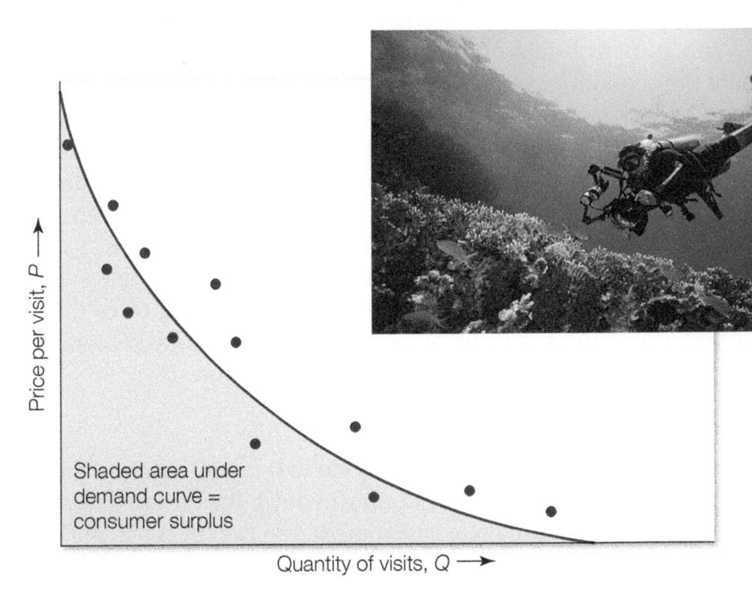

Figure 7.12 The travel cost method generates a demand curve by relating the price per visit to an ecosystem or site (y-axis) to the quantity of visits (x-axis) made by different individuals (data points). The area under the curve quantifies consumer surplus, which is a measure of the economic value of the site to consumers.

price of the good or service increases, which is why the demand curve slopes downward. The area under the demand curve is the consumer surplus—that is, the economic benefit of the site to visitors.

Travel cost methods are not only used to estimate the value of an ecosystem; they are commonly used to determine how people's willingness to pay is influenced by the characteristics of the ecosystem. For example, say we had estimates of consumer surplus (*CS*) from many different coral reefs and wanted to know how much people's willingness to pay to visit a reef is influenced by their desire to see sharks (sightings reported per day) as opposed to the quality of accommodations (measured by online ratings). We could model consumer surplus as a function of the ranking of accommodations, shark sightings per day, and any other factors that are known to drive tourism to coral reefs:

$$CS = b_0 + b_1 \times \textit{accommodation ratings} + b_2 \times \textit{shark sightings per day} \ldots + b_i \times \textit{other factors} \quad (7.4)$$

Assuming we included all of the major factors that influence people's willingness to visit coral reefs, Equation 7.4 could be used to estimate the value of sharks for reef tourism.

HEDONIC PRICING The **hedonic pricing method** uses variation in pricing to determine how people value certain experiences or characteristics that give them pleasure; for example, variation in housing or other real estate prices can be used to estimate the value of local environmental attributes that surround properties (see Table 7.2). The basic premise of hedonic pricing is that the price of a marketed good is a function of its various characteristics, and each of those characteristics contributes some value to the good. For example, the total price of a car reflects a bundle of characteristics of that car, such as its comfort level, style, luxury, or fuel economy. We can estimate the value of each individual characteristic of the car by looking at how people's willingness to pay changes when characteristics of the car change, such as when we add or delete options or when we compare the cost of cars that do versus those that do not have certain characteristics.

Similarly, a typical hedonic price model assumes that the sale price of a house is a function of a bundle of characteristics of that house (**Figure 7.13**). These include transaction characteristics such as the year and season of sale, property characteristics such as the age and size of the house, neighborhood characteristics such as crime rates and quality of the school system, and environmental characteristics such as whether the house is close to a green space or an industrial park. With sufficient data on property sale prices and characteristics, one can estimate the economic benefits or costs attributable to specific aspects of environmental quality (air, water, or noise pollution) and/or environmental amenities (aesthetic views, proximity to recreational sites). The assumption is that, all else being equal, people will pay more for properties that have desirable environmental characteristics, and less for undesirable characteristics. For example, hedonic pricing has been used to estimate the value of water quality by looking at how housing prices vary around lakes that have experienced increasing levels of nutrient pollution.[40,41] It has similarly been used to estimate the costs of air pollution[42] and the value of urban green spaces[43] by comparing property values that vary in distance to an industrial site or forest.

Cost-based techniques

Cost-based techniques are most often used to value the regulating services of ecosystems (see Table 7.2) and are based on estimating the cost of avoiding damages due to lost services (damage cost avoided), the cost of replacing environmental assets after they are lost (replacement cost), or the cost of providing substitute services (substitute cost). These methods are all variants of the

Hedonic Price Model

$$\$_i = b_0 + b_1T_i + b_2P_i + b_3N_i + b_3E_i + \varepsilon_i$$

$\$_i$ - Sale price of property i

T_i - Transaction characteristics (e.g., year and season of sale)

P_i - Property characteristics (e.g., house age, size)

N_i - Neighborhood characteristics (e.g., school quality, crime rate)

E_i - Environmental characteristics (e.g., amount of green space, pollution)

ε_i - Residual unexplained error

Figure 7.13 The hedonic pricing method uses variation in housing or other real estate prices to estimate the value of different characteristics of the property. These characteristics can include local environmental attributes that represent an ecosystem service.

production function method that was described previously. Like the production function method, cost-based techniques ask what impact a change in some ecosystem service, such as a regulating service, would have on the value of goods/services they help to produce. Damage cost avoided, replacement cost, and substitute cost are just measures of the change in value of final goods and services.

The damage cost avoided method uses either the value of property protected or the cost of actions taken to avoid damages, as a measure of the benefits provided by an ecosystem. For example, if a wetland protects adjacent property from a potentially damaging flood, the benefits of flood protection can be estimated by the damages avoided if the flooding does not occur or, alternatively, by the expenditures a property owner must make to protect the property from flooding (e.g., insurance or infrastructure costs of protection).

The replacement cost method estimates the value of an ecosystem or its services by calculating what it would cost to replace the ecosystem if it were lost or damaged. Similarly, the substitute cost method estimates the value of an ecosystem or its services by calculating what it would cost to provide some alternative.

For example, in Chapter 6 (see Box 6.1) we described how officials in New York City weighed the substitution cost of constructing expensive water treatment facilities against the cost of protecting natural watersheds in the Catskill Mountains, where their water is naturally filtered and purified at a much lower cost. Similarly, the cost of building a retaining wall or levee to ensure flood protection services might be used to estimate the substitute cost for flood protection provided by a natural wetland.

Cost-based methods assume that the costs of avoiding damages or replacing natural assets or their services provide useful estimates of the minimum value of these services. In other words, the assumption is that if people incur costs to avoid damages caused by lost ecosystem services, or to replace the services of ecosystems, then those services must be worth at least what people paid to replace them. While this assumption may not always hold, in many cases it may be reasonable to make such assumptions, and measures of damage cost avoided or replacement cost are generally much easier to estimate than people's willingness to pay for certain ecosystem services. The methods can be appropriately applied to a number of important scenarios, such as the following:

- The cost of removing sediment from rivers after storms or stream erosion can be used as a measure of the soil-retaining value of surrounding forests or adjacent wetlands.
- The cost of building levees and seawalls can be used to value the storm protection provided by coastal barrier islands and wetlands.
- The cost of chemical treatment and physical filtration of drinking water can be used as a value for a wetland's or forest's natural water treatment.
- The cost of cleaning up point source pollution of rivers and lakes can be used to measure the value of pollution control improvements.[44]

Stated preference methods

The revealed preference methods described previously estimate the value of ecosystem services by using direct observations that reveal the amount people are willing to pay. However, many ecosystem services are not traded in markets and are not closely related to any marketed goods. For example, many threatened and endangered species are not directly tied to marketed goods; thus, people cannot "reveal" what they are willing to pay for a threatened species through their market purchases or actions. Nevertheless, people may value these species for their option and existence values (see Figure 7.7). In cases like these, stated preference methods can be used.

Stated preference methods utilize questionnaires that directly ask respondents for their **willingness to pay (WTP)** or **willingness to accept compensation (WTAC)**. Stated preference methods offer a direct survey approach for estimating individual or household preferences and, more specifically, WTP or WTAC for changes in the provision of nonmarket goods, which are related to respondents' underlying preferences.[27] Two of the most common stated preference methods are contingent valuation and choice modeling.

CONTINGENT VALUATION **Contingent valuation** is most often used to estimate the non-use values of ecosystems (see Table 7.2), though it can be used to quantify most any type of ecosystem service. The contingent valuation method involves directly asking people, in a survey, their WTP for a specific ecosystem service or, in some cases, their WTAC to give up an ecosystem service (**Figure 7.14**). The method is called *contingent* valuation because people are asked to state their willingness to pay or to accept compensation, contingent on a hypothetical scenario and description of the ecosystem service. The method is referred to as a *stated* preference because it asks people to directly state their values, rather than inferring values from the actual choices they make as the revealed preference methods do.

Contingent valuation is one of the only known ways to assign monetary values to non-use values of the environment—values that do not involve market purchases or even direct participation by a consumer. Such values include everything from the basic life support functions provided by ecosystems, to the enjoyment of a scenic vista or a wilderness experience, to appreciating the option to fish or bird watch in the future, or the right to bequest those options to

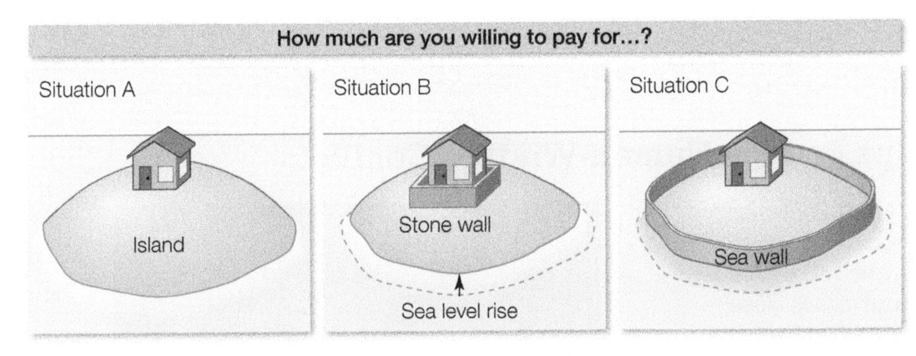

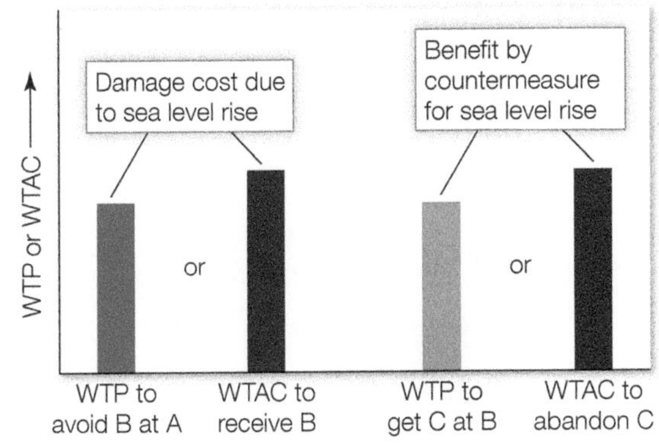

Figure 7.14 Contingent valuation is a stated preference method that involves surveying people to ask how much they would be willing to pay (WTP) for an ecosystem service, or what their willingness is to accept compensation (WTAC) to give up an ecosystem service. (After G. Rincón Polo. 2016. *Air Quality Assessment Workshop, March 16, 2016*. Ecuador: Ministry of Higher Education, Science, Technology and Innovation.)

your grandchildren. It also includes the value people place on simply knowing that giant pandas or whales exist. While it is clear that people value these non-use benefits of nature, such benefits cannot be included in economic analyses unless their dollar values are somehow estimated.

The fact that contingent valuation is based on what people say they would do, as opposed to what people actually do in any real market, gives some economists pause. Some argue the method is misleading because the amounts that people say they are willing to pay (or accept) are often different from what they would actually pay (or accept) in a real market. Despite such concerns, the method has a long history of use by leading economists. For example, when the *Exxon Valdez* struck Prince William Sound's Bligh Reef in 1989 and spilled an estimated 11 million gallons of crude oil, the spill precipitated one of the greatest environmental disasters and wildlife kills in U.S. history. But environmental damage assessment to Prince William Sound was difficult because there was little economic activity in the pristine, undeveloped region. Carson and colleagues[45] performed a contingent valuation analysis for the State of Alaska using a survey of 1599 residents spanning every state in the United States. The median willingness to pay for spill prevention was $31, which was multiplied by the number of U.S. households to produce a damage estimate of $2.8 billion.

The National Oceanic and Atmospheric Administration (NOAA), which was charged with leading the damage assessment, subsequently asked Nobel laureate economist Kenneth Arrow to chair a panel of expert economists to advise NOAA on whether contingent valuation studies were capable of providing sound economic estimates.[46] The panel concluded in its 1993 report that contingent valuation "can produce estimates reliable enough to be the starting point of a judicial process of damage assessment."[47] The *Exxon Valdez* spill went on to become the first case in which contingent valuation was used to quantitatively assess damages for purposes of litigation. Contingent valuation has since been used as the foundation for some of conservation's best success stories (**Box 7.3**).

BOX 7.3 Success Story

Contingent Valuation Helps Reduce Human-Wildlife Conflict

In the South Asian country of Sri Lanka, conflict between humans and elephants kills about 50 people and 150 elephants per year.[53] The conflict occurs when elephants "raid" the crops of poor rural farmers whose land is close to elephant habitat, which is declining because of human population growth and development. The displaced elephants consume large portions of crops, and they damage large areas of farm fields (see Figure). The farmers have few options to manage the elephants and few resources to compensate their losses. In desperation, they turn to poison, traps, and guns to kill the elephants.

In 2016, Sri Lanka's leading insurance company, Ceylinco Insurance, took a bold step to reduce the human-wildlife conflict and enhance elephant conservation.[54] With the help of universities, Ceylinco performed a contingent valuation study in which they surveyed urban residents and asked their willingness to pay additional premiums for life and vehicle insurance, assuming additional premiums would be used to establish a trust fund that would help conserve elephants and minimize conflict with farmers. A substantial proportion of respondents (89%) expressed their willingness to join the proposed insurance plan. Estimates revealed that urban dwellers were willing to pay (in principle) 2012 million Sri Lankan rupees (Rs 2012 million = US$11.5 million) per year in perpetuity for the conservation of wild elephants—a figure that far exceeded the economic value of annual crop and property damage caused by elephants (Rs 1121 million, or US$6.4 million).

Man versus elephant. Villagers try to protect their crops from raiding elephants.

To establish ownership of the plan, farmers were asked to pay a nominal annual fee of Rs 650 (less than US$6). In turn, they were eligible for payments for accidental death, property damage, and crop loss.[55] A commission of 10% was set aside for local government agencies or organizations selected by the Department of Wildlife Conservation to help organize and implement the plan in remote areas.

This story illustrates how a company like Ceylinco Insurance can use its corporate social responsibility to turn a hypothetical market into demand for a new product that helps advance the goals of conservation for both people and wildlife.

CHOICE MODELING The **choice modeling method** is similar to contingent valuation in the sense that it can be used to estimate economic values for virtually any ecosystem service (both use and non-use values). However, it differs from contingent valuation in two ways: First, unlike contingent valuation, choice modeling does not ask people to state their values in dollars. Instead, values are inferred from the hypothetical choices and trade-offs that people make. Second, choice modeling tends to focus more on multidimensional decisions that have multiple outcomes, whereas contingent valuation surveys tend to focus on single variables (see Table 7.2).

The application of choice modeling to valuing multidimensional environmental problems has grown in recent years, and the method is now routinely discussed alongside contingent valuation in the design, analysis, and use of stated preference studies. While there are a number of different approaches that fall under the umbrella of choice modeling, the **choice experiment** has become one of the dominant approaches for measuring the stated preferences for ecosystem goods and services.

In a typical choice experiment, survey respondents are asked to compare and then choose between two groups of ecosystem goods and services, each of which has a different price or cost. The survey usually focuses on tradeoffs among scenarios that have different characteristics; therefore, choice experiments are well suited to aid policy decisions where a set of possible actions might result in different impacts on natural resources or ecosystem services. For example, protecting coastal wetlands may protect against property damage by reducing storm surge and flooding, but it may also decrease the amount of beach habitat available for swimming. A choice experiment can determine which services people value most, either in terms of monetary values, or simply as ranked options.

Benefits transfer method

Benefits transfer is used when it is too expensive or there is too little time available to conduct an original valuation study, yet some measure of economic benefit from ecosystem services is needed. In this case the information gathered from a similar place and condition can be applied as an estimate (see Table 7.2). For example, the economic value of recreational fishing in Florida may be estimated by applying measures of recreational fishing value quantified from a study conducted in South Carolina.

By the nature of **benefit transfer methods**, their use introduces subjectivity and uncertainty into appraisals, which require analysts to make numerous assumptions and judgments in addition to those contained in the original studies. The key question is whether the added subjectivity and uncertainty surrounding the transfer is acceptable and whether the transfer is still, on balance, reliable and informative. Similarity of conditions between the site of the original study and the site to which its results will be applied is key to a good fit. Those conditions include location, type of ecosystem, quality/degradation of the ecosystem, and surrounding human population. The transfer of information will also be most acceptable when the intended action and the action in the original study are the same. It should go without saying that the quality and accuracy of the original study should be investigated before any attempt is made to transfer its results. The people living in the two areas, and especially their willingness to pay, also should be compared, and adjustments should be made for their differences.[27]

One Final Comment

As we end this chapter and Part II of the book, a final comment is warranted. While a disproportionate number of decisions about biodiversity come down to how people perceive the value of nature for themselves and their own good, it is again prudent to point out the challenges and opportunities that exist when distilling nature and its biodiversity down to instrumental value (Box 7.4). People value biodiversity for a wide variety of reasons (see Chapter 5), and any attempt to distill the value of ecosystems and their biodiversity down to a single number or metric that is based on instrumental value (i.e., dollars) is bound to be misleading, and will most certainly underrepresent the value of nature.

But the important point to make is that not all people have the same values. The result is that different people confronted with the same sets of conditions often choose to make different decisions. Biocentrists, ecocentrists, and anthropocentrists are just three examples of differing views. Traditional and Indigenous communities are another, as are foresters, fishers, and farmers. There are often differences between urban and rural people, between young

BOX 7.4 **Challenges & Opportunities**

Taking Anthropocentrism to the Extreme

In November of 2017, Alexander Pyron—an evolutionary biologist and professor of biology at the George Washington University—published an essay in the *Washington Post* that took the anthropocentric view of biodiversity to an extreme.

The Washington Post

We don't need to save endangered species. Extinction is part of evolution.

The only creatures we should go out of our way to protect are *Homo sapiens*.
— *R. Alexander Pyron, November 22, 2017*

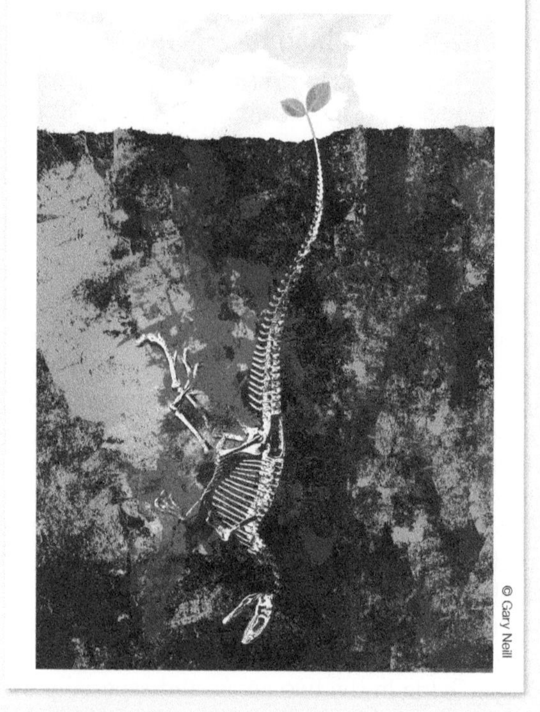

In his essay, Pyron argued that species extinction is a natural process that has occurred throughout the history of life on Earth and that extinct species are eventually replaced by even greater numbers of species later. He wrote:

> *Extinction is the engine of evolution, the mechanism by which natural selection prunes the poorly adapted and allows the hardiest to flourish.... Species constantly go extinct, and every species that is alive today will one day follow suit. There is no such thing as an "endangered species," except for all species.*[78]

Pyron used this logic to further argue that humans should only be concerned about their own fate and focus their attention on using other life-forms solely to help prevent their own extinction:

> *The only reason we should conserve biodiversity is for ourselves, to create a stable future for human beings.... The impulse to conserve for conservation's sake has taken on an unthinking, unsupported, unnecessary urgency.*[78]

Pyron's essay generated thousands of comments from readers of the *Washington Post*, as well as a published response that was coauthored by 3700 people, including Nobel laureates, concerned scientists, and citizens from 88 countries around the world. These individuals expressed their fundamental disagreement with Pyron's views, outlining how his essay was logically flawed and was at odds with scientific facts about extinction and speciation. Importantly, the vast majority of responses also pointed to our moral responsibility as human beings to save the world's biodiversity and to prevent extinction wherever possible. The authors of the response wrote:

> *We cannot condone actions that lead to a future in which our children and grandchildren will only know of rhinos, gorillas and polar bears from picture books. This outcome is not inevitable; it is preventable.*[79]

The backlash to Pyron's essay emphasized that the intrinsic values of biodiversity often motivate more people to conserve biodiversity than do utilitarian values. The challenge for conservation biologists is to balance the variety of things that different stakeholders value about biodiversity. For those who have the tools and methodologies to simultaneously consider values ranging from intrinsic, to relational, to instrumental values, the opportunity to save biodiversity will be great.

and old, rich and poor. People's values also change over time. Because many ecosystem services and much of the world's biodiversity are public goods (or common-pool resources) or lie beyond national jurisdiction, the differences among people's values often need to be negotiated. It is our hope that Part II of this book has provided the concepts and tools needed to negotiate a better outcome for biodiversity.

Summary

1. The field of ecological economics assumes that economic systems are constrained by the same physical laws that all living organisms abide by. Economies can alter the form of natural resources in space or time, but they cannot create nor destroy resources. For this reason, ecological economics must account not only for the value of market and nonmarket goods and services that ecosystems provide to society, but also for the wastes and externalities that are produced by transactions that involve these goods and services.

2. Three decision-making tools are commonly used to consider the value of ecosystem goods and services in conservation projects or policies. Cost-effectiveness analysis (CEA) allows one to determine the most efficient allocation of resources by combining an ecological measure of a conservation program or action (effectiveness) with an economic measure of program input (costs). Cost-benefit analysis (CBA) is the primary decision-making tool used when one wants to maximize the economic benefits of a decision or project while minimizing risk. In contrast, multicriteria decision analysis (MCDA) is designed to weigh the variety of things that people value—both monetary and nonmonetary—and then optimize decision making when there are multiple, conflicting goals that require trade-offs.

3. Total economic value (TEV) is the sum of all economic benefits that ecosystems provide to society. TEV includes both the use and non-use values of ecosystems. Use values include direct use values (consumptive and nonconsumptive use of ecosystems), indirect use values (the regulating services of ecosystems), and option value (future value of ecosystem services). Non-use values include the existence and bequest values of ecosystems.

4. Three general methods exist to estimate the economic value of an ecosystem good or service. Revealed preference methods use direct observations of the amount people are willing to pay for goods or services, in either real (existing) markets or surrogate (implicit) markets or via the cost of replacements. Stated preference methods use surveys to assess people's willingness to pay for goods or services in hypothetical scenarios. Benefits transfer methods estimate the economic value of ecosystem services by transferring information from prior studies completed in other locations and/or contexts.

For Discussion

1. Can you name an environmental market failure or externality that has directly affected you or your family? Who has paid the costs associated with that market failure?

2. Which type of natural resource do you think is more difficult to manage in a sustainable manner—open-access resources or common-pool resources? Why?

3. Say, hypothetically, that you wanted to value the pharmaceuticals that are produced by a tropical rain forest. Describe two methods of valuation you might use to estimate the economic value of this ecosystem service.

4. Option and bequest values both estimate the future value of protecting an ecosystem service. How do these two types of value differ? Can you give an example of each one?

5. Do you think contingent valuation is a useful method for quantifying the economic value of an ecosystem service? If people state on a survey that they are willing to pay $100 per year to save an endangered species from extinction, is this existence value real? Why or why not?

6. Do you consider cost-benefit analysis (CBA) or multicriteria decision analysis (MCDA) to be the more rigorous tool for making conservation decisions? What are the pros and cons of each method? Why might you trust one more than the other?

Suggested Readings

Barbier, E. B. 2011. Pricing nature. *Annual Review of Resource Economics* 3: 337–353. Good overview of the field of ecological economics, as well as the tools of the trade.

Barbier, E. B. 2015. Valuing the storm protection service of estuarine and coastal ecosystems. *Ecosystem Services* 11: 32–38. Demonstrates the damage cost avoided method (called the expected damage function in this paper) for valuing an ecosystems service.

Davies, A. L. et al. 2013. Use of multicriteria decision analysis to address conservation conflicts. *Conservation Biology* 27: 936–944. Summarizes how MCDA can be used to improve conservation decisions.

Fisher, B. et al. 2015. *A Field Guide to Economics for Conservationists.* New York, NY: W.H. Freeman. This easy-to-understand text is a must-read for conservation practitioners.

Jones, T. E. et al. 2017. Assessing the recreational value of world heritage site inscription: A longitudinal travel cost analysis of Mount Fuji climbers. *Tourism Management* 60: 67–78. Shows how the travel cost method can be used to value an ecosystems service.

Lee, C. K. et al. 2010. Preferences and willingness to pay for bird-watching tour and interpretive services using a choice experiment. *Journal of Sustainable Tourism* 18: 695–708. Uses choice experiments to value an ecosystems service.

Liebelt, V. et al. 2018. Hedonic pricing analysis of the influence of urban green spaces onto residential prices: The case of Leipzig, Germany. *European Planning Studies* 26: 133–157. Shows how the hedonic pricing method can be used to value an ecosystems service.

Ostrom, E. 2009. A general framework for analyzing sustainability of social-ecological systems. *Science* 325: 419–422. This culmination of Elinor Ostrom's work conveys her insight into how we might sustainably manage nature and the services it provides to humanity.

Torras, M. 2000. The total economic value of Amazonian deforestation, 1978–1993. *Ecological Economics* 33: 283–297. Provides an example of how to calculate the total economic value (TEV) of an ecosystem.

Visit the
Conservation Biology
Companion Website
oup.com/he/cardinale1e
for videos, exercises, links, and other study resources.

Part III

Threats to Biodiversity

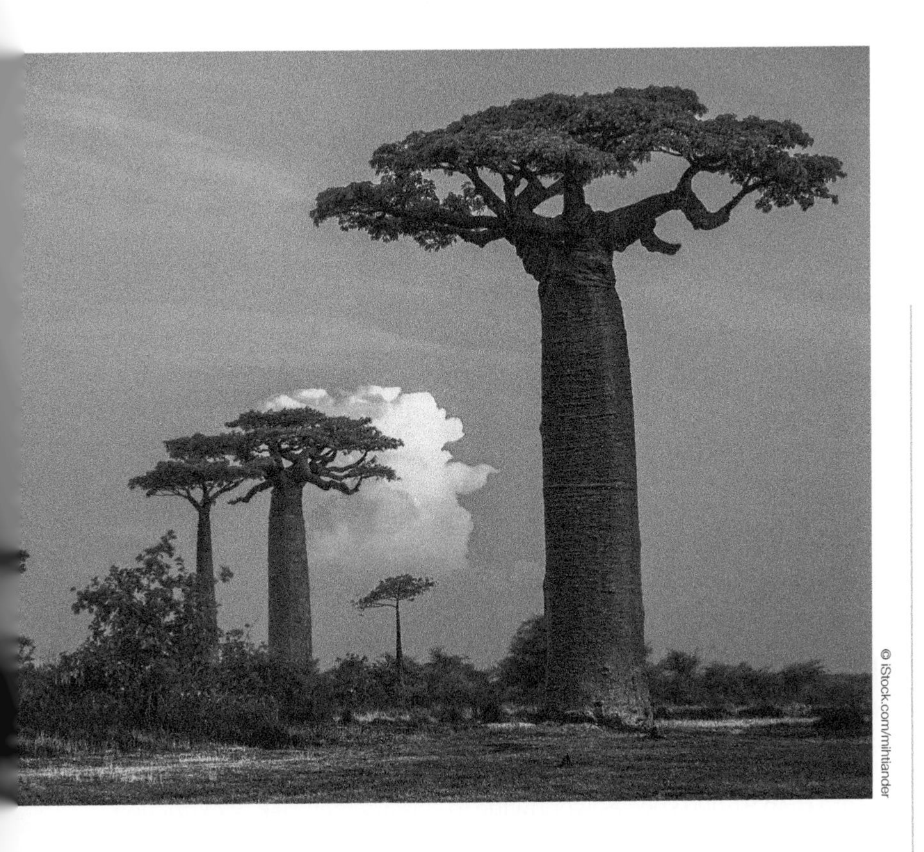
© iStock.com/mihtiander

8

Extinction

The iconic baobab tree faces ever-increasing threats from climate change and habitat loss. The most threatened of the African baobabs is *Adansonia perrieri*, which has just 99 known trees remaining.

Extinction is the loss of a species that occurs with the death of the last individual of that species. Extinction results in the complete loss of the genetic diversity of that species, as well as the evolutionary history that generated that diversity. The meaning of the word *extinct* can vary somewhat, depending on context. A species is considered **globally extinct** when, after a thorough search, no member of the species is found alive anywhere in the world. If individuals of a species remain alive only in captivity or in other human-controlled situations (e.g., zoos, gardens, aquaria), the species is said to be **extinct in the wild**. A species can also be considered locally extinct, or **extirpated**, when it is no longer found in a specific area that it once inhabited but is still found elsewhere in the wild. Some conservation biologists speak of a species as being **ecologically** or **functionally extinct** if it persists at such reduced numbers that its effects on other species and the function of ecosystems are negligible. To maintain a declining species that is headed toward extinction, conservation biologists must identify and mitigate the human activities that affect the stability of its populations and that are driving the species to extinction.

E. O. Wilson, one of the leading advocates of conservation biology, has argued that the most serious aspect of environmental damage is the extinction of species, because once it is extinct, there is no way to get the species or its genetic diversity back again. In this chapter we cover global patterns of species endangerment, overview the primary drivers of extinction, describe local patterns of biodiversity change, and detail the key factors that control extinction risk. Here we do not cover any of the naturally occurring patterns of biodiversity—either through geological time or spatially across the globe—as those were covered in Chapter 4. Instead, we focus this chapter on how human activities are altering modern biodiversity across the planet.

Global Patterns of Endangerment

In order to track rates of global extinction, it is necessary to keep track of species that are known to have already gone extinct, along with the species that are presently threatened with extinction. The primary tool that conservation biologists use to keep track of the global status of extinct, threatened, and endangered species is the International Union for Conservation of Nature's Red List (**Box 8.1**).

The IUCN Red List

The **IUCN Red List** (oup-arc.com/e/cardinale-w8.1) is widely considered to be the world's most comprehensive listing of the global conservation status of threatened and endangered species. Since it was founded in 1964, the Red List has assessed more than 91,000 species worldwide to classify their risk of extinction. The main advantages of the Red List are that it (1) uses clear taxonomic standards to help ensure all species listings adhere to international nomenclature codes, thus avoiding confusion and redundant assessments; (2) standardizes the process for assessing species populations and provides agreed-on, quantitative benchmarks for classifying species by their extinction risk; (3) provides a formal listing and review process to ensure basic levels of record quality and control; and (4) serves as a central repository and distribution center for all records of species assessment, which can be accessed by anyone (**Video 8.1**).

VIDEO 8.1 Watch this video about the IUCN Red List.
oup-arc.com/e/cardinale-v8.1

The assessment process—that is, the process by which one evaluates and generates a species listing for the IUCN Red List—can be initiated by anyone and submitted to the IUCN for consideration. However, the vast majority of assessments on the IUCN Red List are carried out by members of the IUCN Species Survival Commission (SSC), by appointed Red List Authorities (RLAs), Red List Partners, or participants of IUCN-led assessment projects. The SSC is a science-based network of more than 7500 experts who work in universities, government agencies, zoos and botanical gardens, museums, and wildlife reserves in nearly every country in the world. These individuals are chosen because of their biological expertise with particular groups of plants, animals, fungi, or other organisms on the Red List, which qualifies them to provide evaluations. In addition to these authorities, the IUCN partners with a variety of organizations (e.g., the Zoological Society of London, Botanic Gardens Conservation International, BirdLife International, and Conservation International) that expand the taxonomic and biological expertise and that have field biologists who are directly working with threatened and endangered species.

The goal of the Red List assessment process is to classify species into categories that represent their risk of extinction. For species classified into three of these categories—Critically Endangered (CR), Endangered (EN), and Vulnerable (VU)—risk of extinction is considered sufficiently high that they are collectively referred to as threatened species. Assignment of a threatened species to the CR, EN, or VU category is based on an assessment of the species' current population size or geographic range, recent declines in population size or range, consideration of whether the species lives in a limited

BOX 8.1 Conservation in Practice

The IUCN Red List

The IUCN Red List of Threatened Species (also known as the IUCN Red List or Red Data List), founded in 1964, has become the world's most comprehensive inventory of the global conservation status of biological species. The IUCN Red List characterizes the extinction risk for thousands of species and subspecies found in all regions of the world using a set of well-defined biological criteria that are evaluated by teams of expert biologists and reviewers. Several Regional Red Lists are produced by countries or organizations, which use these products to assess the risk of extinction to species within their political management units.

Taxonomic experts, conservationists, and other biologists often work together in teams to conduct reviews to evaluate the population status and trends for individual species. After evaluation, each species is placed into one of eight categories (see Figure). The first category (data deficient) indicates there is insufficient information to evaluate the species status. Those species that do have sufficient data for evaluation are placed in one of seven remaining categories that represent increasing risks of extinction (see below).

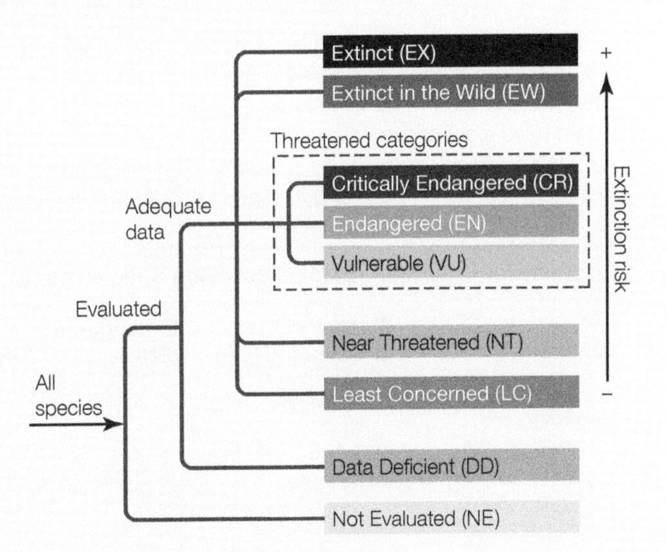

IUCN Red List Categories and Criteria

LEAST CONCERN (LC)	A taxon is of Least Concern when it has been evaluated against the criteria and does not qualify for Critically Endangered, Endangered, Vulnerable or Near Threatened. Widespread and abundant taxa are included in this category.
NEAR THREATENED (NT)	A taxon is Near Threatened when it has been evaluated against the criteria but does not qualify for Critically Endangered, Endangered or Vulnerable now, but is close to qualifying for or is likely to qualify for a threatened category in the near future.
VULNERABLE (VU)	A taxon is Vulnerable when the best available evidence indicates that it meets any of the criteria A to E for Vulnerable (see Table), and it is therefore considered to be facing a high risk of extinction in the wild.
ENDANGERED (EN)	A taxon is Endangered when the best available evidence indicates that it meets any of the criteria A to E for Endangered (see Table), and it is therefore considered to be facing a very high risk of extinction in the wild.
CRITICALLY ENDANGERED (CR)	A taxon is Critically Endangered when the best available evidence indicates that it meets any of the criteria A to E for Critically Endangered (see Table), and it is therefore considered to be facing an extremely high risk of extinction in the wild.
EXTINCT IN THE WILD (EW)	A taxon is Extinct in the Wild when it is known only to survive in cultivation, in captivity or as a naturalized population (or populations) well outside the past range. A taxon is presumed Extinct in the Wild when exhaustive surveys in known and/or expected habitat, at appropriate times (diurnal, seasonal, annual), throughout its historic range have failed to record an individual. Surveys should be over a time frame appropriate to the taxon's life cycle and life form.
EXTINCT (EX)	A taxon is Extinct when there is no reasonable doubt that the last individual has died. A taxon is presumed Extinct when exhaustive surveys in known and/or expected habitat, at appropriate times (diurnal, seasonal, annual), throughout its historic range have failed to record an individual. Surveys should be over a time frame appropriate to the taxon's life cycle and life form.

Source: IUCN. 2012. *IUCN Red List Categories and Criteria: Version 3.1*. Second edition. Gland, Switzerland and Cambridge, UK. https://portals.iucn.org/library/sites/library/files/documents/RL-2001-001-2nd.pdf.

(*Continued on next page*)

BOX 8.1 Conservation in Practice *(continued)*

Species that are placed into the Critically Endangered (CR), Endangered (EN), and Vulnerable (VU) categories are collectively referred to as threatened species. The IUCN system uses a set of five criteria (A–E), each with a set of quantitative values, that are used to determine the severity of the threatened status and to determine whether the species should be designated CR, EN, or VU. Although the IUCN requires that assessments use these quantitative criteria to justify the species' listing, expert opinions regarding what those values may be are both allowed and encouraged whenever actual data are not available.

Overview of Criteria (A–E) for Classifying Species as CR, EN, or VU in the IUCN Red List

Criterion	Critically Endangered (CR)	Endangered (EN)	Vulnerable (VU)	Qualifiers
A.1 Reduction in population size	>90%	>70%	>50%	Over 10 years or 3 generations in the past where causes are reversible, understood, and have ceased
A.2–4 Reduction in population size	>80%	>50%	>30%	Over 10 years or 3 generations in the past, future, or combination, where causes are not reversible, not understood, or ongoing
B.1 Small range (extent of occurrence)	<100 km^2	<5,000 km^2	<20,000 km^2	Plus two of (a) severe fragmentation or few occurrences (CR = 1, EN = 2–5, VU = 6–10), (b) continuing decline, (c) extreme fluctuation
B.2 Small range (area of occupancy)	<10 km^2	<500 km^2	<2,000 km^2	
C Small and declining population	<250	<2,500	<10,000	Mature individuals, plus continuing decline, either over a specific rate in short time periods or with specific population structure or extreme fluctuations
D.1 Very small population	<50	<250	<1,000	Mature individuals
D.2 Very small range	—	—	<20 km^2 or <5 locations	Capable of becoming CR or EX within a very short time
E Quantitative analysis	>10% in 100 years or 3 generations	>20% in 20 years or 5 generations	>50% in 100 years	Estimated extinction risk using quantitative models, e.g., population viability analyses

Source: IUCN. 2012. *IUCN Red List Categories and Criteria, Version 3.1*. Second edition. Gland, Switzerland and Cambridge, UK.

geographic area that is in danger of being destroyed, and quantitative analysis—such as a population viability analysis—that might indicate a high probability of extinction. Each of these criteria has specific definitions that are used by reviewers to determine the severity of the threatened status (Table 8.1; see Box 8.1). Once an assessment is completed, it is subject to review by at least one designee of the appropriate RLA, and the assessment is supposed to be updated every 5 to 10 years.

Suggested Exercise 8.1 Go to the IUCN Red List of Threatened Species to find plants and animals that have been listed as Vulnerable, Endangered, and Critically Endangered in the region that includes your home country. Choose any 10 species you identify and read their assessments to determine what is causing their threatened status. **oup-arc.com/e/cardinale-ex8.1**

TABLE 8.1 Numbers of threatened species by major groups of organisms (1996–2018)

	Estimated number of described species	Number of species evaluated by IUCN	Percentage of described species assessed	Number of threatened species on 2018 IUCN Red List version 2018–2	Percentage of evaluated species threatened
VERTEBRATES					
Mammals	5,692	5,692	100	1,219	21
Birds	11,126	11,126	100	1,492	13
Reptiles	10,793	7,127	66	1,307	18
Amphibians	7,926	6,722	85	2,092	31
Fishes	34,000	16,803	49	2,332	14
Subtotal	69,537	47,470	68	8,442	18
INVERTEBRATES					
Insects	1,000,000	8,037	0.8	1,537	19
Mollusks	85,000	8,627	10	2,195	25
Crustaceans	47,000	3,180	7	733	23
Corals	2,175	864	40	237	27
Arachnids	102,248	324	0.32	182	56
Velvet worms	165	11	7	9	82
Horseshoe crabs	4	4	100	1	25
Others	68,658	839	1.22	146	17
Subtotal	1,305,250	21,886	2	5,040	23
PLANTS					
Mosses	16,236	102	0.6	76	75
Ferns & allies	12,000	558	5	249	45
Gymnosperms	1,052	1,012	96	401	40
Flowering plants	268,000	25,771	10	12,564	49
Green algae	6,050	13	0.2	0	0
Red algae	7,104	58	0.8	9	16
Subtotal	310,442	27,514	9	13,299	48
FUNGI & PROTISTS					
Lichens	17,000	23	0.14	10	43
Mushrooms	31,496	43	0.137	33	77
Brown algae	3,784	15	0.4	6	40
Subtotal	52,280	81	0.15	49	60
TOTAL	**1,737,509**	**96,951**	**6**	**26,840**	**28**

Source: Data from IUCN. 2018. *The IUCN Red List of Threatened Species*. Version 2018-2, updated as of January 22, 2019 at www.iucnredlist.org. NOTE: the data available via this link change as new updates are issued.

In addition to the Red List of Threatened Species, the IUCN has recently started to develop the **Red List of Ecosystems** (RLE, oup-arc.com/e/cardinale-w8.2), which is intended to assist conservation and management efforts that extend beyond single species (**Video 8.2**). The RLE was first envisioned at the fourth IUCN World Conservation Congress (Barcelona, Spain, October 5–14, 2008), at which participants approved a motion to initiate development of a

VIDEO 8.2 View this explanation about the IUCN Red List of Ecosystems.
oup-arc.com/e/cardinale-v8.2

global standard for assessing ecosystems that are at risk of collapse. The RLE was then formally established in 2014 as part of the IUCN's growing tool kit for biodiversity risk assessments. The RLE process is much like that for threatened species, using evaluations to place ecosystems into categories that describe their risk of collapse, which is defined as virtual certainty that the key biotic or abiotic features of the ecosystem will be globally lost and that the characteristic native biota will no longer be sustained. The evaluation process ranks ecosystems that are at increasing risk of collapse, from *vulnerable*, to *endangered*, to *critically endangered*. The IUCN plans to have a complete evaluation of all the world's ecosystems by 2025.

The stated goal of the IUCN Red List is to provide "objective, scientifically-based information on the current status of globally threatened biodiversity." The Red List is widely presented as the most comprehensive resource detailing the global conservation status of plants and animals[1] and one of the most widely used tools available to conservationists worldwide.[2] IUCN products and publications such as the **Red Data Books** and the **Red List Index** provide detailed lists of threatened species by group and country and are routinely used to estimate global rates of extinction,[3] forecast the impacts of environmental change on future extinction rates,[4,5] quantify the most pervasive threats to imperiled species across the globe,[6] and determine how species are responding to conservation efforts.[7]

Despite claims that the IUCN is scientifically based, critics have regularly argued that many, if not most, species assessments on the Red List are not backed by any real data.[8–10] They argue that the assessment and listing processes have been driven more by expert opinion than evaluation of real data. Although the IUCN requires that assessments use quantitative criteria to justify the current status and trends for any given species population, expert opinions regarding what those values may be are both allowed and encouraged whenever actual data are not available. A recent review of the listings for mammal species (perhaps the most studied of all groups) revealed that only a small fraction (12%) of the listings reference any data or scientific study to justify the species classification.[9]

Proponents of the Red List claim that revisions to the listing process in the early 1990s and 2000s led to a more transparent process in which a greater fraction of species listings are based on actual data that are both reliable and verifiable.[1,11] They further argue it is often necessary to rely on the opinion of expert biologists because many species have not been scientifically studied, and the time and money needed to do so is often not available—particularly in developing countries or those in tropical locations where the number of threatened species is greatest. Even so, users of the Red List should keep in mind that much of the database is constructed from the opinions of expert biologists, many of whom work directly with threatened species in the field.

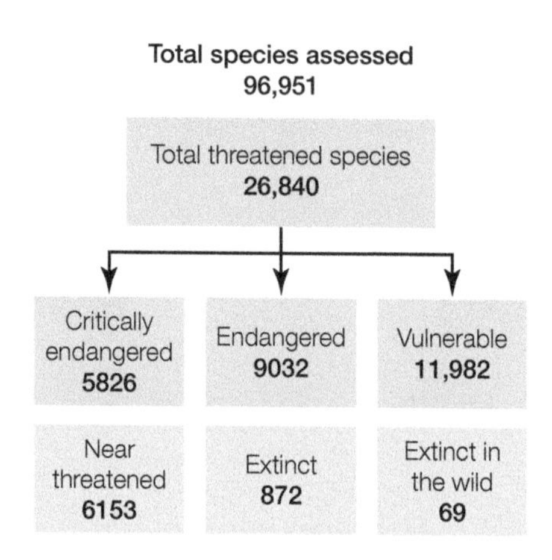

Figure 8.1 A summary of species that comprise the 2017 IUCN Red List of threatened species. (IUCN. 2018. *The IUCN Red List of Threatened Species. Version 2018-2*. www.iucnredlist.org)

To date, expert biologists have evaluated the status and population trends for 96,951 species spanning all regions of the world (see Table 8.1). Assessments have now been performed for almost all mammal and bird species and for the vast majority of amphibians and gymnosperms. Assessments of some groups of vertebrates, such as reptiles and fishes, are only about half complete. Assessments for nearly all invertebrate groups, most plant groups, and most fungi and protists are woefully incomplete, often representing 10% or less of described species.[12]

Of the 96,951 species that have been assessed, 26,840 are currently listed as threatened (**Figure 8.1**). Of these threatened species, almost 6000 are considered critically endangered with an extremely high risk of extinction in the

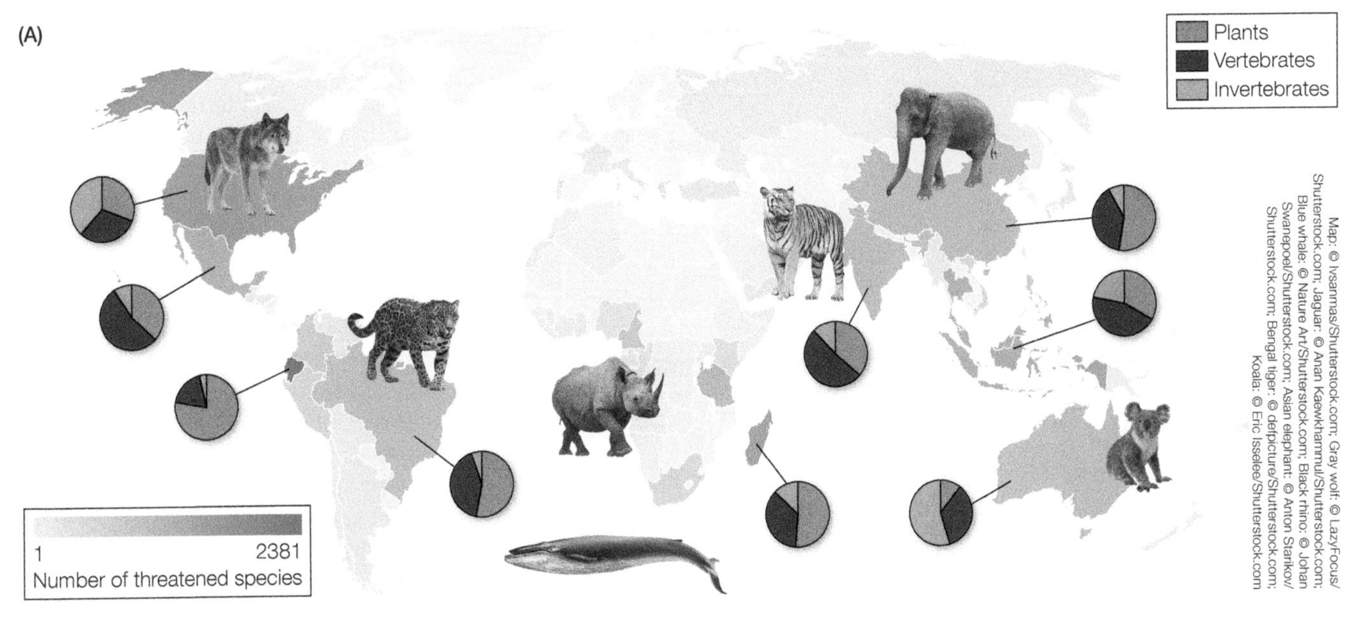

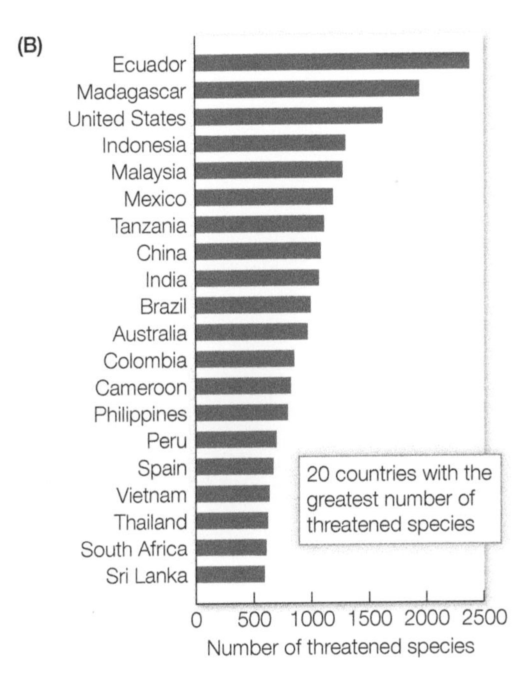

Figure 8.2 Distribution of species threatened with extinction. (A) World map of countries having the highest and lowest numbers of threatened species on the IUCN Red List. Inset pie charts show the proportion of threatened species of plants, vertebrates, and invertebrates. (B) Twenty countries having the greatest numbers of threatened species on the Red List. (From IUCN. 2018. *The IUCN Red List of Threatened Species*. Version 2018-2. Gland, Switzerland: International Union for Conservation of Nature and Natural Resources.)

wild in the immediate future. Another 9032 species are endangered, and 11,982 are vulnerable to extinction. Four out of every ten amphibian and gymnosperm species are currently threatened with extinction. One out of every four mammal species is threatened, with risks concentrated in four countries that each have more than 90 threatened mammal species: Indonesia (201), Madagascar (122), Mexico (99), and India (95). Twenty countries presently have more than 500 threatened species apiece (**Figure 8.2**), with the highest number in Ecuador (2410) followed by Madagascar (1936), the United States (1626), and Indonesia (1434). In most countries around the world, records on the Red List are heavily skewed toward plants and vertebrate animals, with invertebrates very much underrepresented. Because many developing countries have no dedicated efforts for biodiversity monitoring, they are likely to have a higher proportion of species that have not yet been assessed, which may lead to underreporting in many parts of the world.

Estimates of global extinction

At global scales, changes in biodiversity are ultimately regulated by the balance between the rate of formation of new species (i.e., speciation) and the rate of species loss (i.e., extinction). Rates of speciation at the global scale are not particularly well known; however, certain types of human activities are known to accelerate and augment speciation. For example, humans have domesticated 474 animal and 269 plant species over the last 11,000 years, and some of this domestication has resulted in the formation of new species: of the world's 40 most important crop species, 6 to 8 can be considered entirely new.[13]

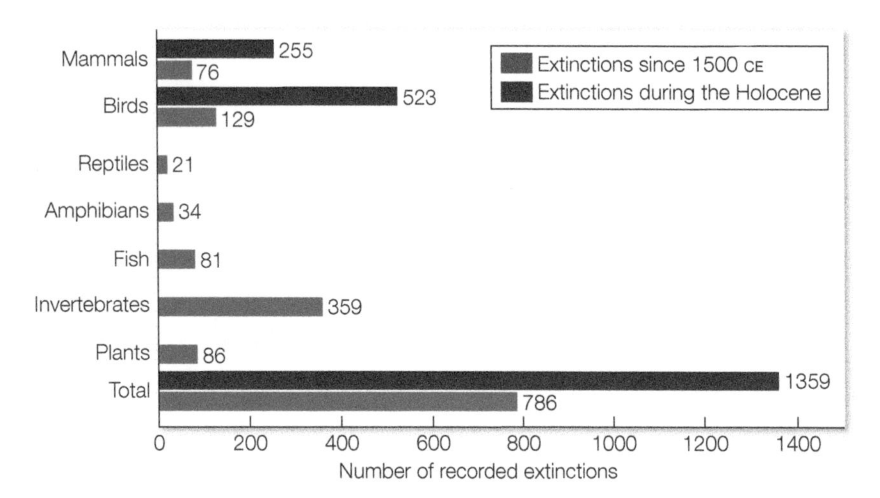

Figure 8.3 Number of recorded animal and plant species extinctions since 1500 CE (blue), and during the Holocene (red). (After J. W. Bull and M. Maron. 2016. *Proc R Soc Lond* [*Biol*] 283: 20160600.)

While human activities can augment speciation, most conservation biologists agree that rates of human-mediated extinction in the modern era are outpacing rates of human-mediated speciation. In turn, the net effect of human activities has been to cause biodiversity to decline at the global scale. Estimates of global-scale extinctions vary slightly depending on the source. The IUCN Red List includes 872 species that are considered extinct (EX) and an additional 69 species that are considered extinct in the wild (EW), with representatives alive only in zoos, gardens, or aquaria (see Figure 8.1). A recent tally of extinct species by Bull and Maron[13] included 1359 known extinctions that have occurred during the Holocene (11,000 years ago to the present) and 786 extinctions documented since 1500 (**Figure 8.3**). Documented extinctions over the past 500 years have included 76 species of mammals, 129 species of birds, 21 species of reptiles, 34 amphibian species, 81 fish species, 359 species of invertebrates (of which 291 were mollusks), 86 plant species, and 1 species of protist.[13–15]

Ceballos and colleagues[16] used listings of extinct species from the 2014 version of the IUCN Red List and estimated that 338 species of vertebrates (mammals, birds, reptiles, amphibians, and fishes) have gone extinct since 1500 (**Figure 8.4**). That estimate increases to 617 extinctions if one also includes those that are extinct in the wild and currently presumed to be extinct (PE). Since 1900, the rate of extinctions has risen considerably (see Figure 8.4), with 198 to 477 vertebrate species having gone globally extinct. Dirzo and colleagues[17] arrived at similar estimates of vertebrate extinctions but emphasized that their numbers were almost certainly underestimates because many extinctions have gone unnoticed and unrecorded. This bias is particularly problematic with less-studied groups of organisms, such as invertebrates.[18] For example, Régnier and colleagues[12] estimated that the actual number of mollusk extinctions is probably double the current number on the IUCN Red List, because they have simply not been recorded and added to a central repository.

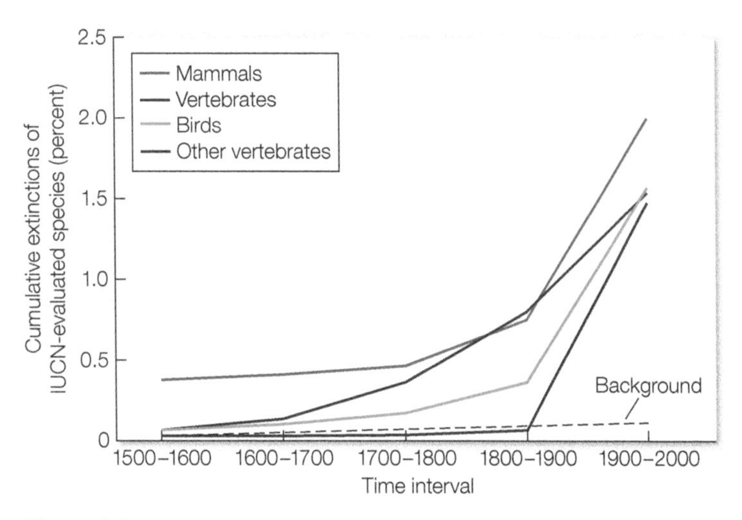

Figure 8.4 Cumulative percentage of vertebrate species that have been recorded as extinct or extinct in the wild by the IUCN (by 2012) over various time intervals. Assessment data are for 5513 mammals (100% of those described); 10,425 birds (100%); other vertebrates, including 4414 reptiles (44%), 6414 amphibians (88%), and 12,457 fishes (38%); and all 39,223 vertebrates combined (59%). The dashed black line represents the number of extinctions expected under a constant standard background rate of 2 extinctions per million species years (E/MSY). (After G. Ceballos et al. 2015. *Sci Adv* 1: e1400253. Reprinted with permission of AAAS. © G. Ceballos et al., some rights reserved; exclusive licensee American Association for the Advancement of Science. Distributed under CC BY-NC 4.0.)

If we use the number of global extinctions that are currently tallied in the IUCN Red List (872 extinctions, and 69 extinctions in the wild), then a sum total of 941 known extinctions of modern taxa represents a relatively small fraction of all species on Earth. Roughly 1.74 million species have been described to date (see Table 8.1 that uses numbers from the IUCN, which is slightly higher than the estimate of 1.24 million given in Chapter 4), and our best estimate is that there may be 8.74 million eukaryotic (multicellular) species on the planet.[19] Given these numbers, the sum total of all extinctions that have taken place over the last 500 years represents just

0.05% of described species (941/1.74 million) and less than 0.01% of predicted species (941/8.74 million). Even if we add in the estimated 26,840 species that are listed as threatened on the IUCN Red List and we assume that all of these will go extinct in the near future, then the prediction for global extinctions still reaches just 1.5% of described species and less than 1% of predicted species richness.

While the total *percentage* of species that have gone globally extinct in the past few centuries is not particularly high, the *rate* at which species are going extinct is a major cause for alarm. Rates of global extinction rates are commonly expressed as extinctions per million species years (E/MSY), which is the number of species expected to go extinct in 1 million years given the available number of species that have the potential to go extinct. The metric is calculated as:

$$E/MSY = \frac{E}{S} \times \frac{10^6}{t} \tag{8.1}$$

where E = the number of documented extinctions, S = the total number of known species, t = the number of observation years used to assess extinctions, and 10^6 = 1 million years.

Based on fossil records, we know that the average species persists for roughly 1 million years. Given this, Pimm and colleagues[20] argued that the "background" rate of extinction throughout Earth's history has been about 1 E/MSY (if a species lives 1 million years, then we expect 1 species to go extinct per million species per year). This widely used benchmark has recently been updated with analyses showing that background rates of extinction in the fossil record are probably closer to 0.1 E/MSY.[21] Therefore, we should expect somewhere between 0.1 and 1.0 species out of a million to go extinct per year, depending on which estimate we use.

Assuming there are 1,736,546 described species (see Table 8.1) and that 941 of those have gone extinct in the past 500 years (see Figure 8.1), then the modern rate of extinction over the last five centuries has been:

$$E/MSY = \frac{935}{1{,}736{,}546} \times \frac{10^6}{500} = 1.077 \tag{8.2}$$

This number means that rates of extinction for all described species in the modern era are occurring at a rate that is 108% to 1077% higher than what is considered normal in the fossil record (1.077 / 1.0 × 100, or 1.077 / 0.1 × 100). Pimm and colleagues[20] summarize data on extinction rates using a narrower time frame of just the past century (**Figure 8.5**). These authors showed that extinction rates are routinely 100 to 1000 times faster than the background rate of extinction in the fossil record and

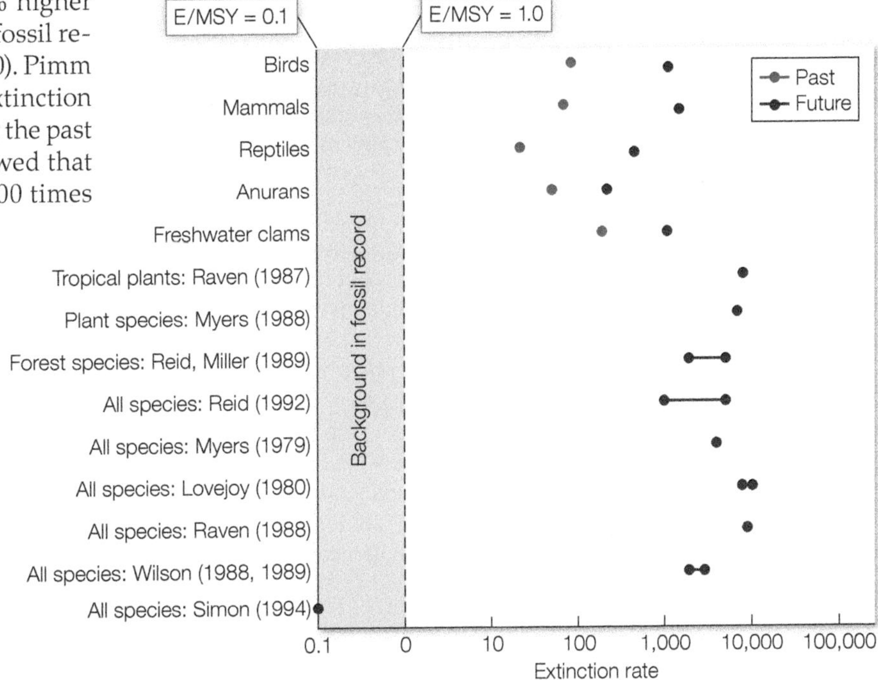

Figure 8.5 A summary of studies (*y*-axis) that have published estimates of modern extinction rates (*x*-axis) for various groups of organisms. Extinction rates are reported in extinctions per million species years (E/MSY). Estimates are given for extinctions since 1900 (blue) and extinctions projected in the near future (red) assuming threatened species are extinct in the next 100 years. (After S. L. Pimm et al. 1995. *Science* 269: 347–350.)

approach 10,000 times faster when species that are currently threatened with extinction are considered in the analyses. Using numbers from Ceballos and colleagues,[16] who estimated that 477 of the 68,574 known vertebrate species have gone extinct since 1900, vertebrate extinctions are occurring at a rate of

BOX 8.2 **Challenges & Opportunities**

Are We Entering the Sixth Mass Extinction?

Many biologists have claimed that Earth is entering the sixth mass extinction in the history of life.[17,108,109] This claim has been repeated in the title of popular books,[110,111] and in news stories and magazine articles that have presented the idea to the public as the general scientific consensus (see Figure A). But despite how it has been presented, surprisingly few scientific papers have directly tested the hypothesis that Earth is entering a mass extinction.

(A)

INDEPENDENT

Oceans on brink of catastrophe: Marine life facing mass extinction within one human generation
– Michael McCarthy, June 20, 2011

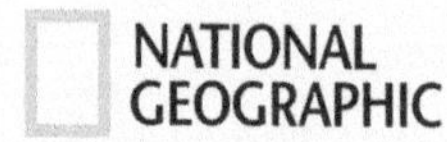

Will humans survive the sixth great extinction?
– Nadia Drake, June 23, 2015

Sixth mass extinction: The era of 'biological annihilation'
– John D. Sutter, July 11, 2017

USA TODAY

Biological annihilation: Earth's 6th mass extinction is underway
– Doyle Rice, July 10, 2017

DISCOVER
MAGAZINE

Species Census: Yes, the 6th Mass Extinction Is Happening Now
– Andrew Moseman, March 2, 2011

The Washington Post

Earth is on brink of a sixth mass extinction scientists say
– Sarah Kaplan, June 22, 2015

Many biologists believe we are presently entering the sixth mass extinction in the history of life on Earth, and this view has been popularized in the public media. But until recently, the scientific evidence to support this view has been lacking.

In 2011, paleontologist Anthony Barnosky and his colleagues published a paper in the journal *Nature* in which they collated and summarized the variety of data needed to directly test whether Earth is indeed entering a mass extinction.[3] Barnosky and colleagues began by clearly defining a mass extinction as an event that leads to 75% or more of species going extinct at a global scale. This benchmark was chosen because it represents the level of species loss that occurred during the previous Big Five mass extinction events (the Ordovician-Silurian, Devonian, Permian-Triassic, Triassic, and Cretaceous-Tertiary) that have occurred in the geologic record (black vertical line in the Figure B).

Next, Barnosky and colleagues used records from the IUCN Red List to compare the magnitude of modern extinction to that of prior mass extinctions. They showed that for most taxonomic groups with sufficient data, the number of documented extinctions represents less than 1% of the known or estimated number of species on the planet, though estimates for select groups like gastropods and bivalves have potentially reached 10%–13% (white numbers and icons in Figure B). Barnosky and colleagues then added the number of species that are presently threatened with extinction to those that have already gone extinct, and they estimated that the percentage of extinct species could soon reach greater than 30%, with values

58.9 E/MSY, which is 59 to 589 times faster than normal. These abnormally high rates of global extinction are what have led many scientists to claim that Earth is entering the sixth mass extinction in the history of life (Box 8.2).

ranging from 14% for bird species (Aves) to 64% for certain groups of plants like the Cycadopsida (black numbers and icons in Figure B). Even so, when compared with the 75% extinction rate that characterized the Big Five, the total percentage of modern species that have gone extinct or may soon become extinct does not meet the criteria for a mass extinction.

Barnosky and colleagues' next step was to compare the rate of modern extinction to the normal background rate of extinction that has been documented in the fossil record. The authors showed that over the past 1000 years, the extinction rate across taxonomic groups has been about 24 E/MSY (extinctions per million species years). Breaking the data into 1-year periods, they showed the maximum extinction rate over the modern period was about 693 E/MSY. These rates are several orders of magnitude faster than what is considered normal and are above rates of extinction that have been associated with prior mass extinctions in the geologic record:

> *Current extinction rates for mammals, amphibians, birds, and reptiles, if calculated over the last 500 years (a conservatively slow rate) are faster than (birds, mammals, amphibians, which have 100% of species assessed) or as fast as (reptiles, uncertain because only 19% of species are assessed) all rates that would have produced the Big Five extinctions over hundreds of thousands or millions of years.*[3]

When the authors projected current rates of extinction into the future, they estimated that terrestrial amphibian, bird, and mammal extinctions are likely to reach the 75% threshold of global disappearance within 240 to 540 years.

Their results led the authors to two main conclusions:

(B)

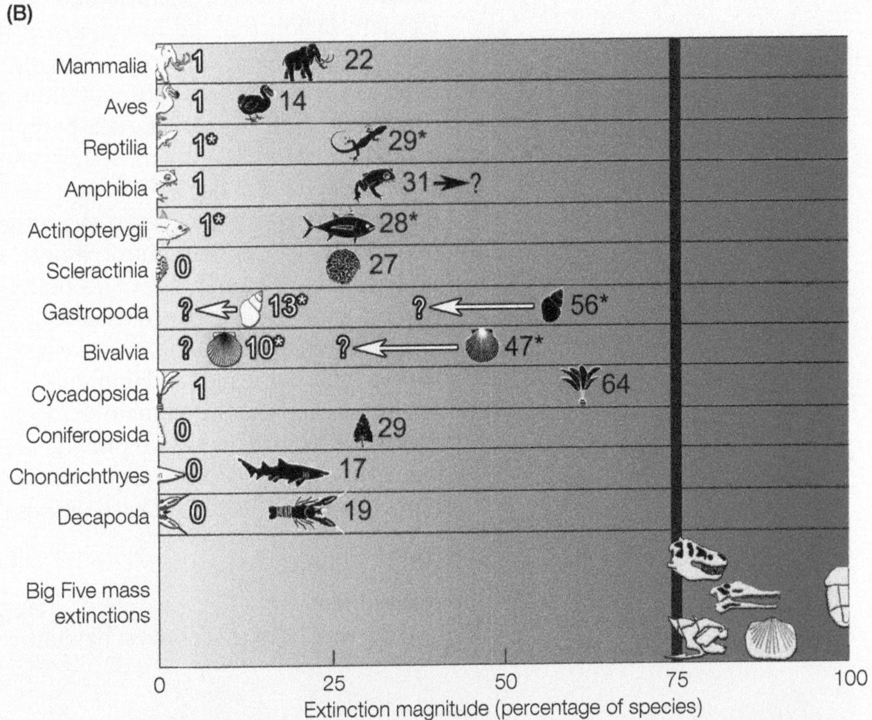

Extinction magnitudes for well-studied taxonomic groups compared with the 75% benchmark used to classify prior mass extinction events. White icons and numbers indicate species extinctions that have occurred over the past 500 years. Black icons add currently threatened species to those that are already known to be extinct. Yellow icons indicate species lost in the Big Five extinctions: Cretaceous-Tertiary, Devonian, Triassic, Ordovician-Silurian, and Permian-Triassic (from left to right). Asterisks indicate taxa for which very few species (less than 3% for gastropods and bivalves) have been assessed; white arrows show where extinction magnitudes are probably inflated (because species perceived to be in peril are often assessed first). (After A. D. Barnosky et al. 2011. *Nature* 471: 51–57.)

(1) Although we're clearly in dangerous territory in terms of extinction rates, we still have time to reverse course, and most species still exist somewhere on Earth and can be saved. However, (2) if we continue on our present course with business as usual, we may well enter the sixth mass extinction event within a time frame of just a few centuries.

Drivers of Extinction

The ultimate cause of most modern extinctions is people. The primary human drivers of extinction are often described qualitatively using the IPAT equation:

$$I = P \times A \times T \quad (8.3)$$

where impact (*I*) on ecosystems and their biodiversity is the product of three factors: human population size (*P*), affluence (*A*), and technology (*T*). The IPAT equation is overly simplistic in both form and assumptions. It does not account for many factors that are known to contribute to extinction, such as social dilemmas (Chapter 7) or human psychology (Chapter 10). It does not account for the many contributions that affluent or technologically advanced societies make to conservation efforts. And it risks perpetuating a myth that human development and pursuit of well-being must ultimately lead to extinction (see the myth of the noble savage, Chapter 3). For these reasons, the IPAT equation is rarely used in practice. Even so, it does serve as a reminder that growth of the human population and key factors that control resource consumption (affluence and technology) underlie many human activities that alter the environment (**Figure 8.6**). These human activities give rise to three major forms of environmental change (habitat loss, overexploitation, and invasive species) that are the largest drivers of biodiversity loss on the planet. Human activities are also producing a fourth form of environmental change (climate change) that has potential to drive much diversity loss in the future. Here, we provide an overview of these proximate drivers of extinction in order to lay the groundwork for the rest of the chapter and to set up more in-depth coverage of each of these causes in Chapters 9–12.

Habitat loss

Habitat loss is actually a suite of changes (habitat conversion, fragmentation, and degradation) that convert habitats from forms that are more usable to forms

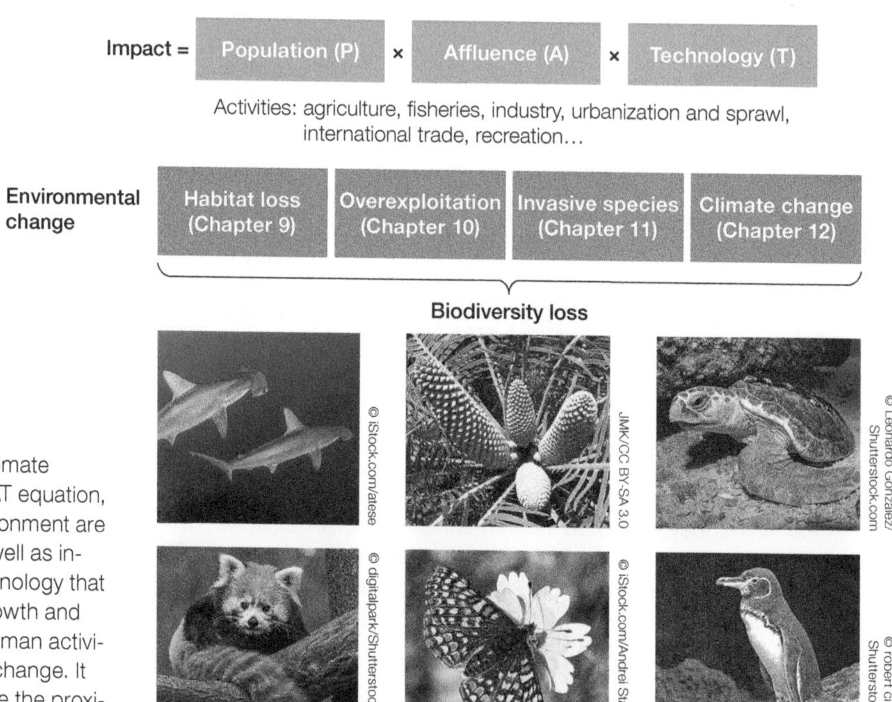

Figure 8.6 Drivers of biodiversity loss. The ultimate causes of biodiversity loss are given by the IPAT equation, which shows how human impacts on the environment are proportional to human population growth, as well as increases in affluence and improvements in technology that underlie resource consumption. Population growth and consumption, in turn, drive a wide variety of human activities that lead to major forms of environmental change. It is these forms of environmental change that are the proximate causes of biodiversity loss on the planet.

that are less usable by nonhuman species. **Habitat conversion** is the transformation of natural or seminatural habitat into human-modified habitat. Conversion typically leads to a biological community in the new habitat type that is less diverse than the community of the original habitat type—for example, as might occur when a native forest or grassland is converted to urban habitat.[22] **Habitat fragmentation** is a term used to describe the process by which habitat loss results in the division of large, continuous habitats into a greater number of smaller patches, each of less total area and isolated from the others by a matrix of dissimilar habitats. Fragmentation leads to a suite of changes in the area, shape, interior-to-edge ratio, connectivity, and microclimate of a habitat patch, making it different from other parts of a landscape. **Habitat degradation** refers to a suite of human-caused changes that convert biologically diverse and complex habitats into simpler forms. These changes include pollution, sedimentation, desertification, and altered disturbance regimes. Habitat loss, fragmentation, and degradation are discussed in Chapter 9.

Overexploitation

Overexploitation, also called **overharvesting**, refers to the harvesting of a renewable resource at a rate that is faster than the resource can be regenerated. Conservation biologists use the term to describe what happens to a species population that is harvested at rates that exceed its population growth rate, which is defined by its natural rate of mortality and reproduction. Harvesting individuals of a species at rates that exceed that natural population growth rate is unsustainable and can lead to the decline and eventual extinction of the species. Overexploitation is a common problem in the fisheries industry, where the United Nations FAO estimates that over 33% of all harvested fish populations in the oceans are overexploited and in decline.[23] Similarly, it is a major problem in the forestry industry, where in many parts of the world, trees are still logged faster than the rate of regrowth and reforestation.[24] Overexploitation is common in a variety of legal trade industries where species are harvested from their natural habitats and sold as pets, ornamentals, or medicines.[25–27] And it is common in illegal trade markets where high demand for animal parts, such as the tusks of elephants,[28] rhinoceros horns,[29] and dorsal fins of sharks,[30] is responsible for a large fraction of species population harvesting. The direct impact of persistent overexploitation is the localized, and ultimately global, extinction of a species. But there are equally dangerous indirect impacts of overexploitation, such as the risk of an **extinction cascade**,[31,32] that is, a series of secondary extinctions that are triggered when an important species like a keystone predator, an ecosystem engineer that provides habitat, or a basal resource species that serves as food for many other taxa goes extinct. Overexploitation can also indirectly lead to changes in the productivity and stability of entire ecosystems.[33,34] Overexploitation is discussed in Chapter 10.

Invasive alien species

Species that invade or are introduced to an area or habitat and become abundant where they do not naturally occur are a significant threat to biodiversity. Invasive species have been implicated in the decline of native species on land,[35] in freshwater habitats,[36] and in the oceans.[37] A review of 329 invasive species in marine ecosystems illustrates how pervasive invasive species have become, concluding that 84% of all marine ecoregions around the globe are now influenced by invasive species that were introduced by international shipping or aquaculture.[37] Invasive species can compromise native species populations through direct interactions (e.g., predation, parasitism, disease, competition, or hybridization), as well as through indirect paths (e.g., by disrupting mutualisms, changing abundances or dynamics of native species, or reducing habitat quality). Invasive species can also disrupt ecosystem services[38] and cause significant economic damage to ecosystems.[39] Emerging infectious diseases are a category of invasive species that are becoming more widespread, have potential to kill entire ecosystems like coral reefs[40] or forests,[41] or kill entire groups of species like bats and frogs. Diseases can have the potential to "spill over" from free-living wildlife populations to humans and affect our health.[42] Invasive alien species are covered in more detail in Chapter 11.

Climate change

While habitat loss, overexploitation, and invasive species are the three main current drivers of extinction, climate change is widely viewed as a major emerging threat to biodiversity worldwide. Anthropogenic climate change represents a suite of alterations to the global environment that are occurring as a result of human combustion of fossil fuels. These alterations include increased concentrations of CO_2 in the atmosphere, increased mean air and water temperatures, greater variation in temperature and precipitation extremes, stronger and more frequent storm events, rising sea levels, and more acidic oceans. Collectively, these alterations are expected to produce rapid biological changes. In response to changing temperature regimes, species are already shifting their distributions toward the poles at a rate of 6.1 km per decade,[43] and continued north/south migrations are going to generate one of the greatest geographic redistributions of species ever recorded on the planet.[44] Even as species move toward a more suitable climate, they will be challenged by **population asynchronies** (interactions between two or more species that become out of phase) that decrease their fitness. Plants will flower days to weeks earlier than their pollinators appear.[43] The hatch of bird chicks will be out of sync with the hatch of their insect prey items.[45,46] Marine organisms may not be able to make their calcified body parts and siliceous shells as changes in pH alter seawater chemistry.[47] The fact that we are already beginning to document these biological changes, and that modern climate change is expected to be at least as fast as the climatic shifts that were responsible for mass extinctions in the geologic record, suggests the effects of anthropogenic climate change on biodiversity will be enormous; yet, most of these effects have yet to be realized. The impacts of climate changes on biodiversity are examined more fully in Chapter 12.

Most threatening factors

Every two years, the World Wildlife Fund (WWF) publishes the **Living Planet Index** (**LPI**) in its ***Living Planet Report***.[48] The LPI is used as an indicator of the state of global biodiversity, based on trends in vertebrate populations of species from around the world that are maintained in the Living Planet Database by the Zoological Society of London. The 2018 LPI was constructed from the trends of over 16,704 populations of birds (1513), mammals (597), fishes (1501), reptiles and amphibians (394) that represented almost 4000 terrestrial, freshwater, and marine species. These time series were gathered from a variety of sources such as journals, online databases, and government reports. The population time-series data were augmented with additional information, including the main threats to populations based on information provided by each data source.

Suggested Exercise 8.2 Go to the World Wildlife Fund's data portal for the Living Planet Index to find records of species in your home country. Go to one of these records and examine how the species population size has changed through time. As a bonus, read about how you can contribute your own monitoring data to the LPI project. **oup-arc.com/e/cardinale-ex8.2**

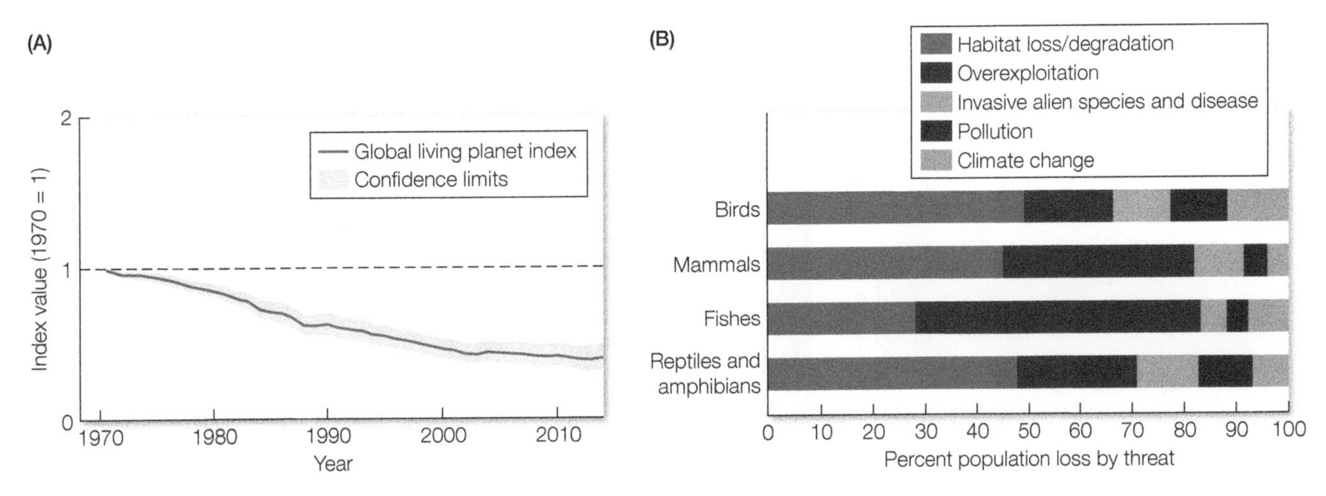

Figure 8.7 (A) The Living Planet Index (LPI), represented on the *y*-axis, is an indicator of the state of global biodiversity, based on trends in more than 16,000 populations that represent 1678 terrestrial, 881 freshwater, and 1353 marine species. The most recent LPI shows that relative to 1970 levels, the average species population has declined by almost 60%. (B) The primary causes of these population declines are habitat loss and degradation, followed by overexploitation. Invasive species and disease are common threats to amphibians and reptiles. (After WWF. 2018. *Living Planet Report 2018: Aiming Higher.* Gland, Switzerland: WWF International.)

Analysis of the 2018 LPI shows that between 1970 and 2014, the average population of mammals, birds, reptiles, amphibians, and fishes on the globe has declined by nearly 60% (**Figure 8.7A**). Species population declines have been especially pronounced in the tropics, with South and Central America suffering an average 89% loss of monitored populations compared to 1970. Freshwater species populations have also declined dramatically, with the Freshwater Index showing an 83% decline since 1970.

Habitat loss and degradation are the most commonly reported threats to species populations (**Figure 8.7B**), followed by overexploitation. Other threats vary in importance according to taxonomic group. Next to habitat loss and degradation, invasive alien species and disease are the most common threats to amphibians and reptiles. The introduction of nonnative rats, cats, and mongooses has had an enormous impact on native reptiles, especially on islands.[49] It is worth noting that while the perceived contribution of climate change to population declines is presently small, that is primarily because climate change effects are only recently being documented, and their largest effects will be mostly realized in the future.

Very often, species face multiple threats, and the interaction among different threats may be the ultimate cause of biodiversity losses. For example, the dodo (*Raphus cucullatus*) went extinct on the island of Mauritius in 1681 because of a combination of human overexploitation and nest predation by introduced cats, dogs, pigs, and rats. Many corals become more susceptible to diseases caused by viral, bacterial, and fungal pathogens when they become stressed physiologically by increases in water temperature (**Figure 8.8**).[50] Another example of threats combining to influence extinction risk occurred in 2015 when a mass mortality event led to the death of more than 200,000 saiga antelopes (*Saiga tatarica tatarica*) in central Kazakhstan, taking them from a healthy population to the brink of extinction in a mere three weeks. The proximate cause of death was hemorrhagic septicemia caused by the bacterium *Pasteurella multocida*. This bacterium is normally harmless to the antelopes and lives as a commensal.

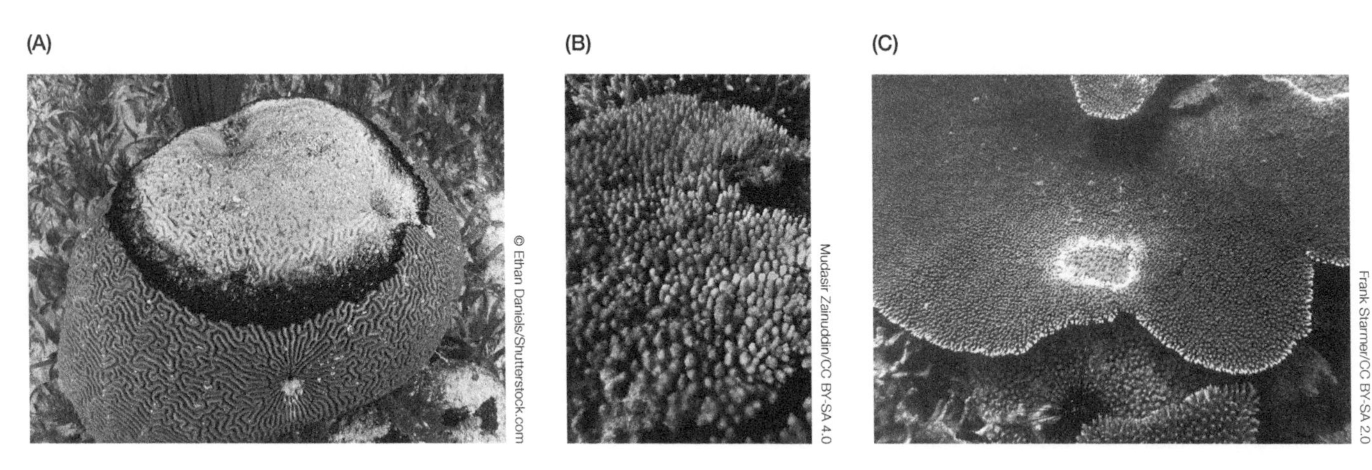

Figure 8.8 Corals are a good example of how multiple stressors can interact to cause population declines. As they respond to increasing water temperatures, many types of coral become physiologically stressed, which can subsequently make them more susceptible to disease due to viral, bacterial, and fungal pathogens. (A) Black band disease is present as dark pigmented areas of tissue on this coral in the Caribbean. (B) An acroporid coral is infected with white band disease, which is caused by a species of *Vibrio* bacteria that destroys the coral tissue. (C) White pox, characterized by circular lesions, is caused by the gram negative enterobacterium *Serratia marcescens.*

However, the strain became deadly after an environmental trigger—a period of unusually high relative humidity and temperature—increased its virulence. This was an example in which climate change resulted in a formerly harmless bacteria suddenly becoming deadly to wildlife.[51]

Local Changes in Biodiversity

At a global scale, changes in biodiversity are controlled by the balance of speciation and extinction. In contrast, biodiversity at subglobal scales is controlled by two opposing processes: (1) the local extirpation of species populations that cause species richness to decline and (2) the addition of new species to a habitat through migration and colonization that cause species richness to increase. The addition of new species can be either natural, such as the migration of species from one habitat to another, or accidental, such as occurs with the introduction of nonnative species.

The commonly held view in conservation biology has long been that local extirpations are leading to widespread declines in biodiversity at most locations on the planet. This view has primarily come from two lines of evidence: analysis of species-area relationships, and direct empirical measurements of biodiversity at various locations across the planet. However, the view has been modified recently by a subset of empirical studies that have shown biodiversity is increasing in certain types of habitats. Below we describe these different studies and their conclusions.

Species-area relationships

The species-area relationship describes a well-known empirical relationship between the area of a habitat and the number of species found within that area. The species-area relationship can be summarized by the formula:

$$S = CA^z \qquad (8.4)$$

where S is species richness (the number of species present), A is habitat area, C is a constant that depends on the unit of measurement used, and the exponent Z is a scaling factor that dictates how species richness changes with area. Some prefer to convert Equation 8.4 to the equation for a line, which is done by taking the log of both sides:

$$log(S) = log(CA^Z) = log(C) + Z \times log(A) \quad (8.5)$$

In this linear form, the exponent Z simply becomes the slope of a linear regression that relates species richness to habitat area. The relationship between S and A is easy to interpret for Equation 8.5, as a 1% change in A leads to a Z% change in S.

Values for C depend on the types of habitat being compared (tropical vs. temperate, dry vs. wet, etc.) and the types of species involved (birds vs. reptiles, etc.). Values of C will be high in groups such as insects that are high in species numbers, and they will be low in groups such as birds that are low in species numbers.[52] Drakare and colleagues[53] summarized results from nearly 800 studies relating species richness to habitat area and showed that Z is typically around 0.27, with a range from 0.15 to 0.35 depending on the group of organisms and ecosystem being studied. These values indicate that S is usually a positive, but decelerating, function of A.

If we have estimates of C and Z, it is straightforward to estimate how species richness S will change with habitat area A. Imagine a simple, hypothetical scenario where $C = 3.13$ and $Z = 0.28$ for fish that inhabit lakes. The species-area relationship would be:

$$S = (3.13)A^{0.28} \text{ or } log(S) = log(3.13) + 0.28 \times log(A) \quad (8.6)$$

Both forms of Equation 8.6 predict that fish richness increases with lake area in a positive, but decelerating, manner (**Figure 8.9**). The log-log form of Equation 8.6 is particularly easy to interpret, as it says that a 1% increase in lake habitat area (A) corresponds to a 0.28% increase in species richness (S).

The species-area relationship has been empirically validated to the point of acceptance by most biologists.[54] For numerous groups of plants and animals, it describes reasonably well the levels of species richness observed in natural systems.

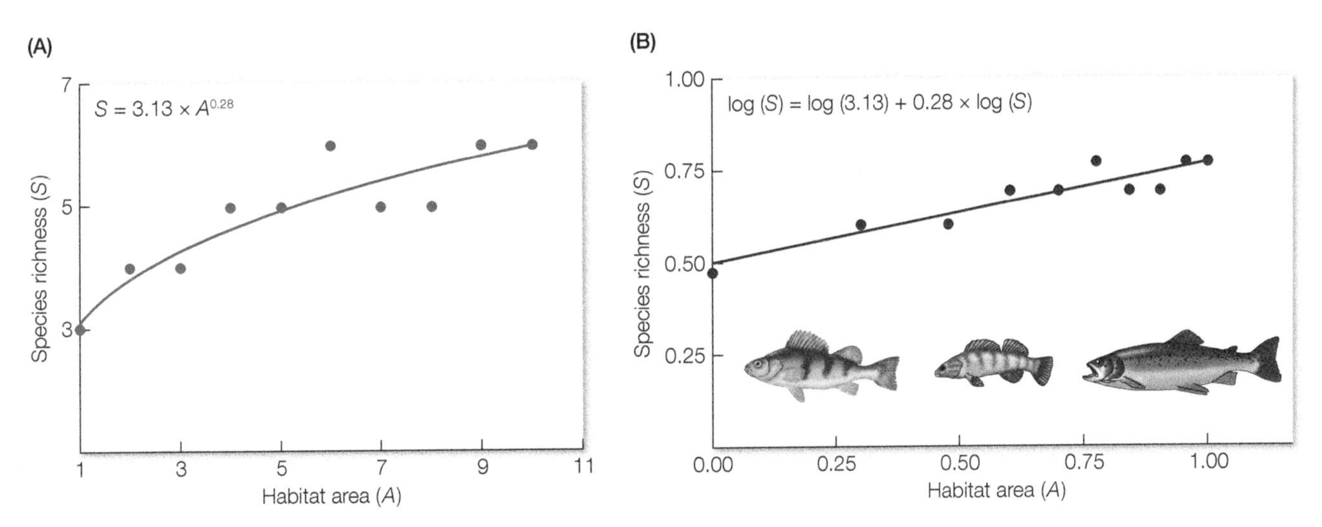

Figure 8.9 A hypothetical species-area relationship relating the species richness (S) of freshwater fish (y-axis) to the surface area (A, in hectares) of lakes (x-axis). (A) The relationship is expressed as the nonlinear power function $S = 3.13 \times A^{0.28}$. (B) The relationship is shown log transformed to the linear function.

The species-area relationship has also been used to predict the number and percentage of species that would go extinct if habitats were destroyed[55,56] and to estimate how many species might be protected by the formation of parks, reserves, or protected areas.[57] To understand how the species-area relationship is used in these situations, consider the analysis by Pimm and Askins,[58] which attempted to predict the number of bird extinctions that have occurred as a result of habitat loss in eastern North American forests. If the original area of the forest, A_O, is reduced to A_n (the area now), then the original number of species, S_O, should decline to S_n (the number of species now) according to the following:

$$\frac{S_n}{S_o} = \frac{CA_n^z}{CA_o^z} = \left(\frac{A_n}{A_o}\right)^z \tag{8.7}$$

The authors assumed that bird diversity scaled with forest habitat by $Z = 0.25$, and they estimated that deforestation since the 1600s had led to a 50% loss of habitat; therefore, $(A_n / A_o)^Z = 0.5^{0.25} = 0.84$. This means that 84% of the estimated 28 endemic bird species were expected to persist through the 50% habitat loss, whereas 16% of the bird species (4.4) were predicted to have gone extinct. Pimm and Askins[58] argued that this estimate matches the four known bird extinctions in eastern North America: passenger pigeon (*Ectopistes migratorius*), Carolina parakeet (*Conuropsis carolinensis*), ivory-billed woodpecker (*Campephilus principalis*), and Bachman's warbler (*Vermivora bachmanii*).

Wilson[59] used similar logic to predict how many species will go extinct as a result of clearing of rain forests. Using the conservative estimate that 1% of the world's rain forests is being destroyed each year, and assuming those forests harbor 5 million species worldwide, Wilson estimated a loss of 10,000 to 15,000 species per year, or 34 species per day. This estimate suggests that 125,000 species should have gone extinct over the 10-year period from 1989 to 1999 as the result of loss of rain forests. Yet, these predictions have not been born out for well-studied groups, such as birds and mammals. For example, there are 17 large areas of the world, including the Amazon Basin, the trans-Himalaya region, northern Canada, and the island of New Guinea, where there have been no recorded extinctions of vertebrates in the last 200 years.[60] There are several potential reasons why extinction rates are lower than predicted by the species-area relationship. One is that there are time delays between impacts on a species, such as through destruction of habitat, and the species' ultimate disappearance. Species-area models make no prediction about how long it will take for species to go extinct. Yet, small populations of some species can persist for decades or centuries in habitat fragments, even though they are no longer viable populations and their ultimate fate is extinction (**Figure 8.10**).[61] For example, it is estimated that 9% of Madagascar's species will eventually go extinct in coming decades and centuries because of the forest destruction that has already occurred.[62] The remaining individuals of species that are doomed to extinction have

Figure 8.10 St. Helena ebony (*Trochetiopsis ebenus*) is endemic to the island of St. Helena in the southern Atlantic Ocean. The wild population has been reduced to just two individuals on the side of a cliff. Given such low numbers, the species will almost certainly go extinct in the wild.

been called "the living dead,"[63] and the eventual extinction of species that will occur because of current habitat destruction is called the **extinction debt**.[64]

A second potential cause of the disparity between predicted and actual extinctions is the assumptions of the species-area model that all habitat areas are equivalent and that areas of habitat are lost at random. In reality, biodiversity is unequally distributed across a landscape, and areas of high species richness are sometimes targeted for species conservation. As a result, a greater percentage of species may be protected than is assumed in the species-area model. Another questionable assumption of the species-area model is that the ecological processes that generate species-area relationships in natural systems also hold true as species go extinct in those systems—that is, that the processes that add species to larger areas are the same as those that lead to extinctions from smaller areas. He and Hubbell[65] have argued that this assumption is rarely true and that species-area models almost always overestimate extinctions as a result. So while empirical models for species-area relationships are a useful tool for estimating extinctions caused by habitat loss and estimating the conservation value of protected areas, and while they may be reliable in some situations, they are not always accurate, and they should be used only as first approximations to be tested with actual empirical data on biodiversity change.

Empirical measures

There are two general types of empirical study that have been used to quantify biodiversity change and species extinctions at local and regional scales: spatial comparisons, and time-series monitoring. Spatial comparisons involve sampling or observing biodiversity at two or more locations and then using the difference in biodiversity between the two sites to estimate the magnitude of human impact. Time-series monitoring (in a biological monitoring program) involves repeatedly sampling biodiversity at the same location over time to determine whether there is any directional change. These two types of study have not always produced the same answer about local biodiversity change.

SPATIAL COMPARISONS Spatial comparisons are usually based on a comparison of biodiversity among two or more sites. At least one of those sites is selected to serve as a "reference" (or control) and is chosen to represent locations that have experienced relatively low human impact on the biological community. The other location(s) are selected as "impact" sites, representing areas that have been subjected to higher levels of human influence. The difference in biodiversity between the reference and impact site is an estimate of changes in biodiversity that have been caused by human activities.

Many empirical studies of local biodiversity change have used spatial comparisons between reference and impact sites.[22,66–70] Perhaps the largest of these, the PREDICTS project (Projecting Responses of Ecological Diversity In Changing Terrestrial Systems, oup-arc.com/e/cardinale-w8.3), has collated more than 3 million data records comparing over 47,000 species at over 26,000 sampling locations.[71] Newbold and colleagues[72] used the PREDICTS database to estimate that local species richness has declined by an average 13.6% across all habitats, and by 76.5% in habitats that are most affected by humans (**Figure 8.11A**). Murphy and Romanuk[66] got comparable numbers in their data synthesis of 327 comparisons of species richness in disturbed and undisturbed habitat, finding that human-mediated disturbances reduce local species richness by an average of 18.3%.

Studies that use spatial comparisons to contrast human-affected sites with reference sites have a number of limitations that are worth noting. Of these, the key challenge is to properly select and pair reference and impact sites. Ideally, they should be comparable in all respects other than human impacts so

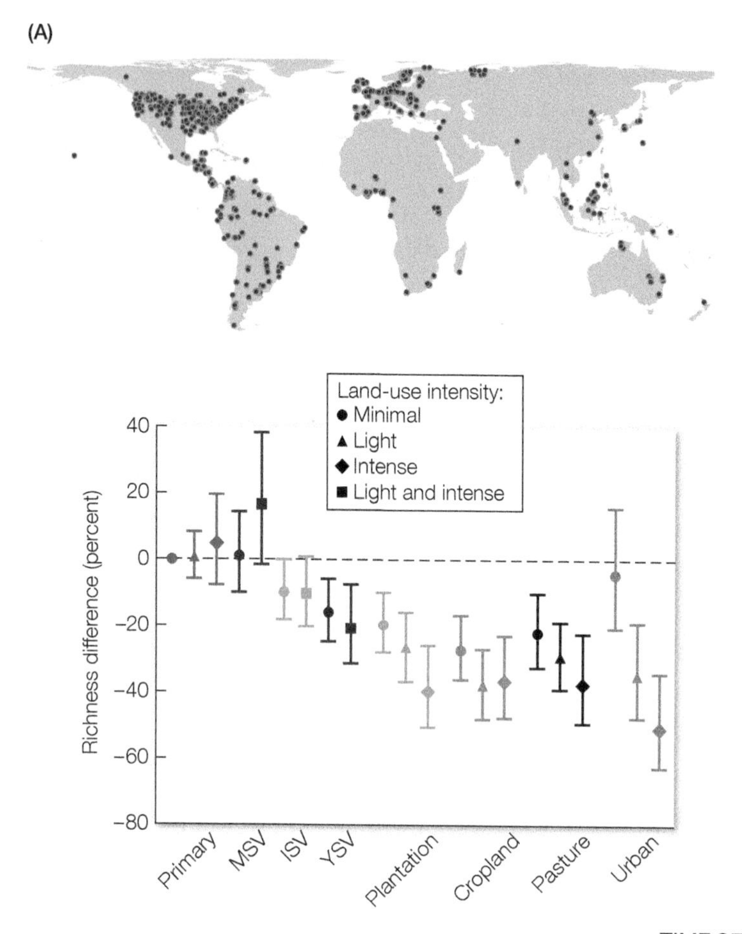

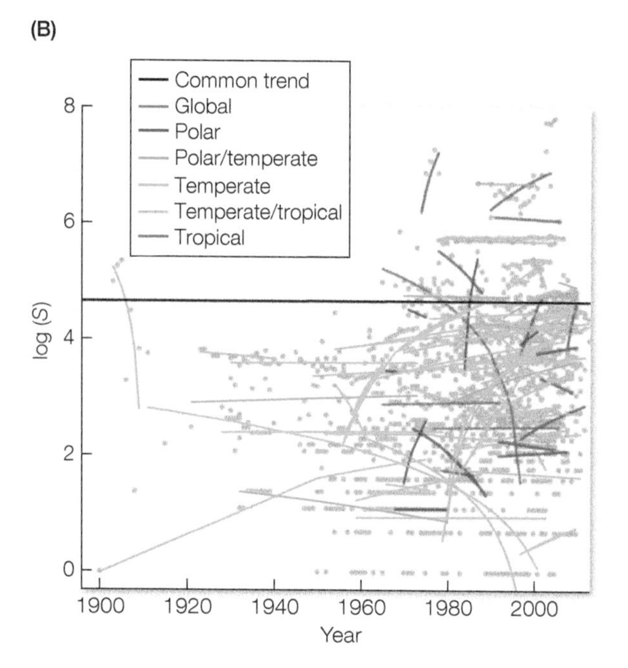

Figure 8.11 Empirically documented changes in local biodiversity. (A) Newbold and colleagues used spatial comparisons to quantify the difference in biodiversity between reference sites and paired sites (locations of paired sites shown as dots in map) that have experienced human disturbance. Results are shown for different habitat types: primary vegetation (Primary); mature secondary vegetation (MSV); intermediate secondary vegetation (ISV); young secondary vegetation (YSV); plantations; croplands; pastures; and urban habitats. (B) Dornelas and colleagues collated results from 100 biological monitoring programs that have collected time-series data on biodiversity (lines of varying colors), most of which were in intact sites that have not been destroyed or converted by human activities. The black line shows the authors' best statistical model, one that has a single slope, but which allows the time series to have differing intercepts. Note the line shows no net change through time. (A, from T. Newbold et al. 2015. *Nature* 520: 45–50; B, from M. Dornelas et al. 2014. *Science* 344: 296–299.)

that any differences in biodiversity can be attributed to the influence of humans. In practice, this is hard to achieve, because it is rare for reference and impact sites to be identical in all aspects of their geology, ecology, biology, and other nonhuman factors that influence biodiversity. Because of this, researchers interested in quantifying biodiversity change also look for other sources of information.

TIME SERIES (BIOLOGICAL MONITORING PROGRAMS) A second type of empirical study is the time-series, or biological monitoring, program. With this approach, a researcher repeatedly samples biodiversity at the same location over time to determine whether there is any directional change. Thousands of biological monitoring programs have been put in place around the world to monitor the populations of many different groups of organisms responding to, or recovering from, a wide variety of human perturbations. Most of these monitoring programs have not been coordinated with each other, have not been uniform in their methods and coverage, and are not easily accessible by scientists who wish to study broad trends in biodiversity across the terrestrial land surface and oceans of the planet.[73–77]

However, there have been a number of recent efforts to collate data from the vast array of biological monitoring programs in order to get a first approximation of local biodiversity change documented from time-series data.[78–80] An interesting result of these summaries is that they have generally failed to detect any systematic reduction in species richness at the locations that have been monitored. Species richness has declined at some locations, stayed constant in some places, but increased in others. Some researchers have speculated that the increases in diversity are due to the introduction of nonnative species, where these introductions have outpaced the local extirpation of native species. However, increased richness could also be driven by the turnover of native species that where once rare or by species' natural migration and colonization of new habitats. Whatever the cause, when locations experiencing increases in species richness are considered alongside those experiencing decreases, the net change in estimates of local biodiversity has centered around zero (**Figure 8.11B**).

Conclusions from syntheses of biological monitoring programs have been questioned on several grounds, including that (1) they mix and match studies of biodiversity responding to human impacts with those of biodiversity recovering from human impacts and (2) they tend to lack clear baselines that tell us whether changes in biodiversity are up or down relative to what is historically normal for a site.[81,82] But in spite of their limitations, time-series data serve to emphasize an important point, which is that we don't necessarily expect biodiversity to be in decline at all locations on the planet. Human activities affect a landscape in a variety of ways, and only some of these are expected to systematically reduce biodiversity. Consider the hypothetical scenario shown in **Figure 8.12**, where a relatively pristine forest habitat that existed at some time in the past has been transformed into a spatial mosaic of human impacts. Some habitats in this mosaic, such as at point 1 where the land has been cleared and prepared for agriculture, have experienced dramatic declines in biodiversity. Some habitats in the mosaic, like that at point 2, are patches of intact forest where biodiversity might be declining because of the effects of fragmentation, or they might be increasing because of colonization by exotic species or by species that thrive on edge habitat. Still other habitats in the mosaic, like the abandoned farm fields at point 3, are recovering from historic disturbance and show an increase in biodiversity as they return toward their original condition.

Spatial comparisons generally tend to focus on habitats that have undergone dramatic conversion or degradation (like those at point 1) and tend to emphasize losses of biodiversity as a result. In contrast, biological monitoring programs often focus on habitats that are still intact (like those at point 2) or that are in recovery after a historic disturbance (like those at point 3). Indeed, few

Figure 8.12 Natural or seminatural landscapes (A) that are transformed by human activities (B) are often a spatial mosaic of habitats, each with unique changes in biodiversity. Habitats that are destroyed or converted to other uses (point 1) tend to experience large declines in biodiversity. Habitat patches (point 2) may experience declines caused by fragmentation or increases in biodiversity due to colonization of novel species. Habitats that are recovering from disturbed conditions, such as the abandoned farm field at point 3, could return to historic conditions with no net loss of species.

biologists have historically monitored biodiversity in farm fields, cities, or other habitats that have been destroyed or converted, though that is now changing. Therefore, it should not be surprising that the different types of studies give differing answers about biodiversity change. However, the two approaches do need to be better coordinated into monitoring and assessment programs that keep track of which types of habitats and human impacts are being studied and of whether biodiversity change is being driven by the loss of native species or the introduction and spread of nonnative species.

Better monitoring and assessment programs

Because local biological monitoring programs are poorly developed in many parts of the world, and the monitoring programs that do exist often lack coordination, there is a great need for the development of more robust biodiversity monitoring and assessment programs. Fortunately, advances in remote sensing, new initiatives to coordinate regional networks, emerging global assessment programs, and more involvement by citizen scientists are helping us move ever closer to the goal of having real-time assessment programs that can detect changes to ecosystems and their biodiversity at local to global scales and then convey the risks associated with those changes to the public and to decision makers.

SATELLITE REMOTE SENSING (SRS) Habitat loss, which is the number one driver of biodiversity loss (see Chapter 9), has become routine monitoring for many ecosystems around the globe as free or low-cost satellite data and aerial images have become widely available in recent years. Long-term satellite programs such as the Polar-orbiting Operational Environmental Satellite (POES) system, Landsat, and NASA's Earth Observing System monitor large geographic areas around the globe. Images from these programs are available to conservation biologists developing land cover databases, and the data can be verified and updated easily and often. Four decades of Landsat images provide historical data that are crucial for comparison of land cover over time and for estimation of rates of change. Instruments and systems with high spatial resolution, such as light detection and ranging (lidar) and synthetic aperture radar (SAR), now provide excellent imagery that make it possible to measure the condition of plants and to determine soil properties.

REGIONAL OBSERVATION AND MONITORING NETWORKS While satellite remote sensing (SRS) now offers high spatial and temporal resolution monitoring, the biological information that can be provided is coarse. For example, SRS cannot monitor changes to biological diversity or the health of ecosystems and ecological processes they perform that provide biological benefits to society. For this type of biological information, one needs a set of instruments to be coordinated with on-the-ground data collection by biologists. Several of these regional monitoring platforms have been initiated in the past few years. In the United States, the National Science Foundation helped establish the National Ecological Observatory Network (NEON). NEON is a continental-scale ecological observation facility designed to collect high-quality, standardized data from 81 field sites (47 terrestrial and 34 aquatic) across the United States (including Alaska, Hawaii, and Puerto Rico). Data collection methods are standardized across sites and include in situ instrumented measurements, field sampling, and airborne remote sensing (Figure 8.13). Field sites are strategically selected to represent the major biomes of North America.

Various regional observation and monitoring networks like NEON are coordinated into a global set of Earth-observing systems by the Group on Earth

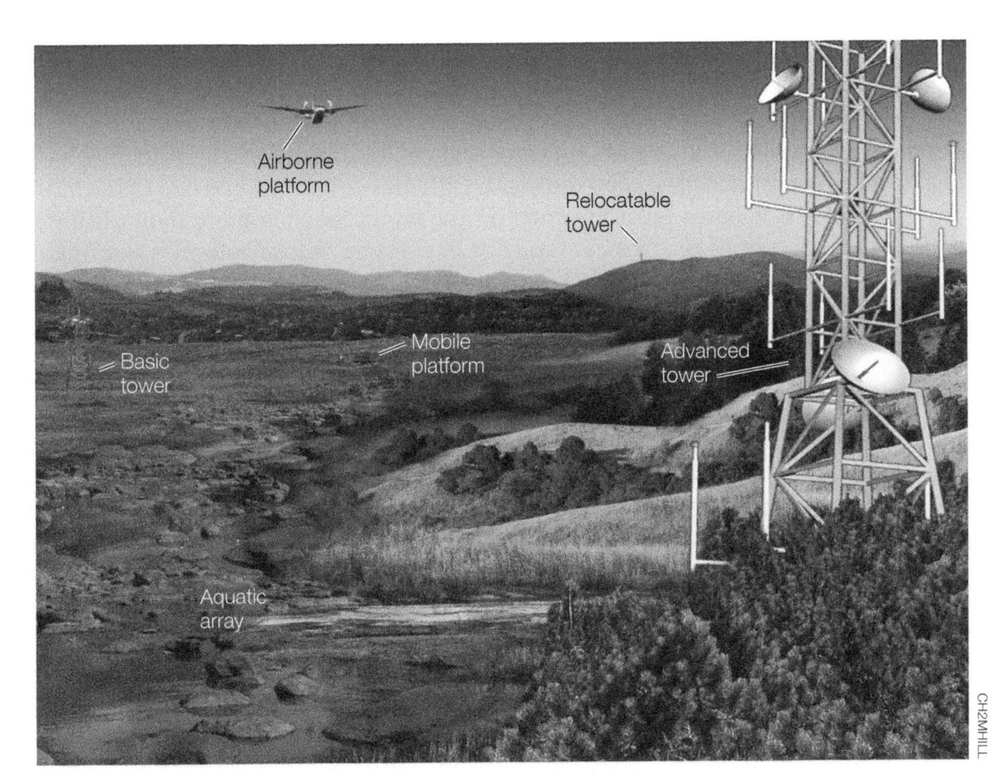

Figure 8.13 Regional observation and monitoring networks like the U.S. National Ecological Observatory Network (NEON) are designed with in situ sensor arrays, advanced instrumentation towers, and airborne platforms that monitor the health of natural ecosystems and the services they provide to society. The Global Earth Observation System of Systems (GEOSS) is an effort by more than 100 national governments to connect the instrumentation from regional networks like NEON to enhance data sharing and promote common standards for monitoring Earth's ecosystems.

Observations (GEO, oup-arc.com/e/cardinale-w8.4). GEO is a partnership of hundreds of government institutions, academic and research institutions, data providers, and businesses that routinely collect physical and biological data on Earth's ecosystems. Partners in GEO have worked together to create the Global Earth Observation System of Systems (GEOSS), which helps connect networks of sensors and monitoring infrastructures from individual partners, promotes the use of common methods so their data are comparable, and then uses agreed-upon standards for data access and sharing so that observations from regional monitoring programs can be joined together and made broadly available at larger scales. At present, there are more than 400 million open-access data resources in GEOSS that have been contributed from more than 150 national and regional providers, including the National Science Foundation the National Aeronautics and Space Administration (NASA) in the United States, and international organizations like the World Meteorological Organization (WMO). The Global Ocean Observing System (GOOS), which is sponsored by the WMO and the United Nations, is a similar effort to coordinate observing and monitoring systems from the global ocean, as well as related analysis and modeling of ocean fields in support of operational oceanography and climate change applications.

Within GEO, there is a subgroup of partners called the GEO Biodiversity Observation Network (GEO BON, oup-arc.com/e/cardinale-w8.5)[83] that specifically coordinates biodiversity observation and monitoring programs around the globe for integration into GEOSS. One of GEO BON's focal projects has been to create a regionally customizable online tool kit for the start-up of regional or national biodiversity observation systems (called BON in a Box; **Video 8.3**). In addition to GEO BON, there are complementary initiatives that are helping to coordinate local and regional observations of biodiversity into more global data sets. The best known of these is the Global Biodiversity Information Facility (GBIF, oup-arc.com/e/cardinale-w8.6).[84] GBIF is an international network and

VIDEO 8.3 Learn about GEO BON's BON in a Box project.
oup-arc.com/e/cardinale-v8.3

research infrastructure funded by the world's governments and aimed at providing anyone, anywhere, open access to data about all types of life on Earth. GBIF provides data-holding institutions around the world with common standards and open-source tools that enable them to share information about where and when species have been recorded. GBIF has developed into one of the most extensive databases on biodiversity in the world, hosting more than a trillion records of species occurrences recorded by everything from museum specimens to geotagged smartphone photos shared by amateur naturalists.

GLOBAL ASSESSMENT PROGRAMS The past five years have seen the formation or reorganization of several international platforms and initiatives that aim to address changes in global biodiversity and ecosystem services.

Recent establishment of the Intergovernmental Science-Policy Platform on Biodiversity and Ecosystem Services (IPBES)[85] has provided a mechanism for undertaking thematic, regional, and global assessments of scientific, indigenous, and local knowledge relating to biodiversity and ecosystem services. For example, IPBES has organized groups of world experts from most countries to perform coordinated reviews on the global status of animal pollinators and the ecosystem services they provide, as well as the threats posed by invasive alien species to biodiversity. These assessments are intended to be informed by regional observation and monitoring networks like those described in the section "Regional observation and monitoring networks" above.

Another important program is Future Earth, a 10-year international research program that aims to advance global sustainability science. Future Earth was launched in 2012 at the United Nations Conference on Sustainable Development (Rio+20). It incorporates several previous scientific programs of relevance to biodiversity and ecosystems (including DIVERSITAS) and is now serving as a high-level mechanism for coordinating and catalyzing research efforts across this broad field.[86]

Assessment programs like IPBES and Future Earth help inform, and meet the stated goals of, intergovernmental agreements or conventions. The best known is the Convention on Biological Diversity (CBD), a multilateral treaty that came into force in 1993, involving 195 member states.[87] Other conventions on biodiversity and ecosystems include the Convention on the Conservation of Migratory Species of Wild Animals; the Convention on International Trade in Endangered Species of Wild Fauna and Flora; the International Treaty on Plant Genetic Resources for Food and Agriculture; the Ramsar Convention on Wetlands of International Importance; the World Heritage Convention; the International Plant Protection Convention; the UN Framework Convention on Climate Change (UNFCCC); and the UN Convention to Combat Desertification.

CITIZEN SCIENCE Perhaps the fastest-growing source of biodiversity data is citizen science, which represents a major conservation success story (**Box 8.3**). Technological advances have changed the engagement of citizens in science over the past 30 years, with the internet reaching large and diverse pools of volunteers who can be engaged quickly to amass data.[88] The growth of personal sensor technology and development of handheld devices like smartphones with built-in global positioning system (GPS) capability and high-definition cameras encourages the nonprofessional to participate in data collection.[89] It has also fueled the development of mobile phone apps that facilitate biodiversity observation and reporting (e.g., iNaturalist, Map of Life, eBird, Pl@ntNet), while connecting people with nature. Citizen involvement in the frontier of biodiversity assessment at the continental scale in North America probably

BOX 8.3 Success Stories

Citizen Science and Biodiversity Monitoring

Biodiversity and the ecosystem functions and services associated with it are undergoing rapid changes worldwide, with potentially serious implications for human well-being and sustainability. Forecasting the changes in biodiversity and preventing undesirable consequences requires monitoring programs that produce records of species locations and abundances in real time and with high spatial resolution.

One of the greatest challenges in conservation biology has been to establish a globally distributed network of monitoring systems that can track changes in biodiversity. While many individual monitoring programs have been established, most have not been coordinated with each other, and the majority have been set up in well-developed countries that have high funding rates for environmental science. Scientists have increasingly called for a global biodiversity monitoring network similar to the worldwide network of meteorological stations that predict changes in weather. But the funding for such an ambitious goal has not materialized.

A major success story in modern conservation biology is the rise of citizen science programs that have begun to establish a global biodiversity monitoring network that tracks species distributions and population sizes. Among these programs, iNaturalist (see Figure) has been called the "standard-bearer."

iNaturalist is a citizen science project and online social network of naturalists, citizen scientists, and biologists that maps and shares observations of organisms across the globe. iNaturalist began as the master's project of Nate Agrin, Jessica Kline, and Ken-ichi Ueda at UC Berkeley; these students envisioned using crowdsourcing as a means to establish a global network of biodiversity observations. Any individual can participate in the crowdsourcing project and serve as an observer. An iNaturalist observer who encounters an organism takes a photo (or audio recording) with a smartphone or other mobile device and uses a mobile app to upload the photo and record the location and time of the observation. Other users of iNaturalist add identifications to each other's observations to confirm or improve the "community identification." Observations are classified as "casual," "needs ID" (needs to be identified), or "research grade" based on the quality of the data provided and the community identification process. In addition to crowdsourcing identifications, iNaturalist uses computer vision, an automated species identification tool that helps users identify organisms from photos, using an artificial intelligence model that has been trained using the

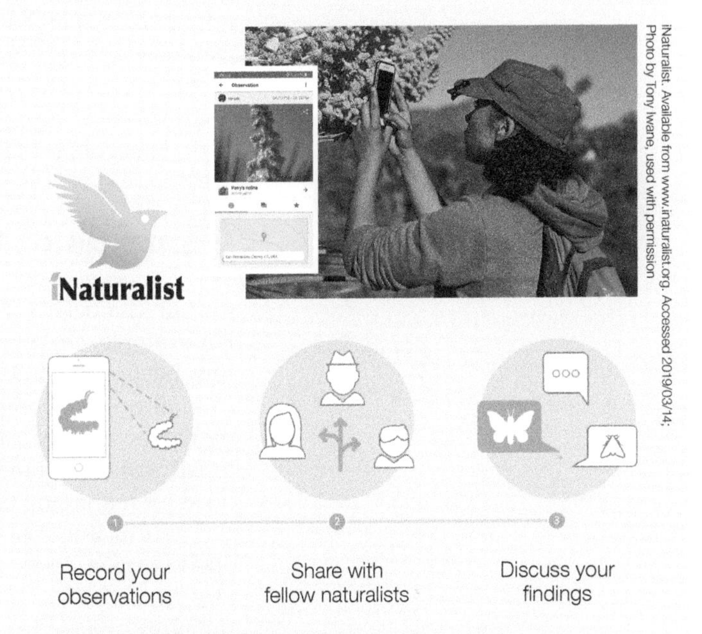

Citizen science programs like iNaturalist, which use crowdsourcing to identify species and collect data on their distribution and population sizes, are helping to establish a much-needed global biodiversity monitoring network.

research grade observations. Research grade observations are ultimately uploaded and incorporated into other online databases such as the IUCN Red List (see Box 8.1) or the Global Biodiversity Information Facility, which are used by scientists to monitor biodiversity.

As of May 2018, iNaturalist had nearly 1 million users who had contributed over 11,200,000 observations of plants, animals, and other organisms worldwide. Users of iNaturalist have used the tool to monitor roadkill, track fishing catches, and document the spread of invasive species. The tool is also routinely used to perform bioblitzes that survey biodiversity of a designated taxon or area, such as the Global Amphibian BioBlitz, which was used to help monitor the occurrence and distribution of the world's amphibian species (oup-arc.com/e/cardinale-w8.7).

iNaturalist is just one example of the extraordinary array of citizen science programs that have emerged to help monitor the world's biodiversity (Map of Life, Great Backyard Bird Count, eBird, NestWatch, eButterfly, Monarch Watch, Pl@ntNet, FrogWatch USA, and others). As the outcome of a master's degree project, it serves as an example showing that individuals at all stages of their career can make a difference. It also serves as a reminder that every individual who loves being in nature—regardless of their biological expertise—can help conserve biodiversity.

began with the National Audubon Society's Christmas Bird Count in the early 1900s and was accelerated by the North American Breeding Bird Survey that was initiated in 1966.[90] At local and regional scales, citizen scientists have monitored pollinators,[91] marine coral reef habitats,[92] wetland plants,[93] mammals by camera-trapping,[94] and threats to biodiversity such as invasive species.[95,96] The growth of citizen science has been accompanied by development of statistical methods to account for sources of uncertainty and to link these observations with data from established monitoring programs.[97,98]

Controls of Extinction Risk

Certain characteristics make some species more prone to extinction than others. For example, a high percentage of seabirds are in danger of extinction because they have low reproductive rates, they form dense breeding aggregations often in small areas where their eggs and nestlings are prone to attack by introduced predators, and their eggs are overharvested by people. By identifying characteristics of extinction-prone species, conservation biologists can anticipate the need for managing populations of vulnerable species.

There is another great need for identifying the characteristics of threatened species: Most threatened species that have been identified so far are also in the most-studied groups, such as birds and mammals, highlighting the point that only when we are knowledgeable about a species can we recognize the dangers it faces. A lack of knowledge about a group of species, such as beetles, ocean fishes, or fungi, should not be taken to mean that the species are not threatened with extinction; rather, a lack of knowledge should be an argument for urgent study of those species. The conservation status of amphibians, for example, was relatively unknown until about 20 years ago, when intensive study revealed that a high proportion of species were in danger of extinction (Box 8.4).

Demographic parameters

Demographics help us understand the size, status, and behavior of populations, and the biological parameters that control demographics have a large control on extinction risk. Here we describe just a few demographic parameters that are linked to extinction risk:

POPULATION SIZE Species that are characterized by small population sizes are often referred to as rare species. Rare species face significantly higher risks of extinction than common species and, as noted by Darwin,[99] the decline of a population to a rare status is often a sign that the species is at risk of extinction.

> To admit that species generally become rare before they become extinct, to feel no surprise at the rarity of the species, and yet to marvel greatly when the species ceases to exist, is much the same as to admit that sickness in the individual is the forerunner of death—to feel no surprise at sickness, but when the sick man dies, to wonder and to suspect that he died of some deed of violence.[99]

A species can be considered rare in any particular location where its population dwindles to a small number of individuals. Small populations are more vulnerable to chance events and natural fluctuations that lead to local extirpations. In addition, small populations can lose the genetic variability needed to adapt to a changing environment or to recover from a new disease or predator.

BOX 8.4 Case in Point

Why Are Frogs and Toads Croaking?

At the First World Congress of Herpetology in 1989 in Canterbury, England, what had previously seemed like casual findings began to take on a disturbing significance: scientists from around the world were seeing a decline in amphibian populations. Frogs, toads, salamanders, and other amphibians that had been common less than two decades before were becoming rare, with some species even going extinct. This led to a call for action, to determine what was happening and what could be done to stop it. In the years since the meeting, hundreds of studies have been published, in addition to dozens of review articles and books. To pull together this vast body of new information, a global amphibian assessment was carried out from 2000 to 2004.[112] This report and a recent international survey by the IUCN show the astonishing conclusion: 43% of amphibian species are declining in numbers, and 36% are threatened with extinction.

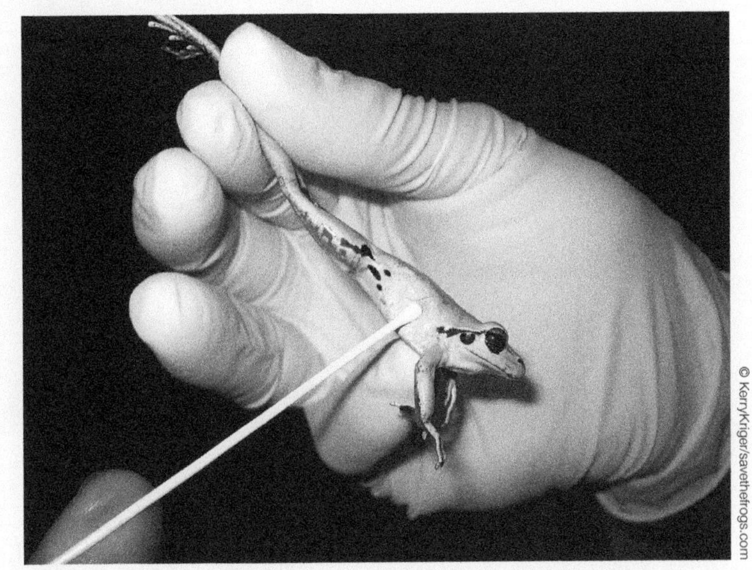

Worldwide studies have shown that amphibians face multiple threats. Infection by a waterborne fungus is a major contributor to amphibian decline. Here a researcher uses a cotton swab on an Australian frog, which will be tested for the presence of the disease-causing chytrid fungus.

These studies demonstrate that amphibians are particularly vulnerable to human disturbance, perhaps because many species require two separate habitats, aquatic and terrestrial, to complete their life cycles. If either habitat is damaged, the species will not be able to reproduce. Amphibians, like many other taxa, are also sensitive to pesticides, chemical pollution, and acid rain. The latter two factors may be particularly dangerous to these animals: chemical pollution and pesticides can easily penetrate the thin epidermis characteristic of amphibians, while slight decreases in pH can destroy eggs and tadpoles.

The loss of wetland habitat is especially important. For instance, the number of farm ponds, a favorite habitat for amphibians in Britain, has declined by 70% over the last 100 years. Introduced predatory fish, drought, unusual climatic events, and increased ultraviolet radiation due to a decrease in the protective ozone layer have subsequently been blamed for the decline of individual species. In many cases, these stress factors have apparently made species susceptible to fatal infections from a waterborne chytrid fungus (*Batrachochytrium dendrobatidis*).[113] The fungus has been implicated in the decline and extinction of 200 amphibian species (see Figure). This deadly fungus may be spread around the world by global commerce, especially trade in live aquatic organisms.

In the past 25 years, a huge effort on the part of the conservation and herpetology communities has provided evidence for the crises facing amphibians and many of the causes. Now that this information is available, people need to develop and implement an effective course of action. Such a conservation program may include protecting wetlands from destruction and pollution, reducing the spread of the harmful fungus, establishing new populations of threatened species at unoccupied sites, and when all else fails, developing captive populations. For less-known species, we cannot develop conservation strategies as long as the reasons for the declines remain unknown or beyond our control.

A species can also be considered rare when it occupies a narrow geographical range. Species that occupy only one or a select few sites in a restricted geographical range are at high risk of extinction because of the potential that the entire range is affected by human activity.[100] For example, the Iberian lynx (*Lynx pardinus*), which formerly occurred across Spain, Portugal, and southern France,

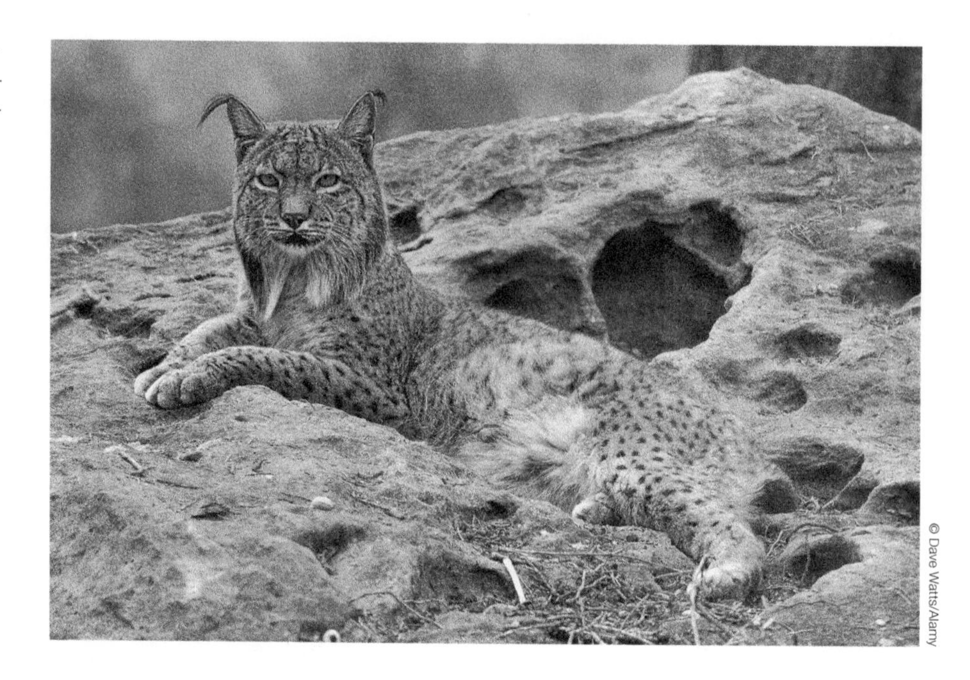

© Dave Watts/Alamy

Figure 8.14 The Iberian lynx (*Lynx pardinus*) occupies highly restricted geographic ranges, with only two breeding populations, in Spain. As a result, the species is the world's most endangered feline.

now has a restricted range with only two breeding populations, both in Spain (**Figure 8.14**). Species with limited ranges are particularly vulnerable to global climate change. For example, many tropical bird species with narrow ranges face increasing threats of extinction due to climate change in the coming decades.[101]

POPULATION DENSITY Population density is distinct from population size and is defined as the size of a population per amount of space that it occupies. Population density controls the influence of density-dependent factors on population growth, and some of these density-dependent factors influence extinction risk. For example, **Allee effects** are a set of phenomena in biology whereby population density influences the fitness of individuals.[102] Common examples include (1) **mate limitation**, whereby sexual reproduction becomes limited when population density gets sufficiently low to reduce the chance of finding a mate; (2) the reduced effectiveness of **cooperative defense**, in which aggregations protect against predators; and (3) the reduced effectiveness of **cooperative feeding**, in which the successful capture of prey is influenced by pack size. In each of these cases, populations that fall below a threshold density face problems with reproduction, defense, or feeding. Density-dependent Allee effects are especially important to consider for the conservation of social animals.

FECUNDITY AND MORTALITY Fecundity is the potential for reproduction of a population, measured by the number of gametes (eggs), seed set, or asexual propagules produced per unit time. Mortality is a measure of the number of deaths in a particular population per unit of time. Together, fecundity and mortality dictate population growth rates. Any population that has low fecundity for a given mortality rate will grow slowly. Similarly, any population that has high mortality for a given fecundity will grow slowly. Species populations that grow slowly (low fecundity or high mortality) are less resilient to human perturbations and are especially vulnerable to overexploitation.[103]

Ecological controls

Certain aspects of a species' ecological niche also influence the risk of extinction. Here we discuss just two: endemism and isolation.

ENDEMISM A species found naturally in a single geographical area and no other place is **endemic** to that location. Endemism is an important ecological factor that determines a species' risk of global extinction. If the populations of an endemic species on Madagascar, or any isolated island, go locally extinct, the species will also become globally extinct. In contrast, mainland species often have many populations distributed over wide areas, so the loss of one population is not catastrophic for a species.

The concept of endemism may seem similar to the properties of rare species that live in a narrow geographical range. But a species may be endemic to a large area and abundant throughout it. And a rare species may be found in only a limited area, or it may be considered rare in only part of its range. For example, the sweetbay magnolia (*Magnolia virginiana*) is reasonably common throughout the southeastern United States, but in the New England region this species is considered rare because it occurs in only one population of 100 individuals in one particular swamp in Magnolia, Massachusetts.

Expansion of an endemic species' geographical distribution that is caused deliberately or accidentally by humans is not considered part of the species' natural distribution. For example, the giant panda (*Ailuropoda melanoleuca*) is endemic to China, even though it now lives in zoos throughout the world (see Chapter 16). Similarly, the black locust tree (*Robinia pseudoacacia*) is native to the eastern and southern United States but has been widely planted elsewhere in North America, Europe, and other temperate regions as a timber tree and an ornamental, and it has spread aggressively into native vegetation in these regions. A species may be endemic to a wide geographical area. For example, the black cherry tree (*Prunus serotina*) is endemic to the Western Hemisphere and is found across North, Central, and South America. Or a species may be endemic to a small geographical area, such as the giant Komodo dragon (*Varanus komodoensis*), which is endemic to just a few small islands in the Indonesian archipelago. Species that occupy small areas because they have only recently evolved from closely related species are designated **neoendemics**; examples include the hundreds of species of cichlid fish that occupy Lake Victoria in East Africa. In contrast, **paleoendemics** are ancient species whose close relatives have all gone extinct; examples include the giant panda and the Indian Ocean coelacanth. All such narrowly distributed endemic species are of concern for their potential to become extinct.

ISOLATION Species populations that occupy isolated geographical units (e.g., remote islands, solitary mountain peaks) often face a higher risk of extinction than populations that are connected. Isolation reduces the probability of colonization by individuals from other nearby populations, which is unfortunate because a population that is not isolated can sometimes benefit from the **rescue effect** (the rescue of declining populations by immigration) if it becomes small, experiences high mortality from a disturbance, or begins to suffer from inbreeding. Isolation, and the probability of colonization by neighboring populations, is one of the greatest determinants of extinction risk for **metapopulations**, that is, local populations that interact spatially with other populations.

Life-history traits

In addition to demographic parameters and ecological controls, extinction risk is influenced by life-history traits that represent inherent properties of a species' biology. Here we discuss several of the most important.

BODY SIZE Large-bodied animals tend to have large individual ranges, low reproductive rates, and high food requirements. In addition, they tend to be hunted and subjected to overexploitation by humans. Top carnivores, especially, are often killed by humans because they compete with humans for wild game, sometimes damage livestock, and are hunted for sport. Within groups of species, often the largest species will be the most prone to extinction; that is, the largest carnivore, the largest lemur, the largest whale (Figure 8.15). In Sri Lanka, for example, the largest species of carnivores—leopards and eagles—and the largest species of herbivores—elephants and deer—are at the greatest risk of extinction. For plants, species with large, short-lived seeds are more vulnerable than are species with smaller, long-lived seeds.[104]

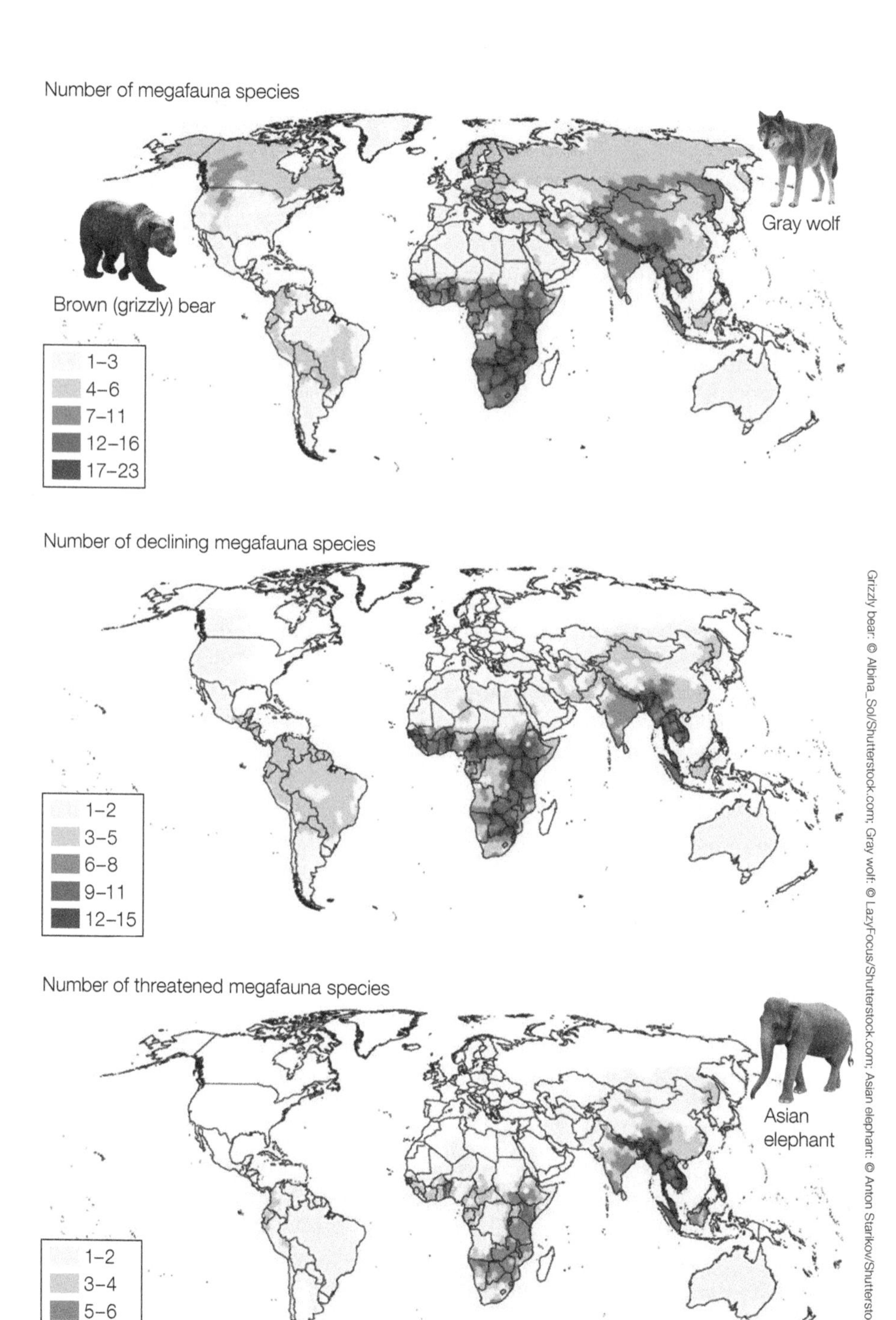

Figure 8.15 The world's largest species at risk. Scientists examined the conservation status of 105 of the world's largest land animals and found that 60% are threatened with extinction. These megafauna (very large animals) are concentrated in sub-Saharan Africa and Southeast Asia and are often keystone species with profound impacts on their ecosystems. (After W. J. Ripple et al. 2016. *BioScience* 66: 807–812.)

DISPERSAL AND MIGRATION Species that are unable to adapt physiologically, genetically, or behaviorally to changing environments must migrate to more suitable habitat or face extinction. The rapid pace of human-induced environmental change often prevents evolutionary adaptation, leaving migration as the only alternative. Species that are unable to cross roads, farmlands, and disturbed habitats are often doomed to extinction as their original habitat becomes affected by fragmentation, pollution, exotic species, and global climate change. Dispersal is important in the aquatic environment as well, where dams, point sources of pollution, channelization, and sedimentation can limit movement. Limited ability to disperse may partially explain why 69% of freshwater mussels and snails in the United States are extinct or threatened with extinction, while only 18% of dragonfly species, which can fly between the aquatic sites needed for their larval stages, face extinction.[105]

Species that migrate seasonally depend on two or more distinct habitat types, and any human activity that disrupts their migration among habitats increases the risk of extinction. For example, where barriers to dispersal are created by roads, fences, or dams between the needed habitats, a species may be unable to complete its life cycle[106] (see Chapter 15). Species that cross international barriers represent a special problem, in that conservation efforts must be coordinated by more than one country. Imagine the difficulties of coordinating conservation and management plans that protect the dispersal routes of Siberian cranes (*Grus leucogeranus*), which migrate 4800 km each year from Russia to India and back, crossing six highly militarized, tense international borders.

AGGREGATION Animals that live or rest in tight groups, especially in enclosed or isolated areas, are at risk from hunters and other predators. For example, some species of bats aggregate in caves during the day, during the time of year when infants are being nursed, or when they hibernate. Many large herbivorous mammals, like caribou and bison, aggregate in herds, and several species of fishes and birds tend to travel in schools or flocks. All of these species are vulnerable to local extinction because of these lifestyles.[107] People who know where and when to find them can, and sometimes do, harvest every individual in the area (see Chapter 10).

Even global species like sea turtles can be threatened with extinction by hunters who overharvest at the few breeding grounds available to them. Plants that aggregate, in groves or along water courses, for instance, are also at risk. Overharvesting of wild nuts or fruits can prevent the growth of seedlings that would continue their species in those areas.

Stochastic processes

The term *stochastic* is used to describe something that is randomly determined. Stochastic processes are random processes that generate variation (or "noise") in the biological parameters that influence a species' population growth. Variations in biological parameters caused by stochastic processes increase the probability of extinction, particularly for species that are rare. Stochastic processes also make it hard for conservation biologists to predict the future state of a population with precision, so population models that quantify a species' viability and survival often use biological parameters drawn at random from potential distributions (see Chapter 13).

There are two types of stochasticity that conservation biologists care about: **Demographic stochasticity** is random fluctuations in population size that occur because the birth and death of each individual in a population is a discrete and probabilistic event. That is, the probability of birth and death varies randomly among individuals in the population, which causes the size of the population to fluctuate through time. Demographic stochasticity can also affect the gender ratio in a population, sometimes causing it to depart from equal numbers of males and females. Demographic stochasticity is particularly important for small populations because it can increase the probability of extinction. **Environmental stochasticity** is unpredictable spatial or temporal fluctuation in a population caused by varying environmental conditions. The environment may be any set of abiotic (e.g., temperature or nutrient availability) or biotic (e.g., predators, competitors) conditions that organisms experience. Environmental stochasticity influences how population abundance fluctuates and affects the fate (e.g., persistence or extinction) of populations. For example, the amount of rainfall might vary greatly from year to year and affect the seed set and reproduction of a rare annual wildflower species.

Random variation in biological parameters caused by demographic or environmental stochasticity makes it more difficult to predict the future size of a population with precision, and it increases the risk of extinction, particularly for rare species (see Chapter 13). This point can be illustrated with simulations of the simplest population growth model—the geometric growth equation:

$$N_{t+1} = (b - d)N_t + N_t \tag{8.8}$$

where N_t is the current population size at time t, N_{t+1} is the population size at one time step in the future, and b and d are the per capita birth and death rates, respectively. If the demographic parameters b and d are fixed (there is only one possible value for each), then population growth is said to be deterministic because there is only one possible value for N_{t+1} at each time step. The population will grow deterministically with no variation (**Figure 8.16**, black line).

If, however, the demographic parameters b and d are variable, chosen at random from a normal distribution to simulate stochastic variation in birth or death rates caused by a fluctuating environment (e.g., years with good or bad food conditions), then population growth is stochastic (see Figure 8.16, gray lines). Each simulation of Equation 8.8 shows the population is more variable through time as b and d change at each time step (note variation in each gray line). Different simulations of Equation 8.8 show different outcomes for the population, even the potential of extinction (see Figure 8.16, red line) caused by stochastic variation. Because of the variation through time, and different possible outcomes, stochastic population models are usually run thousands of times to generate probability distributions showing the chance of survival of a species.

Stochastic population models are a fundamental tool used by biologists to assess the health of species populations and to predict the probability of persistence through time. Once you understand how to use and interpret these models, you will be capable of completing population viability analyses that are used to assess whether species are threatened with extinction (Chapter 13), estimate the effective population size needed for

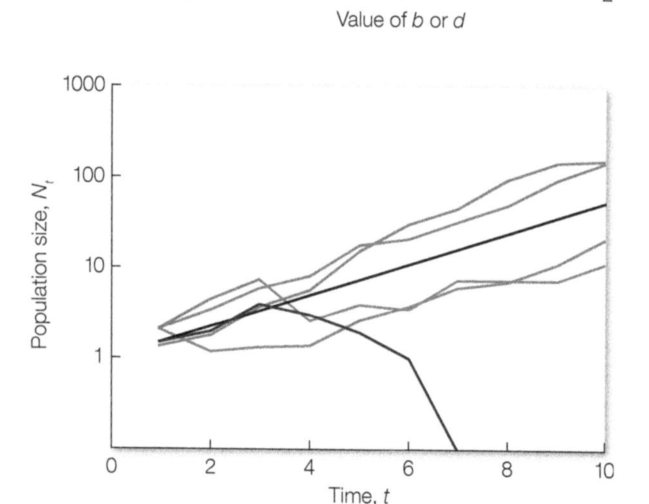

Figure 8.16 Simulations of a simple population model illustrate the impacts of stochasticity on population growth. With no variation in birth (b) or death (d) rates, populations grow deterministically (lower graph, black line). When b and d are selected at random from a normal probability distribution (upper graph), populations may grow faster or slower (gray lines) or even go extinct (lower graph, red line).

a species to persist (Chapter 13), calculate the maximum sustainable yields for species that are harvested (Chapter 10), predict a species response to climate change (Chapter 12), and calculate the probability of a species' persistence in a metapopulation or metacommunity (Chapter 9).

Suggested Exercise 8.3 Practice developing a stochastic population model and use it to predict the impacts of demographic stochasticity on species survivorship, using this online exercise. **oup-arc.com/e/cardinale-ex8.3**

Summary

1. The IUCN Red List is the world's most comprehensive listing of the global conservation status of threatened and endangered species. Of the 96,951 species that have been assessed, the Red List includes records for 941 species that have gone extinct, as well as 26,840 species that are considered threatened with extinction.
2. The total number of species that have gone globally extinct represents less than 1% of known or estimated species for most groups of organisms. However, the documented rate at which species have gone extinct in the last several hundred years, particularly since 1900, is 100 to 10,000 times faster than what is considered normal in the fossil record. This high rate of extinction has led some scientists to claim we are entering the sixth mass extinction.
3. The ultimate causes of biodiversity loss are human population growth and resource consumption. Population growth and consumption drive a variety of human activities that lead to major forms of environmental change like habitat loss, overexploitation, spread of invasive alien species and disease, and climate change. These forms of environmental change are the proximate causes of biodiversity loss on the planet.
4. Species-area relationships have been used to estimate both how many species will be lost when a certain amount of habitat is damaged, and how many species will remain when habitat is conserved by national parks and other protected areas.
5. Spatial comparisons and time-series monitoring show that biodiversity has declined substantially in locations that have been subjected to intense human pressure, such as habitats that have been heavily degraded or converted to human use. In contrast, habitat fragments that have remained intact, and habitats that have experienced less intense human pressure, have shown little decline in biodiversity or, in some instances, increases in biodiversity. The increases suggest that the colonization of habitats by novel species—either naturally or via human introductions—has outpaced local extirpations in some locations.
6. Species that are most vulnerable to extinction have particular characteristics, including small population size/density; low fecundity/high mortality; narrow range and isolation; and select life history traits like large body size, low dispersal, and a propensity to aggregate, and they are subject to demographic and environmental stochasticity.

For Discussion

1. Are we presently in the sixth mass extinction in the history of life on Earth? How does your answer depend on whether you consider the percentage of species that have already gone extinct this century verses the rate at which species are going extinct?
2. Take a moment to look up a well-known threatened species online, such as the Australian koala, the North Atlantic right whale, or the African cheetah (oup-arc.com/e/cardinale-ex8.1). Why are these particular species currently threatened with extinction? Use the IUCN criteria (in Box 8.1) to determine the appropriate conservation category for one or more species.
3. For predicting local extinctions and subglobal changes in biodiversity, common methods include using the species-area relationship, making spatial comparisons, and collecting time-series data from biological

monitoring programs. What are the strengths and weaknesses of each approach? Which of these do you trust most, or least?

4. Animals with certain life history traits, such as large body size, long-distance dispersal, and aggregation behaviors, are generally considered to be more prone to extinction. However, there are certainly exceptions to these generalities. Can you think of species that have declined or persisted despite these trends?

Suggested Readings

Bull, J. W. and Maron, M. 2016. How humans drive speciation as well as extinction. *Proceedings of the Royal Society B: Biological Sciences* 283: 20160600. A balanced summary of how human activities can both increase and decrease biodiversity.

Cardinale, B. J. et al. 2018. Is local biodiversity declining or not? A summary of the debate over analysis of species richness time trends. *Biological Conservation* 219: 175–183. Evaluates the pros and cons of a common line of evidence used to estimate local extinction.

Ceballos, G. et al. 2015. Accelerated modern human–induced species losses: Entering the sixth mass extinction. *Science Advances* 1: e1400253. A great summary of data that have led some to argue we are entering the sixth mass extinction.

De Vos, J. M. et al. 2015. Estimating the normal background rate of species extinction. *Conservation Biology* 29: 452–462. Shows how background rates of extinction are calculated.

He, F. L. and Hubbell, S. P. 2011. Species-area relationships always overestimate extinction rates from habitat loss. *Nature* 473: 368–371. Provides an important warning about the use of species-area relationships for estimating extinctions caused by habitat loss.

Himes Boor, G. K. 2014. A framework for developing objective and measurable recovery criteria for threatened and endangered species. *Conservation Biology* 28: 33–43. The author proposes developing statistical models and gathering data that can be used to determine how conservation measures will reduce the probability of extinction.

Mace, G. M. et al. 2008. Quantification of extinction risk: IUCN's system for classifying threatened species. *Conservation Biology* 22: 1424–1442. Describes the IUCN's system for measuring threats to species.

Murphy, G. E. P. and Romanuk, T. N. 2014. A meta-analysis of declines in local species richness from human disturbances. *Ecology and Evolution* 4: 91–103. Good data synthesis on local biodiversity loss.

Pena, J. C. C. et al. 2014. Assessing the conservation status of species with limited available data and disjunct distribution. *Biological Conservation* 170: 130–136. Scientists must often assess the conservation status of species using incomplete data.

Sekercioglu, C. H. 2012. Promoting community-based bird monitoring in the tropics: Conservation, research, environmental education, capacity-building, and local incomes. *Biological Conservation* 151: 69–73. Nice example of how citizen science can help advance biodiversity-monitoring programs.

Tingley, R. et al. 2013. Life-history traits and extrinsic threats determine extinction risk in New Zealand lizards. *Biological Conservation* 165: 62–67. Certain characteristics make particular species vulnerable to extinction.

Wilcove, D. S. and Wikelski, M. 2008. Going, going, gone: Is animal migration disappearing? *PLOS Biology* 6: 1361–1364. Fences, roads, farms, and other human activities prevent the free movement of migratory species.

Visit the
Conservation Biology
Companion Website
oup.com/he/cardinale1e
for videos, exercises, links, and other study resources.

9

Habitat Loss, Fragmentation, and Degradation

The bay checkerspot butterfly (*Euphydryas editha bayensis*) is endemic to the San Francisco Bay region of California in the United States. Habitat loss and fragmentation have caused the butterfly population to hit record lows, leading to its listing under the U.S. Endangered Species Act.

Every living thing on Earth needs somewhere to live, a place to find food, and room to reproduce. But the habitats that organisms need to live, feed, and reproduce are increasingly being lost and converted to human use, fragmented and disconnected from one another, and degraded by human activities. The dwindling area of fragmented habitat, along with the degradation of what remains, is making it harder for organisms to complete their life cycles. Figuring out how to save biodiversity in these types of landscapes is also the single greatest challenge in conservation biology.

Habitat loss poses the greatest threat to the variety of life on this planet today. Habitat loss refers to complete elimination of habitats, along with their biological communities and ecological functions. It usually results from the conversion of natural or seminatural habitat (e.g., a managed forest) into human-dominated habitat (e.g., a village or parking lot). Habitat loss has been identified as the leading cause of population decay for most wildlife,[1] and it is responsible for nearly 50% of all threatened mammal, bird, and amphibian species on the International Union for Conservation of Nature (IUCN) Red List of Threatened Species.[2] It is also the primary contributor to the decline of many of the world's most diverse ecosystems.[3] Whether it's deforestation in the Amazon, urban development in Asia, or mining in the Arctic, wholesale loss of Earth's habitats is jeopardizing the long-term survival of species.

Habitat fragmentation refers to the process by which larger, continuous habitats become subdivided into a greater number of smaller patches. Even as fragmentation contributes to the loss of habitat area, it also initiates a cascade of changes in the configuration of landscapes that alter the shape, size, and isolation of habitat patches in a landscape—all of which can have distinct impacts on biodiversity. **Habitat degradation** refers to a suite of human activities that make the remaining habitat patches in a landscape less conducive to life, in turn eroding biodiversity. These activities include many forms of pollution (pesticides, heavy metals, nutrients, plastics, personal care products). They also include activities that lead to desertification, erosion, and sedimentation, all of which make habitats less hospitable.

As the human population continues to grow in the coming century, even more of the world's forests, wetlands, streams, grasslands, coral reefs, and other habitats will disappear or be degraded as humans make way for more agriculture, housing, roads, pipelines, and other modes of development. In coming years, conservation biologists will be trying to protect biodiversity in a world that is going to have less habitat, and where the habitats that remain will be more fragmented and degraded. This chapter explains how habitat loss, fragmentation, and degradation affect biodiversity, and it introduces some key concepts and models needed to increase biodiversity's chance of survival in human-dominated landscapes.

Habitat Loss

Habitat loss, or **habitat destruction**, refers to the disappearance of ecosystems and the native organisms that inhabit them.[4] Habitat loss most often occurs as a result of **land-use change**, which is the conversion of one habitat type to another. An example would be when a natural or seminatural habitat like a forest or wetland is converted to a human-dominated ecosystem, such as a palm oil plantation or apartment complex. While human-dominated ecosystems still support various forms of life, the native species populations residing in the original habitat are typically destroyed or displaced by the land-use change, leading to a loss of biodiversity.

Many human societies tend to follow a common sequence of land-use change as they develop (**Figure 9.1**).[5] During the initial human colonization of natural ecosystems, there is a frontier stage when people clear the land of vegetation to gather building materials (e.g., lumber) and to make way for farms. After the frontier clearings, small human settlements are supported by small-scale subsistence agriculture. This is a period of land-use change when the proportion of natural habitat in a landscape drops precipitously. As development progresses, food systems transition into more-intensive agricultural practices, such as large-scale farms for commercial markets, which are associated with irrigation and transportation networks and other major infrastructure support. The increased amount of land dedicated to agriculture, and the intensification of food production systems, results in a surplus of food that can support a growing urban population. As agriculture and urban areas come to dominate the landscape, the proportion of natural habitat declines toward zero. In response, some societies may set aside a small fraction of the landscape as protected or recreational habitat.

Different parts of the world are presently in different stages of land-use transition, depending on their history, as well as social and economic conditions. Furthermore, not all parts of the world move linearly through these

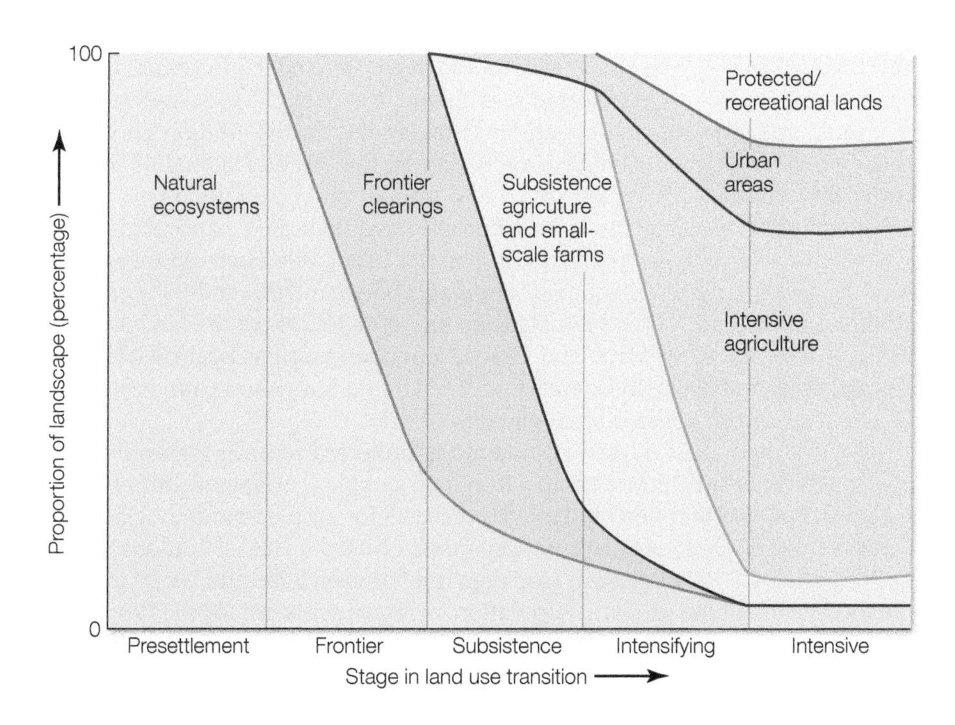

Figure 9.1 An idealized sequence of land-use transitions. As societies progress through demographic and economic transitions, they tend to follow a predictable sequence of land-use change: natural ecosystems that are dominant during pre-settlement give way to frontier clearings, which then transition to subsistence agriculture and small-scale farms, which then give way to intensive urban development, and protected or recreational lands. Various parts of the world are in different stages of these transitions depending on their history, socioeconomic conditions, and ecological histories. Furthermore, some places remain in one stage of land-use transition for a long period of time while others move rapidly between stages. (After R. S. DeFries et al. 2004. *Front Ecol Environ* 2: 249–257.)

transitions—some places remain in one stage for long periods of time while others move rapidly between stages. Nevertheless, Figure 9.1 shows a common sequence of land-use changes that has played out repeatedly in different parts of the world.

Primary drivers of habitat loss

Human activities have already transformed more than 50% of the land surface of the planet,[6] modified two-thirds of the world's water supplies,[7] and altered greater than 40% of the world's oceans.[8] Some of the most prominent human activities that have driven these transformations are described next.

AGRICULTURAL EXPANSION Agricultural habitats presently occupy about 40% of Earth's terrestrial surface, representing the single largest use of land on the planet. Crop and livestock farming have contributed to habitat loss since agrarian societies began; however, land-use change caused by agricultural expansion accelerated greatly during the Industrial Revolution as farming became mechanized and global markets became linked by swifter transportation and commerce.[9] The green revolution that took place between the 1930s and 1960s produced several technological advances that intensified the manner in which crops are grown, including modern irrigation, development and use of pesticides, and the formation of synthetic nitrogen fertilizer. Those technologies helped double global grain production, greatly reducing food shortages in turn. But technological advances also facilitated expansion of agriculture into areas that had previously been uncultivated. Between 1700 and 1950, the amount of land on the planet dedicated to agriculture increased from 2.65 to 12 million km^2—a 450% increase over a period of 250 years.[10]

By the late 1990s, the rate of agricultural expansion began to slow, and even decline in some areas of the world, as a fraction of previously used farms were abandoned. In 2017, a collation of satellite imagery put together by the U.S. Geological Survey (USGS) showed there were 18.7 million km^2 of cropland in the world.[11] In addition to croplands, pastures covered an estimated 33.8 million km^2 (Figure 9.2A).[12] While earlier studies had suggested that China or the United States had the most area dedicated to cropland, the 2017 USGS study[11] showed that India actually ranks first in agricultural land, with 1.80 million km^2 (9.6% of the global cropland area). India is followed by the United States (1.68 million km^2, 8.9%), China (1.65 million km^2, 8.8%), and Russia (1.56 million km^2, 8.3%). Select countries in South Asia and Europe have dedicated particularly high fractions of their total land area to croplands. Croplands made up more than 80% of Moldova, San Marino, and Hungary; between 70% and 80% of Denmark, Ukraine, Ireland, and Bangladesh; and 60% to 70% of the Netherlands, United Kingdom, Gaza Strip, and Czech Republic as well as Spain, Lithuania, Poland, Italy, and India. By comparison, the United States and China each had 18% of their land dedicated to croplands.

By 2050, the global human population is projected to have grown from its present 7.7 billion to 9 or perhaps 10 billion people.[13] Somehow, all of these additional people will need to be fed. Forecasts for crop demand in 2050 vary widely. The Food and Agriculture Organization of the United Nations (FAO) has projected a 56% increase in crop demand between 2006 and 2050[14] to meet the food needs of the growing population. A report prepared for the World Resources Institute adjusted this forecast to account for revised population growth estimates and the need to ensure adequate nutrition in all world regions, arriving at a 69% higher crop demand in 2050 compared with 2006.[15] Using a

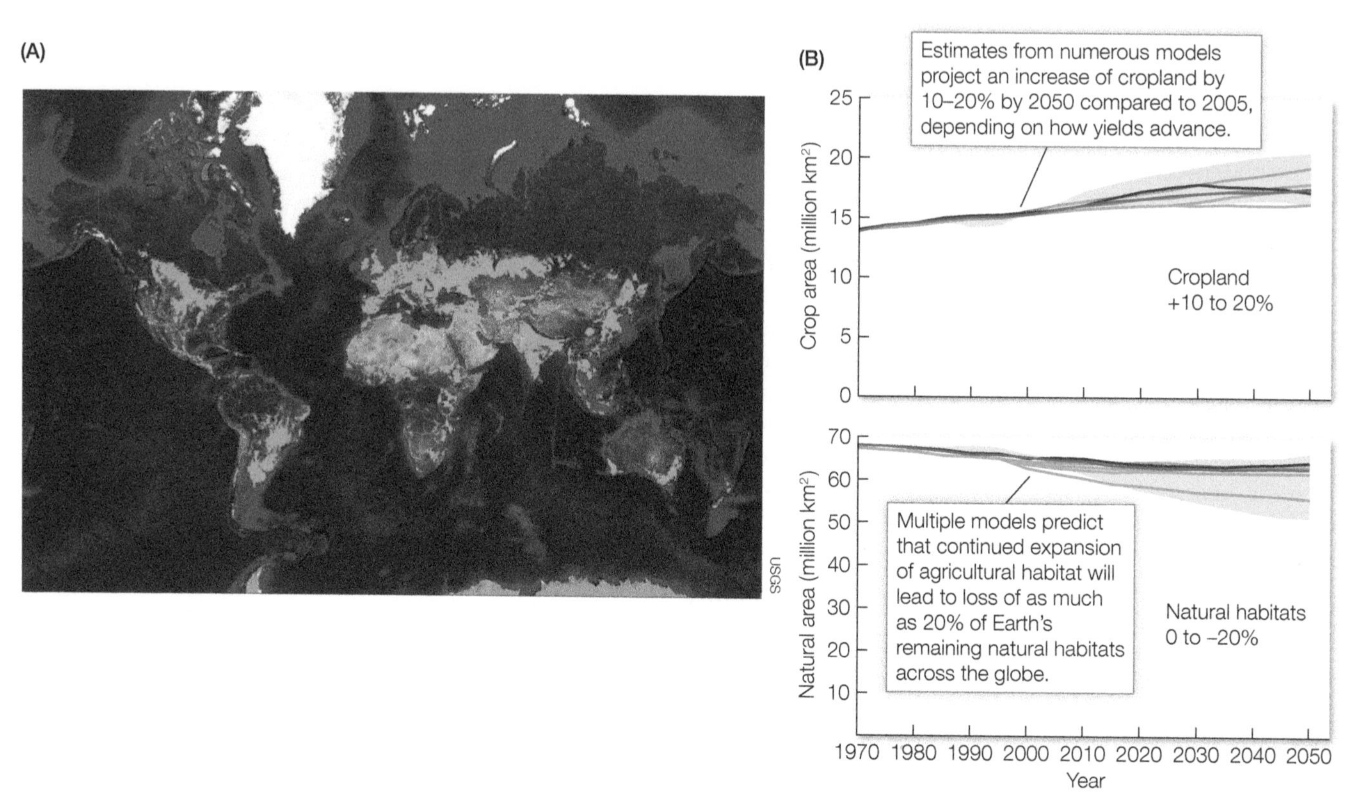

Figure 9.2 Global habitat loss due to expansion of agriculture. (A) As of 2017, the world has 18.7 million km^2 of cropland and 33.8 million km^2 of pasture lands. Altogether, agriculture occupies about 40% of Earth's terrestrial surface, representing the single largest use of land on the planet (green area in figure). (B) Smith and colleagues summarized several integrated assessments of future land-use change (different lines; gray shading indicates 20th to 80th percentiles of literature estimates) and projected that the amount of agricultural land on Earth will need to increase 10% to 20% by 2050 (compared with 2005 levels) to feed the growing human population. This will correspond to a future loss of up to 20% of the natural habitats remaining on Earth. (B, after P. Smith et al. 2010. *Phil Trans Roy Soc Lon B* 365: 2941–2957.)

different methodology, Tilman and colleagues estimated that crop demand will need to grow by a daunting 100% between 2005 and 2050.[16]

In order to meet growing demand for food, we will need to increase either crop yields (the amount harvested per unit of existing land) or the amount of land dedicated to agriculture (**Video 9.1**). There is some optimism that higher yields can be achieved through advances in genetic engineering and through advanced technologies that improve the precision of farming and increase efficient use of limited resources (nutrients, water, etc.). But some scientists believe we have hit the point of diminishing returns on increasing yield. Several major crop-producing regions have seen yields stagnate in recent years.[17,18] Irrigation has only modest potential to expand, as many rivers and aquifers are already fully exploited and subject to competing demands,[15] and the application of more fertilizers and biocides is more likely to increase pollution than yield. Given possible contraints on increased yield, pessimistic estimates suggest that an additional 10 million km^2 or so of land will need to be cleared globally by 2050 to meet demands for crop production.[16] Optimistic estimates project a 10%–20% increase of cropland by 2050 compared with 2005, depending on how yields advance (**Figure 9.2B**).[19,20] The majority of this agricultural expansion is predicted to occur in Latin America, the Caribbean, sub-Saharan Africa, and the Near East and North Africa, where human populations are growing the fastest.

Video 9.1 Watch a video about the challenge of feeding Earth's growing population.
oup-arc.com/e/cardinale-v9.1

Continued expansion of agricultural habitat is expected to cause the loss of as much as 20% of Earth's remaining natural habitats across the globe (see Figure 9.2B).[20] If these predictions come true, then the consequences for biodiversity would be dire. In a study that summarized records of occurrence for 320,924 species of plants and animals at 11,525 sites, Newbold and colleagues[21] showed that local species richness has been reduced by an average of 30% to 40% in pastures and croplands, compared with less-affected reference habitats. Similarly, de Baan and colleagues[22] compared species richness in agricultural habitats to reference habitats and found that local species richness was reduced by a median 60% in annual cropping systems and by 33% in pastures.

Certain types of agricultural practices are disproportionately harmful to biodiversity (see Chapter 15). The cultivation of industrial monocultures—single species of crops or tree plantations that are grown across vast areas of land where they are heavily managed with agrochemicals (fertilizers, pesticides, and herbicides)—actively seeks to eliminate species that might compete with, eat, or damage the focal crop species or that might make harvesting more difficult or expensive. Industrial monocultures are the dominant food production system in many developed nations (think of corn/maize, wheat, soybeans, rice, and palm oil), and they are becoming more prominent through time.

In contrast to industrial monocultures, less-intensive agricultural practices have the potential to be more "friendly" (less damaging) to biodiversity. For example, small-scale local farming of crops within landscapes that also preserve some natural habitat can allow species to persist in the landscape, even when they do not persist locally in the agricultural habitat.[23,24] This is especially true for highly mobile species like birds or mammals, which can disperse across habitat patches in a landscape.[25,26] The contrasting effects of different kinds of agriculture on biodiversity have spawned a vigorous debate about whether conservation in the future will be best achieved by intensifying industrial monocultures while setting aside large reserves for biodiversity conservation or, alternatively, interspersing small-scale farms within landscapes that also have patches of natural habitat (**Box 9.1**).[27,28]

URBANIZATION Approximately 3% of Earth's surface has already been converted to urban habitat. Projections suggest that the number of people living in urban areas will increase to nearly 5 billion by 2030 and that 7 out of every 10 people

BOX 9.1 Challenges & Opportunities

Land Sharing versus Land Sparing

Under the current scenario of rapid human population increase, achieving efficient and productive agricultural land use while conserving biodiversity is a global challenge. There is an ongoing debate about whether land for nature (protected areas) and land for food production (farm and rangelands) should be segregated (land sparing) or integrated on the same land (land-sharing, wildlife-friendly farming).

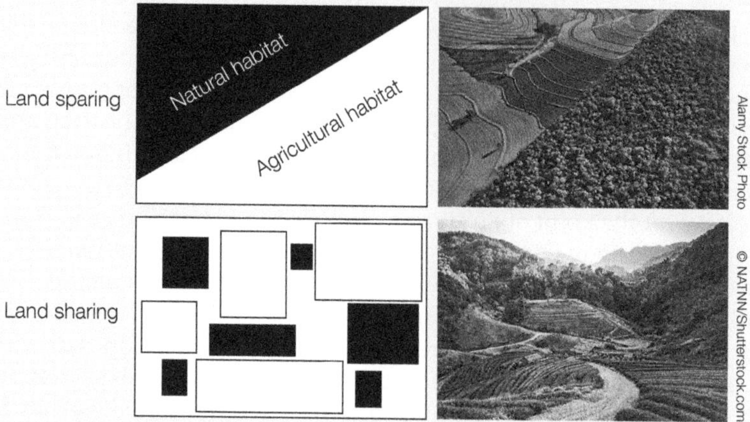

The left panels in this figure depict the distribution of natural habitat (black) and agricultural habitat (white) for land sparing (top) and land sharing (bottom). The right panels give examples of what real landscapes look like under these two alternative land-uses.

LAND SPARING Proponents of the land-sparing option argue that the best way to save biodiversity is to set aside large areas of land as national parks and other protected areas that are separated from potentially damaging human activities like agriculture. To feed the growing human population, those who favor land sparing argue, we should intensify production and increase yields in current agricultural lands rather than expanding food production systems at the expense of natural habitat. Land-sparing strategies are more consistent with "classical" conservation practices, which focused on setting aside large, well-preserved natural areas.

LAND SHARING Proponents of the land-sharing option argue that the best way to conserve biodiversity is to integrate patches of natural habitat into landscapes that are characterized by small-scale farms with low-intensity land-use practices such as selective logging, organic farming, and low-intensity grazing. This is sometimes called wildlife-friendly farming by researchers who demonstrate that farming and conservation can coexist as a spatial mosaic in a heterogeneous landscape. Land-sharing strategies have gained popularity as an alternative to the wasteful and environmentally destructive practices of large-scale commercial farming. Land-sharing practices are already common in many developing countries where landscapes are dominated by smallholder farms and traditional agriculture (see Figure).

The land-sparing versus land-sharing (LSLS) debate gained increased attention after Cambridge University's Ben Phalan and colleagues published a prominent 2011 paper[27] testing the efficacy of these two alternative strategies for protecting bird and tree species diversity across agricultural gradients in Ghana and northern India. Phalan and colleagues found that "more species were negatively affected by agriculture than benefited from it" and that "land sparing is a more promising strategy for minimizing negative impacts of food production, at both current and anticipated future levels of production."

Critics of the paper made three counterarguments: (1) Land sparing fails to account for the fact that smallholder (not large-scale commercial) farms already dominate landscapes in much of the developing world where agriculture will need to increase to feed the growing population.[28] Therefore, smallholder farms are already conducive to, and even practicing, land sharing. (2) Much food produced by large-scale, commercial farming practices is wasted, which creates a larger ecological footprint than is needed for food production. In addition, commercial farming has environmental externalities (e.g., nutrient pollution, greenhouse gases) that degrade biodiversity but that have not been considered in the LSLS debate. (3) Land sharing provides certain ecological benefits, such as higher levels of biological pest control and greater rates of pollination, that often do not exist with large-scale, intensive agricultural practices. These benefits have not been considered in the LSLS debate.

Since Phalan and colleagues' paper was published, a number of data syntheses have shown that one can find support for either land sparing or land sharing as the better option for biodiversity, depending on the scale of measurement and assumptions of the analyses.[137,138] In addition, numerous authors have pointed to additional elements that have been missing from the debate but are important for considering the economic viability, and social acceptability, of different options.[138–141]

Given these arguments, new research on the debate is starting to move beyond a simple dichotomy and beginning to consider when, where, and why each of the two management options might contribute to biodiversity conservation. In her recent critique of the debate,[137] Claire Kremen wrote, "Both large, protected regions and favorable surrounding matrices are needed to promote biodiversity conservation; they work synergistically and are not mutually exclusive." Figuring out how land sharing and sparing can contribute to conservation is both a challenge and opportunity for the next generation of practitioners.

will be living in megacities by 2050 (**Figure 9.3**). As a result, the amount of urban habitat on the planet is expected to triple by 2030, with 70% of the human population living in cities. The rather dramatic shift in the human population from rural environments to cities will be accompanied by substantial growth of urban habitat. Globally, more than 5.87 million km^2 of land has the potential to be converted to urban habitat by 2030, and 20% of this (1.2 million km,^{2} an area about the size of South Africa) is projected to have a very high probability (>75%) of conversion. If all areas with high probability undergo urban land cover transformation, then the current global land area dedicated to built environments will roughly triple.[29]

Nearly half of all urban expansion is expected to occur in Asia.[29] In China, urban growth is expected to create a 180 km urban corridor along the coast from Hangzhou to Shenyang. In India, urban growth is expected to be associated primarily with its seven state capital cities, though high growth may also occur in the Himalayan region where small villages and towns exist. Increased urban land cover is predicted to be highest in Africa, which is expected to grow by 590% compared to 2000.[29] Expansion in Africa will occur primarily in the Nile River region of Egypt, the coast of western Africa on the Gulf of Guinea, the northern shores of Lake Victoria in Kenya and Uganda, the Kano region in northern Nigeria, and the greater Addis Ababa in Ethiopia. In North America where the percentage of the human population living in urban areas already lies at 78%, forecasts predict a doubling of urban land cover by 2030.[29]

(A)

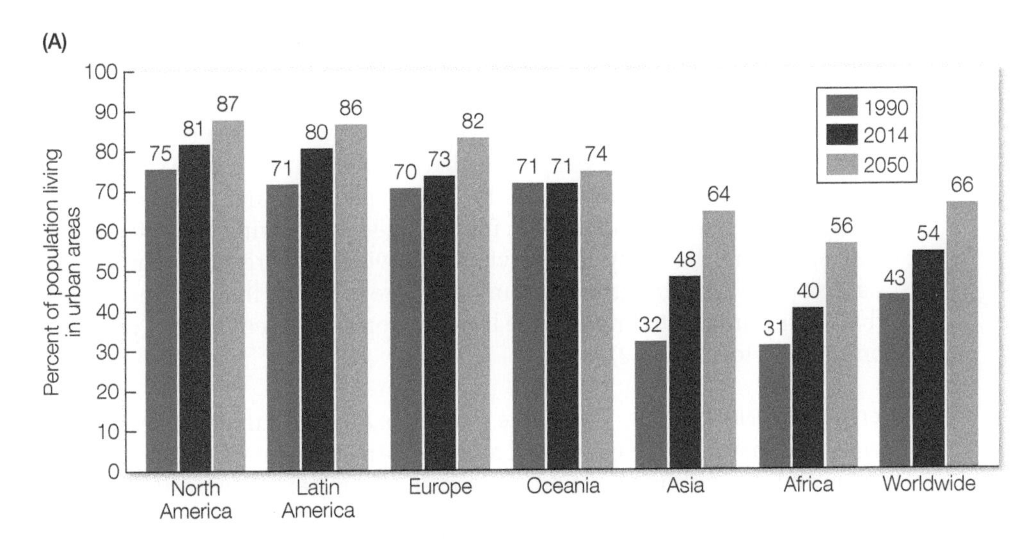

(B)

Figure 9.3 Urbanization. (A) The world is in the middle of a dramatic transition where the majority of the human population is moving into urban habitats. By 2050, almost 7 out of every 10 people will live in a megacity. The large population shifts from rural to urban habitat are taking place in Asia and Africa. (B) Tokyo, Japan, is currently the world's most populous city, with 38 million inhabitants. (A, Data from United Nations. *World Urbanization Prospects: The 2018 Revision*, Online Edition. POP/DB/WUP/Rev.2018/1/F02. CC BY 3.0 IGO.)

Built environments have a range of impacts on biodiversity. These range from relatively small impacts associated with low-density human populations that have minimal infrastructure (e.g., villages) to massive impacts associated with heavily populated epicenters that are filled with human infrastructures (e.g., megacities). The continuum of impacts on biodiversity was quantified by Newbold and colleagues,[21] who found that urban habitats characterized by low land-use intensity had no consistent impact on species richness. While some low-intensity urban environments had declines in richness that were as great as 20%, these were offset by other low-intensity habitats that had increases in richness of up to 15%. However, when the authors looked at urban environments characterized by intense land use, the impacts on species richness were consistently negative, averaging a 55% loss of biodiversity and reaching close to an 80% loss in the most extreme cases. The findings correspond well to those of Aronson and colleagues,[30] who compiled a large data set comparing plant diversity in 110 major cities throughout the world, and bird diversity in 54 cities, with natural areas that were near those cities. Those authors showed that cities have only 8% of the native bird species, and 25% of the native plant species, found in nonurban areas.

Despite their generally negative impacts on biodiversity, particularly in heavily populated cities with intense land-use change, urban environments are by no means biological deserts. (Chapter 15 discusses opportunities for biodiversity conservation in urban habitats). The introduction of agricultural domesticates and ornamental exotic plants can sometimes exceed the loss of native plant species, leading to species enrichment for certain groups of organisms.[31] Furthermore, the field of **urban ecology** has shown that cities often have their own unique sets of flora and fauna,[32] and the biota living in urban environments can provide a suite of ecosystem services to the residents of cities. For example, urban trees improve urban air quality and reduce surface temperatures, which benefit public health.[33] Urban planning and design, as well as the use of green infrastructure, are increasingly engineering cities to enhance biodiversity and the ecosystem services it provides. The rise of urban ecology as a discipline plus the growing number of projects that integrate nature into urban environments to improve the lives of people represent a major conservation success story (**Box 9.2**).

EXTRACTIVE INDUSTRIES Extractive industries are those that remove raw materials from nature; they include oil and gas extraction, mining and quarrying, fishing and aquaculture, hunting, and forestry. Many extractive activities cause habitat degradation (e.g., pollution), which is discussed later in this chapter. Other extractive activities cause overexploitation of renewable resources (e.g., overfishing), which is discussed in Chapter 10. Here we focus more narrowly on how extractive industries directly lead to habitat loss.

Comprehensive estimates of habitat loss caused by extractive industries are difficult to find for the global scale, and when they exist, they are often out of date. However, several publications give an idea of the scope of habitat loss caused by certain activities in more geographically constrained regions. Oil, gas, and mineral extraction now account for an estimated 7% of global deforestation in the subtropics, with increasing exploration and development taking place in the Amazon and Congo Basins. As of 2008, Finer and colleagues reported,[34] national governments had leased an estimated 180 oil and gas blocks covering 688,000 km^2 of the Amazon rain forest habitat (an area nearly the size of Chile) to multinational energy companies for exploration and production. In North America, the land area converted to oil-drilling well pads, roads, and storage facilities between 2000 and 2012 was about 30,000 km^2, equivalent to the area of three Yellowstone National Parks.[35] The development of tar sands infrastructure in the Canadian boreal forest has resulted in the clearing and degradation of over 7500 km^2 of forested areas since the year 2000.[36]

In the United States, mining operations disturb just 0.02% to 0.1% of land.[37] In Canada, a meager 0.01% of land has been used for mining,[38] and in Australia mining sites have disturbed less than 0.26% of total land mass.[39] While the areal coverage of mining activities is very small compared with the main drivers of habitat loss (e.g., agriculture), it is worth noting that mining activities often completely destroy all biodiversity at the local scale of operations. Surface mining of coal or minerals is an example. Open pit mines and strip mines cover modest expanses of land; the Yanacocha gold mine in the Cajamarca region of northern Peru spans 215 km,^{2} the Haerwusu coal mine in China spans 67 km,^{2} and the Garzweiler lignite strip mine in Germany

BOX 9.2 Success Story

The Rise, Fall, and Resurrection of Detroit

During the nineteenth century, the city of Detroit, Michigan became a world epicenter for industry and manufacturing after Henry Ford established the first automotive assembly plant in Detroit. Nicknamed the Motor City, Detroit's expansion of the auto industry fueled a growth spurt that allowed it to become the fourth-largest city in the United States, peaking at nearly 5 million people.

But population growth and heavy industrialization came with an environmental cost. Housing, road, and infrastructure projects sprawled outward from the city center, destroying nearly all green spaces and natural habitat in an area of 3000 km^2. By the 1940s, an estimated 6 million gallons of oil and other petroleum products had been dumped into the Detroit River and River Rouge. By the 1950s, air quality in the Detroit metro area was a chronic health problem.[142] By the 1960s, enough untreated garbage and raw sewage had made their way into surrounding rivers that the Detroit River was declared one of the most polluted rivers in the United States.[143] As this pollution drained into Lake Erie, the lake was declared dead due to rampant hypoxia and fish and wildlife kills. In the 1970s, the entire fishing industry in the St. Clair River, Lake St. Clair, the Detroit River, and Lake Erie had to be closed because of toxic levels of mercury in the water.[143] The rapid rise of Detroit as an economic powerhouse had made it one of the most polluted cities in North America, if not the world.

By the 1970s, Detroit's automotive industry began a slow, painful decline. The 1970s energy crisis and the economic recession of the 1980s led to massive layoffs in the automobile industry, causing unemployment and poverty in Detroit to skyrocket. The city's tax base declined, infrastructure fell into disrepair, and by 2000, Detroit's population had declined by 50%. The 2007–2008 global financial recession was an existential crisis for Detroit as a modern American city. Two of the three major automotive companies (Chrysler and General Motors) became financially insolvent, and in 2012, Detroit became the largest city in U.S. history to file for bankruptcy. Detroit's urban decay left behind 70,000 abandoned buildings, 31,000 empty houses, and 90,000 vacant lots.[144]

After hitting rock bottom, Detroit is now undergoing an economic revival and reinventing itself. Politicians, city planners, designers, and nonprofit groups are now envisioning a cleaner, greener city. Vacant houses are being torn down and replaced with urban green spaces. Urban forestry programs have planted hundreds of thousands of trees in the city. Urban agriculture is booming, and green infrastructure projects are replacing outdated physical infrastructures that were once eyesores.

A noteworthy successes in the "greening" of Detroit has been formation of the Detroit River International Wildlife Refuge (DRIWR). DRIWR is the first international wildlife refuge in North America and represents a joint effort between the United States and Canada. The DRIWR was envisioned in 2000 when Canadian Deputy Prime Minister Herb Grey and U.S. Congressman John Dingell commissioned a group of agencies and prominent individuals to re-envision the future of the Detroit River watershed:

> *In ten years the lower Detroit River ecosystem will be an international conservation region where the health and diversity of wildlife and fish are sustained through protection of existing significant habitats and rehabilitation of degraded ones, and where the resulting ecological, recreational, economic, educational, and quality of life benefits are sustained for present and future generations.*[145]

(Continued on next page)

BOX 9.2 Success Story *(continued)*

This vision was used by Congressman Dingell to introduce legislation creating the DRIWR, which was signed into law by the U.S. president in 2001. Canada responded to the legislation by using a number of its own existing laws to grow the size of the refuge.

Today, the DRIWR consists of 24 km^2 of islands, coastal wetlands, marshes, shoals, and waterfront lands that extend along 48 miles of the Detroit River and western Lake Erie shoreline. Many areas included in the refuge were sites that had been historically degraded by industrial activities, which needed chemical pollution removed, or were abandoned brownfields that had to be restored.

Native animals, including some endangered species, have returned to the area after years of local extinction. The Detroit River is now home to over 30 species of waterfowl, 17 species of raptors, 31 species of shorebirds, 160 species of songbirds, and 117 species of fish that live along or regularly migrate through the Detroit River. The area has breeding populations of reintroduced eagles, ospreys, and peregrine falcons and now supports sport fisheries for lake whitefish, sturgeon, salmon, and walleye. The site has been listed among the Wetlands of International Importance by the 1971 Ramsar Convention to protect wetlands.

While many people of Detroit still live in poverty and urban decay, the Detroit River International Wildlife Refuge is helping to transform an unhealthy, unpleasant part of urban Detroit into a highly valued green space where thousands of residents and tourists now fish, hike, bird watch, and enjoy nature each year.

Then...

U.S. National Archives and Records Administration

© James Fassinger/Alamy Stock Photo

versus now

All images from USFWSmidwest

In the 1950–1970s, Detroit was heavily polluted by industry, and its waterways like the Detroit River were filled with oil, chemicals, and raw sewage (left). Today, the urban environments of Detroit are being reinvented to be cleaner and greener. Much of the Detroit River waterfront has been renovated and revitalized and now supports hundreds of species of fish, birds, and mammals, as well as recreational activities that draw tourists and that provide opportunities for residents of the city.

spans 48 km.[2] But in these areas, mining operations are sufficiently intense that they have left the land devoid of life (**Figure 9.4A**). Some studies have shown that more than 25% of metal mines worldwide are located within 10 km of the boundary of a protected natural area.[40]

A recent advance in fossil fuel extraction is even more environmentally destructive than pit and strip mines. Mountaintop removal (MTR) mining, which has been referred to as "strip mining on steroids," involves blasting the tops of mountains off to reach the coal buried below (**Video 9.2**). Mountaintop removal began in the 1970s as a cheap alternative to underground mining, and it is now used for extracting coal mainly in the Appalachian Mountains of the United

▶ **Video 9.2** Watch this video about mountaintop mining.
oup-arc.com/e/cardinale-v9.2

States, where it is the dominant driver of land use change.[41] The environmental consequences of MTR mining are radical and severe, as the biology of entire landscapes is destroyed in ways that will never allow recovery.

In marine environments, there are several destructive harvesting practices that contribute to habitat loss. Bottom trawling is an industrial fishing method where large nets with heavy weights are dragged across the seafloor, destroying the seafloor habitat in the process. Fishing trawlers drag across an estimated 15 million km^2 of ocean floor each year, an area 150 times greater than the area of forest cleared in the same time period. Bottom trawling can reduce biodiversity by 50% or more,[42] and it takes decades for benthic ecosystems to recover.[43]

Destructive fishing practices also contribute to habitat loss in coral reefs. The use of poisons like cyanide to kill fish is common. But the practice kills coral as well as fish, contributing to the devastation of the reefs of eastern Indonesia, western Pacific countries, and the Philippines—where an estimated 65 tons of cyanide are poured into the sea each year (**Figure 9.4B**).[44] The use of explosives for blast fishing is also on the rise.[45] Explosions produce large craters that destroy 10 to 20 m^2 of the seafloor (**Figure 9.4C**). They kill not only fish, but all of the surrounding fauna and flora. Explosives are inexpensive and easily purchased, and they often come from the mining and building industries.

AQUACULTURE Aquaculture, which now provides more than half of all the world's fish and shellfish, represents one of the fastest-growing food sectors in the world.[46] China accounted for 45.5 million metric tons of aquaculture production in 2014, including over 60% of global fish production from aquaculture. Other major producers were India, Vietnam, Bangladesh, and Egypt. The rapid expansion of aquaculture globally has led to the farming of fish and shellfish becoming a significant contributor to habitat loss, particularly in coastal marine

Figure 9.4 Habitat loss caused by extractive industries. (A) A large surface coal mine in western Australia destroys all local plant and animal life. (B,C) Fishermen use illegal methods of cyanide poisoning and dynamite blasting to collect fish from coral reefs. These methods simultaneously kill the coral habitat. (D) A coastal mangrove forest in Thailand is clear-cut to make way for shrimp farms. (E) A rain forest in the Amazon is slashed and burned to prepare for cattle ranching.

environments and certain types of inland freshwater habitats.[46] The best example of habitat loss caused by aquaculture is the loss of mangrove forests. Mangrove forests are among the most important wetland habitats in tropical areas.[47] Composed of woody plants that are able to tolerate salt water, mangrove forests occupy coastal areas with saline or brackish water, typically where there are muddy bottoms. Mangroves are important breeding and feeding grounds for fish and shellfish, and they are known to have great economic value for their utility for protecting coastal areas from storms and tsunamis (see Chapter 6).[48]

But mangrove habitat is also prime habitat for commercial shrimp and prawn hatcheries, and large areas of mangrove forests have been cleared and transformed into shrimp and prawn aquaculture facilities (**Figure 9.4D**). These transformations have been particularly dramatic in Southeast Asia, where as much as 15% of the mangrove area has been removed for aquaculture. Over half of the world's mangrove ecosystems have already been destroyed, and more are being destroyed every year.[49] Of the vertebrate species that are endemic to mangroves, 40% are now threatened with extinction.[50] There is, however, some good news. Public awareness of the environmental impacts of shrimp farms, and the recognition that mangroves protect coastal villages from storms, is fostering efforts to replant mangrove forests in many parts of Thailand, as well as other regions of Asia.[51]

Mangrove habitats are not the only forest type to be influenced by deforestation. Certain kinds of logging, such as clear-cutting and slash-and-burn practices (**Figure 9.4E**), have contributed to rampant loss of terrestrial forests in some parts of the world. These trends are detailed in the next section, which describes habitat loss by biome.

Habitat loss by biome

The human activities that drive habitat loss, discussed in the previous section, have affected certain biomes on the planet more than others. What follows is a discussion of biomes that have been disproportionately influenced by habitat loss.

FORESTS Over the last 5000 years, the global area of forested habitat has declined by roughly 50% (18 million km^2, an area slightly larger than Russia), due to agricultural expansion and unsustainable harvesting of lumber.[52] Until the late nineteenth century, the highest rates of deforestation were in the world's temperate regions; however, during the late nineteenth century, deforestation began to slow in the temperate and boreal regions of the world, and the geographical distribution of deforestation began to shift.[52]

In western Europe, for example, rates of deforestation declined because of improvements in the productivity of existing agricultural land, increased timber imports from other parts of the world, and the replacement of wood by coal as the main source of fuel. By the end of the twentieth century, the forest area in most of Europe was stable or increasing, with forests covering about one-third of the total land area.[52] Similarly, following two centuries of deforestation, forest area throughout most of North America has been stable since the early twentieth century. Although forest cover in China had fallen to a historical low of less than 10% of the land area by 1949, it had recovered to nearly 20% of the land area by the end of the twentieth century as a result of reforestation programs.[52]

But even as deforestation slowed in temperate regions of the world, it increased in tropical regions of the world during the twentieth century, especially in developing countries. The destruction of tropical rain forests has come to be synonymous with habitat and biodiversity loss. Tropical moist forests occupy 7% of the Earth's land surface but are estimated to contain over 50% of its species.[53,54] In addition to their biodiversity, tropical forests have regional importance in protecting watersheds and moderating climate, local significance as home to numerous

Indigenous cultures, and global importance as sinks to absorb some of the excess carbon dioxide (CO_2) produced by the burning of fossil fuels (see Chapter 6). But tropical areas of the planet are currently facing some of the highest development threats due to rapidly growing populations, conversion of forests to agricultural and grazing lands, and fossil fuel exploration and extraction.

Much of the forest losses in tropical countries has been driven by forest management policies that were established during the period of colonialism. While colonialism was mostly abolished after World War II, many of the original management practices continued to be practiced by newly independent countries. For example, Nigeria lost more than 90% of its forests because of practices that began in the colonial period, which include mechanized logging of forest reserves, establishment of state-owned plantations (e.g., for cocoa and oil palm), and mining.[55] Deforestation in sub-Saharan Africa was, however, lower than other parts of the tropics. In Latin America, forest area declined to about 50% of its original land area by the end of the twentieth century.[56]

During the first two decades of the twenty-first century, loss of tropical forests has continued, even as many other forest types are undergoing **afforestation** (Figure 9.5). In boreal regions, the area of forested habitat has increased over the past decade. In temperate regions, afforestation has continued for the past several decades. These trends can largely be explained by the natural expansion of forest on abandoned agricultural lands, including rangelands, in territories that were part of the former Soviet Union. For example, there was an increase in forest area of 260,000 km^2 on abandoned farmland in Belarus, Kazakhstan, and the Russian Federation.[57]

In contrast to increased forest area in boreal and temperate regions, forests in tropical regions had the highest decrease in forest area of any domain from 2000 to 2010, and it was the only domain to show an increase in agricultural area. An estimated 70,000 km^2 of forest was lost per year over that period in the tropics, and the area of agricultural land increased by 60,000 km^2 per year. More than 60% of the recent loss has occurred in the Neotropics, with Brazil alone

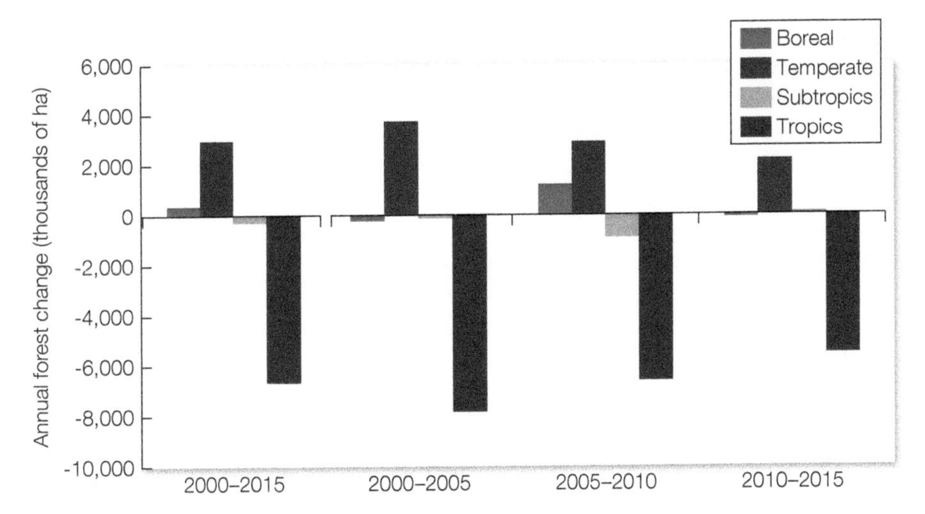

Figure 9.5 Changes in the net annual forest area differ by climatic region. Note that temperate forests have been undergoing afforestation for several decades as they recover from historic overharvesting and as new forests have grown up on abandoned agricultural lands. However, tropical forests (and, to a lesser extent, subtropical forests) continue to experience chronic loss of habitat. (After FAO. 2016. *State of the World's Forests 2016. Forests and Agriculture: Land-Use Challenges and Opportunities*. Rome [www.fao.org/publications/sofo/2016/en/]; data from *Global Forest Resources Assessment 2015*. [www.fao.org/forest-resources-assessment/en]. Reproduced with permission.)

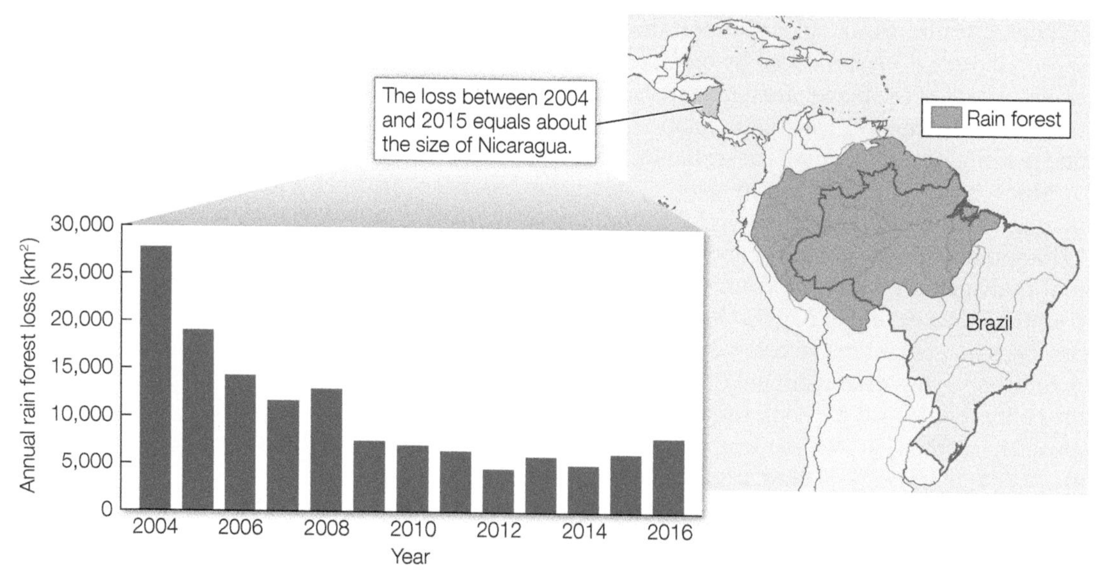

Figure 9.6 With 60% of the Amazon rain forest (green) lying within its borders, Brazil is home to a large fraction of the world's rain forests. But Brazil is also responsible for a large fraction of the world's loss of rain forest habitat. While new policies have helped slow deforestation in Brazil, the country is still losing nearly 8000 km^2 every year as of 2016. (Data from Brazil's National Institute for Space Research [INPE]. www.obt.inpe.br/OBT/assuntos/programas/amazonia/prodes)

accounting for almost half (Figure 9.6). Another third has occurred in Asia, with Indonesia second to Brazil in the absolute rate of forest loss. Africa contributed only 5.4% to the total rain forest area lost, reflecting the current absence of extensive industrial-scale agricultural clearance and commercial logging. If these rates of loss continue, there will be little tropical forest left in the world after this century, except in relatively small national parks and remote, rugged, or infertile areas of the Amazon Basin, Congo Basin, and New Guinea.

Much rain forest destruction results from small-scale cultivation of crops by poor farmers who are often forced to remote forest lands by poverty, or sometimes moved there by government-sponsored resettlement programs[58] (**Video 9.3**). Much of this subsistence farming is termed **shifting cultivation** or **slash-and-burn agriculture**, in which trees are cut down and then burned away. The cleared patches are farmed for two or three seasons, after which soil fertility has usually diminished and soils have eroded to the point where adequate crop production is no longer possible.[59] The patches are then abandoned, and natural vegetation must be cleared in a new patch.

Video 9.3 This video explores the causes and effects of deforestation.
oup-arc.com/e/cardinale-v9.3

Increasingly, the clearing of rain forests by peasant farmers to meet subsistence needs is being dwarfed by the impacts of large landowners and companies that clear land to create pasture for cattle ranching or to plant cash crops. One of the greatest cash crops that is contributing to tropical deforestation today is oil palm. Palm oil is currently the most widely used vegetable oil in the world; it is used in more than half of all packaged goods sold in the supermarket. High consumer demand for products containing palm oil has led to a rapid proliferation of oil palm plantations, particularly in Southeast Asia, Latin America, and Africa, where more than 270,000 km^2 of native forests have been bulldozed or torched to make room for plantations. An additional 2700 km^2 of forests are being converted to palm plantations in countries that are major palm oil exporters.[60] The rapid conversion of native forests to palm plantations is threatening many species with imminent extinction,[61] including the orangutan,[62,63] Borneo elephant,[64] and Sumatran tiger[65] (Box 9.3).

GRASSLANDS Grasslands are a diverse set of habitats that include (1) temperate grasslands in North America (prairies), Europe and Asia (steppe), South America (pampas), and South Africa (veldt); (2) tropical grasslands in areas with seasonal rain and drought, including savannas and their woodlands; and (3) tundra, which occurs around the Arctic Circle above latitudes where trees survive and is dominated by shrubs, sedges, grasses, lichens, and mosses.

Estimates of the areal extent of grasslands vary but generally fall between 40 and 56 million km², or 30% to 40% of the Earth's land area.[66] Some of the highest known rates of habitat loss have been for grasslands, in large part because of their suitability for growing crops like wheat and corn and for conversion to grazing lands.[3] For example, the U.S. Geological Survey estimates that since

BOX 9.3 Case in Point

The Unsustainable Use of Palm Oil

As consumers, many people are unaware of the environmental and social issues that are associated with the palm oil industry. And as with many other environmental problems that affect biodiversity (e.g., overharvesting of fish), they are unaware of how the choices they make each day—choices about what to eat, what to drink, and which products to buy—affect habitat loss, biodiversity, and human rights in other parts of the world.

Palm oil is the most widely traded vegetable oil globally, with demand projected to increase substantially in the future. Palm oil has a high melting point, which makes it semisolid at room temperature. This property, along with the fact that it is one of the least expensive vegetable oils on the market, has contributed to its widespread use in processed foods, as well as personal care and home products. It is estimated that palm oil is used in about half of all packaged products that are sold in a typical supermarket, in items as diverse as frozen pizzas, margarine, cookies, ice cream, body creams, soaps, makeup, candles, and detergents (see oup-arc.com/e/cardinale-w9.1 and Figure A).

In addition to its use in packaged products, palm oil is increasingly used as biofuel. Almost half of the palm oil imported into the European Union is used to meet requirements for the European Renewable Energy Directive (RED), which mandates the blending of biofuels into motor vehicle fuels.[146,147]

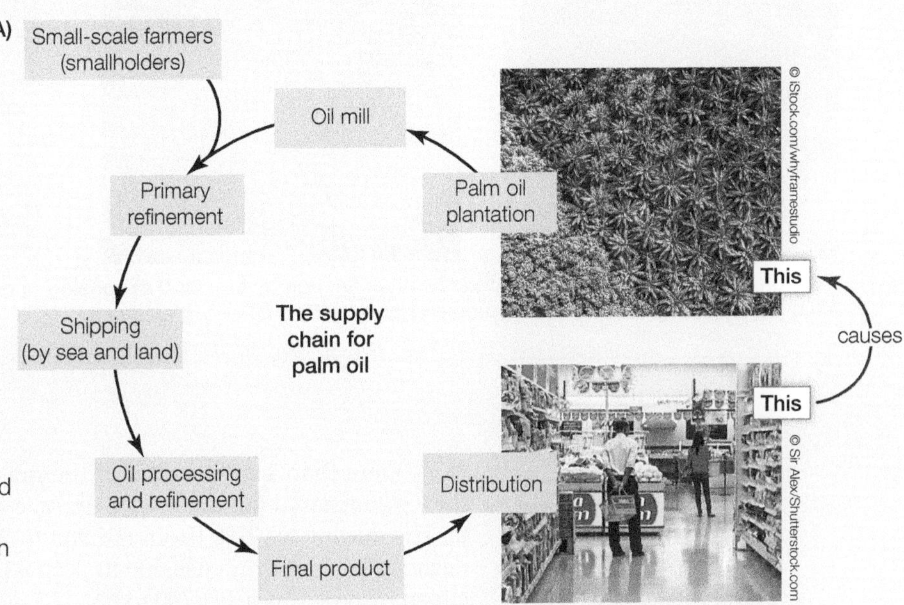

The supply chain for palm oil, which shows how purchasing decisions in developed nations cause tropical deforestation in developing nations.

Palm oil is harvested from the African oil palm (*Elaeis guineensis* Jacq.), which is the world's highest-yielding species for vegetable oil. The African oil palm grows in warm, humid climates such as those that support tropical forests. In response to increasing demand for palm oil, large tracts of rain forest in Southeast Asia, Latin America, and Africa have been bulldozed or logged to make room for oil palm plantations. Between 2000 and 2011, an average of 2700 km² of tropical forests were destroyed and converted to oil palm plantations in tropical countries.[60] Oil palm plantations

(Continued on next page)

BOX 9.3 Case in Point *(continued)*

now cover more than 270,000 km² of the Earth's surface, making it one of the greatest contributors to deforestation in the world.

The rapid loss of tropical forests caused by conversion to palm plantations is putting many species on the brink of extinction.[61] These include such creatures as the orangutan,[62,63] Sumatran tiger,[65] and Borneo elephant.[64] Studies of species richness in oil palm plantations consistently show that they have fewer than half of all vertebrate species that live in primary forests.[61] Analyses that look more broadly across all taxa have found that only 15% of the species found in primary forests also occur in oil palm plantations (see Figures B–D).

In addition to its impacts on biodiversity, the palm oil industry has contributed to environmental injustices and human rights violations; small landholders and Indigenous people who have inhabited and protected the forest for generations have often been brutally driven from their land.[148] In Indonesia, more than 700 land conflicts are related to the palm oil industry.

To encourage environmentally conscious choices, conservation biologists must educate themselves, as well as other people, about the supply chains that link individual decisions to consequences that may occur far away and which may not be immediately obvious.

(B)

(C)

(D)

Examples of species that are threatened with extinction by the rapid expansion of palm oil plantations in Southeast Asia: (B) the orangutan, (C) the Sumatran tiger, and (D) the Borneo elephant.

1830, more than 1 million km² of the grasslands of the western United States have disappeared (an area twice the size of California). Tallgrass prairie habitat has decline by 97% (from 677,300 to 21,548 km²), mixed-grass prairie has declined by 64% (from 628,000 to 225,803 km²), and shortgrass prairie has declined by 66% (from 181,790 to 62,115 km²). Conversion of grasslands to farmlands in western Canada and the United States has left only small remnants of original prairie grasslands. Additional declines are occurring in grasslands in other parts of the world as well, but the rate and extent of these declines is less well documented so is harder to quantify accurately. Oakleaf and colleagues[67] project that an additional 5.98 million km² of grassland habitat will be lost by 2030 to human development.

CORAL REEFS Coral reefs are the "rain forests of the sea" and are some of the most biodiverse and productive ecosystems on Earth. They occupy less than 1% of the ocean floor; yet, they are home to more than a quarter of all marine species: crustaceans, reptiles, seaweeds, bacteria, fungi, and over 4000 species of fish make their homes in coral reefs. Coral reefs provide food and resources for more than 500 million people in over 100 countries and territories and have a global economic value of $375 billion a year.

But coral reefs are in crisis. According to a 2011 report by the World Resources Institute,[68] 25% of the world's coral reefs have already been damaged beyond repair, 75% are at risk from both global and local stressors, and 90% are projected to be in danger of being lost by 2030. Historically, the most severe destruction of coral reefs has taken place in the Philippines, where a staggering 90% of the reefs are dead or dying. In China, coral reefs have declined by 80% over the past 30 years.[69] In Australia's Great Barrier Reef—the world's largest coral ecosystem, composed of roughly 2900 individual reefs that span more than 344,400 km^2—a 2016 heat wave caused the death of one in three corals, including nearly 50% of all corals in the northern third of the reef (**Figure 9.7A**).[70]

While multiple factors have contributed to coral reef habitat loss, the main culprit by far is global climate change.[71] Corals cannot survive if the water temperature is too high, and when stressed by water temperature, corals expel the symbiotic algae living in their tissues, causing the corals to turn white, known as **coral bleaching**. Global warming and rising ocean water temperatures have already led to increases in levels of coral bleaching and mass mortality events across the planet, and these events are predicted to increase in frequency and severity in the coming decades as ocean waters warm even further (**Figure 9.7B**).[72,73] The consequences of global climate change on coral reefs have already been sufficiently severe that several of the world's leading marine ecologists and coral biologists have conceded the extinction of warm water coral reefs as we presently know them and argued that returning reefs to past configurations is no longer an option.[74] Instead, they suggest our best option now is to steer reefs through the Anthropocene in a way that maintains some of their biological functions.

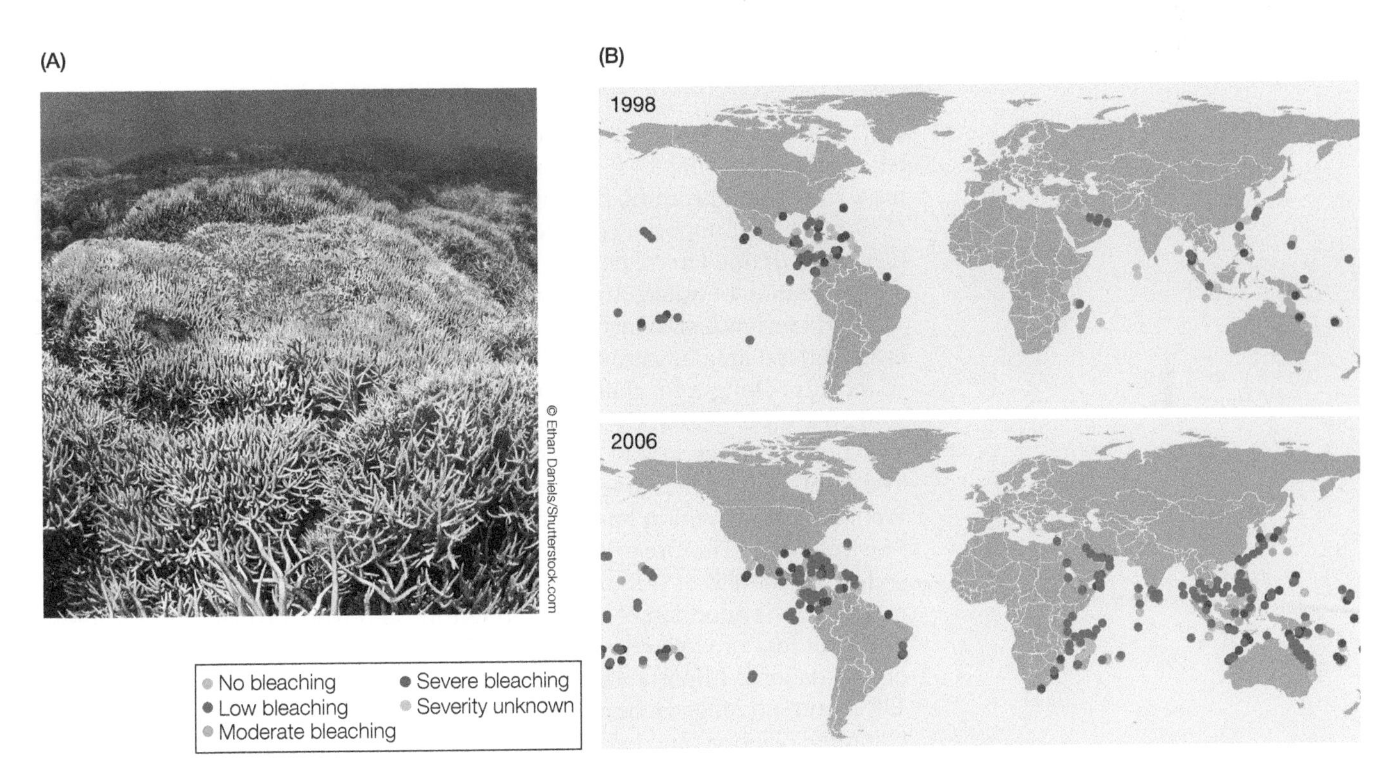

Figure 9.7 Impacts of global climate change on coral reefs. (A) In 2016, rising water temperatures led to heat stress that caused severe bleaching (the loss of symbiotic algae) in Australia's Great Barrier Reef. One in every three corals died during the mass mortality event, and large portions of the reef were killed. (B) The extent and severity of coral bleaching events have been increasing globally over the last decade. Prior to 1998, coral bleaching had been recorded in most of the world's main reef regions, but few reefs had experienced severe bleaching. Since 1998, however, every reef region has experienced severe bleaching, with many areas suffering significant bleaching-induced mortality. (After P. Marshall and H. Schuttenberg. 2006. *A Reef Manager's Guide to Coral Bleaching*. © Commonwealth of Australia [GBRMPA], 2006.)

But even as some have predicted coral reef extinction due to the inevitability of ocean warming, others have continued to focus on more localized factors that contribute to the decline of coral reefs:

- Destructive fishing practices like cyanide fishing, blast or dynamite fishing, and bottom trawling
- Overfishing, particularly of large predators like sharks and herbivorous fish, whose removal alters the food chain and causes trophic cascades that promote overgrowth of corals by algae
- Careless tourism by boaters, divers, anglers, and tour and resort operators who touch reefs, collect coral, drop anchors on reefs, build infrastructures on top of reefs, and empty sewage or other wastes into water surrounding reefs
- Pollution, including industrial waste, sewage, agrochemicals, and oil pollution that poison reefs and contribute to eutrophication
- Sedimentation caused by construction, mining, logging, and farming that sends unnatural amounts of sediment to the ocean, where it can smother corals by depriving them of the light needed to survive

While these local factors contribute to and hasten the loss of coral reef habitat in individual locations, studies have shown that they do not explain the global decline of coral reefs.[71] While taking action to reduce these local threats may slow the extinction of corals and loss of habitat, it is unlikely to offset the ultimate impacts of global climate change. The loss of reef-building corals—at least, as the hard corals exist today—will go down in history as one of conservation biology's greatest failures.[74]

WETLANDS Wetlands that are permanently or seasonally inundated with water are critical habitat for fish, aquatic invertebrates, and birds. They are also a resource for flood control, drinking water, and power production (see Chapter 6). Despite their importance ecologically and economically, wetlands are often filled in or drained to create agricultural or urban habitat, or they are altered by channelization of watercourses and dams.

A 2014 review by Davidson[75] summarized the results of 189 studies of changes in wetland area in various locations around the globe. Davidson found that all records of long-term change in natural wetlands have reported a loss of wetland area since 1700, with rates of loss accelerating almost fourfold since 1900. The global average loss of natural wetland area has been 53.5%, with more loss of inland wetlands (average 60.8%) than coastal wetlands (average 46.4%). Wetland loss by region has been as follows: Europe, 56.3%; North America, 56.0%; Asia, 45.1%; Oceania, 44.3%; and Africa, 43.0%.

The global rate of wetland loss has now slowed to –0.3% per year, driven mainly by the adoption of policies in Europe and North America that commit to no net loss of wetlands. Under the 1971 Ramsar Convention on Wetlands of International Importance, named after the city in Iran where it was signed, 138 countries pledged to include wetland protections in their national land-use planning. National and local laws established by the Ramsar Convention now provide for the management of more than 1300 wetland areas covering over 1 million km^2 (an area roughly the size of Venezuela or Egypt). Even so, the rate of wetland habitat loss continues to be high in Asia, where large-scale and rapid conversion of coastal and inland natural wetlands is continuing. Record keeping on wetland habitat change is especially poor for Africa and the Neotropics; however, the records that exist suggest these areas are losing wetlands at a rate of 1% to 2% of the total area per year.

STREAMS, RIVERS, AND LAKES Freshwater ecosystems make up just 0.01% of the world's water and approximately 0.8% of the Earth's surface. Yet, this tiny fraction of global water supports an estimated 125,000 species,[76] representing 10% of all known organisms and roughly one-third of all known vertebrates. Among freshwater habitats, river, stream, and lake networks are disproportionately important for certain groups of organisms. Freshwater fishes comprise almost 45% of all fishes and 25% of all mollusks on Earth, which is remarkable given how small freshwater ecosystems are in area and volume compared with the world's oceans. Other major groups that depend on freshwater include reptiles, amphibians, insects, plants, and mammals.

But freshwater habitat loss has been pervasive, particularly for streams and rivers. River systems have been fragmented by more than 1 million dams globally, confined by levees, and dredged and straightened for navigation and flood control.[76] During the twentieth century alone, humans constructed more than 45,000 dams that were over 15 m tall to help meet increasing demands for agriculture, hydropower, and drinking water supplies.[77] By 2000, the flow of roughly two-thirds of all of Earth's rivers were regulated, and 48% of river habitats were considered moderately to severely affected by flow regulation.[78] Several of the world's great river systems, including the Ganges–Brahmaputra, Yellow, Nile, and Colorado, stop flowing to the sea during dry periods.[76]

In intensively developed areas, such as Europe and North America, it is not unusual for one-third or more of the freshwater species in a taxonomic group to be extinct or imperiled.[76] Globally, between 10,000 and 20,000 freshwater species are already extinct or imperiled as a result of human alterations to freshwater systems.[79] Major dam construction projects in China (e.g., the Three Gorges Dam on the Yangtze River), throughout the Amazon Basin, and along the Mekong River in Thailand, Laos, and Cambodia are going to exacerbate biodiversity loss significantly by contributing to habitat loss in some of the largest and most diverse freshwater habitats on Earth (**Video 9.4**).

Video 9.4 Watch this video about the social impacts of Brazilian Amazon dam projects.
oup-arc.com/e/cardinale-v9.4

Future development threats

A growing and more affluent human population is expected to increase the demand for natural resources in the coming century, in turn accelerating land-use change and habitat loss. In a 2015 paper, Oakleaf and colleagues[67] offered their projection of threats to natural habitats from the expansion of urban and agricultural habitat, as well as from growth of conventional and unconventional oil and gas, coal, solar, wind, biofuels, and mining development (**Figure 9.8**). Cumulatively, these authors estimated, all threats combined place 19.68 million km^2 of natural land at risk of being developed, which represents 20% of the remaining global total of natural habitat.

At present, one-fifth of all regions have already lost 50% or more of their natural habitats (see Figure 9.8A). Oakleaf and colleagues projected that future development could lead to half of the world's biomes having more than 50% of their natural habitats converted (see Figure 9.8B). Three biomes are likely to experience disproportionately high habitat loss: tropical and subtropical grasslands, savannas, and shrublands (loss of 5.98 million km^2); deserts and xeric shrublands (3.74 million km^2); and tropical and subtropical moist broadleaf forests (3.4 million km^2). After both current and potential future development are accounted for, three biomes are projected to become predominantly human modified: tropical and subtropical dry broadleaf forests (83% of global coverage modified), mangroves (72%), and temperate broadleaf and mixed forests (71%). Clearly, these biomes are in the greatest danger of future biodiversity loss simply due to loss of habitat.

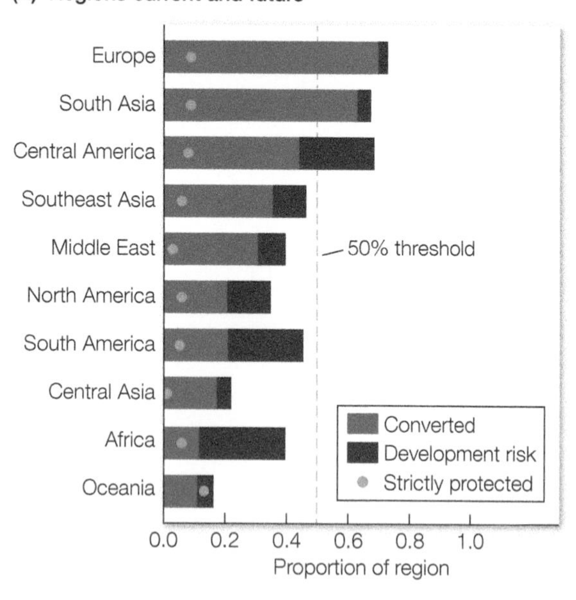

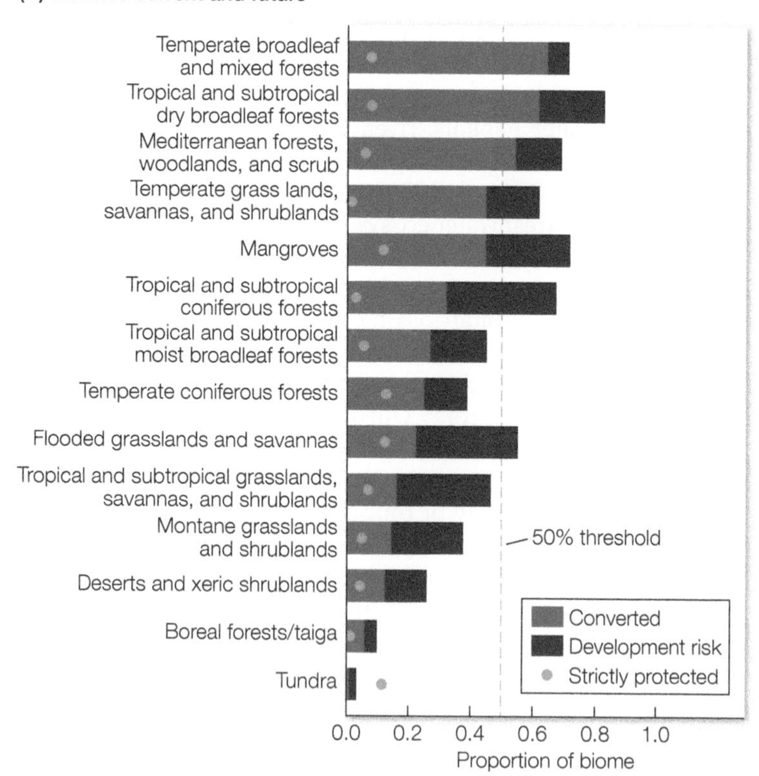

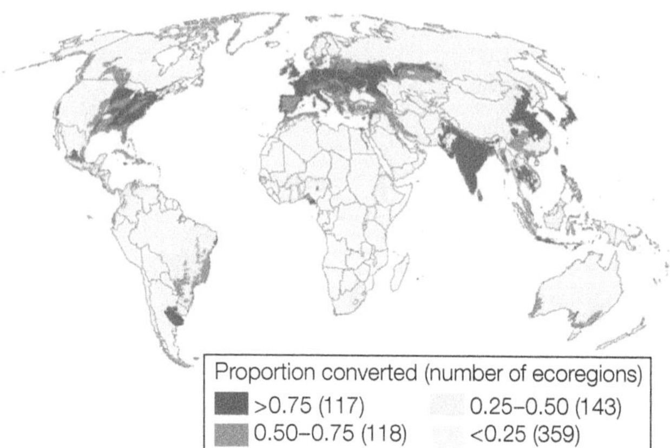

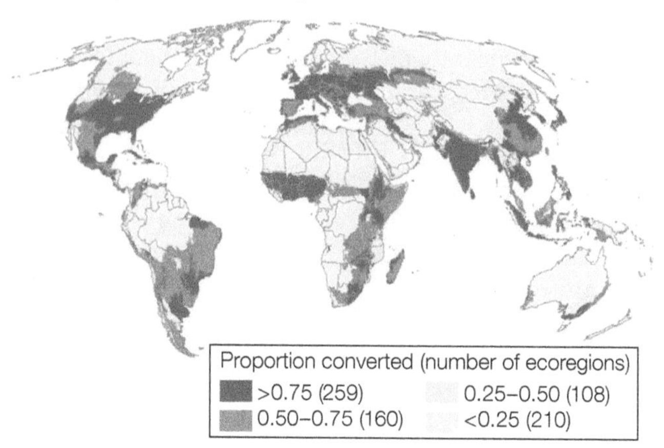

Figure 9.8 Land conversion at present and in the future. The proportion of land in each geopolitical region (A) and biome (B) that is currently converted (blue) and at high risk of future development (red). The blue and red represent total future conversion, with dashed lines showing the 50% threshold. The proportion of strictly protected natural lands (orange dots) is given for comparison. The maps on the bottom show the distribution of terrestrial ecoregions that have 75%, 50%–75%, 25%–50%, and less than 25% of their lands currently converted (C) and potentially converted in the future (D) due to areas with high development risk. (After J. R. Oakleaf et al., 2015. *PLOS ONE* 10: e0138334 CC BY.)

The three regions of the world with the largest amounts of converted habitat already (Central America, Europe, and South Asia) are projected to remain the most affected after future development risk is accounting for (compare Figure 9.8C and D). In marked contrast, Africa and South America, which are currently among the least developed regions of the world, have the most land that is at risk of development (8.18 and 4.32 million km^2, respectively). The amount of converted lands could double for South America and triple for Africa, which places these geographic areas at the greatest risk for future biodiversity loss.

Habitat Fragmentation

Habitat fragmentation is the process by which a larger expanse of habitat is subdivided into smaller pieces, or patches, within a landscape. Fragmentation and habitat loss almost always occur together, and in fact, one consequence of fragmentation is further loss of total habitat area. But habitat fragmentation leads to a suite of changes in a landscape that differ from those of habitat loss and that deserve their own attention in conservation.[80,81] This point is illustrated in Figure 9.9, which shows a hypothetical landscape that is originally composed of just one habitat type (e.g., a native forest) but that is transformed through time into a different habitat type (e.g., an urban habitat). Transformation caused by habitat loss is shown in the upper panels, whereas transformation by habitat fragmentation is shown in the lower panels. Note that while both habitat loss and habitat fragmentation lead to a reduction in the total area of the original habitat in the landscape, fragmentation also leads to a suite of changes in the configuration of the habitat patches that remain in the landscape. Specifically, fragmentation (1) alters the size of habitat patches, (2) changes the number of habitat patches, (3) increases the distance between, and isolation among, habitat patches, and (4) changes the area-to-perimeter ratio of habitat patches, in turn increasing the amount of edge habitat. Each of these four changes to the configuration of habitat patches has its own

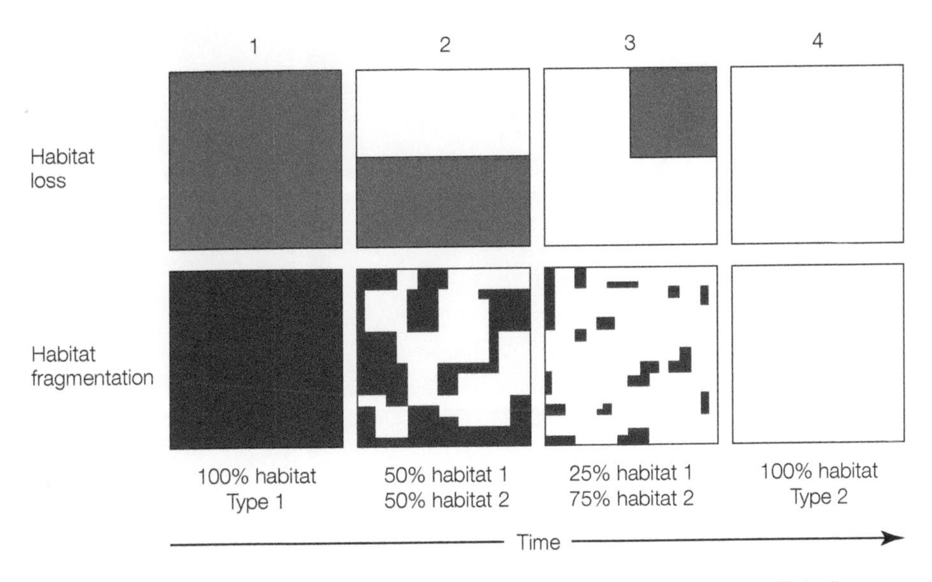

Figure 9.9 The relationship between habitat loss and habitat fragmentation. This diagram depicts a time series showing the loss and fragmentation of habitat from a natural landscape. Between time steps 1 and 4, the landscape is altered from being composed entirely of habitat type 1 (blue or red) to being composed entirely of habitat type 2 (white). Note that while habitat loss and fragmentation are both associated with a loss of area of habitat type 1, during time steps 2 and 3 fragmentation also changes the configuration of patches, including the number of patches in the landscape, the mean size of patches, the distance between and isolation of patches, and the amount of edge habitat in the landscape. (After L. Fahrig. 2003. *Annu Rev Ecol Evol Syst* 34: 487–515.)

biological consequence; collectively, they can generate very different impacts on biodiversity than habitat loss alone can.

Suggested Exercise 9.1 Learn how to quantify different aspects of fragmentation with this online exercise on landscape patterns.
oup-arc.com/e/cardinale-ex9.1

Biological consequences of fragmentation

While a reduction in habitat area is almost always detrimental to biodiversity, changes in the configuration of habitat patches (patch size, number, isolation, and edge effects) can have either a positive or a negative influence on biodiversity. This is true for two reasons: First, several aspects of patch configuration change with habitat area in ways that are nonlinear (**Figure 9.10A**), which implies that habitat loss and patch configuration can affect landscapes differently (i.e., they are not correlated in a 1:1 manner). Second, the different aspects of patch configuration relate to habitat loss in different ways, which implies that habitat fragmentation has potential to generate a much broader array of landscape patterns than will habitat loss alone (**Figure 9.10B**). Some of these landscape patterns can benefit the population size and persistence of certain types of organisms, whereas other landscape patterns can decrease population size and persistence. These contrasting effects are why fragmentation per se can have positive or negative impacts on biodiversity, even as the effects of habitat loss are generally negative.

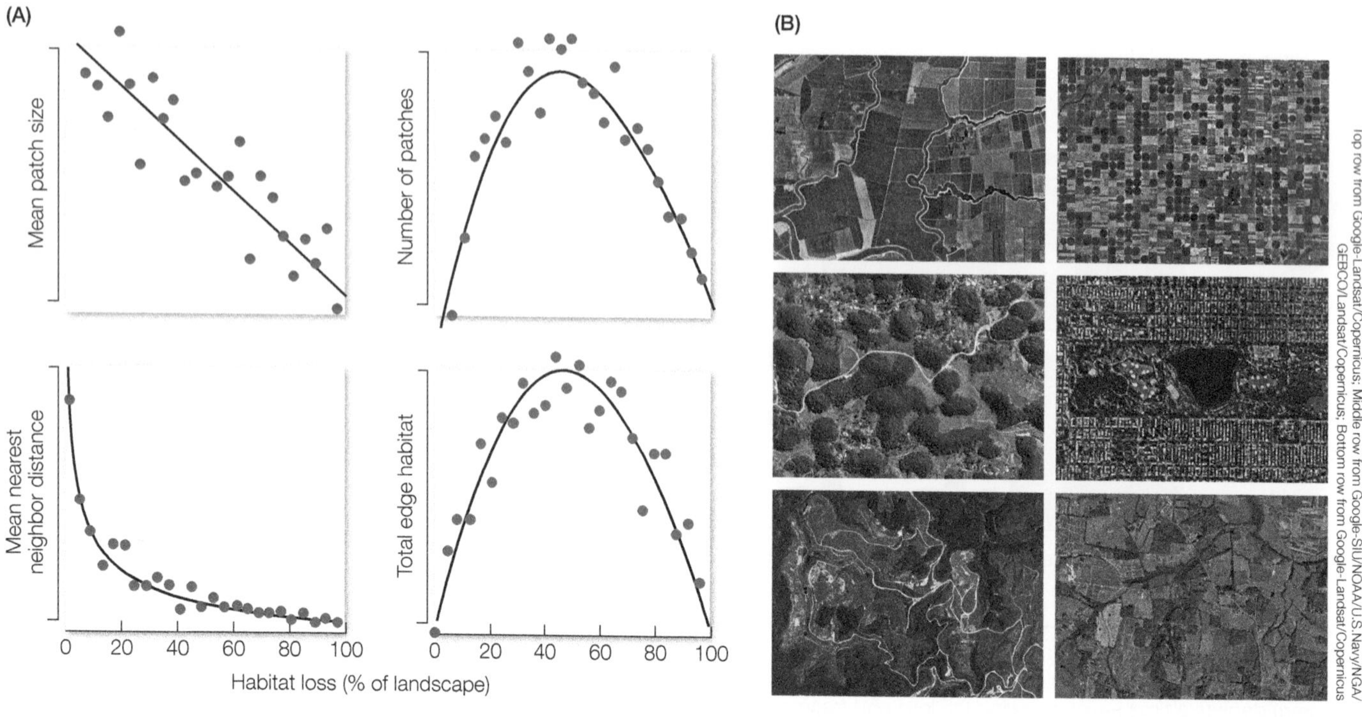

Figure 9.10 Habitat loss and patch configuration. (A) Illustration of how four aspects of patch configuration relate to habitat loss in a landscape. Note that the four aspects of patch configuration change differently as the amount of habitat is reduced, and several change nonlinearly. This variety of effects allows fragmentation to produce a wide variety of patterns in a landscape. To illustrate, consider panel (B), which shows examples of real fragmented landscapes that have different patterns of patch configuration. Note that some landscapes have many small patches, and others have only a few large patches. Some have patches close together; others are far apart. Some have elongated patches with high edge:interior ratios; others are round with low edge:interior ratios. Fragmentation will influence biodiversity very differently in these landscapes. (A, after H. Bird-Jackson and L. Fahrig. 2013. In S. A. Levin [Ed.], *Encyclopedia of Biodiversity* [2nd ed.], p. 54. New York, NY: Academic Press.)

PATCH SIZE AND NUMBER The most obvious changes to a landscape that result from habitat fragmentation are a decrease in the size of patches that remain as an original habitat type is lost. Indeed, the mean size of a patch remaining in a landscape tends to decrease linearly as the original habitat is lost (see Figure 9.10A). In Chapter 8 we introduced species-area relationships, which are well-established empirical relationships showing that species richness tends to decline with the area of a habitat as a power function, $S = CA^Z$, where S is species richness (the number of species present), A is habitat area, C is a constant that depends on the unit of measurement used, and the exponent Z is a scaling factor that dictates how species richness changes with area. More than 800 studies have related species richness to the area of a habitat and shown that Z is typically about 0.27, with a range from 0.15 to 0.35 that depends on the group of organisms and ecosystem being studied.[82]

The values of Z for the species-area relationship indicate that S is a positive, but decelerating, function of A. This means that species losses are relatively small when the initial decline in patch size is minimal. Therefore, the initial stages of habitat fragmentation, which produce only small declines in habitat patch size, have minimal impact on biodiversity. On the other hand, species losses accelerate rapidly as patch sizes get increasingly small. In other words, late stages of fragmentation that generate increasingly small patches have a disproportionately negative effect on biodiversity.

One reason late stages of fragmentation have a disproportionate effect is because species have **extinction thresholds**—that is, thresholds of habitat size that are required for them to persist and complete their life cycles.[83] Many animals require territories of a minimum size to find food, to locate mates, and to breed and rear their young. Many types of plants must disperse a sufficient distance from their parent in order to germinate and grow. Both plants and animals need space to accumulate large enough populations that they don't go extinct because of stochastic forces. The threshold at which a habitat patch size becomes too small to support species populations tends to be higher for large-bodied, mobile organisms that require larger areas to complete their life cycles, so these organisms tend to have higher values of Z in the species-area relationship. The threshold is lower for smaller organisms that require less area or for species that do not need to disperse to complete their life cycles; these organisms tend to have smaller values of Z in the species-area relationship.

In contrast to patch size, which declines linearly with habitat loss, the number of patches that comprise a landscape is often highest at intermediate levels of habitat loss (see Figure 9.10A). When there is no habitat loss, the landscape is composed of one large continuous patch. At the opposite extreme, when the original habitat is entirely lost and the area is converted to a new habitat type, it once again represents a continuous patch (of the alternative habitat type). But in between these two extremes, the number of patches in the landscape is higher. When coupled with trends for patch size, this means that as habitat loss proceeds, the process of fragmentation first generates a few large patches of habitat, then generates a larger number of patches that are intermediate in size, and then dwindles to a small number of natural habitat patches that are small in area.

Obviously, the scenario of having a small number of natural habitat patches of small area is bad for biodiversity. But which of the other two scenarios is better for biodiversity—having a landscape composed of just one or a few large patches, or having a landscape composed of a greater number of smaller patches (assuming they are large enough for a species to persist)? This question fueled an intense debate in conservation biology in the 1970s and 1980s, referred to as the SLOSS debate. *SLOSS* stands for "single large or several small" and refers to two different approaches to land conservation to protect biodiversity in a given region. The "single large" approach favors maintaining one sizable, contiguous land reserve. The "several small" approach favors maintaining many smaller reserves of land whose total areas equal that of a large reserve.

The SLOSS debate began in 1975 after Jared Diamond used the theory of island biogeography (see Chapter 4) to propose some general rules for the design of protected areas.[84,85] One of his propositions was that a single large reserve was always preferable to having several smaller reserves of equal area. This idea became popular among ecologists, was incorporated into many conservation biology textbooks, and was used to guide real world conservation planning. However, the idea that a single large protected area is best for biodiversity was challenged by Daniel Simberloff, who pointed out that the idea relied on the certain assumptions—namely, that smaller reserves contained nested subsets of species found in the larger reserve.[86] If, however, smaller reserves each had unique species, then it would be possible for two small reserves to have more species than a single large reserve.

The SLOSS debate never came to a definitive conclusion and, in fact, it continues to this day. But the debate did help set the stage for fragmentation research to become an important area of study in conservation biology, stimulating an entire generation of new experiments.[87] The debate also made clear that the ultimate effects of patch size and number on biodiversity depend on two things: (1) whether patches in a landscape are sufficiently heterogeneous that they increase diversity across the landscape by adding new species, such as when patches represent novel environments with new niche opportunities, and (2) whether species can disperse among the patches in a landscape, ultimately behaving like a set of populations that are linked by dispersal. The answer to the latter depends on the degree of isolation among patches (discussed at the end of this section), which determines whether species behave like a metapopulation (discussed in "Metapopulations and landscape mosaics" at the end of this chapter).

EDGE EFFECTS **Edge habitats**, also referred to as **ecotones**, mark the transition between two different habitats. One change that occurs to the configuration of patches in a fragmented landscape is an increase in the proportion of edge habitat (see Figure 9.10A). Recent analyses have revealed the extreme magnitude of edge habitat around the globe. For example, a 2015 study by Haddad and colleagues[87] analyzed a high-resolution map of global tree cover to quantify the magnitude of fragmentation in the world's forests. Their analysis showed that 70% of the world's remaining forests are within 1 km of a forest edge, and nearly 20% of the world's forests are within 100 m of an edge. Thus, organisms are living in landscapes composed of an increasing number of small patches that are in close proximity to agricultural, urban, or other human-modified environments.

Edge habitat becomes more prominent in fragmented habitats because the area of a patch declines as the square of patch size, whereas the perimeter of a patch declines linearly with patch size. Consider, for example, the following two patches:

Square patch 1: 16 × 16 km, 256 km^2, 64 km perimeter, perimeter:area = 0.25

Square patch 2: 8 × 8 km, 64 km^2, 32 km perimeter, perimeter:area = 0.5

Note that as the length and width of the square patch declines by 50% (16 to 8 km), the perimeter also declines by 50% (64 to 32 km). Yet, the area of the patch declines by 75% (256 to 64 km^2). As a result, the proportion of edge habitat (perimeter) increases as the patch gets small. And as the proportion of edge habitat increases, the proportion of **interior habitat**, also called **core habitat**, gets smaller as patch size declines (**Figure 9.11**).

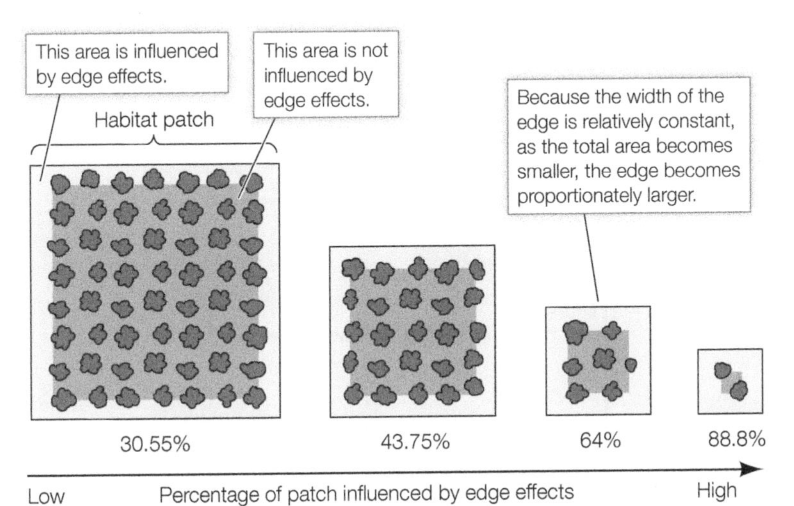

Figure 9.11 As patch sizes are reduced in a landscape, the proportion of core (interior) habitat declines, and the proportion of edge habitat increases. As a result, edge effects have a greater influence on the dynamics of small patches.

The amount of edge habitat is affected not only by patch size, but also by patch shape. Consider three geometric patch shapes: a square patch, a rectangular patch, and a circular patch:

Square patch: 8 × 8 km, 64 km^2, 32 km perimeter, perimeter:area = 0.50

Rectangular patch: 16 × 4 km, 64 km^2, 40 km perimeter, perimeter:area = 0.63

Circular patch: 4.5 km radius, 64 km^2, 28 km perimeter (circumference), perimeter:area = 0.44

Note that all of these three patch configurations have the same total area (64 km^2). Compared with the square and rectangular patches, the circular patch has the lowest perimeter-to-area ratio (0.44), indicating the smallest proportion of edge habitat and highest proportion of core habitat. Circular geometries always have the lowest proportion of edge habitat. In contrast, the rectangular patch—which has elongated sides—has the highest perimeter-to-area ratio (0.63), and thus the greatest proportion of edge habitat and lowest proportion of core habitat. Perimeter-to-area ratios can get even higher for irregular and complex patch geometries, such as polygons that have many projections or invaginations.

The proportion of edge habitat for an individual patch, or collectively for an entire landscape, is important because it influences what are called **edge effects**. The term *edge effects* is used broadly to refer to a suite of physical and biological changes that tend to occur at patch edges. Many of these changes can be detrimental to biodiversity. For example, microclimatic conditions tend be sunnier, drier, and hotter at the edges of forest patches than in interior habitats (**Figure 9.12**).[88] The forest canopy buffers the microclimate of the forest floor, keeping it relatively cool, moist, and shaded during the day and reducing air movement and trapping heat during the night. But at the edge of a forest, the forest floor is exposed to direct sunlight, and the ground becomes hotter during the day. Furthermore, wind dries the soil and increases water loss (evapotranspiration rates) from leaf surfaces, creating drier microclimates with higher fire risk. Without the canopy to reduce heat and moisture loss, the forest edge is also colder at night and less humid, thus inducing greater temperature extremes. Increased wind along the edges of fragments physically damages trees, causing stunted growth or tree falls. This is especially obvious when a fragment first forms, since interior plant species are often not structurally adapted to handle high wind stress. In studies of Amazonian forest fragments, microclimatic changes extended 60 m into the forest interior, and increased tree mortality could be detected within 100 to 300 m of forest edges.[89]

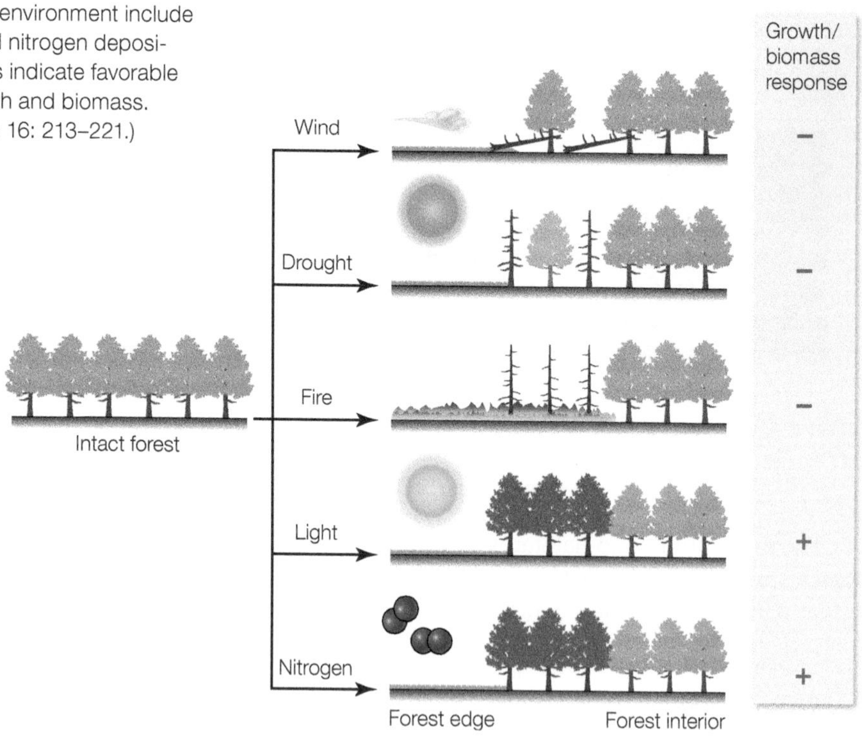

Figure 9.12 Changes in the forest edge microenvironment include exposure to more wind, drought, fire, light, and nitrogen deposition. The positive (+) sign and negative (–) signs indicate favorable and adverse conditions, respectively, for growth and biomass. (After I. A. Smith et al. 2018. *Front Ecol Environ* 16: 213–221.)

Because plant and animal species are often adapted to temperature, humidity, and light levels, changes in microclimatic factors can eliminate many species from forest fragments. Shade-tolerant wildflower species of the temperate forest, late-successional tree species of the tropical forest, and humidity-sensitive animals such as certain insects and amphibians are often eliminated by habitat fragmentation because of altered environmental conditions, which can shift the species composition of biological communities in the remaining patches. Certain types of generalist species that thrive on ecotones between human-dominated and natural habitats can become more common, displacing more-specialized species in the interior habitat. For example, in temperate regions of North America, populations of deer and other generalist herbivores build up in edge areas where plant growth is lush, eventually overgrazing the vegetation and selectively eliminating certain rare and endangered plant species for several kilometers into the forest interior. Omnivorous animals like raccoons, opossums, skunks, and blue jays often have increased population sizes along forest edges where their diets are supplemented by additional foods, including the eggs and nestlings of birds. These aggressive feeders often seek out the nests of interior forest birds, preventing successful reproduction for bird species hundreds of meters from the nearest forest edge.[90] Edges can also become areas of concentrated predation. For example, domestic cats are the predominant predator of birds in fragmented urban landscapes.[91]

In addition to altered rates of herbivory and predation, parasitism and disease can become more prominent at edge habitats. A classic example of edge habitat promoting parasitism involves nest-parasitizing cowbirds, which thrive along the ecotones between agricultural fields and woodlots or forests. Cowbirds use edge habitat as invasion points to fly up to 15 km into forest interiors, where they lay their eggs in the nests of forest songbirds.[92] When the cowbirds hatch, they outcompete the smaller songbird hatchlings for food, usually leading to stunted growth and death of the songbird hatchlings. Habitat fragmentation that leads to increased brood parasitism is partly responsible for the dramatic decline of certain migratory songbird species of North America, such as the cerulean warbler (*Dendroica cerulea*).[93] An example of disease promoted by edge habitat is Lyme disease. Fragmented forest habitats characteristic of suburban development often have high densities of white-footed mice and blacklegged ticks with high rates of infection with *Borrelia burgdorferi*, along with a corresponding increase in Lyme disease in people living in those areas.[94]

Increased edge habitat can also increase the vulnerability of a habitat fragment to invasion by nonnative pest species.[95] Edge habitat serves as the entry point for invasive species, such as aggressive vines that overgrow and kill mature trees. The road edges themselves may represent dispersal routes for invasive species, such as rats and cats. The forest edge represents a high-energy, high-nutrient, disturbed environment in which many pest species of plants and animals can increase in number and then disperse into the interior of the fragment. In general, edge habitats tend to be dominated by species of plants and animals that are distinct from those found in the habitat interior—and often, these species are composed of generalist native and nonnative species.

PATCH ISOLATION Isolation of populations among habitat patches is another consequence of fragmentation. Isolation is a function of both the distance between habitat patches and the matrix that separates those habitat patches (see Figure 9.10A). As the distance between patches grows, dispersal between them becomes increasingly difficult. As the matrix that separates habitat patches becomes increasingly hostile to dispersal, the successful emigration out of patches, and immigration into patches, becomes less likely. Certain kinds of built habitats represent a particularly hostile matrix for organisms. For example, roads, railway lines, fences, and water channels impose significant barriers to dispersal, while extensive croplands or urban developments create hostile environments that make it difficult for certain types of organisms to move among habitat patches.

Patch isolation and barriers to dispersal among patches affect organisms in two primary ways. First, they impede the movement of individuals between the different habitats in a landscape that are required to complete their life cycles. For example, many organisms exhibit **ontogenetic niche shifts**—that is, changes in habitat requirements to complete the different parts of their life cycles. Many insects, and certain types of vertebrates like frogs, spend the juvenile portion of their life cycle in aquatic habitats, and then they spend the adult portion of their life on land. Some species, such as many types of birds, require one type of habitat for foraging and another type of habitat for breeding and nesting. Other species migrate seasonally or annually among habitat types to gather food, obtain water, or breed. When the habitat types needed to complete a life cycle become more isolated, or the matrix that separates those habitats becomes more hostile to dispersal, then a portion of the life cycle can be jeopardized.

Second, patch isolation, and barriers to dispersal among patches, can restrict immigration and emigration to and from patches in ways that reduce population size and increase the risk of extinction in the landscape. When populations become small and isolated from colonization, they become increasingly vulnerable to stochastic variation in demographic parameters (e.g., birth and death rates), as well as environmental stochasticity like floods, fires, or droughts that can eliminate a small local population (see Chapter 8). Small population size also reduces genetic variation within populations, which may lead to inbreeding or genetic drift, making a population more vulnerable to recessive lethal alleles or to changing environmental conditions.

Dispersal of individuals among patches in a landscape not only has consequences for the genetic diversity of populations in habitat patches, but also influences the size and persistence of populations living in a spatially subdivided landscape. The simplest equation to describe the dynamics of a population living in a spatially subdivided landscape is:

$$N_{t+1} = N_t + B_t - D_t + I_t - E_t \quad (9.1)$$

where N_t represents the size or density of a species population in a patch at time t, N_{t+1} is the size of the population one time step later, B_t represents the total number of births in the interval from time t to time t+1, D_t represents the total number of deaths in the same time interval, I_t represents the total number of immigrants in the same time interval, and E_t represents the total number of emigrants in the same time interval.

Birth, immigration, death, and emigration are the four most important parameters in population dynamics (called the BIDE factors). Together, they determine whether a population will survive or go extinct over time. If $B_t + I_t$ (birth and immigration) exceed $D_t + E_t$ (death and emigration), then N_{t+1} will be larger than N_t and the population in a patch will grow. If, however, $D_t + E_t > B_t + I_t$, then $N_{t+1} < N_t$ and the population will decline and go extinct. Dispersal of individuals among the patches in a landscape therefore plays an important role in determining whether the population in any given patch is able to maintain a positive growth rate.

Species-specific responses to fragmentation

The impacts of fragmentation on individuals and populations are often species-specific, which is another reason why fragmentation can have either positive or negative effects on biodiversity, depending on the types of organisms living in the fragmented landscape. For example, certain types of generalist species (e.g., deer) may not be affected by edge habitat and may even thrive on the ecotone between natural and human-dominated habitats. Yet, these same edge habitats can be **ecological traps** for other species, such as songbirds that get exposed to nest parasitism. Ecological traps are areas of low-quality habitat that reduce survival and reproduction.

Highly mobile species, such as species that fly (e.g., birds, bats, flying insects), are often less affected by patch isolation than species that are less mobile (e.g., frogs and beetles). For some species, such as African elephants (*Loxodonta* spp.), that are large and perceive a landscape as more connected, an open field may not be a dispersal barrier as they migrate among habitat patches. Yet, for a species with small bodies and short-range dispersal (e.g., a shrew), or even a larger-bodied species that spends most of its life in the treetops (e.g., many types of primates and marsupials), an open field may represent a barrier to dispersal that will never be crossed. A study in the Amazon by Malcolm[96] revealed distinct responses of similar animals to fragmentation. Two species of opposum—the wooly opossum (*Caluromus sp.*) and the mouse opposum (*Marmosa sp.*)—were tracked by use of radio transmitters to determine if they would travel a gap of 135 to 275 m to reach the fragment on the other side. Mouse opposums were tracked crossing the gap, while the more strictly arboreal wooly opossums would not cross.

Unfortunately, very little information exists on the qualities of suitable dispersal habitat or on barriers for various species. Species- and habitat-specific dispersal studies are essential for gaining a better understanding of fragmentation effects. However, what we do know suggests that human-created structures and habitats—roads, urban areas, agricultural fields, clearcuts—can greatly inhibit movement and negatively affect population viability of many kinds of animals and plants (especially those pollinated or dispersed by animals).

Habitat Degradation

Even when a landscape is unaffected by overt habitat loss or fragmentation, that landscape can be degraded by human activities that reduce its potential to support life. Sometimes the human activities that degrade a landscape are obvious. For example, keeping too many cattle in a prairie and allowing them to graze along streams produces telltale signs of overgrazing and soil erosion that most anyone can recognize. Other types of degradation, such as many forms of chemical pollution, are not visually obvious. This makes them particularly dangerous, as they can go unnoticed and uncorrected.

Pollution

The most widespread form of habitat degradation is pollution, commonly caused by pesticides, herbicides, sewage, fertilizers from agricultural fields, industrial chemicals and wastes, and emissions from factories and automobiles. Pollution is the largest environmental cause of human disease and premature death in the world today. Diseases caused by pollution were responsible for an estimated 9 million premature deaths in 2015 (16% of all deaths worldwide), which is 3 times more deaths than from AIDS, tuberculosis, and malaria combined and 15 times more than from all wars and other forms of violence. In the most severely affected countries, pollution-related disease is responsible for more than one death in four.[97] In addition to its direct toll on human health, many forms of pollution have negative impacts on biodiversity. We cover those forms next.

PESTICIDES The dangers of pesticides were brought to the world's attention in 1962 by Rachel Carson's book *Silent Spring*. Carson described a process known as **biomagnification** (Figure 9.13) through which dichlorodiphenyltrichloroethane (DDT) and other organochlorine pesticides become concentrated at higher

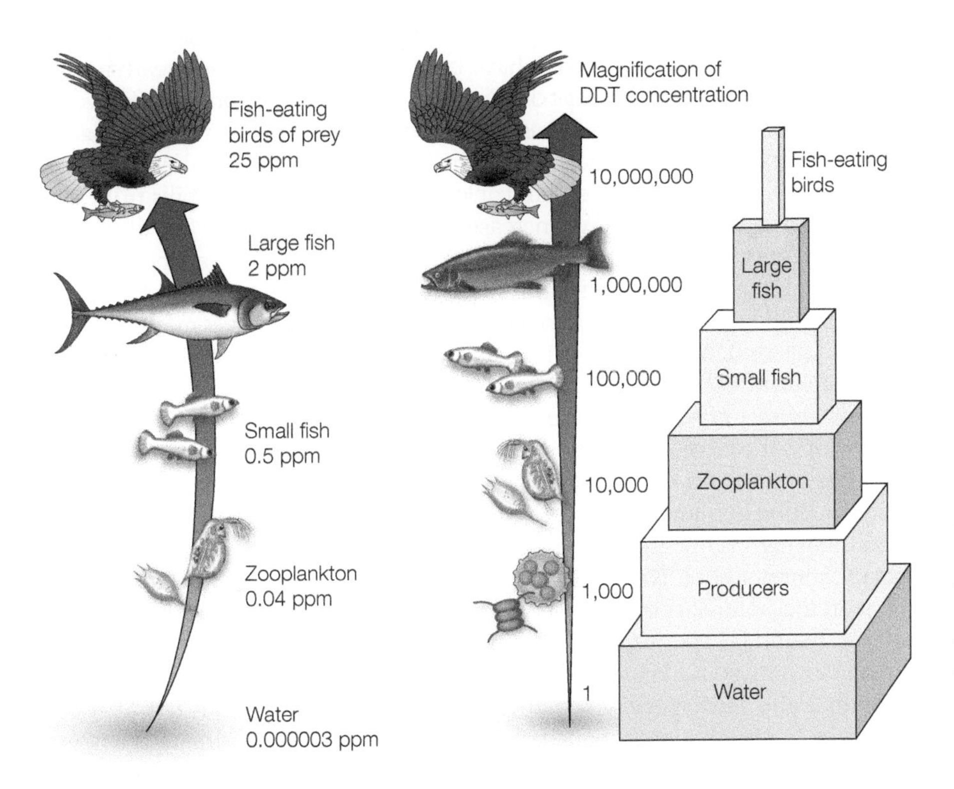

Figure 9.13 Biomagnification of DDT. (Left) The actual concentration of DDT in various components of the food web in parts per million (ppm). (Right) The relative magnification of DDT at successive trophic levels of the food web. (After G. M. Woodwell et al. 1967. *Science* 156: 821–824.)

levels of the food chain. These pesticides, which are used on crops to kill insects and sprayed on water bodies to kill mosquito larvae, were dramatically reducing wildlife populations and threatening many species with extinction. Birds, such as hawks and eagles, that ate large amounts of insects, fish, or other animals from habitats contaminated with DDT suffered reproductive failure due to the thinning and breaking of eggshells. In agricultural areas, beneficial insect species were killed even as the targeted pest species developed resistance to the pesticide. In aquatic habitats, DDT became concentrated in predatory fish and in sea mammals such as dolphins. While DDT has since been banned for use by many industrialized nations, it continues to be used along with other organochlorine compounds for pest control in some countries.

Despite lessons learned from DDT, harmful pesticides with nontarget effects continue to be widely used throughout the world. Recent surveys suggest these chemicals are continuing to contribute to substantial declines in biodiversity. A 2010 publication by Geiger and colleagues[98] summarizes a study of nine large agricultural landscapes spanning eight countries in Europe. The authors found that use of fungicides and insecticides was the one measure of agricultural intensification that had consistent negative effects on biodiversity of native plants and animals. A 2013 study by Beketov and colleagues[99] analyzed the effects of pesticides on insect biodiversity in 63 streams in Germany, France, and Australia. The researchers found there were 42% fewer species in highly contaminated streams than in uncontaminated streams in Europe. Highly contaminated streams in Australia showed a 27% decrease in the number of entire invertebrate families, compared with uncontaminated streams. Moreover, the continuing exposure to harmful pesticides has been implicated in the recent decline of several important groups of organisms, such as bees that serve as pollinators,[100] amphibians that are in widespread decline,[101] and bats that control insect pests.[102]

OIL SPILLS Despite the steps by some governments toward renewable sources of energy, reliance on oil continues to grow in many parts of the world. As oil exploration, extraction, and transportation has increased, the risk of spills has increased. Historically, oil tankers have been responsible for the greatest number of spills and largest volumes of oil released into the environment. However, the number of spills by tankers has decreased over the past 30 years because of improved navigation tools and ship design.[103] In contrast, the number of large oil spills from pipelines has more than tripled since the 1970s, and spills from storage and refinery facilities have increased more than fivefold.[104] The number of major disasters from exploration and production facilities, such as the 2010 failure of the Deepwater Horizon drilling rig that spilled somewhere between 134 and 206 million gallons of oil into the Gulf of Mexico (the largest marine oil spill in world history), have held constant at just under one per year.[104]

Oil spills affect plants and wildlife through direct exposure and toxicity, as well as indirect effects that occur through cleanup programs that degrade the ecosystems they live in. The most obvious form of direct exposure occurs when oil films coat the fur of mammals or the feathers of birds, causing them to lose their function in buoyancy, heat regulation, and flight.[105] Mortality rates are exceedingly high for plants and animals that have been directly exposed to even small amounts of oil (**Figure 9.14A**). Under most circumstances, spilled oil will float on the water surface, in turn, minimizing the direct exposure experienced by most subtidal species. Species with canopies that reach the surface of the water, such as some kelp and seagrass species, represent exceptions. Marine mammal and bird species, which must regularly pass through the air-water interface to breathe, are particularly vulnerable to oil exposure,[105,106] whereas pelagic fish species will have less exposure.[107] In spill disasters in which the oil

(A)

© National Geographic Image Collection/Alamy Stock Photo

(B)

© Pulsar Imagens/Alamy Stock Photo

(C)

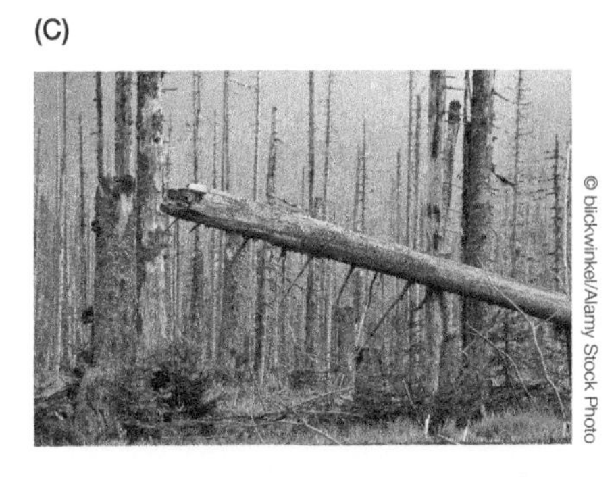

© blickwinkel/Alamy Stock Photo

Figure 9.14 (A) Birds, marine mammals, and many other ocean animals sicken and die when they are covered by oil following spills. This brown pelican was found following the Deepwater Horizon oil spill in the Gulf of Mexico in 2010, which, at 4.9 million barrels, is the largest marine oil spill in history. (B) Coastal hypoxia. Fish in the Rodrigo de Freitas Lagoon in Rio de Janeiro died off as a result of pollution. High nitrogen and phosphorus levels from human activity resulted in excessive algal growth, followed by algal death and decay leading to oxygen levels so low that fish could not survive. (C) Acid rain leads to the death of trees over large expanses of forest area in New England.

floats, the greatest exposure will occur in the intertidal zone, where rising and falling tides bring species in direct contact with the bulk of the spilled oil.

Ingestion or inhalation of oil and/or absorption through the skin can lead to accumulation of contaminants in tissues that cause DNA damage, compromised immune function, and cardiac dysfunction and cause mass mortality of eggs and larvae.[105] Species differ in their physiological responses to an oiling event, because of variation in their morphological and physiological traits. For example, barnacle populations are sometimes resistant to direct exposure to oil,[108] whereas amphipod species generally experience strong and long-lasting population declines in response to oiling.[109] Nonetheless, subtle differences among even closely related species can result in divergent responses to oiling events,[105] though oil spills are generally damaging for most of the exposed species.

Aside from direct exposure and toxicity, oil spills can affect biodiversity through cleanup efforts that degrade habitat. Many procedures are used to clean up oil spilled in marine environments, but all of these procedures further damage marine ecosystems, and in some instances they can actually increase the amount of time required for ecosystems to recover from oil spills.[110] For example, chemical dispersants are often applied to the oil to break it down into small droplets and diffuse it. But the chemical dispersants used to break down oil are themselves toxic,[111–113] and the combination of oil and dispersant can have stronger negative effects on marine species than the oil alone.[105]

TOXIC METALS Human activities such as mining and smelting operations, use of leaded gasoline (still used in some countries), burning of coal for heat and power, and disposal of electronic wastes (e.g., computers, printers, TVs, mobile phones) release large quantities of lead, zinc, mercury, cobalt, and other toxic heavy metals into the environment.[114] Metals can accumulate to levels that are directly toxic to terrestrial and aquatic organisms. There are numerous examples, such as habitats around smelting operations, where the toxic effects of metal poisoning have destroyed most life for kilometers around. In addition to lethal effects, metal poisoning can have sublethal effects that reduce or eliminate species from an ecosystem through increased susceptibility to disease, mortality, and decreased fecundity.

Biomagnification is a serious problem with heavy metals—not only because it increases concentrations of metals to toxic levels in the tissues of the organisms at higher levels of the food chain, but also because it can pose a human health risk.

Aquatic flora and fauna, ranging from algae to invertebrates (oysters, mussels) to fish, can accumulate trace metals in their tissues to levels that are several orders of magnitude (thousands to million of times) above background levels in the aquatic environment. Biomagnification of mercury was responsible for one of the worst environmental disasters of the past century when, in 1956, the Chisso Corporation released industrial wastewater containing highly toxic methylmercury into Minamata Bay and the surrounding Shiranui Sea in southern Japan (see Chapter 2). Local people who ate contaminated shellfish and fish developed neurological disorders, including difficulty seeing, hearing, and swallowing, leading in some cases to convulsions, coma, and eventually 2265 fatalities.

EUTROPHICATION Human sewage, agricultural fertilizers, detergents, and industrial processes often release large amounts of nitrates and phosphates into aquatic systems, initiating the process of **eutrophication**—which is the leading cause of degraded water quality worldwide.[115,116] Humans release as much nitrate into the environment as is produced by all natural processes combined, and the human release of nitrogen is expected to keep increasing as the human population continues to increase. Even small amounts of these nutrients can stimulate plant and animal growth, and high concentrations of nutrients released through human activities often result in thick "blooms" of algae at the surface of ponds and lakes. These algal blooms may be so dense that they outcompete other plankton species and shade bottom-dwelling plant species. As the algal mat becomes thicker, its lower layers sink to the bottom and die. The bacteria and fungi that decompose the dying algae grow in response to this added sustenance and consequently absorb all of the oxygen in the water (**Video 9.5**). Without oxygen, much of the remaining animal life dies off, sometimes visibly in the form of masses of dead fish floating on the water's surface (**Figure 9.14B**). The result is a greatly impoverished and simplified community, a dead zone consisting of only those species tolerant of polluted water and low oxygen levels.

Video 9.5 Watch how eutrophication has led to harmful algal blooms and a state of emergency in Florida.
oup-arc.com/e/cardinale-v9.5

This process of eutrophication has had a particularly dramatic effect on coastal marine systems, generating excessive biochemical oxygen demand that has produced 245,000 km^2 of dead zones in over 400 coastal habitats around the world.[117] These dead zones are especially prominent in bodies of water in confined areas, such as the Gulf of Mexico, the Mediterranean, the North and Baltic Seas in Europe, and the enclosed seas of Japan. In warm tropical waters, eutrophication favors algae, which grow over coral reefs and completely change the biological community. The key to stopping eutrophication and its negative effects is to reduce the release of excess nutrients through improved sewage treatment and better farming practices, including reduced applications of fertilizer and establishment of buffer zones between fields and waterways.

ACID RAIN In the past, people assumed that the atmosphere was so vast that materials they released into the air would be widely dispersed and their effects would be minimal. But today several types of air pollution are sufficiently widespread that they can degrade whole ecosystems. One case in point is acid rain, which is produced when industries like smelting operations and coal- or oil-fired power plants release nitrogen and sulfur oxides into the air. When those chemicals combine with moisture in the atmosphere, they produce nitric and sulfuric acids that become part of cloud systems and, in turn, dramatically lower the pH of rainwater. Acid rain can lead to the weakening and deaths of trees over large expanses of area (**Figure 9.14C**). In addition, acid rain can lower the pH of soils to sufficiently low levels that important cations (K^+, Ca^{2+}) required for plant growth are leached from entire ecosystems, in turn reducing productivity and preventing regrowth of plants.[118]

Acid rain is currently a problem in eastern North America, in central Europe and other parts of Europe, and in China, Korea, and other parts of East Asia. In the United States alone, about 27 million metric tons of nitrogen and sulfur oxides are released into the atmosphere each year, with deposition of acid rain concentrated in the northeastern United States. The heavy reliance of China on high-sulfur coal and the rapid increase in automobile ownership and industrialization in China, India, and East Asia represent serious threats to biodiversity in these regions, with dramatic increases in acid rain and nitrogen deposition predicted over the next 50 years.[119]

Increased acidity alone damages many plant and animal species; as the acidity of water bodies increases, many fish either fail to spawn or die outright. Both increased acidity and water pollution are contributing factors to the dramatic decline of many amphibian populations throughout the world.[120] Most amphibian species depend on bodies of water for at least part of their life cycle, and a decline in water pH causes a corresponding increase in the mortality of eggs and young animals. Acidity also inhibits the microbial process of decomposition, lowering the rate of mineral recycling and ecosystem productivity. Many ponds and lakes in industrialized countries have lost large portions of their animal communities as a result of acid rain. These damaged water bodies are often in supposedly pristine areas hundreds of kilometers from major sources of urban and industrial pollution, such as the North American Rocky Mountains and Scandinavia. While the acidity of rain is decreasing in many areas because of better pollution control, it still remains a serious problem in many developing countries, especially in East and South Asia.

PHARMACEUTICALS AND PERSONAL CARE PRODUCTS (PPCPs) PPCPs include medicines that are used by people or used by agribusiness to boost the growth or health of livestock (e.g., hormones, antibiotics, lipid regulators). They also include substances used by individuals for their personal health or cosmetic reasons (e.g., fragrances, preservatives, disinfectants, sunscreens). PPCPs enter the environment directly when hospitals, households, or industries discharge materials directly into surface waters. Another important route, particularly for pharmaceuticals, is the excretion or improper disposal of medicines into plumbing and sewer systems from patients undergoing treatment. Wastewater treatment plants seldom have the technology needed either to detect the hundreds of thousands of potential chemicals in pharmaceuticals or to reduce their concentrations.

Several studies have shown that PPCPs are now ubiquitous in the environment. For example, a 1999–2000 study led by the U.S. Geological Survey (USGS) performed the first national reconnaissance of "emerging pollutants" in the United States.[121] The study, which covered 142 streams, 55 wells, and 7 effluents in 36 states, found that steroids, nonprescription drugs, insect repellents, and detergents were present in 70% or more of all water bodies (**Figure 9.15A**). Antibiotics, fire retardants, and plasticizers (chemicals used to make plastics) were found in 50% or more. Steroids and plasticizers represented over 20% of the total concentration of chemicals in the water bodies, while detergents (cleaning agents) represented over 30% (**Figure 9.15B**).

The consequences of PPCPs for biodiversity are poorly understood and represent an ongoing area of research. However, limited evidence suggests that certain types of PPCPs could have adverse effects on the physiology, behavior, and reproduction of fish and other animals that ingest these biologically active chemicals.[122] In 2007, Kidd and colleagues[123] reported results of a seven-year experiment in which they exposed an entire lake to low concentrations of 17-ethynylestradiol (an estrogen medication used widely in birth control pills) and then looked at how exposure influenced a common species of fish (the flathead minnow, *Pimephales*

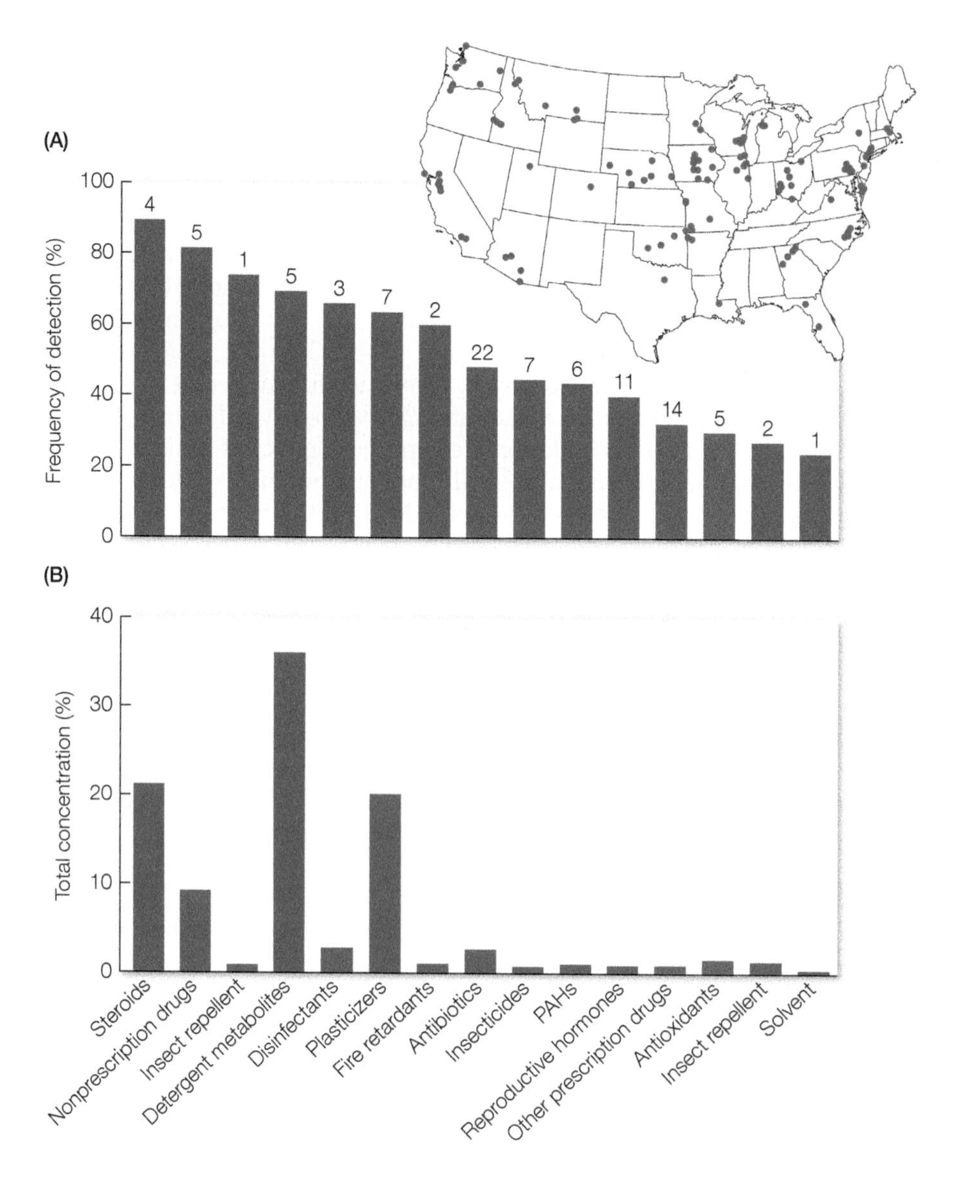

Figure 9.15 Results of a 1999–2000 study led by the USGS that performed the first national reconnaissance of "emerging pollutants" in the United States. The *x*-axis lists the different compounds that were detected. (A) The frequency of detections for each compound, as the percentage of water bodies in which they were detected. The number of compounds in each category is shown above the bar. (B) The total concentration of the compound when detected. (From D. Kolpin et al., 2002. *Environ Sci Technol* 36: 1202–1211.)

promelas). The authors showed that exposure to 17-ethynylestradiol feminized the male fathead minnows by reducing the size of their gonads. Loss of males led to reproductive failure, causing the fathead minnow population to crash.

PLASTICS Plastics are one of the fastest-growing forms of environmental pollution and have the potential to become a major contributor to the loss of wildlife in certain biomes, especially in marine ecosystems. Recent estimates suggest that more than 5 trillion pieces of plastic weighing more than 250,000 tons are afloat in the world's oceans.[124] The harmful impact of plastic pollution through ingestion and entanglement of marine fauna, ranging from zooplankton to cetaceans, seabirds, and marine reptiles, is well documented.[124, 125] Certain types of organisms, like whales, turtles, and seabirds, are particularly vulnerable because they mistake plastic particles and floating bags as food items, and they can ingest large quantities that decrease nutrition and interfere with digestion (**Figure 9.16**).[126] Plastics can also concentrate certain types of organic pollutants and metals that sorb to plastic debris. When ingested, these toxic chemicals can become concentrated at higher levels in the food chain.[127] There is also an increasing recognition of the danger posed by microplastics, which are tiny particles of plastic waste that result from the breakdown of plastic products in the marine environment.

Thailand Dept. of Marine and Coastal Resources/Social Media/via REUTERS

Figure 9.16 In 2018, a small male pilot whale was found near the Thai border with Malaysia, unable to swim or breathe and vomiting plastic bags. After the whale died, a necropsy revealed 80 plastic bags weighing 7.7 kg, which the whale likely mistook for food, clogging its stomach.

Desertification

Desertification is a form of habitat degradation where poor soil management practices lead to soil degradation that causes relatively dry areas of land to become increasingly arid and lose their water bodies, vegetation, and wildlife.[128] Drylands account for about 40% of Earth's land area and are home to more than 2 billion people. Between 6 and 12 million km^2 of drylands (10%–20% of the total) are already heavily degraded, and more than a billion people are under threat from further desertification (**Figure 9.17**).[129] The problems of desertification will become more severe in coming decades with the greater frequency of drought and heat waves associated with climate change, as well as the increasing pressure put on the land by a rising human population. Many areas of the world that currently support grasslands and shrublands are in danger of becoming deserts.

There are several dramatic examples of the devastating impact of desertification on natural landscapes. One example is that of the Aral Sea. Once the fourth-largest inland lake by area in the world, with an area of 67,339 km^2, the Aral Sea became a victim of the Soviet Union's agricultural policies that began in the 1950s. Water from its two river sources—the Amu Darya and Syr Darya—were intentionally diverted for cotton cultivation.[130] As water flow decreased, the Aral Sea lost 74% of its area and 90% of its volume, resulting in ecological damages that had dire human consequences. As the sea began to dry up, salinity rose substantially, causing the once-abundant freshwater fish species to die out. By the 1980s, the fishing industry in the Aralsk district was wiped out, forcing a mass migration of people. Declining groundwater levels and salt accumulation in the soils harmed farming in the region. Large and frequent dust storms carried toxic salts and agricultural chemicals from the seabed, causing acute human health problems, crop failure, and harm to wild and domestic animals. Desiccation altered the regional climate-regulating effect, resulting in more heat waves and droughts that further promoted desertification.[130] The rapid collapse of the Aral Sea ecosystem in just under three decades prompted one of the largest restoration efforts in history.* The Syr Darya Control and Northern Aral Sea (NAS) Project, which is an $86 million restoration

* See *National Geographic*'s report at oup-arc.com/e/cardinale-w9.2.

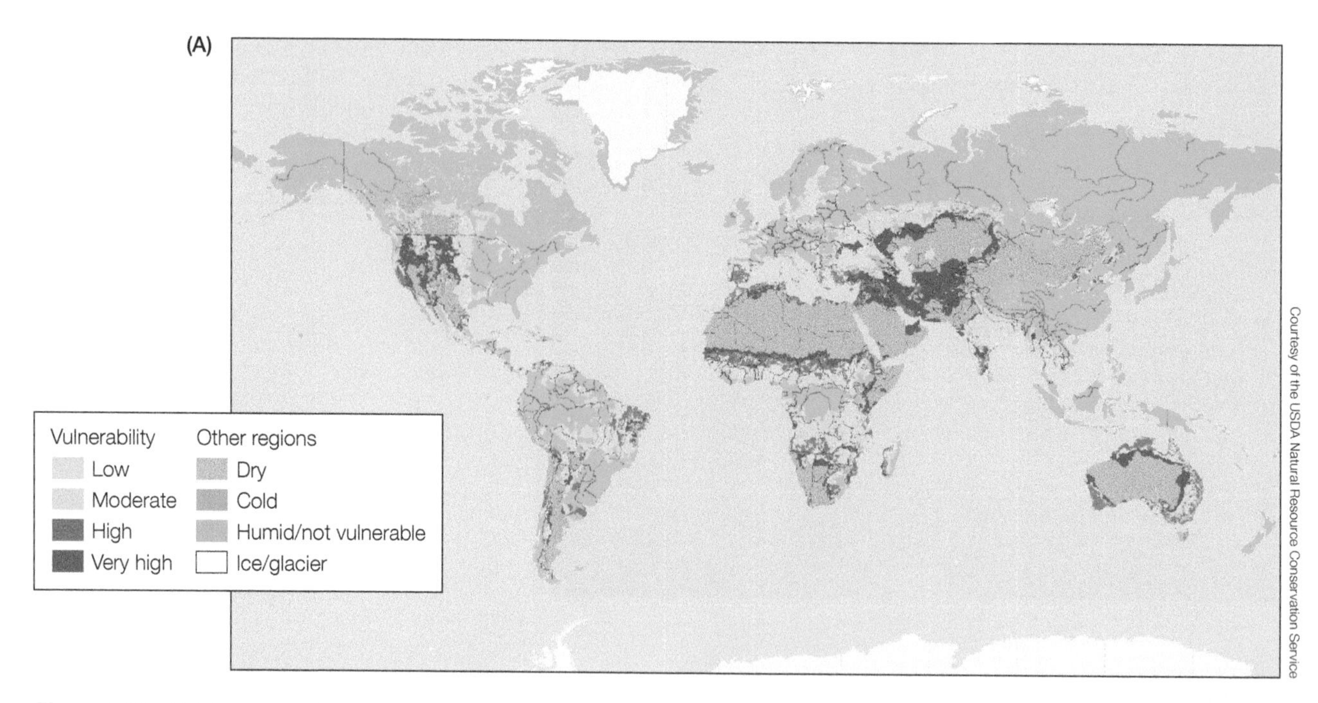

Figure 9.17 (A) Arid areas of the world are experiencing encroaching desertification, with increasing aridity and expanding deserts. The regions shaded in red, orange, and yellow are vulnerable to desertification and are potentially at risk of becoming desert over the next several decades. (B) Deserts are expanding in the Sahel region south of the Sahara as human activities stress semiarid ecosystems.

project partially funded by the World Bank, is working to increase agriculture and fish production in the Syr Darya basin and improve ecological and environmental conditions in the delta area. The project consists of removing major water flow obstructions, directing more water to the sea, and then retaining that water with control structures like dykes, levees, and dams. Water levels rose by 11 feet during the first seven months of the effort, which increased optimism that parts of the once-rich natural habitat and native biodiversity of the Amu Darya and Syr Darya deltas might be restored.

In 1996, the United Nations established the Convention to Combat Desertification for countries experiencing serious drought and/or desertification, particularly in African nations. The convention, which came out of a recommendation of the 1992 Rio Conference's *Agenda 21*, is the only internationally legally binding framework set up to address the problem of desertification. The convention, which has 197 parties, combats desertification and the effects of drought through national action programs that incorporate long-term strategies supported by international cooperation and partnership arrangements.

Erosion and sedimentation

At geologic time scales, rates of continental erosion are roughly balanced by rates of sediment accumulation. But human activities have sharply increased net rates of continental sediment erosion and the displacement of soil to other habitats. Deforestation, overgrazing, and unsustainable agricultural methods have elevated current rates of soil erosion to 6 cm (2.4 inches) per 100 years, which is between 11 and 38 times faster than what has been recorded in the past geologic record.[131] It is estimated that half of all topsoil in the world has been eroded in the last 150 years. Each year, an estimated 20.7 gigatons (Gt) of terrestrial sediments are eroded into the world's rivers. Of this, about 3.6 Gt is retained behind dams in reservoirs, 1.6 Gt remains as bed load in the rivers, 2.9 Gt remains in suspension, and 12.6 Gt flows to the oceans, where it is deposited as sediment at the mouths of rivers in deltas, bays, and estuaries.[131]

Soils that are rich in organic materials and the minerals required for growth of bacteria, fungi, and plants take hundreds if not thousands of years to form. Loss of organic material and minerals reduces soil fertility and productivity. In extreme cases, erosion can cause a landscape to become a barren moonscape with little plant life and, in turn, little food or habitat for other organisms to live, and can leave little value to people.

When soil runs into waterways, sediment in the water can make aquatic ecosystems inhospitable to certain groups of organisms. The breathing gills of many types of invertebrates, including the aquatic nymph stages of groups like stoneflies (Plecoptera), mayflies (Ephemeroptera), and caddisflies (Trichoptera), can become clogged with water-borne sediment, and the animals can suffocate and die. The smothering and suffocation of benthic (mud-dwelling) fauna is a primary cause of reduced biodiversity in rivers and streams.[76] Die-offs of sensitive organisms that cannot tolerate sedimentation are often associated with shifts in the composition of benthic communities toward species that have burrowing lifestyles or that have different modes of respiration and can tolerate low oxygen conditions.

Metapopulations and Landscape Mosaics

Habitat loss, fragmentation, and degradation are causing the landscapes of the world to increasingly look like a patchy, spatial mosaic where some habitats have been lost and converted to human uses, others have been heavily degraded, and those that remain intact are small, fragmented, and disconnected from one another (**Figure 9.18**). Species that survive, reproduce, and complete

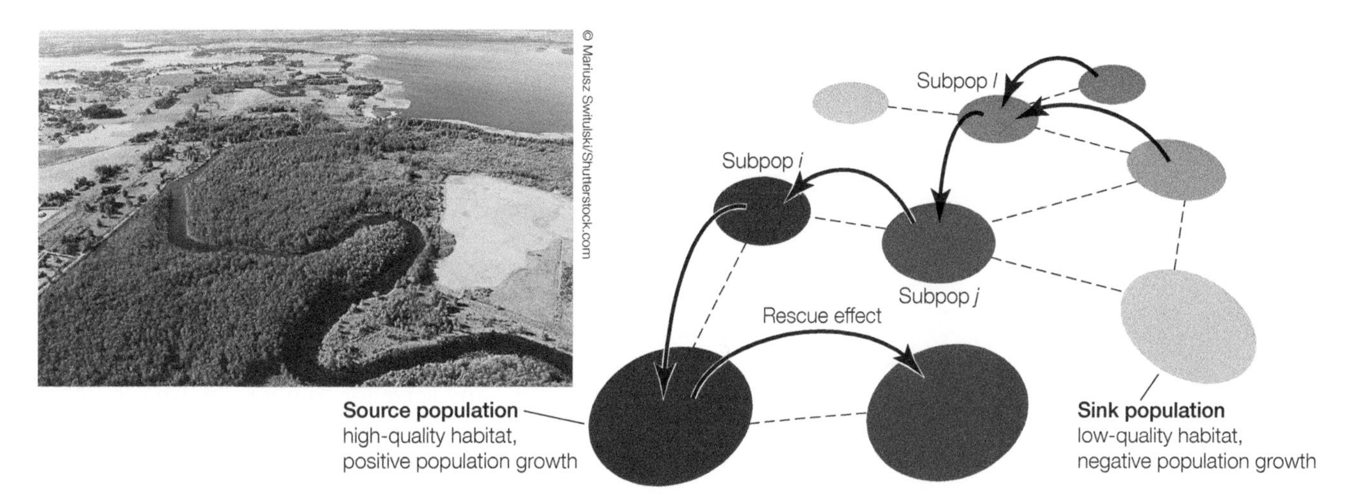

Figure 9.18 Modern landscapes increasingly look like patchy mosaics where some habitats have been lost and converted to human uses, others have been degraded by human activities, and those that remain are small, fragmented, and disconnected from each other. Species populations in landscape mosaics are often subdivided, and behave like a metapopulation where some populations are sources and others are sinks. (After C. Poletto et al. 2013. *PLOS Comp Biol* 9: e1003169. CC BY 4.0.)

their life cycles in these **landscape mosaics** are spatially subdivided. Some parts of the landscape mosaic may have habitat fragments whose populations are separated by barriers to dispersal. On other parts of the landscape mosaic, individuals that occupy subdivided habitat fragments may attempt to migrate across the landscape to other patches. Some of the habitat patches in the landscape mosaic may not have populations at all, perhaps because they are too degraded for the species to persist.

The "classic" metapopulation model

Small, isolated populations in single habitat fragments are highly vulnerable to local extinction. But when small, local populations are connected by occasional movements of individuals among them, a species has a much greater likelihood of survival in the landscape as a whole. A **metapopulation** is a collection of local populations that are spatially separated from each other in individual patches of a landscape but connected to each other by the dispersal and exchange of individuals (see Figure 9.18). The term *metapopulation* was coined by Richard Levins in 1969[132] to describe the population dynamics of insects in agroecosystems. Levins called it "a population of populations." Later, Ilkka Hanski developed the basic concepts, models, and empirical examples of metapopulations in the influential book *Metapopulation Biology*,[133] which helped conservation biologists gauge the prospects of long-term survival for species living in fragmented landscapes.

A metapopulation is generally considered to consist of several spatially distinct populations that exist in a landscape that has areas of suitable habitat that are unoccupied. Individual populations in single suitable habitat patches have finite life spans, and they blink in and out of the landscape as small local populations go extinct as a consequence of inbreeding depression or demographic and environmental stochasticity. But even as local populations go extinct, the metapopulation as a whole can persist through time so long as colonization of unoccupied but suitable habitat patches exceeds the rate of extinction in occupied patches.

A description of how a metapopulation can persist even in the face of local extinctions comes from the classic metapopulation model, which is remarkably simple:

$$\frac{df}{dt} = p_i f(1-f) - p_e f \tag{9.2}$$

In Equation 9.2, f is the fraction of patches occupied by a species in a landscape, which ranges from 0 (none) to 1 (all). The change in the proportion of patches occupied through time, df/dt, is influenced by just two things: p_i is the probability that an unoccupied patch in the landscape will be converted to an occupied patch via immigration, and p_e is the probability that a patch currently occupied will be made vacant by extinction. Note that in this model, p_i is proportional to f, which means there are more colonists produced when more patches are occupied and only unoccupied patches can be colonized at each time step (which is the term $1-f$). In contrast, the final term $p_e f$ means only populations in patches currently occupied can go extinct.

If we set $df/dt = 0$ in Equation 9.2 and solve for equilibrium, we can determine the fraction of patches that would be occupied, on average, for any given probability of immigration or extinction. The equilibrium is given by:

$$\bar{f} = 1 - \frac{p_e}{p_i} \tag{9.3}$$

Equation 9.3 says the frequency of patches occupied in the landscape at equilibrium, $\bar{f}$, is simply influenced by the ratio of extinction to immigration. So long as the probability of immigration to unoccupied patches exceeds the probability of extinction from occupied patches, then the population will persist. Note that the frequency of patches occupied is always less than 1, which means that on average some habitat patches always remain unoccupied. This is true because the probability of extinction can never be zero, and the probability of immigration can never be infinite. But even while individual populations in single habitat patches have finite life spans, the metapopulation as a whole can persist through time because of the exchange of individuals among patches.

The model given by Equation 9.3, which is called the classic (or traditional) metapopulation model, has proven to be a powerful tool for estimating the extinction risk of species in fragmented landscapes and for devising management plans that increase the chance a population will persist. For example, the bay checkerspot butterfly (*Euphydryas editha bayensis*) is endemic to the San Francisco Bay region of California in the United States (**Figure 9.19**). It exists on shallow serpentine soils that are high in magnesium, because this is the habitat where its host plant (the dwarf plantain, *Plantago erecta*) is found. By the 1980s, habitat loss and fragmentation around the San Francisco Bay area caused the butterfly population to hit record lows, which led to it being listed as threatened under the U.S. Endangered Species Act. Paul Erhlich and his colleagues at Stanford University began developing population models of the bay checkerspot butterfly that could be used to guide conservation efforts, and they found that population dynamics of the species throughout the fragmented landscape could be approximated by simple metapopulation models much like that in Equation 9.2.[134] These models were used to estimate the probability of extinction of the species and the number of habitat patches required for the butterfly to persist in the landscape and to determine high-priority habitats for conservation. The classic metapopulation model (Equations 9.2 and 9.3) has since been used to estimate the patch sizes and connectivity thresholds needed to restore butterfly habitat and ensure long-term persistence.[135]

Source-sink dynamics

The classic metapopulation model makes some simplifying assumptions that are not always realistic. For example, the original idea of a metapopulation assumed that all habitat fragments in a landscape were of equal quality. But in real-world landscapes, habitat patches tend to vary in size and quality. In fact, some patches may be so resource poor or afford such bad protection from predators that population growth is negative. Other patches may represent perfect conditions, in which population growth is highly positive.

One improvement to metapopulation models is the integration of source-sink dynamics. A **source population** is defined as a patch that provides a net donation of immigrants to nearby

(A) Geographic range of *Euphydryas editha*

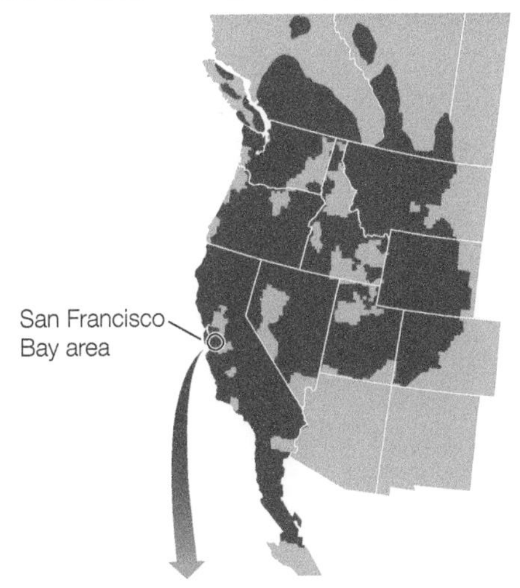

(B) Serpentine outcrops (potential butterfly habitat) in Santa Clara Valley

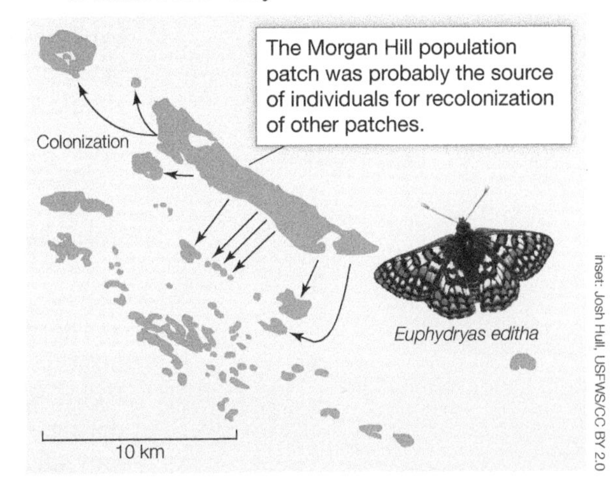

Figure 9.19 (A) The bay checkerspot butterfly (*Euphydryas editha bayensis*) was once distributed throughout the western United States. However, habitat loss, fragmentation, and degradation have severely reduced its population size and restricted its range to the San Francisco Bay region of California. (B) In its current habitat, the bay checkerspot butterfly exists as a metapopulation composed of many distinct subpopulations that inhabit shallow serpentine soils where its host plant (the dwarf plantain, *Plantago erecta*) is found. Biologists have used metapopulation models to estimate the probability of extinction of the species and the number of habitat patches required for the butterfly to persist in the landscape and to determine high-priority habitats for conservation. (A, after R. E. Stanford and P. A. Opler. 1993. *Atlas of Western USA Butterflies, Including Adjacent Parts of Canada and Mexico*. Published by authors. Denver, CO: K. Lotts and T. Naberhaus, coordinators. 2017. Butterflies and Moths of North America, www.butterfliesandmoths.org; B, after S. Harrison et al. 1988. *Am Nat* 132: 360–382.)

patches with lower-quality habitat (see Figure 9.18). A source population should persist indefinitely even in isolation. A **sink population** is defined as a patch that would go exinct if it were not for the constant input of immigrants from nearby source populations. As immigrants from a source population colonize a sink population, or as they recolonize a habitat patch that has been left open by the extinction of another population, they can "rescue" local populations from extinction. **Rescue effects** like these are common in patchy landscapes where subpopulations may blink in and out, going locally extinct in a patch, only to be rescued by colonists from another patch. **Source-sink dynamics** is the term used to describe the immigration and emigration of individuals across a landscape with source and sink populations.

Source-sink dynamics extend the types of models that can be applied to conservation problems. For example, one modification of the classic metapopulation model is called the **mainland-island model** of population dynamics. The mainland-island model, also called a **propagule rain model**, is derived from the theory of island biogeography (see Chapter 4) and describes a scenario where a large mainland population (such as a conservation reserve) provides a source of emigrants that disperse to nearby small populations. The mainland population has a low probability of extinction, whereas the small populations become extinct relatively frequently. Emigration from the mainland supplements the small populations, introduces new genetic material, and allows recolonization should local extinction occur (Box 9.4).

Suggested Exercise 9.2 Learn how to construct both classic and mainland-island metapopulation models in this online exercise. **oup-arc.com/e/cardinale-ex9.2**

The mainland-island model of a metapopulation helped explain the recolonization process of bay checkerspot butterfly in the San Francisco Bay area (see Figure 9.19B). Many subpopulations of this endangered butterfly went extinct during a severe drought that gripped California between 1975 and 1977.[136] The only subpopulation that did not go extinct was the largest one, on Morgan Hill. The extinct subpopulations remained at zero density until 1986, when nine habitat patches were recolonized by immigrants from the Morgan Hill source population. Patches closest to Morgan Hill were most likely to be recolonized because adult butterflies do not fly very far.

Conservation of metapopulations

Metapopulation studies recognize that local populations are dynamic; that is, the locations of populations change over time, and individuals can move between populations and colonize new sites. And sites within the range of the species may be occupied only because they are repeatedly colonized after local extinction occurs. Conservation of species that live as metapopulations must take into account the infrequent colonization events, in which individuals immigrate to unoccupied sites, as well as dispersal of individuals between existing populations that can "rescue" small populations that would otherwise be headed toward extinction. Such rescue effects determine not only how many patches in a landscape are colonized, but also the long-term persistence of a metapopulation in a landscape.

BOX 9.4 Conservation in Practice

The Mainland-Island Metapopulation Model

The classic metapopulation model assumes that each subpopulation in a landscape has the same probability of extinction and that each patch has the same probability of being colonized. While these assumptions are realistic for some metapopulations, they don't apply to all situations.

Many modifications have been made to the classic metapopulation model to extend its usefulness for conservation and management of species that exhibit different spatial configurations in a landscape. One modification is called the *mainland-island model* of population dynamics. The mainland-island model, sometimes referred to as the *propagule rain model*, describes a scenario where a large mainland population (such as in a conservation reserve) provides a source of emigrants that disperse to nearby small populations. The mainland population has a low probability of extinction, whereas the small populations become extinct relatively frequently. Emigration from the mainland supplements the small populations, introduces new genetic material, and allows recolonization should local extinction occur (see Figure).

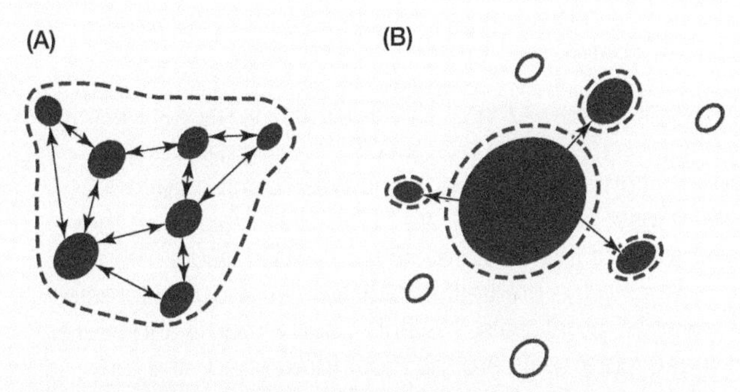

Stylistic representation of two different kinds of metapopulations. Ovals represent habitat patches: filled ovals represent occupied patches; open ovals represent vacant patches. Arrows indicate migration, and dashed lines indicate the boundaries of local populations. (A) Metapopulation model. (B) Mainland-island (or source-sink) metapopulation. (After S. Harrison. 1991. *Biol J Linn Soc* 42: 73–88; S. A. Boorman and P. R. Levitt. 1973. *Theor Pop Biol* 4: 85–128.)

The equation used to describe a mainland-island metapopulation is a simple modification of the classic model:

Classic model

$$\frac{df}{dt} = p_i f(1-f) - p_e f$$

Mainland-island model

$$\frac{df}{dt} = p_i(1-f) - p_e f$$

Where f is the fraction of patches occupied by a species in a landscape (ranges from 0 to 1), p_i is the probability that an unoccupied patch in the landscape will be colonized by immigration, and p_e is the probability that a species will go extinct in a patch that is presently occupied.

Note that for the classic model, the probability of immigration, p_i, is proportional to the number of patches that are occupied, f. In other words, when more patches are occupied, then there are more colonists and a higher probability that an unoccupied patch will be colonized. In contrast, for the mainland-island model, p_i is not proportional to the number of patches occupied, because there is a constant rate of colonization from the mainland. This constant rate of colonists is called the *propagule rain*.

The propagule rain makes the probability of immigration independent of the number of patches that are presently occupied in the landscape.

Although the difference between the classic and mainland-island models is minimal, the change has a big impact on the fraction of patches that are occupied in a landscape at equilibrium:

Classic model equilibrium

$$\bar{f} = 1 - \frac{p_e}{p_i}$$

Mainland-island model equilibrium

$$\bar{f} = \frac{p_i}{p_i + p_e}$$

It just so happens that, for any given probability of immigration and extinction, the equilibrium value for the mainland-island model is greater than the equilibrium value for the classic model (prove this to yourself by solving the equations for any values of p_i and p_e). This means that spatially subdivided populations that have a propagule rain of immigrations from a large, permanent source of colonists will achieve larger population sizes and, in turn, have a higher probability of persistence. That's an important rule of thumb that we'll revisit in Chapter 14 when we discuss the design of nature reserves and the conservation of species populations.

Summary

1. The habitats that organisms need to live, feed, and reproduce are increasingly being lost, fragmented, and degraded by human activities. Figuring out how to save biodiversity in these types of landscapes is the single greatest challenge in conservation biology.

2. The single greatest threat to biodiversity is habitat loss, which is largely driven by agricultural expansion and urbanization. Extractive industries also contribute to habitat loss. Their effects are on smaller scales, but they have some of the most severe localized impacts. Biomes experiencing the greatest habitat loss are tropical forests, grasslands, coral reefs, wetlands, and streams and rivers. Of these, coral reefs are the most threatened, and they may be mostly lost in the coming decades.

3. Habitat fragmentation is the process by which a larger expanse of habitat is subdivided into smaller patches in a landscape. Fragmentation contributes to habitat loss and changes the number, size, isolation, and amount of edge habitat of remaining patches. Changes in patch configuration can produce complex landscape patterns—some of which can benefit certain species, and some of which are detrimental to others. Fragmentation can have positive or negative effects on biodiversity, depending on the landscape pattern and the types of species in the landscape.

4. Habitat degradation refers to a suite of human activities that make the remaining habitat patches in a landscape less conducive to life. These activities include many forms of pollution (pesticides, oil spills, toxic metals, eutrophication, acid rain, pharmaceuticals and personal care products, and plastics), as well as activities that lead to desertification or to erosion and sedimentation.

5. Habitat loss, fragmentation, and degradation are creating landscapes that are patchy, spatial mosaics where some habitats have been lost and converted to human uses, others have been heavily degraded, and those that remain are small, fragmented, and disconnected from one another. The small, isolated populations living in degraded fragments are highly vulnerable to local extinction; however, movement of individuals among patches can enhance the likelihood of survival in the landscape as a whole.

6. A metapopulation is the collection of local populations that are spatially separated from each other in individual patches of a landscape but connected to each other by the dispersal and exchange of individuals. Even as local populations go extinct in individual patches, the metapopulation as a whole can persist through time so long as colonization of unoccupied habitat patches exceeds the rate of extinction in occupied patches. Patches with good habitat and positive population growth rates can serve as sources of immigrants to patches of poor habitat with negative population growth rates, essentially rescuing local populations from extinction.

For Discussion

1. Take a look at the wide variety of products that contain palm oil (oup-arc.com/e/cardinale-w9.1). Could you realistically eliminate these products from your daily life in order to reduce the rate of tropical deforestation? Which products have easy alternatives, and which do not? Consider also soybean products, coffee, and orange juice, which are similarly responsible for much habitat loss.

2. Examine a map of a park, green space, or nature reserve that is close to you (with an internet connection you can use the free program Google Earth [earth.google.com] to find the location). Has the area been fragmented by roads, power lines, or other human constructs? Has fragmentation affected the size of habitats, isolation among them, or amount of edge habitat? If so, how do you think these changes have affected the species in the landscape?

3. Do you tend to agree more with the land-sparing or the land-sharing approach to conservation? Do you see them as mutually exclusive or complementary to each other? If complementary, how would the two approaches be used simultaneously in any given landscape to conserve biodiversity?

4. Do you think we will ever be able to design agricultural or urban habitats that do not cause biodiversity loss? If not, what do you think are the key limitations? If so, what will it take for these human-dominated habitats to be more biodiversity "friendly"?

Suggested Readings

Balmford, A. et al. 2012. What conservationists need to know about farming. *Proceedings of the Royal Society B: Biological Sciences* 279: 2714–2724. Lays out the big topics conservation biologists will need to resolve in order to halt biodiversity loss caused by expansion of agriculture.

Fahrig, L. 2003. Effects of habitat fragmentation on biodiversity. *Annual Review of Ecology, Evolution, and Systematics* 34: 487–515. This classic paper is still an excellent summary of fragmentation.

Haddad, N. M. et al. 2015. Habitat fragmentation and its lasting impact on Earth's ecosystems. *Science Advances* 1: e1500052. Presents results from a global analysis on the state of fragmentation in forests and summarizes the biological impacts of fragmentation from experiments.

Hughes, T. P. et al. 2017. Coral reefs in the Anthropocene. *Nature*, 546: 82–90. Presents a new vision for conservation of coral reefs in the coming decades, which requires radical changes in science, management, and governance.

Laurance, W. F. et al. 2011. The fate of Amazonian forest fragments: A 32-year investigation. *Biological Conservation* 144: 56–67. Results of a large-scale experiment investigating the impacts of tropical forest fragmentation on biodiversity and ecosystem functioning.

McPhearson, T. et al. 2016. Advancing urban ecology toward a science of cities. *BioScience* 66: 198–212. Poses some key questions, conceptual models, and approaches for understanding ecology in urban habitats.

Oakleaf, J. R. et al. 2015. A world at risk: Aggregating development trends to forecast global habitat conversion. *PLOS ONE* 10: e0138334. A comprehensive summary of habitat loss to date, with forecasts for future development that can help prioritize biomes and geographic regions at greatest risk.

Schultz, C. B. and Crone, E. E. 2005. Patch size and connectivity thresholds for butterfly habitat restoration. *Conservation Biology* 19: 887–896. Excellent example of the use of a metapopulation model to guide conservation of a threatened species.

Schuyler, Q. et al. 2014. Global analysis of anthropogenic debris ingestion by sea turtles. *Conservation Biology* 28: 129–139. Worldwide action is needed to reduce the marine debris that is harming sea turtle populations.

Vijay, V. et al. 2016. The impacts of oil palm on recent deforestation and biodiversity loss. *PLOS ONE* 11: e0159668. Enormous areas of tropical forest are being converted to oil palm plantations. This paper reviews the consequences for biodiversity.

Visit the
Conservation Biology
Companion Website
oup.com/he/cardinale1e
for videos, exercises, links, and other study resources.

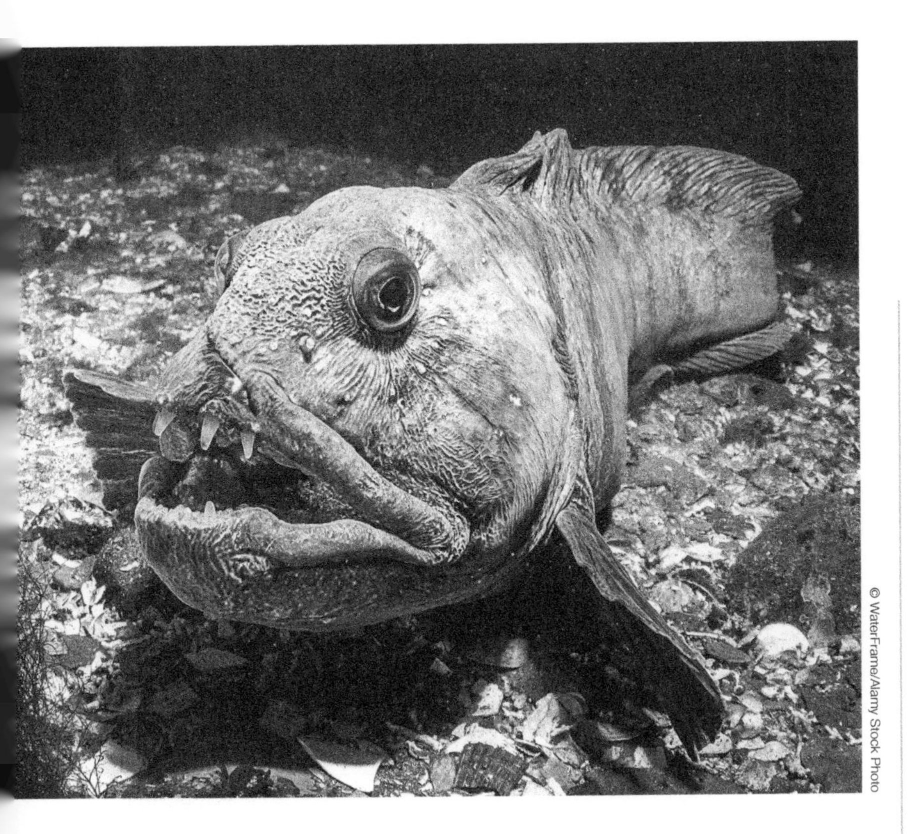
© WaterFrame/Alamy Stock Photo

10 Overexploitation

Anarhichas is a genus of wolffishes native to the northern Atlantic and Pacific oceans. Populations of several species of wolffish have decreased drastically due to overfishing and bycatch.

Overexploitation, also called **overharvesting**, refers to the harvesting of a renewable resource faster than it can be produced. The term applies to resources like populations of wild plants and animals (medicinal plants, game animals, fish), whole ecosystems (e.g., grazing pastures), and resources that are required to sustain life (e.g., water aquifers). While short-term overexploitation often can be corrected, persistent overexploitation can lead to the destruction of the resource.

In a world that seems intent on liquidating natural resources for short-term gain, overexploitation has become the second greatest threat to the survival of the world's flora and fauna. Consider the summary by Maxwell and colleagues,[1] who in 2016 analyzed records of 8688 species that are listed as Threatened or Near-Threatened on the IUCN Red List for groups in which all species have been assessed (**Figure 10.1**). First on the list were activities leading to habitat loss, fragmentation, and degradation (i.e., agricultural activity, urban development, pollution, system modification; see Chapter 9). Overexploitation was the second greatest threat to species extinction. Of the species currently listed as Threatened or Near Threatened, 72% (6241) are overexploited for commerce, recreation, or subsistence.

We begin this chapter with a brief overview of the history of overexploitation by humans, after which we review the

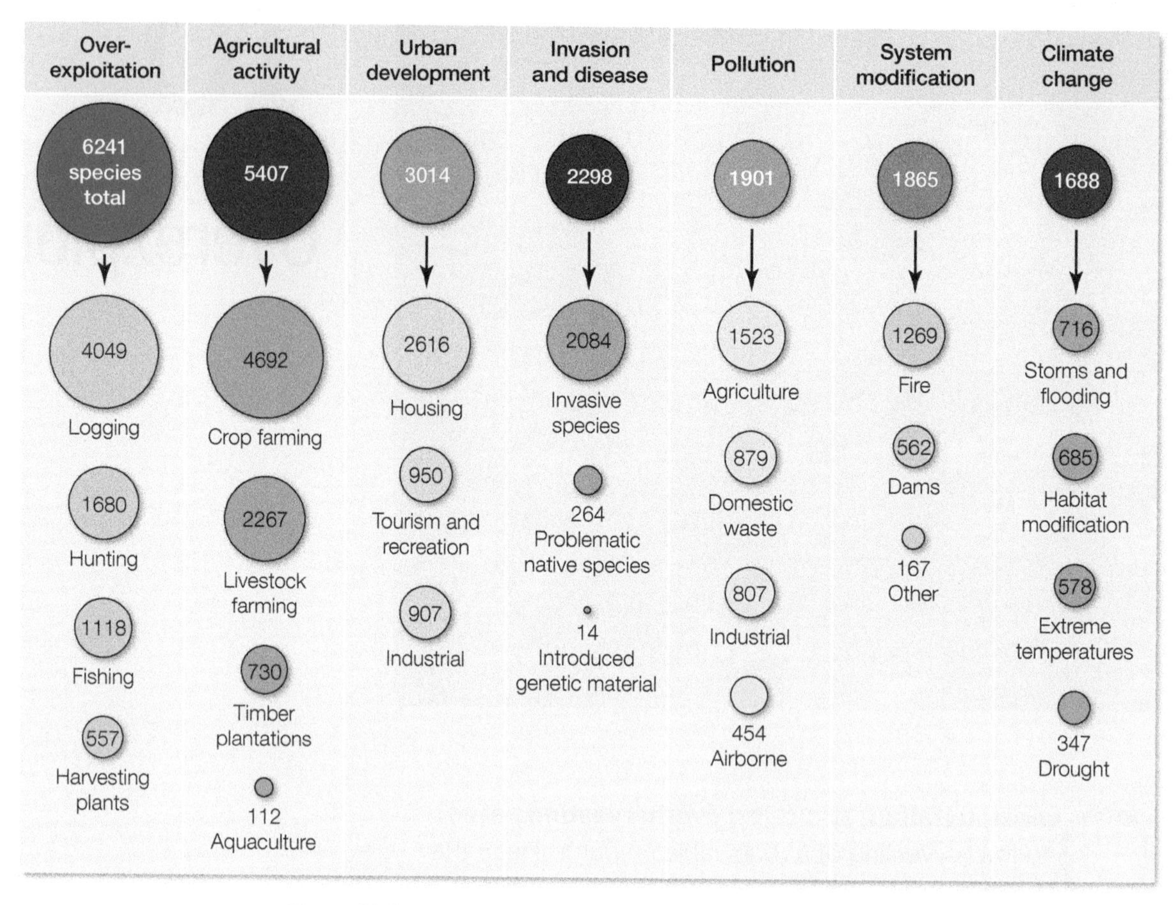

Figure 10.1 Threats of extinction. Maxwell and colleagues analyzed records for 8688 Threatened or Near-Threatened species on the IUCN Red List. After activities related to habitat loss (agricultural activity, urban development, pollution, and system modification), overexploitation was the second most frequent threat of extinction. Circled numbers are the number of species affected, with the top circle giving the total and all others being sized proportional to the subsequent number of species. Note that a given species may be counted more than once; more than 80% of the species analyzed were affected by more than one type of threat. (After S. Maxwell et al. 2016. *Nature* 536: 143–145. Data from IUCN Red List of Threatened Species, Version 2016-1.)

human psychology of overexploitation, pointing out the wide variety of reasons people use wild populations of plants and animals to excess. This is followed by a summary of the major types of overexploitation on both target and nontarget species in a variety of habitats. We end by discussing theories and models for the sustainable harvest of natural resources. By the end of this chapter, students should be well versed in the major causes and forms of overexploitation, and understand the primary tools resource managers use to prevent overexploitation from causing species extinctions.

History of Overexploitation

Overexploitation is not a new phenomenon. Ever since *Homo sapiens* first evolved as a species, humans have hunted and harvested food and other resources needed to survive. Fossil evidence suggests that overexploitation by early hunter-gatherer societies caused the extinction of mammals and birds thousands of years ago (**Figure 10.2**). For example, after the glaciers of the last ice age receded, the mammal

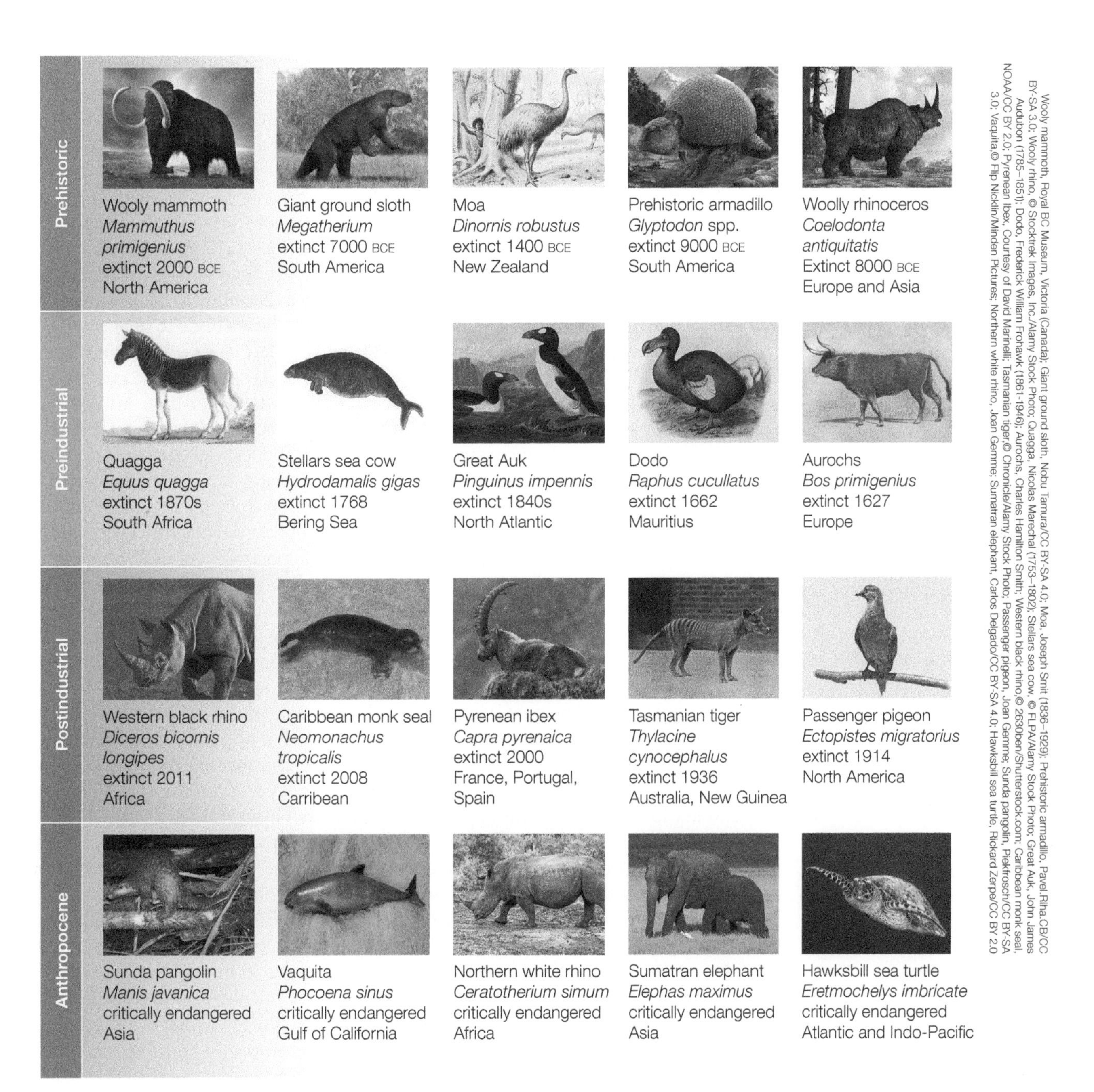

Figure 10.2 A few of the species that have been driven globally extinct by overexploitation, and that are currently threatened by extinction due to overexploitation. While human harvest has driven species extinction since prehistoric times, rates of extinction caused by overharvesting have greatly accelerated in the modern era due to the rapidly growing human population, increased demand, and advanced technologies for killing and harvesting.

fauna of North America diversified into a rich array of antelopes, horses, cheetahs, ground sloths, mammoths, and mastodons that rivaled the large mammal fauna of modern Africa. But more than half of the large mammal biota of the Americas disappeared in a wave of extinction that occurred around 11,000 years ago, near the end of the Pleistocene. Over a period of roughly 1000 years, 34 genera of large mammals became extinct in North America, while 40 more became extinct in South America.[2,3] These included woolly mammoths, Colombian mammoths,

American mastodons, three types of ground sloths, glyptodonts, giant armadillos, several species of horses, four species of pronghorn antelope, three species of camels, giant deer, several species of oxen, and giant bison. This massive die-off of megafauna occurred almost immediately after the arrival of early humans to the Americas,[3,4] which are known to have hunted these species for food.Similar mass extinction events of large-bodied vertebrates following the arrival of humans are documented in the fossil record for Europe, parts of Asia, and Madagascar.[2,5]

The extinction of flightless birds on islands provides a concrete example that early humans harvested species to extinction. On remote islands where birds evolved in absence of predators, they sometimes lost the ability to fly. When early humans arrived on these islands, they found the flightless birds to be easy prey. For example, when the Polynesians (now known as Maoris) arrived in New Zealand around 1200 CE, the islands were populated with 11 species of moas, a group of flightless birds that ranged in size from a turkey to larger than an ostrich. Archaeologists know the Maoris ate moas and their eggs, as numerous archeological sites have been uncovered with piles of bird bones and eggshells around campsites and cooking fires. Within a few hundred years, humans had driven all of the moa species extinct.[6,7] This pattern of extinction was repeated on small islands throughout the Pacific, where scores of bird species went extinct soon after the arrival of humans. On the Hawaiian Islands, for example, 44 of 82 land birds went extinct between the arrival of Polynesians and later arrival of Europeans.[8]

It is sometimes tempting to think that human-caused overexploitation has occurred only with large, somewhat rare mammals like mastodons, or with flightless birds that were isolated on islands and easy to capture. But history has shown that humans have overexploited even the most abundant species (see Figure 10.2). The passenger pigeon (*Ecopistes migratorius*) was once one of the most numerous bird species in the world, with a population of 1 to 5 billion individuals in the early to mid 1800s.[9] But the commercialization of pigeon meat as cheap food led Europeans to hunt the birds on a massive scale for several decades, leading to rapid population declines between 1870 and 1890. The last passenger pigeon died on September 1, 1914 in the Cincinnati Zoo.

The story of the passenger pigeon is not unique. Steller's sea cow (*Hydrodamalis gigas*), the dodo (*Raphus cucullatus*), the great auk (*Pinguinus impennis*), and the Caribbean monk seal (*Neomonachus tropicalis*) are just a few examples of species that once had healthy, vibrant populations, with most spanning large geographic ranges. Yet all have recently been hunted to the point of global extinction by humans. Many species that are now Critically Endangered are on the cusp of extinction because of human overexploitation, which is a more significant threat to biodiversity today than ever before.

Suggested Exercise 10.1 Use Wikipedia's Timeline of Extinctions in the Holocene to determine how many extinctions during the period of modern humans have been caused by overexploitation. **oup-arc.com/e/cardinale-ex10.1**

While people have always hunted and harvested the food and other resources they need to survive, two things are different in the Anthropocene. First, human populations are larger. For most of history, overexploitation was driven by small groups of nomadic people. Today, more than 1 billion people worldwide rely on wild plants for rural subsistence, and populations in some 62 developing countries rely on wildlife for at least one-fifth of their animal protein, including insects, amphibians, reptiles, birds, fish, rodents, and larger mammals.[10] Fish alone provides more than 3 billion people with almost 20% of their average intake of animal protein.[11] The modern scale of exploitation of biological resources is unprecedented in human history. Second, our technologies for hunting, fishing, killing, and harvesting have become far more efficient. Primitive people hunted with blowpipes, spears, and arrows. Today we use high-powered rifles, motorized vehicles and fishing boats, enormous factories, and global distribution networks. Our current ability to harvest nature greatly exceeds nature's ability to supply.

Psychology of Overexploitation

The root causes of overexploitation are diverse. Perhaps the most widely cited causes of overexploitation fall into a category of scenarios called social dilemmas. **Social dilemmas** are situations in which individuals make decisions in their own self-interest that have inferior outcomes compared to what they would have achieved by cooperating with others.[12] Social dilemmas encompass several well-known scenarios that help explain why perfectly rational individuals often make irrational decisions that cause harm to themselves, such as when individuals overexploit a vital biological resource even to their own detriment.

One such scenario is exemplified by the **prisoner's dilemma**, which is a common example from the field of **game theory** that illustrates why two rational individuals might choose not to cooperate with each other even when it is in their best interest to do so. The prisoner's dilemma goes like this: Two members of a criminal gang are arrested and placed in solitary confinement with no means of communicating with each other. Prosecutors lack sufficient evidence to convict the pair on the principal charge, but they have enough evidence to convict each gang member on a lesser charge. Therefore, the prosecutors offer each prisoner a bargain. If each prisoner betrays the other, s/he will only serve two years in prison. If prisoner A betrays B but B remains silent, A will be set free and B will serve three years in prison (and vice versa). If A and B both remain silent, then both of them will only serve one year in prison (on the lesser charge).

Figure 10.3A shows the payoff matrix for the four possible decisions that can be made by the two prisoners. Because betraying a partner offers a greater reward than cooperating with them (i.e., going free versus one year in prison), most

(A) Prisoner's dilemma

Prisoner A payout / Prisoner B payout	Prisoner A stays silent	Prisoner A betrays B
Prisoner B stays silent	−1 −1	0 −3
Prisoner B betrays A	−3 0	−2 −2

(B) Assurance dilemma

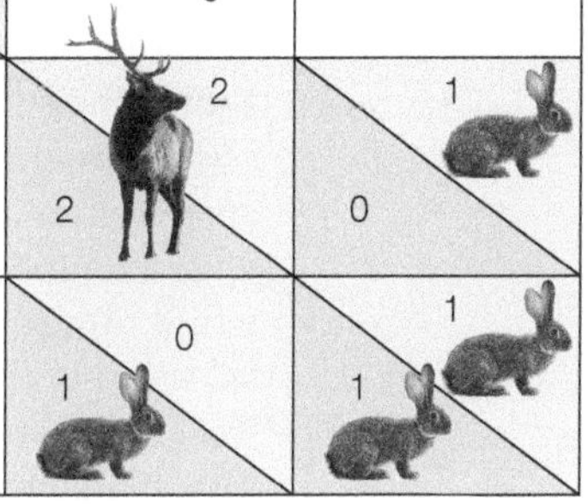

Hunter A payout / Hunter B payout	Hunter A cooperates and hunts a stag	Hunter A defects and hunts a hare
Hunter B cooperates and hunts a stag	2 2	1 0
Hunter B defects and hunts a hare	0 1	1 1

Figure 10.3 Social dilemmas are situations in which individuals make decisions in their own self-interest that have inferior outcomes compared to what they would have achieved by cooperating with others. Social dilemmas are a common explanation for why individuals overexploit natural resources to their own long-term detriment. This figure shows the "payoff matrix" for two common social dilemma scenarios. Numbers in the cells represent outcomes from −3 (worst) to 2 (best).

rational self-interested prisoners will betray the other. But when both prisoners betray each other, they each receive a longer sentence than they would have if they had remained silent and cooperated (two years versus one year in prison). The prisoner's dilemma has been applied to natural resource management to try and explain why rational people fail to cooperate in protecting a natural resource, which can occur when individuals have incomplete information and cannot predict the behavior of the other actor(s) in the interaction.

A variation of the prisoner's dilemma that more directly applies to irrational resource overexploitation is the **assurance dilemma**, a scenario often conveyed with an example called *the stag hunt*. The stag hunt envisions two people hunting for food. Each person has the choice to pursue either a rabbit or a stag. The stag, of course, provides much more food than the rabbit, but it can only be caught if the two hunters cooperate and pursue it together. In contrast, each hunter can capture a rabbit, regardless of what the other hunter does. The rabbit provides less food, but it is sufficient nourishment for one person. Before starting, the two hunters can discuss which strategy to employ, after which, they must decide whether to trust each other, or to defect and hunt alone. In this situation, how should the two hunters behave?

Figure 10.3B shows the payoff matrix for the stag hunt. Cooperation among hunters that leads to capture of a stag is worth more than the individual capture of a rabbit. But despite the obvious benefits of cooperation, rational individuals may still make inferior choices. Although capture of the single stag is the optimal interaction, it requires that both actors trust each other, which also involves taking a risk. If one hunter cheats, the other gets nothing. In contrast, while capture of a rabbit is a suboptimal choice, it is a guaranteed reward and doesn't require trust in another individual. The assurance dilemma has been used to explain why people don't cooperate in their harvest of natural resources, which can occur when the lack of trust generates unacceptable levels of risk. It includes not only mistrust at a personal level (two individuals), but also mistrust between individuals and their governments.

The prisoner's dilemma and assurance dilemma illustrate why rational people sometimes make seemingly irrational decisions because (1) they have incomplete information needed to properly evaluate the short- verses longer-term consequences of their actions; and (2) their perception of risk exceeds their trust in other individuals. Both of these factors can contribute to the **tragedy of the commons** (explained in Chapter 7), which refers to a particular kind of social dilemma where individuals who use a **common-pool resource** (a public good, or common property) act independently and according to their own self-interest; yet, the collective action of many individuals pursuing their own self-interest depletes, spoils, or even destroys the resource, to the detriment of everyone.

While social dilemmas are often the cause overexploitation, much research shows they can be avoided by management actions that promote communication, trust, and cooperation among groups and organizations.[13,14] However, cooperation can be unstable because any collectively produced benefit comes with a temptation to cheat or freeload—that is, to collect a share of the benefit while minimizing one's personal risk and/or contribution. Crucially, natural selection is known to favor freeloading even if it makes everyone, including the freeloader, worse off than they would have been if no one freeloaded. Therefore, management actions must not only encourage communication, trust, and cooperation; such actions must also include negative feedbacks or penalties that discourage freeloading and penalize cheaters.

In addition to social dilemmas, there are numerous other causes of overexploitation. Overexploitation can occur simply because of need, such as when people exhaust resources while trying to meet their subsistence needs for food, water, shelter, or safety. Overexploitation can occur because of ignorance, such as when people lack awareness of the finite limits of a resource or are overly optimistic about the potential for recovery.[15] Overexploitation can occur because of unfounded optimism, such as when individuals deplete a resource while believing it can be replaced by a technological solution or alternative resource. Overexploitation can occur because of a drive for thrill, or the need for conquest, dominance, and status, such as is common among trophy hunters, particularly male hunters.[16,17] And lastly, overexploitation can occur because of psychological spite ("if I can't have it, no one can"), a behavior believed to be uniquely human. Kim Sterelny[18] captured the idea of psychological spite when she reflected on the following passage from Jared Diamond's influential book *Collapse*:[19]

> I have often asked myself, 'What did the Easter Islander who cut down the last palm tree say while he was doing it?' Like modern loggers, did he shout 'Jobs, not trees'? Or 'Technology will solve our problems, never fear, we'll find a substitute for wood'? [19]

Sterelny's response is that it was "much more likely" that the Easter Islander shouted,

> '[A]t least those bastards in the XYZ clan won't get this one.'[18]

The key point is that there are many causes of overexploitation, some of which involve individual human behaviors and others that involve social conflicts of interest among two or more individuals who are engaged in conflicts for a common resource. Conservation biologists who strive to understand which of these human psychologies are driving overexploitation for any given species are in a much better position to formulate effective management strategies.

Types of Overexploitation

The term *overexploitation* is used very broadly to refer to a variety of activities that affect the long-term survivorship of species and natural habitats. Some of these activities are legal, others are not. And most have different socio-economic drivers. Therefore, it's useful to break-down the different forms of exploitation into a few distinct categories.

Commercial exploitation

Commercial overexploitation is the unsustainable harvesting of a natural resource that is driven by the existence of a legal or illegal market. Money is the driving force behind the majority of species extinctions that are caused by overexploitation. Large markets driven by consumer demand—on the scales of hundreds of billions of dollars—exist for certain types of extractive industries (e.g., commercial fishing and logging). Legal and illegal trafficking of wildlife are also multibillion dollar industries. With such large amounts of money to be made, overexploitation of commercially viable species is common, particularly in countries that lack strong regulations and enforcement, or have no incentives for species protection.

Two industries, commercial fishing and commercial logging, are considered to be the "poster children" for overexploitation. We discuss both of these below, focusing in particular on the impact these industries have on species population decay and extinction. We then discuss the legal and illegal wildlife trades that in recent years have become major contributors to the loss of wildlife.

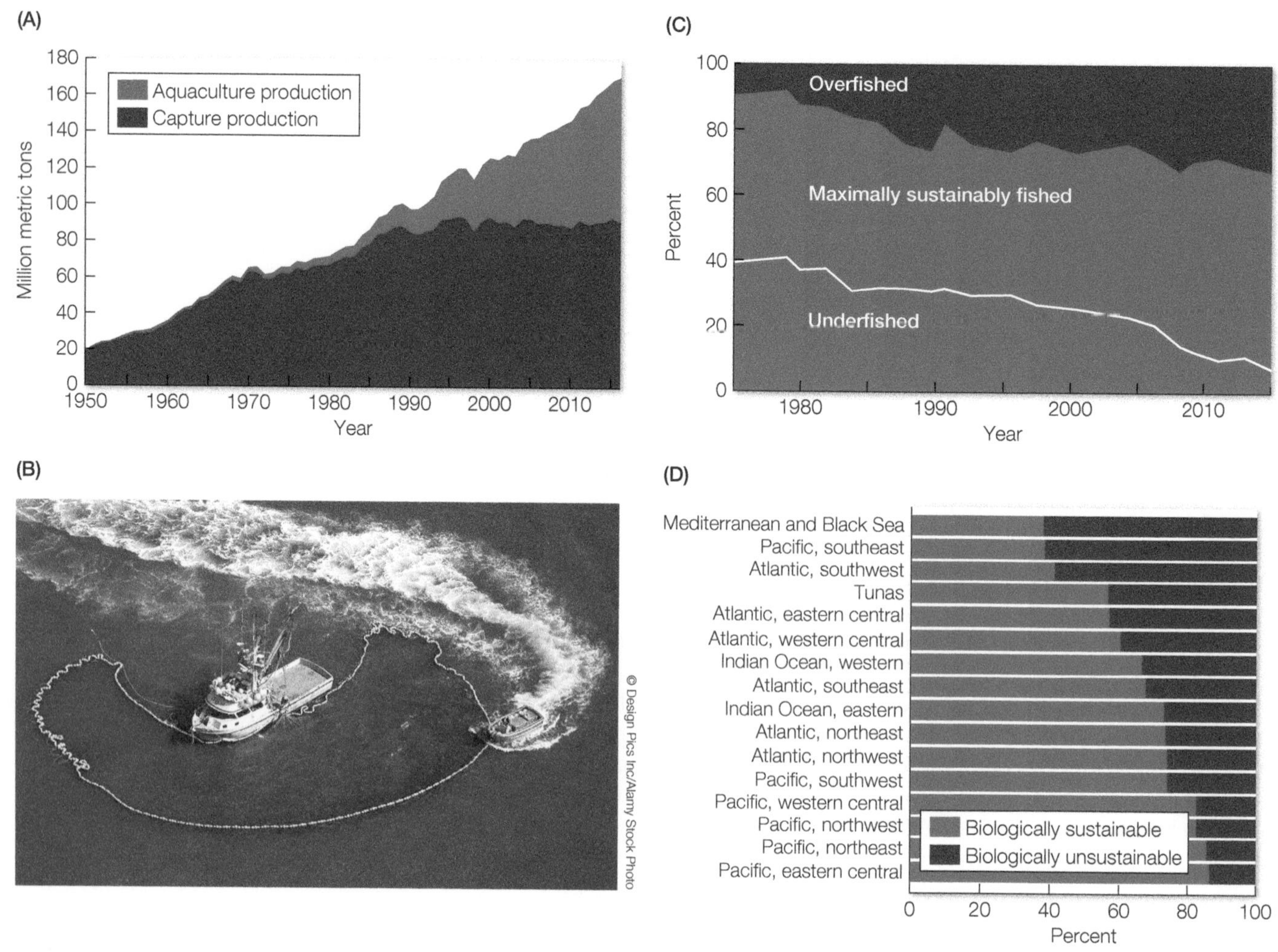

Figure 10.4 Statistics from the Food and Agriculture Organization (FAO) of the United Nations' 2018 State of the World's Fisheries report. (A) Yields for capture production (wild-caught fish) fisheries and for aquaculture. (B) A herring purse seiner with modern equipment is able to catch a massive number of fish in a short time. (C) The percentage of world fisheries the FAO reports as unsustainably overfished, maximally fished, and underfished. (D) The percentage of fisheries in different parts of the world that are sustainably versus unsustainably fished. (After FAO. 2018. *The State of the World's Fisheries and Aquaculture 2018*. Rome.)

FISHING Commercial fishing is a large, lucrative industry worth more than $360 billion worldwide.[20] Global fish production (fish, crustaceans, mollusks, and other aquatic animals) reached an all-time high of 171 million metric tons in 2016 (**Figure 10.4A**), of which 88% was for direct human consumption.[20] Wild-capture fisheries represented 53% of the total, with an estimated sale value of US$131 billion. The other 47%, worth an additional $232 billion, was from aquaculture.

Capture fishery production has remained relatively static since the late 1980s, even as aquaculture has increased to meet global demand. In 2015, fish provided about 3.2 billion people with almost 20% of their average per capita intake of animal protein, with people in developing countries having a higher share of fish protein in their diets than those in developed countries. To provide the human population with fish, the total number of fishing vessels in the world in 2016 was estimated at 4.6 million.[20] Asia's fishing fleet was by far the largest, consisting of 3.5 million vessels and accounting for 75% of the global fleet, followed by Africa (15%), Latin America and the Caribbean (6%), North America (2%), and Europe (2%).

There is abundant evidence that global consumer demand for fish has led to overharvesting and severe population crashes of commercially fished species. A few well-documented examples of fisheries collapse are given here:

- **Atlantic northwest cod (*Gadus morhua*) fishery.**[21] The Newfoundland Grand Banks were once renowned as the world's most productive fishing grounds. The first European explorers described the waters as being so full, one had to just lower a basket into the water and it would come up filled with cod. Throughout the 1950s and 1960s, the bountiful fishery attracted giant trawlers from distant countries. Working around the clock, trawlers could take an unprecedented 200 metric tons of fish *per hour* in their huge nets, and in 1968, the cod catch peaked at 800,000 metric tons. By 1975, the annual catch had fallen by more than 60%. As cod catches declined, factory trawlers used ever more powerful sonar and satellite navigation to target what fish were left. Members of the Canadian government, many of whom owned shares in industrial fishing companies, refused to listen to scientific warnings that the cod fishery was in crisis. But by 1992, when the cod catch reached the lowest point ever measured, the government was forced to close the fishery. In what may be the single biggest lay-off in history, the moratorium put 40,000 people out of work in five Canadian provinces, and required a several-billion-dollar relief package to be disbursed to coastal communities.
- **Peruvian anchoveta (*Engraulis ringens*) fishery.**[22,23] The west coast of South America off Peru and Chile, was once the world's largest fishery. Throughout the 1950s and 1960s, global demand for fish meal, coupled with decline of California's sardine fishery, led to exponential growth in the catch of anchoveta in the Peruvian current. In 1965, the Peruvian government performed its first stock assessment of the fishery, which showed the stock was fully exploited, or nearly so, at an annual catch of 8 million metric tons. But subsequent stock assessments based on more optimistic assumptions suggested that the annual catch could potentially be sustained at 10 million tons, and the quota was raised in 1968. By the early 1970s, the anchoveta fishery appeared to be a model of successful management. But in 1972, overfishing and a severe El Niño event severely depleted the adult stock of anchoveta, and recruitment for the year largely failed. Anchoveta populations plummeted, and fewer than 2 million tons were harvested. Loss of the anchoveta fishery hit Peru's economy hard, and it took the anchoveta population more than two decades to recover.
- **Atlantic herring (*Clupea harengus*) fishery.**[24] Herring are arguably one of the most important fish on the planet, being a staple food of societies dating back to 3000 BCE. In the 1950s, traditional herring fishing techniques were revolutionized by the advent of large boats fitted with sonar, power blocks (mechanized winch used to haul fishing nets), and large nets (Figure 10.4B). Technological advances led to a rapid rise in the herring catch that contributed to the collapse of major stocks throughout the world. One of the most dramatic declines was that of the Atlanto-Scandian stock between Iceland and Norway. By the 1960s, it ranked among the most productive fisheries in the world, producing nearly 2 million metric tons per year. The Icelandic economy was heavily dependent on the fishery, with salting and reduction of herring being the country's most important export industry and a major source of its wealth. But there was no fisheries management authority to limit either catch or effort, and by 1970 the fishery had collapsed from 2 million to

only a few thousand metric tons, and herring populations were on the border of extinction within just over a decade. The sudden closure of the fishery was a shock to many communities in Norway and Iceland.

▶ **Video 10.1** Watch this video about overfishing in some of the richest fishing grounds in the world.
oup-arc.com/e/cardinale-v10.1

These are just a few of the many examples that show how overexploitation can cause serious population decay of aquatic species. Even so, it is worth noting that estimates of modern fisheries overexploitation vary widely. Some of the most pessimistic claims argue that two-thirds of the world's seafood is overfished,[25] or that 80% of the world's fish stocks are either fully exploited, overexploited, or have already collapsed.[26] Some scientists have argued that overexploitation is so severe it is leading to the depletion of all large predatory fish from the oceans, causing the commercial fishing industry to increasingly exploit previously spurned small fish and invertebrates. This process has been referred to as "fishing down the food web"[27] (**Video 10.1**).

In contrast to the more dire predictions, the 2018 State of the World's Fisheries report by the United Nations Food and Agriculture Organization[20] suggests that approximately 33% of the world's fish stocks are currently being fished unsustainably, whereas 60% are being fished at their maximum sustainable rate, and 7% are underfished (**Figure 10.4C**). The proportion of fisheries being harvested sustainably versus unsustainably differs considerably between different parts of the world. In geographic areas where countries have poor regulation and enforcement, such as the Mediterranean and Black Seas, the southeastern Pacific, and the southwestern Atlantic, more than 60% of fisheries are being unsustainably harvested (**Figure 10.4D**). In areas with good regulation and enforcement, such as the northeastern and northwestern Pacific, most fisheries stocks are being harvested sustainably.

Given the widely varying estimates about the state of the world's fish stocks, it is sometimes difficult for someone who isn't a fisheries biologist to understand how widespread the problem is (**Box 10.1**). In addition, it is sometimes hard to understand the frequency with which overexploitation of fisheries leads to extinction.[28–31] While the direct link between overharvesting and extinction risk is well documented for terrestrial species,[32,33] several scientists have noted that humans have caused comparatively few verifiable complete extinctions in the oceans.[28,34] Overfishing has certainly driven numerous species to sufficiently low numbers that they have gone locally extinct, or can be considered functionally extinct in the ecosystem (see Table 1 in Jackson et al.[31] for examples). However, many marine fish species have life-history strategies that are characterized by extremely high fecundity with no parental care, which leads to high mortality in early life stages and survival-to-maturity rates as low as 1:100,000.[35] Such life-history strategies result in a lower probability of population extinction, as only a few female fish are needed to potentially generate millions of juvenile fish.[29]

Figure 10.5 During cod fishing, other species are caught in the nets as bycatch and then discarded into the ocean, usually dead or near-dead.

For many marine species, the greatest impacts are not direct harvest, but rather the indirect effects of commercial fishing through **bycatch**.[36] Many marine vertebrates and invertebrates are caught incidentally as bycatch during fishing operations and are subsequently injured or killed in the process (**Figure 10.5**). Between 25% and 75% of the harvest in fishing operations is dumped back into the sea, almost dead or soon to die. Population declines in skates, rays, and many types of seabirds have all

BOX 10.1 Challenges & Opportunities

No Fish Left, or Let Us Eat Fish?

One of the challenges to understanding the extent of fisheries overexploitation is the wide disparity of estimates that exist regarding the state of marine fisheries. Nowhere has this disparity been more evident than in a debate between marine conservation biologist Boris Worm and fisheries scientist Ray Hilborn.

In 2003, Worm assembled a group of colleagues at the National Center for Ecological Analysis and Synthesis in Santa Barbara, California to summarize how marine biodiversity influences the goods and services that oceans provide to society. One of the first set of analyses from their working group[25] was published in the prestigious journal *Science* and received a huge amount of attention. It made headlines on major news networks and was distributed in widely read newspapers like the *New York Times* and *Los Angeles Times*. The paper received so much publicity because of the authors' rather dire prediction that there would be a "global collapse of all taxa currently fished [in the oceans] by the mid-twenty-first century."[25] Worm and his colleagues had collated data on the proportion of collapsed fisheries in oceans around the world, plotted the proportion of collapsed fisheries that have been recorded through time, and then extrapolated the time-trends to the *x*-axis to predict the date when all fisheries in the ocean would be collapsed (Figure A). The analysis led the authors to an apocalyptic claim that all fisheries in the ocean will be gone by 2048. The claim was widely cited (>3,700 times by October 2018) and used to justify marine conservation initiatives, such as the implementation of marine reserves and protected areas.

Not surprisingly, the claim made by Worm and colleagues was controversial. Perhaps the biggest critic was fisheries biologist Ray Hilborn who, in media interviews, called the study "incredibly sloppy"[115] and "mind-boggling stupid."[116] Along with other fisheries biologists, Hilborn claimed the analyses cherry-picked data, focusing only on fisheries known to be doing poorly while ignoring fisheries that were doing well or recovering after decades of overfishing. In addition, Hilborn and his colleagues, who were trained in the use of quantitative models and techniques for determining the abundance of fish stocks, argued that the methods used by Worm and colleagues were flawed and produced misleading conclusions.

(Continued on next page)

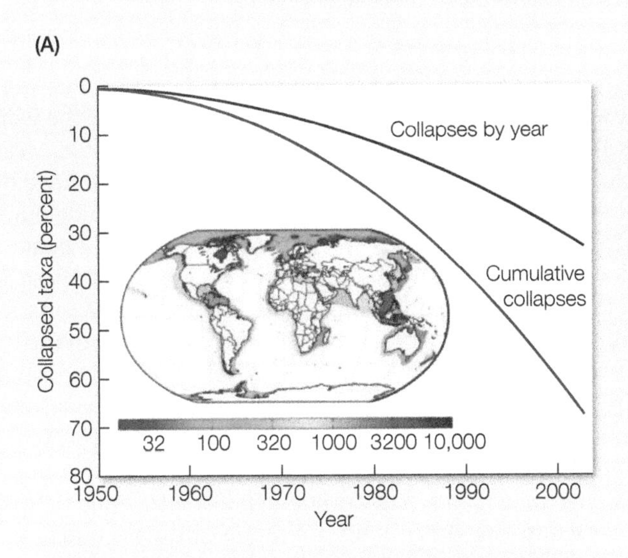

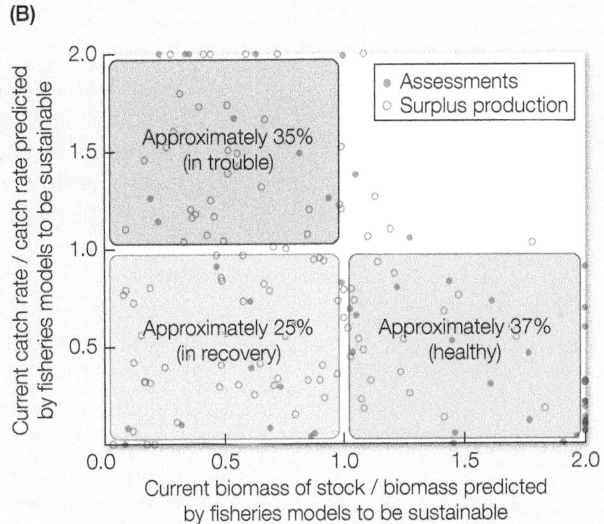

(A) Worm and colleagues[25] collated data from fisheries in 64 large marine ecosystems (LMEs) throughout the world that varied widely in species richness (color coded map, see inset). They then reconstructed trajectories of collapsed fish and invertebrate taxa over the past 50 years by each year (upper line) as well as cumulatively (lower line), and used these historical trends to extrapolate the year at which lines would reach the *x*-axis. The most dire of their forecast (lower line) predicted there would be no fish in the oceans by the year 2048. (B) After their work was criticized, the authors published a new analysis in collaboration with fisheries biologists using stock assessments to characterize the health of populations.[118] That new analysis suggested that more than one-third of fish stocks are healthy (lower right), one-fourth are in recovery from historic overfishing (lower left), and one-third are being managed unsustainably (upper left). (A, after B. Worm et al. 2006. *Science* 314: 787–790; B, based on B. Worm et al. 2009. *Science* 325: 578–585.)

BOX 10.1 Challenges & Opportunities *(continued)*

To their credit, several authors of the original Worm and colleagues paper teamed up with Hilborn and other critics to run a second working group at NCEAS to try and find common ground in the state of marine fisheries. Worm, Hilborn, and colleagues re-analyzed scientific estimates of biomass for 166 fisheries stocks around the world, comparing the current biomass to levels predicted to be sustainable from fisheries management models. In addition, the authors compared current rates of exploitation (fish catch) to rates of growth and recovery of the stocks.

Their re-analysis, which was also published in *Science*[117] but received far less attention, came to very different conclusions about the state of marine fisheries (Figure B). The new analysis showed that 35% of the world's fish stocks are overexploited (current biomass less than that predicted to be sustainable) and in decline, as rates of harvest continue to exceed rates of recovery. These stocks are truly in trouble. In contrast, 25% of the stocks that have been historically overexploited are now in recovery, as management actions have reduced catch rates low enough to allow regeneration. In addition, 37% of the assessed fisheries stocks were considered healthy, with levels of biomass higher than, and catch rates lower than, what is predicted to be sustainable by fisheries management models.

These new analyses led to conclusions that were substantially different than those of the original paper. In fact, the conclusions were sufficiently optimistic that Hilborn went on to write an OpEd piece in the *New York Times* titled "Let Us Eat Fish"[118] (oup-arc.com/e/cardinale-w10.1) In that piece, Hilborn argued that many doomsday predictions about the state of the world's fisheries are inaccurate. He further argued that many underfished stocks could potentially be used as a source of protein to feed the world's growing population, and that harvesting marine fish would be more sustainable, and better for the environment, than expansion of terrestrial agriculture.

Interestingly, even after publishing their results together, Worm and Hilborn continued to disagree about the interpretation of data. Their glass-half-full versus glass-half-empty viewpoints were captured in a widely read blog called the SeaMonster, which also received input from a wide variety of marine biologists and fisheries scientists[119] (oup-arc.com/e/cardinale-w10.2).

While this type of debate has made it challenging for anyone who is not specialized in marine fisheries to understand the state of overexploitation, it has also made clear that some portion of the world's fisheries are in a good shape, and may—with proper management—represent an opportunity for food production in the future.

---

been linked to their wholesale death as bycatch.[37] For example, all of the world's 22 albatross species are threatened with extinction, mostly as a result of bycatch.

In recent decades, however, the commercial fishing industry has made huge strides in reducing bycatch. After a massive public outcry over the huge number of sea turtles and dolphins killed as bycatch by commercial fishing boats, the industry responded with the development of improved nets to reduce these accidental catches. People can now look for "dolphin-safe" labels on tuna cans, indicating that no dolphins were injured or killed during the harvesting of the tuna. In addition, public awareness campaigns are increasingly allowing consumers to choose their sources of seafood more wisely, using their money to reward more environmentally friendly practices. For example, seafood guides like those published (or developed into smartphone apps) by the Monterey Bay Aquarium (oup-arc.com/e/cardinale-w10.3), the Environmental Defense Fund (oup-arc.com/e/cardinale-w10.4), and the WWF (oup-arc.com/e/cardinale-w10.5) allow people to easily choose fish species that are harvested sustainably when shopping or dining at a restaurant.

LOGGING The FAO's 2016 State of World's Forests report[38] showed that, although temperate forests in many developed countries are recovering and accruing after decades of historic logging, a net forest loss of 7 million hectares per year continues to take place in tropical countries, particularly in low-income countries where rural populations are increasing. In Chapter 9 we described how habitat loss and the fragmentation of forests, caused by indiscriminate logging for conversion to agriculture, has impacted biodiversity. But to what extent has overexploitation of species from intact forests contributed to biodiversity loss?

As is true for the commercial fishing industry, there are numerous examples where consumer demand has caused overexploitation, severe population decay, and local extinction of forest species, particularly tree species that are harvested for lumber. Some examples include:

- **Broad leaf mahogany (*Swietenia macrophylla*),**[39,40] also called *caoba* throughout much of Latin America, *mogno* in Brazil, *mara* in Bolivia, and *ahuano* in Ecuador, is a species of tree in the Meliaceae family that is native to Central and South America. It it is one of three species that yields genuine mahogany timber, which is prized for its rich reddish color and technical characteristics that make it sought after for the manufacture of furniture, musical instruments, and other wood products of high quality, beauty, and durability (Figure 10.6A). Mahogany fetches an extraordinarily high price on international markets, which makes it lucrative for loggers to seek out and selectively harvest the species, even in remote wilderness areas. At the height of mahogany extraction in the early 1990s, one cubic meter of export-quality sawn wood was worth about US$700. A typical mill with one band saw was producing an average 4500 m^3 of sawn wood per year, which brought approximately US$3,000,000 on the international market and provided US$800,000 in annual profit.[41] The extraordinarily high value of the wood led to population reductions in Central America that are estimated at greater than 70% since 1950. The species was harvested to such low densities in El Salvador, Costa Rica, and many parts of South America that it is now considered commercially extinct (depleted to such low numbers that harvesting is not profitable).[40] The species survives in cultivated in plantations, including some outside its native range, and in heavily managed forest reserves. Its distribution and sale are regulated by the Convention on International Trade in Endangered Species of Wild Fauna and Flora (CITES; Box 10.2).
- **African cherry (*Prunus africana*).**[42,43] Extracts from the bark of mature African cherry trees of central Africa (also called "red stinkwood" and "wotanto") have long been used by locals in traditional medicine, including treatments for malaria and other fevers (Figure 10.6B). Today there is a high demand for the bark among European pharmaceutical companies, where the extract (whose complex chemical compounds have yet to be synthesized in the laboratory) is used in medications that treat enlarged

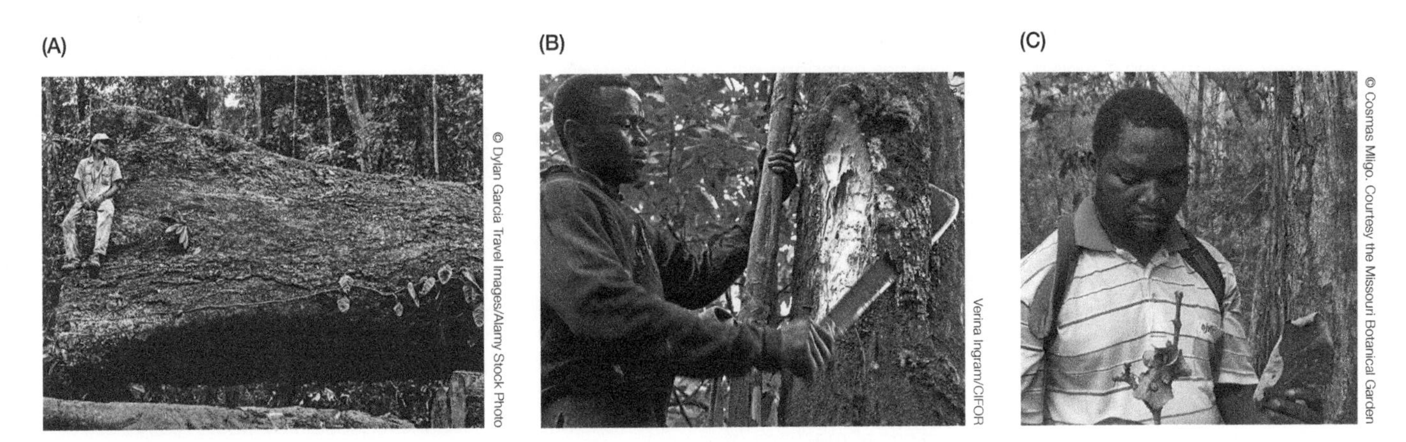

Figure 10.6 Three of the world's most overexploited trees. (A) Broad leaf mahogany (*Swietenia macrophylla*) is highly prized for its timber. (B) The bark of African cherry (*Prunus africana*) is harvested for a medicinal extract. (C) *Karomia gigas* is one of the rarest tree species in the world, in part, due to harvesting by botanists for collection of specimens.

prostates. In the West African country of Gabon, it has been harvested beyond sustainable levels. Before 1985, there was no threat of overexploitation because a French pharmaceutical company had a monopoly on the extract and harvested the tree in a sustainable manner. However, in 1985 the Gabon government ended the monopoly and licensed local businesses to harvest the bark, inciting competition and unsustainable levels of collecting as local villagers vied for the much-needed source of income. Today the company that processes the bark foresees the extinction of natural stocks and has begun growing seedlings, which it distributes to locals interested in starting their own plantations. Such small-scale agroforestry may be the only hope for survival of this tree.

- ***Karomia gigas.***[44] This species in the Lamiaceae family is one of the world's rarest trees. Historically found in Kenya and Tanzania, the species is believed to have become extinct in Kenya after the last known specimen was cut down in the 1970s. The species was also thought to be extinct in Tanzania until, in 2016, botanists from Botanic Gardens Conservation International found six trees growing in a single location in Tanzania (**Figure 10.6C**). They employed local people to guard the trees and report if there

BOX 10.2 Conservation in Practice

CITES: The Convention on International Trade in Endangered Species of Wild Fauna and Flora

The Convention on International Trade in Endangered Species of Wild Fauna and Flora (CITES) is a multilateral treaty to protect endangered plants and animals by regulating unsustainable wildlife trade. It was drafted as a result of a resolution adopted in 1963 at a meeting of members of the International Union for Conservation of Nature (IUCN). The convention was opened for signature in 1973 and CITES entered into force on July 1, 1975. As of 2019 CITES has 183 parties (member countries) that have agreed to regulate the international trade in specimens of wild animals and plants so that it does not threaten the survival of species in the wild. Parties voluntarily adopt their own domestic legislation to implement the CITES treaty, with that legislation requiring four key components: designation of Management and Scientific Authorities; laws prohibiting the trade in violation of CITES; penalties for such trade; laws providing for the confiscation of specimens.

Today, CITES accords varying degrees of protection to over 36,000 species of animals and plants, whether they are traded as live or dead specimens, body parts (such as ivory or leather) or derivatives (such as medicines). Each protected species or population is included in one of three lists, called Appendices, which delineate the extent of risk faced by species in the wild, and thus, the degree of regulation required for trade. Species are proposed for inclusion in or deletion from the Appendices at meetings of the Conference of the Parties (CoP), which are held once every three years.

- Appendix I lists species that are the most endangered among CITES-listed animals and plants. They are presently threatened with extinction, and CITES prohibits international trade in specimens of these species except when the purpose of the import is not commercial, for instance for scientific research. Examples of species listed in Appendix I include the western gorilla (*Gorilla gorilla*), tigers (*Panthera tigris*), manatees (*Sirenia*), and Brazilian rosewood (*Dalbergia nigra*).
- Appendix II lists species that are not necessarily threatened with extinction at present, but may become so unless trade is controlled. A permit to trade these species may be granted if authorities are satisfied that trade will not be detrimental to the species' survival in the wild. Examples of species listed on Appendix II are the great white shark (*Carcharodon carcharias*), the green iguana (*Iguana iguana*), and broad leaf mahogany (*Swietenia macrophylla*).
- Appendix III lists species that are included at the request of a country that already regulates trade in the species but needs cooperation from other countries to prevent unsustainable or illegal exploitation. These species can be traded only with appropriate documentation from the listing country. Examples include the two-toed sloth (*Choloepus hoffmanni*, requested by Costa Rica), the alligator snapping turtle (*Macrochelys temminckii*, requested by the U.S.), and the Manchurian ash (*Fraxinus mandshurica*, requested by the Russian Federation).

were any seeds, which could then be cultivated in a Tanzanian botanical garden. A second site of the species was found in 2017. Today, fewer than 20 individuals of *Karomia gigas* exist across both locations, and the species is assessed as Critically Endangered on the IUCN Red List.

It is hard to know what fraction of the world's forest species have been subjected to increased extinction risk due to overexploitation. In part, this is because broader summaries and comparisons of tree species exploitation often fail to use comparable methods that are reported adequately. Currently, conservation status has only been determined for some 20,000 tree species, which represent perhaps 30% of the world's trees.[44] Of those 20,000, almost half—9600 tree species—may be threatened with extinction. The UK-based Botanic Gardens Conservation International recently joined forces with the IUCN to compile data from more than 500 published sources and create the online database GlobalTreeSearch, which is the world's first authoritative list of global tree species.[44] This effort has summarized data for 60,065 currently living tree species. Of that number, more than half were found to have a highly restricted geographic ranges, often occurring in only a single country.

36,000 species are protected under CITES

30,000 plant species

5,800 animal species

62 proposals under consideration at CoP17, relating to some 500 species

Appendix I

Trade permitted only in exceptional circumstances, 3% of all species

(Example: African rhinoceros, threatened with extinction)

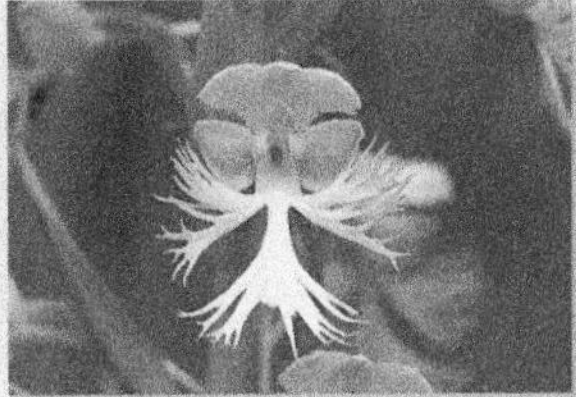

Appendix II

Trade strictly controlled, 97% of all species

(Example: Orchidaceae, at risk of becoming threatened)

Appendix III

Request for assistance in controlling trade of species protected in a specific country

(Example: Chilean toad)

CITES currently protects almost 36,000 species (30,000 plant and 5800 animal species), with an additional 62 proposals being considered that would add another 500 species to international trade regulations for the 183 parties to CITES. Species protected by CITES are placed in one of three Appendices. Species listed in Appendix I are already threatened with imminent extinction. Species listed in Appendix II have trade that is strictly controlled because it could endanger the species. And species listed in Appendix III are special requests by individual parties.

In addition to improved monitoring, an important development taking place in the forestry industry is the rise of sustainable forestry initiatives and certification programs that provide market-based incentives for landowners and operator groups to reduce overharvesting and protect threatened tree species. The adoption of Forest Principles and Agenda 21 at the 1992 Rio Earth Summit formalized for the first time the agreement on sustainable development as part of forest management worldwide. In response to the adoption of sustainable development principles, forest certification programs emerged as a tool to uphold compliance of forest management with sustainable harvesting practices. The most widely used certification programs include the Forest Stewardship Council (FSC, oup-arc.com/e/cardinale-w10.6), the Sustainable Forestry Initiative (SFI, oup-arc.com/e/cardinale-w10.7), and the Program for the Endorsement of Forest Certification (PEFC, oup-arc.com/e/cardinale-w10.8). These programs set standards on timber products, certify and label them as eco-friendly, and establish "chains of custody" for wood products so they can be traced back to the forests of origin.

Forest owners or operator groups can initiate the certification process by requesting a third-party, independent inspection of their forest to see if its management practices meet certification requirements. For most certification programs, these requirements include (1) compliance with international laws and treaties like CITES, and (2) reduction of environmental impacts in order to maintain the ecological function and integrity of the forest. In exchange, the producers have access to a niche market of environmentally conscious consumers who are willing to pay up to 39% more for sustainably harvested products,[45] as well as popular "green building" programs (such as Leadership in Energy and Environmental Design, oup-arc.com/e/cardinale-w10.9) that have their own certification programs. In addition, certification can assist with export/import regulations for countries who are parties in international conservation treaties like CITES (see Box 10.2).

Forest certification schemes have fallen under scrutiny in the past few years for a lack of rigor and evaluation of their certification programs,[46] a lack of transparency in decision-making,[47] for being overly influenced by industry representatives,[48] and for using labels to simply "greenwash" or "launder" trafficking in illegal timber.[48] Some conservation biologists are more optimistic, however, and believe the timber harvest techniques that are encouraged by forest certification programs help maintain biodiversity in forests.[49,50] Major international conservation organizations like The Nature Conservancy and World Wildlife Fund still consider certification programs to be one of the best ways to ensure environmentally responsible, socially beneficial, and economically viable management of forests. And increased competition among the different certification programs is helping to ratchet up standards for certification and accountability.[51]

LEGAL WILDLIFE TRADE The legal trade in wild animal and plant resources and their derivatives is a multibillion dollar international industry. In 2005, the estimated global value of wildlife trade of live plants and animals, as well as wildlife products (excluding timber), was US$202 billion a year.[52] Of this amount, the modern pet trade accounts for the exchange of some 350 million live animals worth about US$20 billion per year.[53]

The estimated volume of legal trade for certain groups of organisms is staggering. Consider, for example, the trade volumes of primates. In the late 1960s and early 1970s, India exported between 20,000 and 50,000 primates per year, while Peru exported an average 30,000 primates per year.[54] Since 1995, China

(31%) and Mauritius (18%) have supplied almost half of all primates traded internationally. The single largest importer of live primates is the United States (26%), followed by Japan (14%) and China (13%). Apart from the live primate trade, there is a significant international trade in dead primates and their body parts. Over the last 30 years, countries have exported 1365 primate bodies (mainly macaques and baboons), 6143 skins (mostly colobus monkeys), and 11,292 skulls (mainly baboons and vervets).[54]

Wild-caught fish are another group with especially large trade volumes. Over 1 billion ornamental fish, comprising more than 4000 freshwater and 1400 marine species, are traded internationally each year.[55] The United States is the largest importer of ornamental fish, importing 10.5 million marine fish in 2005, with most coming from the Philippines, Indonesia, and Sri Lanka.[56] After the U.S. comes Japan, and several European countries that are also large importers of ornamental fish. In addition to ornamental fish, 18 million pieces and 2 million kg of live corals were exported over a 10-year period.[57]

Asian countries are perhaps the greatest exporters of wild flora and fauna. Nijman[57] obtained data on international trade of wildlife for all Southeast Asian nations that are signatories to CITES. In total, 35 million animals (0.3 million butterflies; 16.0 million seahorses; 0.1 million fish; 17.4 million reptiles; 0.4 million mammals; 1.0 million birds) were exported in the period 1998 to 2007, of which 30 million individuals were wild-caught (**Figure 10.7**).

Much of the legal wildlife trade involves species that are abundant, that reproduce fast enough to recover from harvest, or that can be bred in captivity. As such, some have argued that legal wildlife trade has the potential to be a sustainable commercial enterprise that can provide a source of income for millions of producers, as well as an array of goods and products for millions of consumers who depend on wildlife products for medicine, construction materials, food, or cultural experiences. This view is espoused by some conservation organizations like the wildlife trade monitoring network TRAFFIC (oup-arc.com/e/cardinale-w10.10), which is a leading NGO working globally on the trade of wild animals and plants in the context of both biodiversity and sustainable development. TRAFFIC was founded in 1979 as an alliance of the World Wildlife Fund (WWF) and the International Union for the Conservation of Nature (IUCN). The organization's aim is to combat wildlife crime and illegal trafficking while promoting sustainable wildlife trade, with the hope that such equitable trade can both enhance the lives of local people and help to conserve biodiversity.

On the other hand, many species of plants and animals currently face serious threats due to uncontrolled trafficking. The high prices that wildlife products fetch in international markets can drive species decline, degradation of ecosystems, and impoverishment of local livelihoods. Regulations that are poorly enforced or that neglect the stewardship rights of indigenous peoples and communities can undermine local support for conservation, making attempts to counter organized

Figure 10.7 The Chinese water dragon (*Physignathus cocincinus*) is a species of agamid lizard native to China and mainland Southeast Asia. Here, wild-caught individuals are being traded at a rural market, after which they will likely be exported for the pet trade.

crime poaching ineffective. Attempts to tighten enforcement can sometimes be counter-productive, increasing demand and driving prices up.[58]

Suggested Exercise 10.2 Use the CITES database to determine how many species from your home country have trade regulated by this international agreement. **oup-arc.com/e/cardinale-ex10.2**

ILLEGAL WILDLIFE TRADE The total value of all environmental crime is estimated to be between US$91 and US$258 billion per year, which makes it one of the largest crime sectors in the world after drug trafficking (US$344 billion) and counterfeiting (US$288 billion), and on par with human trafficking (US$157 billion).[59] Of the various forms of environmental crime, illegal wildlife trade is estimated to be worth between US$7 and US$23 billion per year, which is on par with illegal mining (US$12–48 billion), forestry crimes and illegal logging (US$9–26 billion), and illegal fishing (US$11–24 billion).[59] Illegal wildlife trade is one of the most profitable crime sectors in the world.

The primary driver of illegal wildlife trade is the extraordinarily high prices that are paid on the black market for rare specimens and their body parts (**Table 10.1**). Tigers (*Panthera tigris*), whose body parts are used in traditional medicine, folk remedies, and increasingly as a status symbol in some Asian cultures, can fetch US$50,000 on the black market. Scent glands from the musk deer, which are used in traditional medicines and the manufacturing of perfumes, can fetch US$250,000 per musk gland. Rhinoceros horns, which are incorrectly thought to be an aphrodisiac or to have medicinal properties, are the world's most valuable animal appendage, valued at US$65,000 per kg (**Video 10.2**). A single bull rhino carrying a 16-kg horn is enough to change the life of a poacher (**Figure 10.8**).

▶ **Video 10.2** Watch this video about the sale of rhino horn. **oup-arc.com/e/cardinale-v10.2**

The illegal wildlife trade is vast in species diversity and geographic scope, with large numbers of wildlife being traded successfully on a regular basis. Rosen and Smith[60] analyzed 12 years of seizure records (1996–2008) compiled by TRAFFIC. Over this period, TRAFFIC Bulletins reported 967 seizures of wildlife and wildlife products originating from more than 100 different

Figure 10.8 Poachers using a high-caliber rifle killed this black rhinoceros (*Diceros bicornis*) in order to harvest its horn. Mistaken beliefs and superstitions about the properties of rhino horns have led to a lucrative black market trade and subsequent widespread slaughter of these animals; only about 5000 black rhinos exist today. This one was killed illegally in South Africa's protected Hluhluwe-Imfolozi Park.

TABLE 10.1 Ten high-value CITES-listed species subject to ongoing illicit international trade

Common name	Scientific name	IUCN Red List status[b]	CITES Appendix	Population trend (from Red List)	Poaching pressure	Estimated retail value (US$)[a]	Price trend
Tiger	*Panthera tigris*	EN	I	Decreasing	Increasing	50,000 / animal	–
Chinese pangolin	*Manis pentadactyla*	CR	II	Decreasing	Increasing	1550 / animal	Increasing
Sunda pangolin	*Manis javanica*	CR	II	Decreasing	Increasing	1550 / animal	Increasing
Musk deer	*Moschus* spp.	EN/VU	I/II	Decreasing	Persistent	250,000 / kg musk	Increasing
Saiga antelope	*Saiga tartarica*	CR	II	Decreasing	Persistent	877 / kg horn	Increasing
Snow Leopard	*Panthera uncia*	VU	I	Decreasing	Persistent	73–1,670 / kg bone, < 10,000 / skin	–
White rhino	*Ceratotherium simum*	NT	I/II	Increasing	Increasing	65,000 / kg horn	Increasing
Asiatic Black bear	*Ursus thibetanus*	VU	I	Decreasing	Persistent	109,700 / kg gall-bladder, 710 / paw	–
African elephant	*Laxodonta africana*	VU	I/II	Increasing	Increasing	6500 / kg ivory	Increasing

Source: D. W. S. Challender and D. C. MacMillan. 2014. *Conserv Lett* 7: 484–494.
[a]Values are rough estimates and may not be representative of general illicit trade markets, which vary considerably by location and by rarity of a species or body part.[114]
[b]EN = Endangered; CR = Critical; VU = Vulnerable; NT = Near-Threatened

countries, with a marked concentration in South Africa and Southeast Asia (**Figure 10.9A**). Mammals and mammal derivatives constituted 51% of all seizures (**Figure 10.9B**). Of these, skins, pelts, and furs, mostly of tigers (49 seizures) and leopards (66 seizures), constituted 26% of mammal product seizures. Ivory constituted an additional 25% of seized mammal products, with more than 42,000 kg of elephant ivory representing about 5400 elephants seized over the 12-year period. More than 192,000 live animals were seized (see Figure 10.9B), 69% of which were reptiles, including more than 72,000 turtles and tortoises. Sixteen seizures included more than 4224 live primates, and an additional 13 contained primate products. Invertebrates, amphibians, reptiles, and fish were frequently confiscated as live specimens (see Figure 10.9B). Seventy-six percent of seizures included species that are listed CITES Appendix I (threatened with extinction) or Appendix II (trade controlled to ensure survival) (**Figure 10.9C**).

Suggested Exercise 10.3 Download the latest volume of the TRAFFIC Bulletin. Using the Seizures and Prosecutions section, determine if there were recent illegal wildlife trade activities recorded in your country.
oup-arc.com/e/cardinale-ex10.3

Illegal wildlife trade is commonly run by organized crime syndicates that are able to evade local and international laws, often by bribing officials and police. These syndicates also create trafficking networks for the sale of body parts, and help distribute weapons to facilitate operations. Regional sectors of these syndicates run the day-to-day poaching and trafficking operations on the ground, hiring and paying poachers and couriers a modest amount for acquiring the animal parts. In parts of Africa and Asia, these regional-level syndicates operate across international boundaries as they fund illegal hunting of high-value wildlife or trapping and smuggling of live animals.

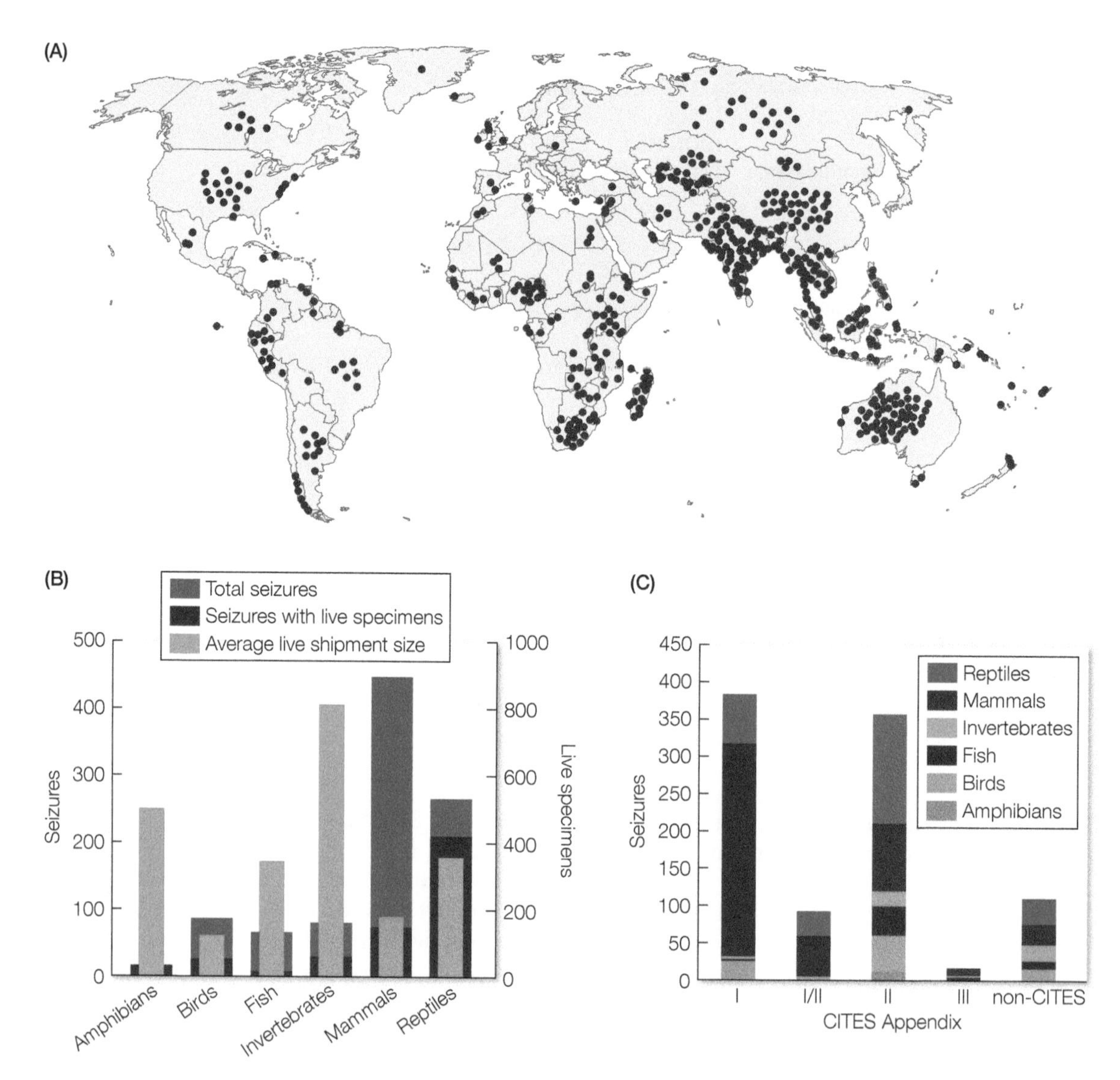

Figure 10.9 TRAFFIC Bulletins report seizures of wildlife and wildlife products. (A) Between 1996 and 2008, there were 967 seizures in over 100 different countries (each dot represents a seizure). (B) The frequency of seizures for various groups of organisms, along with the number and size of shipments containing live specimens. (C) The number of seizures, by organismal group, falling into each of the three CITES Appendices. (After G. E. Rosen and K. F. Smith. 2010. *Ecohealth* 7: 24–32, based on data from TRAFFIC Bulletin.)

The poaching and trade of pangolins (**Figure 10.10A**) exemplifies the successful establishment of trade routes by criminal syndicates. Pangolin meat is considered a delicacy in Asia, their scales are used in traditional medicines, and their skins are processed into leather products. Because of these multiple uses, pangolins are currently the most heavily trafficked wild mammals in the world, with more than a million individuals trafficked between the years 2000 and 2013.[61] During this period, a total of 1270 seizures were reported from 67 countries and territories across 6 continents. A total of 159 unique international trade routes were identified, with an average of 27 new routes identified each year, highlighting a highly mobile transnational trade network with constantly shifting routes to evade authorities (**Figure 10.10B**).

In 2010, CITES founded the International Consortium on Combatting Wildlife Crime (ICCWC), a collaborative effort between the CITES Secretariat,

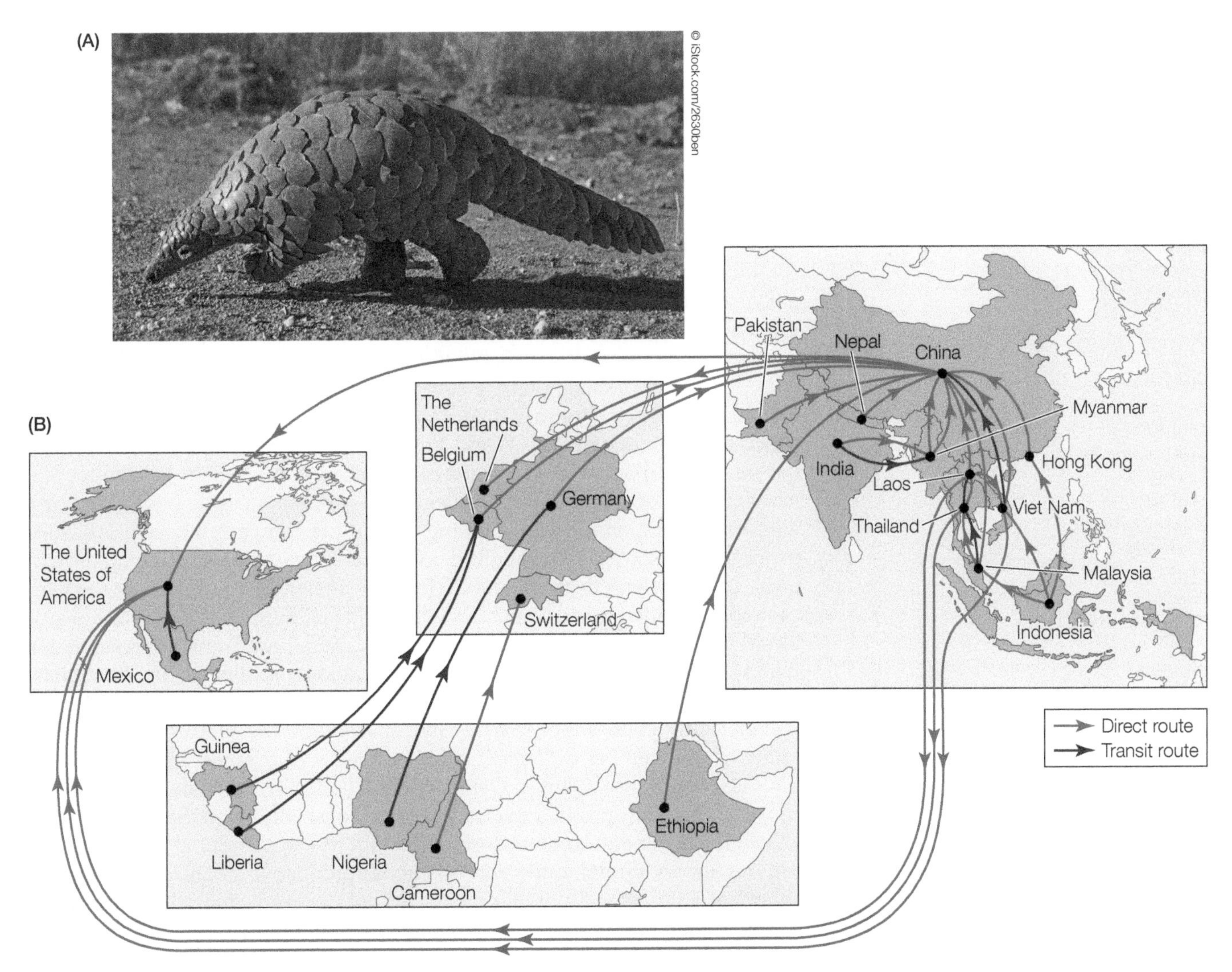

Figure 10.10 (A) Pangolins are currently the most heavily trafficked wild mammals in the world, with more than a million individuals captured and sold each year. (B) The nearly 160 criminal trade routes for pangolins span 67 countries and territories across six continents, and these shift year-to-year as smugglers attempt to evade authorities. (After S. Heinrich et al. 2017. *The Global Trafficking of Pangolins: A Comprehensive Summary of Seizures and Trafficking Routes from 2010–2015*. Selangor, Malaysia: TRAFFIC, Southeast Asia Regional Office.)

INTERPOL, the United Nations Office on Drugs and Crime (UNODC), the World Bank, and the World Customs Organization (WCO). The ICCWC was formed to increase prosecution and punishment for smugglers and poachers as well as increase law enforcement in developing nations. In 2017, a global operation dubbed "Operation Thunderbird" resulted in the identification of nearly 900 suspects and 1300 seizures of illicit wildlife and timber products worth an estimated US$5.1 million. Seizures included:

- More than 300 metric tons of wood and timber
- More than 5200 birds
- More than 2800 reptiles including at least 2000 turtles and tortoises
- 13 wild cats (including six dead bodies)
- More than 3.9 metric tons of pangolin scales

- More than 50 kg of raw and processed ivory
- 25 metric tons of various animal parts, including meat, horns and feathers
- More than 22,000 derivatives and processed products such as medicines, ornaments, and carvings
- 14.3 metric tons of marine wildlife, including 180 dead seahorses

Despite important efforts by organizations the ICCWC, the international community still lags far behind in fighting environmental crimes. The current global resources allocated to organizations that combat transnational wildlife crimes are in the tens of millions of U.S. dollars globally, which is a mere fraction of the resources available to those committing the crimes.[59] Because crime-fighting organizations are severely underfunded and understaffed, the private and public sectors are increasingly teaming up to fight wildlife and forest crime. For example, the World Resources Institute's Global Forest Watch has set up as an online resource that allows local communities to monitor and manage their local forests, report illegal and unsustainable activities, and stop illegal deforestation. Private shipping and airline companies have established policies to deter wildlife smuggling. In China, for example, 17 companies (e.g., EMS, DHL, Federal Express, and SF Express) that account for 95% of the Chinese courier market made a public declaration pledging zero tolerance towards illegal wildlife trade. In June 2016, the International Air Transport Association (IATA) adopted a resolution on the illegal trade in wildlife at the 72nd IATA Annual General Meeting, denouncing illegal trade in wildlife and calling on member airlines to adopt policies that discourage the illegal wildlife trade. Many of the member airlines (e.g., Air Canada, Air France, British Airways, Emirates, Lufthansa, Qantas, Qatar, Virgin Atlantic, Delta, United, and American) now explicitly prohibit the transport of illegal wildlife in their cargo holds.

While greater enforcement is an important part of the global war on poaching, regulatory efforts alone can be counterproductive because reducing the supply of wildlife and wildlife parts without also reducing demand leads to price inflation that encourages even more illicit trade.[58] This is why many conservation practitioners have emphasized that fighting illegal wildlife trade is far more than an enforcement problem. Other strategies must focus on reducing societal demand for illicit products, as is the goal of many public service campaigns that attempt to change public perception about the use of wildlife. Conservationists must focus on providing incentives to protect wildlife against poaching, such as programs that engage communities in wildlife protection efforts, or that convert poachers into protectors of endangered species by providing them with more sustainable and legal forms of livelihood[62] (**Video 10.3**). Still other approaches must use market-based approaches, such as the ranching and farming of wildlife, to flood markets with increased supplies, thus hopefully reducing demand and market prices of illicit wildlife.[63]

▶ **Video 10.3** Watch this video about former egg poachers in Malaysia who are now saving the river terrapin from extinction.
oup-arc.com/e/cardinale-v10.3

Subsistence overexploitation

Unsustainable hunting of **wild meat** (also known as **bushmeat**) for human consumption represents a significant threat to many wild mammals, reptiles, amphibians, and bird, particularly in developing countries where growing populations are faced with food insecurity. For millennia, humans hunted wildlife to survive, and the practice is vital for subsistence in some areas even today. However, the unsustainability of subsistence hunting has accelerated due to growing human populations, increased access to wild animals caused by logging and development of roads, and an increasing tendency for wild meat to be traded commercially. Poaching pressure within many parks and reserves has increased as wildlife populations outside have declined, causing surrounding landscapes to become void of wild mammals.[64]

The harvest and sale of wild meat has become big business in certain parts of the world. An estimated 89,000 metric tons of meat with a market value of approximately US$200 million are harvested annually in the Brazilian Amazon.[65] The volume of the bushmeat trade in West and Central Africa was estimated at 1 to 5 million metric tons per year at the turn of this century,[66] and some estimates suggest as much as 5 million tons of wild meat are consumed across the Congo Basin annually.[67] The United Nations Food and Agriculture Organization (FAO) estimates that between 30% and 80% of protein intake in rural households in central Africa comes from wild meat, making it one of the primary sources of food security for hundreds of millions of rural people.[68]

Ripple and colleagues[69] analyzed records for 4557 of the world's terrestrial mammal species that have been assessed by the IUCN, and 1169 listed as Threatened on the IUCN Red List . They showed that human consumption of wild meat was the primary motivation for hunting in 301 (95%) of those cases, which represents almost 7% of all assessed mammals, and 26% of all Threatened mammals (Figure 10.11). Mammals with the highest percentages of species threatened by

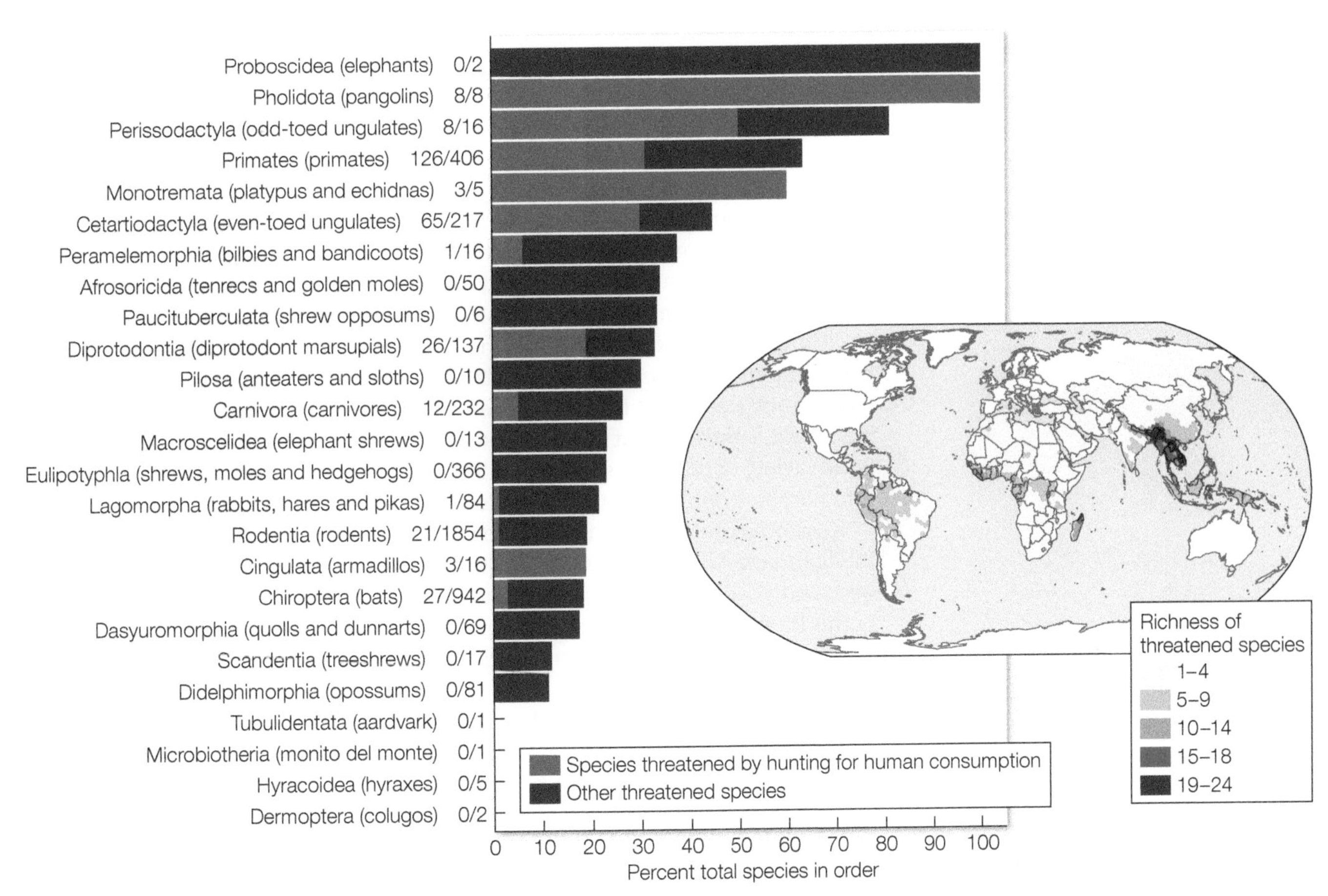

Figure 10.11 The percentage of terrestrial mammal species, by order, that are threatened by hunting for human consumption (indicated in blue along the *x*-axis) verses other human activities (indicated in red). Numbers after the order names on the *y*-axis show the number of species threatened by hunting followed by the total number of species in the order. The map indicates areas of the world where the greatest number of species are threatened by hunting. Note that Elephants are indeed threatened by hunting, but are not shown in blue because they are predominately killed for their ornamental ivory and not for the consumption of meat or medicine. (After W. J. Ripple et al. 2016. *Royal Soc Open Sci* 3: 160498/CC BY 4.0.)

hunting include pangolins (100%), platypus and echidnas (60%), odd-toed ungulates (50%), primates (31%) and even-toed ungulates (30%). Nearly all of the hunting and trapping occurs in developing countries of Africa, South America, and Southeast Asia. There are 113 species threatened by hunting in Southeast Asia (13% of all threatened mammals are east of India and south of China), 91 in Africa (8%), 61 in the rest of Asia (7%), 38 in Latin America (3%) and 32 in Oceania (7%). Notably, all of the 301 mammals that are threatened by subsistence hunting are found in developing countries, underscoring the huge contrast in dietary patterns, wildlife management, and conservation efforts between the developing and developed world.

Hunting and consumption of wild meat has another direct consequence to humans. Wild meat is a primary route for **zoonotic diseases**—diseases that are transmitted from animals to humans. Of the 1415 species of infectious viruses, bacteria, fungi, or parasites that are known to be pathogens to humans, 868 (61%) are zoonotic.[70] Zoonotic diseases represent the majority of emerging infectious diseases (EIDs) in the world, and more than 70% of EIDs stem from contact with wild animals.[71] Disease transmission between wild animals and humans can occur whenever there is direct contact, including wildlife encounters with loggers, poachers, and tourists. But the risks are particularly great for those who handle or eat wild meat.

Zoonotic diseases caused by consuming wild meat are responsible for some of the largest human health epidemics in history. Perhaps the best-known example is HIV and the AIDS epidemic. The human immunodeficiency virus (HIV) is a type of retrovirus whose infection in humans leads to acquired immunodeficiency syndrome (AIDS). AIDS is characterized by progressive failure of the immune system, which in turn leads to opportunistic infections and tumors that rarely affect those with healthy immune systems. According to the World Health Organization, more than 70 million people have been infected with HIV, and 35 million have died (940,000 in 2017 alone), making it one of the deadliest epidemics in world history. HIV evolved from a similar retrovirus in chimpanzees, the simian immunodeficiency virus (SIV). SIV is thought to have first entered the human population early in the twentieth century in the Democratic Republic of Congo[72] after humans came into contact with infected bushmeat.[73] Subsequent work has shown that non-human primates harbor a wide variety of retroviruses beyond SIV, and some of these are actively crossing into human populations that continue to come in contact with wild meat.[73,74]

Zoonotic diseases also include the EIDs Zika and West Nile that first emerged in Uganda; SARS and bird flu in China; and brucellosis in Malta. One of the best-known EIDs that has been linked to human consumption of wild meat is the Ebola virus, for which the primary host is suspected to be fruit bats.[75] Ebola has been transferred from animals to humans at least 30 times between the first recorded outbreak in 1976. The 2014 outbreak in West Africa, which was the largest and deadliest thus far, likely stemmed from a single zoonotic transmission event involving a 2-year-old-boy in the remote Guinean village of Meliandou. The boy fell ill with a fever, black stools, and vomiting,[76] and died 2 days later. Retrospective analysis suggests the child, whose family routinely hunted fruit bats for food, might have been infected by hunting or playing with insectivorous free-tailed bats living in a nearby hollow tree.[76,77]

Despite health organization warnings about the disease risks associated with wild meat, surveys indicate that those who eat wild meat are usually unaware of the risks of zoonotic infection. Because of its crucial and longstanding role as a protein source, particularly in West Africa, efforts to stop the sale and consumption of bushmeat have been difficult to enforce and have met with

(A)

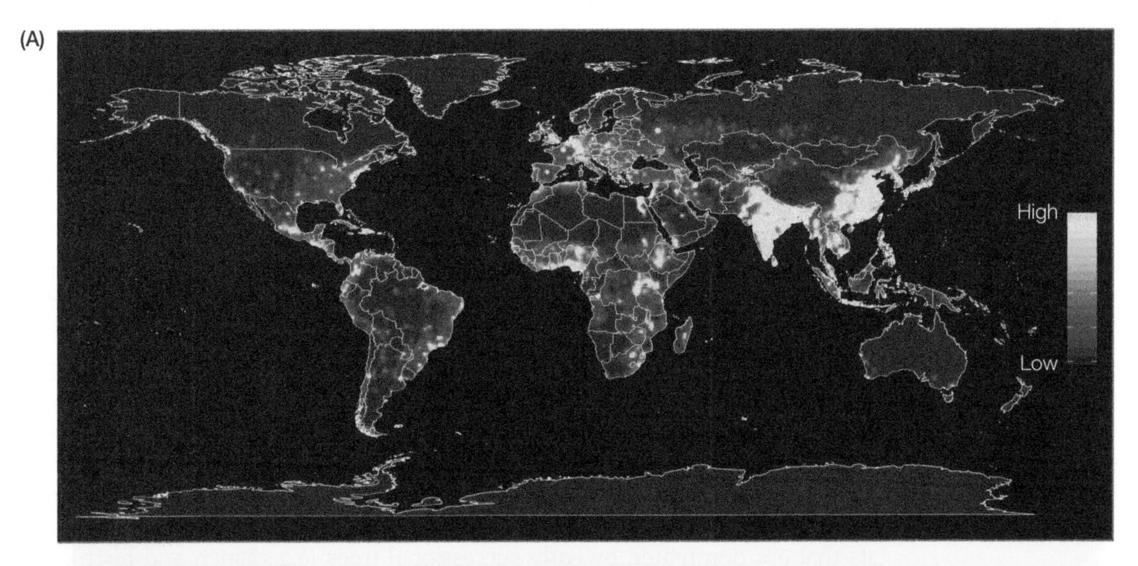

(B)

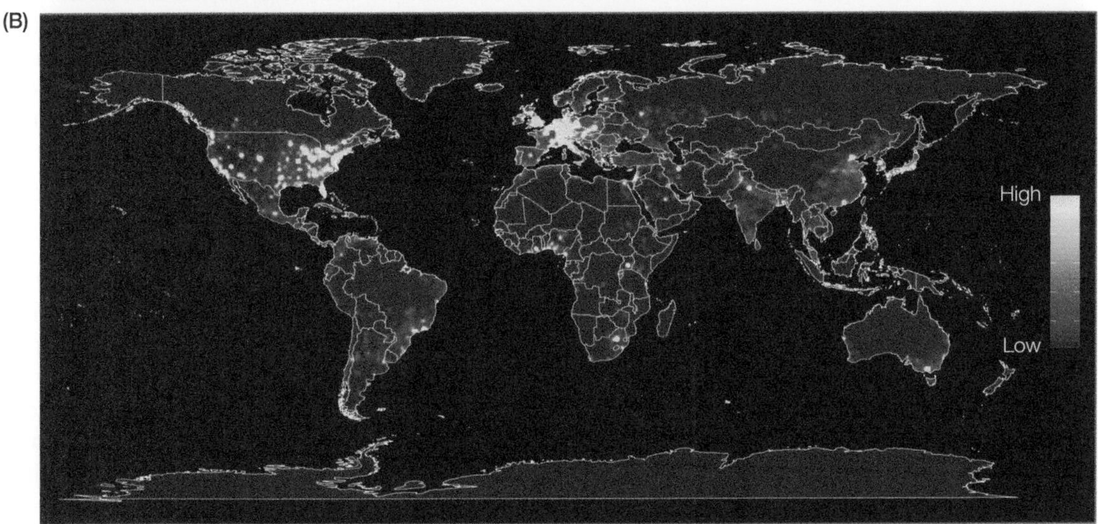

Figure 10.12 Global hotspots of emerging zoonotic disease. (A) Areas of the world where both the reservoirs of pathogens in wild animals and human contact with wild meat is high. (B) However, the risks to human populations are greater in developed countries that host the major transportation hubs, and thus have the greatest potential to spread disease. (From T. Allen et al. 2017. *Nat Comm* 8: 1124/CC BY 4.0.)

suspicion from rural people.[75] Given this situation, most researchers suggest that tropical countries in Asia, Africa, and to a lesser extent South America, will continue to be the source of thousands of new zoonotic EIDs in the decades to come.[78] These are the areas where the reservoirs of pathogens in wild animals is high and contact with wild meat is growing (**Figure 10.12A**). At the same time, the greatest risk to the human population will be in developed countries that host the major transportation hubs, which have the greatest potential to spread disease and create global pandemics (**Figure 10.12B**).

Recreational overexploitation

Hunting and fishing for sport are popular pastimes in many cultures, and participants frequently argue that their sport helps benefit nature conservation. This argument presumes that any harm done by killing individual animals is outweighed by the benefits that hunters and anglers bring to the conservation of nature.

RECREATIONAL HUNTING Hunting is not just a recreational experience. In many places and in many families it is cultural experience, passed down from generation to generation. Recreational hunting can also be a major contributor to local economies. Consider, for example, a report by the Outdoor Industry Association that reported 13.7 million U.S. citizens ages 16 and older went hunting in 2011.[79] These individuals spent an average of 21 days hunting that year, during which, they spent US$34 billion on trips, equipment, and licenses.

Many animals that are targeted in recreational hunting are not threatened with extinction; in fact, some are sufficiently abundant that hunting is used to control their population sizes. However, recreational hunting and conservation tend to conflict in two circumstances. The first is when an ecosystem is managed for game animals in a manner that is detrimental to that ecosystem's biodiversity. One example is management of forested ecosystems for the white-tailed deer (*Odocoileus virginianus*), a highly prized game animal that is hunted throughout much of North America. In parts of North America, such as the U.S. Midwest, management agencies use thinning and clear-cutting to maintain forest habitats that are dominated by early successional trees (e.g., aspen and birch), and that have interspersed open areas of grasses, forbs, legumes and berry-producing shrubs. Such conditions provide food and optimal habitat for deer, and have been used to justify frequent logging. But these land management practices can also promote hyper-herbivory by deer, which causes local extinction of native plants[80,81] and endangers the persistence of certain mammals and songbirds that depend on the development of late-successional forests for habitat.[82]

Recreational hunting and conservation also conflict when the targets for kill are species that are rare or threatened with extinction. Trophy hunting—the killing of big game for a set of horns or tusks, a skin, or a taxidermied body—is a lucrative, profit-driven industry. Individual hunters pay tens of thousands to hundreds of thousands of dollars to purchase the right to legally hunt and kill rare megafauna, such as Africa's "big five": elephant, *Loxodonta* spp.; lion, *Panthera leo*; leopard *Panthera pardus*; rhino *Ceratotherium simum* and *Diceros bicornis*; and Cape buffalo, *Syncerus caffer* (**Figure 10.13**). One of the largest payments for trophy hunting was made in January 2014, when the Dallas Safari Club auctioned a permit to

(A)

(B)

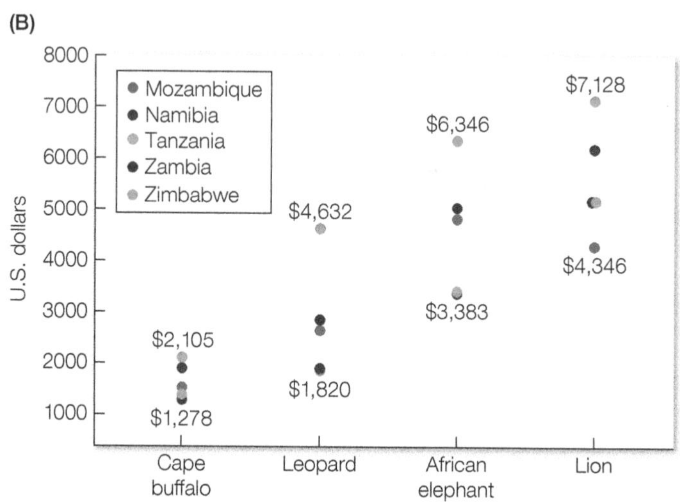

Figure 10.13 (A) Taxidermy exposition room in Namibia. (B) Hunters will pay tens of thousands of dollars to kill a "trophy," with the amount varying by country and animal. Shown here are the minimum prices (US$) for trophy-hunting packages in five African nations in 2011. (B, after P. A. Lindsey et al. 2012. *PLOS ONE* 7: e29332.)

hunt a single individual of the Critically Endangered black rhino (*Diceros bicornis*) in Namibia. The winner of the auction, Corey Knowlton of Dallas, paid $350,000, with the funds going to the Namibia Ministry of Environment and Tourism.

The killing of endangered animals for sport tends to arouse deep emotions and fierce opposition, which became apparent in 2015, when Walter Palmer, an American dentist and recreational game hunter from Minnesota, created an international furor by killing Cecil the Lion. Cecil was a 13-year-old male lion that lived primarily in the Hwange National Park in Zimbabwe. Cecil was the park's best-known animal and a major attraction, identifiable from his black-fringed mane and a GPS tracking collar that he wore as part of study being run by the University of Oxford. Palmer reportedly paid US$50,000 to a professional guide to help him track and kill a lion. In the late afternoon of July 1, the guide and his assistant planted an elephant carcass just outside the park to lure Cecil out of his protected area. That evening, Palmer shot and wounded Cecil with an arrow. The next morning, the hunters tracked the wounded lion to location less than 250 meters from the initial shot, and killed him with a second arrow, 12 hours after initially wounding him.[83] Cecil left behind a pride of 3 females and 7 cubs.

Though Palmer had a permit to hunt and ultimately was not charged with any crime, the killing of such a prominent animal, and the ethical conditions of the hunt (luring an animal outside a protected area, extended suffering during the kill) resulted in international media attention, caused outrage among animal rights groups, prompted criticism from politicians, and drew public condemnation from numerous celebrities.[84] Palmer received hate messages, was subjected to harassment on social media, and had the words "Lion Killer" spray-painted on the garage door of his Florida vacation home. Artists from around the world commemorated Cecil with art and musical compositions. Online petitions signed by over a million people urged Zimbabwe's government to stop issuing hunting permits for endangered animals.

But even as Cecil's killing sparked widespread discussion about the ethics of trophy hunting, it also reinvigorated a debate about the role of hunting in species conservation.[85] Hunters, and some conservation biologists, argue that trophy hunting helps protect species by generating a market that places a value on them that promotes their protection.[86] The market value, they argue, creates economic demand that will be matched by supply as governments, local communities, or private businesses set up and manage protected areas or game farms that cater to hunters. Proponents further argue that the large amount of money spent on trophy hunting improves the livelihoods of local communities. which in turn encourages them to protect the wild animals that are vital to their economy (**Table 10.2**).

But critics argue that the economic benefits are overstated, and that little revenue from trophy hunting stays in local communities. For example, a report by the nonprofit group Economists at Large showed that, while trophy hunting in Africa generates about US$200 million in annual revenue, only 3% of the fees paid for the hunts reach local communities.[87] Instead, the vast majority of revenues go to firms, government agencies, and individuals located internationally or in national capitals. Corruption usually ensures that revenues that accrue with government agencies rarely reach local communities. The authors also point out that trophy hunting is an insignificant portion of tourism revenue (just 1.3%); yet killing of animals diminishes opportunities for nature-based ecotourism (e.g., safaris, photography) that are worth far more to local economies, create more jobs, and are more sustainable.

Even so, some conservation biologists believe the debate over trophy hunting and conservation is more than just an economic argument. The single greatest threat to wildlife, including Africa's "big five," is habitat loss. Demand for

TABLE 10.2 Hunting contributions to biodiversity conservation and national economies in seven sub-Saharan countries

Country	Area covered by game ranches (% of total land area)	Terrestrial protected areas (% of total land area)	Top three most exported trophies (2012)[a]	Annual revenue (US$ million)[b]
South Africa	13.1	6.2	Impala, warthog, greater kudu	68.0
Tanzania	26.4	32.2	Leopard, hippopotamus, elephant	56.3
Botswana	23.0	37.2	Elephant, leopard, lechwe	40.0
Namibia	11.4	43.2	Zebra, chacma baboon, leopard	28.5
Zimbabwe	16.6	27.2	Elephant, leopard, chacma baboon	15.8
Mozambique	10.5	17.6	Nile crocodile, elephant, hippopotamus	5.0
Zambia	21.3	37.8	Lechwe, hippopotamus, leopard	3.6
TOTAL REVENUE	–	–	–	217.2

Source: E. Di Minin et al. 2016. *Trends Ecol Evol* 31: 99–102.
[a]Species designations: chacma baboon, *Papio ursinus*; African elephant, *Loxodonto africana*; hippopotamus, *Hippopotamus amphibius*; impala, *Aepyceros melampus*; greater kudu, *Tragelaphus strepsiceros*; lechwe, *Kobus leche*; leopard, *Panthera pardus*; Nile crocodile, *Crocodylus niloticus*; warthog, *Phacochoerus africanus*; zebra, *Equus* spp.
[b]Data not adjusted for inflation.

trophy hunting helps prompt large areas of land to be set aside as protected areas, and these have been complemented by the creation of new areas such as game ranches that cater to hunters (see Table 10.2). Some feel that the conservation benefits of these protected lands for whole populations of animals far outweigh any detrimental impacts that hunting might have on a relatively small number of individuals.[88] This debate will almost certainly continue for decades to come.

RECREATIONAL FISHING Although the dramatic effects that commercial fishing can have on marine fish stocks are well recognized, the potential role that recreational fishing plays as a threat to aquatic biodiversity has been historically ignored. The IUCN Red List rarely cites recreational angling as a threat to fish species,[89] even though ichthyofauna—particularly in freshwater habitats—are considered to be some of the most threatened organisms on the planet.[90]

But growing evidence suggests that recreational fishing poses a far greater threat to aquatic biodiversity than has been historically appreciated.[89,91] Approximately 11% of the global human population participates in recreational fishing, with participation exceeding 45% in some Scandinavian countries.[92] In certain developed countries, the recreational fishing industry has grown into a multibillion dollar industry[93–96] where individual anglers spend thousands of dollars in licenses, equipment, and travel.[94]

The collective impact of millions of recreation anglers is substantial. Cooke and Cowx estimated that recreational fishing is responsible for the landing of 47 billion fish per year globally, representing a harvest of approximately 11 million metric tonnes.[91] This estimate represents nearly 12% of global fish catch (commercial fishing is approximately 80 million metric tons per year; see Figure 10.4A), which is far greater than previous estimates.[97] In certain fisheries, recreational angling is responsible for an equal or larger portion of fish catch than commercial fishing. For example, Coleman and colleagues estimated that recreational fishing accounts for 38% to 65% of all finfish catch in the Gulf of Mexico, South Atlantic, and Pacific Coast.[98]

The potential impact of recreational angling on fish populations has become increasingly apparent, with numerous studies having implicated recreational angling as the primary cause of population decay for fish. For example, Post and colleagues collated information on the collapse of several prominent freshwater fisheries in Canada and showed that declines of walleye (*Sander vitreus*), northern pike (*Esox lucius*), and numerous species of trout (*Oncorhynchus*, *Salmo*, and *Salvelinus* spp.) were all caused by overfishing by recreational anglers, which was exacerbated by poor management practices that ignored the warning signs of overfishing.[99] Sadovy de Mitcheson and colleagues summarized the conservation status of grouper species (*Epinephelinae* spp.), which collectively represent a billion dollar recreational fishery throughout Southeast Asia, the Caribbean, and coastal Brazil.[100] The authors determined that 20 species (12% of all grouper species) are currently under threat of extinction, including three considered to be Critically Endangered as a result of fishing pressure.

In contrast to commercial fishing, which is concentrated in deeper coastal oceans or large inland water bodies, recreational fisheries have near-exclusive access to most of the world's freshwater ecosystems, as well as the shallow near-shore marine habitats like estuaries, reefs, and mangroves that are important spawning, nursery, or feeding grounds. Aside from their unique access to these habitats, recreational fishing differs in that anglers often specifically target the largest, most fecund individuals that have disproportionate impacts on population growth.[101,102] These large, trophy-size fish are often sacrificed when caught. Of the 1222 fish species that are tracked by the International Game and Fishing Association—the largest organization that maintains and reports the size/weight records of trophy fish—85 of these species are currently threatened with extinction.[103] Because they must usually be transported to a land-based weigh station for certification, catch-and-release of these large, rare trophy fish is uncommon.

Unsustainable recreational fishing is often compensated for by expensive fish stocking programs, or with management interventions that restrict the harvest of threatened species by limiting gear types or applying seasonal or area closures. But attempts to limit fishing are sometimes vehemently opposed by anglers. For example, in the Florida Keys, overfishing has caused the mean size of recreationally caught "trophy" reef fish to decline by 88% over the last 60 years, shifting landings from large predators like groupers (*Epinephelus* spp.) and sharks to small snappers (*Lutjanus* spp. and *Ocyurus chrysurus*).[104] Marine angling groups have joined in a "right to fish" movement that opposes and obstructs conservation efforts to regulate the Keys fishery and instead supports legislation to protect their perceived right to fish in the reef ecosystem. Their political activism lies in stark contrast to that of conservation-minded sportsmen, who have traditionally supported habitat protection and closed degraded reserves both on land and in freshwater.[105]

Despite the anti-conservation behavior of some angler groups, most recreational anglers have been remarkably good at self-regulation, and have been proactive in the conservation of aquatic resources. Unlike trophy hunting, no-kill, catch-and-release sport fishing has become a common environmental ethic among anglers, and it is widely promoted by management agencies and professional organizations.[89,106] In addition, there are many examples of recreational fishers being instrumental in fisheries conservation through active involvement in, or initiation of, conservation projects that protect species from decline. These efforts represent a major conservation success story in the management of recreational fisheries (**Box 10.3**).

BOX 10.3 Success Story

The Taimen Conservation Project

Eurasian giant trout (*Hucho taimen*) are the world's largest salmonids, reaching sizes of over 45 kg (100 lbs) and 1.5 m (5 ft; see the Figure). Also called the taimen, the fish is a slow-growing, long-lived apex predator that is particularly vulnerable to overexploitation. Recreational sport fishing of taimen is common and threatens the species' survival, particularly in areas where recreational fishers use catch-and kill-methods. As the species' range has become constricted, the Eg-Uur River watershed in northern Mongolia has become home to one of the few taimen populations still characterized by large adult fish. This has led to increased pressure by anglers, who travel to the remote location for the opportunity to catch the large, rare fish.

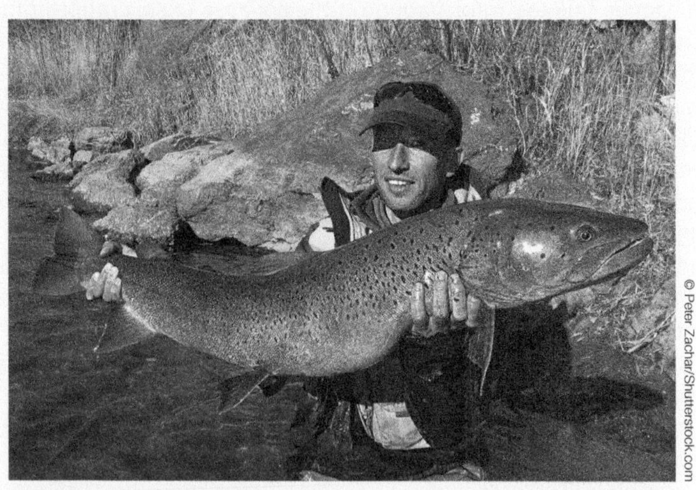

An angler holds a taimen (Eurasian giant trout, *Hucho taimen*), the world's largest salmonid.

To protect taimen, a local Mongolian nongovernmental organization (NGO), the Taimen Conservation Fund,[120] established a conservation project in the Eg-Uur watershed. Using revenues from recreational fishing permits, the project funds local managers, provides infrastructure, and works to facilitate business partnerships between local communities and recreational anglers.

Along with working to increase the perceived economic value of protecting the taimen, one of the project's major goals is to treat the fish as a locally rather than externally managed resource. Locally based management gives communities increased incentive to protect the resource and to prevent overfishing and poaching. The Taimen Conservation Fund sees that fees for catch-and-release ecotourism operations are paid to a community council that oversees river management in the Eg-Uur watershed.

Another goal of the Taimen project has been to develop local revenue sources and local scientific and monitoring capacity to ensure long-term sustainability. To accomplish this goal, recreational anglers contribute in-kind support to scientific and enforcement teams that study and protect the taimen. This support includes boats and equipment, air transportation, fishing expertise, and access to clients' fish for tagging, creel surveys, and spawning fish counts.

By demonstrating that local residents, scientists, and recreational fishers can work together to study and protect a threatened species, the Taimen Project is a conservation success story.

Theory of Sustainable Harvesting

Conservation biologists involved in the management of natural resources have a number of options at their disposal to try and prevent the overexploitation of species. They can help set policies and regulations that limit the take of threatened species. They can use financial or other incentives to encourage individuals and communities to abide by those regulations and policies, as well as disincentives and enforcement for those who do not comply. They can also augment the populations of species in decline in some areas by protecting them in others (e.g., parks, reserves, sanctuaries) that have higher population growth, and which serve as a source of immigrants. But for all of these options, resource managers face a common challenge, which is to maximize a trade-off between those who wish to harvest and use a resource *right now*, versus the need to keep that resource from going extinct, and protecting it for those who might use it in the future. In other words, managers must figure out how much harvesting is sustainable and how much is unsustainable.

To determine if harvests are sustainable, resource managers often rely on population models that are founded in the theory of **maximum sustainable yield (MSY)**. The theory of MSY helps define what a sustainable harvest is, and aims to maintain the population size of a species at a point below the sustainable harvesting level so that the population can continue to reproduce indefinitely. Models based on MSY are widely used in fisheries management, the harvesting of trees from forests, and the management of game species. MSY is also a component of population viability models (PVAs) used to project the survivorship of endangered species (see Chapter 13).

Nearly all models that use MSY are derivatives of simple ecological models of population growth. But the complexity of these models can be varied to increase realism, and to be more specific to the species or group of organisms being managed. In the final section of this chapter, we introduce the theory of maximum sustainable yield as well as two simple models (fixed quota and fixed effort) that make differing assumptions about how a population is harvested.

Sustainable yield

In principle, maximum sustainable yield is the largest yield (or catch) that can be taken from a species' population without depleting that population through time. Two assumptions that underlie nearly all sustainable harvesting models are: (1) that biological populations grow and replace themselves (i.e., they are renewable resources); but that (2) populations cannot grow indefinitely, and must ultimately reach an upper bound of population size the environment can support. This upper bound is called the **carrying capacity**. Given these two assumptions, most MSY models begin with logistic population growth (Figure 10.14A), which in continuous time can be expressed as:

$$N_t = \frac{K}{1+\frac{K-N_0}{N_0}e^{-rt}} \tag{10.1}$$

where N_0 is the initial population size, N_t is the size of the population at time t, K is the carrying capacity, and r is the intrinsic rate of population increase. If we differentiate Equation 10.1 with respect to time, we get an expression

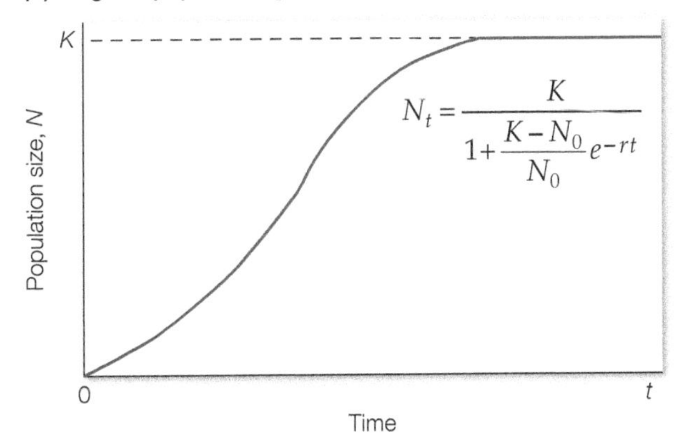

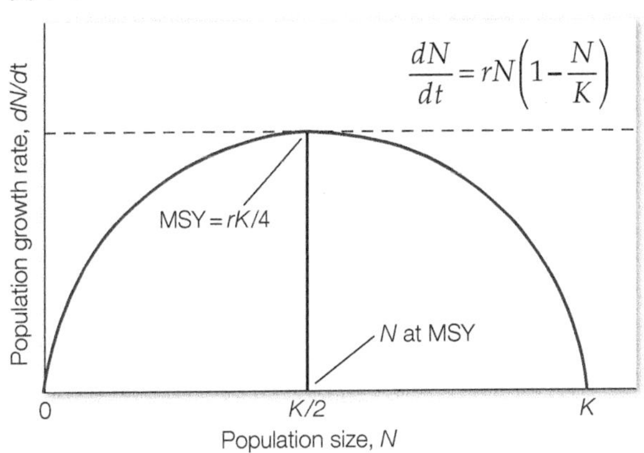

Figure 10.14 Maximum sustainable yield (MSY). (A) The logistic population growth model is the foundation for most estimates of MSY. Populations grow and replace themselves until they reach an upper bound of population size the environment can support, called the carrying capacity. (B) When the rate of population growth, dN/dt, is plotted as a function of population size N, the MSY for the population is equal to $rK/4$, which occurs at a population size equal to half the carrying capacity, or $K/2$.

that predicts how fast the population grows through time in the absence of any exploitation:

$$\frac{dN}{dt} = rN\left(1 - \frac{N}{K}\right) \tag{10.2}$$

Equation 10.2 is the logistic growth equation commonly presented in introductory biology and ecology textbooks. For our purposes here, Equation 10.2 is important because we can plot the species population growth rate, dN/dt, as a function of the population size (**Figure 10.14B**). When the species population size N is small (e.g., point 0 on the x-axis), the population growth rate is very low because there are few individuals reproducing and adding to the population. When N is large and close to the carrying capacity (e.g., point K on x-axis), then population growth rate is again low, because resources have been used up and no new individuals are being added to the population. Importantly, the rate of population growth rate is highest at intermediate population sizes (note the curve in Figure 10.15B is concave down). Specifically, the highest rate of population growth occurs when N is exactly half the carrying capacity, or $N = K/2$.

Note that $K/2$ is the population size N at which growth is maximal—in other words, it is the point on the x-axis of Figure 10.14B where the curve is highest. But what we want to know is the point on the y-axis where the curve is highest. That represents the maximum rate at which new individuals are being added to the population, or dN/dt is greatest. To get that number, we plug $K/2$ into Equation 10.2 for all entries of N, and simplify the equation to get:

$$\frac{dN}{dt_{max}} = \text{MSY} = \frac{rK}{4} \tag{10.3}$$

Equation 10.3 gives the biological definition of sustainability for a species population. It says that the maximum rate at which new individuals are added to a population is one-fourth the product of the intrinsic growth rate times the carrying capacity. If harvesting exceeds the maximum rate at which individuals are added to a population (that is, exploitation rates are greater than the MSY), then the population will not maintain a positive growth rate, and will decline to extinction if not corrected.

Fixed quota (Q) harvesting

Equation 10.2 is the fundamental population model used to estimate MSY. The simplest way to change this to a harvesting model is to subtract a constant Q that represents a fixed number of individuals removed from the population at each time step:

$$\frac{dN}{dt} = rN\left(1 - \frac{N}{K}\right) - Q \tag{10.4}$$

Equation 10.4 is called the **fixed quota** (or **constant harvest**) **model**. The fixed quota model implies a system of exploitation in which the number of individuals removed from a population is both independent of the population size and independent of the amount of effort put into harvesting. The fixed quota model is used to estimate the number of individuals that can be harvested from the population per unit time. To do so, we set $dN/dt = 0$ and then solve Equation 10.4 for Q, which leads to:

$$Q = rN\left(1 - \frac{N}{K}\right) \tag{10.5}$$

The results of the fixed quota model are depicted in **Figure 10.15**, which gives scenarios for three different levels of fixed harvest. If the quota Q exceeds MSY—that is, if $Q > rN(1 - N/K)$—then harvesting exceeds the populations ability to reproduce, and the population will decline toward zero (note arrows showing population trajectory for the "high quota" scenario). This scenario is obviously unsustainable.

The second scenario is one where the quota is set to equal MSY (MSY quota). If the initial population is larger than N_{MSY} (e.g., $K/2$) prior to the start of harvest, then the population will naturally decline to N_{MSY} (note arrows showing population trajectory). This is, in principle, a sustainable harvesting regime. It is a risky management strategy, however, because if the population declines to anything lower than N_{MSY} (by poor monitoring, or due to stochasticity), then the population will be driven toward extinction.

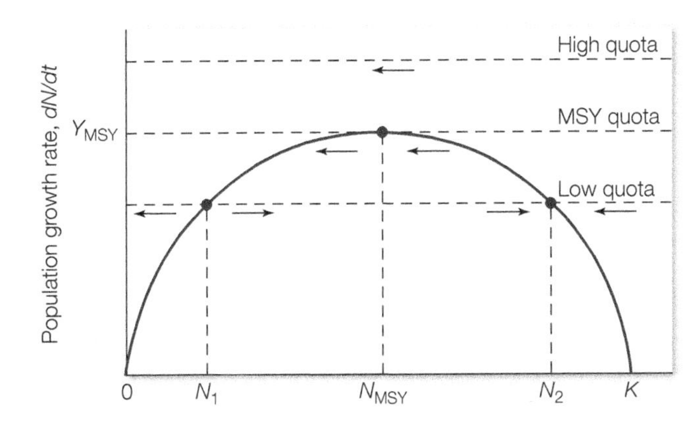

Figure 10.15 Equilibria and population stability for the fixed quota, or constant harvest, model. The arrows indicate the directions of change in population size for each quota. MSY occurs at half the carrying capacity, or $rK/4$. Continued harvest at the "high" quota will drive the population toward zero. The sustainability of the other quotas (MSY, low) depend on where the initial population sizes are.

Lastly, consider the "low quota" example, where $Q < rN(1 - N/K)$. If the population is initially below N_1 (even slightly), then harvesting will cause the population to crash. If, however, the initial population is between N_1 and N_2, then its production will exceed the harvest set by the quota, and the population size will grow to N_2. N_1 is what we call an unstable equilibrium, in the sense that small fluctuations in population size near N_1 can mean the difference between population growth verses extinction. In general, exploitation in that region should be avoided because of the associated uncertainty and risk.

In contrast, N_2 is considered a stable equilibrium. If the initial population size is larger than N_1 but less than N_2, then it will grow to N_2 because production will exceed the rate of harvest. If, however, the initial population is above N_2 then it will also be driven down to N_2 because the quota harvest exceeds the population's rate of production. Therefore, Thus, N_2 is the point of both sustainable and stable harvest of the resource.

Fixed effort (proportional) harvesting

In many ways, it is more sensible to set exploitation targets based on the size of the population, allowing for more harvesting when populations are large and less when they are small. Aside from the sensibility of harvesting according to population size, there is the practical issue that resource managers are often in a better position to regulate the effort that users devote to harvesting. For example, in fisheries, managers often regulate the number of boats that can fish in a certain area, what kinds of gear people can use, or how many days they can fish. In game management, agencies issue limited numbers of hunting licenses and limit the number of days of the hunting season. Mathematically, fixed-effort harvesting can be described as:

$$\frac{dN}{dt} = rN\left(1 - \frac{N}{K}\right) - EN \tag{10.6}$$

Equation 10.6 is called the **fixed effort** (or **proportional harvest**) **model**. In this model, N is the population size, and E is a constant that represents the exploitation rate. The product EN represents the proportion of the population that can be harvested from the population per unit time as:

$$EN = rN\left(1 - \frac{N}{K}\right) \tag{10.7}$$

We can rearrange this equation to find the equilibrium population size for any rate of exploitation, as long as the value of E is below that of r:

$$N = K\left(1 - \frac{E}{r}\right) \tag{10.8}$$

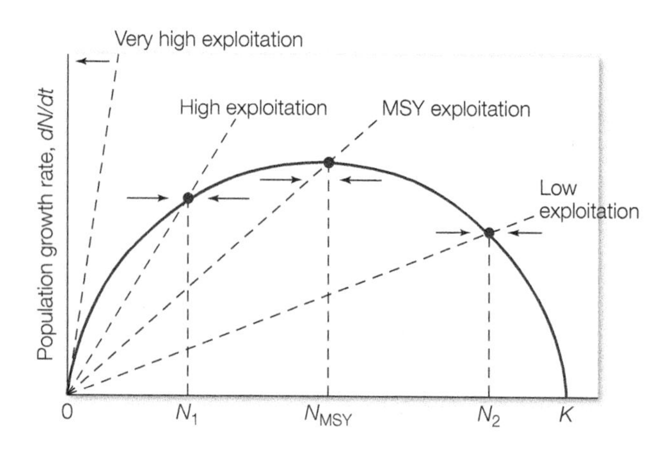

Figure 10.16 Equilibria and population stability for the fixed effort, or proportional harvest, model. Here, a constant fraction of the population is removed at each time step. The dashed lines have slopes corresponding to different rates of exploitation. The arrows indicate the directions of change in population size for each scenario. As with the fixed quota model, MSY occurs at $rK/4$. The model is more stable than the fixed quota model, with only extremely high exploitation rates driving the population toward zero.

The equilibria for Equation 10.8 are shown in **Figure 10.16**

The advantage of fixed effort harvesting is that, as long as the exploitation rate E is below the intrinsic rate of natural increase r, then all equilibria in Figure 10.16 are stable. Whereas the population crashed under the "high" fixed quota scenario (top dashed line in Figure 10.15), in the fixed effort scenario a population that is initially below the high exploitation rate will increase to N_1; if it is above, it will decrease to N_1. The only scenario that is not sustainable is one with a very high exploitation rate—that is, where $EN > r$.

Limitations of MSY models

The theory of maximum sustainable yield has been broadly adopted by many local, federal, and international bodies that manage wildlife, forests, and fisheries.[107] Even so, the use of MSY models in management has been heavy criticized by some biologists on both theoretical and practical grounds.[108] Critiques have pointed out that MSY models like those presented in this chapter (sometimes referred to as "stock surplus" models) are overly simplistic, failing to capture important aspects of species biology and realistic environments. For example, these models fail to account for the size or age structure of populations, as well as key sources of demographic and environmental stochasticity that can alter species population growth rates and carrying capacities, which are not fixed parameters but can vary through time.[107] Failure to account for such biological realities can generate inaccurate predictions for stock assessments. While these issues are important, they are not necessarily limitations. More biologically realistic MSY models that account for age/size structures and demographic/environmental stochasticity have been developed and are increasingly used to manage stocks[109] (see Chapter 13 for examples).But as is true for all models, those with greater reality also tend to be more complex and data-intensive, and the data required to parameterize more realistic models often don't exist.

Another common problem with MSY models is their improper implementation and use. Management authorities sometimes set harvesting levels too high in order to satisfy local business interests, protect jobs, and generate revenue, which damages the resource base.[110] It can also be difficult to coordinate international agreements and to monitor compliance with MSY limits when species migrate across national boundaries and through international waters. And of course illegal harvesting results in additional resource removal that is not accounted for in official records.[111]

Despite these limitations, it is important to note that the concept of MSY as a management tool has morphed through time. MSY is now widely considered to be an undesirable upper limit for harvest that should be avoided, rather than a management target per se.[112,113] It is also important to note that MSY models have evolved with an improved set of assumptions and revisions as more data has become available. These models continue to provide a far superior way to manage fish and wildlife stocks than subjective approaches based on expert opinions and personal judgments. Therefore, students of conservation biology need to learn the fundamentals of MSY models that form the foundation for many management decisions about fish and wildlife stocks.

Suggested Exercise 10.4 Practice using the fixed quota and fixed effort models to estimate maximum sustainable yield with this spreadsheet exercise. **oup-arc.com/e/cardinale-ex10.4**

Summary

1. Humans have always exploited wild plants and animals to obtain the food and materials needed to survive, and this has led to extinctions in the past. However, overexploitation is now the second leading cause of extinction behind habitat loss. The advent of guns, nets, and bulldozers has made it incredibly easy for humans to hunt, fish and harvest, which is threatening the existence even of species that were once abundant.
2. The human psychology of overexploitation is complex. People overexploit for a variety of reasons, including meeting their needs for subsistence, social dilemmas that cause rationale individuals to make irrational decisions that in the long run cause harm to themselves, a lack of awareness about the finite limits of natural resources, an overly optimistic faith in technological solutions, and spite or selfishness. Conservation biologists must understand which of these are driving overexploitation in any given situation in order to come up with effective management strategies.
3. At a global scale, commercial exploitation, such as commercial fishing and logging, is the largest contributor to overexploitation. However, improved regulations, government oversight, and privately-based sustainability certification programs have helped reduce rates of exploitation that threated plants and animals.
4. Wildlife trade, both legal and illegal, poses some of the most immediate threats to biodiversity, with many species now Critically Endangered because of the demand for them and the benefits to suppliers of their capture from wild habitats and subsequent sale in markets. International treaties and regulations like CITES, market-based incentives, and greater engagement of local communities in conservation are all part of a portfolio to reduce the impacts of wildlife trade on endangered species.
5. Subsistence exploitation, including the unsustainable harvest of wild meat, not only endangers wildlife, it endangers the human population due to it being the source of most of the world's emerging infectious diseases. Providing people with stable, alternative sources of food is key to global security for human health.
6. Recreational exploitation is perhaps the most controversial of all forms of overexploitation. Hunters and anglers who kill for sport have clearly contributed to the population decay and endangerment of many terrestrial, freshwater, and marine species. At the same time, the economic value that hunting and fishing brings to conservation can be substantial, and many who practice hunting and fishing are vocal proponents and advocates for survival of their target species.
7. Models of maximum sustainable yield (MSY) are the primary tools that resource managers use to perform stock assessments, and to manage take from populations. They have been adopted by the vast majority of local, federal, and international bodies that manage wildlife, forests, and fishing. Though simple MSY models like the ones presented in this chapter are limited in biological reality and can sometimes make flawed assumptions, they are the basis for more realistic and complex models that are improving quantitative stock predictions.

For Discussion

1. Is it fair and reasonable for people who live in rich countries—many of which have become rich by exhausting their own natural resources—to try and help or encourage people in poor countries to exploit their plants and animals sustainably?
2. Many cases of overexploitation involve not only people's livelihoods, but their very survival. How can we prioritize conservation versus human subsistence, and what mechanisms can be used to reduce conflict between people's essential needs and the needs of nature?
3. Why do you think people trophy hunt? Do you feel that trophy hunting has a net benefit, or net detriment to conservation efforts?
4. Approaches that attempt to reduce illegal wildlife trade include the establishment of government regulations, voluntary international trade agreements, increased policing and enforcement, engaging local communities to become more involved management, and market-based solutions like sustainably harvested certification programs. Which of these approaches do you think works best? Which perform poorly?

5. Is overexploitation a bigger problem on land, freshwater, or in the oceans? Why does overexploitation tend to differ among these habitat types, and how do these differences influence the success of conservation efforts?
6. What are the limitations of quantitative models that help us exploit wild populations more sustainably? What improvements are needed to use these models more safely?

Suggested Readings

Allen, T. et al. 2017. Global hotspots and correlates of emerging zoonotic diseases. *Nature Communications* 8: 1124. Reviews the risk of emerging infectious diseases that arise from human contact with wildlife.

Benitez-Lopez, A. et al. 2017. The impact of hunting on tropical mammal and bird populations. *Science* 356: 180–183. Provides a synthesis of studies that have quantified hunting-induced declines of mammal and bird populations across the tropics.

Biggs, D. et al. 2017. Developing a theory of change for a community-based response to illegal wildlife trade. *Conservation Biology* 31: 5–12. Sets out a conceptual framework to guide efforts to effectively combat illegal wildlife trade through actions at community level.

Cooke, S. J. et al. 2016. Angling for endangered fish: Conservation problem or conservation action? *Fish and Fisheries* 17: 249–265. Presents numerous case studies showing that anglers are effective proponents of fish and habitat conservation, and for endangered species.

Cooney, R. et al. 2017. From poachers to protectors: Engaging local communities in solutions to illegal wildlife trade. *Conservation Letters* 10: 367–374. Lays out a conceptual framework to guide efforts to effectively combat illegal wildlife trade through actions at community level.

Di Minin, E. et al. 2016. Banning trophy hunting will exacerbate biodiversity loss. *Trends in Ecology & Evolution* 31: 99–102. Though controversial, this paper provides an interesting and important counter-point to arguments against trophy hunting.

Jackson, J. B. C. et al. 2001. Historical overfishing and the recent collapse of coastal ecosystems. *Science* 293: 629–638. This classic, highly cited paper reviews select examples of overexploitation leading to population declines of marine organisms.

Naidoo, R. et al. 2011. Effect of biodiversity on economic benefits from communal lands in Namibia. *Journal of Applied Ecology* 48: 310–316. An excellent example of how conservation of biodiversity can provide income for local communities.

Nellemann, C. et al. 2016. *The Rise of Environmental Crime: A Growing Threat to Natural Resources, Peace, Development and Security.* Nairobi, Kenya: United Nations Environment Programme (UNEP). Provides a comprehensive review of the role of criminal syndicates in overexploitation of species.

Ripple, W. J. et al. 2016. Bushmeat hunting and extinction risk to the world's mammals. *Royal Society Open Science* 3: 160498. An outstanding and comprehensive synthesis of the contribution of subsistence hunting to species endangerment.

Visit the
Conservation Biology
Companion Website
oup.com/he/cardinale1e
for videos, exercises, links, and other study resources.

11

Invasive Alien Species

Throughout the history of life, species have spread into new regions of the globe via natural routes of dispersal. While some species have evolved the potential for long-distance migration over short time scales (e.g., migratory birds, planktonic larvae in ocean currents), dispersal for most species is a slow process. In fact, migration rates for many plant species are on the order of just a few meters per year.[1] While occasional long-distance dispersal is possible for all species, even sessile plants,[2] there have historically been very limited possibilities for global transport across entire oceans or continents.

But the historical constraints on dispersal have largely disappeared in the Anthropocene. During any 24-hour period, there is a car, truck, plane, or ship that connects nearly all locations on the planet (**Figure 11.1**). These routes of human transportation are dispersing species faster, farther, and in greater numbers than ever before, obscuring past regional differences in flora and fauna that have persisted for millennia[3–5] (**Video 11.1**). Every region of the world has been affected, even the Antarctic.

One of the consequences of global transportation is an increased probability that pest species will be introduced into new locations where they can harm native biodiversity and the ecosystems that host them. In this chapter, we review the causes and consequences of invasive alien species, as well as the methods to manage and mitigate their impacts.

The signal crayfish (*Pacifastacus leniusculus*) is a North American species that was intentionally introduced to Europe in the 1960s to bolster Scandinavian crayfish fisheries, which were being damaged by a water mold disease. It turns out, the signal crayfish also carries the disease, and it is now an invasive alien species across Europe.

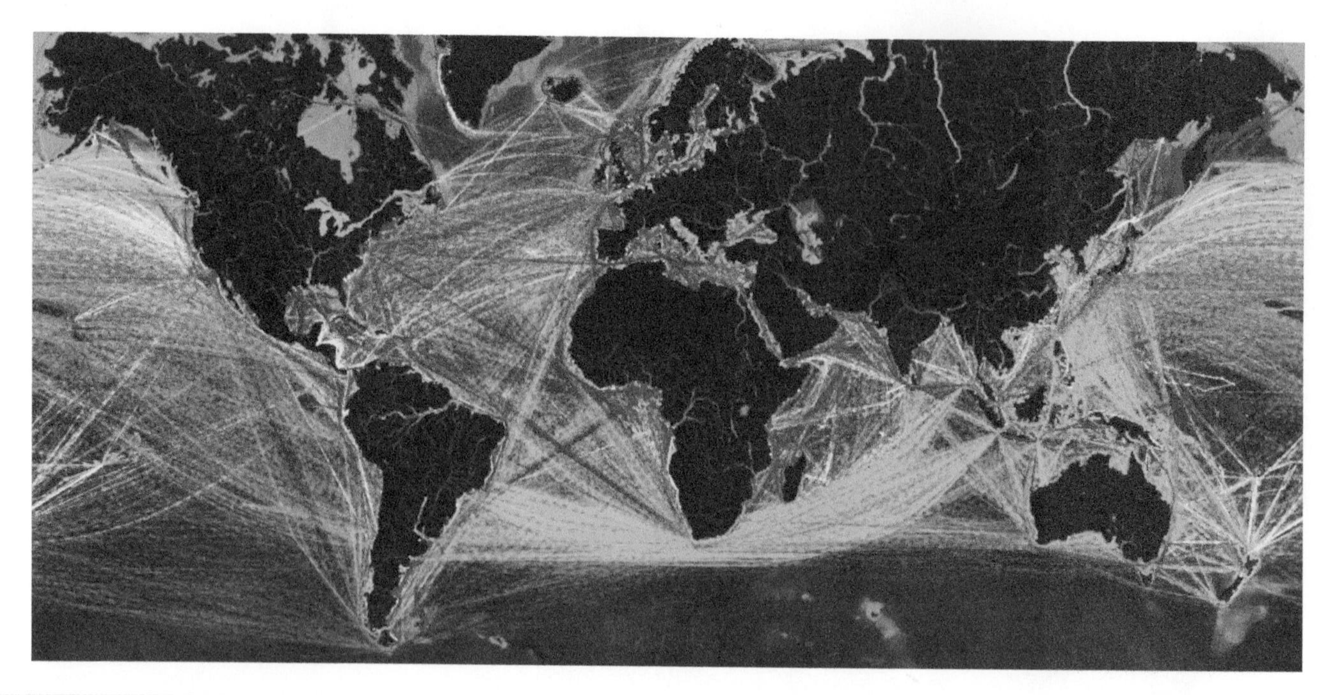

Figure 11.1 This map shows the extraordinary global scale of commercial shipping on a single day, which is but one route of international transportation (colors represent different types of goods). The map demonstrates not only the interconnected nature of today's world, but the potential for species redistribution at a global scale. Color codes: yellow (container ships), blue (dry bulk), red (tanker), green (gas bulk), pink (vehicles). (From www.shipmap.org. Map created by Kiln (www.kiln.digital) based on data from the UCL Energy Institute [UCL EI].)

VIDEO 11.1 Watch the video on global shipping routes to better appreciate the extent of global transport and the potential for invasive species to spread around the world.
oup-arc.com/e/cardinale-v11.1

Overview of the Problem

The literature on invasive alien species is filled with terminology, some of which is used consistently, some of which is not.[6] To be clear about terms used in this chapter, Table 11.1 shows the Convention on Biological Diversity's Glossary of Terms for invasive species, which itself summarizes definitions that are used by major international organizations that work on invasive species and conservation.

A **native species** (or indigenous species) is one that occurs within its natural range (past or present) and that does so in the absence of dispersal that is assisted, either directly or indirectly, by humans. In contrast, an **alien species** (also called exotic, foreign, introduced, nonindigenous, and nonnative species) is one that occurs outside of its natural range (past or present) and beyond its natural dispersal abilities, assisted in some way by human activities. Alien species are distinguished from **invasive alien species (IAS)** based on their impact. *IAS* has become the preferred term to refer to the subset of alien species that become threats to native biodiversity following their introduction. Some organizations extend the definition of *IAS* beyond impacts to biodiversity, and define the term more broadly to include all alien species that have a demonstrable environmental or socioeconomic impact, including impacts on ecological processes and ecosystem services.

The spread of IAS has become a global problem involving nearly every major group of species. Several researchers have recently summarized the scale of IAS distributions, using global and regional databases (Table 11.2).[4,7] For example, Turbelin and colleagues[4] compiled data from the Global Invasive

TABLE 11.1 Convention on Biological Diversity's invasive alien species terminology

Term	Definition	Comment
NATIVE SPECIES Common synonyms: indigenous species	A species, subspecies, or lower taxon occurring within its natural range (past or present) and dispersal potential (i.e. within the range it occupies naturally or could occupy without direct or indirect introduction or care by humans)	UNEP[a] and ICES[b] use nearly identical definitions.
ALIEN SPECIES Common synonyms: exotic, foreign, introduced, nonindigenous, and nonnative species	A species, subspecies, or lower taxon occurring outside of its natural range (past or present) and dispersal potential (i.e. outside the range it occupies naturally or could not occupy without direct or indirect introduction or care by humans) [including] any part, gametes or propagule of such species that might survive and subsequently reproduce	UNEP, CBD,[c] and ICES definitions do not include phrase about dispersal potential. UNEP and ICES definitions do not specify gametes and propagules.
INVASIVE ALIEN SPECIES Common synonyms: alien invasive species	An alien species which becomes established in natural or seminatural ecosystems or habitat, is an agent of change, and threatens native biological diversity	CBD definition does not include phrase about being an agent of change.

Sources: Convention on Biological Diversity's Glossary of Terms for invasive species (www.cbd.int/invasive/terms.shtml); Guidelines for the Prevention of Biodiversity Loss Caused by Alien Invasive Species. 2000. Approved by the IUCN Council February 2000.
[a]UNEP: United Nations Environment Programme World Conservation Monitoring Centre—Glossary of Biodiversity Terms (oup-arc.com/e/cardinale-w11.1).
[b]ICES: International Council for the Exploration of the Sea. Code of Practice on the Introduction and Transfer of Marine Organisms. 2005.
[c]CBD: Convention on Biological Diversity. Decision VI/23 of the Conference of the Parties to the CBD, Annex, footnote to the Introduction.

Species Database of the International Union for Conservation of Nature and Natural Resources (IUCN) and the Centre for Agriculture and Bioscience International's Invasive Species Copendium (CABI ISC). The combined data set spanned 243 countries and overseas territories, with 1517 different IAS species represented (886 terrestrial plants, 222 arthropods, 72 mammals, 66 fishes, 52 aquatic plants, 37 birds, 21 reptiles, 14 amphibians, and 147 other

TABLE 11.2 Examples of databases that track the distribution and spread of invasive alien species

Sponsor/Host	Database	Geographic scope	Taxonomic scope	Link to database	Comment
Centre for Agriculture and Bioscience International (CABI)	Invasive Species Compendium (ISC)	Global	All	oup-arc.com/e/cardinale-w11.2	Peer-reviewed data sheets, images, and maps; a bibliographic database; and full text articles
Invasive Species Specialist Group (ISSG) of the IUCN Species Survival Commission	Global Invasive Species Database (GISD)	Global	All	oup-arc.com/e/cardinale-w11.3	List of "100 of the world's worst" invasive alien species
University of Georgia Center for Invasive Species and Ecosystem Health	Early Detection & Distribution Mapping System (EDDMapS)	United States	All	oup-arc.com/e/cardinale-w11.4	Distribution maps for over 1000 invasive plants in the United States both at county level and point level
European Commission	Delivering Alien Invasive Species Inventories for Europe (DAISIE)	Europe	All	oup-arc.com/e/cardinale-w11.5	Records of >12,000 exotic species found throughout Europe; species searchable by region or name; complete records with detailed maps for widespread invasive species
Dyer et al.[143]	Global Avian Invasion Atlas (GAVIA)	Global	Birds	oup-arc.com/e/cardinale-w11.6	Data repository associated with original paper[143]

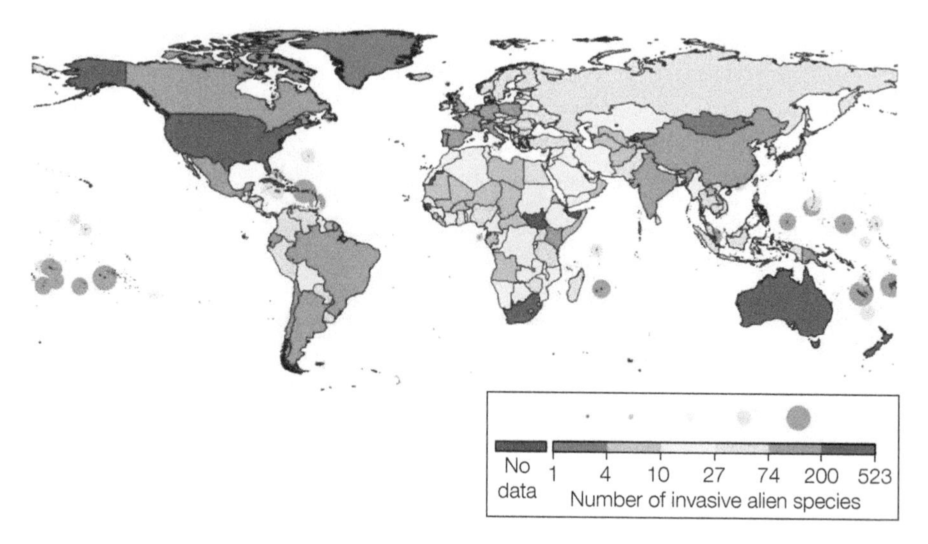

Figure 11.2 A global map of the number of documented invasive alien species (IAS) per country, excluding overseas territories. Note the scale increases logarithmically. To aid visualization of smaller land areas, circles represent island nations with areas < 20,000 km^2. (After A. J. Turbelin et al. 2017. *Global Ecol Biogeogr* 26: 78–92. Data from the Global Invasive Species Database [GISD, 2016], www.issg.org/database [accessed 6 June 2016] and the CABI Invasive Species Compendium [CABI ISC, 2016] http://www.cabi.org/isc/ [accessed 7 January 2016]. CC BY 4.0.)

organisms). **Figure 11.2** shows the number of IAS per country in 2016, which ranged from 1 to 523 species, with a mean of 44. A total of 19 countries have more than 100 IAS. Countries (excluding territories) with the highest numbers of recorded IAS include the United States (523), New Zealand (329), Australia (322), Cuba (318), and South Africa (208). If overseas territories are included, then the top five countries include the United States (1071), France (927), New Zealand (511), Australia (465), and the United Kingdom (463). If countries are standardized by land area (square kilometers), then economically developed countries in the global north (e.g., the United States) continue to rank high in IAS, but they are joined by newly industrialized countries (e.g., China, India, Brazil) and small tropical and subtropical islands (e.g., French Polynesia, Fiji).

To emphasize the sheer number of alien species that are being introduced to new locations and becoming invasive, consider the case of nonnative amphibians and reptiles in the state of Florida in the United States. Florida is somewhat famous for having a large number of IAS. Krysko and colleagues[8] showed that 137 alien species of amphibian and reptile were introduced into Florida between 1863 and 2010. Of these introduced species, 56 had established reproducing populations, including 3 species of frog, 4 species of turtle, 1 crocodilian species, 43 lizard species, and 5 species of snake.

The rate at which new alien species are being introduced to novel ecosystems shows no sign of slowing down. To document the discovery rate of new alien species, Seebens and colleagues[7] compiled data on the 45,813 records documenting the first recorded instances of alien species for 282 regions spanning all continents. Their data included over 16,000 species from well-studied taxa like vascular plants, mammals, insects, birds, and fishes. The spread of alien species, as reflected in the global rate of discovery of all alien species outside their home ranges, remained low between 1500 and 1800, averaging just 7.7 new discoveries per year (**Figure 11.3**). Since 1800, the rate of discovery of alien species outside their home ranges has been increasing exponentially, reaching a maximum of 585 in 1996 (the last year of record in that study), reflecting more than 1.5 new records every

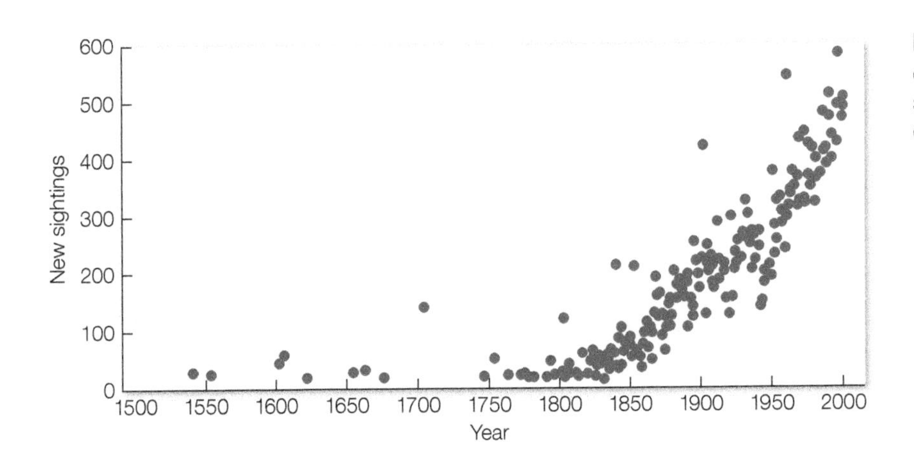

Figure 11.3 Global time trends in the discovery rate (e.g., of first recorded instance) of alien species. (After H. Seebens et al. 2017. *Nat Comm* 8: 14435. CC BY 4.0.)

single day. This increase in the spread of species beyond their native ranges since 1800 is consistent across nearly all taxonomic groups (plants, bryophytes, birds, reptiles, amphibians, insects, mollusks, etc.), with exception of mammals and fishes, whose discovery rates for new alien species have slowed in recent decades.

Suggested Exercise 11.1 Use the advanced search option of the IUCN's Global Invasive Species Database to identify an animal species that has been accidentally released in your country. Once it is identified, read about its distribution, impact, and management options. **oup-arc.com/e/cardinale-ex11.1**

Early and colleagues[5] recently reviewed the risk of further alien species introductions that may occur in the coming decades and concluded that one-sixth of the global land surface of the planet is still highly vulnerable to invasion, including substantial areas within developing economies that harbor biodiversity hotspots. Early and colleagues predict that biological invasion rates will remain high in wealthy countries that are already highly invaded and will increase substantially in developing nations where alien species have been poorly studied, monitored, or recognized. The exponential rise in IAS over the past century and the rather dire warnings for more IAS in the immediate future have led several major international initiatives to prioritize the prevention, eradication, and control of IAS. For example, target 15.8 of Goal 15 of the United Nations' Sustainable Development Goals (SDGs) calls for measures to be implemented by 2020 that prevent the introduction and significantly reduce the impact of IAS on land and water ecosystems and that control or eradicate certain particularly damaging species. Aichi Target 9 of the Convention on Biological Diversity's Strategic Plan for Biodiversity (2011–2020) similarly calls for the prioritization of IAS for prevention, eradication, or control.

The take-home message of this section is that IAS are a global problem spanning all taxonomic groups and geographic locations. They are becoming more prevalent through time, and nearly all governments are recognizing the problems IAS create. Therefore, it is important to understand how they are introduced, become established, and spread.

The Population Biology of IAS

Like all species, alien species must successfully navigate the process of community assembly in order to become established in a local community, where

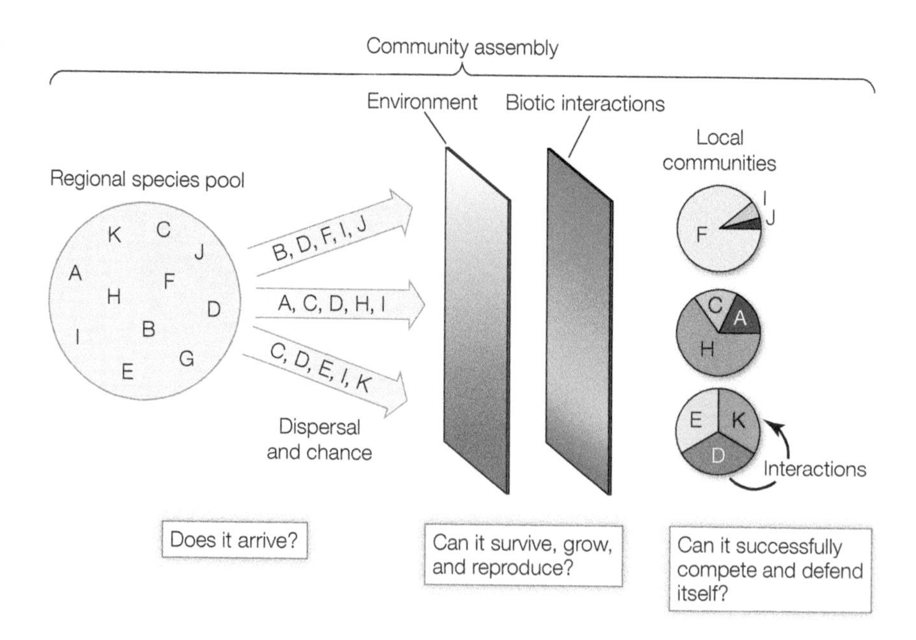

Figure 11.4 Like all species, alien species are subject to the process of community assembly that ultimately determines the membership of local communities. (After J. Hille Ris Lambers et al. 2012. *Ann Rev Ecol Evol Syst* 43: 227–248.)

they may subsequently affect biodiversity, ecological processes, and ecosystem services (**Figure 11.4**). Community assembly involves a suite of processes that operate at a range of spatiotemporal scales. At the largest scales of entire geographic regions, the regional species pool is constrained by historical processes, such as the processes of evolution that generate new species. A subset of the regional species pool, which is influenced both by chance and by limitations to dispersal, becomes available for colonization of a particular site. Thus, the first question is whether a foreign species can overcome chance and **dispersal limitation** to arrive at a new site as an alien colonist.

The subpool of species that arrive as alien colonists to a new site must then pass through two "filters" before they become established and can be considered IAS in a local community. First, they must pass through an environmental (abiotic) filter, which determines whether the species can sustain a population in the new location.[9,10] Successful passage through the environmental filter is a function of the abiotic conditions of the environment—whether the new environment has temperature, soil chemistry, water pH, and so on that are conducive to the survival, growth, and ultimately reproduction of the species. Second, a successful IAS must pass through a biotic filter, proving it is able to compete with native species for limited resources and to successfully defend itself against predators, pathogens, and parasites. Assuming the IAS successfully passes through these two filters, then its introduction may lead to spread and impact.

Conceptualizing invasions through the lens of the community assembly process is useful because it has been used repeatedly to identify factors that constrain alien species and to identify stages at which they might be managed or controlled.[11] We cover these stages next.

Introduction of alien species

There is widespread evidence that the range distributions of many species of herbaceous plants,[12] trees,[13] birds,[14] mammals,[15] insects,[16] microbes,[17] and other taxonomic groups are limited by their dispersal abilities. Therefore, it is not surprising that limitations in species' abilities to disperse and recruit to new sites is the most widely cited constraint on the distribution of alien species.[18–20] A good example of dispersal limitation comes from Capinha and colleagues,[20] who looked at the biogeographic distribution of terrestrial gastropods (e.g., snails and slugs) among 56 different regions of the globe and compared their distributions before and after

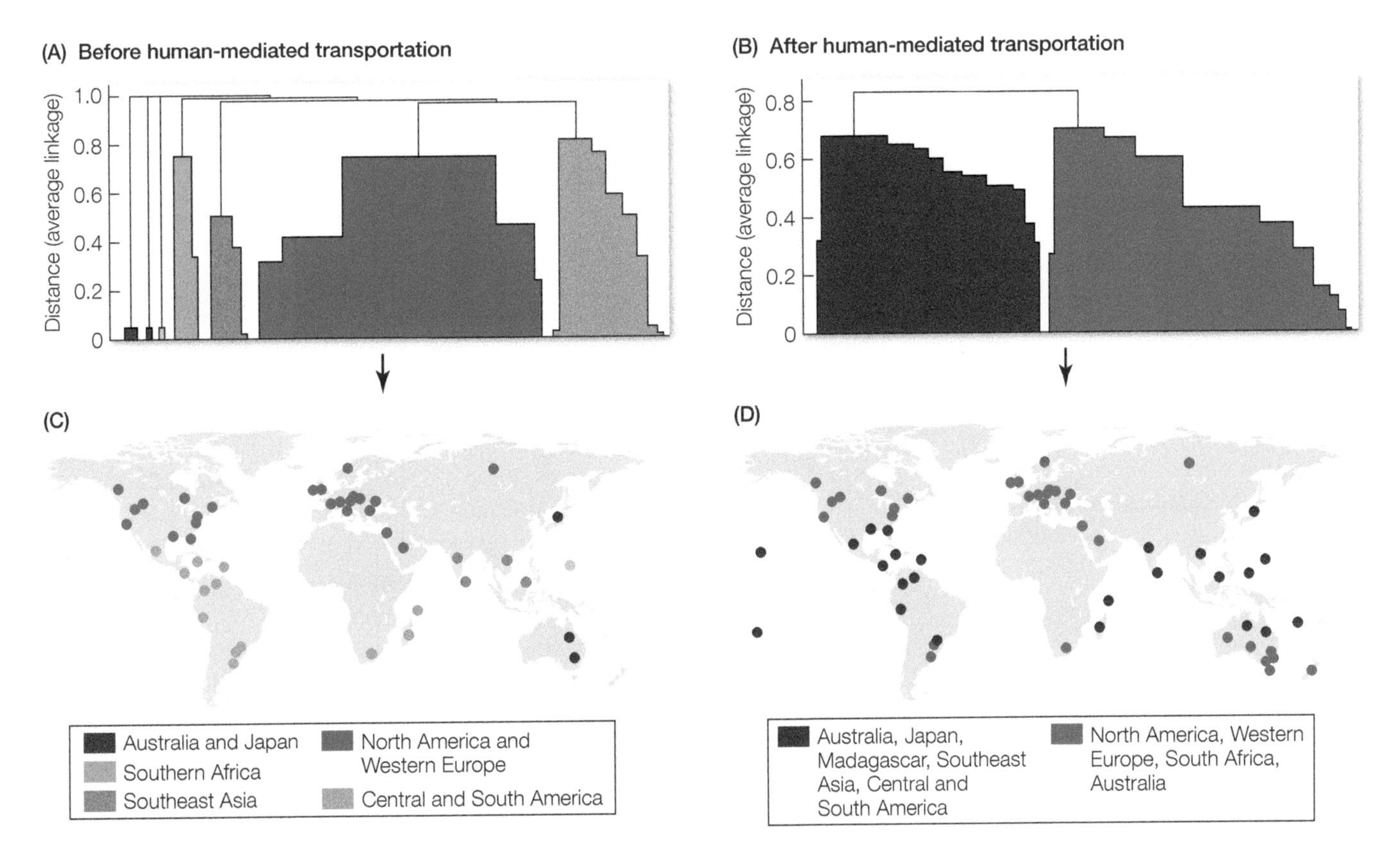

Figure 11.5 Dendrogram and map of the compositional similarities among groups of alien terrestrial gastropods before dispersal by humans (A and C) versus after dispersal by humans (B and D). Note how dispersal by humans has helped gastropods overcome natural dispersal barriers that have historically generated distinct groups of fauna in different geographic locations. (After C. Capinha et al. 2015. *Science* 348: 1248–1251.)

transport by humans. Prior to human transport, the native ranges conformed to well-known biogeographic realms where groups of fauna remained distinct because of natural barriers known to inhibit dispersal (**Figure 11.5**). However, after human transport, gastropod taxa were found to cluster into just two groups. The differences in taxonomic composition between these two groups were partly explained by climate (a temperate vs. tropical group). However, they were also partly explained by commodity trade routes that have allowed species to overcome geographic barriers to dispersal, introducing them into new areas of the globe. Thus, communities of gastropods worldwide have become more similar because the snails and slugs are being anthropogenically introduced to new locations.

Humans help species overcome dispersal limitation by serving as a **transport vector**—either as an unintentional vector (through accidental or inadvertent transport) or through intentional introductions to new locations. Of the 1517 recorded IAS that were reviewed by Turbelin and colleagues,[4] 26% were introduced unintentionally, 39% were introduced intentionally, 22% were spread by both intentional and unintentional introductions, and for 13% there was no information available. Unintentional introductions most often occur when species are transported as "hitchhikers" in human transportation, in planes, trains, boats, cars, etc. International trade is a primary source of introduction of IAS that hitchhike as stowaways or contaminants in goods and packing materials.[21,22] Stowaways in passenger planes are an expanding source of IAS introductions,[23,24] and marine shipping ports are known to be epicenters of invasion.[23,25] Well-known examples of IAS being unintentionally introduced to new ecosystems include the following:

- The three most invasive rat species—black or ship rat (*Rattus rattus*), brown or Norway rat (*R. norvegicus*), and Pacific rat (*R. exulans*)—have spread throughout much of the world as hitchhikers on ships, including to remote islands that they never would have colonized had it not been for human exploration of the world's oceans (**Figure 11.6A**).[26] Invasive alien rats have had dramatic effects on the unique flora and fauna of islands,[26] perhaps most notably by preying on the eggs of a variety of seabird species that use islands as nesting habitat.[27]
- The Eurasian zebra mussel (*Dreissena polymorpha*) is a small freshwater mussel that was originally native to the Black Sea and Caspian Sea in Eurasia. However, the zebra mussel was accidentally introduced to numerous countries in Europe, as well as to the United States and Canada after the planktonic larvae of the species traversed the Atlantic Ocean in the ballast water of ocean-going ships (**Figure 11.6B**; **Video 11.2**).[28] The invasion has been widely publicized and noted for both the speed of the range expansion and the large ecological and economic impacts.[29]
- Giant salvinia, also called kariba weed (*Salvinia molesta*), is a free-floating fern that is native to southeastern Brazil (**Figure 11.6C**). The plant was exported as part of the pet industry to be used in aquaria and garden ponds. From there, it escaped or was deliberately released into the wild. Once in a waterway, the plant spreads rapidly by water movement and transport by contaminated watercraft and wildlife. Giant salvinia was recently voted the world's worst IAS by 650 experts from 63 countries.[30]

VIDEO 11.2 Watch this video about the spread of invasive species.
oup-arc.com/e/cardinale-v11.2

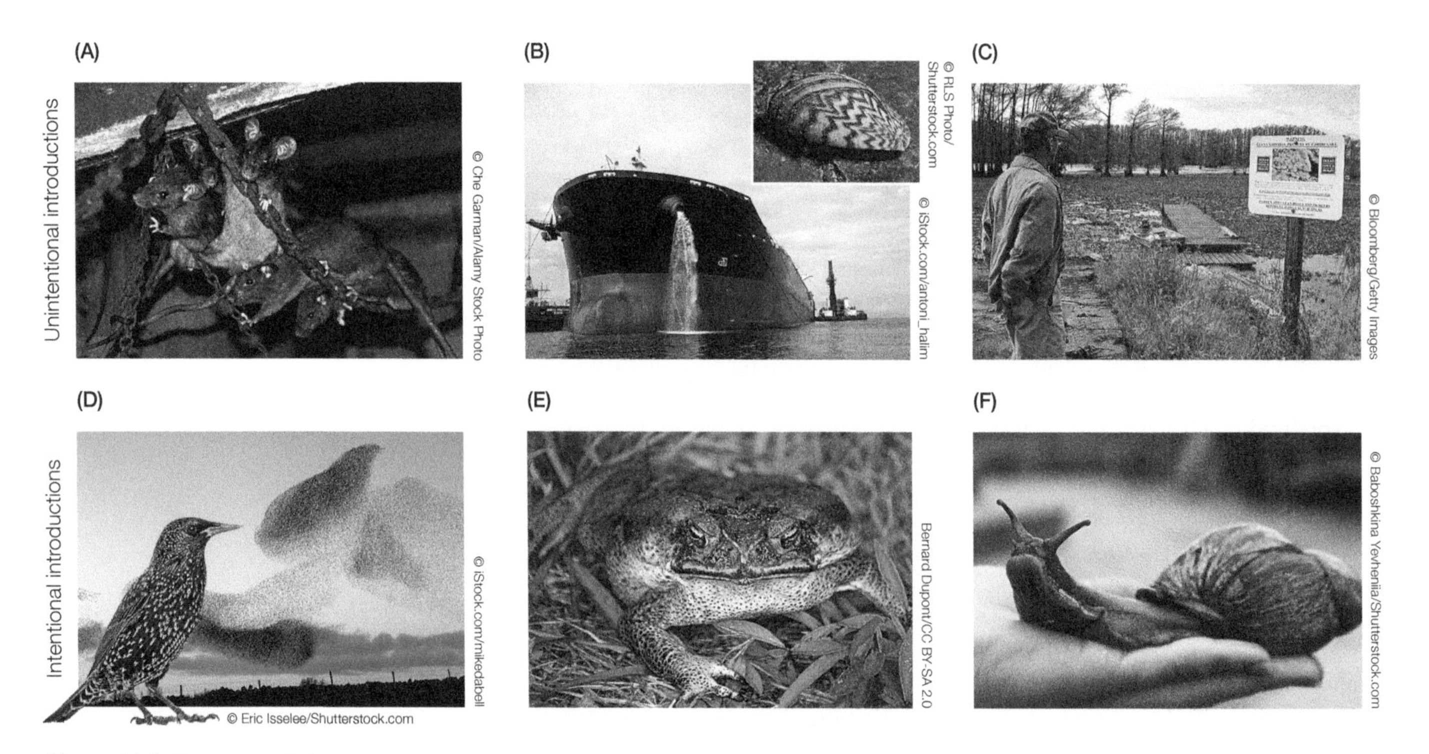

Figure 11.6 Examples of alien invasive species that were unintentionally (A–C) and intentionally (D–F) introduced into novel ecosystems. (A) Rats (*Rattus* spp.) spread throughout much of the world as hitchhikers on ships; (B) the Eurasian zebra mussel (*Dreissena polymorpha*) was introduced to North America in the ballast water of ships; (C) giant salvinia, a free-floating fern dubbed the world's worst IAS, was introduced by the aquarium trade; (D) European starlings (*Sturnus vulgaris*) were introduced to New York City's Central Park in the 1890s by a group who wanted America to have all the birds that Shakespeare ever mentioned in his works; (E) cane toads (*Rhinella marina*) were introduced to Australia to control insect pests of sugarcane crops; (F) the giant African snail (*Lissachatina fulica*), introduced to much of the world through the pet trade and as a food resource, serves as an intermediate host to a nematode that causes meningoencephalitis in humans.

In contrast to unintentional introductions, many species are intentionally transported and introduced to new ecosystems by humans. In some cases, the purpose of the introduction is to establish new wild populations, in which case it is clear that human desires have motivated the introduction. In other cases, species that are intentionally transported into a new region as captive or cultivated populations escape and establish breeding populations. Escaped organisms are included in this category because their initial transport is human motivated.

The dominant pathway for intentional species introductions is horticulture and the nursery trade, which is responsible for about 31% of species that have been introduced outside of their natural geographical ranges.[4] The pet and plant trades are also major sources of animal and plant introductions due to the frequent escape or release of imported species into the wild.[31] The pet and plant trades are also a primary mechanism for the introduction of nonnative insect pests and pathogens.[32] Some well-known examples of IAS that were intentionally introduced by humans include the following:

- European starlings (*Sturnus vulgaris*) are one of the most abundant birds in North America (**Figure 11.6D**). European starlings, which are native to Europe, became established in North America after 100 birds were intentionally set loose in Central Park, New York in the early 1890s. The birds were released by Shakespeare enthusiasts who wanted America to have all the birds that Shakespeare had mentioned in his works. After several tries, the population exploded,[33] and today there are more than 200 million European starlings from Alaska to Mexico.[34]
- Cane toads (*Rhinella marina*) were introduced to Australia from Hawaii in 1935 by the Bureau of Sugar Experiment Stations to control insect pests of the economically valuable sugarcane crop. After release, the toads reproduced rapidly, increasing in population size to well over 200 million individuals, spreading throughout Queensland and ultimately reaching New South Wales and the Northern Territory (**Figure 11.6E**).[35] The unanticipated invasive wave had numerous ecological effects, including the loss of native species that the toads eat and the loss of native predators that die from eating cane toads, which have venom-secreting glands that release bufotoxin when threatened.[36] There is no evidence that introduction of the toads had any impact on the pests they were introduced to control.
- The giant African snail (*Lissachatina fulica*) is native to East Africa, but it has been widely introduced to other parts of the world as a food item and via the pet trade (**Figure 11.6F**). The snail, which thrives in warm, moist climates, has spread to many parts of Africa, China, India, Southeast Asia, the West Indies, the Caribbean, Brazil, and the United States.[35] Introduction of the snail to Florida in the United States traces back to a Miami-based religious leader named Charles Stewart, who used the snail in rituals associated with his practice of the traditional African religion of Ifa Orisha. According to an interview conducted by the *Miami Herald* back in 2010,[37] Stewart would hold the snail over a devotee and then cut the live snail and pour its raw fluid into the mouth of the devotee. The ritual was supposed to cure worshippers of medical problems, but they instead went to authorities complaining it made them violently ill. In response, Stewart's house was raided in January 2010, where authorities found at least 20 of the giant snails in a wooden box kept in his backyard. At the time there was no sign that the snails had gotten loose, but that changed a year later when a family 8 miles from Stewart's home reported that their backyard had become infested. The species is now considered to be one of the top 100 problematic IAS in the world because

of its damage to agricultural crops and native plants. The snail species also serves as an intermediate host of *Angiostrongylus cantonensis*, a nematode worm that can cause deadly meningoencephalitis in people.

Establishment of alien species

The vast majority of alien species die in transport before they are ever introduced into a new ecosystem.[38] Of the individuals that are introduced to a new ecosystem alive, most do not become established in that location.[39] Consider, for example, introduction of the European red deer (*Cervus elaphus*) to New Zealand, which failed to establish a population after 31 intentional introductions were made to establish game for hunting. But on the thirty-second try, the species established a rapidly growing population that spread over the entire South Island, ultimately becoming a serious economic problem.[40] The scenario with the red deer is not uncommon. The percentage of colonists of alien species that survive when introduced to a new location has been reported to be in the range of less than 1% to 24%, with a mean of perhaps 10%.[38] These numbers are probably substantial overestimates, since failed colonization is difficult to monitor, and studies of failed colonization are rare. Nevertheless, they suggest that the vast majority of new colonists of alien species tend to die before they can establish a population. This is why many studies show that the probability of successful establishment of an alien species in a new site is proportional to the number of times the species is introduced[41] and proportional to the **propagule pressure** (the number of colonists released) at each introduction.[42]

So why do the vast majority of alien species fail to establish? As discussed in Chapter 8 (and later in Chapter 13), demographic and environmental stochasticity play a large role in the dynamics of small populations because random variation in the biological parameters that influence population growth increases the probability of extinction. Invasive species are small populations during their initial phases of colonization and establishment. As such, they are subject to all of the same constraints as small populations of native species.[39]

Oftentimes, the explanation is that the location's abiotic characteristics are unsuitable to the organism's physiological needs for survival, growth, and reproduction (the environmental filter, see Figure 11.4). Increasing the number of introductions and/or the number of individuals introduced per event increases the chance that an individual with a suitable genotype and physiology will be able to tolerate the abiotic conditions of the site.

Other times, the failure of alien species to establish is explained by biotic defenses whereby interactions with native species that comprise the resident community preclude survival and growth (the biotic filter, see Figure 11.4). Four hypotheses have been used to explain how biotic interactions can limit the establishment of alien species:

1. The **empty niche hypothesis**[43,44] proposes that the primary way alien species become established in a local community is by exploiting a vacant niche—either using biological resources in a new way that allows them to avoid competition with native species, or by performing a completely new "job" that was previously unfilled by a native species (**Figure 11.7A**). If, for example, an alien species can use a resource that is spatially or temporally distinct from the resources used by native species, then the alien could potentially avoid competition for resources, which would allow it to survive and grow. Levine and colleagues[45] found support for this hypothesis in a **meta-analysis** (i.e., a synthesis of studies that are collated and subjected to collective statistical analysis) of 65 experiments that have quantified the strength of competition between native and alien plant species. The authors showed that competition

with native plants significantly reduced the establishment and performance (biomass, growth, or fecundity) of alien species. In many cases, the reductions were substantial. For example, one of the reviewed studies showed that an average of 455 seedlings of an alien perennial grass (*Anthoxanthum odoratum*) could establish and grow in California grasslands where native perennials were removed with an herbicide. But when the natives were allowed to grow, the alien species could germinate just 5.9 seedlings on average (a reduction of nearly 99%). Such results support the notion that competition is a major contributor to biotic resistance against alien species.

2. The **novel weapons hypothesis**[46] proposes that alien species gain advantage over native plants by expressing allelopathic chemicals *that kill resident plants* (**Figure 11.7B**). Thorpe and colleagues[47] studied *Centaurea maculosa*—a forb of the family Asteraceae that is native to Eurasia and that creates near monocultures in many parts of its invaded range in western North America. The plant produces the root exudate (±)-catechin, which reduces the germination, growth, and survival of native plants. Interestingly, production of (±)-catechin

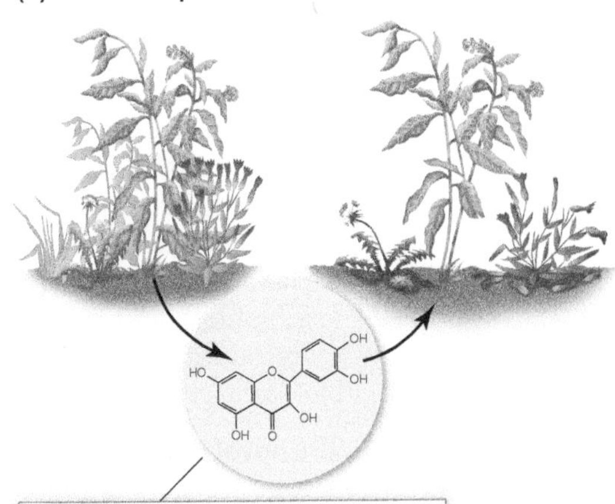

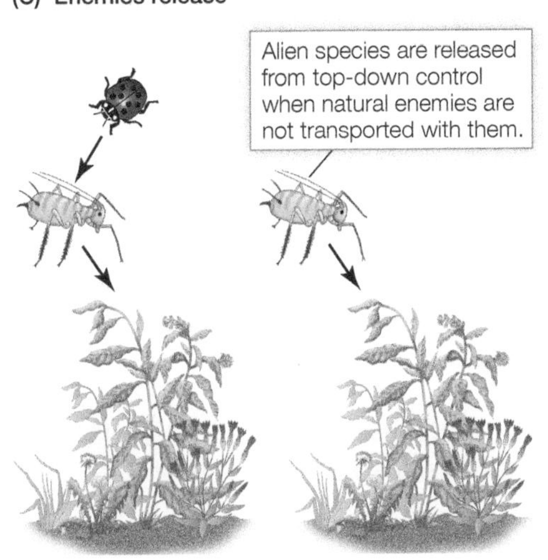

Figure 11.7 Four hypotheses about how biological interactions influence the successful establishment of alien species.

is low in the species' native range but tends to be induced upon introduction to a new habitat. Some data suggest that formation of novel chemicals is a common feature of successful plant invasions. Ni and colleagues[48] found that allelopathic chemicals have contributed to the successful establishment of 25 out of 33 (76%) of the most noxious invasive plants in China. The effects seem to be common in certain groups of plants, such as the Asteraceae.

3. The **enemies release hypothesis**[49] proposes that while populations of alien species tend to be controlled by natural enemies, such as pathogens, predators, and parasitoids, in their native ranges (**Figure 11.7C**), when a small number of individuals of an alien species are introduced to a new location, their natural enemies may not be transported with them. When released from top-down control by natural enemies, the alien species can then establish in the new location. In support of this hypothesis, Torchin and colleagues[50] compared the parasite loads of 26 species of mollusks, crustaceans, fishes, birds, and mammals between their native and invasive ranges. While species averaged 16 types of parasites in their native ranges, only 3 of these were transported with the alien species to their introduced ranges. Also, alien populations were less heavily parasitized, with the frequency of infections and the infection loads per individual significantly reduced. One of the studies reviewed provided a good example of the consequences of parasites being lost.[51] In their native range, European shore crabs (*Carcinus maenas*) have a high prevalence of parasitic castrators (parasites that remove reproductive organs) that reduce fitness and fecundity. But introduced populations of *C. maenas* were not infected with these parasites and had significantly higher fitness than European populations. Enemies release is also a common explanation for the successful establishment of many agricultural pests (e.g., aphids), which is why translocation of natural enemies from native to introduced habitats is one of the most common approaches for the biological control of invasive species, particularly pests of crops.[52,53]

4. The **novel environments hypothesis**[39] proposes that successful invasive species are particularly well adapted for, and successful in, habitats that have been directly modified or disturbed by human activities (**Figure 11.7D**). Plant IAS in many geographic regions are composed primarily of "weeds" that occur in agricultural fields, along roadsides, and around human settlements. Many of the most successful animal IAS have commensal relationships with humans; the housefly (*Musca domestica*), common cockroach (*Periplaneta Americana*), house sparrow (*Passer domesticus*), and house mouse (*Mus musculus*) have all tracked human settlers throughout the world. The novel environments hypothesis differs from others in that it suggests successful invasions have little to do with the resident community of native species, and more to do with a species history of association with humans and human-modified ecosystems. As the world's ecosystems are increasingly modified by humans through the effects of climate change, pollution, logging, agriculture, fishing, and urbanization, native biological communities will become more vulnerable to invasive species.

While none of these four hypotheses apply to all alien species, it is clear that each of them helps explain the successful establishment of at least some alien species. But once established, an alien species must spread and proliferate before it is considered an "invasive" alien species.

Spread of invasive alien species

Once initial colonization and establishment have occurred, the proliferation and spread of most alien invaders follows a similar sequence: (1) there is a lag phase characterized by slow population growth and a low rate of spread, (2) an expansion phase is characterized by exponential population growth and a rapidly expanding invasion front, and (3) the IAS moves to a saturation phase (at carrying capacity), and population growth and the rate of spread both reach plateaus.[54] The lag phase between initial establishment of an alien species and the start of expansion can be quite substantial.[55] For example, a summary of 184 woody plant species introduced to Brandenburg, Germany, showed that the average time between first introduction and the appearance of unplanted seedlings was 147 years.[56] While lags are rarely this extreme, long initial periods of inactivity followed by seemingly sudden changes in invader dynamics are common.[57]

There are several mechanisms that can account for prolonged lag phase. First, by definition, exponential growth is initially slow when densities are low and then accelerates as a population gets larger. There is, therefore, an inherent period of initially slow growth for any biological population. However, the length of slow growth can be exacerbated by several other mechanisms. For example, an Allee effect is a phenomenon whereby the fitness of individuals and growth of a population depend on the population's size or density. Allee effects can occur when animals are unable to find mates at low population densities, or plants are unable to attract pollinators, both of which can inhibit population growth. Alien species may also require local genetic adaptation to environmental conditions before they are able to proliferate, and the rate of adaptation is slow in a small population. Propagules of alien species can also live in a period of dormancy or stasis waiting for a shift in abiotic conditions that favor growth. If the cause of the lag is known, then predicting the time available to establish an eradication and containment plan can be quite helpful to IAS management, as will be discussed later in this chapter.[57]

After the lag phase, a small number of alien species enter a period of rapid growth and spatial expansion to new locations. During this period, there are few constraints on population growth, which is near its maximum exponential potential. The spread of the population to new locations is primarily constrained by (1) the dispersal distance the species can travel, which is usually described by a dispersal kernel that gives the probability distribution of the distance traveled by any single individual; (2) additional human-assisted transport, either accidental or deliberate, that augments the dispersal kernel; and (3) establishment of colonists in new locations, which must pass through the environmental and biotic filters (see Figure 11.4).

Suggested Exercise 11.2 Read and discuss this article about how rearing of invasive alien species for business sales contributes to their spread. **oup-arc.com/e/cardinale-ex11.2**

After the IAS colonizes all potential sites or becomes subject to some type of biological control, then the expansion phase transitions into a saturation phase where growth and spread slow or stop. One of the primary goals for conservation biologists is to predict the rate of spread of an IAS, as well as the species distribution once saturation is reached. One of the key tools we use for making these predictions (reaction-diffusion models) is described in **Box 11.1**, which also gives students the opportunity to practice this tool.

BOX 11.1 Conservation in Practice

Predicting an Invasive Alien Species Wave Front

The most common type of model used to describe the expansion and saturation phase of invasive population dynamics is called a reaction-diffusion model. Reaction-diffusion models have long been used to describe the speed at which waves propagate across the ocean.[154] They were adapted by R. A. Fisher in the 1930s to quantify the speed at which advantageous genes spread through a population.[155] The models were subsequently adapted by Skellam,[156] who used Fisher's equation to model the spread of the invasive alien muskrat (*Ondatra zibethica*) throughout central Europe.

The basic form of a reaction-diffusion model (Fisher's equation) that is used to describe the spread of an invasive species is:[157]

$$\frac{\delta N}{\delta t} = rN\left(1 - \frac{N}{K}\right) + D\frac{\delta^2 N}{\delta x^2}$$

This equation should look familiar. The first part (highlighted in yellow) is simply the equation for logistic population growth, where N is the number of individuals of the population, t is time, r is the intrinsic rate of population growth, and K is the carrying capacity. This part of the equation predicts that a population will initially grow at an exponential rate until the density of individuals N approaches the carrying capacity K, at which time the population saturates and stops growing. The second term in Fisher's equation (highlighted in blue) is new and represents the rate at which individuals in the population move in a single dimension, x. The term D, which stands for diffusivity, is a constant that describes the rate at which the IAS spreads to new locations along the x-coordinate. The term after the D, or $\frac{\delta^2 N}{\delta x^2}$, is a density gradient that represents how the density of the population changes with respect to x. Note that while growth of the IAS is logistic, diffusion (or spread) is directly proportional to the density gradient.

If we want to describe the movement of an IAS along two spatial dimensions (x and y), such as latitude and longitude coordinates, then a simple modification of the equation above is:[158]

$$\frac{\delta N}{\delta t} = rN\left(1 - \frac{N}{K}\right) + D\left(\frac{\delta^2 N}{\delta x^2} + \frac{\delta^2 N}{\delta y^2}\right)$$

This equation simply says that the area occupied by an IAS spreading over a two-dimensional geography will increase as the square of time.

Both equations are generalizable models that can be used to predict the rate of spread of IAS to new locations, as well as the population density of invaders at any location and time (see Figure).

The model has successfully been used to predict the spread of numerous IAS, including Argentine ants (*Linepithema humile*) from South America into the United States;[159] the coypu (the nutria, *Myocastor coypus*) throughout Great Britain;[158] the noxious shrub *Cytisus scoparius* throughout Chile, Australia, and New Zealand;[157] and damaging insect pests of forests.[160] It can easily be modified to account for more realistic biological scenarios, such as populations of alien species that exhibit size or age structure, that experience demographic or environmental stochasticity or that have varying dispersal distances.[157]

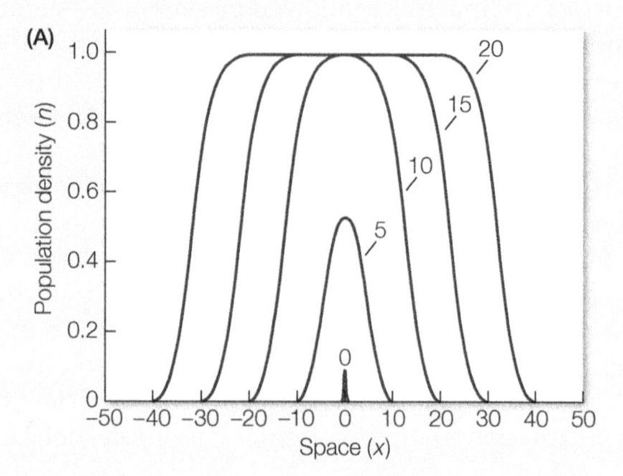

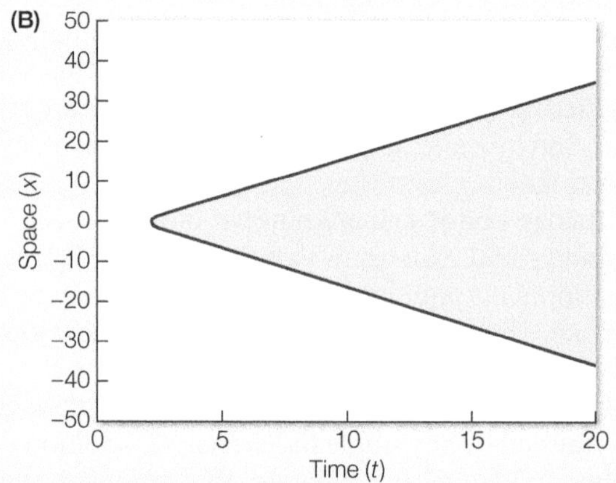

Fisher's equation predicts (A) how the population density of an IAS spreads across a spatial gradient (x-axis) and how the population grows through time at each location (y-axis; curves for t = 0 to 20). The model also predicts (B) the time it takes (x-axis) for an invasive wave front to reach each location (y-axis; gray shows wave front). Fisher's equation was simulated with $r = D = 1$. (After M. G. Neubert and I. M. Parker. 2004. *Risk Anal* 24: 817–831.)

Suggested Exercise 11.3 Model the range expansion of an invasive species, and explore how population growth and emigration rates influence the speed of an invasive wave front.
oup-arc.com/e/cardinale-ex11.3

Impacts of Invasive Alien Species

The population and community-level impacts of IAS are well known, and their impacts on whole ecosystems are becoming clearer. A growing number of studies have estimated the economic costs of IAS as well. Below, we review the impacts of IAS on biodiversity, ecosystems, and economies, including their potential positive values.

Impacts on biodiversity

Ecologists, conservation biologists, and natural resource managers have long believed that invasions by nonnative species are a leading cause of modern species extinctions. This belief is exemplified in a 1998 paper led by David Wilcove—an ecologist with the Environmental Defense Fund at the time, and now a professor at Princeton. In their paper, Wilcove and his colleagues tallied the presumed causes of extinction for many different groups of organisms (e.g., plants, mammals, birds). The tallies were based primarily on qualitative sources of data, such as interviews with taxonomic specialists or from databases that allow threatened species to be listed based on expert opinion. Wilcove and colleagues concluded that, across all taxa, IAS were responsible for 49% of imperiled species, second only to habitat loss and degradation. Estimates were highest for birds (69%) and plants (57%) and lowest for select groups of invertebrates (none to 27%).

The Wilcove et al. paper[58] has become one of the most widely cited sources in the literature for the belief that invasive species are a direct and leading cause of extinction. The views of expert biologists that were tallied in that paper have been supported by certain well-known case studies that have shown how waves of an IAS can drive species extinctions. Perhaps the most famous case study is that of the brown tree snake (*Boiga irregularis*), which is an arboreal (i.e., tree-dwelling) species of colubrid snake that is native to Australia, Indonesia, Papua New Guinea, and a number of islands in Melanesia. Shortly after World War II, the brown tree snake was accidentally transported from its native range in the South Pacific to the snake-free island of Guam.[59] It probably made it to Guam as a stowaway in the cargo of a ship, or possibly on the landing gear of a Guam-bound aircraft. After introduction to Guam, the snake had a lengthy time lag, remaining at low densities for more than a decade. But then, perhaps because it had reached a critical density that allowed it to reproduce more successfully or because it had adapted genetically to the site, the snake began to proliferate rapidly, spreading from its original site of introduction to other sites around the island, where it started to reach high population densities of up to 5000 individuals per square kilometer.[60]

The brown tree snake is a voracious predator of arboreal fauna, including birds, their nestlings, and their eggs. As the snake spread around Guam, the 10 species of forest birds that were known to be native to Guam began disappearing. All of them followed a similar pattern, with population losses spatially tracking the invasive wave of snakes (**Figure 11.8**). Birds first disappeared from southern Guam by the late 1960s. This was followed by progressive loss as the snake spread north through the 1970s and 1980s. By the early 1980s, the only location that still harbored populations of all 10 bird species was a small forest patch on the northern tip of the island, but by 1986 they had disappeared from this area as well.

In total, the brown tree snake is believed to have caused 12 bird species to go extinct (including 3 endemic species) and to have caused the populations of 8 others to decline by more than 90%.[61] In addition to the birds, the brown tree snake consumed large numbers of native reptiles (e.g., anoles, geckos, skinks) and mammals (e.g., rats and bats), radically restructuring the food web on

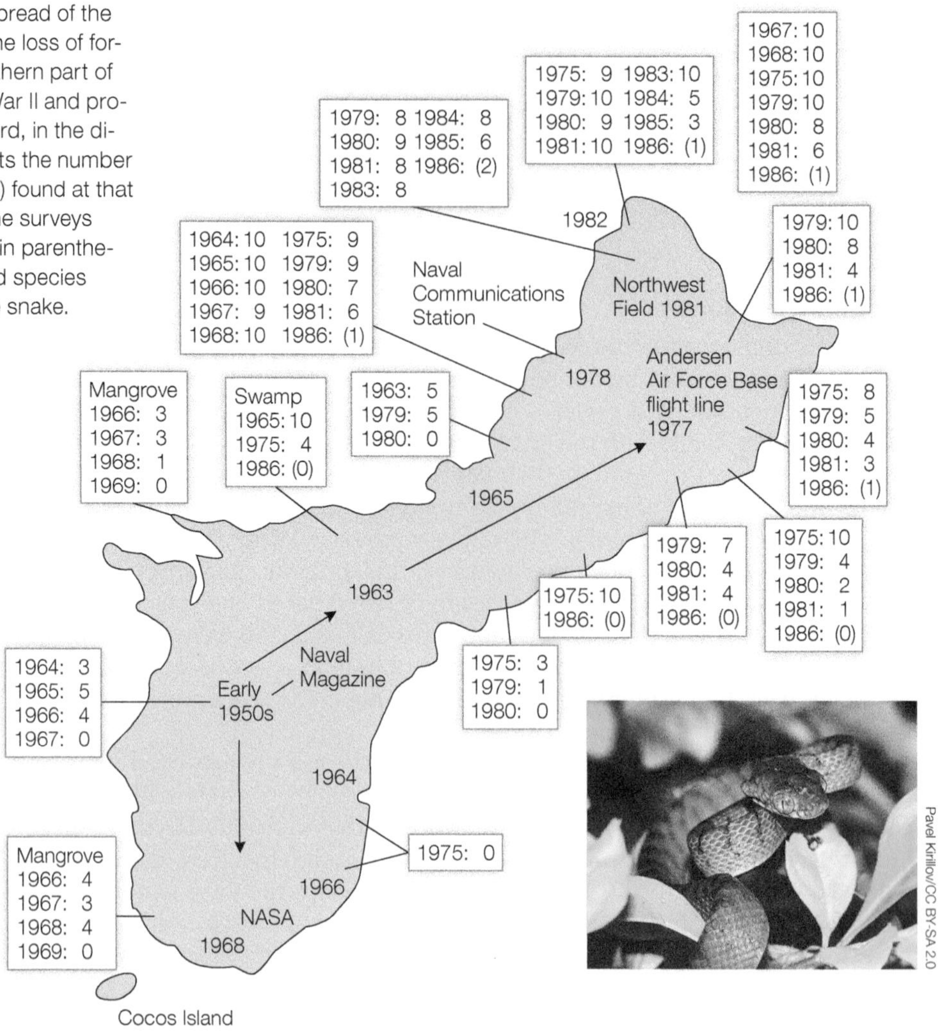

Figure 11.8 Map of Guam showing the spread of the brown tree snake (*Boiga irregularis*) and the loss of forest bird species. *Boiga* colonized the southern part of the island in the early 1950s after World War II and progressively spread northward and southward, in the direction shown by the arrows. Each box lists the number of forest bird species (out of a possible 10) found at that location in the year indicated. For 1986, the surveys were incomplete, and those numbers are in parentheses. Note the decline in the number of bird species that corresponds to the spread of the tree snake. (After J. A. Savidge. 1987. *Ecology* 68: 660–668.)

Guam.[60] The brown tree snake is but one example of an IAS that radically altered the biological composition of native communities. Other well-known examples include the introduced Nile perch (*Lates niloticus*) and their impacts on native cichlids of Lake Victoria,[62,63] zebra mussels (*Dreissena polymorpha*) and their impacts on native mussels throughout the Mississippi River basin,[64,65] and ice-plant (*Mesembryanthemum crystallinum*) and cheat grass (*Bromus tectorum*) in the western United States whose invasions have led to monocultures in many locations.[39]

While case studies like these have helped solidify the idea that IAS are a major driver of species extinctions, this view began to be challenged around the turn of the twenty-first century. Two of the first ecologists to openly question this view were Jessica Gurevitch and Dianna Padilla, who published a paper in 2004 titled "Are Invasive Species a Major Cause of Extinctions?"[66] The first sentence of their abstract reads, "The link between species invasions and the extinction of natives is widely accepted by scientists as well as conservationists, but available data supporting invasion as a cause of extinctions are, in many cases, anecdotal, speculative and based upon limited observation."

After reanalyzing data that were presented in Wilcove et al.[58] and extending analyses further using threats to species reported in the IUCN Red List, Gurevitch and Padilla pointed out that few examples of extinction could unambiguously be linked to IAS. In most records, multiple threats to species were reported, making it hard to determine whether IAS were primary drivers or just minor contributors to the fall of already imperiled species. The authors further pointed out that in most studies, including many "high-profile" case studies of IAS, the evidence linking IAS to extinction was circumstantial or correlative. In no way did Gurevitch and Padilla suggest that IAS are not a contributor to extinction. Rather, the point of their paper was that researchers needed to take a more rigorous, evidence-based approach to understanding the role that invasive species play in extinctions.

Gurevitch and Padilla's paper not only spawned debate,[67] but also prompted a far more critical inquiry into the role of IAS in biodiversity loss. In one line of inquiry, researchers began to amass large data sets comparing various measures of biodiversity (e.g., species richness or evenness) between sites dominated by IAS and nearby reference sites that had yet to be invaded. In addition, they collated a growing number of experimental studies in which researchers had directly manipulated the presence or absence of an alien species in a resident community by, for example, removing alien species from field plots or by adding them to contained greenhouses. Several formal syntheses have recently collated and analyzed the results of these studies:

- Murphy and Romanuk[68] summarized results of 116 observational studies where researchers compared species richness between an invaded site and an uninvaded reference site. In addition, they reviewed 16 controlled experimental studies that manipulated the presence/absence of an IAS. The authors showed that IAS significantly decrease native species richness by an average of 24% and went on to show that the impacts of IAS on biodiversity ranked second only to the impacts of various types of land-use change.
- Vila and colleagues[69] performed a formal meta-analysis of 1041 observational or experimental field studies that described the impacts of 135 different plant IAS on resident species of plants and animals. IAS reduced the growth of resident plant species by 22%, decreased population abundance by 44%, and lowered species diversity (richness or evenness) by 51%. Plant IAS did not significantly decrease animal species richness but did reduce the population abundance of animal species by 17%, on average.
- Doherty and colleagues[70] quantified the number of bird, mammal, and reptile species that are known to be threatened by invasive alien mammalian predators or thought to have gone extinct (since 1500 CE) because of them. In total, 596 threatened and 142 extinct species (total 738) have suffered negative impacts from 30 species of invasive mammalian predators (**Figure 11.9**). Invasive mammalian predators were implicated as a causal

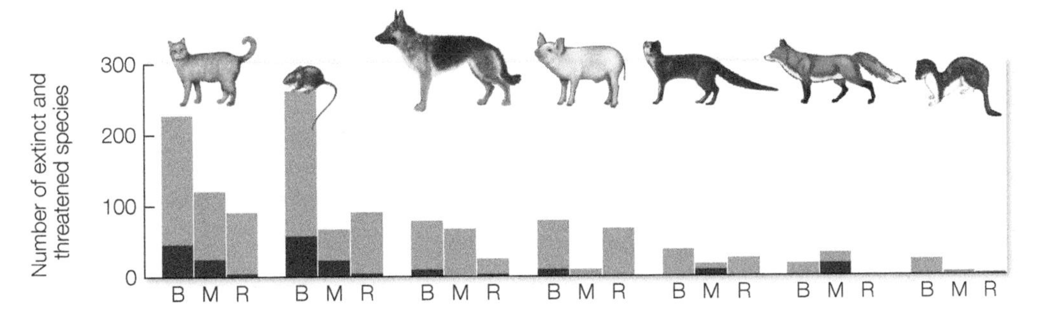

Figure 11.9 This figure shows the number of threatened and extinct bird (B), mammal (M), and reptile (R) species that have been negatively affected by invasive mammalian predators. The gray bars give the total number of extinct and threatened species, while the red bars show number of extinct species (including those that are extinct in the wild). Predators (left to right) are cats, rodents, dogs, pigs, mongoose, red fox, and stoat. (After T. S. Doherty et al. 2016. *Proc Natl Acad Sci USA* 113: 11261–11265.)

factor in the extinction of 87 bird, 45 mammal, and 10 reptile species, which equates to 58% of these groups' contemporary extinctions worldwide. Invasive rodents are linked to the extinction of 75 species (52 bird, 21 mammal, and 2 reptile species; 30% of all extinctions), and cats are linked to 63 extinctions (40, 21, and 2 species, respectively; 26%), whereas red foxes, dogs (*Canis familiaris*), pigs (*Sus scrofa*), and the small Indian mongoose (*Herpestes auropunctatus*) are implicated in 9 to 11 extinctions each. Invasive mammalian predators currently threaten an additional 596 species that are classed by the IUCN Red List as "Vulnerable" (217 species), "Endangered" (223), or "Critically Endangered" (156).

Clearly, recent data syntheses of experimental and observational studies confirm that IAS are a major driver of extinctions of native flora and fauna across a range of ecosystems and that the number of extinctions is substantial. But interestingly, a separate group of data syntheses have reported another trend that influences how IAS affect biodiversity. Numerous studies have now shown that the rates at which alien species are introduced to ecosystems sometimes outpace the rates at which those aliens cause the extinction of native species. In turn, alien invasions can actually cause total biodiversity in some locations to increase rather than decrease. Among the first to note this trend were Sax and colleagues,[71] who tallied the number of native plant and bird species known to occur on remote oceanic islands, as well as the number of alien species that had been introduced to, and become naturalized on, the islands (Table 11.3). The authors showed that while extinctions of native plants on the islands have occurred, the loss of native species has been exceeded by new introductions by a margin of about 2:1, leading to a near doubling in species richness for plants. For birds, each extinction was matched by roughly one new introduced species, causing species richness of birds to remain relatively constant. The authors subsequently showed that the same pattern tends to occur on mainlands.[72]

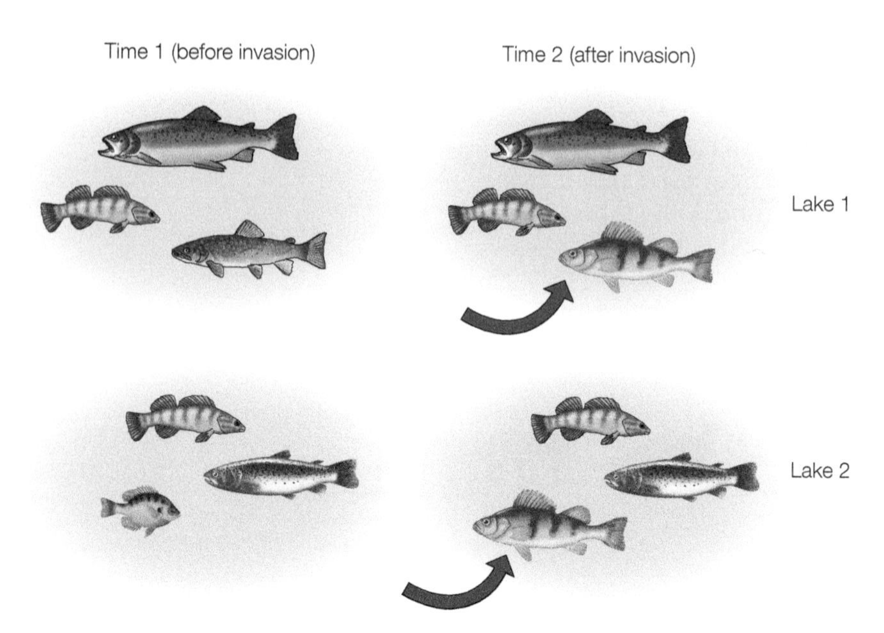

Figure 11.10 An example of how invasive species can cause biotic homogenization. At time 1, only one of three species is shared between the two lakes, giving a compositional similarity of 33%. At time 2, the same IAS has been introduced to both lakes and has caused extinction of one fish species. Two species are now shared among lakes (one native, one IAS), resulting in a compositional similarity of 66%. Thus, invasion has caused the lakes to become less diverse (more similar), even though species richness has remained unchanged.

Data syntheses have also revealed yet another influence of IAS on biodiversity—a reduction in beta-diversity. As discussed in Chapter 3, beta-diversity is a measure of the turnover of ecological communities in space or time, which determines how unique species assemblages are, among any two locations or between two points in time. A growing number of studies have shown that IAS are decreasing beta-diversity by causing communities to have more similar species composition than they have had historically. This process of **biotic homogenization**[73,74] is illustrated in Figure 11.10, which shows how invasion of two lakes by an alien species can cause the two lakes to become more similar (less distinct) in their species composition.

Biotic homogenization often leads to a relatively small number of "winners," usually generalist IAS that are highly

TABLE 11.3 Number of native species (extant and extinct) of vascular plants on oceanic islands, as well as the number of alien species that have become naturalized

Island or archipelago	Class	Native species (extant)	Natives species (extinct)	Naturalized species (extant)	Total extant species
Christmas Island (Indian Ocean)	B	201	2	151	352
Cocos (Keeling) Islands	B	61	1	53	114
Easter Island[a]	D	43	7	68	111
Hawaiian Islands[a]	D	1223	71	1090	2313
Heron Island	B	27	...[b]	25	52
Lord Howe Island	A	219	2	202	421
Mangareva Island	D	85	...[b]	60	145
Nauru Island	D	50	1	85	135
New Zealand Islands[a]	D	2065	3	2069	4134
Norfolk Island	C	157	3	244	401
Northern Line Islands	C	35	0	41	76
Pitcairn Island	C	40	4	40	80
Tristan da Cunha Island	A	70	0	54	124

Source: Data are from Table 1 of D. F. Sax et al. *Am Nat* 160: 766–783.

Note: "All islands are currently inhabited by humans but are split into one of four categories based on their history of human occupation before European contact. Class A islands have no evidence of human occupation or visitation before European contact and are otherwise unlikely to have been visited. Class B islands have evidence of human visitation, or are likely to have been visited, but have no evidence of human settlement and are unlikely to have been settled. Class C islands appear to have had human settlements but were abandoned before European contact. Class D islands had human settlements when European contact occurred."[71]

[a]A fossil record exists.

[b]Insufficient information was available to determine whether any species had become extinct.

flexible in their habitat requirements, that replace a large number of "losers," often unique and specialized native species that are locally adapted to a habitat and have a limited geographic range.[73] The best examples of biotic homogenization come from studies of freshwater habitats. Rahel[75] compared present-day freshwater fish faunas in the United States with those present before European settlement to assess the degree of fish homogenization among U.S. states. He showed that, on average, any two randomly chosen states have 15 more species in common now than before European settlement. The 89 pairs of states that formerly had no species in common now share an average of 25 species. This pattern is largely the result of species introductions associated with fish stocking for recreational purposes (e.g., brown trout, *Salmo trutta;* rainbow trout, *Oncorhynchus mykiss;* and smallmouth bass, *Micropterus dolomieu*) or aquaculture (e.g., common carp, *Cyprinus carpio*).

Dornelas and colleagues recently analyzed 100 time-series data sets where researchers had monitored 35,613 species of mammals, birds, fishes, invertebrates, and plants measured at 430,324 locations around the planet (mostly in oceans). These authors found that in 79 out of 100 time-series data sets, ecological communities were being homogenized at a rate of roughly 10% of their species per decade.[76] Collectively, these analyses show that biotic homogenization of the world's flora and fauna is widespread, and it is happening quickly as species are transported around the globe. Some researchers have begun to project how similarity levels of species assemblages will change in the future if the present patterns of introductions and extinctions continue. Rosenblad and Sax[77] project that while plant assemblages will stabilize at intermediate levels of biotic homogenization, long-term trends will cause some taxonomic assemblages, such as birds, to become highly homogenized. Homogenization is projected to have an even greater impact on assemblages of mammals.[78]

Impacts on ecosystems

Compared with the amount of work that has focused on how IAS influence biodiversity, less work has focused on how IAS affect whole ecosystems. In part, this may be due to a historical perception among biologists that IAS only infrequently affect ecosystem-level processes. For example, some experts on invasive species have noted that the majority of introductions of alien species are benign and cause no significant changes to an ecosystem even after the alien has become established.[79] Some estimates suggest that the fraction of established invaders that affect ecosystem-level process is about 10%,[80,81] but this may be an underestimate. Vila and colleagues[69] found slightly higher numbers when they summarized information for 8628 IAS listed in the European Commission's DAISIE database (Delivering Alien Invasive Species Inventories for Europe). They found that 6% of terrestrial plants (326 of 5789), 14% of terrestrial invertebrates (342 of 2481), and 30% of terrestrial vertebrates (109 of 358) had records of significant ecosystem-level impacts as invaders. Freshwater flora and fauna were more frequently listed as having ecosystems impacts (30% of records) than marine flora and fauna (16%).

While the majority of alien invaders may be benign, the large number of alien species that are being introduced ensures that most ecosystems, including ecosystem processes and services, are likely being altered by at least one IAS.[81] And a growing body of literature shows that when those impacts occur, they can be large. For example, Liao and colleagues[82] summarized the results of 94 experimental manipulations of invasive alien plant species and found that IAS generally increase annual rates of net primary production (ANPP) by an average of 80% (**Figure 11.11**). Increased rates of plant production are also associated with increased storage of carbon and nitrogen in plant shoots and roots and in the soils (soil NH_4, NO_3). Plant IAS can also double the speed at which dead plant materials are decomposed and increase the rates at which essential nutrients such as nitrogen are made biologically available (by nitrification, mineralization). Vila and colleagues[69] found similar results in their summary of 1041 field studies looking at ecosystem-level effects of alien plant species. Plant IAS increased production of biomass by 57% following invasion. They also enhanced the activity of soil microbes, which in turn increased the rate at which nutrients were recycled and made biologically available. These results suggest that plant IAS tend to speed up rates of nutrient cycling and enhance the production of plant biomass, which has a variety of consequences for the global ecosystem and the humans who depend on it.

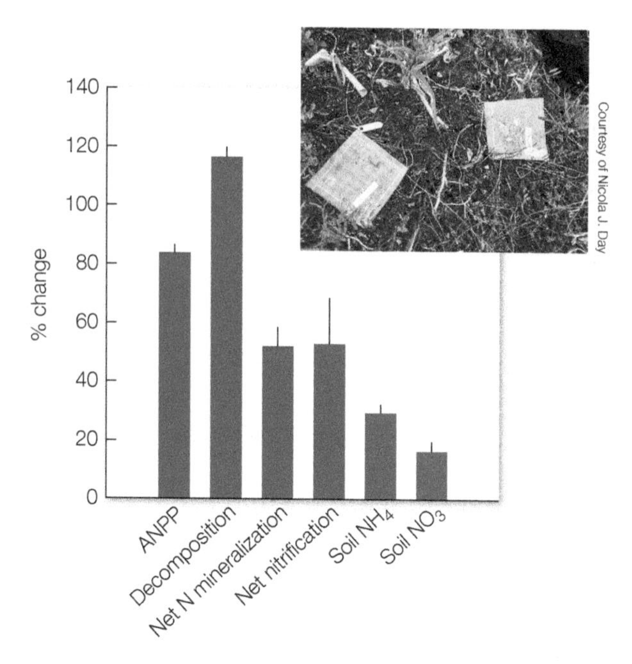

Figure 11.11 Impacts of invasive alien species of plants on ecosystem-level processes. Inset shows litter decomposition bags being measured for mass loss. (Data from C. Z. Liao et al. 2008. *New Phytol* 177: 706–714; graph after D. L. Strayer. 2012. *Ecol Lett* 15: 1199–1210.)

One of the best examples of the ecosystem-level impact of a plant IAS comes from Peter Vitousek's long-term studies of *Myrica faya* in Hawaii. *M. faya* is a nitrogen-fixing shrub that is native to Macaronesia. The shrub was first introduced to the Hawaii Volcanoes National Park about 1961, where it was the only N-fixing plant growing on recent lava flows on the island.[83] This unique functional trait, coupled with the nutrient-poor volcanic soils, gave *M. faya* a growth advantage over other plants. By 1977, the shrub had expanded to more than 600 ha despite intensive efforts to control its distribution. As the N-fixing shrub spread, it pumped nitrogen from the atmosphere into the nutrient-poor volcanic soils. In doing so, it increased the biological availability of nitrogen more than fourfold in some locations, which in turn allowed some plants to grow more luxuriantly and outcompete others. By changing the availability of what had been a limiting resource for millions of years, *M. faya* was able to alter basic ecosystem processes and species interactions.[84]

Some of the most profound impacts of IAS occur not through their direct impacts on ecological processes, but rather through indirect effects that stem from their altering the physical characteristics of the ecosystem itself. Ehrenfeld claimed that alteration of physical habitats by ecosystem engineers and alteration of disturbance regimes that shift ecosystems to a new state are the most common pathways of ecosystem change caused by IAS.[85] One example of an IAS that transformed an ecosystem through engineering is the European periwinkle sea snail (*Littorina littorea*). As the periwinkle spread throughout the coastal landscape of New England, it bulldozed sediments from hard substrates and prevented the growth of algae, transforming much of the coast from mud flats and salt marshes to rocky shores.[86] In doing so, the periwinkle essentially created a new ecosystem that had an entirely different set of ecological functions and a different composition of species.

There are numerous examples of IAS altering disturbance regimes that shift ecosystems to a new state, perhaps the best of which involve plants. After the alien grasses *Schizachyrium condensatum* and *Melinis minutiflora* were introduced to Hawaii, their spread helped increase fire frequencies that were responsible for converting the submontane woodlands into grasslands.[87] Invasion by Australian *Melaleuca quinquenervia* (paperbark) into grass and sedge communities in south Florida led to more intense fires that helped created large expanses of nearly monospecific melaleuca forests.[88] In the Cape Province of South Africa, invasion of areas of treeless fynbos by nonnative, fire-adapted trees like *Acacia*, *Hakea*, and *Pinus* helped convert heathlands into dense forests.[89]

Economic impacts

An increasing number of studies have estimated the economic impacts of IAS. At a national or regional scale, estimates vary by several orders of magnitude; however, nearly all are in the billions of US dollars, sometimes expanding into the tens and even hundreds of billions of dollars (**Table 11.4**). The United States and India have the highest reported economic losses caused by IAS, followed by Brazil, several countries in Southeast Asia, and then China and Canada (**Video 11.3**). The median reported economic loss caused by IAS is 1.4% of a country's gross domestic product (GDP). National estimates like these must be interpreted with caution, as they often suffer from inaccurate numbers and poor accounting methods, and seldom do they use methods that are directly comparable among studies. Nevertheless, the key point is that the economic impacts of IAS tend to be large.

VIDEO 11.3 Watch this report on the impacts of invasive species in the United States.
oup-arc.com/e/cardinale-v11.3

The economic impacts of IAS tend to be concentrated in certain sectors of national and regional economies. For example, Bradshaw and colleages[90] reviewed the global costs associated with invasive insects and showed that impacts are greatest on infrastructure, agriculture, and forestry, with the highest costs occurring in North America and Europe (**Figure 11.12**). One of the best-known impacts of an IAS on physical infrastructure is that of the zebra mussel (*Dreissena polymorpha*), which in 1995 was estimated to be causing $10 million in infrastructure damage per year to nuclear power facilities, sewage and water treatment plants, and industries associated with transportation in Canada and the United States.[91]

In the agricultural sector, Pimentel[92] estimated that crop loss in the United States caused by invasive weeds is $23 billion, and losses caused by invasive

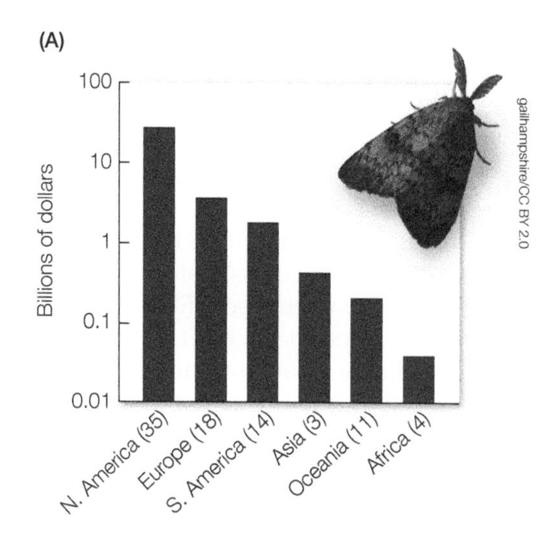

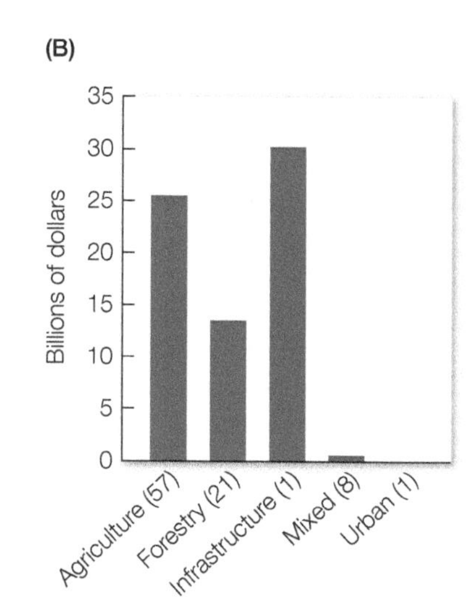

Figure 11.12 Costs of annual goods and services associated with invasive insects for (A) major geographic regions of the world and (B) economic sectors. All costs are expressed as 2014 US dollars. Numbers in parentheses in the *x*-axis labels indicate the number of estimates per category. (After C. J. A. Bradshaw et al. 2016. *Nat Comm* 7: 12986. CC BY 4.0.)

plant pathogens add another $21 billion. Oliveira and colleagues[93] reviewed crop losses caused by insect pests in Brazil and concluded that 24 introduced insect pest species are responsible for a collective 13% of all annual crop losses (US$1.6 out of US$12.0 billion). The Mediterranean fruit fly (*Ceratitis capitata*) is responsible for US$242 million in annual damage to fruit, while the silverleaf whitefly (*Bemisia tabaci*) is responsible for US$714 million in annual crop loses to vegetables.

Paini and colleagues[94] estimated the future threat to agriculture posed by 1297 insect pests and pathogens that have potential to invade, establish in, and damage economically important crops in 124 countries around the world. Large agricultural producers such as China, the United States, India, and Brazil are at greatest financial risk from future invasive species (**Figure 11.13A**). China and the United States also pose the greatest threat to the rest of the world, given their expansive network of agricultural trade partners and

TABLE 11.4 Estimates of economic losses caused by invasive alien species by country or region

Country or region	Billion USD/yr	% of GDP	Reference
United States	121	0.9	D. Pimentel et al. *Agricult Ecosys Environ* (2001)[146]
India	117	12	D. Pimentel et al. *Agricult Ecosys Environ* (2001)[146]
Brazil	47	4.5	D. Pimentel et al. *Agricult Ecosys Environ* (2001)[146]
Southeast Asia	32	2.6	L. T. P. Nghiem et al. *PLOS ONE* (2013)[147]
China	18	< 0.01	H. G. Xu et al. *Biol Invasions* (2006)[148]
Canada	15	0.9	R. I. Colautti et al. *Biol Invasions* (2006)[149]
Australia	7.7	1.0	G. Marbuah et al. *Diversity* (2014)[144]
South Africa	3.9	1.4	D. Pimentel et al. *Agricult Ecosys Environ* (2001)[146]
Sweden	3.3	0.1	I. M. Gren et al. *Ambio* (2009)[150]
United Kingdom	2.1	< 0.01	F. Williams et al. CABI (2010)[151]
Germany	1.9	0.01	F. Reinhardt et al. Federal Environmental Agency. (2003)[152]
New Zealand	1.8	1.9	N. Giera and B. Bell. MAF Biosecurity New Zealand (2009)[153]

Source: Most entries previously summarized in G. Marbuah et al. 2014. *Diversity* 6: 500, and B. D. Hoffmann and L. M. Broadhurst. 2016. *NeoBiota* 31: 1–18; original studies and recently published entries have been added.
Note: All values are in 2008 U.S. dollars (USD).

incumbent pool of invasive species. However, when potential impacts are standardized by GDP, which is an indicator of a country's ability to mitigate impacts (via pest management, crop substitutions, imports, subsidies, etc.), the countries at highest risk are all developing nations, with the top six most at risk all located in sub-Saharan Africa (**Figure 11.13B**).

After infrastructure and agriculture, forestry experiences the third largest economic losses caused by IAS. Aukema and colleagues[95] summarized the impacts of 62 of the most damaging species of phytophagous (feeding on plants) insects that are IAS in U.S. forests. They showed that wood- and phloem-boring insects cause the largest economic impacts by annually inducing nearly $1.7 billion in government expenditures and causing an approximately $830 million loss in residential property values. Among these, the most costly IAS pest has been the emerald ash borer (*Agrilus planipennis*), a wood-boring beetle native to northeastern Asia that was accidentally introduced into the United States and Canada in the 1990s (**Figure 11.14**). In 2002, the beetles were discovered in Canton, Michigan, and then rapidly spread throughout much of the midwestern and eastern United States. The female beetles lay their eggs in bark crevices on ash trees (*Fraxinus* spp.), and the larvae feed underneath the bark, ultimately killing these common trees when infestations get heavy enough. These beetles are responsible for the death of 53 million ash trees throughout the United States and Canada, with an estimated economic damage between 2009 and 2019 of US$11 billion.[96]

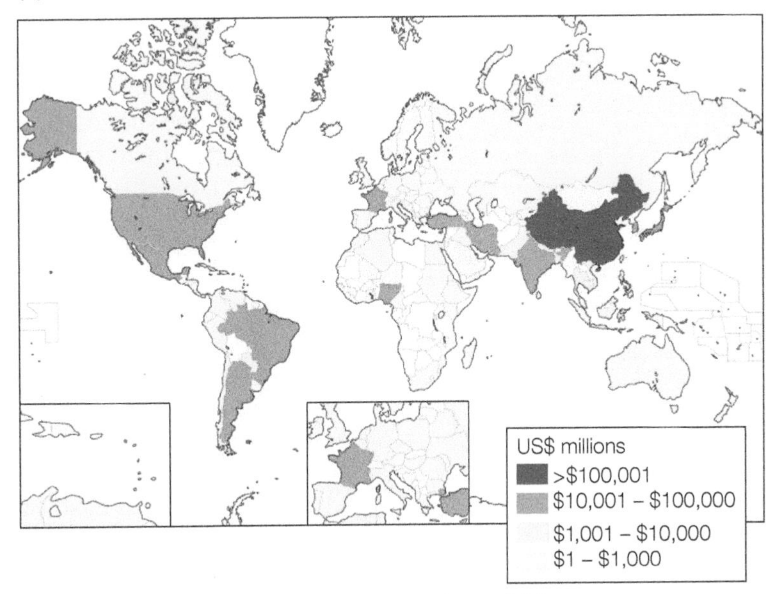

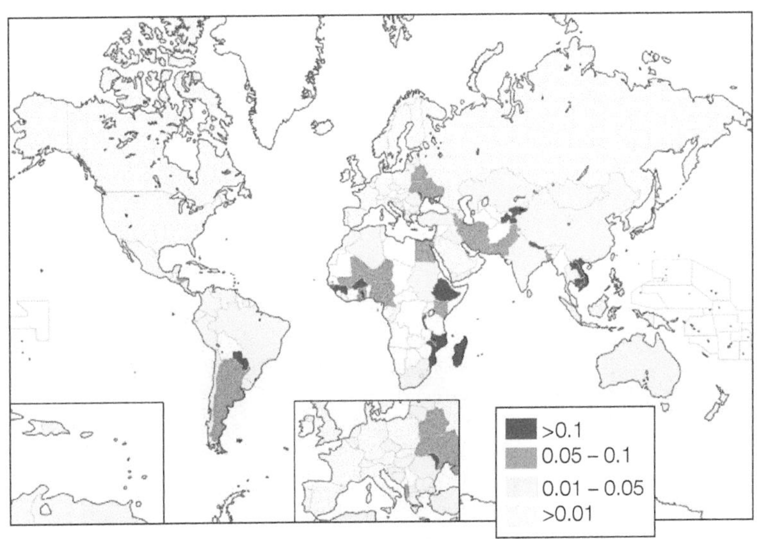

Figure 11.13 World map showing (A) the total cost of invasion (TICt) (in millions of US dollars) to threatened countries and (B) the total cost of invasion to threatened countries as a proportion of GDP. Those countries without color were not included in the analysis. (From D. R. Paini et al. 2016. *Proc Natl Acad Sci USA* 113: 7575–7579.)

Positive values of IAS

While most economic evaluations of IAS have focused on tallying the perceived negative impacts of IAS, often there is little or no consideration of the benefits of IAS (economic or otherwise), which would be needed for a full cost-benefit analysis (see Chapter 7). In fact, many species have been specifically translocated from one country to another because they perform valuable functions.

Indeed, one of the most common motivations for humans to introduce species into new locations is to grow food. The vast majority of food crops eaten by people in developed countries did not originate in those countries, but rather were introduced from other locations and domesticated. For example, nonnative crops and livestock comprise 98% of food eaten in the United States, including such staples as soybeans, wheat, honey from honeybees, and nearly all livestock. Alien species like these have become such common parts of our culture and diet that little thought is given to their geographic origins.

Suggested Exercise 11.4 Use the interactive map of the International Center for Tropical Agriculture to learn about the geographic origins of your favorite food crops. **oup-arc.com/e/cardinale-ex11.4**

Many tree species have been intentionally introduced to forests because of their high growth rates and increased wood production. For example, in Sweden the nonnative lodgepole pine (*Pinus contorta*) grows faster than the native Scots pine (*Pinus sylvestris*), producing up to 36% more wood. Other alien tree species, such as the Sitka spruce (*Picea sitchensis*) and hybridized aspen (*Populus tremula* × *tremuloides*), exceed the productivity of nearly all native species in the region. Their superior productivity can generate higher profit margins, which can be 50% to 100% greater for certain combinations of nonnative species grown as alternatives to the native assemblages of trees.[97]

Many fish that have been introduced to lakes and streams have significant economic value despite their environmental damage. For example, brown trout (*Salmo trutta*) is one of several species of Salmonidae that have been intentionally introduced to freshwater habitats throughout the world, as they are highly prized for the challenge they provide to anglers. Brown trout were introduced to New Zealand in 1867 to promote a sport fishery. Despite significant negative impacts on New Zealand's native galaxiid fishes,[98] New Zealand's brown trout fishing has come to be regarded as one of the world's best, attracting anglers from all over the world. Anglers spend significant amounts of money for the chance to catch trophy trout in scenic locations on New Zealand's South Island.[99] A lucrative fishing industry valued at more than $400 million (New Zealand dollars) per year has developed to service angler demand—complete with fishing guides, accommodation, lodges, shops, and transportation services that all provide jobs, many in areas that once had high unemployment.[100]

Biological control of pest species is yet another area where some IAS are beneficial. Classic approaches to biological control involve the release of a natural enemy (predator, parasite, or pathogen) that is known to control a pest's population in its native habitat, as a means to control the pest in its introduced habitat. Approximately 16% of the introductions of nonnative natural enemies result in complete control of the target pest.[101] One of the best examples is that of the cassava mealybug, *Phenacoccus manihoti*, in Africa. Cassava (*Manihot esculenta*) is a shrub that produces an edible starchy tuberous root. It was introduced to Africa from South America about 300 years ago, and by the middle of the twentieth century, it had become a staple food for more than 200 million people in sub-Saharan Africa. In the 1970s, the cassava mealybug was accidentally introduced from South America, and in the absence of natural enemies, it spread rapidly. By the 1980s, it had become a major pest causing up to 50% crop loss of cassava.[102] Biologists began to search in South America for natural enemies of the cassava mealybug and found a small parasitic

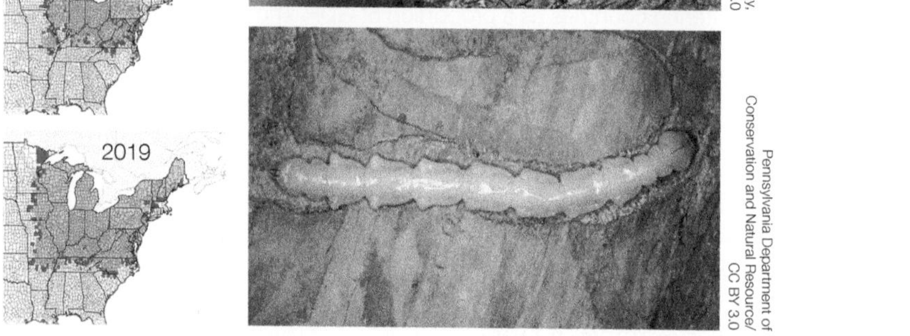

2010
2015
2019

Daniel Herms, The Ohio State University, Bugwood.org/CC BY 3.0

Penn DCNR/CC BY 3.0

Pennsylvania Department of Conservation and Natural Resource/CC BY 3.0

Figure 11.14 The emerald ash borer has been one of the most economically damaging invasive alien species in U.S. forestry. Maps show the rate of spread of the pest from 2010 (top) to 2019 (bottom). Pictures show the bore holes (top) made by the larvae (bottom). (Maps from K. F. Kovacs et al. 2010. *Ecol Econ* 69: 569–578.)

wasp, *Epidinocarsis lopezi*, that appeared to be an effective control agent. With support from the International Fund for Agricultural Development, mass rearing and distribution techniques were developed at the International Institute of Tropical Agriculture in Nigeria. By 1987, the alien species of parasitic wasp was intentionally released and established in 90% of the cassava-growing regions of Africa.[102] Losses caused by mealybug were almost immediately brought under control, and the project soon after earned the World Food Prize for saving millions of lives and billions of dollars across the African continent.

While the examples above in no way diminish the devastating impacts that some IAS have on biodiversity, ecosystem services, and economies, they do make the important point that alien species are frequently used to benefit humans. This is a point that has been increasingly made by some biologists who feel there has been a historical tendency for researchers and managers to demonize all alien species as "bad."[103–105] But the common practice of designating native species as good and introduced species as bad fails to recognize positive contributions nonnative species sometimes make to society, and it fails to recognize that different stakeholders can view an introduced population as either "harmful" or "useful." In fact, some have gone on to suggest that alien species may have an important role to play in conservation and restoration of ecosystems and their biodiversity (Box 11.2; **Video 11.4**; **Audio 11.1**).

VIDEO 11.4 This video examines some surprisingly helpful invasive alien species.
oup-arc.com/e/cardinale-v11.4

AUDIO 11.1 Listen to Mark Davis's *Nature Podcast* about species xenophobia.
oup-arc.com/e/cardinale-a11.1

BOX 11.2 Challenges & Opportunities

Should Species Be Judged by Their Origin, or by What They Do?

In 2011, invasive species biologist Mark Davis and 18 prominent colleagues published a comment about IAS in the prestigious journal *Nature* that set off a firestorm.[79]

In their opening sentence, Davis and colleagues wrote, "Over the past few decades, 'nonnative' species have been vilified for driving beloved 'native' species to extinction and generally polluting 'natural' environments." But Davis and colleagues argued that many of people's perceptions about the threats posed by introduced alien species to ecosystems and their biodiversity are not backed by real data. Rather, the perceptions are based more on an intrinsic value system that views native species as "good" and alien species as "bad." This value system has been perpetuated by the use of military-like metaphors (with terms like *invasive* and *alien*) and exaggerated claims of impending harm that are intended to convey the message that introduced specie are enemies of man and nature.

Davis and colleagues argued that policy and management decisions must begin to take into account the positive values many invaders bring to society (see Figure). In

(Continued on next page)

Monarchs (*Danaus plexippus*) are a threatened species of butterfly that are known for two things—their beauty and their epic migrations from the mountain forests in central Mexico to the United States and Canada. Along the route, monarchs roost on eucalyptus trees (*Eucalyptus obliqua*), which provide the preferred habitat for overwintering butterflies. Eucalyptus trees are invasive alien species in North America. But their contribution to the survival of monarch butterflies is an example of the sometimes beneficial effects of IAS.

BOX 11.2 Challenges & Opportunities *(continued)*

many cases, nonnative species create "novel ecosystems" that provide significant ecosystem services, such as when plants stabilize the ground and provide valuable green cover in disturbed urban environments. Conservation biologists perhaps should incorporate alien species into management plans, rather than try to achieve the often-impossible goal of eradicating them or drastically reducing their abundance. They ended by writing

> *We are not suggesting that conservationists abandon their efforts to mitigate serious problems caused by some introduced species, or that governments should stop trying to prevent potentially harmful species from entering their countries. But we urge conservationists and land managers to organize priorities around whether species are producing benefits or harm to biodiversity, human health, ecological services and economies. Nearly two centuries on from the introduction of the concept of nativeness, it is time for conservationists to focus much more on the functions of species, and much less on where they originated.*[79]

The article by Davis and colleagues was interpreted by some news outlets as a suggestion that invasive species are not bad after all and may even be good. The response to the article was swift. One response, led by invasive species biologist Daniel Simberloff and signed by 141 scientists,[161] argued that Davis and his colleagues had set up a "strawman" argument, falsely claiming that conservation biologists oppose nonnative species per se when, in fact, they have always focused on species that pose negative threats to ecosystems while acknowledging those that provide benefits. The critics went on to suggest that Davis and colleagues had severely downplayed the negative impact of IAS as well as the wealth of evidence that has been used to demonstrate those impacts.

Mark Davis has continued to promote his controversial views, many of which were first published in his 2009 book *Invasion Biology*.[162] Meanwhile, many other experts have used the debate as an opportunity to move beyond some of our historically subjective views about nonnative species, toward more evidence-based descriptions that help distinguish alien species that have net negative, versus net positive, impacts on ecosystems and their biodiversity.[3,104]

Management and Control

Despite the potential positive values of some IAS, it is generally agreed that most IAS—particularly those that represent unintentional releases—generally have negative impacts on biodiversity, ecosystems, and economies. Given this recognition, multiscale (regional, national, and international) management programs have been developed in many parts of the world to reduce the current and future impacts of IAS. For example, the Global Invasive Species Programme (GISP) represents a partnership formed by the Centre for Agriculture and Bioscience International (CABI), the IUCN, the South African National Biodiversity Institute, and The Nature Conservancy to support implementation of Article 8(h) of the Convention on Biological Diversity (prevent introduction of, control or eradicate harmful IAS). One role of the GISP has been to publish a *Global Strategy on Invasive Alien Species* and a *Toolkit of Best Prevention and Management Practices*.[106] GISP's strategy has helped facilitate the development of more regional management plans, such as the European Union's ground-breaking legislation (EU Regulation No. 1143/2015) that sets a common standard for combatting IAS across political jurisdictions at a multinational scale.[107,108] That legislation is an agreement to establish early warning and surveillance systems (Articles 16 and 22), the development of action plans to address priority pathways of introduction (Article 13), and rapid eradications to prevent establishment and long-term mitigation and control mechanisms (Article 17). Legislation like this has potential to build on the growing number of success stories of IAS control efforts that have helped restore ecosystems and their biodiversity (**Box 11.3**)

Figure 11.15 illustrates the process for developing a management plan to control IAS or mitigate their impacts. Development typically begins with **horizon scanning**, which involves a systematic search to identify species that

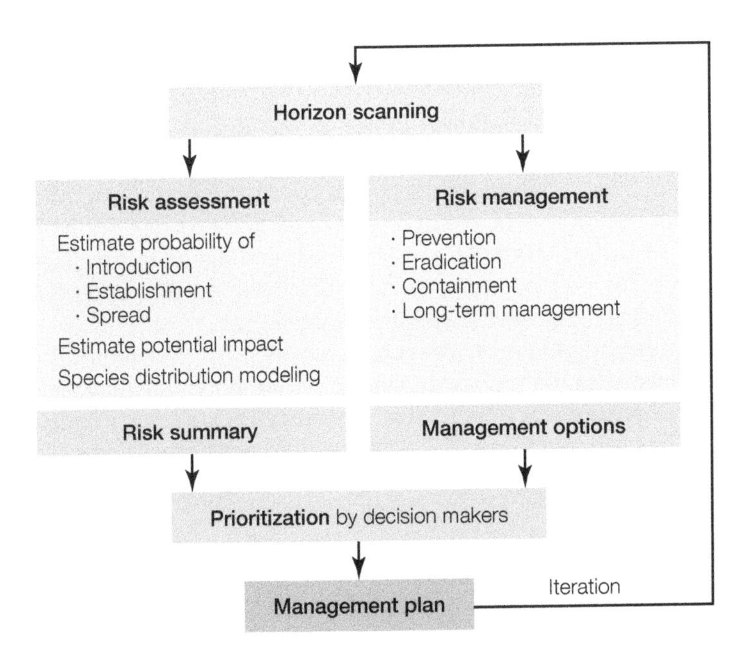

Figure 11.15 Steps for developing an IAS management plan. (Adapted from Great Britain's Nonnative Species Secretariat.)

BOX 11.3 Success Story

To Catch a Goat

About 18 years ago, scientists in Mexico decided it was time to get rid of goats on Isla Guadalupe. Guadalupe Island is a remote Pacific island about 150 miles west of the Baja Peninsula, with fewer than 150 permanent human inhabitants. The island was once lushly vegetated with 150 native species of plants, including endemic varieties of Monterey pine (*Pinus radiata* var. *binata*) and the endemic cypress. Then, in the mid-nineteenth century, goats were released on the island by American and Russian sailors hoping to establish a reliable source of meat.

The goat population exploded, reaching an estimated 100,000 animals by 1870. Overgrazing by the goats led to disappearance of entire plant communities on the island, including several dozen endemic species. Without the vegetation, soils began to erode and dry out, and the island began to take on the appearance of a desert.

Guadalupe cypress (*Hesperocyparis guadalupensis*) on the central highlands of Guadalupe Island.

In a bid to stabilize the island ecosystem and its biodiversity, an NGO called Grupo de Ecología y Conservación de Islas embarked on a campaign to eradicate Guadalupe's goat population with support from the Mexican government. Between 2003 and 2006, more than 10,000 goats were removed. By 2007, the entire goat population had been eradicated.

Within a year, thousands of new seedlings of pine and cypress sprung up all over the island. Without the goats around to eat them, seedlings developed into trees that are now several meters tall and showing healthy growth. In addition, many native shrubs, including some endemic species, have reappeared in large numbers, and several species that were thought to be extinct have been rediscovered. The island is quickly revegetating and reverting back to a functional forested ecosystem (see Figure).

The success of Guadalupe Island is not unique. Jones and colleagues have summarized the results of 251 eradications of invasive alien mammal species on islands and shown that 107 highly threatened bird, mammal, and reptile species on the IUCN Red List (6% being highly threatened species) have likely benefitted from invasive mammal eradications on islands.[163]

Invasive mammal eradications on islands represent a suite of conservation biology's greatest success stories—stories that are sufficiently abundant that they are now being archived at Database of Island Invasive Species Eradications (oup-arc.com/e/cardinale-w11.7).

pose potential threats before they are introduced to a novel ecosystem. The information gathered from horizon scans can come from a variety of sources, including literature reviews, analysis of species pools at the beginning and end of trade or transportation routes, or surveys of experts. Regardless of how potential threats are identified, a key point is that best management practices require that horizon scans be performed on a regular basis so that potential threats can be identified and assessed before they materialize.

After horizon scanning, a good invasive species management plan requires a risk assessment that evaluates the probability and impact of potential alien species, and a list of risk management options that represent the tools, techniques, and policies that can be used to mitigate risk. We describe each of these next.

Risk assessment

Risk assessment is a formal process for determining the probability that an event will occur and the consequences or impact if it does occur. Risk assessments are widely used in fields like public health or toxicology where decision makers must assess the risk of many different types of disease or pollutants in order to prioritize and manage those that pose the greatest risk to society. Risk assessments are often required components of environmental impact assessments, and they have become increasingly common in fields involving biosecurity, such as the field of invasive species biology, where they are used to evaluate risks associated with a multitude of species being introduced (intentionally or accidentally) to a given region.

The typical IAS risk assessment involves five steps (**Figure 11.16**). The first step is to identify potential threats, such as through a horizon-scanning exercise like that discussed in the previous section. The second step is to determine the scope of the potential threat by deciding which stakeholders are likely to be affected by the IAS and to determine how they might be affected. The impacts can be economic, as when infrastructure is damaged (e.g., water intake pipes), or the impacts can be on other aspects of well-being, such as public health, human happiness, or perceptions of the aesthetic value of ecosystems. Listing these impacts explicitly is important because it helps determine how risks are analyzed (step 3) and summarized (step 4). Once these risks are known and quantified, the final step is to review and monitor the threat.

TWO COMPONENTS OF RISK Any analysis of risk involves two components: (1) the probability or likelihood that the threat will become reality and (2) the impact if the threat does become reality. These two components can be formalized as follows:

$$Risk_i = I_i \times P(I_i) \tag{11.1}$$

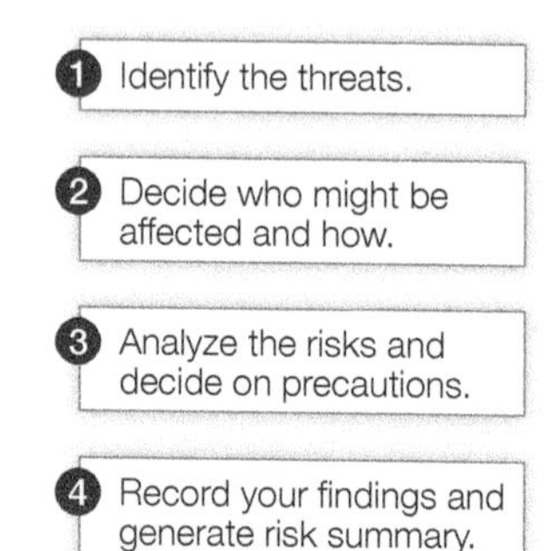

Figure 11.16 Five steps for generating an IAS risk assessment.

In Equation 11.1, I_i is the impact that threat i will have if it becomes reality (in whatever units of impact are relevant), and $P(I_i)$ is the probability that threat i will become reality. For an IAS, the probability that the threat will become a reality depends on the probability of introduction, P_{intro}, the probability of establishment, P_{estab}, and the probability of spread, P_{spread}. We can rewrite Equation 11.1 to represent the product of these three probabilities:

$$Risk_i = I_i\left[P_{intro} \times P_{estab} \times P_{spread}\right] \tag{11.2}$$

Equation 11.2 quantifies the risk associated with any one potentially invasive species threat (species i). If any of the three probabilities (introduction, establishment, or spread) is zero, then risk of that threat is zero. If the probability of introduction is near certain ($P_{intro} \sim 1$), but the probability of establishment or spread is low (e.g., P_{estab} and P_{spread} both = 0.01), then the risk associated with species i is just 0.0001 times the potential impact of species i.

Estimates from Equation 11.2 can be used to generate a risk matrix to help establish priorities for management. To illustrate, **Figure 11.17** gives a hypothetical example of analysis of Equation 11.2 for eight species (A–H) that were identified as potential threats in a horizon scan. From the risk matrix, we would conclude that species G and, to a lesser extent, species C and H pose the greatest risk and warrant the most attention for management actions. In contrast, species A and B and, to a lesser extent, species E all pose comparatively less risk. Risk matrices are widely used to set management priorities. For example, Singh and colleagues[109] used matrices like these to quantify the risk associated with 97 species of parasitic nematodes that pose a biosecurity risk to agriculture in Australia.

In some instances, it may be helpful to estimate the total risk posed by all IAS threats. To do this, we simply sum Equation 11.2 for all IAS identified in a horizon scan, from $i = 1$ to n potential threatening species:

$$Risk_{total} = \sum_{i=1}^{n} I_i \left[P_{intro} \times P_{estab} \times P_{spread} \right] \quad (11.3)$$

Estimates from Equation 11.3 could also be used to generate a risk matrix, except the species presently shown in Figure 11.17 might instead be replaced with different locations (e.g., various national parks) to illustrate which ones are at greatest total risk from alien species.

ESTIMATING PROBABILITIES In order to use Equations 11.1 through 11.3 to generate a risk matrix, one must first estimate the probabilities that an alien species will be introduced, become established, and spread. There are many different ways to estimate these probabilities. Some methods are qualitative methods, such as those that rely on the opinions of experts to estimate probabilities. Expert opinions are often solicited using questionnaires that ask individuals to identify potential alien threats and score their potential impacts and probabilities of introduction, establishment, and spread. Answers are often converted to an ordinal scale (e.g., 1 to 10) that rank the risks posed by different species relative to one another. These rankings can be used to generate a risk matrix that can, in turn, be used by decision makers as they formulate an IAS management plan.

Many regional and national risk assessment tools use qualitative approaches that are based on expert opinion.[110] One advantage of these approaches is that qualitative assessments can be performed rapidly with little to no data. There is also some evidence that qualitative assessments can provide accurate predictions for certain well-studied groups like invasive plants where experts have

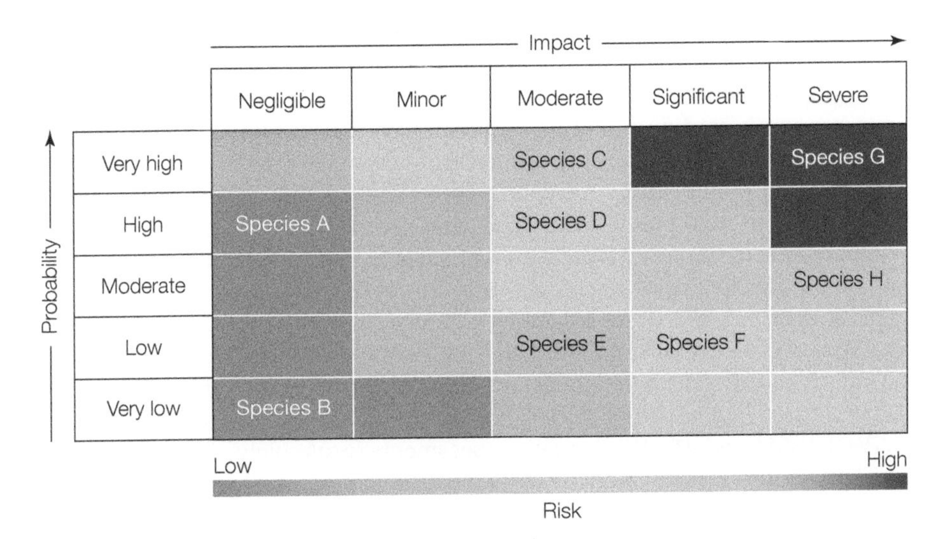

Figure 11.17 The IAS risk matrix. Entries in the cell give estimates of risk from low to high. These are organized according to how likely an IAS is to be introduced, become established, and spread, and by the potential impact the IAS is expected to have after spread.

a long history of case studies to base their opinions on. For example, one of the most widely used risk assessment frameworks for plants is the Australian Weed Risk Assessment System (AWRA), which uses responses to 49 questions to generate a score that classifies species as "low risk—acceptable for import," "possible risk—further evaluation needed," or "high risk—reject import."[111] The accuracy of the AWRA (i.e., its capacity to correctly identify invasive species) has ranged from an average of 80% for weeds rejected[112] to 90% for major invaders correctly identified.[113] Because of its success, the AWRA has been adapted for use in multiple regions of the world outside Australia and New Zealand, for which it was developed.[114]

Of course, the limitation of qualitative methods is that experts can be wrong, and inaccurate predictions are more likely for species or ecosystems that are less well known and studied. Because of this, there has been a push to complement qualitative approaches with more quantitative methods. One quantitative method that is becoming increasingly common is **trait-based risk assessment**. Trait-based risk assessments use measurable life history traits and other biological characteristics (genetic, morphological, etc.) of potential alien species to try to predict their introduction, establishment, and spread. Baker[115,116] published one of the first trait-based risk assessments for plants, defining the "ideal weed" based on a suite of 12 biological characteristics that included rapid growth from seedling to sexual maturity, the ability to reproduce both sexually and asexually, and ability to rapidly adapt to a new environment (e.g., phenotypic plasticity).

Since Baker's analysis of the ideal weed, many subsequent analyses have tried to identify the traits/characteristics that distinguish IAS from native species and that distinguish successful from unsuccessful invaders. For plants, characteristics that are common among successful invaders include *r*-selected life history traits, such as the ability to use recently disturbed habitat, short generation times, high fecundity and growth rates, broad environmental tolerances (e.g., ability to function across a wide range of physical conditions), aggressive competitive ability, and high phenotypic plasticity. In addition to these, studies of vertebrates such as birds and fish suggest that biological characteristics that generate a close association with humans (edibility, aesthetic beauty, desirability as pets, etc.) are common predictors of successful animal IAS.[117,118]

Trait-based risk assessments are not as widely used as expert opinion surveys, in part because there has historically been a paucity of large data sets that summarize the life history traits for many different species. However, this problem has begun to diminish with the advent of several large databases of species traits. The Plant Trait Database (TRY, oup-arc.com/e/cardinale-w11.8) represents a network of plant scientists around the world who have collected over 3 million traits/characteristics on nearly 70,000 plant species. Other databases are available for plants (U.S. Department of Agriculture PLANTS Database [oup-arc.com/e/cardinale-w11.9], and the World Agroforestree Database [oup-arc.com/e/cardinale-w11.10]), and similar data sets have been developed for animals (TraitNet [oup-arc.com/e/cardinale-w11.11], and the Global Biotraits Database [oup-arc.com/e/cardinale-w11.12]). These databases have been increasingly used to predict what traits make some species more invasive.[119]

In addition to improved databases, statistical methods for identifying the traits of IAS are rapidly improving. Algorithms in machine learning are allowing the rapid scan of large databases to identify commonalities among IAS, and improved methods of classification and regression are now allowing researchers to use fewer traits to identify IAS and to take advantage of imperfect and incomplete data.[110]

IMPACT ASSESSMENT While many assessments focus solely on the probability of alien species introduction, establishment, and spread, any true risk assessment requires an estimate of the potential impact of an alien species (the I_i in

Equation 11.3). It is quite possible for the probabilities of introduction, establishment, and spread to be high but for the risk to be low if a species' potential impact is negligible (see Species A in Figure 11.17). This is why most legislation, including the new legislation in the European Union, requires an explicit impact assessment as part of the risk assessment process.[108]

In general, we lack the empirical data needed to quantify the real or known impacts of alien species in a potentially invaded range. In some instances, we can assume the impacts of an alien species introduced to one location or country will be similar to its impacts elsewhere. But in many instances, we do not have historical or spatial comparisons to rely on. Therefore, most protocols for performing an alien species impact assessment are qualitative and rely on expert opinion.[120]

There are four impact assessment protocols currently in widespread use at national and international levels. These include the Generic Impact Scoring System (GISS), the Environmental Impact Classification for Alien Taxa (EICAT), Harmonia+, and the Great Britain Non-native Risk Assessment (GB NNRA). While GISS and EICAT are impact-only protocols, Harmonia+ and GB NNRA are full risk assessment protocols, considering not only IAS impacts, but also the likelihood of introduction and spread.

GISS was first developed and applied to alien mammals in Europe, after which it was refined and used to predict the impacts of alien birds, fishes, and terrestrial and aquatic invertebrates.[121] The latest version has been used to assess the environmental and economic impacts of 349 alien plant and animal species in 12 different categories of ecological and socioeconomic impact. EICAT is a recent derivative of GISS, modified to classify species according to the magnitude of their detrimental environmental impacts, using a broader range of categories that directly correspond to those used by the IUCN Global Invasive Species Database.[122] Harmonia+ assesses the impacts IAS have on infrastructure, plant, animal, and human health and on ecosystems. It was designed to be maximally compliant with international law.[123] GB NNRA was originally developed for use in Great Britain[124] but has recently been updated and modified to be applicable to the whole European Union. It is based on the European and Mediterranean Plant Protection Organization risk assessment framework, which is a set of 18 questions that are widely used in international plant health regulations, focused on potential impacts of IAS on biodiversity and ecosystems.

All four impact assessment protocols have comparable methodologies and scoring scales; however, they differ substantially in their underlying emphases and assumptions. For example, GB NNRA emphasizes the biodiversity and ecosystem-level impacts of IAS, whereas Harmonia+ concentrates more heavily on health and infrastructure. As a result, the four protocols can produce substantially different assessments of risk. For example, Turbé and colleagues[120] evaluated the ability of three of these four assessment protocols to predict impacts associated with two nonnative parrots in Europe (**Figure 11.18**). They found that predictions differed significantly among assessment protocols because they gave differing weights to environmental, economic, and social impacts.

SPECIES DISTRIBUTION MODELING After one has estimated the probabilities and impacts that go into Equations 11.1–11.3, the next step of the risk assessment process is to map the potential spread of IAS to other locations. The primary tool used for this is called a **species distribution model** (**SDM**), or **ecological niche model**. SDMs have become central in efforts to understand the spread of IAS across habitats, including how species shift their range distributions in response to some form of environmental change (e.g., global warming; see Chapter 12). Most SDMs are based on correlations that relate a species current distribution to suites of environmental variables that are used

Figure 11.18 Plot of additive impact scores and confidence for the ring-necked (left) and monk parakeet (right) derived from three independent assessment protocols—the generic impact scoring system (GISS), Harmonia+, and the Great Britain Nonnative Risk Assessment (GB NNRA). Error bar represents ± 1 SE. Red dots indicate results of the consensus assessment. (After A. Turbé et al. 2017. *Divers Distrib* 23: 297–307.)

to forecast their future distribution in space (e.g., an alternate region) or time (e.g., under a climate change scenario). When the predictor variables of these correlations are climatic, the models are called climate envelope models.

A key assumption of many, if not most, SDMs is that a species' realized niche—the range of environmental conditions that constrain its range distribution—is primarily a function of the abiotic environment. For example, the range distribution of a particular plant might be constrained by environmental variables such as temperature, moisture, solar radiation, and soil type. Given these constraints on the current distribution, the suitability of sites that might be colonized by the plant species in the future can be predicted based on the similarity of the environmental conditions in the occupied and the potential sites (**Figure 11.19**).

There are many different programs and computer algorithms that practitioners use to predict the suitability of sites that might be colonized in the future. Elith[125] has written an excellent introduction to the various approaches to SDM methods. Here, we briefly describe just two—CLIMEX and MaxEnt—both of which are commonly used to model the distribution of IAS.[126] CLIMEX (oup-arc.com/e/cardinale-w11.13) is a commercially available software program that was specifically developed for modeling of invasive species distributions and has been adopted worldwide by many management agencies and governmental departments.[125] The primary output of CLIMEX is a mapped prediction of the suitability of a set of grid locations for a given species.

CLIMEX requires location records of a species in its native range and then uses those records to infer the environmental requirements of a species. These environmental requirements are generated from simple statistical functions that estimate a "growth index" (*GI*) describing the potential for population growth during a favorable event or season. For example, a growth index for an IAS plant might be modeled as a multiplicative function of environmental covariates such as temperature (*T*), moisture (*M*), solar radiation (*R*), and soil type (*S*). Each of the predictor variables would be scaled from 0 to 1 to ensure each is given equal weight in contributing to the outcome:

$$GI = T \times M \times R \times S \tag{11.4}$$

CLIMEX then modifies the growth index with a stress index (*SI*) that gives the probability that a population will survive through unfavorable events or seasons. The *GI* and *SI* are then combined into a final "ecoclimatic index" (*EI*) that gives an overall measure of the suitability of locations that can then be presented on a map (e.g., Figure 11.19):

$$EI = GI \times SI \tag{11.5}$$

Another popular program for generating SDMs is MaxEnt (oup-arc.com/e/cardinale-w11.14). MaxEnt is an open-source software package that takes a

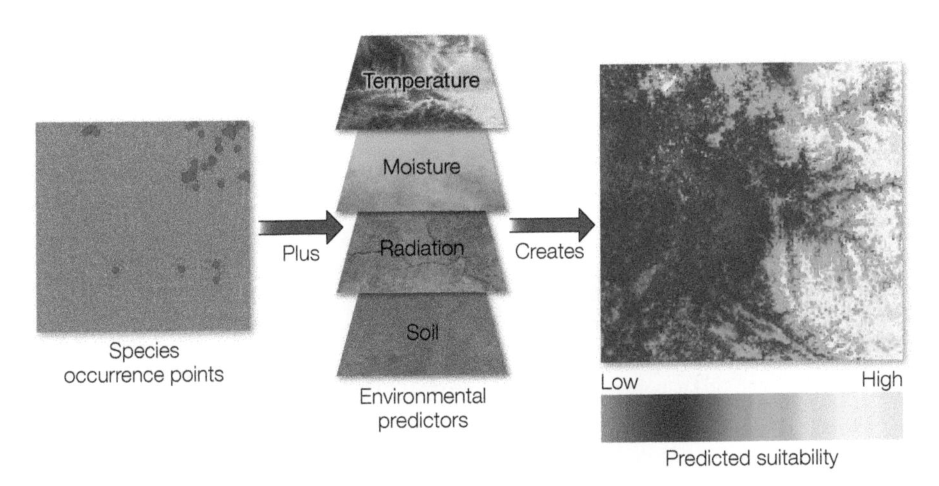

Figure 11.19 A conceptual diagram of how species distribution models (SDMs) work. The species' current distribution (left) is related to abiotic factors such as temperature, moisture, radiation, and soil type (center), using a statistical model. That statistical model is used with maps of those same environmental factors to predict the potential suitability of habitat not yet occupied (right). (After NatureServ.)

list of species presence and absence locations as input, and a set of environmental predictors (e.g., precipitation, temperature) that are overlaid on a user-defined landscape that is divided into grid cells. From this landscape, MaxEnt extracts a sample of background locations that it contrasts against the presence locations. The name MaxEnt stems from the machine learning algorithm the program uses to predict future distributions, which is based on the principle of maximum entropy. The principle of maximum entropy postulates that the "best" model is one that is not only consistent with the data, but which is as similar as possible to prior expectations. Merow and colleagues have written an excellent practical guide for practitioners who would like to use MaxEnt,[127] and several tutorials exist[128] that make it remarkably easy to get started doing SDMs with MaxEnt.

Suggested Exercise 11.5 Go to the MaxEnt website, download the current version of the software, and complete the tutorial available on the website. You'll be running SDMs in no time at all! **oup-arc.com/e/cardinale-ex11.5**

SDM is an important visual tool for communicating risk to policy makers and managers. If a species is detected in a habitat but is unlikely to persist there, costly eradication efforts can be avoided. By contrast, prime habitat for establishment may justify both aggressive surveillance and eradication or control efforts.

Risk management

Risk management represents the suite of tools, techniques, and policies that one can use to mitigate the risks identified in a risk assessment. Decision makers must consider management options and their associated costs alongside the summary of risks in order to prioritize their limited funding and personnel as they develop a final management plan.

The management and control options for IAS that are available to a decision maker depend on what stage an alien species is at in the process of introduction, establishment, and spread (**Figure 11.20**). As a general rule, success rates in managing and controlling alien species are far higher during the early stages

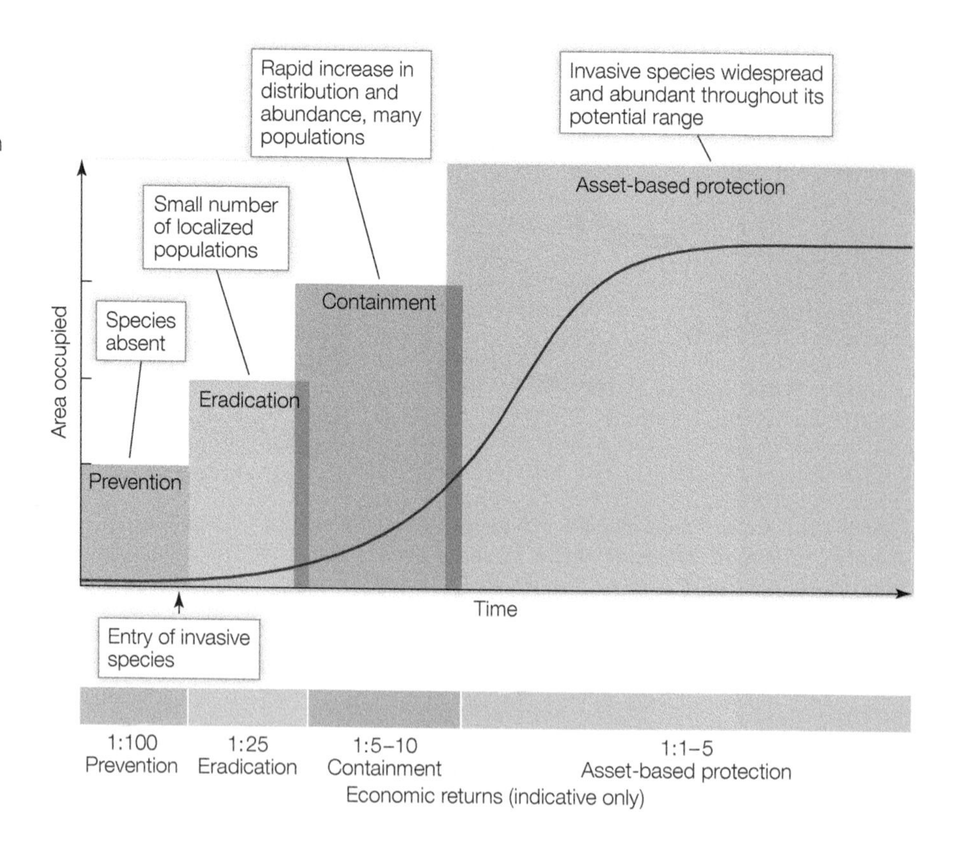

Figure 11.20 Generalized invasion curve showing management options (prevention, eradication, containment, and asset protection) appropriate to each stage of an alien species' population growth (solid line). Bar at bottom shows approximate economic returns as the cost:benefit ratio of dollars spent on management to dollars saved in damages avoided. (© State of Victoria, Department of Jobs, Precincts and Regions. Reproduced with permission.)

of community assembly, where dispersal routes can be controlled, and established populations at low densities can be eradicated. But once an alien species becomes established and begins the process of exponential population growth, the probability of successful management declines precipitously, and economic costs rise substantially.[110,129]

PREVENTION Numerous studies have shown that prevention of colonization by alien species is not only the most successful of all management options, but also the most cost-effective option for minimizing the risks associated with IAS.[130,131] Successful prevention requires good risk assessments that forecast routes of dispersal and help prioritize sites that are at greatest risk of introduction. Once dispersal routes and high-risk sites are identified, management actions can focus on minimizing propagule pressure and the probability of introduction of an alien species (P_{intro}, Equations 11.2 and 11.3).

The primary tools available for minimizing propagule pressure are (1) public engagement and education, (2) trade and transportation policies, and (3) biosecurity screening. The need for public engagement in the management of IAS has been acknowledged by the Convention on Biological Diversity,[132] as well as by national and international strategies like GISP's Global Strategy on Invasive Alien Species[106] and the Invasive Alien Species Strategy for Canada.[133] Establishing collaborations between all stakeholders and codesigning management actions are well known to increase public support for efforts designed to prevent the introduction of alien species.[134] Common mechanisms for increasing public engagement are public workshops and citizen science programs. When active public engagement is not possible, public awareness campaigns can still increase public support for prevention of invasive species. Tools that help raise awareness include advertising, developing websites or printed documents, establishing forums, organizing school talks or public events, creating informative films, and publishing popular science articles.

Good trade and transportation policies are also needed to prevent alien species introductions. International trade policies that consider the risk of alien species are rare, but there are exceptions. For example, the World Trade Organization (WTO) has established certain trade agreements (e.g., the Agreement on the Application of Sanitary and Phytosanitary Measures) that require countries to establish national regulations that protect human, animal, and plant life from risks associated with alien species, pests, and pathogens.[110] Several regional trade agreements, such as the Central American Free Trade Agreement (between the United States and several countries in Central America) or the Comprehensive and Progressive Agreement for Trans-Pacific Partnership (between Australia, Brunei, Canada, Chile, Japan, Malaysia, Mexico, New Zealand, Peru, Singapore, and Vietnam), have explicit provisions requiring countries to monitor and prevent the transportation and introduction of alien species to new locations. These formal trade agreements have been reinforced by a number of international treaties that aim to prevent introductions. One example is the International Convention for the Control and Management of Ships' Ballast Water and Sediments, which requires signatories to ensure their ships have certified management plans for discharge of ballast water and sediments while still far out to sea rather than in coastal zones, to minimize the spread of harmful aquatic organisms.

The final tool for pathway and vector management is biosecurity screening. The goal of biosecurity screening is to identify and intercept alien species before they are introduced into a country. Most developed nations have preborder screening programs that are designed to intercept alien species that are considered high risk, such as those that might threaten agriculture or forestry or have other high economic impacts. Biosecurity screening has proven to be a highly efficient, cost-effective method to prevent invasions. For example, Keller and colleagues[135] showed that the use of a preborder screening and weed risk assessment system in Australia provided net economic benefits of US$1.67 billion over 50 years by allowing authorities to screen out costly invasive species.

ERADICATION AND CONTAINMENT The most effective way to deal with invasive species, short of keeping them out, is to discover them early and attempt to eradicate or contain them before they spread.[136] The success of eradication and containment efforts is highly dependent on the quality of surveillance programs that allow for early detection of an IAS and the ability to mount a rapid response. Surveillance and early detection of IAS are difficult because, by definition, species are rare in the early stages following their introduction, and rare species are often difficult to detect. Traditional biological monitoring programs that rely on periodic sampling and censuses are often inadequate for IAS because they are generally designed to monitor populations of more common and abundant species and are not designed for detection of rare species.

The problem of rare species detection has been partially overcome by increasing the number of people involved in surveillance, and collating data into ever-increasing data sets. This is the approach taken by the Early Detection & Distribution Mapping System (EDDMapS, oup-arc.com/e/cardinale-w11.4), which is a web-based mapping system for documenting the distribution and spread of IAS. EDDMapS combines data records collected by science-based organizations with records submitted by volunteers and citizen science monitoring programs. By increasing the number of people performing surveillance and the area being monitored, managers can use EDDMapS to more quickly identify potential introductions.

In addition to the increase in total effort, early detection systems have been greatly improved by recent advances in high-tech surveillance tools. One such high-tech surveillance tool is environmental DNA (eDNA), which takes advantage of the fact that individual organisms leave traces of DNA in the

> **VIDEO 11.5** Watch this video on detection of invasive species using eDNA.
> **oup-arc.com/e/cardinale-v11.5**

environment through their excretions (e.g., urine or feces) or shedding of skin, scales, hair, or other body parts (**Figure 11.21**). This DNA can be sampled directly from the environment rather than from fluid or tissue from the organism itself. Because the DNA of a species can be detected at remarkably low concentrations in soil and water samples, we don't have to actually see an individual of an IAS to know it is present in its early stages of invasion (**Video 11.5**).

If alien species can be detected early when they are in low abundance, then the chance for eradication (complete elimination of an alien species from an ecosystem) is quite good for certain kinds of species. Success rates tend to be relatively high for large animals such as mammals. Indeed, many successful eradications from islands have been reported for cats, foxes, goats, rats, and other types of mammals.[137] In contrast, there are relatively few reports of successful eradications of invasive alien plants, likely because of the difficulties of removing certain life history stages (e.g., seeds) and body parts (e.g., roots).

The potential for eradication also increases if the alien species is contained within a small area. Rejmanek and Pitcairn[138] reviewed eradication attempts by the California Department of Food and Agriculture involving 18 plant species and 53 separate infestations targeted for eradication. They showed that the likelihood of eradication declined rapidly with an increasing area of infestation. While professional eradication of infestations smaller than 1 hectare were usually possible, eradication of infestations 101–1000 ha in size were successful only 25% of the time. Costs of eradication projects increased dramatically as the size of the infestation increased such that eradication success for species occupying more than 1000 ha was very unlikely, given the extensive cost and resources required for such operations.

Myers and colleagues[139] concluded that six factors are common among successful eradication programs:

1. Because eradication projects are expensive, often costing millions of dollars, resources must be sufficient to fund the program to its conclusion.
2. Eradication programs often require treatments (e.g., poisons, trapping) or regulations (e.g., restricted access) that are controversial and that often cover private lands and multiple jurisdictions. Eradication is only feasible if the lead individual or agency has clear lines of authority to carry out the required procedures across all lands and jurisdictions.

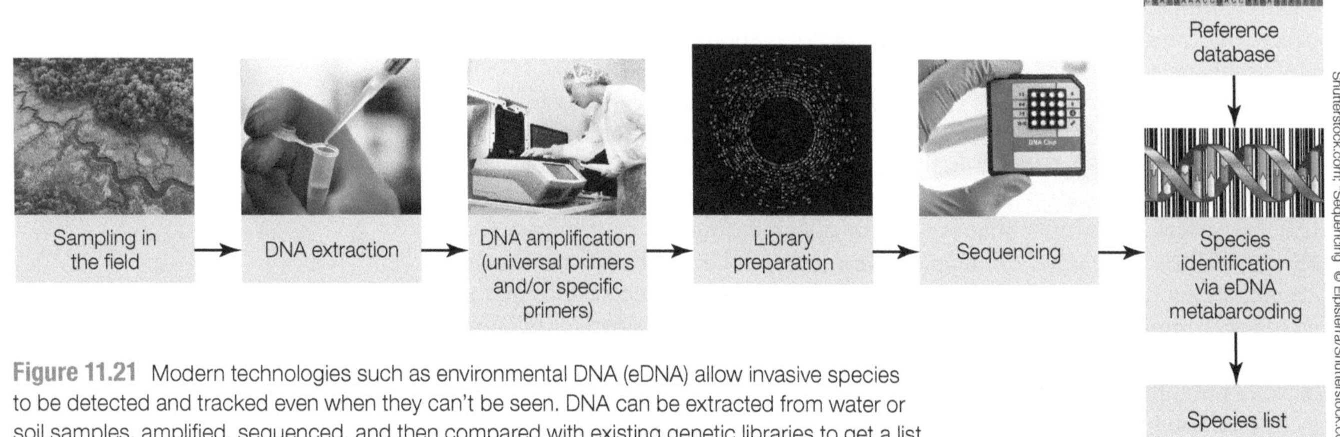

Figure 11.21 Modern technologies such as environmental DNA (eDNA) allow invasive species to be detected and tracked even when they can't be seen. DNA can be extracted from water or soil samples, amplified, sequenced, and then compared with existing genetic libraries to get a list of both native and alien species that are present in a site. (After Aarhus University, Danish Center for Energy and Environment/Environmental DNA Center. envs.au.dk/en/about-the-department/sections/environmental-microbiology-and-biotechnology/environmental-dna-center/)

3. The biology of the target organism must make it susceptible to control procedures. The dispersal ability, reproductive biology, and life history traits of the target alien species will determine the ease of population reduction and its potential for reinvasion. If these are not all considered, eradication will fail.
4. Reinvasion must be prevented. Eradication will only be temporary if the influx of propagules continues.
5. The pest must be detectable at relatively low densities. Easy detection allows residual pockets of individuals to be identified and targeted for treatment.
6. The individual or agency leading the eradication must anticipate collateral damage to other species and the ecosystem and be prepared to mitigate and/or restore the native community and ecosystem processes that are damaged by the eradication effort.

LONG-TERM MANAGEMENT If prevention, eradication, and containment efforts fail, an IAS will spread and increase in population size. At this point, elimination of the species becomes virtually impossible or financially infeasible. Once an IAS reaches the saturation phase (see Figure 11.20 where area occupied reaches a plateau), the only realistic option may be to manage the population in an effort to reduce impacts and protect important assets.

There are three types of control that are commonly used to manage populations of IAS. Mechanical control is the physical removal of individuals from the population, such as by cutting, mowing, or pulling of plants or by trapping and hunting of animals. Mechanical control tends to be labor intensive and has the potential to generate significant amounts of disturbance that, in some instances, can encourage even further establishment by alien species. Therefore, mechanical controls need to be considered carefully and monitored to ensure they do more good than harm.

Chemical control refers to the use of pesticides (herbicides, insecticides, fungicides, animal poisons, etc.) that increase mortality and/or inhibit growth and reproduction of the target organism. A wide variety of chemicals are available for IAS management. These tend to be organism specific. The list of acceptable chemicals often differs among countries as well as individual locations within countries, so local regulations and suppliers must be consulted before chemical controls can be used. While many invasive species simply cannot be controlled without the use of pesticides, the main disadvantage of chemical control that practitioners must consider is the collateral impacts they have on nontarget species and their potential toxic effects on human health.

Biological control programs rely on the use of predators, pathogens, and parasites, collectively called *natural enemies*, to increase mortality of the IAS. Because IAS often do not have biological control agents in their introduced ranges, natural enemies from their native habitats are sometimes transported and introduced into the nonnative habitat. Introducing a second alien species as a means to control another is riddled with risk, the greatest of which is for an introduced natural enemy to attack nontarget hosts and harm native biodiversity. Therefore, biocontrol efforts that use introduced natural enemies are heavily regulated in most countries and tend to require extensive study to demonstrate host specificity and minimal nontarget effects.

It is worth noting that humans often make an excellent biological control agent for IAS, and there are many creative programs that take advantage of humans as natural enemies. For example, the Florida Fish and Wildlife Conservation Commission has hosted Python Challenges, inviting teams of hunters to compete for prize money for the most invasive Burmese pythons caught or killed in the Everglades. The South Florida Water Management District pays hunters $50 per snake kill (oup-arc.com/e/cardinale-w11.15).

Another creative program takes advantage of people's ability to hunt invasive species for food. Joe Roman, a conservation ecologist at the University of Vermont, has written a number of articles in which he has argued that if humans can hunt native species to extinction, as we have since the beginning of humankind (see Chapter 10), then we should be able to deploy our insatiable appetites against invaders.[140] Roman's proposal has caught the attention of several prominent chefs who have been interested in using sustainable food sources to promote stronger environmental ethics at their restaurants. For example, in 2005, Asian American Chef Bun Lai—a pioneer of the sustainable food movement—created an invasive species menu for his family restaurant, Miya's, in New Haven, Connecticut. Miya's became the first sustainable sushi restaurant in the world.

The idea that invasive species can be used as a food source has increased in popularity. In 2010 the National Oceanic and Atmospheric Administration launched its Eat Lionfish campaign to combat the species' invasion of the Caribbean. In 2011 Food & Water Watch hosted an invasive species banquet at the James Beard House in New York City. In 2012, the University of Oregon's Institute for Applied Ecology began an annual invasive species cook-off contest. In 2016, the upscale American grocery chain Whole Foods added lionfish to its shelves and started promoting it as "an invasive species" in the Atlantic Ocean and Caribbean Sea. Websites like invasivore.org and EatTheInvaders.org encourage people to be "invasivores" by promoting home recipes for exotic species (**Video 11.6**).

▶ **VIDEO 11.6** This video shows how eating invasive alien species can become part of management and control efforts.
oup-arc.com/e/cardinale-v11.6

NOVEL ECOSYSTEMS In a select number of cases, all attempts at management and population control of IAS may prove futile, impractical, or financially infeasible, and an ecosystem will become irreversibly dominated by the alien species. In these cases, there may be no other option but to modify expectations and embrace the notion of "novel ecosystems." Novel ecosystems are those systems characterized by species that occur in combinations and relative abundances that have not occurred previously at a given location or biome.[141] Novel ecosystems may be sufficiently altered that it is no longer possible to remove the IAS without collapse of native communities, or without permanent damage to the ecological processes that sustain life in the ecosystem. An example is that of the invasive *Tamarix* (saltcedar) tree. A native of Eurasia and Africa, tamarix forms dense forest stands in the southwestern United States, where it consumes large amounts of precious desert groundwater. However, establishment of tamarix has also created a novel ecosystem that provides scarce habitat for birds, including endangered species like the southwestern willow flycatcher (*Empidonax traillii*). As a result, value judgments must now be made. Do we value the endangered flycatcher more than the native vegetation? Should we remove the tamarix trees? Some would argue that in cases like these, the only reasonable option may be to refocus management goals.[137]

Many alien species that pose threats in their introduced locations are, paradoxically, threatened or endangered in their own native ranges.[142] For example, the yellow-crested cockatoo (*Cacatua sulphurea*) is a Critically Endangered species that has been subject to severe population declines in its native range in eastern Indonesia and East Timor because of international demand for pet birds. Yet, the release of captive birds has led to its introduction to Hong Kong and Singapore, both important trade hubs in Asia that now sustain sizable feral populations of the species. What should we do in such circumstances? Should we attempt to curtail the populations of introduced cockatoos in Hong Kong and Singapore to mitigate the damage they may potentially cause to native species? Or should we protect the cockatoo in its introduced range because of its status as a Critically Endangered species?

Questions about the role of novel ecosystems in conservation are becoming even more important in the face of climate change, which is causing a mass migration as species move to more northerly latitudes and to higher altitudes. Shall we embrace, perhaps even assist, these migrations as species form new assemblages and novel ecosystems that have not previously existed in Earth's history? Or shall we deny species their efforts to move in an effort to persist on a changing planet, in order to protect native communities? These are difficult questions we explore even further in Chapter 12, Climate Change.

Summary

1. The spread of invasive alien species (IAS) has become a global problem involving nearly every major group of species and every country. Controlling and managing IAS is considered a high priority for governments and conservation organizations.
2. Like most species, alien species are limited by dispersal, and their spread is facilitated by human transport. Most individual organisms that are accidentally or intentionally introduced by humans to a new location die before they can establish. But the few that are able to survive in the new abiotic environment and can overcome biological constraints, such as competition and predation, can then proliferate and spread.
3. IAS affect biodiversity, though the impacts differ from what has been historically presumed. There is good evidence that IAS cause population declines and even extinction of native species; however, the rate of alien species introductions sometimes exceeds the rate of native species extinctions, causing the local biodiversity in an ecosystem to increase rather than decrease. The spread of IAS across ecosystems also tends to reduce beta diversity, which is a measure of how unique communities are in space or time.
4. In many case studies, IAS have altered rates of ecosystem processes such as primary production, decomposition, and nutrient cycling. Some of the more dramatic ecosystem-level impacts of alien species occur when an invader acts as an ecosystem engineer to completely alter habitat types and disturbance regimes.
5. The economic impacts of IAS can be exceptionally large, amounting to billions of dollars annually for certain sectors of the economy (e.g., agriculture, forestry, infrastructure) and in certain countries that have many IAS (e.g., countries in North America and Europe). It is, however, worth noting that cost-benefit analyses that quantify both the negative and positive values of IAS are rare, which ignores the fact that some invasive species also provide ecosystem services.
6. Horizon scanning, risk assessment, and risk management are the three components of an IAS management plan. Horizon scanning is a systematic search to identify species that pose potential threats before they are introduced to a novel ecosystem. Risk assessment is the formal process for evaluating the probability and impact of potential alien species. Risk management is a list of tools, techniques, and policies that can be used to mitigate risk.
7. Good risk assessments must estimate the probability of introduction, establishment, and spread of an alien species, as well as the potential impact if the species gets established. There are both qualitative (e.g., expert opinion survey) and quantitative (e.g., trait-based assessment) methods to do this. Good risk assessments also predict the future distribution of an IAS using species distribution models (SDMs).
8. Risk management considers options for IAS that are available to a decision maker to manage and control an IAS, which depend on what stage an alien species is at in the process of introduction, establishment, and spread. As a general rule, success rates in managing and controlling alien species are far higher during the early stages of community assembly, where dispersal routes can be controlled and established populations at low densities can be eradicated. But once an alien species becomes established and begins the process of exponential population growth, the probability of successful management declines precipitously, and economic costs rise substantially.

For Discussion

1. Many studies suggest that invasive alien species (IAS) of plants increase rates of primary production, speed rates of nutrient cycling, and enhance decomposition. In what ways might these impacts of IAS increase or decrease the value of ecosystem services to people at the local and global scales?
2. Which do you think is the more effective tool for preventing the introduction of alien species to a new site: (a) engaging the public in development of a management plan, (b) educating the public about the impacts of invaders and how to prevent their introduction, (c) involving the public in monitoring and eradicating invasive species, or (d) government-instituted trade and transportation policies that control routes of dispersal of alien species?
3. Do you think your country should have stronger biosecurity at its borders, and perhaps even a task force that identifies alien species along transportation routes before they arrive at your borders? What are the pros and cons of strong biosecurity at borders?
4. Considering the four hypotheses that are often used to explain biotic resistance to alien species (empty niche, novel weapons, enemies release, novel environments), how could you manage a natural ecosystem to be more resistant to invasion?
5. Why do you think economic estimates of IAS so seldom consider both the negative and positive value of species? What kind of biases do you think this has generated in the minds of the public? In the minds of policy makers? Or even in the minds of scientists?
6. Think of an invasive alien species that is well known where you live. What would it take in terms of financial resources and personnel to eradicate the species? Is it more realistic to simply control its population size, or has it created a novel ecosystem that can no longer revert back to its original state?

Suggested Readings

Early, R. et al. 2016. Global threats from invasive alien species in the twenty-first century and national response capacities. *Nature Communications* 7: 12485. Many of the threats of IAS that are yet to come and to be dealt with.

Gurevitch, J. and Padilla, D. K. 2004. Are invasive species a major cause of extinctions? *Trends in Ecology & Evolution* 19: 470–474. The common assumption that invasive alien species cause biodiversity loss was questioned, resulting in more rigorous studies.

Marbuah, G. et al. 2014. Economics of harmful invasive species: A review. *Diversity* 6: 500. One of the best and most reliable summaries outlining the economic costs of IAS.

Rosenblad, K. C. and Sax, D. F. 2017. A new framework for investigating biotic homogenization and exploring future trajectories: Oceanic island plant and bird assemblages as a case study. *Ecography* 40: 1040–1049. Predicts the future impacts of invasive alien species on beta diversity.

Schlaepfer, M. A. et al. 2011. The potential conservation value of nonnative species. *Conservation Biology* 25: 428–437. A balanced perspective on the pros and cons of invasive alien species.

Simberloff, D. et al. 2013. Impacts of biological invasions: What's what and the way forward. *Trends in Ecology & Evolution* 28: 58–66. A good summary of modern thinking on IAS, from impacts, to management options, to public perception.

Vila, M. et al. 2011. Ecological impacts of invasive alien plants: A meta-analysis of their effects on species, communities and ecosystems. *Ecology Letters* 14: 702–708. One of the more comprehensive summaries of how IAS affect communities and ecosystems.

Visit the
Conservation Biology
Companion Website
oup.com/he/cardinale1e
for videos, exercises, links, and other study resources.

12

Climate Change

The polar bear (*Ursus maritimus*) is a species that is considered vulnerable to extinction due to the impacts of climate change on its arctic habitat.

If the geologic record tells us anything about the history life, it is that the abundance, distribution, and diversity of species across the globe are all strongly controlled by Earth's climate. Because the rates of most biochemical processes are temperature-dependent, Earth's climate has exerted strong control over the rate at which biodiversity has formed. Indeed, the kinetic effects of temperature on metabolism help explain variation in mutation and speciation rates in the oceans over the past 30 million years, with diversification slowing when the oceans cooled and speeding up when the waters warmed.[1]

Earth's climate has also exerted strong control over species distributions, with taxa expanding or contracting their ranges as they have attempted to survive or bounce back from global cooling or warming events. Indeed, after the range distributions of most tree species contracted during the last ice age (approximately 115,000 to 11,700 years ago), the tree species that were able to survive migrated northward by 450 to 2200 km in less than 10,000 years as Earth warmed and the glaciers receded.[2]

Past climate change has also been associated with large extinction events. During the Paleocene-Eocene Thermal Maxima that occurred 56 million years ago, rising temperatures of the coastal oceans produced widespread hypoxia believed to have caused one of the smaller, but still significant, mass extinctions of marine fauna.[3,4] In general, the number of genera of marine animals has declined whenever Earth's temperatures have risen because rates of extinction have increased faster with rising temperatures than have rates of generation of new taxa.[5]

Given that climate change has radically altered biodiversity throughout Earth's history,[6–8] it should be no surprise that most biologists believe anthropogenic climate change will have great impacts on the abundance, distribution, and diversity of species in the modern era. This chapter reviews how modern anthropogenic climate change impacts biodiversity and how conservation biologists can mitigate these effects. This chapter does not discuss a wide range of direct effects that climate change has on people (e.g., water availability, fire frequencies, natural disasters); we instead limit our scope to the impacts of climate change on biodiversity and the ecosystem services organisms provide to people. The chapter begins with a brief review of anthropogenic climate change, after which we review the predicted impacts of climate change on biodiversity, the impacts that have been documented to date, and the management strategies available to practitioners to mitigate the impacts of climate change.

Anthropogenic Climate Change

Anthropogenic climate change is a general term that refers to a suite of abiotic variables that are changing simultaneously across the planet as a result of human activities (**Figure 12.1**). Each year, thousands of scientific papers are published that advance our understanding of how the averages, ranges, variability, frequency, predictability, and seasonality of the abiotic variables that regulate Earth's climate are changing across the globe. The main governing body charged with summarizing this literature is the Intergovernmental

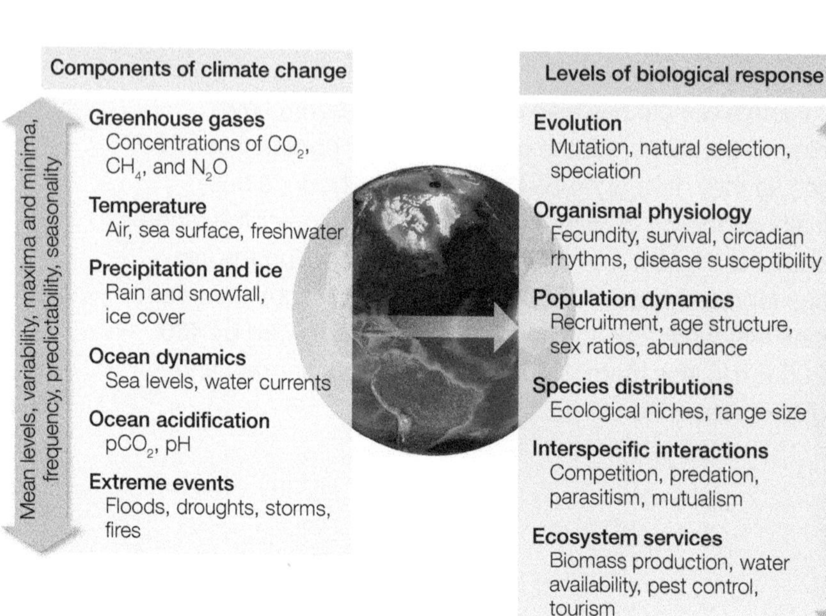

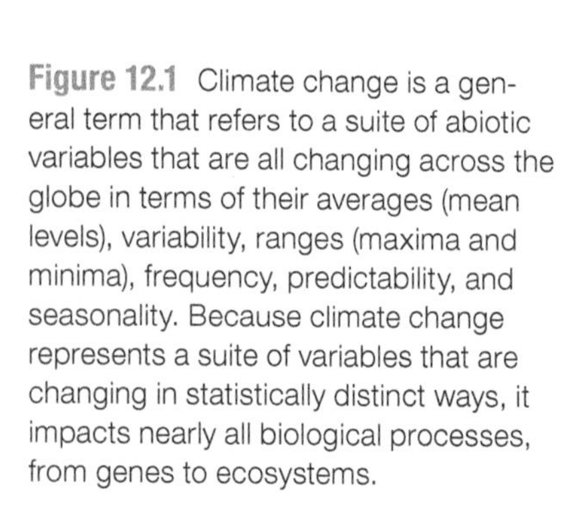

Figure 12.1 Climate change is a general term that refers to a suite of abiotic variables that are all changing across the globe in terms of their averages (mean levels), variability, ranges (maxima and minima), frequency, predictability, and seasonality. Because climate change represents a suite of variables that are changing in statistically distinct ways, it impacts nearly all biological processes, from genes to ecosystems.

Panel on Climate Change (IPCC). The IPCC was established in 1988 by the World Meteorological Organization (WMO) and the United Nations Environment Programme (UNEP). With membership open to all members of the WMO and UN, the IPCC is charged with periodically publishing objective, science-based summaries of climate change, as well as its natural, political, and economic impacts and risks. Reports published by the IPCC contribute to work of the United Nations Framework Convention on Climate Change (UNFCCC), which is the main international treaty on climate change and is designed to "stabilize greenhouse gas concentrations in the atmosphere at a level that would prevent dangerous anthropogenic (human-induced) interference with the climate system."

The most recent Synthesis Report was published in 2014 and represents the IPCC's Fifth Assessment Report[9] (AR5; the Sixth Assessment is due in 2022). The Fifth Assessment Report was written by 831 experts from disciplines that included meteorology, physics, oceanography, statistics, engineering, ecology, the social sciences, and economics. These experts were nominated by their home countries and contributed to writing and reviewing the report on a voluntary basis. They summarized the results of 9200 peer-reviewed scientific studies and concluded that:

> ... human influence on the climate system is clear and growing, with impacts observed across all continents and oceans. Many of the observed changes since the 1950s are unprecedented over decades to millennia. The IPCC is now 95 percent certain that humans are the main cause of current global warming. In addition, the SYR [Synthesis Report] finds that the more human activities disrupt the climate, the greater the risks of severe, pervasive and irreversible impacts for people and ecosystems, and long-lasting changes in all components of the climate system.[9]

The ultimate cause of climate change is human emission of greenhouse gasses (GHGs) like carbon dioxide (CO_2, emitted from fossil fuel combustion and deforestation), methane (CH_4, emitted from agricultural activities, waste management, and biomass burning), and nitrogen oxide (N_2O, emitted from agricultural activities like fertilizer use) into the atmosphere. Concentrations of CO_2, CH_4, and N_2O have increased by 40%, 150%, and 20%, respectively, since the mid-1700s.[9] The concentration of CO_2 in the atmosphere, which is the single largest contributor to **radiative forcing**—the difference between sunlight absorbed by the Earth and energy radiated back to space—reached 415 parts per million (ppm) by May 2019, a level that is unprecedented in at least the last 800,000 years of Earth's history (**Figure 12.2A**).[10]

As GHGs accumulate in the atmosphere, they trap solar radiation, which in turn increases the temperature of the troposphere (the lowest region of the atmosphere, extending 6–10 km above Earth's surface), the land, the oceans, as well as freshwater lakes, streams, and wetlands (**Video 12.1**). Multiple independent analyses of the modern instrumental temperature record (i.e., roughly the past 150 years, when thermometers have been widely used) have shown that the combined surface air and sea surface temperatures averaged over the globe have increased by approximately 1°C above preindustrial levels (circa 1850–1900; **Figure 12.2B**).[9] Reconstructions of historical temperature records based on climate proxies (e.g., gas bubbles in ice cores, growth of tree rings, isotopic composition of snow, corals, and stalactites) suggest that the second half of the twentieth century had higher average temperatures than nearly any period in the last 1500 years.[11] The majority of the additional energy stored in Earth's climate system has been absorbed by the oceans, where the upper 75 m of surface water warmed

VIDEO 12.1 Watch a video about the causes and effects of climate change.
oup-arc.com/e/cardinale-v12.1

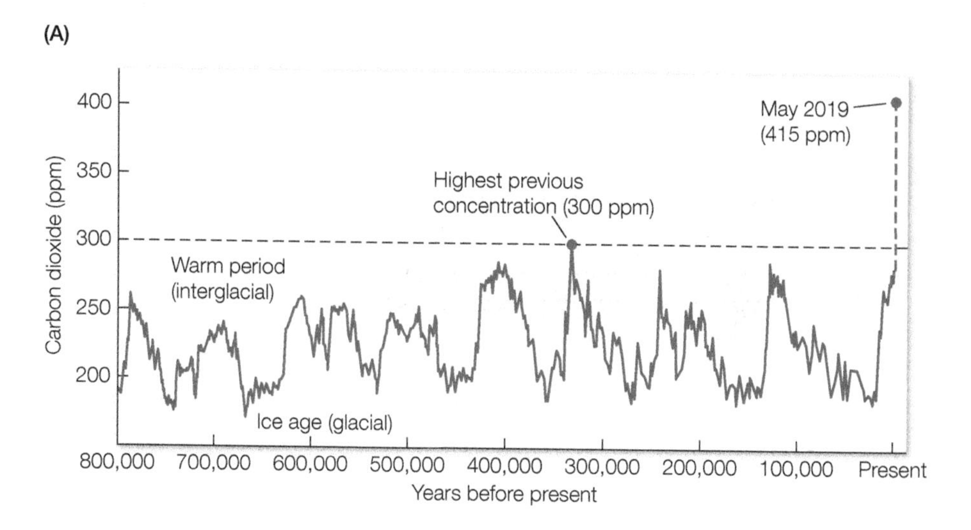

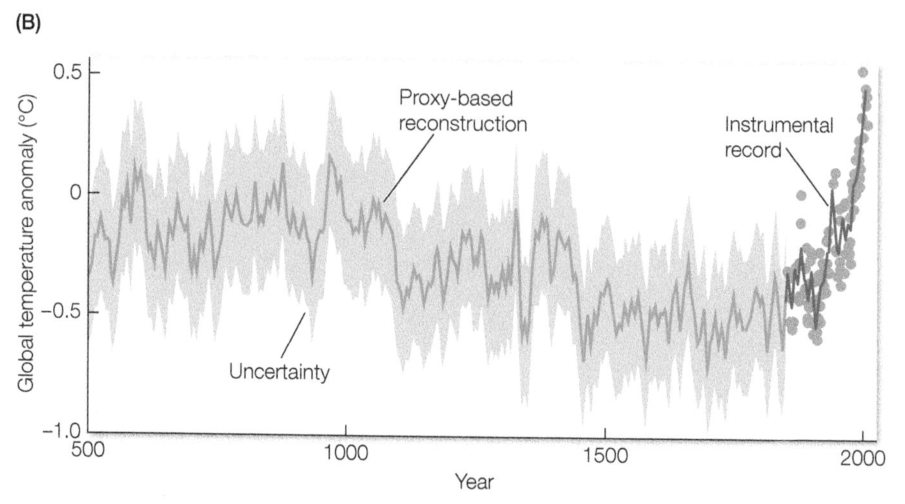

Figure 12.2 (A) Atmospheric carbon dioxide concentrations in parts per million (ppm) for the past 800,000 years. Peaks and valleys in CO_2 levels track the coming and going of ice ages (low CO_2 levels) and warmer interglacial periods (higher levels). Throughout these cycles, atmospheric CO_2 was never higher than 300 ppm (dashed line). However, in May 2019, concentrations of CO_2 reached a record high of 415.0 ppm (blue dot). (B) Temperature histories from paleoclimate data (orange line) compared to the history based on modern instruments (blue line). Note that global temperature is higher now than it has been in the past 1500 years, and possibly longer. (A and B after NASA. 2010. How is Today's Warming Different from the Past? [https://earthobservatory.nasa.gov/features/GlobalWarming/page3.php]. A, based on EPICA Dome C data, in D. Lüthi et al. 2008. *Nature* 453: 379–382; B, adapted from M. E. Mann et al. 2008. *Proc Natl Acad Sc* 105: 13252–13257. © 2008 National Academy of Sciences, USA.)

by 0.11°C per decade over the period 1971 to 2010.[9] Warming has been unequally distributed across the globe, with certain high-latitude locations in the Northern hemisphere and select equatorial regions warming considerably more than the global average (up to 2.5°C; **Figure 12.3A**). If warming continues at the current rate, average global temperatures are projected to reach 1.5°C to 2.0°C higher than preindustrial levels sometime between 2030 and 2052.[12]

Increased air and water temperatures have led to cascading impacts on many other components of Earth's climate. Numerous aspects of the global water cycle have been altered. For example, increases in atmospheric moisture have been associated with global-scale shifts in patterns of precipitation, including intensification of heavy precipitation events over land regions (**Figure 12.3B**). The Earth's **cryosphere** (the frozen-water portion of Earth's system) has been altered with reduced ice caps, receding glaciers, and declining sea ice and snow. The annual mean extent of Arctic sea ice decreased by 3.5%–4.1% per decade from 1979 to 2012 (**Figure 12.3C**), and the extent of Northern hemisphere snow cover has decreased by about 1.6% per decade since the mid-twentieth century.

Between 1901 and 2010, global mean sea level rose by 19 cm (**Figure 12.3D**), a rate of sea level rise larger than the mean rate documented during the previous two millennia.[9] Taken together, glacier mass loss and ocean thermal expansion from warming explain about 75% of the observed global mean sea level rise.[9] Some areas of the world have experienced higher sea level rise and others

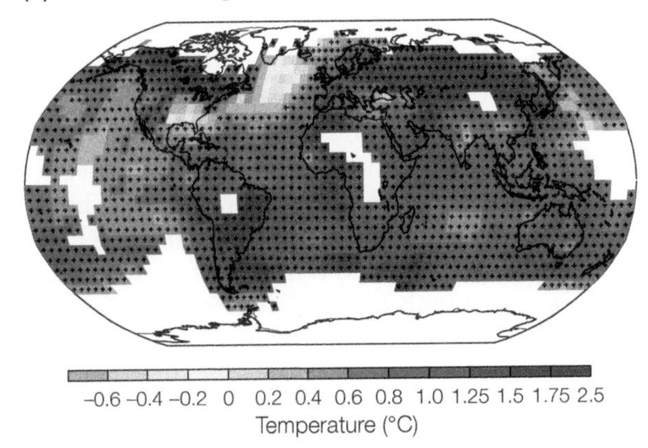

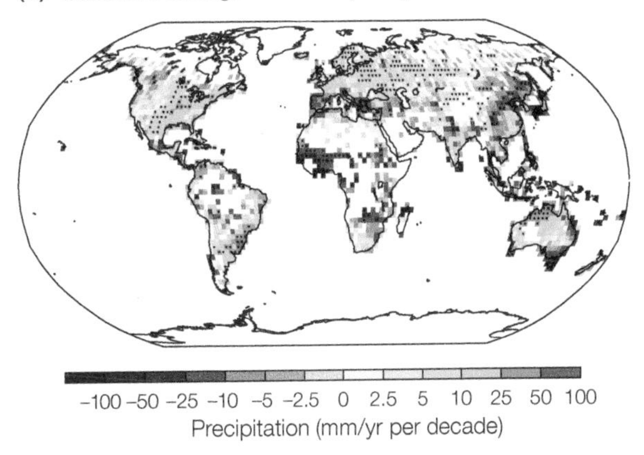

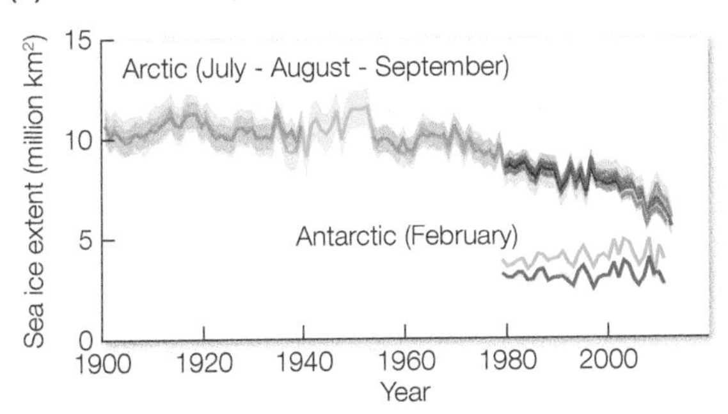

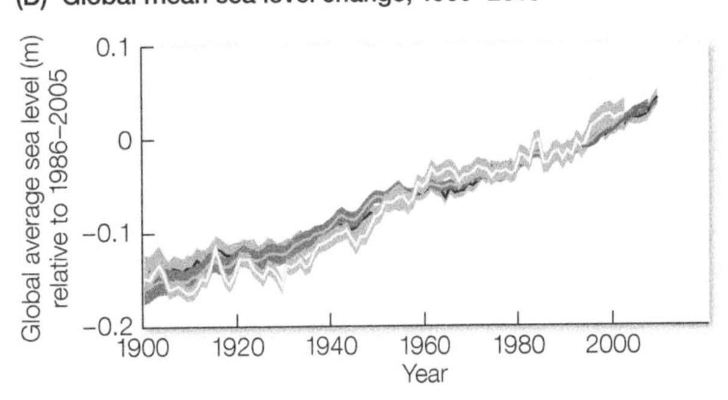

Figure 12.3 Documented changes in (A) Earth surface temperature from 1901 to 2012; (B) annual precipitation over land per year from 1951 to 2010; (C) Arctic (July to September average) and Antarctic (February) sea ice extent. (D) Changes in global mean sea level relative to the 1986–2005 mean of the longest running data set, and with all data sets aligned to have the same value in 1993 (the first year of satellite altimetry data). Colored lines in C and D indicate different data sets. (From IPCC. 2014. *Climate Change 2014: Synthesis Report. Contribution of Working Groups I, II and III to the Fifth Assessment Report of the Intergovernmental Panel on Climate Change* [Core Writing Team, R. K. Pachauri and L. A. Meyer (eds.)]. Geneva, Switzerland: IPCC.)

lower, due to fluctuations in ocean circulation. For example, since 1993, regional rates of sea level rise for the Western Pacific are up to three times greater than the global mean, while those for much of the Eastern Pacific are near zero or negative.[9] Since the beginning of the industrial era, oceanic uptake of carbon dioxide from the atmosphere has increased the partial pressure of CO_2 (pCO_2) in ocean waters, leading to acidification of the ocean. Indeed, the pH (a measure of hydrogen ion concentration) of ocean surface water has decreased by 0.1, corresponding to a 26% increase in acidity.[9]

Changes in many extreme weather events have been observed since 1950, and some of these have been linked to climate change. Extreme temperature events (heat waves and cold spells) have become more frequent and extreme.[9] The number of heavy precipitation events has increased significantly in several regions of the world.[12] Some analyses predict that weather-related natural disasters like forest fires,[13] floods,[14] and cyclones[15] may increase in frequency and magnitude as a result of climate change, though most papers acknowledge a high degree of uncertainty and variability among the predictions of different climate models.

The key point here is that climate change involves many different abiotic variables that are all changing simultaneously in terms of their averages, ranges, variability, frequency, predictability, and seasonality (see Figure 12.1, left panel). Because climate change represents a suite of variables that are changing in statistically distinct ways, this form of human-induced environmental change has the potential to alter almost every aspect of biodiversity at levels ranging from genes to species to communities to whole landscapes (see Figure 12.1, right panel). Although few documented extinctions have been shown to be caused by climate change, there is reason to believe its impacts are just beginning and that climate change may become a dominant contributor to biodiversity loss in the near future. A 2019 report from the Intergovernmental Panel on Biodiversity and Ecosystem Services[16] concluded that as many as 1 million plant and animal species are threatened with extinction in the next few decades, with climate change ranking as one of the top five drivers.

Indeed, models predict that the impacts of modern climate change on biodiversity are likely to be substantial, and empirical evidence gathered to date suggests that biological impacts of climate change have already begun and are on their way to equaling or exceeding model predictions. In the next two sections of this chapter, we detail climate models and their predictions that forecast the impacts of climate change on diversity, and then compare model predictions to biological changes that have been observed and documented with real empirical evidence. We end with thoughts on how to manage and mitigate the biological impacts of climate change on biodiversity.

Predicted Response to Climate Change

As the climate changes, species may find themselves maladapted to the new conditions in their native habitat. If these new environmental conditions move outside of a species' **climatic niche** (i.e., the set of climatic conditions to which the species is adapted), then the species must change if it is to persist. Persistence can be achieved by responding along any of three axes (**Figure 12.4**).[17] Species can change their *phenotype*, altering their morphology, physiology, or behavior to become better suited to the new local conditions. Alternatively, species can change their *phenology*, altering the timing of life-cycle events like reproduction or migration in order to adjust important biological activities to be better suited to the environment.[18] Lastly, species can shift their *spatial distributions* and move to more appropriate environmental conditions.[19] These responses can arise either by **phenotypic plasticity** in those individuals whose genotypes have the ability to express different phenotypes under different environmental conditions, or via genetic adaptation, whereby intraspecific variation within the population's gene pool allows natural selection to weed out less fit phenotypes.[20,21]

While there are ways for species to adapt and cope with climate change, the key question is whether their responses will be adequate to counter the speed and magnitude of modern climate change. Several methods and models are used to predict whether species will be able to adapt and persist in a changing climate.

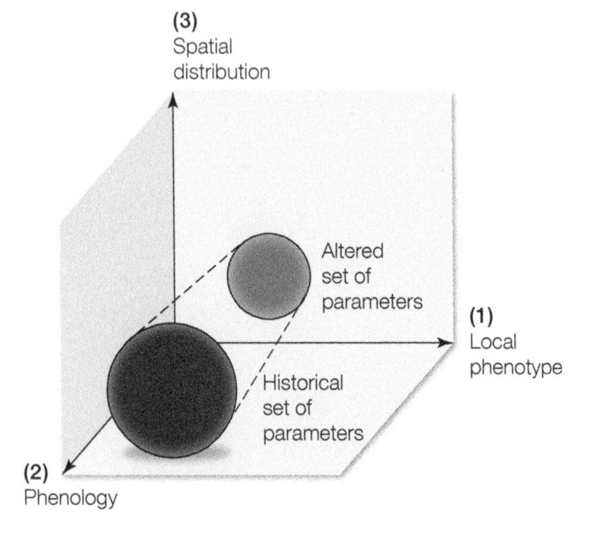

Figure 12.4 Potential responses that can allow a species to cope with climate change. (1) Change in local phenotype: alteration of the species' morphology, physiology, or behavior to be more adapted to local environmental conditions. (2) Change in phenology: adjusting life cycle events to match new climatic conditions. (3) Change in spatial distribution: populations disperse to areas with more suitable habitat, or change location on a microhabitat scale. (After C. Bellard et al. 2012. *Ecol Lett* 15: 365–377.)

Methods and models

Most approaches that predict the response of species, communities, or whole ecosystems to climate change are based on modeling studies, which generally have five steps in common (Figure 12.5). First, nearly all modeling studies begin with one or more scenarios that predict the extent to which the variables that affect Earth's climate are likely to change in the future. Scenarios like the **representative concentration pathways (RCPs)** adopted by the IPCC for its Fifth Assessment Report (AR5) give four trajectories for future greenhouse gas concentrations, with scenarios that differ based on assumptions about how societies and governments will respond to climate change. Those scenarios result in differing predictions about the range of potential increases in radiative forcing that will drive increases in average global temperature, with predictions ranging from 1.0°C to 3.7°C by 2100. These differing assumptions and predictions mean that projections of biological responses to climate change can yield contrasting results, depending on the scenarios used. Nevertheless, the scenarios bound the range of potential temperature increases and biological responses that are likely to occur in the coming century.

Once a scenario or set of scenarios that project emissions and radiative forcing has been decided on, **general circulation models (GCMs,** also called **global climate models)** are used to forecast the mean, range, and variation for future climatic variables like temperature and precipitation (step 2 in Figure 12.5). There are both atmospheric GCMs (AGCMs) and ocean GCMs (OGCMs), which are often coupled to form atmosphere-ocean general circulation models (AOGCMs). Many AOGCMs have been developed by different academic institutions or government agencies. Some examples include the Canadian Regional Climate Model (CanRCM4) developed by the Canadian Center for Climate Modelling and Analysis (CCCMA); the Community Earth System Model (CESM) developed by the National Center for Atmospheric Research in the U.S.; and the United Kingdom's Hadley Centre Climate Model (HadCM3).

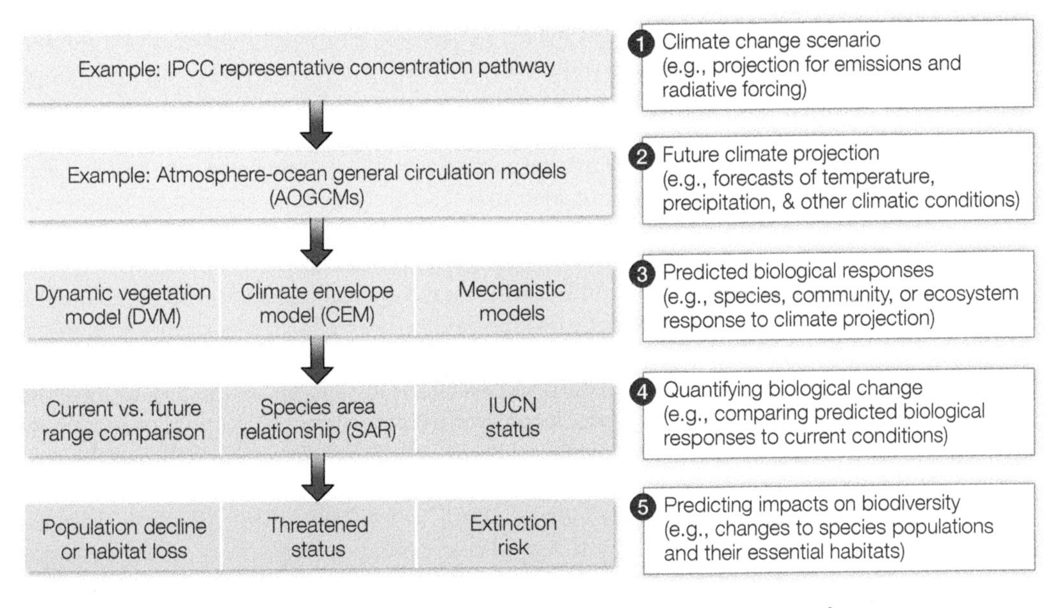

Figure 12.5 Five general steps for analyses that attempt to predict the response of species, communities, and ecosystems to climate change.

VIDEO 12.2 Watch this video about how the climate is modeled.
oup-arc.com/e/cardinale-v12.2

Though models differ in assumptions and construction, they operate similarly by dividing Earth's atmosphere, land, and oceans into three-dimensional grids of varying sizes, and then running differential equations within each grid to simulate changes in physical processes like heat transfer, radiation, relative humidity, wind, and surface hydrology. The IPCC uses the predictions of several different AOGCMs to forecast the range of future climatic conditions[22] (**Video 12.2**).

The next step is to predict biological responses to climate change (step 3 in Figure 12.5). Several kinds of models are used to make these predictions. One of the key differences among these models lies in the level of complexity at which they treat biodiversity. **Dynamic vegetation models (DVMs)** are somewhat crude in the sense that that focus primarily on plant responses to climate change. Plant species are grouped into coarse plant functional types that have similar physiological and structural properties related to climate change variables (e.g., leaf surface area, thermal regulation, water stress tolerance). DVMs use time series of climate data (e.g., temperature, precipitation, number of days with sunshine) to explain the current distribution of plant functional types, and those relationships are subsequently used to project shifts in entire vegetation zones and biomes at a regional or global scale.[17]

Climatic envelope models (CEMs, also referred to as *bioclimatic envelope* or *environmental niche* models) are more refined than DVMs, and focus on predicting the response of individual species to climate change. At their core, CEMs are simply a form of species distribution models (SDMs, introduced in Chapter 11), but where the main predictors of a species' distribution are climate-related variables. CEMs begin by developing response curves that describe relationships between a species' current distribution and variables like temperature, precipitation, and soil moisture. These relationships are then used to describe a species climatic niche—that is, the set of climatic conditions in which the species is currently found. Assuming the climatic niche is fixed and is the primary control over the species' distribution, one can compare the current climatic niche to that for future climate conditions predicted from a GCM.

Once a DVM or CEM is used to define a climatic niche, biological responses to climate change by species, communities, or ecosystems can be quantified by comparing current conditions to future conditions (steps 4 and 5 in Figure 12.5). One of the most common ways to do this is to compare the current range distribution of the species to the projected distribution for a given climate change scenario, and then predict population declines as a function of the amount of suitable habitat loss. Species-area relationships (see Chapter 8) can also be used to predict loss of diversity from trophic groups like plants as a result of the loss of suitable habitat. One can also use IUCN criteria for classifying species or habitats as vulnerable, endangered, or critically endangered (see Box 8.1) based on the amount of contraction in species range distributions or habitat loss.

An example of a CEM comes from Tanaka and colleagues,[23] who predicted the future suitability of habitat for five species of *Abies* trees (*A. nephrolepis, A. mariesii, A. veitchii, A. homolepis,* and *A. firma*). These fir species are native to Japan, China, North Korea, South Korea, and southeastern Russia, and they rank among the most abundant species in the forests of these regions. The authors used climate projections from GCMs developed by the Canadian Centre for Climate Modelling and Analysis (CCCMA) and Australia's Commonwealth Scientific and Industrial Research Organization (CSIRO) to forecast future climate conditions throughout Japan and South Asia. Projections were based on greenhouse gas emission scenario A1B, a scenario used by the Fourth IPCC Assessment. The authors then developed a CEM to explain the current climatic niche of *Abies* using four climatic variables: (1) the minimum temperature of the coldest month; (2) the mean temperature of the warmest quarter; (3) the precipitation level in the warmest quarter; and (4) the precipitation level in the coldest quarter. They then overlaid the current climate niche onto a map of future climate conditions to estimate the amount of suitable habitat projected to exist in 2070 to 2099. Their results indicated that three *Abies* species (*A. mariesii*, *A. veitchii*, and *A. homolepis*) that occur mainly on high mountains in Japan are likely to lose large extents of habitat, and may be completely replaced by other plant species. Projections for *A. nephrolepis* (Khingan fir) in China predicted this common fir species may gain some suitable habitat area in the north, but that its current distribution in the lowland habitats of northeastern China would largely become unsuitable in the near future (**Figure 12.6**). Thus, this species was deemed as vulnerable to potential extinction by climate change unless it shifts its spatial distribution northwards.

DVMs and CEMs have a number of limitations that limit the accuracy of their predictions. First, they focus almost exclusively on just one axis of the biological response to climate change—namely, the ability of organisms to spatially shift their distributions as they track optimal climatic conditions (see Figure 12.4). Most DVMs and CEMs have not considered the possibility that species populations will alter their local phenotypes or phenologies, which has potential to give misleading predictions about biological change. Second, DVMs and CEMs are phenomenological models in the sense that they simply represent statistical descriptions of the current range distribution of species, communities, or ecosystems. The models are not based on any real biological mechanisms that constrain the distributions of species or their populations. Given these limitations, there is currently a push to develop more mechanistic models that can provide improved predictions on biological responses to climate change. These models not only consider spatial redistribution through dispersal, they consider local changes

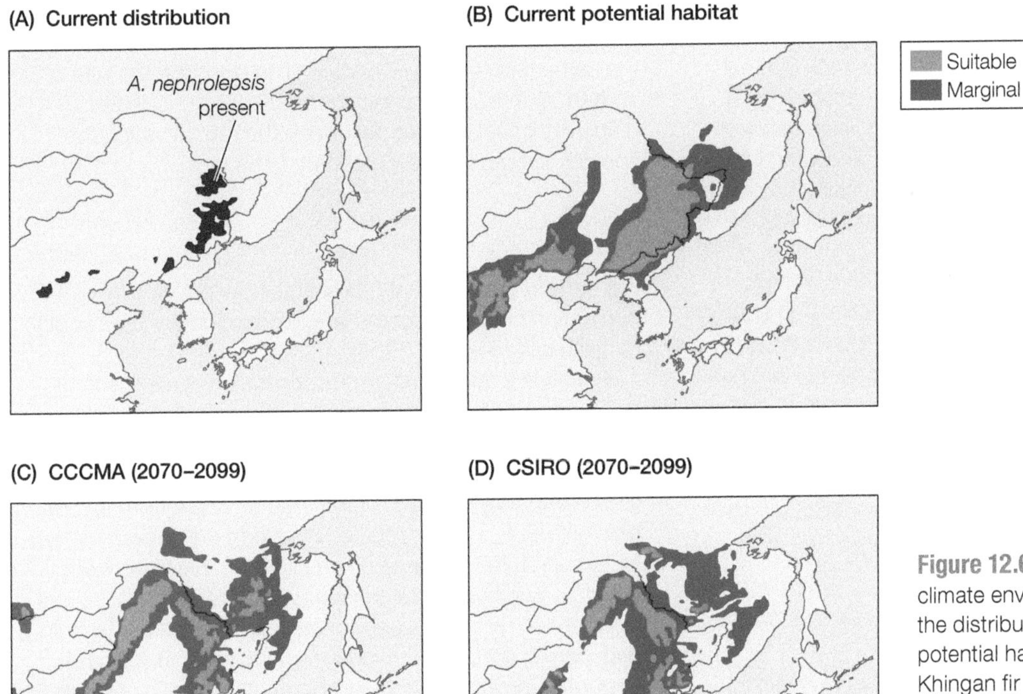

Figure 12.6 Example of results from climate envelope models (CEMs). Map of the distribution in China (A) and predicted potential habitats in East Asia (B) for Khingan fir (*Abies nephrolepis*) under the current climate. (C,D) Predicted changes in potential habitat for this species under two future climate scenarios (CCCMA and CSIRO) for 2070 to 2099. The maps are depicted at approximately 20-km resolution. (From N. Tanaka et al. 2012. *Proc Environ Sci* 13: 455–466.)

in species physiology, demography, species interactions, and even evolutionary responses through genetic adaptation (Box 12.1). More mechanistic models will almost certainly give more robust predictions about the biological responses to climate change, but they require more time and information to develop.

Current forecasts

Studies that attempt to predict the impact of anthropogenic climate change on biodiversity vary widely in their assumptions and methods. As a result, they are vary widely in predictions, suggesting anywhere from 0 to 58% of species could become extinct as a result of climate change.[24] For example, Thomas and colleagues[25] used CEMs to forecast the distributions and amount of suitable habitat

BOX 12.1 **Conservation in Practice**

Toward Mechanistic Climate Envelope Models

As anthropogenic climate change accelerates, one of the most urgent needs is to develop accurate predictions of species responses that can help guide the protection and management of Earth's biodiversity. Nearly all forecasts of biological responses to climate change come from some form of climate envelope models (CEMs), which estimate a species' current climatic niche and then project that niche onto a map of future climatic conditions.

But current models are crude tools with limited accuracy. Most focus exclusively on the ability of organisms to spatially shift their distributions as they track optimal climatic conditions. They leave out important biological information about species demographics, dispersal abilities, physiology, potential for adaptation, and interactions with other species in a community. In addition, they are not based on real biological mechanisms that involve survivorship and reproduction.

There is currently a push to develop more biologically realistic models that provide better predictions on species responses to climate change. Models being developed now generally account for six biological components (left panel of the Figure) that influence a species response to climate change. These include:

- Physiology: Temperature-dependent population growth
- Demography: Explicit models of population dynamics
- Evolution: The ability to adapt through natural selection and gene flow
- Dispersal: Realistic models of dispersal distances and rates
- Species interactions: Interactions like competition or predation that influence survivorship and reproduction
- Environment: Spatial climatic gradients and rates of abiotic (including climate) change

CEMs accounting for these six components are modeled using a set of differential equations (detailed in the Figure) that track the population dynamics, movement, and physiological or evolutionary response of species. Equation 1 in the figure tracks changes in the size of a population by modeling population dynamics (shaded in purple) and the rate of dispersal (green). Population dynamics for a species i can be represented by any function g_i (geometric, exponential, logistic growth equations or their derivatives). The rate of dispersal of the species is the same term used in Chapter 11 to describe the rate of spread of an invasive species (see Box 11.1). D is the diffusivity constant that describes the rate at which the species spreads to new locations. The term

$$\frac{\partial^2 N}{\partial x^2}$$

is a density gradient that represents how the density of the population changes along a single coordinate x (e.g., south to north latitudes).

Equation 2 takes the term for population dynamics from Equation 1 and converts fitness (species reproduction and growth) at each location x into two parts. The first part (in light purple) accounts for species physiology by modeling population growth as a function of temperature. The maximum possible growth (r_{max}) is discounted by a term that measures how far temperatures (TC) are from the species thermal growth optima z, which is then standardized by a term that accounts for selection on thermal performance ω. The second term (in dark purple) accounts for species interactions. In this case, population growth is discounted to account for competition among species (α) for biologically limited resources.

Equation 3 allows the species thermal growth optima (z in Equation 2) to evolve and increase fitness. The first term in Equation 3 allows directional selection to eliminate for 1103 plant and animal species by 2050 under three scenarios of climate change: a scenario of minimum expected change (0.8°C –1.7°C), a mid-range scenario (1.8°C –2.0°C), and a scenario of maximum expected change (greater than 2.0°C). The authors then used range comparisons, species-area relationships (SARs), and IUCN Red List criteria (see Figure 12.5, step 4) to forecast species extinctions. Based on their modeling efforts, the authors projected that 34%–58% (minimum to maximum scenario) of species are committed to extinction for the given climate change scenarios if they are unable to disperse to new locations, while 11–33% are committed to extinction if they can disperse to future areas that are within their current climatic niche. For certain groups of organisms, the predicted estimates of extinction were exceptionally high, such

individuals from the population that have growth optima with low fitness. The rate of selection is proportional to the amount of genetic variation in the population V and the rate at which variation is lost in small populations q. The second term in Equation 3 accounts for the addition of new genetic variation to the population through gene flow that occurs when individuals from other locations x enter the local population. These "immigrant" individuals influence the distribution of traits that can be selected on by adding new variants with different thermal optima z.

While more mechanistic and biologically realistic models will certainly provide greater accuracy when predicting species' responses to climate change, they require more time to develop and more information to properly parameterize the terms. The challenge for future practitioners is to find the balance between model accuracy and complexity that is needed to protect biodiversity in the face of anthropogenic climate change.

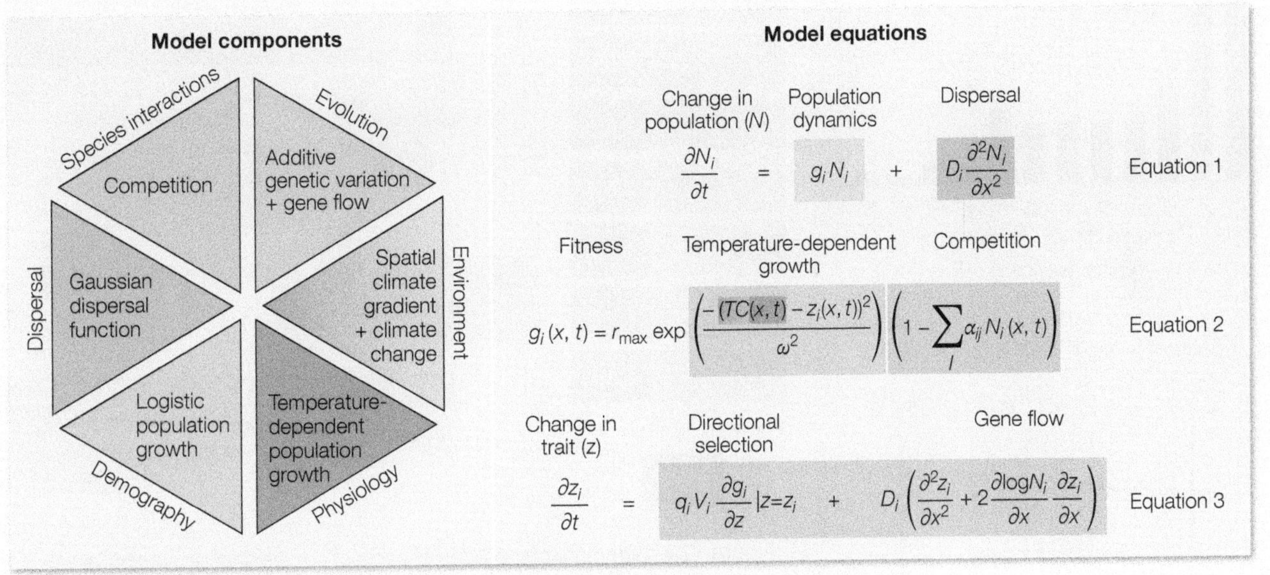

A generalized mechanistic climate envelope model (left) that accounts for at least six biological components of a species' response to climate change. The six components are matched by color to their representation in equations on the right. N = population size, i = species i (1 to I), t = time, g = fitness, D = diffusion constant for dispersal, x = spatial position along a single dimension, r_{max} = maximum intrinsic population growth rate, T = temperature, C = temperature change, z = temperature optimum for each species, ω = width of selection on thermal performance trait, α = strength of competition among species i and j, q = reduction in genetic variance at low population abundances, V = additive genetic variation in thermal optimum. (After M. C. Urban et al. 2016. *Science* 353: aad8466.).

as the projected 87% of Amazonian plants, or 48% of European birds projected to go extinct under scenarios of maximum climate change.

In contrast to more dire predictions about climate change, other studies have suggested that far fewer species are likely to be driven extinct by climate change if we make different assumptions. For example, Foden and colleagues[26] argued that CEMs overestimate the fraction of species of birds, amphibians, and corals that are likely to be driven extinct because they generally fail to consider the potential for species to adapt to new local environmental conditions. The authors argued that when adaptive traits are considered, just 6%–9% of bird species, 11%–15% of amphibian species, and 6%–9% of coral species are both vulnerable to climate change and presently threatened with extinction. Warren and colleagues[27] further showed that if reductions in CO_2 emissions allow more time for species to redistribute themselves, then the proportion of species that will lose large fractions of their current climatic range will be greatly reduced, limiting extinctions even further.

Given widely varying forecasts, Urban[24] recently summarized 131 published predictions of biodiversity response to climate change using a meta-analysis. The majority of predictions included in Urban's analyses used DVMs or CEMs to predict suitable habitat under future climate scenarios. A smaller number of predictions came from mechanistic models of physiology or demography (15%), from species-area relationships (5%), or were just expert opinions (4%). Across all 131 published studies, the mean prediction is that 8% of species are likely to be driven extinct by climate change (**Figure 12.7A**). Estimates of extinction ranged from

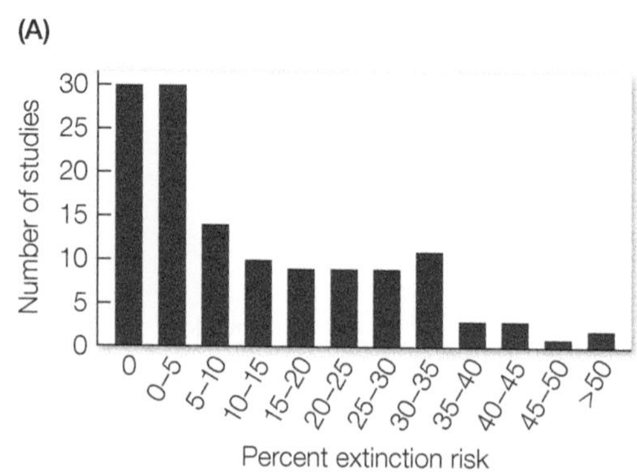

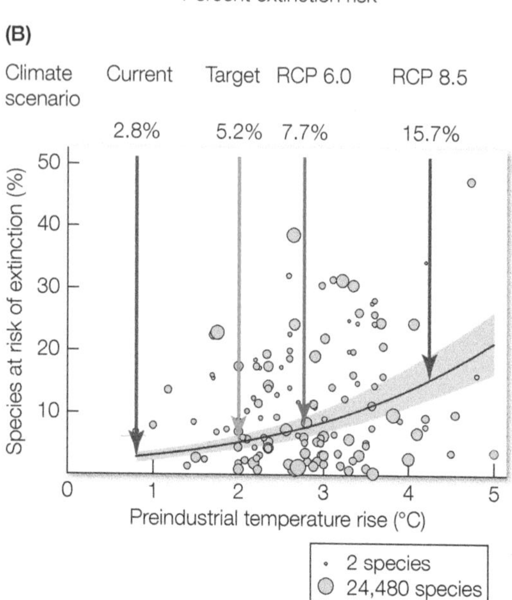

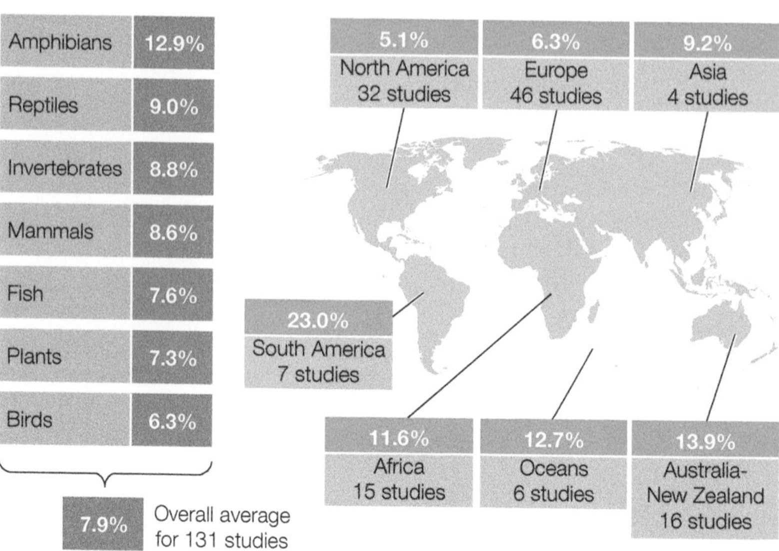

Figure 12.7 Summary of predicted impacts of climate change on biodiversity. (A) Percentage of species at risk of extinction based on a synthesis of the predictions from 131 independent studies. The mean extinction risk from all studies is 7.9% with a 95% confidence interval of 6.2% to 9.8%. (B) The predicted extinction risk from climate change depends on the future climate scenarios: the current postindustrial temperature rise of 0.8°C, the policy target of 2°C, and RCPs 6.0 and 8.5. The gray band indicates 95% confidence intervals for extinction risk. Circles are proportional to the number of species considered in each study. (C) Predicted extinction rates by taxonomic group (left) and geographic region (right). (A–C after M. C. Urban. 2015. *Science* 348: 571–573. C, also after R. McSweeney. 2015. *Carbon Brief* 30 April 2015. London: Carbon Brief Ltd. www.carbonbrief.org/climate-change-threatens-one-in-six-species-with-extinction-study-finds.)

2.8% of species already committed to extinction by postindustrial temperature increase, to 5.2% committed to extinction under the current international policy target of a 2°C postindustrial rise, to a high of 15.7% species extinctions for RCP 8.5 (**Figure 12.7B**). The 15.7% estimate represents the IPCC "business-as-usual" pathway of CO_2 emissions, which is expected to increase global temperatures by 4.5°C.[24] Another primary factor that influenced predictions was the threshold of habitat loss researchers assumed would commit a species to extinction. For example, assuming that a species would go extinct after 80% of their habitat was lost more than tripled the predicted extinction risk (from 5% to 15%) compared to those studies that assumed 100% habitat loss was required for extinction.

Urban's survey[24] also found that predicted extinction rates differ significantly among different taxonomic groups and for different regions of the globe (**Figure 12.7C**). The highest risk of extinction is projected for amphibians (12.9%) and reptiles (9%), while birds (6.3%) and plants (7.3%) are predicted to have the lowest extinction risk. North America and Europe are expected to have the lowest percentage of extinctions due to climate change (5.1% and 6.3%, respectively), whereas South America (23%) and Australia and New Zealand (13.9%) are expected to have the highest due to their diverse assemblages of endemic species, many of which have small range sizes. While 60% of projections made to date focus on North America and Europe, many other regions have been poorly studied. For example, estimates of biodiversity loss in Asia stem from just four published studies.

Pearson and colleagues[28] used machine-learning algorithms to identify the most important predictors of extinction risk in CEMs that have been developed for 36 amphibian and reptile species. The most important variables driving extinction risk in these models were a mixture of spatial and demographic factors (**Figure 12.8**). The total area presently occupied by a species was consistently the most important predictor of future extinction risk from climate change, with extinction risk increasing precipitously for populations that occupy less than 3000 km^2. Population size was the second most important predictor, with extinction risk increasing precipitously for populations less than 8000 individuals. These two factors were likely important because they are correlated with the breadth of climatic and habitat conditions under which a species currently exists. The third and fourth most important predictors were the spatial correlation in climatic conditions among spatially subdivided populations, and the species' generation length (time, in years, from birth to sexual maturity and reproduction). Compared to spatial and demographic variables, recent trends in population size, area occupied, and fragmentation of habitat were weak predictors of extinction risk due to climate change. This finding suggests that biological impacts of climate change observed in the recent past cannot necessarily be used to predict future impacts as climate change accelerates.

While most predictions of the effects of climate change on biodiversity have focused on terrestrial environments, CEMs have also been used to predict effects on marine biodiversity.[29,30] For example, Cheung and colleagues[29]

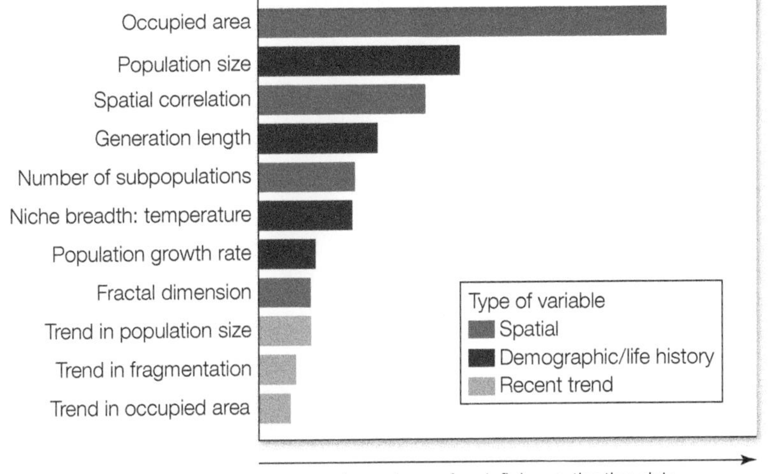

Figure 12.8 Ranked importance of the top predictors of extinction risk due to climate change based on CEMs for amphibian and reptile species. Predictors are categorized as spatial (blue), demographic/life history traits (red), and recent historical trends (yellow). (After R. G. Pearson et al. 2014. *Nat Clim Change* 4: 217–221.)

used CEMs to project the future range distributions of 1066 species of fish and marine invertebrates. The authors predicted that climate change will generate numerous local extinctions in the subpolar regions of the globe, as well as in the tropics and semi-enclosed seas. Their models predicted that polar species, which have particularly narrow temperature tolerances, are at high risk of extinction due to shrinking of their low-latitude range boundaries that will occur as oceans warm. However, the models also predicted the simultaneous poleward expansion of a large number of species as warming increases high-latitude range limits. Therefore, local loss of polar species is expected to occur alongside introduction of many new species, possibly generating as much as 60% turnover of the present biodiversity and potentially leading to increases in polar species richness.

Documented Responses to Climate Change

In the previous section, we described how biodiversity is expected to respond to climate changes based mostly on the predictions of models that forecast biological responses to climate scenarios for some 30 to 70 years in the future (2050 to 2100 are common time scales). In this section, we describe the empirically documented evidence for biological responses to climate change that have already happened.

We begin by noting that documented global extinctions caused by climate change are surprisingly rare. For example, only 33 of 872 (3.7%) of the global extinctions recorded on the IUCN Red List cite climate change as one of the potential causes of these extinctions, and evidence linking them directly to climate change is tenuous at best. In fact, the first clear evidence of global extinction driven by climate change occurred in 2019, when the Bramble Cay melomys (*Melomys rubicola*) was declared extinct due to climate-induced inundation of its island habitat off the coast of Australia (**Box 12.2**).[31]

While there is scant evidence to date for global extinction, there is abundant evidence that climate change has already caused ecological changes that are threatening the persistence of entire groups of organisms.[16,32–34] Here we review the impacts of climate change on population decay and local extirpations, altered species phenologies that are reducing fitness, and wholesale distribution shifts in communities and biomes.

Population decay and local extirpation

There is substantial evidence that climate change has already led to narrowing of species geographic range distributions and contributed to population declines that are threatening persistence. MacLean and Wilson[32] reviewed 74 studies that provided 130 empirical observations of how species have already responded to climate change in terms of changes in their geographic ranges and population sizes. Using the IUCN Red List criteria, the authors converted estimates of range constriction and population decay into estimates of extinction risk, which allowed them to compare increases in extinction risk that have already been observed to those predicted by modeling studies over comparable time periods. The authors showed that observed extinction risk has increased by 14% due to climate change, while the predicted increase from modeling studies is 10% (**Figure 12.9**). The point here is that the models are coming close to predicting actual population declines and range contractions of real species.

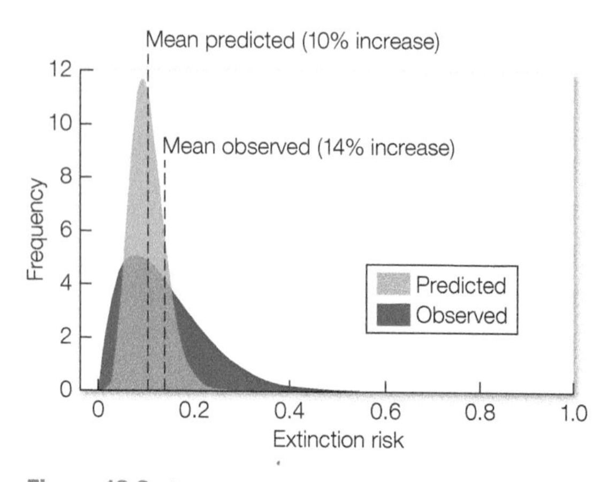

Figure 12.9 Observed versus predicted extinction risk caused by climate change. Frequency distributions comparing the observed increases in extinction risk caused by climate change to the risk predicted by models that forecast range constrictions and population declines of comparable taxa. This comparison indicates that current models can be quite accurate in predicting actual extinction risk. (After I. M. D. Maclean and R. J. Wilson. 2011. *Proc Nat Acad Sci* 108: 12337–12342.)

There is also substantial evidence that climate change has led to local extirpation of certain species throughout the world. Cahill and colleagues[33] compiled a brief list of examples of local extirpations that

BOX 12.2 Case in Point

Meet the First Species to Go Extinct Because of Anthropogenic Climate Change

On a small island off the northern tip of the Great Barrier Reef, a small rodent called the Bramble Cay melomys (*Melomys rubicola*) once made its living by dining on plants and occasional turtle eggs. Just 15 cm (6 inches) long, this small, cute mammal was recognizable from its prehensile tail that is covered in scales (see the Figure).

The melomys' downfall was its narrow habitat distribution, limited to a single 3.6-ha island called Bramble Cay. Bramble Cay is a grassy sand cay (island) that sits less than 3 meters (10 feet) above sea level. As sea levels rose around the Great Barrier reef, storms and high tides caused extensive flooding that not only drowned and swept away the animals, it killed the vegetation that melomys used for food and shelter.

The melomys was last seen in 2009 by fishermen. In 2016, Australian scientists reported that the Bramble cay melomys had likely gone extinct. The Australian government declared the species extinct in 2019,[34] and this declaration was confirmed by the IUCN two days later.

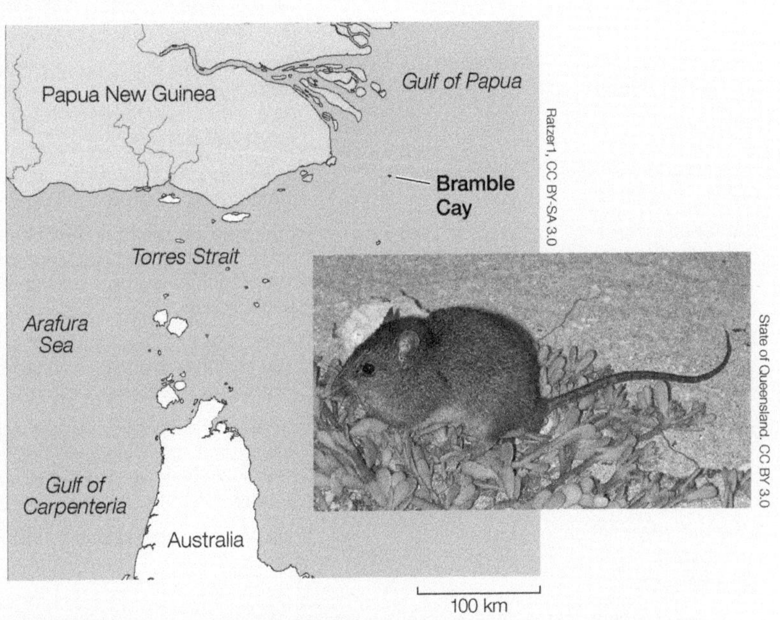

The Bramble Cay melomys (*Melomys rubicola*) is the first known global extinction caused by climate change. The species was declared extinct after rising oceans repeatedly flooded its lone island habitat on Bramble Cay.

The Bramble cay melomys is believed to be the first mammal to be driven globally extinct by rising sea levels, and it may well be the first of all species to succumb to anthropogenic climate change.

have been attributed to various aspects of climate change (Table 12.1). In some instances, species have gone locally extinct because they have been unable to tolerate changing temperature extremes. For example, Sinervo and colleagues[35] monitored populations of 48 species of *Sceloporus* lizards at 200 sites in Mexico and showed the genus had gone locally extinct from 12% of sites due to new temperature extremes that extended beyond the lizards' thermal niche. In other instances, extirpations have occurred because climate change caused habitat loss. For example, Trape[36] estimated that more than a third of the local populations of Adrar mountain fishes of Mauritania have been extirpated as climate-induced drought has led to perennial water bodies drying up. In still other instances, species have gone locally extinct because climate change has altered species interactions. Durance and Ormerod[37] studied *Planaria* (invertebrate flatworms) in Welsh streams in the United Kingdom and showed that increasing water temperatures caused by a large North Atlantic Oscillation (a weather phenomenon related to atmospheric pressure at sea level) intensified competition among planarians for limited resources, causing less competitive flatworm species to go locally extinct from streams. Thus, there is strong evidence that climate change not only leads to range contractions and population decay, but that it causes local extirpations.

TABLE 12.1 Examples of local extinctions caused by climate change

Species	Location	Proximate cause of local extinction	Reference
American pika (*Ochotona princeps*)	Great Basin region, USA	Limited physiological tolerance to temperature extremes (both high and low)	E. A. Beever et al. 2010. *Ecol Appl* 20: 164;[97] E. A. Beever et al. 2011. *Glob Change Biol* 17: 164.[98]
Planarian (*Crenobia alpina*)	Wales, UK	Loss of prey resources and increased competition with other planarians as a result of increasing stream temperatures	I. Durance and S. J. Ormerod 2010. *J. N. Am Benth Soc* 29: 1367[37]
Desert bighorn sheep (*Ovis canadensis*)	California, USA	Decreased precipitation led to altered plant community (food source)	C. W. Epps et al. 2004. *Cons Biol* 18: 102[99]
Edith checkerspot butterfly (*Euphydryas editha bayensis*)	San Francisco Bay area, USA	Increase in variability of precipitation corresponding with reduction of temporal overlap between larvae and host plants	J. F. McLaughlin et al. 2002. *Proc Nat Acad Sci* 99: 6070[43]
Clown goby (*Gobiodon* sp. *A*)	New Britain, Papua New Guinea	Destruction of obligate coral habitat due to coral bleaching caused by increasing water temperatures	P. L. Munday 2004. *Global Change Biology* 10: 1642[100]
Spiny lizards (48 species from the genus *Sceloporus*)	Mexico	Increased maximum air temperature approaches physiological limit, seemingly causing decreased surface activity during the reproductive season	B. Sinervo et al. 2010. *Science* 328: 894[35]
Adrar mountain fishes (multiple genera and species)	Mauritania	Loss of water bodies due to drought	S. Trape 2009. *PLOS ONE* 4: e4400[36]

Source: From A. E. Cahill et al. 2013. *Proceedings of the Royal Society B* 280: 9.[32]

Geographic range shifts

One of the strongest signals of biotic change from anthropogenic global warming is the induction of geographic range shifts as species adjust their spatial distributions to track new environmental conditions (**Video 12.3**). A pattern of range shifts (generally poleward, or upward in elevation) has been documented in hundreds of plant and animal species.[38,39] Parmesan and Yohe[38] reviewed range boundaries for 99 species of birds, butterflies, and alpine herbs and showed that the majority of species distributions had shifted toward the poles, at an average rate of 6.1 km per decade. Hickling and colleagues[40] analyzed changes in the northern range boundaries of 329 species in Britain representing a variety of taxonomic groups (e.g., millipedes, spiders, butterflies, beetles, dragonflies, fish, reptiles, amphibians, birds, and mammals). Of the 329 species considered, 84% (275) showed clear evidence of northward range shifts between the years 1960 and 2000.

One of the earliest and best studied examples of a species range shift driven by climate change is that of Edith's checkerspot butterfly (*Euphydryas editha*), a North American species with a historical distribution concentrated in northern Mexico and the southern United States, but extending up the west coast of North America into Canada (**Figure 12.10A**). Parmesan[41] monitored populations of the butterfly in the 1990s and showed the species had exhibited significant latitudinal and altitudinal shifts in its distribution. Much of the range shift was driven by high rates of local extirpation of populations in more southerly portions of the species' range and at low elevations. Populations were four times more likely to go extinct in Mexico than in Canada, and populations above 2400 m in elevation were more persistent than those at lower elevations (**Figure 12.10B**). Subsequent studies showed that extirpation of local populations was being driven largely

VIDEO 12.3 Watch the U.S. National Park Service's video on species range shifts in Sequoia-Kings Canyon National Park.
oup-arc.com/e/cardinale-v12.3

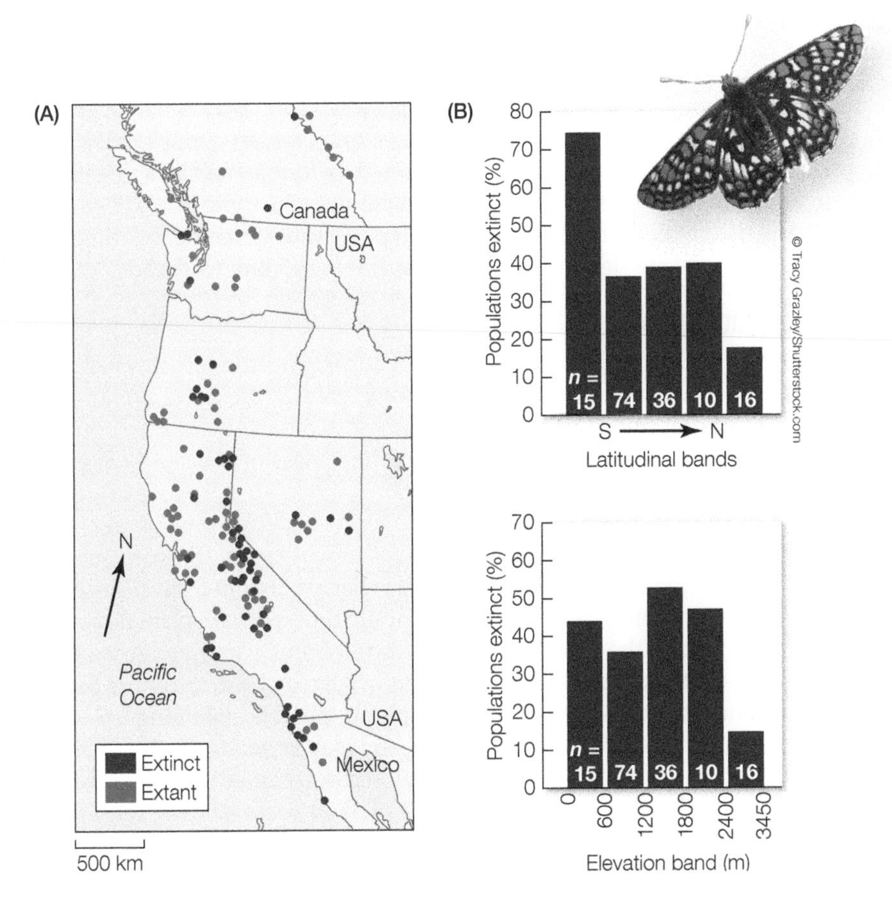

Figure 12.10 Geographic range shifts in the Edith Checkerspot butterfly (*Euphydryas editha*). (A) A map of the distribution of the butterfly showing locations of extinct (red) versus extant (blue) populations. (B) Percent population extinctions as a function of latitude (top) and elevation (bottom). (After C. Parmesan. 1996. *Nature* 382: 765–766.)

by climate change, particularly by changes in precipitation and snowpack that reduced the availability of the species' host plant.[42,43]

There are now many well-documented examples of climate induced range shifts in marine environments as well (**Figure 12.11**).[44,45] For example, Poloczanska and colleagues[46] reviewed 1735 biological responses of marine organisms to climate change and found that more than 80% responded to climate change in the predicted direction. Range shifts of marine organisms were caused by both **cold-edge expansion** (expansion into regions beyond the species' former "cold"

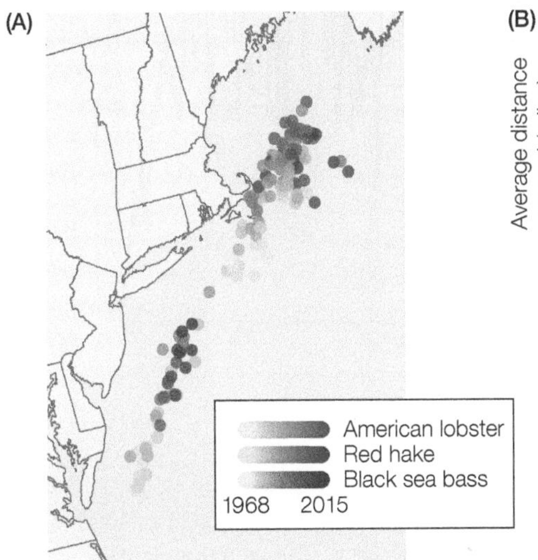

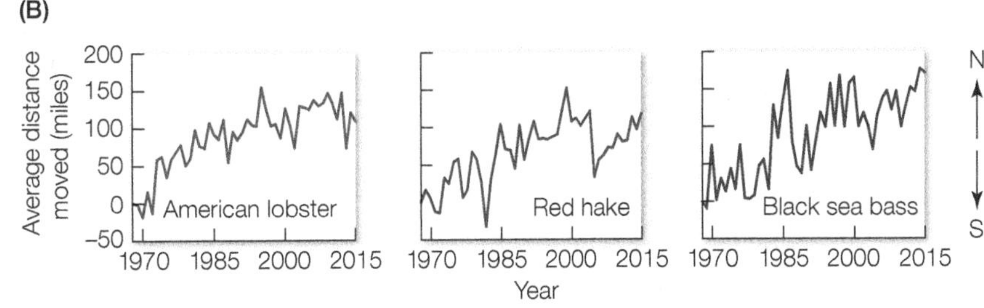

Figure 12.11 Range shifts in marine habitats. (A) Changes in the locations of populations of a crustacean (lobster) and two fish species (hake and sea bass) along the northeast coast of the United States from 1968 to 2015. (B) Average distance moved (in miles) over the same timeframe. (From U.S. Environmental Protection Agency [EPA]. 2016. *Climate Change Indicators in the United States*. www.epa.gov/climate-indicators. Data from National Oceanic and Atmospheric Administration [NOAA] and Rutgers University. 2016. OceanAdapt. oceanadapt.rutgers.edu.)

limits; also called leading-edge expansion) and **warm-edge contraction** (inability to remain in areas that have become too warm). The mean rate of expansion at the leading (cold) edge of ranges for marine species was 72 km per decade, which is nearly an order of magnitude faster than rates reported for terrestrial species. The fastest leading-edge expansions were documented for highly mobile or dispersive pelagic organisms like phytoplankton (470 km per decade), bony fish (278 km per decade), and invertebrate zooplankton (142 km per decade). Trailing (warm-edge) range contractions were significantly slower, averaging 15 km per decade.

Altered phenologies

Another well-documented biological signal of anthropogenic climate change is the alteration of species phenologies. **Phenology** refers to life-cycle events (e.g., reproduction, migration) that plants and animals exhibit seasonally, annually, or interannually. Climate change can impact the fitness of species by altering the timing of these life cycle events.[18]

One of the best examples of climate change altering a species' phenology comes from observations of cherry trees (*Prunus jamasakura*) in Japan. In Kyoto, records have been kept on the date of celebration of the cherry blossom festival since the ninth century. This record gives the longest running time-series of plant flowering date from any place in the world. Primack and colleagues[47] analyzed more than 1200 years of the timing of flowing of cherry trees in Japan. Over the entire period of record, the first day of flowering has been highly variable, spanning a 6-week range from as early as late March to as late as early May (Figure 12.12). By the 1980s and early 1990s, average flowering times had become earlier than at any time previously during the entire flowering record of 1200 years. During the period from 1971 to 2000, plants flowered an average of 7 days earlier in comparison to the average of all previous records. Reconstructed temperature records show that earlier flowering dates correspond to increased regional temperatures.

Many additional studies show that plants around the world have altered the timing of their flowering as a result of climate change.[18,48] Amano and colleagues[49] reviewed the timing of first flowering dates for 405 plant species in the

(A)

From R. B. Primack et al. 2009. *Bio Cons* 142: 1943–1949. Photo by Hiroyoshi Higuchi

(B)

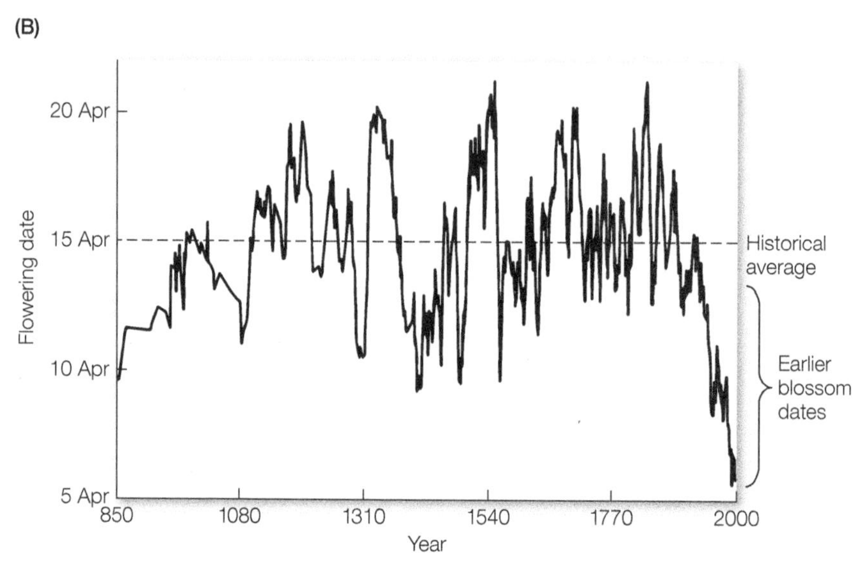

Figure 12.12 Cherry blossoms in Japan. (A) People enjoying the cherry blossom festival in Ueono Park, a popular spot in the center of Tokyo. (B) Ten-year running averages of cherry blossoming dates in Kyoto. Each point represents the average of 1 year and the 5 flowering dates earlier and later than that date. (B, after R. B. Primack et al. 2009. *Bio Cons* 142: 1943–1949. Data from Yasuyuki Aono, personal communication.)

United Kingdom from 1753 to 2008. The authors then converted flowering dates to a single community-level "index" that displays how the average timing of reproduction has changed for plants in the U.K. Analyses showed that flowering dates in the most recent 25 years have been 2.2 to 12.7 days earlier than any other 25-year period since 1760. Early spring flowering dates were closely correlated with February–April temperatures in central England, with flowering occurring 5.0 days earlier for every 1°C increase in temperature. Many other aspects of plant phenology, such as the date of leaf unfolding and fruit ripening, are shifting to earlier days in the year, while the date of fruit harvesting and leaf fall are shifting to later dates as growing seasons are extended by climate change.[48,50]

Phenological shifts have been documented for many types of animals as well. The appearance and day of first flight of butterflies has advanced in the U.S.,[51] the U.K.,[52] and Spain,[53] in some locations occurring a month earlier than these events have historically occurred.[51] The timing of reproduction in amphibians has advanced by several weeks per decade as a result of increasing daytime and nighttime temperatures.[54,55] Dates of egg-laying in some birds have advanced by several days per decade, while others have delayed their long-distance migration to breeding grounds due to later cues received in their overwintering habitat.[56–58] Cohen and colleagues[59] summarized more than 1000 time-series of how animal phenology responds to climate change for insects, mammals, reptiles, and birds spanning 5 continents. The authors showed that animals have, on average, advanced the timing of their major life cycle events by 3 days per decade, with most of these advances explained by increasing temperatures.

By themselves, changing phenologies are not necessarily a threat to biodiversity. However, they become a threat when such changes reduce the fitness of species and their populations. One way this can happen is when changing phenologies lead to population asynchronies.[60,61] **Population asynchronies** occur when interactions between two or more species become out of phase, leading to a reduction in fitness of at least one of those species. For example, Visser and colleagues[62] studied how the timing of reproduction in the Dutch population of great tits (*Parus major*) responded to climate change (Figure 12.13). The authors showed that date of egg-laying by the birds was not influenced by climate change over a 24-year period. However, one of their primary food sources—caterpillars—responded strongly to climate change, hatching sooner each year, causing peak caterpillar biomass to come earlier in the season. Because the caterpillars hatched sooner than the birds, the birds' reproduction became out of sync with the

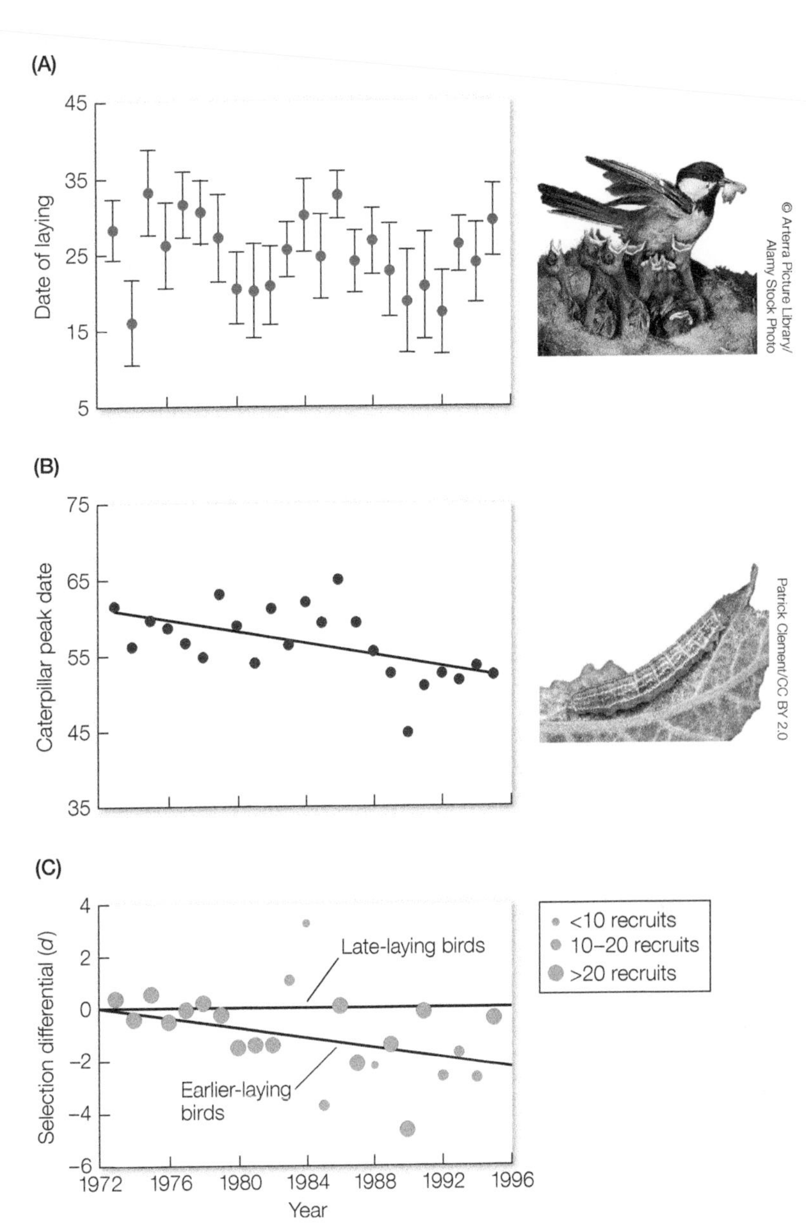

Figure 12.13 Example of a population asynchrony. (A) Over time, the egg laying phenology of great tits has not changed in response to climate change. Numbers on the *y*-axis indicate the number of days eggs have been laid after April 1. (B) However, the timing of reproduction of the bird's primary food source—caterpillars—*has* responded to climate change, advancing by several days per decade (relative to April 1). (C) This asynchrony has led to selective pressure on the birds to shift the timing of their reproduction. In this study, based on the number of surviving offspring (recruits) per female, a negative selection differential indicates a selective advantage for earlier-laying birds. (After M. E. Visser et al. 1998. *Proc Roy Soc B* 265: 1867–1870.)

timing of reproduction in their food source, leading to higher mortality of chicks and increased selective pressure for earlier egg-laying.

Population asynchronies appear to be common among predators and their prey, among competitors, and among species involved in mutualistic relationships.[33,63] For example, Pyke and colleagues[64] studied plants and their pollinators in the Rocky Mountains of Colorado over a 33-year period. While flowering times of most plant species shifted toward earlier in the season, bumblebee phenology generally did not change over this time period, resulting in asynchrony between the availability of pollinators and the timing of flowering by host plants. Reduced synchrony led to declines in the abundance of bumblebee populations.

Biome regime shifts and alternative states

Most studies to date have focused on the effects of climate change on individual organisms or on species and their populations, emphasizing effects on the phenology and physiology of organisms, as well the distribution and range shifts of species. But some of the most dramatic effects of climate change occur at larger biological scales, where whole biomes shift or change states, leading to the regional loss of entire communities of organisms.[60]

There are numerous examples of climate-induced shifts in entire biomes. One of the best studied examples is the expansion of woody vegetation into the Arctic tundra. Over the last 150 years, warming has led to northward expansion of woody plant species (e.g., alder, birch, willow) into high-latitude ecosystems,[65] and woody vegetation has replaced large areas of herbaceous vegetation. Peñuelas and Boada[66] documented a similar climate-induced biome shift in the Montseny mountains of Spain (**Figure 12.14**). This region was historically characterized by beech (*Fagus sylvatica*) forests in mid-elevations (1200–1400 m) and heather (*Calluna vulgaris*) heathlands and grasslands in the mid- to high elevations (1400–1700 m). But since the mid 1900s, the high-elevation heathland habitat has been replaced by beech (*Fagus sylvatica*) as the forest has undergone an altitudinal shift. The mid-elevation heathland and grassland systems have been replaced by Mediterranean forests of broad leaf oak (*Quercus ilex*). These biome shifts have led to nearly complete loss of the local heathland habitat and its biological community.

There are also many examples of climate-induced state changes to biomes. For example, in parts of Africa, Australia, and South America, altered precipitation and fire frequencies driven by climate change are shifting ecosystems from forests to savannas.[67] In oceans, warming waters have led to increased frequency of coral bleaching events, as well as population declines of major reef-forming species in entire regions of the globe (e.g., staghorn and elkhorn corals from the Caribbean).[68] Some coral researchers argue these trends are causing a phase shift in coral reefs to alternate states characterized by the loss of reef-forming cnidarians (a diverse group of aquatic invertebrates that includes corals) and overgrowth by macroalgae.[69]

One of the most dramatic examples of a regime shift and rapid transition to an alternative state occurred in Australia starting around 2011.[70] Much of the western coast of Australia, from Kalbarri to Cape Leeuwin, had been lined with kelp forests. But after decades of ocean warming, a set of extreme marine heat waves that began in 2011 forced a 100-kilometer range contraction of extensive kelp forests. The kelp forests were replaced by persistent seaweed turfs as well as invertebrates, corals, and fishes that were characteristic of subtropical and tropical waters (**Figure 12.15**). This community-wide shift altered key ecological processes (e.g., changes in herbivory and competitive interactions) that enhanced the persistence of the tropical community and suppressed the recovery of kelp forests.

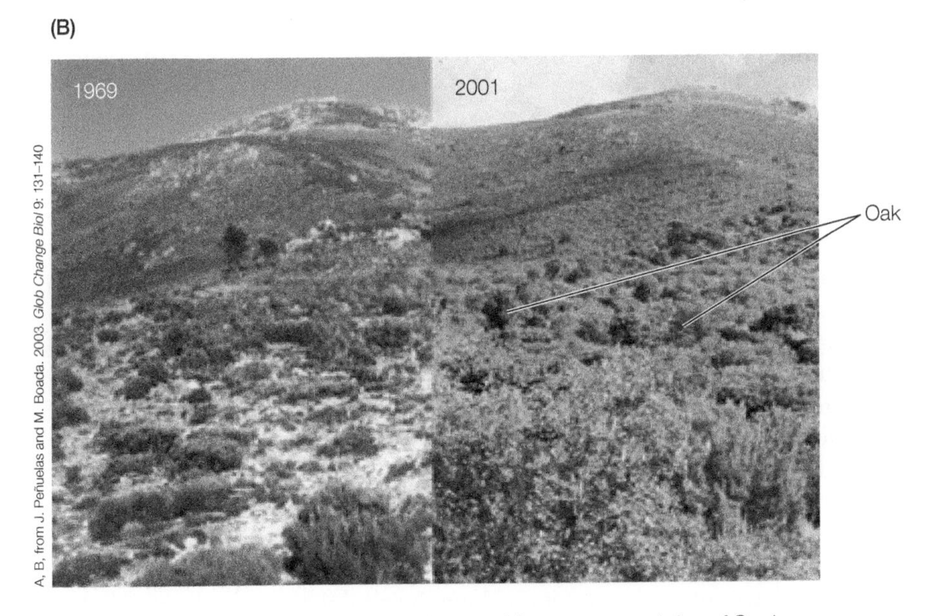

A, B, from J. Peñuelas and M. Boada. 2003. *Glob Change Biol* 9: 131–140

Figure 12.14 Climate-driven biome shift in the Montseny mountains of Spain. (A) High-elevation (1700 m) heathland dominated by heather (*Calluna vulgaris*) is being replaced by a beech (*Fagus sylvatica*) forest that has undergone an altitudinal shift. (B) Mid-elevation (1400 m) heathland and grassland are being replaced by a Mediterranean forest of broad leaf oak (*Quercus ilex*).

Change in ecosystem services

Biological changes induced by anthropogenic climate change are expected to alter the goods and services that ecosystems provide to humanity.[71] Although changes to many ecosystem services have yet to be realized because the shuffling and redistribution of life on Earth has yet to fully play out, there is a growing list of examples of ecosystem services that have already been altered by climate change. In 2017, Pecl and colleagues[72] collated 30 examples of climate-driven changes in species distributions that have impacted human well-being,

(A)

(B)

(C)

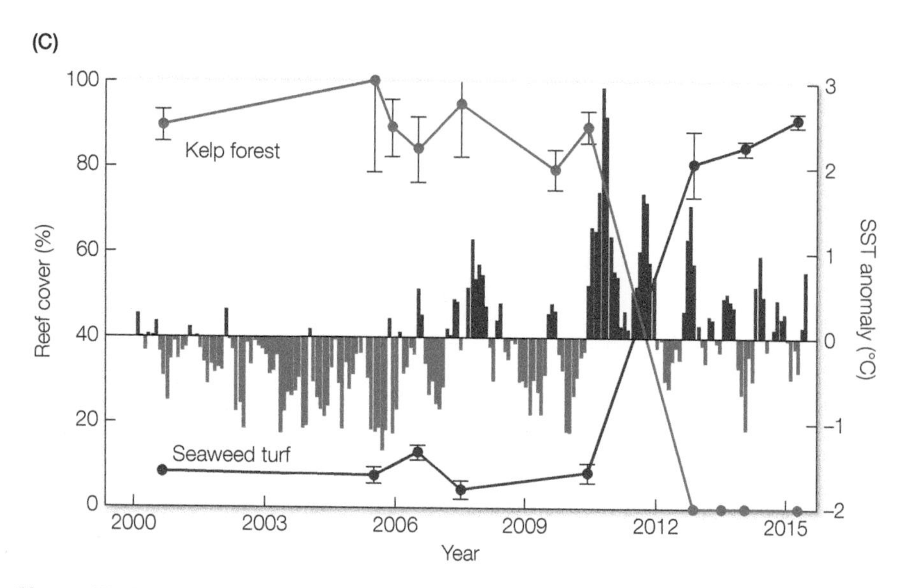

A, B, from T. Wernberg et al. 2016. *Science* 353: 169–172

Figure 12.15 Climate-induced regime shift. Kelp forests (A) were dense in Kalbarri, along the western coast of Australia, until 2011, when the kelps disappeared from hundreds of kilometers of coastline and were replaced by seaweed turfs (B). (C) The shift from kelp forest cover (blue trace) to seaweed turf cover (red trace) coincided with sea surface temperature (SST) anomalies (blue and red bars) that showed abnormally warm ocean temperatures from 2011 to 2013. (From T. Wernberg et al. 2016. *Science* 353: 169–172.)

ecosystem health, and feedbacks that have reinforced the rate of climate change. Eleven of these are shown in **Figure 12.16**.

Some of the best-documented examples of climate change altering biology in ways that impact human well-being are the spread of human diseases and vectors of disease. Siraj and colleagues[73] showed that warming in the highlands of Africa and South America has allowed mosquitoes carrying protozoan parasites that cause malaria to expand their range into higher altitudes. Densely populated areas in Colombia and Ethiopia that were once disease-free are now inundated with mosquitos. This has led to pronounced increases in the number of cases of malaria, one of the most deadly vector-borne diseases in the world (see Figure 12.16, case 3).

Food security is another example of climate change altering biology in ways that impact human well-being. Kerr and colleagues[74] showed that many species of bumblebees, which are key pollinators for many food crops, have experienced high levels of local extirpation along their southern range limits. Yet bee populations have failed to track changing temperatures and seasonality in their northern range limits, which has led to substantial range losses for bee species across North America and Europe (see Figure 12.16, case 2). An example from the oceans comes from skipjack tuna (*Katsuwonus pelamis*), which supports an important fishery in the Western and Central Pacific Ocean (WCPO). Several island countries in the WCPO depend heavily on skipjack for economic development, government revenue, and food security. But skipjack populations have shifted in response to warming waters, becoming less abundant in Western Pacific but more abundant in the Eastern Pacific as the species has shifted its regional distribution. This shift is benefiting some Eastern Pacific countries (e.g., French Polynesia, Kiribati) while hurting some Western Pacific countries (e.g., Papua New Guinea, Federated States of Micronesia). Management and governance of the tuna fishery has become complicated and created tensions among governments as the species has shifted across political boundaries, transitioning from one economic exclusive zone to another (see Figure 12.16, case 4).[75]

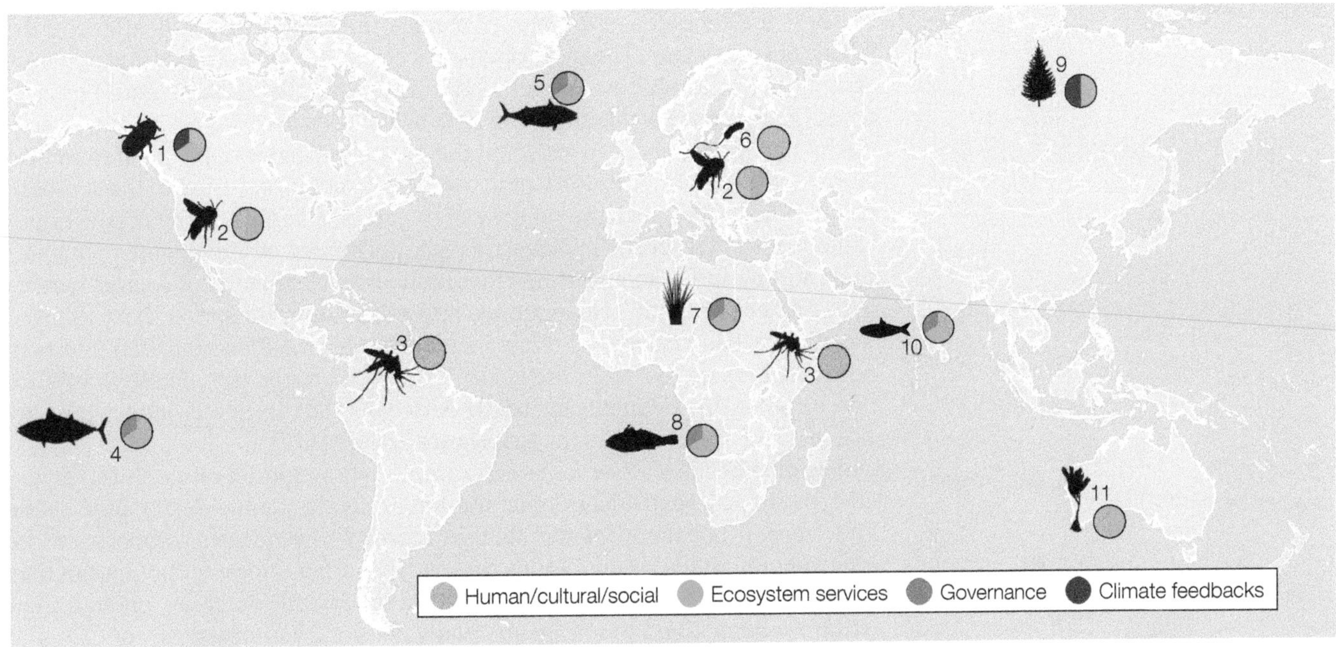

1. Bark beetles 

Northward and elevational shift of bark beetles in North America driven by warming climate. The combined effects of increasing temperatures and droughts predispose trees to defoliators and to bark beetles, thus contributing to the severity of pest outbreaks, which in turn may impact climate through increasing fuel loads and fire frequency at high latitudes.

2. Bumblebees

Southern range contraction and elevational shift for southern species of bumblebees in North America and Europe due to climate change. While species have experienced significant losses from equatorward range boundaries, there has been no corresponding expansion of range limits northwards for these important pollinators. Shifts to higher elevations have been restricted to southern species.

3. Malaria

Upslope shift in malaria distributions. The median elevation of malaria cases has increased in warmer years in both Ethiopia and western Colombia. In Ethiopia, high-elevation locations previously free of Malaria are now within the viable range for this disease. In Colombia, temperatures have fluctuated without a consistent trend of warming, and Malaria cases at high elevations have fluctuated very closely with the temperature change.

4. Skipjack tuna 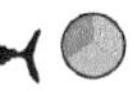

Skipjack tuna is projected to become less abundant in western, and more abundant in eastern, areas of the Western and Central Pacific Ocean (WCPO). Several Pacific Island countries in the WCPO depend heavily on skipjack tuna for economic development, government revenue, and food security.

5. Mackerel 

Expanded distribution of mackerel into Icelandic waters in the recent warm period since 1996. This expansion initially supported a bycatch fishery, which then developed into a direct fishery within the Icelandic EEZ, increasing from about 1700 tonnes in 2006 to about 120,000 tonnes in 2009 and 2010. Negotiations over new quotas for mackerel were key to discussions of Iceland and the Faeroe Islands joining the EU.

6. *Vibrio* 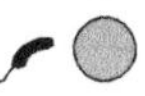

Unexpected emergence of *Vibrio* infections, a bacterial waterborne disease, in northern Europe. Changes in sea surface temperature in the Baltic are thought to be responsible.

7. Sahel vegetation 

Changes in Sahel vegetation. Over past decades, vegetation in the Sahel region has changed, affecting the livelihood and culture of people in the region. Pastoralists have increasing difficulties finding dry-season grazing areas for their livestock and suffer from lack of security of tenure over land and resources.

8. Coastal fish 

Poleward range shift in the coastal fish species, Argyrosomus coronus, from Angola into Namibia. This shift crosses Economic Exclusive Zones, complicating fishery management, particularly in light of a lack of congruence in the fisheries policies between nations.

9. Arctic vegetation 

Altered distribution, composition, and density of terrestrial vegetation in the Arctic, driven by climate warming, through both increasing average temperatures and a longer growing season. These changes in vegetation affect the albedo, vegetation biomass, and evapotranspiration, exacerbating climate warming.

10. Oil sardine 

Northward shift in the range of the oil sardine. Historically, the sardine had a restricted distribution between 8°N to 14°N, but in the past two decades, it has increased in abundance to the north: the region 14°N–20°N now makes up 15% of the catch. The range shift of the species is a boon for coastal fishing communities in this region in India.

11. Kelp, fish, and invertebrates

Range contraction of 100 km in kelp forests and other habitat-forming seaweeds in Western Australia. Increases in warm-water fish and invertebrates associated with ocean warming, leading to increased herbivory, loss of kelp and replacement of by seaweed turf (eastern and western Australia). These changes in ecosystem structure could impact Australia's most valuable single-species fishery (rock lobster).

Figure 12.16 Several examples of how climate-driven changes in species distributions have already begun to impact human physical, social, or cultural well-being (yellow), ecosystem services (green), governance systems (orange), and feedbacks that reinforce the rate of climate change (blue). (After G. T. Pecl et al. 2017. *Science* 355: eaai9214.)

There are numerous examples of climate-induced range shifts affecting the health of ecosystems. The most obvious occur when shifts in the distribution of agricultural or forest pest species lead to invasions that damage entire ecosystems. One example is the mountain pine beetle (*Dendroctonus ponderosae*), a species of bark beetle that is considered the most destructive of all forest insects in North America.[76] Prior to 1970, pine beetle infestations occurred in the southern half of British Columbia, the southwestern part of Alberta, and the northwestern United States. The beetle's northern range was limited by lethal winter temperatures and by spring temperatures that limited development. But shorter winters with higher minimum temperatures, and earlier onset of spring, have allowed the pine beetle to expand its range into lodgepole pine (*Pinus contorta*) forests in northern Canada that were previously unexposed to the pest. These forests are now experiencing frequent pine beetle outbreaks and high levels of tree mortality as a result of these infestations (see Figure 12.16, case 1).

The pine beetle also serves as an example of how species range shifts can exert ecosystem-level feedbacks that reinforce climate change. As the pine beetle kills trees, it both reduces the ability of forests to sequester carbon from the atmosphere, and releases CO_2 to the atmosphere as trees decay. Researchers have estimated that during the most recent pine beetle outbreak, enough trees will have been killed to release the equivalent of 270 megatonnes of CO_2 into the atmosphere from Canadian forests,[77] an amount sufficient to convert the forest from a net carbon sink to a large net carbon source.

Managing Effects of Climate Change on Biodiversity

Traditional biodiversity conservation has focused on protecting and managing systems to maintain their current state. Often this involves preserving habitat borders and population boundaries in order to conserve existing species and habitats as if they are in steady state. But with climate change, the threats to biodiversity are a moving target, and the approaches and locations that best conserve biodiversity today may not be the best approaches or locations in the future.[78] Therefore, conservation biologists must use a variety of management strategies that maximize flexibility, opportunity, and adaptability for species in the coming century.

Fortunately, many freely available tools can help practitioners identify key management options (**Table 12.2**). Nearly every practitioner nowadays can access projections of climate change at global to regional scales using free datasets like those published by WorldClim, which offers global climate layers (gridded climate data) for ecological modeling. Numerous tools are also available to determine which species are most vulnerable to climate change. For example, the U.S. Department of Agriculture's SAVS program (A System for Assessing Vulnerability of Species) allows users to answer questions about a species' current habitat requirements, physiology, phenology, and biotic interactions using an online survey that then produces a species-specific assessment of vulnerability to climate change.

Suggested Exercise 12.1 Use the SAVS (A System for Assessing Vulnerability of Species) online survey tool to complete a hypothetical climate vulnerability assessment for your favorite species. Don't worry about having all the relevant information—the goal is to learn how the assessment works and what information is needed. **oup-arc.com/e/cardinale-ex12.1.**

TABLE 12.2 Examples of freely available tools that can be used to answer key management questions about the effects of climate change on species and their habitats

Question	Tool	Scale	Variables	Source
How is climate projected to change?	WorldClim	Global	Temperature; precipitation; solar radiation; wind speed; water vapor pressure; other bioclimatic variables	oup-arc.com/e/cardinale-w12.1
	ClimateData	Regional (U.S.)	Temperature; precipitation	oup-arc.com/e/cardinale-w12.2
How vulnerable is the species to climate change?	System for Assessing Vulnerability of Species (SAVS)	Global	Projected habitat change; life-history traits	oup-arc.com/e/cardinale-w12.3
	Climate Change Vulnerability Index	Regional (U.S.)	Sensitivity to changes in temperature, precipitation, and landscape change; life-history traits	oup-arc.com/e/cardinale-w12.4
How is the species range likely to change?	MaxEnt	User-defined	Species occurrence; user's choice of environmental and climatic variables	oup-arc.com/e/cardinale-w12.5
	Climate Change Atlas	Regional (U.S., terrestrial)	Species occurrence; temperature; precipitation; elevation	oup-arc.com/e/cardinale-w12.6
	OceanAdapt	Regional (U.S., marine)	Species occurrence; latitude; depth; temperature	oup-arc.com/e/cardinale-w12.7
How will management options affect species persistence?	NatureServe Vista	User-defined	User's choice	oup-arc.com/e/cardinale-w12.8
	Marxan	User-defined	User's choice	oup-arc.com/e/cardinale-w12.9
How will climate change affect the connectivity of populations?	Circuitscape and Linkage Mapper	User-defined	Resistance to movement; user's choice of environmental and climate variables.	oup-arc.com/e/cardinale-w12.10

Source: Adapted from B. Abrahms et al. 2017. *Biodivers Conserv* 26: 2277.

Many free software programs are available for users to model species climatic niches and predict future range distributions that may result from climate change. One of the most widely used is MaxEnt, an open-source software package for modeling species' climatic niches and distributions by applying a machine-learning technique called maximum entropy. Programs like Marxan and NatureServe's Vista subsequently allow users to optimize selection of sites for conservation, and to design networks of reserves to accomplish management goals under different scenarios. Other programs, like Circuitscape, can be used to visualize species' range shifts and movements across large landscapes under changing climate conditions. These tools can help practitioners plan for certain management goals such as those described next.

Establishing refugia and safe havens

The fossil record shows that species ranges of plants and animals have routinely expanded and contracted as climate has changed throughout the Quarternary period (2.6 to 0.004 Mya). There is extensive evidence that during past climatic shifts, many species populations were able to survive in **refugia**—limited geographic areas that avoided the worst impacts of climate change. The term was first used to describe places of limited spatial extent to which populations retracted during the Last Glacial Maximum (LGM).[79] After the glaciers receded, those refugia served as the source of new populations that subsequently expanded from the refugia into the surrounding landscape.

Given that refugia have been a primary means by which biodiversity has survived during periods of past climate change,[80] there is much interest in identifying refugia that might be used as safe havens for biodiversity during anthropogenic climate change.[81] Of course, the concept of refugia was originally developed to describe the shift of species toward the Equator as they sought habitat that would allow them to survive periods of global cooling and glaciation. The concept must now be modified for the opposite scenario as species move poleward or upward in elevation to avoid global warming.

While the application is different, the tools used to identify climate refugia are the same as those described earlier in the chapter, in the section "Methods and Models." The general approach is to use climate envelope models (CEMs) to describe species' modern distributions as a function of climatic variables and then to search for the same climatic niche using the projections of global climate models (GCMs) that forecast future climate conditions. These areas can then be used to prioritize areas for conservation. For example, Li and colleagues[82] used records of the distribution of snow leopards (*Panthera uncia*) to construct a CEM that accurately predicted the historic distribution of the species since the last glaciation, as well as the current range of the species across all countries in Asia. They then projected results of the CEM onto forecasts of a GCM to identify habitat that is likely to be available for snow leopards by 2070 (**Figure 12.17**). The authors identified 1.1 million km^2 of habitat composed of 1317 patches that are likely to be within the snow leopard's climatic niche by 2070. Among these were three patches of more than 50,000 km^2 in the high-mountain environments of Asia that would be large enough to support viable populations. These analyses help make clear which habitats must be protected if the snow leopard is to have refugia that allow the species to persist in the face of future climate change.

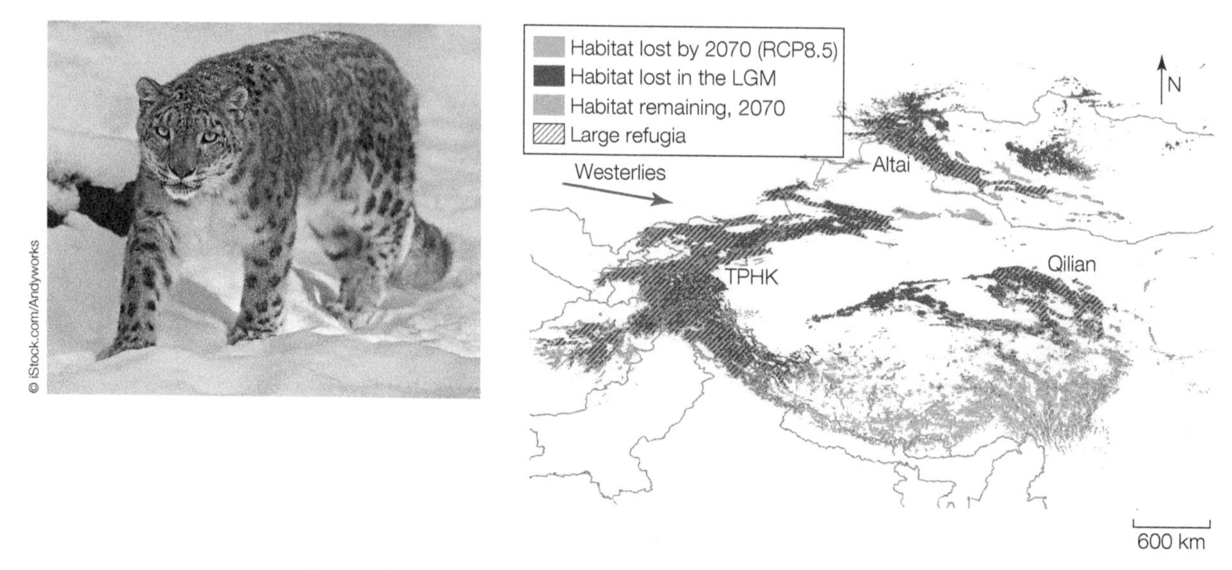

Figure 12.17 Climate envelope models (CEMs) have been used to describe the modern climatic niche of snow leopards (*Panthera uncia*). When researchers overlaid the results of CEMs onto projections of global climate models (GCMs), they were able to identify habitat already lost during the Last Glacial Maximum (blue), habitat that will be lost by 2070 (green), habitat that should persist (orange), and large climate refugia (hatched) that should exist in 2070, and which could allow the species to persist if conserved. (After J. Li et al. 2016. *Biol Conserv* 203: 188–196.)

Optimizing migration pathways

Because species will need to migrate to find future suitable habitat, conservation biologists must also focus on maximizing opportunities for dispersal. A good example of a conservation initiative that maximizes opportunities for dispersal is the Yellowstone to Yukon (Y2Y) project. The Y2Y conservation initiative is a joint effort by Canadian and U.S. conservation organizations to connect a 3200-km long corridor of habitat along the Rocky Mountains of North America (Figure 12.18). The project is designed to allow north–south migration pathways for large migratory species (e.g., American bison, *Bison bison*; grizzly bears, *Ursus arctos*), and for species that are shifting their geographic distributions northward in response to climate change (e.g., boreal caribou, *Rangifer tarandus caribou*). The project began in 1993 when Harvey Locke, a lawyer and environmentalist, had the idea to develop a wildlife corridor so that mammals with large territory sizes and home ranges could move safely among their habitats. After being formally established in 1997, a small staff and a large number of volunteers engaged with scientists, conservation groups, landowners, government agencies, Indigenous people's governments and communities, and businesses to stitch together a landscape of protected habitats that now traverses 2 countries, 5 U.S. states, 2 Canadian provinces, 2 Canadian

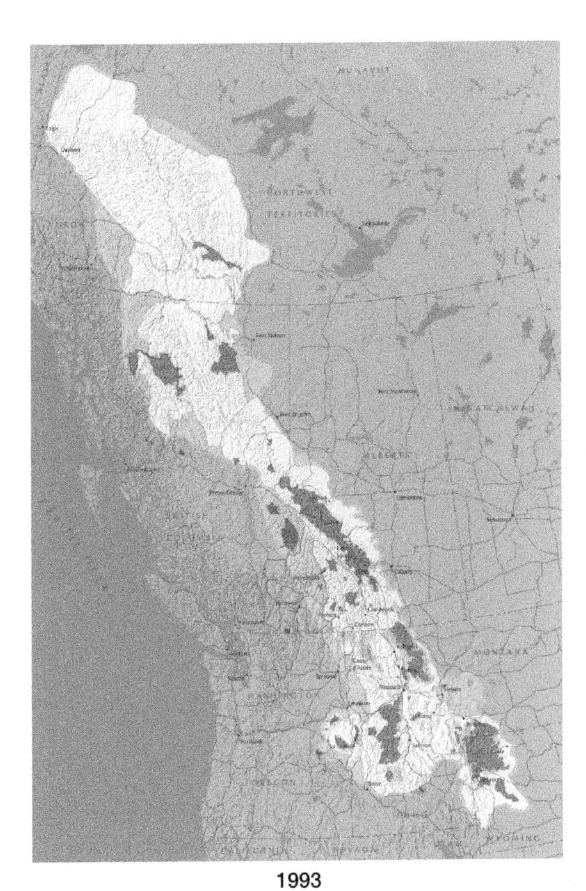

1993

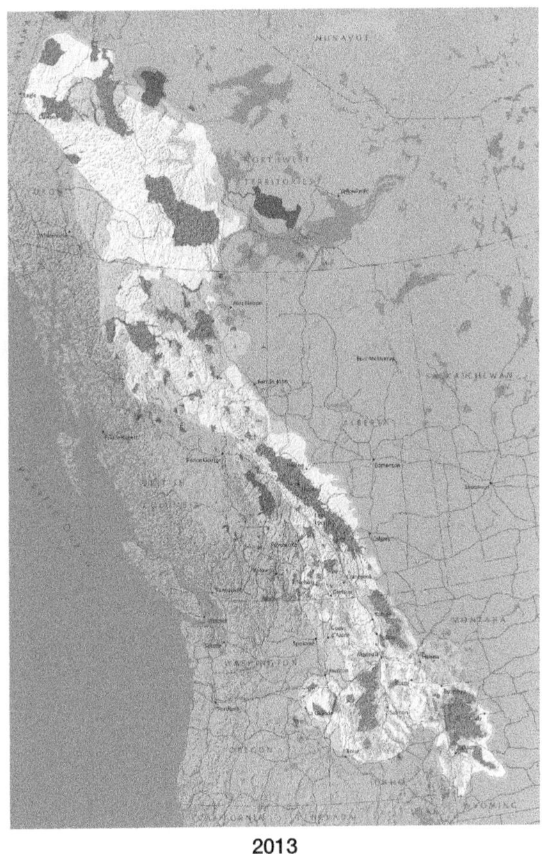

2013

Figure 12.18 The Yellowstone to Yukon (Y2Y) Conservation Initiative is a project designed to allow north-south and east-west migration pathways between protected areas for large organisms (e.g., American bison, boreal caribou) that migrate seasonally, or that are shifting distributions in response to climate change. The two years show before and after the Y2Y initiative was established. Green areas indicate protected lands, such as provincial parks and wilderness preserves. Yellow areas indicate other conservation designations, including but not limited to Provincial Natural Areas, Special Management Zones, and Restricted Use Wilderness Areas. (Images courtesy of Yellowstone to Yukon Conservation Initiative.)

territories, the reservation or traditional lands of more than 30 Native American and First Nation governments, and numerous government land management agencies. The Y2Y project has been named by the IUCN-World Conservation Union as one of the world's leading conservation initiatives. In Chapter 15, we will discuss how to include corridors and stepping stones in the design of reserve networks like that of the Y2Y in order to give species the best possible chance to disperse naturally to new locations.

One of the greatest challenges and controversies in conservation biology is what to do with species that have poor dispersal abilities, or that cannot disperse to new habitats fast enough to overcome the impact of climate change. A strategy that we need to consider is to transplant isolated populations of rare and endangered species to new localities at higher elevations and closer to the poles, where they can survive and thrive. This has been termed **assisted colonization** (or **managed relocation**). There is considerable debate within the conservation community about whether assisted colonization represents a valid strategy or whether it is too problematic because of the potential for transplanted species to become invasive in their new ranges (Box 12.3).

Building evolutionary resilience

For many groups of organisms, changing spatial distributions will not be an option, either because they are too sessile, or because the rate of climate change will outpace the speed at which their populations can migrate. For these species, the only option for long-term persistence may be to maximize their potential for evolutionary resilience. Ecologists and evolutionary biologists now recognize that evolution can occur far more rapidly than once thought—often within just a few generations. There is already evidence that many species of plants and animals with short generation times (e.g., most invertebrates, annual plants, and algae) have exhibited evolutionary responses to climate change with adaptations that have improved their physiological tolerance to new temperatures or levels of CO_2.[83] There is also limited evidence that, among some larger vertebrates, genetic adaptations have arisen that allow their phenologies to better match locally changing climatic conditions.[83] These limited examples offer hope that some portion of the world's biodiversity may be able to adapt to climate change within their current geographic ranges.

Sgro and colleagues[84] described several steps that conservation biologists can take to maximize the potential for evolutionary resilience. These measures include:

- *Increase population sizes* Genetic variation increases with population size (i.e., more individuals translates to more genetic diversity). Therefore, the larger a population size, the more mutation and recombination can generate new genetic diversity and the greater the chance that new traits will arise—some of which may allow for favorable adaptations to climate change.
- *Maintain adaptive potential in target genes and traits* Certain biological traits, such as those associated with physiological response to temperatures, are likely to give species the greatest chance of resilience to climate change. Identifying the genes that underlie these traits, and then intentionally maintaining heterozygosity of these genes, may help maximize the potential for evolutionary resilience.
- *Identify and protect evolutionary hotspots and refugia* Searching out and protecting geographic areas that are hotspots (areas of high diversity) of mitochondrial or nuclear DNA, or that have unique sequences of DNA, can help maximize the potential for evolutionary responses to climate change.

BOX 12.3 Challenges & Opportunities

Assisted Colonization: A Key Tool for Conservation, or Pandora's Box?

The western swamp turtle (*Pseudemydura umbrina*) is a short-necked freshwater turtle that is critically endangered in its native habitat of Western Australia due to habitat loss and impacts by alien invasive species. To add insult to injury, the species is being pushed even closer to extinction by anthopogenic climate change. Warming is drying up their swampy habitat, and that which remains is becoming so warm that scientists believe the waters will soon exceed the turtles' thermal tolerance.[101]

Given concern about its pending extinction, a team of researchers from the University of Western Australia has relocated around a dozen captive-bred turtles to two sites, about 250 km outside of their known range north and south of Perth. Although the new habitat is colder than in the native range, scientists believe the temperatures should be optimal in as little as 50 years.

Courtesy of Dr. Nicola Mitchell

The tiny western swamp turtle (*Pseudemydura umbrina*), critically endangered due to habitat loss and alien invasive species, is now also threatened by anthropogenic climate change.

Attempts to conserve the western swamp turtle illustrate a dilemma that conservation biologists face for many species. Should we intentionally move these species to new habitats, assisting their migration and colonization to ensure their long-term survival? Or should we stand by and watch them go extinct?

Opponents of assisted colonization argue that the tool is a Pandora's box.[102] Moving a species to nonnative habitat is the equivalent of intentionally introducing an alien species. Alien species that proliferate in population size (which is the goal of assisted colonization) have the potential to impact other species in the native habitat, potentially even driving the other species extinct. Saving one species at the potential expense of others, some argue, is not a good strategy for dealing with climate change.

In contrast, proponents of assisted colonization argue that this may be the only way to save the many species facing almost certain extinction by climate change, but which don't have the ability to adapt locally or disperse quickly to new locations.[103] They further argue that, with appropriate scientific studies, the risks of cascading impacts by alien invasive species can be minimized or controlled.[104]

Debate over the pros and cons of assisted colonization (also called managed relocation) has been ongoing for well over a decade[105,106] and is unlikely to be resolved any time soon. Assisted colonization will represent a challenge for future practitioners in the sense that they will need to mitigate the risks and unintended consequences associated with moving species beyond their native ranges. However, it also represents an opportunity. If done properly and with the appropriate risk analyses (see Chapter 11 on risk assessment for alien invasive species), then assisted colonization may be one of the key tools we use to save species from climate change.

- *Increase adaptability to future environments by translocating genes* Taking individuals from the warmest parts of a species' climate niche and introducing their genetic material to populations near the centroid could introduce genetic variation that allows populations to adapt more readily to future climate change while remaining within their current range distribution.
- *Identify species with low adaptive potential* Some species may have low genetic diversity for traits involved in maintaining their spatial distributions, or for traits that may be required to tolerate climate change in their native habitats. Identifying species that have little potential for adaptation can help biologists devise alternative plans before it is too late.

Using biodiversity for climate mitigation

Conservation of biodiversity may be useful as a tool to help minimize or even reverse the effects of anthropogenic climate change. Common mitigation strategies use reforestation, afforestation (i.e., planting trees on lands not previously forested), and ecosystem restoration to increase carbon sequestration from the atmosphere by plants. Certain kinds of plants, notably trees, have the potential to store large amounts of carbon in their woody tissues, and that storage can extend for centuries and even millennia if tough woody tissue is eventually buried underground, forming soil carbon stores.

Many programs have been developed to encourage conservation of ecosystems and their biodiversity for the purpose of mitigating atmospheric increases in CO_2. International treaties like the 1997 Kyoto Protocol have been adopted by nations that have agreed to reduce their emission of greenhouses gases, in part, by conserving existing forests and restoring areas that have been deforested. Carbon emission trading markets that allow firms or individuals to buy and sell carbon have been established by many countries to help meet emission caps put in place by the Kyoto Protocol.[85] The United Nations REDD program (Reducing Emissions from Deforestation and Degradation) has created financial incentives for developing countries to reduce emissions from forested lands and invest in low-carbon paths to sustainable development.

Conservation of plant biodiversity can help maximize carbon sequestration and storage in ecosystems. Cardinale and colleagues[86] summarized results of nearly 500 experiments that have manipulated the species richness of plants in a variety of ecosystems (e.g., grasslands, forests, lakes, streams). They showed that, on average, the most diverse plant communities produced one and a half times more biomass and carbon sequestration than less diverse communities. Studies have shown that plant biodiversity has an even greater impact on carbon sequestration in natural ecosystems, where the most diverse natural grasslands produce up to seven times more, and the most diverse forests produce up to four times more biomass than their less diverse counterparts (after controlling for abiotic drivers of productivity).[87–89] Based on estimates of how grassland plant diversity impacts carbon storage, Hungate and colleagues[90] predicted that restoring plant biodiversity across 12 million ha of grassland prairies in the U.S. could increase carbon storage by 17 teragrams of carbon per year, which would be worth US$2.3 billion on the social carbon market. Clearly, natural ecosystems and their biodiversity can be a tool to reduce atmospheric concentrations of CO_2, and to mitigate the future impacts of anthropogenic climate change.

Another example comes from the loss of sea otters along the western coast of North America. As sea otters (*Enhydra lutris*) have been killed for fur, their loss has led to increased densities of one of their preferred food sources—sea urchins (*Strongylocentrotus* spp.). As sea urchin densities have increased, they have eaten the stipes (basal supports) of giant kelp (order Laminariales), which forms underwater forests that rank among the most productive on Earth. Conservation of sea otters increases predation on sea urchins and has a cascading effect on kelp that increases carbon storage by 4 to 8 teragrams. At 2012 prices (US$47 per ton of C stored), this stored C would be valued at US$205 million–$408 million on the European Carbon Exchange.[91]

Developing green infrastructure

Natural ecosystems and their biodiversity can also be managed in ways that help society adapt to the most severe impacts of climate change. **Green infrastructure** (or **natural infrastructure**) is the collection of genes, species, and biological communities that regulate the services natural and managed ecosystems provide to people (ecosystem services; see Chapter 6). These services include the provision of goods (e.g., food, water, building materials) or services (e.g., water purification, climate regulation, or natural disaster mitigation). When natural infrastructure is damaged, neglected, or poorly managed, critical functions and services can be lost, and people's health, safety and prosperity can be affected.

Green infrastructure is likely to play an important role in human adaptation to climate change, particularly in the area of natural disaster mitigation. Natural disasters are expected to increase in the coming century as droughts, floods, heat waves, and intense storms become more common. Impacts from extreme weather like these can have devastating social, economic, and political consequences. Humanitarian crises can develop from disruptions to food production and availability, power outages, contaminated or diminished water supplies, and spread of disease. The most vulnerable areas are often the most populated, which increases the likelihood of major disruptions from natural disasters.

Green infrastructure can help protect societies from the worst effects of natural disasters. For example, flooding is one of the most common natural disasters in the world, killing thousands of people each year. The incidence of flooding has increased globally from the 1950s to the present, in part due to larger storm events that are associated with climate change.[92,93] But there is substantial evidence that green infrastructure can help mitigate the social impacts of floods (**Figure 12.19**). For example, mangrove forests reduced the death toll caused by coastal flooding after a massive cyclone hit India.[94] Coastal wetlands along the east coast of the United States helped thwart US$625 million worth of property damage during

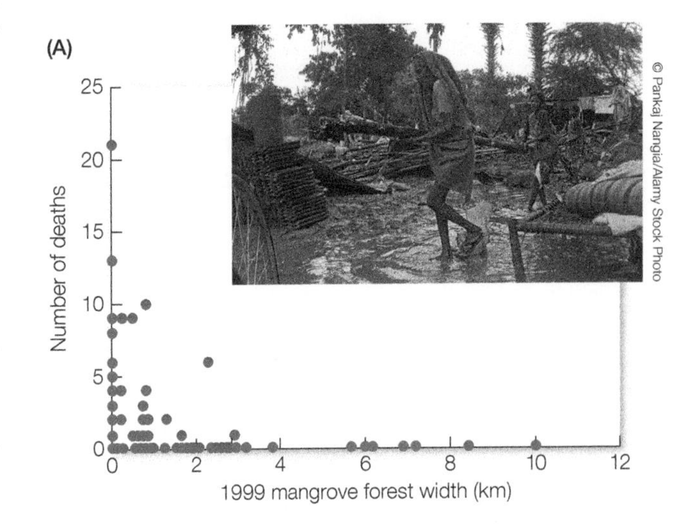

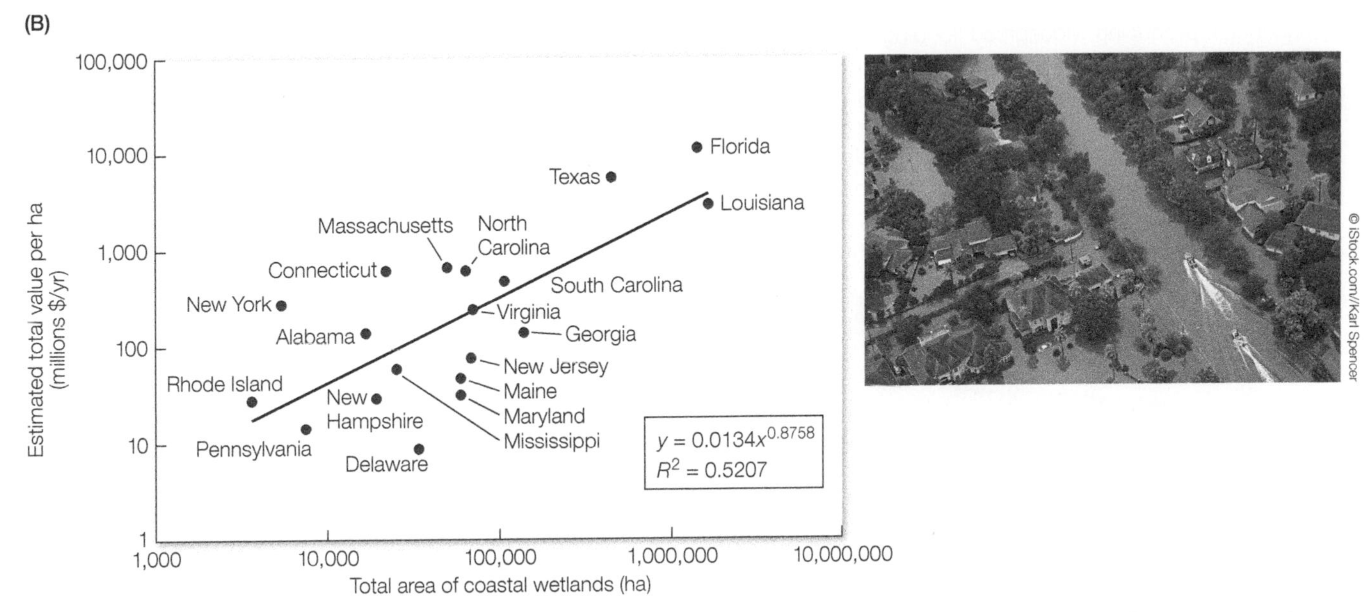

Figure 12.19 Natural infrastructure can help mitigate natural disasters like floods, which are increasing in frequency and magnitude due to anthropogenic climate change. (A) The number of deaths following the 1999 "super-cyclone" that hit Orissa, India was substantially lower (near zero, in fact) in areas where mangrove forests provided extensive flood protection. (B) Property damage costs are substantially reduced in coastal cities that protect wetlands, since wetlands serve as a "sponge" that can protect people from the worst impacts of flooding. This graph shows the total value of coastal wetlands per hectare for avoiding property damage after storms (A, after S. Das and J. R. Vincent. 2009. *Proc Nat Acad Sci* 106: 7357–7360; B, after R. Costanza et al. 2008. *Ambio* 37: 241–248. © The Royal Swedish Academy of Sciences.)

2012's Hurricane Sandy.[95] Arkema and colleagues estimated that wetlands and other natural habitats (e.g., coral reefs, dunes, kelp forests) protect about 67% of the U.S. coastline, protecting millions of people (and more than US$4 billion in property) that live within 1 km of the coast that is vulnerable to storms.[96]

Statistics like these have led to increased recognition that green infrastructure can help protect people from the worst impacts of climate change. Conservation of wetlands, floodplains, and coastal habitat has become a priority in many regions of the globe, where conservation is used as a management strategy to improve coastal resilience and adaptation to storms, and to reduce flooding of developed areas. Many green infrastructure projects now explicitly incorporate biodiversity into their design, which represents a success for conservation (Box 12.4).

BOX 12.4 Success Story

China's Sponge City Program: Using Green Infrastructure to Mitigate Natural Disasters

Rapid urbanization is one of the most dramatic changes in human demographics over the last century. Nowhere is this demographic shift more striking than in China, where rates of urbanization (the percentage of the population living in urban rather than rural areas) increased from 12% to 50% between 1952 and 2011.[107]

Like most countries that undergo rapid development, China's construction of major urban areas led to vast habitat loss of vegetated areas like forests, grasslands, and wetlands. But the loss of these pervious habitats and their replacement by impervious surfaces had dramatic consequences, producing widespread flooding in urban areas—a problem that has grown worse as climate change has produced larger and larger precipitation events. In 2012, a severe flood in Beijing wreaked havoc on the city's transportation systems, and in 2016 floods overwhelmed drainage systems in Wuhan, Nanjing, and Tianjin (see the Figure).

In 2013, China embarked on a bold new initiative to combat urban flooding. The Sponge City Program (SCP) seeks to replace large areas of impervious surfaces with green infrastructure like wetlands and "rain gardens" that serve as water storage and provide flood control for major cities. The program has an ambitious goal: by 2020, 80% of urban areas are charged with absorbing at least 70% of rainwater. In turn, the rainwater captured and stored by these natural "sponges" is recycled and reused for city functions (e.g., street cleaning, firefighting) or stored to maintain the city's water supply during shortages.[108]

In 2015, China selected 16 cities to serve as pilot studies, awarding the equivalent of US$59–88 million annually for three years to fund construction of green infrastructure projects.[109] Wuhan, which was declared one of China's first sponge cities, has now completed 228 projects that have developed green infrastructure in more than 38 km^2 of public spaces, schools and residential areas. These areas not only serve their intended function of flood control, they provide habitat for plants and wildlife, recreational areas for people, and a healthier overall environment.[110]

(A)

(B)

(A) Due to rapid urbanization and the spread of impervious surfaces, many large cities in China experience massive flooding during rainstorms. (B) China's Sponge City Program is a bold initiative to replace impervious surfaces with green infrastructure that serves as water storage and flood control.

Summary

1. Anthropogenic climate change is the result of human activities that release greenhouse gases into the atmosphere. It involves many abiotic variables (gas concentrations, temperature, precipitation, ocean acidification, and extreme events) that are all changing simultaneously in terms of their averages, ranges, variability, frequency, predictability, and seasonality.
2. Modeling studies predict that as much as 8% of species could go extinct globally if CO_2 emissions continue to increase as they presently are. Species with smaller population sizes and/or more narrow geographic ranges or climatic niches are at greatest risk.
3. While few species have yet to be driven globally extinct, the biotic signatures of anthropogenic climate change are pervasive. Population decay caused by climate change has led to a higher number of threatened species than was predicted by models. Species on land and in the oceans have exhibited rapid range shifts toward the poles and upward in elevation. Phenological shifts have led to population asynchronies that have reduced the fitness of interacting species, and entire biomes have exhibited regime shifts to alternative states.
4. Climate-driven changes in species distributions have already begun to alter ecosystem services that impact human well-being, ecosystem health, and feedbacks that reinforce the rate of climate change. These changes include shifts in human disease, emergence of pest species of agriculture and forested ecosystems, and changes in species that are vital to food security.
5. Conserving biodiversity in the face of climate change requires management strategies that maximize flexibility, opportunity, and adaptability for species in the coming century. Management options include establishing climate refugia, optimizing migration pathways, building evolutionary resilience, utilizing biodiversity for climate mitigation, and developing natural infrastructure.

For Discussion

1. Very few global species extinctions have been linked to anthropogenic climate change. Does this influence your thinking about the seriousness of climate change?
2. Most biological models used to predict species responses to climate change are unrealistic in the sense that they do not account for the potential of species to adapt locally to changing conditions. Do you think this invalidates the predictions of these models?
3. Which forms of biological change do you think pose the greatest risk to biodiversity over the next century: population decay, local extirpations, geographic range shifts, altered phenologies, or biome regime shifts and alternative states?
4. Should Conservation biologists put more time, energy, and money into (1) protecting areas that presently have high biodiversity, but are at risk of habitat loss or overexploitation; or (2) protecting areas that will serve as climate refugia later this century?
5. Do you think assisted colonization is a wise strategy for conserving species faced with extinction by climate change? Why or why not?

Suggested Readings

Abrahms, B. et al. 2017. Managing biodiversity under climate change: challenges, frameworks, and tools for adaptation. *Biodiversity and Conservation* 26: 2277–2293. Reviews management options for biodiversity conservation under different scenarios of climate change.

Cohen, J. M. et al. 2018. A global synthesis of animal phenological responses to climate change. *Nature Climate Change* 8: 224–228. Review of how climate change is disrupting the timing of migration and breeding of animals.

Hughes, T. P. et al. 2017. Coral reefs in the Anthropocene. *Nature* 546: 82–90. Coral reefs face special challenges with climate change, which are reviewed in this paper.

Pecl, G. T. et al. 2017. Biodiversity redistribution under climate change: Impacts on ecosystems and human well-being. *Science* 355: eaai9214. Outstanding summary of how climate change is altering ecosystem services, with many useful examples.

Pinsky, M. L. et al. 2018. Preparing ocean governance for species on the move. *Science* 360: 1189–1191. Describes challenges and poses solutions for managing fisheries as species distributions shift with climate change.

Poloczanska, E. S. et al. 2013. Global imprint of climate change on marine life. *Nature Climate Change* 3: 919–925. Describes how marine life is responding to climate change.

Urban, M. C. et al. 2016. Improving the forecast for biodiversity under climate change. *Science* 353: aad8466. Great review of how models can be improved to give more accurate predictions about biological responses to climate change.

Visit the
Conservation Biology
Companion Website
oup.com/he/cardinale1e
for videos, exercises, links, and other study resources.

Part IV

Approaches to Conservation

© SuperStock/Alamy Stock Photo

13

Species-Level Conservation

The snow leopard (*Panthera uncia*) is a large cat native to the mountainous regions of Asia. It is listed as Vulnerable on the IUCN Red List, with a global population of less than 10,000 individuals.

The field of conservation biology has historically focused on protecting and improving the status of species and their populations. The reason for focusing on conservation at the levels of species and their populations is that species are tangible, observable, and often relatable to the general public, which can result in greater public support and funding. Efforts have targeted single populations of species like those that are rare, threatened, or endangered. This might include discrete, small populations of species like snow leopards (*Panthera uncia*), hyacinth macaws (*Anodorhynchus hyacinthinus*), or giant barred frogs (*Mixophyes iteratus*). Efforts have also targeted metapopulations, which are a "population of populations" (see Chapter 9).

There is a long precedent for targeting species and their populations for conservation, a precedent that has been reinforced by legislation and policy at national and international levels. For example, the U.S. Endangered Species Act (ESA) passed overwhelmingly in 1973 and is a powerful piece of legislation that focuses on protecting and recovering species deemed to be threatened or endangered. The ESA has led to a strong focus on species by federal agencies in the U.S. and has resulted in numerous successes, like the delisting of island foxes (*Urocyon littoralis*), a species endemic to the Channel Islands of California that declined to near extinction in the 1990s but which has now largely recovered (**Box 13.1**).

BOX 13.1 Success Story

Record-Breaking Recovery: Island Foxes of California

The Channel Islands are a collection of small islands not far off the coast of southern California. They include eight islands that altogether cover an area about 900 km^2. As is often the case with island ecosystems, the Channel Islands harbor several endemic species of plants and animals. One such species is the island fox (*Urocyon littoralis*), a small fox closely related to the gray fox (*U. cinereoargenteus*) of mainland North America. Island foxes are thought to have diverged from gray foxes after they reached the Channel Islands over 10,000 years ago. These foxes occur on six of the eight Channel Islands, including those owned and managed by the U.S. National Park Service, U.S. Navy, The Nature Conservancy, and Santa Catalina Island Conservancy. A unique subspecies occurs on each island (see Figure A).

Until recent times, island foxes lived largely free of natural predators and threats from humans. In the early 1990s, however, researchers began to notice rapid declines of foxes on some islands. Between 1994 and 2001, the population declined more than 95% on both San Miguel and Santa Cruz Islands, and more than 99% on Santa Rosa Island (see Figure B). Not long after, more than 90% of the fox population disappeared from Santa Catalina Island.

Why did these populations decline so drastically? The main driver in the northern islands (San Miguel, Santa Rosa, and Santa Cruz) was the arrival of golden eagles (*Aquila chrysaetos*) from the mainland.[83] Once golden eagles reached the islands, they found an abundance of food, mainly feral pigs that had been brought to the islands in the nineteenth century. They also found an absence of competitors like bald eagles (*Haliaeetus leucocephalus*), which were extirpated from the islands in the 1960s due to pesticide poisoning (DDT). Golden eagles flourished in the islands and quickly switched to preying on island foxes

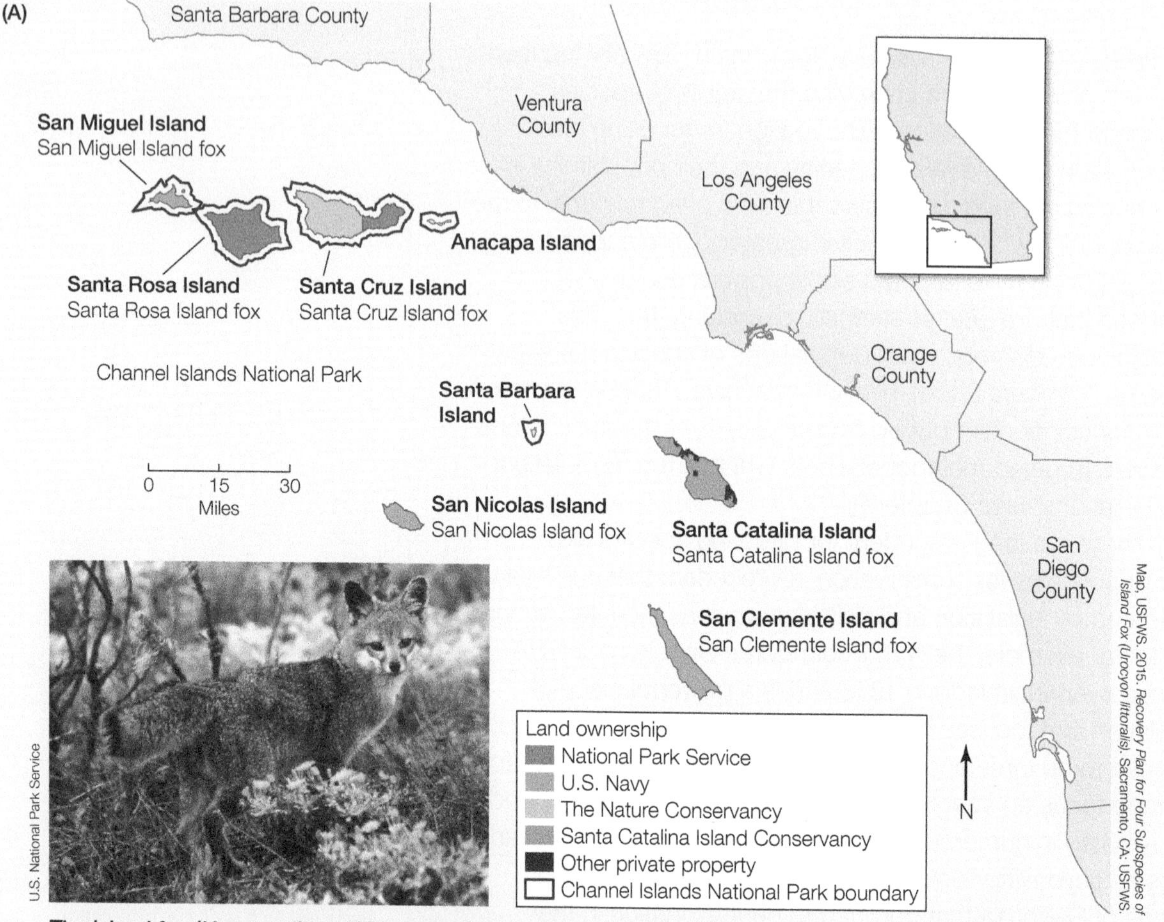

The island fox (*Urocyon littoralis*) is a small canid endemic to the Channel Islands of southern California. Island foxes occur on six of the eight Channel Islands, including San Miguel, Santa Rosa, and Santa Cruz in the north, and Santa Catalina, San Nicolas, and San Clemente in the south.

that were unaccustomed to predators. Hyperpredation by golden eagles decimated fox populations. On Santa Catalina Island, the population declined for another reason: an outbreak of canine distemper virus that appears to have originated in a stowaway raccoon (*Procyon lotor*) that made it to the island from the mainland.[84]

In response to declines, the U.S. Fish and Wildlife Service listed four of the six subspecies of island fox as Endangered in 2004. This listing led to a recovery plan, and conservationists took several actions to address the threats and bring populations back to pre-eagle levels. Efforts aimed to reduce golden eagles included capturing the birds and relocating them back to the mainland and eradicating feral pigs. Other efforts included the reintroduction of bald eagles, which actively defend and exclude golden eagles from their territories. A captive breeding program also began on several islands, which resulted in the release of hundreds of foxes into the wild to supplement populations, and a vaccination program was instituted to reduce the threat from another potential disease outbreak.

Today, as a result of these efforts, island fox numbers have increased considerably. The population of island foxes as of 2015 was greater than 700 on San Miguel, greater than 1200 on Santa Rosa, greater than 2100 on Santa Cruz, and greater than 1800 on Santa Catalina. All populations except the one on Santa Rosa are at or above pre-decline numbers. On some islands, like San Miguel, the population is even thought to have returned to carrying capacity—a remarkable achievement, given that the population there had reached as low as 15 individuals. Furthermore, population viability analyses suggest that extinction risk has been less than 5% for San Miguel, Santa Cruz, and Santa Catalina since 2008, and the same for Santa Rosa since 2011.

In response to the successful recovery of the island fox, the U.S. Fish and Wildlife Service removed three of the subspecies from the Endangered Species list and lowered the status of the remaining subspecies from Endangered to Threatened in 2016.[85] The delisting of island foxes represents the fastest recovery of any mammal ever listed under the U.S. Endangered Species Act. The success of the recovery may be attributed to interdisciplinary, out-of-the-box thinking to address the problems that faced island foxes; the use of science and data to drive solutions; and teamwork by federal, state, municipal, and nonprofit organizations coming together to implement solutions. Recovery also benefited from an abundance of intact habitat (of which there was just as much as in pre-eagle times), the near-complete protection of that habitat by partner organizations, and the scientific expertise of other partners like the Institute for Wildlife Studies, which had been studying the island foxes since before golden eagles arrived (**Video 13.1**).

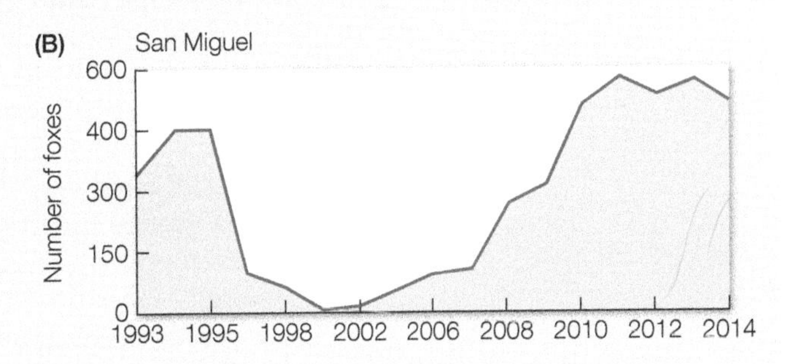

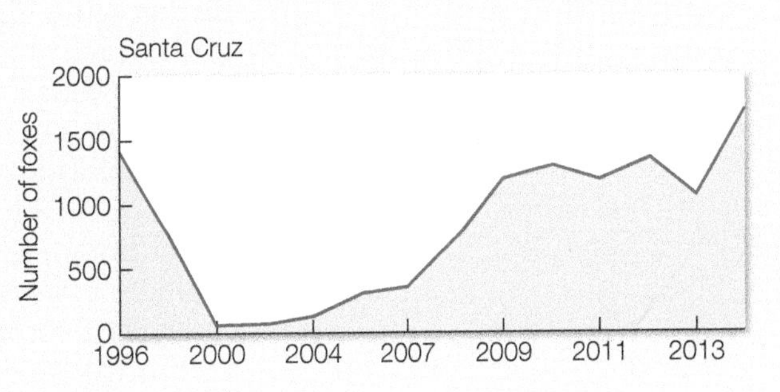

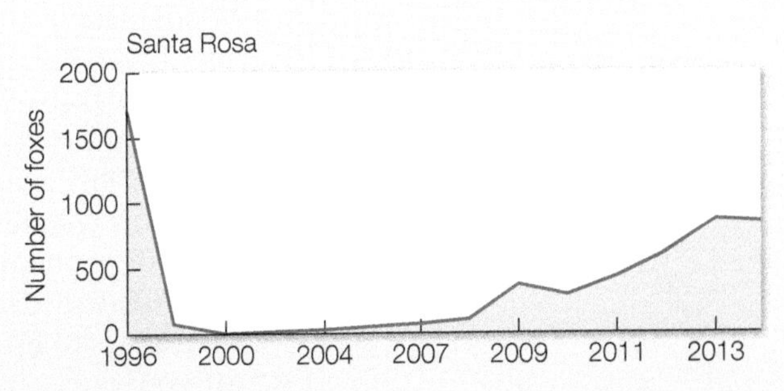

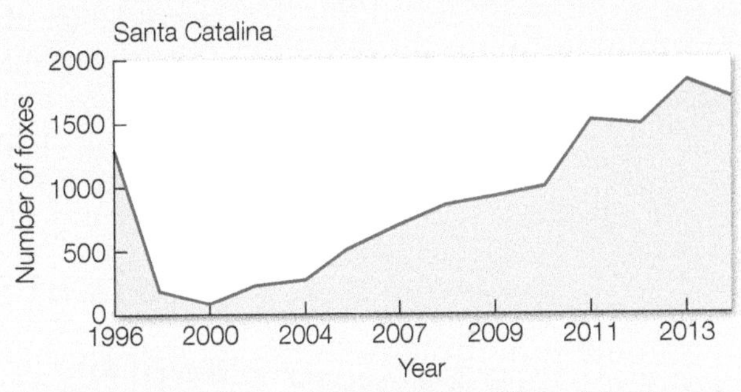

Population estimates of island foxes on four of California's Channel Islands. Each island has a unique subspecies, all four of which were listed as Endangered under the U.S. Endangered Species Act in 2004 due to sharp population declines mainly from golden eagle predation and disease. (Courtesy of Friends of the Island Fox [www.islandfox.org] from data presented at June 16–17, 2015 Island Fox Conservation Working Group.)

Video 13.1 Watch an overview of the efforts to restore island foxes and the ecosystems on the Channel Islands.
oup-arc.com/e/cardinale-v13.1

At the international level, the continued focus on species and populations has been reinforced by organizations such as the International Union for Conservation of Nature (IUCN; see Chapter 8). The IUCN Species Survival Commission, a science-based network of more than 7500 experts from across the world, supports 140 specialist groups, Red List authorities, and task forces. Specialist groups, in particular, are taxon-specific and range from plants (e.g., the Cactus and Succulent Plants Specialist Group) to large vertebrates (e.g., the Bear Specialist Group). Much of the work of specialist groups involves evaluating the conservation status of species, and these evaluations can feed into environmental legislation in many countries, as well as into development projects funded by organizations such as the World Bank (see Chapter 17). Large, broadly supported international conventions further reinforce species-focused conservation. The Convention on International Trade in Endangered Species of Wild Fauna and Flora (CITES, oup-arc.com/e/cardinale-w13.1) is an international agreement between governments that came into force in 1975 and regulates the international trade in plants and animals to ensure that trade does not threaten their survival (see Chapter 10). CITES lists species in three appendices depending on the degree of threat from international trade. Approximately 5800 species of animals and 30,000 species of plants are listed on appendices.

The historic focus on species has led to a profusion of tools and approaches that further reinforce species- and population-level conservation. Tools such as population models and population viability analyses that predict the state and dynamics of populations have supported the recovery of species ranging from koalas to elephants. Species-level management activities such as the captive breeding of threatened species and their reintroduction into the wild (see Chapter 16) have advanced and underpin many species conservation programs.

In this chapter, we will focus mainly on issues related to the conservation of small populations, which are a common target for species-level conservation. We will discuss the challenges that small populations face, then cover some ways to estimate population size and model how we expect populations to grow or change over time. Estimating and predicting population size can be crucial for planning conservation actions and making decisions to protect a given population. We will then introduce the concept of minimal population size and use a simple model to conduct a population viability analysis that aims to evaluate a population's long-term viability. Finally, we will discuss the trade-offs of any species-level conservation effort.

Goals of Species-Level Conservation

The primary goal of species-level conservation is to ensure the persistence of a species. At the population level, conservation focuses on ensuring that a given population is large enough to be viable or sustainable over time. This means that there are enough individuals in the population that survive and reproduce at rates that allow the population to maintain a positive growth rate over the long term. It also often means that the population is genetically diverse, which can allow species to better adapt to uncertain conditions in the future and, in some cases, help avoid problems associated with low diversity such as reductions in the biological fitness of individuals. When a species has gone extinct in a particular area, sustaining a viable population can mean restoring a population through actions like the reintroduction of individuals sourced from zoos and aquaria, and in some cases, wild populations from elsewhere, which we will cover in Chapter 16.

Species-level conservation can also have goals related to the broader conservation of ecosystems. Conservation actions targeting one species can result in benefits for other species, populations, and even ecological processes in an ecosystem. Such species are often called **surrogate species**—those that provide an efficient means for achieving broader conservation goals.[1] There are two main types of surrogate species. The first is a **flagship species**, which is a species used to attract the attention of the public, usually about the plight of a given ecosystem or region. Examples of flagship species include jaguars (*Panthera onca*), used to bring attention to the threats facing tropical ecosystems in Central America, and golden lion tamarins (*Manorina melanophrys*), which have brought attention to the Atlantic Forest in Brazil (an extremely biodiverse ecosystem that has been highly fragmented and currently covers less than 8% of its historic distribution). Another example is the giant panda (*Ailuropoda melanoleuca*), used in the logo by the World Wildlife Fund to draw attention to global issues around the state of wildlife and ecosystems. Flagship species tend to be charismatic and grab the attention of the general public.

The second type of surrogate is an **umbrella species**, which is one that provides a protective "umbrella" that benefits other species.[2] In other words, by virtue of protecting an umbrella species, the species living underneath the umbrella of that species benefit. An example is the wildebeest (*Connochaetes taurinus*), which has famously large herds that exhibit seasonal migrations in East Africa. In the mid-1900s, the annual range of migratory wildebeest was used to delineate the boundaries of the Serengeti and Ngorongoro protected areas in Tanzania.[3] The rich diversity of species living within these areas subsequently received protection. Umbrella species tend to have large home ranges (i.e., they cast a large protective shadow). However, not all species with large home ranges act as effective umbrella species. For example, a study[4] of black rhinoceros (*Diceros bicornis*)—a large-bodied herbivore that occupies very large home ranges—indicated that the species is probably not an effective umbrella for other herbivores, such as giraffe (*Giraffa camelopardalis*), springbok (*Antidorcas marsupialis*), and ostrich (*Struthio camelus*). The area used by rhinos in the study was insufficient to ensure other herbivores reached reasonable population sizes (at least 250 individuals). Care must be taken in the selection of umbrella species, making sure such a species truly serves broader conservation goals.

Challenges to Conserving Small Populations

Small populations are often a focus of species-level conservation efforts because they are especially vulnerable to local extinction. Populations of many species have been driven to small numbers due to the impacts of human activities. Habitat conversion and fragmentation can cause populations to become small and isolated.[5] Recovering and increasing the size of small populations can be accomplished by addressing the root causes of a given population's decline. For example, Ethiopian wolves (*Canis simensis*) are one of the rarest canids in the world, with a global population that varies between 300 and 500 individuals. In the past, outbreaks of rabies transmitted by domestic dogs periodically reduced the number of wolves to even smaller numbers. A domestic dog vaccination program initiated by the Ethiopian Wolf Conservation Programme, coupled with a science-driven wolf vaccination plan when an outbreak occurs, has been successful at reducing rabies outbreaks.[6]

Even in circumstances in which the causes of population decline can be removed or considerably reduced, as with the case of Ethiopian wolves, small populations still face three challenges that could lead to extinction. These include:

- Demographic stochasticity caused by random variations in birth and death rates.
- Environmental stochasticity caused by variation in the biological and physical environment.
- Loss of genetic variability due to the effects of genetic drift and inbreeding.

Demographic stochasticity

As introduced in Chapter 8, **demographic stochasticity** represents random variation in the birth and death rates of individuals, and it can have a magnified effect on small populations. Each individual in a population has an average chance, or probability, of surviving and reproducing given the conditions of the environment. However, not every individual will survive and reproduce at the average rate; some survive and reproduce at higher rates, others at lower rates. There is variation, and whether a given individual is above or below the average can be due to random chance. In large populations, this variation has little impact on population size, but in small populations, demographic stochasticity can cause population declines and increase extinction risk. As a simplistic illustration, if the average probability of a given individual surviving from one year to the next is 50% (as in flipping a coin, heads = survive, tails = die), a small population of, say, 3 individuals stands a good chance of going extinct (i.e., the probability that 3 coin flips in a row result in 3 tails is high, $0.5^3 = 0.125$). In a large population, say 1000 individuals, the likelihood of all individuals dying and the population going extinct (i.e., 1000 sequential coin flips resulting in tails) is vanishingly low (0.5^{1000}).

Demographic stochasticity can also lead to deviations in sex ratio in small populations, which can result in a further decrease in population size. Such unequal sex ratios are discussed later in the chapter, but for current purposes, imagine a small population of a monogamous bird species. The population has 4 birds and includes 2 male-female pairs, and each female produces 2 offspring per year (with an even chance that an offspring will be male or female). In the next year, the pairs would produce 4 offspring, but there is a 12.5% (1-in-8) chance that all the offspring will be the same sex—in which case no eggs will be laid to produce the following generation. There is also a 50% (8-in-16) chance that there will be either 3 males and 1 female, or 3 females and 1 male in the next generation, in which case only one pair of birds will mate, reducing recruitment into the population, and skewing the sex ratio. This scenario is illustrated by the now-extinct dusky seaside sparrow (*Ammodramus maritimus nigrescens*); the last 5 individuals were males, so there was no opportunity to establish a captive breeding program. Such demographic effects are also seen in the Spanish imperial eagle (*Aquila adalberti*). When the population is large, only mature birds breed; immature birds are more likely to breed when the population is small. Immature birds of this species are more likely to produce predominantly male offspring, contributing to further population decline and increasing the probability of local extinction[7] (**Figure 13.1**). For these eagles, a management strategy involving supplemental feeding was able to increase population size and restore the sex ratio. An imbalanced sex ratio in the kakapo (*Strigops habroptila*), a large flightless parrot species that now exists in a very small population on three offshore islands in New Zealand, has also created challenges to increasing population size.[8] Although the current population is about 150 individuals, in the 1990s, the population was 20 females and 34 males, and a breeding program resulted in 9 chicks, of which only 2 were female.

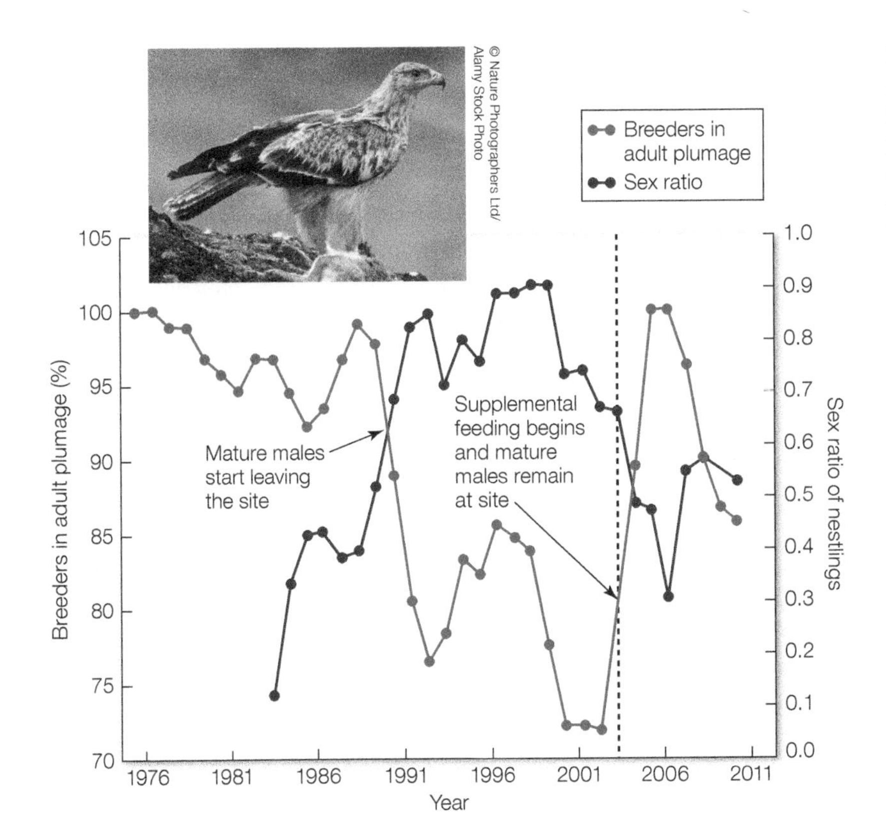

Figure 13.1 Population changes in Spanish imperial eagles (*Aquila adalberti*). Small population size resulted in changes in the sex ratio of eagles in the population partly due to demographic stochasticity. The population initially declined due to food scarcity, as mature birds left the population to find food elsewhere. Subsequent matings with immature males resulted in mainly male offspring, causing a change in sex ratio and population declines. Supplemental feeding led to mature males remaining at the site to breed and a return to the original sex ratio. (After M. Ferrer et al. 2009. *Conserv Biol* 23: 1017–1025 with updates by M. Ferrer.)

In the bird studies above, the influence of demographic stochasticity on sex ratio is an example of what is referred to as an **Allee effect**. An Allee effect, named after ecologist W. C. Allee (1885–1955), is a population response that is premised on the notion that the biological fitness of individuals is positively correlated with population size (or density) in some circumstances.[9,10] In simpler terms, fitness tracks population size: it increases as the population grows and density increases. However, if population size (or population density) falls below a particular critical threshold, fitness may then follow and decrease, in some cases to the point of causing the population to crash. Demographic stochasticity is one of several underlying mechanisms that can lead to an Allee effect by altering the sex ratio to such an extent that individuals simply have trouble finding mates as the density of one of the sexes is too low, and so the population declines.

Some other mechanisms that cause Allee effects include loss of **heterozygosity** due to inbreeding as population size declines (discussed later in this chapter), and reduction in cooperative interactions when there are fewer individuals.[9,10] African wild dogs (*Lycaon pictus*) provide an example of the latter. Wild dogs live in packs and are obligate cooperative breeders, with pack members all assisting in raising litters. If pack size falls below a critical threshold number, Allee effects could reduce fitness as packs will be less able to raise pups, have lower hunting success, and be less effective at defending against predators. Ultimately, these consequences could lead to population declines and increased risk of extinction.

Environmental stochasticity

Another form of variation introduced in Chapter 8 is environmental stochasticity. **Environmental stochasticity** is variation in the biotic and abiotic conditions of a landscape that affects the birth and survival rates of a population of a given species. This includes variation in conditions like the amount of food, the number of predators and competitors, and the weather and climatic conditions that a species experiences. Unlike demographic stochasticity, which acts on individuals, environmental stochasticity acts on populations. For example, imagine a population of lions that has an average birth rate and survival rate across all individuals. Variation in prey abundance from year to year can result in variation in those average rates, which in turn can result in changes in population size from one year to the next. In years with high prey abundance, birth and survival may be higher than average, whereas in years with low prey abundance, these rates may be lower than average. In small populations, such variation can increase the risk of extinction.[11]

As an example of environmental variation and its potential impact on small populations, imagine a rabbit population of 100 individuals in which the average birthrate is 0.2 and an average of 20 rabbits are eaten each year by foxes. On average, the population will maintain its numbers at exactly 100 individuals, with 20 rabbits born each year and 20 rabbits eaten each year. However, if there are 3 successive years in which the foxes eat 40 rabbits per year, the population size will decline to 80 rabbits, 56 rabbits, and 27 rabbits in years 1, 2, and 3, respectively. If these high-predation years are then followed by 3 years with no fox predation, the rabbit population will increase from 27 to 33, 39, and 47 individuals in years 4, 5, and 6. Even though the average rate of predation (20 rabbits per year) is the same over this 6-year period, variation in year-to-year predation rates caused the rabbit population size to decline by more than 50%. At a population size of 47 individuals, the rabbit population may very well go extinct within the next 5 to 10 years when subjected to the average rate of 20 rabbits eaten by foxes per year (and the high degree of variation in predation).

Natural catastrophes represent a form of environmental stochasticity that often occur at unpredictable intervals. Catastrophes include events like droughts, storms, earthquakes, and fires, along with cyclical die-offs of the surrounding biological community, which can cause dramatic fluctuations in population levels. Natural catastrophes can kill part of a population or even eliminate an entire population from an area. Numerous examples exist of die-offs in populations of large mammals; in many cases, 70% to 90% of the population dies.[12] For a wide range of vertebrates, the probability of a catastrophe is about 0.15, or 15% per generation.[13] Even though the probability of a natural catastrophe in any one year is low, over the course of decades and centuries, natural catastrophes have a high likelihood of occurring.

The interaction between population size and environmental variation was demonstrated using the biennial herb garlic mustard (*Alliaria petiolata*), an invasive plant in the United States, as an experimental subject.[14] Populations of various sizes were assigned at random to one of two conditions—either to be left alone as controls, or to be experimentally eradicated by removing every flowering plant in each of the 4 years of the study (removal of all plants being an example of an extreme environmental event). Overall, the probability of an experimental population's going extinct over the 4-year period was 43% for small populations (initially fewer than 10 individuals), 9% for medium-sized populations (10–50 individuals), and 7% for large populations (more than 50 individuals). For control populations, the probability of going extinct for small, medium, and large populations was 11%, 0%, and 0%, respectively. Large numbers of dormant seeds in the soil apparently allowed most experimental populations to persist even when every flowering plant was removed in four successive years. However, small populations were far more susceptible to extinction than large populations.

Reintroduced populations also demonstrate the magnified impact that environmental stochasticity can have on small populations. Przewalski's horse (*Equus ferus przewalskii*) once occurred across much of Eurasia, but went extinct in the wild in the late 1960s.[15] Conservation efforts led to the reintroduction of

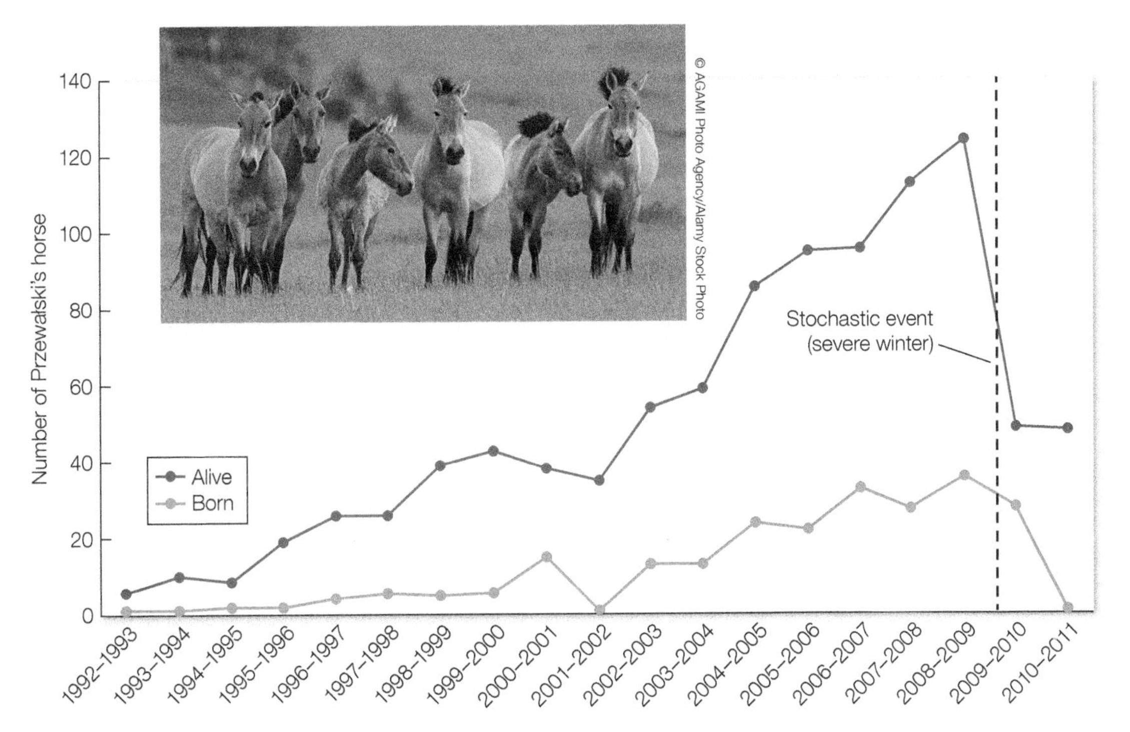

Figure 13.2 Population size and number of newborn Przewalski's horse (*Equus ferus przewalskii*) in a reintroduced population in Mongolia. A stochastic event (the severe winter of 2009/2010) caused a 60% decline in the population, and only one newborn was recruited into the population. (After P. Kaczensky, P. et al. 2011. *PLOS ONE* 6:e28057.)

the species at a small number of sites in Mongolia and China. One reintroduced population in the Mongolian Gobi Desert grew steadily at an average rate of 12% per year (since 2002) and had reached 138 animals by 2009 (**Figure 13.2**). A severe winter in 2009/2010—the most extreme in the previous 50 years in Mongolia—caused the population to decline 60% from December 2009 to April 2010.[16] Only one foal was born in 2010. The substantial decline and impact on reproduction reflects the impact that stochastic events can have on small, isolated populations. In this case, creating more reintroduction sites with spatially dispersed populations represents a potential strategy for reducing the effects of stochasticity and improving the persistence of the species.

Studies of other reintroduced ungulates show similar results. Saltz and colleagues[17] examined the effects of environmental variation on reproduction success of a wild ass (*Equus hemionus*) population reintroduced in Israel's Makhtesh Ramon Nature Reserve, using data from 1985 to 1999. Simulations of the population showed that increased variation in weather conditions as expected under global climate change (see Chapter 12) resulted in 30% higher variation in population size compared to a no-change scenario. Minor die-offs (≥15%) following droughts also increased the probability of population extinction nearly tenfold. Results of the study suggested that environmental stochasticity due to changing climate conditions poses a threat to the species.

Loss of genetic variability

A population's ability to adapt to a changing environment depends on **genetic variability**, which is a measure of the degree of genetic variation among individuals in a population. Genetic variability is a result of individuals' having different **alleles**—that is, different forms of the same gene (see Chapter 3).

Individuals with certain alleles or combinations of alleles may have the characteristics needed to survive and reproduce under new conditions that may emerge in the future.[18,19] Conserving the diversity of alleles in a population is therefore important for the long-term persistence of a species.

In small populations, allele frequencies may change significantly from one generation to the next simply because of chance—based on which individuals survive to sexual maturity, mate, and leave offspring. This random process of changes in allele frequency is known as **genetic drift** (see Chapter 4), and it is a separate process from changes in allele frequency caused by natural selection or other evolutionary forces.[20] When an allele occurs at a low frequency in a small population, it has a significant probability of being lost in each generation. For example, if a rare allele occurs in 5% of all the genes present (the gene pool) in a population of 1000 individuals, then 100 copies of the allele are present (1000 individuals × 2 alleles per individual × 0.05 allele frequency = 100), and the allele will probably remain in the population for many generations (assuming that individuals are mating randomly and no other evolutionary forces are acting on the population). By contrast, in a population of 10 individuals, only 1 copy of the allele is present (10 individuals × 2 alleles per individual × 0.05 allele frequency = 1). In this circumstance, with only one individual in the population carrying the rare allele, it is quite possible it will be lost from the population in the next generation or two simply by chance. The effects of genetic drift are magnified for small populations, and can result in alleles being lost or expunged from a population by chance alone. The loss of genetic diversity may increase the probability of extinction.[21]

Genetic diversity can also be lost through **inbreeding**, which is when two individuals that mate are more similar in their genotypes than if they were paired randomly. Inbreeding is essentially nonrandom mating between relatives, and the consequence, from a genetic perspective, is a loss of heterozygosity in the population. In other words, it results in fewer heterozygous genotypes and more homozygous genotypes among individuals in the population in subsequent generations. In very small populations, inbreeding occurs when no other mates are available, leading to individuals mating with their offspring, siblings, and cousins (and self-fertilization in hermaphroditic species).

In severe cases, the loss of genetic variability due to inbreeding can result in **inbreeding depression**, a condition that occurs when an individual receives two identical copies of a defective allele from each of its parents. Inbreeding depression essentially "depresses" the biological fitness of individuals, and is characterized by higher mortality of offspring, fewer offspring, or offspring that are weak or sterile or have low mating success.[21] For example, in a study of 38 species of mammals, Ralls and colleagues[22] showed that mortality was 33% higher in the offspring of parent-offspring or full-sibling pairs than in the offspring of unrelated parents. Higher rates of mortality result in even fewer individuals in the next generation, leading to a downward spiral of more pronounced inbreeding depression and further declines in the population.

The effects of inbreeding are often deleterious in small populations in the wild. A study that examined over 150 datasets across multiple taxa (birds, mammals, poikilotherms, and plants) found that 53% showed inbreeding to be detrimental.[23] The scarlet gilia (*Ipomopsis aggregata*) provides an example. Plants that come from inbred populations with fewer than 100 individuals produce smaller seeds with a lower rate of seed germination and exhibit greater susceptibility to environmental stress than do plants from larger populations.[24] In a second study, Bouzat and colleagues[25] examined isolated small populations of greater prairie chickens (*Tympanuchus cupido pinnatus*) in Illinois. These populations were showing the effects of declining genetic variation and inbreeding depression, including lowered fertility and lowered rates of egg hatching. However, when individuals from large, genetically diverse populations were released among the small populations, egg viability was restored and the populations began to increase in numbers. This result demonstrates the importance of maintaining genetic variation in existing populations and of restoring genetic variation in genetically impoverished populations as a conservation strategy. Inbreeding can also be problematic for managing small captive populations in ex situ facilities like zoos and aquariums, which we will explore in greater detail in Chapter 16.

Effective Population Size

Genetic management frequently involves actions that try to maintain or restore genetic diversity in a population to counteract the effects of genetic drift and inbreeding and promote evolutionary flexibility. Efforts often focus on adding more genetic material to the population. This can occur by reducing the isolation of a population by increasing the natural movement of animals (i.e., their genes) from elsewhere into the population. It can also occur through more direct actions, like translocating individuals into the population from elsewhere. Deciding on any action to maintain or restore genetic diversity can be informed by some measure of the genetic nature of the population. One key measure used by conservation biologists is **effective population size**, which is essentially the number of breeding individuals in a population (or the number of individuals that contribute genes equally to the next generation).

Determining effective population size

Sewall Wright[26] first proposed the concept of effective population size, symbolized N_e, to describe changes in genetic diversity in a population over time. Wright used the following equation to show the impact of breeding individuals on the original heterozygosity (i.e., the proportion of individuals with 2 different allele forms of the gene) remaining after each generation (H):

$$H = 1 - \frac{1}{2N_e} \tag{13.1}$$

According to this equation, a population of 50 breeding individuals would retain 99% of its original heterozygosity after 1 generation:

$$H = 1 - \frac{1}{100} = 1.00 - 0.01 = 0.99 \tag{13.2}$$

The proportion of heterozygosity remaining after t generations (H_t) decreases over time according to the equation:

$$H_t = H^t \tag{13.3}$$

For a population of 50 animals, then, the remaining heterozygosity would be 98% after 2 generations (0.99 × 0.99), 97% after 3 generations, and 90% after 10 generations. A population of 10 individuals would retain 95% of its original heterozygosity after 1 generation, 90% after 2 generations, 86% after 3 generations, and 60% after 10 generations (**Figure 13.3**). This formula demonstrates that significant losses of genetic variability can occur in isolated small populations, and also shows how the number of breeding individuals affects genetic diversity.

Effective population size is a simple measure that captures the genetic nature of a population and can be used as a benchmark for management actions. This is especially important for small populations, but the measure can also reveal the genetic state of larger populations. For example, consider a population of 1000

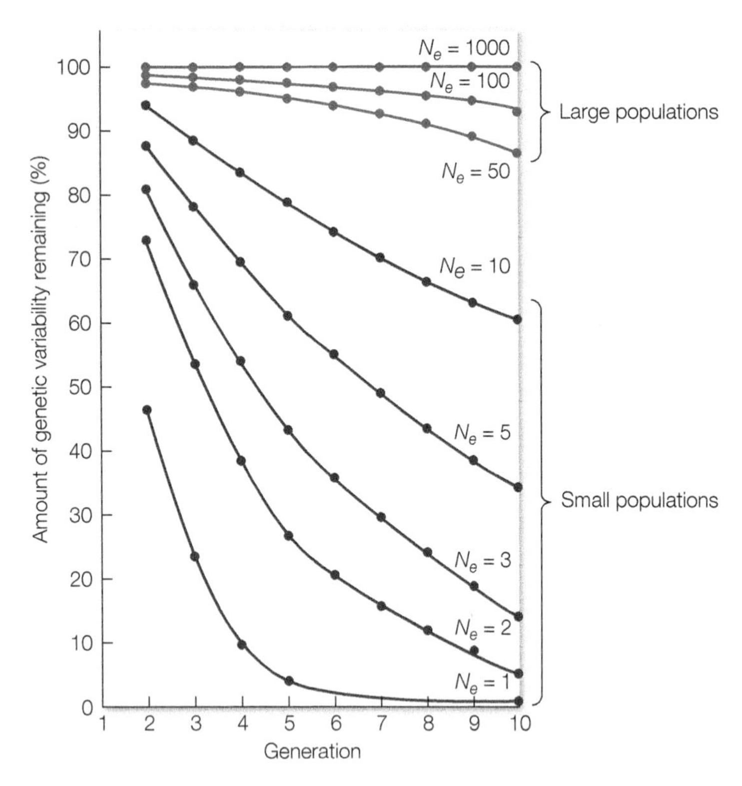

Figure 13.3 Genetic variability is lost randomly through genetic drift. This graph shows the average percentage of genetic variability remaining after 10 generations in theoretical populations of various effective population sizes (N_e). (After G. C. Meffe and C. R. Carroll 1997. *Principles of Conservation Biology*, 2nd ed. Sunderland, MA: Oxford University Press/Sinauer.)

alligators with 990 immature animals and only 10 mature breeding animals—5 males and 5 females. In this case, the effective population size is 10, not 1000. For a rare oak species, there might be 20 mature trees, 500 saplings, and 2000 seedlings, resulting in a population size of 2520 but an effective population size of only 20. In both of these cases, the populations appear large from a demographic perspective, which can give the impression that the populations are in good shape. However, the low effective population size indicates that from a genetic perspective, these populations are very small because there are so few "breeders." If maintaining genetic diversity is an important management goal, then this lack of breeders is concerning, as the seeming large populations are actually very small from a genetic perspective.

Effective population size can be influenced by several factors, including the sex ratio among individuals in the population, variation in reproductive output, and fluctuations in population size.[27,28] Understanding how these factors impact N_e can be important for structuring management actions.

Unequal sex ratio

A population may consist of unequal numbers of males and females due to chance, selective mortality, or the harvesting of only one sex by people (e.g., the selective hunting of male elephants or male white-tailed deer). If, for example, a population of a goose species that is monogamous (with one male and one female forming a long-lasting pair bond) consists of 20 males and 6 females, then only 12 individuals—6 males and 6 females—will mate, and the effective population size would be 12, not 26.

The effect of unequal numbers of breeding males and females on N_e can be described by the equation:

$$N_e = \frac{4N_m N_f}{N_m + N_f} \tag{13.4}$$

where N_m and N_f are the numbers of adult breeding males and breeding females, respectively, in the population. In general, as the sex ratio of breeding individuals becomes increasingly unequal, the ratio of the effective population size to the number of breeding individuals (N_e/N) also goes down (**Figure 13.4**). This occurs because only a few individuals of one sex are making a disproportionately large contribution to the genetic makeup of the next generation, rather than the equal contribution found in monogamous mating systems. In the case of Asian elephants (*Elephas maximus*), for example, males are hunted by poachers for their tusks at the Periyar Tiger Reserve in India.[29] In 1997, there were 1166 elephants, of which 709 were adults. Of these adults, 704 were female and 5 were male. If all of these elephants were breeding, this would result in an effective population size of only 20, using Equation 13.4. Poaching of saiga antelope (*Saiga tatarica*) also targets males, which have horns that are valued in traditional Asian medicine. Selective mortality of males in saiga populations has dramatically altered the sex ratio, and consequently effective population size, contributing to large scale declines in the species.[30] In unhunted populations, the proportion of males is reported to be 0.20

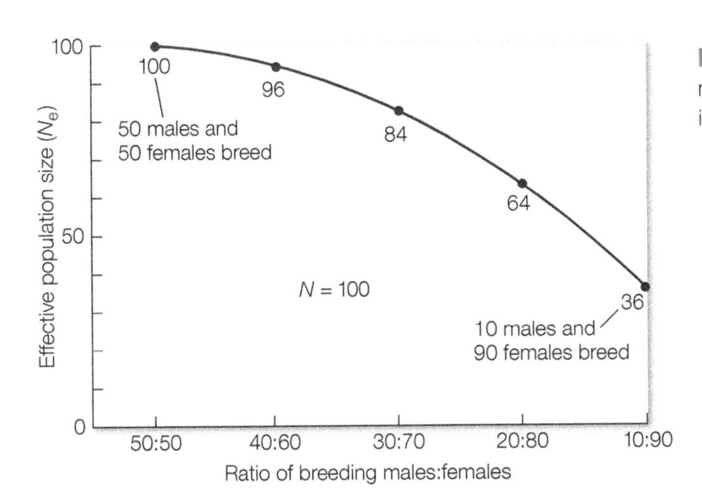

Figure 13.4 The effective population size (N_e) declines when the number of males and females in a breeding population (N) of 100 individuals is increasingly unequal.

to 0.25, but hunted populations can experience proportions as low as 0.02, which appears to also impact female fecundity and calf survival.[31]

Variation in reproductive output

In many species, the number of offspring varies substantially among individuals. This phenomenon is particularly true of highly fecund species, such as plants and fish,[20] where many or even most individuals produce a few offspring while others produce large numbers. This is also observed in polygynous species like elephant seals (*Mirounga* spp.), where only a small number of dominant males breed; and some social animals like African wild dogs, which live in packs that typically have only one breeding pair. In cases of unequal reproductive output among individuals, N_e can be estimated with the following equation:

$$N_e = \frac{k(N \times k - 1)}{V_k + k(k-1)} \tag{13.5}$$

Here, N is the number of individuals, k is the mean number of offspring per individual, and V_k is a measure of the variance in the number of surviving offspring from parent to parent. In cases where variance is large, which reflects a population with some individuals producing far more offspring than others, effective population size will be smaller. As an example, imagine a population consisting of 100 individuals. Monitoring of the population over time reveals that $k = 2$ and $V_k = 3$. Using the equation above, $N_e = 80$ individuals. If the variance were slightly higher—say, 4—then N_e would be smaller, at 66 individuals.

Unequal production of offspring leads to a substantial reduction in N_e because few individuals in a given generation will be disproportionately represented in the gene pool of the next generation. In general, the greater the variation in reproductive output, the more the effective population size is lowered. For a variety of species in the wild, Frankham[18] estimated that variation in offspring number reduces effective population size by 54%. In many annual plant populations that consist of large numbers of tiny plants producing one or a few seeds and a few gigantic individuals producing thousands of seeds, N_e could be reduced even more.

Population fluctuations and bottlenecks

Population size can also vary dramatically from generation to generation. Some small mammals such as mice, voles, squirrels, and chipmunks fluctuate periodically in response to food availability.[32,33] Other species may experience large changes in population size over time. For example, plains bison (*Bison bison*

bison) in North America experienced a significant reduction in population size due to overhunting, declining from tens of millions to a few hundred in the mid-1800s.[34] In extreme fluctuations, the effective population size is somewhere between the lowest and the highest numbers of breeding individuals. The effective population size can be calculated over a period of t years using the number of individuals N breeding in any one year:

$$N_e = \frac{t}{\left(\frac{1}{N_1} + \frac{1}{N_2} + \cdots \frac{1}{N_t}\right)} \quad (13.6)$$

Consider a butterfly population, monitored for 5 years, that has 10, 20, 100, 20, and 10 breeding individuals in the successive years. In this case:

$$N_e = \frac{5}{\left(\frac{1}{10} + \frac{1}{20} + \frac{1}{100} + \frac{1}{20} + \frac{1}{10}\right)} \quad (13.7)$$

The effective population size over the course of 5 years is 16, which is above the lowest population size (10), but well below both the maximum (100) and the average (32) population sizes. You may recognize this equation as the harmonic mean, which is commonly used to calculate an average from a set of rate data. For example, if you traveled 30 miles per hour in one direction to your destination and 60 miles per hour on the way back, your average rate of speed would be calculated as 2/(1/30 + 1/60) = 40 miles per hour—not 45 mph as estimated by an arithmetic mean, which we commonly use to calculate an average. The harmonic mean in the case of effective population size is essentially treating population sizes as though they are rates for periods of time. Throughout the first year in the example above, the population size was 10, then in the second year it was 20, and so forth through all 5 years. The average size given these changes represents N_e.

Suggested Exercise 13.1 Practice calculating and interpreting the effective population size for a fluctuating population. **oup-arc.com/e/cardinale-ex13.1**

Effective population size tends to be influenced more by the years in which the population has the smallest numbers. A single year of drastically reduced population numbers will substantially lower the value of N_e. This principle applies to a phenomenon known as a **population bottleneck**, which occurs when a population is greatly reduced in size and loses rare alleles if no individuals possessing those alleles survive and reproduce.[27] With fewer alleles present and a decline in heterozygosity, the overall fitness of the individuals in the population may decline.

A special category of bottleneck, known as the **founder effect**, occurs when a few individuals leave one population and establish a new population. The new population often has less genetic variability than the original (larger) population. For example, the Swedish wolf population was established by only 5 individuals, a very small gene pool.[35] If a population is fragmented by human activities, each of the resulting small subpopulations may lose genetic variation and go extinct. Such is the fate of many fish populations fragmented by dams.[36] Bottlenecks can also occur when captive populations are established using relatively few individuals.

The lions (*Panthera leo*) of Ngorongoro Crater in Tanzania provide a well-studied example of a population bottleneck.[37] The lion population in the crater consisted of 60 to 75 individuals until an outbreak of biting flies in 1962 reduced the population to 9 females and 1 male (**Figure 13.5**). Two years later, 7 additional males immigrated to the crater; there has been no further immigration since

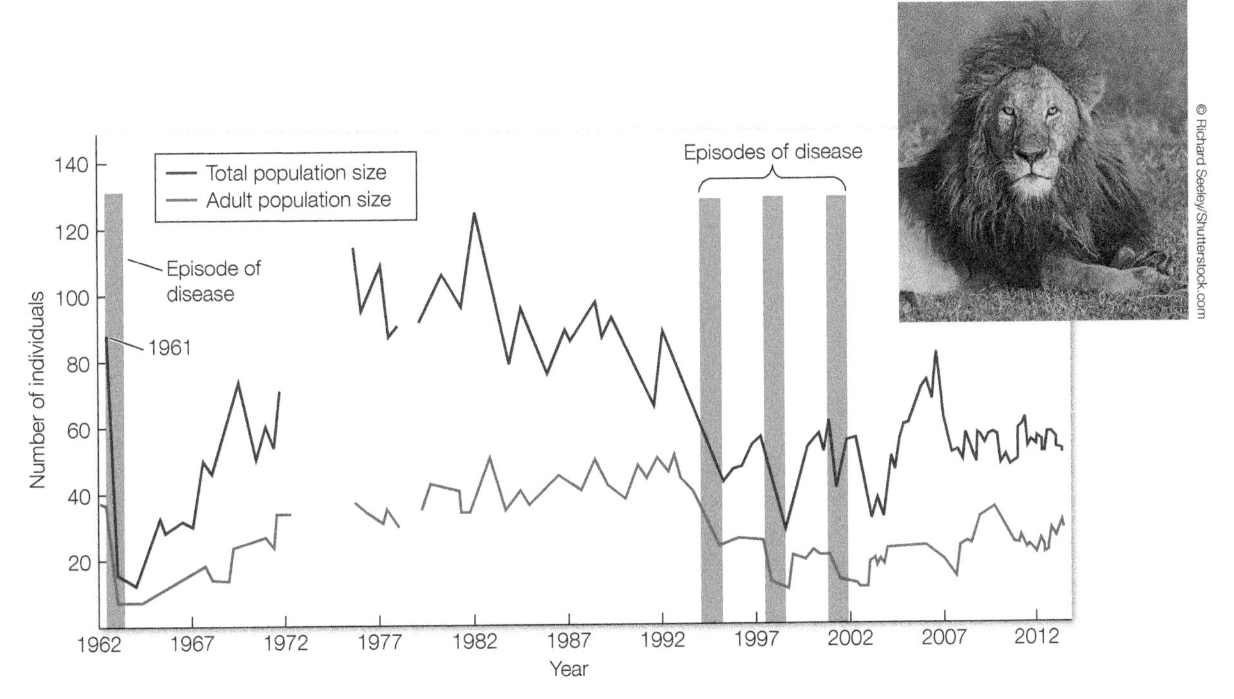

Figure 13.5 The Ngorongoro Crater lion population in Tanzania consisted of about 90 individuals in 1961 before it crashed in 1962. Since that time, the population reached a peak of 125 individuals in 1983 before collapsing to 34 individuals (fewer than 20 of which were adults). Small population size, an isolated location, lack of immigration since 1964, and disease have contributed to the loss of genetic variability caused by a population bottleneck. A lack of census data for certain years is the cause of gaps in the lines. The four orange bars represent episodes of disease outbreak. (After L. Munson et al. 2008. *PLOS ONE* 3: e2545., with updates from C. Packer.)

that time. The small number of founders, the isolation of the population, and the variation in reproductive success among individuals have apparently created a population bottleneck, leading to inbreeding depression. In comparison with the large Serengeti lion population nearby, the crater lions show reduced genetic variability, high levels of sperm abnormalities, reduced reproductive rates, increased cub mortality, and higher rates of infection.[37] After reaching a peak of 125 animals in the 1980s, the population has declined again. By 2003, the population had dropped to 34 animals following an outbreak of canine distemper virus that spread from domestic dogs kept by people living just outside the crater area. More recently, the population size has been around 70 animals.

Population bottlenecks do not always lead to reduced heterozygosity. If the population expands rapidly after a temporary bottleneck, average heterozygosity in the population may be restored even though the number of alleles present is severely reduced. An example of this phenomenon is the high level of heterozygosity found in the greater one-horned rhinoceros (*Rhinoceros unicornis*) in Nepal's Chitwan National Park, even after the population passed through a bottleneck. Population size in the park declined from 800 individuals to fewer than 100 individuals, and fewer than 30 were breeding. With an effective population size of 30 individuals for one generation, the population would have lost only 1.7% of its heterozygosity after one generation. As a result of strict protection of the species by park guards, the population recovered to 400 individuals and lost little genetic diversity. The Mauritius kestrel (*Falco punctatus*) represents an even more extreme case, with a long population decline that resulted in only one breeding pair remaining in 1974. An intensive conservation program has allowed the population to recover to about 300 adult birds today.

A study comparing the present birds with preserved museum specimens and kestrels living elsewhere has found that the Mauritius kestrel lost only about half of its genetic variation after passing through this bottleneck.[38]

Extinction Vortices

The smaller a population becomes, the more vulnerable it is to further demographic variation, environmental variation, and genetic factors that tend to lower reproduction, increase mortality rates, and so reduce population size even more, driving the population ever downward to extinction (Figure 13.6). This tendency of small populations to decline toward extinction has been likened to a vortex, a whirling mass of gas or liquid spiraling inward in which the closer an object gets to the center, the faster it moves. The outcome of an **extinction vortex** is the local extinction of the species. Once caught in such a vortex, it is difficult for a species to resist the pull toward extinction.[39]

As one example, a decrease in an Iberian lynx (*Lynx pardinus*) population caused by loss of its rabbit prey may cause inbreeding depression and associated higher mortality rates, further decreasing population size.[39] Decreased population size may then result in biased sex ratios and the inability to find mates (an Allee effect), leading to an even lower population size. The smaller population is then more vulnerable to further population reduction, loss of heterozygosity, and eventual extinction caused by unusual environmental events. Another example is the heath hen (*Tympanuchus cupido cupido*), a now-extinct subspecies of the greater prairie chicken that was once common in sandy scrub-oak plains in the northeastern United States.[40,41] By the early 1900s, overhunting and habitat destruction had reduced the subspecies to a single population of 50 on Martha's Vineyard. Despite habitat improvements that increased the population over the next few years, the remaining population experienced subsequent declines due to stochastic events including fire, weather, and a pulse of predation by goshawks. Inbreeding depression then occurred and led to lower reproduction, which was followed by an outbreak of a poultry disease that reduced the population to 2 females and 11 males. The last heath hen died in 1932.

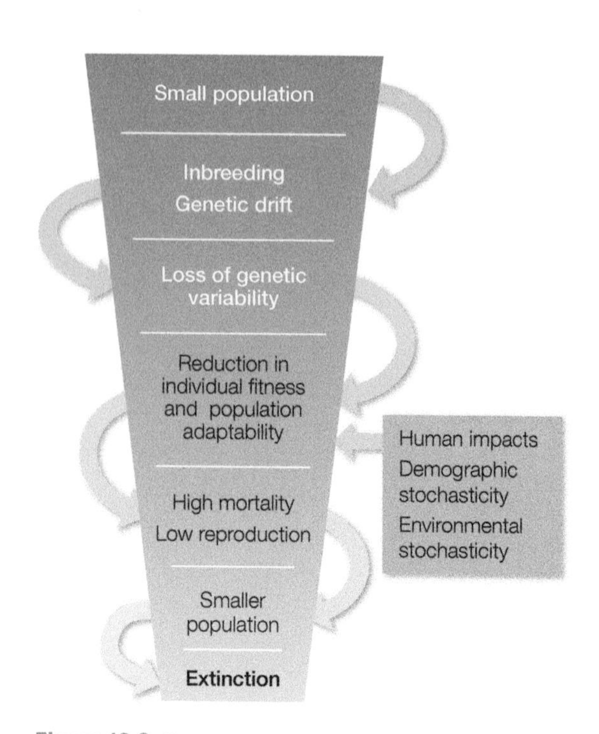

Figure 13.6 Once a population drops below a certain size, it can enter an extinction vortex in which the factors that affect small populations tend to drive its size progressively lower. This downward spiral can lead to local extinction of the species.

The case of the heath hen highlights an important implication of the extinction vortex: addressing the original cause of population decline may not be sufficient to recover a threatened population. Even though hunting and habitat destruction were minimized in the early 1900s, the problems associated with small population size ultimately doomed the heath hen population. This was also the case with the greater prairie chicken population in Illinois described earlier. The original population of over 1 million prairie chickens declined to below 50 following the arrival of European settlers, with a decline in fertility and hatchability. Habitat restoration of the prairie landscape, reversing one of the major original causes of decline, failed to help the population recover. The Illinois prairie chicken population began to grow only after it was outcrossed to populations from other states to reverse inbreeding depression.

As both prairie chicken examples illustrate, once a population has declined to a very small size, it will most likely go extinct unless unusual and highly favorable conditions allow the population size to increase.[42] Such populations often require a careful program of population and habitat management to increase population growth rate and allow the population to escape from the harmful effects of small population size.

Estimating Population Size

Planning conservation actions for a population, whether large or small, requires information on the size of the population. Although population size is a simple and basic demographic measure, it can be challenging to estimate. For species that are easy to detect and occur in a discrete location—perhaps the number of giraffes in a fenced game reserve—it may be relatively easy. But for species that are cryptic, secretive, and difficult to detect, or that occur at low density across large areas, it can be especially difficult (and costly). Next we will cover two basic approaches to estimating population size: conducting a census, and using capture-mark-recapture techniques. The latter approach in particular provides a foundation for more sophisticated approaches that rely on sampling data to estimate the number of individuals in an area.

Census

A **census** is simply a count of the number of individuals present in a population, and can be used for organisms that are easy to detect and not particularly mobile. By repeating a census over successive time intervals, biologists can determine whether a population's numbers are stable, increasing, or decreasing. In one example of a monitoring study, censuses of Hawaiian monk seal (*Monachus schauinslandi*) populations on the beaches of several islands in the Kure Atoll of the South Pacific revealed a decline from almost 100 adults in the 1950s to fewer than 14 in the late 1960s (**Figure 13.7**). The number of seal pups similarly declined during this period. On the basis of these trends, the Hawaiian monk seal was declared endangered in 1976 under the U.S. Endangered Species Act.[43] Subsequent conservation efforts reversed the trend, but only for some populations. The Tern Island population, for example, increased after the Coast Guard station there was closed in 1979, but has substantially declined since the 1990s because of high juvenile mortality.

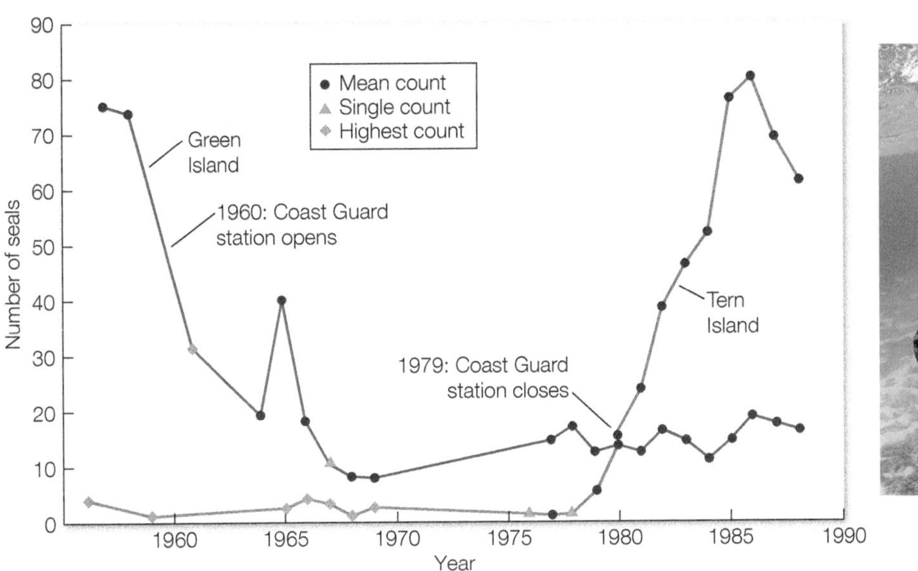

Figure 13.7 Censusing the Hawaiian monk seal (*Monachus schauinslandi*) populations on Green Island, Kure Atoll and Tern Island, French Frigate Shoals revealed that this species was in danger of extinction. Population counts were plotted from a single count, the mean of several counts, or the maximum of several counts. Seal populations declined when a Coast Guard station was opened on Green Island in 1960 due to disturbance by people and dogs; the population increased on Tern Island after the closing of a Coast Guard station in 1979. (After T. Gerrodette and W. G. Gilmartin. 1990. *Conserv Biol* 4: 423–430.)

One major disadvantage of censuses, and other surveying methods, is that they are often not repeated at later dates due to lack of funding, changing priorities, lack of time, the records getting lost or forgotten, or poor documentation as to what was done in the previous census. In addition, a census is meant to be a complete count of individuals in a population, but there are often uncertainties in count data that can lead to bias. For example, censuses of species that are difficult to detect (e.g., whales underwater, or evasive tigers) can result in incomplete counts, as researchers may miss some individuals during a census. There are sampling techniques that help overcome issues of detection probability (**Box 13.2**), and care should be taken to ensure that census data are collected and analyzed to reduce potential sources of bias.

In addition to censusing populations, censuses of a biological community can be conducted to determine what species are currently present in a locality. Censuses conducted over a wide area can help to determine the range of a species and its areas of local abundance. A comparison of current with past censuses can highlight species that have been lost and changes in species ranges. The most extensive censuses of this kind have been carried out in the British Isles by large numbers of local amateur naturalists supervised by professional societies. The most detailed mapping efforts have involved recording the presence or absence of plants, lichens, and birds in a mosaic of 10-km squares covering the British Isles. The Biological Records Centre (BRC) at Wallingford in the United Kingdom maintains and analyzes the 4.5 million distribution records, which contain information on 16,000 species. One part of these efforts involved the Botanical Society in the British Isles Monitoring Scheme, under which 1600 volunteers intensively sampled the British Isles in 1987/1988. During that timeframe the volunteers collected approximately one million records of all plant species occurrences.[44] When these data were compared with detailed censuses from 1930 to 1960, it was found that numerous species of grassland, heathland, aquatic, and swamp habitats had declined in frequency, while introduced weed species had increased (**Figure 13.8**).

Another example of large-scale censusing is the National Audubon Society's Christmas Bird Count in North America. The Christmas Bird Count began over a century ago and is an annual, early-winter bird census (of both species and number) conducted by thousands of volunteers for one day around Christmastime. In 2018/2019, the 119th Christmas Bird Count conducted more than 2500 counts and recorded over 48 million birds.

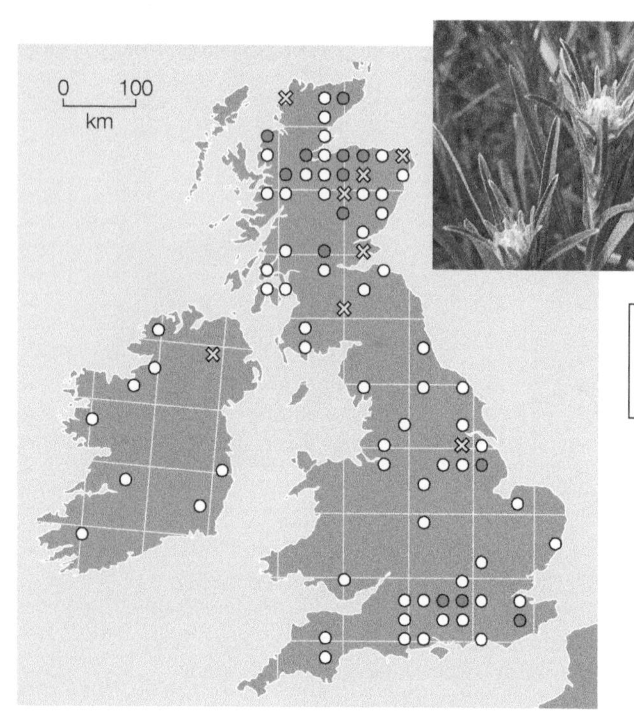

○ Historical population absent
● Historical population persists
⊗ New population

Figure 13.8 The Botanical Society of the British Isles Monitoring Scheme has documented the decline in the heath cudweed (*Gnaphalium sylvaticum*), a perennial herb covered by silvery hairs. Many populations present from 1930–1960 were no longer present in the period from 1987–1988 (white dots), particularly in Ireland and England. Some populations, mostly in Scotland, persisted during this interval (orange dots), and there were a few new populations (yellow Xs). (After T. C. G. Rich and E. R. Woodruff. 1996. *Biol Conserv* 75: 217–229.)

BOX 13.2 Conservation in Practice

Estimating Animal Density: The Distance Sampling Approach

How many animals occur in a given area? This is a central question that is often asked in conservation biology, and it is not always easy to answer. Counting animals and precisely estimating their density can be challenging because our ability to detect them is imperfect. Imagine driving along a line transect and counting zebras. You will most likely be able to count just about every animal near your line, but the farther away animals are from the line, the harder it will be to detect them. Some may be obscured by vegetation, others may be lying down and less visible, and some may just be too far away to be easily identified. The problem of imperfect detection can lead to substantial bias when estimating density.

Distance sampling is a commonly used approach for estimating density that accounts for this issue of detectability.[86] It involves traveling along a transect and measuring the perpendicular distance to each animal you detect. The distances are used to estimate the proportion of animals detected and not detected as a basis for calculating density. Let's look at a simple example to see how this approach works. Say you are estimating zebra density along a 1000-m transect and have measured the perpendicular distance from the transect line to every zebra detected, then plotted those data (see the Figure). You can see that the number of detections declines with increasing distance, which shows the detection problem. Presumably there were animals at these different distances, but they were not spotted. How can this be accounted for?

The first important point is that you assume the probability of detection is 100% on the centerline. In other words, you assume that you will always be able to detect a zebra on the centerline, which makes intuitive sense for many species. So for zebra, you recorded 35 individuals on the centerline. With this assumption, if you had perfect detection, you would expect to record 35 individuals not only on the centerline, but also in each one of the distance bins. To calculate density, you need to calculate the proportion of individuals detected out of the total. The proportion of individuals detected is 0.52 (or 52%), which is the average proportion detected across all 7 bins.

If you multiply the average proportion of individuals detected by the furthest distance from the transect (*w*), that gives you a new value called the effective strip half-width (μ). In this case, 0.52 × 70 m = 36.4 m. This means that you effectively surveyed an area (or rectangular strip) with a half-width (one side) of 36.4 m (the area of your total rectangular strip would then be 1000 m × 72.8 m). More specifically, it means that if you drew a vertical line on your graph at 36.4 m, the blue area to the right of the line would fill in the red area to the left of the line. In other words, it is

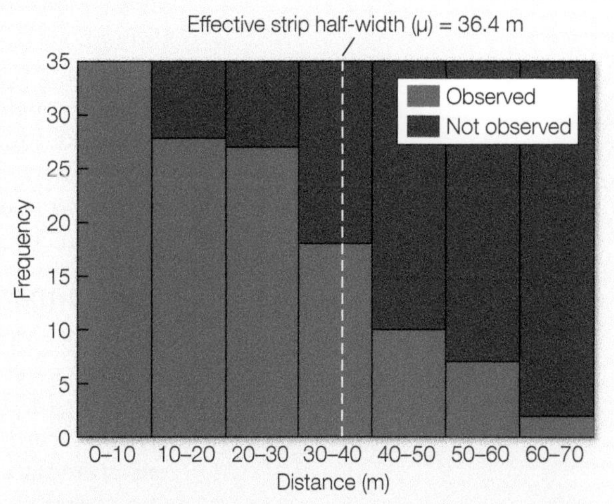

A graph showing the number (or frequency) of animals detected in different distance "bins," each 10 m from the centerline of a transect survey. For each animal detected during the survey, the perpendicular distance from the line was measured. These results are from a hypothetical example of a zebra survey, but are typical for transect (or circular plot) surveys for many other species.

like you are conducting a transect in which everything was detected within 36.4 m of the line. Now you can calculate density: number of individuals / (transect length × μ × 2) = 127 zebra / (1000 m × 36.4 m x 2) = 0.00174 zebra/m^2, or 17.4 zebra/ha.

The distance sampling approach goes one step further to improve the density estimate. The next step basically involves fitting a line to the distance data (it is actually the data transformed into probabilities). The advantage of fitting a line is that it provides a continuous curve, so the data are not treated in discrete, chunky distance bins (like 0–10 m), and allows for different detection probabilities on the centerline (values other than 100%, which might occur under some circumstances). This allows for a better approximation of the blue area or the proportion of individuals detected out of the total.

The distance sampling approach has been used to estimate density of many species, from gazelles to whales to snakes, lizards, and songbirds. The approach has helped overcome the issue of imperfect detection and led to more precise measures of density that ultimately provide better information to conservationists.

Suggested Exercise 13.2 Learn how to conduct a full analysis of distance sampling data in this free online course. **oup-arc.com/e/cardinale-ex13.2**

The Christmas Bird Count is an example of a successful community science program that has yielded long-term monitoring data across a large geographic extent. These data have helped evaluate historic and future trends in bird populations that can inform conservation planning. Analysis of data from the Christmas Bird Count combined with data from the North American Breeding Bird Survey—another long-term program that monitors bird populations along survey routes throughout the United States and Canada—showed that 314 species (53%) are projected to lose more than half of their current geographic range across three scenarios of climate change by the end of the century.[45] Loss is expected to occur without range expansion for some species (126) and with opportunities to colonize new replacement ranges for other species (188). Results demonstrate the climate sensitivity of species and indicate the need for conservation actions that accommodate shrinking and shifting geographic ranges.

Capture-mark-recapture techniques

Another common approach used to estimate the abundance or density of animals in a given area is **capture-mark-recapture (CMR)**. This approach involves capturing individuals and marking them in one period, then capturing individuals and recording the number of previously marked animals (i.e., those that are recaptured) in a second period. Abundance is then estimated based on the ratio of recaptured animals to total animals captured in the second period.

To show how this works, consider estimating the population size of a species of jerboa (a desert rodent) in a protected area. Begin by recognizing that the value of interest, which is unknown at this point, is N representing total population size. You decide to capture jerboas by using a grid of small box-traps set across the study area. During the first sampling period, you capture 34 jerboas, and each one is marked with a small ink mark on its belly fur. Then a day later, you set traps again and this time capture 40 jerboas, of which 25 are marked from the previous trapping period. Population size can be estimated using the simple formula:

$$\frac{n_1}{N} = \frac{m_2}{n_2} \tag{13.8}$$

where n_1 is the total number of animals captured in the first period (34), n_2 is the total number of animals captured in the second period (40), and m_2 is the total number of recaptured animals (25) in the second period. Equation 13.8 can then be rearranged to solve for the unknown, N:

$$N = \frac{(n_1 \times n_2)}{m_2} \tag{13.9}$$

Using Equation 13.9, population size is calculated as 54 jerboas.

An important point is that CMR is essentially creating a subpopulation of marked animals within the larger unknown population. You know the size of the subpopulation because all animals have marks from the first capture period. When you capture animals in the second period, the ratio of the number of recaptured animals in the subpopulation to the total number of animals in the subpopulation (right side of Equation 13.8) should be equivalent to the ratio of the number of animals initially captured and marked in the first period to the total unmarked population, N (left side of Equation 13.8).

This particular CMR technique is called the Lincoln-Petersen estimator. The Lincoln-Petersen is considered the first capture-mark-recapture estimator of animal abundance.[46] It is simple, intuitive, still widely used, and meant for **closed populations.** In other words, it assumes that the population under consideration is "closed" in the sense that there are no animals entering or leaving the population through immigration, emigration, birth, or death (during the period of sampling). This assumption may hold true for short sampling periods, like trapping jerboas over two 1-day periods, but for longer-term studies that might sample once a year over several years, it may not. Fortunately, there are other estimators for **open populations**, which take into account animals entering or leaving the population. One common estimator is the Jolly-Seber model, which tracks the capture history of each marked animal and uses these data to estimate capture probabilities and survival probabilities.[47,48] By accounting for the chance of animals getting captured and surviving from one sampling period to the next, the Jolly-Seber model provides a more precise estimate of population size in open populations.

The CMR approach has been used to estimate the population size of many species ranging from field mice to elephants, and expanded to include noninvasive sampling and more advanced computation techniques to improve the precision of abundance estimates.[49] Capture methods include live traps like box or cage traps (for mammals) and mist nets (for birds and bats); camera traps (for species with recognizable marks, like stripes or spots); and even DNA sequenced from scats, hair, or other biological samples (Box 13.3). Marking methods are just as variable and include permanent markers like tattoos, ear punches, and toe clips; semi-permanent markers like radio collars, ear/flipper tags, and leg rings; and temporary markers like dyes, paints, and powders. Any study involving the capture and marking of individuals should have their methods approved by an institutional committee that reviews procedures for ethics and standards of animal care.

Suggested Exercise 13.3 Use a spreadsheet to calculate animal abundance with the Lincoln-Peterson estimator and assess the accuracy of abundance estimates using a Monte Carlo simulation.
oup-arc.com/e/cardinale-ex13.3

BOX 13.3 **Case in Point**

Detector Dogs Support Research and Conservation Efforts

Fundamental information about species in need of conservation can be challenging to obtain. Basic information such as the number of animals in a population or how individuals are distributed across a landscape can be difficult to estimate, and yet is often critical for conservation planning and decision-making. This is especially the case for species that occur in low density, live in areas that are hard for humans to access, or are cryptic or secretive in nature. Obtaining necessary information on abundance, distribution, and measures of animal health can be costly in terms of time, money, and resources.

One solution to overcoming the problem of detecting species has been the use of specially trained domestic dogs. Dogs have an extraordinary sense of smell; many are able to scent up to 100 million times more odorants than people and have the capacity to detect certain compounds in concentrations as low as 500 parts per trillion.[87] Detector dogs have been deployed in many contexts, including finding people in search and rescue missions, and in narcotics and bomb detection.[88] Researchers are now using them as a tool for finding species that are in need of conservation but are difficult to detect, and the dogs have been highly effective.

Consider the San Joaquin kit fox (*Vulpes macrotis mutica*), a native of the Central Valley of California, that is listed as Threatened by the state and Endangered under the U.S. Endangered Species act, mainly due to habitat loss. Kit foxes are nocturnal, largely solitary, and occur in low density in arid grassland and shrubland environments. Researchers trained dogs to identify the odor of kit fox scat, then walked transects through a large study area to find scats.[89] These working dogs found 1298 scats, which were then genetically tested to verify they were from the correct species. Remarkably, the dogs had identified kit fox scats with 100% accuracy. This is particularly impressive given that scats were often dry and desiccated from the desert environment, and that several other carnivores occur in the region, including coyotes (*Canis latrans*), skunks (*Mephitis mephitis*), and badgers (*Taxidea taxus*). An assessment of effort also revealed that dogs found about four times as many scats as an experienced person searching for scats visually. Genetic

(*Continued on next page*)

BOX 13.3 Case in Point *(continued)*

information from the scats provided information on the number of individuals in the population that can be used to estimate population characteristics.

Dogs have even been used to find animals living in marine environments. For example, researchers have used dogs positioned on the bow of a research boat to find fecal samples of whales. The dogs are trained to give cues that are used to orient the direction of travel (see the Figure). Once located and retrieved, fecal samples (which are understandably difficult to find using conventional survey approaches) provide valuable data on identity, health, diet, reproduction, and physiological stress on individuals (among other factors). During three years of a study[90] of highly endangered North Atlantic right whales (*Eubalaena glacialis*), detector dogs helped researchers locate nearly 100 fecal samples, many of them found in areas where the human crew did not even observe whales. This rate of fecal collection was more than four times higher than opportunistic methods, and dogs detected samples from as far away as one nautical mile.

The photo depicts Tucker, the Center for Conservation Biology scat detection dog, locating whale scat from Southern Resident Killer Whales on the Salish Sea in Washington State. Scats are genotyped to determine the whale's identity and sex. Reproductive, stress and nutrition hormones extracted from the samples indicate physiological health as well as pregnancy occurrence and loss. Toxins extracted from these samples indicate how nutritional stress interacts with toxin loads, leading to spontaneous abortion of nutritionally compromised whales.

Detector dogs have expanded researchers' ability to find signs of many species, including black-footed ferrets (*Mustela nigripes*), grizzly bears (*Ursus arctos*), bush dogs (*Speothos venaticus*), and tigers (*Panthera tigris*). Quantitative assessments of detection indicate that in some cases dogs find up to 15 times more samples than people.[91] Samples can be used to document presence, estimate population characteristics like abundance (e.g., with CMR techniques), sex ratio, and individual relatedness using genetic analyses. They also allow estimates of physiological measures of health and reproductive status by analyzing for hormones and metabolites. Detector dogs have also been used to find invasive species to assist in eradication efforts, and diseased animals to help in the surveillance of infectious diseases. Detector dogs provide researchers with a noninvasive means of collecting species information, offering an alternative to more invasive techniques like capturing and marking individuals.

Models of Population Size

The conservation of populations can be aided by models of how we expect a given population to grow or change over time. A model is simply a mathematical way of representing some phenomenon (which in this case is the growth or decline of a population) and is usually developed from a set of population data (often from censuses or CMR estimates). Models inherently have limitations because they are simplified representations and usually based on a sample of information, but try to make use of the best available information. George Box (1919–2013), a renowned British statistician, famously wrote that "all models are wrong, but some are useful." While this statement points out that models are merely approximations of the "truth" about some phenomenon, conservation biologists often invest in developing models that allow us to make quantitative predictions about population change. These models are especially useful tools for exploring (or simulating) how we expect changes in a landscape to impact populations, whether from habitat fragmentation, increasing periodicity of weather events, or some other type of environmental change. In this section, we introduce three basic models of population growth, including the geometric growth model, exponential growth model, and logistic growth model; and some of the important fundamental elements of population modeling. These models provide a foundation for building more complex models.

Geometric growth model

Nearly all population models are based on four primary parameters that govern growth. These parameters include births, immigrants, deaths, and emigrants, and are referred to as the **BIDE parameters** (introduced in Chapter 9). The number of births and immigrants that move into the population add individuals to a population, and the number of deaths and emigrants out of the population reduce individuals in a population. The rates of these parameters determine the size of a population at a given time point. A simple model of growth expressed in terms of the BIDE parameters is:

$$N_{t+1} = N_t + B_t + I_t - D_t - E_t \tag{13.10}$$

Here the size of the population is N and the t refers to time period (say, a year, for example). The size of the population one year from now (N_{t+1}) is expected to be the size of the population now plus the number of births and immigrants, minus the number of deaths and emigrants. This simple mathematical model is often distilled even further into a model that focuses just on births and deaths. In other words, it assumes that the population is "closed" or does not experience immigration or emigration (fortunately these parameters can be added back in to the model in cases where a population is considered "open"). This basic population model is called the **geometric growth model** and looks like this:

$$N_{t+1} = N_t + RN_t \tag{13.11}$$

Here you'll notice a new term, R, which is the **geometric rate of increase**. It is basically a measure of growth rate and is estimated as $(b - d)$, or the rate of birth (per individual) minus the rate of death (per individual). To put this into context, imagine that we observed a population of 100 tigers and recorded 50 births and 10 deaths in a given year. The per individual or per capita birth rate would be 0.5 (= 50/100) and death rate would be 0.10 (= 10/100). R in this case would be 0.4 (= 0.5 – 0.1). An R value greater than zero indicates that population experiences positive growth and an R value less than zero indicates that the population will decline (i.e., there are more deaths than births).

The geometric growth model is simple and says that the size of the population in the next time step (e.g., 1 year) is the size of the population now plus some measure of growth, which could be positive or negative. In the tiger example above, the geometric growth model could be used to predict the size of the population in the next time period: 100 tigers now + (0.4 growth rate × 100 tigers now) = 140 tigers. It could then also be used to predict the population in the next time period after that: 140 + (0.4 × 140) = 196, and so forth into the future. If we carried out this calculation for many years, the geometric growth model would produce a pattern of growth that is exponential. In other words, population size starts to increase slowly, then increases rapidly (Figure 13.9).

The geometric growth model has a few key properties. First, it shows a pattern of growth that is unconstrained; that is, birth and death rates are constant, which implies an unlimited supply of resources like food and habitat. Second, it is considered a **discrete-time model**. This means that time is being treated in discrete packets, whether years, seasons, or some other time unit, and that growth only occurs once in each period. Because of this, the geometric

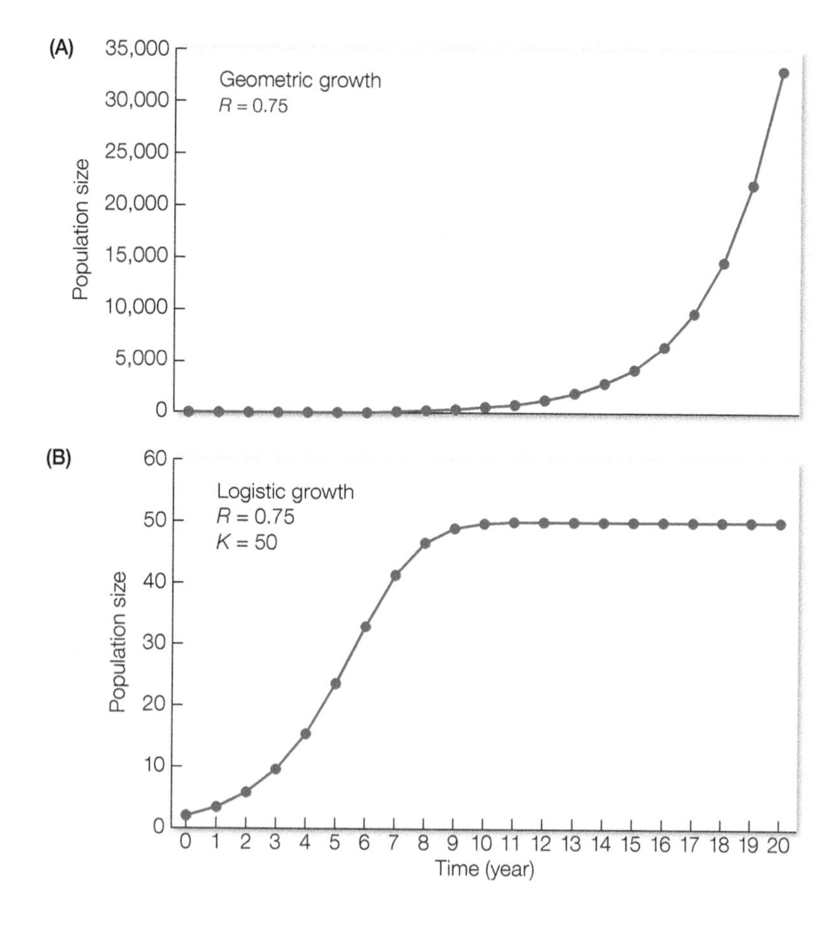

Figure 13.9 Population growth calculated using the geometric growth model (A) and logistic growth model (B). The geometric growth model shows a pattern of exponential growth that is unconstrained, as birth and death rates are constant. The logistic growth model introduces a carrying capacity (*K*, the maximum number of individuals an area can support) to account for the effect of intraspecific competition in the population. Both models are discrete-time models (with year as the time unit) and have the same growth rate (*R*).

growth model is usually best suited to a species that breeds once at a particular time of the year, like those occurring in more temperate environments. Consider caribou (*Rangifer tarandus*) in polar regions of North America. They breed and produce offspring in a pulse, once per year, in the spring months.

The geometric growth model can also be rewritten in another form:

$$N_{t+1} = \lambda N_t \qquad (13.12)$$

Lambda (λ) is called the **finite rate of increase** and is based on R. It is simply: $R + 1$, so in the tiger example above, λ is 1.4 (= 0.4 + 1), indicating a growth rate of 40%. Both R and λ essentially represent the same measure of growth, and more advanced population models may use one or the other depending on the mathematical structure of the model equation. The value of λ can also be written as:

$$\lambda = \frac{N_{t+1}}{N_t} \qquad (13.13)$$

Getting back to tigers, λ can also be calculated using this equation: (140 in Year 1) / (100 in Year 0) = 1.4; and (196 in Year 2) / (140 in Year 1) = 1.4. In other words, if we took a census of a population each year as part of a monitoring population, we could estimate a growth rate for each year. This growth rate could form the basis of a model to predict the size of the population at some time point in the future. For example, the geometric growth model was used as a basis for projecting population size of rare boreal toads (*Anaxyrus boreas boreas*) at sites in the U.S. Rocky Mountains.[50] Boreal toads have experienced range-wide declines, mainly due to amphibian chytrid fungus (*Batrachochytrium dendrobatidis*), and population projections across 20 years explored the potential impacts of the incursion of chytrid into disease-free populations. The results provide information that can help plan for the management of these populations and mitigation of chytrid impacts.

Exponential growth model

The **exponential growth model** is very similar to the geometric growth model. It assumes constant birth and death rates and results in unconstrained growth over time. However, there is one key difference and that is in the way it treats time. The exponential growth model is a **continuous-time model**, which means that growth does not occur in discrete periods (as you would expect from caribou or other seasonal breeders), but rather occurs continuously over time. This model is better suited to a species that breeds continuously throughout the year. Examples might include insects, birds, or small mammals in the tropics that breed year-round. The exponential growth model has also been used to describe and predict growth of human populations. The exponential growth model is typically written as:

$$N_t = N_0 e^{rt} \tag{13.14}$$

This equation states that the population size at some given time point (t) is predicted by multiplying the starting population size (N_0, size at time zero) by e raised to the power of $r \times t$. Let's dissect this further: the constant e is Euler's number, which is an irrational number (2.71828...), and r is called the **instantaneous** (or **intrinsic**) **rate of increase** and is the same as R (calculated as $b - d$). The lowercase letter is used to indicate that it is representing continuous growth; the uppercase R indicates discrete growth.

Let's now consider an example. Imagine that you have been studying a population of poison dart frogs in the Amazon and estimated the rate of growth r as 0.1. Your starting population is 100 frogs and you would like to project forward to what you expect the population size will be in 5 years. Applying the exponential model $N_5 = 100 \times 2.71828^{0.1 \times 5}$, the result is 165 frogs.

One property of the exponential growth model is that it produces a pattern of growth that is faster than the geometric growth model, even if the same growth rate value is used. This is due to the nature of continuous growth. In the exponential growth model, growth compounds (or occurs) continuously over the course of a time period (like a year), whereas in the geometric growth model, growth compounds (or is applied) only once. The exponential growth model has been used to describe and predict population size for many species, ranging from those that are rare to those that are harvested, and can be particularly helpful for understanding the invasion of non-native species, which often grow exponentially following a lag period after arrival[51,52] (see Chapter 11).

Logistic growth model

The geometric and exponential growth models represent unconstrained growth. These types of models may be appropriate for some species and circumstances, such as how we expect an invasive species to grow in the absence of any competitors or predators, or how we expect a population of a newly reintroduced species to grow. However, in most cases, there will be constraints on growth. The **logistic growth model** builds on the geometric and exponential models by adding a term that accounts for the constraint of carrying capacity. (As we saw in Chapter 10, **carrying capacity**, symbolized ***K***, is the maximum number of individuals that an area can support given the resources available.) For example, there are only so many deer that a forest patch can support. When deer exceed this level, individual fitness is expected to decline because there are simply too many individuals for the amount of food and other resources available. This makes intuitive sense, and the underlying effect that the logistic model addresses is the impact of **intraspecific competition** (i.e., competition from other members of the species) on population growth. The expectation is that as a population approaches carrying capacity, competition from conspecifics intensifies and leads to a reduction in birth rate and increase in death rate. Let's have a look at the model structure:

$$N_{t+1} = N_t + RN_t\left(\frac{K - N_t}{K}\right) \tag{13.15}$$

This is a discrete-time version of the model, and says that the size of the population at the next time period (N_{t+1}) can be estimated as the current population size (N_t) plus some measure of growth (RN_t) that is tempered by the effect of K or carrying capacity. You will notice that the first part of this equation looks just like the geometric growth model. The actual measure of growth, RN_t, though, is now multiplied by an expression that adjusts growth according to how close the population is to carrying capacity.

To demonstrate the tempering effect of carrying capacity in the model, consider an example of wild boar in a large forested patch. If the carrying capacity is estimated to be 1000 boar and there are currently only 10 boar in the area, then the carrying capacity element in Equation 13.15 would be a number that is close to 1: $(K - N_t)/K = (1000 - 10) / 1000 = 0.99$. When this value is multiplied by the measure of how many animals are added to the population (the RN_t part of the equation), then the population should grow in a geometric fashion. In other words, there is little constraint from competition at this low population level because there is very little competition for resources.

Now consider a different scenario. Imagine that the boar population has grown to a point near carrying capacity—say, 950 animals. The carrying capacity part of the equation now becomes a very small number: $(1000 - 950) / 1000 = 0.05$. When this value is multiplied by RN_t, then the growth we expect under the geometric model is severely reduced. It is tempered by the fact that the population is just about at carrying capacity.

The logistic growth model reflects a pattern of growth that starts out by following the geometric growth model pattern. The population begins to grow slowly, then increases rapidly in an exponential manner. Under the logistic growth model, growth begins to decline and eventually becomes zero when carrying capacity is reached (see Figure 13.9). This pattern of growth is referred to as **density-dependent growth**, because the rate of growth at any given time point is a function of how many animals are currently there. The logistic growth model can also be expressed as a continuous-time model, which may be better suited for some species and circumstances.

The logistic growth model is a commonly used model in population biology because it includes the effect of carrying capacity, which probably makes the model more realistic than exponential models of growth for many species. For example, it is often use as a basis for setting harvest quotas or bag limits for hunted species (see Chapter 10). The idea being that harvest can be maximized by harvesting the number of animals at the fastest period of population growth, which is the region where growth is exponential, just before it starts to reduce due to the carrying capacity effect. The logistic growth model has also been used in other circumstances, including how we expect reintroduced populations to grow, and simulations of how changes in carrying capacity due to landscape change will affect growth.

Age- and stage-based models

The geometric, exponential, and logistic growth models provide a means of predicting the size of a population. They are relatively simplistic, as they are based on rates of birth and death that are averaged among all individuals in a population in order to produce a single number. While helpful, these models may miss critical aspects of how populations change that could be valuable for addressing a species conservation problem. More complex models separate out growth by meaningful divisions such as age or stage.

- **Age-based models** estimate growth parameters by age classes because birth and survival tend to differ among animals at different ages. For example, among moose (*Alces alces*), newly born animals experience high mortality in their first year and do not reproduce. Yearlings and 2-year-olds will have higher survival rates and begin reproducing, and older adults will have even higher survival rates, but birth rates will senesce or reduce over time.
- **Stage-based models** estimate growth parameters by meaningful life-stages, which may not necessarily be associated with age. An example is insect developmental stages like larva, pupa, and adult. Sometimes it may be more appropriate to use size to define a stage. For example, seedlings, small-sized individuals, medium-sized individuals, and large-sized individuals may be meaningful categories for a tree species (assuming these divisions differ with respect to survival and reproduction).

As you can imagine, estimating population growth over time can become complicated when considering different rates among ages or stages. Different ages/stages have different rates, and over time, individuals advance from one age/stage to another. Keeping track of these changes over time often requires charting populations using a matrix of values. Age- and stage-based models can predict the total population size at a given time point, but they can also predict the number of individuals in each age or stage, which provides a more detailed picture of the structure of the population.[53]

Matrix models are particularly appropriate for age- and stage-based modeling. To illustrate how a matrix model works, consider a population that has age structure (Figure 13.10). We will focus on a species that produces offspring once per year (in a pulse) and at the same time each year. The species only lives for 4 years, and we will assume 4 age classes. (Age class 1 = 0–1 year old, 2 = 1–2 years old, 3 = 2–3 years old, and 4 = 3–4 years old). Animals in each age class have a distinct probability of surviving (P) to the next age class. For example, newborns or those in the first age class may have a 70% chance of surviving ($P_1 = 0.70$) beyond their first birthday and into the second age class. If there were 100 individuals in the first age class, on average, 70 would survive to the second. Those in the second age class have a different probability (P_2) of making it into the next age class, and those in the third age class have yet a different probability (P_3) of making it into the final age class. Similarly, each age class has a different birth rate. The first age class produces no offspring, but each year the other ages produce offspring that all enter the first age class. Birth rates are often expressed as *fertilities* (noted as F) in matrix models because they are adjusted by survival rates.[54] To explain, for an animal that has just entered the second age class, in order to produce offspring that will be born in the next time step, it has to both reproduce and survive to make it to the next time step (or age class). So its fertility (F_2 in this case) is the average birth rate for that age class multiplied by the survival rate for making it into the next age class (P_2).

Survival rates and fertilities can be arranged in a matrix, and used to predict the size of each age class in the next time period. A **Leslie matrix**, named after the biologist P. H. Leslie, is commonly used.[55] In our example, we construct a matrix that is four rows by four columns (because we have four age classes) that includes our fertility and survival values (see Figure 13.10). The values in each row are used to estimate the size of an age class population in the next time step. For example, in row 1, the number of age class 1 individuals in the next time step will be the number of offspring produced by each age class that year

(A) Life cycle

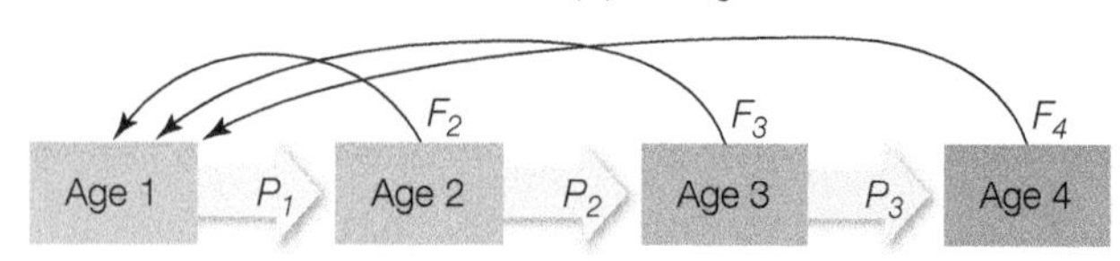

Survival probability (P) to the next age class

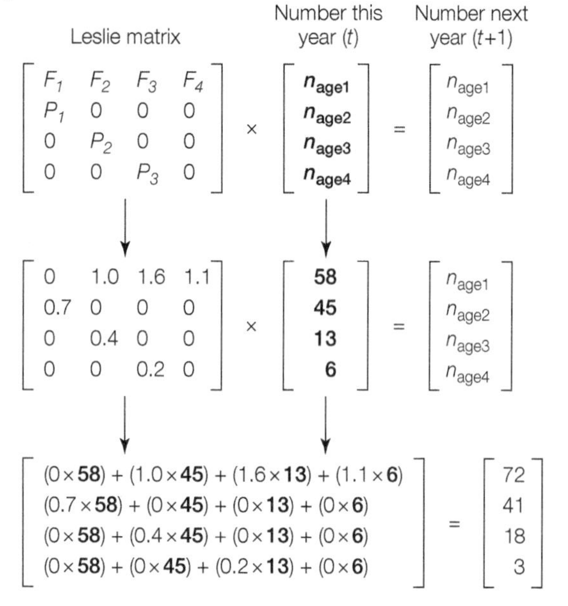

(C) Population size and growth rate

Population size, $N_{t+1} = (72 + 41 + 18 + 3) = 134$

Growth rate, $\lambda_t = N_{t+1}/N_t = (72 + 41 + 18 + 3)/(58 + 45 + 13 + 6) = 1.098$

Figure 13.10 Using a Leslie matrix to estimate the size and growth rate of a population when there is age structure. (A) In this example, the species lives for 4 years and ages are divided into classes (1 = 0–1 year; 2 = 1–2 years; 3 = 2–3 years; and 4 = 3–4 years). Animals in each class have a probability *P* of surviving to the next class. Animals in each age class except the first produce offspring at a rate *F*; these offspring then enter the first age class. (B,C) Because survival and fecundity differ by age class, there is age structure in this population. If we estimate these parameters and census the population in the current year, a Leslie matrix can be used to estimate the number of individuals in each class in the next time step. Total population size is then the sum of all animals in each age class. Population estimates can also be used to calculate the growth rate for the population.

(i.e., each fertility value F_1, F_2, F_3, and F_4). The number of offspring produced is a function of how many individuals there are in each age class (fertility × number of individuals = number of offspring). In row 2, the number of age class 2 individuals in the next time step is just the number of age class 1 individuals that survive to make it into age class 2, so you will see P_1 in the first column and zero values in the next three columns (obviously, older individuals can't go back into younger age classes). The same is repeated for rows 3 and 4.

If we census the population in our starting year and know how many individuals are in each age class at time *t*, then we can simply multiply these values across our row values in the matrix to estimate the number of animals we expect in each age class in the next time step (see Figure 13.10B). The values can be summed to calculate total population size, and a growth rate like λ can be estimated as the population size divided by the population size from the census in the previous year (see Figure 13.10C). This approach can then be used to project the size of the population in the next time step, the step after that, and so forth into the future.

In addition to a Leslie matrix, age or stage-structured populations are often modeled using a Lefkovitch matrix. A Lefkovitch matrix differs from a Leslie matrix by allowing some individuals to remain in the same class and others to move on.[56] Named after biologist L. P. Lefkovitch, this form of a matrix model essentially adds terms for the probability of remaining in a stage/size class into a Leslie matrix.

Separating growth into age/stage components can be helpful for simulating the impacts of changes in the growth of different ages/stages. An example of the effectiveness at simulating these types of changes to address a species conservation problem involves efforts to reduce declines of marine turtles. Crowder and colleagues[57] developed a stage-based model of loggerhead sea turtles (*Caretta caretta*), which are listed under the U.S. Endangered Species Act. Loggerhead turtles face a number of threats from people, including loss and degradation of beach nesting habitat due to coastal development, and fisheries that can ensnare turtles in nets meant for other species. The researchers explored how simulated increases in the survival of five different life stages—egg/hatchlings, small juveniles, large juveniles, subadults, adults—impacted overall population growth. Their results indicated that

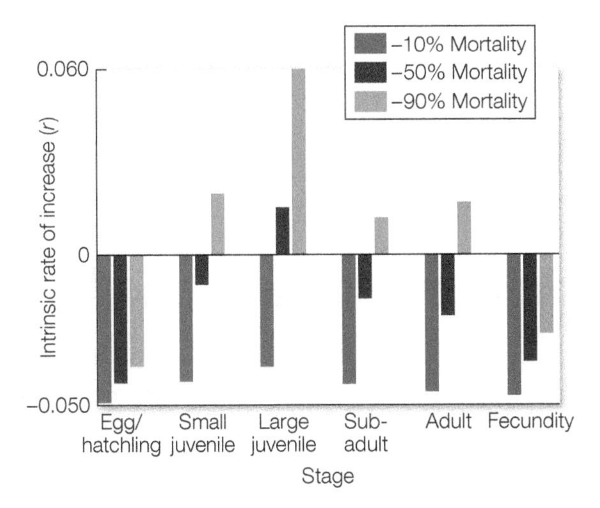

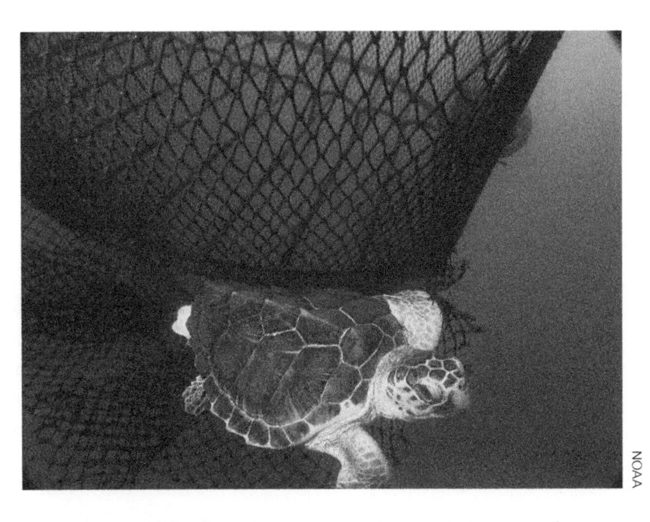

Figure 13.11 A stage-based model of loggerhead turtles (*Caretta caretta*) was used to simulate of the effects of 10%, 50%, and 90% reductions in mortality on the intrinsic population growth rate (r) of different life stages. The results revealed that efforts to reduce mortality of large juveniles had the greatest effect on the population. A main source of mortality for turtles in this age class is entrapment in fishing nets. These results led to more widespread use of escape hatches in nets, known as TEDs (turtle excluder devices). (After L. B. Crowder et al. 1994. *Ecol Appl* 4:437–445.)

population growth was most affected by the survival of large juveniles followed by small juveniles, over other stages (including egg/hatchlings, which had been the target of previous conservation efforts; **Figure 13.11**).[57] The study led to a shift in focus to better protect the two juvenile age classes, which are mainly threatened by trawling nets for shrimp. More specifically, the study led to more widespread use of turtle excluder devices (TEDs) in nets; these devices act as escape hatches for ensnared turtles.

Stochastic models

Population models, including the geometric, exponential, logistic, and age/stage based models, can be modified to incorporate variation in parameters, which may help improve their ability to predict. In the examples above, we treated parameters like R, λ, and birth and death rates as constants. They reflect average values and do not change from time period to time period. We call a model with fixed values a **deterministic model**. Each value is fixed and determined, and these values are always used to make predictions. An alternative is to build a **stochastic model** that accounts for variation in the model parameters (among time points). For example, we know that birth rates will change from year to year even in a stable population, so rather than use a fixed, average birth rate in a model, a stochastic model would use a value drawn from a distribution of birth rates collected over time (or from a theoretical distribution).

For example, imagine that we recorded birth rates of a gazelle species for several years. From these data, we could estimate a mean birth rate and a measure of the spread of data that went into that mean, like a standard deviation. In our population model, if we were to use it to predict population size in the next time period, we would not use the deterministic mean value, but rather draw a value from the distribution of values recorded. To predict the following year, we would then not use the same value, but a new random value from the distribution, and so forth for subsequent years. The goal of a stochastic model is to build more realism into a model in an effort to make predicted estimates of population size as close to the "truth" as possible.

Population models, whether deterministic or stochastic, geometric/exponential or logistic, age-or stage-based, allow conservation biologists to describe how a population grows and to predict how a population will change in the future. A key question that models can help answer is: How large does a population need to be in order ensure that it will be viable at some point in the future? This question is often critical for conserving small populations that have few individuals. In the next section, we will cover the concept of a minimum viable population size, then introduce a technique for estimating whether a population is in fact viable.

Predicting Population Persistence

In a groundbreaking paper in 1981, Mark Shaffer[58] defined the number of individuals necessary to ensure the long-term survival of a species as the **minimum viable population (MVP)**:

> A minimum viable population for any given species in any given habitat is the smallest isolated population having a 99% chance of remaining extant for 1000 years despite the foreseeable effects of demographic, environmental, and genetic stochasticity, and natural catastrophes.[58]

In other words, the MVP is the smallest population size predicted to have a very high chance of persisting for the foreseeable future. Shaffer emphasized the tentative nature of this definition, saying that the survival probabilities could be set at 95%, 99%, or any other percentage and that the timeframe might similarly be adjusted, for example to 100 or 500 years. The key point is that the MVP size allows a quantitative estimate to be made of how large a population must be to ensure long-term survival.

Shaffer compared MVP protection efforts to flood control.[58] It is not sufficient, he pointed out, to use average annual rainfall as a guideline when planning flood control systems and developing regulations for building on wetlands. Instead, we must plan for extreme situations of high rainfall and severe flooding, which may occur only once every 50 or 100 years. In protecting natural systems, we understand that certain catastrophic events, such as hurricanes, earthquakes, forest fires, epidemics, and die-offs of food items, may occur at even greater intervals. To plan for the long-term protection of endangered species, we must provide for their survival not only in average years, but also in exceptionally harsh years.

Minimum population size

An accurate estimate of the minimum population size for a species' survival requires both a detailed demographic study of the species' populations and an analysis of their environments. This can be expensive and require months, or even years, of research. Analyses of over 200 species for which adequate data were available (mainly vertebrates) indicated that most MVP values for long time periods fall in the range of 3000 to 5000 individuals, with a median of 4000.[59,60] In general, protecting a larger population increases the chance of the population persisting for a longer period of time (**Figure 13.12A**). For species with extremely variable population sizes, such as certain invertebrates and annual plants, protecting a population of about 10,000 individuals might be the ideal strategy.

Unfortunately, many species, particularly endangered species, have population sizes much smaller than these recommended minimums. For instance, half of 23 isolated elephant populations remaining in West Africa have fewer than 200 individuals, a number considered to be inadequate for long-term survival of the population.[61] The wolf population on Isle Royale, Michigan—a long, narrow island approximately 535 km^2—has been fluctuating around 20 individuals since the early 1980s, and declined to 2 individuals in 2016.[62,63] The Isle Royale example clearly

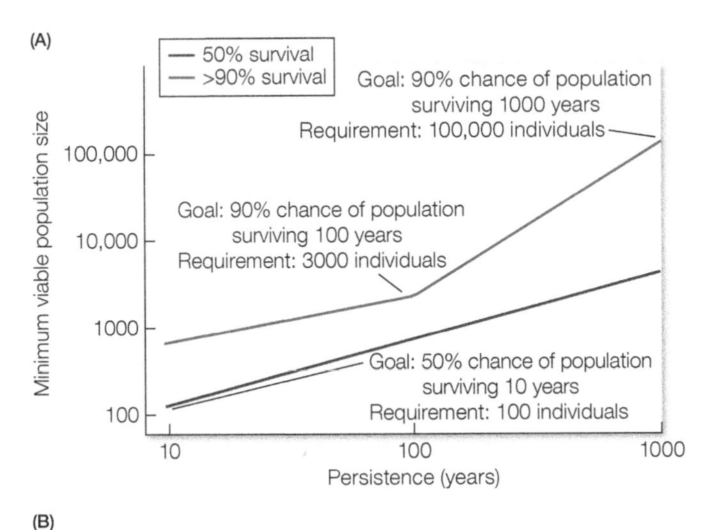

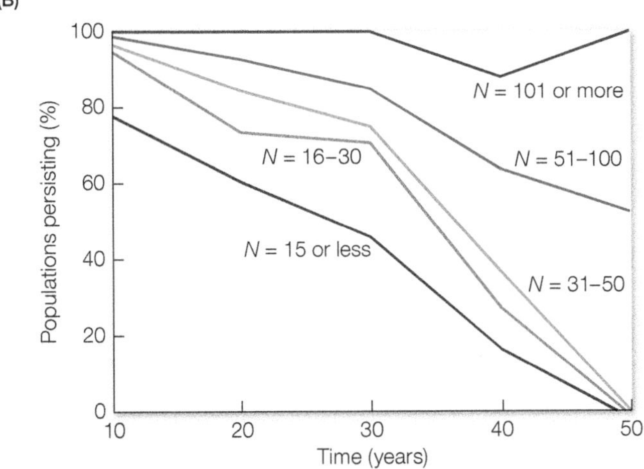

Figure 13.12 (A) If the goal is persistence for a greater number of years, then a larger minimum viable population (MVP) size is needed. This point is illustrated by two lines: one showing the minimum viable population size (*y*-axis) required to ensure a 50% chance of survival (red line), and a second showing the size needed to ensure a 90% chance of survival (blue line). These log scale values were derived from data on changes in population size and persistence of 1198 different species. (B) The relationship between initial population size (*N*) of bighorn sheep and the percentage of populations that persisted over time. Almost all populations with more than 100 sheep persisted beyond 50 years, while populations with fewer than 50 individuals died out within 50 years. Not included are small populations that were actively managed and augmented by the release of additional animals. (A, after L. W. Traill et al. 2010. *Biol Conserv* 143: 28–34. B, after J. Berger. 1990. *Conserv Biol* 4: 91–98.)

demonstrates the perils of small populations, and efforts have since been taken to translocate wolves to the island from other parts of the United States and Canada.

Field studies confirm that small populations are most likely to decline and go extinct.[64] One of the best-documented studies of MVP size tracked the persistence of 120 bighorn sheep (*Ovis canadensis*) populations (some of which have been followed for 70 years) in the deserts of the southwestern U.S.[65,66] The striking observation is that all (100%) of the unmanaged populations with fewer than 50 individuals went extinct within 50 years, while virtually all of the populations with more than 100 individuals persisted within the same time period (Figure 13.12B). No single cause was evident for most of the populations that died out; rather, a wide variety of factors appear to be responsible for the extinctions. Thus, an empirical estimate of MVP for bighorn sheep, is at least 100 individuals. Unmanaged populations below 50 could not maintain their numbers, even in the short term. Additional research suggests that bighorn sheep populations have a greater chance of persisting when they occupy large habitats (which allow populations to increase in size) that are more than 23 km from domestic sheep (a source of disease).[67] However, despite the factors hindering the survival of small populations of bighorn sheep, habitat management by government agencies and the release of additional animals have allowed some small populations to persist that might otherwise have gone extinct. Other long-lived species, such as turtles and trees, can often persist for extended periods in small populations.[68]

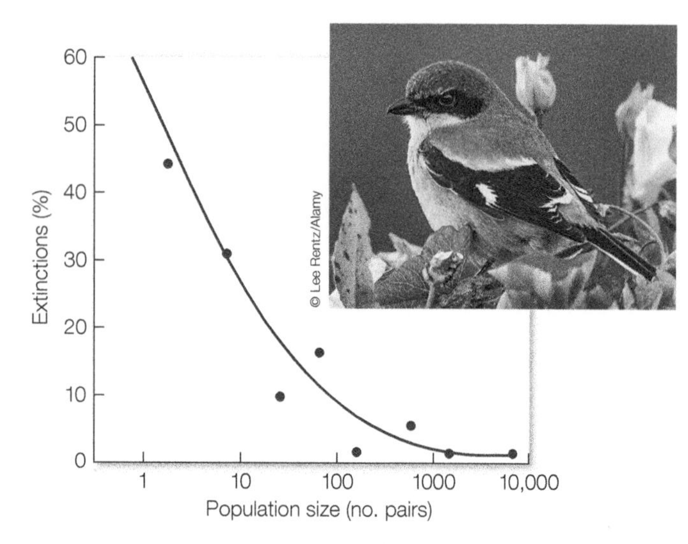

Figure 13.13 Extinction rates of bird species like the loggerhead shrike, *Lanius ludovicianus anthonyi* (shown above), on the Channel Islands of southern California. Each dot represents the extinction percentage of all the species in that population size class; extinction rate decreases as the size of the population increases. Populations with fewer than 10 breeding pairs had an overall 39% probability of extinction over 80 years. Populations of between 10 and 100 pairs averaged about 10% probability of extinction, and populations of over 100 pairs had a very low probability of extinction. (After H. L. Jones and J. M. Diamond. 1976. *Condor* 76: 526–549.)

Field evidence from long-term studies of birds on the Channel Islands off the California coast supports the fact that large populations are needed to ensure population persistence; only bird populations with more than 100 breeding pairs had a greater than 90% chance of surviving for 80 years (**Figure 13.13**). In spite of most evidence to the contrary, however, small populations sometimes do prevail; many bird populations have survived for 80 years with 10 or fewer breeding pairs. Of course, birds are especially mobile and can readily recolonize areas following local extinction. Less mobile species do not have this ability.

Once an MVP size has been established for a species, the **minimum dynamic area (MDA)**—the area of suitable habitat necessary for maintaining the minimum viable population—can be estimated by studying the home range size of individuals and colonies of species.[69,70] It has been estimated that reserves in Africa of 100 km^2 to 1000 km^2 are needed to maintain many small mammal populations. To preserve populations of large carnivores such as lions, reserves of 10,000 km^2 are needed.

Population viability analysis

Estimates of the minimum viable population size and minimum dynamic area required for a population to persist can be calculated using a **population viability analysis**, or **PVA**. A PVA can be thought of as a risk assessment, as it aims to estimate the probability that a population will go extinct at some time point in the future. A basic PVA requires (1) an estimate of the current population size and (2) a model for estimating how a population will change over time. Any of the population models we covered earlier in this chapter can be used, although researchers favor stochastic models that help account for potential variation in the parameters of a model (like birth and survival rates). When projecting how a population is expected to change in the future, variation is undoubtedly going to occur and the model should try to account for this. Let's walk through a simple PVA using a basic geometric growth model to demonstrate how it could be used to estimate the probability of extinction among African wild dogs (*Lycaon pictus*), a species that was mentioned earlier in the chapter.

African wild dogs historically occurred through much of sub-Saharan Africa, but today have been reduced to 39 discrete subpopulations that are estimated to range in size from 2 to 276 mature individuals.[71] Most populations are small and declining. Imagine that we have been monitoring one of those discrete populations of wild dogs for the past 30 years, estimating the population size each year. Today there are 100 wild dogs in the population, but since our monitoring began, the population has fluctuated considerably. For each year of monitoring, we can calculate a growth rate, λ, which you might recall from earlier is estimated as N_{t+1}/N_t, or the population size in one year (or any specified time period) divided by the population size in the previous year. This gives us a collection of λ values that when plotted may look like a normal, bell-shaped distribution. The distribution is defined by the mean (or the peak) and the width is determined by the standard deviation of the values.

We will use a basic geometric growth model ($N_{t+1} = \lambda N_t$) to predict the size of the population each year for the next 50 years. At the end of the 50 years, we will evaluate whether the remaining population is viable. To do this, we start with our current population size, which is 100, then multiply it by a λ growth rate value.

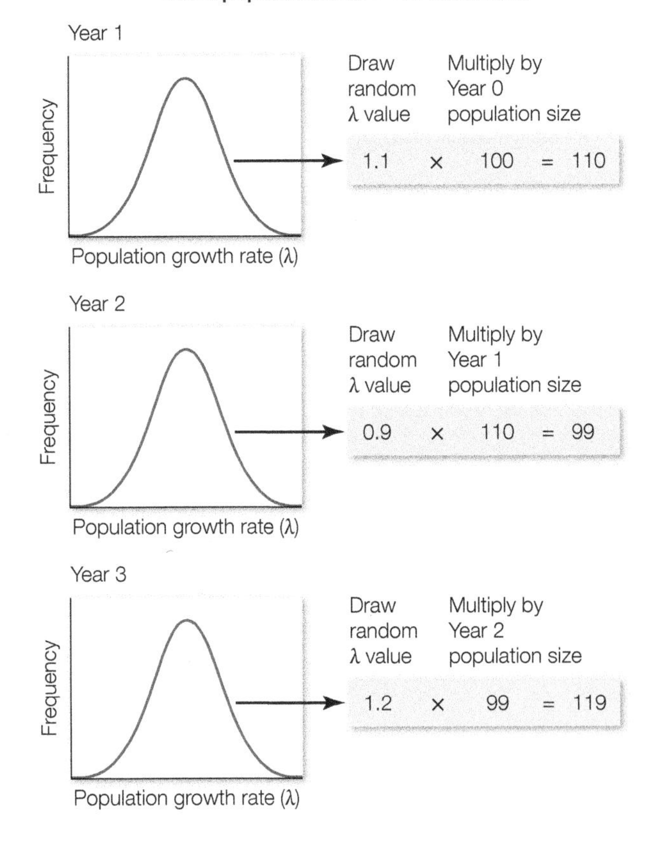

Figure 13.14 Projecting a population forward in a basic PVA using a geometric growth model; each projection is referred to as a trial. Starting with an initial population size, estimating the population size in the next step involves drawing a random population growth rate value (λ) from a distribution of values from past data, then multiplying this value by the initial population size. Estimating population size in the next year involves drawing a new λ value and multiplying it by the Year 1 population size. This procedure is repeated year after year until the final time period of the analysis is reached.

Rather than use a deterministic value like the average λ recorded across our 30 years of monitoring, we will randomly select a value from our distribution of values to build in stochasticity. We do not know what the actual growth rate will be for the next year, but we are assuming that it would be similar to what we observed in the past. If we draw a random value from our distribution of values, we are essentially making a reasonable guess at what the value of the population growth rate will be. Let's say we draw a λ value of 1.1 from a normal distribution of growth rates (**Figure 13.14**). When we apply this to our model, our estimate of population size in Year 1 is 110 (1.1×100 wild dogs). Now we move on to Year 2. To predict population size, we draw a new random value from the distribution of λ values and use it in our model. Let's say the value is 0.9. We multiply 0.9 times the population size in Year 1 to estimate the Year 2 population ($0.9 \times 110 = 99$). We follow this procedure year after year until we reach 50 years, at which point we make a judgement about viability. Are there any animals left after 50 years? Are there enough individuals to constitute a *viable* population (i.e., are there at least 2 individuals or some other minimum threshold required for viability)?

We just conducted a single projection of the population over 50 years. This projection is referred to as a trial, and a PVA typically involves conducting not just one, but many trials (e.g., 1000 or more). Each trial will give a different projection because our growth rate values change with each yearly population size estimate (**Figure 13.15**). In the end, we are left with a collection of outcomes. If we conduct 1000 trials, we will have 1000 estimates of the population size after 50 years. Some of the estimates may be zero (i.e., the species is predicted to go extinct), whereas others may be high (depending on your growth rate data). This collection of values can be used to estimate the probability of extinction. If we had 318 trials that ended with no individuals left, then the chance of extinction is 318/1000, or 38%. The inverse represents the probability of being viable, which in this case is 62%. We may use this value to inform how we manage the population now. For example, we could simulate the effect of increasing the population through a reinforcement program. If we brought in 50 new individuals, our starting population would jump from 100 to 150 wild dogs. The PVA would then give us the ability to evaluate the effect of this management action on the broader population. The PVA could also give us the ability to evaluate the effect of periodic landscape changes on the population. We could, for example, simulate the impacts of 10-year droughts on a population by reducing growth by a certain amount at 10-year intervals in the analysis, which ultimately will affect estimates of viability.

Suggested Exercise 13.4 Develop a simple population viability analysis using a spreadsheet. **oup-arc.com/e/cardinale-ex13.4**

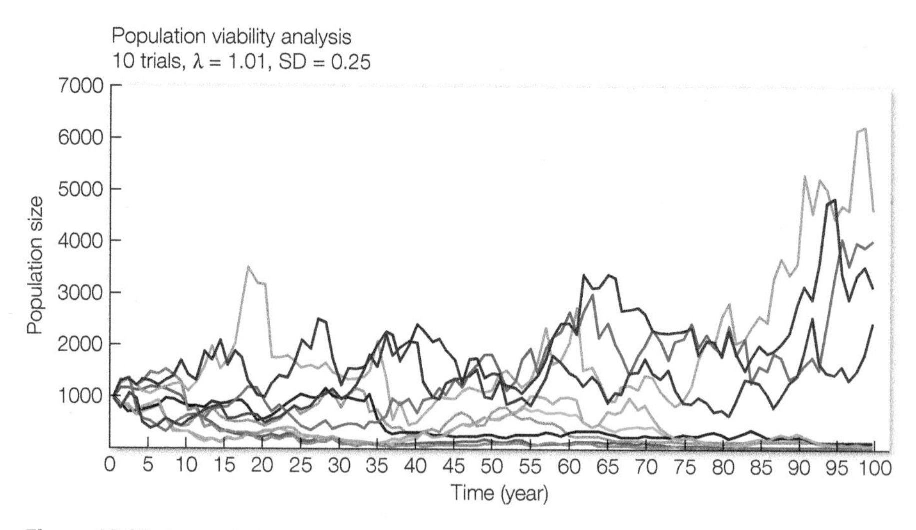

Figure 13.15 A population viability analysis based on 10 population projections, or trials. The probability of viability in this case is the number of trials that end with more than 2 individuals out of the total number of trials = 4/10, or 40%. PVAs usually conduct many more than 10 trials.

Population viability analyses can certainly incorporate more complexity. The wild dog example above involved a simple population model. More sophisticated models may include parameters (along with distributions) for birth and death rates by different age classes or life stages. Future projections would then draw from multiple distributions in order to estimate population size.

It is important to recognize that PVA models are only as good as the data that goes into developing them, and they rely on the assumption that data collected in the past capture all of the variability that the population is expected to exhibit in the future.[72] Models should be used with some degree of caution and a clear understanding of their assumptions and limitations.[73] Attempts to validate models should also be conducted. This might include pitting the model predictions against real data (perhaps from the past or from populations elsewhere), or confirming the model's ability to accurately predict the future by comparing predicted population estimates to incoming monitoring data. PVAs also require a clear understanding of the ecology of the species, the threats it faces, and its demographic characteristics. Here are some examples of PVAs that have been used in real conservation scenarios:

- **Hawaiian stilt.** The Hawaiian stilt (*Himantopus mexicanus knudseni*) is an endangered endemic bird of the Hawaiian Islands. Hunting and coastal development reduced the number of birds to 200 by the early 1940s, but protection has allowed recovery to the present population size of around 1600 individuals.[74,75] The goal of government protection efforts is to allow the population to increase to 2000 birds. A PVA was made of the species' ability to have a 95% chance of persisting for the next 100 years. Models treated the stilts as either one continuous population or 6 subpopulations inhabiting individual islands. Given the stilts' current positive growth under existing conditions, the models predicted that stilt numbers would increase until they occupied all available habitat, but that they would show a rapid decline if nesting failure and mortality rates of first-year birds exceeded 70%, or if the mortality rate of adults

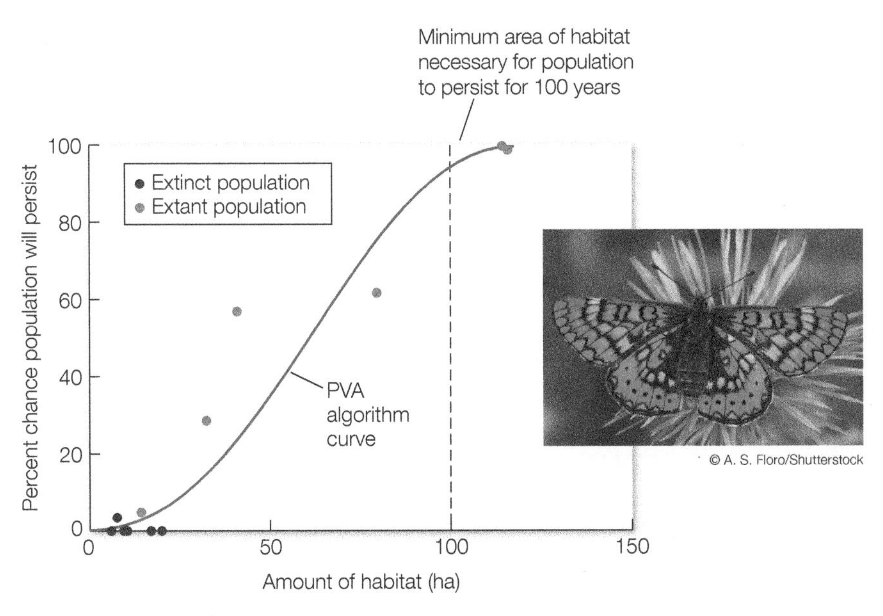

Figure 13.16 Population viability analyses predict that it takes 100 ha of habitat to ensure (at 95% likelihood) the persistence of a marsh fritillary butterfly (*Euphydryas aurinia*) population for 100 years. All of the extinct populations occupied areas much smaller than 100 ha. Four of the six extant populations occupy areas smaller than 100 ha and are predicted to go extinct unless their habitat is increased. (After C. R. Bulman et al. 2007. *Ecol Appl* 17: 1460–1473.)

increased above 30% per year. Keeping mortality rates below these levels would require the control of exotic predators, protection of existing wetland, and restoration of additional wetland habitat.

- **Marsh fritillary butterfly.** The marsh fritillary butterfly (*Euphydryas aurinia*) is declining in abundance in the United Kingdom, where it occupies lightly grazed grasslands. The average area occupied by the six extant populations studied is larger than the average area formerly occupied by six extinct populations. A PVA showed that an area of at least 100 ha is necessary to ensure a 95% probability that a population will persist for 100 years (**Figure 13.16**). Only two of the extant populations encompass areas of this magnitude. The other four populations face a high probability of extinction unless the habitat is enlarged in area and managed to encourage the growth of food plants.[76]
- **Leadbeater's possum.** The most complete PVA ever undertaken is probably that of Leadbeater's possum (*Gymnobelideus leadbeateri*), an endangered arboreal marsupial inhabiting a rare type of eucalyptus forest in southeastern Australia.[77] Populations of this species were predicted to decline by more than 90% over the 20 to 30 years following the PVA because logging and wildfires remove the large trees with cavities that the possums need for their dens. Population models based on extensive field research, including the spatial distribution of habitat patches and dispersal corridors, the abundance of tree cavities, and forest dynamics, informed forest management decisions by government agencies. Unfortunately, continued logging and severe wildfires have resulted in a substantial decline in the numbers of Leadbeater's possum from a high of approximately 7500 individuals to only 1500 today—an 80% decline, in line with the dire predictions made decades earlier.[78]

Conservation Trade-Offs

In this chapter, we have focused on the conservation of single species, and especially of small populations, which are frequently a target of conservation efforts. The broader goal of species-level conservation is to ensure the long-term persistence of populations, and we covered a number of scientific tools that are available to estimate population characteristics, which can aid in the development of conservation actions or decisions. However, it is important to recognize that whether conservation efforts target populations, groups of populations, or are aimed at achieving broader ecosystem benefits, any conservation effort will have trade-offs that might be ecological, economic, or social.

Consider efforts to restore wolf (*Canis lupus*) populations in North America and Europe. Actions that result in increases in wolf abundance, like the reintroduction or translocation of individuals into a population, may be good from a population recovery perspective. However, they may also result in higher rates of livestock depredation and negative economic impacts to local people. Similarly, these actions may lead to greater genetic diversity in populations (again, good for the population), but could be expensive and take funding away from other species in need. Trade-offs can be complex and challenging to address when developing conservation actions.

One means of addressing the complexity of trade-offs is to think about species conservation actions as a decision-making problem. In other words, an action taken is really a decision that is usually made from a suite of other potential actions. When viewed through the lens of decision-making, principles of formal decision-making science can be used to help visualize the problem and account for trade-offs. As you'll recall, **structured decision making** (or SDM for short) is a commonly used framework for the management of natural resources[79] (see Box 2.3). SDM aims to provide a transparent and data driven means of evaluating potential decisions and is rooted in the PrOACT approach that we introduced in Chapter 2. PrOACT stands for Problem, Objectives, Alternatives, Consequences, and Trade-offs and can be operationalized in many ways, some of which we covered earlier such as multicriteria decision analysis (see Chapter 7).

Another common approach is to use a simple decision-making tool—a consequences table—that incorporates the elements of PrOACT and helps identify the trade-offs of potential decisions, which can be helpful for addressing a species-level conservation problem. Consider an example involving bighorn sheep (*Ovis canadensis*) in Montana. Bighorn sheep in this region face an ongoing threat from disease (like pneumonia) spread from domestic livestock. Mitchell and colleagues[80] used the PrOACT approach to address the problem of how best to reduce a potential outbreak of disease in two bighorn populations. They used a SMART (Simple Multi-Attribute Rating Technique) table to organize their objectives, alternatives, consequences, and assessment of trade-offs (**Table 13.1**). In this study, objectives including maximizing the probability of persistence of bighorn, minimizing costs, and maximizing opportunities for bighorn viewing and hunting. Objectives included a set of aggressive proactive actions (like fencing domestic sheep herds to reduce interactions with bighorn, and increasing harvest of bighorn males that range widely during the rut), moderate proactive actions (like communicating with livestock producers about minimizing contact between domestic sheep and bighorn), and simply being reactive to an outbreak of disease. Consequences of each alternative on each objective were generated from predictive models, data, and expert opinion. The

(A)

(B) Petty Creek

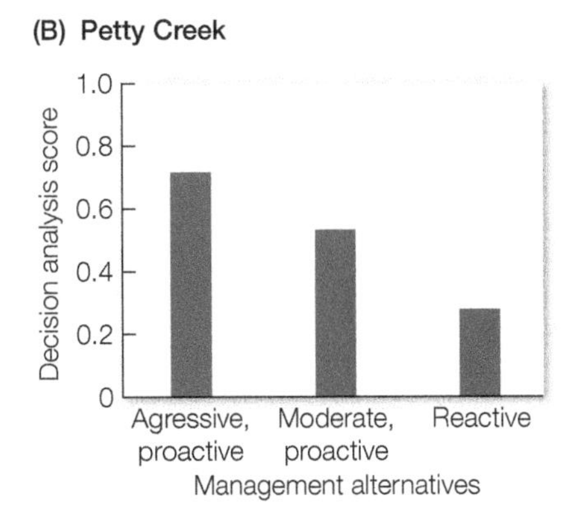

(C) Missouri Breaks

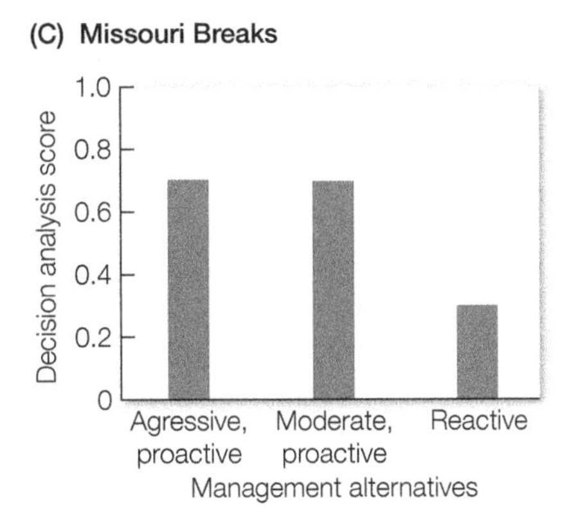

Figure 13.17 Scores for three management alternatives for two populations of bighorn sheep in Montana, calculated using a decision-making framework (see Table 13.1). (From M. S. Mitchell et al. 2013. *Wild Soc Bull* 37: 107–114.)

consequence data across objectives were then used to score each alternative, which they plotted in a simple bar graph (**Figure 13.17**). The graph shows how alternatives compared to each other, which can be helpful to decision-makers. The results indicated that aggressive actions were best for one population, while either aggressive or moderate actions were best for the other population. Being reactive was the worst scoring alternative, indicating that proactive management of both populations would be best. In other words, doing nothing to prevent an outbreak of disease was the least effective option for achieving the objectives associated with the problem.

This simple SMART table has advantages. It requires that the problem, objectives, and alternatives are explicitly articulated. These basic elements of a species-level decision problem can sometimes be overlooked or vaguely defined, which can limit the effectiveness of any resulting conservation actions. The table also makes use of data to drive the evaluation of trade-offs of potential decisions, and in this case, also included weights that allowed for some objectives to carry more importance than others in the analysis (see Table 13.1). Weights can come from a variety of sources such as the opinions of decision-makers or key stakeholders. Another advantage is that the table allows for the consideration of objectives from multiple disciplines (e.g., ecological, economic, and social), and thereby provides a means of integrating interdisciplinary elements of a species conservation problem. Conservation problems are rarely confined to a single discipline and the most effective solutions usually arise when considering multiple disciplines or dimensions of a problem.[81] Several other types of frameworks (like SDM) and associate tools (like a consequences table) also exist to support decision-making and help improve conservation outcomes.[82]

TABLE 13.1 Results of a decision-making analysis using a 6-objective framework to evaluate and compare 3 management alternatives for reducing a potential outbreak of disease in a bighorn sheep population[a]

Management alternative	Objective: Probability of persistence Goal: maximize Weight: 0.21	Operating costs Goal: minimize Weight: 0.15	Personnel costs Goal: minimize Weight: 0.14	Crisis response costs Goal: minimize Weight: 0.19	Viewing opportunity Goal: maximize Weight: 0.15	Hunting opportunity Goal: maximize Weight: 0.18	Final score[b]
AGGRESSIVE							
Expected outcome	0.9	105	220	8	0.9	190	
Normalized score	1.00	0.00	0.00	1.00	1.00	1.00	
Normalized score × wt.	0.21	0.00	0.00	0.19	0.15	0.18	**0.72**
MODERATE							
Expected outcome	0.8	100	170	16	0.8	135	
Normalized score	0.80	0.05	0.23	0.80	0.80	0.35	
Normalized score × wt.	0.17	0.01	0.03	0.15	0.12	0.06	**0.53**
REACTIVE							
Expected outcome	0.4	0	0	48	0.4	105	
Normalized score	0.00	1.00	1.00	0.00	0.00	0.00	
Normalized score × wt.	0.00	0.15	0.14	0.00	0.00	0.00	**0.28**

Source: M. S. Mitchell et al. 2013. *Wild Soc Bull* 37: 107–114.
[a]The consequence of each alternative decision on each objective was estimated from models, data, and expert opinion. These values were normalized by objective (i.e., scaled from 0 to 1.0, with 0 being the worst and 1.0 being the best; if the goal for an objective is to maximize, then the lowest value among the alternatives becomes 0 and the highest value becomes 1.0; if the goal is to minimize, the lowest value is the best and becomes 1.0 and the highest value becomes 0). Finally, normalized scores were multiplied by weights and summed across objectives to result in a final score for each alternative management decision. The higher the score, the better the decision.
[b]Equations for normalizing values in an objective: 1) if the goal is to maximize, then: (value – min)/(max – min), 2) if the goal is to minimize, then: (value – max)/(min – max). For example, the normalized value for the Moderate alternative in the Probability of Persistence objective = (0.8 – 0.4)/(0.9 – 0.4) = 0.8.

Summary

1. Species-level conservation efforts often target populations, groups of populations (like a metapopulation), and the restoration of locally or globally extinct species. Species focused conservation can also achieve broader conservation goals, such as when surrogate species (flagship or umbrella species) help protect entire communities or ecosystems.

2. Small populations face three main challenges that can increase extinction risk: (1) demographic stochasticity, which refers to random variation in average birth and death rates among individuals, (2) environmental stochasticity, which refers to random variation in the biological and physical environment, and (3) loss of genetic variability due to genetic drift and inbreeding, which can have fitness consequences to individuals and limit adaptability.

3. Management of populations often focuses on increasing genetic variation. Effective population size (N_e) reflects the genetic nature of a population and generally refers to the number of breeding individuals in a population. N_e is typically far less than the actual number of individuals and can be influenced by several factors including sex ratio in the breeding population, reproductive output among individuals, and population changes over time. Low effective population size can be concerning as it reflects low genetic diversity.

Summary (continued)

4. Population size is an important demographic characteristic. Approaches to estimating population size include conducting a census or complete count of individuals in a population or capture-mark-recapture. The latter involves estimating population size by creating a marked subsample in the broader population.
5. Models that describe population growth can be developed from population data. Models allow conservation biologists to predict how populations are expected growth over time and explore how landscape changes affect populations. Three fundamental population models include the geometric growth model, exponential growth model, and logistic growth model. More advanced models include age- and stage-based models.
6. Population viability analysis (PVA) is a common approach that uses a population model to estimate whether a given population will be viable at some point in the future. PVAs can help determine the minimum viable population (MVP) size and be used to explore the effects of conservation actions and landscape changes on viability.
7. Actions targeting species-level conservation issues are bound to have ecological, economic, and social trade-offs. The structured decision making approach provides a framework for evaluating a suite of potential conservation actions or decisions around a species-level problem that account for these trade-offs.

For Discussion

1. Use the IUCN Red List of Threatened Species (oup-arc.com/e/cardinale-w13.2) to find a Critically Endangered species that exists only in one or a few small populations. Carefully read the species profile. How vulnerable are the remaining populations to the effects of genetic drift, inbreeding, demographic stochasticity, and environmental stochasticity? What types of conservation actions (if any) have been taken to reduce the risk of extinction from these and other effects (see the Conservation Actions section of the species profile)?
2. Effective population size (N_e) is a measure of the genetic nature of a population and influenced by the sex ratio of individuals in the population, reproductive output of individuals in the population, and fluctuations in population size over time. We presented equations to calculate N_e, one for each of these influences. How might you calculate N_e for a population that experiences effects from all three influences?
3. Population viability analysis can be a helpful tool for species conservation. However, there is always uncertainty when projecting a population into the future. What are the sources of uncertainty in a PVA? How might you account for uncertainties in future events, like natural disasters or other catastrophes, that could severely impact a population? Can genetic factors be incorporated into a PVA?
4. Consider the case of island fox recovery in the Channel Islands (see Box 13.1). Managers took several actions to help island foxes including removing pigs and golden eagles, and reintroducing bald eagles. These actions were clearly meant to improve the status of the island fox population, but there were bound to be other trade-offs and consequences. What do you suppose some of these trade-offs were? What other types of consequences do you think managers kept in mind when implementing these actions? How might you apply a decision-making framework to this problem?

Suggested Readings

Courchamp, F. L. et al. 2008. *Allee Effects in Ecology and Conservation.* Oxford, UK: Oxford University Press. A comprehensive review of Allee effects, mechanisms that lead to the effects, and their relevance to conservation biology.

Crowder, L. B. et al. 1994. Predicting the impact of turtle excluder devices on loggerhead sea turtle populations. *Ecological Applications* 4: 437–445. Demonstrates the use of a stage-based model around a species conservation problem.

Gotelli, N. J. 2008. *A Primer of Ecology*, 4th ed. Sunderland, MA: Oxford University Press/Sinauer. An excellent mathematical overview of models of population growth (e.g., exponential, logistic, age/stage/size).

Kaczensky, P., et al. 2011. The danger of having all your eggs in one basket: Winter crash of the re-introduced Przewalski's horses in the Mongolian Gobi. *PLOS ONE* 6: e28057. Provides a vivid example of the perils of small populations and impacts of stochastic events.

Lacy, R. C. 2019. Lessons from 30 years of population viability analysis of wildlife populations. *Zoo Biology* 38: 67–77. An overview of lessons learned from the use of population viability analysis in the context of in situ and ex situ wildlife conservation.

Langham, G. M. et al. 2015. Conservation status of North American birds in the face of future climate change. *PLOS One* 10(9): e0135350. Shows the use of Christmas Bird Count and Breeding Bird Survey census data to evaluate the impacts of climate change on 588 species.

Lindberg, M. S. 2012. A review of designs for capture-mark-recapture studies in discrete time. *Journal of Ornithology* 152: 355–370. Covers the history of capture-mark-recapture designs, techniques, and considerations.

Runge, M. C. et al. 2013. Structured decision making. Pp. 51–72 in *Wildlife Management and Conservation: Contemporary Principles and Practices*, P. R. Krausman and J. W. Cain, (Eds.). Baltimore, MD: Johns Hopkins University Press. Provides an overview of the principles of the structured decision making framework.

Schwartz, M. W., et al. 2018. Decision support frameworks and tools for conservation. *Conservation Letters* 11: 1–12. Provides a balanced overview of five common frameworks and associated tools used to support conservation decision-making.

Shaffer, M. L. 1981. Minimum population sizes for species conservation. *BioScience* 31: 131–134. A classic paper that introduces the concept of minimum population size.

Wang, J. et al. 2016. Prediction and estimation of effective population size. *Heredity* 117: 193–206. A comprehensive overview of various calculations of effective population size and relevance of the concept to the field of conservation genetics.

Visit the
Conservation Biology
Companion Website
oup.com/he/cardinale1e
for videos, exercises, links, and other study resources.

© imageBROKER/Alamy Stock Photo

14

Community and Ecosystem Conservation

In Chapter 13 we described species-level approaches to conservation, which have arguably been the mainstay of the discipline. A historical focus on the species level of biodiversity has been driven by the fact that species conservation has legal precedent (e.g., CITES, the U.S. Endangered Species Act), well-established tools (e.g., population viability analysis), and has enjoyed relatively strong public support and funding (e.g., flagship species).

But conservation efforts aimed at higher levels of the biological hierarchy, such as that of entire biological communities or whole ecosystems, have the potential to protect even greater amounts of biodiversity, often with greater success and lower cost. Some of the tools for community- and ecosystem-level conservation have historical precedent and success, while others are rapidly developing. And although it is sometimes hard to gather public support for conservation of less-tangible concepts like biological communities or ecosystems, the now widespread acceptance of ecosystem services and the indirect benefits that habitats have for human well-being are becoming increasingly engrained in the minds of society.

This endangered green sea turtle (*Chelonia mydas*) rests on a sandy seabed in Australia's Great Barrier Reef, a UNESCO World Heritage Site.

The primary tool used for conservation of whole communities and ecosystems is the **protected area**. A protected area is a clearly defined geographical space that is recognized, dedicated, and managed to achieve long-term conservation of nature and its associated ecosystem services. The concept of protected areas has been around for centuries, dating back to the sacred *fengshui* forests established during the Song Dynasty (960–1279 CE) in China, and the royal game reserves of medieval England and China's Qing Dynasty (1644–1911). But establishment of protected areas for the purpose of conserving whole ecosystems and their biodiversity is a relatively new concept that began with establishment of national parks like the Bogd Khan Mountain nature reserve in Mongolia (1783) and Yellowstone in the U.S. (1872). While these natural areas were originally set aside and protected for their religious value (Bogd Khan Mountain) and scenic beauty (Yellowstone), formation of national parks was a turning point in conservation that was soon after imitated by countries around the world, and ultimately inspired the formation of Marine Protected Areas (see Chapter 2). By the mid-1900s, it was clear that protected areas were the primary pathway to conserving biodiversity at the level of entire communities and ecosystems.

We begin this chapter by reviewing the classification and global status of protected areas. We then review the approaches used to select protected areas, as well as the effectiveness of protected areas for conservation. When readers combine the information presented in this chapter with that in the next, they will understand the strengths and limitations of different types of protected areas, how to select and establish new protected areas, and how to create networks of protected areas and manage them alongside unprotected habitats for the benefit of biodiversity in entire landscapes.

Classification of Protected Areas

While the concept of protected areas began with formation of national parks, protected areas now include a variety of habitat types established and maintained by governments, Indigenous peoples and traditional societies, private individuals, conservation organizations, and research institutions. Organizations and individuals protect areas for many different reasons, and often conservation of biodiversity is just one of those. But habitats protected for other uses (e.g., recreation, tourism, forestry, grazing) often have value for biodiversity conservation.

The World Commission on Protected Areas (WCPA) is responsible for classification of the world's protected areas. The WCPA is one of six commissions of the International Union for Conservation of Nature (IUCN). Administered by the IUCN's Global Protected Areas Program, the WCPA is composed of ca. 2400 world experts on protected areas who represent 140 countries. With help from WCPA, the IUCN has developed a six-category system for classifying protected areas around the world (Table 14.1; a list of protected areas by category is maintained in the World Database on Protected Areas at oup-arc.com/e/cardinale-w14.1).

Strict nature reserves and wilderness areas (Category I)

Category I is the highest level of protection recognized by the WCPA, and is reserved for the most stringently protected natural habitats on Earth. Category I includes two subcategories—strict nature reserves (Category Ia); and wilderness areas (Category Ib)—that are created and managed for protection of large, unspoiled areas of wilderness. The primary purpose of Category I areas is to

TABLE 14.1 Types of protected areas recognized by the International Union for Conservation of Nature (IUCN)

IUCN Management Category	Description
Ia Strict nature reserve	Category Ia comprises strictly protected areas set aside to protect biodiversity and also possibly geologic/geomorphological features, where human visitation, use, and impacts are strictly controlled and limited to ensure protection of the conservation values. Such protected areas can serve as indispensable reference areas for scientific research and monitoring
Ib Wilderness area	Category Ib protected areas are usually large, unmodified or slightly modified areas, retaining their natural character and influence without permanent or significant human habitation, which are protected and managed so as to preserve their natural condition
II National park	Category II protected areas are large natural or near-natural areas set aside to protect large-scale ecological processes, along with the complement of species and ecosystems characteristic of the area, which also provide a foundation for environmentally and culturally compatible, spiritual, scientific, educational, recreational, and visitor opportunities
III Natural monument or feature	Category III protected areas are set aside to protect a specific natural monument, which can be a landform, sea mount, submarine cavern, geological feature such as a cave or even a living feature such as an ancient grove. They are generally quite small protected areas and often have high visitor value
IV Habitat/species management area	Category IV protected areas aim to protect particular species or habitats and management reflects this priority. Many Category IV protected areas will need regular, active interventions to address the requirements of particular species or to maintain habitats, but this is not a requirement of the category.
V Protected landscape/seascape	A protected area where the interaction of people and nature over time has produced an area of distinct character with significant, ecological, biological, cultural, and scenic value; and where safeguarding the integrity of this interaction is vital to protecting and sustaining the area and its associated nature conservation and other values.
VI Protected area with sustainable use of natural resources	Category VI protected areas conserve ecosystems and habitats together with associated cultural values and traditional natural resource management systems. They are generally large, with most of the area in a natural condition, where a proportion is under sustainable natural resource management and where low-level non-industrial use of natural resources compatible with nature conservation is seen as one of the main aims of the area.

Source: N. Dudley (Ed.) 2013. *Guidelines for Applying Protected Area Management Categories.* Gland, Switzerland: IUCN.
Note: A list of protected areas by category is maintained in the World Database on Protected Areas at protectedplanet.net/

protect biodiversity and maintain evolutionary and ecosystem processes, and they are often managed for scientific research and environmental monitoring. Category I areas are generally free of direct human intervention, have little to no human infrastructure, do not allow mechanized transportation, prohibit extraction of natural resources, and have minimal recreation. However, these areas do allow Indigenous people to maintain their lifestyle and traditional forms of ecosystem management. The U.S. Papahānaumokuākea Marine National Monument is the largest Category Ia protected area in the world (see Box 2.1, Chapter 2). It encompasses 1,510,000 km^2 of ocean waters, including ten atolls of the Northwestern Hawaiian Islands. The monument supports more than 7000 species, at least 25% of which are endemic. The largest Category Ib protected area is The Queen Maud Gulf Migratory Bird Sanctuary. As Canada's largest federally owned protected area, The Queen Maud covers 61,765 km^2, and harbors the largest population of migratory geese in the world. Under terms of the Ramsar Convention, an international, intergovernmental treaty dedicated to the conservation of wetlands worldwide, this area was designated as a wetland of international importance in 1982.

National parks (Category II)

National parks tend to be large areas similar in size to Wilderness Areas. While direct exploitation is usually excluded in national parks, management of these protected areas often focuses on multiple objectives, such as balancing conservation alongside human tourism and recreation. Achieving the dual mandate of conservation and opportunities for recreation can be complicated, since the latter objective requires certain types of development and supporting infrastructure, and is often accompanied by the economic development of surrounding communities. An example of a Category II protected area is the 32,000-ha Tubbataha Reef Marine Park of the Philippines. This park is comprised of two atolls that protect marine resources considered to be of great importance for sustaining the region's fisheries. The park also supports a vibrant recreational diving industry. Other examples of Category II national parks are Yellowstone National Park (U.S.), Banff National Park (Canada), Serengeti National Park (Tanzania), and Fiordland National Park (New Zealand). It is worth noting that protected areas under this category often include designations like *game reserves* and *nature reserves* that are not necessarily labeled as national parks in their home country; the Niassa Nature Reserve in Mozambique is an example.

Natural monuments (Category III)

Category III protected areas include national monuments and are established to protect natural geological or geomorphological features (e.g., mountains, waterfalls, cliffs, caves, fossil beds, sand dunes, rock formations), culturally influenced natural features (e.g., cave dwellings), natural cultural sites (e.g., sacred groves, springs, waterfalls, mountains, or sea coves), or cultural sites with their associated ecology (e.g., archeological sites with natural features). Category III sites are generally small in area, but they contribute to conservation of local biodiversity associated with the particular features. The Giant Sequoia National Monument in California, U.S.A., is an example of a natural monument with significant biological features that are the focus of conservation.

Habitat/species management area (Category IV)

Category IV protected areas are established to protect or restore flora or fauna of international, national, or local importance, including resident or migratory fauna and the habitats they require. Category IV sites require management and intervention in order to maintain their biodiversity value, and they are often sites of environmental monitoring and other scientific research. In Papua New Guinea, for example, the Baiyer River Sanctuary is located in an area that is largely under cultivation for coffee and tea. But the sanctuary is closely monitored to protect the habitat of wildlife that includes large populations of birds of paradise.

Protected landscape/seascape (Category V)

Category V protected areas are established to protect whole biological communities and ecosystems across an entire body of land or sea. Category V areas often accommodate human developments and infrastructure for multiple uses, including ecotourism, fishing, and farming. Management objectives focus on safeguarding the tradition of this interaction, which may involve protecting traditional land uses like low-intensity grazing by domestic animals, and maintaining traditional social and cultural values, including both natural and cultural features of significance. An example is the Mount Emei and Leshan Giant Buddha in Sichuan, China, one of the four holy lands of Chinese Buddhism and site of a 71-m statue of the Buddha that was carved into a mountain peak in the eighth century. The site's status protects some 2000 people, including Buddhist monks and nuns living in temples and monasteries. Mount Emei also harbors a number of endemic and globally threatened species of plants and animals.

Managed resource protected area (Category VI)

Category VI protected areas are generally large areas that maintain much of the landscape in a natural condition. However, a proportion of the protected area is associated with non-industrial, sustainable harvesting and use of natural resources. These are seen as areas where human activities of relatively low-intensity are compatible with nature conservation. The Ngorogoro Crater Conservation Area of northern Tanzania is an example. The crater, one of the largest inactive calderas in the world, has been used for centuries by the pastoral Maasai people for cattle grazing, and was originally developed as a conservation area to benefit the Maasai. However, the area is also home to one of Africa's largest aggregations of wildlife, including a relict population of the black rhinoceros (*Diceros bicornis*), and it is now also recognized for its vital conservation value. Examples of Category VI protected areas in the U.S. include the 769,000 km^2 of public forests that are managed by the U.S. National Forest Service, as well as the 672,262 km^2 of public land that is managed by the Bureau of Land Management, but leased to ranchers for grazing of livestock.

Global Status of Protected Areas

In 2010, parties to the United Nations Convention on Biological Diversity (CBD) adopted the **Strategic Plan for Biodiversity** along with its 20 **Aichi Biodiversity Targets** (oup-arc.com/e/cardinale-w14.2) that were designed to slow biodiversity loss and speed international efforts in conservation.[1] Among these, **Aichi Target 11** set out goals for protected areas:

> By 2020, at least 17 percent of terrestrial and inland water, and 10 percent of coastal and marine areas, especially areas of particular importance for biodiversity and ecosystem services, are conserved through effectively and equitably managed, ecologically representative and well connected systems of protected areas and other effective area-based conservation measures, and integrated into the wider landscapes and seascapes.

In 2015, almost immediately after the Aichi Biodiversity Targets were formulated, members of the United Nations adopted the **2030 Agenda for Sustainable Development**, a blueprint for achieving human well-being and equality while transitioning to sustainable use of the planet's natural resources (see Chapter 17). The 2030 agenda was formed on 17 **Sustainable Development Goals** (**SDGs**), of which SDGs 14 and 15 repeated calls for increased protection of terrestrial and marine ecosystems.

The CBD's Strategic Plan for Biodiversity and the UN's Agenda for Sustainable Development represent two of the most important commitments ever made by governments to biodiversity conservation, and the global network of protected areas was deemed an important mechanism for achieving these commitments. Every two years, the United Nations Environment World Conservation Monitoring Centre (UNEP-WCMC), International Union for Conservation of Nature (IUCN), and, more recently, the National Geographic Society (NGS) publish the Protected Planet Report, which assesses the global trends and status of protected areas. The report uses data from the World Database on Protected Areas (WDPA),[2] which compiles information annually and makes it publicly available online on the Protected Planet platform (oup-arc.com/e/cardinale-w14.1). The most recent Protected Planet Report was published in 2018.[3]

Suggested Exercise 14.1 Visit Protected Planet and the World Database on Protected Areas (WDPA). Examine the site's many resources to learn about and visualize global efforts to protect terrestrial and marine areas. **oup-arc.com/e/cardinale-ex14.1**

There are currently 245,449 protected areas recorded in the WDPA. Most of these areas are terrestrial, and collectively they represent about 20 million km^2 of land, equivalent to 15% of the earth's land surface (**Figure 14.1A,B**). Marine protected areas, which number around 11,000, (oup-arc.com/e/cardinale-w14.3), cover more than 26 million km^2 of the earth and represent 8% of the world's oceans. Marine areas under national jurisdiction (Exclusive Economic Zones within 200 nautical miles of coastlines) have significantly more protection (17%) than Areas Beyond National Jurisdiction (>200 nautical miles from the coast), for which only 1% is protected.

The total area designated as marine protected areas (MPAs) within Exclusive Economic Zones has increased at a rapid rate over the last two decades. Assuming current trends continue, the goal to protect 10% of coastal and marine areas (Aichi Target 11, SDG 14) will be achieved by 2020. In contrast to marine environments, the formation of new terrestrial protected areas has slowed since 2012, and has fallen short of international commitments. However, information on future land commitments reported to the CBD Secretariat suggest that coverage of terrestrial protected areas will rise over the next two years, possibly allowing international commitments to be met by 2020.

Key biodiversity areas (KBAs) are defined as "sites contributing significantly to the global persistence of biodiversity."[3] Around 15,000 KBAs have been identified in terrestrial, marine, and freshwater ecosystems to date. Protected area coverage of KBAs is used by the CBD as one of the measures to track progress towards Aichi Biodiversity Target 11, and as an indicator for the United Nations Sustainable Development Goals. As of 2018, 21% of KBAs were completely covered by protected areas, while 35% of KBAs had no protection (**Figure 14.1C,D**). On average, 47% of each terrestrial, 44% of each freshwater, and 16% of each marine KBAs are within protected areas. While protected area coverage of KBAs in marine areas had tripled during the period 2010 and 2018 (5% to 16%), there was slower progress in the inclusion of terrestrial and freshwater KBAs into the global collection.

Figure 14.2 shows the percentage of land, marine coastal zones, and freshwater that are covered by protected areas for different countries. Some countries have been very progressive at protecting their natural resources. Three countries have protected more than half their land (New Caledonia, Venezuela, and Slovenia), and 25 countries have protected at least one-third or

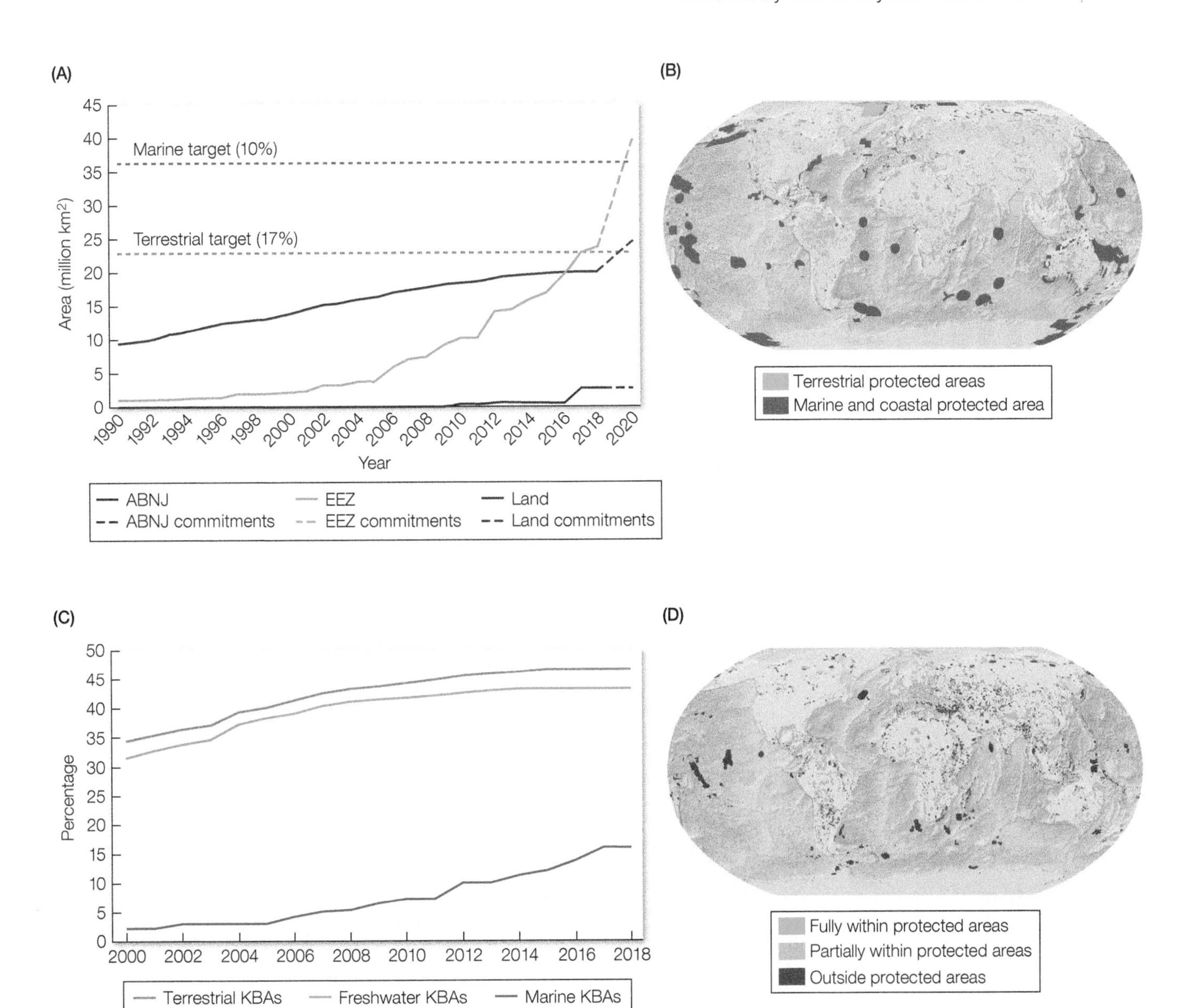

Figure 14.1 Global status of protected areas. (A) Growth in protected area coverage on land and in the ocean between 1990 and 2018. Projected growth to 2020 is based on future commitments from countries and territories. (EEZ, exclusive economic zones; ABNJ, areas beyond national jurisdiction) (B) Map showing the spatial distribution of world's protected areas. (C) Growth in the percent coverage of terrestrial, freshwater, and marine Key Biodiversity Areas (KBAs) since 2000. (D) Map showing KBAs that are fully within, partially within, or outside of protected areas on land and in the ocean (within EEZ). (A–D after UNEP-WCMC, IUCN and NGS [2018]. *Protected Planet Report 2018.* Cambridge UK: UNEP-WCMC, IUCN and NGS; based on data in UNEP-WCMC and IUCN. 2018. *Protected Planet: The World Database on Protected Areas [WDPA].* July 2018 version, Cambridge, UK: UNEP-WCMC and IUCN; A, unpublished data from CBD Secretariat; C and D, data compiled by BirdLife International and IUCN.)

more. Five countries (Slovenia, Monaco, New Caledonia, Saint Martin, and Palau) have protected more than 50% of their marine coastal zones and important freshwater habitats, while 12 countries have protected at least a third of these aquatic regions.

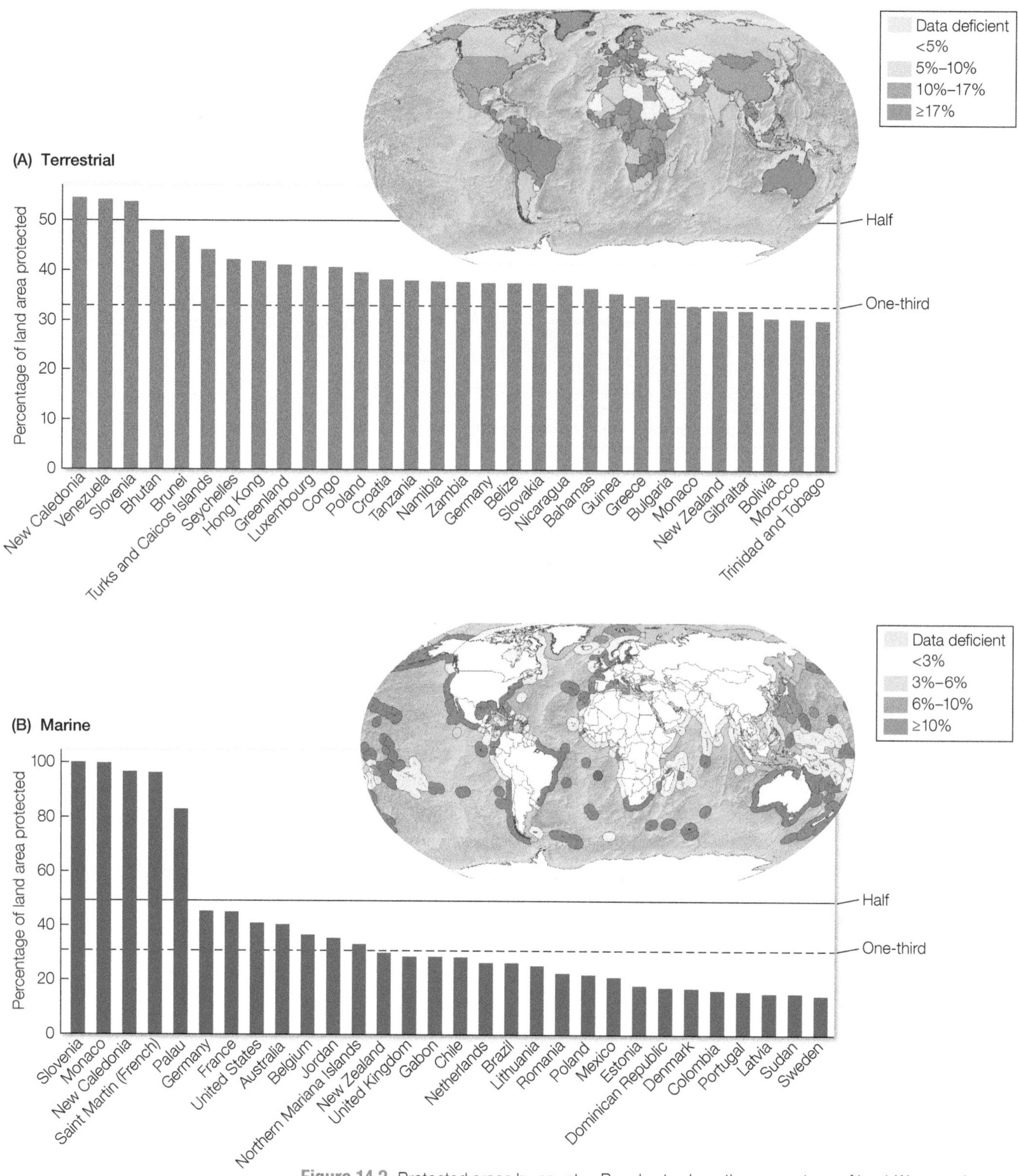

Figure 14.2 Protected areas by country. Bar charts show the percentage of land (A) or marine (B) habitat that is in protected areas for the top 25 countries. Maps display proportion of protected area for all countries. (Maps from UNEP-WCMC, IUCN and NGS [2018]. *Protected Planet Report 2018*. Cambridge UK: UNEP-WCMC, IUCN and NGS; based on data in UNEP-WCMC and IUCN. 2018. *Protected Planet: The World Database on Protected Areas [WDPA]*. July 2018 version, Cambridge, UK: UNEP-WCMC and IUCN. Graphs based on 2017 data from ourworldindata.org.)

Approaches for Choosing Protected Areas

The species approach to conservation (see Chapter 13) often requires the protection of habitats that are needed for a species to persist. As such, protected areas are sometimes chosen based on the habitat requirements of an individual species. For example, Wood Buffalo National Park, Canada's largest park (approximately 45,000 km^2), was created in 1922 to protect the last remaining herds of bison (*Bison bison*) in northern Canada, and happens also to be one of the last remaining natural nesting areas for whooping cranes (*Grus americana*). Mountain Zebra National Park in South Africa was established in 1937 to protect endangered Cape mountain zebra (*Equus zebra*).

However, there are at least five additional approaches for choosing protected areas that focus on conservation of higher levels of biodiversity, such as whole communities or ecosystems. These approaches differ in their emphases about what aspects of biodiversity are most important to protect, but they have potential to be complementary when used in combination.

Hotspots of biodiversity

The first approach is to protect hotspots of biodiversity. A **biodiversity hotspot** is a geographic area that has particularly high levels of biodiversity and which is also highly threatened by human activities (**Video 14.1**). The concept was developed by British environmentalist Norman Myers, who wrote about ten tropical forest hotspots that had exceptional concentrations of endemic species, but which were undergoing major loss of habitat.[4] A subsequent analysis added eight additional hotspots, including four from Mediterranean-type ecosystems.[5] Myers and colleagues then completed a systematic update and analysis of hotspots using two strict criteria. First, a hotspot had to contain at least 1500 vascular plants as endemics (>0.5% of the world's total). Second, it had to have lost 70% or more of its original vegetation. Based on these two criteria, Myers and colleagues published a seminal paper[6] in the journal *Nature* showing that 44% of all species of vascular plants and 35% of all species in four vertebrate groups (mammals, birds, reptiles, and amphibians) are confined to 25 hotspots of biodiversity that comprise just 1.4% of the Earth's land surface. Subsequent analyses showed that up to two-thirds of all threatened plant species and over half of all threatened terrestrial vertebrates are endemic to these 25 hotspots.[7] The implication was that a large fraction of the world's species, particularly those threatened with extinction, could be conserved by protecting a few relatively small geographic areas.

VIDEO 14.1 Watch this video on biodiversity hotspots.
oup-arc.com/e/cardinale-v14.1

After these original publications, the concept of biodiversity hotspots quickly evolved to incorporate other geographic locations and additional types of biodiversity. Analyses now have revealed a set of 34 hotspots that collectively hold 50% or more of vascular plants and 42% of terrestrial vertebrates in just 3.4 million km^2, which is 2.3% of the world's land area (**Figure 14.3**).[8] Many hotspots represent isolated areas of tropical rain forest found in places such as the Atlantic coast of West Africa, Brazil, and the tropical Andes. The Andes alone harbor more than 30,000 plant species, 1728 bird species, 569 mammal species, 610 reptile species, and 1155 amphibian species that persist in tropical forests and high-altitude grasslands on about 0.3% of the Earth's total land surface.

The concept of hotspots has been used to prioritize the conservation of marine organisms as well. For example, Lucifora and colleagues[9] studied the geographic distribution of 507 species of shark and identified areas of high endemism off the coasts of Japan, Taiwan, the east and west coasts of Australia, southeast Africa, southeast Brazil, and the southeast United States. The hotspot

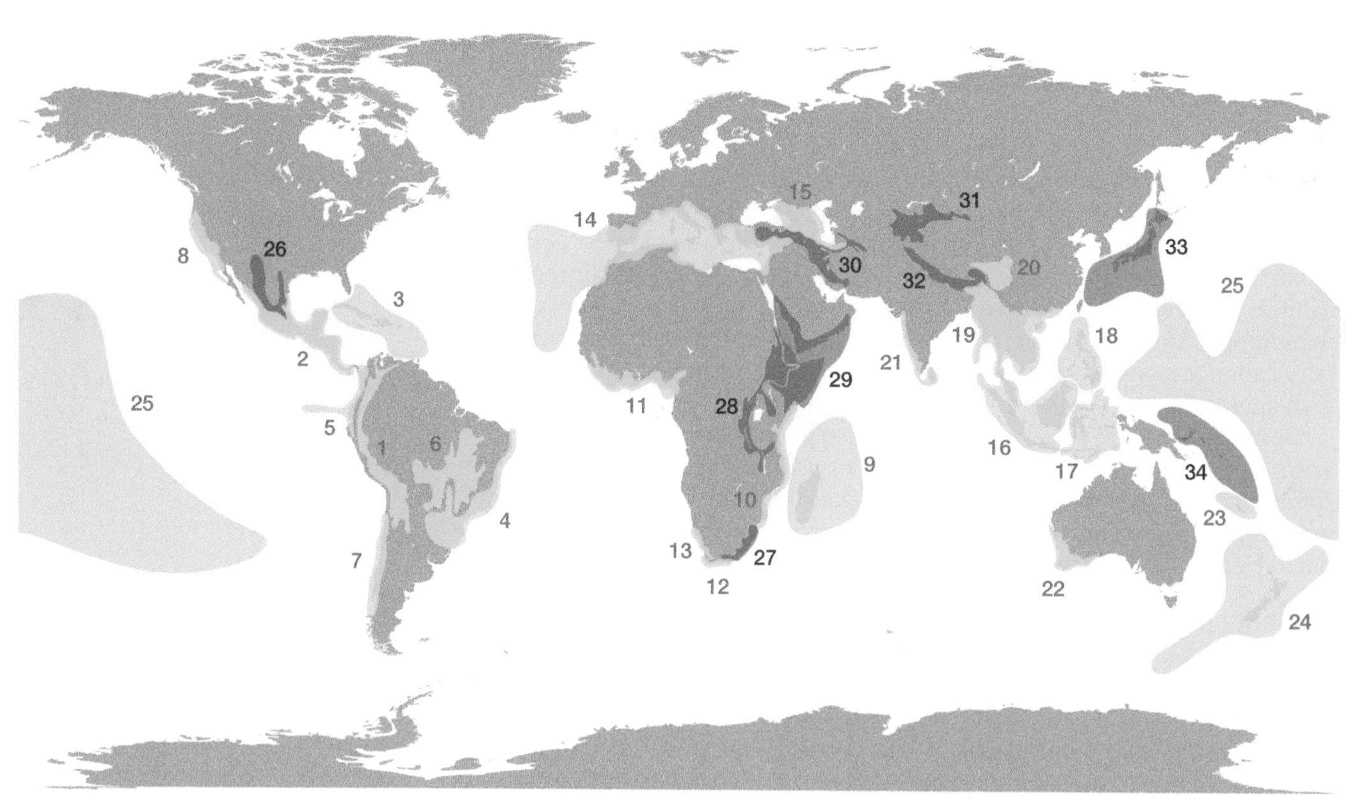

1. The Tropical Andes
2. Mesoamerica
3. The Caribbean Islands
4. The Atlantic Forest
5. Tumbes-Chocó-Magdalena
6. The Cerrado
7. Chilean Winter Rainfall-Valdivian Forests
8. The California Floristic Province
9. Madagascar and the Indian Ocean Islands
10. The Coastal Forests of Eastern Africa
11. The Guinean Forests of West Africa
12. The Cape Floristic Region
13. The Succulent Karoo
14. The Mediterranean Basin
15. The Caucasus
16. Sundaland
17. Wallacea
18. The Philippines
19. Indo-Burma
20. The Mountains of Southwest China
21. Western Ghats and Sri Lanka
22. Southwest Australia
23. New Caledonia
24. New Zealand
25. Polynesia and Micronesia
26. The Madrean Pine-Oak Woodlands
27. Maputaland-Pondoland-Albany
28. The Eastern Afromontane
29. The Horn of Africa
30. The Irano-Anatolian
31. The Mountains of Central Asia
32. Eastern Himalaya
33. Japan
34. East Melanesian Islands

Figure 14.3 The 25 original hotspots of biodiversity (green) proposed by Myers and colleagues in 2000, as well as 9 additional hotspots that have been added since 2000 (blue). (After N. Myers et al. 2000. *Nature* 403: 853–858. Modified and updated by R. A. Mittermeier et al. 2011. In F. E. Zachos and J. C. Habel [Eds.], *Biodiversity Hotspots: Distribution and Protection of Conservation Priority Areas*. London, UK: Springer. Map: Wikimedia/Biodiversity Hotspots.svg/Ninjatacoshell/CC BY-SA 3.0.)

concept has been extended to other levels of biodiversity, such as the identification of hotspots of genetic diversity that might be used to conserve the evolutionary history of species and their potential for adaptation.[10] Biodiversity hotspots have also been predicted to house a large fraction of undescribed plant species; thus, it has been proposed they are a primary means to conserve species that have yet to be discovered.[11]

Since its conceptualization in 1988,[4] the hotspot approach has generated a considerable amount of enthusiasm and funding, and a number of organizations have used the approach as a way to prioritize their conservation efforts. For example, Conservation International, a large nonprofit organization based in the U.S., invested heavily in development of the hotspot concept, ultimately developing its own maps of 36 hotspots to guide its contribution to the protection of biodiversity. BirdLife International, a global partnership of organizations focused on bird conservation, has used the hotspot approach to identify Important

Bird Areas (IBAs) that represent geographic locations in terrestrial, freshwater, and marine environments with large concentrations of bird species that have restricted ranges. The organization's website (oup-arc.com/e/cardinale-w14.4) presently lists about 12,000 IBAs in more than 200 countries, and they aspire to engage a global network of partners in the protection of 15,000 IBAs covering 10 million km^2, equal to 7% of the world's land surface.

Analysis of IBAs has provided one of the few concrete estimates of how much it might cost to protect biodiversity using the hotspot approach. McCarthy and colleagues[12] tallied the cost of protecting 4445 globally threatened bird species in 11,731 IBAs. They estimated that the total cost worldwide would be US $7 billion annually (2012 dollars), of which US $1.6 billion (22%) would be required in low-income countries. For perspective, the operating cost of the Large Hadron Collider, a single intergovernmental physics project to find a Higgs-Boson particle, is approximately $1 billion per year,[13] and Jeff Bezos, the CEO of Amazon, made US$39.2 billion in 2018 alone.[14]

The biodiversity hotspot approach to selection of protected areas has received some criticism. Several scientists have pointed out that the delineation of hotspots is heavily dependent on one's definition of biodiversity and the measure used, and that these definitions and measures don't always lead to the same conclusions about which areas to protect.[15] Others have argued that, with its sole focus on protecting areas of high species richness, the hotspot approach fails to consider other conservation goals, such as the need to protect crucial but naturally low-diversity ecosystems (e.g., seagrass beds, open oceans),[16] or the need to protect ecosystem services that are important to people.[17] Therefore, the hotspot approach represents but one way to prioritize the selection of protected areas.

Ecoregions

An alternative approach to the selection of protected areas is the ecoregion approach. Ecoregions were introduced in Chapter 4, where an ecoregion was defined as "a unit of land containing a distinct assemblage of natural communities and species, with boundaries that approximate the original extent of natural communities prior to major land-use change."[18] In 2001, scientists at the World Wildlife Fund (WWF) classified Earth's land-based biomes into 867 terrestrial ecoregions that represent naturally occurring boundaries for terrestrial species and their ecosystems. Comparable maps of ecoregions have been developed for ocean ecosystems based on biogeographical distributions of marine species and data on major oceanic currents.[19] Development of freshwater ecoregions has lagged behind, as some argue freshwater is already incorporated into terrestrial ecoregion classifications.[20]

Proponents of the ecoregion approach argue that these units make biological sense for conservation because they represent naturally occurring biological boundaries that likely existed prior to human domination of the planet.[18–20] The ecoregion approach also avoids certain criticisms of the hotspot approach, particularly in that it seeks to protect biodiversity in all types of ecosystems, not just those that are the most diverse. Several major conservation organizations have adopted the ecoregion approach as a means to prioritize their conservation efforts. For example, WWF uses their Global 200 Program to guide conservation projects in each of 142 terrestrial, 53 freshwater, and 43 marine ecoregions of the world that are chosen for their species richness, endemism, taxonomic uniqueness, unusual ecological or evolutionary phenomena, and global rarity.[21] The Nature Conservancy (TNC) bases part of its conservation planning in North America on a set of 81 fine-scale ecoregions, and has developed similar ecoregion-based plans in Latin America, the Caribbean, and the Asia-Pacific region.

Suggested Exercise 14.2 Use the Ecoregions2017 dataset to find five imperiled ecoregions in Africa. **oup-arc.com/e/cardinale-ex14.2**

Dinerstein and colleagues[22] recently used the Ecoregions2017 dataset (oup-arc.com/e/cardinale-w14.5) to assess the status of all 476 forested and 370 non-forested terrestrial ecoregions of the world. A primary goal of their assessment was to measure progress on Aichi Target 11, which called for 17% of terrestrial areas to be protected by 2020. **Figure 14.4** shows data for ecoregions aggregated into their corresponding biomes. The figure shows the percent of historical biome area that has already been converted to other habitat, that remains but is unprotected, and which is currently designated as IUCN categories I-IV protected areas. Only two biomes—flooded grasslands and savannas, and montane grasslands and shrublands— have met Aichi Target 11. The remaining nine biomes fall short of Target 11, and some (e.g., deserts and xeric shrublands; temperate grasslands; savannas and shrublands) fall exceptionally short, with only 4–6% of their historical areal coverage currently protected.

A second goal of the Dinerstein and colleagues[22] assessment was to measure progress toward goals of the Nature Needs Half initiative (oup-arc.com/e/cardinale-w14.6). Nature Needs Half is an international coalition of scientists, conservationists, nonprofits, and public officials who have set a bold goal to get 50% of each of the world's ecoregions into protected areas by 2030 (**Box 14.1**). Nature Needs Half is perhaps the most visionary conservation initiative yet seen. If successful, it could turn the tide in favor of conserving Earth's biodiversity and life support systems, as well as transform society's relationship with nature.

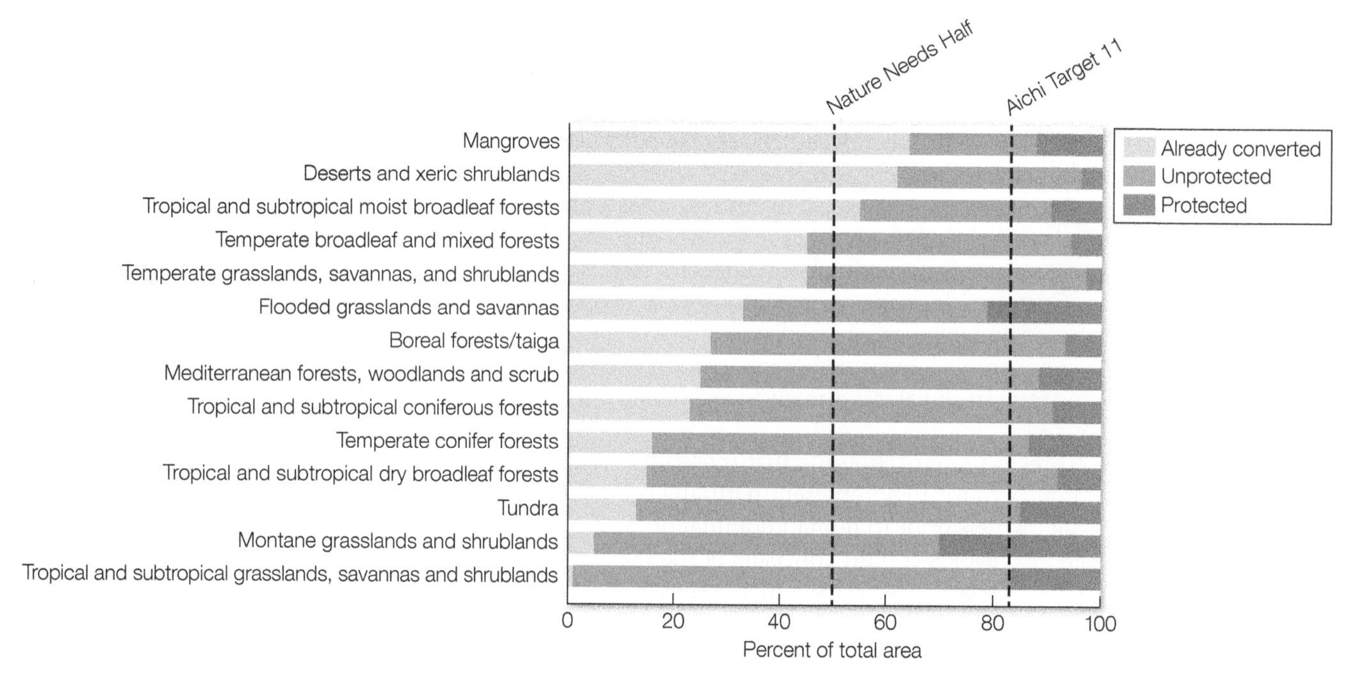

Figure 14.4 Protected areas by biome. The graph shows the percentage of historical biome area that has already been converted to other habitat (grey), or which is currently designated as IUCN Categories I–IV protected areas (green), or presently unprotected (orange). Vertical lines show the Aichi Biodiversity Target 11, which calls for 17% of land areas to be protected by 2020 (100% – 17% = 83%); and the Nature Needs Half Initiative, which calls for 50% of land areas to be protected by 2030. (After E. Dinerstein et al. 2017. *BioScience* 67: 534–545, and J. R. Oakleaf et al. 2015. *PLOS ONE* 10: e0138334.)

BOX 14.1 Case in Point

Conserving Half: A Bold Agenda for Conservation

How much land and water do we need to set aside to conserve biodiversity? One answer to this question was given by the Convention on Biological Diversity's (CBD) Aichi Target 11, which set a goal to protect at least 17% of terrestrial and inland waters, and 10% of coastal and marine areas.

Aichi Target 11 is certainly achievable in the near future (see Figure 14.1). But many conservation biologists argue that policy-based targets for protected areas are not established using the best available science, and are likely to be woefully insufficient for achieving global conservation of biodiversity.[71–73]

Two initiatives—the WILD Foundation's Nature Needs Half initiative, and the E. O. Wilson Biodiversity Foundation's Half-Earth Project—both seek to set aside half of the planet's land and water for the conservation of biodiversity.

In contrast to policy-driven targets, scientific studies and reviews tend to suggest that higher fractions of Earth must be protected to achieve representation of species and ecosystems. Noss and colleagues[71] summarized the results of 17 studies that have estimated the percentage of terrestrial regions required to meet conservation goals. They found results ranging from less than 10% to nearly 75%; however, the estimates from studies that used real empirical data were consistently higher, with a median of around 50%. After reviewing 144 marine studies, O'Leary and colleagues[74] came to a similar conclusion—i.e., that conservation of the world's oceans requires that approximately 50% of the sea be protected.

The convergence of different analyses has led several peer-reviewed papers to propose that 50% of the planet must be set aside as protected areas in order to conserve the world's biodiversity.[22,71,74,75] The idea was the subject of E. O. Wilson's book *Half-Earth: Our Planet's Fight for Life*,[73] and has become the focus of major conservation initiatives, such as the Nature Needs Half and the Half-Earth projects.

The **Nature Needs Half Project** (oup-arc.com/e/cardinale-w14.6) is an international coalition of scientists, conservationists, nonprofits, and public officials working together to protect at least half of all nature on Earth—land and water—by 2030 in order to support the existence of nature and the services it provides (**Video 14.2**). Nature Needs Half is an idea put forth by the WILD Foundation at the 9th World Wilderness Congress in Mérida, Mexico, and meant to be taken up as a common banner by people and organizations endeavoring to protect wild places. The project has been endorsed by several leading conservationists, including marine biologist Sylvia Earle and primatologist Jane Goodall. Since its inception, WILD has been collecting and conducting case studies of places around the world that have, or are on track to achieve, at least 50% protection.

The **Half-Earth Project** (oup-arc.com/e/cardinale-w14.7) is an initiative being led by the E. O. Wilson Biodiversity Foundation to protect half the land and sea in order to manage sufficient habitat to reverse the species extinction crisis and ensure the long-term health of our planet. The project fosters this goal by mapping priorities for place-based conservation to maximize species survival; advancing education and public engagement initiatives; and partnering with decision-makers at the local, regional and global levels to bring science-drive species information to the conservation management processes to achieve the conservation of half the Earth for biodiversity (oup-arc.com/e/cardinale-w14.8).

Video 14.2 Watch this video about the Nature Needs Half initiative.
oup-arc.com/e/cardinale-v14.2

Political and economic boundaries

Although selecting protected areas using the ecoregion approach overcomes certain criticisms of the hotspot approach, it has its own limitations. In particular, the scale at which ecoregions are delineated and defined is often arbitrary, and ecoregions defined at large scales can become impractical to manage when they cross multiple political jurisdictions. Because of the difficulties of getting multiple governments to cooperate and coordinate their conservation activities, it can sometimes be more effective to plan conservation at the level of countries, provinces, or states. For better or worse, planning efforts by governments are likely to follow geopolitical boundaries rather than natural biological boundaries. Organizations that use political boundaries to prioritize biodiversity conservation can better align with the plans and actions of government agencies.

Many government agencies have also moved towards selecting their conservation and management units based on regional planning for large economic regions. For example, the Gulf of Mexico Alliance (oup-arc.com/e/cardinale-w14.9) was formed in 2004 to assess and balance the competing demands on the Gulf in hopes of achieving smarter resource use and public planning.[17] The result of this alliance is that governors from the U.S. states of Florida, Alabama, Mississippi, Louisiana, and Texas work together on actions that seek to reduce nutrient pollution, restore habitats, manage for coastal resilience, and invest in environmental education. The Gulf Alliance is successful, in part, because it represents a region with a shared cultural identity.

Ecosystem services

Another approach is to base selection of protected areas on their delivery of ecosystem services.[23] While the ecoregion approach focuses on protecting whole communities and ecosystems, the ecosystem services approach focuses on protecting the biological functions of nature, particularly those that underlie the delivery of important goods and services. For example, we often want to protect natural systems to protect water quality, prevent soil erosion, or mitigate against storms. While hotspots or ecoregions may maximize both biodiversity and ecosystems services, this is not always the case.

Take, for example, recent work by Xu and colleagues,[24] who performed a nationwide assessment of China to quantify (1) the habitat of threatened species, and (2) the potential for protected areas to deliver important ecosystem services (**Figure 14.5**). Important areas for water retention, soil retention, and

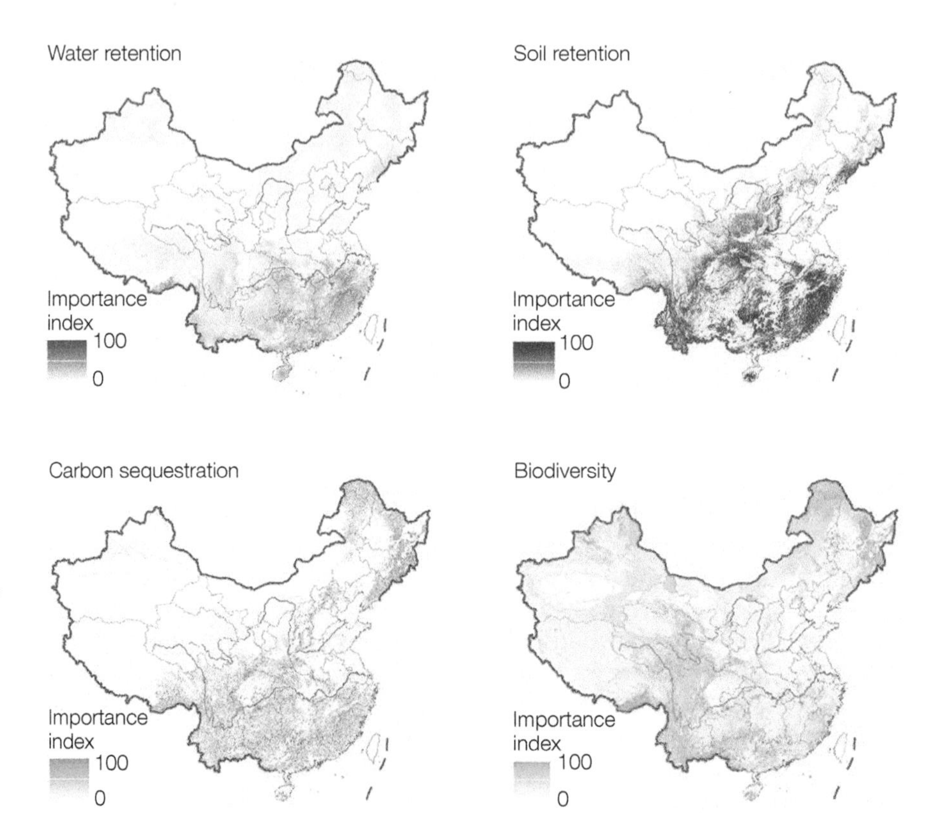

Figure 14.5 The ecosystem services approach in China. Important habitat areas for water retention, soil retention, and carbon sequestration were distributed mainly in places with forests, shrubs, and wetlands in southern and northeastern China. In contrast, much of the biodiversity (mainly of mammals and birds) was concentrated in the southwestern (Tibet and Yunnan) or northern (Inner Mongolia) China, which showed low overlap with areas of high ecosystem services. (From W. Xu et al. 2017. *Proc Nat Acad Sci* 114: 1601–1606.)

carbon sequestration were mainly in places with forests, shrubs, and wetlands in southern and northeastern China. These locations included the Khingan, Changbai Mountains, and Loess Plateau in the north region; and the Qinling-Ba Mountains, Nanling Mountains, and Jiangnan Hills in the south region. While there was some overlap between ecosystem services and the protection of biodiversity, particularly for plants, much of the mammalian and bird diversity of China was concentrated in Qinghai-Tibetan region, or in areas of Mongolia that showed low overlap with areas of high water/soil retention or carbon sequestration. This example illustrates how a desire to protect ecosystem services can lead to distinct but potentially complementary goals of protecting biodiversity. It also illustrates that protecting ecosystem services that serve people may not be the best way to preserve mammalian and avian diversity. Ecosystem services is probably the least often used method of choosing areas for protection; however, we would expect this approach to become more common as our understanding of ecosystem services increases.

Areas of cultural importance

A final approach is to select protected areas according to their cultural importance or significance. The primary means of obtaining protection of such areas is through the United Nations Educational, Scientific, and Cultural Organization (UNESCO). UNESCO runs two programs that help set aside protected areas as either Biosphere Reserves or World Heritage Sites. Biosphere Reserves are selected to be part of the World Network of Biosphere Reserves and are managed under UNESCO's Man and the Biosphere Programme (MAB). MAB was launched in 1971 as an intergovernmental program to study the scientific basis for improving relationships between people and their environments (**Video 14.3**). MAB sites are considered "learning places" for sustainability, using interdisciplinary approaches to balance economic development with environmentally sustainable approaches. The primary benefit offered by the MAB program is the support, planning, and implementation of research and training programs that benefit from the technical assistance and scientific advice of an international community. As of May 2019, the program includes 686 sites in 122 countries.

Video 14.3 Watch this video about Man and the Biosphere reserves.
oup-arc.com/e/cardinale-v14.3

An example of a UNESCO MAB site is the 8000 km^2 Dja Biosphere Reserve located between the forests of southern Nigeria and the forests of the Congo Basin and encircled by the Dja River in Cameroon (**Figure 14.6A**). The reserve is characterized by dense evergreen rainforest that hosts more than 100 mammal species, including chimpanzee (*Pan troglodytes*), giant pangolin (*Manis gigantea*), elephant (*Loxodonta africana cyclotis*), leopard (*Panthera pardus*) and the critically endangered western lowland gorilla (*Gorilla gorilla*). About 4000 people live in the reserve's core area, including seminomadic groups like the Baka pygmies, who hunt within the reserve using traditional methods. An additional 40,000 people inhabit the surrounding areas, which places pressure on the reserve for timber exploitation, harvest of plants for domestic or pharmaceutical use, and the hunting of bushmeat. The reserve is managed by the Dja Conservation Services, which receives sustained support from cooperating international partners. Local anti-poaching strategies have been developed, and there are regular patrols in the forest and around the Reserve who cooperate with forestry operators for monitoring illegal activity. Community education and communication are conducted in collaboration with 19 village vigilance committees who prioritize issues on poaching, collection of forest data, and the code of laws and procedures needed to sustain the reserve.

UNESCO World Heritage Sites are cultural, historic, natural, or scientific landmarks of significant collective importance to humanity. Countries that have signed the 1972 World Heritage Convention, pledging to protect their

(A)

© Earwig

(B)

© nouseforname/Shutterstock.com

Figure 14.6 Protected areas of cultural importance. (A) the Dja Biosphere Reserve in Cameroon is a UNESCO Man and the Biosphere Reserve selected as a "learning place" for balancing economic development with environmentally sustainable activities. (B) The Galápagos Islands of Ecuador are a UNESCO World Heritage Site, selected for protection based on their biological importance for humanity.

natural and cultural heritage, can submit nominations for properties on their territory to be considered for inclusion in UNESCO's World Heritage List. Nominations are evaluated by the International Council on Monuments and Sites and the World Conservation Union, which assess proposals based on ten criteria, six cultural and four natural. The four natural criteria, which are the ones most often used to establish protected areas, are:

1. The site contains superlative natural phenomena or areas of exceptional natural beauty and aesthetic importance.
2. The site is an outstanding example representing major stages of Earth's history, including the record of life, significant ongoing geological processes in the development of landforms, or significant geomorphic or physiographic features.
3. The site is an outstanding example representing significant ongoing ecological and biological processes in the evolution and development of terrestrial, freshwater, coastal, and marine ecosystems, and communities of plants and animals.
4. The site contains the most important and significant natural habitats for in situ conservation of biological diversity, including those containing threatened species of outstanding universal value from the point of view of science or conservation.[25]

Locations selected as UNESCO World Heritage Sites are protected under international treaties. As of May 2019, a total of 1092 World Heritage Sites (845 cultural, 209 natural, and 38 mixed properties) exist across 167 countries. Italy, with 54 sites, has the most of any country, followed by China (53), Spain (47), France (44), Germany (44), India (37), and Mexico (35). Some examples of UNESCO World Heritage Sites that are important for biodiversity conservation are the Serengeti National Park in Tanzania, the Great Barrier Reef in Australia, Białowieża Forest in Belarus and Poland, Pantanal Conservation Area in Brazil, Sichuan Giant Panda Sanctuaries in China, and Ecuador's Galápagos Islands (**Figure 14.6B**).

Effectiveness of Protected Areas

Countries across the globe are continuing to invest resources into acquiring and managing protected areas, believing they are the backbone of biodiversity conservation, and that they deliver other benefits to society. But with global biodiversity continuing to decline, it is important that we critically evaluate the extent to which reserves really do protect biodiversity and deliver benefits to people. Critical evaluations are also essential if we are to develop best management practices and optimize our use of limited resources.

Terrestrial protected areas

A number of assessments have recently quantified the effectiveness of terrestrial protected areas for reducing habitat loss, maintaining species population sizes, and protecting species richness. Most of these have shown that protected areas significantly reduce habitat loss. For example, Geldmann and colleagues[26] reviewed 76 studies that used satellite remote sensing or aerial photos to evaluate the impacts of protected areas on habitat cover. Sixty-two of the 76 studies (82%) found habitat loss to be higher outside of protected areas than inside. Habitat loss outside of protected areas ranged from 1.25 to 22.7 times greater than habitat loss inside (average 5.4 times greater). Some studies found the reduction in habitat loss to be less than expected due to continued pressure on natural resources inside protected areas. Cuenca and colleagues[27] evaluated the impact of protected areas on deforestation of Ecuador's tropical Andean forest between 1990 and 2008 and found that protection reduced deforestation in only 6% of the protected forests, which would have been deforested had they not been protected. Andam and colleagues[28] found similar results in Costa Rica, where approximately 10% of forests in a network of protected areas would have been deforested had they not been under protection.

Joppa and colleagues[29] provided a striking example of the potential for protected areas to reduce habitat loss in areas that are under intense pressure for development and extraction. These authors used the Global Land Cover 2000 (GLC2000) dataset to quantify the cover of tropical moist forests at progressively larger distances inside and outside of protected areas in the Amazon, Congo, South American Atlantic Coast, and West Africa. The authors found that protected areas of forest in the Amazon and Congo differed little in cover from surroundings outside the protected areas, which also retained high levels of forest cover because (at the time) they were undeveloped and inaccessible. However, along the South American Atlantic Coast and in West Africa, where much larger fractions of tropical forests had already been destroyed, there were sharp boundaries in forest cover at the edges of protected areas where forest declined precipitously, sometimes representing up to 80% habitat loss (**Figure 14.7**).

The majority of assessments performed to date further suggest that protected areas maintain significantly higher population sizes of species and higher levels of biodiversity. Geldmann and colleagues[26] reviewed 42 studies that evaluated impacts of protected areas on species populations. Protection and management had positive effects in 31 of the 42 studies. In 12 of these, species populations did exhibit declines in protected areas, but the declines were smaller than what had occurred in unprotected areas, suggesting that there was a benefit. Coetzee and colleagues[30] compared 861 measures of (1) the abundances of individual species, (2) the abundance of all species in a community, and (3) levels of species richness measured inside and outside of protected areas. They found that

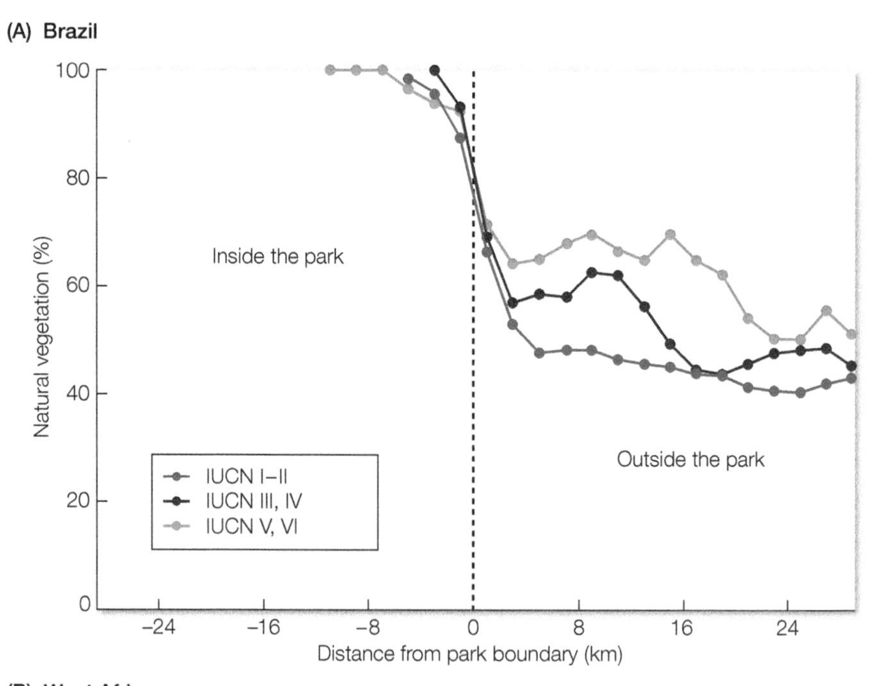

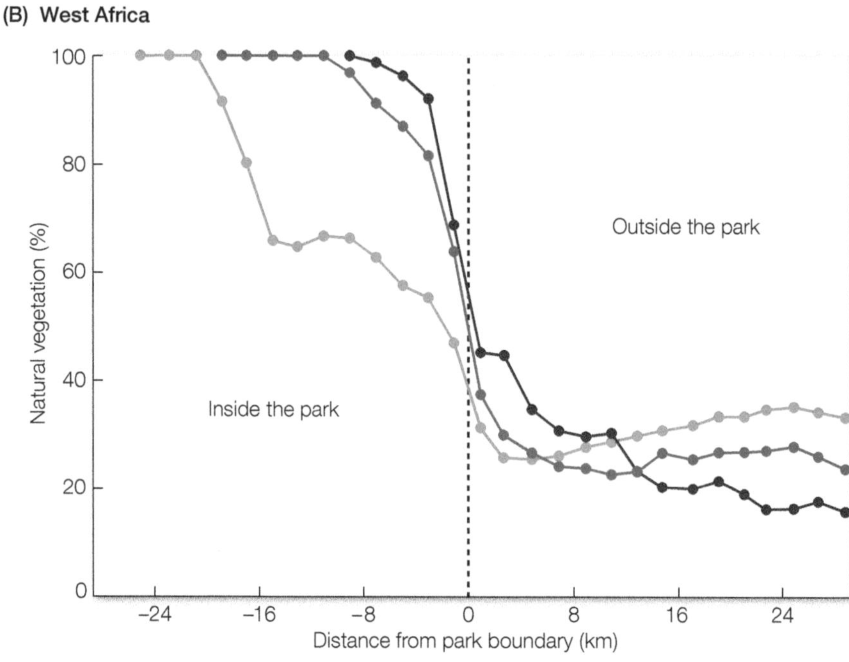

Figure 14.7 National parks and other protected areas have been able to prevent damage to the natural forests in (A) the Atlantic coast forests of Brazil and (B) West Africa. In the Atlantic coast forest, there is a sharp boundary, with intact forest inside the protected areas and around 50% intact forest outside the protected area. For protected areas in West Africa, there is considerable forest degradation within 16 km of the park boundary, particularly for IUCN categories V and VI (see Table 14.1), which are mainly forest reserves. (After L. N. Joppa et al. 2008. *Proc Nat Acad Sci* 105: 6673–6678. © 2008 National Academy of Sciences, USA.)

all three measures were significantly higher inside protected areas than outside. The benefits of protected areas were significantly positive for all categories of threatened species, with exception of critically endangered species that have exceptionally low population sizes.

In 2016, Gray and colleagues[31] published what is perhaps the most comprehensive assessment of terrestrial protected areas to date. Using the PREDICTS database (Projecting Responses of Ecological Diversity In Changing Terrestrial Systems, oup-arc.com/e/cardinale-w14.10), the authors compared the abundance of 13,669 species of vertebrates, invertebrates, and plants from sites that were surveyed both inside and outside of 359 terrestrial protected

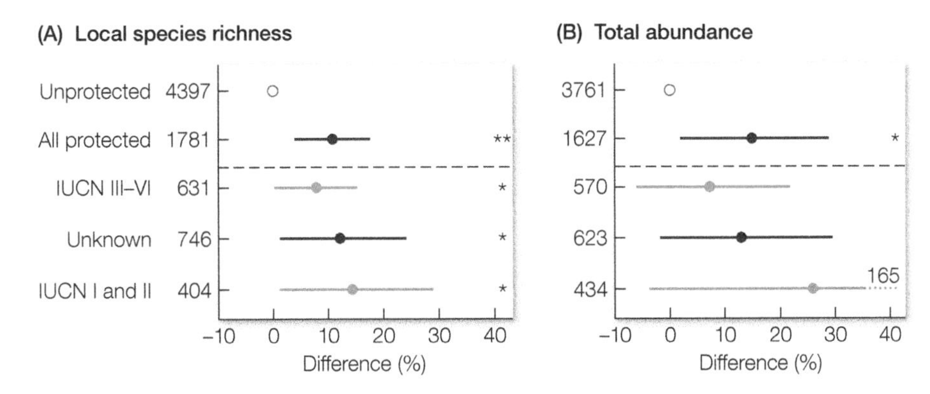

Figure 14.8 Effectiveness of terrestrial protected areas on measures of (A) local species richness, and (B) total abundance at sites inside (filled circles) relative to sites outside protected areas (open circles). Estimates are given for protected areas in different IUCN management categories, and unknown (missing IUCN category). Bars indicate 95% confidence intervals; **, $P < 0.01$; *, $P < 0.05$. The number of sampled sites in each category is shown on the *y*-axis next to categories. (After C. L. Gray et al. 2016. *Nat Comm* 7: 12306. CC BY 4.0.)

areas in 48 countries and 101 ecoregions. On average, protected sites had 14% more individuals and 11% more species than unprotected counterparts (**Figure 14.8**). However, when measures of species richness were rarified (*rarification* was described in Chapter 3) to account for difference in abundance, protected areas did not show consistently higher levels of species richness or higher levels of endemism. Such results suggest that the benefits of protected areas for biodiversity are primarily due to their maintenance of higher population size, particularly of rare species, rather than the hosting of unique species that are not found outside their area.

Marine protected areas

Marine protected areas (MPAs) have been embraced by many nations around the world, and by high-level international bodies, as an important mechanism to achieve the Convention on Biodiversity's Aichi Targets and the United Nation's Sustainable Development Goals. While most MPAs are small (around 50% are less than 10 km^2), recent trends to establish large-scale MPAs have produced some of conservation's greatest success stories (**Box 14.2**).

While terrestrial protected areas have been used in conservation for hundreds of years, MPAs are a comparatively new conservation tool, with most established in the latter part of the twentieth century.[32] As such, less data have accumulated to evaluate the effectiveness of MPAs compared to terrestrial counterparts, and there is more debate regarding their impact on conservation.[33] Nevertheless, an increasing number of assessments suggest that MPAs have benefits for conservation that are similar to those of their terrestrial counterparts

For example, studies suggest that MPAs reduce the loss of important marine habitats. Selig and Bruno[34] compiled a database of 8534 live coral cover surveys from 1969–2006 to compare annual changes in coral cover inside 310 MPAs to changes in unprotected areas. They showed that after an MPA is established, coral cover continues to decline for several years before losses slow, stabilize, and begin to recover (**Video 14.4**). After stabilization, cover increases in MPAs even as unprotected areas continue to experience losses, and the benefits of an MPA for coral growth accelerate the longer the protected area has been established. For the most recent year in their dataset (2004–2005), the authors

Video 14.4 Watch this video on coral habitat loss.
oup-arc.com/e/cardinale-v14.4

BOX 14.2 Success Story

Innovative Partnerships Protect One of Earth's Last Remaining Marine Wilderness Areas

As one of the world's last intact coral archipelago ecosystems, the Phoenix Islands, located within the Republic of Kiribati in the central Pacific Ocean, represent an ocean wilderness with immense biodiversity. The eight Phoenix Islands boast 120 species of coral, more than 500 species of reef fish, and vast bird populations. Unlike most other coral ecosystems in the world today, those of Kiribati are healthy and sport an abundance of marine species that is seldom encountered anywhere else on earth.

But the remoteness and isolation of Kiribati's Phoenix Islands were not enough to protect this ocean paradise. Kiribati's reef ecosystems were being degraded by climate change, invasive alien species, and commercial fishing and overharvesting. Much of this fishing was legal, because Kiribati used the sale of licenses to fishing fleets from South Korea, Japan, China and the U.S. to generate income from its territorial waters. However, much was unregulated fishing from nations that illegally entered Kiribati's exclusive economic zone to harvest from its rich, diverse waters.

But what could be done to stop the reefs from being degraded? Kiribati's only viable natural resource is its 3.5 million km^2 of coastal waters (an area roughly the size of India). Kiribati has a limited economic market that is disconnected from other parts of the world, and its primary source of income for its government was the sale of fishing licenses.

In January of 2008, the Republic of Kiribati entered into a unique partnership with the nonprofit environmental organization Conservation International (CI) and the New England Aquarium. Together, the three partners formed the Phoenix Island Protected Area (PIPA). With support from CI's Global Conservation Fund, the PIPA partners also established the PIPA Conservation Trust, a nongovernmental organization (NGO) established under the laws of the Republic of Kiribati, and governed by a board of directors appointed jointly by the partners.

The primary objective of the PIPA Conservation Trust was to provide long-term, sustainable financing for the conservation effort, which was accomplished by establishing a modest US$5 million endowment fund. The annual interest on this endowment was used to cover (1) the annual costs associated with managing PIPA, and (2) payments to the government of Kiribati for ensuring that exploitation of all or part of PIPA remains limited or prohibited. This financial arrangement ensured that, in exchange for protecting the reef ecosystem, Kiribati would not lose money from fishing revenue and, in fact, would get a long-term, sustainable source of income.

On January 30, 2009, the Republic of Kiribati submitted an application for the Phoenix Islands Protected Area to be considered for the United Nations Educational, Scientific and Cultural Organization (UNESCO) World Heritage List. This was the first nomination submitted by Kiribati since they ratified the Convention in 2000. On August 1, 2010 at the 34th session of the World Heritage Committee in Brasília, Brazil, the decision was made to inscribe PIPA onto the World Heritage List, making it the largest and deepest World Heritage site in the world.

Kanton Island is the northernmost and largest of the eight islands in the Republic of Kiribati's Phoenix Islands Protected Area (inset).

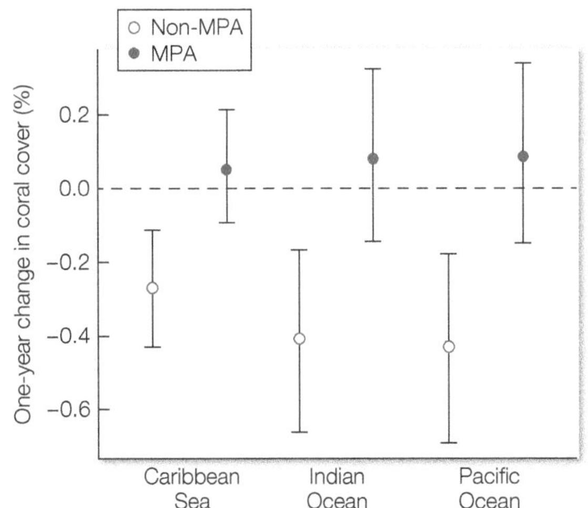

Figure 14.9 Change in coral cover from 2004 to 2005 inside and outside of MPAs. Reefs protected in MPAs had slightly positive changes in percent coral cover, although not significantly different from zero (dashed line). Percent coral cover was obtained by back-transforming predictions from a logit model. Error bars show 95% confidence intervals. (After E. R. Selig and J. F. Bruno. 2010. *PLOS ONE* 5: e9278. CC BY 4.0.)

showed that coral cover within MPAs increased by 0.05% in the Caribbean and 0.08% in the Pacific and Indian Oceans (**Figure 14.9**). In that same time period when coral was showing slow recovery within the MPAs, average declines on unprotected reefs ranged from −0.27% in the Caribbean to −0.41% and −0.43% in the Indian and Pacific Oceans, respectively. Such results have been confirmed by others, such as Strain and colleagues,[35] who studied 30 MPAs in tropical ecoregions and found that well-enforced no-take MPAs that were more than 10 years old had 1.08 to 1.19 times higher total coral cover than sites open to fishing, and 1 to 2 times more massive corals (i.e., large, slow-growing ball or boulder-shaped corals).

MPAs also appear to maintain higher population sizes and higher biodiversity of many species, such as fish species that are often the targets of conservation efforts. Gill and colleagues compiled data from 218 MPAs to compare fish biomass inside MPAs to fish biomass in a statistically matched control site (e.g., pre-establishment and/or outside MPA). The authors documented positive responses to protection in 71% of the MPAs. On average, fish biomass was 1.6 times higher in MPAs than in matched non-MPA areas, and positive responses were observed across almost all geographies and habitats (**Figure 14.10**). Topor and colleagues[36] showed that MPAs have nearly twice the local fish species richness as unprotected areas in the Atlantic, Indian, and Pacific Oceans.

MPAs can be particularly effective for protecting threatened species. For example, White and colleagues[37] assessed the capacity of large MPAs to conserve grey reef sharks (*Carcharhinus amblyrhynchos*), which are listed as Near Threatened by the IUCN. Using a combination of conventional tags, satellite tags, and vessel tracking technology, the authors showed that the 54,000-km^2 Palmyra Atoll National Wildlife Refuge in the central Pacific Ocean provides substantial protection for grey reef sharks. Two-thirds of satellite-tracked sharks remained within the MPA boundary for the duration of the study, and satellite detections of commercial fishing vessels identified no fishing effort within the refuge despite significant effort beyond the MPA perimeter, suggesting that large MPAs can effectively benefit reef sharks if properly enforced.

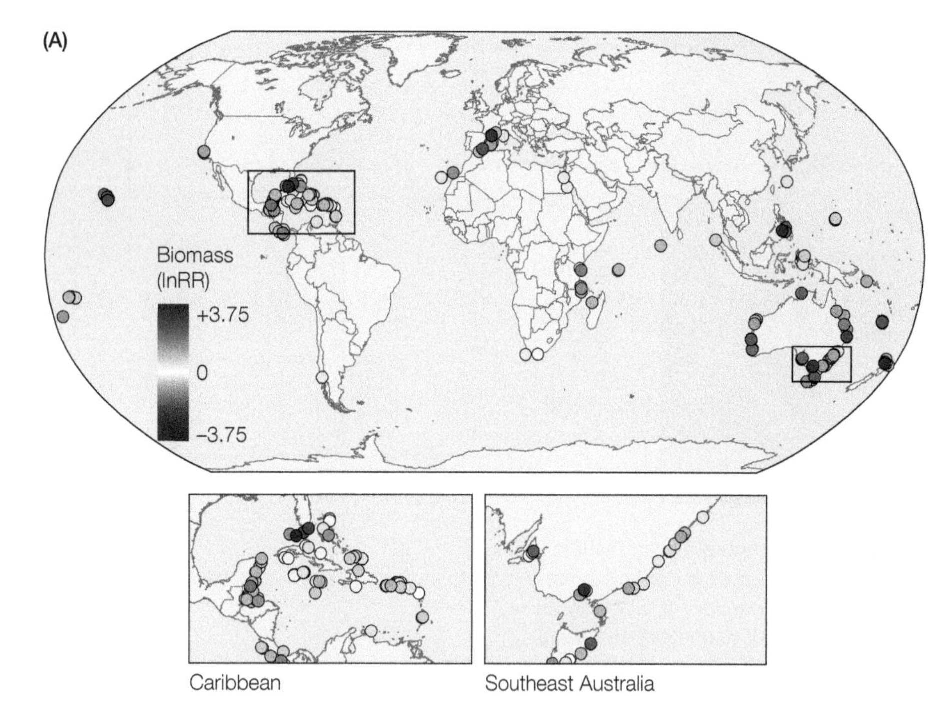

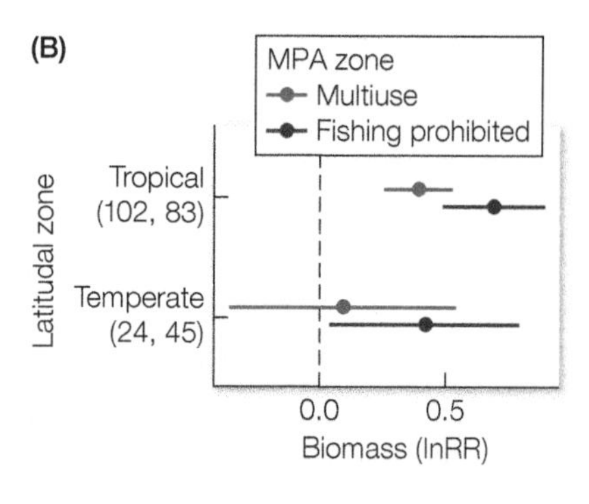

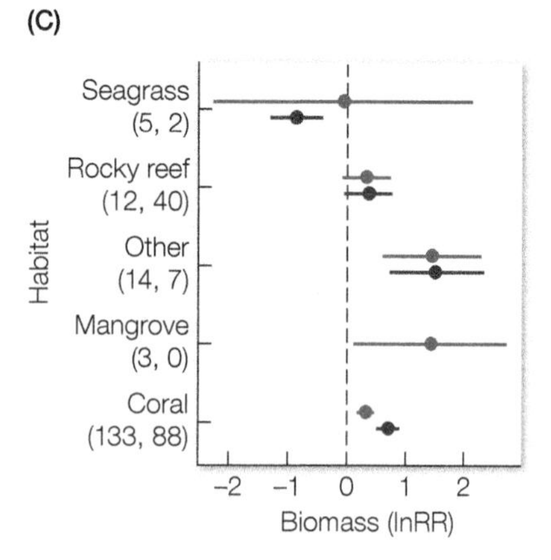

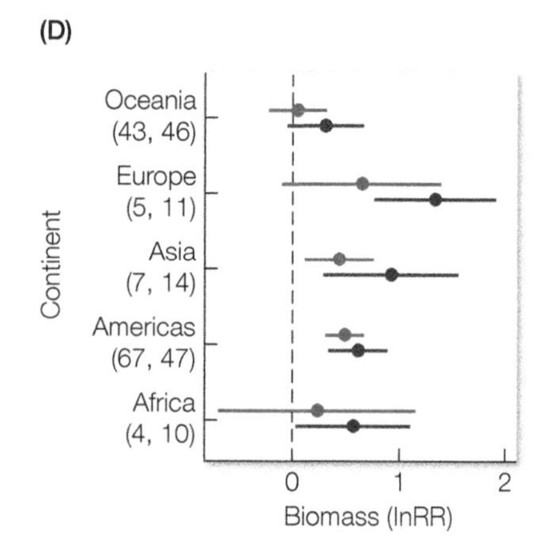

Figure 14.10 MPA effects on fish populations (biomass) measured as the natural log response ratio of mean fish biomass per unit area inside an MPA relative to mean fish biomass in a statistically matched control site (e.g. outside the MPA). (A) Global variation in fish biomass for 218 MPAs. Positive response ratios (blue) indicate MPAs with greater biomass inside the MPA relative to that in matched non-MPA areas. Negative values are in red. (B–D) Mean response ratios (dot) and 95% confidence interval (error bars) for multiuse areas (blue) and areas where fishing is prohibited (red) for 218 MPAs shown by latitudinal zone (B), habitat type (C), and continental region (D). Values in parentheses on the *y*-axes indicate the number of MPAs/zones that are multiuse and those where fishing is prohibited, respectively. RR = response ratio. (After D. A. Gill et al. 2017. *Nature* 543: 665. Base map sourced from B. Sandvik, World Borders Dataset. thematicmapping.org/downloads/world_borders.php [2016].)

Impacts on Human Well-Being

While the conservation benefits of protected areas for biodiversity are well established, the social, political, and economic impacts of protected areas on people have long been a matter of debate. Many conservation biologists have pointed out the positive benefits of protected areas for human well-being. In Chapter 6 we presented data showing that protected areas can be worth millions of dollars in income and serve as a major source of employment for local communities (see Table 6.1). For example, in the United States, more than 250 million people visit national parks, state parks, wildlife refuges, and other protected public lands each year.[38] These visitors generate economic activity worth about US$1 trillion per year, which produces more than 9 million jobs.[39] The total contribution from hunting, fishing, wildlife viewing, and the "human-powered" recreations

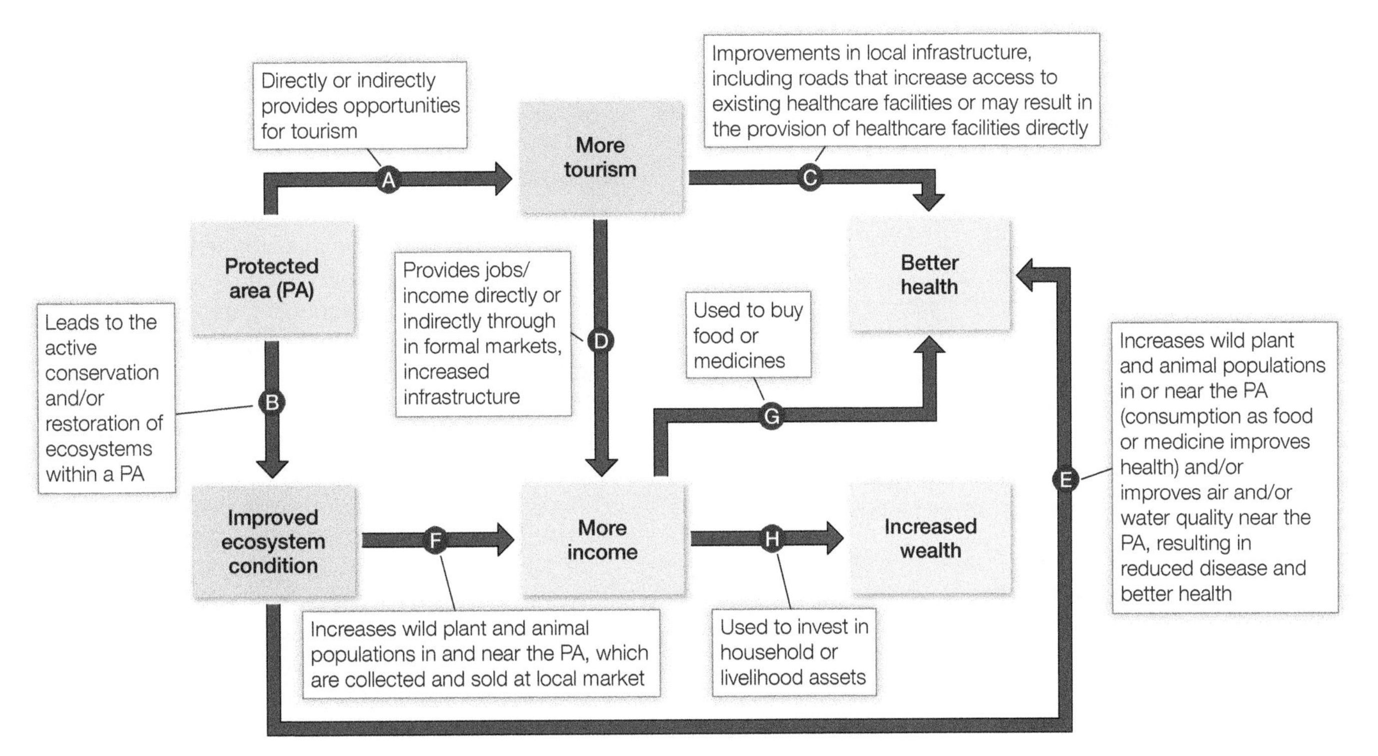

Figure 14.11 Potential mechanisms by which protected areas impact the health and prosperity of nearby people. Individual pathways can be combined to conceptualize an impact mechanism. For example, pathway ADG suggests that protected areas can lead to better health outcomes via income gains from PA-related tourism employment, income that is then spent on improving children's health. (After R. Naidoo et al. 2019. *Sci Adv* 5: eaav3006.)

(hiking, camping, skiing, and bicycling) in the U.S. is worth an additional $730 billion a year,[39] and the economic value of outdoor recreational sales (gear and trips combined) adds another $325 billion per year. These values far exceed the annual returns from pharmaceutical and medicine manufacturing ($162 billion), legal services ($253 billion), and power generation and supply ($283 billion).

In developing nations, it has been proposed that protected areas can improve the health and prosperity of local people, either through the extraction, sale, and use of products from the protected area, or as a result of economic activity when protected areas are managed for ecotourism (e.g., safaris) or other forms of recreation (e.g., hunting and fishing) (**Figure 14.11**).[40] Recent analyses confirm that there are economic benefits of protected areas that also translate to improved human health. Naidoo and colleagues[41] summarized data on the socioeconomic conditions of more than 87,000 children in over 60,000 households that were situated either close to or far from about 600 protected area in 34 developing countries. They showed that households located near protected areas with tourism had 17% higher wealth levels and a 16% lower likelihood of poverty than comparable households that were far from protected areas. This translated into greater food security and health benefits for children under the age of five, who also had 10% higher height-for-age scores and were 13% less likely to have stunted growth than children living far from protected areas (**Figure 14.12**). Importantly, while strict protected areas showed no socioeconomic benefits to local people, multiple-use protected areas managed simultaneously for conservation and sustainable resource use improved people's health and welfare compared to those who were not located near a protected area.

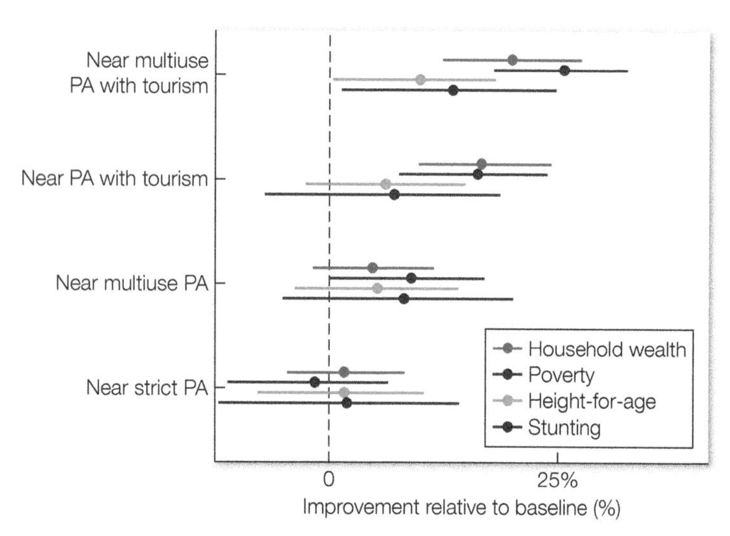

Figure 14.12 Impacts of protected areas on the socioeconomic condition of local communities. Values on the *x*-axis show improvements in four socioeconomic indicators (household wealth, likelihood of poverty, height-for-age scores, and likelihood of stunted growth) for communities that are close to a protected area relative to a baseline of communities that are not (*x*-axis). The *y*-axis shows values for protected areas with four different management regimes (strict PA, multiuse PA, PA with tourism, and multiuse PA with tourism). Data points give means and 95% confidence intervals. (After R. Naidoo et al. 2019. *Sci Adv* 5: eaav3006.)

While there is abundant evidence that protected areas can benefit people's health and prosperity after they are established, there is also abundant evidence that the process by which protected areas are established often creates great hardship for people. There are numerous examples where the creation of a protected area has led to the displacement of local communities and loss of their livelihoods.[42,43] For example, when Yellowstone National Park was established in 1872, U.S. military camps were set up to forcefully evict the Shoshone and Crow Native American tribes from the land, despite the fact they had lived in and used the area for centuries before Europeans arrived.[44] In 1947, during British colonial rule in Africa, Amboseli National Park was created across the Kenya–Tanzania border. After the ecosystem was declared a national reserve, Indigenous groups like the Maasai—nomadic herders who had grazed their cattle in Amboseli for many generations—were prohibited from using the land for grazing of the livestock.

Of course displacement of people as a result of conservation projects has negative consequences for human well-being. Conflict and violence are common, particularly when decisions are made top-down by governments and force is used to remove people from protected areas.[43,45] Studies have documented significant loss of livelihood and agricultural incomes, as well as indirect losses caused by lost access to areas set aside for conservation.[42,46,47] Displaced people have historically been those who are poor, or who are part of groups that lack strong political representation. Rarely have displaced people been compensated for their losses and, in many cases, their displacement has not been legally recognized, leading to even further political disempowerment.[42,43]

People who are displaced by the establishment of protected areas have been called **conservation refugees**.[48] No systematic records are kept for conservation refugees. However, Geisler[49] estimated that as the number of IUCN categorized protected areas grew from fewer than 1000 in 1950 to more than 29,000 by 2002, the 8.5 million km^2 of land placed into protection (an area the size of the continental U.S. plus half of Alaska) may have displaced as many as 136 million people. Others have estimated that 14 million people have been displaced in Africa alone.[50] The potential for future injustices at the hands of conservation is sobering when one considers that more than one-fifth of the world's population inhabits the planet's major hotspots of biodiversity, which are precisely the areas that many conservationists argue need to be protected from people[51] (**Video 14.5**).

Video 14.5 Watch a video about creating an equity framework for protected areas.
oup-arc.com/e/cardinale-v14.5

Given the jaded history of protected area creation, it should not be surprising that many people have a negative perception of protected areas.[52] Some feel they or their relatives have been treated unfairly during the formation of protected areas. Others see protected areas as a challenge to their property and sovereignty rights. Still others view protected areas as a limitation on their livelihoods and economic development. In Thailand, for example, surveys of public attitudes toward Marine Protected Areas suggest a belief that MPAs undermine people's ability to make a living by limiting access to fishing and agricultural opportunities; yet in fact they have negligible benefits for fishing or tourism.[53] Attitudes like these have sometimes generated a public backlash against protected areas that has led to governments taking steps to downsize or reduce the number of protected areas. Mascia and colleagues[54] documented

543 examples in 57 countries that were practicing what they called **protected areas downgrading, downsizing, and degazettement (PADDD)**, a response to social pressure from local communities who were angry they had not been fairly compensated for the loss of their lands and livelihoods, or from those that argue greater access to and use of natural resources is needed for economic development. In Brazil, for example, 73,000 km^2 were affected by PADDD events documented between 1981 and 2012, and of these, 52,000 km^2 were downsized or degazetted.[55] The United States ranks among counties with the highest rates of PADDD, having withdrawn 17 Biological Reserves from the UNESCO Man and the Biosphere Programme in 2017 alone.[56]

Clearly, the impacts of protected areas on the social, political, and economic well-being of people has a mixed history. It is precisely this mixed history that requires us to understand what factors best predict the failure verses success of protected areas.

Predictors of Success

Routine monitoring of the threats facing protected areas, and evaluation of their effectiveness in achieving conservation goals are recognized as essential components of maintaining successful biodiversity conservation strategies. There have been numerous calls for coordinated national and regional assessments of protected area effectiveness, including a call by the Convention on Biological Diversity (CBD) at the 2010 Conference of the Parties (COP10) to expand and institutionalize management effectiveness assessments internationally.

Protected area management effectiveness (PAME) evaluations are the primary tool conservation biologists use to evaluate the effectiveness of protected areas. Many different methods for PAME evaluation have been developed to meet specific regional goals, or specific criteria of conservation organizations. However, most PAME methodologies share at least four common objectives: (1) to analyze the range of threats facing a protected area; (2) to identify the most pressing management issues; (3) to assess how the protected area is functioning and performing; and (4) to suggest corrective steps need to improve management effectiveness.

PAME assessments are consolidated into the Global Database on Protected Area Management Effectiveness (GD-PAME, oup-arc.com/e/cardinale-w14.11), which contains thousands of assessments of protected areas collated from around the world that are translated into a common reporting format, allowing the cross analysis of information collected by different assessment methodologies. GD-PAME started as a research database at the University of Queensland in 2006 under a program that was jointly funded by WWF and The Nature Conservancy (TNC). It is now maintained as a project between the United Nations Environment Program (UNEP) and the International Union for Conservation of Nature (IUCN).

PAME evaluations have recently been summarized and analyzed to quantify the effectiveness of protected areas in accomplishing their goals, and to identify key predictors of success. Those reviews suggest at least three factors are key predictors of successful conservation in protected areas: effective management, community involvement, and the integration of social goals with conservation goals.

Effective management

Recent analyses of PAME evaluations suggest that only 22% of protected areas have sound management,[57] and some experts estimate that only half of all tropical reserves are effective.[58] Declines in plant and animal abundance have been documented inside many protected areas,[57,59] and the effectiveness of protection is being compromised by increased human pressures and inadequate government support in places as diverse as Latin America, Africa, and Asia.[60–62]

Schulze and colleagues[63] used PAME evaluations to summarize the primary threats reported for 1961 terrestrial protected areas in 149 countries. Unsustainable and illegal hunting was the most commonly reported threat to successful conservation, and occurred in 61% of all protected areas. Hunting was followed by disturbance from recreational activities occurring in 55% of protected areas, and natural system modifications from fire or its suppression in 49%. In developing countries, threats were linked to overexploitation for resource extraction, while negative impacts from recreational activities dominated in developed countries.

Given these findings, it should not be surprising that one of the strongest predictors of effective management in protected areas was enforcement and compliance with regulations. Leverington and colleagues[57] compiled data from 4092 PAME evaluations for terrestrial protected areas and showed that some of the strongest correlates of success included natural resource protection activities and adequacy of law enforcement. Table 14.2 shows the top five factors correlated with overall success, of success with conservation of biodiversity, and success with social goals. The greatest limitations to adequate protection and enforcement included: (1) inadequate financial support, particularly in developing countries that lacked the infrastructure, equipment, and facilities needed to execute conservation plans; and (2) lack of clear legal authority to regulate and enforce, including legal establishment of areas, legislation and boundary marking that allow for effective governance.

Enforcement and compliance also show up as top predictors of success in evaluations of MPAs. For example, Edgar and colleagues[64] evaluated factors

TABLE 14.2 Predictors of the success of terrestrial protected areas[a]

Predictor	Correlation to overall success	Correlation to conservation goals	Correlation to social goals
Adequacy of infrastructure, equipment, and facilities	0.70		
Effectiveness of administration	0.70		
Natural resource and cultural protection activities	0.67		
Communication program	0.67		0.33
Adequacy of law enforcement	0.67	0.38	
Research and monitoring		0.36	0.25
Management effectiveness evaluations			0.29
Community and stakeholder involvement			0.32
Constraint or support by external political and civil environment		0.42	
Appropriate program of community assistance			0.29
Achievement of work set program		0.36	
Skill level of staff/partners		0.43	

Source: F. Leverington et al. 2010. *Environ Manage* 46: 685–698.
[a]The table shows the top five factors correlated with overall success, success with conservation of biodiversity, and success with social goals.

contributing to the success of 87 MPAs throughout the world. While MPAs often fail to reach their full potential as a consequence of factors like illegal harvesting, these authors showed that success of conservation goals increases substantially with strong regulation and enforcement (**Figure 14.13**). When no-take regulations were weak or not enforced, the total biomass of all fish, biomass of large (>25 cm length) fish, species richness of fish, and species richness of large fish were all at or below values that would be predicted from global models of fished coastlines. In other words, MPAs with weak regulations or enforcement were no better than, and sometimes even worse than, areas with no protection.

However, MPAs that had stronger regulations and higher levels of enforcement had substantially higher fish biomass and species richness, sometimes double or triple that of more poorly managed areas. Increases in biomass and species richness were particularly prominent for large species like jacks, grouper, and sharks, which showed some of the greatest benefits from effective management and enforcement of MPAs. In their own review of MPAs, Gill and colleagues[65] further noted that regulations and enforcement rank among the strongest predictors of success. However, they cited the lack of adequate funding for personnel, monitoring, and policing activities as one of the greatest limitations to the success of MPAs.

Stakeholder and community involvement

Successful conservation efforts using protected areas require more than just effective ecological management and enforcement. They also require public support by local communities, politicians, and businesses—all of whom may

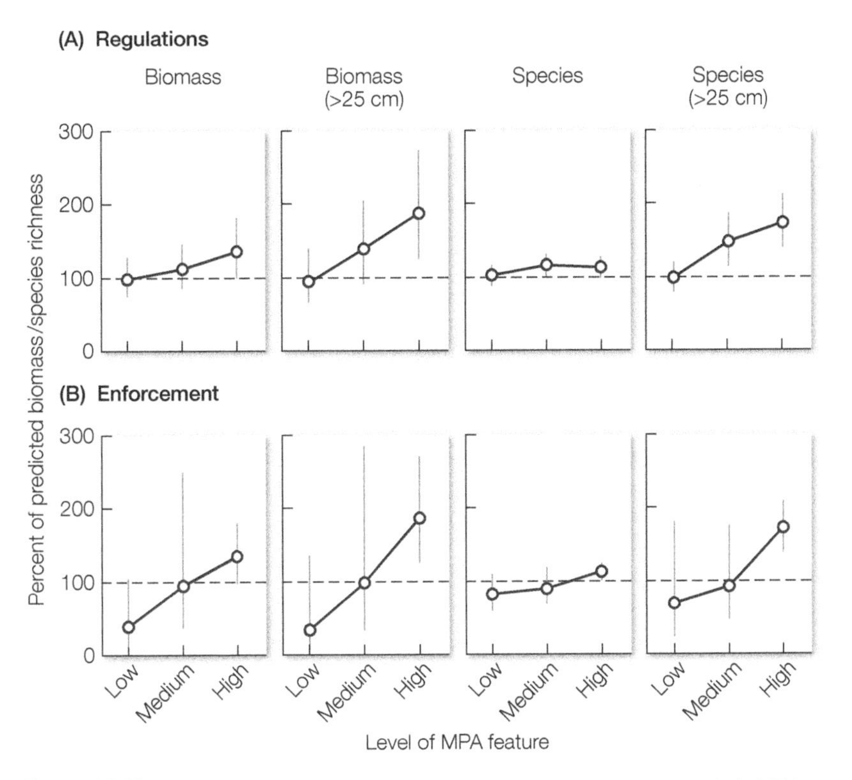

Figure 14.13 Effects of regulation and enforcement on conservation goals in MPAs. The *x*-axis shows increasing levels of regulation (A) or enforcement (B) from low to high. The *y*-axis shows total fish biomass, biomass of fish longer than 25 cm, total species richness, and richness of fish larger than 25 cm relative to that predicted for a corresponding unprotected area of ocean habitat. (After G. J. Edgar et al. 2014. *Nature* 506: 216–220.)

have different vested interests in the protected area. With the collaboration and support of various stakeholders, and the involvement of local community members, the long-term success of a protected area is far more likely than without it.

Several reviews of PAME evaluations show that involvement of communities and stakeholders in the planning, formation, and management of a protected area, as well as effective communication programs that keep local communities and stakeholders connected to protected area managers, rank among the top predictors of protected area success (see Table 14.2).[57] In their review of factors contributing to the success and failure of Marine Protected Areas around the world, Giakoumi and colleagues[66] found that stakeholder engagement was the single most important factor affecting MPA success, and equally, its absence, was the most important predictor of failure to achieve conservation goals.[66] Stakeholder and community involvement enhances the success of protected areas by:

- Creating a mechanism to recognize and discuss different motivations and needs, which are often divergent and must be balanced.
- Encouraging clear conservation goals and providing a means of educating the public as to why those goals are important.
- Facilitating transparency in decision-making, and empowering parties by giving them a chance to have input to the decision-making process.
- Engaging people in the monitoring, management, and oversight of the protected area, thus giving them a vested interest in its success.[67]

Given the importance of community and stakeholder engagement, governments, NGOs, and natural resource managers have moved away from top-down regulatory styles, instead using styles that feature stronger, more diverse partnerships, and collaborations between management agencies and local communities. These collaborations often offer citizens a far greater role in the oversight and management of protected areas, as well allowing them a share of power and participation in political and policy decisions. Unfortunately, in their analysis of 4092 PAME evaluations from 100 countries, Leverington and colleagues[57] concluded that communication and community involvement programs are generally inadequate. In all regions studied, these factors score poorly. Frequently, reports mentioned that the staff who help manage protected areas have inadequate skills and training in communication and community involvement. This underscores the need for practitioners who have the tools to balance competing desires, such as the use of structured decision-making (Chapter 2), cost-benefit analysis (Chapter 7), and multicriteria decision analysis (Chapter 7). Training in the "human dimensions" of natural resource management has become very important in many state and federal management organizations (for an example, see oup-arc.com/e/cardinale-w14.12).

Integrated social development and conservation goals

A growing body of research suggests that positive socioeconomic outcomes of protected areas are some of the strongest predictors of conservation success. Oldekop and colleagues[68] summarized results of 171 published studies of 165 protected areas to assess how socioeconomic factors that influence the well-being of local communities (e.g., displacement, conflict, monetary or other impacts on livelihoods) relate to the reporting of positive versus negative outcomes on biodiversity conservation. The authors reported that positive socioeconomic outcomes were strongly associated with positive conservation outcomes, whereas negative conservation outcomes were often associated with negative socioeconomic outcomes. Sustainable-use protected areas (IUCN Category VI)

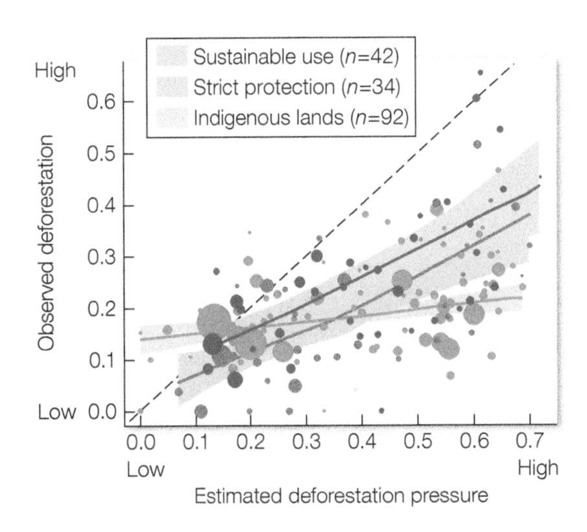

Figure 14.14 Observed levels of deforestation in protected areas as a function of their estimated deforestation pressure (x and y- axes are in arbitrary units, but increase from left to right, and bottom to top). Each point on the graph represents a set of forested protected areas that were matched to unprotected areas as controls (i.e., that were biologically similar). Circle size corresponds to the number of forest areas included in the study, and data points below the dashed diagonal line represent protected areas that avoided deforestation relative to unprotected controls. Regression lines (solid colored lines with 95% confidence intervals shaded) show the success of different types of protected areas in avoiding deforestation. Sustainable use protected areas (blue) appear to have had the same impact on deforestation as strict protected areas (green), as indicated by the fact the lines are parallel. However, protected areas occupied by Indigenous people (orange) were the most effective at reducing deforestation, as evident from the low values on the y-axis and shallow slope. (From C. Nolte et al. 2013. *Proc Nat Acad Sci* 110: 4956–4961.

were more likely to report overall positive socioeconomic outcomes than more strictly protected ones (IUCN Category I), and protected areas in which local people experienced positive cultural outcomes and fewer negative impacts on their livelihood had the highest reported success.

Numerous studies have also shown that conservation efforts can be just as effective in multiuse protected areas as in strict protected areas. For example, Nolte and colleagues[69] studied how different approaches to land management influenced rates of deforestation in Brazil for forests being subjected to varying rates of deforestation. Areas that were managed for sustainable use of forest products (e.g., timber) had the same impact on rates of deforestation as areas that were strictly protected (note the parallel blue and green lines in **Figure 14.14**, which indicate comparable effects on deforestation). Areas occupied and managed by Indigenous peoples were even more effective at reducing deforestation than protected areas. A likely explanation for these results is that community involvement in local resource management reduced rates of illegal logging, and protection from illegal logging outweighed any use of forest resources by local communities.

There are other examples as well. Nelson and Chomitz[70] found that multiuse protected areas in Latin America and Asia were more effective than strict protected areas at reducing the incidence of fires used for deforestation; Joppa and colleagues[29] found negligible differences in forest cover for protected areas in different IUCN management categories in the Amazon, Congo, South American Atlantic Coast, and West Africa; Gray and colleagues[31] found no significant difference in measures of species richness or population abundance for protected areas assigned to different IUCN Categories.

Studies that fail to find conservation benefits of strict protected areas challenge the way practitioners have historically thought about conservation. Many, particularly those from the natural sciences, have been taught that socioeconomic development and habitat use by humans are, in fact, the very reason biodiversity loss is such a problem in the Anthropocene. This perspective is not wrong (see Chapters 9–12), but it is incomplete. While there are many situations where people have the propensity to destroy nature, there are also situations where people are compelled to protect nature, such as when their health, happiness, and well-being are tied to the conservation and management of their local ecosystems and biodiversity. When people become connected to their land and water, it is possible for the goals of conservation to coexist with sustainable use of natural resources. One of conservation biology's greatest challenges in the coming century, as well as a great opportunity, is to determine how best to balance conservation with sustainable development (**Box 14.3**).

Establishment of protected areas has been ethically troubling when conservation efforts have resulted in the displacement of communities and have had disadvantageous social outcomes for local people. Such events can create a social backlash that undermines the goals of conservation when it leads to protected areas being downgraded, downsized, and degazetted (PADDD).

BOX 14.3 Challenges & Opportunities

Fences and Fines ... or ... Integrated Conservation and Development?

Conservation of biodiversity has often been viewed as being at odds with social and economic development. Many believe that you can develop land and oceans and use their natural resources for the good of people, or you can conserve the land and oceans for the good of biodiversity, but you can't do both. This viewpoint underlies many twentieth century approaches to conservation that sought to keep people and nature separated from one another, and that used punitive management practices relying on regulations and enforcement. Separation with punitive management has been called the **fences-and-fines** approach to conservation.

There is certainly evidence that physical separation, regulations, and enforcement can be an effective way to conserve biodiversity.[65,69] But there is increasing recognition that the fences-and-fines approach has limits, and can sometimes create a societal backlash that undermines the goals of conservation. Exclusionary tactics can be ethically troubling when conservation efforts result in the displacement of communities, or lead to disadvantageous social outcomes for local people. Regulation and enforcement, when used too heavily, can alienate people and turn them against conservation efforts. A growing body of research also suggests that multiuse and sustainable use areas can be as effective in protecting biodiversity as strictly protected areas when local communities become invested in their success.[29,31,70]

A frequently advocated strategy for conservation is the use of **integrated conservation and development projects** (**ICDPs**; see Figure). ICDPs seek to meet the goals of biodiversity conservation while at the same time dealing with the social and economic needs of communities who might otherwise threaten biodiversity. ICDPs work to benefit local communities and Indigenous populations through programs that encourage economic sharing (e.g., transfer of money from tourism), the creation of jobs (e.g., park management), and the integration of agricultural habitat (e.g., food security). In doing so, ICDPs try to address the underlying cause(s) of biodiversity loss (e.g., poverty that leads to habitat loss or overexploitation), and to encourage protection of the area through shared management and governance, and by promoting economic development.

Of course, the ICDP approach to conservation has its own limitations. Local people and their living practices comprise the greatest threat to biodiversity conservation, and there is no guarantee that improved living conditions will reduce human pressure on the natural resources of an area. There is, in fact, a fine balance where economic development taken too far can be counterproductive to the original goal of conservation. The great challenge and opportunity for future practitioners is to determine the right balance between conservation and development when establishing an ICDP.

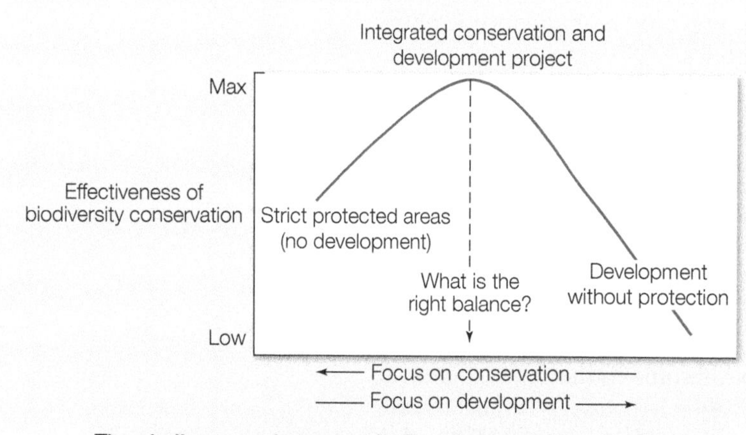

The challenge and opportunity for integrated conservation and development projects is to determine the right balance between conservation and development foci.

© ephotocorp/Alamy Stock Photo

Integrated conservation and development projects that seek to meet the goals of biodiversity conservation while at the same time dealing with the social and economic needs of people are being increasingly adopted as a best management practice for conservation efforts.

Summary

1. The primary tool used for conservation of whole communities and ecosystems is the protected area. A protected area is a clearly defined geographical space that is recognized, dedicated, and managed to achieve long-term conservation of nature and its associated ecosystem services.
2. With help from the World Commission on Protected Areas, the International Union for Conservation of Nature categorizes protected areas into six types: (I) strict nature reserves and wilderness areas, (II) national parks, (III) national monuments, (IV) habitat/species management areas, (V) protected landscape/seascape, and (VI) managed resource protected area. The IUCN Categories I–IV are often referred to as areas where biodiversity is strictly protected. Categories V–VI are referred to as multiple-use areas.
3. Roughly 15% of the earth's land surface is currently managed as a protected area (IUCN Categories I–VI), which is just short of the Conservation of Biological Diversity's Aichi Target 11 to have 17% of land and freshwater habitat in protected areas by 2020. The total areal coverage of marine protected areas (MPAs) within exclusive economic zones has increased at a rapid rate over the last two decades, and currently stands at 7.5% of the world's oceans. Assuming current trends continue, the Aichi goal to protect 10% of coastal and marine areas will be achieved by 2020.
4. There are five common approaches for choosing areas to protect entire biological communities and ecosystems. The hotspot approach chooses geographic areas that host the highest levels of biodiversity and the greatest risk of habitat loss. The ecoregion approach prioritizes conservation of whole ecosystems using boundaries that approximate the original extent of natural communities prior to major land-use change. A third approach bases selection of conservation areas on political or economic boundaries. The fourth is to base selection of protected areas on their delivery of ecosystem services. The final approach is to select protected areas according to their cultural importance for inclusion in the United Nations Educational, Scientific and Cultural Organization's (UNESCO) biosphere reserves or world heritage sites.
5. Both terrestrial and marine protected areas are effective tools for preventing habitat loss, minimizing overexploitation, and protecting biodiversity. However, there is considerable variation in the success of individual protected areas in achieving conservation goals. The best predictors of success are effective management through regulation and enforcement; strong stakeholder and community engagement programs; and integrated social development and conservation goals.
6. Establishment of protected areas has, at times, been ethically troubling when conservation efforts have resulted in the displacement of communities, and led to disadvantageous social outcomes for local people. Such events can create a social backlash that undermines the goals of conservation through protected areas downgrading, downsizing, and degazettement (PADDD). Integrated conservation and development projects that seek to meet the goals of biodiversity conservation while, at the same time, dealing with the social and economic needs of people are being increasingly adopted as a best management practice for conservation efforts.

For Discussion

1. Which approach(es) to the selection of protected areas do you think work(s) best: biodiversity hotspots, the ecoregion approach, political or economic boundaries, ecosystem services, or areas of cultural importance?
2. How much of Earth's land and ocean should be set aside to conserve biodiversity? Is the 17% (land) and 10% (ocean) proposed by Aichi Target 11 enough? Is the 50% proposed by Nature Needs Half too much, or too little?

3. In your experience, are the goals of conservation compatible with the goals of socioeconomic development? Is the idea of integrated conservation and development projects a viable strategy for the future of conservation, or just wishful thinking?

4. PAME evaluations clearly show that regulations and enforcement are key predictors of successful conservation in protected areas. At the same time, many studies also show that the fences and fines approach to conservation has eroded support for conservation efforts. How can we balance the need for proper management and oversight of protected areas with the need to build public support and involvement in those areas?

Suggested Readings

Dinerstein, E. et al. 2017. An ecoregion-based approach to protecting half the terrestrial realm. *BioScience* 67: 534–545. Assesses progress on the Nature Needs Half initiative, which calls for protection of 50% of the terrestrial biosphere.

Dowie, M. 2011. *Conservation Refugees: The Hundred-year Conflict between Global Conservation and Native Peoples.* Cambridge, MA: MIT press. A book filled with cautionary tales about how the establishment of protected areas have negatively influence peoples—tales we should learn from, and avoid repeating.

Gray, C. L. et al. 2016. Local biodiversity is higher inside than outside terrestrial protected areas worldwide. *Nature Communications* 7: 12306. A comprehensive assessment of the effectiveness of protected areas in conserving biodiversity.

Naidoo, R. et al. 2019. Evaluating the impacts of protected areas on human well-being across the developing world. *Science Advances* 5: eaav3006. Uses data from more than 600 protected areas to show how living either close to or far from a protected area influences aspects of human well-being.

Oldekop, J. A.et al. 2016. A global assessment of the social and conservation outcomes of protected areas. *Conservation Biology* 30: 133–141. Shows the relationship between a protected areas conservation and socioeconomic outcomes.

Schulze, K. et al. 2018. An assessment of threats to terrestrial protected areas. *Conservation Letters* 11: e12435. Reviews the threats primary facing 1961 protected areas in 149 countries.

Strain, E. M. A. et al. 2019. A global assessment of the direct and indirect benefits of marine protected areas for coral reef conservation. *Diversity and Distributions* 25: 9–20. Looks at 30 MPAs around the world to determine how reserve age, size, isolation, and enforcement influence protected area success.

Visit the
Conservation Biology
Companion Website
oup.com/he/cardinale1e
for videos, exercises, links, and other study resources.

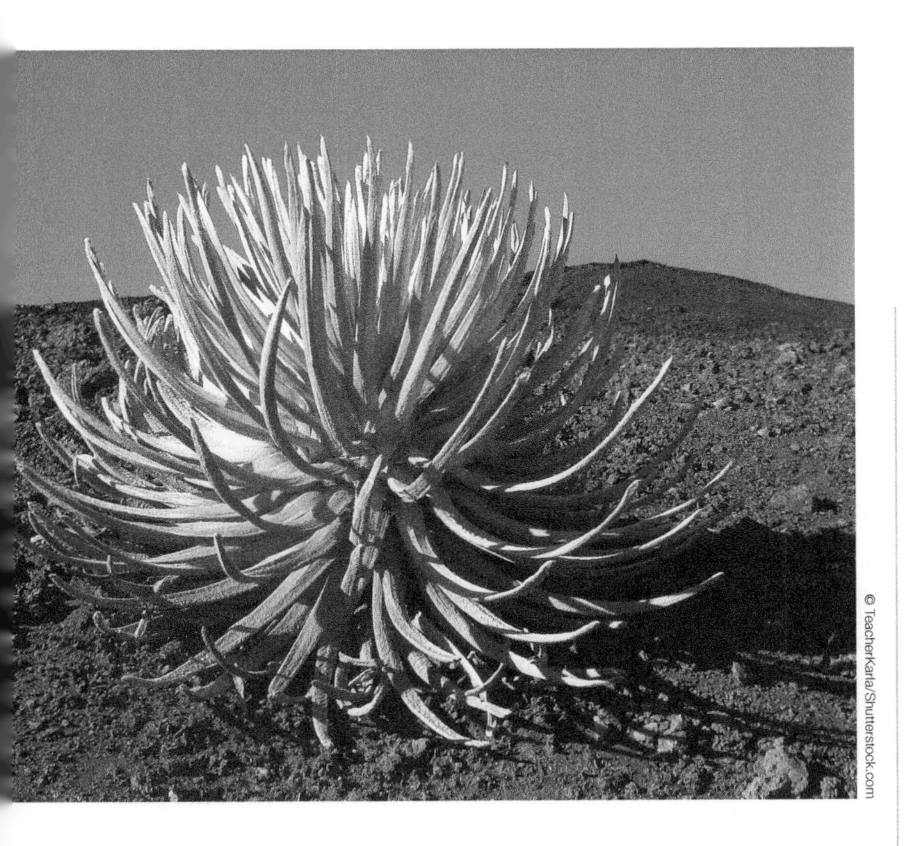

15 Landscape-Scale Conservation

Many threatened silver sword plants (genus *Argyroxiphium*), like this one in the Haleakala National Park in Hawaii, survive because of conservation reserve networks.

In Chapter 13, we covered the tools and techniques used in the conservation of single species, particularly focusing on the challenges of conserving species that have been driven to small population sizes as a result of human activities. In Chapter 14, we covered approaches used to conserve entire biological communities and ecosystems, focusing primarily on the use of protected areas, which have been a cornerstone of conservation efforts for over a century. Here in Chapter 15, we turn to an even larger spatial scale of conservation, that of entire landscapes.

A **landscape** is a large geographic region that spans multiple biological communities and ecosystems and that includes different habitat types (Figure 15.1). Because of habitat loss, fragmentation, and degradation (see Chapter 9), landscapes throughout the world are increasingly patchy spatial mosaics where natural habitats have been lost and converted to human uses (e.g., agriculture, urban), others have been heavily degraded (e.g., industrial), and the natural habitats that remain are small, fragmented, and disconnected from one another. Landscapes like these are referred to as **landscape mosaics** or **patch mosaics**.

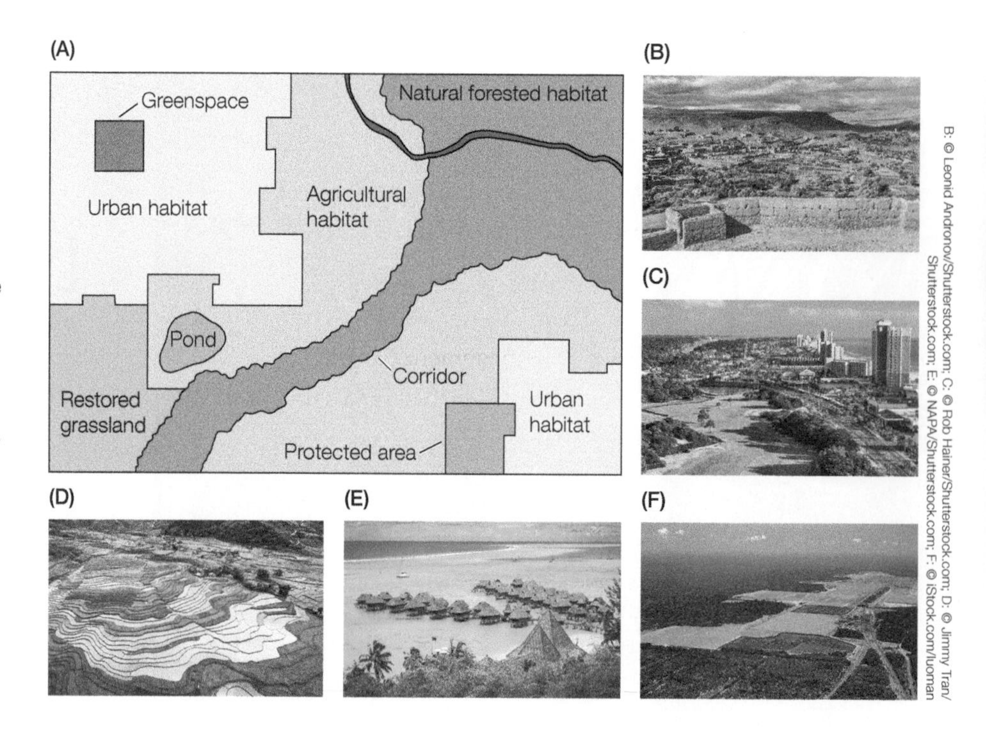

Figure 15.1 Conservation in landscape mosaics. (A) Most present-day landscapes are patchy mosaics composed of many different habitat types. Some of these habitat types are more conducive to conservation (e.g., protected areas) than others (e.g., urban or agricultural habitat). But practitioners must use all habitat types to maximize conservation of biodiversity while ensuring human well-being at the scale of whole landscapes. (B–F) Some examples of landscape mosaics. (B) Ait Ben Haddou village in Morocco, North Africa. (C) Aerial view of Panama City Beach, Florida, U.S.A. (D) Aerial view of terraced rice fields in water season, Mu Cang Chai, Vietnam. (E) Aerial view of bungalows in a luxury lagoon resort, Moorea, French Polynesia. (F) Aerial view of Amazonian rain forest, showing logging roads and deforestation. (A, after G. Bentrup. 2008. *Conservation Buffers: Design Guidelines for Buffers, Corridors, and Greenways*. Gen. Tech. Rep. SRS-109. Asheville, NC: USDA, Forest Service, Southern Research Station.)

Within landscape mosaics, some habitat types are more conducive to conservation of biodiversity than others. There are several reasons why conservation biologists must work with this heterogeneity to maximize conservation of biodiversity at the scale of whole landscapes. First, conservation efforts in natural habitats or protected areas ultimately succeed or fail based on our ability to manage what is happening in the surrounding landscape. For example, analyses of mammalian extinctions in national parks in the United States, Canada, and West Africa have all shown that the pressures exerted by increased human density, and the proportion of modified habitat surrounding parks are often better predictors of extinction than the size or other characteristics of the protected areas themselves.[1–4] Understanding the landscape context of protected areas and working with the appropriate stakeholders (e.g., policy makers, regional planners, landscape architects) to develop projections of changes in human population density and habitat types becomes a necessary part of creating effective conservation plans within protected areas.[4,5]

The second reason conservation biologists must work to protect biodiversity within the context of landscape mosaics is a simple one: species often do not stay put. Many species move large distances across different habitats. This can occur because of large home ranges or territory sizes (e.g., bears, mountain lions, tigers) or because of migration for purposes of feeding or breeding (e.g., wildebeest, buffalo, breeding birds). While some of the habitats in a landscape mosaic may be inhospitable for mobile species (e.g., wolves that cross farmland or pasture are sometimes shot and killed by ranchers), other habitats may be suitable (e.g., birds living and breeding in urban habitat), and still other habitats may be suboptimal but temporarily useful (e.g., foxes moving through farm fields to get to forest habitat). Species that are highly mobile must be managed at a landscape scale where the sizes of conservation units more closely match the natural population sizes, migration patterns, and biological requirements of the species.[6,7]

Lastly, landscape-scale conservation is important because conservation biologists must protect species while also ensuring a good quality of life for people. As discussed in Chapter 14, conservation efforts that produce undue social and economic hardships on communities often fail, and even lead to backlashes against conservation. However, when efforts integrate social development and conservation goals, people often feel more compelled to protect nature because their health, happiness, and well-being become intimately tied to the conservation and management of their local ecosystems and biodiversity. Integrating social development and conservation is inherently a landscape-scale process that requires establishing spatial mosaics that integrate human needs with those of other organisms occupying the landscape.

In this chapter, we discuss (1) the creation of networks of protected areas in a landscape mosaic, (2) the incorporation of unprotected habitats into the landscape network in order to complement the goals of conservation, and (3) the measurement, monitoring, and management the landscape to achieve broad-scale conservation goals.

Creating Networks of Protected Areas

The backbone for conservation at a landscape scale is the **protected area network** (also called a **reserve network**). Few things ensure more long-term success in conservation than establishing a collection of protected areas that represent all scales in the biological hierarchy (genes, species, communities, and ecosystems), and which are connected within a landscape and managed collectively. In this section, we discuss how new protected areas are established, and we cover criteria for selecting new protected areas, methods for filling gaps in networks and how to connect protected areas using corridors and stepping-stones.

Establishing new protected areas

New protected areas are established by national governments, local governments, corporations, land trusts, and individuals. National governments play the largest role in establishing new protected areas through their creation of national parks, wilderness areas, biological reserves, and mixed-use public lands (e.g., national forests). Countries that are member parties to the United Nations Convention on Biological Diversity (CBD) are required to develop **national biodiversity strategy and action plans (NBSAPs)** for the conservation and sustainable use of biodiversity.[8] NBSAPs must lay out a country's plan for accomplishing goals in the CBD's Strategic Plan for Biodiversity 2011–2020, which include the 20 Aichi Biodiversity targets (see Chapter 14).[9] As you may recall from Chapter 14, Aichi Target 11 calls for 17% or more of terrestrial land and inland water, and 10% or more of coastal and marine areas to be conserved and managed as a network of well-connected protected areas that are integrated into the broader landscapes and seascapes. As of 2019, 190 of the 196 (97%) signatories to the CBD have developed and submitted national NBSAPs.[9]

Suggested Exercise 15.1 Find your home country's national biodiversity strategy and action plan, and learn how it plans to accomplish Aichi Biodiversity Target 11. **oup-arc.com/e/cardinale-ex15.1**

In some countries, especially Western nations where land and water rights are afforded to corporations and individuals who own the land, protected areas can be established privately. Protected areas under the governance of private entities are known as privately protected areas (PPAs).[10] PPAs have gained considerable attention in recent years as one of the fastest growing sectors of conservation, and are quickly becoming one of the primary ways to meet the Aichi Biodiversity Targets. For example, in the lower 48 U.S. states, more land has been set aside as PPAs than is currently protected within the U.S. National Park System.[11,12] The U.S. National Conservation Easement Database (NCED), which compiles records from land trusts and public agencies throughout the U.S, now includes more than 158,000 conservation easements totaling 109,414 million km^2 (oup-arc.com/e/cardinale-w15.1). That is an area equivalent to 12 times the size of Yellowstone National Park!

PPAs are generally established in one of two ways. The first is the purchase of land by private corporations or individuals, who then set aside that land as a biological reserve but maintain ownership of the reserve. In some instances, these reserves are established for profit. Examples include the many Private Concessions and Game Reserves that have been set up around Kruger National Park in Africa (e.g., the Imbali, Lukimbi, and Nwanetsi Concessions). These areas are not open to the public, but are reserved for paying guests who stay at the camps and lodges for safaris or wildlife hunting. In other instances, the private reserves are nonprofit ventures established primarily for the purpose of biodiversity conservation. For example, Pumalin Park is a 3000 km^2 tropical forest reserve in southern Chile that was purchased for $30 million by wealthy foreign philanthropists who used it in the creation of a network of 11 wilderness parks across Chile and Argentina.[11]

The second way a PPA can be formed is through **conservation easements** or **covenants**. Easements and covenants are voluntary legal agreements that a landholder enters into with another party, such as a government agency or land trust (described in the next paragraph). The agreement places some restrictions on how the property will be used in the future in order to ensure some value for conservation, but the original owner can choose to maintain certain options or activities. For example, an easement may preclude use of the land for housing development while at the same time allowing the landowner and his/her descendants to build their personal houses and maintain certain types of activities (e.g., ranching, lumber extraction). In exchange for a conservation easement/covenant, the landholder may receive tax credits or deductions, and may be shielded from certain legal liabilities. Easements are typically made in perpetuity such that if a property is sold or passed on to relatives, the easement remains with the land. In contrast, covenants are usually not permanent, and may be voided when the property changes ownership.

The other party involved in a conservation easement or covenant is sometimes a government agency that manages the land according to the agreement, as part of its portfolio of public land and protected areas. Oftentimes, however, the other party is a **land trust** (or **land conservancy**), which is a private, nonprofit conservation organization that, as part of its mission, assists in land acquisitions through conservation easements and covenants, and then stewards those lands as part of their own portfolio of protected areas. In the U.S., there are more than 1700 land trusts that engage in land acquisition through easements, including large, well-known conservation organizations like The Nature Conservancy and the Land Trust Alliance.

Though a relatively new conservation tool, easements and covenants have become increasingly popular, particularly among individual landowners who want to see the natural resources on their property preserved for future generations. The tool is now being widely used by land trusts in Australia, Canada, Chile, Costa Rica, Ecuador, Mexico, New Zealand, Paraguay, the United Kingdom, and many other countries. In addition, land trusts like The Nature Conservancy have begun exploring the development of marine conservation agreements that can be used in ways that are similar to easements and covenants in countries where marine areas are under private governance.[13]

Increasingly, federal governments, local governments, charitable organizations, and corporations are forming partnerships in which they work together to establish and integrate new protected areas into larger, coordinated conservation networks. For example, in 2018, Prime Minister Justin Trudeau established the **Canada Nature Fund**, which dedicated $500 million in federal funding over 5 years, and leveraged an additional $500 million through partnership support from foundations, provinces, territories, corporate and not-for-profit sectors, and Indigenous peoples. The anticipated $1 billion is dedicated to establishment of protected and conserved areas that will support **Pathway to Canada Target 1**, a nationwide initiative to achieve Canada's international commitment to the Aichi Biodiversity Targets. As part of this plan, Canada is working with Indigenous peoples to create and legally recognize **Indigenous Protected and Conserved Areas** (IPCAs). IPCAs are lands where Indigenous peoples and their governments play the primary role in protecting and conserving ecosystems while building sustainable local economies. IPCAs not only help Canada's federal government meet its international obligation to the CBD, they also play a small part in reparations to correct historical injustices imposed on indigenous peoples who were displaced from their native land. As such, IPCAs represent a major success story, and an important paradigm for linking conservation, Indigenous culture, and human rights (Box 15.1).[14]

Criteria used to select and prioritize new protected areas

Because the resources needed to establish new protected areas are always limited, organizations and individuals who participate in the establishment of new protected areas must prioritize their efforts. There has been much debate about what factors should be used to prioritize the selection of new protected areas. Some of this debate has focused on desirable properties of individual protected areas that are expected to maximize conservation benefits (Figure 15.2A). Other

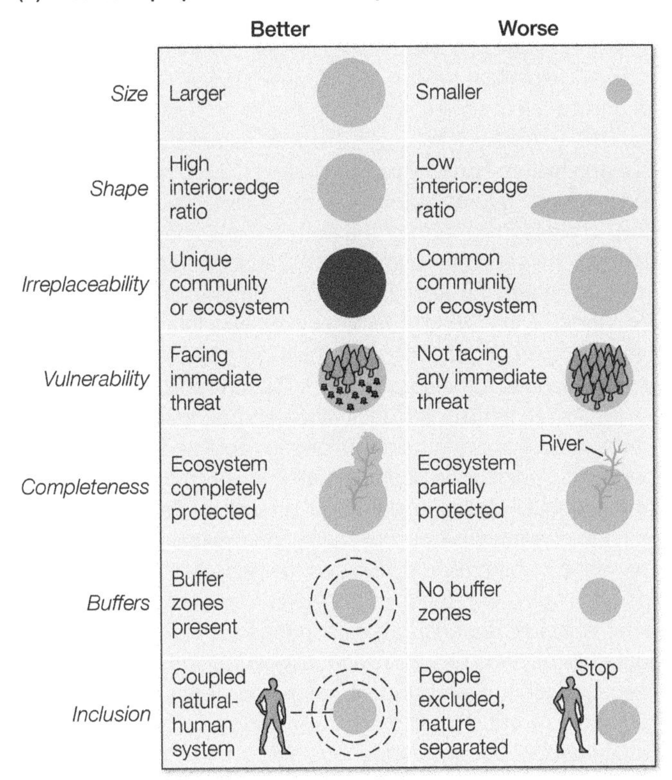

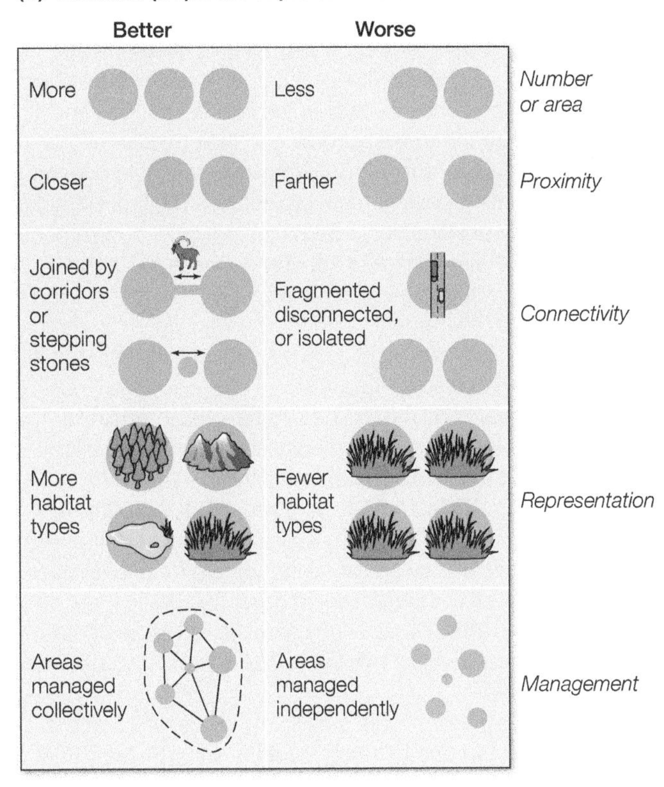

Figure 15.2 General guidelines for the design of protected area networks. (A) Desirable properties of individual protected areas that maximize biodiversity conservation. (B) Desirable properties of networks of protected areas that optimize biodiversity conservation across a landscape. (After C. L. Shafer. 1997. In M. W. Schwartz [Ed.], *Conservation in Highly Fragmented Landscapes*, pp. 345–378. Chapman and Hall: New York, based on C. L. Shafer. 1994. In E. A. Cook and H. N. van Lier [Eds.], *Landscape Planning and Ecological Networks*, pp. 201–223. Elsevier: Amsterdam.)

BOX 15.1 Success Story

Indigenous Protected and Conserved Areas: The Right to Decide

Indigenous peoples have generations of experience with sustainable wildlife management and conservation practices that incorporate people as a key component in a coupled natural-human ecosystem. Tragically, in many countries throughout the world, Indigenous peoples have been subjected to systematic social injustices that include displacement from their native land and forced relocation, policies to eliminate Indigenous governments and social institutions, and often total disregard for Indigenous rights. These injustices have led to poor relationships and mistrust of government institutions. But some governments are now making efforts to reconcile the historical injustices, broken treaties and promises, and the exclusion of Indigenous peoples from decisions about how their traditional territories are managed.

Te Urewera National Park is an Indigenous Protected and Conserved Area in New Zealand that is managed by the Tūhoe peoples.

One small step in the reconciliation process has been the formation of Indigenous Protected and Conserved Areas (IPCAs). IPCAs aim to safeguard Indigenous peoples' rights and secure a space where communities can practice their traditional ways of life, while also maintaining biodiversity. IPCAs establish formal agreements between Indigenous groups and national governments. First Nations are given the legal right to exercise free prior and informed consent—that is, the right to manage and steward their territories according to their traditional practices and values, and the right to give or withhold consent for government development projects like mineral extraction and energy exploration. In exchange, national governments can use ICPA agreements not only a means to rebuild relationships with Indigenous peoples, but also to meet their international agreements with the Convention on Biological Diversity (CBD) to protect 17% of their land in protected areas (Aichi Biodiversity Target 11).

ICPAs were first established in the 1998 in Australia when the Nepabunna community and South Australian Aboriginal Lands Trust established the Nantawarrina Indigenous Protected Area.[91] Between 1998 and 2007, Aboriginal Australians and Torres Strait Islanders contributed two-thirds of all new additions to Australia's National Reserve System.[92] Australia would never be able to meet its promises to the CBD if its Indigenous peoples had not been prepared to allocate more than 270,000 km^2 of their land to conservation.

Australia's ICPAs were sufficiently successful that the idea quickly spread to other countries. In 2014, New Zealand's federal government returned Te Urewera, a former national park, to the Tūhoe peoples. This park became its own legal entity, where Tūhoe are the sole decision makers (see Figure). In 2017, a ruling by the Supreme Court of Canada gave the Tsilhqot'in (pronounced Chil-co-tin) legal title to their traditional territories in British Columbia.[93] The Dasiqox Tribal Park, which protects 3120 km^2 of culturally and ecologically important habitat, was granted to the Tsilhqot'in to govern and manage according to their indigenous laws. In 2018, Prime Minister Justin Trudeau established the Canada Nature Fund, which dedicated C$500 million in federal funding to establish even more protected and conserved areas in support of the Pathway to Canada Target 1, a nationwide initiative to achieve Canada's international commitment to the Aichi Biodiversity Targets. The Canadian Indigenous Circle of Experts (ICE) was formed to develop recommendations for establishing a system of IPCAs across Canada to help accomplish Canada's Target 1.[93]

IPCAs are a major success story, and represent one of the fastest growing contributions to protected area networks. Indigenous peoples make up less than 5% of the global population, yet about a quarter of all land outside of Antarctica (approximately 38 million km^2) is managed or controlled by Indigenous peoples[94] (this includes some two-thirds of the world's most remote and least inhabited regions). It may come as a shock to learn that 40% of lands listed by national governments as being managed for conservation are Indigenous lands.[94] IPCAs represent an important paradigm linking conservation, traditional culture, and indigenous rights, and they build partnerships with people who have thousands of years of experience with sustainable land management and biodiversity conservation.

parts have focused on the properties that are desired in protected area networks (or reserve networks), which represent collections of protected areas throughout a landscape (Figure 15.2B). Properties of both the individual protected areas and the network as a whole influence the total conservation benefits achieved in a landscape. Therefore, it is worth laying out some general guidelines that are routinely used to select new protected areas, and to design protected area networks. Here we summarize 10 such guidelines:

1. **Protected areas should be selected/designed to be as large as possible.** Because population size is the single best predictor of extinction risk, and population size is directly proportional to geographic area, large protected areas are usually better at achieving conservation goals than small geographic areas.[3,15,16]
2. **Individual protected areas should be selected/designed to minimize edge effects.** Protected areas with geometries that maximize interior to edge ratios will minimize edge effects that are detrimental to many species and ecosystem services[16] (see Chapter 9).
3. **Areas that have irreplaceable and/or vulnerable biological features should receive high priority for protection.** Some areas have unique biological features, such as rare or threatened species, communities, or ecosystems that cannot be conserved by the inclusion of other sites in a landscape. Because these aspects of biodiversity could be permanently lost from a landscape if not conserved, these areas are almost always given higher priority for conservation than areas with more common biological features.[17] Similarly, areas that are particularly vulnerable, such as habitats facing immediate threat of destruction, are often given higher priority for conservation than those that are not facing an immediate threat.
4. **Protected areas should completely encompass the conservation unit.** Areas that offer complete protection of the species, biological community, or ecosystem that is the focus of conservation efforts have a greater chance of success than areas that only partially encompass the conservation unit.
5. **Whenever possible, protected areas should be designed with buffer zones.** Buffer zones create a gradient of protection around a protected area that enhances conservation goals. Human activities and resource use in buffer zones are usually restricted, but to a lesser extent than in the core of the protected area. The transitional space established by buffer zones serves as multiuse habitat that serves to "soften" any impact that people might have on biodiversity in the core, thus providing a gradient of protection.
6. **Humans should be included in, not excluded from, the design of protected areas.** As discussed in Chapter 14, excluding people from protected areas often has long-term negative impacts on success, particularly when injustices (e.g., displacement) and negative impacts on human well-being (e.g., loss of resources or livelihoods) erode community support for conservation efforts. There is growing evidence that protected areas are more successful over the long term when designed as integrated conservation and development projects that seek to meet the goals of conservation while, at the same time, dealing with the social and economic needs of people. There is, of course, a balance that requires regulations and oversight on the use of natural

resources to ensure protected areas remain effective. Buffer zones (point 5) can help achieve this balance by managing for multi-use habitat at the exterior, while also protecting the core.

7. **Networks should be designed to maximize the number and/or areas of protected areas in a landscape.** As a general rule, landscape mosaics that have a great number of protected areas, or more total area protected, are better at accomplishing conservation goals than landscapes that have fewer numbers or area. It is, however, important to note that the relationship between the amount of area protected and the amount of biodiversity protected is often non-linear, and gives diminishing returns. For example, the species-area relationship (SAR) is an empirical generality that shows how species richness S increases as a power function of habitat area A raised to a scaling constant Z ($S = CA^Z$; see Chapter 8). This power function predicts that that the marginal returns for conservation (e.g., the number of species protected per area conserved) are large when the amount of protected area is small (**Figure 15.3**). But protected areas give diminishing marginal returns (fewer species conserved per protected area added) when there is an increasing number of them, or when total area under protection increases.

An example of diminishing returns comes from González-Maya and colleagues,[18] who studied the National Protected Areas System (NPAS) of Costa Rica, which currently includes more than 190 protected areas. The initial establishment of protected areas in the 1950s and 1960s led to substantial increases in the protection of mammal species, with approximately 82% of native mammal species represented in at least one protected area by 1975 (**Figure 15.4**). After 1975, however, the addition of new protected areas to the NPAS did not increase the number of mammal species represented, which plateaued at just over 200 species and remained constant through 2014. But unlike the number of species represented, the percentage of mammal species with geographic ranges that overlapped protected areas continued to increase through 2000, after which it also plateaued and has remained constant as new protected areas have been added to the NPAS. These results highlight the fact that there is not necessarily

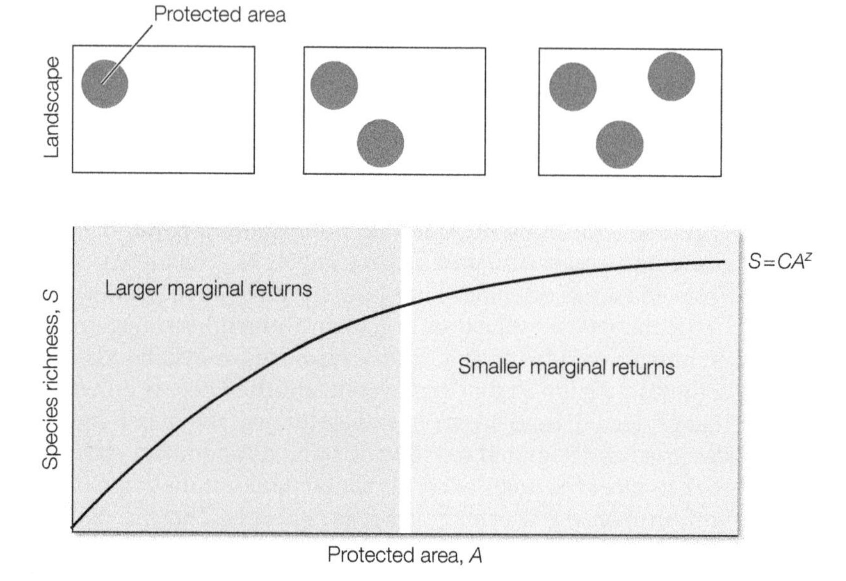

Figure 15.3 The species-area relationship predicts that species richness S will increase with protected area A to the power of Z, as $S = CA^Z$. This function suggests that species richness increases rapidly as new protected areas are added to a landscape, initially giving large marginal returns (number of species per additional area). However, as more protected areas are added, the marginal returns diminish, and fewer species are added per additional area.

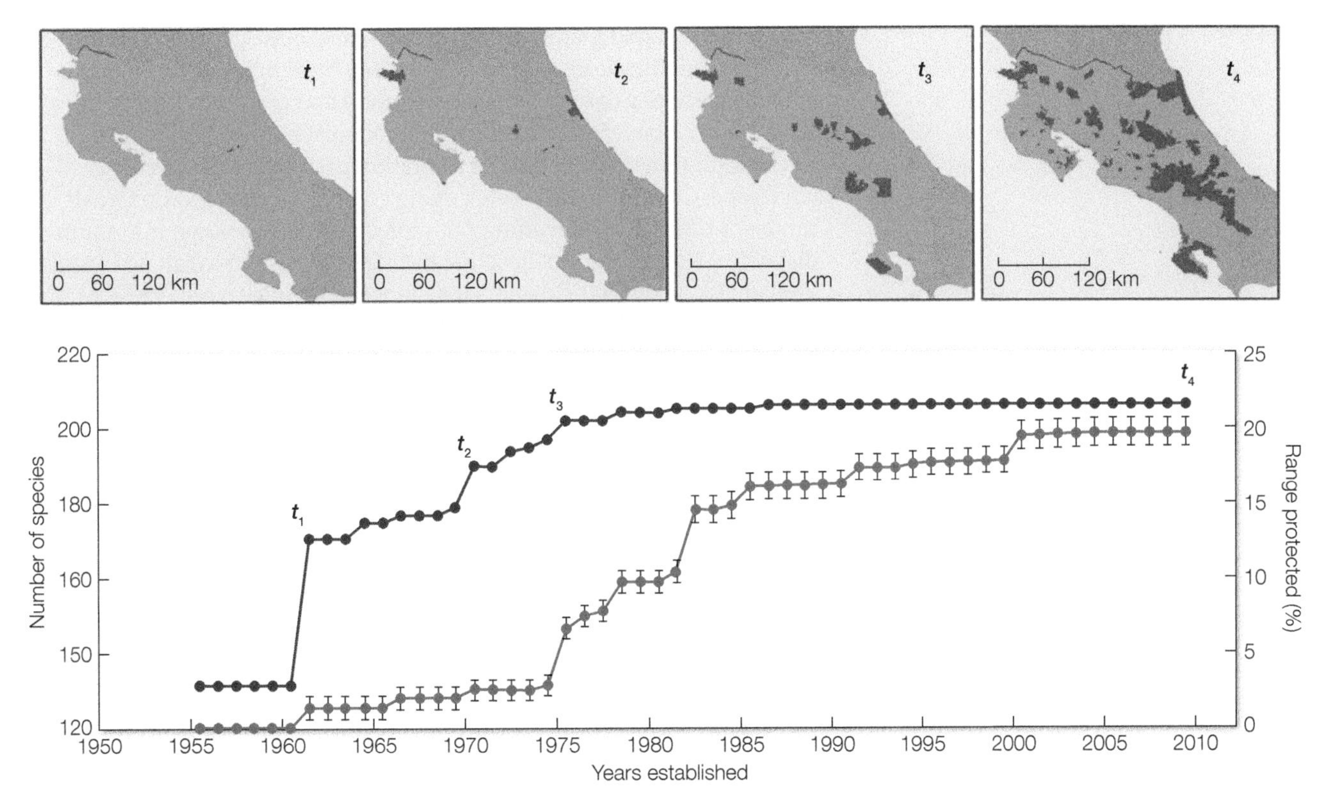

Figure 15.4 Nonlinear relationship between amount of protected area in a landscape and biodiversity. The chronology of protected areas established as part of Costa Rica's National Protected Areas System is shown on the *x*-axis. The number of mammal species represented within the protected areas is shown in the left-hand *y*-axis (red line). The percentage of species ranges that are within protected areas is shown on the right-hand *y*-axis (blue line represents means ± one standard error for various species). Maps show the locations of protected areas at four time points (t_1 = 1960, t_2 = 1970, t_3 = 1975, and t_4 = 2009. (After J. F. González-Maya et al. 2015. *PLOS ONE* 10: e0124480/CC BY 4.0.)

a linear relationship between the area protected in a landscape and the outcome for conservation. Because of this, the practitioner must decide whether the marginal returns on biodiversity are worth the purchase and management costs of adding new protected areas to a network.

8. **Networks should be designed so that the linked protected areas are in close proximity and well connected to allow for optimal dispersal of the focal organisms.** Optimizing conservation at the landscape scale requires that connected protected areas function biologically as a whole. This is necessary to allow individuals to disperse across a conservation landscape, which is important for several reasons. First, dispersal is often needed for species to maintain their genetic diversity and viable population sizes (see Chapter 13) that allow for persistence in metapopulations (see Chapter 9). Second, large, mobile species often require more space than afforded by an individual protected area in order to complete their life cycles or meet their biological requirements (territory size, breeding and feeding habitats). Third, as climate change alter the biogeographic distributions of species (see Chapter 12), migration will be required for species to find new and hospitable habitats.

In some situations, having a well-connected network of protected areas can run counter to other conservation goals, such as the need to minimize certain types of risk. Such risks, like those posed by catastrophes (floods, fires, hurricanes, disease, alien invasive species) or illegal activities (e.g., habitat destruction, hunting), can be elevated when protected areas are in close proximity and connected.[19] In these instances, the benefits of minimizing risk must be weighed against the biological benefits of having migration pathways through the landscape. Often the trade-offs among competing conservation goals can be optimized by spreading out protected areas by some minimum distance—that is, the smallest distance needed to achieve an acceptable level of risk, such as the dispersal distance of an invasive species or disease vector, or the maximum extent of a potential fire or flood.

9. **Protected area networks should be designed to conserve the full range of biodiversity in a landscape.**[20] Each element of biodiversity, whether it be at the level of genes, species, communities, or ecosystems, should be represented in at least one protected area in the network.[21] Accomplishing this may be difficult in practice since, even with the best designs, protected area networks typically cover only a fraction of the landscape and a subsample of its biodiversity. Nevertheless, conservation biologists often prioritize areas with biological features that are underrepresented in a network compared to their representation in a landscape.

10. **All protected areas in a network should be managed collectively as a whole.** In order to realize the full potential of conservation at a landscape scale, the protected areas in a network need to be managed consistently and with a coordinated plan. If the protected areas are managed inconsistently through time, or are managed for differing goals, then management plans at one protected area might conflict with the plans of another. Because protected area networks often span multiple jurisdictions and stakeholder groups who have vested interest in how natural resources are managed across a landscape, management styles that are collaborative and based on principles of **co-design** (or **participatory design**) are needed to ensure that all stakeholders participate in the development of a coordinated management plan.[22] We discuss such management styles later in this chapter.

Suggested Exercise 15.2 Use this online exercise to practice designing a protected area network that conserves multiple species that have different habitat requirements. **oup-arc.com/e/cardinale-ex15.2**

In addition to these ten guidelines for prioritizing new protected areas and designing networks, it is important that a protected area network be adequate to ensure the persistence of biodiversity over the long term. There is little benefit in adding additional protected areas to a network if they will not be maintained and managed now and into the future. A practitioner must consider practical issues that may limit success, such as whether there are sufficient funds, political will, and community support not only to acquire and protect lands, but also to regulate and manage the protected areas.

In addition, practitioners must deal with the fact that many protected areas are established not based on the guidelines given above, but rather because the land to be protected has no immediate value, represents areas where few people live, and/or is considered unsuitable or too remote for agriculture, logging, urban development, or other profitable human activities. Other protected areas are added to networks as unplanned, opportunistic gifts. While these types of protected areas often lead to suboptimal network designs,[23] practitioners must still incorporate them into their conservation management plans, and seek to augment and improve those plans with coordinated efforts that use clear rationale and methodologies for selection and addition of new areas.

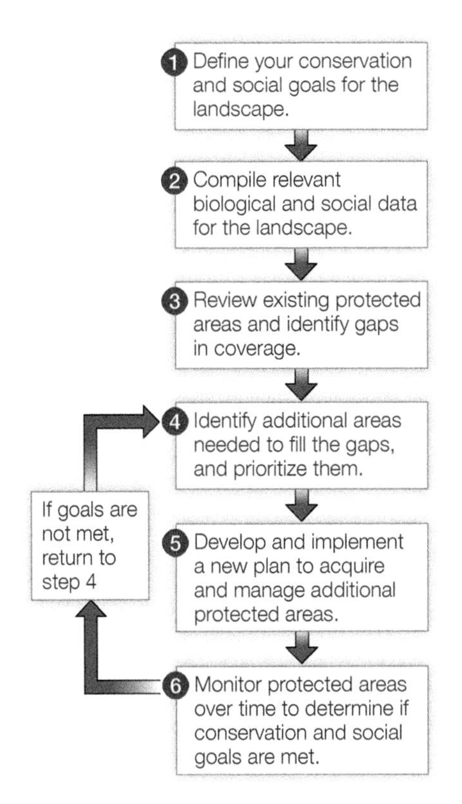

Figure 15.5 The six steps of a gap analysis.

Tools used to optimize selection of protected areas

Rigorous tools, techniques, and software packages are available to help practitioners select protected areas and optimize the design of protected area networks. Most of these tools, techniques, and software packages begin with a gap analysis. A **gap analysis** is an assessment of how close an existing network of protected areas comes to meeting a set of defined conservation and social goals, followed by a process of identifying and prioritizing additional protected areas that are required to achieve those goals. The gap analysis process generally requires six steps (**Figure 15.5**):

1. **Define conservation and social goals for the landscape.** Conservation goals could, for example, include the amount of area to be protected for each ecosystem in the landscape, or the population sizes of rare and threatened species to be protected. Social goals might account for the proportion of landscape dedicated to agricultural, urban, and protected habitats, as well as protection of socially important areas (e.g., important recreational habitat, areas of high human density or particular economic importance).
2. **Compile relevant biological and social data for the landscape.** Biological data includes geographic information about the presence and distribution of species, ecosystems, and physical features of the landscape, which are sometimes referred to as conservation units. Social data might include information on land use and habitat types, human densities, and economic factors.
3. **Review existing protected areas and identify gaps in coverage.** In this step, currently protected areas are reviewed and summarized to determine what goals have already been achieved in the landscape, and which goals have not been achieved with current levels of protection, such as species or biological communities not currently protected (the "gaps").
4. **Identify unprotected areas that "fill the gaps" and use an appropriate decision-making algorithms to prioritize them.** Decision-making algorithms are simply formal ways for practitioners to define their priorities for selection (**Box 15.2**).
5. **Develop and implement a new plan to acquire and manage additional protected areas.** The additional areas are reviewed in more detail and, if appropriate and practical, they are protected in some way (often by being directly purchased or designated as national parks). Management plans are then developed and implemented.
6. **Monitor new protected areas to determine whether they are meeting their stated goals.** If not, the management plan can be changed or, possibly, additional areas can be acquired to meet the goals.

Steps 2–4 are usually completed using software programs that have been specifically designed to aid practitioners in the design of protected area networks. Many of these software programs rely on geographic information systems (GIS), as a framework for gathering, managing, and analyzing spatial

BOX 15.2 Conservation in Practice

Decision-Making Algorithms

How do you prioritize which new areas to add to a protected area network? While gap analysis lays out a process by which new areas can be selected, decision-making algorithms used by various gap analysis software programs ultimately determine how areas are selected and prioritized.[95] In order to perform gap analyses appropriately, practitioners must understand what various decision-making algorithms do, and what goals they are set up to accomplish.

There are many different decision-making algorithms; however, two are particularly relevant because they optimize classic problems faced when designing a protected area network. These are (1) the minimum set problem and (2) the maximal coverage problem. These problems are described next, along with the mathematical functions that software programs try to optimize. After describing the problems, we briefly discuss some other types of algorithms that are used for optimization.

Minimum set algorithm

The minimum set problem assumes there is a feature, or set of features, that you want to protect. These features could be anything (species, biological communities, entire ecosystems). But for purposes of this example, let us say you want to conserve a set of threatened species. There are a set of conservation units available to you, which we will call *sites*. Your goal is to find a set of sites that will protect all of the threatened species at the lowest possible cost (e.g., money).

Assume that i is an index for different species you wish to protect (e.g., $i = 1$ to S species), j is an index for the sites that are available (e.g., $j = 1$ to N sites), and c_j is the cost associated with each site j (e.g., \$ per site). The goal of the minimum set algorithm is to minimize the following expression:

$$min \sum_{j \in J} c_j x_j$$

subject to the constraint:

$$\sum_{j \in J} a_{ij} x_j > r_i$$

In both equations, x_j is an index that keeps track of whether or not site j is selected (0 if site j is not selected, 1 if it is selected). The term a_{ij} represent an $i \times j$ matrix of all possible species i by sites j. The term r_i in the second equation tallies whether the objective has been met, meaning that a chosen site j has a threatened species i (e.g., 1 site for each i). When the condition of this equation is met—that is, when a site containing a threatened species is found—then the first equation adds that site to the running tally if it also helps to minimize the total cost.

Maximal coverage algorithm

The maximal coverage problem assumes that you have a fixed amount of total money T available to you. The goal now is to purchase the set of sites that will maximize the number of protected species while staying below the total available funding, T. To accomplish this goal, the maximal coverage algorithm attempts to maximize the following expression:

$$max \sum_{j \in J} r_i$$

subject to the constraint:

$$\sum_{j \in J} c_j x_j \leq T$$

The second equation sums up the set of sites x_j that allow costs c_j to stay below the total T, while the first maximizes the number of species i that are found in those sites.

Other types of algorithms

In addition to the minimum set and maximum coverage algorithms, there are dozens of additional mathematical functions that practitioners can choose in order to accomplish their specific goals, such as selecting new areas that maximize protected area size or connectivity in a network, or that minimize edge effects. Regardless of what function is used to accomplish a given goal, nearly all software programs use one of three types of algorithms to optimize (maximize or minimize) those functions. While a full discussion of algorithms and how they work is an advanced topic and beyond the scope of this text, readers should be at least broadly familiar with three common types:

1. **Iterative heuristic algorithms** step through a list of sites, choosing the best site at each step according to the rules given by the functions. These algorithms implicitly consider complementarity among sites, as the contribution of unselected sites to meeting the conservation objectives is recalculated each time a site is added to the system of protected areas.

2. **Greedy heuristic algorithms** add sites to a protected area network by sequentially selecting the set of sites that add the most unprotected species to the already-selected set. Each step in these "greedy" heuristics maximizes progress toward some desired objective, such as maximizing the number of threatened species protected.

3. **Global heuristic algorithms**, like the simulated annealing algorithm used in popular software programs like Marxan, begin by generating a random selection of protected areas. One site is randomly chosen for addition or removal during each iteration (the site may or may not be in the protected system already). With each addition or removal, the algorithm evaluates the change in value to the reserve system. If negative changes are accepted early in the process, the algorithm allows the system to move temporarily through suboptimal solutions. The advantage of allowing both negative and positive changes is that it helps the algorithm avoid becoming "trapped" in a local optimum, which may or may not be the best solution globally (i.e., across all possible choice of different sites).

data. By digitizing the dates and times of occurrence, as well as the *x*-, *y*-, and *z*-coordinates (i.e., longitude, latitude, and elevation) of spatial information, GIS allows spatial data to be mapped, analyzed, and compared among different sources of information. **Figure 15.6** shows how GIS can be used to facilitate gap analysis by overlaying maps that represent different sources of information. Spatial information on habitat types in a landscape can be mapped onto the distribution of threatened animal species in order to show the relationships between habitat types and species distributions. These maps can subsequently be overlaid onto a map of existing protected areas, which can then be used to identify gaps in the protection of species—simply, areas where species and their distributions are not protected.

Spatial information can be stored in any number of GIS file formats. Some of the more common formats use vectors to represent geospatial data as points, lines, or polygons:

- **Geodatabase** is the format recommended by the Environmental Systems Research Institute (ESRI) for use in its popular GIS software ArcGIS.
- **Shapefile** originated with ESRI's old ArcView software and remains a common format used by many software programs. The Shapefile format is actually made up of at least three files with the same name but different extensions: *.shp, *.shx, and *.dbf.
- **KML** stands for Keyhole Markup Language and is the file format used by Google to view geographic data in Google Earth.
- **OpenStreetMap** is a crowdsourcing GIS data project that uses either XML-based file formats or smaller, more efficient formats called Protocolbuffer Binary Formats (PBF).

Spatial information stored in any of these formats is readily available on the internet and can be downloaded and used in software packages for conservation planning. (For a link to a list of some of the more commonly used sources of GIS data go to oup-arc.com/e/cardinale-w15.2). In addition to existing files, it is relatively straightforward to gather spatial information from surveys, censuses, aerial photographs, and/or satellite imagery and then convert those data to any of the above GIS file formats for use in specific conservation projects.

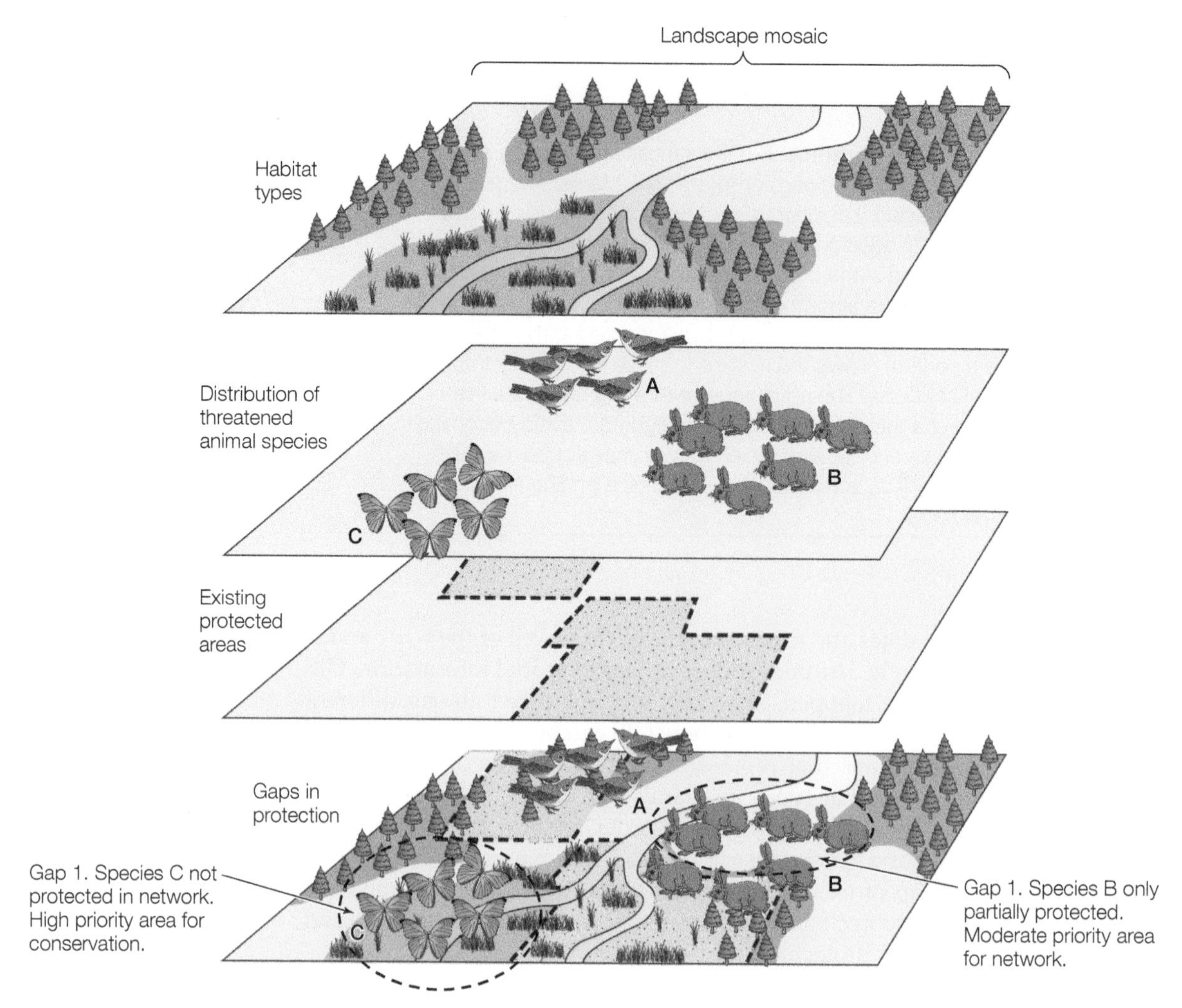

Figure 15.6 Geographic information systems (GIS) provide a method for integrating a wide variety of data for analysis and display on maps. In this example, vegetation types, distributions of endangered animal species, and protected areas are overlapped to identify areas that need additional protection. The overlapped maps show that the distribution of species A is predominantly in a protected area, species B is only partially protected, and species C is found entirely outside of the protected areas. Establishing a new protected area to include the range of species C would be the highest priority. (After J. M. Scott et al. 1991. In D. J. Decker et al. [Eds.], *Challenges in the Conservation of Biological Resources: A Practitioner's Guide.* Boulder, CO: Westview Press.)

Conservation planners often supplement their use of GIS with additional software that allows them to use decision-making algorithms or other planning tools to sort through biophysical, social, and economic data to identify gaps and optimal designs for protected area networks. Table 15.1 lists some common conservation planning software programs. **Marxan** is perhaps the most widely used program for designing protected area networks in landscapes. The software was initially developed by a PhD student named Ian Ball, who was working with Hugh Possingham, now chief scientist for The Nature Conservancy. Once Marxan was developed, the Great Barrier Reef Marine Park Authority in Australia used the software to help design the expansion of no-take areas in a reef system that covers 344,000 km^2 (almost the size of California) and consists of about 17,000 planning units.[24] The Authority wanted to identify the most cost-effective way to increase the area of no-take zones so that roughly 20% of each ecosystem type or feature was

TABLE 15.1 Examples of commonly used conservation planning software programs

Software	Description	Tutorials
Marxan oup-arc.com/e/cardinale-w15.3	Marxan assists systematic reserve design on conservation planning, using stochastic optimization routines (Simulated Annealing) to prioritize protected areas to achieve a set of conservation goals with reasonable optimality. Marxan is the most widely used reserve planning software, and has been used extensively by The Nature Conservancy and the World Wildlife Fund.	oup-arc.com/e/cardinale-w15.4
CLUZ oup-arc.com/e/cardinale-w15.5	The Conservation Land-Use Zoning (CLUZ) software is a freely available plug-in for QGIS (an open source GIS application) that allows users to design protected area networks and other conservation landscapes and seascapes. It can be used for on-screen planning and also acts as a link for the Marxan conservation planning software.	oup-arc.com/e/cardinale-w15.6 and oup-arc.com/e/cardinale-w15.7
Prioritizr oup-arc.com/e/cardinale-w15.8	Prioritizr, a package developed for the open source software *R*, uses integer linear programming (ILP) to build and solve conservation planning problems. It supports a broad range of objectives, constraints, and penalties that can be used to customize conservation planning problems. In contrast to algorithms conventionally used to solve conservation problems, the exact algorithms used by Prioritizr are guaranteed to find optimal solutions.	oup-arc.com/e/cardinale-w15.9 and oup-arc.com/e/cardinale-w15.10
Zonation oup-arc.com/e/cardinale-w15.11	Zonation is a freely available decision-support software for land use planning that includes applications for spatial conservation planning and ecological impact avoidance. The program operates using spatial data about any aspect of biodiversity species, habitats, and ecosystem services, along with data on various costs and threats. It is able to factor in information about uncertainty and ecological factors such as connectivity, and can prioritize many dimensions of biodiversity.	oup-arc.com/e/cardinale-w15.12
Miradi oup-arc.com/e/cardinale-w15.13	Miradi is a subscription-based software program that helps practitioners design, manage, and monitor networks of protected areas. Miradi was co-designed by members of The Conservation Measures Partnership (CMP) that includes Conservation International, The Nature Conservancy, World Wildlife Fund, and the U.S. Fish and Wildlife Service.	oup-arc.com/e/cardinale-w15.14

protected within a no-take zone, while at the same time preserving important biological processes such as larval dispersal, and also accommodating commercial fishing, recreational needs, and political and financial realities. They used Marxan to help work through the data, and then followed an iterative process in which they discussed the model outputs with managers and other stakeholders to assess their practicality, refined the models, and continued through another cycle.[25] In the end, they expanded the amount of no-take area from 5% of the reef system to 33%.

Besides Marxan, there are numerous other software programs that can incorporate spatial information on biodiversity (e.g., species distributions) with data on natural habitats (e.g., vegetation or soil types), climate (e.g., precipitation or temperature), patterns of land use (e.g., human developments and zoning), or any other factor that one can obtain spatial information for. Most of these programs rely on decision-making algorithms (see Box 15.2) to identify gaps in protected areas, and to identify new areas for protection that will accomplish conservation goals and/or optimize the trade-offs among potentially competing goals. Output is usually in the form of maps showing areas that should be protected to fulfill a variety of conservation goals, whether the goal is to protect irreplaceable locations, target species, biodiversity hotspots, or representative ecosystem types.

(A)

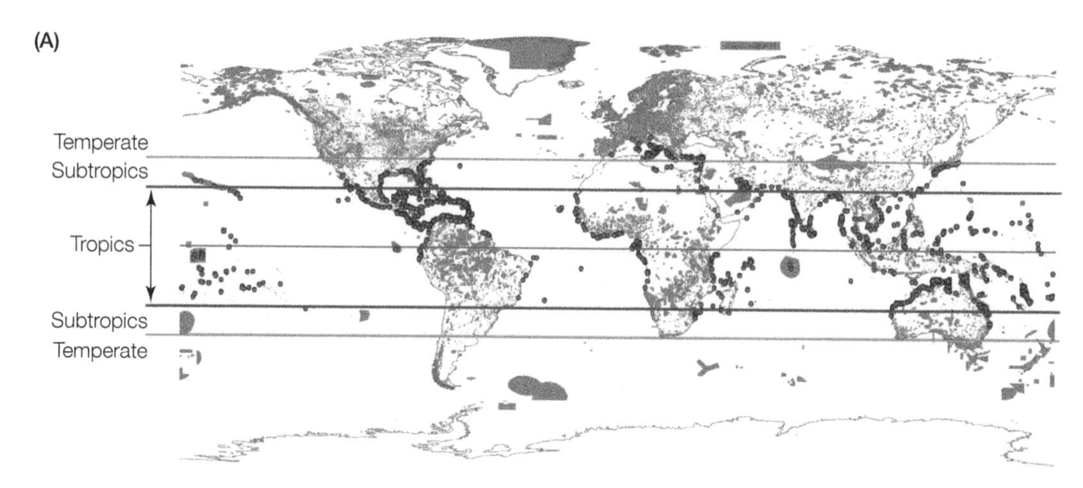

(B)

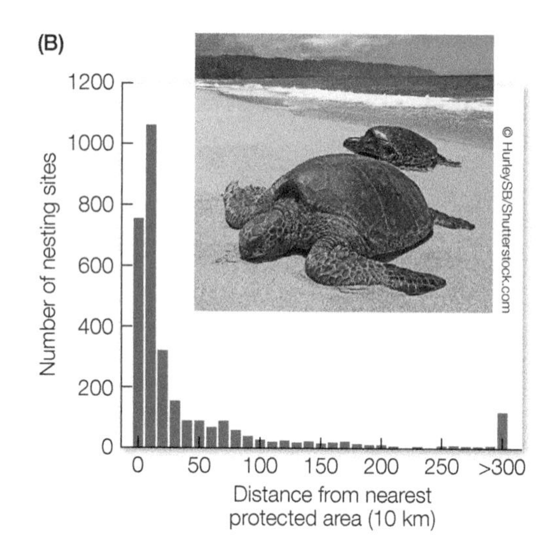

Figure 15.7 A gap analysis for sea turtles. (A) Sea turtle nesting sites (red circles) are almost entirely located in the tropics and subtropics, and comparatively few are located within protected areas (green shading). (B) Number of sea turtle nesting sites located within and varying distances from protected areas using 10-km grids. (From A. D. Mazaris et al. 2014. *Biol Conserv* 173: 17–23.)

Many countries have completed gap analyses at national scales. For example, the Gap Analysis Program (GAP) is a nationwide program in the United States that is directed by the U.S. Geological Survey and implemented in coordination with state and regional programs. The program began in the 1980s when Michael Scott, a professor at the University of Idaho who helped pioneer gap analysis for wildlife conservation, collected data on the geographic distributions of finches throughout Hawaii, and then overlaid those data onto a map of protected areas. He found little overlap between finch distributions and the protected areas, revealing large gaps in their conservation.[26,27] Results of his analysis led to creation of the Hakaiau Forest National Wildlife Refuge, one of Hawaii's most species-rich habitats. Given its success in Hawaii, the U.S. Geological Survey launched GAP nationwide in 1989 with the goal of providing geographically explicit information on the distribution of native vertebrate species, their habitat preferences, and their management status in order to determine gaps in biodiversity protection. GAP has produced national land cover and protected areas datasets for more than 2000 species of mammals, birds, reptiles, and amphibians in the U.S.

Gap analyses have also been used to identify holes in conservation at international and global scales. For example, in 2004 Rodriquez and colleagues[28,29] published a set of papers in which they overlaid distribution maps for 11,633 terrestrial mammal, bird, turtle, and amphibian species onto the global distribution of protected areas maintained by the World Database on Protected Areas. Their analysis revealed that 1424 species (about 12% of those analyzed) had distributions that were not covered by any protected area on Earth, and an additional 12% were only partially covered in areas that were small (less than 1000 ha) or that were in multiuse protected areas (IUCN category V and VI; see Chapter 14).

Global gap analyses have also been used to assess the conservation status of particular groups of organisms that are threatened.[30,31] For example, Mazaris and colleagues[30] used a global dataset of 2991 sea turtle nesting sites to determine the extent to which the global network of protected areas covers the egg-laying grounds of seven threatened sea turtle species (loggerhead, *Caretta caretta*; green, *Chelonia mydas*; hawksbill, *Eretmochelys imbricata*; Kemp's ridley, *Lepidochelys kempii*; olive ridley, *Lepidochelys olivacea*; flatback, *Natator depressus*; and leatherback, *Dermochelys coriacea*). Their analyses showed that protected areas (mostly marine protected areas) cover just 25% of all sea turtle nesting sites (**Figure 15.7A**), with the remaining 75% (2240) being located at varying distances from protected areas (**Figure 15.7B**). Nesting sites used by loggerhead and flatback turtles receive the highest protection, with about 35% of all sites located within existing protected areas; from 20% to 27% are protected for the other five species. More than 80% of sea turtle nesting sites (2459) occur in tropical countries that have developing economies, which afford far less protection (22% of these sites are in protected areas) than nesting sites covered by protected areas in developed countries (38%). Gap analyses like these make clear where priorities lie for establishing additional protected areas.

Connecting Individual Protected Areas into Networks

Optimizing conservation at a landscape scale requires that protected areas are well connected and function as a whole. Connectivity is required for the persistence of animals that have large territories or range sizes, species that migrate to complete their life histories, or subdivided populations that persist in a landscape as a metapopulation. Therefore, as protected areas are selected and prioritized for incorporation into a network, it is important to consider how they will be connected to each other.

The two primary tools that conservation biologists use to connect protected areas into a network are habitat corridors and stepping stones (**Figure 15.8**). A **habitat corridor** is a continuous strip of protected habitat that runs between and connects two protected areas.[32,33] **Stepping stones** are discontinuous pieces of habitat that fill in gaps between two protected areas and help facilitate movement.[34] Habitat corridors and stepping stones allow plants and animals to disperse from one protected area to another, which can be important for maintaining genetic diversity and viable population sizes of species that persist in a landscape as metapopulations. Corridors and stepping stones also help conserve populations of migratory animals. Several case studies are useful for illustrating the concept and practical applications of habitat corridors and stepping stones.

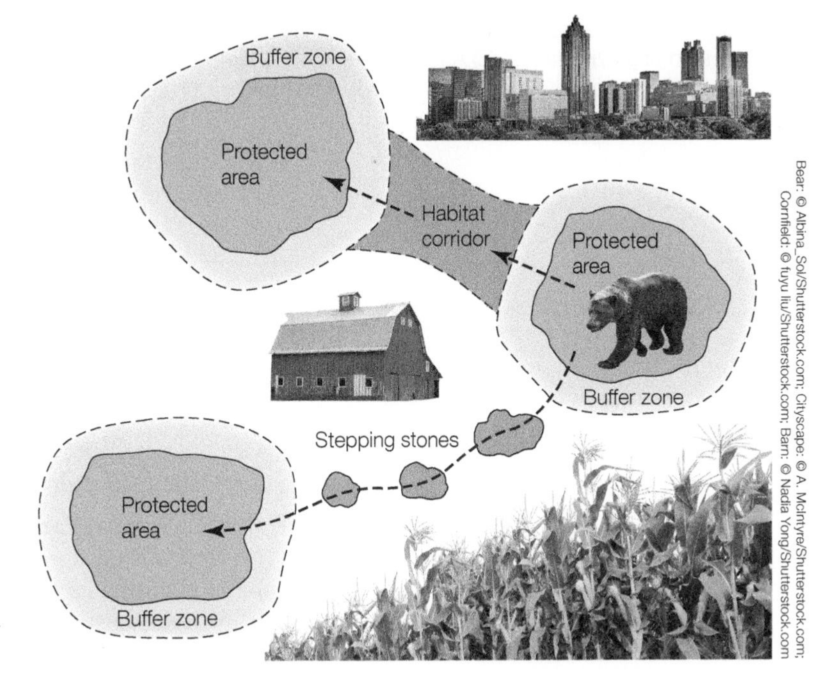

Figure 15.8 Corridors and stepping stones are the primary tools used to connect protected areas in a landscape network.

(A)

© Hemis/Alamy Stock Photo

(B)

© iStock.com/pics-xl

(C)

From R. G. Seidler et al. 2018. *Glob Ecol Conserv* 15: e00416

(D)

Parks Australia/CC BY 3.0 AU

Figure 15.9 Examples of corridors to facilitate wildlife dispersal. (A) In Kenya and Tanzania, wildebeest herds migrate annually through the Maasai Mara and Serengeti reserve corridor in search of grazing lands. (B) A fenced-off highway directs wildlife to an overpass that allows them to safely disperse through Banff National Park, Canada. (C) A highway overpass (on the left in this photo), part of the Path of the Pronghorn wildlife corridor in North America. (D) Red crabs use a tunnel to cross under a road safely during their annual migrations on Christmas Island, New Zealand.

- **Maasai Mara and Serengeti.** Two protected areas, the Serengeti National Park in Tanzania and the Maasai Mara National Reserve in Kenya, together cover 16,260 km^2 of area that help protect the migration route of roughly 1.2 million wildebeests (*Connochaetes* spp.), zebras (*Equus* spp.), and gazelle (*Gazelle* spp.) that migrate annually from short-grass plains in the southern Serengeti to the grasslands and savannas of the Maasai Mara. Virtually the entire migration occurs within the protected areas—the only mammal migration in Africa, and one of the only ones in the world, that is completely protected (**Figure 15.9A**).[35] As a result, wildebeests are one of the only migratory mammals not to have declined in abundance in recent years, even as populations in unprotected areas have declined.
- **Banff National Park.** In many areas of the world, roads are the single greatest obstruction to animal movement through a landscape.[36,37] In Banff National Park, the Canadian government has worked to overcome these obstructions by building a variety of underpasses and overpasses across the four-lane Trans-Canada Highway (**Figure 15.9B**). These structures are combined with fences along the road that direct wildlife

movement and force animals to cross the highway at specific locations. Thirteen recent structures were evaluated for their use by wildlife, which was measured by camera traps and animal tracks left in raked beds of soil.[38,39] Certain mammals, such as grizzly bears, wolves, elks, and deer, used wide overpasses, whereas narrow underpasses were favored by black bears and cougars. Cougars favored crossings with vegetation cover, but grizzly bears, elk, and deer preferred more open habitats. These results demonstrate that a mixture of crossing types and associated vegetation covers are needed to allow connectivity across road barriers. An added benefit of these passageways is that road collisions involving deer, elk, and other large mammals declined by 96% after fences, overpasses, and underpasses were installed, thus saving lives and money.

- **Pronghorn in North America.** Pronghorn (*Antilocapra americana*) represent one of North America's last surviving endemic ungulates. These animals exhibit one of the longest terrestrial migrations of species in North America. Each autumn when snows begin to fall, herds of 300 to 500 animals depart their summer feeding grounds in the high valleys of Grand Teton National Park and begin to migrate 200 km south, to their winter feeding grounds in the Green River Basin of Wyoming.[40] Females make the migration while pregnant, giving birth in the spring (commonly to twins), and raising the offspring over the summer. The pronghorn's historical migration route has been compromised by numerous obstructions, including highways, housing subdivisions, and ranches with extensive fencing. In response, government agencies like the U.S. Forest and Park Service, nongovernmental organizations like the Wildlife Conservation Society, and private citizen groups joined together to create the Path of the Pronghorn (POP).[41] POP is a wildlife corridor that connects the summer and winter feeding grounds of the pronghorn, designed to allow them to complete their annual migration. The corridor was formed by (1) negotiating migration-friendly management practices with the cooperation of private landowners, (2) constructing overpasses and underpasses that allow pronghorn to move safely across highways, and (3) constructing new fences to direct herds around obstacles (**Figure 15.9C**).[42] Recent research shows that these actions have been largely successful. Pronghorn prefer the overpasses to underpasses, and appear to have modified their migration to use the new, safer route.
- **Christmas Island crab tunnels.** Not all wildlife corridors are designed for large mammals (**Video 15.1**). Christmas Island, Australia, is home to the extraordinary annual migration by red crabs (*Gecarcoidea natalis*). The crabs spend most of their lives in the island's interior forests. But once a year, tens of millions of the crabs migrate to the ocean in order to breed and lay eggs. During this migration, individual red crabs must cross as many as four roads to get to their breeding grounds and then back to forest. Large numbers of red crabs are crushed by vehicles, and they sometimes cause accidents when their tough exoskeletons puncture tires. To ensure the safety of both humans and crabs, local park rangers create aluminum barriers called "crab fences" that funnel the crabs toward underpasses called "crab grids" so they can safely cross the roads (**Figure 15.9D**). Furthermore, 5-m-high bridges have been constructed that help crabs crawl over the roads.

Video 15.1 Watch this video of red crab migration.
oup-arc.com/e/cardinale-v15.1

Other well-known national or international corridor projects include the Terai Arc Landscape, a wildlife corridor in Nepal that allows threatened species like the Asian elephant (*Elephas maximus*), Indian rhinoceros (*Rhinoceros unicornis*), and Bengal tiger (*Panthera tigris tigris*) to migrate through 11 protected areas; the European Green Belt, a natural strip of forests and parks that extends more than 8,500 km connecting the Baltic Sea to the Adriatic Sea, which replaced wire fences and landmines from the Iron Curtain era; and the Zona Protectora Las Tablas, a 77 km^2 forest corridor that allows 75 species of birds to migrate between the two large conservation areas in Costa Rica, the Braulio Carrillo National Park and La Selva Biological Station.

Despite these examples of well-connected networks of protected areas, recent studies suggest there is much room for improvement. Saura and colleagues[43] recently assessed progress towards Aichi Biodiversity Target 11, which not only calls for 17% of terrestrial and inland water to be protected by 2020, but also calls for those protected areas to be well connected in a global network. After developing an index of connectivity based on graph theory (which they called the Protected Connected, or ProtConn, index), Saura and colleagues quantified the percentage of Earth's land surface that is both protected, and connected for species dispersal distances ranging from less than 1 km to 100 km. The authors found that only 8.5% of protected areas could be considered connected for organisms able to disperse less than 1 km, and only 11.7% could be considered connected for organisms with dispersal abilities of 100 km (**Figure 15.10**). Such results show that, while the number of protected areas has increased globally, most of those protected areas are not well connected, and the vast majority of species would not be able to disperse among them naturally.

Although the idea of corridors is intuitively appealing, some have warned that corridors can have unintended negative consequences for conservation efforts.[44,45] In particular, some of the same mechanisms that facilitate species

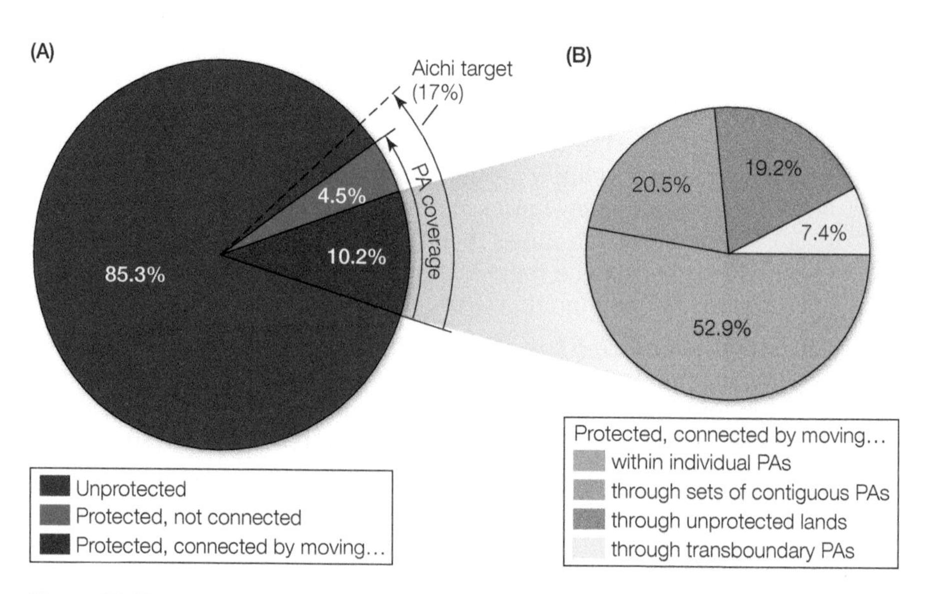

Figure 15.10 Aichi Target 11 calls for 17% of terrestrial land and inland waters to be protected by 2020, and to be well connected in a global network. (A) The percentage of terrestrial land and inland water that has been protected to date (blue + dark purple = 14.7%). Dark purple represents the percentage of protected area that can be considered connected for a species with a maximum dispersal distance of 30 km (10.2%). (B) For the 10.2% of protected areas that are connected, the pie chart at right shows the percentage of area that a species with a 30-km dispersal distance could potentially access. Most protected area (52.9%) is accessible by moving only within individual protected areas. Additional protected area could potentially be reached by moving among contiguous protected areas (20.5%), by moving through unprotected lands to get to new protected areas (19.2%), or by moving through transboundary protected areas (7.4%). (After S. Saura et al. 2017. *Ecol Indic* 76: 144–158.)

dispersal may also increase the spread and impact of predators and pathogens, foster edge effects (see Chapter 9), increase invasion by alien species, facilitate the spread of disturbances such as fire, and/or increase biological asynchrony (see Chapter 12) in ways that reduce species persistence. While examples of these negative effects of corridors exist, a recent review by Haddad and colleagues[46] of 33 studies of corridor effects found no consistent evidence that habitat corridors have unintended negative consequences on wildlife. The authors also found that negative impacts were small when compared to the benefits of facilitating dispersal and maintaining or even increasing species diversity.

Incorporating unprotected areas into conservation plans

Even though establishing a network of protected areas is essential for species conservation, there are several reasons why it would be shortsighted to rely solely on protected areas for conservation at the landscape scale. First, 85% of the world's land and 92% of the world's oceans are currently unprotected under any type of conservation (see Chapter 14). Even the most optimistic projections suggest that the vast majority of Earth's land and water will remain unprotected in the future. It would be unwise to ignore the largest areas of land and water when developing comprehensive conservation strategies. Second, a large fraction of the world's species are not covered by protected areas. In the U.S., for example, 60% of species that are globally rare or listed under the U.S. Endangered Species Act occur on private lands.[47] And finally, many unprotected areas can serve as viable habitat for species, even if those habitats are suboptimal. For all these reasons, conservation efforts at a landscape scale must focus on both protected and unprotected areas.

The ecosystems that make up a landscape span a continuum from natural, uncontrolled wilderness areas that are optimal habitat for many native plants and wildlife, to completely artificial industrial and urban ecosystems that are conducive to only those species that can thrive in close association with humans (Figure 15.11). Between these two extremes of the continuum, there are many ecosystem types that still harbor some of their original biota. These unprotected areas can add value to conservation efforts. For example:

- Forests that are either selectively logged on a long cutting cycle or are cut down for farming using traditional shifting cultivation methods may still contain a considerable percentage of their original biota, and can maintain some of their original ecosystem services.[48,49] In Malaysia, for example, most forest bird species are still found in rain forests 30 years after selective logging has occurred, because undisturbed forest is available nearby to act as a source of colonists.[50] In African tropical forests, gorillas, chimpanzees, and elephants can tolerate selective logging and other land uses that involve low levels of disturbance if hunting levels are controlled by active anti-poaching patrols.[51] In Indonesia, 75% of the remaining orangutans live outside of protected forests, often in logged forests and tree plantations.[52]
- Government-owned lands that are protected and managed for purposes other than biodiversity protection often include habitat that can contribute to conservation. For example, watersheds adjacent to metropolitan water supplies are often protected to maintain water quality. Security zones surrounding government military reservations preserve large areas of natural habitat. The U.S. Department of Defense manages more than 110,000 km^2, much of it undeveloped, containing around 420 threatened and endangered species of plants and animals. The White Sands Missile Range in New Mexico is 10,000 km^2, about the same size as Yellowstone National Park. While sections of military reservations are often damaged by military activities, much of the habitat remains as an undeveloped buffer zone with restricted access.

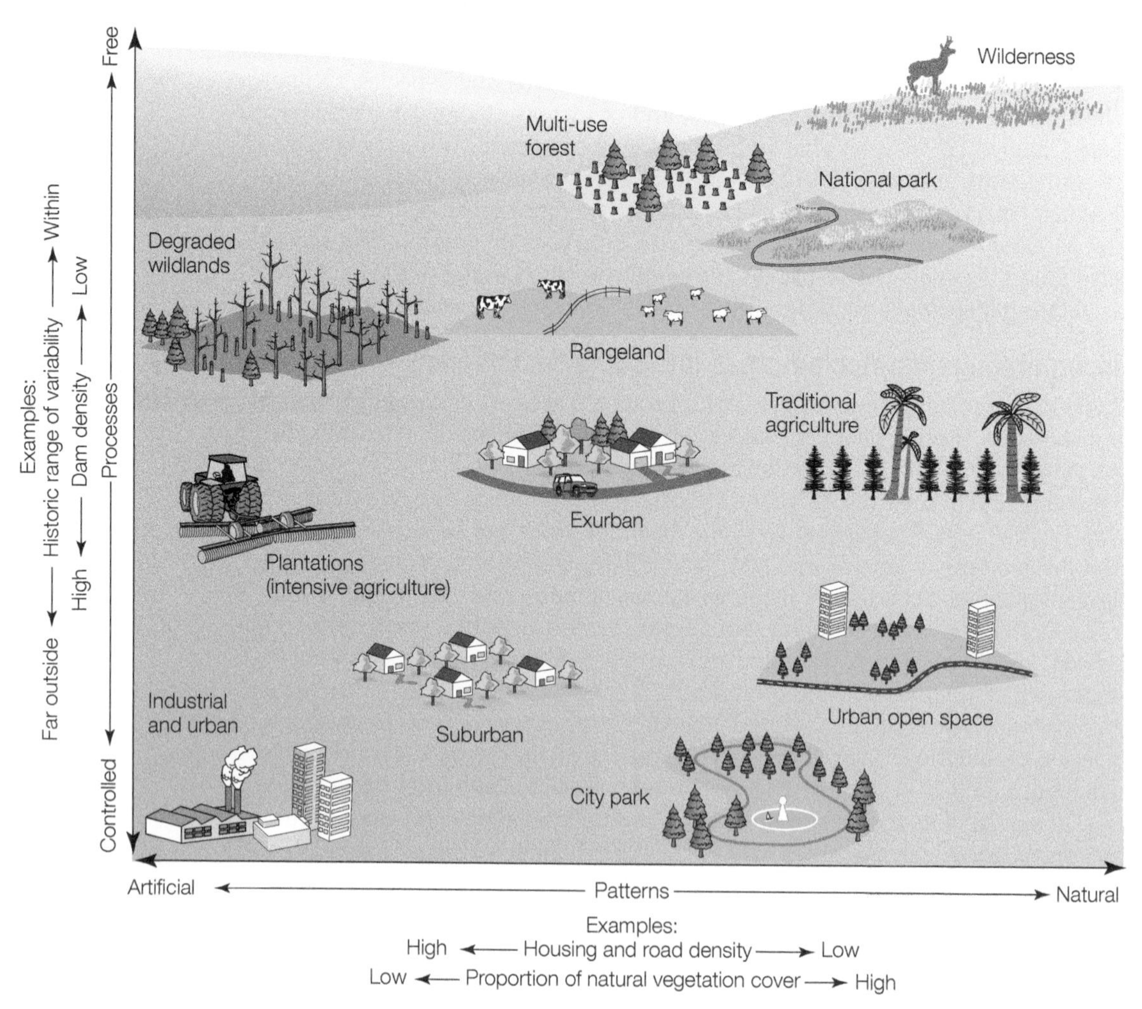

Figure 15.11 The land use continuum. Ecosystems in modern landscapes span a continuum from natural, uncontrolled wilderness areas that are optimal habitat for many native plants and wildlife (upper right), to artificial industrial and urban ecosystems that are conducive only to those species that thrive when in close association with humans (bottom left). In between these two extremes, there are many ecosystem types and unprotected areas that harbor some of their original biota, and which can add value to biodiversity conservation efforts. (After D. M. Theobald 2004. *Front Ecol Environ* 2: 139–144.)

- In many parts of the world, wealthy individuals have acquired large tracts of land for their personal estates and for private hunting. These estates are frequently used at low intensity, often in a deliberate attempt by the landowners to maintain large wildlife populations. Some estates in Europe preserve unique old-growth forests that have been owned and protected by royal families for hundreds of years. Such privately owned lands, whether owned by individuals, families, corporations, or tribal groups, often contain important aspects of biodiversity. Strategies that encourage private landowners and government land managers to protect rare species and ecosystems are obviously essential to the long-term conservation of biodiversity.

In almost every country, there are examples of threatened species or ecosystems that exist primarily or exclusively on unprotected public lands or on lands that are privately owned. For example, the Florida panther (*Puma concolor coryi*) is an endangered species of mountain lion that is protected by the U.S. Endangered Species Act. While conservation areas and public land have been established to help protect the species, radio telemetry studies have shown that the

(A)

(B)

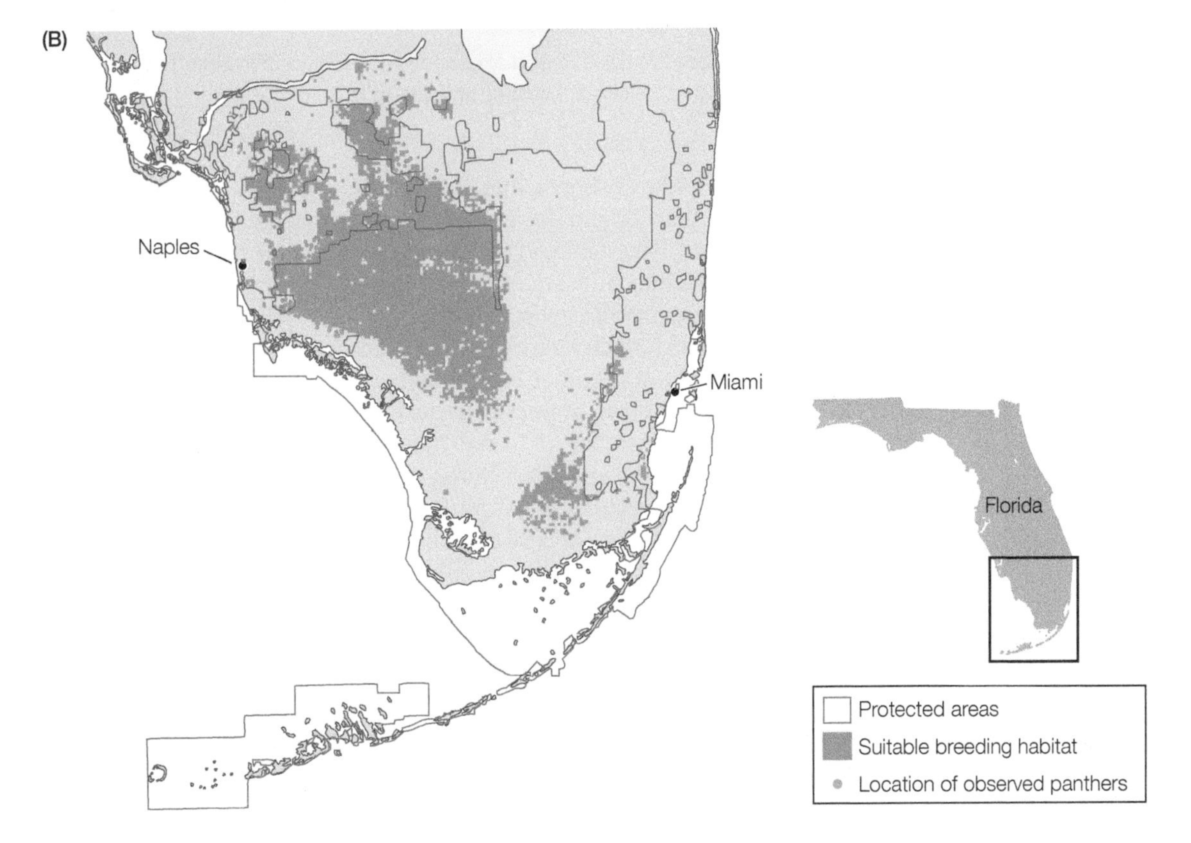

Figure 15.12 The value of unprotected lands for conservation. (A) The Florida panther (*Puma concolor coryi*) moves between public and private lands in South Florida. (B) The orange dots represent radio telemetry records of 17 collared panthers from 2017–2018. (B, after Florida Fish and Wildlife Conservation Commission. 2018. *Annual Report on the Research and Management of Florida Panthers: 2017–2018.* Fish and Wildlife Research Institute & Division of Habitat and Species Conservation, Naples, FL, USA; R. A. Frakes et al. 2015. *PLOS ONE* 10: e0133044/CC0; Florida Natural Areas Inventory. Florida Conservation Lands, April 2019. https://www.fnai.org/webmaps/ConLandsMap/.)

panthers have large territory sizes and home ranges that extend well beyond protected areas; thus, they spend significant portions of their life on private land (**Figure 15.12**). Successful conservation of this species requires managers to incorporate both protected and unprotected lands into their conservation plan.

Clearly, strategies for reconciling human needs and conservation interests in unprotected areas are critical to the success of conservation.[53] A crucial component of conservation strategies must be the protection of biodiversity inside and outside—immediately adjacent to and away from—protected areas.[54,55] Here we discuss opportunities for biodiversity conservation in urban, agricultural, and mixed-use habitats.

Conservation opportunities in urban habitat

Cities rank among the worst habitats for biodiversity, with fewer than half of species that are native to a landscape able to persist within an urban environment (see Chapter 9). Even so, many native species have adapted to cities and built environments, and learned to thrive in urban habitats like green spaces, urban streams and ponds, golf courses, and among buildings, bridges, and other infrastructure.[56] In addition to the persistence of native species, introduction of domesticated agricultural and ornamental exotic plants has, in some cities, exceeded the loss of native plant species, causing urban environments to have enriched taxonomic diversity for certain groups of organisms (though these pools are often disproportionately composed of alien species).[57] Thus, built environments often have their own unique set of plants and animals,[58] and these species can provide a suite of ecosystem services to residents, such as improvements in air quality and reduced surface temperatures by trees that benefit public health.[59] Therefore, this makes cities worthy of inclusion in landscape conservation planning.

Suggested Exercise 15.3 Read about the recovery of the peregrine falcon in New York City. **oup-arc.com/e/cardinale-ex15.3**

Numerous studies have shown that certain kinds of cities are more "friendly" to biodiversity than others. In a comprehensive analysis of biodiversity in urban environments, Newbold and colleagues[60] found that cities characterized by high human densities and intense land use (i.e., high proportions of land use dedicated to physical infrastructure) exhibited up to 80% loss of species. But urban environments characterized by lower human densities and less intense land use had far lower, sometimes even undetectable impacts on biodiversity.

Several practices in cities appear to reduce species loss in a landscape, or even improve the persistence of some native species. Incorporating green spaces (e.g., parks, woods, and grasslands) into city design, replacing physical infrastructure with green infrastructure (e.g., using rain gardens and natural wetlands for water drainage and storage instead of pipelines), and increased use of sustainable architecture can all enhance biodiversity conservation at a landscape scale (**Figure 15.13A–C**).[61,62] Many studies have shown that urban green spaces can offer viable habitat for certain native plants, insects, amphibians, reptiles, birds, and occasionally mammals to complete their life cycles. In addition, urban green spaces can be used as critical corridors and stepping stones as species migrate among protected areas or mixed-use habitat where they have higher survivorship and reproduction. For example, Beninde and colleagues[63] performed a meta-analysis of biodiversity in 75 cities around the world to determine how characteristics of urban green spaces affect the biodiversity of a wide variety of plants and animals. The authors showed that biodiversity generally increases when urban habitats (1) have a greater total area of green space; (2) have a greater richness of habitat types in the green spaces; (3) have more total ground cover of herbs, shrubs, and trees; (4) are less intensively managed; and (5) are in close proximity to water (**Figure 15.13D**).

Results like those in Figure 15.13D suggest it is possible for cities to be more friendly (or at least less hostile) to biodiversity, and that good examples exist of cities around the world that are on the right track. In addition to their benefits for biodiversity, urban green spaces and green infrastructure and architecture offer other benefits that make them attractive to people and indirectly promote conservation. For example, in Chapter 6 we described some of the well-documented ecosystem services that green spaces provide, which include improved physical and mental health—part of which is due to outdoor recreational opportunities that improve fitness, and part to improvements in air and water quality.[64,65] Case studies also show that urban green spaces are sufficiently attractive that people will pay higher prices for housing located next to them, which can increase income to land owners who rent property,[66] or revenues of local governments that collect property taxes.[67,68]

(A) (B) (C)

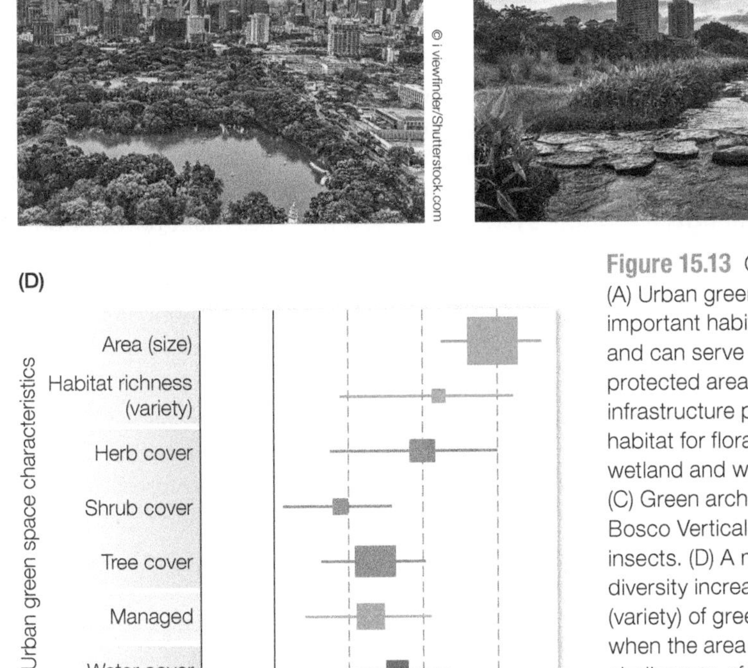

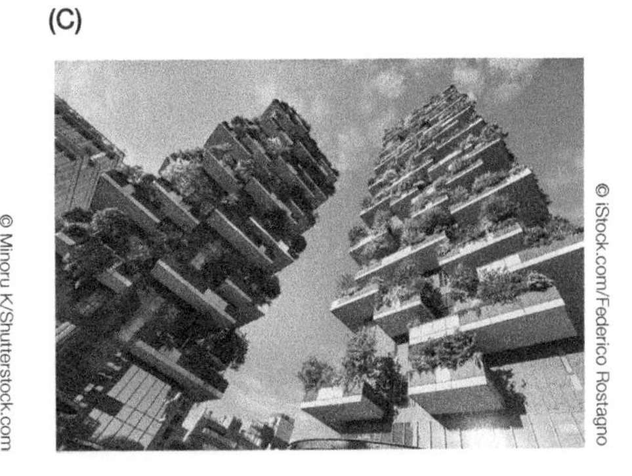

Figure 15.13 Opportunities for biodiversity conservation in urban habitats. (A) Urban green spaces like Lumpini Park in Bangkok, Thailand serve as important habitat for plants and wildlife that have adapted to urban habitat, and can serve as stepping stones for other species as they migrate among protected areas or multiuse habitats throughout a landscape. (B) A green infrastructure project at Bishan-Ang Mo Kio Park in Singapore offers more habitat for flora and fauna than physical infrastructure. For example, natural wetland and water courses rather than pipelines can be used for drainage. (C) Green architecture, such as used on these two residential towers called Bosco Verticale in Milan, Italy can serve as migratory habitat for birds and insects. (D) A meta-analysis of 75 cities around the world shows that biodiversity increases when urban habitats have a greater area and richness (variety) of green spaces, with more cover of herbs, shrubs, and trees; and when the area is less intensively managed (less physical intervention, including use of pesticide, etc.) and close to water. Data points are the mean of all cities ± one standard error. (D, after J. Beninde et al. 2015. *Ecol Lett* 18: 581–592.)

Even as we make built environments friendlier to biodiversity, it is important to recognize the potential negative consequences of integrating people with nature so that we can try to mitigate those consequences. Increasing the presence of wild animals in the urban landscape can have serious consequences for both animals and humans. Transmission of disease and other potentially harmful direct interactions among people, domestic animals, and wild animals is a major concern. For example, as woodland areas and mountain canyons become urbanized or suburbanized, people tend to create yards and gardens that attract deer. Deer bring with them a host of problems: they can carry ticks that transmit illnesses to humans, such as Lyme disease and Rocky Mountain spotted fever; they are a significant potential road hazard; and males can become aggressive toward humans during mating season. In some areas, deer that live within developments also attract predators such as cougars, increasing the potential for human–wildlife conflicts for a scarce and ecologically important top carnivore.

Understanding the ecology, ecosystem processes, and human use of a location is critical for implementing policies to promote conservation in urban areas. Conservation biologists and policy makers have a number of tools they can use to achieve their urban conservation aims. These tools include: (1) zoning and building regulations; (2) endangered species and clean air and water protections; (3) the planning and improvement of roads and culverts; (4) education and training programs; (5) partnerships with nonprofits, garden clubs, and other community organizations; and (6) financial and other types of incentives for pro-conservation behaviors. Deciding on the proper tools, though, requires good information on ecology and complex urban human–natural systems and knowledge of how best to motivate people to behave in conservation-friendly ways. These areas of research are growing and beginning to provide insights that are improving urban conservation.

Conservation opportunities in agricultural habitat

Like urban habitat, agricultural habitat generally houses less biodiversity than more natural systems (see Chapter 9). However, there are a wide range of farming systems that are used throughout the world, and some of these systems are far more conducive to conservation than others. Garibaldi and colleagues[69] recently summarized farming systems into several types that are commonly mentioned in the scientific literature:

- **Conventional farming** is characterized by large fields of monoculture crops that are managed with heavy machinery and external inputs like synthetic fertilizers and pesticides.
- **Sustainably intensified farming** uses practices from agroforestry, conservation agriculture, and biological pest control, to promote favorable ecological interactions that produce a low-input, resource-conserving agroecosystems.
- **Organic farming** was traditionally characterized by low-input, small-scale, diversified farms. More recently, certified organic farming prohibits the use of most synthetic inputs and genetically modified organisms (GMOs), while allowing organic fertilizers and pesticides.
- **Diversified farming** integrates several crops and/or animals in the production system to promote agrobiodiversity, ecosystem services, and reduced need for external inputs.
- **Agroecological farming** emphasizes integration of farms into a surrounding landscape that conserves and manages biodiversity to enhance ecological processes that support crop production, biological pest control, nutrient cycling, and pollination.

Garibaldi and colleagues[69] then summarized some of the characteristics of these farming systems that influence the conservation of biodiversity (**Table 15.2**). Because of its reliance on monocultures and intensive management practices to maintain and harvest monocultured crops (e.g., pesticides, fertilizers, heavy machinery), conventional farming—particularly as it is practiced in highly developed countries—is the least desirable form of agriculture for biodiversity conservation in a landscape. Sustainably intensified farming systems, which still rely heavily on monocultures but seek to reduce intensive management by promoting beneficial ecological interactions like biological pest control, are slightly better than conventional farming systems.

Organic farming, diversified farming, and agroecological farming systems are friendlier to biodiversity than alternatives. All three of these farming systems seek to diversify the crops and/or livestock that are farmed, thus moving away from the intensive management of monocultures. All three systems also seek to increase biodiversity of non-farmed species. This is often accomplished by increasing the heterogeneity of habitats around the farms, such as by integrating patches of woodland into farm fields, leaving riparian buffer strips along streams, and incorporating hedgerows along edges. A key goal of encouraging non-farmed biodiversity is to take advantage of the ecosystem services provided by biodiversity, such as pollination of crop plants, biological control of herbivorous pests and fungal diseases, and the retention and natural fertilization of soils.

Agroecological farming differs from organic and diversified systems in that it emphasizes integration of farms into a broader landscape that is designed to conserve and manage natural resources sustainably. Agroecological food systems tend to focus on development of small, diverse farms that are interspersed among other habitat types, and which use low levels of inputs to ensure the long-term balance between food production and the sustainability of natural resources. They often emphasize local production of food for local communities.

There is good evidence that agricultural practices that diversify crops and livestock, which manage for habitat heterogeneity to promote non-farmed biodiversity, and that take advantage of naturally occurring ecosystem services to reduce their intensity of management, do in fact promote biodiversity in landscapes. Rahmann[70] reviewed 396 scientific papers that studied whether organic farming practices have any advantages for biodiversity compared to conventional farming systems (**Figure 15.14A**). Of these papers, 327 (83%) concluded that biodiversity is higher in organic compared to conventional farms. Only 3% of papers concluded that organic farms have less biodiversity, and 14% concluded that patterns were unclear or that levels of biodiversity were indistinguishable between farming systems. The results were consistent for every group of organisms considered,

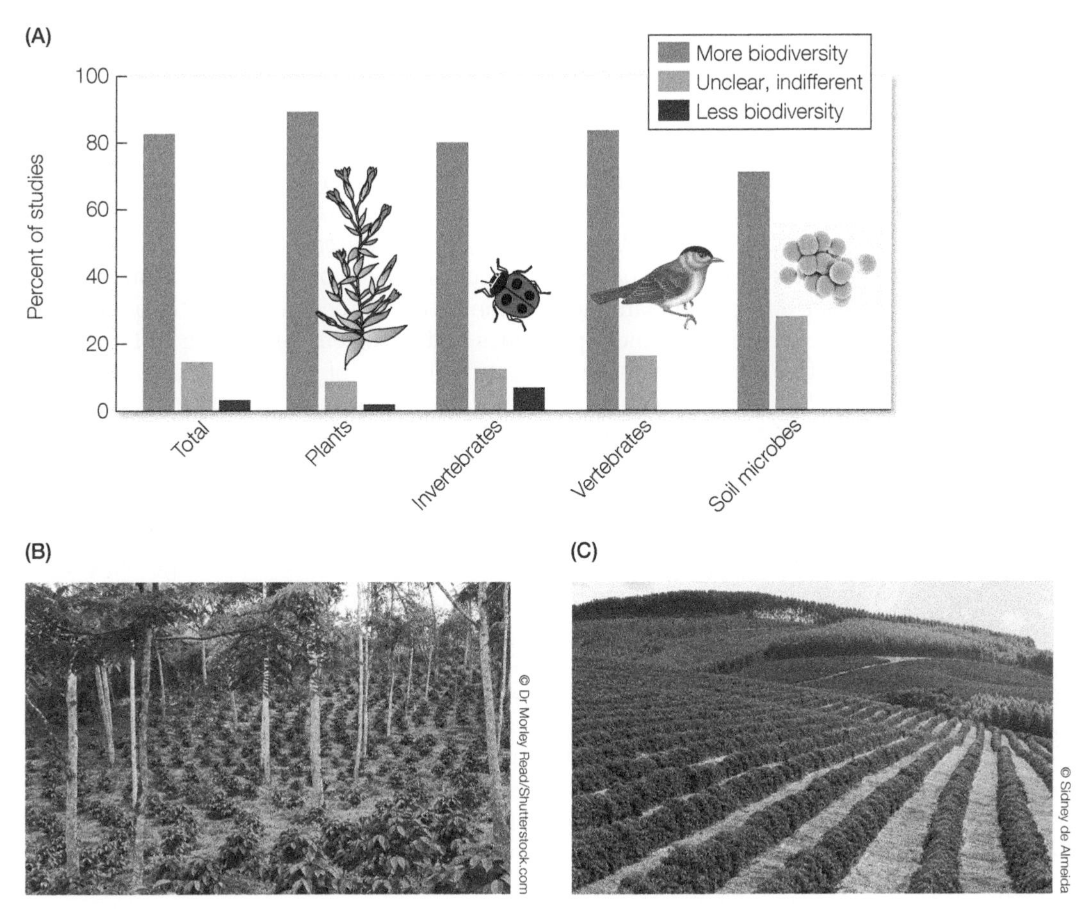

Figure 15.14 Opportunities for biodiversity conservation in agricultural habitat. (A) A summary of studies that have compared biodiversity in organic farms to that in conventional farms. The *y*-axis shows the percentage of studies that have found greater (green bars), less (red bars), or indistinguishable (yellow) levels of biodiversity for several groups of organisms. (B,C) A notable example of biodiversity conservation in agricultural systems come from tropical countries that grow coffee. (B) A shade-grown organic coffee plantation in Ecuador. The coffee is grown under a diverse canopy of trees that provides forest structure for birds, insects, and other animals. (C) A conventional coffee plantation (low-diversity monoculture) in Brazil offers little habitat value for other species. (A based on data in G. Rahmann. 2011. *Agricult Forest Res* 3: 189–208.)

TABLE 15.2 Similarities and differences among five farming systems that are commonly referred to in the scientific literature

Farming System	Uses synthetic inputs	Uses GMOs	Uses crop and livestock diversity	Encourages non-farmed biodiversity	Encourages spatial heterogeneity	Provides and uses ecosystem services	Integrates into landscape-scale conservation
Conventional farming	Very often	Very often	Sometimes	Rarely	Rarely	Rarely	Almost never
Sustainably intensified farming	Sometimes	Sometimes	Rarely	Sometimes	Sometimes	Sometimes	Almost never
Organic farming	Rarely	Almost never	Sometimes	Sometimes	Sometimes	Very often	Sometimes
Diversified farming	Rarely	Rarely	Very often	Very often	Very often	Very often	Sometimes
Agroecological farming	Rarely	Almost never	Very often	Very often	Very often	Very often	Very often

Source: Modified from Table 1 in L. A. Garibaldi et al. 2017. *Trends Ecol Evol* 32: 68–80.

with the majority of studies showing that the biodiversity of plants, invertebrates, vertebrates, and soil microbes is higher in less intense farming systems.

A notable example of preserving biodiversity in an agricultural setting comes from tropical countries and their traditional shade coffee plantations, in which coffee is grown under a wide variety of shade trees, often as many as 40 tree species per farm (**Figure 15.14B-C**).[71] In northern Latin America alone, shade coffee plantations cover 27,000 km^2. These plantations have structural complexity created by multiple vegetation layers and a diversity of birds and insects comparable to adjacent natural forest, and they represent a rich repository of biodiversity.[72] The presence of such coffee plantations can also potentially slow the pace of deforestation.[73] Unfortunately, in many areas, the spread of the fungal disease coffee leaf rust has encouraged conversion of shade plantations to high-yielding sun coffee plantations without shade trees, which incorporate coffee varieties that require more pesticides and fertilizers (see Figure 15.14B). These sun coffee plantations have only a tiny fraction of the species diversity found in shade coffee areas and are far more prone to water runoff and soil erosion. Therefore, programs are being developed to encourage and subsidize farmers to maintain their shade-grown coffee plantations and to market the product at a premium price as "environmentally friendly" shade-grown coffee. But let the buyer beware: there are currently no uniform standards for shade coffee, and some coffee marketed as "environmentally friendly, shade-grown coffee" may actually be sun coffee grown among only a few small, interspersed trees. Shade-grown chocolate and other tropical tree crops share similar issues.[74]

New government programs that compensate farmers and rural landowners directly for protecting elements of biodiversity are sometimes called payments for ecosystem services (PES) programs (described in Chapter 6). In Florida, for example, private conservation organizations and government agencies compensate ranchers for maintaining unimproved pastures that are rich in wildlife, such as native wet prairie species. Such programs allow farmers an option other than changing to intensive agriculture (**Video 15.2**), which would result in declines in biodiversity.[75,76]

Video 15.2 The award winning documentary *Biggest Little Farm* could change your views on modern agriculture. Start by watching the trailer.
oup-arc.com/e/cardinale-v15.2

Conservation opportunities in mixed-use habitat

In many countries, large parcels of government-owned land are designated as multiple use habitat—that is, habitats managed to provide a variety of goods and services. An emerging and important area of research involves the development of innovative ways to reconcile competing claims on land use, such as logging, grazing, mining, species conservation, and tourism. This requires careful analyses and considerations of the trade-offs of pursuing alternative development options concerning both environmental and socioeconomic priorities. A different approach is to use regulations, the legal system, and political pressure to prevent government-approved activities on public lands if these activities threaten the survival of endangered species.

In the U.S., the Bureau of Land Management oversees more than 1.1 million km², including 83% of the state of Nevada and large amounts of Utah, Wyoming, Oregon, Idaho, and other western states (**Figure 15.15**). National forests cover over

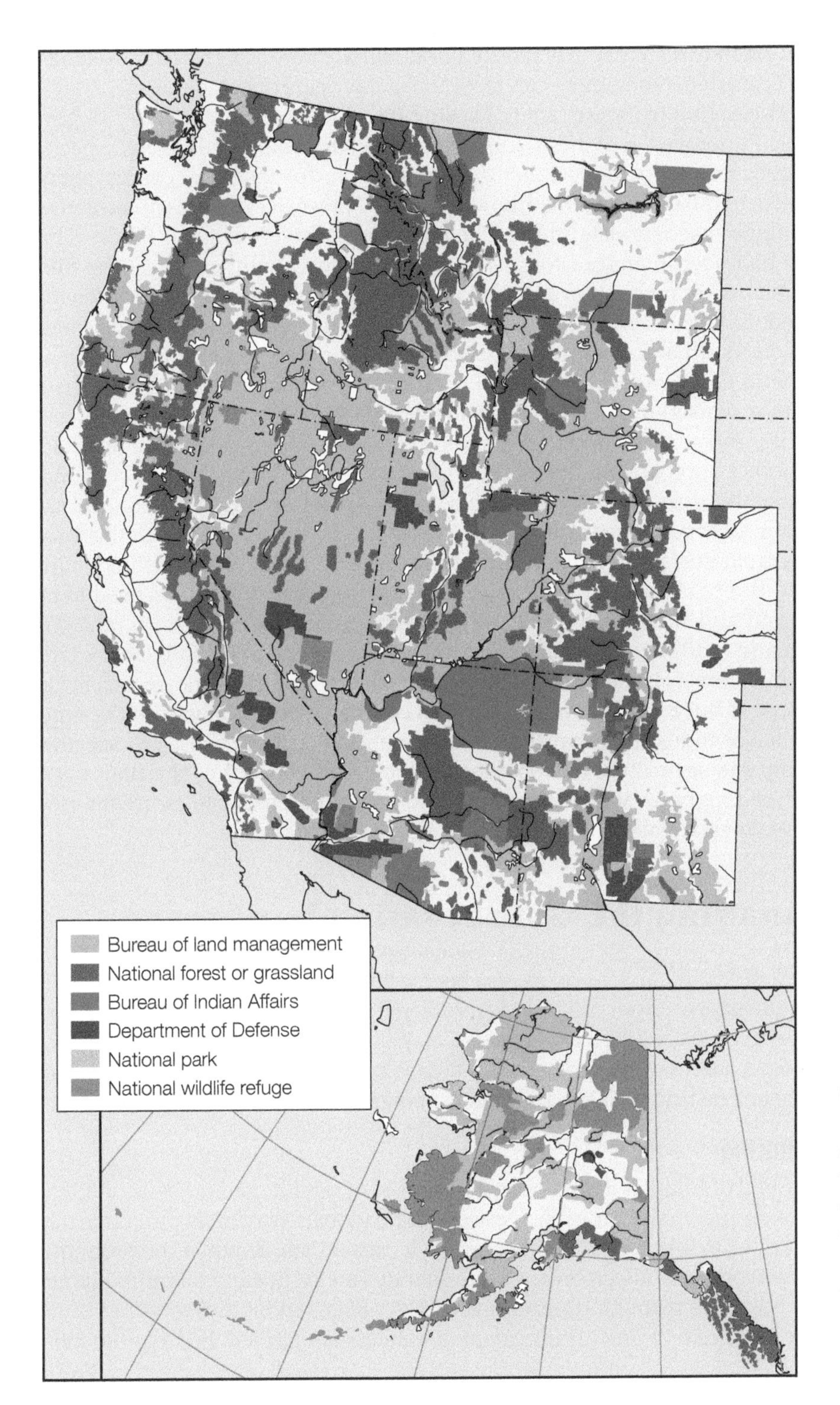

Figure 15.15 Conservation opportunities in mixed-use habitat. In Alaska and the western states, agencies of the U.S. government own the majority of the land, including some truly enormous blocks of land. The management of this multiple use land increasingly incorporates the protection of biodiversity as a major objective. (Data from U.S. Department of the Interior/U.S. Geological Survey. 2017. The National Map. nationalmap.gov/small_scale/printable/fedlands.html.)

830,000 km^2, including much of the Rocky Mountains, the Cascade Range, the Sierra Nevada, the Appalachian Mountains, and the southern coast of Alaska. In the past, these lands have been managed for logging, mining, grazing, wildlife, and recreation. The challenge is that often each one of these activities is managed in isolation from the others, and their cumulative effects may threaten biodiversity. Increasingly, multiple-use lands also are being valued and managed for their ability to protect species, biological communities, and ecosystem services.[77] The U.S. Endangered Species Act of 1973 and similar laws, such as the 1976 National Forest Management Act, require landowners, including government agencies, to avoid activities that threaten listed species. One such activity is overgrazing by cattle; when cattle grazing is reduced or eliminated on overgrazed rangelands, these ecosystems sometimes can recover in a few years or decades.[78]

Conservation biologists are now using laws and court systems to halt government-approved activities that threaten the survival of endangered species on public lands. In the late 1980s in Wisconsin, for instance, conservation-oriented botanists questioned how the U.S. Forest Service was interpreting its multiple-use mandate in the Chequamegon-Nicolet National Forest. This forest had been managed for a wide variety of uses by the U.S. Forest Service, but timber production and managing deer populations for hunting tended to predominate, leading to the decline of many migrant songbirds and forest wildflowers.[79] Lawsuits involving scientists and conservation groups such as the Sierra Club compelled the U.S. Forest Service to increase the emphasis it placed on conserving biodiversity, though finding a balance between logging, hunting, and conservation that everyone can accept has proved to be difficult.

Another approach has been to define standards of best practices such that resource use does not harm biodiversity. The Forest Stewardship Council has been one of several organizations that promote the certification of timber produced from sustainably managed forests. For the Forest Stewardship Council and comparable organizations to grant certification, the forests need to be managed and monitored for their long-term environmental health, and the rights and well-being of local people and workers need to be protected. Certification of forests is increasing rapidly in many areas of the world, especially in response to buyers in Europe who request certified wood products. At the same time, major industrial organizations representing such industries as logging, mining, and agriculture are lobbying for their own versions of alternative certification programs, which generally have lower requirements for monitoring and weaker standards for judging practices to be sustainable.

Managing the Conservation Landscape

After a protected area network has been established, and once opportunities for biodiversity conservation in unprotected areas have been integrated into conservation plans, the next step is to manage the conservation landscape for long-term success. Managing the landscape over the long term requires monitoring, forecasting with models, and ecosystem-based management plans.

Monitoring

At the end of Chapter 8, we detailed a number of satellite remote sensing (SRS) programs, regional observation networks, and global assessment programs that are used to monitor biodiversity and track extinctions. Some of the programs and networks are also relevant here, as they can be used to quantify changing landscapes through time, and to track changes in biodiversity that occur as the composition and arrangement of habitats are altered. Perhaps the most important programs for monitoring at the scale of landscapes are the satellite data and aerial imagery that have become widely available in recent decades. Regional to global land cover products derived from long-term satellite missions like Landsat, the Terra and Aqua Earth Observing System satellites, and the Polar-orbiting Operational Environmental Satellite (POES) series are widely accessible. They offer a relatively inexpensive means for obtaining high spatial resolution (pixel size that can get to less than 10 m^2) images of landscapes that are updated on short time scales (daily to biweekly). Data from these satellite images are even available for free through programs like Google Earth, which provides spatial data online in KML format for GIS applications. Some of these programs have made long-term datasets available, such as the archive of terrestrial satellite imagery from the Landsat system, which spans more than four decades at spatial resolutions of 15–82 m, allowing for historical comparisons and estimates of rates of change.

In addition to satellite remote sensing, aerial surveys of landscapes have become increasingly common and are often available in the records of local or regional governments (e.g., city and regional planning commissions). The rise of unmanned aerial vehicle (UAV) technology, in which amateur operators use unmanned drones and gliders to get aerial photographs, now allows landscape photographs to be collected on a routine basis at very low cost.

Once satellite images or aerial photographs of a landscape have been collected, it is relatively straightforward to digitize the images, delineate different habitat types in that image (e.g., tracing them manually on a computer, or using image recognition software to automate the process). After habitats on a digitized image have been delineated, it is possible to calculate several metrics that quantify important landscape patterns (Figure 15.16). The simplest—but perhaps most

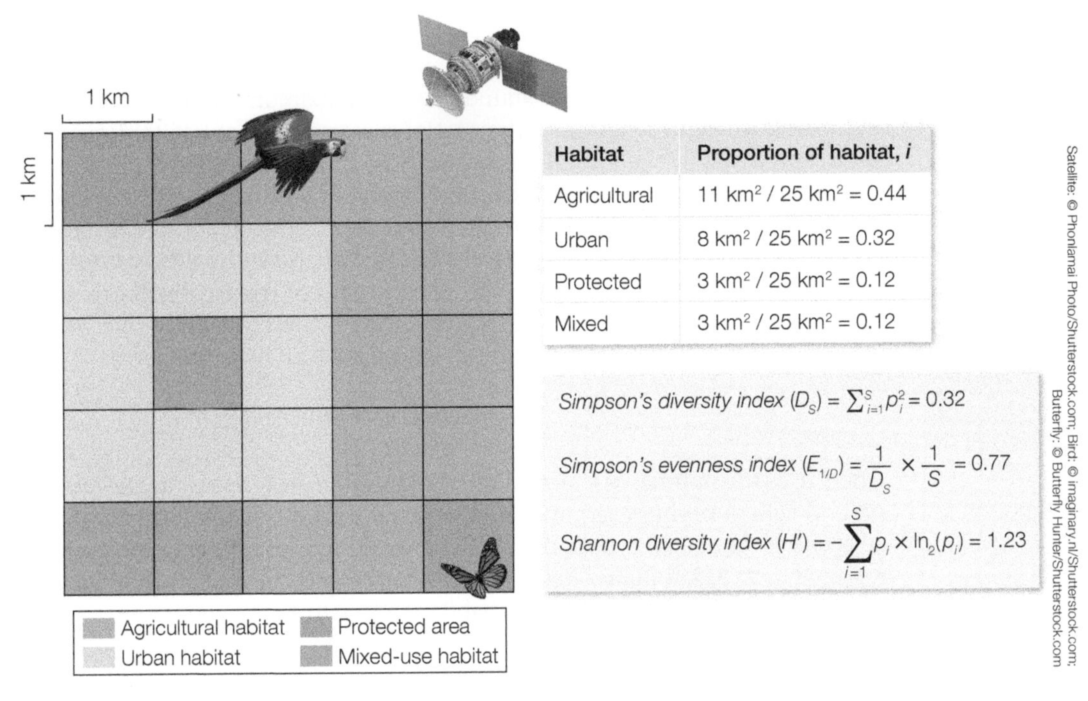

Habitat	Proportion of habitat, *i*
Agricultural	11 km^2 / 25 km^2 = 0.44
Urban	8 km^2 / 25 km^2 = 0.32
Protected	3 km^2 / 25 km^2 = 0.12
Mixed	3 km^2 / 25 km^2 = 0.12

Figure 15.16 Metrics of landscape pattern that can be monitored through time. Satellite images and aerial photographs of a landscape can be digitized and delineated into quadrats that are classified according to the type of habitat present. After this is done, various metrics of can be used to quantify important landscape patterns, such as the proportion of each habitat type, as well as the diversity and evenness of all habitat types across the landscape.

important—metric is the proportion p of each habitat type i in the landscape. For example, the proportion of protected habitat in a landscape can be calculated as:

$$p_i = \frac{a_i}{A} \quad (15.1)$$

where A is the total area of the landscape, and a_i is the total area of protected habitat (i.e., summed across all protected areas in the landscape). Using these proportions, we can use the metrics described in Chapter 3 for quantifying species diversity to quantify the diversity (or heterogeneity) of habitat types in a landscape:

Simpson's diversity index:

$$D_S = \sum_{i=1}^{S} p_i^2 \quad (15.2)$$

Simpson's evenness index:

$$(E_{1/D}) = \frac{1}{D_S} \times \frac{1}{S} \quad (15.3)$$

Shannon diversity index:

$$(H') = -\sum_{i=1}^{S} p_i \times \ln_2(p_i) \quad (15.4)$$

where p_i is the proportion of each habitat i in the landscape, and S as used here is the "richness" of habitat types (i.e., the number of different habitat types in the landscape). Simpson's diversity index ranges from 0 and 1, with values of D_S close to 1 representing a landscape with the highest possible diversity given S habitat types, and values close to 0 indicating low diversity. Values of Simpson's evenness index $E_{1/D}$ also range from 0 to 1, with 1 being complete evenness of habitats in a landscape, and 0 being total dominance of one habitat type. The maximum value of the Shannon index H' is ln(S), which occurs when every habitat type in the landscape has the same relative abundance (i.e., maximum evenness).

Often it is useful to quantify the spatial configuration of habitat types in a landscape—that is, the spatial positions of the different habitats in relation to each other. For example, if we know that protected areas do poorly at conserving biodiversity when adjacent to cities, but fare better when placed next to mixed-use habitat, then we might want to calculate how often the patch types occur together in a landscape so that we can manage the combinations properly. The *probability of adjacency* ($q_{i,j}$), which gives the probability that habitat type i is adjacent to habitat type j in the landscape, can be expressed as:

$$(q_{i,j}) = \frac{n_{i,j}}{n_i} \quad (15.5)$$

where $n_{i,j}$ is the number of times habitat types i and j are adjacent to one another on grid cells superimposed on a landscape map, and n_i is the total number of adjacencies for habitat type i on the map. Probabilities of adjacency are typically reported in an $N \times N$ matrix (referred to as the **Q** matrix) that show the adjacency of all habitat types to all others. Probabilities close to 1 indicate that two habitat types commonly occur together in the landscape, whereas probabilities close to 0 indicate that two habitat types are never adjacent. One thing that is important to keep in mind for analyses like these is that the numbers obtained will depend on the grain of the grid—that is, the resolution of the grid size in area. One might look at small grid sizes (e.g., 1 km^2) for organisms like butterflies. But this will give different answers than the larger grid sizes (e.g., 100 km^2) chosen for characterizing the landscape for birds. So it is important to think carefully about what organisms and grid sizes are being summarized by the **Q** matrix.

Contagion (C) uses values of $q_{i,j}$ from the **Q** matrix to compute an index describing the degree of clumping in a landscape. The index is calculated as:

$$C = \frac{\sum_i \sum_j \left[\left(p_i \times q_{i,j} \right) \times ln \left(p_i \times q_{i,j} \right) \right]}{C_{max}} \quad (15.6)$$

where $q_{i,j}$ are the adjacency probabilities, and $C_{max} = 2 \times \ln(S)$, which gives the maximum value of the index for a landscape with S habitat types. Values for contagion range from 0 to 1. High values indicate that habitat types are "clumped" together in the landscape, whereas low values indicate that habitat types tend to be spread out across the landscape. Contagion can be useful, for example, if one wants to keep protected areas close together so as to enhance connectivity.

While some of the landscape metrics can easily be calculated by hand (e.g., proportion of habitat types), others require serious computation. As such, these and other metrics of landscape pattern are usually calculated with spatial analysis software. There are many different software packages that can be used to quantify landscape patterns. (For a link to a list of software packages used to quantify landscape patterns go to oup-arc.com/e/cardinale-w15.15.) Some of the more commonly used in Conservation biology are ArcGIS, R-Analysis of Spatial Data, and FRAGSTATS.

Suggested Exercise 15.4 Learn how to use FRAGSTATS to quantify landscape patterns with this tutorial. **oup-arc.com/e/cardinale-ex15.4**

In addition to monitoring habitats and their configuration in a landscape, we must monitor biodiversity in the landscape to ensure long-term success of conservation efforts. Practitioners can set up their own monitoring programs tailored to their specific needs. However, depending on the goals and level of sampling required, new monitoring programs can sometimes be expensive and time-consuming. Therefore, it is worth noting that in previous chapters, we discussed a variety of existing tools, programs, and websites that make biodiversity data publicly available. These include:

- Regional monitoring networks like the GEO Biodiversity Observation Network (GEO BON; oup-arc.com/e/cardinale-w15.16)[80] that coordinates biodiversity observation and monitoring programs around the globe for integration into the Global Earth Observation System of Systems (GEOSS).
- Global and regional biodiversity assessment programs like the Intergovernmental Platform for Biodiversity and Ecosystem Services (IPBES),[81] which undertakes thematic, regional and global assessments of biodiversity and ecosystem services.
- Citizen science programs that monitor biodiversity for certain groups of organisms (e.g., the Audubon Society's Christmas Bird Count) or that report on the spatial location of observed species through mobile apps like iNaturalist, Map of Life, eBird, and PlantNet.
- Scientific surveys and censuses to record the occupancy of species in different habitats in a landscape (e.g., camera trapping) or that organize "bioblitzes" for intensive sampling of biodiversity at particular locations on single dates.

Modeling

It is somewhat rare for the monitoring of landscape habitats and their species to be coupled with modeling efforts that can forecast potential problems with biodiversity in advance. But when it is possible to do so, using monitoring data in a spatial model to predict pending changes in a species population size or distribution can be a powerful tool for managing biodiversity at a landscape scale. Many different kinds of spatial models can be applied at a landscape scale, and some of these have been covered in earlier chapters. For example, the metapopulation models covered in Chapter 9 are often used to predict changes in **patch occupancy**—that is, the proportion of habitat patches in a landscape that will occupied at some time in the future. If one has monitoring data on the presence/absence of a species from each habitat patch in a landscape at the present time (called patch occupancy), and estimates of the rate at which the species colonize new patches of habitat in the landscape, then you can predict patch occupancy in the near future (see Box 9.4, Equations 1 and 2) and predict whether the species will persist in the landscape (see Box 9.4, Equations 3 and 4).

A good example of the use of metapopulation models in landscape scale conservation comes from the work of Karanth et al.[82] These authors used sign surveys (i.e., evidence of tracks or scat) to monitor the presence/absence of tigers (*Panthera tigris*) in different habitats throughout a 38,000-km^2 landscape in India called the Malenad-Mysore Tiger Landscape (MMTL; **Figure 15.17**). Data from this monitoring effort were used to parameterize a metapopulation model, which was then simulated to determine what characteristics of the different

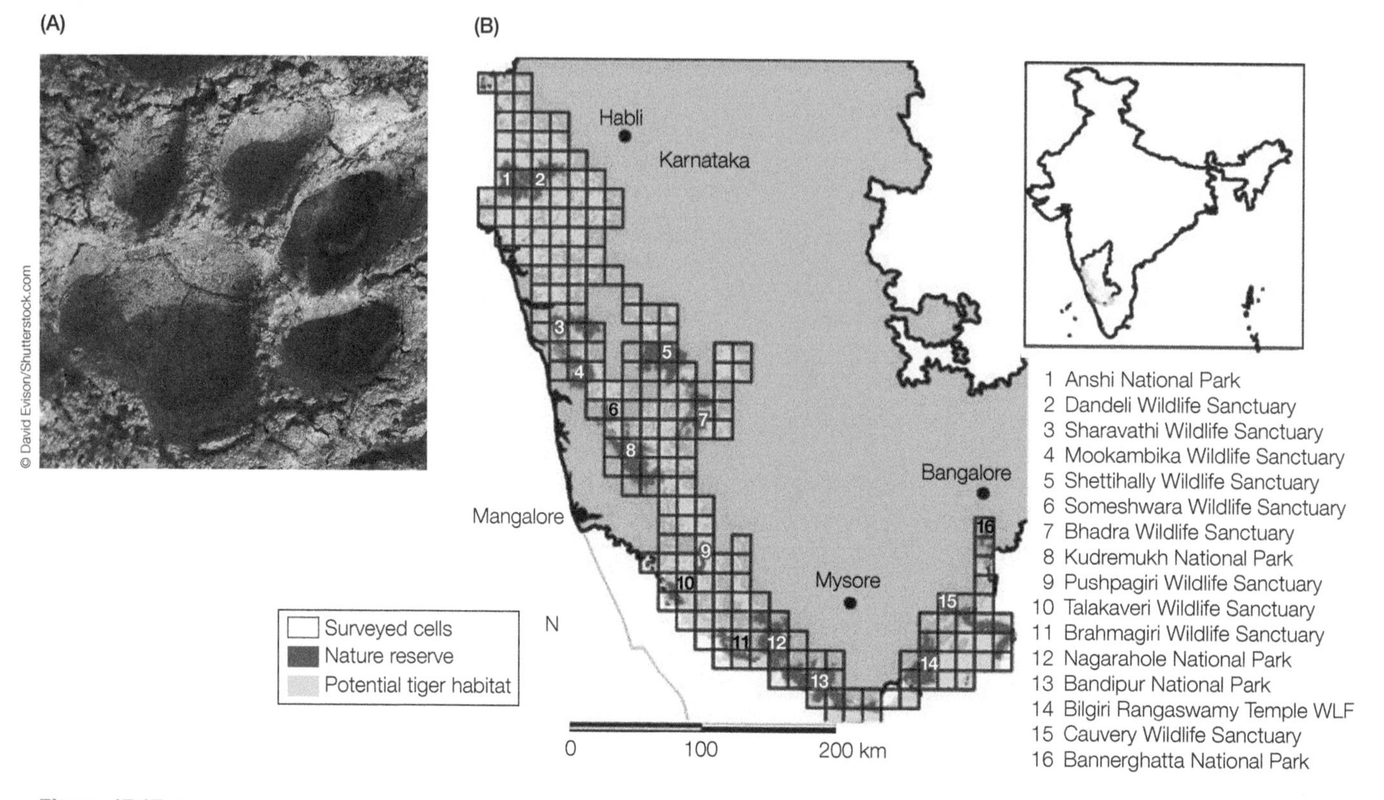

Figure 15.17 Landscape-scale models and monitoring. (A) To ensure the persistence of tigers (*Panthera tigris*) in the 38,000 km^2 Malendad-Mysore Tiger Landscape in India, researchers used "sign" surveys (e.g., presence of tracks or scat) to parameterize a metapopulation model that predicted the species' distribution across the landscape. (B) Their model allowed them to identify key source habitats (i.e., where tiger reproduction exceeds deaths), as well as areas where human activities, such as hunting of ungulate prey, are harming tiger populations. (B from K. U. Karanth et al. 2011. *J Appl Ecol* 48: 1048–1056.)

habitats in the MMTL influenced the distribution of tigers. Two important results came out of the analyses. First, depletion of ungulate prey by human hunters was a key predictor of declining tiger populations, which suggested that new management strategies and restrictions on hunting would be needed for long-term persistence of tigers in the MMTL. Second, the analyses showed that persistence of the tiger population was dependent on a few effectively protected source populations in the MMTL where levels of reproduction were above replacement levels. As such, their monitoring and models identified the most important parts of the landscape for protection that were required for long-term persistence. Similar use of metapopulation models and analyses of landscape persistence have been applied to conservation of African lions (*Panthera leo*) in Kenya and Tanzania,[83] southern ground hornbills in protected areas of South Africa,[84] frogs and salamanders in the northeastern United States[85], the persistence of fish populations in marine protected areas in Papua New Guinea,[86] and the persistence of big game species (rhino, elephants, cheetahs, leopards, lions, and wild dogs) in KwaZulu-Natal, South Africa.[87]

In addition to metapopulation models, species distribution modeling (SDM), mentioned in Chapter 11, is a type of spatial model that is used to predict the spread of invasive species. A specialized type of SDM called climate envelope models (CEMs) was introduced in Chapter 12 as a way to predict the future distribution of species under climate change based on their current habitat requirements. SDMs and CEMs are also used to predict whether protected area networks are sufficient to protect species distributions and population sizes in a landscape. For example, Kaky and Gilbert[88] used SDM to confirm that the network of protected areas established throughout Egypt was sufficient to conserve representatives of 121 species of native medicinal plants. Sundblad and colleagues[89] used SDM to assess the amount of protection offered to marine fish species in the Natura 2000 network, a 30,000-km^2 archipelago in the Baltic Sea. They concluded that the network of marine reserves in this landscape did not include a sufficient variety of habitats to protect the fish fauna; however, they were able to identify additional areas that could be added to the network to offer complete protection.

Lastly, in Chapter 13, we covered population viability analyses (PVAs), which are models of population growth (geometric, exponential, and logistic) that are modified to account for demographic stochasticity and the probability of extreme events in order to predict the probability of a species persistence through time. PVAs are sometimes turned into spatially explicit models by overlaying a grid or lattice onto a digitized landscape, running a PVA model for each grid, and then simulating those runs to predict the probability a species will persist in the landscape as whole. These models can even account for dispersal across the landscape by establishing rules for how individuals are exchanged between adjacent grids. An example comes from Bonnot and colleagues,[90] who used landscape-based PVAs to show how two migratory songbirds (the prairie warbler *Dendroica discolor* and the wood thrush *Hylocichla mustelina*) will be impacted by land use change in the Central Hardwoods Bird Conservation Region of the U.S. Midwest. In particular, the authors showed that restoration and afforestation of certain habitats could ensure long-term persistence of the two bird populations in the landscape.

Ecosystem-based management

Monitoring changes in species and their habitats, and modeling those changes in order to predict negative events before they happen, are just the first steps in landscape-scale conservation. Conservation must ultimately use these data and model predictions to establish effective management plans. Management at a landscape scale can be particularly difficult for two reasons. First, given their large spatial scale and inclusion of many habitat types, landscapes often span multiple jurisdictions (local authorities, regional and national government agencies, etc.), as well as many different stakeholder groups (private land owners, land trusts, corporations, etc.) that sometimes have divergent ideas about how natural resources should be managed. These jurisdictions and divergent interests must somehow be coordinated and steered toward a common conservation goal. Second, conservation at the scale of whole landscapes is generally ridden with more uncertainty and unpredictability than conservation efforts undertaken at smaller scales. There are simply more variables that can influence natural resources at large scales, and a greater chance for unanticipated problems to crop up. Because of this, management strategies used at a landscape must be particularly flexible and adaptive, and able to respond quickly to unanticipated events.

Ecosystem-based management (**EBM**, or just **ecosystem management**) is the term most often used to describe management of natural resources at a landscape scale. Ecosystem-based management is an integrated management approach that recognizes the full array of interactions within an ecosystem, including humans, rather than considering single issues, species, or ecosystem services in isolation. The term *ecosystem* is somewhat of a misnomer in EBM, since EBM is typically used to refer to management of entire landscapes that are composed of many different habitat or ecosystem types—both natural and anthropogenic (Figure 15.18). Regardless, EBM has several core characteristics:

- **EBM is place-based, and the geographic units of management are defined by ecological criteria.** EBM is inherently linked to a particular landscape, and the units of management within the landscape are defined by the natural boundaries of ecosystems that compose the landscape. Because ecosystems are dynamic—their boundaries shift (e.g., a river moves), they experience natural disturbances (e.g., a coastline can experience a hurricane)—management units may have "fuzzy" or imprecise boundaries, or boundaries that change through time as ecosystems in the landscape are altered. Regardless, the key point here is that EMB as a management strategy focuses on natural boundaries that are defined by biological processes, rather than by political boundaries that often do not correspond to biological processes.
- **A key goal of EBM is to achieve environmental, social, and economic sustainability.** EBM as a management strategy seeks to do more than just conserve biodiversity in a landscape. It seeks to conserve biodiversity while balancing that goal with other, sometimes competing, goals in the landscape. Therefore, EBM uses the best available science to try and achieve all three pillars of sustainability: (1) environmental sustainability, which requires that natural resources are not extracted faster than they are produced to ensure they are available for future generations; (2) social sustainability, which focuses on alleviating human inequality, social injustices, and poverty; and (3) economic sustainability, which requires economic models that ensure a fair distribution and efficient allocation of resources, and which allows jobs to be created and businesses to be profitable (see Chapter 17). The key point here is that EBM recognizes people and nature as a socioecological system (sometimes called a "coupled natural-human system") where human needs must be balanced with needs of the environment.
- **EBM is cross-sectoral, considering interactions between various sectors of human activity.** In addition to considering landscapes as socioecological systems, EBM recognizes that effective management must consider interactions between various sectors of human activity in a landscape (e.g. farming, fishing, energy development). This differs from most

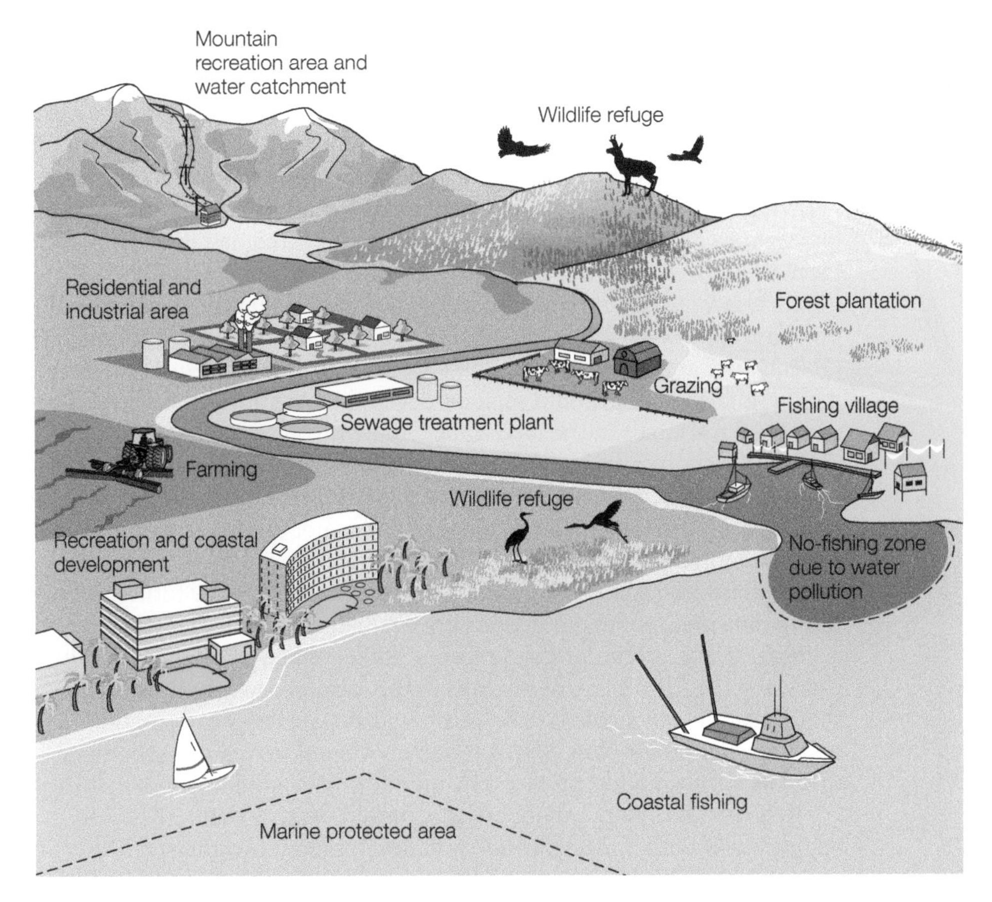

Figure 15.18 Ecosystem-based management (EBM) involves bringing together all of the stakeholders that affect a large ecosystem and receive benefits from it. In this case, a watershed needs to be managed for a wide variety of purposes, many of which influence each other. (After WRI/IUCN/UNEP. 1992. *Global Biodiversity Strategy: Guidelines for Action to Save, Study, and Use Earth's Biotic Wealth Sustainably and Equitably.* Washington, DC: World Resources Institute.)

types of resource management (whether by national, local, or tribal governments), which usually focus on a single sector and is often called for by a statute (local or national or both) specific to that sector. In contrast, EBM attempts to manage the cumulative impacts of different sectors — not just their impacts on the environment, but their impacts on each other.

- **EBM is proactive and works to optimize inherent tradeoffs in the management of natural resources among various sectors.** EBM recognizes that trade-offs are inherent in the use and management of natural resources and human activities. Rarely is it possible to optimize all activities at once without some trade-off in uses and goals. For example, increased energy development might result in some loss or degradation of habitats; yet it may be necessary to meet a nation's energy demand. Many traditional management practices do not deal explicitly with trade-offs among sectors, but with EBM trade-offs are made transparent and become part of the planning process. This ensures that all who have a stake in the sustainable use of natural resources have the opportunity to engage and resolve issues proactively, and to understand the consequences of their decisions.
- **EBM is inclusive and collaborative, encouraging participation from all stakeholders.** EBM strives to be inclusive and collaborative at all stages of the process. It uses principles of co-design (or participatory design) to ensure that all stakeholders in the landscape whose lives depend on its ecosystems have an opportunity to give input and/or participate in the development of a coordinated management plan for sustainable use of resources.[22] A diverse mix of collaborators and stakeholders is required not only to ensure that all perspectives and all livelihoods are

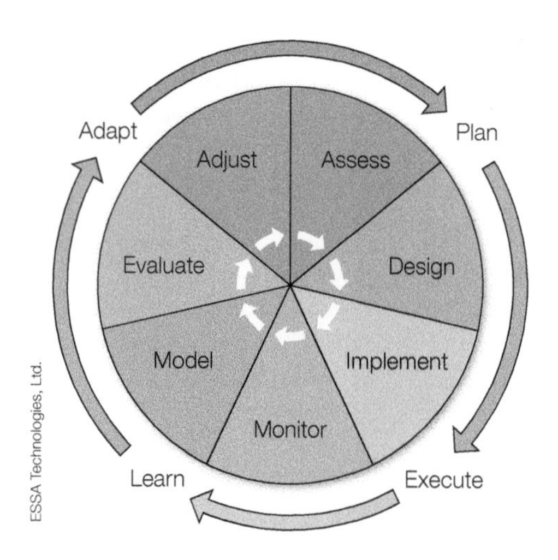

Figure 15.19 Adaptive management is an iterative process for making decisions in the face of uncertainty. The goal is to be able to take corrective action if monitoring data show, or models predict, that the goals of the plan will not be met.

considered before taking action, but also to help build acceptance of the plan needed to achieve success. Collaboration shows stakeholders they are part of a solution that is best achieved by working together.

- **EBM is adaptive and flexible, and uses monitoring and modeling results to make course corrections. Adaptive management (AM)**, also known as **adaptive resource management (ARM)**, is an essential part of managing a landscape effectively. AM is a structured, iterative process for making decisions in the face of uncertainty, with the aim of taking corrective action if monitoring data and models predict a pending problem. It involves four sequential steps: (1) planning, (2) executing, (3) learning, and (4) adapting (**Figure 15.19**). Step 1 begins with an assessment and design of an EBM plan co-designed to achieve sustainable use of natural resources in the landscape. Step 2 involves implementing the plan and then monitoring the natural resources in the landscape to see if the goals of the plan are achieved (e.g., that threatened species persist, fisheries are managed sustainably, and water quality is maintained above a defined threshold). Step 3 uses the monitoring data to predict potential undesirable changes that might occur in the near future. Monitoring data and model predictions are evaluated on a regular basis, such as might occur at regular meetings of a stakeholder group charged with executing the EBM plan, or perhaps in the form of town-hall meetings that allow all stakeholders to be given updates and new information on progress of their plan. Step 4 requires evaluation and adjustment of the plan if information provided by monitoring or models show that the goals of the EBM plan are not being achieved. At this point, the cycle begins anew with a reassessment and development of a revised EMB plan. Adaptive management requires people and institutions to be flexible and responsive to new information and experiences. The process is most likely to succeed when there are clear linkages among information, actions, and results and a strong climate of trust and mutual respect among partners.

EBM has been used by many government agencies, nongovernmental organizations, and local communities since the 1990s as their primary management strategy for conserving biodiversity. One example is the **community-based natural resource management (CBNRM)** programs in Africa, where governments grant local landowners and communal groups the authority to manage wildlife in their landscapes in ways that help build sustainable economies and improve people's livelihoods (**Box 15.3**). Another example is the U.S. National Oceanic and Atmospheric Administration (NOAA), whose National Marine Fisheries Service is charged with stewardship of marine resources. Historically, the NOAA fisheries service focused on developing single-species fishery management plans that were limited in geographic scope and focused entirely on natural systems (**Figure 15.20**). Realizing that those models were insufficient to describe the dynamics of species that were part of a complex coastal or open ocean comprised of different ecosystems and broader food webs, NOAA began developing fisheries ecosystem plans that were multispecies models embedded in ecosystems that had multiple habitats, and that were dynamic—changing through time. But these models and management plans still focused entirely on managing natural systems, with little consideration of human social systems. Therefore, NOAA's Marine Fisheries Service has recently evolved toward developing regional ocean plans for eight large marine landscapes in U.S. coastal oceans. Each landscape is managed with an EBM plan that was co-designed with regional partners and stakeholders, and which include cross-sectoral and

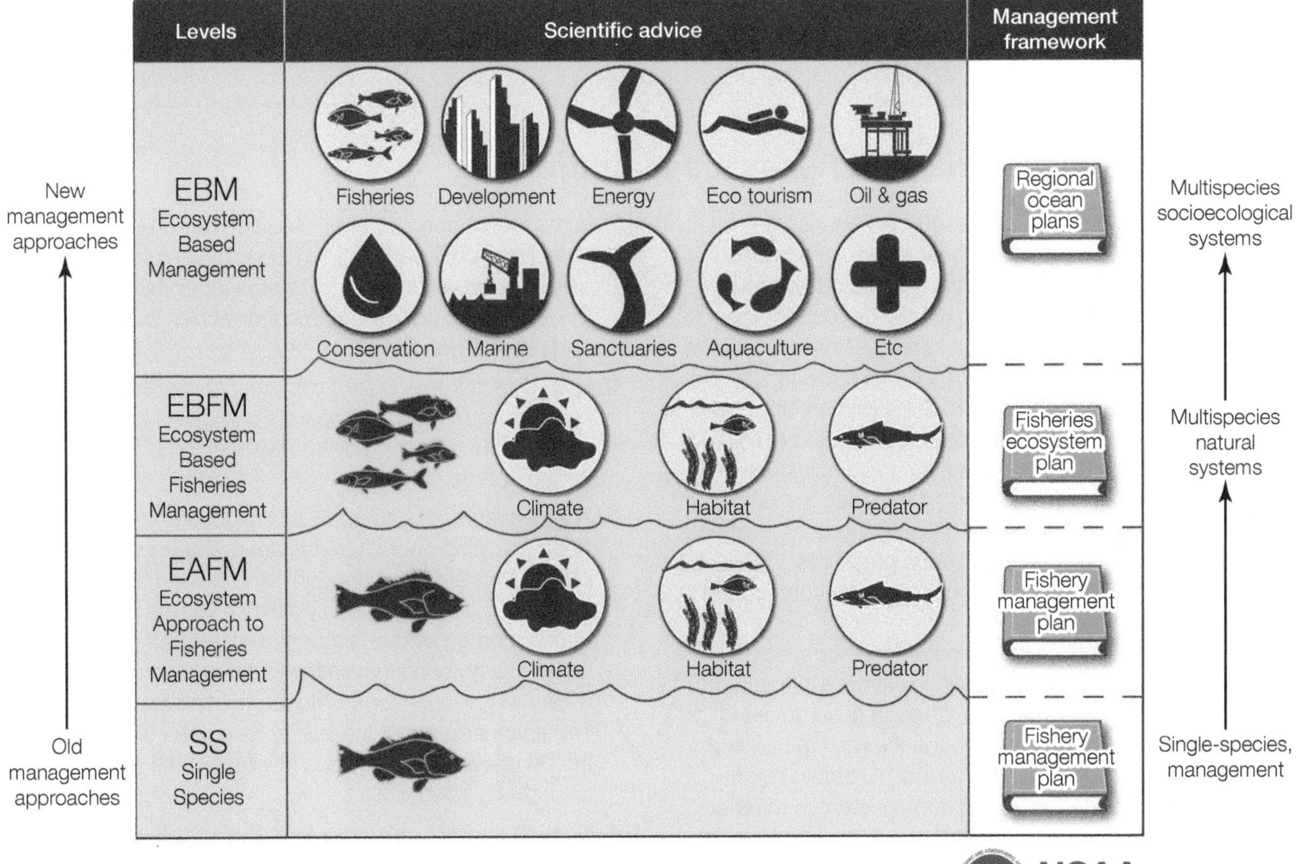

Figure 15.20 An example of the shift to ecosystem-based management is the evolving management strategy of the U.S. National Oceanic and Atmospheric Administration (NOAA), whose National Marine Fisheries Service has evolved from single-species fishery management plans to management of entire socioecological systems with of many different stakeholders. (Courtesy of NOAA Fisheries.)

cross-agency consideration of impacts on fisheries, along with development of management measures to address those impacts.

Landscape-Scale Challenges

Methods for designing optimal protected area networks, integrating unprotected areas into a conservation plan, and effectively managing whole landscapes are constantly being improved by the publication of new research and case studies of successes and failures. It is, however, important to recognize there is no "one size fits all" strategy for conservation planning at a landscape scale. And there are no definitive tools, techniques, or models that can ensure success in all cases. Successful conservation at the landscape scale depends as much, if not more, on social science and interactions with people as it does on natural science and understanding the needs of threatened species or other aspects of biodiversity. The greatest challenge for practitioners engaged in landscape-scale conservation is to simultaneously master the tools and techniques from natural science that conservation biologists have traditionally used to protect biodiversity at smaller scales (see Chapters 13, 14, and 16), as well as the tools and techniques from social science that allow them to meet the needs and desires of different stakeholders (see Chapters 5, 6, and 7). Only in combination can we achieve biodiversity conservation in a socio-ecological system.

BOX 15.3 Case in Point

Community-Based Natural Resource Management

Community-Based Natural Resource Management (CBNRM) programs in Africa represent an approach to conservation in which local landowners and communal groups are given the authority to manage and profit from the wildlife on their own property. In many African countries, wildlife both inside and outside of national parks is managed by government officials, often with no input from the local people, who gain little or no economic benefit from the wildlife on their own land and have no incentive to protect the wildlife. By changing the management system to CBNRM, government officials and conservation organizations hope to counterbalance pressures threatening local wildlife while simultaneously contributing to rural economic development.

One of the most ambitious programs for local communities managing wildlife is found in Namibia in southern Africa.[96] Beginning in 1996, the Namibian government granted traditional communal groups the right to use and manage the wildlife on their own lands. To obtain these rights, a group needed to form a management committee and determine the boundaries of its land. The government then designated the group as a "community conservancy," which has four benefits:

1. The conservancy can form joint ventures with tour operators, with 5% to 10% of the gross earnings paid to the conservancy. A certain number of the employees in the tourist operation are hired from among the communal group. Revenues from the joint ventures are used to train and pay game guards (again hired from the communal group) who monitor wildlife populations and prevent poaching.
2. Using funds from the joint ventures, conservancy members can build and operate campsites for tourist groups, thus providing direct revenue and employment for the communal group.
3. The conservancy can apply to the government for a trophy-hunting quota. Quotas are granted if wildlife populations are large enough, as indicated by routine monitoring and population viability analyses. This quota can then be sold or auctioned off to professional hunters, who bring in wealthy foreign tourists willing to pay a high price for an African hunting experience. The entire amount of the trophy fees goes directly to the conservancy, regardless of whether the animals are actually killed.
4. Once the conservancy has formed a wildlife management plan, four species of wildlife—gemsbok, springbok, kudu, and warthog—can be hunted for subsistence. In practice, the hunting is often done by game guards and professional hunters, and the meat is distributed to everyone in the community.

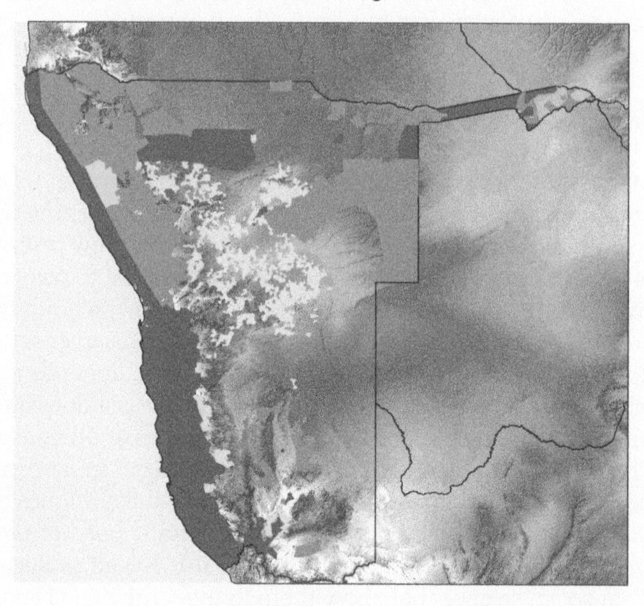

(A) Location of communal conservancies and protected areas across Namibia. (B) Total cash income (fees paid to conservancies as well as resident wages from operations) and in-kind benefits (game meat and fringe benefits provided to employees by private sector) of Namibia's Community-Based Natural Resource Management (CBNRM) program. (C) Trends in wildlife population sizes over the course of the CBNRM program. (From MET/NACSO. 2018. The state of community conservation in Namibia: A review of communal conservancies, community forests and other CBNRM activities [Annual Report 2017]. MET/NACSO, Windhoek.)

As of 2017, the Namibian Association of CBNRM Support Organizations (NACSO) reported there are 86 registered conservancies in Namibia covering 166,045 km² of land (20% of Namibia's land surface) that are inhabited by 212,092 people (see Figure). Conservancies in the CBNRM program are generating increasing revenues from their wildlife operations, which they are using to build more tourist facilities, erect communal structures such as schools, distribute money to their members, and even establish bank accounts. The communal management system seems to be having positive conservation effects on many species, with populations of large herbivores and predators increasing since the start of the CBNRM program

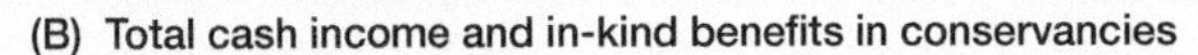
(B) Total cash income and in-kind benefits in conservancies

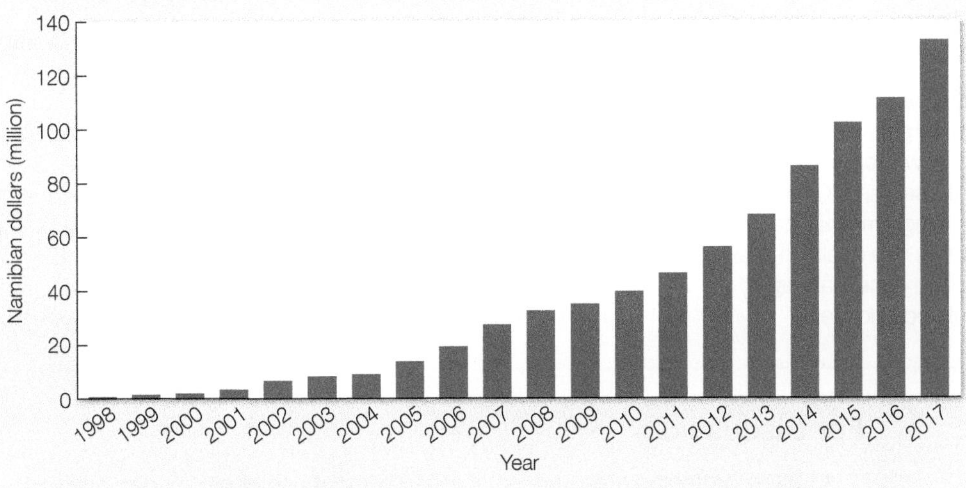

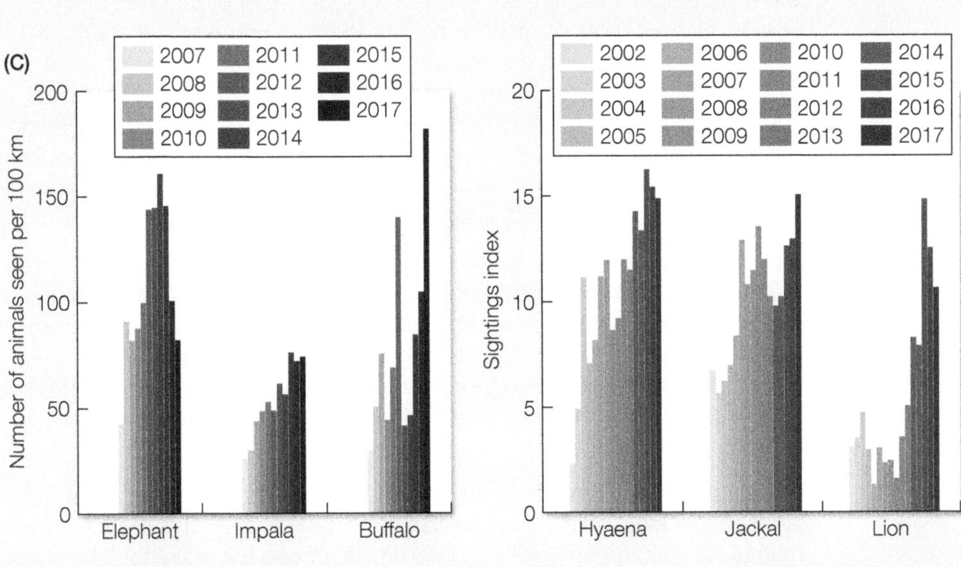

Summary

1. Conservation at a landscape scale is achieved by establishing networks of protected areas. Historically, national governments have played the largest role in setting up protected area networks through their creation of national park systems, wilderness areas, biological reserves, and mixed-use public lands. Today, Privately Protected Areas (PPAs) and Indigenous Protected and Conserved Areas (IPCAs) are also making large contributions to protected area networks.
2. When designing protected area networks, individual protected areas should be as large as possible, minimize edge effects, prioritize protection of irreplaceable or vulnerable biological features, encompass the entire conservation unit, incorporate buffers, and include humans in the design. Networks should maximize the number or area of protected areas in a landscape, ensure protected areas are well connected, cover the full range of biodiversity in the landscape, and manage protected areas collectively as a whole.
3. Gap analysis is a widely used process to identify areas in a landscape where conservation goals are not being met, and to select and prioritize new protected areas for inclusion in a network to cover those gaps. Gap analysis is typically done using software programs that can perform spatial analysis, and which use decision-making algorithms to prioritize the selection of new areas for protection according to user-defined priorities.
4. Optimizing conservation at a landscape scale requires that protected areas are well connected and function collectively as a whole. A habitat corridor is a continuous strip of protected habitat that runs between and connects two protected areas. Stepping stones are discontinuous pieces of habitat that help fill gaps between two protected areas and help facilitate the movement of species. Habitat corridors and stepping stones allow plants and animals to disperse from one protected area to another, which is important for maintaining genetic diversity and viable population sizes.
5. Protected area networks alone are often not enough to conserve biodiversity at a landscape scale. Management plans must also consider conservation opportunities in urban, agricultural, and mixed-use habitats, which often can provide valuable habitat for species of concern and which provide ecosystem services for people.
6. Managing the conservation landscape requires routine monitoring of land use change and biodiversity, and developing models to forecast changes before they happen. Ecosystem-based management uses monitoring and modeling in an adaptive approach to make corrections as needed to improve outcomes. Ecosystem-based management is collaborative and inclusive of all who have a stake in natural resource management in the landscape.

For Discussion

1. Many protected areas are established without use of proper gap analysis, or even without clear conservation goals in mind. For example, tracts of land that are considered useless for other purposes are often set aside to meet international requirements for Aichi Biodiversity Target 11. Would you invest your time and money into improving these areas and integrating them into your plan? Or would you focus your time and energy on acquiring other areas that are chosen more carefully and that better meet your goals? Explain what you think are the pros and cons of each approach.
2. It often takes a long time, sometimes many decades, to set up a protected area network that effectively conserves biodiversity at the landscape scale. Therefore, once a network is set up, it should remain intact for a long time. But there are always pressures to convert protected areas into other uses such as development, mining, or energy extraction. How would you go about making sure that the protected areas in a network remain protected over the long term?
3. Ecosystem-based management (EMB) is a terrific way to manage a landscape if stakeholders cooperate, but it can fall apart if there is not trust, mutual respect, and a spirit of compromise among the stakeholder groups. What could you do to foster trust, respect, and compromise when trying to implement EBM? What would you do if stakeholders were unable to cooperate with each other?

Suggested Readings

Aronson, M. F. J. et al. 2017. Biodiversity in the city: Key challenges for urban green space management. *Frontiers in Ecology and the Environment* 15: 189–196. Describes approaches for, and challenges to, maximizing biodiversity in urban green spaces.

Bingham, H. et al. 2017. Privately protected areas: Advances and challenges in guidance, policy and documentation. *Parks* 23: 13–28. Describes national policies and reporting of privately protected areas, which represent the fastest growing component of protected area networks.

Cumming, G. S. and Allen, C. R. 2017. Protected areas as social-ecological systems: Perspectives from resilience and social-ecological systems theory. *Ecological Applications* 27: 1709–1717. This editorial discusses use of social-ecological approaches to understand the resilience and sustainability of protected areas.

Dolrenry, S. et al. 2014. A metapopulation approach to African lion (*Panthera leo*) conservation. *PLoS One* 9: e88081. Gives the first metapopulation analysis of lions and shows how spatially realistic models can help prevent local extirpations.

Garibaldi, L. A. et al. 2017. Farming approaches for greater biodiversity, livelihoods, and food security. *Trends in Ecology & Evolution* 32: 68–80. A good review that shows how alternative farming practices can be friendlier to biodiversity while still providing high yields and profits.

Garnett, S. T. et al. 2018. A spatial overview of the global importance of Indigenous lands for conservation. *Nature Sustainability* 1: 369–374. Reviews the scale and location of lands managed by Indigenous peoples, as well as their contribution to global conservation and climate agreements.

Goettsch, B., Duran, A. P. and Gaston, K. J. 2019. Global gap analysis of cactus species and priority sites for their conservation. *Conservation Biology* 33: 369–376. Excellent example of how to complete a gap analysis for a threatened group of species.

Miller, D. A. and Grant, E. H. C. 2015. Estimating occupancy dynamics for large-scale monitoring networks: Amphibian breeding occupancy across protected areas in the northeast United States. *Ecology and Evolution* 5: 4735–4746. Shows how monitoring and patch-occupancy models can be used to identify parts of the landscape most needed for long-term species persistence.

Visit the
Conservation Biology
Companion Website
oup.com/he/cardinale1e
for videos, exercises, links, and other study resources.

© leungchopan/Shutterstock.com

16

Ex Situ Conservation

Megafauna like the endangered whale shark (*Rhincodon typus*) are a major attraction at marine aquariums.

One of the most important goals of conservation is to preserve all levels of biodiversity (genetic, species, community, and ecosystem) in the natural environment where organisms have evolved and are adapted. Conservation of biodiversity in the natural environment is called **in situ (on-site) conservation** (Figure 16.1). In situ conservation was covered in Chapters 13–15 where we described the tools and techniques that conservation biologists use to protect populations, conserve whole communities and ecosystems, and collectively manage protected areas and biological reserves at the scale of entire landscapes.

While in situ conservation gives biodiversity the greatest chance of long-term persistence, there are many instances where the sole reliance on in situ conservation is insufficient to protect biodiversity. Even with good in situ management and protection strategies in place, species populations may still decline and go extinct in the wild for any of number of reasons, including too much habitat destruction, fragmentation and degradation, climate change, impacts of invasive species or disease, or severe overexploitation. And sometimes the remnant populations that are conserved in sanctuaries, parks, protected areas, and reserves are too small to maintain genetic variation or to overcome the demographic and environmental stochasticity that increases the probability of extinction. In these instances, the only way to prevent a species and all of

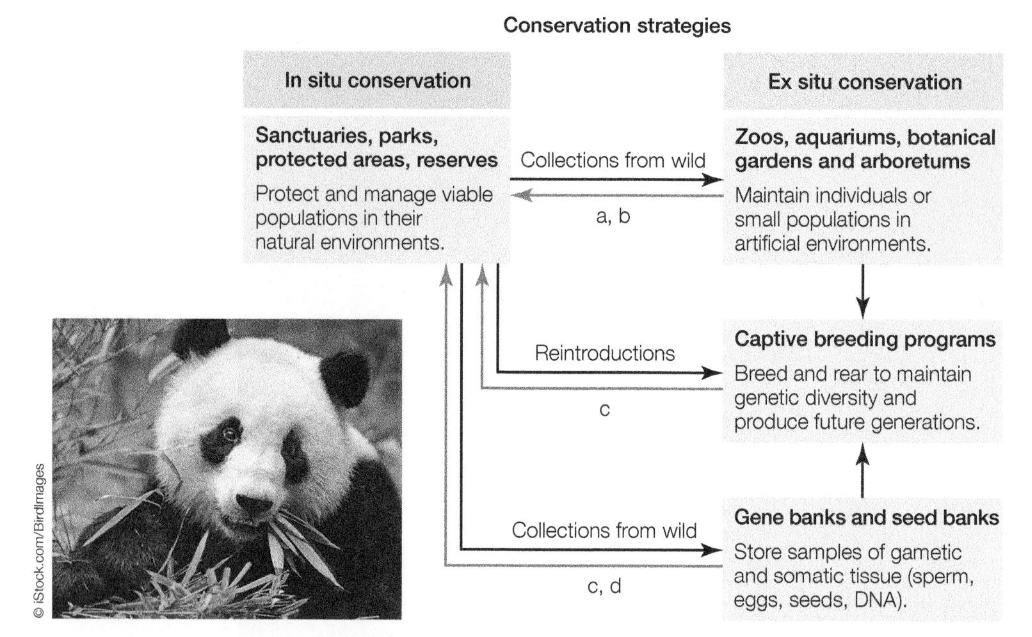

Figure 16.1 The complementary roles of in situ (on-site) and ex situ (off-site) conservation. Black arrows illustrate the direct routes by which in situ and ex situ conservation efforts benefit each other with alternative conservation strategies. Collections from the wild that are maintained in artificial environments under human supervision can serve as short-term buffers against extinction and as material for captive-breeding programs that allow species and their genes to be reintroduced to the wild. Gray arrows show additional indirect benefits of ex situ facilities for in situ conservation efforts: (a) outreach and education activities at ex situ facilities can raise public awareness in ways that encourage in situ conservation; (b) some revenues generated at ex situ facilities are used to fund in situ conservation; (c) captive-bred individuals offer the opportunity for research, reintroductions, and reinforcements that can improve the success of in situ conservation efforts; (d) seed banks are used to restore native populations of plants. While no single species conforms to this full, idealized model of conservation, the giant panda is an example where numerous elements from both in situ and ex situ conservation were used to save the species from extinction. (After N. Maxted. 2001. In S. A. Levin [Ed.], *Encyclopedia of Biodiversity*, pp. 683–696. San Diego, CA: Academic Press.)

its genetic diversity from going extinct may be to maintain individuals in artificial environments.[1]

Ex situ (off-site) conservation is the protection and management of biodiversity in artificial, human-built environments (see Figure 16.1). Ex situ conservation allows wild-caught individuals to be maintained under close human supervision in facilities like zoos, aquariums, botanical gardens, or arboretums. Collections of individuals that are kept alive in artificial environments can serve as short-term buffers against extinction. They also allow for captive individuals to be bred and reared so they can be reintroduced to the wild to reestablish or augment populations (**Box 16.1**).

In addition to maintaining live plants and animals, we can collect samples of sperm, eggs, seeds, or DNA from organisms and store them in gene or seed banks to ensure that a species' gametic and somatic tissues are preserved. Collections of gametic and somatic tissues allow us to use biomedical technologies to improve the success of captive-breeding programs (e.g., artificial insemination) and provide opportunities for use of new technologies that may one day help resurrect species from extinction (e.g., cloning).

Ex situ conservation has a controversial history that has led to several criticisms and negative stereotypes, some of which are discussed in this chapter. But ex situ

BOX 16.1 Case in Point

A Comprehensive Conservation Plan Saved the Giant Panda

Efforts to save the giant panda (*Ailuropoda melanoleuca*), a worldwide symbol of wildlife conservation for more than half a century, are finally paying off: the iconic black-and-white bear is no longer endangered, upgraded in 2016 by the IUCN from "endangered" to "vulnerable."[79] But it took a decades-long mix of in situ and ex situ conservation efforts to save this unique bear from extinction.

Giant pandas live mainly in high forests in the mountains of central China, where they subsist almost entirely on bamboo. The panda's problems began decades ago as agricultural expansion and deforestation of their mountain habitat led to substantial habitat loss for the species. Roads and human settlements fragmented their few remaining habitats, making it difficult for pandas to migrate in search of bamboo. To make matters worse, giant pandas were a target of poaching, killed for their valued skins and furs.[80] The reclusive nature of the animals made it difficult to get accurate censuses; however, estimates in the 1980s suggested fewer than 1000 pandas were still alive,[81] so in 1988 the species was listed as endangered by the IUCN.

Giant panda populations in the wild have been on the rise, however, for nearly two decades, up to 1864 animals during the last national census in 2014.[82] Their success can be attributed to several factors: First, the Chinese government moved quickly to protect the giant panda's habitat. A logging ban declared at the end of 1998 put most remaining habitat off-limits to commercial logging, and Chinese authorities established 67 giant panda reserves for in situ conservation.

Second, new policies were established to reduce poaching. In 1983, the Chinese government enacted wildlife protection laws that increased protections for the giant panda. And in 1984, the species was listed in Appendix I of CITES, which prohibited trade of the species or its products internationally.

Last, the Chinese government established an ex situ population of giant pandas that included a captive-breeding program (see Figure). Initial success rates of the breeding program were low. Little was known about the reproductive biology of adults, which limited the number of successful matings. And poor understanding of the nutritional needs and care of newborn cubs led to high infant mortality. But study of captive population biology and the increased use of artificial insemination greatly improved success. Today, the Wolong Giant Panda Center in Sichuan Province, China, produces over two dozen cubs per year, some of which are released into the wild to augment giant panda populations (**Video 16.1**).

The giant panda provides a case in point for how in situ and ex situ conservation efforts can complement one another in a comprehensive plan to save a threatened species.

A captive-bred giant panda being reintroduced to the mountains of central China.

conservation often means the difference between persistence and extinction.[2] There are, in fact, 69 plant and animal species that are listed by the International Union for Conservation of Nature (IUCN) as extinct in the wild but which persist in captive colonies in ex situ facilities. Without these captive colonies, these species would have gone globally extinct, forever lost with all of their beauty, genetic variation, and potential ecological functions. This is why some have compared ex situ facilities to Noah's Ark, the biblical boat that served as a last refuge that allowed species to survive a short-term catastrophe (40-day flood). Similarly, ex situ conservation facilities provide species a short-term refuge from extinction until catastrophes in their natural habitat can subside or be overcome.

In addition to serving as a short-term refuge from extinction, ex situ conservation facilities have the potential to indirectly benefit in situ conservation efforts

VIDEO 16.1 Watch a video about the challenges of raising giant pandas.
oup-arc.com/e/cardinale-v16.1

(see Figure 16.1). Research on captive populations can provide insight into the basic biology, physiology, animal behavior, and genetics of the species through research studies that would not be possible on wild animals. Results of these studies can suggest new conservation strategies for in situ populations. Similarly, the ease of access to individual animals in captivity allows scientists to develop and test relevant technologies (e.g., radio collars) that enhance the study and preservation of the species in the wild. Long-term, self-sustaining ex situ populations can also reduce the need to collect individuals from the wild for display and research. Captive-bred individuals on display can help educate the public about the need to preserve the species in the wild. Zoos, aquariums, botanical gardens, and the people who visit them may contribute money to in situ conservation programs. In addition, ex situ programs can be used to develop new products that can potentially generate funds from profits or licensing fees to protect species in the wild.

In this chapter, we describe the role that ex situ conservation plays in local, regional, and global conservation strategies. We begin by describing the major types of ex situ conservation facilities and what they contribute to conservation. We then describe captive-breeding programs that serve as the source of reintroductions of species to the wild.

Ex Situ Conservation Facilities

The most common types of ex situ conservation facilities currently in use are zoos, aquariums, botanical gardens, and gene and seed banks. In this section, we'll examine each of these facilities and describe its role in conservation programs.

Zoos and aquariums

A **zoo**, also called an **animal park** or **zoological park**, is a facility in which animals are housed in human-built enclosures and displayed to the public and in which they may also breed. An **aquarium** is the aquatic counterpart of a zoo, housing living aquatic animal specimens for public viewing. The concept of a zoo has changed radically through time (Figure 16.2). While the display of captive wild animals is ancient, the practice of managing whole collections of captive animals for the purpose of display and entertainment began with **royal menageries**. Royal menageries were large collections of exotic animals that were kept by aristocrats or royal courts within palace gardens for the purpose of entertainment. Menageries were widely used throughout the Middle Ages (fifth to fifteenth centuries) to display wealth and power by European royals. For example, the eighth-century Roman emperor Charlemagne built three menageries in his palaces in Germany (Aachen and Ingelheim) and the Netherlands (Nijmegen), where he housed elephants, monkeys, lions, bears, camels, and exotic birds that had been given to him as gifts from rulers of Africa and Asia.[3]

The most prominent collection of exotic animals in medieval England was the Tower Menagerie in London. Established in 1204 by King John, the Tower Menagerie persisted as the royal collection for more than six centuries, acquiring a large assortment of exotic animals that were given to kings and queens as gifts.[4] These included elephants, lions, leopards, and a "white" bear—presumably a polar bear that was muzzled and chained so that it could be kept on display in the courtyard. During the seventeenth century, King Louis XIV of France established two large royal menageries—one at Vincennes, where "ferocious" beasts like lions, tigers, and leopards were used to entertain nobles and visiting dignitaries with animal fights,[5] and a more elaborate menagerie at the palace of Versailles, where animals housed in small cages were used as decorations to augment the baroque style architecture.[6]

Duncan from Nottingham, UK/CC BY 2.0

© Hulton Archive/Stringer/Getty Images

© Dominic Johnson/Minden Pictures

© Jochen Tack/Alamy Stock Photo

5th to 18th centuries
Royal menageries
Goal: Private exhibition and entertainment
Theme: Symbols of wealth and power
Displays: Royal gardens

18th to 20th centuries
Traveling menageries
Goal: Public exhibition and entertainment
Theme: Strange and diverse collections
Displays: Cages

20th century
Zoological parks
Goal: Living museum
Theme: Animal behavior, species management
Displays: Dioramas

21st century
Conservation centers
Goal: Environmental resource center
Theme: Ecosystems, holistic conservation
Displays: Immersion exhibits

Figure 16.2 The evolution of zoos and aquariums. Zoos and aquariums began as menageries that put animals on display, first for royalty, and then for the general public. In the twentieth century they evolved into zoological parks that had a dual purpose in entertainment and research. In 2015, the World Zoo Organization and the Captive Breeding Specialist Group of IUCN published the World Zoo Conservation Strategy, which called on zoos and aquariums to evolve yet again, to morph into conservation centers of the twenty-first century that address sustainable relationships between humans and nature and that explain the values of ecosystems and the importance of conserving biological diversity. (After IUDZG/CBSG [IUCN/SSC]. 1993. *Executive Summary. The World Zoo Conservation Strategy; The Role of the Zoos and Aquaria of the World in Global Conservation.* Brookfield, IL: Chicago Zoological Society. © 1993 IUDZG—The World Zoo Organization; R. Barongi et al. 2015. *Committing to Conservation: The World Zoo and Aquarium Conservation Strategy.* Gland, Switzerland: WAZA Executive Office.)

Royal menageries were not generally open for public viewing. But as word spread about the strange creatures held in royal gardens, demand for public showing of animals was met by the emergence of **traveling menageries**. Traveling menageries were touring groups of showmen and animal handlers who visited towns and cities with collections of common and exotic animals. Traveling collections of animals first appeared in England about 1703. Perhaps the largest and most well known was Wombwell's Traveling Menagerie that, by its peak, touted more than 600 animals on exhibit. Traveling menageries appeared in 1710 in America,[7] where they were soon combined with circuses that sold both animal and human performances as entertainment. The Ringling Bros. and Barnum & Bailey Circus, which ran in one form or another as one of the largest traveling circuses in America for 146 years (1871–2017), originally advertised their shows as the "World's Greatest Menagerie."[8] But stories of poor animal welfare and violations of animal care laws ultimately led to intense public pressure to eliminate animal performances from circuses. Thirty-four countries have now banned the use of exotic animals in circus performances, as have many large cities and local governments.[9]

Even though traveling menageries persisted into the twentieth century, the general concept of zoos and aquariums began to change in two ways during the eighteenth century. First, royal menageries were transitioned into zoological parks for the purpose of public viewing (see Figure 16.2). The first zoological park was the Tiergarten Schönbrunn in Vienna, Austria, which was originally constructed as an imperial menagerie in Schönbrunn Palace in 1752 but was made accessible to the public in 1765.[10] Second, zoological parks began to embrace the scientific study of animals as part of their mission. The world's oldest scientific zoological park, the London Zoo, opened in 1828. Initially known as the Gardens and Menagerie of the Zoological Society of London, the zoo was intended to be used as a collection of animals for scientific study.[4] But it was opened to the public in 1847 to try to augment funding. The first public aquarium was opened as part of the London Zoo shortly after, in 1853.

Throughout the twentieth century, many types of ex situ facilities were developed to manage collections of terrestrial and aquatic animal species so that they were easier to view and study than in nature. These facilities included not only zoos and aquariums, but also game reserves, aviaries, safari parks, rescue centers, and sanctuaries. While there are now thousands if not tens of thousands of these facilities around the globe, only a fraction of them are accredited by organizations that govern and oversee regulations for the care of wild animals in captivity. The World Association of Zoos and Aquariums (WAZA, oup-arc.com/e/cardinale-w16.1), based in Barcelona, Spain, is the prominent umbrella organization of zoos and aquariums worldwide. WAZA supports and guides its partner organizations in animal care and welfare, conservation of biodiversity, and environmental education.

WAZA is composed of approximately 400 leading zoos, aquariums, and affiliated organizations (e.g., zoo veterinarians and zoo educators) from around the world. In addition, roughly 1300 zoos and aquariums are members through regional associations that help coordinate conservation and educational goals and provide accreditation in specific geographic areas of the world. These include regional associations like the Pan-African Association of Zoos and Aquaria (oup-arc.com/e/cardinale-w16.2), the Latin American Zoo and Aquarium Association (oup-arc.com/e/cardinale-w16.3), the South East Asian Zoos and Aquariums Association (oup-arc.com/e/cardinale-w16.4), the Zoo and Aquarium Association for Australia and New Zealand (oup-arc.com/e/cardinale-w16.5), the European Association of Zoos and Aquaria (oup-arc.com/e/cardinale-w16.6), and the Association of Zoos and Aquariums (oup-arc.com/e/cardinale-w16.7) that primarily represents North America.

Suggested Exercise 16.1 Read this summary of how zoos and aquariums earn accreditation from major accrediting organizations. **oup-arc.com/e/cardinale-ex16.1**

In addition to WAZA and its regional associations that coordinate conservation and educational goals and oversee accreditation processes, there are numerous organizations that help coordinate the activities and biological collections of zoos and aquariums. One of these is **Species360** (oup-arc.com/e/cardinale-w16.8), which is an international nonprofit organization that maintains an online database of wild animals that are under human care in more than 1100 zoos and aquariums spanning 96 countries. Species360 provides its members with data collection and management software called ZIMS—the Zoological Information Management System (oup-arc.com/e/cardinale-w16.9)—which is designed to provide a comprehensive, global source of information on animals that are maintained in zoos, as well as the environments they are housed in, to better serve animal management and conservation goals.

According to ZIMS, modern zoos and aquariums presently hold about 10 million individual animals representing more than 22,000 species. These collections are disproportionately dedicated to large vertebrates—especially mammals—since these species are of greatest interest to the general public. However, zoos and aquariums have increasingly directed more of their attention and efforts to the housing and management of smaller-bodied species such as insects, amphibians, and reptiles, which are less expensive to maintain in large numbers than large-bodied mammals such as giant pandas, elephants, and rhinos. In addition, displays of colorful species (e.g., frogs) and those that can be maintained as large populations (e.g., butterflies) have proven to be increasingly popular to the public and are appealing avenues to teach about the diversity of life.

Conde and colleagues[11,12] provided the most recently published estimate of how many species of conservation concern are housed in zoos and aquariums. The authors summarized records from the Species360 database (which was then called the International Species Information System, ISIS) and cross-referenced species in that database to their status on the IUCN Red List. While zoos and aquariums harbor just 1.8% of the world's 1.24 million described eukaryotic species,[13] they hold a disproportionately large number of threatened species (**Figure 16.3**, upper panels). Roughly one in seven species (15%) threatened with extinction (categorized by the IUCN as vulnerable, endangered, or critically endangered) is held in a zoo or aquarium. This includes 19%–24% of threatened mammal species, 9%–17% of threatened bird species, 28%–51% of threatened reptile species, and 2%–4% of threatened amphibian species.

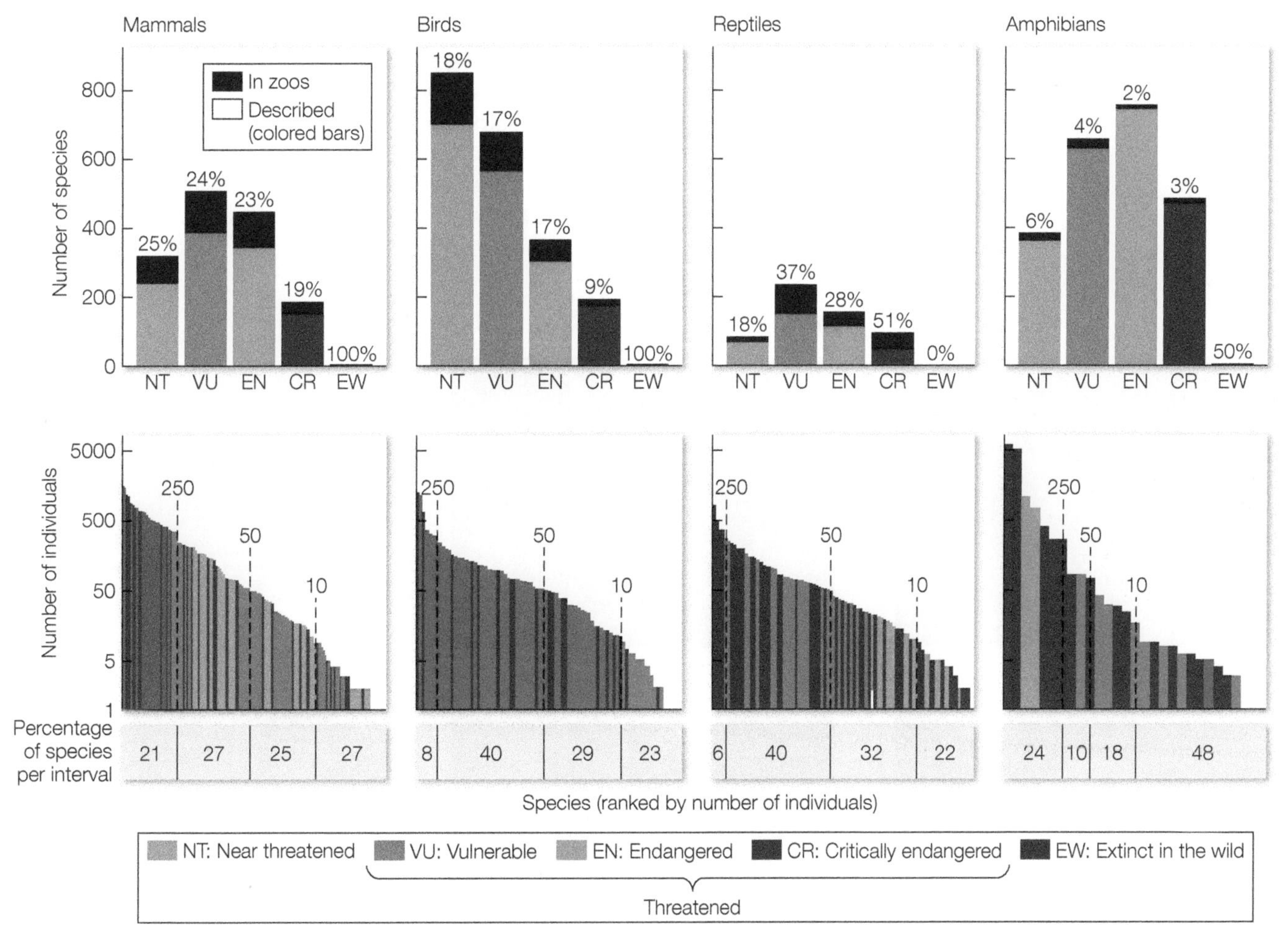

Figure 16.3 Threatened animal species in captivity. Approximately 22,000 species are held in zoos and aquariums, representing about 1.8% of global biodiversity but which represent 4%–25% of species of concern, depending on the taxonomic group. The top panels in this figure show the number of species that had been globally described at the time of this study (colored bars) compared with the number that were held in zoos (black bars) for each IUCN status. The bottom panels give frequency distributions showing how many individuals are held in zoos—for mammals (142 species), birds (83 species), reptiles (90 species), and amphibians (29 species). The vertical dashed lines give boundaries for the percentage of species (*x*-axis) that have 10, 50, and 250 individuals represented in zoos (*y*-axis). (After D. A. Conde et al. 2011. *Science* 331: 1390–1391.)

In addition to the number of species threatened with extinction, zoos and aquariums hold the only surviving individuals for numerous species of mammals, birds, and amphibians that are considered to be extinct in the wild, including the following:

- The Panamanian golden frog (*Atelopus zeteki*) is a poisonous species of toad that historically inhabited streams along the Codillera cloud forests of west central Panama (**Figure 16.4A**). The species is considered to be one of the most beautiful frogs in Panama and is a national symbol. Even so, the golden frog began vanishing from its high mountain forests in the late 1990s, possibly because of an outbreak of fungal disease. The species was found for the last time in the wild in 2006 when David Attenborough hosted the BBC's series *Life in Cold Blood*. Fortunately, before being declared extinct in the wild in 2007, a small population was taken into captivity for safekeeping, and several zoos are now collaborating on a conservation project to keep the species safe from extinction.[14]
- The Micronesian or Guam kingfisher (*Todiramphus cinnamominus*) is a brilliantly colored, medium-sized species of kingfisher native to the U.S. territory of Guam (**Figure 16.4B**). The species was extirpated from its native habitat on the island after the accidental introduction of the brown tree snake, which is a voracious predator of birds, their eggs, and their nestlings. The kingfisher now persists only as a captive population of less than 200 birds (as of 2017) in U.S. mainland and Guam breeding facilities.[15]

Figure 16.4 Examples of animal species (top panels) and plant species (bottom panels) that have been declared "extinct in the wild" but which still exist as captive populations in ex situ conservation facilities: (A) the Panamanian golden frog (*Atelopus zeteki*), (B) the Guam kingfisher (*Todiramphus cinnamominus*), (C) the Socorro isopod (*Thermosphaeroma thermophilum*), (D) the Franklin tree (*Franklinia alatamaha*), (E) Wood's cycad (*Encephalartos woodii*), (F) brome of the Ardennes (*Bromus bromoideus*).

- The Socorro isopod (*Thermosphaeroma thermophilum*) is a crustacean that was endemic to the Sedillo thermal spring located in Socorro county, New Mexico (Figure 16.4C). When the spring was diverted to the city of Socorro for drinking water, the isopods persisted in three small populations in a horse trough and two bathing pools before going extinct in 1988. Fortunately, before they went extinct, captive populations were established at New Mexico Tech's biology department and New Mexico's Department of Game and Fish, as well as the Albuquerque Biological Park and the Minnesota Zoo.[16]

While twentieth-century zoos and aquariums serve as a temporary refuge for many threatened species, they have struggled to move beyond the poor reputations and negative stereotypes that stem from their beginnings as menageries that often mistreated their animals. One current criticism is that exhibit sizes remain far too small to maintain proper animal care and well-being, often orders of magnitude smaller than the minimum home range sizes of animals in the wild. This can lead to abnormally high levels of animal stress and mortality. For example, Clubb and Mason[17] studied 35 species of large carnivores in captivity and showed that the larger their wild home range was, the higher their infant death rate was in captivity, and the higher was the frequency of stereotypy pacing that indicates animal stress (Figure 16.5). Among the worst off were polar bears, which have a home range of 1000 km^2 or more, which is 1 million times bigger than their average enclosure in a zoo. Polar bears paced approximately 25% of the day in captive enclosures and had infant mortality rates of 65%. Films like the 2013 documentary *Blackfish*, which described the stress experienced by captive orcas (killer whales) held in public aquariums, have fueled much public debate about the ethics of holding animals in captivity and have even provoked boycotts that have caused ex situ facilities like SeaWorld in San Diego, California, to close their orca exhibits.

The management of surplus animals has also become an ethical lightning rod for zoos and aquariums. Many ex situ facilities do not have the space to house animals that breed in captivity. In the absence of programs to release excess animals into the wild, they must be euthanized. This problem was

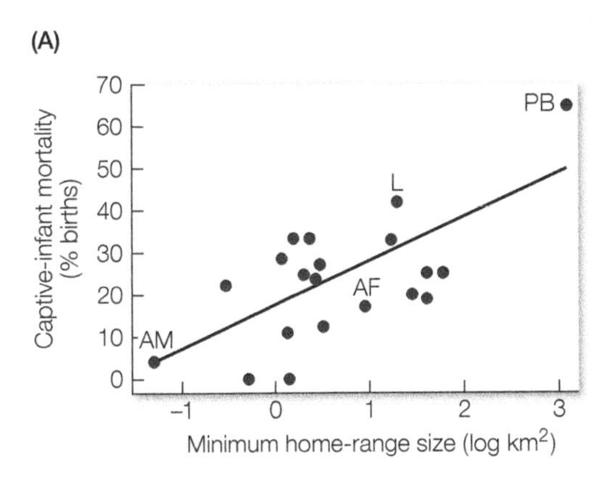

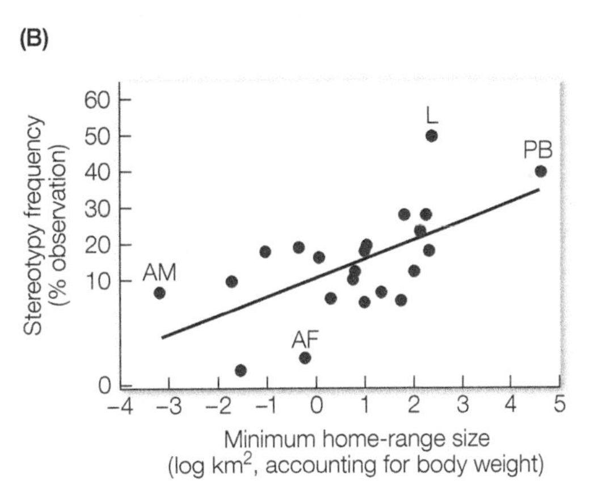

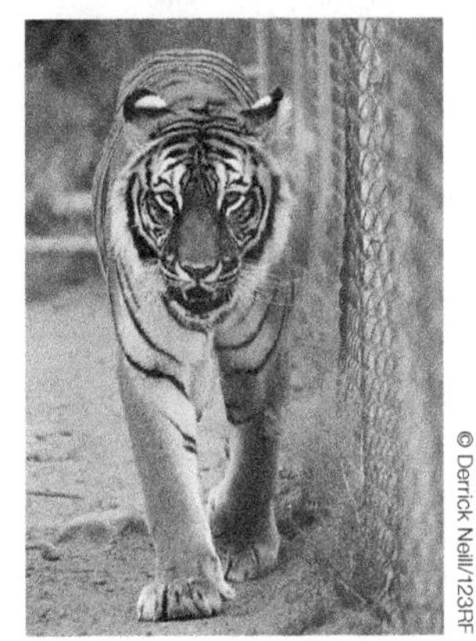

Figure 16.5 Measures of animal welfare for large carnivores held in captivity. These plots show (A) infant mortality and (B) stereotypic pacing in captive populations as functions of the carnivores' minimum home-range sizes in the wild. Highlighted species: AF, arctic fox (*Alopex lagopus*); PB, polar bear (*Ursus maritimus*); AM, American mink (*Mustela vison*); L, lion (*Panthera leo*). Note that values on the *x*-axes differ because fitted values used in B also incorporate animal body weight. (After R. Clubb and G. Mason. 2003. *Nature* 425: 473–474.)

made public in 2014 when the Copenhagen Zoo killed a healthy male giraffe named Marius. Because his genotype was considered too common in zoos, he was not sought out for breeding, and the zoo thought the expense of keeping him was too great. Instead, they euthanized him and opened his necropsy to public viewing, describing it as educational, and then fed his remains to other zoo animals. The incident raised much concern among the public and animal ethics organizations.

In addition to ethical concerns about animal welfare, some have questioned the contributions that twentieth-century zoos and aquariums make to animal conservation. Critics argue that less than 3% of the budgets of accredited zoos go toward conservation efforts[18] and that caring for animals in captivity is far more expensive than is in situ conservation. One study showed that the cost of maintaining African elephants and black rhinos in zoos is 50 times greater than protecting the same number of animals in East African national parks, suggesting that in situ conservation is a better use of funds.[19] Critics also argue that population sizes of threatened or extinct species maintained in zoos and aquariums are too small to save species from extinction. Nearly a quarter of all threatened species housed in zoos and aquariums are represented by 10 or fewer individuals, and more than half of all threatened species are represented by fewer than 50 individuals (see Figure 16.3, lower panels).

In response, advocates for ex situ facilities have noted that the world community of zoos and aquariums has historically been the third-largest contributor to wildlife conservation efforts, ranking only behind major conservation organizations like The Nature Conservancy and the World Wildlife Fund (**Figure 16.6**).[20] In addition, zoos and aquariums are working to shed historically negative stereotypes and practices and reinvent themselves as twenty-first-century conservation centers (see Figure 16.2). In its 2015 report *Committing to Conservation: The World Zoo and Aquarium Conservation Strategy*,[21] WAZA outlined its plan for zoos and aquariums around the world to evolve, from simply being public exhibits of exotic animals, to environmental resource centers of biodiversity conservation (**Video 16.2**). WAZA called on its members to improve public education efforts to focus more on the causes and consequences of biodiversity loss in the wild and to make research and in situ conservation of wild populations a greater part of their financial portfolios. Many zoos and aquariums have already started this transition by rebranding themselves as "bioparks" and "wildlife conservation parks."[22] They have begun to change the way animals are displayed, using immersion exhibits that better mimic native habitats in order to improve animal health and educate the public about the plight of endangered species and their ecosystems.

VIDEO 16.2 Watch a video on the future of zoos.
oup-arc.com/e/cardinale-v16.2

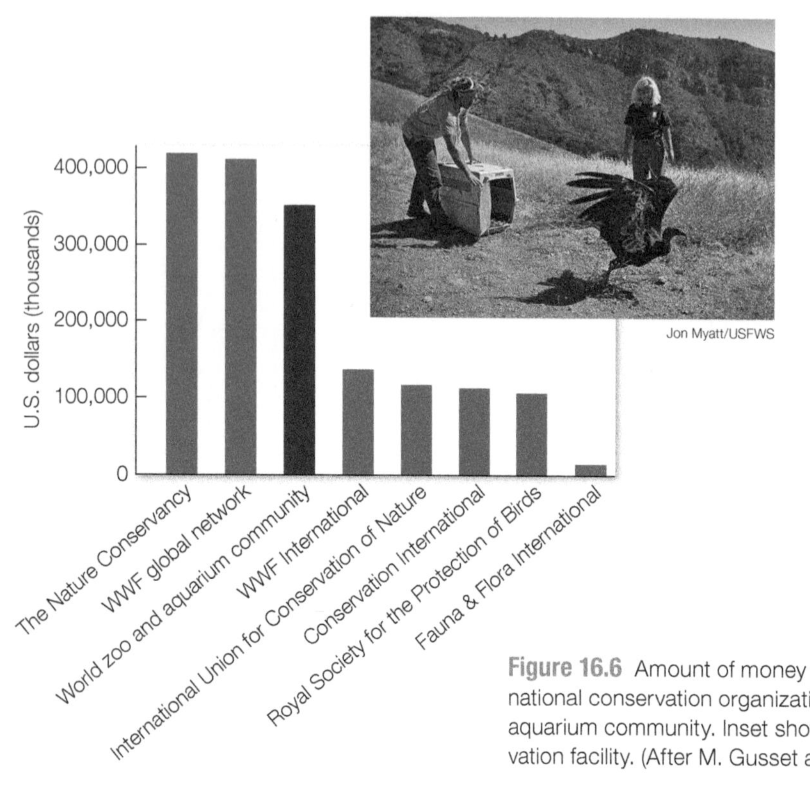

Figure 16.6 Amount of money spent in 2008 on wildlife conservation by major international conservation organizations compared with that spent by the world zoo and aquarium community. Inset shows a condor being released from an ex situ conservation facility. (After M. Gusset and G. Dick. 2011. *Zoo Biol* 30: 566–569.)

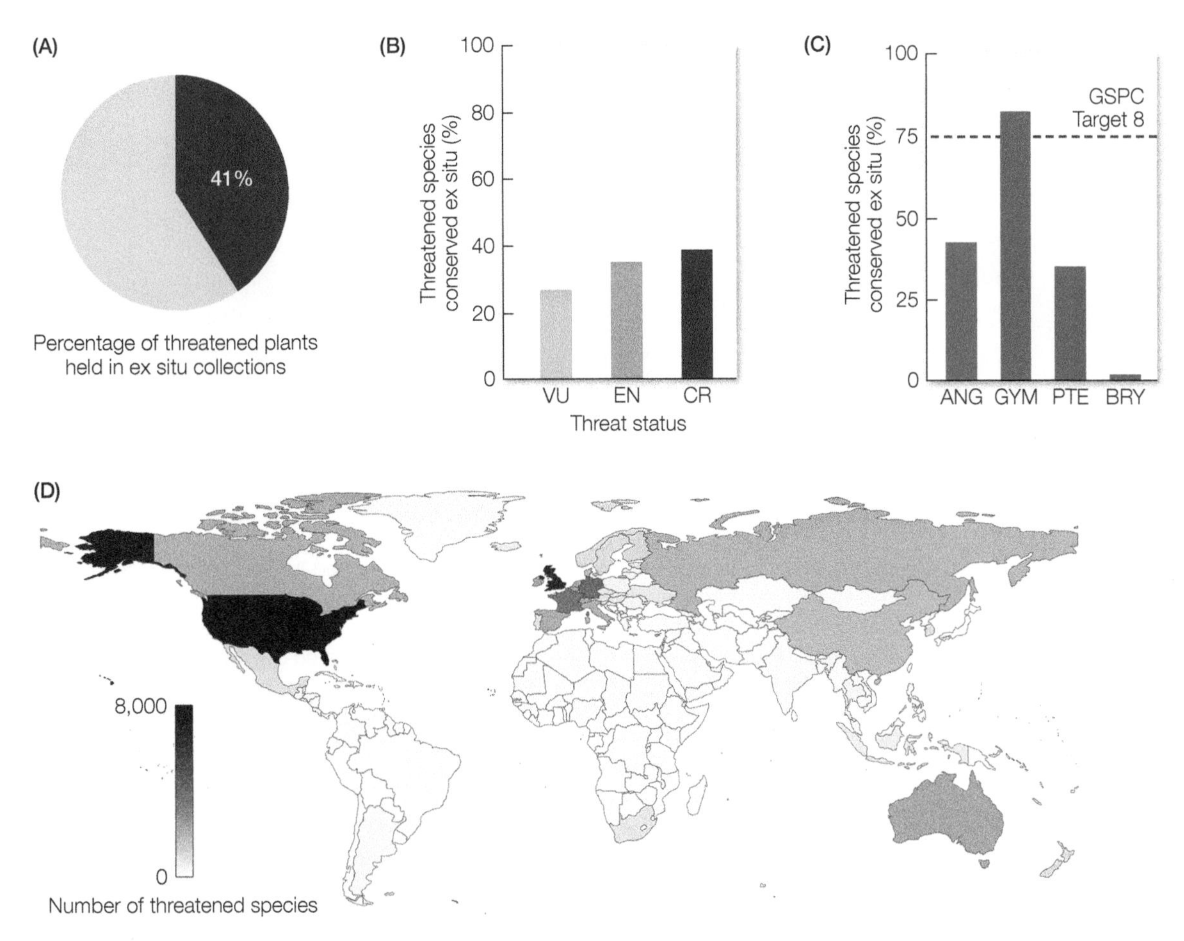

Figure 16.7 Threatened plants in botanical gardens. Botanical gardens hold representatives of 30% of all plant species and 59% of all genera. Figures here show (A) the percentage of threatened plant species that are conserved in ex situ facilities; (B) the percentage of plant species in IUCN threatened categories that are held ex situ (VU = vulnerable to extinction, EN = endangered, CR = critically endangered); (C) the percentage of plant species that are conserved ex situ, by different plant groups (ANG = angiosperms, GYM = gymnosperms, PTE = pteridophytes, BRY = bryophytes); (D) the number of threatened plant species held in ex situ facilities in different countries around the world. (After R. Mounce et al. 2017. *Nature Plants* 3: 795–802.)

English botanist David Aplin discovered preserved seeds in collections of Belgium's National Botanic Garden. Since 2009 the garden has been able to germinate seeds and grow plants.[28]

As is true with zoos and aquariums, there is a worldwide bias in the distribution of botanical gardens and availability of associated digitized collections (**Figure 16.7D**). While the majority of species live wild in the tropics, more than 90% of archived species are in ex situ collections in the Northern Hemisphere. The United States and Great Britain host the most botanical gardens and, in turn, the most threatened plant species. Several regional efforts in both countries are coordinating plant conservation efforts among gardens. For example, the Center for Plant Conservation (oup-arc.com/e/cardinale-w16.16) in the United States maintains a collection of more than 1400 of America's most imperiled native plants through a coordinated effort among its network of 47 participating institutions.

Despite the geographic biases, plants are often easier to translocate and maintain in captivity than animals, because they have similar basic needs for light, water, soils, and minerals that can be readily supplied in greenhouses and gardens. Since plants do not move or require territories, they often can be grown in high densities, which helps reduce problems associated with small

(A)

(B)

Figure 16.8 Gene banks and seed banks. (A) The Plant and Animal Genetic Resources Preservation center, located in Fort Collins, Colorado, is the U.S. Department of Agriculture's National Laboratory for Genetic Resources Preservation. (B) The Svalbard Global Seed Vault is an important and secure seed collection, located deep inside a mountain on the Svalbard archipelago, halfway between mainland Norway and the North Pole.

population sizes. Many plant species readily produce seeds on their own, which can be collected, stored, and later germinated to produce more plants.

Gene banks and seed banks

Gene banks are biorepositories that preserve genetic material (Figure 16.8). For animals, gene banks collect and hold tissues, cells, gametes (sperm and egg cells), and DNA of endangered animals in cryogenic freezers. Sometimes referred to as **frozen zoos**, these facilities can store samples for long periods of time. The samples can be used later to assist with reproduction through technologies such as **artificial insemination** (introduction of a male's sperm into a female's cervix or uterine cavity to try to achieve pregnancy) and **in vitro fertilization** (fertilization of an egg with sperm outside the body, followed by implantation of the embryo into the uterus). Such technologies have become more common over the past several decades, and there are numerous examples of their being used to impregnate living females of threatened animal species. In addition, gene banks serve as an important source of material for emerging technologies like **cloning** (the process of creating a new multicellular organism through asexual reproduction of cells).

The idea of the frozen zoo originated in the 1970s with geneticist and conservation biologist Kurt Benirschke.[29] Benirschke worked at the San Diego Zoo, where he established a collection of frozen cells and reproductive material from dozens of highly endangered species. In the 1970s when he began his collection, there were no artificial reproductive technologies available to make use of the collection. But Benirschke believed the technologies would eventually be developed, so he proceeded to create a gene bank that has now grown into one of the largest in the world, containing more than 10,000 living cell cultures, oocytes, sperm, and embryos representing nearly 1000 taxa (oup-arc.com/e/cardinale-w16.17). This collection contains many precious samples, including those of the po'ouli (black-faced honeycreeper, *Melamprosops phaeosoma*), a species of passerine bird endemic to Hawaii that is now thought to be globally extinct (Figure 16.9).[30]

Figure 16.9 The po'ouli, or black-faced honeycreeper (*Melamprosops phaeosoma*), is a passerine bird endemic to Hawai'i that is now thought to be extinct. Cell cultures of the species are stored cryogenically at the San Diego Zoo's Beckman Center for Conservation Research. These samples provide some potential opportunity for the species to be resurrected from the dead once technologies allow.

By the 1990s, frozen zoos had been established in many developed countries around the world that had active wildlife conservation programs (**Video 16.3**). For example, in 1995 Australia established the Animal Gene Storage and Resource Centre of Australia (AGSRCA) as a joint venture between Monash University and the Zoological Parks Board of New South Wales, which included the Taronga Zoo at Sydney and the Western Plains Zoo at Dubbo. The AGSRCA maintains a cryopreservation facility at Monash University in Melbourne that holds over 100 different species, including critically endangered species like the northern hairy-nosed wombat (*Lasiorhinus krefftii*) and the black rhinoceros (*Diceros bicornis michaeli*).[31]

> **VIDEO 16.3** Watch a video on how frozen zoos could be endangered species' best hope for survival.
> **oup-arc.com/e/cardinale-v16.3**

The umbrella organization that coordinates activities and global collaboration among gene banks is called the Frozen Ark Project (oup-arc.com/e/cardinale-w16.18), which is a nonprofit charity run out of the University of Nottingham in the United Kingdom. It facilitates the conservation of tissues, cells, and DNA from endangered animals; provides a portal where information can be accessed about samples that are stored and available from different gene banks around the world; and helps establish best practices for the collection and storage of the tissues, cells, gametes (sperm and egg cells), and DNA of endangered animals. Today, the Frozen Ark Consortium consists of 27 members who run 22 storage facilities in 12 countries, including the United Kingdom, United States, Australia, New Zealand, Germany, India, Korea, South Africa, Norway, and Ireland. The consortium includes the Natural History Museum in London, the American Museum of Natural History, the San Diego Zoo, and the Australian Frozen Zoo.

For plants, preservation of species and genetic variation is most often accomplished by stocking the seeds in a specialized gene bank called a **seed bank**. Seeds of most plant species can be stored for long periods of time in cold, dry conditions where metabolism is slowed. Seeds can be germinated later to produce new plants. Because seeds of large numbers of important or rare species can be stored in small spaces with minimal supervision at a low cost, seed banks are particularly well suited to ex situ conservation efforts.

There are about 370 institutions from 74 different countries around the world that collect and bank seeds of wild plant species (**Figure 16.10**).[32] The majority of

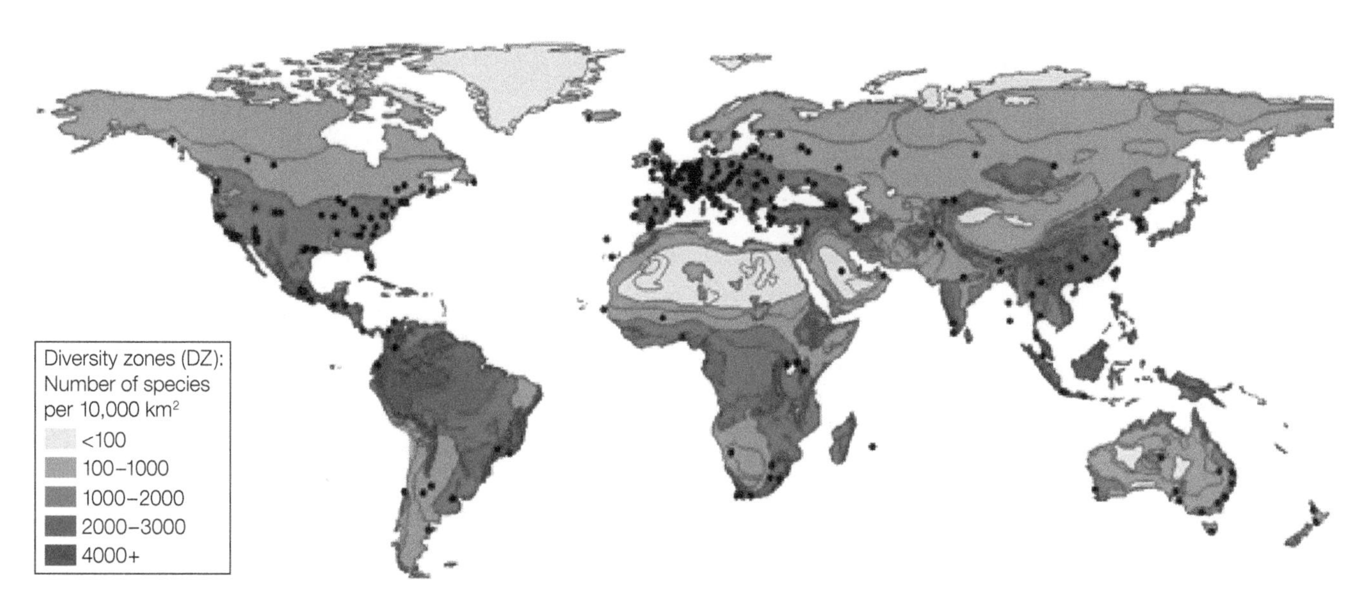

Figure 16.10 The location of seed-banking institutions (black dots) in relation to plant biodiversity (colors, with legend at left). (After K. O'Donnell and S. Sharrock. 2017. *Plant Diversity* 39: 373–378; G. Kier et al. 2005. *J Biogeogr* 32: 1107–1116.)

these (316) are botanical gardens and arboretums, but they also include specialized seed banks (e.g., the National Tree Seed Centre in Canada) and university research facilities (e.g., Universidad Politécnica de Madrid Seed Bank). The United States, Australia, and France have more than 20 institutions devoted to seed conservation. However, most countries have only one or two institutions involved, and countries in geographic regions like South America, Central Africa, and Southeast Asia that have particularly high plant diversity tend to have little or no seed-banking capabilities.

Seed-banking institutions and networks have developed numerous goals at the national, regional, and global levels to help achieve the GSPC's Target 8 to conserve 75% or more of threatened plant species in ex situ collections by 2020 (Table 16.1). O'Donnell and Sharrock[32] used the BGCI databases (GardenSearch, PlantSearch, ThreatSearch, and GlobalTreeSearch) to analyze the contribution seed banks have made toward Target 8. They showed that 56,987 distinct plant taxa are now represented in seed banks. Of the species that have had conservation assessments completed at the global or national level, 34% (9696) are considered threatened with extinction. A total of 6881 tree species from 166 countries are stored in seed bank collections. More than half (3562) of these tree species are single-country endemics. Only 32 countries have more than 20% of their endemic tree species conserved in seed banks, and just five countries have 75% or more.

One of the world's largest banks is the Millennium Seed Bank project of the Royal Botanic Gardens, Kew (oup-arc.com/e/cardinale-w16.19), which has a goal of conserving seeds of 25% of the world's plant species by 2020. The U.S. Department of Agriculture's National Center for Genetic Resources Preservation in Fort Collins, Colorado, currently stores seed samples from 10,000 plant species (oup-arc.com/e/cardinale-w16.20). The Institute of Crop Germplasm Resources in Beijing, China, has over 400,000 seed collections. Perhaps the most famous seed bank is the Svalbard Global Seed Vault in Norway, where

TABLE 16.1 Examples of global, regional, and national level seed bank targets

2020 targets	Main implementer	Facilitators
GLOBAL		
Double the number of threatened species in seed banks	BGCI's Global Seed Conservation Challenge (GSCC)	200 GSCC member botanical gardens
25% of world's bankable species conserved	RBG Kew's MSBP	MSB Partnership institutions
REGIONAL: MULTI COUNTRY		
500 vascular plant species	The Alpine Seed Conservation and Research Network	5 plant science institutions in 4 countries (France, Switzerland, Italy, Austria)
NATIONAL		
60% of Korea's native plant species	Korea National Arboretum	12 botanic garden and plant conservation institutions
75% of Australia's threatened species	Australian Seed Bank Partnership	71 organizations including botanic gardens, nature reserves and universities
10,000 [of] China's native taxa	Germplasm Bank of Wild Species Kunming Institute of Botany, Chinese Academy of Science	
REGIONAL: SUB COUNTRY		
100% of the California flora	California Plant Rescue Project	Conservation organizations, botanic gardens and seed banks
75% of the regions threatened species conserved in seed banks or living collections	New England Wild Flower Society	

Source: K. O'Donnell and S. Sharrock. 2017. *Plant Divers* 39: 373–378.

968,000 seed samples, originating from almost every country in the world, are stored below permafrost (see Figure 16.8B). The Svalbard seed vault has the capacity to store 4.5 million varieties of crops with an average of 500 seeds each, for a maximum storage capacity of 2.5 billion seeds.

Gene banks and seed banks are particularly important for the agricultural industry as a means to preserve the genetic diversity of agricultural crops, livestock, and their wild relatives. Resistance to pests and disease is often found in only one variety of a crop that is grown in one small area of the world, known as a **landrace**, or in a wild relative. Preserving the genetic variability represented by landraces is crucial to the agricultural industry's ability to maintain high productivity of modern crops and to their ability to respond to changing environmental conditions such as climate change.[33] Agricultural researchers comb the world for landraces of major food crops that can be stored and later hybridized with modern varieties in crop improvement programs. Varietals of many of the world's major food crops such as rice, wheat, corn (maize), oats, millet, potatoes, sorghum, soybeans, and other legumes are intensively collected. These efforts have led to some large collections, such as that of the International Rice Research Institute in the Philippines, which maintains 127,000 accessions of rice and wild relatives, and the International Maize and Wheat Improvement Center in Mexico, which holds 28,000 samples of maize and 150,000 samples of wheat. Researchers are racing to preserve genetic variability before traditional farmers around the world abandon their diverse local crop varieties in favor of standard, high-yielding varieties.[34] For example, Sri Lankan farmers, who grew over 2000 varieties of rice until the late 1950s, have now switched to using about five high-yielding varieties. Many of these accessions are collated in **Genesys** (oup-arc.com/e/cardinale-w16.21), an online data portal that contains records of over 4 million germplasm accessions from 460 institutions around the world to facilitate the access to, and use of, accessions in ex situ gene banks.

Comparable collection efforts exist for important domesticated animals. The Center for Genetic Resources, the Netherlands (CGN), maintains gene bank collections for more than 100 breeds of cattle, pigs, horses, sheep, goats, and poultry to conserve rare domestic animal breeds and to encourage animal breeders to back up their commercial varieties in the gene bank. In the United States, Congress approved the National Genetic Resources Program (NGRP) in 1990. NGRP gathers, studies, and preserves germplasm from all agriculturally important species and makes their resources available to scientists through the Germplasm Resources Information Network (GRIN) web server. The program is run by the U.S. Department of Agriculture's Agricultural Research Service.

A major controversy in the development of gene banks, particularly agricultural seed banks, revolves around who owns and controls the genetic resources.[35] In the past, international seed banks freely collected samples, often disproportionately from developing countries, and then provided those samples to research agencies and companies, often from developed countries that had funding to develop new commercial varieties through sophisticated breeding programs and field trials. The resulting seeds were sold at high prices to maximize profits, which often totaled hundreds of millions of dollars a year; yet, the countries from which the original seeds were collected did not receive any profit or royalties from this activity and even had to pay for seeds of the improved crop.

The problem of ownership of genetic resources was addressed in 2010 by the **Nagoya Protocol** (**Video 16.4**). The Nagoya Protocol was a supplementary agreement to the 1992 Convention on Biological Diversity (CBD) that sought to establish an international protocol for the fair and equitable sharing of benefits that arise from the access and use of genetic resources. The agreement, which has been ratified by 113 UN member states and the European Union, sets forth

VIDEO 16.4 Watch a video about implementation of the Nagoya Protocol in the Caribbean.
oup-arc.com/e/cardinale-v16.4

a general framework for sharing the financial benefits of genetic resources fairly, and it gives incentives to countries that preserve biological diversity. The policy recommendations include the following:

- Countries have the right to control access to their biological diversity and should be paid for its use.
- Countries have a responsibility to inventory their biological diversity and protect it.
- Collectors must have permission to collect samples from the host country, the local community, and individual landowners.
- As much as possible, research, breeding, processing, and production of new varieties should take place in the countries where the biological resources occur.
- The financial benefits, new products, and new varieties should be shared fairly with countries that contributed genetic resources used in the final product.

Many countries, international agencies, conservation organizations, and corporations are presently developing the financial and legal mechanisms to carry out the Nagoya Protocol. For example, there is a growing number of preferential trade agreements (PTAs) between nations that benefit the country that is the source of the genetic resource. They give that country "preferential access" and define the monetary share that it will receive from the sale and use of the resource. Some of these PTAs even incorporate language and intent from the Nagoya Protocol.[36] As just one example, in article 16.5 of the free trade agreement between the Republic of Columbia and the Republic of Korea, both parties agreed to adopt the Nagoya Protocol for the sharing and use of genetic resources, as well as the innovations and practices of Indigenous communities.

Contributions to In Situ Conservation

Aside from serving as biological repositories of threatened and endangered species, ex situ facilities play several additional roles that assist with in situ conservation efforts. Here we discuss three contributions that ex situ facilities make to in situ conservation, including captive-breeding programs, reintroduction and reinforcement programs, and public education and engagement.

Captive-breeding programs

Captive breeding is the process of maintaining reproducing populations of plants and animals in artificial environments. Zoos, aquariums, botanical gardens, and other ex situ conservation facilities are good candidates for developing captive-breeding populations of rare and endangered species because (1) they have the required experience, staff, and facilities to provide animal care and veterinary medicine for reproducing animals; (2) they engage with universities, government agencies, and conservation organizations in research designed to identify conditions that enhance reproductive success of threatened species; and (3) they have developed many of the tools and technologies that are needed to establish and house breeding colonies of animals.[37,38]

Captive-breeding programs have benefited from a number of artificial reproductive technologies that have greatly advanced over the past several decades.[39,40] Below are a few of the reproductive tools and techniques that have been used to increase the reproductive output of threatened animal species that do not reproduce well in captivity.

CROSS FOSTERING is a technique used in animal husbandry, animal science, and conservation whereby offspring are removed from their biological parents at birth and raised by surrogates. Surrogate parents, often from a closely related species, can sometimes be used to raise offspring of a rare species in order to raise reproductive rates. For example, many bird species, such as the bald eagle (*Haliaeetus leucocephalus*), normally lay only one clutch of eggs per year. But if biologists remove this first clutch of eggs, the female bird will lay and raise an additional second clutch. If the first clutch of eggs is given to another bird of a related species, then two clutches of eggs will be produced per year for each rare female. This technique, known as **double clutching**, can double the number of offspring one female of a rare species can produce. **Artificial incubation**, where eggs are incubated and hatched in a device that simulates avian incubation by keeping eggs warm at the correct temperature, is an alternative to cross fostering that works for some species.

ARTIFICIAL INSEMINATION is a procedure in which sperm are collected from a male animal and then introduced into a female animal's cervix or uterine cavity in order to achieve a pregnancy. Artificial insemination is a useful procedure for animal species that lose interest in mating when they are held in captivity. It is also useful when a zoo or aquarium has just one or a few individuals of a rare species, such as the giant panda (see Box 16.1). In these circumstances, the female can be artificially inseminated when she becomes ready to breed, either as part of her natural cycle or because she was induced with hormones. The procedure is similar to that which occurs in a human fertility clinic: Biochemical tracking of hormonal levels in urine and feces can be used to determine the timing of sexual receptivity in the female. Sperm is collected directly from a living male or is taken from a frozen sample stored in a gene bank, and it is introduced into the cervix or uterine cavity of the receptive female. While artificial insemination has become routine with agricultural animals such as cattle, the techniques differ and must be worked out for each species individually in conservation breeding programs. Artificial insemination is used increasingly for maintaining genetic diversity, as the sperm of distantly related males held at other facilities can be used to inseminate females. For example, the gene pool of cheetahs (*Acinonyx jubatus*) in North American zoos has been augmented by inseminating females with semen collected from wild-born Namibian males.[41] Transporting semen is preferable to removing endangered species from the wild and is much more cost-effective than transferring males between ex situ facilities. Reproductive biologists have even developed the ability to sort sperm cells by sex for certain species.[42] Using this new technology, ex situ facilities can maintain more ideal sex ratios needed to promote breeding and captive management.

IN VITRO FERTILIZATION is the fertilization of an egg with sperm outside the body, followed by the transfer and implantation of the viable embryo into a uterus. Embryo transfer has been accomplished successfully in a few rare and endangered animals, such as the bongo (*Tragelaphus eurycerus*), the Indian bison or gaur (*Bos gaurus*), tiger (*Panthera tigris*), ocelot (*Leopardus pardalis*), and Przewalski's horse (*Equus ferus przewalskii*). Superovulation, or production of multiple eggs, is sometimes induced using fertility drugs. The extra eggs are surgically collected, fertilized with sperm, and laparoscopically implanted into surrogate mothers, sometimes of related common species. The surrogate mothers carry the offspring to term and then gives birth (Figure 16.11). In the future, this technology may be used to increase the reproductive output of certain rare species.

Figure 16.11 A young bongo, an endangered species, stands with its surrogate mother, an eland (*Taurotragus oryx*), at the Cincinnati Zoo. This bongo calf was produced by embryo transfer.

CLONING Cutting-edge medical and veterinary technologies are being used to develop new approaches for some species that are difficult to breed in captivity, or that are already extinct.[43] These include cloning individuals from single cells (when only one or a few individuals remain) and cross-species hybridization (when the remaining members of a species cannot breed among themselves). The success rate for cloning of wild animals is generally less than 1%, and most cloned animals suffer lethal malformations.[44] For this reason, the technology is not generally viewed as a viable option for conservation at present. Even so, there have been a limited number of successes. For example, in 2001, the first cloned animal of the threatened gaur (Indian bison, *Bos gaurus*) was born at Trans Ova Genetics in Sioux Center, Iowa. The calf was brought to term by a domestic cow (*Bos taurus*) that served as the surrogate mother. Though healthy at birth, the calf died within 48 hours from dysentery, most likely unrelated to cloning. In 2003, an endangered banteng (*Bos javanicus*) was successfully cloned,[45] which was followed by three African wildcats, cloned from a thawed frozen embryo, that subsequently produced an additional litter of kittens.[46] Such successes have provided hope that cloning techniques might one day be used to saved threatened species, or resurrect those that are extinct. The process of using cloning to create an organism that is either a member of or resembles an extinct species has been referred to as **de-extinction**, **resurrection biology**, or **species revivalism**. De-extinction is a controversial topic that poses both great challenges and great opportunities for the field of conservation biology (Box 16.2).

Many of the aforementioned technologies were pioneered in human medicine. But others are novel methods that were developed at facilities that have a long history of captive breeding of animals. Some of these facilities are the San Diego Zoo Institute for Conservation Research (oup-arc.com/e/cardinale-w16.22); the Audubon Nature Institute's Species Survival Center in New Orleans (oup-arc.com/e/cardinale-w16.23); the Durrell Wildlife Conservation Trust in the United Kingdom (oup-arc.com/e/cardinale-w16.24); and the Smithsonian Conservation Biology Institute in Virginia (oup-arc.com/e/cardinale-w16.25), which is part of the Conservation Centers for Species Survival network (oup-arc.com/e/cardinale-w16.26) that links zoos with the U.S. Fish and Wildlife Service and other organizations to breed endangered species. In addition, many fish-breeding techniques were originally developed by fisheries biologists for large-scale stocking operations involving trout, bass, salmon, and other commercial species. Other techniques were discovered in the aquarium pet trade, when dealers attempted to propagate tropical fish for sale. These techniques are now being applied to endangered freshwater fauna. Programs for breeding endangered marine fishes and coral species are still in an early stage, but both public and private groups are making impressive efforts to unlock the secrets of propagating some of the more difficult species. Commercial production levels have been achieved for numerous species, and home aquarists can now expect fishes, corals, and other creatures to have been raised in captivity or be certified as having been sustainably collected from the wild.

The use of captive-breeding programs in species recovery has grown enormously in recent years, and these programs have the potential to play a crucial role in the short-term conservation of species.[47] Even so, captive-breeding programs do not provide a viable long-term solution to extinction, because they suffer numerous limitations, discussed next.

MAINTENANCE OF GENETIC DIVERSITY Captive populations are usually established using a relatively small number of individuals that are transferred from the wild to ex situ breeding facilities. Because the populations are small, they are particularly susceptible to genetic bottlenecks and founder effects when the

BOX 16.2 Challenges & Opportunities

Resurrection Science: Should Extinct Species Be Brought Back to Life?

The idea of using medical technology to bring species back from extinction is not new. Medical scientists have recreated viruses and other pathogens in the past to study their behavior in hopes of preventing new disease outbreaks. But it wasn't until 2009 that scientists brought an extinct mammal back to life—the Pyrenean ibex or bucardo (*Capra pyrenaica pyrenaica*), the last of which had died in 2000.[83] The process involved fusing frozen bucardo cells with goat egg cells from which the nuclei had been removed and then implanting the resulting cells into a domestic goat as a surrogate mother (see Figure). However, even though the embryo developed, the cloned ibex was badly deformed and died within minutes of birth.

In recent years, however, genomics and synthetic biology have made huge advances, and the field of "de-extinction" or "resurrection science" is gaining new momentum as a potential tool for conservation.[84,85] The process of bringing a species back from extinction requires DNA, ideally a complete genome. This limits the candidate organisms to those that have not been extinct so long that their genetic material has degraded. The genome of the extinct species, perhaps obtained from tissue preserved in a freezer, buried under a glacier, or preserved deep in an anoxic bog or oceanic mud, is then compared with a closely related living species—for example, the extinct passenger pigeon and a currently living band-tailed pigeon—and the differences are identified. At that point, the living species' DNA and cells can potentially be modified and used to create stem cells and germ cells with the DNA of the extinct species. These germ cells can then be implanted into eggs, which are then transferred to the wombs of living relatives. At the end of the gestation period, the result could be an extinct species brought back to life. This process has recently been successfully used to create early-stage embryos of an extinct gastric-brooding frog (*Rheobatrachus silus*), but they soon died.

The technologies needed to resurrect recently extinct plants, animals, and other species may soon be perfected. And those technologies could potentially be applied to the resurrection of ancient species, using tissue from permafrost to resurrect the mastodon, or bits of DNA from fossilized bones to de-extinct the saber-toothed tiger. But while it may be possible to bring extinct species back to life, should we do it? There are numerous ethical questions associated with de-extinction.[86] For example, if the re-created passenger pigeons were brought up by another pigeon species and lacked the flocking, feeding, and other behaviors of extinct passenger pigeons, should they really be considered passenger pigeons? Would resurrected passenger pigeons function in an ecosystem in the same ways as past passenger pigeons? What is the conservation value of bringing back a species if its habitat no longer exists?

How will de-extinction programs affect and interact with existing conservation efforts? Will the ability to bring species back to life reinforce a technocentric attitude toward wildlife and wild places—and convince us that we can wash away the conservation failures of the past? Will the ability to reverse conservation failures reduce the incentive to preserve species in the wild? Will research and efforts with de-extinction divert funding from research and conservation projects aimed at preserving existing species and ecosystems?

The questions that these new techniques raise are thought provoking and require thoughtful discussions. De-extinction and resurrection science represent a great challenge—and great opportunity for conservation.

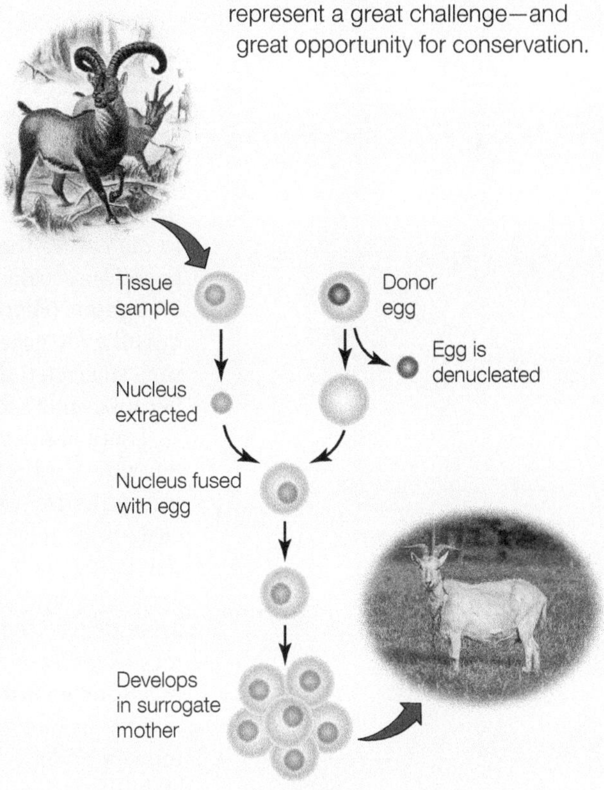

The process used to clone the Pyrenean ibex in 2009. The tissue culture had been taken from the last living female Pyrenean ibex, named Celia. The egg cell was taken from a goat (*Capra hircus*), and the egg's nucleus was removed to ensure the offspring was purely Pyrenean ibex. The egg was then implanted into a surrogate goat mother for development. (After 15ldavenport. CC BY-SA 4.0.)

transferred individuals have only a fraction of the genetic heterozygosity that is represented in the original population. The limited amount of genetic diversity in a founding population can then be reduced further by genetic drift to the point that releasing genetically depauperate animals back into a wild population can be detrimental. Therefore, captive-breeding programs must work to establish populations that have a sufficiently large number of individuals to be representative of the genetic variation of the original wild population.[48] To prevent genetic drift, ex situ populations of at least several hundred and preferably several thousand individuals need to be maintained. Because of space limitations, few zoos and aquariums can accommodate such large numbers, particularly for larger animal species.

INBREEDING AND OUTBREEDING In Chapter 13 we described inbreeding and outbreeding, which are common problems in small populations. Inbreeding leads to **inbreeding depression**, which is a loss of biological fitness due to lower reproductive and/or survival rates of offspring. Outbreeding occurs when organisms mate with genetically unrelated individuals. Outbreeding depression is a reduction in fitness similar to inbreeding depression. However, it is due to different mechanisms, such as sexual incompatibility or chromosomal differences between the individuals, or due to offspring no longer being adapted to their environment.

Because zoos and aquariums often house very small populations of animals, inbreeding and outbreeding depression are particular challenges for ex situ facilities. Historically, captive populations held in zoos and aquariums have exhibited extensive inbreeding due to failure to maintain genetic heterozygosity. However, outbreeding depression has occurred in some instances. One example is the orangutan. Prior to the 1980s, Bornean orangutans captured from Malaysia were often paired in captivity with orangutans captured from Sumatra. As the individuals were allowed to interbreed, it was noted that the hybrid orangutans tended to die earlier than their nonhybrid counterparts. After the 1980s, new tools in genetics revealed that the Bornean orangutan (*Pongo pygmaeus*) and Sumatran orangutan (*Pongo abelii*) were actually two genetically distinct species. Subsequently, the government of Indonesia and the Association of Zoos and Aquariums suggested that zoos stop interbreeding Sumatran and Bornean orangutans because outbreeding depression was further endangering the two species.[49]

There has been a concerted effort by ex situ facility managers to monitor the amount of inbreeding and outbreeding in captive populations (**Box 16.3**) and to minimize impacts on fitness by choosing genetically compatible matches when assigning mates.[50] Most modern accredited zoos and aquariums now use global databases, such as the Zoological Information Management System (ZIMS) that is collated and managed by Species360 to track the genetic lineages of captive animals. Studbooks that document the pedigree and entire demographic history of each animal managed within AZA member institutions are used as part of a **Species Survival Plan**, or **SSP**, that helps maintain healthy and genetically diverse animal populations. SSPs establish breeding goals and management recommendations to achieve the maximum genetic diversity and demographic stability for a species, given transfer and space constraints. AZA members currently have about 500 SSP programs for captive populations around the world.

ADAPTATION Ex situ facilities can exert natural selection on captive populations, causing them to undergo genetic adaptation to their artificial environments.[51,52] Plants or animals that have become genetically adapted to an artificial environment can have reduced rates of survival and reproduction when introduced back into the wild.

BOX 16.3 Conservation in Practice

Measuring Inbreeding and Outbreeding

Managers of captive-breeding programs need to monitor the amount of heterozygosity in an animal population, and keep track of the amount of inbreeding and outbreeding, in order to avoid fitness depression in offspring.

In Chapter 3, we showed how one can measure the frequency of alleles and genetic heterozygosity in a population. Assuming one has information on allele frequencies and heterozygosity, the F-statistic to quantify inbreeding and outbreeding can be calculated as:

$$F = \frac{H_0 - H}{H_0}$$

where H is the observed level of heterozygosity in the population, and H_0 is the amount of heterozygosity that is expected if a large population is mating randomly. H_0 must be calculated using the Hardy-Weinberg equilibrium model, which is:

$$p^2 + 2pq + q^2 = 1$$

where p is the frequency of the dominant allele in the population, q is the frequency of the recessive allele, p^2 is the frequency of the homozygous dominant genotype, q^2 is the frequency of the homozygous recessive phenotype, and $2pq$ is the frequency of the heterozygous genotype.

The second equation is often represented by a Punnett's square showing all possible combinations of sperm and eggs in a population that can be used to generate progeny (see Figure). There is one combination of sperm and egg that results in a homozygous dominant genotype (pp, or p^2), one that leads to the homozygous recessive (qq, or q^2), and two combinations that produce a heterozygous genotype ($2pq$). The expected level of heterozygosity, H_0, in a large randomly mating population that is behaving according to Hardy-Weinberg equilibrium is $2pq$.

Now consider an example. Assume you have measured the frequencies of p and q in a captive population and found them to be 0.4 and 0.6, respectively ($p + q$ must equal 1). Based on this second equation, you would expect

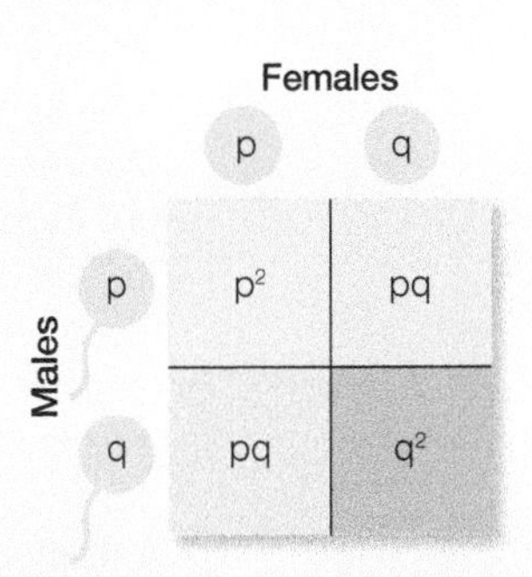

A Punnett's square showing all possible combinations of a dominant allele *p* and recessive allele *q* during random mating of males and females in a population.

$p^2 = 0.16$ or 16% homozygous dominant genotypes, $q^2 = 0.36$ or 36% homozygous recessive genotypes, and $2pq = 0.48$ or 48% heterozygous genotypes in the population. Now assume you measured the actual frequency of heterozygotes in the population and found them to be 0.47. The F-statistic would be:

$$F = \frac{0.48 - 0.47}{0.48} = 0.02$$

The equation above is close to zero because the observed value of heterozygosity, H, is similar to the expected value, H_0, from Hardy-Weinberg equilibrium. This means the population is not experiencing inbreeding or outbreeding. However, if $H < H_0$, then F gets large (approaching a maximum of 1), indicating that the population is experiencing inbreeding. If $H > H_0$, then F becomes negative (approaching a minimum of –1), indicating the population is experiencing outbreeding. This last equation is how one monitors inbreeding and outbreeding in a captive population.

Suggested Exercise 16.3 Now practice how to minimize inbreeding and outbreeding depression with the online exercise. **oup-arc.com/e/cardinale-ex16.3**

Lewis and Thomas[53] provided an example of maladaptive traits related to dispersal and reproduction developing in a culture of the large white (or cabbage) butterfly, *Pieris brassicae* (L.), that had been held in captivity for 100–150 generations. Individuals in the captive culture were heavier and had smaller wings with lower wing aspect ratios than their wild counterparts. Such traits would make it more difficult for the butterflies to fly, disperse, and evade predators if released back into the wild. Similar evolutionary changes can affect vertebrates, as well as invertebrates, that are reared in ex situ conservation programs, decreasing the likelihood that these species can be reestablished in the wild. The single best approach in order to minimize genetic adaptation is to reduce the number of generations that a species spends in captivity.[51]

BEHAVIORAL MALADAPTATION Captive breeding can generate changes in animal behavior that are maladaptive when individuals are reintroduced to the wild. Animals raised in captivity are commonly less capable of hunting and foraging for food and less able to avoid predators when released into the wild, which leads to starvation and death. This maladaptation is most common in social animals, such as primates, that teach their offspring what to do, and what not to do, to survive. For example, captive-raised golden lion tamarins often die after being released in the wild because they have not learned to climb and properly forage for food. This leads to continuing population declines, despite reintroduction, as the species are unable to produce viable offspring. Released animals also tend to display more risk-taking behavior.[54] For example, salmon bred in captivity fail to show the behaviors needed to avoid predators, and they experience higher mortality than their wild counterparts. However, salmon that were reared in an environment enriched with natural predators showed less risk-taking behaviors and were more likely to survive.[55] Providing appropriate training for captive animals may be needed before they can be released in the wild.

COST AND CONTINUITY Ex situ programs are very expensive, often considerably more expensive than other conservation efforts. As mentioned earlier in this chapter in the section on zoos and aquariums, one study showed that the cost of maintaining large animals in zoos (e.g., African elephants and black rhinos) is up to 50 times greater than the cost of protecting the same number of individuals in national parks.[19] In addition to their high cost, ex situ programs sometimes lack continuity in funding and are subject to political instabilities and shifting priorities of funding agencies and organizations. The high cost and instability limit the effectiveness of ex situ facilities as a single, long-term strategy for conservation. However, for smaller animals and many plant species that are less expensive to house and care for, or for plant or animal species for which habitat preservation and management is prohibitively expensive, ex situ conservation can be more effective than attempting to sustain a wild population.

Reintroduction and reinforcement

One goal of captive-breeding programs is to reestablish wild populations of rare and endangered species or increase the size of existing populations. Three basic approaches are used with plants and animals. **Reintroduction** (or reestablishment) involves releasing captive-bred or wild-caught individuals into an ecologically suitable site within their historical range where the species no longer occurs. The main objectives of a reintroduction program are to create a new population in its original environment and to help restore a damaged ecosystem and allow species to regain their ecological and evolutionary roles within their native ecosystem. **Reinforcement** can be thought of as restocking or augmentation, which involves releasing individuals into an existing population to increase its size and genetic diversity. These released individuals may be raised in captivity or may be wild individuals collected elsewhere. The release of greater prairie chickens (*Tympanuchus cupido pinnatus*) is an example of a successful reinforcement program to restore genetic variation in small, isolated populations. Reintroduction and reinforcement programs are distinct from **introduction** programs (also called **assisted colonization** or **translocation**, see Chapter 12). Reintroduction and reinforcement involve restoring populations to their original habitats, whereas introduction involves moving organisms outside their historical range because the original range is no longer suitable.

Summaries of reintroduction and reinforcement programs suggest that success rates tend to be low.[56,57] For example, a survey of more than 400 releases of short-lived fish species into wild habitats of the western United States showed a success rate of about 26%.[58] An extensive survey of practitioners estimated the success rate of plant reintroductions into the wild is less than 33%.[59] These estimates may, in fact, be high, given that scientists tend to publish only the results of successful projects, and many projects, particularly those that failed, are often not published or are poorly documented.[60] In addition to low success rates, establishing new populations is often expensive and requires a long-term commitment. The programs to capture, raise, monitor, and release California condors, peregrine falcons, and black-footed ferrets, for example, have cost millions of dollars and have required years of work. When the animals involved are long-lived, the program may have to continue for many years before its outcome is known.[61]

Brichieri-Colombi and colleagues[62] recently summarized the contribution of ex situ captive-breeding programs to 279 efforts to reintroduce animal populations in the wild. These authors showed that contributions by captive-breeding programs have increased over the last several decades and account for 162 (58%) of the 279 animal species reintroductions. Even so, animals bred for release by zoos represented just 14% of all animal species for which ex situ reintroductions were published, and they represented only 25% of all animal species that were bred for releases occurring in North America. Zoos were disproportionately important to reintroduction of certain groups, such as amphibians, for which they account for 42% of captive release efforts. Botanical gardens don't fare much better, with only 25% having any type of in situ conservation program and only 12% involved in plant restoration efforts (Table 16.2). These results suggest that ex situ conservation facilities could do far better at contributing to in situ conservation efforts than they currently do.

Despite limitations and challenges, there are numerous case studies showing that successful reintroduction and reinforcement of threatened plant and

TABLE 16.2 The percentage of botanical gardens providing facilities and activities related to ecological restoration

Facilities and activities	Estimated percentage of gardens
FACILITIES	
Computerized plant record systems	38
Herbarium	26
Seed bank	19
CONSERVATION WORK	
Conservation program	26
Ex situ program	25
Reintroduction program	16
RESEARCH PROGRAMS	
Horticulture	25
Conservation biology	21
Ecosystem conservation	16
Restoration ecology	12

Source: K. A. Hardwick et al. 2011. *Conserv Biol* 25: 265–275. Data from Botanic Gardens Conservation International. 2010. GardenSearch database as of June 2010. BGCI: UK.

animal species can be done, and such efforts that have saved species from extinction represent some of conservation biology's greatest success stories (Box 16.4). Several techniques exist to help maximize the chance of success by helping individuals make successful transitions to their new homes when they are reintroduced or reinforced. We discuss some below.

BOX 16.4 Success Story

Reintroduction Programs That Saved Species from Extinction

Summaries of captive-breeding and reintroduction programs suggest that such efforts are costly and tend to have low success rates. But there are numerous examples showing that successful reintroduction and reinforcement of threatened species can be done and that, when done correctly, they can bring species back from the brink of extinction. Here are three success stories:

- Atlantic puffins (*Fratercula arctica*) were virtually eliminated from the Maine coast by overharvesting of both eggs and the puffins themselves for their meat. In 1973, Project Puffin began a program to reintroduce puffins to Eastern Egg Rock Island off the Maine coast. Over a 13-year period, researchers performed a soft release of more than 900 chicks into artificial burrows while supplying them with a diet of fish and vitamins. Aggressive gulls had to be regularly chased from the island. After a season of growing, fledged chicks left the island for the open ocean. To encourage them to return, researchers set up puffin decoys on the island so it appeared that the island was occupied by an active puffin colony. Puffins began to return to the island in 1977, and as of 2017, there were 172 breeding pairs.[87]
- Red wolves (*Canis rufus*) have been reestablished in the Alligator River National Wildlife Refuge in northeastern North Carolina through the release of 42 captive-born animals starting in 1987. Over 100 animals currently occupy about 700,000 hectares of private and government land, including a military base. Animals in the program have produced pups and established packs and survive by hunting deer, raccoons, rabbits, and rodents.[88] Even though the red wolf recovery program appears to be successful, many landowners remain unwilling to accept the presence of wolves on their land. At this point, the greatest threat to the species may be the loss of integrity of its gene pool and obscuring of species boundaries by mating and hybridization with coyotes.
- The Big Bend gambusia (*Gambusia gaigei*), also called the Big Bend mosquitofish, is a small fish originally known from two small springs in Texas. One population was eliminated when one spring dried up in 1954; at the same time, the second population began to decline rapidly when its spring was diverted to create an artificial fishing pond, and by 1960 it too had disappeared. In the interim, however, two females and one male had been taken from the artificial pond to establish a captive-breeding program. A combination of captive breeding and releases into new artificial ponds in Big Bend National Park helped this vulnerable species survive a series of droughts and invasions by exotic fish. The species is now reestablished in two spring pools with a population size of several thousand individuals, and the natural flow of the spring is mandated under the management plan for this protected species.[89] A captive population is still maintained in a fish hatchery in New Mexico, however, in case the wild population declines again.

(A)

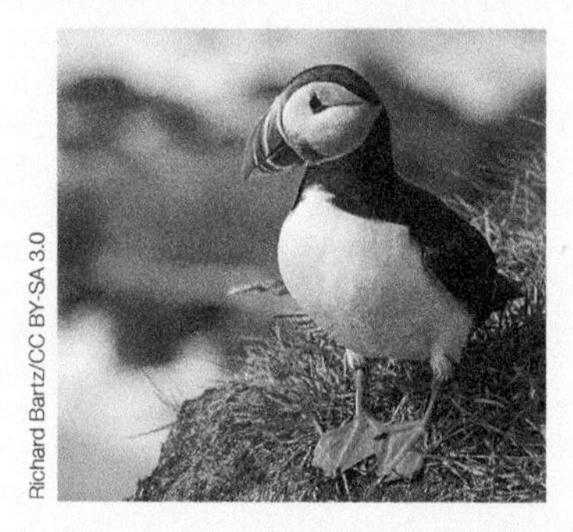

Richard Bartz/CC BY-SA 3.0

(B)

B. Bartel/USFWS/CC BY 2.0

(C)

USFWS

Three species with successful captive-breeding and reintroduction programs: (A) Atlantic puffins (*Fratercula arctica*) off the coast of Maine, (B) red wolves (*Canis rufus*) in the Alligator River National Wildlife Refuge, (C) Big Bend mosquitofish (*Gambusia gaigei*) from springs in Texas.

(A)

(B)

Figure 16.12 Examples of headstarting. (A) Seedlings of native rain forest trees ready to be planted back into their natural habitat in eastern Madagascar. These seedlings will have a higher chance of survival in the wild compared with trees reintroduced directly from seeds. (B) Newly hatched baby turtles are reared at a conservation facility before being released into the ocean. The rearing allows the turtles time to grow, giving them a higher chance of success in the wild.

HEADSTARTING Headstarting is a technique in which plants or animals are raised in captivity during their vulnerable young stages, after which they are released into the wild (**Figure 16.12**). For plants, the chance of success can be increased by germinating seeds in controlled environments so that the young plants can be grown in protected conditions. Plants are then transplanted into the wild after they are past the fragile seedling stage. For animals, headstarting usually involves hatching and/or rearing young under human care in a highly controlled environment where major causes of juvenile mortality (e.g., starvation or predation) can be controlled or eliminated. This approach has been used extensively with egg-laying species such as sea turtles, birds, fishes, and amphibians; hatchlings are protected and fed during their vulnerable early stages, and the young are then released into the wild or raised in captivity. The release of sea turtle hatchlings produced from eggs collected in the wild and raised in nearby hatcheries is an example of headstarting.

SOFT RELEASE Soft-release methods, which involve caring for plants and animals as they acclimate to a new site, can be used to improve the chance of success of organisms released into an unfamiliar environment. For example, to ensure successful establishment of reintroduced plants, caging or fences are often used to exclude herbivores, fertilization is used to augment mineral nutrients to ensure fast growth, and existing vegetation is sometimes removed by controlled burns to reduce competition during critical early periods of growth (**Figure 16.13**).[63,64] Animals may need to be fed and sheltered at their release point until they are able to subsist on their own, or they may need to be caged temporarily at the release point and introduced gradually, once they become familiar with the sights, sounds, smells, and layout of the area. Even when animals appear to have enough food to survive, supplemental feeding may help increase reproduction and allow the population to increase and persist. Intervention may be necessary if animals experience unexpected hardships, such as during episodes of unanticipated drought or low food supply.[65] Outbreaks of diseases and pests may need to be monitored and dealt with. While a decision must be made about whether it is better to give temporary help to the species or to force the individuals to survive on their own, much evidence suggests that soft-release methods often increase the rate of success.[66]

(A)

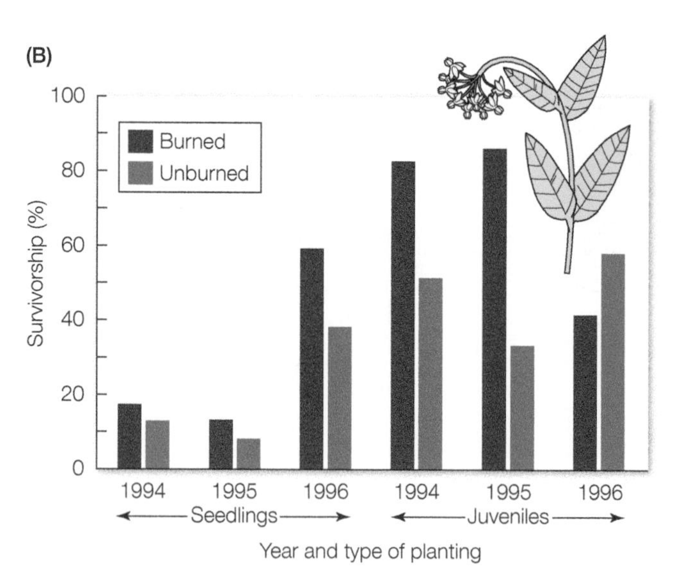

Figure 16.13 (A) Several methods are being used to create new populations of rare wildflower species on U.S. Forest Service land in South Carolina. Seeds are being planted in a pine forest from which the oak understory has been burned away. Wire cages will be placed over some plantings to determine whether excluding rabbits, deer, and other animals will help in plant establishment. (B) Introduced seedlings and juvenile plants of Mead's milkweed are evaluated in a reintroduction experiment. Survivorship is greater for older juvenile plants than for seedlings and greater in burned habitat than in unburned habitat. Seedling survivorship was greatest in 1996, a year with high rainfall. (B after M. L. Bowles et al. 1998. *Ann Mo Bot Gard* 85: 110–125.)

BEHAVIORAL TRAINING Many animals born in the wild learn behaviors that are important for their survival, including how to forage or hunt, what to eat, how to avoid predators, and how to interact socially with members of the same species. But as mentioned earlier in the Captive-Breeding Programs section, when animals like mammals and birds are raised in captivity, they do not learn how to appropriately search for natural food items, they do not learn antipredator behaviors, and their social behaviors often become distorted and detrimental to their survival.[60] To overcome these behavioral problems, captive-raised mammals and birds may require extensive training before and after their release into the wild.[67,68] For example, captive-bred California condor hatchlings (*Gymnogyps californianus*) were originally unable to learn normal social bonds with other condors because they had imprinted on their human keepers (**Video 16.5**). Newly hatched condors are now fed with condor puppets and kept from seeing human visitors so they learn to identify with their own species rather than a foster species or humans (**Figure 16.14**). However, even with such training, when captive-raised condors were released into the wild in protected areas, they often congregated around buildings, causing damage and frightening people. To break this association, condors are now being captive-reared in enclosed outdoor areas without any buildings.

When captive-bred animals are released into the wild as part of a reinforcement program, developing social relationships with wild animals is crucial to their success. To foster such interactions, one

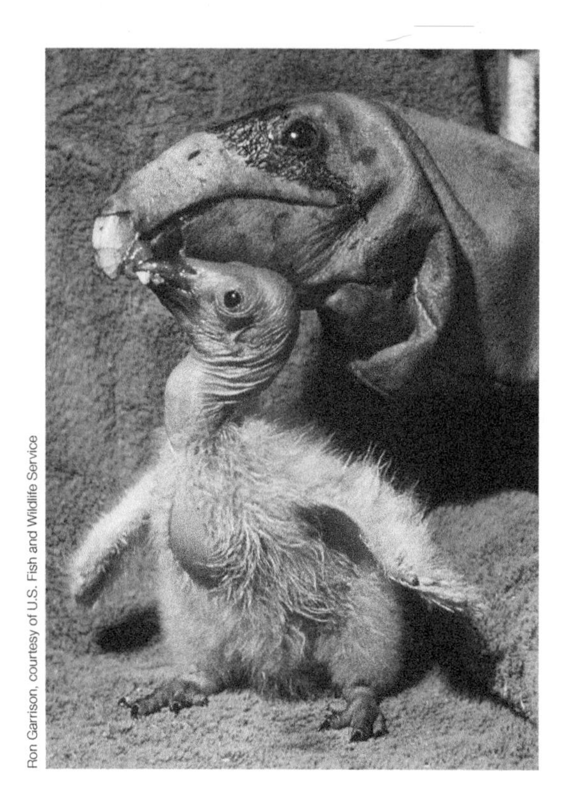

Figure 16.14 California condor chicks raised in captivity are fed by researchers using puppets that look like adult birds. Conservation biologists hope that minimizing human contact with the birds will improve their chances of survival when they are returned to the wild.

behavioral training technique employs wild individuals as "instructors" for captive individuals of the same species. For example, wild golden lion tamarins are caught and held with captive-bred tamarins so that the captive-bred tamarins will learn appropriate behavior from the wild ones. After they form social groups, they are released together in protected forests in Brazil. These captive-reared animals then gain some knowledge of food items and potential danger by watching the wild animals in their group.[69]

VIDEO 16.5 Watch a video about a California condor release program.
oup-arc.com/e/cardinale-v16.5

Public education and engagement

While ex situ facilities have much opportunity for public education and engagement, there has been considerable debate about whether zoos, aquariums, and botanical gardens are actually effective at, and dedicated to, teaching people about biodiversity or conservation. Certainly, in the early years of royal and traveling menageries, public education and engagement, if practiced at all, were not focused on conservation. In contrast, most modern zoos and aquariums now include public education and engagement in conservation as part of their mission. Consider, for example, the stated missions of these organizations:[70]

- World Association of Zoos and Aquariums (WAZA): "The educational role is to interpret living collections to attract, inspire and enable people from all walks of life to act positively for conservation."
- Association of Zoos and Aquariums (AZA): "Facilitate multi-institutional conservation education, outreach, and collaborations that activate the public to connect with and take personal action to conserve wildlife and wild habitats."
- European Association of Zoos and Aquaria (EAZA): "To create an urgent awareness among the many millions of European zoo visitors of the fact that the long-term survival of a thriving human population on earth is fully dependent on the rapid development of sustainability on a global scale. And, through the creation of this awareness, to evoke individual and collective political action aiming at reaching global sustainable levels of all human activities within the next three to five decades."
- Pan-African Association of Zoos and Aquaria (PAAZA): "The education message should be well defined and holistically presented in terms of the integrated conservation approach of the institution."
- South East Asian Zoos Association (ESAZA): "The vision of the South East Asian Zoos Association is that its member zoos utilize their animal collections for the primary purposes of educating our public by imparting messages on the urgent need for environmental conservation in a manner that upholds the respect and dignity of the wild animal."

Some of these organizations even have education committees who support conservation education within the member institutions and help zoo and aquarium staff to apply the current best practices. They help staff to develop educational lectures and displays, promotional materials (e.g., videos, pamphlets, and brochures), and exhibits of ambassador animals with whom people can make personal connections. Some of these educational programs have been highly successful, such as the Monterey Bay Aquarium's Seafood Watch (oup-arc.com/e/cardinale-w16.27), which produces pocket guides that help diners and consumers choose fish species that are harvested sustainably.

There is no doubt that zoos and aquariums represent themselves as ambassadors of conservation education. And their own studies tend to support that view. For example, one of the most widely cited studies supporting the efficacy of public education by zoos was supported and published by WAZA itself.[71] The study claimed to be the largest and most international study of zoo and aquarium visitors

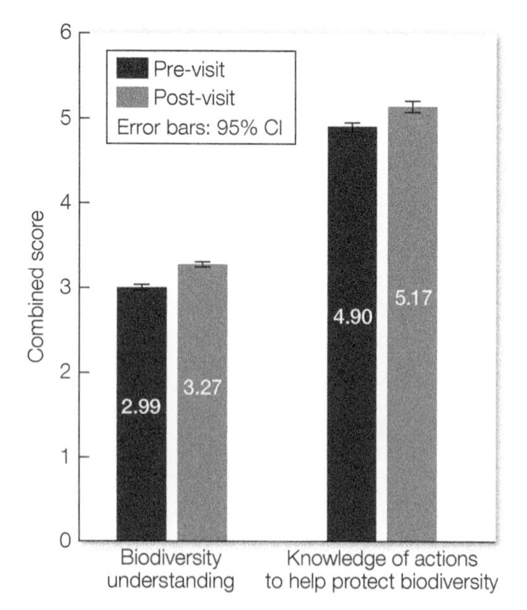

Figure 16.15 A survey of visitors to zoos and aquariums suggests that as a result of their visit, people have a better understanding of biodiversity (left) and actions that can be taken to protect biodiversity (right). (After A. Moss et al. 2015. *Conserv Biol* 29: 537–544.)

ever conducted. The researchers used a pre- and post-visit survey design to evaluate biodiversity understanding and knowledge of actions to help protect biodiversity among zoo and aquarium visitors. Key results of this study are shown in **Figure 16.15**, which suggests that people tend to have a slightly better understanding of biodiversity and the actions that can be taken to protect biodiversity after they visit a zoo or aquarium. This same graph has been published multiple times in multiple journals,[72,73] and the results have been widely used by zoo and aquarium advocates to justify their educational contributions to conservation. Yet, results of the WAZA study contrast with those of other studies, such as that of Jensen,[74] who surveyed 2800 children after guided and unguided visits to the London Zoo and found that 62% showed no change in knowledge regarding facts about animals or environmental conservation. Mellish and colleagues reviewed 48 peer-reviewed studies of the effectiveness of zoo and aquarium conservation programs published between 1998 and 2016 and found that 83% of the studies were based on flawed methodologies (e.g., lack of proper controls, biased survey designs, failure to account for confounding variables), a finding that is sufficient to cause doubt about the data and the conclusions drawn from them.[75]

While there is ongoing debate about the effectiveness of conservation education, there can be little doubt that ex situ facilities have great potential for public education and engagement and much opportunity to help visitors form connections to animals that influence people's environmental values and attitudes.[76,77] Ex situ conservation facilities host large numbers of visitors each year. WAZA reports that more than 700 million visitors annually pass through the gates of zoos and aquariums. If accurate, this would mean that zoos and aquariums of the world engage more people than all of the world's major sports leagues combined (**Figure 16.16**). The AZA, which is the umbrella accreditation organization for zoos and aquariums in North America, reports 181 million annual visitors, with two out of three adults visiting a zoo with a child and with half of adults visiting aquariums with a child (oup-arc.com/e/cardinale-w16.28). These estimates exceed the estimated annual attendance at all games of the National Football League, National Basketball Association, National Hockey League, and Major League Baseball combined. Therefore, ex situ facilities have an unrivaled opportunity to educate the general public about biodiversity and the plight of threatened species and to engage people in discussions that lead to pro-conservation behaviors.[21]

In addition to the large number of people who visit zoos and aquariums each year, there is some evidence to suggest that people trust zoos and aquariums to provide reliable information. For example, a Roper poll in 1992 found that the three most trusted sources of information about wildlife and conservation were *National Geographic*, Jacques Cousteau, and aquariums and zoos.[78] This level of public trust is partly why, in 2015, WAZA called on its worldwide membership of zoos and aquariums to evolve yet again, morphing into conservation centers of the twenty-first century that address sustainable relationships between humans and nature and that explain the values of ecosystems and the importance of conserving biological diversity (see Figure 16.2).

In addition to their educational programs, many types of ex situ conservation facilities perform a number of important public service activities. For example, many aquariums have animal rescue units of personnel who respond to public requests for assistance in handling animals that are stranded on beaches or disoriented in shallow waters. Zoos, aquariums, and botanical gardens often have citizen science programs that assist with reintroduction and reinforcement programs. For example, Zoo Atlanta partnered its captive-breeding program for golden lion

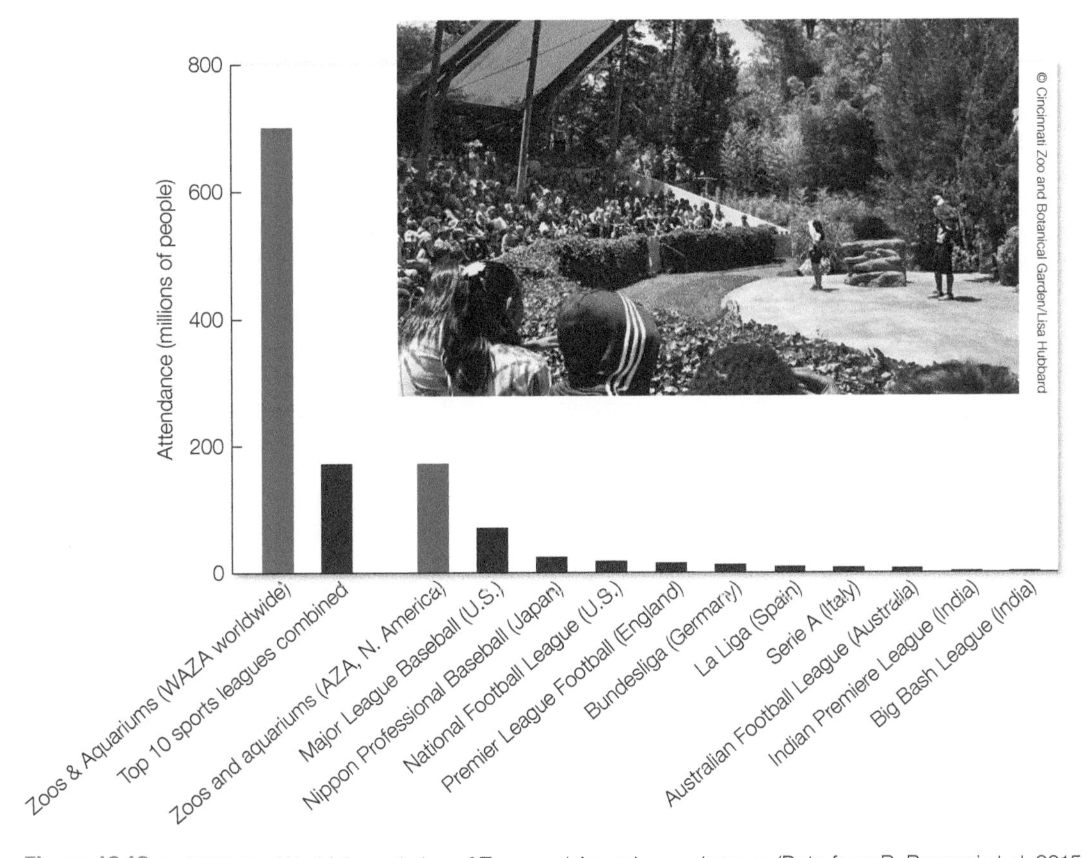

Figure 16.16 In 2015, the World Association of Zoos and Aquariums (WAZA) reported an annual visitation to zoos and aquariums of more than 700 million people. That was four times more people than attended all games during the 2017–2018 seasons of the world's top 10 sporting leagues combined. In 2019, the Association of Zoos and Aquariums in North America reported annual attendance that was higher than the 2017–2018 attendance for any single sports league. (Data from R. Barongi et al. 2015. *Committing to Conservation: The World Zoo and Aquarium Conservation Strategy.* Gland, Switzerland: WAZA Executive Office; Association of Zoos and Aquariums. 2019. www.aza.org/partnerships-visitor-demographics; Wikipedia. Accessed 2019. List of sports attendance figures. www.wikipedia.org/wiki/List_of_sports_attendance_figures.)

tamarins (*Leontopithecus rosalia*) with NGOs that use volunteers to help reintroduce the species in Brazil and work with local landowners to ensure their survival. Some aquariums have established hatcheries where large numbers of baby sea turtles can be raised and later released back into the wild. Such programs are often of great interest to the public and become a source of pride for the community.

Final Thoughts on Ex Situ Conservation

As more of the world's land and water become dominated by human activities, ex situ facilities will play an increasingly important role in the conservation of species in the wild. Captive facilities can serve as short-term refuges for threatened species. In the cases of highly endangered species, captive-breeding programs can be used to reestablish viable populations in the wild, once threats have been identified and controlled. Ex situ facilities also have great potential to foster in situ conservation efforts with public education and engagement activities that encourage pro-conservation behaviors. However, ex situ conservation is by no means a substitute for in situ conservation, nor does it provide viable long-term solutions for the conservation of threatened species. But ex situ conservation can buy valuable time for species that are on the verge of extinction, and it can play a key role by augmenting the goals of in situ conservation efforts.

Summary

1. While in situ (on-site) conservation gives biodiversity the greatest chance of long-term persistence, there are many instances where species must be protected in artificial environments like zoos, aquariums, botanical gardens, and arboretums. Such ex situ (off-site) conservation can serve as a short-term refuge against extinction and be an important part of a broader conservation plan.

2. Zoos and aquariums have evolved considerably through time. They began as menageries that put exotic animals on display for the purpose of entertainment, but in the twentieth century they morphed into zoological gardens with a dual purpose of entertainment and research. The World Association of Zoos and Aquariums (WAZA) has recently developed a new twenty-first-century mission that pushes zoos and aquariums to become holistic conservation, research, and public education centers.

3. According to the Zoological Information Management System (ZIMS), zoos and aquariums hold 22,000 species, including 15% of species threatened with extinction and several species that are now extinct in the wild. Even so, zoos and aquariums still struggle with poor animal care and welfare, and population sizes in zoos and aquariums are often too small to prevent the loss of genetic variation and to overcome demographic stochasticity. Also, animals in captivity are prone to disease and loss of their wild behavior, making them unsuitable for being returned to the wild.

4. Botanical gardens and arboretums trace back to royal gardens and the curation of medicinal plants by sixteenth-century universities. Botanic Gardens Conservation International (BGCI), which serves as the umbrella organization for the world's modern botanical gardens, suggests there are more than 3500 botanical gardens worldwide and that these host more than 500,000 species of plants. This global network is over halfway toward achieving its goal to protect 75% or more of threatened plant species in ex situ collections by 2020.

5. Gene banks, often referred to as frozen zoos, for animals, and seed banks, for plants, are biorepositories that collect and preserve tissues, cells, gametes (sperm and egg cells), and DNA of endangered plants and animals in cryogenic freezers. Seed collections, like those in the Svalbard Global Seed Vault in Norway, protect sources of genetic variation that are important for plant conservation programs and the improvement of agricultural crops. Collections of animal gametes, like those in the Frozen Ark Project at the University of Nottingham in the United Kingdom, provide greater opportunities for captive breeding of animals in zoos.

6. Many ex situ facilities engage in reintroduction and restoration programs that contribute back to in situ conservation efforts by establishing new wild populations. The chances of success are improved when animals are given a soft release that allows them to gradually adjust to their new environment and when animals are given behavioral training before release. Such programs have a limited record of success, so there is room for improvement.

7. In addition to serving as refuges from extinction, ex situ conservation facilities have high numbers of visitors and high public trust, giving them a tremendous opportunity to encourage pro-conservation behaviors through public education and engagement. There is currently mixed evidence regarding how effective and dedicated zoos and aquariums are to this mission; however, the World Association of Zoos and Aquariums (WAZA) has emphasized public education and engagement as a primary role of twenty-first-century facilities.

For Discussion

1. How effective are ex situ facilities like zoos, aquariums, and botanical gardens in helping to protect the world's biodiversity? Have zoos overcome the negative stereotypes associated with their early beginnings as menageries? If not, what do they need to do in the future to convince people that they are providing proper animal care and can contribute effectively to conservation efforts?

2. How much of an ex situ facility's resources should be devoted to conservation efforts in order for the institution to announce that it is a conservation organization? What sorts of conservation activities are appropriate for each institution?

3. Like zoos and aquariums, botanical gardens and arboretums are disproportionately located in North America and Europe, despite the fact that most plant biodiversity resides in the tropics. The transport of live tropical

plants or seeds has potential to spread invasive alien species (IAS), and indeed, the horticultural industry is a large contributor to IAS. How do we protect plants in ex situ facilities while still protecting native habitats from IAS?

4. Ex situ conservation, and particularly captive-breeding programs, tend to be very expensive. Personnel in charge of these programs usually have limited resources and must make hard choices. If you were in charge of captive-breeding programs at a zoo or aquarium, at what point would you give up on a species and say its survival in captivity is not worth the cost?
5. Cloning technologies may soon allow scientists to resurrect species that have already gone extinct, like the passenger pigeon, woolly mammoth, or flightless moa. Should we bring these species back to life? What benefits, consequences, and ethical considerations should be weighed in that decision?
6. Based on your experience visiting zoos, botanical gardens, and aquariums, do you think they do a good job of public education about biodiversity, threatened species, and the need for conservation of wild populations? If not, what could they do better?

Suggested Readings

Barongi, R. et al. 2015. *Committing to Conservation: The World Zoo and Aquarium Conservation Strategy.* Gland, Switzerland: World Association of Zoos and Aquariums. This report by WAZA outlines the organization's vision for twenty-first-century zoos and aquariums to become holistic conservation centers.

Brichieri-Colombi, T. A. et al. 2019. Limited contributions of released animals from zoos to North American conservation translocations. *Conservation Biology* 33: 33–39. This review highlights the good work of many ex situ facilities toward in situ conservation but also shows the global community of zoos, aquariums, and botanical gardens needs to do better.

Conde, D. A. et al. 2013. Zoos through the lens of the IUCN Red List: A global metapopulation approach to support conservation breeding programs. *PLOS ONE* 8: e80311. Really cool paper that shows how management of the global network of zoos as a metapopulation can enhance the persistence of captive populations.

Fraser, D. J. 2008. How well can captive breeding programs conserve biodiversity? A review of salmonids. *Evolutionary Applications* 1: 535–586. This widely cited paper reviews the potential for captive breeding programs to maintain genetic diversity, through the lens of salmonid fishes.

Minteer, B. A. et al. (Eds.). 2018. *The Ark and Beyond: The Evolution of Zoo and Aquarium Conservation.* Chicago, IL: University of Chicago Press. This new book is the best available summary of the evolution of zoos and aquariums and their role in biodiversity conservation.

Mounce, R. et al. 2017. Ex situ conservation of plant diversity in the world's botanic gardens. *Nature Plants* 3: 795–802. This is the most comprehensive summary of the contribution that botanical gardens make to biodiversity conservation of plants.

Visit the
Conservation Biology
Companion Website
oup.com/he/cardinale1e
for videos, exercises, links, and other study resources.

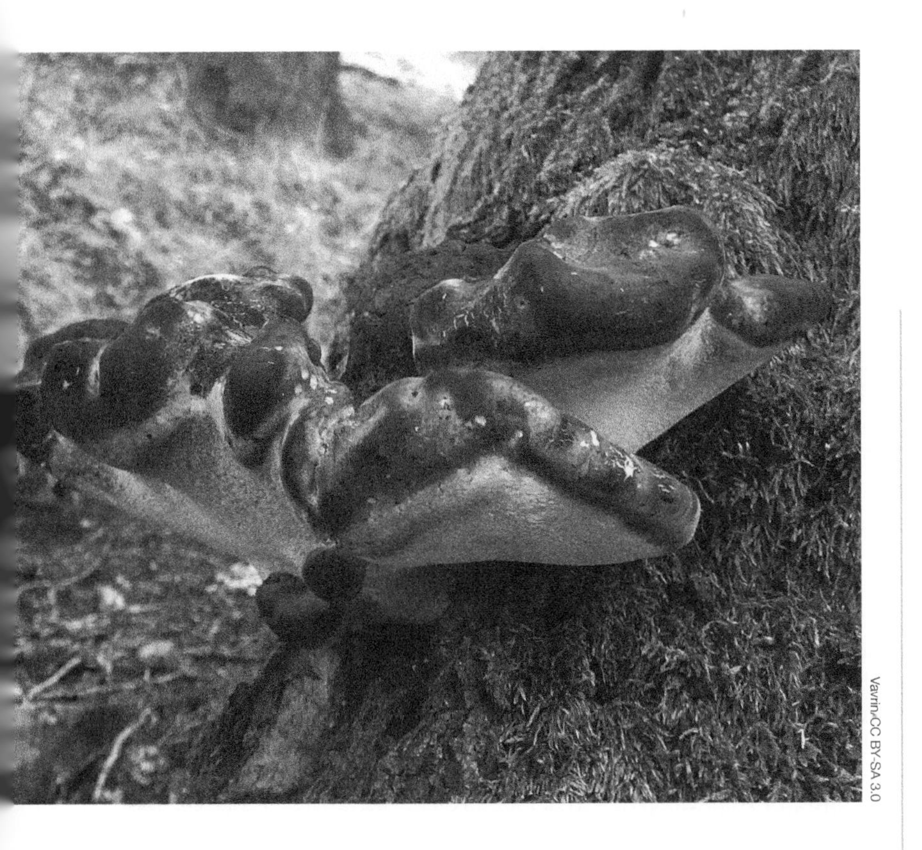

17

Conservation and Sustainable Development

The rare oak polypore, *Piptoporus quercinus*, is one of five fungi accorded the legal protection by the UK's 1981 Wildlife and Countryside Act. It fruits exclusively on oak (*Quercus* spp.) in old growth wood pastures that are increasingly rare habitats in southern England.

Conservation biology began as a crisis discipline in the early 1980s following a period of alarming environmental degradation (see Chapter 2). The Industrial Revolution resulted in unprecedented economic growth that appeared limitless up through the early and mid-1900s. Rapid economic growth allowed some countries to prosper and led to the notion that humans have the right to dominate the natural world in pursuit of maximum economic production. However, it eventually became apparent that natural resources were not limitless and that economic growth had consequences. By the mid-1900s, the gap between rich and poor societies widened as the benefits of the new world economic system flowed mainly to industrialized nations.[1] The world's population more than tripled from 1800 to 1970, which resulted in increased levels of consumption and concern over the impacts of demand on natural resources. Pollution of air and water, a byproduct of economic growth, also became apparent and led to the realization that unconstrained industrialized capitalism had trade-offs that could in fact threaten many aspects of human well-being.

By the early 1970s, environmental concerns had become an important international issue that launched a new era focused on growth that was more sustainable and would minimize the negative trade-offs with the environment. The Club of Rome, an organization of prominent figures in society including politicians, diplomats, scientists, economists, and business leaders, commissioned a landmark study on economic and human population growth. The study, published in 1972 as *The Limits to Growth,* warned that the Earth's resources are finite and that over-exploitation of those resources could harm much of the human population.[2] In that same year, the United Nations held the first Conference on the Human Environment in Stockholm, Sweden that focused on the emerging ecological crisis, and represented the first of several related international conferences. The conference (also known as the Stockholm Conference) acknowledged that human progress and prosperity required a more sustainable form of economic growth that also strives to minimize environmental damage.[3] Sustainability of natural resources became a theme of several of the principles adopted by the conference.

A new paradigm of sustainable development subsequently emerged and gained popularity in the 1980s, contributing to the founding of the field of conservation biology (see Chapter 2). It led to the **World Commission on Environment and Development (WCED)**, a group of 22 people from developed and developing countries convened by the United Nations to identify long-term environmental strategies for the international community. Commonly known as the **Brundtland Commission** (after Gro Harlem Brundtland, former Prime Minister of Norway and chair of the commission), the WCED issued its report to the UN in 1987 entitled *Our Common Future,* which formalized the goals of sustainability and launched a new global focus on sustainable development.[4]

In this chapter, we provide an overview of sustainable development since that time. We begin by defining sustainable development and the challenges it presents for conservation. We then describe some of the major initiatives and conventions that guide sustainable development at the international level, cover sources of funding for sustainable development activities, and describe some examples of sustainable development projects. We end by describing areas that could improve the success of sustainable development in the future.

What Is Sustainable Development?

Sustainable development has been defined in several ways. Perhaps the most widely used definition comes from the Brundtland Commission report to the United Nations:

> Sustainable development is development that meets the needs of the present without compromising the ability of future generations to meet their own needs.[4]

This definition, while general in scope, reflected the themes of the report and the Commission's work—namely, that meeting the basic needs of people and advancing progress and prosperity worldwide requires economic growth. However, the commission argued that economic growth should be carried out in a sustainable way, such that any activities now do not compromise the well-being of people in future generations.

The Brundtland definition of sustainable development led to a profusion of discussion and discourse around a more precise description of what the concept really means, especially in practice. In a general sense, *sustainable* means to survive and persist, and *development* means to expand and become more advanced (toward a better state).[5] Sustainable development combines these ideas and seeks to achieve development that strikes a balance between three dimensions of sustainability: economic, societal, and environmental. These three dimensions are commonly referred to as the pillars of sustainability, or the triple bottom line, and sustainable development is often viewed as the intersection of all three (Figure 17.1).[6,7] A sustainable development activity, then, involves a compromise between the three pillars that aims to achieve a balanced outcome. For example, logging a forest for timber at such a rate that it exhausts all trees in a short timeframe may yield the maximum profit immediately, but is not sustainable, as it causes the loss of the forest and any economic, social, or environmental benefits it provides for future generations. Logging at a lower rate that allows the forest to regenerate, along with restoration activities like tree planting, allows for economic growth from the sale of timber (although less than the maximum-profit scenario) while ensuring that the forest is available and ecologically intact (although not in an entirely pristine state) for future generations.

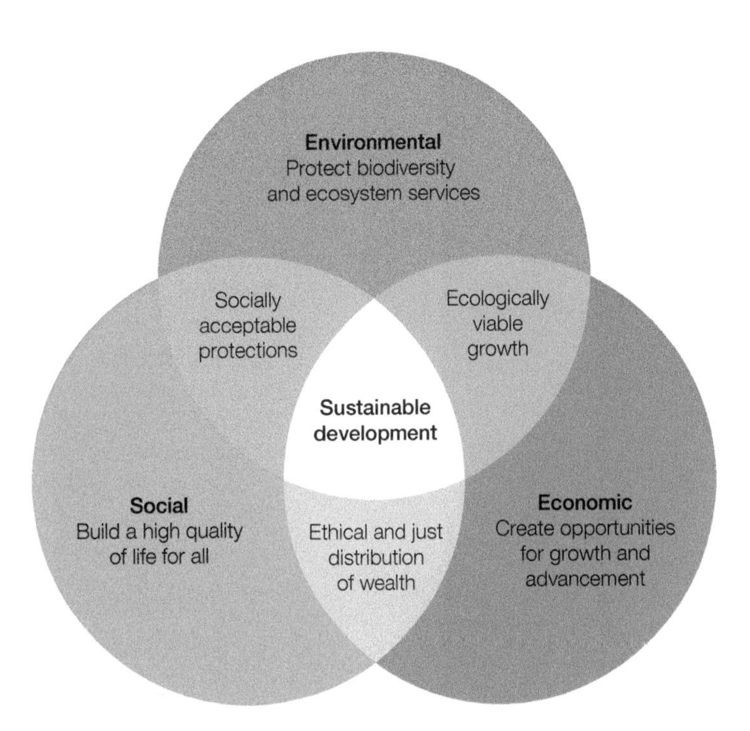

Figure 17.1 Sustainability is defined along environmental, economic, and social dimensions. Sustainable development represents the intersection of all three. (After E. B. Barbier. 1987. *Environ Conserv* 14: 101–110, and miscellaneous sources.)

The term *sustainable development* is considered by some to be an oxymoron because of the nature of compromise between the dimensions of sustainability.[8,9] The very idea that economic growth or development is "sustainable"—that is, that it can potentially continue indefinitely across generations—seems to contradict the idea that resources are finite and ultimately limited. Others[8] have argued that sustainable development should not be taken as a literal term, but rather viewed as an ideal that seeks to integrate and reconcile three imperatives:

1. The *ecological imperative* to stay within the biophysical carrying capacity of the planet
2. The *economic imperative* to provide an adequate material standard of living for all
3. The *social imperative* to provide systems of governance that propagate the values that people want to live by[10]

Taking another approach, Holden and colleagues[11] suggested that sustainable development is better defined as a set of constraints on human activities, and proposed a different model based on three moral imperatives: satisfying human needs, ensuring social equity, and respecting environmental limits.

While the concept of sustainable development has been refined since its inception, a clear and operational definition remains plagued by questions related to the timeframe and nature of the system under consideration.[12] For example, "sustainability" can be interpreted to mean "maintenance forever"—but nothing lasts forever, because economic, social, and environmental systems are dynamic and change over time. What is an appropriate timescale to view sustainability? Certainly efforts should be intergenerational (see Chapter 2), but to what extent? Similarly, what is the specific system of interest? The intersection of the three pillars of sustainability essentially reflects the system or target of development actions, but defining the characteristics of that intersection remains challenging.

International Efforts to Achieve Sustainable Development

Despite the challenges of defining sustainable development, following the Brundtland Report, the international community largely embraced a global emphasis on the concept that continues today. International efforts, often formalized through bodies like the United Nations, have invested in large-scale initiatives to better define the goals of sustainable development, and to encourage funding of activities and projects at various scales (local, national, regional, and global) meant to advance society in sustainable ways. Beginning in the 1990s, a series of landmark international meetings and conventions galvanized support for sustainable development and led to specific actions to improve the lives of people and the state of the environment. In this section, we describe the role of the United Nations in sustainable development and key international conventions along with their protocols and agreements that guide sustainable development worldwide.

The United Nations

The **United Nations (UN)** is an international, intergovernmental organization founded in 1945 to maintain international peace and security, develop friendly relations among nations, achieve international cooperation solving problems, and act as a center for harmonizing the actions of nations.[13] The organization consists of 193 member states (countries) and has been a driving force behind large-scale international sustainable development. Much of the efforts toward sustainable development began in 1992 at the **Earth Summit** in Rio de Janeiro, Brazil (also known as the UN Conference on Environment and Development, or UNCED) that built on the momentum of the Brundtland report. The Earth Summit brought together representatives from over 170 countries, including heads of state, leaders of the UN, and individuals from major conservation organizations, specialized agencies and intergovernmental organizations, and from groups representing religions and Indigenous peoples.[14] The Summit's purpose was to discuss ways of combining increased protection of the environment with sustainable economic development in less wealthy countries.

The Earth Summit successfully heightened awareness of the seriousness of the environmental crisis by placing the issue at the center of the world's attention. The conference also emphasized the connection between the drive to protect the environment and the need to alleviate poverty in the developing world—goals that require increased levels of financial assistance from developed countries. At the time, many people in developing countries believed that economic progress could only come from rapidly exploiting natural resources to spur development and reduce poverty. While this strategy provides a short-term gain, it is costly over the long term. At the Earth Summit, the developed countries in attendance collectively agreed to assist developing countries of the world in the long-term goals of protecting their biodiversity and the global environment and providing for their people through sustainable development.

The Earth Summit resulted in three major non-legally binding documents aimed at guiding future sustainable development. The first was **Agenda 21**, a comprehensive outline of actions to be taken globally, nationally, and locally by organizations of the United Nations system, governments, and major groups in every area in which humans impact the environment.[15] The four sections of Agenda 21 effectively constitute an action plan for guiding sustainable development in the twenty-first century. The first section focuses on social and economic dimensions of sustainable development, and discusses combatting poverty, changing resource use patterns, promoting human health and community development, and integrating the ideas of environmental protection and sustainable development into decision-making. The second section focuses on the conservation and management of resources for development, and covers a range of topics related to protecting and managing natural resources and biodiversity, as well as the reduction of pollution and managing biotechnology. The third focuses on strengthening the role of major groups in development, including women, children, and Indigenous communities, and the fourth focuses on mechanisms for implementing the agenda. Each section includes specific objectives and activities for their implementation.

The second non-legally binding document was the **Rio Declaration on Environment and Development**, a document that included 27 principles to guide the actions of both wealthy and poor nations on issues of the environment and development.[16] The right of nations to utilize their own resources for economic and social development is recognized, provided the environments of other nations are not harmed in the process. The declaration affirms the "polluter pays" principle, in which companies and governments take financial responsibility for the environmental damage that they cause. The declaration also declares that "States shall cooperate in a spirit of global partnership to conserve, protect and restore the health and integrity of the Earth's ecosystem."[15]

The third document was the Principles for a Global Consensus on the Management, Conservation and Sustainable Development of All Types of Forests, usually referred to as the **Rio Forest Principles**.[17] This document recognized the value of forests and provided a set of 15 principles for sustainable forest management. The principles include a statement recognizing that countries have a sovereign and inalienable right to utilize, manage, and develop their forests, and that forest resources and forest lands should be sustainably managed to meet the social, economic, ecological, cultural, and spiritual needs of present and future generations. Other principles recognize the role of forests in maintaining ecological processes and provide recommendations on national forest policies, forest use and restoration activities, financial resources for forest conservation, capacity building for sustainable forest management, trade in forest products, and controlling pollution that affects forest health.

The Millennium Summit held in 2000 at United Nations Headquarters in New York City represented another major milestone in the evolution of the sustainable development concept. The Summit, which included 149 heads of state and government and high-ranking officials from over 40 other countries, led to the elaboration of eight **Millennium Development Goals** (**MDGs**) to be achieved by 2015.[18] These eight goals were:

1. To eradicate extreme poverty and hunger
2. To achieve universal primary education
3. To promote gender equality and empower women
4. To reduce child mortality
5. To improve maternal health
6. To combat HIV/AIDS, malaria, and other diseases
7. To ensure environmental sustainability
8. To develop a global partnership for development

A review of the MGDs in 2015 indicated that the program was successful in some areas.[19] One major success was the substantial reduction in poverty: the number of people living in extreme poverty declined by more than 50%, from 1.9 billion in 1990 to 836 million in 2015. However, many inequalities remained, which led to an effort to establish a new set of sustainable development goals to build on the momentum of the MDGs.

Events that followed the Millennium Summit included the 2002 World Summit on Sustainable Development in Johannesburg, South Africa that reaffirmed the global community's commitments to poverty eradication and protection of the environment, and the 2012 United Nations Conference on Sustainable Development (Rio+20) in Brazil that launched a process to develop a new set of sustainable development goals. These efforts culminated in the **2030 Agenda for Sustainable Development** that was adopted by all United Nations member states in 2015.[20] The core of the 2030 Agenda included 17 ambitious **Sustainable Development Goals** (**SDGs**) that built on the MDGs (which concluded in 2015) and aimed to complete what they did not achieve. The SDGs are expansive and include improving health and education, reducing inequality, tackling climate change, and working to preserve oceans and forests (Table 17.1). Each SDG includes a set of targets, for which progress can be measured by monitoring a set of indicators. For example:

- SDG 15: *Life on Land* aims to "protect, restore and promote sustainable use of terrestrial ecosystems, sustainably manage forests, combat desertification, and halt and reverse land degradation and halt biodiversity loss."[20]
- Within SDG 15 are 12 targets, such as Target 15.7, to "take urgent action to end poaching and trafficking of protected species of flora and fauna and address both demand and supply of illegal wildlife products."[20]
- Progress toward the targets can be assessed by indicators such as the "proportion of traded wildlife that was poached or illicitly trafficked"[65] (indicator 15.17.1).

Video 17.1 Watch this short video that provides an overview of the dimensions of sustainable development.
oup-arc.com/e/cardinale-v17.1

The use of targets and indicators provides a means for evaluating progress and level of success at meeting each SDG. In total, SDGs include 169 targets and 242 indicators. The UN Division for Sustainable Development Goals supports the implementation, evaluation, outreach, and advocacy of the SDGs (**Video 17.1**).

International conventions

The 1992 Earth Summit in Rio de Janeiro led to three legally binding conventions on biodiversity, climate change, and desertification. All three were eventually ratified. Although these conventions have been discussed in various parts of earlier chapters, we cover them collectively here because they have been a driving force at mobilizing the world around issues of human welfare, biodiversity, and climate change within the framework of sustainable development and are still in effect today.

CONVENTION ON BIOLOGICAL DIVERSITY Threats to biodiversity emerged as an important global issue in the 1980s, which led to the development of the **Convention on Biological Diversity** (**CBD**). The CBD is a legally binding, multilateral treaty that came into force at the end of 1993 and has 196 Parties (countries)[21] that are obligated to implement its provisions. The CBD has three main goals: (1) conservation of biodiversity, (2) sustainable use of biodiversity, and (3) fair and equitable sharing of the benefits arising from the use of genetic resources.

The CBD recognizes the intrinsic value of biological diversity and its importance for maintaining life-sustaining systems, and that the conservation of biodiversity is a common concern of humankind. It acknowledges that biodiversity is being significantly reduced by human activities while recognizing that states have sovereign rights over their own biological resources and that they are responsible for sustainable use of those resources. The Convention

TABLE 17.1 United Nations Sustainable Development Goals (SDGs) from the 2030 Agenda for Sustainable Development

Sustainable Development Goal (SDG)	Description
1. No Poverty	End poverty in all its forms everywhere.
2. Zero Hunger	End hunger, achieve food security and improved nutrition, and promote sustainable agriculture.
3. Good Health and Well-Being	Ensure healthy lives and promote well-being for all at all ages.
4. Quality Education	Ensure inclusive and equitable quality education and promote lifelong learning opportunities for all.
5. Gender Equality	Achieve gender equality and empower all women and girls.
6. Clean Water and Sanitation	Ensure availability and sustainable management of water and sanitation for all.
7. Affordable and Clean Energy	Ensure access to affordable, reliable, and sustainable modern energy for all.
8. Decent Work and Economic Growth	Promote sustained, inclusive and sustainable economic growth, full and productive employment, and decent work for all.
9. Industry, Innovation and Infrastructure	Build resilient infrastructure, promote inclusive and sustainable industrialization, and foster innovation.
10. Reduced Inequalities	Reduce inequality within and among countries.
11. Sustainable Cities and Communities	Make cities and human settlements inclusive, safe, resilient, and sustainable.
12. Responsible Production and Consumption	Ensure sustainable consumption and production patterns.
13. Climate Action	Take urgent action to combat climate change and its impacts.
14. Life below Water	Conserve and sustainably use the oceans, seas, and marine resources for sustainable development.
15. Life on Land	Protect, restore, and promote sustainable use of terrestrial ecosystems, sustainably manage forests, combat desertification, halt and reverse land degradation, and halt biodiversity loss.
16. Peace, Justice and Strong Institutions	Promote peaceful and inclusive societies for sustainable development, provide access to justice for all and build effective, accountable and inclusive institutions at all levels.
17. Partnerships for the Goals	Strengthen the means of implementation and revitalize the global partnership for sustainable development.

Source: United Nations General Assembly. 2015. *Transforming Our World: the 2030 Agenda for Sustainable Development.* United Nations, A/RES/70/1. Icons © United Nations 2030 Agenda for Sustainable Development (https://sustainabledevelopment.un.org/). © United Nations 2015. Reprinted with the permission of the United Nations. The content of this publication has not been approved by the United Nations and does not reflect the views of the United Nations or its officials or Member States.

covers biodiversity at multiple levels, including ecosystems, species, and genetics, and includes provisions outlined in 42 articles. Among the provisions are requirements for countries to develop national strategies/action plans for the conservation and sustainable use of biodiversity; these plans represent the main tool for implementing the convention and support the mainstreaming of biodiversity into policies of other economic sectors such as agriculture, forestry, and fisheries.[22] Provisions also call on countries to engage in specific conservation actions including (among many others) establishing protected areas, restoring degraded ecosystems, and adopting measures to recover threatened species.

Advances in the field of genetics since the turn of the century led to two protocols to the CBD. The Cartegena Protocol on Biosafety, which entered into force in 2003, aims to ensure the safe handling, transport, and use of living modified organisms that may adversely affect biological diversity and human health. The Nagoya Protocol on Access and Benefit-sharing, which entered into force in 2014, focuses on sharing the benefits from the utilization of genetic resources in a fair and equitable way.

At the tenth meeting of the CBD held in Nagoya, Aichi Prefecture, Japan, The Strategic Plan for Biodiversity 2011–2020 was adopted.[23] The Plan provided a framework for biodiversity conservation for all countries, and included a set of 20 targets referred to as the **Aichi Biodiversity Targets** (Table 17.2). Targets were organized around five strategic goals or themes. Some targets are broad such as:

> *Target 1*: By 2020, at the latest, people are aware of the values of biodiversity and the steps they can take to conserve and use it sustainably.[23]

Other targets are more specific:

> *Target 11*: By 2020, at least 17 percent of terrestrial and inland water areas, and 10 percent of coastal and marine areas, especially areas of particular importance for biodiversity and ecosystem services, are conserved through effectively and equitably managed, ecologically representative and well connected systems of protected areas and other effective area-based conservation measures, and integrated into the wider landscapes and seascapes.[23]

> *Target 15*: By 2020, ecosystem resilience and the contribution of biodiversity to carbon stocks has been enhanced, through conservation and restoration, including restoration of at least 15 percent of degraded ecosystems, thereby contributing to climate change mitigation and adaptation and to combating desertification.[23]

The CBD provides guidance for setting national targets and indicators, and developing actions per each target. Guidance is meant to help countries revise and update national biodiversity strategies and action plans. The 2018 CBD Conference of the Parties in Sharm El-Sheikh, Egypt acknowledged that most of the Aichi Biodiversity Targets are not on track to be achieved by 2020 and called for actions to accelerate progress on the targets.

UNITED NATIONS FRAMEWORK CONVENTION ON CLIMATE CHANGE Climate change has the potential to become a major disruption to global biodiversity and ecological processes, services, and health (see Chapter 12). Recognizing the need to address growing concerns about climate change and its impacts on the natural world and ultimately human well-being, the United Nations adopted the legally binding Framework Convention on Climate Change, or UNFCCC, at the Earth Summit. The UNFCCC entered into force in 1994 and has 197 Parties.[24] The UNFCCC broadly aims to prevent dangerous human interference with the climate system of the planet. This agreement requires countries to reduce their emissions

TABLE 17.2 Aichi Biodiversity Targets along with their broader strategic goals that are part of the Convention on Biological Diversity Strategic Plan for Biodiversity 2011–2020

Strategic goal		Target
A. Address the underlying causes of biodiversity loss by mainstreaming biodiversity across government and society.	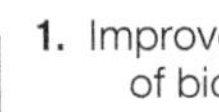	**1.** Improve awareness of values of biodiversity.
		**2.** Integrate biodiversity values into development.
		**3.** Reform incentives to support biodiversity.
		**4.** Implement plans for sustainability.
B. Reduce the direct pressures on biodiversity and promote sustainable use.		**5.** Reduce loss of all natural habitats.
		**6.** Harvest fish and invertebrates sustainably.
		**7.** Manage agriculture, aquaculture, forestry sustainably.
		**8.** Reduce pollution.
		**9.** Control and manage invasive alien species.
		**10.** Minimize climate change impacts on coral reefs.
C. Improve the status of biodiversity by safeguarding ecosystems, species, and genetic diversity. 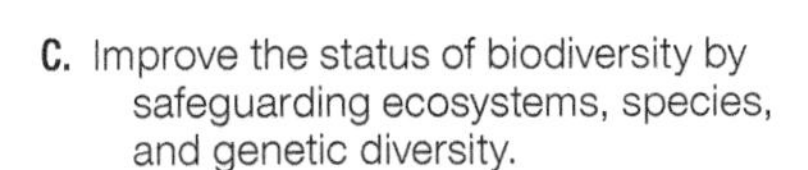		**11.** Protect important areas for biodiversity and ecosystem services.
		**12.** Prevent extinction of threatened species.
		**13.** Maintain genetic diversity.
D. Enhance the benefits to all from biodiversity and ecosystem services.		**14.** Restore and safeguard ecosystem services.
		**15.** Improve ecosystem resilience and carbon storage.
		**16.** Implement the Nagoya Protocol.
E. Enhance implementation through participatory planning, knowledge management and capacity building.		**17.** Implement national biodiversity strategy and action plan.
		**18.** Protect traditional knowledge.
		**19.** Share science knowledge and technologies.
		**20.** Mobilize financial resources for the Strategic Plan for Biodiversity.

Source: Adapted from Conference of the Parties to the Convention on Biological Diversity. 2010. *The Strategic Plan for Biodiversity 2011–2020 and the Aichi Biodiversity Targets*. United Nations Environment Programme, UNEP/CBD/COP/DEC/X/2. For a complete description of targets, see www.cbd.int/sp/. Icons © United Nations Convention on Biological Diversity.

of CO_2 and other greenhouse gases and to make regular reports to the United Nations on their progress. While specific emission limits were not decided on, the Convention states that greenhouse gases should be stabilized at levels that will not interfere with Earth's climate. The agreement puts the onus on industrialized developed nations (and those in transition) to do the most to cut emissions. Industrialized nations further agreed to support climate change activities in developing countries by providing financial support for climate change action.

The UNFCCC has a key protocol, called the Kyoto Protocol, which was adopted at a meeting in Kyoto, Japan in 1997 and entered into force in 2005.[25] The Protocol created binding emission reduction targets for developed countries and included two commitment periods, 2008–2012 and 2013–2020. During the first commitment period, 36 industrialized countries (and the European Union) pledged to reduce emissions of six greenhouse gases to specific levels (5% below 1990 levels). An Adaption Fund was created to help finance adaptation projects and programs in developing countries that are parties to the Protocol. The second commitment period includes conditions established under the Doha Amendment (adopted in 2012), under which the parties committed to reduced greenhouse gas emissions by at least 18% below 1990 levels. The Doha Amendment will enter into force once it has been formally accepted by 144 Parties; as of July 2019, 130 Parties have formally accepted the amendment.

The UNFCCC also has a key agreement, the Paris Agreement, that was adopted by Parties to the Convention at a meeting in Paris, France in 2015.[26] The Paris Agreement is separate from the Kyoto Protocol, but builds on the UNFCCC by bringing all nations into combatting climate change and adapting to its effects. The main goals of the Agreement are (1) to keep global temperature rise well below 2°C above preindustrial levels, and (2) to pursue efforts to limit increases to 1.5°C this century. The Agreement calls for regularly reported Nationally Determined Contributions from each country to monitor progress toward achieving climate goals. The Paris Agreement entered into force in 2016 and as of 2019 has been ratified by 185 of 197 Parties to the Convention.

The UNFCCC framework along with the Kyoto Protocol and the more recent Paris Agreement represent major global efforts to reduce greenhouse gas emissions and the impacts of climate change. These efforts are directly tied to the UN Sustainable Development Goals to achieve broader, sustainable development for all people in all countries. Despite widespread support for the UNFCCC, global CO_2 levels continue to rise.

CONVENTION TO COMBAT DESERTIFICATION Habitat degradation represents a significant concern for biodiversity, and is caused by several factors. One major cause is desertification, which is driven by poor soil management practices, climate change, and land use pressure from a growing human population (see Chapter 9). Recognizing the need to address the expansion of desertification, the United Nations adopted the Convention to Combat Desertification (UNCCD) in 1994. The UNCCD came into force in 1996 and has 197 Parties.[27] The UNCCD recognizes that land degradation and desertification in dryland regions is a major environmental, social, and economic problem. The Convention is legally binding and contains provisions that aim to combat desertification and reduce the effects of drought in arid, semiarid, and dry subhumid areas, where many of the most vulnerable ecosystems and people occur. The UNCCD calls for parties to create national action programs that identify both the factors contributing to desertification and practical measures for combatting and mitigating desertification and drought.

The UNCCD developed a strategic plan for 2008–2018 that outlined steps for reversing and preventing desertification and land degradation. The more recent UNCCD 2018–2030 Strategic Framework built on the first strategic plan and aims to achieve "land degradation neutrality" by restoring productivity of degraded land, improving the livelihoods of more than 1.3 billion people, and reducing the impacts of drought on vulnerable human populations.

Funding for Sustainable Development

Sustainable development activities, initiatives, and programs, whether aimed at reducing poverty or protecting land and ocean biodiversity, require funding for implementation. At the international level, working toward achieving the UN SDGs or those outlined in the three conventions governing sustainable development rely on large sources of funding. Sources of funding typically include loans from multinational banks and loans or grants that come from other countries and nonprofit organizations. In this section, we describe four sources of funding for large-scale sustainable development activities.

The World Bank

The **World Bank** is a global enterprise that funds sustainable development activities, many of which are congruent with several of the UN Sustainable Development Goals. Established in 1944, the World Bank has 189 member countries and is a multilateral development bank composed of five institutions, collectively referred to as the World Bank Group. These institutions include the IBRD (International Bank for Reconstruction and Development), IDA (International Development Association), IFC (International Finance Corporation), MIGA (Multilateral Investment Guarantee Agency), and ICSID (International Center for Settlement of Investments). The primary mission of the World Bank is to end extreme poverty, with the goal of reducing the share of the global population that lives in extreme poverty to 3%, and promote shared prosperity, with the goal of increasing the incomes of the poorest 40% of people in every country.

Much of the World Bank's role in sustainable development is providing funding for large-scale development projects. Many of these projects are not directly related to conservation, but are meant to achieve the organization's mission on poverty and prosperity. Among the five institutions of the World Bank, the most relevant to conservation and sustainable development are the IBRD, IDA, and IFC. IBRD is considered the world's largest development bank (owned by member countries) and provides loans and related financial products and services to the governments of medium- and low-income countries. The IBRD raises most of its funds in the world's financial markets and has provided more than US$500 billion in loans to alleviate poverty around the world since 1946, with its shareholder governments paying-in about US$14 billion in capital. The IDA complements the IBRD and provides zero- or low-interest loans and grants to low-income countries for programs that boost economic growth, reduce inequalities, and improve living conditions. The IFC offers financial products and services exclusively for the private sector in developing countries. Examples include large-scale conventional lending for the development of mining, agriculture, and energy projects. Other projects include providing a microloan system for small businesses and entrepreneurs, and funding green construction projects to help countries meet emissions goals under the Paris Climate Agreement.

Projects funded by the World Bank and its institutions require that recipients comply with standards for managing social and environmental risks. Different standards exist for different institutions. An example is the World Bank's Environmental and Social Standards (ESSs), which include a standard (ESS6) related to biodiversity conservation with responsibilities for (1) protecting and conserving biodiversity, (2) mitigating biodiversity impacts, (3) promoting sustainable management, and (4) supporting livelihoods of local communities.[28] ESS6 includes conditions for mitigating biodiversity and habitat loss, and protecting resources for Critically Endangered and Endangered species as defined by the IUCN Red List (see Chapter 8). Another example is the IFC Environmental and Social Performance Standards, which have some similarities to the ESSs and have been adopted by many of the world's largest private lending institutions (i.e., the Equator Principles Financial Institutions). Performance Standard 6 (PS6): Biodiversity Conservation and Sustainable Management of Living Resources[29] in particular provides standards for protecting and conserving biodiversity and renewable natural resources for development projects. Given the magnitude of World Bank and IFC activities around the world, including the development of large infrastructure projects like mines, dams, water treatment plans, and agricultural projects, these standards are vital for minimizing the impacts of such development on people and the environment.

The World Bank was an active participant in the development of the UN SDGs and has made several commitments related to the 2030 Agenda for Sustainable Development. Among them are commitments to funding for development activities and climate financing. It also continues to be one of the largest international financiers of biodiversity conservation. From 2006 to 2016, their portfolio included 241 biodiversity-related projects worth over US$1.35 billion that ranged from support for protected areas to promoting nature-based tourism to combatting wildlife crime.

The World Bank maintains a publicly available dataset of World Development Indicators (oup-arc.com/e/cardinale-w17.1) that contains current global development data at regional, national, and global scales. Indicator data relate to demographics, human health, economics, and the environment. These data are helpful for visualizing trends and informing decisions and actions around development goals. For example, data on levels of extreme poverty over time provides monitoring information that can help inform priority setting and evaluate the effectiveness of sustainable development activities (**Figure 17.2**).

Aside from the World Bank, which is large in scope and aligned with the United Nations, other multilateral development banks fund development projects. Most of these banks are governed by major developed countries.

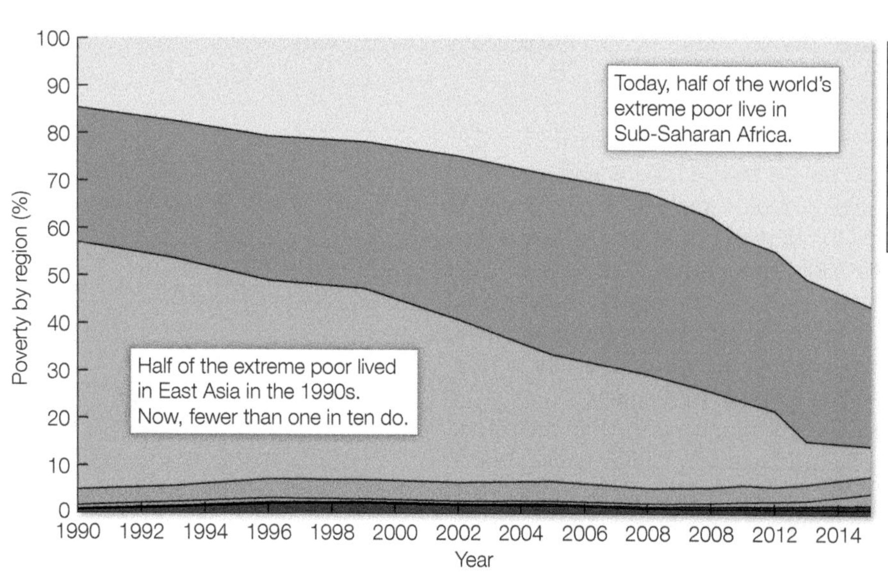

Figure 17.2 Percentage of the world's people living in extreme poverty by region from 1990 to 2015. "Extreme poverty" is defined as those living on less than US$1.90 per day. (© World Bank Development Indicators. [SI.POV.DDAY; SP.POP.TOTL]/CC BY 4.0.)

Some examples include the Inter-American Development Bank, Asian Development Bank, African Development Bank, Caribbean Development Bank, and Islamic Development Bank. Other international financial institutions, such as the European Investment Bank—the lending arm of the European Union—and European Bank for Reconstruction and Development, also finance large development projects. Many of these banks have their own performance standards, including those related to biodiversity conservation.

Global Environment Facility

The **Global Environment Facility** (**GEF**) was created at the 1992 Earth Summit in Rio de Janeiro to help address the planet's most pressing environmental problems. The GEF is an international partnership of 183 countries, international institutions, civil society organizations, and private sector enterprises, and provides funding for environmental projects, many of which link to the United Nations SDGs. Since its inception, the GEF has provided over US$18.1 billion in grants and mobilized an additional US$94.2 billion in co-financing for more than 4500 projects in 170 countries. Some notable achievements include projects related to:

- **Protected areas:** Investments in more than 3300 protected areas covering over 8.6 million km^2.
- **GHG emission reduction:** Support for 940 climate-change mitigation projects expected to contribute 8.4 billion metric tons of direct and indirect GHG emission reductions over time.
- **Water resources:** Sustainable management of 43 transboundary river basins in 84 countries, and improved governance of one-third of the world's large marine ecosystems.

The GEF provides funds to developing countries and countries with economies in transition for environmental projects to assist in meeting international conventions and agreements, including the CBD, UNFCCC, and UNCCD. Funding is provided to government agencies, civil society organizations, private sector companies, and research institutions in recipient countries. Projects are currently supported in a set of focal areas: biodiversity, climate-change mitigation, land degradation, international waters, and chemicals and waste. Programs within those areas focus on food systems, land use, and restoration; sustainable cities; and sustainable forest management. The GEF receives funding principally through the GEF Trust Fund from 39 donor countries (both developed and developing). The Trust Fund is replenished every four years by donor countries, and administered by the World Bank, which serves as the GEF Trustee. The Trust Fund has steadily grown from US$1 billion at its inception to US$4.43 billion in the 2014–2018 replenishment cycle (Figure 17.3).

The GEF is not a direct funding mechanism for the UN SDGs, but its focal areas align with many SDG goals. For example, GEF's US$900 million food security program in Africa operates in 12 countries and funds actions to safeguard natural resources (land, water, soils, trees, and genetic resources) that underpin food and nutrition security. The program focuses on smallholder farmers, who are responsible for most of the food production in the region, and targets nearly 3 million households and 100,000 km^2 of land. Activities vary by country, but include strengthening institutional frameworks that promote integrated approaches to agriculture and practices that improve soil health and sustainable agricultural production. The program supports five of the SDGs, including the elimination of poverty (SDG 1), gender equity (SDG 5), decent work and economic growth (SDG 8), responsible consumption and production (SDG 12), and climate action (SDG 13).

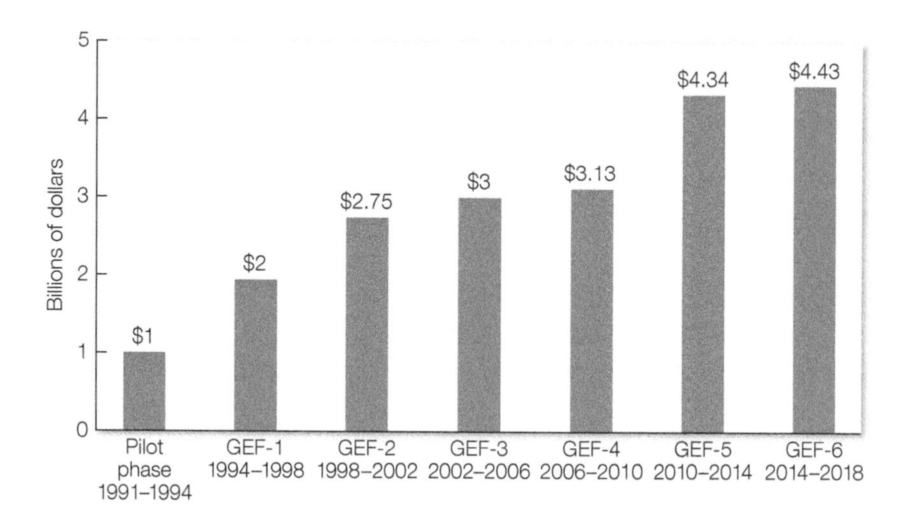

Figure 17.3 Global Environment Facility (GEF) funding provided by countries in 4-year replenishment cycles. Funding shown in US$ (billions). Funding for environment projects has steadily increased since the inception of the GEF in the early 1990s. (© Global Environment Facility, https://www.thegef.org/about/funding.)

Individual governments

Individual governments often contribute to international sustainable development projects. Many contribute to the United Nations to support their efforts and honor commitments to UN conventions, protocols, and agreements. Individual governments may also contribute directly to a country for a given project by making a bilateral contribution (i.e., from one country to another, unlike multilateral, which refers to funding from multiple countries distributed through an organization like the UN or World Bank). Bilateral funding is often provided by developed nations as part of their foreign aid or assistance programs. Examples of aid agencies include the Agence française de développement (AFD), Deutsche Gesellschaft für Internationale Zusammenarbeit GmbH (GIZ), Japan International Cooperation Agency (JICA), Norwegian Agency for Development Cooperation (Norad), the United Kingdom's Department for International Development (DFID), and United States Agency for International Development (USAID). According to 2018 data from the international Organisation for Economic Co-operation and Development, considering the total gross amount contributed bilaterally as foreign aid, the top-ranking countries include the United States (US$34.26 billion), Germany (US$24.99 billion), and the United Kingdom (US$19.40 billion). (China may exceed all three, partly due to investments in the Belt and Road Initiative described in **Box 17.1**, but does not officially disclose aid information.) However, when scaling aid to the relative size of a country (using GDP as a measure), rankings differ (**Figure 17.4**). The Center for Global Development, an independent organization missioned to reduce global poverty and improve lives through economic research, measures aid based on both the quantity and quality of aid delivered. When accounting for quality and quantity in 2018, the top-ranking nations included Luxembourg, Denmark, Sweden, New Zealand, and the Netherlands. The Center's more comprehensive Commitment to Development Index, which accounts for the quality and quantity of contributions across seven categories, ranks Sweden first, reflecting their dedication to policies that benefit people and the environment. One of the categories relates to environment, which is scored based on three subcomponents: a country's environmental policies on global climate, sustainable fisheries, and biodiversity and ecosystems. Sweden ranks second among other countries (behind the Slovak Republic) for having low greenhouse gas emissions and large reductions in emissions over a previous 10-year period. The country could improve its environment score by reducing large subsidies to its fishing industry, raising gasoline taxes, and better reporting on participation in biodiversity treaties.

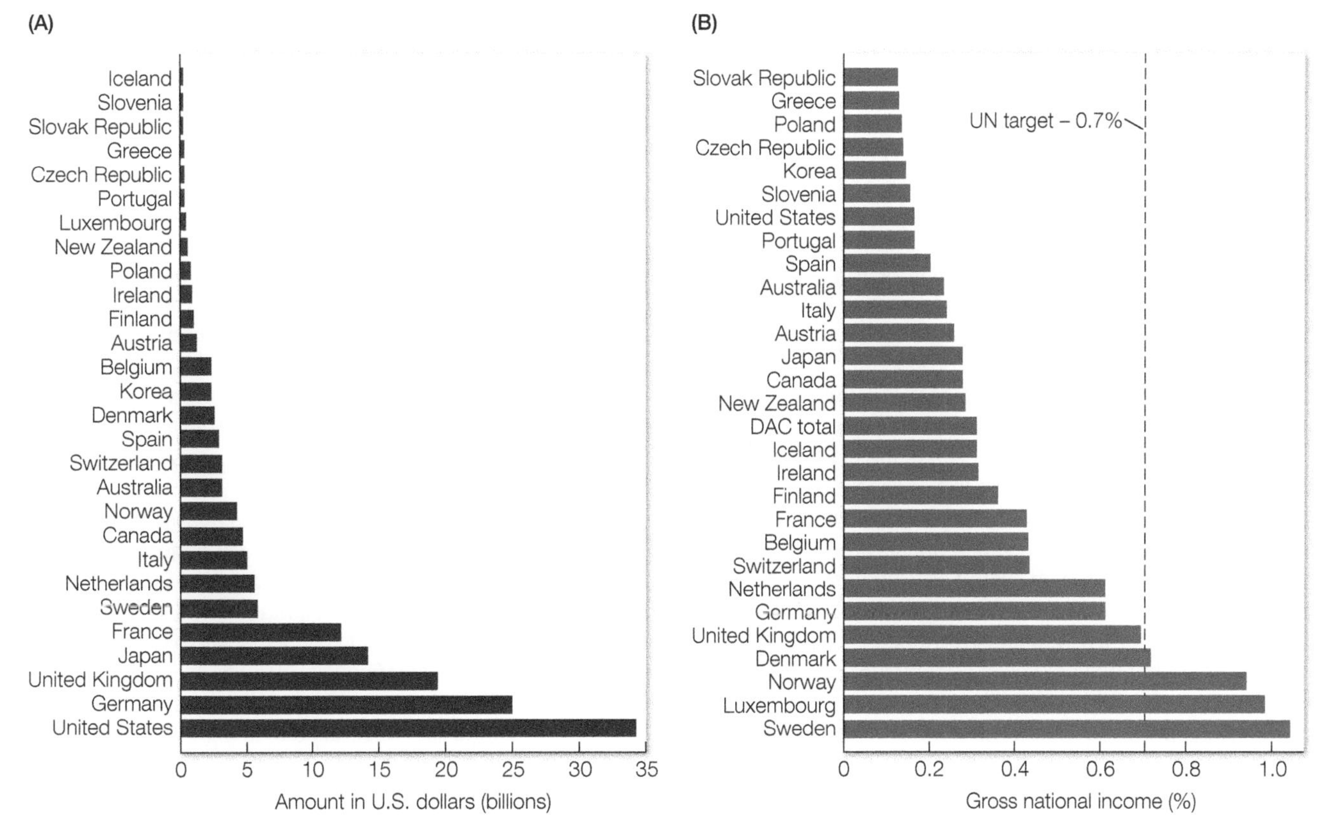

Figure 17.4 Development aid provided by countries in 2018 by total gross amount (A) and amount as a percentage of the country's gross national income (GNI) (B). The UN target of 0.7% of GNI is indicated by the dashed line. DAC refers to the Development Assistance Committee of the Organisation for Economic Co-operation and Development, a group of the 30 largest providers of aid (including those in the graph) plus the European Union and Hungary. (© Organisation for Economic Co-operation and Development [OECD].)

Is foreign aid effective at achieving sustainable development? This question is challenging to answer. There are certainly examples of sustainable development projects funded by foreign aid broadly deemed as successful, and many that are not.[30] Large amounts of foreign aid have resulted in negative consequences for some countries. Among the consequences can be creating a system of donor dependency, in which governments become dependent on aid flows that reduce incentives to find real solutions to problems, lead to wasteful spending, and weaken domestic governance. It has also resulted in corruption in some cases, with aid being used for other purposes or diverted to private individuals. Aid can constitute a large component of some economies (it accounts for over half the GDP of some countries), and some analyses suggest that it has done little to reduce poverty and stimulate positive economic growth.[31] Controversy exists over the impacts of aid, but it is important to recognize that bilateral development aid ultimately is meant to serve the interests of the donor country.[32] Advancing global development priorities, a common goal of aid, is therefore tempered by national interests.

Suggested Exercise 17.1 Read about how the Center for Global Development ranks countries for its Commitment to Development Index, then select a country to view trends across development categories and the country report. How has this country scored on its contribution to promoting a sustainable environment over time? **oup-arc.com/e/cardinale-ex17.1**

BOX 17.1 Challenges & Opportunities

China's Belt and Road Initiative

In 2013, China launched the Belt and Road Initiative (BRI), an ambitious plan to improve trade, connectivity, and connection across Asia, Europe, and Africa. The plan calls for the development of overland trade routes (the "Belt") and maritime trade routes (the "Road"), and has the potential to substantially change the nature of global trade. The BRI trade routes cross through 64 countries that together account for one-third of the global economy, two-thirds of the global population, and three-quarters of the world's known energy reserves (see the Figure). The initiative mainly involves investing in the construction of infrastructure, including roads, bridges, railways, airports, ports, and energy developments. It effectively connects China to other regions of the world and increases their access to global goods, services, and markets, but it is also meant to improve lives and livelihoods by investing in infrastructure that can help build economies and lead to shared prosperity (**Video 17.2**).

China supports the United Nations 2030 Agenda for Sustainable Development, and the BRI will address several Sustainable Development Goals, such as ending poverty (SDG 1) and supporting industry, innovation, and infrastructure (SDG 9). Although there is evidence that the BRI has resulted in positive impacts on sustainable development in some areas,[54] there have been criticisms, especially with respect the potential environmental impacts of BRI projects.[55] For example, an analysis by the World Wildlife Fund indicates that proposed BRI terrestrial routes overlap with the ranges of 265 threatened species (including 39 Critically Endangered species and 81 Endangered species), 1739 Important Bird Areas or Key Biodiversity Areas, and 46 biodiversity hotpots or Global 200 regions.[56] In addition, 32% of the total area of all protected areas in countries crossed by BRI routes is potentially affected. Furthermore, the development and use of infrastructure for transcontinental trade (including oil and gas) could increase levels of greenhouse gas emissions. Other risks of the BRI include the social consequences of large debt taken on by countries for development projects, as well as negative impacts on vulnerable people.

The potential for environmental impacts has led to calls for China to set a new standard for how BRI projects affect the environment.[55] This might include using Strategic Environmental and Social Assessments (SESAs) in line with environmental standards held for projects within China or requiring governments and financial institutions involved in BRI projects to follow environmentally sustainable performance standards, like those used by the International Finance Corporation (IFC). More specific actions include *avoiding* vulnerable environments, *reducing* impacts through mitigation activities, *restoring* damaged environments, and *offsetting* environmental losses.[57] Investment in green infrastructure and renewable energy would also contribute to positive sustainable outcomes.

One challenge to sustainable development the BRI presents is to ensure that the project results in growth that is not exclusively economic. Sustainability is defined along

Non-governmental organizations

Sustainable development activities are also undertaken by **non-governmental organizations (NGOs)**. These nonprofit organizations are mission-driven, and often have the capacity to be more nimble and effective than governments or multilateral organizations like the UN, which are inherently more complex and bureaucratic. NGOs, which include foundations, provide direct funding to governments, other organizations, the private sector, and individuals for sustainable development projects.[33]

Several large conservation NGOs are actively involved in funding and implementing sustainable development projects, often in collaboration with governments and the private sector (**Box 17.2**). Examples include organizations such as the World Wildlife Fund, The Nature Conservancy, Conservation International, the Wildlife Conservation Society, and Birdlife International. NGOs have emerged as leading sources of conservation funding, raising money from membership dues, donations from individuals, sponsorships from corporations, and grants from foundations and international development banks. The largest nature-focused NGO based in the United States is The Nature Conservancy, which in 2018 had an annual revenue of US$1.144

three dimensions: economic, environmental, and societal. If the sole focus is on economic growth and expansion, then the initiative essentially represents business as usual, and may very well result in negative impacts to environment and society. A second challenge is to ensure that development now is sustainable in the long term (i.e., intergenerational) and does not compromise equal access and opportunities for future generations. A third challenge is to ensure that the BRI will result in growth and prosperity that is not only sustainable, but shared among the countries involved.

The BRI will have significant economic impacts, and China has an opportunity to engage in full sustainable development that also results in long-term benefits to people and environment. Concerted alignment of the BRI with UN Sustainable Development Goals presents opportunities to set new standards for global trade and partnership, and bring the initiative in line with China's President Xi Jinping's vision (June 23, 2016) for a "green, healthy, intelligent, and peaceful" Silk Road of the future.

▶ **Video 17.2** Watch a video that provides an overview of China's Belt and Road Initiative.
oup-arc.com/e/cardinale-v17.2

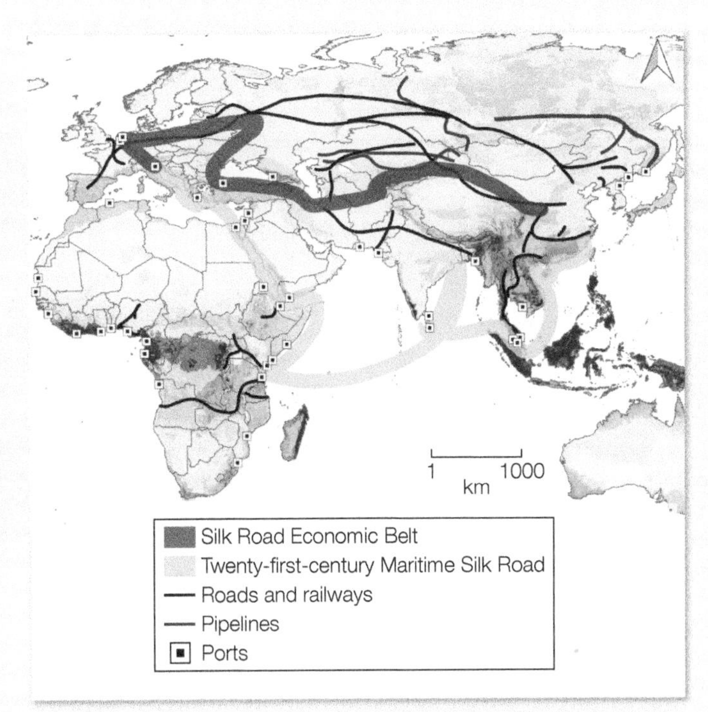

Overland and maritime trade routes proposed for development through China's Belt and Road Initiative. The "Silk Road" references China's historic network of trade routes that were active from the second century BCE right up until the dawn of the Industrial Revolution. (From F. Ascensão et al. 2018. *Nat Sustain* 1: 206–209. Infrastructure mapping is based on infographics from the Mercator Institute for China Studies [MERICS].)

billion, ranking twentieth with respect to revenue among all non-profit organizations across all sectors in the U.S. The Nature Conservancy was founded in 1951 and works in 72 countries across six continents, and has protected over 480,000 km^2 of land. Despite their role in advancing sustainable development projects, it is important to recognize that NGOs can also face issues (like government agencies) related to fluctuating budgets and shifting priorities, corporate influences, and large administrative costs.

Nonprofit organizations often partner with multilateral and bilateral funding agencies to engage in conservation and development activities. One mechanism is through the establishment of **National Environmental Funds (NEFs)**. NEFs are typically set up as conservation trust funds or foundations in which a board of trustees—composed of representatives of the host government, conservation organizations, and donor agencies—allocates the annual income from an endowment to support inadequately funded government departments and nongovernment conservation organizations and activities. NEFs have been established in over 50 developing countries with funds contributed by developed countries and major organizations such as the World Bank, Global Environment Facility, and World Wildlife Fund.

BOX 17.2 Success Story

Snow Leopard Enterprises

The snow leopard (*Panthera uncia*) is a rare large cat species that occurs through the mountainous regions of Asia. The IUCN lists the snow leopard as a threatened species due to population declines.[58] One of the causes of declines is conflict with people, many of whom rely on raising livestock for their livelihoods and are impoverished. Snow leopards will kill livestock, especially in areas where prey populations have declined, and are subsequently killed by people. The loss of even a single goat, sheep, or other livestock can result in financial hardship for many living in the region.

In an effort to increase snow leopard numbers, better protect snow leopard habitat, and support the livelihoods of rural people and communities, the Snow Leopard Trust (SLT), a U.S.-based nonprofit organization, launched Snow Leopard Enterprises (**Video 17.3**). This initiative partners with communities, mainly women, and facilitates a business enterprise in which participants use the raw wool from their livestock to make handmade products such as felt rugs, ornaments, and clothing (see the Figure). SLT purchases these products at fair prices and sells them to a worldwide market. Communities that participate must sign a conservation agreement that requires each person to protect snow leopards and wild prey living in the area from poaching. If no poaching occurs, a financial bonus is provided each year to the community.

Snow Leopard Enterprises now works with more than 400 women in 40 communities in 4 countries, including Mongolia, Kyrgyzstan, Pakistan, and India. The program has generated over US$1 million in revenue, and has increased local incomes up to 40% in some communities. Revenue has also been invested in snow leopard conservation programs that focus on combatting poaching, offsetting livestock losses and preventing depredation, improving livestock health, and environmental education.

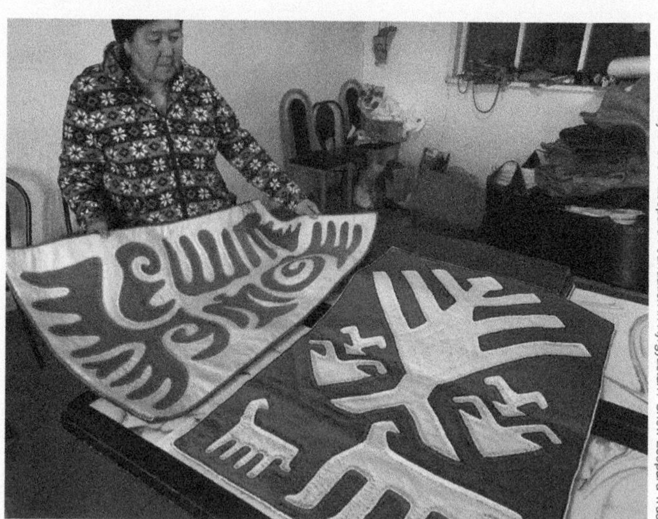

Courtesy of Snow Leopard Foundation in Kyrgyzstan / Snow Leopard Trust

Colorful rugs made by Snow Leopard Enterprises artisans in Ak Shiyrak, Kyrgyzstan.

Snow Leopard Enterprises is an example of an approach to sustainable development that involves partnerships between nonprofit organizations, communities, and businesses and centers around improving the status of snow leopards and their habitats, and the livelihoods and well-being of people. It aims to break the cycle of poverty for many rural communities, empower families, and create incentives to protect wildlife and ecosystems. In 2011, Snow Leopard Enterprises won a BBC World Challenge Award honoring projects that have shown enterprise and innovation at a grass-roots level, resulting in positive change.

▶ **Video 17.3** Watch a video about a Snow Leopard Enterprises project.
oup-arc.com/e/cardinale-v17.3

The Bhutan Trust Fund for Environmental Conservation (BTF, oup-arc.com/e/cardinale-w17.2) was one of the first NEFs and launched in 1992 as a collaborative venture between the Royal Government of Bhutan, United Nations Development Program, and World Wildlife Fund. The BTF established an endowment of US$20 million, and donors to the trust fund include World Wildlife Fund, GEF, and the governments of Bhutan, Denmark, Finland, the Netherlands, Norway, and Sweden. The fund supports field programs in three themes: biodiversity conservation, ecosystem management, and social well-being. Since inception, it has funded over 225 projects that include scientific studies of flora and fauna, protected area management, climate change adaptation, and community capacity building among others.

Another example of a NEF is the Latin American and Caribbean Network of Environmental Funds (RedLAC, oup-arc.com/e/cardinale-w17.3). Established in 1999, RedLAC is a network of 19 countries in Mesoamerica, South America, and the Caribbean and administers 53 environmental funds in the region in support of biodiversity conservation and sustainable development. Members have managed over 5800 projects and supported over 900 protected areas.

Examples of Sustainable Development Projects

Sustainable development projects occur at a variety of scales, from small local projects to large, transboundary international projects. There are many examples of successful projects, although it is important to recognize that measuring success can be challenging, and any sustainable development initiative will have trade-offs. In this section, we briefly describe four large-scale sustainable development projects focusing on ecotourism, wildlife trade, fisheries management, and climate change.

Suggested Exercise 17.2 Use the World Bank and GEF databases to learn about sustainable development projects in your home country. **oup-arc.com/e/cardinale-ex17.2**

Ecotourism and biodiversity

Costa Rica has emerged as a global leader in nature-based tourism. Ecotourism, commonly defined as "responsible travel to natural areas that conserves the environment and improves the wellbeing of local people"[34] has become a substantial component of the Costa Rican economy that relies on the country's rich biodiversity. Despite covering only 0.03% of Earth's land surface, Costa Rica harbors approximately 5% of global biodiversity, including at least 854 bird species, 237 mammal species, 423 reptile and amphibian species, 11,467 plant species, and 66,000 insect species.[35] The high level of biodiversity attracts ecotourism, as well as related areas of adventure-based tourism, volunteer tourism, and sustainable tourism. Despite periods of extensive deforestation (reaching around 600 km^2 per year) and accumulation of debt in the 1970s and 1980s, investments in conservation by government, nongovernment organizations (NGOs), and the private sector resulted in a vibrant ecotourism economy that has embraced biodiversity conservation.[36] In 2017, 2.3 million tourists visited Costa Rica, most to protected areas, which contributed 12.9% of GDP (US$7.5 billion) and supported 1 in 8 jobs in the country.[37] The development and rise of ecotourism mainly occurred from 1970 to 2000; tourist arrivals increased from 155,000 in 1970 to 435,000 in 1990 to 1.1 million in 2000, with tourism revenues increasing from US$21 million to US$1.15 billion during the same period.[36]

Today, a national effort called the Forever Costa Rica Program aims to conserve the country's terrestrial and marine biodiversity in perpetuity and support the well-being of the people of Costa Rica. In 2010, the Costa Rican government along with other foundations and organizations (like The Nature Conservancy) raised funds (US$56 million) to create two trust funds: the Forever Costa Rica Irrevocable Trust, and the Second Debt Swap for Nature. The latter involved an agreement with the United States to commit to projects for conservation and sustainable use of tropical forests in exchange for the cancellation of part of its debt (an approach known as a **debt-for-nature swap**). The Forever Costa Rica Association (Asociación Costa Rica por Siempre, oup-arc.com/e/cardinale-w17.4) was then created to manage and administer the funds and implement conservation and development projects in partnership with the government, private sector, and civil society (it is a member of RedLAC). The Forever Costa Rica Program is part of the activities of the Association and helps Costa Rica fulfill its agreements under the Convention on Biological Diversity. The Association is also part of Pacífico, an alliance of four environmental funds that seeks sustainable financing for conservation actions along the Tropical Eastern Central Pacific region. The Forever Wild Program has implemented projects that improved trails and tourism infrastructure in protected areas, developed protection strategies for protected areas, reclaimed wetlands, and strengthened marine protected area monitoring and management of fishing rights and invasive species.

Illegal wildlife trade

The illegal wildlife trade is valued at between US$7 and US$23 billion worldwide, behind trade in narcotics, human trafficking, and weapons (see Chapter 10). Trade in animals and their parts, including skins, bones, and organs, has severely impacted populations of many species. Much of the illegal trade has been driven by traditional medicine markets in Asia, and especially affected species from Africa and Asia. Black rhinoceros (*Diceros bicornis*) are Critically Endangered, largely because their horns are perceived to have medical qualities and bring large profits that attract poachers. African elephants are poached for their tusks (ivory) that contributed to their widespread decline, but efforts to reduce poaching and demand such as a ban on domestic ivory trade and processing in China have helped increase populations in some areas (**Figure 17.5**; **Box 17.3**).[38,39] Many other species have experienced significant declines from poaching, including tigers and other large cats, primates, birds, and marine

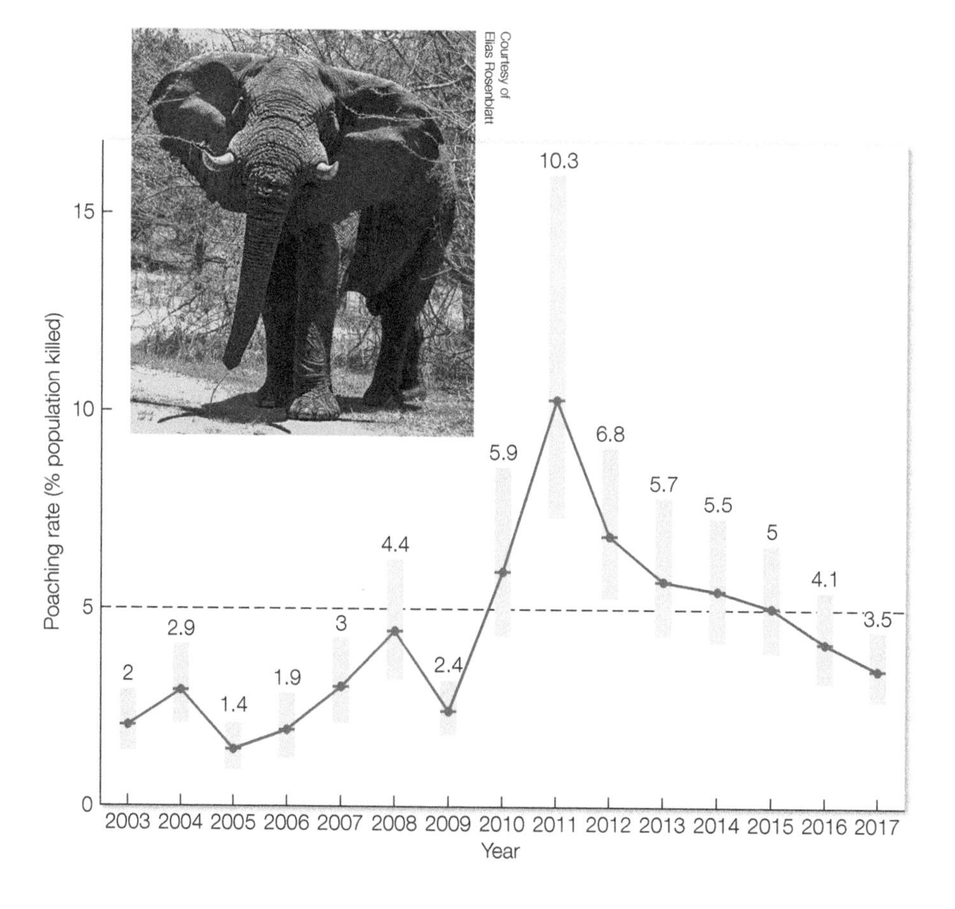

Figure 17.5 Trends in poaching rates of African elephants (*Loxodonta africana*). The data take into account a 3% natural mortality, and the dashed line at 5% represents an average growth rate of a large, well-established population. Poaching rates above this line are thought to result in net population losses. Boxes represent 90% confidence intervals. (Data from CITES Monitoring the Illegal Killing of Elephants Programme [MIKE], 18th meeting of the Conference of the Parties, Colombo, Sri Lanka, CoP18 Doc. 69.2.)

BOX 17.3 Case in Point

Should the Ban on Elephant Ivory Trade Be Lifted?

The African elephant (*Loxodonta africana*) is a majestic and iconic species that experienced large-scale declines in the 1900s mainly due to overhunting and the ivory trade. International outcry over declines led to a 1989 ban on the sale of ivory through the Convention on International Trade in Endangered Species of Wild Fauna and Flora (CITES; see Chapter 10). The ban, together with aggressive conservation actions by governments, nonprofits, and other international bodies, led to recovery of elephant populations in some areas, particularly in southern Africa.

© Denis Farrell/AP/Shutterstock

Stockpile of elephant tusks (ivory) in South Africa.

Increased populations and improvement in elephant protections led some countries to petition CITES to lift the ban on ivory sales to allow for the sale of stockpiled ivory (collected from natural deaths, confiscations, and seizures; see the Figure). This ignited a fierce debate over the risks of lifting the ban (**Video 17.4**). Some countries argued that permitting the sale of ivory would allow them to generate revenue that could be reinvested in elephant conservation and management. Critics argued that it would kick-start a demand for ivory that could lead to higher levels of poaching and further population declines. CITES did approve two sales in response to petitions. In 1999, Botswana, Namibia, and Zimbabwe auctioned 50 metric tons of tusks to Japanese dealers for US$5 million; then, in 2008, Botswana, Namibia, South Africa, and Zimbabwe sold 102 metric tons to Japan and China for US$15.4.[59] To allow these sales, CITES downlisted elephants from Appendix I (the highest form of protection) to Appendix II for each country (all other countries remained at Appendix I). CITES then agreed to a 9-year moratorium on international ivory sales. Tanzania and Zambia subsequently petitioned—unsuccessfully—to sell their ivory stockpiles[60].

At the 2019 Africa Wildlife Economy Summit in Zimbabwe, convened by the UN Environment Programme and African Union, leaders of Zimbabwe, Botswana, Zambia, Angola, and Namibia renewed calls to again lift the ban on ivory sales so they could sell stockpiled ivory. Leaders argued that good governance has led to larger elephant populations and that revenue from ivory sales would allow for better management, including combatting both poaching and high levels of elephant-human conflict (from crop loss, destruction of property, and threats to personal safety). Critics believe lifting the ban could fuel even greater levels of poaching, not only in southern Africa, but elsewhere on the continent. Further declines could lead to ecological impacts on savannah and forest ecosystems, where elephants are a keystone species, and potentially could result in local, regional, or even global extinction.

Should the ban be temporarily lifted? Do the benefits outweigh the risks? Are efforts better spent stopping the ivory trade (and demand) rather than fueling it? Who should govern the use and trade of wildlife? These questions will undoubtedly be debated at future CITES meetings, potentially leading to changes in the governance of trade for other species.

Elephants are an important and valued part of wildlife tourism, which represent a substantial element of some economies, including those of southern Africa. An analysis of the value of elephants to tourism indicated that the species alone accounts for approximately US$25 million each year, an amount based on the loss of elephants to poaching.[61] Meeting United Nations Sustainable Development Goals in several African countries will require managing the complex relationships between elephants, tourism, and the livelihoods of rural communities.

Video 17.4 Watch a video on the ivory ban, temporary sales of ivory, and scientific efforts to track illegal poaching.
oup-arc.com/e/cardinale-v17.4

life. Pangolins (8 species) are the most illegally traded group with over 1 million poached between 2000 and 2013; 159 unique international trade routes for illegal pangolin trafficking have been identified (see Figure 10.10).[40]

In response to the crisis of illegal wildlife trade, the World Bank, with GEF funding, launched the Global Wildlife Program (GWP), more formally called the Global Partnership on Wildlife Conservation and Crime Prevention for Sustainable Development. The GWP is a global partnership, including UN programs, government agencies, and non-profit organizations that promotes wildlife conservation and sustainable development by combatting illegal wildlife trafficking. The program received US$213 million in funding and is engaged in projects in 32 countries in Africa, Asia, and Latin America. Activities are aimed at conserving wildlife and habitats, promoting wildlife-based economies, combatting wildlife crime, and reducing demand.

More specific actions include enhancing anti-poaching and intelligence operations, increasing and better managing protected areas, integrated land-use planning, and providing opportunities for development through tourism and other agriculture, forestry, and natural resource projects. They have supported projects in places like Gorongosa National Park and Niassa National Reserve in Mozambique, large wilderness areas and hotspots for poaching. Poachers have killed half of Mozambique's elephants in the past 7 years, with just 10,000 remaining today. Efforts in this region have involved promoting the value of wildlife, strengthening law enforcement, establishing protected areas, restoring habitats for wildlife, and generating sustainable rural livelihoods (**Video 17.5**). The GWP reported in 2019 that under the program, 26.6 million ha of protected areas have been created or are under improved management for conservation; 2.7 million ha of landscapes are under improved practices; 11.4 million metric tons of CO_2 have been sequestered; and the number of beneficiaries has reached 490,000.

Video 17.5 See what efforts are helping to combat elephant poaching in the Niassa region as part of the Global Wildlife Program.
oup-arc.com/e/cardinale-v17.5

Fisheries

Worldwide, 100 million tons of marine fish are harvested annually and together with freshwater fish provide over 3 billion people, many from developing countries, with an important source of protein and micronutrients (see Chapter 6).[41] Coastal marine fisheries or those lying within Exclusive Economic Zones (offshore areas, defined under the 1982 UN Convention on the Law of the Sea, where individual governments have exclusive rights to fishing and other marine resources) contribute 85% of the 80 million metric tons per year generated by marine fisheries.[42] However, coastal fisheries face significant threats from overexploitation. Overexploitation of marine ecosystems directly impacts people by affecting food security, human health and well-being, and jobs and livelihoods. Overharvesting of commercial fish also results in higher levels of bycatch (incidentally captured fish and marine life that are discarded), which further erodes biodiversity and damages ecosystems (see Chapter 10).

The Coastal Fisheries Initiative (CFI) is a GEF supported program that aims to address large-scale human and environmental issues surrounding coastal marine fisheries. The CFI works in six countries with large coastal fisheries, including Indonesia, Ecuador, Peru, Cabo Verde, Côte d'Ivoire, and Senegal. The Initiative focuses on three areas:

1. *Policy*, by incorporating environmental and economic factors into fisheries policies and regulations
2. *Capacity*, by improving the ability of nations and communities to manage fisheries sustainably
3. *Partnerships*, by promoting public-private partnerships to improve the sustainability of supply chains

Efforts at better managing fisheries also aim to increase greenhouse gas sequestration through better protection of salt marshes, sea grass beds, and mangroves that absorb and store carbon. The CFI links with the GEF's Large Marine Ecosystems (LME) initiative that invests in the sustainable management and governance of 23 of the world's 66 large marine ecosystems in collaboration with 124 countries. Both the CFI and LME directly relate to achieving the UN SDG 14: *Life below Water,* which aims to conserve and sustainably use the oceans, seas and marine resources for sustainable development.

Climate change

Tackling climate change has been a major focus of international sustainable development, and numerous projects aim to reduce greenhouse gas emissions to meet targets of the Paris Agreement. Climate change is accelerating: atmospheric carbon dioxide has reached 415 parts per million, global temperature has risen 1°C (1.8°F) since 1880, the annual mean extent of Arctic sea ice decreased 3.5% to 4.1% per decade from 1979 to 2012, global mean sea level rose by 19 cm between 1979 and 2020, and ocean acidification has increased 26% since the Industrial Revolution (see Chapter 12).[43] Five of the warmest years on record have occurred since 2010. Climate change is already having clear impacts on biodiversity and the lives of people, and is predicted to become the largest contributor to the loss of biodiversity and ecosystem services in the future. Investment in reducing greenhouse gas emissions is urgently needed, especially as only 57 countries of the Paris Agreement (31% of those that ratified it) are on track to meet their targets.

Investment in renewable energy represents one means of reducing fossil fuel emissions. A large project in Morocco is underway, constructing one of the world's largest concentrated solar power plants. Concentrated solar involves using a field of mirrors to focus the sun's energy on a small area to heat a fluid that produces steam, which drives turbines that generate electricity; the fluid also heats molten salts in storage tanks that remain hot enough to generate steam during nighttime. When fully completed, the Noor Ouarzazate complex is expected to produce over 500 megawatts and provide power to 1.1 million people—about the size of the population living in the city of Amsterdam or Islamabad (**Figure 17.6**). It is also expected to reduced carbon emissions by 760,000 tons per year—roughly equivalent to the amount of CO_2

Figure 17.6 Noor Ouarzazate concentrated solar power facility, Morocco. Funded in part by the World Bank and the Global Environment Facility, this project is one of the largest concentrated solar power plants in the world and is expected to reduce carbon emissions by 760,000 tons per year, thus furthering the climate-change goals of the United Nations.

emitted by 165,000 passenger cars per year. The solar plant is important for achieving energy independence, as Morocco is a large energy importer, relying on foreign sources for 97% of its energy. Concentrated solar plants have the capacity to produce large amounts of renewable energy, but are costly. This project, with funding from the World Bank, GEF, and other development banks, hopes to demonstrate the viability of this form of power generation to catalyze other projects around the world.

The Future Success of Sustainable Development

Sustainable development efforts at a global level have been a focus of the international community and resulted in some remarkable achievements, such as a global reduction in extreme poverty. However, challenges lay ahead as many of the problems faced by society are deepening and becoming more complex. The United Nations 2030 Agenda for Sustainable Development has provided a set of goals, targets, and indicators to improve the lives of people without sacrificing the environment. Successfully reaching these goals by 2030 will require continued investment and funding from the international community. The UN Conference on Trade and Development (UNCTAD) estimated that, on average, US$3.9 trillion is needed each year until 2030 to achieve the 17 SDGs in developing countries alone; only 36% is being met by current public investment plans, leaving a US$2.5 trillion per year gap in investment.[44]

Other groups suggest that the costs are even higher. The Council on Foreign Relations—a nonprofit think tank specializing in U.S. foreign policy and international affairs—suggests that the cost needed to achieve all SDGs is probably between US$90 and US$120 trillion (**Figure 17.7**). However, this estimate largely excludes costs related to biodiversity conservation, especially with respect to achieving SDG 14: *Life below Water* and SDG 15: *Life on Land*. According to the United Nations Development Programme, it will cost between US$70 and US$160 million per year to achieve sustainable forest management on a global scale—representing just one component of SDG 15—and an additional US$150 to US$440 billion per year to stop the loss of biodiversity by 2050. Similarly, achieving ocean protection (Target 14.5 under SDG 14) is estimated to cost US$21 billion per year. Numerous financing options can be used to fund biodiversity conservation, ranging from biodiversity offsets to payments for ecosystem services to crowdfunding.

Suggested Exercise 17.3 Explore 20 financing options suggested by the United Nations Development Programme for funding biodiversity conservation activities that support UN Sustainable Development Goal 15, Life on Land. **oup-arc.com/e/cardinale-ex17.3**

Reaching the SDGs by 2030 will also require active leadership and engagement, especially by the developed world. While most countries support the UN SDGs and the conventions that promote sustainable development, several countries have failed to ratify or support some of the key protocols and agreements. The United States—the second largest emitter of greenhouse gas emissions (behind China)—signed the Kyoto Protocol in 1998, but due to perceived economic concerns it was never ratified by the U.S. Senate to become legally binding. Similarly, the United States announced a withdrawal from the Paris Agreement in 2017, again citing that it would be economically disadvantageous to the country.

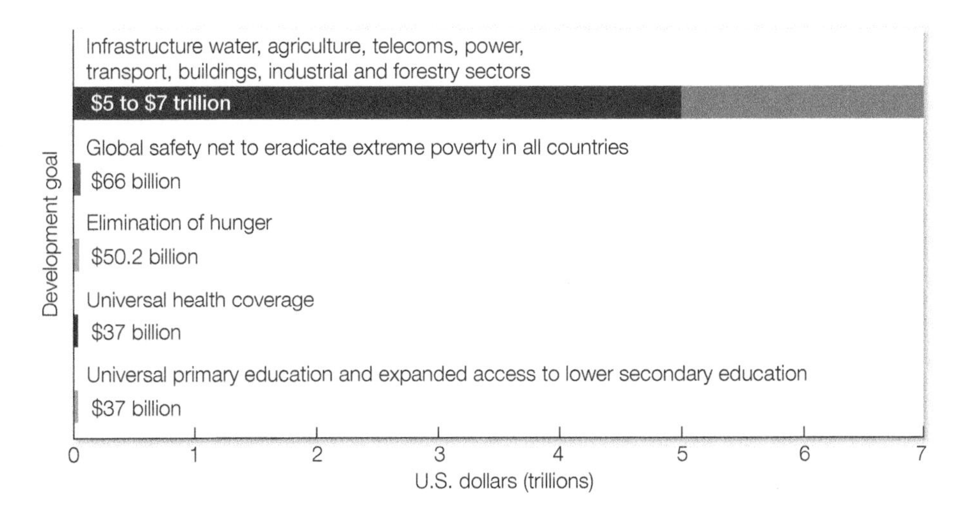

Figure 17.7 Annual costs to achieve the United Nations Sustainable Development Goals estimated by the Council on Foreign Relations. These estimates largely exclude costs associated with biodiversity conservation, which could be several hundred billion US$ more per year, according the United Nations Development Programme. (After United Nations Sustainable Development Goals 2015, by the Council on Foreign Relations. Reprinted with permission.)

The success of sustainable development, regardless of scale, could benefit from better measures of the interconnectedness and interactions among the economic, social, and environmental components of projects. Any sustainable development action will have trade-offs, and better describing those trade-offs can lead to more efficient and effective outcomes. For example, the UN SDGs comprise 17 goals, each with their own set of targets and indicators (see Table 17.1). While targets and indicators are helpful for addressing individual goals, no guidance was given on the interconnectedness between goals, which creates challenges for designing, monitoring, and evaluating sustainable development projects.[45] Addressing one goal may have negative effects (trade-offs) or positive effects (synergies) on other goals. For example, efforts to tackle SDG 2: *Zero Hunger* may involve promoting agricultural development for food production. However, with agricultural development comes a trade-off with SDG 15: *Life on Land*, which aims to protect biodiversity and limit development activities. Synergies also exist. Efforts to promote SDG 7: *Affordable and Clean Energy* in the form of renewables like solar and wind, will reduce greenhouse gas emissions by replacing fossil fuels contributing to SDG 13: *Climate Action*, which then contributes to reductions in ocean acidification from carbon pollution contributing to SDG 14: *Life below Water.*

Developing frameworks and ways of visualizing trade-offs and synergies can inform the creation of policies and actions that maximize human and environmental well-being. In an effort to provide a starting point for evaluating trade-offs, Nillson and colleagues[46] developed a simple rubric that scores the interactions between goals (which could also be applied to targets, indicators, and other policy actions). The rubric scores the interaction between two SDGs on a scale from +3, representing an "indivisible interaction," to –3, representing a "cancelling interaction" (0 indicates no interaction, positive or negative). For example, SDG 2: *Zero Hunger* interacts with other goals such as eradicating poverty (SDG 1), promoting health (SDG 3), and education for all (SDG 4). Addressing hunger is indivisible from addressing poverty (+3) and health (+3), and reinforces education (+2) as children can concentrate and perform better in school; not addressing hunger would "counteract" education (–2), as children have to help provide food rather than attend school.[46] Scoring interactions supports decision-making that minimizes negative interactions and enhances positive ones.

Other efforts have expanded on the simple rubric to provide a more nuanced view of interactions such as mapping more complex interactions.[47] Using Sweden as an example, Weitz and colleagues[48] developed a set of 34 indicators related to SDGs. They then created 34 × 34 matrix that contained pairwise interactions scored using the rubric. Values summed across rows and columns

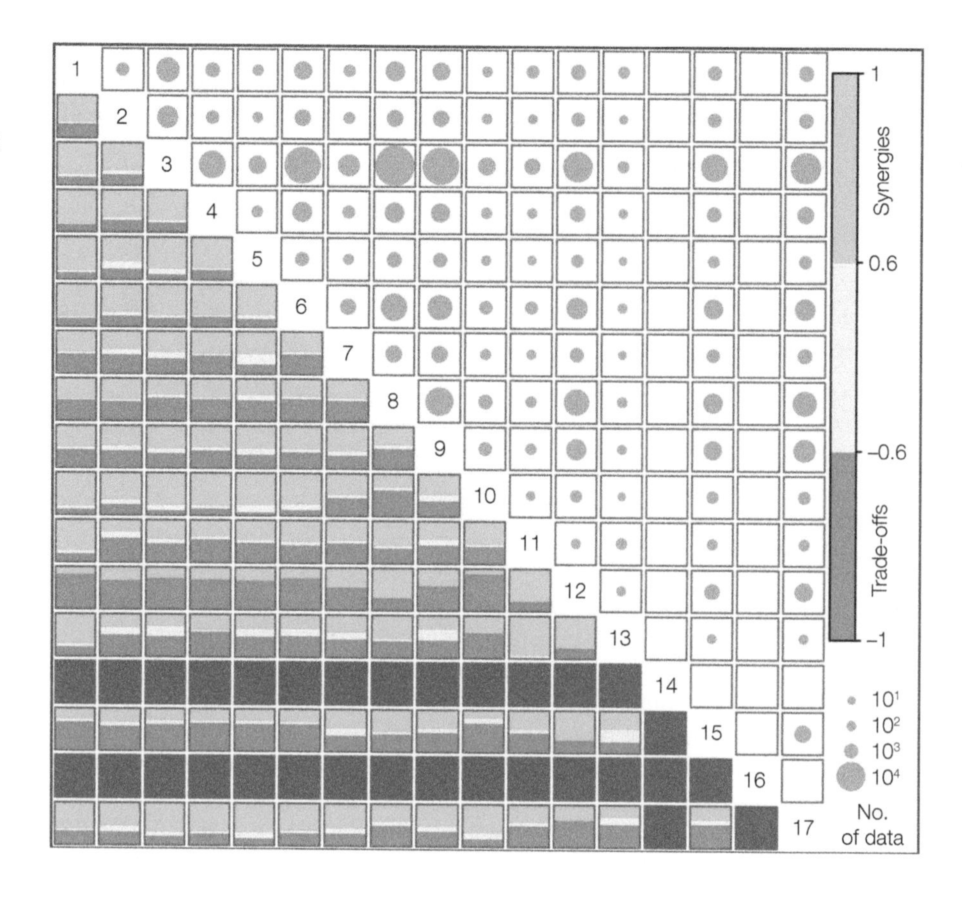

Figure 17.8 Synergies and trade-offs between UN Sustainable Development Goals (numbered 1–17 along the diagonal; see Table 17.1) based on official SDG data on 122 indicators across 227 countries from 1983 to 2016. Synergies (green) represent the proportion of positive correlations between SDG pairs across the dataset, while trade-offs (orange) represent the proportion of negative correlations between SDG pairs. Yellow indicates nonclassified relationships, and gray bars represent insufficient data. The sizes of the circles at right indicate the number of data pairs in each correlation. Notice that SDG 1: *No Poverty* had largely synergetic relationships with most other SDGs, and SDG 12: *Responsible Consumption and Production* and SDG 15: *Life on Land* mostly resulted in trade-offs with other SDGs. (From P. Pradhan et al. 2017. *Earth's Future* 5:1169–1179/CC BY-NC-ND 4.0.)

provide a measure of each target's net influence on other targets and whether they are strongly influenced by progress in other targets.[48] Network analysis was performed to graphically show the strength and magnitude of positive and negative interactions among targets. In another approach, Pradhan and colleagues[49] used data on 122 SDG indicators across 227 countries from 1983 to 2016 and calculated correlations between pairs of SDG values across time. They classified a significant positive correlation between a pair as a *synergy* and a significant negative correlation between a pair as a trade-off. The results revealed relationships between SDGs and indicated that SDG 1: *No Poverty* had a synergetic relationship with most of the other goals, whereas SDG 12: *Responsible Consumption and Production* and SDG 15: *Life on Land* were mostly associated with trade-offs (**Figure 17.8**). SDG 15 trade-offs in particular reflect the conflict between traditional patterns of unsustainable economic growth and the need for environmental sustainability. For example, converting land for agriculture to meet SDG 2: *Zero Hunger* or building infrastructure to meet SDG 9: *Industry, Innovation and Infrastructure* have historically been at odds with the protection of terrestrial ecosystems under SDG 15.

Synergies and trade-offs can also be aggregated into single metrics that capture the interconnectedness between economic development, society, and environment. Gross Domestic Product (GDP) is measure of economic activity, often used as a measure of a country's overall performance. However, it was never designed to account for social, economic, or environmental welfare and represents a misleading measure of success that has ignored inequality and environmental damage in many countries.[45,50] Other more comprehensive measures of progress exist that can be used to better evaluate progress toward SDGs and relationships between them.

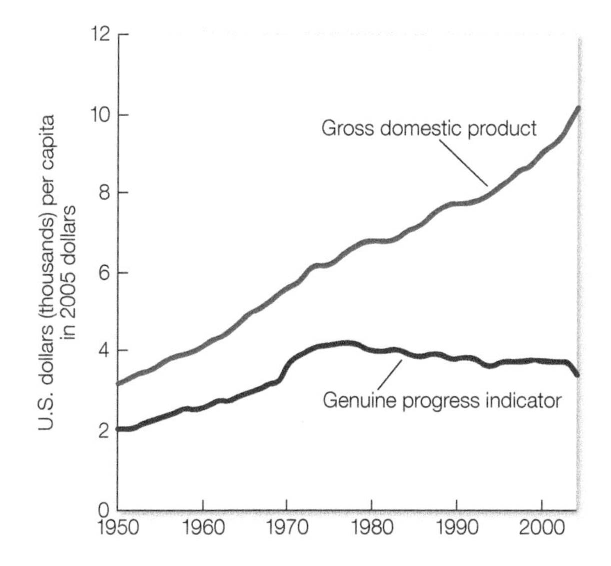

Figure 17.9 Genuine progress indicator (GPI) and gross domestic product (GDP) estimates based on data aggregated from 17 countries. GDP is an economic measure based on market transactions that is often mistakenly used as a measure of well-being. The GPI is an alternative, more holistic measure of well-being that includes economic spending, but also takes into account over 20 additional factors such as income distribution, environmental costs, crime, and pollution. Although GDP has increased consistently since 1950, GPI has not. (After I. Kubiszewksi et al. 2013. *Ecol Econ* 93: 57–68.)

Alternative measures can be classified in three groups:[50]

1. *Adjusted economic measures* modify economic measures by incorporating social and environmental factors. The Genuine Progress Indicator (GPI) is an example that takes economic data on spending, but then makes adjustments for 26 additional factors of well-being (**Figure 17.9**).[51] Nine of those factors relate to the state of environment and include measures of the costs of water and air pollution, the loss of forest and wetland cover, and long-term damage from climate change.[52] The GPI combines economic, social, and environmental factors into a single index value.
2. *Subjective measures of well-being* are based on surveys of the opinions of people. The Bhutan Gross National Happiness index (oup-arc.com/e/cardinale-w17.5) is an example that asks people how they feel in nine domains such as psychological well-being, health, education, governance, and ecological diversity and resilience. The "ecological diversity and resilience" domain includes data on pro-environmental beliefs and behaviors and perceptions of environmental issues faced by people, such as pollution and wildlife conflict. Data from all domains are combined into an index value that ranges from 0 to 1 (the 2015 index was 0.756, up 1.8% since 2010).[53]
3. *Composite measures* of several indicators combine multiple subjective and objective measures into one. Examples include the Happy Planet Index (oup-arc.com/e/cardinale-w17.6), the Better Life Index (oup-arc.com/e/cardinale-w17.7), and the Social Progress Index (oup-arc.com/e/cardinale-w17.8; **Box 17.4**).

Each of these indices includes environmental dimensions. Measures across all groups provide a means of better capturing well-being by accounting for economic, social, and environmental conditions. They also provide a foundation for the development of integrated dynamic models that incorporate the full range of variables that affect the SDGs and overall human and ecosystem well-being.[45] Integrated models allow for predictions about the performance of sustainable development activities.

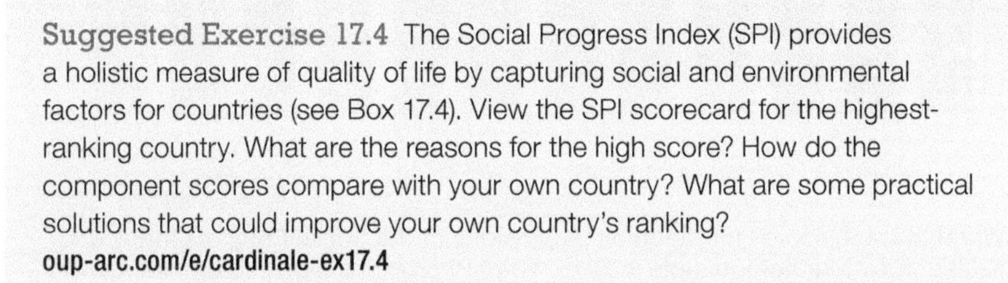
Suggested Exercise 17.4 The Social Progress Index (SPI) provides a holistic measure of quality of life by capturing social and environmental factors for countries (see Box 17.4). View the SPI scorecard for the highest-ranking country. What are the reasons for the high score? How do the component scores compare with your own country? What are some practical solutions that could improve your own country's ranking?
oup-arc.com/e/cardinale-ex17.4

BOX 17.4 Conservation in Practice

The Social Progress Index

Sustainable development aims to improve the lives of people now and into the future, but actually estimating the quality of life for a society is challenging. Often, simple economic measures are used, such as the gross domestic product (GDP). However, GDP only reflects a part of what defines quality of life, which has led some governments to find better measures that capture the state of society.[50,62] Research since the 1970s has produced new measures and indices of progress, success, and quality of life.[63] One in particular that is easy to interpret and understand is the Social Progress Imperative's Social Progress Index (SPI, oup-arc.com/e/cardinale-w17.8), launched in 2013.

The SPI is based on information in three dimensions, each with a set of four associated components (see the Figure):

1. **Basic Human Needs:** Nutrition and basic medical care, water and sanitation, shelter, and personal safety
2. **Foundations of Well-being:** Access to basic knowledge, access to information and communications, health and wellness, and environmental quality
3. **Opportunity:** Personal rights, personal freedom and choice, inclusiveness, and access to advanced education

The SPI is calculated from 51 indicators across these dimensions and aggregated into a score that ranges from 0 (worst) to 100 (best).[64] For example, "environmental quality" is one of the four components of the Foundations of Well-being dimension and includes four indicators: (1) outdoor air pollution-attributable deaths; (2) wastewater treatment; (3) greenhouse gas emissions; and (4) biome protection.

Biome protection in particular incorporates the role of biodiversity in well-being. It is calculated as the percentage of biomes (i.e., naturally occurring communities of flora and fauna) in protected areas. Data come from a variety of reliable sources, and no economic indicators or subjective measures of "happiness" or "satisfaction" are included in the calculation. The nonprofit Social Progress Imperative calculates the SPI for each country and publishes a scorecard regularly. In 2018, the global average score was 63.46, with Norway being the highest at 90.26, and the Central African Republic being the lowest at 26.01. If the world were a single country, its 63.46 average ranking would lie somewhere between Botswana and the Philippines. Across dimensions, the world scores highest in Basic Human Needs and lowest in Opportunity (see the Figure), indicating that societies are generally able to provide the fundamental needs of people, but struggle to create opportunities for all citizens.

The SPI provides an intuitive measure of societal success that can be used as a benchmark for evaluating the progress of sustainable development activities. It complements GDP and incorporates social and environmental measures that better capture the lives of people, which ultimately are the target of large-scale sustainable development initiatives like the United Nations Sustainable Development Goals. While biodiversity conservation could be better represented in the index, the SPI provides one measure of the balance of important elements of sustainability.

Video 17.6 View a TED talk about the Social Progress Index by economist Michael Green.
oup-arc.com/e/cardinale-v17.6

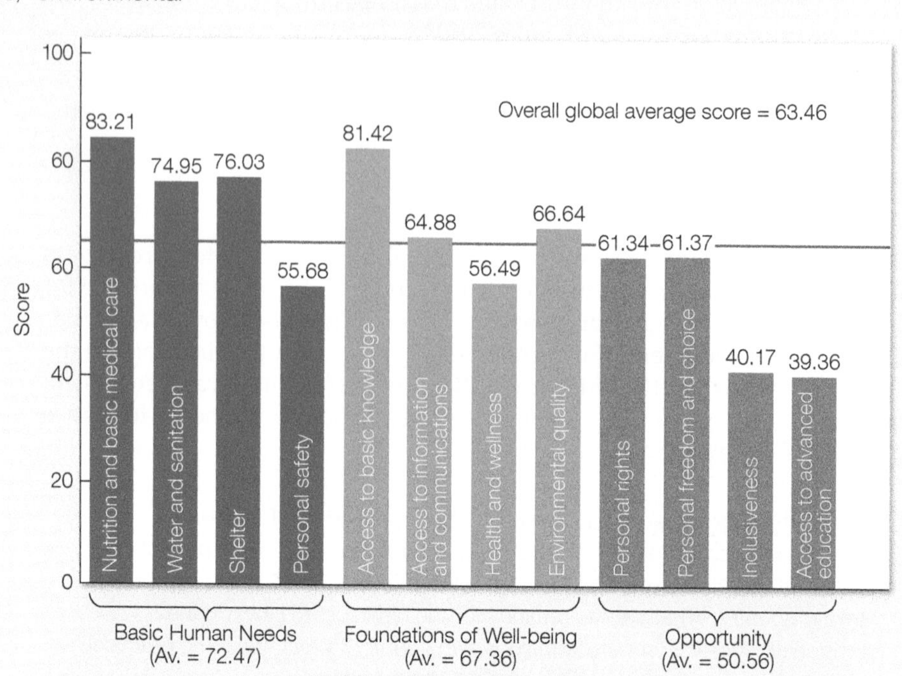

World average scores for the three dimensions of the Social Progress Index and each of their four components in 2018. The SPI provides a measure of the quality of life on a scale from 0 (worst) to 100 (best). (After Social Progress Imperative, www.socialprogress.org.)

Lastly, the success of sustainable development can be enhanced by identifying a vision for a future world.[45] What would the world look like if all UN SDGs were achieved? What would the conditions be like both for individuals and collectively among communities and countries? These questions get to the heart of the matter: what sustainable development is trying to achieve, and how we define the intersection between the three pillars of sustainability (see Figure 17.1). Articulating this vision provides a foundation for creating more effective programs and initiatives and building broader public consensus that can lead to more successful outcomes for people and the environment.

Concluding Remarks

As we end this book, it is worth returning to where we first began. In Chapter 1, we reviewed the state of the planet, including the state of both humans and of non-human life. We emphasized that one of the greatest challenges for conservation in the Anthropocene will be to protect the magnificent variety of life that is the most striking feature of our planet, while also meeting the needs of people. This final chapter has returned full circle, back to that same point by emphasizing how conservation must be interwoven with sustainable development if we are to save the planet's biodiversity.

Throughout this book, we have introduced you to a variety of concepts, tools, techniques, models, and skills that will be needed to address the most pressing questions in conservation, to seek new and innovative solutions to the problem of biodiversity loss, and to help society progress in smart, sensible, and balanced ways. You are now better equipped than your predecessors. You are more interdisciplinary with your approaches. You have success stories to tell. The momentum of society is behind you, and there is reason to be optimistic that we can succeed.

Indeed, you are now prepared to become the next generation of champions for conservation (see Figure 1.10). If together we succeed in saving even a fraction of the world's biodiversity, then our brief time on Earth will have been meaningful and well spent. If we fail, then at least we will be remembered by our children as moral ancestors who dedicated ourselves to a higher purpose.

Summary

1. Sustainable development is often defined as "development that meets the needs of the present without compromising the ability of future generations to meet their own needs."[44] Sustainable development lies at the intersection of the three pillars of sustainability: economic, societal, and environmental. Describing the characteristics of that intersection can be challenging, and has led to different interpretations of the concept.

2. The United Nations has been a leader in advancing large-scale, international sustainable development. The UN Earth Summit in 1992 resulted in three key non-legally binding documents (Agenda 21, Rio Declaration, Rio Forest Principles) and three key legally binding conventions (Convention on Biological Diversity, Framework Convention on Climate Change, and Convention to Combat Desertification) that launched a global movement for sustainable development.

3. Funding for large-scale sustainable development projects typically come from multilateral development banks, bilateral assistance programs, and nonprofit organizations in the form of loans and grants. The World Bank Group, composed of five institutions, is a large multilateral organization that supports many development projects around the world. The Global Environment Facility (GEF) receives funding from donor countries and provides grants for environmentally focused sustainable development projects.

4. The future success of sustainable development depends on several factors, including additional funding, continued engagement by the international community, better methods for quantifying the trade-offs, synergies, and interconnectedness of sustainable development goals, and a more holistic vision for the future.

For Discussion

1. Meeting the United Nations Sustainable Development Goals by 2030 will require unprecedented amounts of funding, far more than has been committed. Yet problems like biodiversity loss and climate change, among many others, are urgent and need bold action now. Why do you think many countries are reluctant to contribute funds to support the UN SDGs and other large-scale international efforts? What can be done to change this?
2. Despite large investments in biodiversity conservation under the umbrella of sustainable development since the early 1990s, nearly 26,000 species are still listed as Threatened on the IUCN Red List, and the average species population has declined by 58% since 1970. Why has the international community been less effective at reducing threats to biodiversity than to other areas like poverty alleviation? Is it only about funding, or are there other reasons? Is continued investment in biodiversity conservation worth it?
3. Costa Rica has been a global leader in ecotourism and biodiversity conservation, attracting over 2 million visitors each year. Why has Costa Rica been so successful at conserving their natural resources and developing a tourist economy relative to other tropical countries with similar biodiversity? Is ecotourism a good model for sustainable development?
4. Sustainable development is considered by some to be an oxymoron. Do you agree with this assessment? Can economic growth really occur without damaging elements of human society and environment?

Visit the
Conservation Biology
Companion Website
oup.com/he/cardinale1e
for videos, exercises, links, and other study resources.

Suggested Readings

Ascensão, F., et al. 2018. Environmental challenges for the Belt and Road Initiative. *Nature Sustainability* 1: 206–209. Provides recommendations for managing the environmental and social risks of China's terrestrial and maritime BRI projects.

Costanza, R., et al. 2014. Time to leave GDP behind. *Nature* 505: 283–285. Discusses how the GDP is a misleading measure of national success and describes alternative metrics that account for human and environmental well-being.

Du Pisani, J. 2006. Sustainable development: Historical roots of the concept. *Environmental Sciences* 3: 83–96. Provides an overview of the origin and history of the concept of sustainable development.

Jones, G. and Spadafora, A. 2017. Creating ecotourism in Costa Rica, 1970–2000. *Enterprise & Society* 18: 146–183. A comprehensive overview of the history of the development of ecotourism in Costa Rica.

Naidoo, R. et al. 2016. Estimating economic losses to tourism in Africa from the illegal killing of elephants. *Nature Communications* 7: 13379. An analysis that describes the economic value of the loss of elephants to society due to poaching.

Nilsson, M. et al. 2016. Map the interactions of sustainable development goals. *Nature* 534: 320–322. Introduces a simple rubric for scoring the interactions between UN sustainability goals and targets that is useful for priority setting.

Robinson, J. 2004. Squaring the circle? Some thoughts on the idea of sustainable development. *Ecological Economics* 48: 369–384. Gives an overview of the concept of sustainable development, the issues and criticisms surrounding it, and alternative ways of viewing sustainability.

United Nations General Assembly. 1992. *Rio Declaration on Environment and Development.* United Nations, A/CONF.151/26 (Vol. I). A key document from the Earth Summit that led to a global emphasis on sustainable development.

United Nations General Assembly. 2015. *Transforming our World: The 2030 Agenda for Sustainable Development.* United Nations, A/RES/70/1. Another key document that is a blueprint for sustainable development that includes the UN Sustainable Development Goals.

Glossary

Numbers in brackets indicate the chapter(s) in which the term is defined.

A

adaptive management (AM) A structured, iterative process for making decisions in the face of uncertainty, with the aim of taking corrective action if monitoring data and models predict a pending problem. Also called *adaptive resource management* (ARM). [15]

adaptive radiation The evolutionary diversification of a group of organisms into forms that fill different ecological niches in an environment. [4]

adaptive resource management (ARM) *See adaptive management.*

afforestation The establishment of a forest or stand of trees (forestation) in an area where there was no previous tree cover. [9]

age-based models Population growth models that estimate growth parameters by age classes. Compare with *stage-based models*. [13]

Agenda 21 One of three non-binding agreements that came out of the 1992 *Earth Summit*. The agenda outlines a comprehensive set of actions for guiding sustainable development in the twenty-first century, including social, economic, conservation, and political aspects of sustainable development. [17]

agroecological farming Farming practices that integrate farms into a surrounding landscape that conserves and manages biodiversity to enhance ecological processes that support crop production, biological pest control, nutrient cycling, and pollination. [15]

ahimsa Ethical concept of nonviolence and kindness to all living beings, preeminent in the religious traditions of Hinduism, Buddhism, and Jainism. [5]

Aichi Biodiversity Targets Twenty specific goals to slow biodiversity loss and speed international efforts in conservation. Set out in 2010 by the United Nations Convention on Biodiversity's Strategic Plan for Biodiversity, the goals are built around five themes and specify a target date of 2020 for their achievement. [14, 15, 17]

Aichi Target 11 Sets out the goal of protecting "at least 17 percent of terrestrial and inland water, and 10 percent of coastal and marine areas" in managed and connected areas "integrated into the wider landscapes and seascapes." [14]

alien species A species, subspecies, or lower taxon occurring outside of its natural range (past or present) and dispersal potential (i.e., outside the range it occupies naturally or could not occupy without direct or indirect introduction or care by humans) [including] any part, gametes or propagule of such species that might survive and subsequently reproduce Also called *exotic, foreign, introduced, non-indigenous,* and *nonnative species*. Compare with *invasive alien species*. [11]

Allee effects A set of phenomena in biology that cause the biological fitness of individuals to be positively correlated with population size (or density). Common examples include mate limitation, cooperative defense, and cooperative feeding. [8, 13]

allele frequency The proportional representation of all of the alleles in a population. [3]

alleles Different forms (i.e., differing DNA sequences) of the same gene—sequences of DNA that are found at the same gene *locus* (physical, fixed location where a gene sits on a chromosome) but that code for the production of different proteins. [3, 13]

alpha (α) diversity Any measure of biodiversity (e.g., the number of different genes or species) at the scale of a local community or specific location. See also *beta diversity, gamma diversity*. [3]

animal park See *zoo*.

Anthropocene Unofficial term for the present geological period in which human beings have come to dominate Earth's ecosystems and their biophysical processes. [1]

anthropocentrism Worldview that considers human beings to be the primary holders of moral standing and views nature and the environment primarily in terms of their value and benefit to humans. Compare with *human exceptionalism, technocentrism*. [5]

anthropogenic climate change A general term that refers to a suite of abiotic variables that are changing simultaneously across the planet as a result of human activities that release greenhouse gasses into the atmosphere. [12]

aquarium The aquatic counterpart of a zoo, housing living aquatic animals for public viewing and conservation research. [16]

arboretum A specialized botanical garden dedicated to collections of trees and other woody plants. [16]

artificial incubation Incubation in a device that simulates avian incubation by keeping eggs warm at the correct temperature. [16]

artificial insemination Human assisted introduction of a male's sperm into a female's cervix or uterine cavity to try to achieve pregnancy. Often used to increase reproductive output of endangered species. [16]

assisted colonization A conservation strategy that focuses on transplanting threatened species, especially those with poor dispersal abilities, to new sites at higher elevations or closer to the poles in order to help them survive climate change. Also called *managed relocation*. See *introduction*. [12]

assurance dilemma A common example from the field of *game theory* that illustrates why a lack of trust causes two rational individuals to not cooperate with each other even when it is in their best interest to do so. Often presented as a "stag hunt" scenario of two hunters searching for food. [10]

asynchrony See *population asynchrony*.

autotrophs See *primary producers*.

B

beneficiary value See *bequest value*.

benefit transfer method An economic analysis that estimates the benefit of an ecosystem good or service in a specific situation based on information gathered from a similar place and condition (rather than conducting a unique study of the new situation). [7]

bequest value A subset of *existence value*. The value that an individual places on ensuring the non-use of certain natural resources so that they will be available for the enjoyment and well-being of future generations. Also known as *beneficiary value*. Compare with *option value*. [5, 7]

beta (β) diversity The turnover of biodiversity across local communities in a region, or through time in a local community. See also *alpha diversity; gamma diversity*. [3]

BIDE parameters Stands for "*b*irth, *i*mmigration, *d*eath, *e*migration," which are key parameters that influence a population's size and growth through time. [13]

binomial nomenclature The unique two-part Latin name taxonomists bestow on a species, such as *Acer rubrum* (red maple), *Canis lupus* (gray wolf) or *Homo sapiens* (human). The first name is that of the genus, the second name identifies the species (*Canis lupus*, gray wolf; *Canis simensis*, Ethiopian wolf). [3]

biocentrism Worldview that extends inherent value and consideration to all living beings, and holds that humans hold no particular moral superiority over other organisms. Compare with *anthropocentrism; ecocentrism*. [5]

biodiversity hotspot A defined geographic area that has particularly high levels of biodiversity and which is also highly threatened by human activities. [14]

biodiversity The variety of life on Earth, including all of its genes, populations, species, communities, and biomes. Also known as *biological diversity*. [3]

biological asynchrony See *population asynchrony*. [8, 12]

biological community A collection of species that occupy and interact in a particular location. [3]

biological diversity See *biodiversity*.

biological species concept The concept that defines a *species* as a group of individuals that can potentially breed among themselves in the wild and that do not breed with individuals of other groups. Compare with *evolutionary species concept, morphological species concept*. [3]

biomagnification Process whereby toxins become more concentrated in the tissues of organisms at higher levels in the food chain. [9]

biophilia The innate tendency by humans to seek connections with nature and other nonhuman forms of life. [5]

biotic homogenization The process by which invasive alien species replace native flora and fauna, in turn causing ecosystems to lose their biological uniqueness. [1, 11]

botanical garden An ex situ facility that is dedicated to the collection, cultivation, and preservation of plants. Compare with *arboretum*. [16]

Brundtland Commission Common name (from its chairman, Gro Harlem Brundtland) for the UN's World Commission on Environment and Development (WCED). The Commission's 1987 report, *Our Common Future*, launched a global focus on sustainable development. [17]

bushmeat Wildmeat, or game meat obtained from non-domesticated mammals, reptiles, amphibians and birds that are hunted for food. [10]

bycatch Animals, including marine mammals, sea turtles, and fish that are not the target of commercial fishing, but which are caught and killed unintentionally during large-scale fishing operations. [10]

C

Canada Nature Fund A Canadian government fund (enhanced with support from nongovernmental sources) formed in 2018 to spearhead the establishment and integration of protected areas into a nationwide protected network. [15]

capture-mark-recapture (CMR) An approach to estimating a species' abundance that involves capturing individuals and marking them in one timeframe, then capturing individuals and recording the number of previously marked animals in a later timeframe. Abundance is estimated based on the ratio of recaptured animals to total animals captured in the second timeframe. [13]

carrying capacity (*K*) The maximum number of individuals (or biomass) of a species that an area can support given the limiting resources available. [1, 10, 13]

census A count of the number of individuals in a population; for example, the Audubon Society's Christmas Bird Count. [13]

charismatic megafauna Large animals like whales, pandas, or elephants that elicit strong emotional responses in people, who may then express a willingness to pay for their preservation despite receiving no direct benefit from their existence. [7]

choice experiments A method of estimating the value of ecosystem goods and services that asks respondents to choose between scenarios with two groups of items, each of which has a different cost. Choice experiments focus on trade-offs among varying characteristics and impacts in the different scenarios. [7]

choice modeling method A method of estimating the value of ecosystem goods and services in terms of trade-offs among hypothetical choices that have multiple possible outcomes. [7]

class Unit of taxonomic classification into which related *orders* of species are grouped. [3]

climatic envelope models (CEMs) A specialized form of *species distribution model* in which the main predictors of a species' distribution are climate-related variables. CEMs describe relationships between a species' current distribution and climatic variables like temperature and precipitation to describe the species' *climatic niche*, which can be compared to that for future climate conditions predicted from a general circulation model. [12]

climatic niche The set of climatic conditions to which a species is adapted. [12]

cloning The process of creating a new multicellular organism through asexual reproduction of cells. [16]

closed population A population in which no individuals are entering or leaving the population through birth, immigration, death, or emigration (the *BIDE parameters*). Compare with *open populations*. [13]

co-design The process of including all relevant stakeholders in the discussion and development of coordinated management plans for sustainable use of natural resources. Also called *participatory design*. [15]

cold-edge expansion Expansion of a species' range into regions that were once beyond its "cold" limits, but which are now accessible as a result of climate warming. Also called *leading-edge expansion*. [12]

common-pool resources Natural resources that are owned by national, regional, or local governments as public goods, or by communal groups as common property. [7, 10]

community and ecosystem diversity The different biological communities and their associated ecosystems that make up whole landscapes. [3]

community structure The organization of a community, including (though not limited to) its trophic structure, species composition, and food web complexity. [3]

community-based natural resource management programs (CBNRMs) Programs where governments grant local landowners and communal groups the authority to manage wildlife in their landscapes in ways that help build sustainable economies and improve people's livelihoods. CBNRMs are particularly common in Africa. [15]

competitive exclusion principle The principle that no two species can coexist when using the same resource, at the same time, in the same location, because one of the competing species will exclude the other. [4]

Conference on Biological Diversity See *Earth Summit*.

connectance A measure of network complexity that quantifies the proportion of all possible interactions between species in an interaction matrix that are actually realized. Connectance (C) is equal to L/S^2, where L is the total number of links in the matrix and S is the total number of species. See also *linkage density*. [3]

conservation easements A voluntary legal agreement that a landholder enters into with another party, such as a government agency or land trust, that places some restrictions on how the property will be used in the future in order to ensure some value for conservation, but which maintains certain options or activities for the original owner. Easements are typically made in perpetuity, and may have legal and/or tax benefits for the original owner. [15]

conservation refugees A term applied to people who are displaced from their lands and homes by the establishment of protected areas. A continuing and serious problem, since as much as 20 percent of the world's population live in hotspots of biodiversity that some conservationists argue need urgent protection. [14]

consumer surplus The difference between the total amount that consumers are willing to pay for a product or service (indicated by the demand curve) and the total amount they actually do pay (the market price). Compare with *economic surplus*. [7]

consumptive use value Direct use value assigned to goods that are collected and consumed locally. Compare with *productive use value*. [5, 7]

contingent valuation A method of estimating the value of ecosystem goods and services that asks respondents to state their willingness to pay for (or to accept compensation for doing without) an ecosystem service, based on a hypothetical scenario and description of the service. [7]

continuous-time model Mathematical model for population growth where reproduction occurs continuously, without regard to seasons or other discrete time periods. Compare with *discrete-time model*. [13]

Convention on Biological Diversity (CBD) A legally binding, multilateral treaty that came into force at the end of 1993 and has 196 Parties (countries) that are obligated to implement its provisions. The CBD has three main goals: (1) conservation of biodiversity, (2) sustainable use of biodiversity, and (3) fair and equitable sharing of the benefits arising from the use of genetic resources. See also *Aichi Biodiversity Targets*; *Earth Summit*. [14, 17]

conventional farming High-production farming based on monoculture crops (often genetically modified crops, GMOs) that are managed with heavy machinery and external inputs like synthetic fertilizers and pesticides. [15]

cooperative defense When aggregations of animals defend each other against shared predators. [8]

cooperative feeding Foraging or hunting in which groups of animals work together and share the food they acquire. [8]

coral bleaching Effect when warming water causes corals to expel the symbiotic algae living in their tissues, so that the corals turn white and lose their biological functioning. [9]

core habitat See *interior habitat*.

cost-based techniques A set of economic valuation methods that are based on estimating the cost of avoiding damages due to lost services (damage cost avoided), the cost of replacing environmental assets after they are lost (replacement cost), or the cost of providing substitute services (substitute cost). [7]

covenants Similar to *conservation easements*, but usually are not permanent and are void if a protected property changes ownership.

crisis discipline Refers to a field like conservation biology (or cancer research or foreign policy) in which urgent issues require practitioners to take action even in the absence of complete information. [2]

cryosphere The frozen-water portion of Earth's ecosystems. [12]

cryptic biodiversity The existence of undescribed but genetically distinct species that look similar to, and consequently have been wrongly classified as, a described species. [3]

cultural services Nonmaterial benefits people obtain from ecosystems. [6]

D

de-extinction The process of using cloning to create an organism that is a member of, or resembles, an extinct species. Also called *resurrection biology* or *species revivalism*. [16]

debt-for-nature swap Agreement in which a developing country agrees to fund conservation activities in exchange for the cancellation of some of its debt owed to a developed nation. [17]

decomposers Organisms that obtain their nutrients and energy from the waste products (detritus) and dead tissues of other organisms. Also called *detritivores*. [3]

deep ecology An environmental movement based on the premise that the living world as a whole should be regarded as having the inalienable right to live and flourish, independent of its instrumental values for human use. The movement's philosophy emphasizes biodiversity protection, personal lifestyle changes, and working toward political change. [5]

demographic stochasticity Random fluctuations in population size that occur because the birth and death of each individual in a population is a discrete and probabilistic event. Small populations are particularly susceptible to declines in population size caused by stochasticity. Also called *demographic variation*. Compare with *environmental stochasticity*. [8, 13]

demographic transition The historical shift in demographics from a population characterized by high birth rates and high infant death rates in societies with minimal technology, education (especially of women), and economic development, to population demographics characterized by low birth rates and low death rates in societies with advanced technology, higher education, and economic development. [1]

density-dependent growth A pattern of population growth in which the growth rate at any given time is dependent on the current size of the population. [13]

desertification The process by which once fertile land becomes a desert, usually due to drought, deforestation, or poor land management practices (e.g., intensive agriculture) that alter the water budget. [1, 9]

deterministic model A population model in which parameters such as birth and death rates are fixed, leading to only one possible outcome for population size. Compare with *stochastic model.* [13]

detritivores See *decomposers.*

dilution effect A biological mechanism that reduces the risk of infection by zoonotic pathogens (pathogens transmitted from animals to human) by reducing the prevalence of the pathogen among the animal hosts that transmit the disease. [6]

direct use value The economic or social value of ecosystem goods or services that are used directly by individuals. These include consumptive uses (e.g., harvesting goods) and non-consumptive uses (e.g., recreation). [5, 7]

discounting An economic tool used to estimate the present value of costs or benefits that might be incurred in the future. It determines the discounted value, or present value. [7]

discrete-time model Population growth model that treats time in discrete packets, whether years, seasons, or some other time unit, with population growth (reproduction) occurring only once in each period. Compare with *continuous-time model.* [13]

dispersal limitation Limits to the dispersal ability of an organism, which constrains a species population growth and geographic distribution. [11]

diversified farming Farming practices that integrate several crops and/or animals in the production system to promote agrobiodiversity, ecosystem services, and reduced need for external inputs. [15]

double clutching The technique of removing the first clutch of eggs from a rare female bird so that she will lay and raise an additional clutch while her first clutch is raised by another bird of a related species. Two clutches of the rare species' eggs are therefore produced each season. [16]

dynamic vegetation models (DVMs) Models that use time series of climate data (e.g., temperature, precipitation, number of days with sunshine) to explain the current distribution of plant functional types, and those relationships are subsequently used to project shifts in entire vegetation zones and biomes at a regional or global scale. [12]

E

Earth Summit Formally called the United Nations Conference on Environment and Development, or UNCED. Held in Rio de Janeiro in 1992, the Earth Summit brought together representatives from over 170 countries, including heads of state, leaders of the UN, and individuals from major conservation organizations, specialized agencies and intergovernmental organizations, and from groups representing religions and indigenous peoples to discuss combining increased protection of the environment with sustainable economic development. [17]

ecocentrism Wholistic worldview that considers the Earth as a biophysical system that includes humans, nonhuman life, and the sum of all the physical and biological processes that are required to sustain life. Compare with *anthropocentrism; biocentrism.* [5]

ecological economics A subdiscipline of the field of economics, which quantifies the instrumental value of nature and its biodiversity. [5, 7]

ecological efficiency The percentage of energy captured at one trophic level that is transferred to the next highest trophic level in a food web. As a "rule of thumb," about 10% of the energy produced by a lower trophic level tends to become available in the next trophic level; efficiency varies, however, among ecosystems and organism types. [3]

ecological niche models See *species distribution models.*

ecological traps Areas of low-quality habitat that reduce survival and reproduction. [9]

ecologically (functionally) extinct Refers to a species that has been so reduced in numbers that it no longer has a significant ecological impact on the biological community or ecosystem in which it resides. [8]

economic surplus The net economic benefit of a good or service—that is, the sum of *consumer surplus* and *producer surplus.* [7]

ecosystem disservice An ecological process (e.g., supporting service of ecosystems) that has a negative impact on something that humans value (e.g., primary productivity by algae that overgrows and kills a coral reef). [6]

ecosystem engineers Species that create new physical habitat or that extensively modify existing environments through their biological activities. [3]

ecosystem management See *ecosystem-based management.*

ecosystem services Nature's contributions to people; the plethora of benefits that humans receive from both natural and managed ecosystems, including things like the production of consumable goods (e.g., production of fish from oceans, or wood from forests), nonconsumable services (e.g., protection from natural disasters like floods, or biological control of pests and disease), and cultural services (e.g., recreation, or ecotourism). [5]

ecosystem-based management An environmental management approach, usually applied to management of natural resources at the scale of whole landscapes or ecoregions, that recognizes the full array of interactions within an ecosystem, including humans, rather than considering single issues, species, or ecosystem services in isolation. [2, 15]

ecosystem A biological community together with its associated physical and chemical environment. [3]

ecotones See *edge habitats.*

ecotourism A type of recreational tourism that involves people visiting places and spending money wholly or in part to experience and enjoy natural ecosystems and their biodiversity (such as rain forests, African savannas, coral reefs, deserts, the Galápagos Islands, and the Everglades), or to view particular "flagship" species (such as elephants, on safari trips). [6]

edge effects A suite of physical and biological changes that tend to occur at patch edges. Many of these changes, such as altered microclimates and increased prevalence of parasitism or disease, can be detrimental to biodiversity. [9]

edge habitats Areas of transition between different habitat types. Also known as *ecotones*, edge regions are often disturbed habitat whose area increases with habitat fragmentation. See also *ecotones.* [9]

effective population size (N_e) The number of breeding individuals in a population (or the number of individuals that contribute genes to the next generation). [13]

empty niche hypothesis Proposes that the primary way alien species become established in a new community is by exploiting a vacant niche—either by using biological resources in a new way, or by performing a completely new role that was previously unfilled by a native species. [11]

endemic Organisms that are native to and occur only in a given location. [4, 8]

enemies release hypothesis Proposes that while populations of alien species tend to be controlled by natural enemies, such as pathogens, predators, and parasitoids, in their native ranges, when a small number of individuals of an alien species are introduced to a new location, their natural enemies may not be transported with them. [11]

environmental ethics A branch of philosophy that studies the foundation of environmental values as well as societal attitudes, actions, and policies to protect and sustain biodiversity and ecological systems. [5]

environmental justice The fair treatment of all people regardless of race, color, national origin, or income in respect to the development, implementation, and enforcement of environmental laws, regulations, and policies. The environmental justice social movement seeks to empower poor and politically marginalized people, who are often members of minority groups, to protect their own environments and well-being. [5]

environmental stochasticity Unpredictable spatial or temporal fluctuation in a population's demographic rates (birth and death) that are caused by varying environmental conditions. Compare with *demographic stochasticity*. [8, 13]

eutrophication When a body of water becomes overly enriched with nutrients like nitrogen and phosphorus, frequently due to runoff from the land, and subsequently causes a dense growth of plant life and death of animal life from lack of oxygen. [9]

evapotranspiration Sum total transfer of liquid water to atmospheric water vapor via (1) evaporation of water from terrestrial and aquatic surfaces and (2) plant transpiration. In plant transpiration, the water plants take up from the soil travels through the plant and is released into the atmosphere as vapor through stomata in the leaves. [6]

evolutionary species concept The concept that defines a *species* as a group of individuals that share similarities of their DNA (e.g., 95% overlap in nucleotide sequence), and hence their evolutionary past. This is the most recent of several approaches to recognizing and classifying species. Compare with *biological species concept, morphological species concept*. [3]

evolutionary-ecological land ethic One of three environmental philosophies that shaped conservation policies in the United States and elsewhere. This conservation ethic , promulgated by Aldo Leopold, is based on the belief that the most important goal of conservation is to maintain the health of natural ecosystems and the ecological processes they perform. Espouses a middle ground between exploitation and total human control over nature on the one hand, and complete preservation of wilderness with no human presence or activity on the other. See also *preservationist ethic,* and *resource conservation ethic*. [2]

ex situ conservation Off-site conservation. Preservation of living individuals in artificial, human-built environments like zoos, aquariums, and botanical gardens. Also applies to captive-breeding programs and to the collection and preservation of genetic materials (see *frozen zoos; seed banks*). [16]

existence value the benefit people receive from knowing that a particular environmental resource, endangered species, or any other organism or thing exists (e.g., a species, wilderness area, or ecosystem such as tropical rainforest). Quantified by the amount people are willing to pay to protect or preserve species, habitats, and other natural resources from being irreparably harmed or completely lost. Compare with *intrinsic value*. [5, 7]

exotic species See *alien species*.

exponential growth model A continuous-time mathematical model for population growth based on the *instantaneous rate of increase* (*r*). Compare with *geometric growth model; logistic growth model*. [13]

externalities Side effects or consequences of an industrial or commercial activity that affect other parties without those effects being reflected in the cost of the goods or services involved. Externalities lead to market failures where the profits of a transaction benefit the business, but certain costs of the transaction are paid by individuals or groups of individuals that were not involved in the transaction. [6, 7]

extinct in the wild Occurs when a species is no longer found in the wild, although individuals may remain alive in ex situ facilities like zoos, botanical gardens, or other artificial environments. [8]

extinction The loss of a species that occurs with the death of the last individual of the species, resulting in the loss of that species' genetic diversity and ending its evolutionary history. See also *ecologically extinct; extinct in the wild; extirpated; globally extinct*. [8]

extinction cascade A series of linked extinctions whereby the extinction of a key species triggers the extinction of other species. [3, 8]

extinction debt The future extinction of species due to events that have occurred in the past. The extinction debt occurs because of time delays between impacts on a species, such as destruction of habitat, and the species' ultimate disappearance. [8]

extinction thresholds Thresholds of habitat size that are required for species to persist and complete their life cycles. [9]

extinction vortex The tendency of small populations to spiral toward extinction at an ever-increasing rate as it becomes more and more vulnerable to the effects of demographic and environmental variation and reduced heterozygosity. [13]

extirpation The local extinction of a species population from a biological community, even as populations of the species still exist elsewhere in other local communities. [1, 8]

F

family Unit of taxonomic classification into which related genera are grouped. [3]

fences-and-fines An approach to conservation that involves delimited separation of protected areas from people ("fences") and the implementation and enforcement of protective regulations ("fines"). [14]

finite rate of increase (λ) The geometric rate of increase + 1; (= R + 1). [13]

fixed effort (proportional harvest) model A mathematical model for determining maximum sustainable yield when the number of individuals harvested from a population is directly proportional to the size of the population. [10]

fixed quota (constant harvest) model A mathematical model for determining maximum sustainable yield when the number of individuals harvested from a population is a constant—both independent of the population size and independent of the amount of effort put into harvesting. [10]

flagship species A species used to attract the attention and concern of the public, often invoked to raise awareness of the plight of a given ecosystem or region. Flagship species are usually appealing and charismatic to humans (e.g., elephants, jaguars, giant pandas). [13]

food web A network of feeding relationships among species interacting as a community. [3]

founder effect A specific type of *genetic bottleneck* that occurs when a small number of individuals breaks away and becomes isolated from a larger population, thus founding a colony that has less genetic diversity than the original population. This can occur when a population becomes fragmented by human activities. [4, 13]

frozen zoos Gene banks that collect and store tissues, cells, gametes (sperm and egg cells), and DNA of endangered animals in cryogenic freezers. [16]

functionally extinct See *ecologically extinct.*

G

game theory The study of mathematical models of strategic interaction in between rational decision-makers. See *assurance dilemma; prisoner's dilemma.* [10]

gamma (γ) diversity Any measure of biodiversity (e.g., the number of different genes or species) at the scale of a whole geographical region (landscape). See also *alpha diversity; beta diversity.* [3]

gap analysis An assessment of how close an existing network of protected areas comes to meeting a set of defined conservation and social goals, followed by a process of identifying "gaps" and prioritizing additional protected areas that are required to achieve those goals. [15]

gene A sequence of nucleotides (DNA sequence) that are located at a specific point on a chromosome (the *locus*), and which code for a specific protein. The gene is the unit of heredity that is transferred from a parent to offspring, and which determines some characteristic of the offspring. [3]

gene pool The total array of genes and alleles present in a population or subpopulation. [3]

general circulation models (GCMs) Models that forecast the mean, range, and variation for future climatic variables like temperature and precipitation based on predictions of variables like radiative forcing and greenhouse gas levels. Also called *global climate models.* [12]

Genesys An online data portal that contains records of germplasm accessions from institutions around the world to facilitate the access to, and use of, accessions in ex situ gene banks. [16]

genetic bottleneck An extreme reduction in a population's genetic diversity that occurs when a population's size is extremely reduced, for example following a natural disaster or outbreak of disease. Compare with *founder effect.* [4]

genetic diversity The genetic variation found within species, both among individuals within single populations and among geographically distinct populations. See also *heterozygosity* and *polymorphism.* [3]

genetic drift The change in allele frequencies from one generation to the next, resulting from random chance rather than being driven by natural selection or other evolutionary constraints. Can have significant effects on small populations, in which an allele that occurs at a low frequency has a significant probability of being lost in each generation. [13]

genetic variability *See heterozygosity.*

genotype The particular combination of alleles that an individual possesses. [3]

genus (plural, *genera*) Unit of taxonomic classification that includes one or more species. In *binomial nomenclature,* the first term identifies the genus. [3]

Geodatabase A GIS file format recommended by the Environmental Systems Research Institute (ESRI) for use in its popular GIS software ArcGIS. [15]

geoengineering The deliberate large-scale manipulation of Earth's land, atmosphere, and oceans in an attempt to counteract climate change and maintain a habitable climate for humans (e.g., cloud seeding, orbiting "sunshades," enhanced carbon sequestration). [1]

geometric growth model A discrete time mathematical model for population growth based on the *geometric rate of increase (R).* Compare with *exponential growth model; logistic growth model.* [13]

geometric rate of increase (*R*) Population growth rate measured as ($b - d$), or the rate of birth (per individual) minus the rate of death (per individual) during a given timeframe (often 1 year). Compare with *instantaneous rate of increase (r).* [13]

global climate models (GCMs) See *general circulation models.* [12]

Global Environment Facility (GEF) Created at the 1992 *Earth Summit,* the GEF is an international partnership of 183 countries, international institutions, civil society organizations, and private sector enterprises, and provides funding for environmental projects, many of which link to the United Nations Sustainable Development Goals. [17]

globally extinct When, after a thorough search and investigation, it is determined that no individuals of a species are alive anywhere in the world. [8]

green infrastructure See *natural infrastructure.*

gross primary production (GPP) Total biomass production by autotrophs. It is controlled by the rate of photosynthesis, which dictates the mass of inorganic CO_2 that is captured from the air and water and converted into organic carbon in tissues of the primary producers. Compare with *net primary production.* [6]

H

habitat conversion See *habitat loss.*

habitat corridor A continuous strip of protected habitat that runs between and connects two protected areas to facilitate dispersal. Compare with *stepping stones.* [15]

habitat degradation Refers to a suite of human activities that make the habitats in a landscape less conducive to life, in turn eroding biodiversity. These activities include many forms of pollution (pesticides, heavy metals, nutrients, plastics, personal care products), as well as activities that lead to desertification, erosion, and sedimentation, all of which make habitats less hospitable. [8, 9]

habitat destruction See *habitat degradation.*

habitat fragmentation The process by which a larger expanse of habitat is subdivided into smaller pieces, or patches, within a landscape. Fragmentation leads to an overall reduction of habitat area as well as changes in patch shape, size, interior-to-edge ratios, connectivity, microclimates, and other factors that can alter biodiversity in a myriad of ways. [8, 9]

habitat loss Complete elimination of habitats, along with their biological communities and ecological functions, due to the conversion of natural or semi-natural habitat (e.g., a managed forest) into human-dominated habitat (e.g., a village or parking lot). [8, 9]

Half-Earth Project An ambitious initiative to protect one-half of Earth's lands and seas in order to manage sufficient habitat to reverse the rate of species extinctions and ensure the long-term health of the planet. [14]

hedonic pricing method An economic analysis that uses variation in the pricing of homes or other real estate to determine how people value certain attributes of the property and its surrounding characteristics. For example, variation in housing or other real estate prices can be used to estimate the value of local environmental attributes that surround the properties. [7]

herbivores Organisms that eat plants or other photosynthetic organisms. Also called *primary consumers*. [3]

heterozygosity The presence of more than one allele of the same gene in an individual or population. Heterozygosity is quantifiable as the proportion H of gene loci at which the average individual in the population has two or more alleles. Also called *genetic variability*. [1, 3, 13]

heterozygous Condition of an individual having two different forms (i.e., two *alleles*) of a gene on homologous chromosomes. [3]

higher consumers (2°, 3°) Predators that kill and eat other animals. Secondary consumers (carnivores) feed on herbivores, and tertiary consumers feed on other carnivores. See also *omnivores*. [3]

homozygous Condition of an individual having two identical alleles of the same gene on homologous chromosomes. [3]

horizon scanning A procedure for gathering information to identify potential threats from invasive species before they are introduced to a novel ecosystem. Usually the first step in forming a management plan for invasive species. [11]

human exceptionalism An extreme form of *anthropocentrism*, especially as promulgated by certain Christian groups who assert the complete moral and biological superiority and rights of humans (whom they believe to be made in God's image) over all other life forms. Compare with *biocentrism; ecocentrism; technocentrism*. Also called *human supremacism*. [5]

human supremacism See *human exceptionalism*.

hybrid Intermediate offspring resulting from mating between individuals of two different species. [3]

I

immobilization The process within nutrient cycles by which biologically essential elements (e.g., nitrogen, phosphorus, and potassium) in their soluble inorganic forms are taken up and reused by plants and microbes for new growth. Compare with *mineralization*. [6]

in situ conservation On-site conservation. Preservation of natural communities, ecosystems, landscapes, and populations of endangered species in their natural habitats, such as with management of protected areas. [16]

in vitro fertilization Fertilization of an egg by sperm outside the body, followed by implantation of the embryo into the uterus of an animal. [16]

inbreeding Nonrandom mating between relatives (i.e., two individuals that are more similar in their genotypes than a random mating pair would be). Results in a loss of heterozygosity. [13]

inbreeding depression The loss of heterozygosity from inbreeding. Often results in reduced biological fitness, characterized by higher mortality of offspring, fewer offspring, or offspring that are weak or sterile. [13, 16]

Indigenous Protected and Conserved Areas (IPCAs) Lands where Indigenous peoples and their legally recognized governments play the primary role in managing, protecting and conserving their ecosystems while building sustainable local economies. IPCAs aim to safeguard Indigenous peoples' rights and secure a space where communities can practice their traditional ways of life, while also maintaining biodiversity. [15]

indirect use value The economic or social value of ecosystem goods or services that people derive from nature without the need to harvest, consume, or destroy the resource (e.g., flood control). [5, 7]

Industrial Revolution The period in the eighteenth and nineteenth centuries during which technology proliferated and predominantly agrarian, rural societies in Europe and America shifted and became industrial and urban. [1]

instantaneous (intrinsic) rate of increase (*r*) Population growth rate measured as $(b - d)$, or the rate of birth (per individual) minus the rate of death (per individual), but calculated for growth that is constant and continuous through time rather than for growth in discrete time units. Compare with *geometric rate of increase (R)*. [13]

instrumental value The value that a thing has in helping us get something else we want. Something has instrumental value if it serves as a means to an end (e.g., currency has intrinsic value because it allows us to purchase things we want). Nature and its biodiversity have instrumental value in the form of the goods and services they provide to humanity. Also called *utilitarian value*. Compare with *instrinsic value*. [5]

insurance value See *option value*.

integrated conservation and development projects (ICDPs) A conservation strategy that seeks to meet the goals of biodiversity conservation while dealing with the social and economic needs of communities who might otherwise threaten biodiversity. [14]

intergenerational sustainability A Chinese philosophy, dating to around the eleventh century CE, of maintaining sacred landscapes and natural systems so that they are preserved for future generations. See also *seventh generation principle*. [2]

interior habitat The habitat of a patch minus its edge habitat. The proportion of interior habitat decreases as patch size declines. Also called *core habitat*. [9]

intraspecific competition Competition among individuals of the same species. [13]

intrinsic value The inherent value that a thing (e.g., a species or ecosystem) possesses simply because of its existence, and which is independent of its instrumental value to human society. See also *instrumental value, relational value*. [2]

introduction A program that moves individuals into an area outside their historical range because their original range is no longer suitable. Also called *assisted colonization* or *translocation*. Compare with *reinforcement; reintroduction*. [16]

invasive alien species (IAS) An alien species which becomes established in natural or seminatural ecosystems or habitat, is an agent of change, and threatens native biological diversity Also called *alien invasive species*. [11]

IUCN Red List A comprehensive and standardized listing of the global conservation status of more than 90,000 threatened and endangered species. Compiled by the International Union for the Conservation of Nature (IUCN). [8]

K

keystone species A species that has a disproportionate impact (relative to its numbers or biomass) on the organization of a biological community. Loss of a keystone species can lead to a broader loss of biodiversity. [3]

kingdom Large unit of taxonomic classification; for example, the kingdom Animalia includes all animals, while Plantae contains all green plants. [3]

KML A GIS file format that uses Keyhole Markup Language to view spatial information in Google Earth. [15]

L

land conservancy See *land trust.*

land trust A private, nonprofit conservation organization that, as part of its mission, assists in land acquisitions through conservation easements and covenants, and then stewards those lands as part of their own portfolio of protected areas. [15]

land-use change The conversion of one habitat type to another, usually human-engendered such as when a natural forest is converted to a suburban housing development, or a natural grassland is converted to agricultural production. The single greatest cause of habitat loss. [9]

landrace A variety of food crop that has unique genetic characteristics and is cultivated in a very small area of the world. Usually refers to a variety (or its wild relative) that gave rise to the hybrids that dominate modern monocultures. Landrace genomes harbor variability that is crucial to maintaining the high productivity of today's agriculture in the face of climate change and the evolution of new and/or resistant strains of pests and diseases. [16]

landscape A large geographic region that spans multiple biological communities and ecosystems and includes different habitat types. [15]

landscape mosaics Heterogeneous landscapes that are patchy and subdivided, where some habitats have been lost and converted to human uses, others have been heavily degraded, and those that remain intact are small, fragmented, and disconnected from one another. Also called *patch mosaics.* [9, 15]

law of demand A rule that states that if all other factors remain equal, people's demand for a good or service will decrease as the price of the good or service increases. [7]

leading-edge expansion See *cold-edge expansion.* [12]

Leslie matrix A discrete-time, age-based matrix model for estimating the growth of populations with age or stage classes. [13]

linkage density A measure of network complexity that quantifies the number of interactions each species has in a food web. Linkage density (D) represents the average number of interactions between species as $D = L/S$, where S is the total number of species in an interaction matrix, and L is the total number of links in the matrix. See also *connectance.* [3]

Living Planet Index (LPI) An indicator of the state of global biodiversity, based on trends in 16,700 vertebrate populations of species from around the world. These population trends are maintained in the Living Planet Database and published in the *Living Planet Report* by the Zoological Society of London. [1, 8]

Living Planet Report A comprehensive report on global biodiversity, including the Living Planet Index, published every two years by the World Wildlife Federation (WWF). [8]

locally extinct See *extirpated.*

locus Plural, *loci.* See *gene.*

logistic growth model Mathematical model for density-dependent population growth that builds on the *geometric growth model* and *exponential growth model* by adding a term that accounts for the constraint of *carrying capacity.* [13]

love it to death syndrome A danger of ecotourism whereby overly abundant and assertive tourists themselves become the source of environmental destruction to the site they are visiting. [6]

M

mainland–island model A modification of the classic metapopulation model in which a large mainland population with a low probability of extinction serves as a source population, providing immigrants to nearby small populations. Also called the *propagule rain model.* [9]

managed relocation See *assisted colonization.* [12]

manifest destiny In the U.S. in the nineteenth century, the widely held belief that it was destined and inevitable for Americans to move west and occupy North America from coast to coast, and to become prosperous as they did so. [2]

market failure A situation where the profits of a business transaction benefit the business, but certain costs of the transaction are paid by individuals or groups of individuals that were not involved in the transaction. Most commonly occur when externalities cause resources to be misallocated, allowing a few individuals or businesses to benefit at the expense of the larger society. [7]

market pricing method An economic analysis that gives the value of either the quantity or quality of an ecosystem good or service that is bought and sold in a commercial market. Determined by the laws of supply and demand. [7]

Marxan A widely used software program for designing protected area networks. [15]

mass mortality event (MME) A rapid, catastrophic die-off of organisms that kills more than 90% of a population in a short time. [1]

mate limitation When sexual reproduction becomes limited because population density gets sufficiently low to reduce the chance of finding a mate. [8]

matrix model An array of numerical data arranged in columns and rows. [13]

maximum sustainable yield (MSY) The maximum level at which a natural resource can be routinely harvested or exploited without long-term depletion. MSY calculations are widely used in managing fisheries, timber forests, and game species. [2, 10]

meta-analysis Synthesis of information from a large number of studies whose data and results are collated and subjected to collective statistical analysis. [11]

metapopulation A collection of local populations that are spatially separated from each other in the individual patches of a landscape, but are connected to each other by the dispersal and exchange of individuals. [8, 9]

Millennium Development Goals (MDGs) A set of 8 goals for sustainable development that arose out of the United Nation's Millennium Summit of 2000. Has been superseded by the *2030 Agenda for Sustainable Development* and its *Sustainable Development Goals.* [17]

mineralization The process within nutrient cycles by which biologically essential elements (e.g., nitrogen, phosphorus, and potassium) from organic tissue are returned to their soluble inorganic forms, which can subsequently be used by plants and microbes. Compare with *immobilization.* [6]

minimum dynamic area (MDA) The area of habitat required to maintain a minimum viable population of a species. [13]

minimum viable population (MVP) The smallest population size predicted to have a very high chance (e.g., 95%) of persisting for the foreseeable future (e.g., 100 years). [13]

morphological species concept The concept that a group of individuals is recognized as a *species* because they are morphologically, physiologically, or biochemically distinct from other groups. Compare with *biological species concept, evolutionary species concept.* [3]

morphospecies A group of individuals that, based on the level of similarity of their morphological characteristics, are probably a distinct species. [3]

mutational meltdown The accumulation of harmful mutations in a small population, which leads to loss of fitness and decline of the population size, which then leads to further accumulation of deleterious mutations. [4]

N

Nagoya Protocol A supplementary agreement to the 1992 UN Convention on Biological Diversity (CBD) that sought to establish an international protocol for the fair and equitable sharing of benefits that arise from the access and use of genetic resources. [16]

national biodiversity strategy and action plans (NBSAPs) Plans that describe how each of the 196 nations that are parties to the *United Nations Convention on Biodiversity* (CBD) will accomplish goals in the CBD's Strategic Plan for Biodiversity 2011–2020, which includes the 20 Aichi Biodiversity targets. [15]

National Environmental Funds (NEFs) Conservation trust funds or foundations in which a board of trustees—composed of representatives of the host government, conservation organizations, and donor agencies—allocates the annual income from an endowment to support inadequately funded government departments and nongovernment conservation organizations and activities. [17]

native species A species, subspecies, or lower taxon occurring within its natural range (past or present) and dispersal potential (i.e., within the range it occupies naturally or could occupy without direct or indirect introduction or care by humans). Also called *indigenous species.* [11]

natural infrastructure The collection of genes, species, and biological communities that comprise the natural and managed ecosystems on which people depend, and which provide goods and services to society. Also called *green infrastructure.* [1, 12]

Nature Needs Half Project An international coalition of scientists, conservationists, nonprofits, and public officials working together to protect at least half of all nature on Earth—land and water—by 2030 in order to support the existence of nature and the services it provides. [14]

neoendemic A species that occupies a small area because it has only recently evolved from a closely related species. [8]

net primary production (NPP) The difference between gross primary production (GPP) and the organic carbon lost as heat during respiration (R) by autotrophs; NPP = GPP – R. [6]

niche differentiation Different species either use different resources or use the same resource differently in space or time, thus reducing competition for resources among different species. Also called *niche partitioning,* or niche complementarity. [4]

niche partitioning See *niche differentiation.* [6]

noble savages The romanticized ideal of prehistoric and/or primitive cultures living in harmony with nature, managing their natural resources sustainably, and being free from the corrupting influence of modern civilization and technology. [2]

non-use value The value that people and governments assign to ecosystem goods and services even if they never have and never will use them. See *bequest value; existence value.* [5]

nonconsumptive use value Value assigned to an ecosystem service that is not extracted but still used and valued, such as scenic beauty or the use of a river for transportation. [7]

nongovernment organizations (NGOs) A nonprofit organization that operates independently of any government, whose purpose is to address a social or political issue. Among the major NGOs involved in conservation and sustainable development activities are the World Wildlife Fund, the Nature Conservancy, and Conservation International. [17].

novel environments hypothesis Proposes that successful invasive species are particularly well adapted for, and successful in, habitats that have been directly modified or disturbed by human activities. [11]

novel weapons hypothesis Proposes that alien plant species gain advantage over native plants by expressing allelopathic chemicals that can kill the native vegetation. [11]

omnivores Animals that eat food from more than one trophic level, for example, eating both plants and other animals. [3]

ontogenetic niche shifts Changes in species' habitat requirements as they complete the different parts of their life cycles (e.g., a tadpole's need for an aquatic environment while the adult frog can live on land). [9]

open populations A population whose size potentially changes through time as individuals a leave or enter the population. Compare with *closed population.* [13]

open-access resources Natural resources (such as water and air) that are collectively owned by society at large and available for everyone to use. Compare with *common-pool resources.* [7, 10]

OpenStreetMap A crowdsourcing GIS data project that uses either XML-based file formats or smaller, more efficient formats called Protocol Buffer Binary Formats. [15]

option value The value placed on an individual's willingness to pay for maintaining or preserving an ecosystem service so that it can be used for benefit at some time in the future. Also called *insurance value.* [5, 7]

order Unit of taxonomic classification; an order includes one or more related *families.* [3]

organic farming Farming practices that generally prohibit the use of genetically modified organisms (GMOs) and synthetic inputs (chemical fertilizations and pesticides) while allowing organic fertilizers and pesticides. [15]

overexploitation The harvesting of a natural resource at a rate faster than it can be regenerated, resulting in decline or loss of the resource. The term applies to resources like populations of wild plants and animals (medicinal plants, game animals, fish), whole ecosystems (e.g., grazing pastures), and resources that are required to sustain life (e.g., aquifers). Also called *overharvesting.* [8, 10]

overharvesting See *overexploitation.*

P

PADDD (protected areas downgrading, downsizing, and degazettement) "Unprotecting" conservation and other protected areas in response to social pressure from local communities that are angry they had not been fairly compensated for the loss of their lands and livelihoods, or from those that argue greater access to and use of natural resources is needed for economic development. [14]

paleoendemic An ancient species that has a narrow geographical range and is not closely related to extant species. [8]

parasites Organisms that live in or on a host organism, feeding on its tissues or body fluids but without necessarily killing it (e.g., ticks, tapeworms). [3]

participatory design See *co-design.*

patch mosaics See *landscape mosaics.*

patch occupancy The proportion of habitat patches in a landscape that are occupied by a species at a given time. Metapopulation models are used to predict changes in patch occupancy across a landscape. [15]

pathogens Organisms that attack and significantly damage the cells of another organism, often resulting in the death of the host organism (e.g., rusts, which are fungi that damage many crop plants; and *Plasmodium,* the protist responsible for malaria in humans.) [3]

Pathway to Canada Target 1 Canada's nation-wide initiative to develop a plan to achieve its international biodiversity commitment to conserve at least 17 percent of its land and freshwater by 2020 through a coordinated network of protected areas. [15]

phenogram Diagram depicting relationships among species based on their overall similarity or dissimilarity in functional traits, without regard to the evolutionary history of the species. Compare with *phylogenetic tree.* [3]

phenology Life-cycle events (e.g., reproduction, migration) that plants and animals exhibit seasonally, annually, or interannually. Climate change can impact the fitness of species by altering the timing of phenological events. [12]

phenotype The morphological, physiological, anatomical, and biochemical characteristics of an individual that result from the expression of its genotype in a particular environment. [3]

phenotypic plasticity The ability of one genotype to produce more than one phenotype when exposed to different environments. [12]

phenotypic variation The different morphological, physiological, and biochemical characteristics of individuals in a population. Phenotypic variation is the result both of *genetic diversity* and the environmental conditions under which an individual's genotype is expressed. [3]

phylogenetic diversity A measure of biodiversity that incorporates phylogenetic differences among species—that is, that considers the amount of genetic differentiation between two or more species that has occurred since their divergence from a common ancestor. [3]

phylogenetic tree A branching diagram showing the evolutionary relationships among species based on similarities and differences in their genetic characteristics. Species on the same branch of a tree have a common ancestor, and species that are closer together have a more recent common ancestor than those that are farther apart on the tree. [3]

phylum (plural, *phyla*) Large unit of taxonomic classification containing related *classes* of species. [3]

polymorphism A common measure of genetic diversity in a population. It is often quantified as the fraction of gene loci in which alternative alleles of a gene occur (i.e., that are polymorphic). [3]

population A geographically defined group of individuals of the same species that mate and otherwise interact with one another. Compare with *metapopulation.* [3]

population asynchrony Occurs when populations fluctuate through time in a manner that is out-of-phase. Asynchrony can maintain the temporal stability of an ecosystems service by ensuring that decreases in one population and the services it provides are compensated by increases in another population that can maintain those services. [6, 8, 12]

population bottleneck Occurs when a population is greatly reduced in size due to stochastic or anthropogenic events. The loss of rare alleles and general reduction in heterozygosity usually reduces overall fitness of the population. Compare with *founder effect.* [13]

population decay A chronic decline in a species' populations size and geographic range that increase its risk of extinction. [1]

population viability analysis (PVA) A risk-assessment process aimed at determining the minimum viable population (MVP) and minimum dynamic area (MDA) required to conserve a threatened species. [13]

portfolio effect Occurs when each species responds differently to changing conditions through time, allowing the total amount of an ecosystem service delivered by the community to remain stable through time. [6]

positive interactions Interactions among species that allow diverse communities to be more efficient and productive. [6]

predators See *higher consumers.*

preservationist ethic One of three environmental philosophies that shaped conservation policies in the United States and elsewhere. This conservation ethic is based on a belief that large, unmodified areas of wilderness should be conserved for their intrinsic value. Its most prominent proponent was John Muir, who was instrumental in designating several national parks in the U.S. See also *resource conservation ethic,* and *evolutionary-ecological land ethic.* [2]

primary (1°) consumers See *herbivores.*

primary (1°) producers Green plants, algae (including seaweed), cyanobacteria, and photosynthetic protists that can produce their own food (energy) directly from the sun via photosynthesis. Also known as *autotrophs.* [3]

primary productivity The production of biomass by autotrophs (primary producers), such as green plants, algae, and cyanobacteria, which obtain their energy directly from the sun via photosynthesis. [6]

prisoner's dilemma A common example of a social dilemma from the field of *game theory* that illustrates why imperfect information causes two rational individuals to not cooperate with each other even when it is in their best interest to do so. [10]

producer surplus The difference between the total amount earned from a product or service (price × quantity sold) and the cost of production. Compare with *economic surplus.* [7]

production function method An economic analysis that estimates the value of an ecosystem good or service by modeling the values of the good or service as a function of the inputs that control it (e.g., the contribution of a mangrove wetland ecosystem to a shrimping industry). [7]

productive use value Direct use value assigned to products that are sold in markets. Compare with *consumptive use value*. [5]

propagule pressure The number of colonizing individuals of a species arriving to a new location. [11]

propagule rain model See *mainland–island model*. [9]

protected area A clearly defined geographical space that is recognized, dedicated, and managed to achieve long-term conservation of nature, including biodiversity and the many goods and services associates with the world's different ecosystems. Protected areas include a variety of nature reserves, wilderness areas, national parks and monuments, habitat/species management areas, protected landscape and seascapes, and mixed use habitats that promote sustainable use of natural resources. Protected areas are established and maintained by governments, indigenous societies, private individuals, conservation organizations, and research institutions. [14]

protected area network A collection of protected areas that represent all scales in the biological hierarchy (genes, species, communities, and ecosystems), and which are connected within a landscape and managed collectively. Also called a *reserve network*. [15]

provisioning services Products people obtain from ecosystems. [6]

R

radiative forcing The difference between sunlight absorbed by the Earth and energy radiated back to space. [12]

rapid evolution Changes in allele frequencies that lead to adaptive responses of organisms, or even formation of new species, on short time-scale of one, or just a few generations. [4]

Red Data Books Compilations of lists of endangered species (Red Lists) by taxonomic group and country prepared by the IUCN and other conservation organizations. [8]

Red List Index An indicator of the changing state of global biodiversity. It defines the conservation status of major species groups, and measures trends in extinction risk over time. [8]

Red List of Ecosystems A listing of ecosystems that are at increasing risk of collapse, from vulnerable, to endangered, to critically endangered. Being developed by the IUCN, which plans to have a comprehensive evaluation of all the world's ecosystems by 2025. [8]

refugia Term used to describe the equatorward areas to which some boreal species were able to retreat and thus survive the climate changes associated with the Last Glacial Maximum. Now being applied to poleward shifts that may enable species to survive the current warming climate. [12]

regulating services Benefits people obtain from the regulation of ecosystem processes, which helps reduce harmful variation and provide a form of insurance to human well-being. [6]

reinforcement Release of individuals into an existing population to increase its size and genetic diversity. Compare with *introduction; reintroduction*. [16]

reintroduction The release of captive-bred or wild-caught individuals into ecologically and historically suitable sites where the species no longer occurs. Compare with *introduction; reinforcement*. [16]

relational values Value of relationships among people, between people and nonhuman organisms, and between people and the land that enhance one's personal or cultural identity, social responsibility, and emotional well-being. Rooted in psychology and philosophy, these values are an important aspect of preserving ecosystems and biodiversity. [5]

representative concentration pathways (RCPs) A greenhouse gas concentration trajectory adopted by the IPCC for its fifth Assessment Report in 2014. Four scenarios of future levels of radiative forcing and greenhouse gases are based on different predictions of societal and government action on controlling greenhouse gas levels and are then used to produce *general circulation models*. [12]

rescue effect The rescue of a local population from extinction by the influx of immigrants from another location that has a positive population growth rate. [8, 9]

reserve network See *protected area network*.

resource conservation ethic One of three environmental philosophies that shaped conservation policies in the United States and elsewhere. This conservation ethic is based on the belief that natural resources should be used for the greatest good of the largest number of people for the longest time. Gifford Pinchot codified its three principles: (1) fair distribution of resources between present and future generations; (2) efficient use of limited resources; and (3) scientific management of resources based on the best available data. See also *preservationist ethic*, and *evolutionary-ecological land ethic*. [2]

resurrection biology See *de-extinction*.

revealed preference methods A set of economic analyses that estimate the value of ecosystem goods or services by using direct observations of the amount people are willing to pay for them. Compare with *stated preference methods*. [7]

Rio Declaration on Environment and Development One of three non-legally binding agreements arising from the 1992 Earth Summit, this document included 27 principles to guide the actions of both wealthy and poor nations on issues of the environment and development. [17]

Rio Forest Principles One of three non-legally binding agreements arising from the 1992 Earth Summit, this document recognized the value of forests and provided a set of 15 principles for sustainable forest management. [17]

risk assessment The formal process for determining the probability that an event will occur and the consequences or impact if it does occur. Compare with *risk management*. [11]

risk management The suite of tools, techniques, and policies that can be used to mitigate the risks identified in a risk assessment. [11]

romantic transcendentalism Philosophy that arose in North America, notably with Ralph Waldo Emerson and Henry David Thoreau, viewing nature as imbued with the divine and thus a temple that provides spiritual nourishment as well as material needs. Importantly, these philosophers believed that nature required protection from the impacts of industrialism and other manifestations of human greed. [2]

royal menageries Large collections of exotic animals historically kept by aristocrats or royal courts within palace gardens for the purpose of entertainment and as a display of wealth and power. [16]

S

secondary extinction An extinction linked to the prior extinction of another species. A series of such extinctions triggered by the loss of a keystone species is an *extinction cascade*. [3]

seed bank An ex situ storage facility for seeds collected from wild and cultivated plants, for use in conservation and agricultural programs. [16]

selection effect A biological mechanism that can improve ecosystem services. When a community consists of a larger variety of species, one or a select few species are likely to be extremely efficient and come to dominate the ecological processes that generate ecosystem services. [6]

seventh generation principle A philosophy, attributed to the Iroquois Confederacy of North America, that the decisions we make today should guarantee a sustainable world seven generations into the future. See also *intergenerational sustainability*. [2]

Shapefile A GIS file format used by the Environmental Systems Research Institute (ESRI) ArcView software. The Shapefile format is made up of files with the same name but different extensions. [15]

shifting cultivation Farming method in which farmers clear land by cutting down and/or burning trees, plant crops for a few years, and then abandon the site when soil fertility declines. Also called *slash-and-burn agriculture*. [9]

sink population A local population in a low-quality habitat that has a negative population growth rate, and which would go extinct if not for immigration from a source population. [9]

slash-and-burn agriculture See *shifting cultivation*.

social dilemmas Situations in which individuals make decisions in their own self-interest that have inferior outcomes compared to what they would have achieved by cooperating with others. See *assurance dilemma, game theory, prisoner's dilemma, tragedy of the commons*. [10]

source population A local population in a high-quality habitat that has a positive population growth rate, and which serves as a source of colonists to other local populations. [9]

source–sink dynamics Term used to describe the immigration and emigration of individuals across a landscape with source and sink populations. [9]

spatial turnover The change in biodiversity among different communities or ecosystems across a geographical landscape. See also *temporal turnover*. [3]

species The base unit of classification and taxonomic rank of an organism. Species represent fundamental units of evolution, and are the primary targets of much conservation legislation. Recognizing and defining a species can be approached in several ways. See *biological species concept, evolutionary species concept, morphological species concept*. [3]

Species360 An international nonprofit organization that maintains an online database of wild animals that are under human care. [16]

species distribution models (SDMs) Predictive habitat distribution models, and range mapping that use computer algorithms to predict the distribution of a species across geographic space and time using environmental data. SDMs are often used to predict how species (including alien species) shift their range distributions in response to some form of environmental change (e.g., global warming). When the predictor variables of these correlations are climatic, the models are called *climate envelope models*. Also called *ecological niche models*. [11]

species diversity The variety of species that comprise a biological community; the collection of species that occupy and interact in a particular location. [3]

species revivalism See *de-extinction*.

species richness (S) The number of unique species observed in an ecological community or other sampling space such as an ecosystem. [3]

Species Survival Plan (SSP) A plan that establishes breeding goals and management recommendations to achieve the maximum genetic diversity and demographic stability for a species managed in ex situ facilities. [16]

stage-based models Population growth models that estimate growth parameters by a meaningful life-stage, which may or may not be associated with age (e.g., insect developmental stages like larva vs. adult, or small- vs. large-sized trees). Compare with *age-based models*. [13]

stated preference methods A set of economic analyses that are used to estimate the value of ecosystem goods and services by directly surveying consumers to ask about their *willingness to pay* (WTP) for the good or service, or about their *willingness to accept compensation* (WTAC) for doing without them. [7]

stepping stones Discontinuous pieces of habitat that fill in gaps between two protected areas and help facilitate dispersal. Compare with *habitat corridor*. [15]

stochastic model A population model that accounts for variation in parameters such as birth and death rates by randomly drawing values from probability distributions. Compare with *deterministic model*. [13]

Strategic Plan for Biodiversity (SPD) A 10-year plan adopted in 2010 by parties to the United Nations Convention on Biological Diversity (CBD) giving a framework for action by all countries and stakeholders to safeguard biodiversity and the benefits it provides to people. The SPD contains 20 goals called the *Aichi Biodiversity Targets*. [14]

structured decision making (SDM) An approach for systematic and organized analysis of natural resource management decisions that integrates objective evidence-based views with subjective value-based views. [2, 13]

supporting services Ecological processes that control the functioning of ecosystems and production of all other services. [6]

surrogate market techniques A set of economic analyses that are used to estimate the value of ecosystem goods and services that do not have a real-market value, but which can be estimated from "surrogates"—products or services, such as travel and housing costs, that are quantified in real markets. See *travel cost method; hedonic pricing method*. [7]

surrogate species Species that provide a means for achieving broader conservation goals. The two main types of surrogate species are *flagship species* and *umbrella species*. [13]

sustainable development Development, economic or otherwise, that meets the needs of the present generation without compromising the ability of future generations to meet their own needs. To qualify as sustainable development, projects must balance three pillars of development: economic, social, and environmental. [5, 17]

Sustainable Development Goals (SDGs) Seventeen broad social, humanitarian, and ecological goals that expand on the eight *Millennium Development Goals* and aim to complete what they did not achieve The SDGs encompass goals for improving health and education; reducing poverty, hunger and inequality; tackling climate change; and working to preserve life on land and in the sea. Each SDG includes a set of specific targets, for which progress can be measured by monitoring a set of indicators. [14, 17]

sustainable forest management The management of forests so that they provide for present timber needs while being maintained at population levels that will sustain their use by future generations. [2]

sustainably intensified farming Farming that uses practices from agroforestry, conservation agriculture, and biological pest control to promote favorable ecological interactions that produce a low-input, resource-conserving agroecosystems. [15]

T

2030 Agenda for Sustainable Development A United Nations blueprint, formulated in 2015, for achieving human well-being and equality while transitioning to sustainable use of the planet's natural resources. Articulates 17 *Sustainable Development Goals.* [14, 17]

taxonomists Scientists involved in the identification and classification of species. [3]

taxonomy The science of identifying and classifying living things. [3]

technocentrism An extreme form of *anthropocentrism* holding that humans should aspire to dominate nature through technology, with the goal of controlling the global environment to suit human needs and maximize human prosperity. [5]

temporal turnover The change in biodiversity in a community or across a landscape between any two points in time. [3]

theory of island biogeography A theory proposing that the number of species on any island is determined by a balance between the rate at which new species colonize the island, and the rate at which populations of species that are already established become extinct. The theory is often used in conservation to predict how habitat destruction will influence species diversity, and to design protected area networks. [4]

tragedy of the commons A social dilemma where individuals who use a shared resource act independently and according to their own self-interests, yet the collective action of many individuals pursuing their own self-interests causes them to deplete or spoil the resource, to the detriment of all. [7, 10]

trailing-edge contraction See *warm-edge contraction.* [12]

trait-based risk assessment Assessments of potential alien species that use measurable life history traits and other biological characteristics (genetic, morphological, etc.) to try to predict their introduction, establishment, and spread. [11]

translocation See *introduction.*

transport vector The means by which an alien species is either intentionally or accidentally transported to a new location. Humans are the transport vector for most invasive species. [11]

travel cost method An economic analysis that quantifies the ecosystem services of visited sites, based on the travel and time costs that individuals incur to visit the sites. [7]

traveling menageries Touring groups of showmen and animal handlers who historically visited towns and cities with collections of common and exotic animals for the purpose of public entertainment. [16]

trophic cascade An ecological phenomenon triggered by the addition or removal of top predators to a food web, which then leads to reciprocal changes in the relative populations of predators and their prey. Trophic cascades can produce powerful indirect interactions that can alter the properties of entire ecosystems. [3]

trophic pyramid A graphical depiction of the amount of biomass in, or energy available to, each trophic level in a community. The amount of energy available for reproduction and growth usually declines with each successive trophic level, so the effect looks like that of a pyramid with primary producers forming a large base that supports sequentially smaller levels of consumers. [3]

U

umbrella species A species whose existence benefits other species living under its protective "umbrella." Protecting an umbrella species (such as wildebeest in East Africa) benefits many other, interconnected species. [13]

United Nations (UN) An international, intergovernmental organization of 193 countries founded in 1945 to help maintain international peace, foster friendly relations among nations, and achieve international cooperation. [17]

urban ecology The scientific study of the relation of living organisms with each other and their surroundings in the context of an urban environment. [9]

utilitarian value See *instrumental value.*

V

voluntary transaction A monetary transaction that takes place only when it is considered beneficial by both parties involved. [7]

W

warm-edge contraction Contraction of a species' historic geographic range because regions that were once within that range have become too warm. Also called *trailing-edge contraction.* [12]

wild meat See *bushmeat.*

willingness to accept compensation (WTAC) Response elicited in stated preference methods of estimating the value of ecosystem goods and services, as participants are asked how much they would be willing to accept as compensation for doing without a service. [7]

willingness to pay (WTP) Response elicited in stated preference methods of estimating the value of ecosystem services, as participants are asked how much they would be willing to pay for a service or suite of services. [7]

World Bank An international financial institution that provides loans and grants to the governments of poorer countries for the purpose of pursuing capital projects. Among other activities, the World Bank finances certain types of global development projects that are intended to be congruent with the UN's *Sustainable Development Goals.* [17]

World Commission on Environment and Development (WCED) See *Brundtland Commission.* [17]

World Ocean Assessment (WOA) A regular process set up in 2004 by the United Nations General Assembly to review the environmental, economic, and social aspects of the world's oceans and seas. [1]

Z

zoo A facility in which animals are housed in human-built enclosures and displayed to the public and in which they may also breed. Also called *animal park* or *zoological park.* [16]

zoological park See *zoo.*

zoonotic diseases Infectious diseases transmitted to humans by animals. [10]

References

Chapter 1

1. Galor, O. and Weil, D. N. 2000. Population, technology, and growth: From Malthusian stagnation to the demographic transition and beyond. *American Economic Review* 90: 806–828.
2. Daily, G. C. and Ehrlich, P. R. 1992. Population, sustainability, and Earth's carrying capacity. *BioScience* 42: 761–771.
3. Meadows, D. H., Randers, J. and Meadows, D. L. 2004. *The Limits to Growth: The 30-Year Update*. White River Junction, VT: Chelsea Green.
4. Cohen, J. E. 1995. Population growth and Earth's human carrying capacity. *Science* 269: 341.
5. Roser, M. 2017. *Child mortality*, https://ourworldindata.org/child-mortality/.
6. MacDorman, M. F., Mathews, T., Mohangoo, A. D. and Zeitlin, J. 2014. *International Comparisons of Infant Mortality and Related Factors: United States and Europe, 2010*. Hyattsville, MD: National Center for Health Statistics.
7. Roser, M. 2017. *Economic growth*, https://ourworldindata.org/economic-growth.
8. Roser, M. and Ritchie, H. 2017. *Food per person*, https://ourworldindata.org/food-per-person.
9. FAO. 2015. *Statistical Pocketbook: World Food and Agriculture*. Rome, Italy: United Nations.
10. Nejat, P., Jomehzadeh, F., Taheri, M. M. et al. 2015. A global review of energy consumption, CO_2 emissions and policy in the residential sector (with an overview of the top ten CO_2 emitting countries). *Renewable & Sustainable Energy Reviews* 43: 843–862.
11. World Bank. 2018. *Energy Use Per Capita*. Washington, DC: World Development Indicators.
12. Royal Society. 2012. *People and the Planet*. London, UK: Royal Society.
13. Vitousek, P. M., Mooney, H. A., Lubchenco, J. and Melillo, J. M. 1997. Human domination of Earth's ecosystems. *Science* 277: 494–499.
14. Millennium Ecosystem Assessment. 2005. *Ecosystems and Human Well-being: Biodiversity Synthesis*. Washington, DC: World Resources Institute.
15. CIESIN. 2011. Global rural-urban mapping project, version 1 (GRUMPv1): Urban extents grid. Palisades, NY: Center for International Earth Science Information Network, Columbia University.
16. Millennium Ecosystem Assessment. 2005. *Ecosystems and Human Well-being: Synthesis*. Washington, DC: Island Press.
17. EPA. 2017. *Report on the Environment*. Washington, DC: U.S. Environmental Protection Agency.
18. MEP. 2013. *Soil Pollution and Human Health*. China: Ministry of Environmental Protection.
19. Bestelmeyer, B. T., Okin, G. S., Duniway, M. C. et al. 2015. Desertification, land use, and the transformation of global drylands. *Frontiers in Ecology and the Environment* 13: 28–36.
20. Geist, H. 2017. *The Causes and Progression of Desertification*. London, UK: Routledge.
21. Haddad, N. M., Brudvig, L. A., Clobert, J. et al. 2015. Habitat fragmentation and its lasting impact on Earth's ecosystems. *Science Advances* 1: e1500052.
22. Dunlap, R. E., McCright, A. M. and Yarosh, J. H. 2016. The political divide on climate change: Partisan polarization widens in the US. *Environment: Science and Policy for Sustainable Development* 58: 4–23.
23. National Research Council. 2011. *Advancing the Science of Climate Change*. Washington, DC: National Academies Press.
24. IPCC. 2014. *Climate Change 2014: Synthesis Report*. Geneva, Switzerland: Intergovernmental Panel on Climate Change.
25. IPCC. 2013. *Climate Change 2013: The Physical Science Basis*. New York, NY: Cambridge University Press.
26. Parmesan, C. and Yohe, G. 2003. A globally coherent fingerprint of climate change impacts across natural systems. *Nature* 421: 37–42.
27. Pecl, G. T., Araujo, M. B., Bell, J. D. et al. 2017. Biodiversity redistribution under climate change: Impacts on ecosystems and human well-being. *Science* 355: 9.
28. Visser, M. E. and Both, C. 2005. Shifts in phenology due to global climate change: The need for a yardstick. *Proceedings of the Royal Society B: Biological Sciences* 272: 2561–2569.
29. Thomas, C. D., Cameron, A., Green, R. E. et al. 2004. Extinction risk from climate change. *Nature* 427: 145–148.
30. United Nations. 2017. *The First Global Integrated Marine Assessment*. New York, NY: Cambridge University Press.
31. Diaz, R. J. and Rosenberg, R. 2008. Spreading dead zones and consequences for marine ecosystems. *Science* 321: 926–929.
32. Halpern, B. S., Frazier, M., Potapenko, J. et al. 2015. Spatial and temporal changes in cumulative human impacts on the world's ocean. *Nature Communications* 6: 7615.
33. Edgar, G. J., Stuart-Smith, R. D., Willis, T. J. et al. 2014. Global conservation outcomes depend on marine protected areas with five key features. *Nature* 506: 216–220.
34. Postel, S. L., Daily, G. C. and Ehrlich, P. R. 1996. Human appropriation of renewable fresh water. *Science* 271: 785–788.
35. World Commission on Dams. 2000. *Dams and Development: A New Framework for Decision-making*. Report No. 1853837989. London, UK: Earthscan.
36. Grill, G., Lehner, B., Lumsdon, A. E. et al. 2015. An index-based framework for assessing patterns and trends in river fragmentation and flow regulation by global dams at multiple scales. *Environmental Research Letters* 10: 15.
37. Davidson, N. C. 2014. How much wetland has the world lost? Long-term and recent trends in global wetland area. *Marine and Freshwater Research* 65: 934–941.
38. Micklin, P. P. 1988. Desiccation of the Aral Sea: A water management disaster in the Soviet Union. *Science* 241: 1170.
39. Kotlyakov, V. M. 1991. The Aral Sea basin: A critical environmental zone. *Environment: Science and Policy for Sustainable Development* 33: 4–38.

40. Bull, J. W. and Maron, M. 2016. How humans drive speciation as well as extinction. *Proceedings of the Royal Society B: Biological Sciences* 283: 20160600.
41. Pimm, S. L., Jenkins, C. N., Abell, R. et al. 2014. The biodiversity of species and their rates of extinction, distribution, and protection. *Science* 344: 1246752.
42. Mora, C., Tittensor, D. P., Adl, S. et al. 2011. How many species are there on Earth and in the ocean? *PLOS Biology* 9: e1001127.
43. Pereira, H. M., Leadley, P. W., Proenca, V. et al. 2010. Scenarios for global biodiversity in the 21st century. *Science* 330: 1496–1501.
44. Baillie, J. E. M., Hilton-Taylor, C. and Stuart, S. N. E. 2004. *2004 IUCN Red List of Threatened Species: A Global Species Assessment.* Gland, Switzerland: International Union for Conservation of Nature.
45. Hudson, L. N., Newbold, T., Contu, S. et al. 2014. The PREDICTS database: A global database of how local terrestrial biodiversity responds to human impacts. *Ecology and Evolution* 4: 4701–4735.
46. Hudson, L. N., Newbold, T., Contu, S. et al. 2017. The database of the PREDICTS (Projecting Responses of Ecological Diversity in Changing Terrestrial Systems) project. *Ecology and Evolution* 7: 145–188.
47. Newbold, T., Hudson, L. N., Hill, S. L. L. et al. 2015. Global effects of land use on local terrestrial biodiversity. *Nature* 520: 45–50.
48. Vellend, M., Baeten, L., Myers-Smith, I. H. et al. 2013. Global meta-analysis reveals no net change in local-scale plant biodiversity over time. *Proceedings of the National Academy of Sciences* 110: 19456–19459.
49. Dornelas, M., Gotelli, N. J., McGill, B. et al. 2014. Assemblage time series reveal biodiversity change but not systematic loss. *Science* 344: 296–299.
50. McRae, L., Freeman, R. and Marconi, V. 2016. *The Living Planet Report.* Gland, Switzerland: World Wildlife Fund International.
51. McRae, L., Deinet, S. and Freeman, R. 2017. The diversity-weighted Living Planet Index: Controlling for taxonomic bias in a global biodiversity indicator. *PLOS ONE* 12: 20.
52. Ceballos, G., Ehrlich, P. R. and Dirzo, R. 2017. Biological annihilation via the ongoing sixth mass extinction signaled by vertebrate population losses and declines. *Proceedings of the National Academy of Sciences* 114: E6089–E6096.
53. Hallmann, C. A., Sorg, M., Jongejans, E. et al. 2017. More than 75 percent decline over 27 years in total flying insect biomass in protected areas. *PLOS ONE* 12: e0185809.
54. Fey, S. B., Siepielski, A. M., Nusslé, S. et al. 2015. Recent shifts in the occurrence, cause, and magnitude of animal mass mortality events. *Proceedings of the National Academy of Sciences* 112: 1083–1088.
55. Frankham, R., Ballou, J. D., Dudash, M. R. et al. 2012. Implications of different species concepts for conserving biodiversity. *Biological Conservation* 153: 25–31.
56. Willoughby, J. R., Sundaram, M., Wijayawardena, B. K. et al. 2015. The reduction of genetic diversity in threatened vertebrates and new recommendations regarding IUCN conservation rankings. *Biological Conservation* 191: 495–503.
57. McGill, B. J., Dornelas, M., Gotelli, N. J. and Magurran, A. E. 2015. Fifteen forms of biodiversity trend in the Anthropocene. *Trends in Ecology & Evolution* 30: 104–113.
58. Olden, J. D., Poff, N. L., Douglas, M. R. et al. 2004. Ecological and evolutionary consequences of biotic homogenization. *Trends in Ecology & Evolution* 19: 18–24.
59. McKinney, M. L. and Lockwood, J. L. 1999. Biotic homogenization: A few winners replacing many losers in the next mass extinction. *Trends in Ecology & Evolution* 14: 450–453.
60. Rahel, F. J. 2000. Homogenization of fish faunas across the United States. *Science* 288: 854–856.
61. Crutzen, P. J. 2006. The "Anthropocene." In E. Ehlers and T. Kraft (Eds.), *Earth System Science in the Anthropocene.* New York, NY: Springer.
62. Pimm, S. L. 2001. *The World According to Pimm: A Scientist Audits the Earth.* New York, NY: McGraw-Hill.
63. Pope Francis. 2015. *Laudato si' ("Praise Be to You"): On Care for Our Common Home.* Vatican City: The Vatican.
64. International Islamic Climate Change Symposium. 2015. *Islamic Declaration on Global Climate Change.* Birmingham, UK: Islamic Foundation for Ecology and Environmental Sciences.
65. Pace, M. L. and Groffmann, P. M. 1998. *Successes, Limitations, and Frontiers in Ecosystem Science.* New York, NY: Springer.

Chapter 2

1. Bartlett, L. J., Williams, D. R., Prescott, G. W. et al. 2016. Robustness despite uncertainty: Regional climate data reveal the dominant role of humans in explaining global extinctions of Late Quaternary megafauna. *Ecography* 39: 152–161.
2. Alroy, J. 2001. A multispecies overkill simulation of the end-Pleistocene megafaunal mass extinction. *Science* 292: 1893–1896.
3. Van Dyke, F. 2008. The history and distinctions of conservation biology. Chapter 1, pages 1–28, in *Conservation Biology: Foundations, Concepts, Applications* (2nd ed.). Dordrecht, The Netherlands: Springer.
4. Hames, R. 2007. The ecologically noble savage debate. *Annual Review of Anthropology* 36: 177–190.
5. Hitzhusen, G. E. and Tucker, M. E. 2013. The potential of religion for Earth stewardship. *Frontiers in Ecology and the Environment* 11: 368–376.
6. A. Chiari. 1989. In *Proceedings of the Symposium on Biosphere Reserves, Fourth World Wilderness Congress, September 14–17, 1987, YMCA at the Rockies, Estes Park, Colorado, USA.* W. P. Gregg et al. (Eds.), p. 235. Atlanta, GA: U.S. Department of the Interior, National Park Service, https://archive.org/stream/fourthworldwilde00greg/fourthworldwilde00greg_djvu.txt
7. Coggins, C., Chevrier, J., Dwyer, M. et al. 2012. Village fengshui forests of southern China: Culture, history, and conservation status. *ASIANetwork Exchange* 19: 52–67.
8. Coggins, C. 2003. *The Tiger and the Pangolin: Nature, Culture, and Conservation in China.* Honolulu, HI: University of Hawaii Press.
9. Clarkson, L., Morrissette, V. and Regallet, G. 1992. *Our Responsibility to the Seventh Generation: Indigenous Peoples and Sustainable Development.* Winnipeg, Canada: International Institute for Sustainable Development.
10. Diamond, J. M. 2005. *Collapse: How Societies Choose to Fail or Succeed.* New York, NY: Viking.
11. Redford, K. H. 1992. The empty forest. *BioScience* 42: 412–422.
12. Ellingson, T. J. 2001. *The Myth of the Noble Savage.* Berkeley, CA: University of California Press.
13. Redford, K. H. 1991. The ecologically noble savage. *Cultural Survival Quarterly* 15: 46–48.
14. Young, C. R. 1979. *The Royal Forests of Medieval England.* Leicester, UK: Leicester University Press.
15. Grant, R. K. J. 1991. *The Royal Forests of England.* Wolfeboro Falls, NH: Alan Sutton.
16. Loyn, H. R. 1991. *Anglo-Saxon England and the Norman Conquest* (2nd ed). New York, NY: Longman.
17. Weatherford, J. M. 2004. *Genghis Khan and the Making of the Modern World.* New York, NY: Crown.
18. Young, C. R. 1978. Conservation policies in the royal forests of medieval England. *Albion: A Quarterly Journal Concerned with British Studies* 10: 95–103.
19. Barton, G. 2002. *Empire Forestry and the Origins of Environmentalism.* Cambridge, UK: Cambridge University Press.
20. Stebbing, E. P. and Champion, H. G. 1922. *The Forests of India.* London, UK: J. Lane.

21. Bobiec, A. 2002. Białowieża Primeval Forest. *International Journal of Wilderness* 8: 33.
22. Andrian, G. and Tufano, M. 2015. Biosphere reserves and protected areas: A liaison dangereuse or a mutually beneficial relationship? In R. Gambino and A. Peano (Eds.), *Nature Policies and Landscape Policies: Urban and Landscape Perspective*. Cham, Switzerland: Springer.
23. Haines, A. L. 1974. *Yellowstone National Park: Its exploration and establishment*. Washington, DC: U.S. Department of the Interior, National Park Service.
24. Foster, D. R. and Aber, J. D. 2006. *Forests in Time: The Environmental Consequences of 1,000 Years of Change in New England*. New Haven, CT: Yale University Press.
25. Ray, A. J. 1998. *Indians in the Fur Trade: Their Role as Trappers, Hunters, and Middlemen in the Lands Southwest of Hudson Bay, 1660–1870*. Toronto, Canada: University of Toronto Press.
26. Emerson, R. W. 1836. *Nature*. Boston, MA: J. Munroe and Company.
27. Thoreau, H. D. 1854. *Walden; or, Life in the Woods*. Boston, MA: Ticknor and Fields.
28. Marsh, G. P. 1864. *Man and Nature; or, Physical Geography as Modified by Human Action*. New York, NY: Charles Scribner.
29. Meine, C. 2013. Conservation movement, historical. In S. A. Levin (Ed.), *Encyclopedia of Biodiversity*. Waltham, MA: Academic Press.
30. Muir, J. 1912. *The Yosemite*. New York, NY: Century Company.
31. Muir, J. and Badé, W. F. 1916. *A Thousand Mile Walk to the Gulf*. Boston, MA: Houghton.
32. Pinchot, G. 1947. *Breaking New Ground*. New York, NY: Harcourt.
33. Callicott, J. B. 1990. Whither conservation ethics? *Conservation Biology* 4: 15–20.
34. Guha, R. 2000. *Environmentalism: A Global History*. New York, NY: Longman.
35. Leopold, A. 1939. A biotic view of land. *Journal of Forestry* 37: 727–730.
36. Leopold, A. C. 2004. Living with the land ethic. *BioScience* 54: 149–154.
37. Ehrlich, P. R. 1968. *The Population Bomb*. New York, NY: Ballantine.
38. Meadows, D. H. 1972. *The Limits to Growth: A Report for the Club of Rome's Project on the Predicament of Mankind*. Report No. 0876631650. New York, NY: Universe.
39. Meine, C., Soulé, M. and Noss, R. F. 2006. "A mission-driven discipline": The growth of conservation biology. *Conservation Biology* 20: 631–651.
40. Soulé, M. 1987. History of the Society for Conservation Biology: How and why we got here. *Conservation Biology* 1: 4–5.
41. Soulé, M. E. 1985. What is conservation biology? *BioScience* 35: 727–734.
42. Ehrlich, P. and Walker, B. 1998. Rivets and redundancy. *BioScience* 48: 387.
43. Lawton, J. H. and Brown, V. K. 1993. Redundancy in ecosystems. In E. D. Schulze and H. A. Mooney (Eds.), *Biodiversity and Ecosystem Function*. New York, NY: Springer-Verlag.
44. Polis, G. A. and Strong, D. R. 1996. Food web complexity and community dynamics. *American Naturalist* 147: 813–846.
45. Kareiva, P., Marvier, M. and Lalasz, R. 2013. Conservation in the Anthropocene. *Breakthrough Journal* 2: 1–6.
46. Jones, H. P. and Schmitz, O. J. 2009. Rapid recovery of damaged ecosystems. *PLOS ONE* 4: e5653.
47. Scheffer, M. and Carpenter, S. R. 2003. Catastrophic regime shifts in ecosystems: Linking theory to observation. *Trends in Ecology & Evolution* 18: 648–656.
48. Schroder, A., Persson, L. and De Roos, A. M. 2005. Direct experimental evidence for alternative stable states: A review. *Oikos* 110: 3–19.
49. Redford, K. and Sanjayan, M. A. 2003. Retiring Cassandra. *Conservation Biology* 17: 1473–1474.
50. Kareiva, P. and Marvier, M. 2012. What is conservation science? *BioScience* 62: 962–969.
51. Doak, D. F., Bakker, V. J., Goldstein, B. E. and Hale, B. 2014. What is the future of conservation? *Trends in Ecology & Evolution* 29: 77–81.
52. Noss, R., Nash, R., Paquet, P. and Soulé, M. 2013. Humanity's domination of nature is part of the problem: A response to Kareiva and Marvier. *BioScience* 63: 241–242.
53. Cafaro, P. and Primack, R. 2014. Species extinction is a great moral wrong. *Biological Conservation* 170: 1–2.
54. National Research Council. 2001. *Marine Protected Areas: Tools for Sustaining Ocean Ecosystem*. Washington, DC: National Academies Press.
55. IUCN. 1976. *Proceedings of an International Conference on Marine Parks and Reserves*. Morges, Switzerland: International Union for Conservation of Nature and Natural Reserves.
56. Salm, R. V., Clark, J. R. and Siirila, E. I. 2000. *Marine and Coastal Protected Areas: A Guide for Planners and Managers* (3rd ed.). Washington, DC: International Union for Conservation of Nature and Natural Reserves (IUCN).
57. Wilhelm, T. A., Sheppard, C. R. C., Sheppard, A. L. S. et al. 2014. Large marine protected areas: Advantages and challenges of going big. *Aquatic Conservation: Marine and Freshwater Ecosystems* 24: 24–30.
58. Leenhardt, P., Cazalet, B., Salvat, B. et al. 2013. The rise of large-scale marine protected areas: Conservation or geopolitics? *Ocean & Coastal Management* 85: 112–118.

Chapter 3

1. Wilson, E. O. 1988. *Biodiversity*. Washington, DC: National Academies Press.
2. Sarkar, S. 2005. *Biodiversity and Environmental Philosophy: An Introduction*. New York, NY: Cambridge University Press.
3. Millennium Ecosystem Assessment. 2005. *Ecosystems and Human Well-being: Biodiversity Synthesis*. Washington, DC: World Resources Institute.
4. Cardinale, B. J., Duffy, J. E., Gonzalez, A. et al. 2012. Biodiversity loss and its impact on humanity. *Nature* 486: 59–67.
5. Laikre, L., Allendorf, F. W., Aroner, L. C. et al. 2010. Neglect of genetic diversity in implementation of the Convention on Biological Diversity. *Conservation Biology* 24: 86–88.
6. Ayala, F. J. 1982. *Population and Evolutionary Genetics: A Primer*. Menlo Park, CA: Benjamin Cummings.
7. Norden, N., Chave, J., Belbenoit, P. et al. 2009. Interspecific variation in seedling responses to seed limitation and habitat conditions for 14 Neotropical woody species. *Journal of Ecology* 97: 186–197.
8. Bickford, D., Lohman, D. J., Sodhi, N. S. et al. 2007. Cryptic species as a window on diversity and conservation. *Trends in Ecology & Evolution* 22: 148–155.
9. Frankham, R., Ballou, J. D., Dudash, M. R. et al. 2012. Implications of different species concepts for conserving biodiversity. *Biological Conservation* 153: 25–31.
10. Hebert, P. D. N., Penton, E. H., Burns, J. M. et al. 2004. Ten species in one: DNA barcoding reveals cryptic species in the Neotropical skipper butterfly *Astraptes fulgerator*. *Proceedings of the National Academy of Sciences* 101: 14812–14817.
11. Seidel, R. A., Lang, B. K., and Berg, D. J. 2009. Phylogeographic analysis reveals multiple cryptic species of amphipods (Crustacea: Amphipoda) in Chihuahuan Desert springs. *Biological Conservation* 142: 2303–2313.
12. Ryan, M. E., Johnson, J. R., Fitzpatrick, B. M. et al. 2013. Lethal effects of water quality on threatened California salamanders but not on co-occurring hybrid salamanders. *Conservation Biology* 27: 95–102.

13. Taberlet, P., Coissac, E., Pompanon, F. et al. 2012. Towards next-generation biodiversity assessment using DNA metabarcoding. *Molecular Ecology* 21: 2045–2050.
14. Brower, J., Zar, J. and von Ende, C. 1998. *Field and Laboratory Methods for General Ecology* (4th ed.). Dubuque, IA: WCB McGraw-Hill.
15. Maurer, B. A. and McGill, B. J. 2011. Measurement of species diversity. In A. E. Magurran and B. J. McGill (Eds.), *Biological Diversity: Frontiers in Measurement and Assessment*. New York, NY: Oxford University Press.
16. Gotelli, N. J. and Colwell, R. K. 2001. Quantifying biodiversity: Procedures and pitfalls in the measurement and comparison of species richness. *Ecology Letters* 4: 379–391.
17. Colwell, R. K., Chao, A., Gotelli, N. J. et al. 2012. Models and estimators linking individual-based and sample-based rarefaction, extrapolation and comparison of assemblages. *Journal of Plant Ecology* 5: 3–21.
18. Gotelli, N. J. and Chao, A. 2013. Measuring and estimating species richness, species diversity, and biotic similarity from sampling data. In S. A. Levin (Ed.), *Encyclopedia of Biodiversity* (2nd ed.). Waltham, MA: Academic Press.
19. Simpson, E. H. 1949. Measurement of diversity. *Nature* 163: 688–688.
20. Mourier, J., Maynard, J., Parravicini, V. et al. 2016. Extreme inverted trophic pyramid of reef sharks supported by spawning groupers. *Current Biology* 26: 2011–2016.
21. Reed, D., Rassweiler, A. and Arkema, K. 2009. Density derived estimates of standing crop and net primary production in the giant kelp *Macrocystis pyrifera*. *Marine Biology* 156: 2077–2083.
22. McClanahan, T. R. and Branch, G. 2008. *Food Webs and the Dynamics of Marine Reefs*. New York, NY: Oxford University Press.
23. Schaal, G., Riera, P. and Leroux, C. 2009. Trophic significance of the kelp *Laminaria digitata* (Lamour) for the associated food web: A between-sites comparison. *Estuarine Coastal and Shelf Science* 85: 565–572.
24. Watson, J. and Estes, J. A. 2011. Stability, resilience, and phase shifts in rocky subtidal communities along the west coast of Vancouver Island, Canada. *Ecological Monographs* 81: 215–239.
25. Estes, J. A., Terborgh, J., Brashares, J. S. et al. 2011. Trophic downgrading of planet Earth. *Science* 333: 301–306.
26. Heleno, R., Devoto, M. and Pocock, M. 2012. Connectance of species interaction networks and conservation value: Is it any good to be well connected? *Ecological Indicators* 14: 7–10.
27. Dunne, J. A., Williams, R. J. and Martinez, N. D. 2002. Network structure and biodiversity loss in food webs: Robustness increases with connectance. *Ecology Letters* 5: 558–567.
28. Kouhei, M. and Morimasa, T. 2017. Structure of the food web including the endangered lycaenid butterfly *Shijimiaeoides divinus asonis* (Lepidoptera: Lycaenidae). *Entomological Science* 20: 224–234.
29. Vangergen, A. J., Woodcock, B. A., Heard, M. S. and Chapman, D. S. 2017. Network size, structure and mutualism dependence affect the propensity for plant-pollinator extinction cascades. *Functional Ecology* 31: 1285–1293.
30. de Visser, S. N., Freymann, B. P. and Han, O. 2011. The Serengeti food web: Empirical quantification and analysis of topological changes under increasing human impact. *Journal of Animal Ecology* 80: 484–494.
31. McDonald-Madden, E., Sabbadin, R., Game, E. T. et al. 2016. Using food-web theory to conserve ecosystems. *Nature Communications* 7: 10245.
32. Albert, C. H., Rayfield, B., Dumitru, M. and Gonzalez, A. 2017. Applying network theory to prioritize multispecies habitat networks that are robust to climate and land-use change. *Conservation Biology* 31: 1383–1396.
33. Gravel, D., Canard, E., Guichard, F. and Mouquet, N. 2011. Persistence increases with diversity and connectance in trophic metacommunities. *PLOS ONE* 6: e19374.
34. Pires, M. M. 2017. Rewilding ecological communities and rewiring ecological networks. *Perspectives in Ecology and Conservation* 15: 257–265.
35. Jones, C. G., Lawton, J. H. and Shachak, M. 1994. Organisms as ecosystem engineers. *Oikos* 69: 373–386.
36. Wright, J. P., Jones, C. G. and Flecker, A. S. 2002. An ecosystem engineer, the beaver, increases species richness at the landscape scale. *Oecologia* 132: 96–101.
37. Power, M. E., Tilman, D., Estes, J. A. et al. 1996. Challenges in the quest for keystones. *BioScience* 46: 609–620.
38. Paine, R. T. 1966. Food web complexity and species diversity. *American Naturalist* 100: 65–75.
39. Nunez-Iturri, G., Olsson, O. and Howe, H. F. 2008. Hunting reduces recruitment of primate-dispersed trees in Amazonian Peru. *Biological Conservation* 141: 1536–1546.
40. Law, A., Gaywood, M. J., Jones, K. C. et al. 2017. Using ecosystem engineers as tools in habitat restoration and rewilding: Beaver and wetlands. *Science of the Total Environment* 605–606: 1021–1030.
41. Beavers will return to the Forest of Dean for the first time in 400 years. December 8, 2017. https://www.bbc.com/news/uk-england-gloucestershire-42278606.
42. Whittaker, R. H. 1960. Vegetation of the Siskiyou Mountains, Oregon and California. *Ecological Monographs* 30: 279–338.
43. Whittaker, R. H. 1972. Evolution and measurement of species diversity. *Taxon* 21: 213–251.
44. Dornelas, M., Gotelli, N. J., McGill, B. et al. 2014. Assemblage time series reveal biodiversity change but not systematic loss. *Science* 344: 296–299.

Chapter 4

1. Saal, A. E., Hauri, E. H., Lo Cascio, M. et al. 2008. Volatile content of lunar volcanic glasses and the presence of water in the moon's interior. *Nature* 454: 192–195.
2. Carr, M. H., Belton, M. J. S., Chapman, C. R. et al. 1998. Evidence for a subsurface ocean on Europa. *Nature* 391: 363–365.
3. Malin, M. C. and Edgett, K. S. 2000. Evidence for recent groundwater seepage and surface runoff on Mars. *Science* 288: 2330–2335.
4. Borucki, W. J., Koch, D. G., Basri, G. et al. 2011. Characteristics of planetary candidates observed by Kepler II: Analysis of the first four months of data. *Astrophysical Journal* 736: 1–22.
5. Baker, V. R. 2008. Planetary landscape systems: A limitless frontier. *Earth Surface Processes and Landforms* 33: 1341–1353.
6. Dietrich, W. E. and Perron, J. T. 2006. The search for a topographic signature of life. *Nature* 439: 411–418.
7. Dalrymple, G. B. 1991. *The Age of the Earth*. Stanford, CA: Stanford University Press.
8. Noffke, N., Christian, D., Wacey, D. and Hazen, R. M. 2013. Microbially induced sedimentary structures recording an ancient ecosystem in the ca. 3.48 billion-year-old dresser formation, Pilbara, Western Australia. *Astrobiology* 13: 1103–1124.
9. Nutman, A. P., Bennett, V. C., Friend, C. R. L. et al. 2016. Rapid emergence of life shown by discovery of 3,700-million-year-old microbial structures. *Nature* 537: 535–538.
10. Dodd, M. S., Papineau, D., Grenne, T. et al. 2017. Evidence for early life in Earth's oldest hydrothermal vent precipitates. *Nature* 543: 60–64.
11. Fox, D. 2016. What sparked the Cambrian explosion? *Nature* 530: 268–270.
12. May, R. M., Lawton, J. H. and Stork, S. E. 1995. Assessing extinction rates. In J. H. Lawton and R. M. May (Eds.), *Extinction Rates*. New York, NY: Oxford University Press.

13. Pimm, S. L., Russell, G. J., Gittleman, J. L. and Brooks, T. M. 1995. The future of biodiversity. *Science* 269: 347–350.
14. Barnosky, A. D., Matzke, N., Tomiya, S. et al. 2011. Has the Earth's sixth mass extinction already arrived? *Nature* 471: 51–57.
15. Alroy, J. 1996. Constant extinction, constrained diversification, and uncoordinated stasis in North American mammals. *Palaeogeography, Palaeoclimatology, Palaeoecology* 127: 285–311.
16. Harnik, P. G., Lotze, H. K., Anderson, S. C. et al. 2012. Extinctions in ancient and modern seas. *Trends in Ecology & Evolution* 27: 608–617.
17. De Vos, J. M., Joppa, L. N., Gittleman, J. L. et al. 2015. Estimating the normal background rate of species extinction. *Conservation Biology* 29: 452–462.
18. Christie, M., Holland, S. M. and Bush, A. M. 2013. Contrasting the ecological and taxonomic consequences of extinction. *Paleobiology* 39: 538–559.
19. Solé, R. V. and Newman, M. 2002. *Encyclopedia of Global Environmental Change*, Vol. 2, pages 297–391. Hoboken, NJ: John Wiley & Sons.
20. Caplan, M. L. and Bustin, R. M. 1999. Devonian-Carboniferous Hangenberg mass extinction event, widespread organic-rich mudrock and anoxia: Causes and consequences. *Palaeogeography, Palaeoclimatology, Palaeoecology* 148: 187–207.
21. McGhee, G. R. 1996. *The Late Devonian Mass Extinction: The Frasnian/Famennian Crisis.* New York, NY: Columbia University Press.
22. Benton, M. J. 2003. *When Life Nearly Died: The Greatest Mass Extinction of All Time.* London, UK: Thames & Hudson.
23. Sahney, S. and Benton, M. J. 2008. Recovery from the most profound mass extinction of all time. *Proceedings of the Royal Society B: Biological Sciences* 275: 759–765.
24. Labandeira, C. and Sepkoski, J. 1993. Insect diversity in the fossil record. *Science* 261: 310–315.
25. Benton, M. J. and Twitchett, R. J. 2003. How to kill (almost) all life: The end-Permian extinction event. *Trends in Ecology & Evolution* 18: 358–365.
26. Kirchner, J. W. and Weil, A. 2000. Delayed biological recovery from extinctions throughout the fossil record. *Nature* 404: 177–180.
27. Ceballos, G., Ehrlich, P. R. and Dirzo, R. 2017. Biological annihilation via the ongoing sixth mass extinction signaled by vertebrate population losses and declines. *Proceedings of the National Academy of Sciences* 114: E6089–E6096.
28. Costello, M. J., May, R. M. and Stork, N. E. 2013. Can we name Earth's species before they go extinct? *Science* 339: 413–416.
29. Mora, C., Tittensor, D. P., Adl, S. et al. 2011. How many species are there on Earth and in the ocean? *PLOS Biology* 9: e1001127.
30. Joppa, L. N., Roberts, D. L. and Pimm, S. L. 2011. The population ecology and social behaviour of taxonomists. *Trends in Ecology & Evolution* 26: 551–553.
31. May, R. M. 2010. Tropical arthropod species, more or less? *Science* 329: 41–42.
32. Erwin, T. L. 1982. Tropical forests: Their richness in Coleoptera and other arthropod species. *Coleopterists Bulletin* 36: 74–75.
33. Gaston, K. J. and Spicer, J. I. 1998. *Biodiversity: An Introduction.* Malden, MA: Blackwell Science.
34. Locey, K. J. and Lennon, J. T. 2016. Scaling laws predict global microbial diversity. *Proceedings of the National Academy of Sciences* 113: 5970–5975.
35. Holdridge, L. R. 1967. *Life Zone Ecology.* San Jose, CR: Tropical Science Center.
36. Whittaker, R. H. 1962. Classification of natural communities. *Botanical Review* 28: 1–239.
37. Whittaker, R. H. 1967. Gradient analysis of vegetation. *Biological Reviews* 42: 207–264.
38. Olson, D. M., Dinerstein, E., Wikramanayake, E. D. et al. 2001. Terrestrial ecoregions of the world: A new map of life on Earth. *Bioscience* 51: 933–938.
39. Corlett, R. and Primack, R. B. 2011. *Tropical Rain Forests: An Ecological and Biogeographical Comparison* (2nd ed). Hoboken, NJ: Wiley-Blackwell.
40. Grassle, J. F. 2001. Marine ecosystems. In S. A. Levin (Ed.), *Encyclopedia of Biodiversity.* San Diego, CA: Academic Press.
41. Spalding, M., Ravilious, C. and Green, E. P. 2001. *World Atlas of Coral Reefs.* Berkeley, CA: University of California Press.
42. Strayer, D. L. and Dudgeon, D. 2010. Freshwater biodiversity conservation: Recent progress and future challenges. *Journal of the North American Benthological Society* 29: 344–358.
43. Mooney, H. A. 1988. Lessons from Mediterranean-climate regions. In E. O. Wilson (Ed.), *Biodiversity.* Washington, DC: The National Academies Press.
44. Myers, N., Mittermeier, R. A., Mittermeier C. G. et al. 2000. Biodiversity hotspots for conservation priorities. *Nature* 403: 853–858.
45. Willig, M. R., Kaufman, D. M. and Stevens, R. D. 2003. Latitudinal gradients of biodiversity: Pattern, process, scale, and synthesis. *Annual Review of Ecology, Evolution, and Systematics* 34: 273–309.
46. Tittensor, D. P., Mora, C., Jetz, W. et al. 2010. Global patterns and predictors of marine biodiversity across taxa. *Nature* 466: 1098–1101.
47. Rombouts, I., Beaugrand, G., Ibañez, F. et al. 2009. Global latitudinal variations in marine copepod diversity and environmental factors. *Proceedings of the Royal Society B: Biological Sciences* 276: 3053–3062.
48. Morey, D. F. 1994. The early evolution of the domestic dog. *American Scientist* 82: 336–347.
49. Lange, K. E. 2002. Wolf to woof: The evolution of dogs. *National Geographic* Vol. 201: 2
50. Lenski, R. E. and Travisano, M. 1994. Dynamics of adaptation and diversification: A 10,000-generation experiment with bacterial populations. *Proceedings of the National Academy of Sciences* 91: 6808–6814.
51. Alyokhin, A., Baker, M., Mota-Sanchez, D. et al. 2008. Colorado potato beetle resistance to insecticides. *American Journal of Potato Research* 85: 395–413.
52. Masel, J. 2011. Genetic drift. *Current Biology* 21: R837–R838.
53. Nei, M., Maruyama, T. and Chakraborty, R. 1975. The bottleneck effect and genetic variability in populations. *Evolution* 29: 1–10.
54. Barton, N. H. and Charlesworth, B. 1984. Genetic revolutions, founder effects, and speciation. *Annual Review of Ecology and Systematics* 15: 133–164.
55. Coyne, J. A. and Orr, H. A. 2004. *Speciation.* Sunderland, MA: Oxford University Press/Sinauer.
56. Yoshida, T., Jones, L. E., Ellner, S. P. et al. 2003. Rapid evolution drives ecological dynamics in a predator-prey system. *Nature* 424: 303.
57. Bassar, R. D., Marshall, M. C., López-Sepulcre, A. et al. 2010. Local adaptation in Trinidadian guppies alters ecosystem processes. *Proceedings of the National Academy of Sciences* 107: 3616–3621.
58. Beckerman, A. P., Childs, D. Z. and Bergland, A. O. 2016. Eco-evolutionary biology: Feeding and feedback loops. *Current Biology* 26: R161–R164.
59. Schoener, T. W. 2011. The newest synthesis: Understanding the interplay of evolutionary and ecological dynamics. *Science* 331: 426–429.
60. Armstrong, R. A. and McGehee, R. 1980. Competitive exclusion. *American Naturalist* 115: 151–170.
61. Gause, G. F. 1934. *The Struggle for Existence.* Baltimore, MD: Williams and Wilkins.

62. Connell, J. H. 1979. Intermediate-disturbance hypothesis. *Science* 204: 1345.
63. Chesson, P. and Huntly, N. 1997. The roles of harsh and fluctuating conditions in the dynamics of ecological communities. *American Naturalist* 150: 519–553.
64. Mackey, R. L. and Currie, D. J. 2001. The diversity-disturbance relationship: Is it generally strong and peaked? *Ecology* 82: 3479–3492.
65. Fox, J. W. 2013. The intermediate disturbance hypothesis should be abandoned. *Trends in Ecology & Evolution* 28: 86–92.
66. Paine, R. T. 1966. Food web complexity and species diversity. *American Naturalist* 100: 65–75.
67. Yu, D. W. and Wilson, H. B. 2001. The competition-colonization trade-off is dead: Long live the competition-colonization trade-off. *American Naturalist* 158: 49–63.
68. MacArthur, R. H. and Wilson, E. O. 1967. *The Theory of Island Biogeography*. Princeton, NJ: Princeton University Press.
69. Inoue, J. G., M. Miya, B. Venkatesh, and M. Nishida. 2005. The mitochondrial genome of Indonesian coelacanth *Latimeria menadoensis* (Sarcopterygii: Coelacanthiformes) and divergence time estimation between the two coelacanths. *Gene* 11: 227–235.

Chapter 5

1. Pascual, U., Balvanera, P., Díaz, S. et al. 2017. Valuing nature's contributions to people: The IPBES approach. *Current Opinion in Environmental Sustainability* 26–27: 7–16.
2. Chan, K. M. A., Balvanera, P., Benessaiah, K. et al. 2016. Opinion: Why protect nature? Rethinking values and the environment. *Proceedings of the National Academy of Sciences* 113: 1462–1465.
3. Diaz, S., Demissew, S., Carabias, J. et al. 2015. The IPBES Conceptual Framework: Connecting nature and people. *Current Opinion in Environmental Sustainability* 14: 1–16.
4. UN General Assembly. 1972. *United Nations Conference on the Human Environment*. Stockholm, Sweden: United Nations.
5. IUCN. 1980. *World Conservation Strategy: Living Resource Conservation for Sustainable Development*. Report No. ISBN 2-88032-104-2. Gland, Switzerland: International Union for Conservation of Nature and Natural Resources (IUCN).
6. Earth Charter Initiative. 2000. *The Earth Charter*. San Jose, Costa Rica: University for Peace. http://earthcharter.org/discover/the-earth-charter/
7. UN. 2012. *The Future We Want*. Rio de Janiero, Brazil: United Nations.
8. UN. 2015. *Transforming Our World*. New York, NY: United Nations.
9. Vidal, J. 2011. Bolivia enshrines natural world's rights with equal status for Mother Earth. April 10. https://www.theguardian.com/environment/2011/apr/10/bolivia-enshrines-natural-worlds-rights.
10. Soulé, M. E. 1985. What is conservation biology? *BioScience* 35: 727–734.
11. Carr, R. 1976. *English Fox Hunting: A History*. London, UK: Weidenfeld and Nicolson.
12. Burns, L., Edwards, D., Marsh, J. et al. 2000. *The Final Report of the Committee of Inquiry into Hunting with Dogs in England and Wales*. London, UK: The Stationery Office.
13. Hitzhusen, G. E. and Tucker, M. E. 2013. The potential of religion for Earth stewardship. *Frontiers in Ecology and the Environment* 11: 368–376.
14. World Wildlife Fund–India. 1999. *Religion and Conservation* (1st ed.). New Delhi, India: Full Circle.
15. Kao, G. Y. 2014. Creaturely solidarity: Rethinking human-nonhuman relations. *Journal of Religious Ethics* 42: 743–768.
16. Hodges, J. 2005. *Ethics, Morals and Law Relative to Animals*. Wageningen, The Netherlands: Wageningen Academic Publishers.
17. Raymond, C. M., Singh, G. G., Benessaiah, K. et al. 2013. Ecosystem services and beyond: Using multiple metaphors to understand human-environment relationships. *BioScience* 63: 536–546.
18. Piccolo, J. J. 2017. Intrinsic values in nature: Objective good or simply half of an unhelpful dichotomy? *Journal for Nature Conservation* 37: 8–11.
19. Daily, G. C. 1997. *Nature's Services: Societal Dependence on Natural Ecosystems*. Washington, DC: Island Press.
20. Barbier, E. B. 2011. Pricing nature. *Annual Review of Resource Economics* 3: 337–353.
21. Barbier, E. B. 2007. Valuing ecosystem services as productive inputs. *Economic Policy*: 178–229.
22. Fisher, B., Naidoo, R. and Ricketts, T. H. 2015. *A Field Guide to Economics for Conservationists*. New York, NY: W. H. Freeman & Company.
23. Barbier, E. 2011. *Capitalizing on Nature: Ecosystems as Natural Assets*. Cambridge, UK: Cambridge University Press.
24. Walsh, R. G., Loomis, J. B. and Gillman, R. A. 1984. Valuing option, existence, and bequest demands for wilderness. *Land Economics* 60: 14–29.
25. Klain, S. C., Olmsted, P., Chan, K. M. A. and Satterfield, T. 2017. Relational values resonate broadly and differently than intrinsic or instrumental values, or the New Ecological Paradigm. *PLOS ONE* 12: e0183962.
26. Ryan, R. M. and Deci, E. L. 2001. On happiness and human potentials: A review of research on hedonic and eudaimonic well-being. *Annual Review of Psychology* 52: 141–166.
27. Pope Francis. 2015. *Laudato si' ("Praise Be to You"): On Care for Our Common Home*. Vatican City: The Vatican.
28. CRT. 2015. *The Case for Responsible Travel: Trends & Statistics 2015*. Washington, DC: Center for Responsible Travel.
29. Strietska-Ilina, O., Hofmann, C., Haro, M. D. and Jeon, S. 2011. *Skills for Green Jobs: A Global View*. Geneva, Switzerland: International Labour Office.
30. OIA. 2012. *The Outdoor Recreation Economy*. Boulder, CO: Outdoor Industry Association.
31. Comley, V. and Mackintosh, C. 2014. *The Economic Impact of Outdoor Recreation in the UK: The Evidence*. London, UK: Sport and Recreation Alliance.
32. Arlinghaus, R., Tillner, R. and Bork, M. 2015. Explaining participation rates in recreational fishing across industrialised countries. *Fisheries Management and Ecology* 22: 45–55.
33. Verschuuren, B. 2010. *Sacred Natural Sites: Conserving Nature and Culture*. Washington, DC: Earthscan.
34. Gould, R. K., Klain, S. C., Ardoin, N. M. et al. 2015. A protocol for eliciting nonmaterial values through a cultural ecosystem services frame. *Conservation Biology* 29: 575–586.
35. Berghoefer, U., Rozzi, R. and Jax, K. 2010. Many eyes on nature: Diverse perspectives in the Cape Horn Biosphere Reserve and their relevance for conservation. *Ecology and Society* 15: 34.
36. Robinson, J. G. 2011. Ethical pluralism, pragmatism, and sustainability in conservation practice. *Biological Conservation* 144: 958–965.
37. Ogden, L., Heynen, N., Oslender, U. et al. 2013. Global assemblages, resilience, and Earth Stewardship in the Anthropocene. *Frontiers in Ecology and the Environment* 11: 341–347.
38. Fromm, E. 1964. *The Heart of Man, Its Genius for Good and Evil*. New York, NY: Harper & Row.
39. Wilson, E. O. 1984. *Biophilia*. Cambridge, MA: Harvard University Press.
40. Pergams, O. R. W. and Zaradic, P. A. 2006. Is love of nature in the US becoming love of electronic media? 16-year downtrend in national park visits explained by watching movies, playing video games, internet use, and oil prices. *Journal of Environmental Management* 80: 387–393.

41. Mayer, F. S. and Frantz, C. M. 2004. The connectedness to nature scale: A measure of individuals' feeling in community with nature. *Journal of Environmental Psychology* 24: 503–515.

42. Stinson, J. 2017. Re-creating wilderness 2.0: Or getting back to work in a virtual nature. *Geoforum* 79: 174–187.

43. Fuller, R. A., Irvine, K. N., Devine-Wright, P. et al. 2007. Psychological benefits of greenspace increase with biodiversity. *Biology Letters* 3: 390–394.

44. Luck, G. W., Davidson, P., Boxall, D. and Smallbone, L. 2011. Relations between urban bird and plant communities and human well-being and connection to nature. *Conservation Biology* 25: 816–826.

45. Attfield, R. 2003. *Environmental Ethics: An Overview for the Twenty-First Century*. Cambridge, UK: Polity Press.

46. Minteer, B. and Collins, J. 2008. From environmental to ecological ethics: Toward a practical ethics for ecologists and conservationists. *Science and Engineering Ethics* 14: 483–501.

47. White, L. 1967. The historical roots of our ecologic crisis. *Science* 155: 1203–1207.

48. Whitney, E. 2015. Lynn White Jr.'s "The historical roots of our ecologic crisis" after 50 years. *History Compass* 13: 396–410.

49. Taylor, B., Van Wieren, G. and Zaleha, B. D. 2016. Lynn White Jr. and the greening-of-religion hypothesis. *Conservation Biology* 30: 1000–1009.

50. Hoffman, A. and Sandelands, L. 2005. Getting right with nature: Anthropocentrism, ecocentrism, and theocentrism. *Organization & Environment* 18: 141–162.

51. Curry, P. 2011. *Ecological Ethics: An Introduction* (2nd ed.). Cambridge, UK: Polity Press.

52. Chapple, C. K. and Tucker, M. E. 2000. *Hinduism and Ecology: The Intersection of Earth, Sky, and Water.* Cambridge, MA: Harvard University Press.

53. Narayan, R. and Kumar, J. 2003. *Ecology and Religion: Ecological Concepts in Hinduism, Buddhism, Jainism, Islam, Christianity and Sikhism*. Muzaffarpur, India: Deep & Deep.

54. Palmer, M. and Finlay, V. 2003. *Faith in Conservation: New Approaches to Religions and the Enviornment.* Washington, DC: World Bank, pp. 77–82.

55. Schweitzer, A. 1969. *Reverence for Life*. New York, NY: Harper & Row.

56. Singer, P. 1975. *Animal Liberation: A New Ethics for Our Treatment of Animals*. New York, NY: Random House.

57. Taylor, P. W. 1986. *Respect for Nature: A Theory of Environmental Ethics*. Princeton, NJ: Princeton University Press.

58. Watson, R. A. 1983. A critique of anti-anthropocentric biocentrism. *Environmental Ethics* 5: 245–256.

59. Callicott, J. B. 1989. *In Defense of the Land Ethic: Essays in Environmental Philosophy*. Albany, NY: State University of New York Press.

60. Naess, A. 1973. Shallow and deep, long-range ecology movement: Summary. *Inquiry: An Interdisciplinary Journal of Philosophy* 16: 95–100.

61. Capra, F. 1997. *The Web of Life: A New Scientific Understanding of Living Systems.* Parklands, RSA: Random House South Africa.

62. Worm, B., Davis, B., Kettemer, L. et al. 2013. Global catches, exploitation rates, and rebuilding options for sharks. *Marine Policy* 40: 194–204.

63. Dulvy, N. K., Fowler, S. L., Musick, J. A. et al. 2014. Extinction risk and conservation of the world's sharks and rays. *eLife* 3: e00590.

64. Hinckley, S. 2016. How Yao Ming appeased the Chinese appetite for shark fin soup. June 2. https://www.csmonitor.com/World/Global-News/2016/0602/How-Yao-Ming-appeased-the-Chinese-appetite-for-shark-fin-soup.

65. Denyer, S. 2013. In China, victory for wildlife conservation as citizens persuaded to give up shark fin soup. October 19. https://www.washingtonpost.com/world/in-china-victory-for-wildlife-conservation-as-citizens-persuaded-to-give-up-shark-fin-soup/2013/10/19/e8181326-3646-11e3-89db-8002ba99b894_story.html?noredirect=on&utm_term=.23a0d9da6bb1.

Chapter 6

1. Cohen, J. E. and Tilman, D. 1996. Biosphere 2 and biodiversity: The lessons so far. *Science* 274: 1150–1151.

2. Millennium Ecosystem Assessment. 2005. *Ecosystems and Human Well-being: Synthesis*. Washington, DC: Island Press.

3. Daily, G. C., Alexander, S., Ehrlich, P. R. et al. 1997. *Ecosystem Services: Benefits Supplied to Human Societies by Natural Ecosystems*. Washington, DC: Ecological Society of America.

4. Gómez-Baggethun, E., de Groot, R., Lomas, P. L. and Montes, C. 2010. The history of ecosystem services in economic theory and practice: From early notions to markets and payment schemes. *Ecological Economics* 69: 1209–1218.

5. Willis, A. J. 1997. Forum—The ecosystem: An evolving concept viewed historically. *Functional Ecology* 11: 268–271.

6. Odum, E. P. 1953. *Fundamentals of Ecology*. Philadelphia, PA: Saunders.

7. Bormann, F. H. and Likens, G. E. 1979. Catastrophic disturbance and the steady-state in northern hardwood forests. *American Scientist* 67: 660–669.

8. Muth, R. F. 1965. Review. Scarcity and Growth: The Economics of Natural Resource Availability. In H. J. Barnett and C. Morse. *Economic Development and Cultural Change* 14: 113–117.

9. Trucost. 2013. *Natural Capital at Risk: The Top 100 Externalities of Business*. (For TEEB: The Economics of Ecosystems and Biodiversity.) London, UK: Trucost.

10. Costanza, R. 2003. *The early history of Ecological Economics and the International Society for Ecological Economics (ISEE).* International Society for Ecological Economics Internet Encyclopaedia of Ecological Economics. https://www.researchgate.net/publication/237547562.

11. Lele, S., Springate-Baginski, O., Lakerveld, R. et al. 2013. Ecosystem services: Origins, contributions, pitfalls, and alternatives. *Conservation and Society* 11: 343–358.

12. Wilson, C. L. and Matthews, W. H. 1972. *Man's Impact on the Global Environment: Assessment and Recommendations for Action* (5th ed.). Williamstown, MA: MIT Press.

13. Schumacher, E. F. 1973. *Small Is Beautiful; Economics As If People Mattered*. New York, NY: Harper & Row.

14. Ehrlich, P. R. and Mooney, H. A. 1983. Extinction, substitution, and ecosystem services. *BioScience* 33: 248–254.

15. Daily, G. C. 1997. *Nature's Services: Societal Dependence on Natural Ecosystems*. Washington, DC: Island Press.

16. Costanza, R., dArge, R., deGroot, R. et al. 1997. The value of the world's ecosystem services and natural capital. *Nature* 387: 253–260.

17. Chan, K. M. A., Balvanera, P., Benessaiah, K. et al. 2016. Opinion: Why protect nature? Rethinking values and the environment. *Proceedings of the National Academy of Sciences* 113: 1462–1465.

18. Jones, C. G., Lawton, J. H. and Shachak, M. 1994. Organisms as ecosystem engineers. *Oikos* 69: 373–386.

19. Sterner, R. W. and Elser, J. J. 2002. *Ecological Stoichiometry: The Biology of Elements from Molecules to the Biosphere*. Princeton, NJ: Princeton University Press.

20. Power, M. E., Tilman, D., Estes, J. A. et al. 1996. Challenges in the quest for keystones. *BioScience* 46: 609–620.

21. Schulze, E. D. and Mooney, H. A. 1993. *Biodiversity and Ecosystem Function*. New York, NY: Springer-Verlag.

22. Heywood, V. H., Watson, R. T. and United Nations Environment Programme.1995. *Global Biodiversity Assessment*. New York, NY: Cambridge University Press.

23. Loreau, M., Arroyo, M., Dirzo, R. et al. 2002. *DIVERSITAS Science Plan*.

24. Loreau, M., Naeem, S., Inchausti, P. et al. 2001. Biodiversity and ecosystem functioning: Current knowledge and future challenges. *Science* 294: 804–808.
25. Naeem, S. 2002. Ecosystem consequences of biodiversity loss: The evolution of a paradigm. *Ecology* 83: 1537–1552.
26. Braat, L. C. and de Groot, R. 2012. The ecosystem services agenda: Bridging the worlds of natural science and economics, conservation and development, and public and private policy. *Ecosystem Services* 1: 4–15.
27. Salzman, J. 2005. Creating markets for ecosystem services: Notes from the field. *New York University Law Review* 80: 870–961.
28. Larigauderie, A. 2015. The Intergovernmental Platform on Biodiversity and Ecosystem Services (IPBES): A call to action. *Gaia—Ecological Perspectives for Science and Society* 24: 73–73.
29. Chapin, F. S. and Mooney, H. 2002. *Principles of Terrestrial Ecosystem Ecology*. New York, NY: Springer.
30. Naiman, R. J., Bilby, R. E., Schindler, D. E. and Helfield, J. M. 2002. Pacific salmon, nutrients, and the dynamics of freshwater and riparian ecosystems. *Ecosystems* 5: 399–417.
31. Loftas, T. and Ross, J. 1995. *Dimensions of Need: An Atlas of Food and Agriculture*. Rome, Italy: Food and Agriculture Organization of the United Nations.
32. FAO. 2016. *The State of Food and Agriculture 2016*. Rome, Italy: United Nations.
33. Foley, J. A., Ramankutty, N., Brauman, K. A. et al. 2011. Solutions for a cultivated planet. *Nature* 478: 337–342.
34. Chivian, E. and Bernstein, A. 2008. *Sustaining Life: How Human Health Depends on Biodiversity*. New York, NY: Oxford University Press.
35. FAO. 2015. *Statistical Pocketbook: World Food and Agriculture*. Rome, Italy: United Nations.
36. FAO. 2016. *The State of the World's Fisheries and Aquaculture 2016*. Rome, Italy: United Nations.
37. Schlesinger, W. H. and Jasechko, S. 2014. Transpiration in the global water cycle. *Agricultural and Forest Meteorology* 189–190: 115–117.
38. Huntington, T. G. 2006. Evidence for intensification of the global water cycle: Review and synthesis. *Journal of Hydrology* 319: 83–95.
39. Brown, A. E., Zhang, L., McMahon, T. A. et al. 2005. A review of paired catchment studies for determining changes in water yield resulting from alterations in vegetation. *Journal of Hydrology* 310: 28–61.
40. Cui, X., Graf, H. F., Langmann, B. et al. 2007. Hydrological impacts of deforestation on the Southeast Tibetan Plateau. *Earth Interactions* 11: 1–18.
41. Zhang, P., Shao, G., Zhao, G. et al. 2000. China's forest policy for the 21st century. *Science* 288: 2135–2136.
42. Corlett, R. and Primack, R. B. 2011. *Tropical Rain Forests: An Ecological and Biogeographical Comparison* (2nd ed.). Hoboken, NJ: Wiley-Blackwell.
43. FAO. 2016. *State of the World's Forests 2016*. Rome, Italy: United Nations.
44. FAO. 2016. *2015 Global Forest Products Facts and Figures*. Rome, Italy: United Nations.
45. Crichton, R. R. 2012. Biomineralisation. Chapter 19, in *Biological Inorganic Chemistry* (2nd ed.). Oxford, UK: Elsevier.
46. Newsome, L., Morris, K. and Lloyd, J. R. 2014. The biogeochemistry and bioremediation of uranium and other priority radionuclides. *Chemical Geology* 363: 164–184.
47. Shanley, P. and Luz, L. 2003. The impacts of forest degradation on medicinal plant use and implications for health care in eastern Amazonia. *BioScience* 53: 573–584.
48. Dias, D. A., S. Urban and Roessner, U. 2012. A historical overview of natural products in drug discovery. *Metabolites* 2: 303–336.
49. Seca, A. M. L. and Pinto, D. C. G. A. 2018. Plant secondary metabolites as anticancer agents: Successes in clinical trials and therapeutic application. *International Journal of Molecular Sciences* 19: 263.
50. Holland, P. M., Abramson, R. D., Watson, R. and Gelfand, D. H. 1991. Detection of specific polymerase chain-reaction product by utilizing the 5′–3′ exonuclease activity of *Thermus aquaticus* DNA-polymerase. *Proceedings of the National Academy of Sciences* 88: 7276–7280.
51. West, P. C., Narisma, G. T., Barford, C. C. et al. 2011. An alternative approach for quantifying climate regulation by ecosystems. *Frontiers in Ecology and the Environment* 9: 126–133.
52. McKinley, D. C., Ryan, M. G., Birdsey, R. A. et al. 2011. A synthesis of current knowledge on forests and carbon storage in the United States. *Ecological Applications* 21: 1902–1924.
53. Hungate, B. A., Barbier, E. B., Ando, A. W. et al. 2017. The economic value of grassland species for carbon storage. *Science Advances* 3: 1–8.
54. Nazareno, A. G. and Laurance, W. F. 2015. Brazil's drought: Beware deforestation. *Science* 347: 1427.
55. Marengo, J. A. and Espinoza, J. C. 2016. Extreme seasonal droughts and floods in Amazonia: Causes, trends and impacts. *International Journal of Climatology* 36: 1033–1050.
56. Bagley, J. E., Desai, A. R., Harding, K. J. et al. 2014. Drought and deforestation: Has land cover change influenced recent precipitation extremes in the Amazon? *Journal of Climate* 27: 345–361.
57. Simpson, J. R. and McPherson, E. G. 1996. Potential of tree shade for reducing residential energy use in California. *Journal of Arboriculture* 22: 10–18.
58. Huang, Y., Akbari, H., Taha, H. and Rosenfeld, A. H. 1987. The potential of vegetation in reducing summer cooling loads in residential buildings. *Journal of Climate and Applied Meteorology* 26: 1103–1116.
59. Nikoofard, S., Ugursal, V. I. and Beausoleil-Morrison, I. 2011. Effect of external shading on household energy requirement for heating and cooling in Canada. *Energy and Buildings* 43: 1627–1635.
60. Allen, D. C., Cardinale, B. J. and Wynn-Thompson, T. 2014. Integrating ecological principles into interdisciplinary ecogeoscience research. *BioScience* 64: 444–454.
61. Lovas, S. M. and Torum, A. 2001. Effect of the kelp *Laminaria hyperborea* upon sand dune erosion and water particle velocities. *Coastal Engineering* 44: 37–63.
62. Albertson, L. K. and Allen, D. C. 2015. Meta-analysis: Abundance, behavior, and hydraulic energy shape biotic effects on sediment transport in streams. *Ecology* 96: 1329–1339.
63. Albertson, L. K., Sklar, L. S., Pontau, P. et al. 2014. A mechanistic model linking insect (Hydropsychidae) silk nets to incipient sediment motion in gravel-bedded streams. *Journal of Geophysical Research—Earth Surface* 119: 1833–1852.
64. Lin, Q. and Wang, Y. 2018. Spatial and temporal analysis of a fatal landslide inventory in China from 1950 to 2016. *Landslides* 15: 2357–2372.
65. Losey, J. E. and Vaughan, M. 2006. The economic value of ecological services provided by insects. *BioScience* 56: 311–323.
66. Wanger, T. C., Darras, K., Bumrungsri, S. et al. 2014. Bat pest control contributes to food security in Thailand. *Biological Conservation* 171: 220–223.
67. Maas, B., Clough, Y. and Tscharntke, T. 2013. Bats and birds increase crop yield in tropical agroforestry landscapes. *Ecology Letters* 16: 1480–1487.
68. Boyles, J. G., Cryan, P. M., McCracken, G. F. and Kunz, T. H. 2011. Economic importance of bats in agriculture. *Science* 332: 41–42.
69. Kamareddine, L. 2012. The biological control of the malaria vector. *Toxins* 4: 748–767.

70. Barbier, E. B., Koch, E. W., Silliman, B. R. et al. 2008. Coastal ecosystem-based management with nonlinear ecological functions and values. *Science* 319: 321–323.

71. Ewel, K. C. 2010. Appreciating tropical coastal wetlands from a landscape perspective. *Frontiers in Ecology and the Environment* 8: 20–26.

72. Narayan, S., Beck, M. W., Wilson, P. et al. 2017. The value of coastal wetlands for flood damage reduction in the northeastern USA. *Scientific Reports* 7: 9463.

73. Arkema, K. K., Guannel, G., Verutes, G. et al. 2013. Coastal habitats shield people and property from sea-level rise and storms. *Nature Climate Change* 3: 913.

74. Ollerton, J., Winfree, R. and Tarrant, S. 2011. How many flowering plants are pollinated by animals? *Oikos* 120: 321–326.

75. Klein, A. M., Vaissiere, B. E., Cane, J. H. et al. 2007. Importance of pollinators in changing landscapes for world crops. *Proceedings of the Royal Society B: Biological Sciences* 274: 303–313.

76. Aizen, M. A., Garibaldi, L. A., Cunningham, S. A. and Klein, A. M. 2009. How much does agriculture depend on pollinators? Lessons from long-term trends in crop production. *Annals of Botany* 103: 1579–1588.

77. Hanley, N., Breeze, T. D., Ellis, C. and Goulson, D. 2015. Measuring the economic value of pollination services: Principles, evidence and knowledge gaps. *Ecosystem Services* 14: 124–132.

78. Garibaldi, L. A., Steffan-Dewenter, I., Winfree, R. et al. 2013. Wild pollinators enhance fruit set of crops regardless of honey bee abundance. *Science* 339: 1608–1611.

79. Ellis, A. M., Myers, S. S. and Ricketts, T. H. 2015. Do pollinators contribute to nutritional health? *PLOS ONE* 10: 17.

80. Chaplin-Kramer, R., Dombeck, E., Gerber, J. et al. 2014. Global malnutrition overlaps with pollinator-dependent micronutrient production. *Proceedings of the Royal Society B: Biological Sciences* 281: 7.

81. Egoh, B., Rouget, M., Reyers, B. et al. 2007. Integrating ecosystem services into conservation assessments: A review. *Ecological Economics* 63: 714–721.

82. Wolff, S., Schulp, C. J. E. and Verburg, P. H. 2015. Mapping ecosystem services demand: A review of current research and future perspectives. *Ecological Indicators* 55: 159–171.

83. Daniel, T. C., Muhar, A., Arnberger, A. et al. 2012. Contributions of cultural services to the ecosystem services agenda. *Proceedings of the National Academy of Sciences* 109: 8812–8819.

84. Chan, K. M. A., Satterfield, T. and Goldstein, J. 2012. Rethinking ecosystem services to better address and navigate cultural values. *Ecological Economics* 74: 8–18.

85. Fish, R., Church, A. and Winter, M. 2016. Conceptualising cultural ecosystem services: A novel framework for research and critical engagement. *Ecosystem Services* 21: 208–217.

86. Hølleland, H., Skrede, J. and Holmgaard, S. B. 2017. Cultural heritage and ecosystem services: A literature review. *Conservation and Management of Archaeological Sites* 19: 210–237.

87. Klain, S. C., Olmsted, P., Chan, K. M. A. and Satterfield, T. 2017. Relational values resonate broadly and differently than intrinsic or instrumental values, or the New Ecological Paradigm. *PLOS ONE* 12: e0183962.

88. Buckley, R. 2009. Parks and tourism. *PLOS Biology* 7: 2.

89. Siikamaki, J. 2011. Contributions of the US state park system to nature recreation. *Proceedings of the National Academy of Sciences* 108: 14031–14036.

90. Southwick. 2011. *The Economics Associated with Outdoor Recreation, Natural Resources Conservation and Historic Preservation in the United States.* (Prepared for the National Fish and Wildlife Foundation.) Fernandina Beach, FL: Southwick.

91. Power, T. M. and Barrett, R. N. 2001. *Post-cowboy Economics: Pay and Prosperity in the New American West.* Washington, DC: Island Press.

92. Pascoe, S., Doshi, A., Dell, Q. et al. 2014. Economic value of recreational fishing in Moreton Bay and the potential impact of the marine park rezoning. *Tourism Management* 41: 53–63.

93. Gurluk, S. and Rehber, E. 2008. A travel cost study to estimate recreational value for a bird refuge at Lake Manyas, Turkey. *Journal of Environmental Management* 88: 1350–1360.

94. Chakraborty, K. and Keith, J. E. 2000. Estimating the recreation demand and economic value of mountain biking in Moab, Utah: An application of count data models. *Journal of Environmental Planning and Management* 43: 461–469.

95. Balmford, A., Beresford, J., Green, J. et al. 2009. A global perspective on trends in nature-based tourism. *PLOS Biology* 7: 6.

96. Center for Responsible Travel. 2015. *The Case for Responsible Travel: Trends & Statistics 2015.* Washington, DC: CRT.

97. Barquero, M. 2016. *País logra récords en divisas y visitantes por el turismo [Country achieves record tourism visitors and earnings].* La Nación (in Spanish).

98. Giese, M. 1996. Effects of human activity on Adélie penguin *Pygoscelis adeliae* breeding success. *Biological Conservation* 75: 157–164.

99. Keniger, L., Gaston, K., Irvine, K. and Fuller, R. 2013. What are the benefits of interacting with nature? *International Journal of Environmental Research and Public Health* 10: 913.

100. Frumkin, H., Bratman, G. N., Breslow, S. J. et al. 2017. Nature contact and human health: A research agenda. *Environmental Health Perspectives* 125: 18.

101. Donovan, G. H., Butry, D. T., Michael, Y. L. et al. 2013. The relationship between trees and human health: Evidence from the spread of the emerald ash borer. *American Journal of Preventive Medicine* 44: 139–145.

102. Fuller, R. A., Irvine, K. N., Devine-Wright, P. et al. 2007. Psychological benefits of greenspace increase with biodiversity. *Biology Letters* 3: 390–394.

103. Selhub, E. M. and Logan, A. C. 2012. *Your Brain on Nature: The Science of Nature's Influence on Your Health, Happiness and Vitality.* Mississauga, Ontario: John Wiley & Sons Canada.

104. Barbier, E. B. 2011. Pricing nature. *Annual Review of Resource Economics* 3: 337–353.

105. Wallace, K. J. 2007. Classification of ecosystem services: Problems and solutions. *Biological Conservation* 139: 235–246.

106. La Notte, A., D'Amato, D., Makinen, H. et al. 2017. Ecosystem services classification: A systems ecology perspective of the cascade framework. *Ecological Indicators* 74: 392–402.

107. Gomez-Baggethun, E. and Barton, D. N. 2013. Classifying and valuing ecosystem services for urban planning. *Ecological Economics* 86: 235–245.

108. Diaz, S., Demissew, S., Carabias, J. et al. 2015. The IPBES Conceptual Framework: Connecting nature and people. *Current Opinion in Environmental Sustainability* 14: 1–16.

109. Díaz, S., Pascual, U., Stenseke, M. et al. 2018. Assessing nature's contributions to people. *Science* 359: 270–272.

110. Pascual, U., Balvanera, P., Díaz, S. et al. 2017. Valuing nature's contributions to people: The IPBES approach. *Current Opinion in Environmental Sustainability* 26–27: 7–16.

111. Agarwala, M., Atkinson, G., Fry, B. et al. 2014. Assessing the relationship between human well-being and ecosystem services: A review of frameworks. *Conservation and Society* 12: 437–449.

112. de Groot, R. S., Wilson, M. A. and Boumans, R. M. J. 2002. A typology for the classification, description and valuation of ecosystem functions, goods and services. *Ecological Economics* 41: 393–408.

113. Maier, D. S. and Feest, A. 2016. The IPBES Conceptual Framework: An unhelpful start. *Journal of Agricultural & Environmental Ethics* 29: 327–347.

114. Elser, J., Dobberfuhl, D., MacKay, N. and Schampel, J. 1996. Organism size, life history, and N:P stoichiometry. *BioScience* 46: 674–684.

115. Elser, J. J., Elser, M. M., Mackay, N. A. and Carpenter, S. R. 1988. Zooplankton-mediated transitions between N-limited and P-limited algal growth. *Limnology and Oceanography* 33: 1–14.

116. Ripple, W. J., Estes, J. A., Beschta, R. L. et al. 2014. Status and ecological effects of the world's largest carnivores. *Science* 343: 1241484.

117. Fortin, D., Beyer, H. L., Boyce, M. S. et al. 2005. Wolves influence elk movements: Behavior shapes a trophic cascade in Yellowstone National Park. *Ecology* 86: 1320–1330.

118. Ripple, W. J., Beschta, R. L. and Painter, L. E. 2015. Trophic cascades from wolves to alders in Yellowstone. *Forest Ecology and Management* 354: 254–260.

119. Smith, D. W. and Tyers, D. B. 2012. The history and current status and distribution of beavers in Yellowstone National Park. *Northwest Science* 86: 276–288.

120. Beschta, R. L. and Ripple, W. J. 2015. Divergent patterns of riparian cottonwood recovery after the return of wolves in Yellowstone, USA. *Ecohydrology* 8: 58–66.

121. Beschta, R. L. and Ripple, W. J. 2016. Riparian vegetation recovery in Yellowstone: The first two decades after wolf reintroduction. *Biological Conservation* 198: 93–103.

122. David Mech, L. 2012. Is science in danger of sanctifying the wolf? *Biological Conservation* 150: 143–149.

123. Ford, A. T. and Goheen, J. R. 2015. Trophic cascades by large carnivores: A case for strong inference and mechanism. *Trends in Ecology & Evolution* 30: 725–735.

124. Nock, C. A., Paquette, A., Follett, M. et al. 2013. Effects of urbanization on tree species functional diversity in eastern North America. *Ecosystems* 16: 1487–1497.

125. Lodge, A. G., Whitfeld, T. J. S., Roth, A. M. and Reich, P. B. 2018. Invasive plants in Minnesota are "joining the locals": A trait-based analysis. *Journal of Vegetation Science* 29: 746–755.

126. Naeem, S. and Bunker, D. E. 2009. *TraitNet: Furthering biodiversity research through the curation of discovery, and sharing of species trait data.* In S. Naeem et al. (Eds.), *Biodiversity, Ecosystem Functioning, and Human Wellbeing: An Ecological and Economic Perspective.* Oxford, UK: Oxford University Press.

127. Curtis, C. A. and Bradley, B. A. 2016. Plant distribution data show broader climatic limits than expert-based climatic tolerance estimates. *PLOS ONE* 11: e0166407.

128. Tilman, D., Reich, P. B., Knops, J. et al. 2001. Diversity and productivity in a long-term grassland experiment. *Science* 294: 843–845.

129. Tilman, D., Reich, P. B. and Knops, J. M. H. 2006. Biodiversity and ecosystem stability in a decade-long grassland experiment. *Nature* 441: 629–632.

130. Cardinale, B. J., Duffy, J. E., Gonzalez, A. et al. 2012. Biodiversity loss and its impact on humanity. *Nature* 486: 59–67.

131. Kirwan, L., Luscher, A., Sebastia, M. T. et al. 2007. Evenness drives consistent diversity effects in intensive grassland systems across 28 European sites. *Journal of Ecology* 95: 530–539.

132. Hughes, A. R., Inouye, B. D., Johnson, M. T. J. et al. 2008. Ecological consequences of genetic diversity. *Ecology Letters* 11: 609–623.

133. Balvanera, P., Pfisterer, A. B., Buchmann, N. et al. 2006. Quantifying the evidence for biodiversity effects on ecosystem functioning and services. *Ecology Letters* 9: 1146–1156.

134. Cardinale, B. J., Srivastava, D. S., Duffy, J. E. et al. 2006. Effects of biodiversity on the functioning of trophic groups and ecosystems. *Nature* 443: 989–992.

135. Gross, K., Cardinale, B. J., Fox, J. W. et al. 2014. Species richness and the temporal stability of biomass production: A new analysis of recent biodiversity experiments. *The American Naturalist* 183: 1–12.

136. Griffin, J. N., O'Gorman, E., Emmerson, M. C. et al. 2009. *Biodiversity and the stability of ecosystem functioning.* In S. Naeem et al. (Eds.), *Biodiversity and Human Impacts.* Oxford, UK: Oxford University Press.

137. Hooper, D. U., Adair, E. C., Cardinale, B. J. et al. 2012. A global synthesis reveals biodiversity loss as a major driver of ecosystem change. *Nature* 486: 105-U129.

138. Tilman, D., Reich, P. B. and Isbell, F. 2012. Biodiversity impacts ecosystem productivity as much as resources, disturbance, or herbivory. *Proceedings of the National Academy of Sciences* 109: 10394–10397.

139. Duffy, E., Godwin, C. and Cardinale, B. J. 2017. Biodiversity effects in the wild are common and as strong as key drivers of productivity. *Nature* 549: 261–264.

140. Cardinale, B. J. 2011. Biodiversity improves water quality through niche partitioning. *Nature* 472: 86-U113.

141. Losey, J. E. and Denno, R. F. 1998. Positive predator-predator interactions: Enhanced predation rates and synergistic suppression of aphid populations. *Ecology* 79: 2143–2152.

142. Losey, J. E. and Denno, R. F. 1998. The escape response of pea aphids to foliar-foraging predators: Factors affecting dropping behaviour. *Ecological Entomology* 23: 53–61.

143. Denoth, M., Frid, L. and Myers, J. H. 2002. Multiple agents in biological control: Improving the odds? *Biological Control* 24: 20–30.

144. Huang, Z. Y. X., Yu, Y., Van Langevelde, F. and De Boer, W. F. 2017. Does the dilution effect generally occur in animal diseases? *Parasitology* 144: 823–826.

145. Ferraguti, M., Martínez-de la Puente, J., Bensch, S. et al. 2018. Ecological determinants of avian malaria infections: An integrative analysis at landscape, mosquito and vertebrate community levels. *Journal of Animal Ecology* 87: 727–740.

146. Liu, X., Lyu, S., Zhou, S. and Bradshaw, C. J. A. 2016. Warming and fertilization alter the dilution effect of host diversity on disease severity. *Ecology* 97: 1680–1689.

147. Doak, D. F., Bigger, D., Harding, E. K. et al. 1998. The statistical inevitability of stability-diversity relationships in community ecology. *American Naturalist* 151: 264–276.

148. Newell, R. G., Pizer, W. A. and Raimi, D. 2013. Carbon markets 15 years after Kyoto: Lessons learned, new challenges. *Journal of Economic Perspectives* 27: 123–146.

149. Sullivan, P., Hellerstein, D., Hansen, L. et al. 2004. *The Conservation Reserve Program: Economic Implications for Rural America.* Washington, DC: U.S. Department of Agriculture.

150. Kolinjivadi, V. and Sunderland, T. 2012. A review of two payment schemes for watershed services from China and Vietnam: The interface of government control and PES theory. *Ecology and Society* 17: 10.

151. Goldman-Benner, R., Benitez, S., Boucher, T. et al. 2012. Water funds and PES: Practice learns from theory and theory can learn from practice. *Oryx* 46: 55–63.

152. Benitez, S., Calvache, A. and Veiga, F. 2015. Water funds as a tool for urban water provision and watershed conservation in Latin America. In I. Aguilar-Barajas et al. (Eds.), *Water and Cities in Latin America.* London, UK: Routledge.

153. Rissman, A. R., Lozier, L., Comendant, T. et al. 2007. Conservation easements: Biodiversity protection and private use. *Conservation Biology* 21: 709–718.

154. Chan, K. M. A., Anderson, E., Chapman, M. et al. 2017. Payments for ecosystem services: Rife with problems and potential for transformation towards sustainability. *Ecological Economics* 140: 110–122.

155. Carver, E. and Caudill, J. 2007. *Banking on Nature 2006: The Economic Benefits to Local Communities of National Wildlife Refuge Visitation.* Washington, DC: U.S. Fish and Wildlife Service.

156. Knoche, S. and Lupi, F. 2012. The economic value of publicly accessible deer hunting land. *The Journal of Wildlife Management* 76: 462–470.

157. Connelly, N. A., Brown, T. L. and Brown, J. W. 2007. Measuring the net economic value of recreational boating as water levels fluctuate. *Journal of the American Water Resources Association* 43: 1016–1023.

158. Bourne, Jr., J. K. 2004. Gone with the Water. *National Geographic* 206: 88–105.

159. Potts, S. G., Biesmeijer, J. C., Kremen, C. et al. 2010. Global pollinator declines: Trends, impacts and drivers. *Trends in Ecology & Evolution* 25: 345–353.

160. Hallmann, C. A., Sorg, M., Jongejans, E. et al. 2017. More than 75 percent decline over 27 years in total flying insect biomass in protected areas. *PLOS ONE* 12: e0185809.

161. Cameron, S. A., Lozier, J. D., Strange, J. P. et al. 2011. Patterns of widespread decline in North American bumble bees. *Proceedings of the National Academy of Sciences* 108: 662–667.

162. van Engelsdorp, D., Traynor, K. S., Andree, M. et al. 2017. Colony collapse disorder (CCD) and bee age impact honey bee pathophysiology. *PLOS ONE* 12: e0179535.

163. Gallai, N., Salles, J. M., Settele, J. and Vaissiere, B. E. 2009. Economic valuation of the vulnerability of world agriculture confronted with pollinator decline. *Ecological Economics* 68: 810–821.

164. Ostfeld, R. S. and Keesing, F. 2000. Biodiversity and disease risk: The case of Lyme disease. *Conservation Biology* 14: 722–728.

165. Keesing, F., Belden, L. K., Daszak, P. et al. 2010. Impacts of biodiversity on the emergence and transmission of infectious diseases. *Nature* 468: 647–652.

166. Dobson, A., Cattadori, I., Holt, R. D. et al. 2006. Sacred cows and sympathetic squirrels: The importance of biological diversity to human health. *PLOS Medicine* 3: 714–718.

167. Ostfeld, R. S. 2009. Biodiversity loss and the rise of zoonotic pathogens. *Clinical Microbiology and Infection* 15: 40–43.

168. Johnson, P. T. J., Ostfeld, R. S. and Keesing, F. 2015. Frontiers in research on biodiversity and disease. *Ecology Letters* 18: 1119–1133.

169. Wood, C. L., Lafferty, K. D., DeLeo, G. et al. 2014. Does biodiversity protect humans against infectious disease? *Ecology* 95: 817–832.

Chapter 7

1. Redford, K. H. and Adams, W. M. 2009. Payment for ecosystem services and the challenge of saving nature. *Conservation Biology* 23: 785–787.

2. Rebitzer, G., Ekvall, T., Frischknecht, R. et al. 2004. Life cycle assessment Part 1: Framework, goal and scope definition, inventory analysis, and applications. *Environment International* 30: 701–720.

3. Millennium Ecosystem Assessment. 2005. *Ecosystems and Human Well-being: Synthesis*. Washington, DC: World Resources Institute.

4. Smith, A. 1909. *An Inquiry into the Nature and Causes of the Wealth of Nations*. New York, NY: P.F. Collier & Sons.

5. Abson, D. J. and Termansen, M. 2011. Valuing ecosystem services in terms of ecological risks and returns. *Conservation Biology* 25: 250–258.

6. Gleick, P. H. 2010. *Bottled and Sold: The Story Behind Our Obsession with Bottled Water*. Washington, DC: Island Press.

7. Landrigan, P. J., Fuller, R., Acosta, N. J. R. et al. The *Lancet* Commission on pollution and health. *Lancet* 391: 462–512.

8. Hardin, G. 1968. The tragedy of the commons. *Science* 162: 1243–1248.

9. Lloyd, W. F. 1832. *Two lectures on the checks to population, delivered before the University of Oxford, in Michaelmas term*. Oxford, UK: J.H. Parker.

10. McWhinnie, S. F. 2009. The tragedy of the commons in international fisheries: An empirical examination. *Journal of Environmental Economics and Management* 57: 321–333.

11. Ostrom, E. 2015. *Governing the Commons*. Cambridge, UK: Cambridge University Press.

12. Ostrom, E. 2009. A general framework for analyzing sustainability of social-ecological systems. *Science* 325: 419–422.

13. Bateman, I. J., Harwood, A. R., Mace, G. M. et al. 2013. Bringing ecosystem services into economic decision-making: Land use in the United Kingdom. *Science* 341: 45–50.

14. Lant, C. L., Ruhl, J. B. and Kraft, S. E. 2008. The tragedy of ecosystem services. *BioScience* 58: 969–974.

15. Hughey, K. F. D., Cullen, R. and Moran, E. 2003. Integrating economics into priority setting and evaluation in conservation management. *Conservation Biology* 17: 93–103.

16. Weinstein, M. C. and Stason, W. B. 1977. Foundations of cost-effectiveness analysis for health and medical practices. *New England Journal of Medicine* 296: 716–721.

17. Montgomery, C. A., Brown, G. M. and Adams, D. M. 1994. The marginal cost of species preservation: The northern spotted owl. *Journal of Environmental Economics and Management* 26: 111–128.

18. Laycock, H., Moran, D., Smart, J. et al. 2009. Evaluating the cost-effectiveness of conservation: The UK Biodiversity Action Plan. *Biological Conservation* 142: 3120–3127.

19. Boardman, A. E., Greenburg, D. H., Vining, A. R. and Weimer, D. L. 2018. *Cost-Benefit Analysis: Concepts and Practice* (4th ed.). New York, NY: Cambridge University Press.

20. Campbell, H. F. and Brown, R. P. 2003. *Benefit-Cost Analysis: Financial and Economic Appraisal Using Spreadsheets*. Cambridge, UK: Cambridge University Press.

21. Davies, A. L., Bryce, R. and Redpath, S. M. 2013. Use of multicriteria decision analysis to address conservation conflicts. *Conservation Biology* 27: 936–944.

22. Ackermann, F. 2008. *Critique of Cost-Benefit Analysis, and Alternative Approaches to Decision-Making. A Report for Friends of the Earth England, Wales and Northern Ireland*. London, UK: Friends of the Earth.

23. Barbier, E. B. 2011. Pricing nature. *Annual Review of Resource Economics* 3: 337–353.

24. Heal, G. M., Barbier, E. B., Boyle, K. J. et al. 2005. *Valuing Ecosystem Services: Toward Better Environmental Decision Making*. Washington, DC: National Academies Press.

25. Plottu, E. and Plottu, B. 2007. The concept of Total Economic Value of environment: A reconsideration within a hierarchical rationality. *Ecological Economics* 61: 52–61.

26. Hansjürgens, B., Schröter-Schlaack, C., Berghöfer, A. and Lienhoop, N. 2017. Justifying social values of nature: Economic reasoning beyond self-interested preferences. *Ecosystem Services* 23: 9–17.

27. Pearce, D. W., Atkinson, G. and Mourato, S. 2006. *Cost-benefit analysis and the environment: Recent developments*. Paris, France: Organisation for Economic Co-operation and Development.

28. Davidar, P., Arjunan, M. and Puyravaud, J.-P. 2008. Why do local households harvest forest products? A case study from the southern Western Ghats, India. *Biological Conservation* 141: 1876–1884.

29. Balick, M. J. and Cox, P. A. 1996. *Plants, People, and Culture: The Science of Ethnobotany*. New York, NY: Scientific American Library.

30. Barbier, E. B. 2007. Valuing ecosystem services as productive inputs. *Economic Policy* 22: 178–229.

31. Smith, V. K. 1983. Option value: A conceptual overview. *Southern Economic Journal* 49: 654–668.

32. Pascual, U., Balvanera, P., Díaz, S. et al. 2017. Valuing nature's contributions to people: The IPBES approach. *Current Opinion in Environmental Sustainability* 26–27: 7–16.

33. Groom, M. J., Meffe, G. K., Carroll, C. R. and Meffe, G. K. 2006. *Principles of Conservation Biology* (3rd ed.). Sunderland, MA: Oxford University Press/Sinauer.
34. Costanza, R., dArge, R., deGroot, R. et al. 1997. The value of the world's ecosystem services and natural capital. *Nature* 387: 253–260.
35. Costanza, R., de Groot, R., Sutton, P. et al. 2014. Changes in the global value of ecosystem services. *Global Environmental Change* 26: 152–158.
36. Adger, W. N., Brown, K., Raffaello, C. and Moran, D. 1995. Total economic value of forests in Mexico. *Ambio* 24: 286–296.
37. Sina, I., Maryunani, Batoro, J. and Harahab, N. 2017. Analysis of total economic value of ecosystem mangrove forest in the coastal zone Pulokerto village district of Kraton Pasuruan Regency. *International Journal of Ecosystem* 7: 1–10.
38. TEEB. 2010. *The Economics of Ecosystems and Biodiversity Ecological and Economic Foundations*. Washington, DC: Earthscan.
39. Emerton, L., Baig, S. and Saleem, M. 2009. *Valuing Biodiversity: The Economic Case for Biodiversity Conservation in the Maldives.* Male, Maldives: Ministry of Housing, Transport and Environment, Government of Maldives.
40. Boyle, K. J., Poor, P. J. and Taylor, L. O. 1999. Estimating the demand for protecting freshwater lakes from eutrophication. *American Journal of Agricultural Economics* 81: 1118–1122.
41. Walsh, P. J., Milon, J. W. and Scrogin, D. O. 2011. The spatial extent of water quality benefits in urban housing markets. *Land Economics* 87: 628–644.
42. Smith, V. K. and Huang, J. C. 1995. Can markets value air quality? A meta-analysis of hedonic property value models. *Journal of Political Economy* 103: 209–227.
43. Tyrväinen, L. 1997. The amenity value of the urban forest: An application of the hedonic pricing method. *Landscape and Urban Planning* 37: 211–222.
44. FAO Technical Report 2014: *Optimising the production of goods and services by Mediterranean forests in a context of global changes.*
45. Carson, R. T., Mitchell, R. C., Hanemann, W. M. et al. 1992. *A Contingent Valuation Study of Lost Passive Use Values Resulting from the Exxon Valdez Oil Spill: A Report to the Attorney General of the State of Alaska*. Juneau, AK: State of Alaska Attorney General's Office.
46. Hackett, S. and Dissanayake, S. T. 2014. *Environmental and Natural Resources Economics: Theory, Policy, and the Sustainable Society* (4th ed.). New York, NY: Routledge.
47. Arrow, K., Solow, R., Portney, P. et al. 1993. *Report of the NOAA Panel on Contingent Valuation*. Washington, DC: National Oceanographic and Atmospheric Administration.
48. Sairam, R., Chennareddy, S., Parani, M. et al. 2005. OBPC Symposium: Maize 2004 & beyond—Plant regeneration, gene discovery, and genetic engineering of plants for crop improvement. *In Vitro Cellular & Developmental Biology – Plant* 41: 411–423.
49. Nabhan, G. P. 2009. *Where Our Food Comes From: Retracing Nikolay Vavilov's Quest to End Famine*. Washington, DC: Island Press.
50. Frisvold, G. B., Sullivan, J. and Raneses, A. 2003. Genetic improvements in major US crops: The size and distribution of benefits. *Agricultural Economics* 28: 109–119.
51. Leopold, A. 1972. *Round River: From the Journals of Aldo Leopold.* New York, NY: Oxford Univeristy Press.
52. King, D. M. and Mazzotta, M. J. 2000. *Ecosystem valuation*, www.ecosystemvaluation.org/default.htm.
53. Bandara, R. 2010. *Human-elephant conflict mitigation through insurance scheme, Sri Lanka*, http://www.teebweb.org/resources/case-studies/.
54. Ceylinco. 2006. Ceylinco introduces Hasthi Uvaduru Rakawaranaya. Sri Lanka: Ceylinco Insurance Company. http://www.prem-online.org/archive/14/doc/Press%20realease%20Ceylinco.pdf.
55. ten Brink, P. 2011. *The Economics of Ecosystems and Biodiversity in National and International Policy Making*. Washington, DC: Earthscan.
56. Lem, A., Bjørndal, T. and Lappo, A. 2014. *Economic Analysis of Supply and Demand for Food up to 2030. Special Focus on Fish and Fishery Products*. Rome, Italy: FAO.
57. Masiero, M., Pettenella, D. and Secco, L. 2016. From failure to value: Economic valuation for a selected set of products and services from Mediterranean forests. *Forest Systems* 25: e051.
58. Ricketts, T. H., Daily, G. C., Ehrlich, P. R. and Michener, C. D. 2004. Economic value of tropical forest to coffee production. *Proceedings of the National Academy of Sciences* 101: 12579–12582.
59. Ahmed, M., Umali, G. M., Chong, C. K. et al. 2007. Valuing recreational and conservation benefits of coral reefs: The case of Bolinao, Philippines. *Ocean & Coastal Management* 50: 103–118.
60. Shrestha, R. K., Seidl, A. F. and Moraes, A. S. 2002. Value of recreational fishing in the Brazilian Pantanal: A travel cost analysis using count data models. *Ecological Economics* 42: 289–299.
61. Jones, T. E., Yang, Y. and Yamamoto, K. 2017. Assessing the recreational value of world heritage site inscription: A longitudinal travel cost analysis of Mount Fuji climbers. *Tourism Management* 60: 67–78.
62. Mueller, H., Hamilton, D. P. and Doole, G. J. 2016. Evaluating services and damage costs of degradation of a major lake ecosystem. *Ecosystem Services* 22: 370–380.
63. Liebelt, V., Bartke, S. and Schwarz, N. 2018. Hedonic pricing analysis of the influence of urban green spaces onto residential prices: The case of Leipzig, Germany. *European Planning Studies* 26: 133–157.
64. Benson, E. D., Hansen, J. L., Schwartz, A. L. and Smersh, G. T. 1998. Pricing residential amenities: The value of a view. *Journal of Real Estate Finance and Economics* 16: 55–73.
65. Barbier, E. B. 2015. Valuing the storm protection service of estuarine and coastal ecosystems. *Ecosystem Services* 11: 32–38.
66. Reyns, N., Casaer, J., De Smet, L. et al. 2018. Cost-benefit analysis for invasive species control: The case of greater Canada goose *Branta canadensis* in Flanders (northern Belgium). *Peerj* 6: 29.
67. Grossmann, M. 2012. Economic value of the nutrient retention function of restored floodplain wetlands in the Elbe River basin. *Ecological Economics* 83: 108–117.
68. Barth, N. C. and Doll, P. 2016. Assessing the ecosystem service flood protection of a riparian forest by applying a cascade approach. *Ecosystem Services* 21: 39–52.
69. Kotchen, M. J. and Reiling, S. D. 2000. Environmental attitudes, motivations, and contingent valuation of nonuse values: A case study involving endangered species. *Ecological Economics* 32: 93–107.
70. Carson, R. T., Mitchell, R. C., Hanemann, M. et al. 2003. Contingent valuation and lost passive use: Damages from the Exxon Valdez oil spill. *Environmental & Resource Economics* 25: 257–286.
71. Lee, C. K., Lee, J. H., Kim, T. K. and Mjelde, J. W. 2010. Preferences and willingness to pay for bird-watching tour and interpretive services using a choice experiment. *Journal of Sustainable Tourism* 18: 695–708.
72. Birol, E., Karousakis, K. and Koundouri, P. 2006. Using a choice experiment to account for preference heterogeneity in wetland attributes: The case of Cheimaditida wetland in Greece. *Ecological Economics* 60: 145–156.
73. Amuakwa-Mensah, F., Barenbold, R. and Riemer, O. 2018. Deriving a benefit transfer function for threatened and endangered species in interaction with their level of charisma. *Environments* 5: 18.
74. Barton, D. N. 2002. The transferability of benefit transfer: Contingent valuation of water quality improvements in Costa Rica. *Ecological Economics* 42: 147–164.

75. Pak, M., Turker, M. F. and Ozturk, A. 2010. Total economic value of forest resources in Turkey. *African Journal of Agricultural Research* 5: 1908–1916.

76. Torras, M. 2000. The total economic value of Amazonian deforestation, 1978–1993. *Ecological Economics* 33: 283–297.

77. Hernandez, A., Caballero, R., Leon, M. A. et al. 2014. Multi-criteria decision modeling for environmental assessment: An estimation of Total Economic Value in protected natural areas. *International Journal of Environmental Research* 8: 551–560.

78. Pyron, R. A. 2017. We don't need to save endangered species: Extinction is part of evolution. https://www.washingtonpost.com/outlook/we-dont-need-to-save-endangered-species-extinction-is-part-of-evolution/2017/11/21/57fc5658-cdb4-11e7-a1a3-0d1e45a6de3d_story.html?noredirect=on&utm_term=.ef6200bb1884.

79. Opinion. 2017. We must protect biodiversity. https://www.washingtonpost.com/opinions/2017/12/15/53e6147c-e0f7-11e7-b2e9-8c636f076c76_story.html?utm_term=.bd492db32fe1.

Chapter 8

1. Rodrigues, A., Pilgrim, J., Lamoreux, J. et al. 2006. The value of the IUCN Red List for conservation. *Trends in Ecology & Evolution* 21: 71–76.

2. Gardenfors, U., Hilton-Taylor, C., Mace, G. M. and Rodriguez, J. P. 2001. The application of IUCN Red List criteria at regional levels. *Conservation Biology* 15: 1206–1212.

3. Barnosky, A. D., Matzke, N., Tomiya, S. et al. 2011. Has the Earth's sixth mass extinction already arrived? *Nature* 471: 51–57.

4. Thuiller, W., Lavorel, S., Araujo, M. B. et al. 2005. Climate change threats to plant diversity in Europe. *Proceedings of the National Academy of Sciences* 102: 8245–8250.

5. Thomas, C. D., Cameron, A., Green, R. E. et al. 2004. Extinction risk from climate change. *Nature* 427: 145–148.

6. Wilcove, D. S., Rothstein, D., Dubow, J. et al. 1998. Quantifying threats to imperiled species in the United States. *BioScience* 48: 607–615.

7. Lacher, T. E., Boitani, L. and da Fonseca, G. A. B. 2012. The IUCN global assessments: Partnerships, collaboration and data sharing for biodiversity science and policy. *Conservation Letters* 5: 327–333.

8. Rodrigues, A. S. L., Pilgrim, J. D., Lamoreux, J. F. et al. 2006. The value of the IUCN Red List for conservation. *Trends in Ecology & Evolution* 21: 71–76.

9. Nolan, M. P., Jacobson, N. D., McClung, T. A. and Cardinale, B. J. 2018. What scientific data are being used to justify the listing of endangered mammals on the IUCN's Red List? *Journal of Biodiversity and Endangered Species* 103. doi: 10.29011/IJBES-103.100003

10. Mrosovsky, N. 1997. IUCN's credibility critically endangered: Commentary. *Nature* 389: 436–436.

11. Butchart, S. H. M. and Bird, J. P. 2010. Data deficient birds on the IUCN Red List: What don't we know and why does it matter? *Biological Conservation* 143: 239–247.

12. Régnier, C., Fontaine, B. and Bouchet, P. 2009. Not knowing, not recording, not listing: Numerous unnoticed mollusk extinctions. *Conservation Biology* 23: 1214–1221.

13. Bull, J. W. and Maron, M. 2016. How humans drive speciation as well as extinction. *Proceedings of the Royal Society B: Biological Sciences* 283: 20160600.

14. Monastersky, R. 2014. Life: A status report. *Nature* 516: 159–159.

15. Baillie, J. E. M., Hilton-Taylor, C. and Stuart, S. N. 2004. *2004 IUCN Red List of Threatened Species: A Global Species Assessment.* Gland, Switzerland, and Cambridge, UK: IUCN.

16. Ceballos, G., Ehrlich, P. R., Barnosky, A. D. et al. 2015. Accelerated modern human–induced species losses: Entering the sixth mass extinction. *Science Advances* 1: e1400253.

17. Dirzo, R., Young, H. S., Galetti, M. et al. 2014. Defaunation in the Anthropocene. *Science* 345: 401–406.

18. Regnier, C., Achaz, G., Lambert, A. et al. 2015. Mass extinction in poorly known taxa. *Proceedings of the National Academy of Sciences* 112: 7761–7766.

19. Mora, C., Tittensor, D. P., Adl, S. et al. 2011. How many species are there on Earth and in the ocean? *PLOS Biology* 9: e1001127.

20. Pimm, S. L., Russell, G. J., Gittleman, J. L. and Brooks, T. M. 1995. The future of biodiversity. *Science* 269: 347–350.

21. De Vos, J. M., Joppa, L. N., Gittleman, J. L. et al. 2015. Estimating the normal background rate of species extinction. *Conservation Biology* 29: 452–462.

22. Aronson, M. F. J., La Sorte, F. A., Nilon, C. H. et al. 2014. A global analysis of the impacts of urbanization on bird and plant diversity reveals key anthropogenic drivers. *Proceedings of the Royal Society B: Biological Sciences* 281: 20133330.

23. FAO. 2016. *The State of the World's Fisheries and Aquaculture 2016.* Rome, Italy: United Nations.

24. FAO. 2016. *State of the World's Forests 2016.* Rome, Italy: United Nations.

25. Maceda-Veiga, A., Dominguez-Dominguez, O., Escribano-Alacid, J. and Lyons, J. 2016. The aquarium hobby: Can sinners become saints in freshwater fish conservation? *Fish and Fisheries* 17: 860–874.

26. Harris, J. B. C., Tingley, M. W., Hua, F. Y. et al. 2017. Measuring the impact of the pet trade on Indonesian birds. *Conservation Biology* 31: 394–405.

27. Sharma, N. and Kala, C. P. 2018. Harvesting and management of medicinal and aromatic plants in the Himalaya. *Journal of Applied Research on Medicinal and Aromatic Plants* 8: 1–9.

28. Wittemyer, G., Northrup, J. M., Blanc, J. et al. 2014. Illegal killing for ivory drives global decline in African elephants. *Proceedings of the National Academy of Sciences* 111: 13117–13121.

29. Bending, Z. J. 2018. Improving conservation outcomes: Understanding scientific, historical and cultural dimensions of the illicit trade in rhinoceros horn. *Environment and History* 24: 149–186.

30. Worm, B., Davis, B., Kettemer, L. et al. 2013. Global catches, exploitation rates, and rebuilding options for sharks. *Marine Policy* 40: 194–204.

31. Roopnarine, P. D. 2006. Extinction cascades and catastrophe in ancient food webs. *Paleobiology* 32: 1–19.

32. Donohue, I., Petchey, O. L., Kefi, S. et al. 2017. Loss of predator species, not intermediate consumers, triggers rapid and dramatic extinction cascades. *Global Change Biology* 23: 2962–2972.

33. Pauly, D., Christensen, V., Dalsgaard, J. et al. 1998. Fishing down marine food webs. *Science* 279: 860–863.

34. Estes, J. A., Terborgh, J., Brashares, J. S. et al. 2011. Trophic downgrading of planet Earth. *Science* 333: 301–306.

35. Clavero, M., Brotons, L., Pons, P. and Sol, D. 2009. Prominent role of invasive species in avian biodiversity loss. *Biological Conservation* 142: 2043–2049.

36. Cambray, J. A. 2003. Impact on indigenous species biodiversity caused by the globalisation of alien recreational freshwater fisheries. *Hydrobiologia* 500: 217–230.

37. Molnar, J. L., Gamboa, R. L., Revenga, C. and Spalding, M. D. 2008. Assessing the global threat of invasive species to marine biodiversity. *Frontiers in Ecology and the Environment* 6: 485–492.

38. Katsanevakis, S., Wallentinus, I., Zenetos, A. et al. 2014. Impacts of invasive alien marine species on ecosystem services and biodiversity: A pan-European review. *Aquatic Invasions* 9: 391–423.

39. Pimentel, D., Zuniga, R. and Morrison, D. 2005. Update on the environmental and economic costs associated with alien-invasive species in the United States. *Ecological Economics* 52: 273–288.

40. Thurber, R. V., Payet, J. P., Thurber, A. R. and Correa, A. M. S. 2017. Virus-host interactions and their roles in coral reef health and disease. *Nature Reviews Microbiology* 15: 12.

41. Freer-Smith, P. H. and Webber, J. F. 2017. Tree pests and diseases: The threat to biodiversity and the delivery of ecosystem services. *Biodiversity and Conservation* 26: 3167–3181.

42. Daszak, P., Cunningham, A. A. and Hyatt, A. D. 2000. Emerging infectious diseases of wildlife: Threats to biodiversity and human health. *Science* 287: 443–449.

43. Parmesan, C. and Yohe, G. 2003. A globally coherent fingerprint of climate change impacts across natural systems. *Nature* 421: 37–42.

44. Pecl, G. T., Araujo, M. B., Bell, J. D. et al. 2017. Biodiversity redistribution under climate change: Impacts on ecosystems and human well-being. *Science* 355: 9.

45. Visser, M. E., van Noordwijk, A. J., Tinbergen, J. M. and Lessells, C. M. 1998. Warmer springs lead to mistimed reproduction in great tits (*Parus major*). *Proceedings of the Royal Society B: Biological Sciences* 265: 1867–1870.

46. Visser, M. E. and Both, C. 2005. Shifts in phenology due to global climate change: The need for a yardstick. *Proceedings of the Royal Society B: Biological Sciences* 272: 2561–2569.

47. Widdicombe, S. and Spicer, J. I. 2008. Predicting the impact of ocean acidification on benthic biodiversity: What can animal physiology tell us? *Journal of Experimental Marine Biology and Ecology* 366: 187–197.

48. World Wildlife Fund (WWF). 2016. *Living Planet Report 2016: Risk and Resilience in a New Era*. Gland, Switzerland: WWF International.

49. Gibbons, J. W., Scott, D. E., Ryan, T. J. et al. 2000. The global decline of reptiles, déjà vu amphibians. *BioScience* 50: 653–666.

50. Harvell, C. D., Kim, K., Burkholder, J. M. et al. 1999. Emerging marine diseases: Climate links and anthropogenic factors. *Science* 285: 1505–1510.

51. Kock, R. A., Orynbayev, M., Robinson, S. et al. 2018. Saigas on the brink: Multidisciplinary analysis of the factors influencing mass mortality events. *Science Advances* 4: eaao2314.

52. Connor, E. F. and McCoy, E. D. 2001. Species-area relationships. In S. A. Levin (Ed.), *Encyclopedia of Biodiversity*. San Diego, CA: Academic Press.

53. Drakare, S., Lennon, J. J. and Hillebrand, H. 2006. The imprint of the geographical, evolutionary and ecological context on species-area relationships. *Ecology Letters* 9: 215–227.

54. Lomolino, M. V. 2000. Ecology's most general, yet protean pattern: The species-area relationship. *Journal of Biogeography* 27: 17–26.

55. Brooks, T. M., Pimm, S. L. and Collar, N. J. 1997. Deforestation predicts the number of threatened birds in insular southeast Asia. *Conservation Biology* 11: 382–394.

56. Castelletta, M., Thiollay, J. M. and Sodhi, N. S. 2005. The effects of extreme forest fragmentation on the bird community of Singapore Island. *Biological Conservation* 121: 135–155.

57. Chittaro, P. M., Kaplan, I. C., Keller, A. and Levin, P. S. 2010. Trade-offs between species conservation and the size of marine protected areas. *Conservation Biology* 24: 197–206.

58. Pimm, S. L. and Askins, R. A. 1995. Forest losses predict bird extinctions in eastern North America. *Proceedings of the National Academy of Sciences* 92: 9343–9347.

59. Wilson, E. O. 1989. Threats to biodiversity. *Scientific American* 261: 108–116.

60. Sanjayan, M., Samberg, L. H., Boucher, T. and Newby, J. 2012. Intact faunal assemblages in the modern era. *Conservation Biology* 26: 724–730.

61. Sharma, N., Madhusudan, M. D. and Sinha, A. 2014. Local and landscape correlates of primate distribution and persistence in the remnant lowland rainforests of the Upper Brahmaputra Valley, northeastern India. *Conservation Biology* 28: 95–106.

62. Allnutt, T. F., Ferrier, S., Manion, G. et al. 2008. A method for quantifying biodiversity loss and its application to a 50-year record of deforestation across Madagascar. *Conservation Letters* 1: 173–181.

63. Janzen, D. H. 2001. Latent extinctions: The living dead. In S. A. Levin (Ed.), *Encyclopedia of Biodiversity*. San Diego, CA: Academic Press.

64. Tilman, D., May, R. M., Lehman, C. L. and Nowak, M. A. 1994. Habitat destruction and the extinction debt. *Nature* 371: 65–66.

65. He, F. L. and Hubbell, S. P. 2011. Species-area relationships always overestimate extinction rates from habitat loss. *Nature* 473: 368–371.

66. Murphy, G. E. P. and Romanuk, T. N. 2014. A meta-analysis of declines in local species richness from human disturbances. *Ecology and Evolution* 4: 91–103.

67. Moreno-Mateos, D., Barbier, E. B., Jones, P. C. et al. 2017. Anthropogenic ecosystem disturbance and the recovery debt. *Nature Communications* 8: 6.

68. Benayas, J. M. R., Newton, A. C., Diaz, A. and Bullock, J. M. 2009. Enhancement of biodiversity and ecosystem services by ecological restoration: A meta-analysis. *Science* 325: 1121–1124.

69. Alroy, J. 2017. Effects of habitat disturbance on tropical forest biodiversity. *Proceedings of the National Academy of Sciences* 114: 6056–6061.

70. Gerstner, K., Dormann, C. F., Stein, A. et al. 2014. Effects of land use on plant diversity: A global meta-analysis. *Journal of Applied Ecology* 51: 1690–1700.

71. Hudson, L. N., Newbold, T., Contu, S. et al. 2014. The PREDICTS database: A global database of how local terrestrial biodiversity responds to human impacts. *Ecology and Evolution* 4: 4701–4735.

72. Newbold, T., Hudson, L. N., Hill, S. L. L. et al. 2015. Global effects of land use on local terrestrial biodiversity. *Nature* 520: 45–50.

73. Sheil, D. 2001. Conservation and biodiversity monitoring in the tropics: Realities, priorities, and distractions. *Conservation Biology* 15: 1179–1182.

74. Green, R. E., Balmford, A., Crane, P. R. et al. 2005. A framework for improved monitoring of biodiversity: Responses to the World Summit on Sustainable Development. *Conservation Biology* 19: 56–65.

75. Henry, P. Y., Lengyel, S., Nowicki, P. et al. 2008. Integrating ongoing biodiversity monitoring: Potential benefits and methods. *Biodiversity and Conservation* 17: 3357–3382.

76. Duarte, C. M., Cebrian, J. and Marba, N. 1992. Uncertainty of detecting sea change. *Nature* 356: 190.

77. Pereira, H. M. and Cooper, D. H. 2006. Towards the global monitoring of biodiversity change. *Trends in Ecology & Evolution* 21: 123–129.

78. Vellend, M., Baeten, L., Myers-Smith, I. H. et al. 2013. Global meta-analysis reveals no net change in local-scale plant biodiversity over time. *Proceedings of the National Academy of Sciences* 110: 19456–19459.

79. Dornelas, M., Gotelli, N. J., McGill, B. et al. 2014. Assemblage time series reveal biodiversity change but not systematic loss. *Science* 344: 296–299.

80. Elahi, R., O'Connor, M. I., Byrnes, J. E. K. et al. 2015. Recent trends in local-scale marine biodiversity reflect community structure and human impacts. *Current Biology* 25: 1938–1943.

81. Gonzalez, A., Cardinale, B. J., Allington, G. R. H. et al. 2016. Estimating local biodiversity change: A critique of papers claiming no net loss of local diversity. *Ecology* 97: 1949–1960.

82. Cardinale, B. J., Gonzalez, A., Allington, G. R. H. and Loreau, M. 2018. Is local biodiversity declining or not? A summary of the debate over analysis of species richness time trends. *Biological Conservation* 219: 175–183.

83. Scholes, R. J., Walters, M., Turak, E. et al. 2012. Building a global observing system for biodiversity. *Current Opinion in Environmental Sustainability* 4: 139–146.

84. Edwards, J. L. 2004. Research and societal benefits of the Global Biodiversity Information Facility. *BioScience* 54: 485–486.

85. Larigauderie, A. 2015. The Intergovernmental Platform on Biodiversity and Ecosystem Services (IPBES): A call to action. *Gaia—Ecological Perspectives for Science and Society* 24: 73–73.

86. Rockstrom, J. 2016. Future Earth. *Science* 351: 319–319.

87. Bragdon, S. 1996. The Convention on Biological Diversity. *Global Environmental Change—Human and Policy Dimensions* 6: 177–179.

88. Bonney, R., Shirk, J. L., Phillips, T. B. et al. 2014. Next steps for citizen science. *Science* 343: 1436–1437.

89. Dickinson, J. L., Shirk, J., Bonter, D. et al. 2012. The current state of citizen science as a tool for ecological research and public engagement. *Frontiers in Ecology and the Environment* 10: 291–297.

90. Miller-Rushing, A., Primack, R. and Bonney, R. 2012. The history of public participation in ecological research. *Frontiers in Ecology and the Environment* 10: 285–290.

91. Roy, H. E., Baxter, E., Saunders, A. and Pocock, M. J. O. 2016. Focal plant observations as a standardised method for pollinator monitoring: Opportunities and limitations for mass participation citizen science. *PLOS ONE* 11: e0150794.

92. Raoult, V., David, P. A., Dupont, S. F. et al. 2016. GoPros™ as an underwater photogrammetry tool for citizen science. *PeerJ* 4: e1960.

93. Buldrini, F., Simoncelli, A., Accordi, S. et al. 2015. Ten years of citizen science data collection of wetland plants in an urban protected area. *Acta Botanica Gallica* 162: 365–373.

94. McShea, W. J., Forrester, T., Costello, R. et al. 2016. Volunteer-run cameras as distributed sensors for macrosystem mammal research. *Landscape Ecology* 31: 55–66.

95. Delaney, D. G., Sperling, C. D., Adams, C. S. and Leung, B. 2008. Marine invasive species: validation of citizen science and implications for national monitoring networks. *Biological Invasions* 10: 117–128.

96. Crall, A. W., Jarnevich, C. S., Young, N. E. et al. 2015. Citizen science contributes to our knowledge of invasive plant species distributions. *Biological Invasions* 17: 2415–2427.

97. Pagel, J., Anderson, B. J., O'Hara, R. B. et al. 2014. Quantifying range-wide variation in population trends from local abundance surveys and widespread opportunistic occurrence records. *Methods in Ecology and Evolution* 5: 751–760.

98. Isaac, N. J. B., van Strien, A. J., August, T. A. et al. 2014. Statistics for citizen science: Extracting signals of change from noisy ecological data. *Methods in Ecology and Evolution* 5: 1052–1060.

99. Darwin, C. 1859. *On the Origin of Species by Means of Natural Selection, or, The Preservation of Favoured Races in the Struggle for Life.* London, UK: John Murray.

100. Hanna, E. and Cardillo, M. 2013. A comparison of current and reconstructed historic geographic range sizes as predictors of extinction risk in Australian mammals. *Biological Conservation* 158: 196–204.

101. Sekercioglu, C. H. 2012. Promoting community-based bird monitoring in the tropics: Conservation, research, environmental education, capacity-building, and local incomes. *Biological Conservation* 151: 69–73.

102. Stephens, P. A. and Sutherland, W. J. 1999. Consequences of the Allee effect for behaviour, ecology and conservation. *Trends in Ecology & Evolution* 14: 401–405.

103. Denney, N. H., Jennings, S. and Reynolds, J. D. 2002. Life-history correlates of maximum population growth rates in marine fishes. *Proceedings of the Royal Society B: Biological Sciences* 269: 2229–2237.

104. Kolb, A. and Diekmann, M. 2005. Effects of life-history traits on responses of plant species to forest fragmentation. *Conservation Biology* 19: 929–938.

105. Stein, B. A., Kutner, L. S. and Adams, J. S. 2000. *Precious Heritage: The Status of Biodiversity in the United States.* New York, NY: Oxford University Press.

106. Wilcove, D. S. and Wikelski, M. 2008. Going, going, gone: Is animal migration disappearing? *PLOS Biology* 6: 1361–1364.

107. Reed, J. M. 1999. The role of behavior in recent avian extinctions and endangerments. *Conservation Biology* 13: 232–241.

108. Ceballos, G., García, A. and Ehrlich, P. R. 2010. The sixth extinction crisis: Loss of animal populations and species. *Journal of Cosmology* 8: 31.

109. Wake, D. B. and Vredenburg, V. T. 2008. Are we in the midst of the sixth mass extinction? A view from the world of amphibians. *Proceedings of the National Academy of Sciences* 105: 11466–11473.

110. Leakey, R. E. and Lewin, R. 1996. *The Sixth Extinction: Patterns of Life and the Future of Humankind.* New York, NY: Doubleday & Company.

111. Kolbert, E. 2014. *The Sixth Extinction: An Unnatural History.* New York, NY: Henry Holt and Company.

112. Stuart, S. N., Chanson, J. S., Cox, N. A. et al. 2004. Status and trends of amphibian declines and extinctions worldwide. *Science* 306: 1783–1786.

113. Murray, K. A., Skerratt, L. F., Speare, R. and McCallum, H. 2009. Impact and dynamics of disease in species threatened by the amphibian chytrid fungus, *Batrachochytrium dendrobatidis*. *Conservation Biology* 23: 1242–1252.

Figure 8.5 original sources: P. H. Raven. 1987. In *Botanic Gardens and the World Conservation Strategy*. London: Academic Press; N. Myers. 1988. *Environmentalist* 8: 187; W. V. Reid and K. R. Miller. 1989. *Keeping Options Alive: The Scientific Basis for Conserving Biodiversity*. Washington, DC: World Resources Institute; M. V. Reid. 1992. In *Tropical Deforestation and Species Extinction*. London, UK: Chapman & Hall; N. Myers 1979. *The Sinking Ark: A New Look at the Problem of Disappearing Species.* Oxford, UK: Pergamon Press; T. E. Lovejoy. 1980. *The Global 2000 Report to the President, vol. 2.* Washington, DC: Council on Environmental Quality; P. H. Raven. 1988. In *Biodiversity*. Washington, DC: National Academy Press; E. O. Wilson. 1988. in *Biodiversity*. Washington, DC: National Academy Press, and E. O. Wilson. 1989. *Sci Am* 261: 108; N. Myers and J. L. Simon. 1994. *Scarcity or Abundance*. New York, NY: Norton.

Chapter 9

1. World Wide Fund for Nature. 2016. *Living Planet Report 2016. Risk and Resilience in a New Era.* Gland, Switzerland: WWF International.

2. IUCN. 2018. *The IUCN Red List of Threatened Species.* Gland, Switzerland: International Union for Conservation of Nature and Natural Resources.

3. Laverty, M. F. and Gibbs, J. P. 2007. Ecosystem loss and fragmentation. *Lessons in Conservation* 1: 72–96.

4. Laurance, W. F. 2010. Habitat destruction: Death by a thousand cuts. In N. S. Sodhi and P. R. Ehrlich (Eds.), *Conservation Biology for All.* New York, NY: Oxford University Press.

5. Foley, J. A., DeFries, R., Asner, G. P. et al. 2005. Global consequences of land use. *Science* 309: 570–574.

6. Hooke, R. L. and Martin-Duque, J. F. 2012. Land transformation by humans: A review. *GSA Today* 22: 4–10.

7. Vörösmarty, C. J., McIntyre, P. B., Gessner, M. O. et al. 2010. Global threats to human water security and river biodiversity. *Nature* 467: 555.

8. Halpern, B. S., Walbridge, S., Selkoe, K. A. et al. 2008. A global map of human impact on marine ecosystems. *Science* 319: 948–952.

9. Groom, M. J. and Vynne, C. H. 2006. Habitat degradation and loss. In M. J. Groom, G. K. Meffe, C. R. Carrol, and G. K. Meffe (Eds.), *Principles of Conservation Biology* (3rd ed.). Sunderland, MA: Oxford University Press/Sinauer.
10. Wood, S., Sebastian, K. and Scherr, S. J. 2000. *Pilot Assessment of Global Ecosystems: Agroecosystems*. Washington, DC: International Food Policy Research Institute and World Resources Institute.
11. Congalton, G. R., Yadav, K., McDonnell, K. et al. 2017. NASA Making Earth System Data Records for Use in Research Environments (MEaSUREs) Global Food Security-Support Analysis Data (GFSAD) cropland extent 2015 validation global 30 m V001 [Data set]. Sioux Falls, SD: NASA EOSDIS Land Processes DAAC.
12. Foley, J. A., Ramankutty, N., Brauman, K. A. et al. 2011. Solutions for a cultivated planet. *Nature* 478: 337–342.
13. Gerland, P., Raftery, A. E., Sevcikova, H. et al. 2014. World population stabilization unlikely this century. *Science* 346: 234–237.
14. Alexandratos, N. and Bruinsma, J. 2012. *World Agriculture Towards 2030/2050: The 2012 Revision*. Rome, Italy: Food and Agriculture Organization of the United Nations.
15. Searchinger, T., Hanson, C., Ranganathan, J. et al. 2013. *Creating a Sustainable Food Future: Interim Findings*. Washington, DC: World Resources Institute.
16. Tilman, D., Balzer, C., Hill, J. and Befort, B. L. 2011. Global food demand and the sustainable intensification of agriculture. *Proceedings of the National Academy of Sciences* 108: 20260–20264.
17. Grassini, P., Eskridge, K. M. and Cassman, K. G. 2013. Distinguishing between yield advances and yield plateaus in historical crop production trends. *Nature Communications* 4: 2918.
18. Fischer, T., Byerlee, D. and Edmeades, G. 2014. *Crop Yields and Global Food Security: Will Yield Increase Continue to Feed the World?* Canberra: Australian Centre for International Agricultural Research.
19. Schmitz, C., van Meijl, H., Page, K. et al. 2014. Land-use change trajectories up to 2050: Insights from a global agro-economic model comparison. *Agricultural Economics* 45: 69–84.
20. Smith, P., Gregory, P. J., van Vuuren, D. et al. 2010. Competition for land. *Philosophical Transactions of the Royal Society B: Biological Sciences* 365: 2941–2957.
21. Newbold, T., Hudson, L. N., Hill, S. L. L. et al. 2015. Global effects of land use on local terrestrial biodiversity. *Nature* 520: 45–50.
22. de Baan, L., Alkemade, R. and Koellner, T. 2013. Land use impacts on biodiversity in LCA: A global approach. *The International Journal of Life Cycle Assessment* 18: 1216–1230.
23. Vandermeer, J. and Perfecto, I. 2005. The future of farming and conservation. *Science* 308: 1257–1257.
24. Perfecto, I., Vandermeer, J. H. and Wright, A. L. 2009. *Nature's Matrix: Linking Agriculture, Conservation and Food Sovereignty*. Sterling, VA: Earthscan.
25. Gonthier, D. J., Ennis, K. K., Farinas, S. et al. 2014. Biodiversity conservation in agriculture requires a multi-scale approach. *Proceedings of the Royal Society B: Biological Sciences* 281: 8.
26. Karp, D. S., Ziv, G., Zook, J. et al. 2011. Resilience and stability in bird guilds across tropical countryside. *Proceedings of the National Academy of Sciences* 108: 21134–21139.
27. Phalan, B., Onial, M., Balmford, A. and Green, R. E. 2011. Reconciling food production and biodiversity conservation: Land sharing and land sparing compared. *Science* 333: 1289–1291.
28. Tscharntke, T., Clough, Y., Wanger, T. C. et al. 2012. Global food security, biodiversity conservation and the future of agricultural intensification. *Biological Conservation* 151: 53–59.
29. Seto, K. C., Güneralp, B. and Hutyra, L. R. 2012. Global forecasts of urban expansion to 2030 and direct impacts on biodiversity and carbon pools. *Proceedings of the National Academy of Sciences* 109: 16083–16088.
30. Aronson, M. F. J., La Sorte, F. A., Nilon, C. H. et al. 2014. A global analysis of the impacts of urbanization on bird and plant diversity reveals key anthropogenic drivers. *Proceedings of the Royal Society B: Biological Sciences* 281: 20133330.
31. Ellis, E. C., Antill, E. C. and Kreft, H. 2012. All is not loss: Plant biodiversity in the Anthropocene. *PLOS ONE* 7: e30535.
32. McPhearson, T., Pickett, S. T. A., Grimm, N. B. et al. 2016. Advancing urban ecology toward a science of cities. *BioScience* 66: 198–212.
33. Kremer, P., Hamstead, Z., Haase, D. et al. 2016. Key insights for the future of urban ecosystem services research. *Ecology and Society* 21: 29.
34. Finer, M., Jenkins, C. N., Pimm, S. L. et al. 2008. Oil and gas projects in the western Amazon: Threats to wilderness, biodiversity, and indigenous peoples. *PLOS ONE* 3: e2932.
35. Allred, B. W., Smith, W. K., Twidwell, D. et al. 2015. Ecosystem services lost to oil and gas in North America. *Science* 348: 401–402.
36. Petersen, R., Sizer, N. and Lee, P. 2014. *Tar Sands Threaten World's Largest Boreal Forest*. Washington, DC: World Resources Institute.
37. Arbogast, B. F., Knepper, D. H. and Langer, W. H. 2000. *The Human Factor in Mining Reclamation*. Circular 1191. Denver, CO: U.S. Geological Survey.
38. EPC. 1998. Background Paper on Land Access, Protected Areas and Sustainable Development. Canada: Natural Resources Canada.
39. MCA. 2010. *The Australian Minerals Industry and the Australian Economy*. Canberra: Minerals Council of Australia.
40. Durán, A. P., Rauch, J. and Gaston, K. J. 2013. Global spatial coincidence between protected areas and metal mining activities. *Biological Conservation* 160: 272–278.
41. Palmer, M. A., Bernhardt, E. S., Schlesinger, W. H. et al. 2010. Mountaintop mining consequences. *Science* 327: 148–149.
42. Pusceddu, A., Bianchelli, S., Martín, J. et al. 2014. Chronic and intensive bottom trawling impairs deep-sea biodiversity and ecosystem functioning. *Proceedings of the National Academy of Sciences* 111: 8861–8866.
43. Jones, J. B. 1992. Environmental impact of trawling on the seabed: A review. *New Zealand Journal of Marine and Freshwater Research* 26: 59–67.
44. Calado, R., Leal, M. C., Vaz, M. C. M. et al. 2014. Caught in the act: How the U.S. Lacey Act can hamper the fight against cyanide fishing in tropical coral reefs. *Conservation Letters* 7: 561–564.
45. Chan, A. and Hodgson, P. A. 2017. A systematic analysis of blast fishing in South-East Asia and possible solutions. *2017 IEEE Underwater Technology (UT)*. https://ieeexplore.ieee.org/document/7890330.
46. FAO. 2016. *The State of the World's Fisheries and Aquaculture 2016*. Rome, Italy: United Nations.
47. Polidoro, B. A., Carpenter, K. E., Collins, L. et al. 2010. The loss of species: Mangrove extinction risk and geographic areas of global concern. *PLOS ONE* 5: e10095.
48. Barbier, E. B. 2016. The protective service of mangrove ecosystems: A review of valuation methods. *Marine Pollution Bulletin* 109: 676–681.
49. Martinuzzi, S., Gould, W. A., Lugo, A. E. and Medina, E. 2009. Conversion and recovery of Puerto Rican mangroves: 200 years of change. *Forest Ecology and Management* 257: 75–84.
50. Luther, D. A. and Greenberg, R. 2009. Mangroves: A global perspective on the evolution and conservation of their terrestrial vertebrates. *BioScience* 59: 602–612.

51. Barbier, E. B. 2008. In the wake of tsunami: Lessons learned from the household decision to replant mangroves in Thailand. *Resource and Energy Economics* 30: 229–249.
52. FAO. 2016. *State of the World's Forests 2016*. Rome, Italy: United Nations.
53. Corlett, R. and Primack, R. B. 2011. *Tropical Rain Forests: An Ecological and Biogeographical Comparison* (2nd ed.). Hoboken, NJ: Wiley-Blackwell.
54. Bradshaw, C. J. A., Sodhi, N. S. and Brook, B. W. 2009. Tropical turmoil: A biodiversity tragedy in progress. *Frontiers in Ecology and the Environment* 7: 79–87.
55. Enuoh, O. and Bisong, F. 2015. Colonial forest policies and tropical deforestation: The case of cross river state, Nigeria. *Open Journal of Forestry* 5: 66–79.
56. Rudel, T. K. 2013. The national determinants of deforestation in sub-Saharan Africa. *Philosophical Transactions of the Royal Society B: Biological Sciences* 368: 20120405.
57. Lambin, E. F. and Meyfroidt, P. 2011. Global land use change, economic globalization, and the looming land scarcity. *Proceedings of the National Academy of Sciences* 108: 3465–3472.
58. Peres, C. A. and Schneider, M. 2012. Subsidized agricultural resettlements as drivers of tropical deforestation. *Biological Conservation* 151: 65–68.
59. Phua, M.-H., Tsuyuki, S., Furuya, N. and Lee, J. S. 2008. Detecting deforestation with a spectral change detection approach using multitemporal Landsat data: A case study of Kinabalu Park, Sabah, Malaysia. *Journal of Environmental Management* 88: 784–795.
60. Vijay, V., Pimm, S. L., Jenkins, C. N. and Smith, S. J. 2016. The impacts of oil palm on recent deforestation and biodiversity loss. *PLOS ONE* 11: e0159668.
61. Fitzherbert, E. B., Struebig, M. J., Morel, A. et al. 2008. How will oil palm expansion affect biodiversity? *Trends in Ecology & Evolution* 23: 538–545.
62. Ancrenaz, M., Oram, F., Ambu, L. et al. 2014. Of *Pongo*, palms and perceptions: A multidisciplinary assessment of Bornean orang-utans *Pongo pygmaeus* in an oil palm context. *Oryx* 49: 465–472.
63. Linder, J. M. and Palkovitz, R. E. 2016. The threat of industrial oil palm expansion to primates and their habitats. In M. Waller (Ed.), *Developments in Primatology: Progress and Prospects*. Cham, Switzerland: Springer.
64. Evans, L. J., Asner, G. P. and Goossens, B. 2018. Protected area management priorities crucial for the future of Bornean elephants. *Biological Conservation* 221: 365–373.
65. Luskin, M. S., Albert, W. R. and Tobler, M. W. 2017. Sumatran tiger survival threatened by deforestation despite increasing densities in parks. *Nature Communications* 8: 1783.
66. Suttie, J. M., Reynolds, S. G. and Batello, C. 2005. *Grasslands of the World*. Rome, Italy: Food and Agriculture Organization of the United Nations.
67. Oakleaf, J. R., Kennedy, C. M., Baruch-Mordo, S. et al. 2015. A world at risk: Aggregating development trends to forecast global habitat conversion. *PLOS ONE* 10: e0138334.
68. Burke, L., Reytar, K., Spalding, M. and Perry, A. 2011. *Reefs at Risk: Revisited*. Washington, DC: World Resources Institute (WRI).
69. Hughes, T. P., Hui, H. and Yount, M. 2013. The wicked problem of China's disappearing coral reefs. *Conservation Biology* 27: 261–269.
70. Hughes, T. P., Kerry, J. T., Baird, A. H. et al. 2018. Global warming transforms coral reef assemblages. *Nature* 556: 492–496.
71. Bruno, J. F. and Valdivia, A. 2016. Coral reef degradation is not correlated with local human population density. *Scientific Reports* 6: 29778.
72. Heron, S. F., Maynard, J. A., van Hooidonk, R. and Eakin, C. M. 2016. Warming trends and bleaching stress of the world's coral reefs 1985–2012. *Scientific Reports* 6: 38402.
73. Marshall, P. and Schuttenberg, H. 2006. *A Reef Manager's Guide to Coral Bleaching*. Townsville, Australia: Great Barrier Reef Marine Park Authority.
74. Hughes, T. P., Barnes, M. L., Bellwood, D. R. et al. 2017. Coral reefs in the Anthropocene. *Nature* 546: 82–90.
75. Davidson, N. C. 2014. How much wetland has the world lost? Long-term and recent trends in global wetland area. *Marine and Freshwater Research* 65: 934–941.
76. Strayer, D. L. and Dudgeon, D. 2010. Freshwater biodiversity conservation: Recent progress and future challenges. *Journal of the North American Benthological Society* 29: 344–358.
77. World Commission on Dams. 2000. *Dams and Development: A New Framework for Decision-making*. Report No. 1853837989. London, UK: Earthscan.
78. Grill, G., Lehner, B., Lumsdon, A. E. et al. 2015. An index-based framework for assessing patterns and trends in river fragmentation and flow regulation by global dams at multiple scales. *Environmental Research Letters* 10: 15.
79. Strayer, D. L. 2006. Challenges for freshwater invertebrate conservation. *Journal of the North American Benthological Society* 25: 271–287.
80. Fahrig, L. 2003. Effects of habitat fragmentation on biodiversity. *Annual Review of Ecology Evolution and Systematics* 34: 487–515.
81. Bird-Jackson, H. and Fahrig, L. 2013. Habitat loss and fragmentation. In S. A. Levin (Ed.), *Encyclopedia of Biodiversity* (2nd ed.). New York, NY: Academic Press.
82. Drakare, S., Lennon, J. J. and Hillebrand, H. 2006. The imprint of the geographical, evolutionary and ecological context on species-area relationships. *Ecology Letters* 9: 215–227.
83. Fahrig, L. 2002. Effect of habitat fragmentation on the extinction threshold: A synthesis. *Ecological Applications* 12: 346–353.
84. Diamond, J. M. 1975. The island dilemma: Lessons of modern biogeographic studies for the design of natural reserves. *Biological Conservation* 7: 129–146.
85. MacArthur, R. H. and Wilson, E. O. 1967. *The Theory of Island Biogeography*. Princeton, NJ: Princeton University Press.
86. Simberloff, D. S. and Abele, L. G. 1976. Island biogeography theory and conservation practice. *Science* 191: 285–286.
87. Haddad, N. M., Brudvig, L. A., Clobert, J. et al. 2015. Habitat fragmentation and its lasting impact on Earth's ecosystems. *Science Advances* 1: e1500052.
88. Smith, I. A., Hutyra, L. R., Reinmann, A. B. et al. 2018. Piecing together the fragments: Elucidating edge effects on forest carbon dynamics. *Frontiers in Ecology and the Environment* 16: 213–221.
89. Laurance, W. F., Lovejoy, T. E., Vasconcelos, H. L. et al. 2002. Ecosystem decay of Amazonian forest fragments: A 22-year investigation. *Conservation Biology* 16: 605–618.
90. Lampila, P., Monkkonen, M. and Desrochers, A. 2005. Demographic responses by birds to forest fragmentation. *Conservation Biology* 19: 1537–1546.
91. Loss, S. R. and Marra, P. P. 2017. Population impacts of free-ranging domestic cats on mainland vertebrates. *Frontiers in Ecology and the Environment* 15: 502–509.
92. Lloyd, P., Martin, T. E., Redmond, R. L. et al. 2005. Linking demographic effects of habitat fragmentation across landscapes to continental source-sink dynamics. *Ecological Applications* 15: 1504–1514.
93. Valiela, I. and Martinetto, P. 2007. Changes in bird abundance in eastern North America: Urban sprawl and global footprint? *BioScience* 57: 360–370.
94. Allan, B. F., Keesing, F. and Ostfeld, R. S. 2003. Effect of forest fragmentation on Lyme disease risk. *Conservation Biology* 17: 267–272.
95. Flory, S. L. and Clay, K. 2009. Effects of roads and forest successional age on experimental plant invasions. *Biological Conservation* 142: 2531–2537.

96. Malcolm, J. 1998. Fragments of the forest: High roads to oblivion. *Natural History* July: 46–49.
97. Landrigan, P. J., Fuller, R., Acosta, N. J. R. et al. 2018. The Lancet Commission on pollution and health. *The Lancet* 391: 462–512.
98. Geiger, F., Bengtsson, J., Berendse, F. et al. 2010. Persistent negative effects of pesticides on biodiversity and biological control potential on European farmland. *Basic and Applied Ecology* 11: 97–105.
99. Beketov, M. A., Kefford, B. J., Schäfer, R. B. and Liess, M. 2013. Pesticides reduce regional biodiversity of stream invertebrates. *Proceedings of the National Academy of Sciences* 110: 11039–11043.
100. Goulson, D. 2013. An overview of the environmental risks posed by neonicotinoid insecticides. *Journal of Applied Ecology* 50: 977–987.
101. Brühl, C. A., Schmidt, T., Pieper, S. and Alscher, A. 2013. Terrestrial pesticide exposure of amphibians: An underestimated cause of global decline? *Scientific Reports* 3: 1135.
102. Bayat, S., Geiser, F., Kristiansen, P. and Wilson, S. C. 2014. Organic contaminants in bats: Trends and new issues. *Environment International* 63: 40–52.
103. ITOPF. 2018. *Oil Tanker Spill Statistics 2017.* London, UK: The International Tanker Owners Pollution Federation.
104. Eckle, P., Burgherr, P. and Michaux, E. 2012. Risk of large oil spills: A statistical analysis in the aftermath of Deepwater Horizon. *Environmental Science & Technology* 46: 13002–13008.
105. Chang, S. E., Stone, J., Demes, K. and Piscitelli, M. 2014. Consequences of oil spills: A review and framework for informing planning. *Ecology and Society* 19: 26.
106. Peterson, C. H., Rice, S. D., Short, J. W. et al. 2003. Long-term ecosystem response to the Exxon *Valdez* oil spill. *Science* 302: 2082–2086.
107. Paine, R. T., Ruesink, J. L., Sun, A. et al. 1996. Trouble on oiled waters: Lessons from the Exxon *Valdez* oil spill. *Annual Review of Ecology and Systematics* 27: 197–235.
108. Nelson-Smith, A. 1971. The problem of oil pollution of the sea. *Advances in Marine Biology* 8: 215–306.
109. Stephen, C. J., Thomas, A. D., Richard, O. S. and Arny, B. 1999. Exxon *Valdez* oil spill: Impacts and recovery in the soft-bottom benthic community in and adjacent to eelgrass beds. *Marine Ecology Progress Series* 185: 59–83.
110. Foster, M. S., Tarpley, J. A. and Dearn, S. L. 1990. To clean or not to clean: The rationale, methods, and consequences of removing oil from temperate shores. *Northwest Environmental Journal* 6: 105–120.
111. Judson, R. S., Martin, M. T., Reif, D. M. et al. 2010. Analysis of eight oil spill dispersants using rapid, in vitro tests for endocrine and other biological activity. *Environmental Science & Technology* 44: 5979–5985.
112. Sriram, K., Lin, G. X., Jefferson, A. M. et al. 2011. Neurotoxicity following acute inhalation exposure to the oil dispersant COREXIT EC9500A. *Journal of Toxicology and Environmental Health, Part A* 74: 1405–1418.
113. Castranova, V. 2011. Bioactivity of oil dispersant used in the Deepwater Horizon cleanup operation. *Journal of Toxicology and Environmental Health, Part A* 74: 1367–1367.
114. Wuana, R. A. and Okieimen, F. E. 2011. Heavy metals in contaminated soils: A review of sources, chemistry, risks and best available strategies for remediation. *ISRN Ecology* 2011, Article 402647.
115. Carpenter, S. R., Caraco, N. F., Correll, D. L. et al. 1998. Nonpoint pollution of surface waters with phosphorus and nitrogen. *Ecological Applications* 8: 559–568.
116. Dodds, W. K. 2006. Eutrophication and trophic state in rivers and streams. *Limnology and Oceanography* 51: 671–680.
117. Diaz, R. J. and Rosenberg, R. 2008. Spreading dead zones and consequences for marine ecosystems. *Science* 321: 926–929.
118. Likens, G. E., Driscoll, C. T. and Buso, D. C. 1996. Long-term effects of acid rain: Response and recovery of a forest ecosystem. *Science* 272: 244–246.
119. Larssen, T., Lydersen, E., Tang, D. et al. 2006. Acid rain in China. *Environmental Science & Technology* 40: 418–425.
120. Singh, A. and Agrawal, M. 2007. Acid rain and its ecological consequences. *Journal of Environmental Biology* 29: 15.
121. Glassmeyer, S. T., Furlong, E. T., Kolpin, D. W. et al. 2017. Nationwide reconnaissance of contaminants of emerging concern in source and treated drinking waters of the United States. *Science of the Total Environment* 581–582: 909–922.
122. Brodin, T., Fick, J., Jonsson, M. and Klaminder, J. 2013. Dilute concentrations of a psychiatric drug alter behavior of fish from natural populations. *Science* 339: 814–815.
123. Kidd, K. A., Blanchfield, P. J., Mills, K. H. et al. 2007. Collapse of a fish population after exposure to a synthetic estrogen. *Proceedings of the National Academy of Sciences* 104: 8897–8901.
124. Eriksen, M., Lebreton, L. C. M., Carson, H. S. et al. 2014. Plastic pollution in the world's oceans: More than 5 trillion plastic pieces weighing over 250,000 tons afloat at sea. *PLOS ONE* 9: e111913.
125. Gregory, M. R. 2009. Environmental implications of plastic debris in marine settings—entanglement, ingestion, smothering, hangers-on, hitch-hiking and alien invasions. *Philosophical Transactions of the Royal Society B: Biological Sciences* 364: 2013–2025.
126. van Franeker, J. A. and Law, K. L. 2015. Seabirds, gyres and global trends in plastic pollution. *Environmental Pollution* 203: 89–96.
127. Seltenrich, N. 2015. New link in the food chain? Marine plastic pollution and seafood safety. *Environmental Health Perspectives* 123: A34–A41.
128. Bestelmeyer, B. T., Okin, G. S., Duniway, M. C. et al. 2015. Desertification, land use, and the transformation of global drylands. *Frontiers in Ecology and the Environment* 13: 28–36.
129. Geist, H. 2017. *The Causes and Progression of Desertification.* Hants, UK: Ashgate.
130. Micklin, P. 2007. The Aral Sea disaster. *Annual Review of Earth and Planetary Sciences* 35: 47–72.
131. Wilkinson, B. H. and McElroy, B. J. 2007. The impact of humans on continental erosion and sedimentation. *GSA Bulletin* 119: 140–156.
132. Levins, R. 1969. Some demographic and genetic consequences of environmental heterogeneity for biological control. *Bulletin of the Entomological Society of America* 15: 237–240.
133. Hanski, I. 1999. *Metapopulation Biology.* Oxford, UK: Oxford University Press.
134. Harrison, S., Murphy, D. D. and Ehrlich, P. R. 1988. Distribution of the bay checkerspot butterfly, *Euphydryas editha bayensis*: Evidence for a metapopulation model. *American Naturalist* 132: 360–382.
135. Schultz, C. B. and Crone, E. E. 2005. Patch size and connectivity thresholds for butterfly habitat restoration. *Conservation Biology* 19: 887–896.
136. Ehrlich, P. R., Murphy, D. D., Singer, M. C. et al. 1980. Extinction, reduction, stability and increase: The responses of checkerspot butterfly (*Euphyryas*) populations to the California drought. *Oecologia* 46: 101–105.
137. Claire, K. 2015. Reframing the land-sparing/land-sharing debate for biodiversity conservation. *Annals of the New York Academy of Sciences* 1355: 52–76.
138. Grau, R., Kuemmerle, T. and Macchi, L. 2013. Beyond "land sparing versus land sharing": Environmental heterogeneity, globalization and the balance between agricultural production and nature conservation. *Current Opinion in Environmental Sustainability* 5: 477–483.
139. Bennett, E. M. 2017. Changing the agriculture and environment conversation. *Nature Ecology and Evolution* 1: 0018.

140. Salles, J.-M., Teillard, F., Tichit, M. and Zanella, M. 2017. Land sparing versus land sharing: An economist's perspective. *Regional Environmental Change* 17: 1455–1465.

141. Fischer, J., Abson, D. J., Butsic, V. et al. 2014. Land sparing versus land sharing: Moving forward. *Conservation Letters* 7: 149–157.

142. McMillan, A. and Foley, G. 2014. A history of air quality management. In E. Taylor and A. McMillan (Eds.), *Air Quality Management*. Dordrecht, The Netherlands: Springer.

143. Hartig, J. 2010. A brief environmental history of the Detroit River and western Lake Erie since the 1940s. *Great Lakes Echo.* April 5.

144. Binelli, M. 2012. How Detroit became the world capital of staring at abandoned old buildings. *The New York Times Magazine*. November 9.

145. Metropolitan Affairs Coalition. 2000. *A Conservation Vision for the Lower Detroit River Ecosystem*. Detroit, MI: Metropolitan Affairs Coalition. https://www.fws.gov/refuge/detroit_river/about/refuge_history.html

146. Oosterveer, P. 2015. Promoting sustainable palm oil: Viewed from a global networks and flows perspective. *Journal of Cleaner Production* 107: 146–153.

147. Saikkonen, L., Ollikainen, M. and Lankoski, J. 2014. Imported palm oil for biofuels in the EU: Profitability, greenhouse gas emissions and social welfare effects. *Biomass and Bioenergy* 68: 7–23.

148. Colchester, M. and Chao, S. 2011. *Oil Palm Expansion in South East Asia: Trends and Implications for Local Communities and Indigenous Peoples*. Moreton-in-Marsh, UK: Forest Peoples Programme.

Chapter 10

1. Maxwell, S., Fuller, R. A., Brooks, T. M. and Watson, J. E. M. 2016. The ravages of guns, nets and bulldozers. *Nature* 536: 143–145.

2. Martin, P. S. 1984. Prehistoric overkill: The global model. In P.S. Martin and R. G. Klein (Eds.), *Quaternary Extinctions*. Tucson, AZ: University of Arizona Press.

3. Martin, P. S. and Steadman, D. W. 1999. Prehistoric extinctions on islands and continents. In R. D. E. MacPhee (Eds.), *Extinctions in Near Time: Causes, Contexts, and Consequences*. New York, NY: Kluwer.

4. Alroy, J. 2001. A multispecies overkill simulation of the end-Pleistocene megafaunal mass extinction. *Science* 292: 1893–1896.

5. Hunter, M. L. and Gibbs, J. 2007. *Fundamentals of Conservation Biology*, 3rd ed. Malden, MA: Blackwell Publishing.

6. Steadman, D. W. and Martin, P. S. 2003. The late Quaternary extinction and future resurrection of birds on Pacific islands. *Earth-Science Reviews* 61: 133–147.

7. Allentoft, M. E., Heller, R., Oskam, C. L. et al. 2014. Extinct New Zealand megafauna were not in decline before human colonization. *Proceedings of the National Academy of Sciences* 111: 4922–4927.

8. Olson, S. L. and James, H. F. 1984. The role of Polynesians in the extinction of the avifauna of the Hawaiian Islands. In P.S. Martin and R. G. Klein (Eds.), *Quaternary Extinctions*. Tucson, AZ: University of Arizona Press.

9. Schorger, A. W. 1955. *The Passenger Pigeon, Its Natural History and Extinction*. Madison, WI: University of Wisconsin Press.

10. Loftas, T. and Ross, J. 1995. *Dimensions of Need: An Atlas of Food and Agriculture*. Rome: Food and Agriculture Organization of the United Nations.

11. FAO. 2015. *Statistical Pocketbook: World Food and Agriculture*. Rome: Food and Agriculture Organization of the United Nations.

12. Van Lange, P. A. M., Joireman, J., Parks, C. D. and Van Dijk, E. 2013. The psychology of social dilemmas: A review. *Organizational Behavior and Human Decision Processes* 120: 125–141.

13. Ostrom, E. 2015. *Governing the Commons*. Cambridge, UK: Cambridge University Press.

14. Ostrom, E. 2009. A general framework for analyzing sustainability of social-ecological systems. *Science* 325: 419–422.

15. Jager, W., Janssen, M. A. and Vlek, C. A. J. 2002. How uncertaintly stimulates over-harvesting in a resource dilemma: Three process explanations. *Journal of Environmental Psychology* 22: 247–263.

16. Darimont, C. T., Codding, B. F. and Hawkes, K. 2017. Why men trophy hunt. *Biology Letters* 13: 20160909.

17. Oleson, J. C. and Henry, B. C. 2009. Relations among need for power, affect and attitudes toward animal cruelty. *Anthrozoös* 22: 255–265.

18. Sterelny, K. 2006. The perverse primate. In A. Grafen and M. Ridley (Eds.), *Richard Dawkins: How a Scientist Changed the Way We Think*. New York, NY: Oxford University Press.

19. Diamond, J. M. 2005. Collapse: How Societies Choose to Fail or Succeed. New York, NY: Viking.

20. FAO. 2018. The State of the World's Fisheries and Aquaculture 2018. Rome, Italy.

21. Harris, M. 1998. Lament for an Ocean: The Collapse of the Atlantic Cod Fishery. McClelland & Stewart: Toronto, Ont.

22. Idyll, C. P. 1973. The Anchovy crisis. *Scientific American* 228: 22–29.

23. Clark, W. The lessons of the Peruvian anchoveta fishery.

24. Sigurdsson, T. 2006. *The Collapse of the Atlanto-Scandian Herring Fishery: Effects on the Icelandic Economy*. (International Institute of Fisheries Economics & Trade) Portsmouth, UK.

25. Worm, B., Barbier, E. B., Beaumont, N. et al. 2006. Impacts of biodiversity loss on ocean ecosystem services. *Science* 314: 787–790.

26. Mora, C., Myers, R. A., Coll, M. et al. 2009. Management effectiveness of the World's marine fisheries. *PLOS Biology* 7: e1000131.

27. Pauly, D., Christensen, V., Dalsgaard, J. et al. 1998. Fishing down marine food webs. *Science* 279: 860–863.

28. McCauley, D. J., Pinsky, M. L., Palumbi, S. R. et al. 2015. Marine defaunation: Animal loss in the global ocean. *Science* 347.

29. Le Pape, O., Bonhommeau, S., Nieblas, A.-E. and Fromentin, J.-M. 2017. Overfishing causes frequent fish population collapses but rare extinctions. *Proceedings of the National Academy of Sciences* 114: E6274–E6274.

30. Burgess, M. G., Costello, C., Fredston-Hermann, A. et al. 2017. Range contraction enables harvesting to extinction. *Proceedings of the National Academy of Sciences* 114: 3945–3950.

31. Jackson, J. B. C., Kirby, M. X., Berger, W. H. et al. 2001. Historical overfishing and the recent collapse of coastal ecosystems. *Science* 293: 629–638.

32. Benitez-Lopez, A., Alkemade, R., Schipper, A. M. et al. 2017. The impact of hunting on tropical mammal and bird populations. *Science* 356: 180–183.

33. Darimont, C. T., Fox, C. H., Bryan, H. M. and Reimchen, T. E. 2015. The unique ecology of human predators. *Science* 349: 858–860.

34. Dulvy, N. K., Sadovy, Y. and Reynolds, J. D. 2003. Extinction vulnerability in marine populations. *Fish and Fisheries* 4: 25–64.

35. Le Pape, O. and Bonhommeau, S. 2015. The food limitation hypothesis for juvenile marine fish. *Fish and Fisheries* 16: 373–398.

36. Lewison, R. L., Crowder, L. B., Read, A. J. and Freeman, S. A. 2004. Understanding impacts of fisheries bycatch on marine megafauna. *Trends in Ecology & Evolution* 19: 598–604.

37. Zydelis, R., Small, C. and French, G. 2013. The incidental catch of seabirds in gillnet fisheries: A global review. *Biological Conservation* 162: 76–88.

38. FAO. 2016. State of the World's Forests 2016. Rome, Italy.

39. Rodan, B. D. and Campbell, F. T. 1996. CITES and the sustainable management of *Swietenia macrophylla King. Botanical Journal of the Linnean Society* 122: 83–87.
40. Lamb, F. B. 1966. *Mahogany of Tropical America: Its Ecology and Management.* Ann Arbor, MI: University of Michigan Press.
41. Veríssimo, A., Barreto, P., Tarifa, R. and Uhl, C. 1995. Extraction of a high-value natural resource in Amazonia: the case of mahogany. *Forest Ecology and Management* 72: 39–60.
42. Stewart, K. M. 2003. The African cherry (Prunus africana): Can lessons be learned from an overexploited medicinal tree? *Journal of Ethnopharmacology* 89: 3–13.
43. Jimu, L. 2011. Threats and conservation strategies for the African cherry (*Prunus africana*) in its natural range- A review *Journal of Ecology and the Natural Environment* 3: 118–130.
44. Beech, E., Rivers, M., Oldfield, S. and Smith, P. P. 2017. GlobalTreeSearch: The first complete global database of tree species and country distributions. *Journal of Sustainable Forestry* 36: 454–489.
45. Cai, Z. and Aguilar, F. X. 2013. Meta-analysis of consumer's willingness-to-pay premiums for certified wood products. *Journal of Forest Economics* 19: 15–31.
46. van der Ven, H. and Cashore, B. 2018. Forest certification: the challenge of measuring impacts. *Current Opinion in Environmental Sustainability* 32: 104–111.
47. Auld, G. and Gulbrandsen, L. H. 2010. Transparency in nonstate certification: Consequences for accountability and transparency. *Global Environmental Politics* 10: 97–199.
48. Contiff, R. 2018. *Greenwashed Timber: How Sustainable Forest Certification Has Failed.* YaleEnvironment360. New Haven, CT: Yale School of Forestry & Environmental Studies.
49. Bicknell, J. E., Struebig, M. J., Edwards, D. P. and Davies, Z. G. 2014. Improved timber harvest techniques maintain biodiversity in tropical forests. *Current Biology* 24: R1119-R1120.
50. Dasgupta, S. 2017. *Does Forest Certification Really Work?* Mongabay series: Conservation Effectiveness. Menlo Park, CA: Mongabay.
51. Overdevest, C. 2010. Comparing forest certification schemes: the case of ratcheting standards in the forest sector. *Socio-Economic Review* 8: 47–76.
52. Engler, M. and Parry-Jones, R. 2007. Opportunity or threat: The role of the European Union in global wildlife trade. Brussels, Belgium: TRAFFIC Europe.
53. Karesh, W., A. Cook, R., Gilbert, M. and Newcomb, J. 2007. Implications of wildlife trade on the movement of Avian Influenza and other infectious diseases. *Journal of Wildlife Diseases* 43: S55–S59.
54. Nijman, V., Nekaris, K. A. I., Donati, G. et al. 2011. Primate conservation: measuring and mitigating trade in primates. *Endangered Species Research* 13: 159–161.
55. Whittington, R. J. and Chong, R. 2007. Global trade in ornamental fish from an Australian perspective: The case for revised import risk analysis and management strategies. *Preventive Veterinary Medicine* 81: 92–116.
56. Rhyne, A. L., Tlusty, M. F., Schofield, P. J. et al. 2012. Revealing the appetite of the marine aquarium fish trade: The volume and biodiversity of fish imported into the United States. *PLOS ONE* 7: e35808.
57. Nijman, V. 2010. An overview of international wildlife trade from Southeast Asia. *Biodiversity Conservation* 19: 1101.
58. Challender, D. W. S. and MacMillan, D. C. 2014. Poaching is more than an Enforcement Problem. *Conservation Letters* 7: 484–494.
59. Nellemann, C., R. Henriksen, A. Kreilhuber et al. 2016. *The Rise of Environmental Crime: A Growing Threat to Natural Resources, Peace, Development and Security.* Nairobi, Kenya: United Nations Environment Programme (UNEP).
60. Rosen, G. E. and Smith, K. F. 2010. Summarizing the evidence on the international trade in illegal wildlife. *Ecohealth* 7: 24–32.
61. Heinrich, S., Wittman, T. A., Ross, J. V. et al. 2017. *The Global Trafficking of Pangolins: A Comprehensive Summary of Seizures and Trafficking Routes from 2010–2015.* Selangor, Malaysia: Petaling Jaya Southeast Asia Regional Office.
62. Cooney, R., Roe, D., Dublin, H. et al. 2017. From poachers to protectors: Engaging local communities in solutions to illegal wildlife trade. *Conservation Letters* 10: 367–374.
63. Biggs, D., Cooney, R., Roe, D. et al. 2017. Developing a theory of change for a community-based response to illegal wildlife trade. *Conservation Biology* 31: 5–12.
64. Ripple, W. J., Newsome, T. M., Wolf, C. et al. 2015. Collapse of the world's largest herbivores. *Science Advances* 1: e1400103.
65. Peres, C. A. 2000. Effects of subsistence hunting on vertebrate community structure in Amazonian forests. *Conservation Biology* 14: 240–253.
66. Davies, G. 2002. Bushmeat and international development. *Conservation Biology* 16: 587–589.
67. Wilkie, D. S. and Carpenter, J. F. 1999. Bushmeat hunting in the Congo Basin: An assessment of impacts and options for mitigation. *Biodiversity and Conservation* 8: 927.
68. Nasi, R., Taber, A. and Van Vliet, N. 2011. Empty forests, empty stomachs? Bushmeat and livelihoods in the Congo and Amazon Basins. *The International Forestry Review* 13: 355–368.
69. Ripple, W. J., Abernethy, K., Betts, M. G. et al. 2016. Bushmeat hunting and extinction risk to the world's mammals. *Royal Society Open Science* 3: 160498.
70. Taylor, L. H., Latham, S. M. and Woolhouse, M. E. 2001. Risk factors for human disease emergence. *Philos Trans R Soc Lond B Biol Sci* 356: 983–989.
71. Jones, K. E., Patel, N. G., Levy, M. A. et al. 2008. Global trends in emerging infectious diseases. *Nature* 451: 990–993.
72. Faria, N. R., Rambaut, A., Suchard, M. A. et al. 2014. The early spread and epidemic ignition of HIV-1 in human populations. *Science* 346: 56–61.
73. Hahn, B. H., Shaw, G. M., De Cock, K. M. and Sharp, P. M. 2000. AIDS as a zoonosis: Scientific and public health implications. *Science* 287: 607–614.
74. Wolfe, N. D., Switzer, W. M., Carr, J. K. et al. 2004. Naturally acquired simian retrovirus infections in central African hunters. *Lancet* 363: 932–937.
75. FAO. 2014. FAO warns of fruit bat risk in West African Ebola epidemic. http://www.fao.org/news/story/en/item/239123/icode/.
76. Marí Saéz, A., Weiss, S., Nowak, K. et al. 2015. Investigating the zoonotic origin of the West African Ebola epidemic. *EMBO Molecular Medicine* 7: 17–23.
77. WHO. 2014. Ground zero in Guinea: The Ebola outbreak smoulders—undetected—for more than 3 months. http://www.who.int/csr/disease/ebola/ebola-6-months/guinea/en/.
78. Allen, T., Murray, K. A., Zambrana-Torrelio, C. et al. 2017. Global hotspots and correlates of emerging zoonotic diseases. *Nature Communications* 8: 1124.
79. OIA. 2012. *The Outdoor Recreation Economy.* Boulder, CO: Outdoor Industry Association.
80. Horsley, S. B., Stout, S. L. and deCalesta, D. S. 2003. White-tailed deer impact on the vegetation dynamics of a northern hardwood forest. *Ecological Applications* 13: 98–118.
81. Rooney, T. P. and Waller, D. M. 2003. Direct and indirect effects of white-tailed deer in forest ecosystems. *Forest Ecology and Management* 181: 165–176.
82. DeCalesta, D. S. 1994. Effect of white-tailed deer on songbirds within managed forests in Pennsylvania. *The Journal of Wildlife Management*: 711–718.

83. Watch, W. 2018. An inside look at Cecil the lion's final hours. *National Geographic*. Washington, DC: National Geographic Society.
84. Capecchi, C. and Rogers, K. 2015. Killer of Cecil the lion finds out that he is a target now, of internet vigilantism. New York, NY: New York Times.
85. Nelson, M. P., Bruskotter, J. T., Vucetich, J. A. and Chapron, G. 2016. Emotions and the ethics of consequence in conservation decisions: Lessons from Cecil the lion. *Conservation Letters* 9: 302–306.
86. Naidoo, R., Weaver, L. C., Stuart-Hill, G. and Tagg, J. 2011. Effect of biodiversity on economic benefits from communal lands in Namibia. *Journal of Applied Ecology* 48: 310–316.
87. Large, E. a. 2013. The $200 Million Question: How Much Does Trophy Hunting Really Contribute to African Communities? Melbourne, Australia: African Lion Coalition.
88. Di Minin, E., Leader-Williams, N. and Bradshaw, C. J. A. 2016. Banning trophy hunting will exacerbate biodiversity loss. *Trends in Ecology & Evolution* 31: 99–102.
89. Cooke, S. J., Hogan, Z. S., Butcher, P. A. et al. 2016. Angling for endangered fish: Conservation problem or conservation action? *Fish and Fisheries* 17: 249–265.
90. Ricciardi, A. and Rasmussen, J. B. 1999. Extinction rates of North American freshwater fauna. *Conservation Biology* 13: 1220–1222.
91. Cooke, S. J. and Cowx, I. G. 2004. The role of recreational fishing in global fish crises. *BioScience* 54: 857–859.
92. Arlinghaus, R. and Cooke, S. J. 2009. Recreational fishing: Socio-economic importance, conservation issues and management challenges. In B. Dickson, J. Hutton, and B. Adams (Eds.), *Recreational Hunting, Conservation and Rural Livelihoods: Science and Practice*. Oxford, UK: Blackwell Publishing.
93. Aiken, R. 2006. *Net Economic Values of Wildlife-Associated Recreation: Addendum to the 2006 National Survey of Fishing, Hunting and Wildlife-Associate Recreation*. US Fish and Wildlife Service Report 5.
94. Shrestha, R. K., Seidl, A. F. and Moraes, A. S. 2002. Value of recreational fishing in the Brazilian Pantanal: A travel cost analysis using count data models. *Ecological Economics* 42: 289–299.
95. Comley, V. and Mackintosh, C. 2014. *The Economic Impact of Outdoor Recreation in the UK: The Evidence*. London, UK: Sport & Recreation Alliance.
96. Southwick. 2011. *The Economics Associated with Outdoor Recreation, Natural Resources Conservation and Historic Preservation in the United States*. Fenandina Beach, FL: Southwick Associates.
97. Council, N. R. 1999. *Sustaining Marine Fisheries*. Washington, DC: The National Academies Press.
98. Coleman, F. C., Figueira, W. F., Ueland, J. S. and Crowder, L. B. 2004. The impact of United States recreational fisheries on marine fish populations. *Science* 305: 1958–1960.
99. Post, J. R., Sullivan, M., Cox, S. et al. 2002. Canada's recreational fisheries: The invisible collapse? *Fisheries* 27: 6–17.
100. Sadovy de Mitcheson, Y., Craig, M. T., Bertoncini, A. A. et al. 2013. Fishing groupers towards extinction: a global assessment of threats and extinction risks in a billion dollar fishery. *Fish and Fisheries* 14: 119–136.
101. Sutter, D. A. H., Suski, C. D., Philipp, D. P. et al. 2012. Recreational fishing selectively captures individuals with the highest fitness potential. *Proceedings of the National Academy of Sciences* 109: 20960–20965.
102. Birkeland, C. and Dayton, P. K. 2005. The importance in fishery management of leaving the big ones. *Trends in Ecology & Evolution* 20: 356–358.
103. Shiffman, D. S., Gallagher, A. J., Wester, J. et al. 2014. Trophy fishing for species threatened with extinction: A way forward building on a history of conservation. *Marine Policy* 50: 318–322.
104. McClenachan, L. 2009. Documenting Loss of Large Trophy Fish from the Florida Keys with Historical Photographs. *Conservation Biology* 23: 636–643.
105. McClenachan, L. 2013. Recreation and the "Right to Fish" Movement: Anglers and Ecological Degradation in the Florida Keys. *Environmental History* 18: 76–87.
106. Cowx, I. G., Arlinghaus, R. and Cooke, S. J. 2010. Harmonizing recreational fisheries and conservation objectives for aquatic biodiversity in inland waters. *Journal of Fish Biology* 76: 2194–2215.
107. Tsikliras, A. C. and Froese, R. 2018. Maximum Sustainable Yield. In Sven Erik Jørgensen and Brian D. Fath (Eds.), *Encyclopedia of Ecology* (2nd ed.). Amsterdam, the Netherlands: Elsevier.
108. Larkin, P. A. 1977. An epitaph for the concept of maximum sustained yield. *Transactions of the American Fisheries Society* 106: 1–11.
109. Maunder, M. N. and Piner, K. R. 2015. Contemporary fisheries stock assessment: Many issues still remain. *Ices Journal of Marine Science* 72: 7–18.
110. Dichmont, C. M., Pascoe, S., Kompas, T. et al. 2010. On implementing maximum economic yield in commercial fisheries. *Proceedings of the National Academy of Sciences* 107: 16–21.
111. Rudd, M. B. and Branch, T. A. 2017. Does unreported catch lead to overfishing? *Fish and Fisheries* 18: 313–323.
112. Mace. 2001. A new role for MSY in single-species and ecosystem approaches to fisheries stock assessment and management. *Fish and Fisheries* 2: 2–32.
113. Hilborn, R. 2007. Defining success in fisheries and conflicts in objectives. *Marine Policy* 31: 153–158.
114. Nijman, V. 2014. Fact or fiction? Be prudent and accurate when attaching monetary value to threatened wildlife (a comment to Douglas and Alie 2014). *Biological Conservation* 179: 148–149.
115. CBC News. 2006. Will still be plenty of fish in sea, despite gloomy study: scientist. December 8.
116. Bernton, H. 2006. Will seafood nets be empty? Grim outlook draws skeptics. *Seattle Times* November 3.
117. Worm, B., Hilborn, R., Baum, J. K. et al. 2009. Rebuilding global fisheries. *Science* 325: 578–585.
118. Hilborn, R. 2011. Let Us Eat Fish. *New York Times* Op-Ed. April 15. https://www.nytimes.com/2011/04/15/opinion/15hilborn.html
119. SeaMonster. 2011. http://theseamonster.net/2011/05/forum-on-fish-food-and-people/
120. Granek, E. F., Madin, E. M. P., Brown, M. A. et al. 2008. Engaging recreational fishers in management and conservation: Global case studies. *Conservation Biology* 22: 1125–1134.

Chapter 11

1. Higgins, S. I. and Richardson, D. M. 1999. Predicting plant migration rates in a changing world: The role of long-distance dispersal. *The American Naturalist* 153: 464–475.
2. Cain, M. L., Milligan, B. G. and Strand, A. E. 2000. Long-distance seed dispersal in plant populations. *American Journal of Botany* 87: 1217–1227.
3. Simberloff, D., Martin, J.-L., Genovesi, P. et al. 2013. Impacts of biological invasions: What's what and the way forward. *Trends in Ecology & Evolution* 28: 58–66.
4. Turbelin, A. J., Malamud, B. D. and Francis, R. A. 2017. Mapping the global state of invasive alien species: Patterns of invasion and policy responses. *Global Ecology and Biogeography* 26: 78–92.
5. Early, R., Bradley, B. A., Dukes, J. S. et al. 2016. Global threats from invasive alien species in the twenty-first century and national response capacities. *Nature Communications* 7: 12485.

6. Colautti, R. I. and MacIsaac, H. J. 2004. A neutral terminology to define "invasive" species. *Diversity and Distributions* 10: 135–141.
7. Seebens, H., Blackburn, T. M., Dyer, E. E. et al. 2017. No saturation in the accumulation of alien species worldwide. *Nature Communications* 8: 14435.
8. Krysko, K. L., Burgess, J. P., Rochford, M. R. et al. 2011. Verified non-indigenous amphibians and reptiles in Florida from 1863 through 2010: Outlining the invasion process and identifying invasion pathways and stages. *Zootaxa* 3028: 1–64.
9. Belyea, L. R. and Lancaster, J. 1999. Assembly rules within a contingent ecology. *Oikos* 86: 402–416.
10. Hille Ris Lambers, J., Adler, P. B., Harpole, W. S. et al. 2012. Rethinking community assembly through the lens of coexistence theory. *Annual Review of Ecology, Evolution, and Systematics* 43: 227–248.
11. Vander Zanden, M. J. V., Olden, J. D., Thorne, J. H. and Mandrak, N. E. 2004. Predicting occurrences and impacts of smallmouth bass introductions in temperate lakes. *Ecological Applications* 14: 132–148.
12. Turnbull, L. A., Crawley, M. J. and Rees, M. 2000. Are plant populations seed-limited? A review of seed sowing experiments. *Oikos* 88: 225–238.
13. Hubbell, S., Foster, R., O'Brien, S. et al. 1999. Light-gap disturbances, recruitment limitation, and tree diversity in a neotropical forest. *Science* 283: 554–557.
14. Moore, R. P., Robinson, W. D., Lovette, I. J. and Robinson, T. R. 2008. Experimental evidence for extreme dispersal limitation in tropical forest birds. *Ecology Letters* 11: 960–968.
15. Munguía, M., Townsend Peterson, A. and Sánchez-Cordero, V. 2008. Dispersal limitation and geographical distributions of mammal species. *Journal of Biogeography* 35: 1879–1887.
16. Thompson, R. and Townsend, C. 2006. A truce with neutral theory: Local deterministic factors, species traits and dispersal limitation together determine patterns of diversity in stream invertebrates. *Journal of Animal Ecology* 75: 476–484.
17. Peay, K. G., Garbelotto, M. and Bruns, T. D. 2010. Evidence of dispersal limitation in soil microorganisms: Isolation reduces species richness on mycorrhizal tree islands. *Ecology* 91: 3631–3640.
18. Siemann, E. and Rogers, W. E. 2003. Herbivory, disease, recruitment limitation, and success of alien and native tree species. *Ecology* 84: 1489–1505.
19. Sullivan, M. J. P., Davies, R. G., Reino, L. and Franco, A. M. A. 2012. Using dispersal information to model the species-environment relationship of spreading nonnative species. *Methods in Ecology and Evolution* 3: 870–879.
20. Capinha, C., Essl, F., Seebens, H. et al. 2015. The dispersal of alien species redefines biogeography in the Anthropocene. *Science* 348: 1248–1251.
21. Bradley, B. A., Blumenthal, D. M., Early, R. et al. 2012. Global change, global trade, and the next wave of plant invasions. *Frontiers in Ecology and the Environment* 10: 20–28.
22. Levine, J. M. and D'Antonio, C. M. 2003. Forecasting biological invasions with increasing international trade. *Conservation Biology* 17: 322–326.
23. McCullough, D. G., Work, T. T., Cavey, J. F. et al. 2006. Interceptions of nonindigenous plant pests at US ports of entry and border crossings over a 17-year period. *Biological Invasions* 8: 611–630.
24. Hulme, P. E., Bacher, S., Kenis, M. et al. 2008. Grasping at the routes of biological invasions: A framework for integrating pathways into policy. *Journal of Applied Ecology* 45: 403–414.
25. Bellard, C., Leroy, B., Thuiller W. et al. 2016. Major drivers of invasion risks throughout the world. *Ecosphere* 7: 14.
26. Harper, G. A. and Bunbury, N. 2015. Invasive rats on tropical islands: Their population biology and impacts on native species. *Global Ecology and Conservation* 3: 607–627.
27. Jones, H. P., Tershy, B. R., Zavaleta, E. S. et al. 2008. Severity of the effects of invasive rats on seabirds: A global review. *Conservation Biology* 22: 16–26.
28. Johnson, L. E. and Padilla, D. K. 1996. Geographic spread of exotic species: Ecological lessons and opportunities from the invasion of the zebra mussel *Dreissena polymorpha*. *Biological Conservation* 78: 23–33.
29. Ludyanskiy, M. L., McDonald, D. and Macneill, D. 1993. Impact of the zebra mussel, a bivalve invader: *Dreissena polymorpha* is rapidly colonizing hard surfaces throughout waterways in the United States and Canada. *BioScience* 43: 533–544.
30. Courchamp, F. 2013. Monster fern makes IUCN invader list. *Nature* 498: 37.
31. Reichard, S. H. and White, P. 2001. Horticulture as a pathway of invasive plant introductions in the United States. *BioScience* 51: 103–113.
32. Smith, K. F., Behrens, M., Schloegel, L. M. et al. 2009. Reducing the risks of the wildlife trade. *Science* 324: 594–595.
33. Gup, T. 1990. *100 years of the starling.* September 1. www.nytimes.com/1990/09/01/opinion/100-years-of-the-starling.html.
34. Link, R. 2004. *Living with Wildlife: Starlings*. Olympia, WA: Washington Department of Fish and Wildlife.
35. Thiengo, S. C., Faraco, F. A., Salgado, N. C. et al. 2007. Rapid spread of an invasive snail in South America: The giant African snail, *Achatina fulica*, in Brasil. *Biological Invasions* 9: 693–702.
36. Shine, R. 2010. The ecological impact of invasive cane toads (*Bufo marinus*) in Australia. *Quarterly Review of Biology* 85: 253–291.
37. Ovalle, D. 2010. Giant African snails smuggled into Florida for use in religious ritual, authorities say. March 10. www.sun-sentinel.com/news/fl-xpm-2010-03-11-fl-illegal-snails-santeria-20100310-story.html.
38. Lodge, D. M. 1993. Biological invasions: Lessons for ecology. *Trends in Ecology & Evolution* 8: 133–137.
39. Sax, D. F. and Brown, J. H. 2000. The paradox of invasion. *Global Ecology and Biogeography* 9: 363–371.
40. Clarke, C. 1971. Liberations and dispersal of red deer in northern South Island districts. *New Zealand Journal of Forestry Science* 1: 194–207.
41. Kolar, C. S. and Lodge, D. M. 2001. Progress in invasion biology: Predicting invaders. *Trends in Ecology & Evolution* 16: 199–204.
42. Lockwood, J. L., Cassey, P. and Blackburn, T. 2005. The role of propagule pressure in explaining species invasions. *Trends in Ecology & Evolution* 20: 223–228.
43. Levine, J. M., Vila, M., D'Antonio, C. M. et al. 2003. Mechanisms underlying the impacts of exotic plant invasions. *Proceedings of the Royal Society B: Biological Sciences* 270: 775–781.
44. Vila, M. and Weiner, J. 2004. Are invasive plant species better competitors than native plant species? Evidence from pair-wise experiments. *Oikos* 105: 229–238.
45. Levine, J. M., Adler, P. B. and Yelenik, S. G. 2004. A meta-analysis of biotic resistance to exotic plant invasions. *Ecology Letters* 7: 975–989.
46. Callaway, R. M. and Ridenour, W. M. 2004. Novel weapons: Invasive success and the evolution of increased competitive ability. *Frontiers in Ecology and the Environment* 2: 436–443.
47. Thorpe, A. S., Thelen, G. C., Diaconu, A. and Callaway, R. M. 2009. Root exudate is allelopathic in invaded community but not in native community: Field evidence for the novel weapons hypothesis. *Journal of Ecology* 97: 641–645.
48. Ni, G., Zhao, P., Huang, Q. et al. 2012. Exploring the novel weapons hypothesis with invasive plant species in China. *Allelopathy Journal* 29: 199–214.
49. Keane, R. M. and Crawley, M. J. 2002. Exotic plant invasions and the enemy release hypothesis. *Trends in Ecology & Evolution* 17: 164–170.

50. Torchin, M. E., Lafferty, K. D., Dobson, A. P. et al. 2003. Introduced species and their missing parasites. *Nature* 421: 628–630.

51. Torchin, M. E., Lafferty, K. D. and Kuris, A. M. 2001. Release from parasites as natural enemies: Increased performance of a globally introduced marine crab. *Biological Invasions* 3: 333–345.

52. Hajek, A. E. and Eilenberg, J. 2018. *Natural Enemies: An Introduction to Biological Control* (2nd ed.). Cambridge, UK: Cambridge University Press.

53. Hajek, A. E., Hurley, B. P., Kenis, M. et al. 2016. Exotic biological control agents: A solution or contribution to arthropod invasions? *Biological Invasions* 18: 953–969.

54. Arim, M., Abades, S. R., Neill, P. E. et al. 2006. Spread dynamics of invasive species. *Proceedings of the National Academy of Sciences* 103: 374–378.

55. Crooks, J. A. and Soulé, M. E. 1999. Lag times in population explosions of invasive species: Causes and implications. In O. T. Sandlund, P. J. Schei and A. Viken (Eds.), *Invasive Species and Biodiversity Management*. Dordrecht, The Netherlands: Kluwer Academic Press.

56. Kowarik, I. 1995. Time lags in biological invasions with regard to the success and failure of alien species. In P. Pyšek, K. Prach, M. Rejmánek and M. Wade (Eds.), *Plant Invasions: General Aspects and Special Problems*. Amsterdam, The Netherlands: SPB Academic Publishing.

57. Crooks, J. 2005. Lag times and exotic species: The ecology and management of biological invasions in slow-motion. *Ecoscience* 12: 316–329.

58. Wilcove, D. S., Rothstein, D., Dubow, J. et al. 1998. Quantifying threats to imperiled species in the United States. *BioScience* 48: 607–615.

59. Fritts, T. H. and Rodda, G. H. 1998. The role of introduced species in the degradation of island ecosystems: A case history of Guam. *Annual Review of Ecology and Systematics* 29: 113–140.

60. Wonham, M. 2006. Species invasions. In M. J. Groom, G. K. Meffe and C. R. Carroll (Eds.), *Principles of Conservation Biology* (3rd ed.). Sunderland, MA: Oxford University Press/Sinauer.

61. Wiles, G. J., Bart, J., Beck, R. E. and Aguon, C. F. 2003. Impacts of the brown tree snake: Patterns of decline and species persistence in Guam's avifauna. *Conservation Biology* 17: 1350–1360.

62. Witte, F., Msuku, B. S., Wanink, J. H. et al. 2000. Recovery of cichlid species in Lake Victoria: An examination of factors leading to differential extinction. *Reviews in Fish Biology and Fisheries* 10: 233–241.

63. Aloo, P. A., Njiru, J., Balirwa, J. S. and Nyamweya, C. S. 2017. Impacts of nile perch, *Lates niloticus*, introduction on the ecology, economy and conservation of Lake Victoria, East Africa. *Lakes & Reservoirs: Science, Policy and Management for Sustainable Use* 22: 320–333.

64. Ricciardi, A., Neves, R. J. and Rasmussen, J. B. 1998. Impending extinctions of North American freshwater mussels (Unionoida) following the zebra mussel (*Dreissena polymorpha*) invasion. *Journal of Animal Ecology* 67: 613–619.

65. Ricciardi, A. 2003. Predicting the impacts of an introduced species from its invasion history: An empirical approach applied to zebra mussel invasions. *Freshwater Biology* 48: 972–981.

66. Gurevitch, J. and Padilla, D. K. 2004. Are invasive species a major cause of extinctions? *Trends in Ecology & Evolution* 19: 470–474.

67. Clavero, M. and Garcia-Berthou, E. 2005. Invasive species are a leading cause of animal extinctions. *Trends in Ecology & Evolution* 20: 110–110.

68. Murphy, G. E. P. and Romanuk, T. N. 2014. A meta-analysis of declines in local species richness from human disturbances. *Ecology and Evolution* 4: 91–103.

69. Vilà, M., Basnou, C., Pyšek, P. et al. 2010. How well do we understand the impacts of alien species on ecosystem services? A pan-European, cross-taxa assessment. *Frontiers in Ecology and the Environment* 8: 135–144.

70. Doherty, T. S., Glen, A. S., Nimmo, D. G. et al. 2016. Invasive predators and global biodiversity loss. *Proceedings of the National Academy of Sciences* 113: 11261–11265.

71. Sax, D. F., Gaines, S. D. and Brown, J. H. 2002. Species invasions exceed extinctions on islands worldwide: A comparative study of plants and birds. *American Naturalist* 160: 766–783.

72. Sax, D. F. and Gaines, S. D. 2003. Species diversity: From global decreases to local increases. *Trends in Ecology & Evolution* 18: 561–566.

73. McKinney, M. L. and Lockwood, J. L. 1999. Biotic homogenization: A few winners replacing many losers in the next mass extinction. *Trends in Ecology & Evolution* 14: 450–453.

74. Olden, J. D. 2006. Biotic homogenization: A new research agenda for conservation biogeography. *Journal of Biogeography* 33: 2027–2039.

75. Rahel, F. J. 2000. Homogenization of fish faunas across the United States. *Science* 288: 854–856.

76. Dornelas, M., Gotelli, N. J., McGill, B. et al. 2014. Assemblage time series reveal biodiversity change but not systematic loss. *Science* 344: 296–299.

77. Rosenblad, K. C. and Sax, D. F. 2017. A new framework for investigating biotic homogenization and exploring future trajectories: Oceanic island plant and bird assemblages as a case study. *Ecography* 40: 1040–1049.

78. Longman, E. K., Rosenblad, K. and Sax, D. F. 2018. Extreme homogenization: The past, present and future of mammal assemblages on islands. *Global Ecology and Biogeography* 27: 77-95.

79. Davis, M., Chew, M. K., Hobbs, R. J. et al. 2011. Don't judge species on their origins. *Nature* 474: 153–154.

80. Ruiz, G. M., Fofonoff, P., Hines, A. H. and Grosholz, E. D. 1999. Non-indigenous species as stressors in estuarine and marine communities: Assessing invasion impacts and interactions. *Limnology and Oceanography* 44: 950–972.

81. Strayer, D. L. 2012. Eight questions about invasions and ecosystem functioning. *Ecology Letters* 15: 1199–1210.

82. Liao, C. Z., Peng, R. H., Luo, Y. Q. et al. 2008. Altered ecosystem carbon and nitrogen cycles by plant invasion: A meta-analysis. *New Phytologist* 177: 706–714.

83. Vitousek, P. M. 1990. Biological invasion by *Myrica faya* in Hawaii: Plant demography, nitrogen fixation, ecosystem effects. *Oikos* 57: 7–13.

84. Vitousek, P. M. and Walker, L. R. 1989. Biological invasion by *Myrica faya* in Hawai'i: Plant demography, nitrogen fixation, ecosystem effects. *Ecological Monographs* 59: 247–265.

85. Ehrenfeld, J. G. 2010. Ecosystem consequences of biological invasions. *Annual Review of Ecology, Evolution, and Systematics* 41: 59–80.

86. Bertness, M. D. 1984. Habitat and community modification by an introduced herbivorous snail. *Ecology* 65: 370–381.

87. D'Antonio, C. M. and Vitousek, P. M. 1992. Biological invasions by exotic grasses, the grass fire cycle, and global change. *Annual Review of Ecology and Systematics* 23: 63–87.

88. Serbesoff-King, K. 2003. Melaleuca in Florida: A literature review on the taxonomy, distribution, biology, ecology, economic importance and control measures. *Journal of Aquatic Plant Management* 41: 98–112.

89. Richardson, D. M., MacDonald, I. A. W., Hoffmann, J. H. and Henderson, L. 1997. Alien plant invasions. In R. M. Cowling , D. M. Richardson and S. M. Pierce (Eds.), *Vegetation of Southern Africa*. Cambridge, UK: Cambridge University Press.

90. Bradshaw, C. J. A., Leroy, B., Bellard, C. et al. 2016. Massive yet grossly underestimated global costs of invasive insects. *Nature Communications* 7: 12986.
91. O'Neill, C. R., Jr. 1997. Economic impact of zebra mussels: Results of the 1995 National Zebra Mussel Information Clearinghouse Study. *Great Lakes Research Review* 3: 35–44.
92. Pimentel, D. 2009. Invasive plants: Their role in species extinctions and economic losses to agriculture in the USA. In Inderjit (Eds.), *Management of Invasive Weeds* (Invading Nature–Springer Series in Invasion Ecology, Vol. 5). Dordrecht, The Netherlands: Springer.
93. Oliveira, C. M., Auad, A. M., Mendes, S. M. and Frizzas, M. R. 2013. Economic impact of exotic insect pests in Brazilian agriculture. *Journal of Applied Entomology* 137: 1–15.
94. Paini, D. R., Sheppard, A. W., Cook, D. C. et al. 2016. Global threat to agriculture from invasive species. *Proceedings of the National Academy of Sciences* 113: 7575–7579.
95. Aukema, J. E., Leung, B., Kovacs, K. et al. 2011. Economic impacts of nonnative forest insects in the continental United States. *PLOS ONE* 6: e24587.
96. Kovacs, K. F., Haight, R. G., McCullough, D. G. et al. 2010. Cost of potential emerald ash borer damage in U.S. communities, 2009–2019. *Ecological Economics* 69: 569–578.
97. Kjær, E. D., Lobo, A. and Myking, T. 2014. The role of exotic tree species in Nordic forestry. *Scandinavian Journal of Forest Research* 29: 323–332.
98. McIntosh, A. R., McHugh, P. A., Dunn, N. R. et al. 2010. The impact of trout on galaxiid fishes in New Zealand. *New Zealand Journal of Ecology* 34: 195–206.
99. Townsend, C. R. 1996. Invasion biology and ecological impacts of brown trout *Salmo trutta* in New Zealand. *Biological Conservation* 78: 13–22.
100. Fish & Game. 2019. Trout facts. fishandgame.org.nz/threat-to-trout/trout-facts/.
101. Messing, R. H. and Wright, M. G. 2006. Biological control of invasive species: Solution or pollution? *Frontiers in Ecology and the Environment* 4: 132–140.
102. Norgaard, R. B. 1988. The biological control of cassava mealybug in Africa. *American Journal of Agricultural Economics* 70: 366–371.
103. D'Antonio, C. and Meyerson, L. A. 2002. Exotic plant species as problems and solutions in ecological restoration: A synthesis. *Restoration Ecology* 10: 703–713.
104. Schlaepfer, M. A., Sax, D. F. and Olden, J. D. 2011. The potential conservation value of nonnative species. *Conservation Biology* 25: 428–437.
105. Schlaepfer, M. A., Sax, D. F. and Olden, J. D. 2012. Toward a more balanced view of nonnative species. *Conservation Biology* 26: 1156–1158.
106. McNeely, J. A., Mooney, H. A., Neville, L. E. et al. 2001. *Global strategy on invasive alien species.* Gland, Switzerland: IUCN.
107. Genovesi, P. and Shine, C. 2004. *European Strategy on Invasive Alien Species.* Strasbourg, France: Council of Europe.
108. Tollington, S., Turbé, A., Rabitsch, W. et al. 2017. Making the EU legislation on invasive species a conservation success. *Conservation Letters* 10: 112–120.
109. Singh, S. K., Ash, G. J. and Hodda, M. 2015. Keeping "one step ahead" of invasive species: Using an integrated framework to screen and target species for detailed biosecurity risk assessment. *Biological Invasions* 17: 1069–1086.
110. Lodge, D. M., Simonin, P. W., Burgiel, S. W. et al. 2016. Risk analysis and bioeconomics of invasive species to inform policy and management. *Annual Review of Environment and Resources* 41: 453–488.
111. Pheloung, P. C., Williams, P. A. and Halloy, S. R. 1999. A weed risk assessment model for use as a biosecurity tool evaluating plant introductions. *Journal of Environmental Management* 57: 239–251.
112. Weber, J., Dane Panetta, F., Virtue, J. and Pheloung, P. 2009. An analysis of assessment outcomes from eight years' operation of the Australian border weed risk assessment system. *Journal of Environmental Management* 90: 798–807.
113. Gordon, D. R., Onderdonk, D. A., Fox, A. M. and Stocker, R. K. 2008. Consistent accuracy of the Australian weed risk assessment system across varied geographies. *Diversity and Distributions* 14: 234–242.
114. Kumschick, S. and Richardson, D. M. 2013. Species-based risk assessments for biological invasions: Advances and challenges. *Diversity and Distributions* 19: 1095–1105.
115. Baker, H. G. 1991. The continuing evolution of weeds. *Economic Botany* 45: 445–449.
116. Baker, H. G. 1965. Characteristics and modes of origin of weeds. In H. G. Baker and G. L. Stebbins (Eds.), *The Genetics of Colonizing Species.* New York, NY: Academic Press.
117. Newsome, A. E. and Noble, I. R. 1986. Ecological and physiological characters of invading species. In R. H. Groves and J. J. Burdon (Eds.), *Ecology of Biological Invasions.* Cambridge, UK: Cambridge University Press.
118. Moyle, P. B. 1986. Fish introductions into North America: Patterns and ecological impact. In H. A. Mooney and J. A. Drake (Eds.), *Ecology of Biological Invasions of North America and Hawaii.* New York, NY: Springer.
119. Lodge, A. G., Whitfeld, T. J. S., Roth, A. M. and Reich, P. B. 2018. Invasive plants in Minnesota are "joining the locals": A trait-based analysis. *Journal of Vegetation Science* 29: 746–755.
120. Turbé, A., Strubbe, D., Mori, E. et al. 2017. Assessing the assessments: Evaluation of four impact assessment protocols for invasive alien species. *Diversity and Distributions* 23: 297–307.
121. Nentwig, W., Bacher, S., Pysek, P. et al. 2016. The generic impact scoring system (GISS): A standardized tool to quantify the impacts of alien species. *Environmental Monitoring and Assessment* 188: 13.
122. Evans, T., Kumschick, S. and Blackburn, T. M. 2016. Application of the Environmental Impact Classification for Alien Taxa (EICAT) to a global assessment of alien bird impacts. *Diversity and Distributions* 22: 919–931.
123. D'Hondt, B., Vanderhoeven, S., Roelandt, S. et al. 2015. Harmonia (+) and Pandora (+): Risk screening tools for potentially invasive plants, animals and their pathogens. *Biological Invasions* 17: 1869–1883.
124. Baker, R. H. A., Black, R. and Copp, G. H. 2008. The UK risk assessment scheme for all nonnative species. In W. Rabitsch, F. Essl and F. Klingensten (Eds.), *Biological invasions—from ecology to conservation.* NeoBiota 7.
125. Elith, J. 2015. Predicting distributions of invasive species. arxiv.org/abs/1312.0851.
126. Byeon, D.-H., Jung, S. and Lee, W.-H. 2018. Review of CLIMEX and MaxEnt for studying species distribution in South Korea. *Journal of Asia-Pacific Biodiversity* 11: 325–333.
127. Merow, C., Smith, M. J. and Silander, J. A. 2013. A practical guide to MaxEnt for modeling species' distributions: What it does, and why inputs and settings matter. *Ecography* 36: 1058–1069.
128. Phillips, S. 2010. *A Brief Tutorial on MaxEnt.* biodiversityinformatics.amnh.org/open_source/maxent/Maxent_tutorial2017.pdf.
129. GAO. 2015. *Aquatic Invasive Species.* Washington, DC: U.S. Government Accountability Office.
130. Lodge, D. M., Williams, S., MacIsaac, H. J. et al. 2006. Biological invasions: Recommendations for U.S. policy and management. *Ecological Applications* 16: 2035–2054.
131. Leung, B., Lodge, D. M., Finnoff, D. et al. 2002. An ounce of prevention or a pound of cure: Bioeconomic risk analysis of invasive species. *Proceedings of the Royal Society B: Biological Sciences* 269: 2407–2413.

132. Convention on Biological Diversity (CBD). 2014. *Pathways of Introduction of Invasive Species, Their Prioritization and Management*. Montreal, Quebec: Secretariat of the Convention on Biological Diversity.
133. Environment Canada. 2004. *An Invasive Alien Species Strategy for Canada*. Ottawa, Ontario: Government of Canada.
134. Novoa, A., Dehnen-Schmutz, K., Fried, J. and Vimercati, G. 2017. Does public awareness increase support for invasive species management? Promising evidence across taxa and landscape types. *Biological Invasions* 19: 3691–3705.
135. Keller, R. P., Lodge, D. M. and Finnoff, D. C. 2007. Risk assessment for invasive species produces net bioeconomic benefits. *Proceedings of the National Academy of Sciences* 104: 203–207.
136. Simberloff, D. 2003. How much information on population biology is needed to manage introduced species? *Conservation Biology* 17: 83–92.
137. Pyšek, P. and Richardson, D. M. 2010. Invasive species, environmental change and management, and health. *Annual Review of Environment and Resources* 35: 25–55.
138. Rejmanek, M. and Pitcairn, M. J. 2002. When is eradication of exotic pest plants a realistic goal? In C. R. Veitch and M. N. Clout (Eds.), *Turning the Tide: The Eradication of Invasive Species*. Gland, Switzerland: Cambridge University Press.
139. Myers, J. H., Simberloff, D., Kuris, A. M. and Carey, J. R. 2000. Eradication revisited: Dealing with exotic species. *Trends in Ecology & Evolution* 15: 316–320.
140. Snyder, M. 2017. Can we really eat invasive species into submission: The tale of a giant Amazon fish reveals the promise and peril of invasiorism. www.scientificamerican.com/article/can-we-really-eat-invasive-species-into-submission/.
141. Hobbs, R. J., Arico, S., Aronson, J. et al. 2006. Novel ecosystems: Theoretical and management aspects of the new ecological world order. *Global Ecology and Biogeography* 15: 1–7.
142. Gibson, L. and Yong, D. L. 2017. Saving two birds with one stone: Solving the quandary of introduced, threatened species. *Frontiers in Ecology and the Environment* 15: 35–41.
143. Dyer, E. E., Redding, D. W. and Blackburn, T. M. 2017. The global avian invasions atlas, a database of alien bird distributions worldwide. *Scientific Data* 4: 170041.
144. Marbuah, G., Gren, I. M. and McKie, B. 2014. Economics of harmful invasive species: A review. *Diversity* 6: 500.
145. Hoffmann, B. D. and Broadhurst, L. M. 2016. The economic cost of managing invasive species in Australia. *NeoBiota* 31.
146. Pimentel, D., McNair, S., Janecka, J. et al. 2001. Economic and environmental threats of alien plant, animal, and microbe invasions. *Agriculture, Ecosystems & Environment* 84: 1–20.
147. Nghiem, L. T. P., Soliman, T., Yeo, D. C. J. et al. 2013. Economic and environmental impacts of harmful non-indigenous species in Southeast Asia. *PLOS ONE* 8: e71255.
148. Xu, H. G., Ding, H., Li, M. Y. et al. 2006. The distribution and economic losses of alien species invasion to China. *Biological Invasions* 8: 1495–1500.
149. Colautti, R. I., Bailey, S. A., Van Overdijk, C. D. et al. 2006. Characterised and projected costs of nonindigenous species in Canada. *Biological Invasions* 8: 45–59.
150. Gren, I.-M., Isacs, L. and Carlsson, M. 2009. Costs of alien invasive species in Sweden. *Ambio* 38: 135–140.
151. Williams, F., Eschen, R., Harris, A. et al. 2010. *The Economic Cost of Invasive Non-native Species on Great Britain*. (Proj. No. VM10066: 1–99.) Wallingford, UK: CABI.
152. Reinhardt, F., Herle, M., Bastiansenn, F. and Streit, B. 2003. *Economic Impact of the Spread of Alien Species in Germany*. Berlin, Germany: Federal Environmental Agency.
153. Giera, N. and Bell, B. 2009. *Economic Costs of Pests to New Zealand*. (Technical Paper No. 2009/31.) Wellington, NZ: MAF Biosecurity New Zealand.
154. Kendall, D. G. 2008. A form of wave propagation associated with the equation of heat conduction. *Mathematical Proceedings of the Cambridge Philosophical Society* 44: 591–594.
155. Fisher, R. A. 1937. The wave of advance of advantageous genes. *Annals of Eugenics* 7: 355–369.
156. Skellam, J. G. 1951. Random dispersal in theoretical populations. *Biometrika* 38: 196–218.
157. Neubert, M. G. and Parker, I. M. 2004. Projecting rates of spread for invasive species. *Risk Analysis* 24: 817–831.
158. Reeves, S. A. and Usher, M. B. 1989. Application of a diffusion model to the spread of an invasive species: The coypu in Great Britain. *Ecological Modelling* 47: 217–232.
159. Shryock, K. A., Brown, S. L., Sanders, N. J. and Burroughs, E. 2008. A reaction-diffusion equation modeling the invasion of the Argentine ant population, *Linepithema humile*, at Jasper Ridge Biological Station. *Natural Resource Modeling* 21: 330–342.
160. Roques, L., Auger-Rozenberg, M.-A. and Roques, A. 2008. Modelling the impact of an invasive insect via reaction-diffusion. *Mathematical Biosciences* 216: 47–55.
161. Simberloff, D. 2011. Non-natives: 141 scientists object. *Nature* 475: 36–36.
162. Davis, M. A. 2009. *Invasion Biology*. Oxford, UK: Oxford University Press.
163. Jones, H. P., Holmes, N. D., Butchart, S. H. M. et al. 2016. Invasive mammal eradication on islands results in substantial conservation gains. *Proceedings of the National Academy of Sciences* 113: 4033–4038.

Figure 11.2: Maps were generated in R (v.3.2.2) using the rworldmap package. Map projection lines and projections are from the Natural Earth (2016) data (v.1.4.0) at a scale of 1:110 and use the geographical coordinate system (projection) WGS84.

Chapter 12

1. Allen, A. P., Gillooly, J. F., Savage, V. M. and Brown, J. H. 2006. Kinetic effects of temperature on rates of genetic divergence and speciation. *Proceedings of the National Academy of Sciences* 103: 9130–9135.
2. Davis, M. B. and Shaw, R. G. 2001. Range shifts and adaptive responses to Quaternary climate change. *Science* 292: 673–679.
3. McInerney, F. A. and Wing, S. L. 2011. The Paleocene-Eocene thermal maximum: A perturbation of carbon cycle, climate, and biosphere with implications for the future. *Annual Review of Earth and Planetary Sciences* 39: 489–516.
4. Penn, J. L., Deutsch, C., Payne, J. L. and Sperling, E. A. 2018. Temperature-dependent hypoxia explains biogeography and severity of end-Permian marine mass extinction. *Science* 362: eaat1327.
5. Mayhew, P., J., Jenkins, G. B. and Benton, T. G. 2008. A long-term association between global temperature and biodiversity, origination, and extinction in the fossil record. *Proceedings of the Royal Society B* 275: 47–53.
6. Crowley, T. J. and North, G. R. 1988. Abrupt climate change and extinction events in Earth history. *Science* 240: 996–1002.
7. Kump, L. 2018. Climate change and marine mass extinction. *Science* 362: 1113–1114.
8. Nogues-Bravo, D., Rodriguez-Sanchez, F., Orsini, L. et al. 2018. Cracking the code of biodiversity responses to past climate change. *Trends in Ecology & Evolution* 33: 765–776.
9. IPCC. 2014. *Climate Change 2014: Synthesis Report* (Fifth Assessment Report, AR5). Intergovernmental Panel on Climate Change: Geneva, Switzerland.
10. Lüthi, D., Le Floch, M., Bereiter, B. et al. 2008. High-resolution carbon dioxide concentration record 650,000–800,000 years before present. *Nature* 453: 379.
11. Mann, M. E., Zhang, Z., Hughes, M. K. et al. 2008. Proxy-based reconstructions of hemispheric and global surface temperature variations over the past two millennia. *Proceedings of the National Academy of Sciences* 105: 13252–13257.

12. IPCC. 2018. *Summary for Policymakers*. Geneva, Switzerland: World Meteorological Organization.
13. Flannigan, M. D., Stocks, B. J. and Wotton, B. M. 2000. Climate change and forest fires. *Science of the Total Environment* 262: 221–229.
14. Hirabayashi, Y., Mahendran, R., Koirala, S. et al. 2013. Global flood risk under climate change. *Nature Climate Change* 3: 816.
15. Knutson, T. R., McBride, J. L., Chan, J. et al. 2010. Tropical cyclones and climate change. *Nature Geoscience* 3: 157.
16. Díaz, S., Settele, J., Brondizio, E. et al. 2019. *Summary for Policymakers of the Global Assessment Report on Biodiversity and Ecosystem Services of the Intergovernmental Science-Policy Platform on Biodiversity and Ecosystem Services*. United Nations.
17. Bellard, C., Bertelsmeier, C., Leadley, P. et al. 2012. Impacts of climate change on the future of biodiversity. *Ecology Letters* 15: 365–377.
18. Anderson, J. T., Inouye, D. W., McKinney, A. M. et al. 2012. Phenotypic plasticity and adaptive evolution contribute to advancing flowering phenology in response to climate change. *Proceedings of the Royal Society B* 279: 3843–3852.
19. Parmesan, C., Ryrholm, N., Stefanescu, C. et al. 1999. Poleward shifts in geographical ranges of butterfly species associated with regional warming. *Nature* 399: 579.
20. Charmantier, A., McCleery, R. H., Cole, L. R. et al. 2008. Adaptive phenotypic plasticity in response to climate change in a wild bird population. *Science* 320: 800–803.
21. Merila, J. and Hendry, A. P. 2014. Climate change, adaptation, and phenotypic plasticity: The problem and the evidence. *Evolutionary Applications* 7: 1–14.
22. Mechoso, C. R. and Arakawa, A. 2015. *Numerical models: General circulation models*. In G. R. North, J. Pyle, and F. Zhang (Eds.), *Encyclopedia of Atmospheric Sciences* (2nd ed.). Oxford, UK: Academic Press.
23. Tanaka, N., Nakao, K., Tsuyama, I. et al. 2012. Predicting the impact of climate change on potential habitats of fir (*Abies*) species in Japan and on the East Asian continent. *Procedia Environmental Sciences* 13: 455–466.
24. Urban, M. C. 2015. Accelerating extinction risk from climate change. *Science* 348: 571–573.
25. Thomas, C. D., Cameron, A., Green, R. E. et al. 2004. Extinction risk from climate change. *Nature* 427: 145–148.
26. Foden, W. B., Butchart, S. H. M., Stuart, S. N. et al. 2013. Identifying the world's most climate change vulnerable species: A systematic trait-based assessment of all birds, amphibians, and corals. *PLOS ONE* 8: e65427.
27. Warren, R., VanDerWal, J., Price, J. et al. 2013. Quantifying the benefit of early climate change mitigation in avoiding biodiversity loss. *Nature Climate Change* 3: 678.
28. Pearson, R. G., Stanton, J. C., Shoemaker, K. T. et al. 2014. Life history and spatial traits predict extinction risk due to climate change. *Nature Climate Change* 4: 217–221.
29. Cheung, W. W. L., Lam, V. W. Y., Sarmiento, J. L. et al. 2009. Projecting global marine biodiversity impacts under climate change scenarios. *Fish and Fisheries* 10: 235–251.
30. Jones, M. C. and Cheung, W. W. L. 2015. Multi-model ensemble projections of climate change effects on global marine biodiversity. *Ices Journal of Marine Science* 72: 741–752.
31. Gynther, I., N. Waller and Leung, L. K. 2016. *Confirmation of the extinction of the Bramble Cay melomys* Melomys rubicola *on Bramble Cay, Torres Strait: Results and conclusions from a comprehensive survey in August–September 2014*. Brisbane, Australia: Department of Environment and Heritage Protection.
32. Maclean, I. M. D. and Wilson, R. J. 2011. Recent ecological responses to climate change support predictions of high extinction risk. *Proceedings of the National Academy of Sciences* 108: 12337–12342.
33. Cahill, A. E., Aiello-Lammens, M. E., Fisher-Reid, M. C. et al. 2013. How does climate change cause extinction? *Proceedings of the Royal Society B* 280: 9.
34. Parmesan, C. 2006. Ecological and evolutionary responses to recent climate change. *Annual Review of Ecology Evolution and Systematics* 37: 637–669.
35. Sinervo, B., Méndez-de-la-Cruz, F., Miles, D. B. et al. 2010. Erosion of lizard diversity by climate change and altered thermal niches. *Science* 328: 894–899.
36. Trape, S. 2009. Impact of climate change on the relict tropical fish fauna of Central Sahara: Threat for the survival of Adrar Mountains Fishes, Mauritania. *PLOS ONE* 4: e4400.
37. Durance, I. and Ormerod, S. J. 2010. Evidence for the role of climate in the local extinction of a cool-water triclad. *Journal of the North American Benthological Society* 29: 1367–1378, 1312.
38. Parmesan, C. and Yohe, G. 2003. A globally coherent fingerprint of climate change impacts across natural systems. *Nature* 421: 37–42.
39. Thomas, C. D. 2010. Climate, climate change and range boundaries. *Diversity and Distributions* 16: 488–495.
40. Hickling, R., Roy, D. B., Hill, J. K. et al. 2006. The distributions of a wide range of taxonomic groups are expanding polewards. *Global Change Biology* 12: 450–455.
41. Parmesan, C. 1996. Climate and species' range. *Nature* 382: 765–766.
42. Ehrlich, P. R., Murphy, D. D., Singer, M. C. et al. 1980. Extinction, reduction, stability and increase: The responses of checkerspot butterfly (*Euphdyryas*) populations to the California drought. *Oecologia* 46: 101–105.
43. McLaughlin, J. F., Hellmann, J. J., Boggs, C. L. and Ehrlich, P. R. 2002. Climate change hastens population extinctions. *Proceedings of the National Academy of Sciences* 99: 6070–6074.
44. Pinsky, M. L., Worm, B., Fogarty, M. J. et al. 2013. Marine taxa track local climate velocities. *Science* 341: 1239–1242.
45. Barton, A. D., Irwin, A. J., Finkel, Z. V. and Stock, C. A. 2016. Anthropogenic climate change drives shift and shuffle in North Atlantic phytoplankton communities. *Proceedings of the National Academy of Sciences* 113: 2964–2969.
46. Poloczanska, E. S., Brown, C. J., Sydeman, W. J. et al. 2013. Global imprint of climate change on marine life. *Nature Climate Change* 3: 919.
47. Primack, R. B., Higuchi, H. and Miller-Rushing, A. J. 2009. The impact of climate change on cherry trees and other species in Japan. *Biological Conservation* 142: 1943–1949.
48. Gordo, O. and Sanz, J. J. 2010. Impact of climate change on plant phenology in Mediterranean ecosystems. *Global Change Biology* 16: 1082–1106.
49. Amano, T., Smithers Richard, J., Sparks Tim, H. and Sutherland William, J. 2010. A 250-year index of first flowering dates and its response to temperature changes. *Proceedings of the Royal Society B* 277: 2451–2457.
50. Ge, Q., Wang, H., Rutishauser, T. and Dai, J. 2015. Phenological response to climate change in China: A meta-analysis. *Global Change Biology* 21: 265–274.
51. Forister, M. L. and Shapiro, A. M. 2003. Climatic trends and advancing spring flight of butterflies in lowland California. *Global Change Biology* 9: 1130–1135.
52. Roy, D. B. and Sparks, T. H. 2000. Phenology of British butterflies and climate change. *Global Change Biology* 6: 407–416.
53. Stefanescu, C., Peñuelas, J. and Filella, I. 2003. Effects of climatic change on the phenology of butterflies in the northwest Mediterranean Basin. *Global Change Biology* 9: 1494–1506.
54. Gibbs, J. P. and Breisch, A. R. 2001. Climate warming and calling phenology of frogs near Ithaca, New York, 1900–1999. *Conservation Biology* 15: 1175–1178.
55. Beebee, T. J. C. 1995. Amphibian breeding and climate. *Nature* 374: 219–220.

56. Bókony, V., Barta, Z. and Végvári, Z. 2019. Changing migratory behaviors and climatic responsiveness in birds. *Frontiers in Ecology and Evolution* 7.

57. Crick, H. Q. P., Dudley, C., Glue, D. E. and Thomson, D. L. 1997. UK birds are laying eggs earlier. *Nature* 388: 526–526.

58. Brown, J. L., Li, S. H. and Bhagabati, N. 1999. Long-term trend toward earlier breeding in an American bird: a response to global warming? *Proceedings of the National Academy of Sciences* 96: 5565–5569.

59. Cohen, J. M., Lajeunesse, M. J. and Rohr, J. R. 2018. A global synthesis of animal phenological responses to climate change. *Nature Climate Change* 8: 224–228.

60. Walther, G.-R. 2010. Community and ecosystem responses to recent climate change. *Philosophical Transactions of the Royal Society B* 365: 2019–2024.

61. Renner, S. S. and Zohner, C. M. 2018. Climate change and phenological mismatch in trophic interactions among plants, insects, and vertebrates. *Annual Review of Ecology, Evolution, and Systematics* 49: 165–182.

62. Visser, M. E., van Noordwijk, A. J., Tinbergen, J. M. and Lessells, C. M. 1998. Warmer springs lead to mistimed reproduction in great tits (*Parus major*). *Proceedings of the Royal Society B* 265: 1867–1870.

63. Visser, M. E. and Both, C. 2005. Shifts in phenology due to global climate change: The need for a yardstick. *Proceedings of the Royal Society B* 272: 2561–2569.

64. Pyke, G. H., Thomson, J. D., Inouye, D. W. and Miller, T. J. 2016. Effects of climate change on phenologies and distributions of bumble bees and the plants they visit. *Ecosphere* 7: 19.

65. Sturm, M., Racine, C. and Tape, K. 2001. Increasing shrub abundance in the Arctic. *Nature* 411: 546–547.

66. Peñuelas, J. and Boada, M. 2003. A global change-induced biome shift in the Montseny mountains (northeast Spain). *Global Change Biology* 9: 131–140.

67. Staver, A. C., Archibald, S. and Levin, S. A. 2011. The global extent and determinants of savanna and forest as alternative biome states. *Science* 334: 230–232.

68. Greene, C. H. and Pershing, A. J. 2007. Climate drives sea change. *Science* 315: 1084–1085.

69. Hughes, T. P., Barnes, M. L., Bellwood, D. R. et al. 2017. Coral reefs in the Anthropocene. *Nature* 546: 82–90.

70. Wernberg, T., Bennett, S., Babcock, R. C. et al. 2016. Climate-driven regime shift of a temperate marine ecosystem. *Science* 353: 169–172.

71. Mooney, H., Larigauderie, A., Cesario, M. et al. 2009. Biodiversity, climate change, and ecosystem services. *Current Opinion in Environmental Sustainability* 1: 46–54.

72. Pecl, G. T., Araújo, M. B., Bell, J. D. et al. 2017. Biodiversity redistribution under climate change: Impacts on ecosystems and human well-being. *Science* 355: eaai9214.

73. Siraj, A. S., Santos-Vega, M., Bouma, M. J. et al. 2014. Altitudinal changes in malaria incidence in highlands of Ethiopia and Colombia. *Science* 343: 1154–1158.

74. Kerr, J. T., Pindar, A., Galpern, P. et al. 2015. Climate change impacts on bumblebees converge across continents. *Science* 349: 177–180.

75. Pinsky, M. L., Reygondeau, G., Caddell, R. et al. 2018. Preparing ocean governance for species on the move. *Science* 360: 1189–1191.

76. Cudmore, T. J., Björklund, N., Carroll, A. L. and Staffan Lindgren, B. 2010. Climate change and range expansion of an aggressive bark beetle: Evidence of higher beetle reproduction in naïve host tree populations. *Journal of Applied Ecology* 47: 1036–1043.

77. Kurz, W. A., Dymond, C. C., Stinson, G. et al. 2008. Mountain pine beetle and forest carbon feedback to climate change. *Nature* 452: 987.

78. Abrahms, B., DiPietro, D., Graffis, A. and Hollander, A. 2017. Managing biodiversity under climate change: Challenges, frameworks, and tools for adaptation. *Biodiversity and Conservation* 26: 2277–2293.

79. Provan, J. and Bennett, K. D. 2008. Phylogeographic insights into cryptic glacial refugia. *Trends in Ecology & Evolution* 23: 564–571.

80. Cain, S. A. 1944. *Foundations of Plant Geography*. New York, London: Harper & Brothers.

81. Keppel, G., Van Niel, K. P., Wardell-Johnson, G. W. et al. 2012. Refugia: Identifying and understanding safe havens for biodiversity under climate change. *Global Ecology and Biogeography* 21: 393–404.

82. Li, J., McCarthy, T. M., Wang, H. et al. 2016. Climate refugia of snow leopards in High Asia. *Biological Conservation* 203: 188–196.

83. Reusch, T. B. H. and Wood, T. E. 2007. Molecular ecology of global change. *Molecular Ecology* 16: 3973–3992.

84. Sgro, C. M., Lowe, A. J. and Hoffmann, A. A. 2011. Building evolutionary resilience for conserving biodiversity under climate change. *Evolutionary Applications* 4: 326–337.

85. Newell, R. G., Pizer, W. A. and Raimi, D. 2013. Carbon markets 15 years after Kyoto: Lessons learned, new challenges. *Journal of Economic Perspectives* 27: 123–146.

86. Cardinale, B. J., Matulich, K. L., Hooper, D. U. et al. 2011. The functional role of producer diversity in ecosystems. *American Journal of Botany* 98: 572–592.

87. Duffy, E., Godwin, C. and Cardinale, B. J. 2017. Biodiversity effects in the wild are common and as strong as key drivers of productivity. *Nature* 549: 261–264.

88. Liang, J. J., Crowther, T. W., Picard, N. et al. 2016. Positive biodiversity-productivity relationship predominant in global forests. *Science* 354: 12.

89. Maestre, F. T., Quero, J. L., Gotelli, N. J. et al. 2012. Plant species richness and ecosystem multifunctionality in global drylands. *Science* 335: 214–218.

90. Hungate, B. A., Barbier, E. B., Ando, A. W. et al. 2017. The economic value of grassland species for carbon storage. *Science Advances* 3: 1–8.

91. Wilmers, C. C., Estes, J. A., Edwards, M. et al. 2012. Do trophic cascades affect the storage and flux of atmospheric carbon? An analysis of sea otters and kelp forests. *Frontiers in Ecology and the Environment* 10: 409–415.

92. Barbier, E. B., Koch, E. W., Silliman, B. R. et al. 2008. Coastal ecosystem-based management with nonlinear ecological functions and values. *Science* 319: 321–323.

93. Ewel, K. C. 2010. Appreciating tropical coastal wetlands from a landscape perspective. *Frontiers in Ecology and the Environment* 8: 20–26.

94. Das, S. and Vincent, J. R. 2009. Mangroves protected villages and reduced death toll during Indian super cyclone. *Proceedings of the National Academy of Sciences* 106: 7357–7360.

95. Narayan, S., Beck, M. W., Wilson, P. et al. 2017. The value of coastal wetlands for flood damage reduction in the northeastern USA. *Scientific Reports* 7: 9463.

96. Arkema, K. K., Guannel, G., Verutes, G. et al. 2013. Coastal habitats shield people and property from sea-level rise and storms. *Nature Climate Change* 3: 913.

97. Beever, E. A., Ray, C., Mote, P. W. and Wilkening, J. L. 2010. Testing alternative models of climate-mediated extirpations. *Ecological Applications* 20: 164–178.

98. Beever, E. A., Ray, C., Wilkening, J. L. et al. 2011. Contemporary climate change alters the pace and drivers of extinction. *Global Change Biology* 17: 2054–2070.

99. Epps, C. W., McCullough, D. R., Wehausen, J. D. et al. 2004. Effects of climate change on population persistence of desert-dwelling mountain sheep in California. *Conservation Biology* 18: 102–113.

100. Munday, P. L. 2004. Habitat loss, resource specialization, and extinction on coral reefs. *Global Change Biology* 10: 1642–1647.

101. Arnall, S. G., Kuchling, G. and Mitchell, N. J. 2015. A thermal profile of metabolic performance in the rare Australian chelid, *Pseudemydura umbrina*. *Australian Journal of Zoology* 62: 448–453.

102. Ricciardi, A. and Simberloff, D. 2009. Assisted colonization is not a viable conservation strategy. *Trends in Ecology & Evolution* 24: 248–253.

103. Hoegh-Guldberg, O., Hughes, L., McIntyre, S. et al. 2008. Assisted colonization and rapid climate change. *Science* 321: 345–346.

104. Mueller, J. M. and Hellmann, J. J. 2008. An assessment of invasion risk from assisted migration. *Conservation Biology* 22: 562–567.

105. McLachlan, J. S., Hellmann, J. J. and Schwartz, M. W. 2007. A framework for debate of assisted migration in an era of climate change. *Conservation Biology* 21: 297–302.

106. Olden, J. D., Kennard, M. J., Lawler, J. J. and Poff, N. L. 2011. Challenges and opportunities in implementing managed relocation for conservation of freshwater species. *Conservation Biology* 25: 40–47.

107. Jia, H., Yao, H., Tang, Y. et al. 2015. LID-BMPs planning for urban runoff control and the case study in China. *Journal of Environmental Management* 149: 65–76.

108. Dong, G., Weng, B., Qin, T. et al. 2018. The impact of the construction of sponge cities on the surface runoff in watersheds, China. *Advances in Meteorology* 2018: 9.

109. Mei, C., Liu, J., Wang, H. et al. 2018. Integrated assessments of green infrastructure for flood mitigation to support robust decision-making for sponge city construction in an urbanized watershed. *Science of the Total Environment* 639: 1394–1407.

110. Zevenbergen, C., Fu, D. and Pathirana, A. 2018. Transitioning to sponge cities: Challenges and opportunities to address urban water problems in China. *Water* 10: 1230.

Chapter 13

1. Caro, T. M. and G. O'Doherty. 1999. On the use of surrogate species in conservation biology. *Conservation Biology* 13: 805–814.

2. Roberge, J.-M. and P. Angelstram. 2004. Usefulness of the umbrella species concept as a conservation tool. *Conservation Biology* 18: 76–85.

3. Grzimek, B. and M. Grzimek. 1973. *Serengeti Shall Not Die*. Ballantine Books, New York, New York, USA.

4. Berger, J. 1997. Population constraints associated with the use of black rhinos as an umbrella species for desert herbivores. *Conservation Biology* 11: 69–78.

5. Fahrig, L. 2003. Effects of habitat fragmentation on biodiversity. *Annual Review of Ecology, Evolution, and Systematics* 34: 487–515.

6. Haydon, D. T., et al. 2006. Low-coverage vaccination strategies for the conservation of endangered species. *Nature* 443: 692–695.

7. Ferrer, M., I. Newton, and M. Pandolfi. 2009. Small populations and offspring sex-ratio deviations in eagles. *Conservation Biology* 23: 1017–1025.

8. Clout, M. N. and D. V. Merton. 1998. Saving the Kakapo: The conservation of the world's most peculiar parrot. *Bird Conservation International* 8: 281–296.

9. Courchamp, F., T. Clutton-Brock, and B. Grenfell. 1999. Inverse density dependence and the Allee effect. *Trends in Ecology and Evolution* 14: 405–410.

10. Courchamp, F., L. Berec, and J. Gascoigne. 2008. *Allee Effects in Ecology and Conservation*. Oxford, UK: Oxford University Press.

11. Lande, R. 1993. Risks of population extinction from demographic and environmental stochasticity and random catastrophes. *American Naturalist* 142: 911–927.

12. Young, T. P. 1994. Natural die-offs of large mammals: Implications for conservation. *Conservation Biology* 8: 410–418.

13. Reed, D. H., E. H. Lowe, D. A. Briscoe, and R. Frankham. 2003. Fitness and adaptability in a novel environment: effect of inbreeding, prior environment, and lineage. *Evolution* 57: 1822–1828.

14. Drayton, B. and R. B. Primack. 1999. Experimental extinction of garlic mustard (*Alliaria petiolata*) populations: implications for weed science and conservation biology. *Biological Invasions* 1: 159–167.

15. King, S. R.B., L. Boyd, W. Zimmermann, and B. E. Kendall. 2015. *Equus ferus spp. przewalskii*. IUCN Red List of Threatened Species 2015: e.T7961A97205530.

16. Kaczensky, P., et al. 2011. The danger of having all your eggs in one basket: Winter crash of the re-introduced Przewalski's horses in the Mongolian Gobi. *PLOS ONE* 6(12): e28057.

17. Saltz, D., D. I. Rubenstein, and G. C. White. 2006. The impact of increased environmental stochasticity due to climate change on the dynamics of Asiatic wild ass. *Conservation Biology* 20: 1402–1409.

18. Frankham, R. 1995. Effective population size/adult population size ratios in wildlife: A review. *Genetical Research* 66: 95–107.

19. Allendorf, F. W., P. A. Hohenlohe, and G. Luikart. 2010. Genomics and the future of conservation genetics. *Nature Reviews Genetics* 11: 697–709.

20. Hedrick, P. 2005. Large variance in reproductive success and the N_e/N ratio. *Evolution* 59: 1596–1599.

21. Frankham, R., J. D. Ballou, and D. A. Briscoe. 2009. *Introduction to Conservation Genetics*. 2nd ed. Cambridge, UK: Cambridge University Press.

22. Ralls, K., J. D. Ballou, and A. Templeton. 1988. Estimates of lethal equivalents and the cost of inbreeding in mammals. *Conservation Biology* 2: 185–193.

23. Crnokrak, P. and D. A. Roff. 1999. Inbreeding depression in the wild. *Heredity* 83: 260–270.

24. Heschel, M. S. and K. N. Paige. 1995. Inbreeding depression, environmental stress and population size variation in scarlet gilia (*Ipomopsis aggregata*). *Conservation Biology* 9: 126–133.

25. Bouzat, J. L., et al. 2008. Beyond the beneficial effects of translocations as an effective tool for the genetic restoration of isolated populations. *Conservation Genetics* 10: 191–201.

26. Wright, S. 1931. Evolution in Mendelian populations. *Genetics* 16: 97–159.

27. Jamieson, I. G. 2011. Founder effects, inbreeding, and loss of genetic diversity in four avian reintroduction programs. *Conservation Biology* 25: 115–123.

28. Wang, J., E. Santiago, and A. Caballero. 2016. Prediction and estimation of effective population size. *Heredity* 117: 193–206.

29. Ramakrishnan, U., J. A. Santosh, U. Ramakrishnan, and R. Sukumar. 1998. The population and conservation status of Asian elephants in the Periyar Tiger Reserve, southern India. *Current Science India* 74: 110–113.

30. IUCN SSC Antelope Specialist Group. 2018. *Saiga tatarica*. IUCN Red List of Threatened Species e.T19832A50194357:

31. Milner-Gulland, E. J., et al. 2003. Reproductive collapse in saiga antelope harems. *Nature* 422: 135.

32. McShea, W. J. 2000. The influence of acorn crops on annual variation in rodent and bird populations. *Ecology* 81: 228–238.

33. Krebs, C. J., K. Cowcill, R. Boonstra, and A. J. Kenney. 2010. Do changes in berry crops drive population fluctuations in small rodents in the southwestern Yukon? *Journal of Mammalogy* 91: 500–509.

34. Freese, C. H., et al. 2007. Second chance for the Plains bison. *Biological Conservation* 136: 175–184.

35. Laikre, L., M. Jansson, F. W. Allendorf, S. Jakobsson, and N. Ryman. 2013. Hunting effects on favourable conservation status of highly inbred Swedish wolves. *Conservation Biology* 27: 248–253.

36. Wofford, J. E.B., R. E. Gresswell, and M. A. Banks. 2005. Influence of barriers to movement on within-watershed genetic variation of coastal cutthroat trout. *Ecological Applications* 15: 628–637.

37. Munson, L., et al. 2008. Climate extremes promote fatal co-infections during canine distemper epidemics in African lions. *PLoS ONE* 3(6): e2545.

38. Ewing, R. R., et al. 2008. Inbreeding and loss of genetic variation in a reintroduced population of Mauritius kestrel. *Conservation Biology* 22: 395–404.

39. Palomares, F., et al. 2012. Possible extinction vortex for a population of Iberian lynx on the verge of extirpation. *Conservation Biology* 26: 689–697.

40. Simberloff, D. 1988. The contribution of population and community biology to conservation science. *Annual Review of Ecology and Systematics* 19: 473–511.

41. Hunter, W. C., D. A. Buehler, R. A. Canterbury, J. L. Confer, and P. B. Hamel. 2001. Conservation of disturbance-dependent birds in eastern North America. *Wildlife Society Bulletin* 29: 440–455.

42. Schrott, G. R., K. A. With, and A. W. King. 2005. Demographic limitations on the ability of habitat restoration to rescue declining population. *Conservation Biology* 19: 1181–1193.

43. Baker, J. D. and P. M. Thompson. 2007. Temporal and spatial variation in age-specific survival rates of a long-lived mammal, the Hawaiian monk seal. *Proceedings of the Royal Society of London Series B* 274: 407–415.

44. Rich, T. C.G. 2006. Floristic changes in vascular plants in the British Isles: Geographical and temporal variation in botanical activity 1836–1988. *Botanical Journal of the Linnean Society* 152: 303–330.

45. Langham, G. M., J. G. Schuetz, T. Distler, C. U. Soykan, and C. Wilsey. 2015. Conservation status of North American birds in the face of future climate change. *PLOS ONE* 10(9): e0135350.

46. Lincoln, F. C. 1930. *Calculating Waterfowl Abundance on The Basis of Banding Returns*. Circular 118. Washington, DC: U. S. Department of Agriculture.

47. Jolly, G. M. 1965. Explicit estimates from capture-recapture data with both death and immigration-stochastic model. *Biometrika* 52: 225–247.

48. Seber, G. A.F. 1965. A note on the multiple-recapture census. *Biometrika* 52: 249–259.

49. Lindberg, M. S. 2012. A review of designs for capture-mark-recapture studies in discrete time. *Journal of Ornithology* 152: 355–370.

50. Lambert, B. A., R. A. Schorr, S. C. Schneider, and E. Muths. 2016. Influence of demography and environment on persistence of toad populations. *Journal of Wildlife Management* 80: 1256–1266.

51. Crooks, J. A. 2005. Lag times and exotic species: The ecology and management of biological invasions in slow-motion. *Ecoscience* 12: 316–329.

52. Sakai, A. K., et al. 2001. The population biology of invasive species. *Annual Review of Ecology and Systematics* 32: 305–332.

53. Donovan, T. M. and C. Welden. 2002. *Spreadsheet Exercises in Ecology and Evolution*. Sunderland, MA: Oxford University Press/Sinauer.

54. Gotelli, N. J. 2008. *A Primer of Ecology*. 4th ed. Sunderland, MA: Oxford University Press/Sinauer.

55. Leslie, P. H. 1945. On the use of matrices in certain population mathematics. *Biometrika* 33: 183–212.

56. Lefkovitch, L. P. 1965. The study of population growth in organisms grouped by stages. *Biometrics* 21: 1–18.

57. Crowder, L. B., D. T. Crouse, S. S. Heppell, and T. H. Martin. 1994. Predicting the impact of turtle excluder devices on loggerhead sea turtle populations. *Ecological Applications* 4: 437–445.

58. Shaffer, M. L. 1981. Minimum population sizes for species conservation *BioScience* 31: 131–134.

59. Traill, L. W., B. W. Brook, R. R. Frankham, and C. J.A. Bradshaw. 2010. Pragmatic population viability targets in a rapidly changing world. *Biological Conservation* 143: 28–34.

60. Flather, C. H., G. D. Hayward, S. R. Beissinger, and P. A. Stephens. 2011. Minimum viable populations: Is there a "magic number" for conservation practitioners? *Trends in Ecology and Evolution* 26: 307–316.

61. Bouche, P., I. Douglas-Hamilton, G. Wittemyer, and A. J. Nianogo. 2011. Will elephants soon disappear from West African savannahs? *PLOS ONE* 6(6): e20619.

62. Mlot, C. 2013. Are Isle Royale's wolves chasing extinction? *Science* 340: 919–921.

63. Hedrick, P. W., J. A. Robinson, R. O. Peterson, and J. A. Vucetich. 2019. Genetics and extinction and the example of Isle Royale wolves. *Animal Conservation* 22: 302–309.

64. Grouios, C. P. and L. L. Manne. 2009. Utility of measuring abundance versus consistent occupancy in predicting biodiversity persistence. *Conservation Biology* 23: 1260–1269.

65. Berger, J. 1990. Persistence of different-sized populations: An empirical assessment of rapid extinctions in bighorn sheep. *Conservation Biology* 4: 91–98.

66. Berger, J. 1999. Intervention and persistence in small populations of bighorn sheep. *Conservation Biology* 13: 432–435.

67. Singer, F. J., L. C. Zeigenfuss, and L. Spicer. 2001. Role of patch size, disease, and movement in rapid extinction of bighorn sheep. *Conservation Biology* 15: 1347–1354.

68. Shoemaker, K. T., A. R. Breisch, J. W. Jaycox, and J. P. Gibbs. 2013. Reexamining the minimum viable population concept for long-lived species. *Conservation Biology* 27: 542–551.

69. Thiollay, J. M. 1989. Area requirements for the conservation of rainforest raptors and game birds in French Guiana. *Conservation Biology* 3: 128–137.

70. Pe'er, G., et al. 2014. Toward better application of minimum area requirements in conservation planning. *Biological Conservation* 170: 92–102.

71. Woodroffe, R. and C. Sillero-Zubiri. 2012. *Lycaon pictus*. IUCN Red List of Threatened Species 2012: e.T12436A16711116.

72. Lacy, R. C. 2019. Lessons from 30 years of population viability analysis of wildlife populations. *Zoo Biology* 38: 67–77.

73. Jakalaniemi, A., H. Postila, and J. Tuomi. 2013. Accuracy of short-term demographic data in projecting long-term fate of populations. *Conservation Biology* 27: 552–559.

74. Reed, J. M., C. S. Elphick, A. F. Zuur, E. N. Ieno, and G. M. Smith. 2007. Time series analysis of Hawaiian waterbirds. In *Analysis of Ecological Data*, A. F. Zuur, E. N. Ieno, and G. M. Smith, (Eds.). The Netherlands: Springer-Verlag.

75. Reed, J. M., C. S. Elphick, E. N. Ieno, and A. F. Zuur. 2011. Long-term population trends of endangered Hawaiian waterbirds. *Population Ecology* 53: 473–481.

76. Bulman, C. R., et al. 2007. Minimum viable metapopulation size, extinction debt, and the conservation of declining species. *Ecological Applications* 17: 1460–1473.

77. Lindenmayer, D. B. 2000. Factors at multiple scales affecting distribution patterns and their implications for animal conservation: Leadbeater's possum as a case study. *Biodiversity and Conservation* 9: 15–35.

78. Lindenmayer, D. B., et al. 2012. A major shift to the retention approach for forestry can help resolve some global forest sustainability issues. *Conservation Letters* 5: 421–431.

79. Runge, M. C., J. B. Grand, and M. S. Mitchell. 2013. Structured decision making. In *Wildlife Management and Conservation: Contemporary Principles and Practices*, P. R. Krausman and J. W. Cain, (Eds.). Baltimore, MD: Johns Hopkins University Press.

80. Mitchell, M. S., et al. 2013. Using structured decision making to manage disease risk for Montana wildlife. *Wildlife Society Bulletin* 37: 107–114.
81. Dick, M., A. M. Rous, V. M. Nguyen, and S. J. Cooke. 2016. Necessary but challenging: Multiple disciplinary approaches to solving conservation problems. *Facets* 1: 67–82.
82. Schwartz, M. W., et al. 2018. Decision support frameworks and tools for conservation. *Conservation Letters* 11: 1–12.
83. Roemer, G. W., T. J. Coonan, D. K. Garcelon, J. Bascompte, and L. Laughrin. 2001. Feral pigs facilitate hyperpredation by golden eagles and indirectly cause the decline of island fox. *Animal Conservation* 4: 307–318.
84. Timm, S. F., et al. 2009. A suspected canine distemper epidemic as the cause of a catastrophic decline in Santa Catalina island foxes (*Urocyon littoralis catalinae*). *Journal of Wildlife Diseases* 45: 333–343.
85. United States Department of the Interior. 2016. *Endangered and threatened wildlife and plants: Removing the San Miguel island fox, Santa Rosa island fox, and Santa Cruz island fox from the federal list of endangered and threatened wildlife, and reclassifying the Santa Catalina island fox from endangered to threatened*. Docket ID: FWS-R8-ES-2015–0170. Washington, DC: United States Fish and Wildlife Service.
86. Buckland, S. T., et al. 2001. Introduction to Distance Sampling Estimating Abundance of Biological Populations. Oxford, UK: Oxford University Press.
87. Dahlgren, D. K., et al. 2012. Use of dogs in wildlife research and management. In *Wildlife Techniques Manual*, N. Silvy, (Ed.). Washington, DC: The Wildlife Society.
88. Wasser, S. K., et al. 2004. Scat detection dogs in wildife research and management: Application to grizzly and black bears in the Yellowhead Ecosystem, Alberta, Canada. *Canadian Journal of Zoology* 82: 475–492.
89. Smith, D., et al. 2003. Detection and accuracy rates of dogs trained to find scats of San Joaquin kit foxes (*Vulpes macrotis mutica*). *Animal Conservation* 6: 339–346.
90. Rolland, R. M., et al. 2006. Faecal sampling using detection dogs to study reproduction and health in North Atlantic right whales (*Eubalaena glacialis*). *Journal of Cetacean Research and Management* 8: 121–125.
91. Long, R. A., T. M. Donovan, P. Mackay, W. J. Zielinski, and J. S. Buzas. 2007. Effectiveness of scat detection dogs for detecting forest carnivores. *Journal of Wildlife Management* 71: 2007–2017.

Chapter 14

1. Visconti, P., Butchart, S. H. M., Brooks, T. M. et al. 2019. Protected area targets post-2020. *Science* 364: 239–241.
2. Bingham, H. C., Juffe Bignoli, D., Lewis, E. et al. 2019. Sixty years of tracking conservation progress using the World Database on Protected Areas. *Nature Ecology & Evolution* 3: 737–743.
3. UNEP-WCMC, IUCN and NGS. 2018. *Protected Planet Report 2018*. Cambridge, UK.
4. Myers, N. 1988. Threatened biotas: "Hot spots" in tropical forests. *The Environmentalist* 8: 187–208.
5. Myers, N. 1990. The biodiversity challenge: Expanded hot-spots analysis. *The Environmentalist* 10: 243–256.
6. Myers, N., Mittermeier, R. A., Mittermeier, C. G. et al. 2000. Biodiversity hotspots for conservation priorities. *Nature* 403: 853–858.
7. Brooks, T. M., Mittermeier, R. A., Mittermeier, C. G. et al. 2002. Habitat loss and extinction in the hotspots of biodiversity. *Conservation Biology* 16: 909–923.
8. Mittermeier, R. A., Turner, W. R., Larsen, F. W. et al. 2011. Global biodiversity conservation: the critical role of hotspots. In F. E. Zachos and J. C. Habel (Eds.), *Biodiversity Hotspots: Distribution and Protection of Conservation Priority Areas*. London, UK: Springer.
9. Lucifora, L. O., V. B. García and Worm, B. 2011. Global diversity hotspots and conservation priorities for sharks. *PLOS ONE* 6: e19356.
10. Sechrest, W., Brooks, T. M., da Fonseca, G. A. B. et al. 2002. Hotspots and the conservation of evolutionary history. *Proceedings of the National Academy of Sciences* 99: 2067–2071.
11. Joppa, L. N., Roberts, D. L., Myers, N. and Pimm, S. L. 2011. Biodiversity hotspots house most undiscovered plant species. *Proceedings of the National Academy of Sciences* 108: 13171–13176.
12. McCarthy, D. P., Donald, P. F., Scharlemann, J. P. W. et al. 2012. Financial costs of meeting global biodiversity conservation targets: Current spending and unmet needs. *Science* 338: 946–949.
13. Knapp, A. 2012. *How much does it cost to find a Higgs Boson?* Forbes. Jersey City, NJ: Randall Lane.
14. Tuttle, B. 2018. Jeff Bezos is already $40 billion richer this year. *Money* Magazine. New York, NY: Meredith Corporation.
15. Possingham, H. P. and Wilson, K. A. 2005. Turning up the heat on hotspots. *Nature* 436: 919–920.
16. Kareiva, P. and Marvier, M. 2003. Conserving biodiversity coldspots: Recent calls to direct conservation funding to the world's biodiversity hotspots may be bad investment advice. *American Scientist* 91: 344–351.
17. Kareiva, P. and Marvier, M. 2017. *Conservation Science: Balancing the Needs of People and Nature*, 2nd ed. New York, NY: W. H. Freeman.
18. Olson, D. M., Dinerstein, E., Wikramanayake, E. D. et al. 2001. Terrestrial ecoregions of the world: A new map of life on Earth. *BioScience* 51: 933–938.
19. Spalding, M. D., Fox, H. E., Allen, G. R. et al. 2007. Marine ecoregions of the world: A bioregionalization of coastal and shelf areas. *BioScience* 57: 573–583.
20. Abell, R., Thieme, M. L., Revenga, C. et al. 2008. Freshwater ecoregions of the world: A new map of biogeographic units for freshwater biodiversity conservation. *BioScience* 58: 403–414.
21. Olson, D. M. and Dinerstein, E. 2002. The Global 200: Priority ecoregions for global conservation. *Annals of the Missouri Botanical Garden* 89: 199–224.
22. Dinerstein, E., Olson, D., Joshi, A. et al. 2017. An ecoregion-based approach to protecting half the terrestrial realm. *BioScience* 67: 534–545.
23. Tallis, H. and Polasky, S. 2009. Mapping and valuing ecosystem services as an approach for conservation and natural-resource management. *Annals of the New York Academy of Sciences* 1162: 265–283.
24. Xu, W., Xiao, Y., Zhang, J. et al. 2017. Strengthening protected areas for biodiversity and ecosystem services in China. *Proceedings of the National Academy of Sciences* 114: 1601–1606.
25. UNESCO. 2012. Operational Guidelines for the Implementation of the World Heritage Convention. Paris, France: UNESCO World Heritage Centre.
26. Geldmann, J., Barnes, M., Coad, L. et al. 2013. Effectiveness of terrestrial protected areas in reducing habitat loss and population declines. *Biological Conservation* 161: 230–238.
27. Cuenca, P., Arriagada, R. and Echeverría, C. 2016. How much deforestation do protected areas avoid in tropical Andean landscapes? *Environmental Science & Policy* 56: 56–66.
28. Andam, K. S., Ferraro, P. J., Pfaff, A. et al. 2008. Measuring the effectiveness of protected area networks in reducing deforestation. *Proceedings of the National Academy of Sciences* 105: 16089–16094.
29. Joppa, L. N., Loarie, S. R. and Pimm, S. L. 2008. On the protection of protected areas. *Proceedings of the National Academy of Sciences* 105: 6673–6678.
30. Coetzee, B. W. T., Gaston, K. J. and Chown, S. L. 2014. Local scale comparisons of biodiversity as a test for global protected area ecological performance: A meta-analysis. *PLOS ONE* 9: e105824.

31. Gray, C. L., Hill, S. L. L., Newbold, T. et al. 2016. Local biodiversity is higher inside than outside terrestrial protected areas worldwide. *Nature Communications* 7: 12306.
32. Council, N. R. 2001. *Marine Protected Areas: Tools for Sustaining Ocean Ecosystems*. The National Academies Press: Washington, DC.
33. Pendleton, L. H., Ahmadia, G. N., Browman, H. I. et al. 2017. Debating the effectiveness of marine protected areas. *ICES Journal of Marine Science* 75: 1156–1159.
34. Selig, E. R. and Bruno, J. F. 2010. A global analysis of the effectiveness of marine protected areas in preventing coral loss. *PLOS ONE* 5: e9278.
35. Strain, E. M. A., Edgar, G. J., Ceccarelli, D. et al. 2019. A global assessment of the direct and indirect benefits of marine protected areas for coral reef conservation. *Diversity and Distributions* 25: 9–20.
36. Topor, Z. M., Rasher, D. B., Duffy, J. E. and Brandl, S. J. Marine protected areas enhance coral reef functioning by promoting fish biodiversity. *Conservation Letters*: e12638.
37. White, T. D., Carlisle, A. B., Kroodsma, D. A. et al. 2017. Assessing the effectiveness of a large marine protected area for reef shark conservation. *Biological Conservation* 207: 64–71.
38. Siikamaki, J. 2011. Contributions of the U.S. state park system to nature recreation. *Proceedings of the National Academy of Sciences* 108: 14031–14036.
39. Southwick. 2011. *The economics associated with outdoor recreation, natural resources conservation and historic preservation in the United States*. Fenandina Beach, FL: Southwick Associates.
40. Naidoo, R., Weaver, L. C., Stuart-Hill, G. and Tagg, J. 2011. Effect of biodiversity on economic benefits from communal lands in Namibia. *Journal of Applied Ecology* 48: 310–316.
41. Naidoo, R., Gerkey, D., Hole, D. et al. 2019. Evaluating the impacts of protected areas on human well-being across the developing world. *Science Advances* 5: eaav3006.
42. Agrawal, A. and Redford, K. 2009. Conservation and displacement: An overview. *Conservation and Society* 7: 1–10.
43. West, P., Igoe, J. and Brockington, D. 2006. Parks and peoples: The social impact of protected areas. *Annual Review of Anthropology* 35: 251–277.
44. Spence, M. D. 1999. *Dispossessing the Wilderness: Indian Removal and the Making of the National Parks*. New York, NY: Oxford University Press.
45. West, P. and Brockington, D. 2006. An anthropological perspective on some unexpected consequences of protected areas. *Conservation Biology* 20: 609–616.
46. McLean, J. and Stræde, S. 2003. Conservation, relocation, and the paradigms of park and people management: A case study of Padampur Villages and the Royal Chitwan National Park, Nepal. *Society & Natural Resources* 16: 509–526.
47. Rao, K. S., Maikhuri, R. K., Nautiyal, S. and Saxena, K. G. 2002. Crop damage and livestock depredation by wildlife: a case study from Nanda Devi Biosphere Reserve, India. *Journal of Environmental Management* 66: 317–327.
48. Dowie, M. 2011. Conservation Refugees: The Hundred-Year Conflict Between Global Conservation and Native Peoples. Cambridge, MA: MIT Press.
49. Geisler, C. 2003. A new kind of trouble: Evictions in Eden. *International Social Science Journal* 55: 69–78.
50. Dowie, M. 2005. *Conservation refugees* Orion online. November–December 1–12.
51. Cincotta, R. P., Wisnewski, J. and Engelman, R. 2000. Human population in the biodiversity hotspots. *Nature* 404: 990–992.
52. Newmark, W. D., Leonard, N. L., Sariko, H. I. and Gamassa, D.-G. M. 1993. Conservation attitudes of local people living adjacent to five protected areas in Tanzania. *Biological Conservation* 63: 177–183.
53. Bennett, N. J. and Dearden, P. 2014. Why local people do not support conservation: Community perceptions of marine protected area livelihood impacts, governance and management in Thailand. *Marine Policy* 44: 107–116.
54. Mascia, M. B. and Pailler, S. 2011. Protected area downgrading, downsizing, and degazettement (PADDD) and its conservation implications. *Conservation Letters* 4: 9–20.
55. Bernard, E., L. A. O. Penna and Araujo, E. 2014. Downgrading, downsizing, degazettement, and reclassification of protected areas in Brazil. *Conservation Biology* 28: 939–950.
56. Smith, C. and Greshko, M. 2018. UN announces 23 new nature reserves while U.S. removes 17. *National Geographic*. New York City, NY: National Geographic Partners.
57. Leverington, F., Costa, K. L., Pavese, H. et al. 2010. A global analysis of protected area management effectiveness. *Environmental Management* 46: 685–698.
58. Laurance, W. F., Carolina Useche, D., Rendeiro, J. et al. 2012. Averting biodiversity collapse in tropical forest protected areas. *Nature* 489: 290.
59. Craigie, I. D., Baillie, J. E. M., Balmford, A. et al. 2010. Large mammal population declines in Africa's protected areas. *Biological Conservation* 143: 2221–2228.
60. Watson, J. E. M., Dudley, N., Segan, D. B. and Hockings, M. 2014. The performance and potential of protected areas. *Nature* 515: 67.
61. Scheffer, M., Barrett, S., Carpenter, S. R. et al. 2015. Creating a safe operating space for iconic ecosystems. *Science* 347: 1317–1319.
62. Geldmann, J., Joppa, L. N. and Burgess, N. D. 2014. Mapping change in human pressure globally on land and within protected areas. *Conservation Biology* 28: 1604–1616.
63. Schulze, K., Knights, K., Coad, L. et al. 2018. An assessment of threats to terrestrial protected areas. *Conservation Letters* 11: e12435.
64. Edgar, G. J., Stuart-Smith, R. D., Willis, T. J. et al. 2014. Global conservation outcomes depend on marine protected areas with five key features. *Nature* 506: 216.
65. Gill, D. A., Mascia, M. B., Ahmadia, G. N. et al. 2017. Capacity shortfalls hinder the performance of marine protected areas globally. *Nature* 543: 665.
66. Giakoumi, S., McGowan, J., Mills, M. et al. 2018. Revisiting "success" and "failure" of marine protected areas: A conservation scientist perspective. *Frontiers in Marine Science* 5.
67. Dovers, S., S. Feary, A. Martin et al. 2015. Engagement and participation in protected area management: who, why, how and when? In G. L. Worboys et al. (Eds.), *Protected Area Governance and Management*. Canberra, Australia: ANU Press.
68. Oldekop, J. A., Holmes, G., Harris, W. E. and Evans, K. L. 2016. A global assessment of the social and conservation outcomes of protected areas. *Conservation Biology* 30: 133–141.
69. Nolte, C., Agrawal, A., Silvius, K. M. and Soares-Filho, B. S. 2013. Governance regime and location influence avoided deforestation success of protected areas in the Brazilian Amazon. *Proceedings of the National Academy of Sciences* 110: 4956–4961.
70. Nelson, A. and Chomitz, K. M. 2011. Effectiveness of strict vs. multiple use protected areas in reducing tropical forest fires: A global analysis using matching methods. *PLOS ONE* 6: e22722.
71. Noss, R. F., A. P. Dobson, R. Baldwin et al. 2012. Bolder thinking for conservation. *Conservation Biology* 26: 1–4.
72. Butchart, S. H. M., M. Di Marco and Watson, J. E. M. 2016. Formulating smart commitments on biodiversity: Lessons from the Aichi Targets. *Conservation Letters* 9: 457–468.
73. Wilson, E. O. 2016. *Half-Earth: Our Planet's Fight for Life*. New York, NY: W.W. Norton & Company.
74. O'Leary, B. C., Winther-Janson, M., Bainbridge, J. M. et al. 2016. Effective coverage targets for ocean protection. *Conservation Letters* 9: 398–404.

75. Locke, H. 2013. Nature needs half: A necessary and hopeful new agenda for protected areas. *Parks* 19.2.

Chapter 15

1. Brashares, J. S., Arcese, P. and Sam, M. K. 2001. Human demography and reserve size predict wildlife extinction in West Africa. *Proceedings of the Royal Society of London Series B* 268: 2473–2478.
2. Harcourt, A. H., Parks, S. A. and Woodroffe, R. 2001. Human density as an influence on species/area relationships: double jeopardy for small African reserves? *Biodiversity and Conservation* 10: 1011–1026.
3. Parks, S. A. and Harcourt, A. H. 2002. Reserve size, local human density, and mammalian extinctions in US protected areas. *Conservation Biology* 16: 800–808.
4. Wiersma, Y. F., Nudds, T. D. and Rivard, D. H. 2004. Models to distinguish effects of landscape patterns and human population pressures associated with species loss in Canadian national parks. *Landscape Ecology* 19: 773–786.
5. Cumming, G. S. and Allen, C. R. 2017. Protected areas as social-ecological systems: Perspectives from resilience and social-ecological systems theory. *Ecological Applications* 27: 1709–1717.
6. Wikramanayake, E., Dinerstein, E., Seidensticker, J. et al. 2011. A landscape-based conservation strategy to double the wild tiger population. *Conservation Letters* 4: 219–227.
7. Runge, C. A., Watson, J. E. M., Butchart, S. H. M. et al. 2015. Protected areas and global conservation of migratory birds. *Science* 350: 1255–1258.
8. Rands, M. R., Adams, W. M., Bennun, L. et al. 2010. Biodiversity conservation: Challenges beyond 2010. *Science* 329: 1298–1303.
9. Tittensor, D. P., Walpole, M., Hill, S. L. L. et al. 2014. A mid-term analysis of progress toward international biodiversity targets. *Science* 346: 241–244.
10. Bingham, H., Fitzsimons, J. A., Redford, K. H. et al. 2017. Privately protected areas: Advances and challenges in guidance, policy and documentation. *Parks* 23: 13–28.
11. Kareiva, P. and Marvier, M. 2017. *Conservation Science: Balancing the Needs of People and Nature* (2nd ed.). New York, NY: W. H. Freeman.
12. Christensen, J., Rempel, J. and Burr, J. 2011. *Land trusts thrive despite, and because of, the Great Recession.* High Country News. Paonia, CO: Paul Larmer.
13. Stolton, S., Redford, K. H. and Dudley, N. 2014. *The Futures of Privately Protected Areas.* Gland, Switzerland: IUCN.
14. Stevens, S. 2014. Indigenous Peoples, National Parks, and Protected Areas: A New Paradigm Linking Conservation, Culture, and Rights. Tucson, AZ: University of Arizona Press.
15. Maiorano, L., Falcucci, A. and Boitani, L. 2008. Size-dependent resistance of protected areas to land-use change. *Proceedings of the Royal Society B* 275: 1297–1304.
16. Woodroffe, R. and Ginsberg, J. R. 1998. Edge effects and the extinction of populations inside protected areas. *Science* 280: 2126–2128.
17. Brooks, T. M., Mittermeier, R. A., da Fonseca, G. A. B. et al. 2006. Global biodiversity conservation priorities. *Science* 313: 58–61.
18. González-Maya, J. F., Víquez-R, L. R., Belant, J. L. and Ceballos, G. 2015. Effectiveness of protected areas for representing species and populations of terrestrial mammals in Costa Rica. *PLoS One* 10: e0124480.
19. Albers, H. J., Busby, G. M., Hamaide, B. et al. 2016. Spatially correlated risk in nature reserve site selection. *PLOS ONE* 11: e0146023.
20. Aycrigg, J. L., Davidson, A., Svancara, L. K. et al. 2013. Representation of ecological systems within the protected areas network of the continental United States. *PLOS ONE* 8: e54689.
21. Rouget, M., Richardson, D. M. and Cowling, R. M. 2003. The current configuration of protected areas in the Cape Floristic Region, South Africa: Reservation bias and representation of biodiversity patterns and processes. *Biological Conservation* 112: 129–145.
22. Redpath, S. M., Young, J., Evely, A. et al. 2013. Understanding and managing conservation conflicts. *Trends in Ecology & Evolution* 28: 100–109.
23. Maron, M., Rhodes, J. R. and Gibbons, P. 2013. Calculating the benefit of conservation actions. *Conservation Letters* 6: 359–367.
24. Day, J., Fernandes, L., Lewis, A. et al. 2002. The representative areas program for protecting biodiversity in the Great Barrier Reef World Heritage Area. *Proceedings of the Ninth International Coral Reef Symposium, Bali, 23–27 October 2000.*
25. Fernandes, L., Day, J., Lewis, A. et al. 2005. Establishing representative no-take areas in the Great Barrier Reef: Large-scale implementation of theory on marine protected areas. *Conservation Biology* 19: 1733–1744.
26. Scott, J. M., Csuti, B., Jacobi, J. D. and Estes, J. E. 1987. Species richness. *BioScience* 37: 782–788.
27. Scott, J. M., Davis, F., Csuti, B. et al. 1993. Gap analysis: A geographic approach to protection of biological diversity. *Wildlife Monographs*: 3–41.
28. Rodrigues, A. S. L., Akçakaya, H. R., Andelman, S. J. et al. 2004. Global gap analysis: Priority regions for expanding the global protected-area network. *BioScience* 54: 1092–1100.
29. Rodrigues, A. S. L., Andelman, S. J., Bakarr, M. I. et al. 2004. Effectiveness of the global protected area network in representing species diversity. *Nature* 428: 640–643.
30. Mazaris, A. D., Almpanidou, V., Wallace, B. P. et al. 2014. A global gap analysis of sea turtle protection coverage. *Biological Conservation* 173: 17–23.
31. Goettsch, B., Duran, A. P. and Gaston, K. J. 2019. Global gap analysis of cactus species and priority sites for their conservation. *Conservation Biology* 33: 369–376.
32. Beier, P., Spencer, W., Baldwin, R. F. and McRae, B. H. 2011. Toward best practices for developing regional connectivity maps. *Conservation Biology* 25: 879–892.
33. Magrach, A., Larrinaga, A. R. and Santamaria, L. 2012. Effects of matrix characteristics and interpatch distance on functional connectivity in fragmented temperate rainforests. *Conservation Biology* 26: 238–247.
34. Saura, S., Bodin, Ö. and Fortin, M.-J. 2014. Stepping stones are crucial for species' long-distance dispersal and range expansion through habitat networks. *Journal of Applied Ecology* 51: 171–182.
35. Harris, G., Thirgood, S., Hopcraft, J. G. C. et al. 2009. Global decline in aggregated migrations of large terrestrial mammals. *Endangered Species Research* 7: 55–76.
36. Feist, B. E., Buhle, E. R., Baldwin, D. H. et al. 2017. Roads to ruin: Conservation threats to a sentinel species across an urban gradient. *Ecological Applications* 27: 2382–2396.
37. Barber, C. P., Cochrane, M. A., Souza, C. M. and Laurance, W. F. 2014. Roads, deforestation, and the mitigating effect of protected areas in the Amazon. *Biological Conservation* 177: 203–209.
38. Ford, A. T., Clevenger, A. P. and Bennett, A. 2009. Comparison of methods of monitoring wildlife crossing-structures on highways. *The Journal of Wildlife Management* 73: 1213–1222.
39. Sawaya, M. A., Clevenger, A. P. and Kalinowski, S. T. 2013. Demographic connectivity for ursid populations at wildlife crossing structures in Banff National Park. *Conservation Biology* 27: 721–730.
40. Cohn, J. P. 2010. A narrow path for pronghorns. *BioScience* 60: 480.

41. Roth, A. 2018. *Epic Yellowstone Migrations Gain New Bipartisan Protections*. National Geographic. Washington, DC: National Geographic Partners.

42. Berger, J. and Cain, S. L. 2014. Moving beyond science to protect a mammalian migration corridor. *Conservation Biology* 28: 1142–1150.

43. Saura, S., Bastin, L., Battistella, L. et al. 2017. Protected areas in the world's ecoregions: How well connected are they? *Ecological Indicators* 76: 144–158.

44. Simberloff, D., Farr, J. A., Cox, J. and Mehlman, D. W. 1992. Movement corridors: Conservation bargains or poor investments? *Conservation Biology* 6: 493–504.

45. Orrock, J. L. and Damschen, E. I. 2005. Corridors cause differential seed predation. *Ecological Applications* 15: 793–798.

46. Haddad, N. M., Brudvig, L. A., Damschen, E. I. et al. 2014. Potential negative ecological effects of corridors. *Conservation Biology* 28: 1178–1187.

47. Robles, M. D., Flather, C. H., Stein, S. M. et al. 2008. The geography of private forests that support at-risk species in the conterminous United States. *Frontiers in Ecology and the Environment* 6: 301–307.

48. Adum, G. B., Eichhorn, M. P., Oduro, W. et al. 2013. Two-stage recovery of amphibian assemblages following selective logging of tropical forests. *Conservation Biology* 27: 354–363.

49. MacKay, A., Allard, M. and Villard, M.-A. 2014. Capacity of older plantations to host bird assemblages of naturally regenerated conifer forests: A test at stand and landscape levels. *Biological Conservation* 170: 110–119.

50. Peh, K. S. H., Jong, J. d., Sodhi, N. S. et al. 2005. Lowland rainforest avifauna and human disturbance: persistence of primary forest birds in selectively logged forests and mixed-rural habitats of southern Peninsular Malaysia. *Biological Conservation* 123: 489–505.

51. Stokes, E. J., Strindberg, S., Bakabana, P. C. et al. 2010. Monitoring great ape and elephant abundance at large spatial scales: Measuring effectiveness of a conservation landscape. *PLOS ONE* 5: e10294.

52. Meijaard, E., Albar, G., Nardiyono et al. 2010. Unexpected ecological resilience in Bornean orangutans and implications for pulp and paper plantation management. *PLOS ONE* 5: e12813.

53. Cox, R. L. and Underwood, E. C. 2011. The importance of conserving biodiversity outside of protected areas in Mediterranean ecosystems. *PLOS ONE* 6: e14508.

54. Hansen, A. J., Davis, C. R., Piekielek, N. et al. 2011. Delineating the ecosystems containing protected areas for monitoring and management. *BioScience* 61: 363–373.

55. Troupin, D. and Carmel, Y. 2014. Can agro-ecosystems efficiently complement protected area networks? *Biological Conservation* 169: 158–166.

56. Meffert, P. J. and Dziock, F. 2012. What determines occurrence of threatened bird species on urban wastelands? *Biological Conservation* 153: 87–96.

57. Ellis, E. C., Antill, E. C. and Kreft, H. 2012. All is not loss: Plant biodiversity in the Anthropocene. *PLOS ONE* 7: e30535.

58. McPhearson, T., Pickett, S. T. A., Grimm, N. B. et al. 2016. Advancing urban ecology toward a science of cities. *BioScience* 66: 198–212.

59. Kremer, P., Hamstead, Z., Haase, D. et al. 2016. Key insights for the future of urban ecosystem services research. *Ecology and Society* 21.

60. Newbold, T., Hudson, L. N., Hill, S. L. L. et al. 2015. Global effects of land use on local terrestrial biodiversity. *Nature* 520: 45–50.

61. Goddard, M. A., Dougill, A. J. and Benton, T. G. 2010. Scaling up from gardens: Biodiversity conservation in urban environments. *Trends in Ecology & Evolution* 25: 90–98.

62. Aronson, M. F. J., Lepczyk, C. A., Evans, K. L. et al. 2017. Biodiversity in the city: Key challenges for urban green space management. *Frontiers in Ecology and the Environment* 15: 189–196.

63. Beninde, J., Veith, M. and Hochkirch, A. 2015. Biodiversity in cities needs space: A meta-analysis of factors determining intra-urban biodiversity variation. *Ecology Letters* 18: 581–592.

64. Frumkin, H., Bratman, G. N., Breslow, S. J. et al. 2017. Nature contact and human health: A research agenda. *Environmental Health Perspectives* 125: 18.

65. Keniger, L., Gaston, K., Irvine, K. and Fuller, R. 2013. What are the benefits of interacting with nature? *International Journal of Environmental Research and Public Health* 10: 913.

66. Czembrowski, P. and Kronenberg, J. 2016. Hedonic pricing and different urban green space types and sizes: Insights into the discussion on valuing ecosystem services. *Landscape and Urban Planning* 146: 11–19.

67. Liebelt, V., Bartke, S. and Schwarz, N. 2018. Hedonic pricing analysis of the influence of urban green spaces onto residential prices: The case of Leipzig, Germany. *European Planning Studies* 26: 133–157.

68. Brander, L. M. and Koetse, M. J. 2011. The value of urban open space: Meta-analyses of contingent valuation and hedonic pricing results. *Journal of Environmental Management* 92: 2763–2773.

69. Garibaldi, L. A., Gemmill-Herren, B., D'Annolfo, R. et al. 2017. Farming approaches for greater biodiversity, livelihoods, and food security. *Trends in Ecology & Evolution* 32: 68–80.

70. Rahmann, G. 2011. Biodiversity and organic farming: What do we know? *Agriculture and Forstery Research* 3: 189–208.

71. Philpott, S. M., Bichier, P., Rice, R. A. and Greenberg, R. 2008. Biodiversity conservation, yield, and alternative products in coffee agroecosystems in Sumatra, Indonesia. *Biodiversity and Conservation* 17: 1805–1820.

72. Vandermeer, J., Perfecto, I. and Philpott, S. 2010. Ecological complexity and pest control in organic coffee production: Uncovering an autonomous ecosystem service. *BioScience* 60: 527–537.

73. Hylander, K., Nemomissa, S., Delrue, J. and Enkosa, W. 2013. Effects of coffee management on deforestation rates and forest integrity. *Conservation Biology* 27: 1031–1040.

74. Waldron, A., Justicia, R., Smith, L. and Sanchez, M. 2012. Conservation through Chocolate: A win-win for biodiversity and farmers in Ecuador's lowland tropics. *Conservation Letters* 5: 213–221.

75. Bohlen, P. J., Lynch, S., Shabman, L. et al. 2009. Paying for environmental services from agricultural lands: An example from the northern Everglades. *Frontiers in Ecology and the Environment* 7: 46–55.

76. Jordan, N. and Warner, K. D. 2010. Enhancing the multifunctionality of US agriculture. *BioScience* 60: 60–66.

77. Kemp, D. R., Guodong, H., Xiangyang, H. et al. 2013. Innovative grassland management systems for environmental and livelihood benefits. *Proceedings of the National Academy of Sciences* 110: 8369–8374.

78. Earnst, S. L., Dobkin, D. S. and Ballard, J. A. 2012. Changes in avian and plant communities of aspen woodlands over 12 years after livestock removal in the Northwestern Great Basin. *Conservation Biology* 26: 862–872.

79. Rooney, T. P., Wiegmann, S. M., Rogers, D. A. and Waller, D. M. 2004. Biotic impoverishment and homogenization in unfragmented forest understory communities. *Conservation Biology* 18: 787–798.

80. Scholes, R. J., Walters, M., Turak, E. et al. 2012. Building a global observing system for biodiversity. *Current Opinion in Environmental Sustainability* 4: 139–146.

81. Larigauderie, A. 2015. The Intergovernmental Platform on Biodiversity and Ecosystem Services (IPBES): A call to action. *Gaia-Ecological Perspectives for Science and Society* 24: 73–73.

82. Karanth, K. U., Gopalaswamy, A. M., Kumar, N. S. et al. 2011. Monitoring carnivore populations at the landscape scale: Occupancy modelling of tigers from sign surveys. *Journal of Applied Ecology* 48: 1048–1056.
83. Dolrenry, S., Stenglein, J., Hazzah, L. et al. 2014. A metapopulation approach to African lion (*Panthera leo*) conservation. *PLOS ONE* 9: e88081.
84. Broms, K. M., Johnson, D. S., Altwegg, R. and Conquest, L. L. 2014. Spatial occupancy models applied to atlas data show Southern Ground Hornbills strongly depend on protected areas. *Ecological Applications* 24: 363–374.
85. Miller, D. A. and Grant, E. H. C. 2015. Estimating occupancy dynamics for large-scale monitoring networks: Amphibian breeding occupancy across protected areas in the northeast United States. *Ecology and Evolution* 5: 4735–4746.
86. Planes, S., Jones, G. P. and Thorrold, S. R. 2009. Larval dispersal connects fish populations in a network of marine protected areas. *Proceedings of the National Academy of Sciences* 106: 5693–5697.
87. Di Minin, E., Hunter, L. T. B., Balme, G. A. et al. 2013. Creating larger and better connected protected areas enhances the persistence of big game species in the Maputaland-Pondoland-Albany biodiversity hotspot. *PLOS ONE* 8: e71788.
88. Kaky, E. and Gilbert, F. 2016. Using species distribution models to assess the importance of Egypt's protected areas for the conservation of medicinal plants. *Journal of Arid Environments* 135: 140–146.
89. Sundblad, G., Bergström, U. and Sandström, A. 2011. Ecological coherence of marine protected area networks: a spatial assessment using species distribution models. *Journal of Applied Ecology* 48: 112–120.
90. Bonnot, T. W., Thompson, F. R., Millspaugh, J. J. and Jones-Farrand, D. T. 2013. Landscape-based population viability models demonstrate importance of strategic conservation planning for birds. *Biological Conservation* 165: 104–114.
91. Muller, S. 2003. Towards decolonisation of Australia's protected area management: The Nantawarrina Indigenous Protected Area experience. *Australian Geographical Studies* 41: 29–43.
92. DEWR. 2017. *Growing Up Strong: The First 10 Years of Indigenous Protected Areas in Australia*. Canberra, Australia: Australian Government, Department of the Environment and Water Resources.
93. Plotkin, R. 2018. *Tribal Parks and Indigenous Protected and Conserved Areas: Lessons from B.C. Examples*. Vancouver, BC: David Suzuki Foundation.
94. Garnett, S. T., Burgess, N. D., Fa, J. E. et al. 2018. A spatial overview of the global importance of Indigenous lands for conservation. *Nature Sustainability* 1: 369–374.
95. Cabeza, M., Moilanen, A. and Possingham, H. P. 2004. Metapopulation dynamics and reserve network design. In Ilkka Hanski and Oscar E. Gaggiotti (Eds.), *Ecology, Genetics and Evolution of Metapopulations*. Waltham, MA: Academic Press.
96. Schumann, M., Watson, L. H. and Schumann, B. D. 2008. Attitudes of Namibian commercial farmers toward large carnivores: The influence of conservancy membership. *African Journal of Wildlife Research* 38: 123–132.

Chapter 16

1. Bowkett, A. E. 2009. Recent captive breeding proposals and the return of the ark concept to global species conservation. *Conservation Biology* 23: 773–776.
2. Redford, K. H., Jensen, D. B. and Breheny, J. J. 2012. Integrating the captive and the wild. *Science* 338: 1157–1158.
3. Fisher, J. 1966. *Zoos of the World: The Story of Animals in Captivity*. London, UK: Aldus.
4. Blunt, W. 1976. *The Ark in the Park: The Zoo in the Nineteenth Century*. London, UK: Hamilton.
5. Robbins, L. E. 2002. *Elephant Slaves and Pampered Parrots: Exotic Animals in Eighteenth-Century Paris*. Baltimore, MD: Johns Hopkins University Press.
6. Baratay, E. and Hardouin-Fugier, E. 2004. *Zoo: A History of Zoological Gardens in the West*. Chicago, IL: University of Chicago Press.
7. Kisling, V. N., Jr. 2001. *Zoo and Aquarium History: Ancient Animal Collections to Zoological Gardens*. Boca Raton, FL: CRC Press.
8. Hancocks, D. 2001. *A Different Nature: The Paradoxical World of Zoos and their Uncertain Future*. Berkeley, CA: University of California Press.
9. Daly, N. 2017. Why all of America's circus animals could soon be free. *National Geographic* May 22.
10. Ash, M. G. 2008. *Humans, Animals and Zoos: The Tiergarten Schönbrunn in an International Comparison from the 18th Century to the Present*. Vienna, Austria: Böhlau Verlag Wien.
11. Conde, D. A., Colchero, F., Gusset, M. et al. 2013. Zoos through the lens of the IUCN Red List: A global metapopulation approach to support conservation breeding programs. *PLOS ONE* 8: e80311.
12. Conde, D. A., Flesness, N., Colchero, F. et al. 2011. An emerging role of zoos to conserve biodiversity. *Science* 331: 1390–1391.
13. Mora, C., Tittensor, D. P., Adl, S. et al. 2011. How many species are there on Earth and in the ocean? *PLOS Biology* 9: e1001127.
14. Becker, M. H., Richards-Zawacki, C. L., Gratwicke, B. and Belden, L. K. 2014. The effect of captivity on the cutaneous bacterial community of the critically endangered Panamanian golden frog (*Atelopus zeteki*). *Biological Conservation* 176: 199–206.
15. USFW. 2004. Draft revised recovery plan for the Sihek or Guam Micronesian Kingfisher (*Halcyon cinnamomina cinnamomina*). Portland, OR: U.S. Fish & Wildlife Service.
16. NMDGF. 2006. *Threatened and endangered species of New Mexico: Final biennial review and recommendations*. Santa Fe, NM: New Mexico Department of Game and Fish.
17. Clubb, R. and Mason, G. 2003. Captivity effects on wide-ranging carnivores. *Nature* 425: 473–474.
18. Fravel, L. 2003. Critics question zoos' commitment to conservation. *National Geographic* Novermber 13.
19. Leader-Williams, N. 1990. Black rhinos and African elephants: Lessons for conservation funding. *Oryx* 24: 23–29.
20. Gusset, M. and Dick, G. 2011. The global reach of zoos and aquariums in visitor numbers and conservation expenditures. *Zoo Biology* 30: 566–569.
21. Barongi, R., Fisken, F. A., Parker, M. and Gusset, M. 2015. *Committing to Conservation: The World Zoo and Aquarium Conservation Strategy*. Gland, Switzerland: WAZA.
22. Maple, T., McManamon, R. and Stevens, E. 1995. Defining the good zoo: Animal care, maintenance, and welfare. In *Zoo and Aquarium Biology and Conservation Series*, pages 219–234. Washington, DC: Smithsonian Institution Press.
23. Spencer, R. and Cross, R. 2016. The origins of botanic gardens and their relation to plant science, with special reference to horticultural botany and cultivated plant taxonomy. *Muelleria* 35: 43–93.
24. Mounce, R., Smith, P. and Brockington, S. 2017. Ex situ conservation of plant diversity in the world's botanic gardens. *Nature Plants* 3: 795–802.
25. Dirr, M. A. 1990. *Manual of Woody Landscape Plants: Their Identification, Ornamental Characteristics, Culture, Propagation and Uses* (4th ed.). Champaign, IL: Stipes.
26. Bartram's Garden. 2019. Franklinia Tree. bartramsgarden.org/explore-bartrams/franklinia-tree/.
27. Jones, D. L. 2002. *Cycads of the World: Ancient Plants in Today's Landscape* (2nd ed.). Washington, DC: Smithsonian Institution Press.
28. Ainouche, M. L. and Bayer, R. J. 1997. On the origins of the tetraploid *Bromus* species (section Bromus, Poaceae): Insights

from internal transcribed spacer sequences of nuclear ribosomal DNA. *Genome* 40: 730–743.

29. Williams, S. 2018. Conservation biologist and placenta expert Kurt Benirschke dies. *The Scientist*, September 14.
30. Butchart, S. H. M., Lowe, S., Martin, R. W. et al. 2018. Which bird species have gone extinct? A novel quantitative classification approach. *Biological Conservation* 227: 9–18.
31. Gunn, I. M. 2007. Problems in developing a gene bank for endangered species: Does the human race even care? *World Small Animal Veterinary Association World Congress Proceedings*. www.vin.com/apputil/content/defaultadv1.aspx?id=3860691&pid=11242&print=1.
32. O'Donnell, K. and Sharrock, S. 2017. The contribution of botanic gardens to ex situ conservation through seed banking. *Plant Diversity* 39: 373–378.
33. Banga, S. S. and Kang, M. S. 2014. Developing climate-resilient crops. *Journal of Crop Improvement* 28: 57–87.
34. Altieri, M. A. 2004. Linking ecologists and traditional farmers in the search for sustainable agriculture. *Frontiers in Ecology and the Environment* 2: 35–42.
35. Brush, S. B. 2007. Farmers' rights and protection of traditional agricultural knowledge. *World Development* 35: 1499–1514.
36. Morin, J. F. and Gauquelin, M. 2016. *Trade Agreements as Vectors for the Nagoya Protocol's Implementation*. Ontario, Canada: Centre for International Governance Innovation.
37. Wildt, D. E., Comizzoli, P., Pukazhenthi, B. and Songsasen, N. 2010. Lessons from biodiversity: The value of nontraditional species to advance reproductive science, conservation, and human health. *Molecular Reproduction and Development* 77: 397–409.
38. Zimmermann, A., Hatchwell, M., Dickie, L. A. and West, C. 2007. *Zoos in the 21st Century: Catalysts for Conservation?* Cambridge, UK: Cambridge University Press.
39. Fraser, D. J. 2008. How well can captive breeding programs conserve biodiversity? A review of salmonids. *Evolutionary Applications* 1: 535–586.
40. Silla, A. J. and Byrne, P. G. 2019. The role of reproductive technologies in amphibian conservation breeding programs. In H. A. Lewin and R. M. Roberts (Eds.), *Annual Review of Animal Biosciences*, Vol. 7. Palo Alto, CA: Annual Reviews.
41. Comizzoli, P., Crosier, A., Songsasen, N. et al. 2009. Advances in reproductive science for wild carnivore conservation. *Reproduction in Domestic Animals* 44: 47–52.
42. Behr, B., Rath, D., Mueller, P. et al. 2009. Feasibility of sex-sorting sperm from the white and the black rhinoceros (*Ceratotherium simum, Diceros bicornis*). *Theriogenology* 72: 353–364.
43. Holt, W. and Lloyd, R. 2009. Artificial insemination for the propagation of CANDES: The reality! *Theriogenology* 71: 228–235.
44. Jabr, F. 2013. Will cloning ever save endangered animals? *Scientific American*, March 11.
45. Holden, C. 2003. Banteng cloned. *Science* 364(6446).
46. Black, R. 2005. First kittens for cloned wildcats. August 22. news.bbc.co.uk/2/hi/science/nature/4172688.stm.
47. Snyder, N. F. R., Derrickson, S. R., Beissinger, S. R. et al. 1996. Limitations of captive breeding in endangered species recovery. *Conservation Biology* 10: 338–348.
48. Robert, A. 2009. Captive breeding genetics and reintroduction success. *Biological Conservation* 142: 2915–2922.
49. Banes, G. L., Galdikas, B. M. F. and Vigilant, L. 2016. Reintroduction of confiscated and displaced mammals risks outbreeding and introgression in natural populations, as evidenced by orangutans of divergent subspecies. *Scientific Reports* 6: 22026.
50. Pelletier, F., Réale, D., Watters, J. et al. 2009. Value of captive populations for quantitative genetics research. *Trends in Ecology & Evolution* 24: 263–270.
51. Williams, S. E. and Hoffman, E. A. 2009. Minimizing genetic adaptation in captive breeding programs: A review. *Biological Conservation* 142: 2388–2400.
52. Frankham, R. 2008. Genetic adaptation to captivity in species conservation programs. *Molecular Ecology* 17: 325–333.
53. Lewis, O. T. and Thomas, C. D. 2001. Adaptations to captivity in the butterfly *Pieris brassicae* (L.) and the implications for ex situ conservation. *Journal of Insect Conservation* 5: 55–63.
54. McPhee, M. E. 2004. Generations in captivity increases behavioral variance: Considerations for captive breeding and reintroduction programs. *Biological Conservation* 115: 71–77.
55. Roberts, L. J., Taylor, J. and Garcia de Leaniz, C. 2011. Environmental enrichment reduces maladaptive risk-taking behavior in salmon reared for conservation. *Biological Conservation* 144: 1972–1979.
56. Griffith, B., Scott, J. M., Carpenter, J. W. and Reed, C. 1989. Translocation as a species conservation tool: Status and strategy. *Science* 245: 477–480.
57. Fischer, J. and Lindenmayer, D. B. 2000. An assessment of the published results of animal relocations. *Biological Conservation* 96: 1–11.
58. Hendrickson, D. A. and Brooks, J. E. 1991. *Transplanting short-lived fishes in North American deserts: Review, assessment and recommendations*. In W. L. Minckley and J. E. Deacon (Eds.), *Battle against Extinction: Native Fish Management in the American West*. Tucson, AZ: University of Arizona Press.
59. Godefroid, S., Piazza, C., Rossi, G. et al. 2011. How successful are plant species reintroductions? *Biological Conservation* 144: 672–682.
60. White, T. H., Collar, N. J., Moorhouse, R. J. et al. 2012. Psittacine reintroductions: Common denominators of success. *Biological Conservation* 148: 106–115.
61. Grenier, M. B., McDonald, D. B. and Buskirk, S. W. 2007. Rapid population growth of a critically endangered carnivore. *Science* 317: 779–779.
62. Brichieri-Colombi, T. A., Lloyd, N. A., McPherson, J. M. and Moehrenschlager, A. 2019. Limited contributions of released animals from zoos to North American conservation translocations. *Conservation Biology* 33: 33–39.
63. Bowles, M. L., McBride, J. L. and Betz, R. F. 1998. Management and restoration ecology of the federal threatened Mead's milkweed, *Asclepias meadii* (Asclepiadaceae). *Annals of the Missouri Botanical Garden* 85: 110–125.
64. Donath, T. W., Bissels, S., Holzel, N. and Otte, A. 2007. Large scale application of diaspore transfer with plant material in restoration practice: Impact of seed and microsite limitation. *Biological Conservation* 138: 224–234.
65. Blanco, G., Lemus, J. A. and Garcia-Montijano, M. 2011. When conservation management becomes contraindicated: Impact of food supplementation on health of endangered wildlife. *Ecological Applications* 21: 2469–2477.
66. Harrington, L. A., Moehrenschlager, A., Gelling, M. et al. 2013. Conflicting and complementary ethics of animal welfare considerations in reintroductions. *Conservation Biology* 27: 486–500.
67. Buchholz, R. 2007. Behavioural biology: An effective and relevant conservation tool. *Trends in Ecology & Evolution* 22: 401–407.
68. Nicholson, T. E., Mayer, K. A., Staedler, M. M. and Johnson, A. B. 2007. Effects of rearing methods on survival of released free-ranging juvenile southern sea otters. *Biological Conservation* 138: 313–320.
69. Brightsmith, D., Hilburn, J., del Campo, A. et al. 2005. The use of hand-raised psittacines for reintroduction: A case study of scarlet macaws (*Ara macao*) in Peru and Costa Rica. *Biological Conservation* 121: 465–472.
70. Moss, A. and Esson, M. 2013. The educational claims of zoos: Where do we go from here? *Zoo Biology* 32: 13–18.

71. Moss, A., Jensen, E. and Gusset, M. 2014. *A Global Evaluation of Biodiversity Literacy in Zoo and Aquarium Visitors*. Gland, Switzerland: World Association of Zoos and Aquariums.
72. Moss, A., Jensen, E. and Gusset, M. 2015. Evaluating the contribution of zoos and aquariums to Aichi Biodiversity Target 1. *Conservation Biology* 29: 537–544.
73. Moss, A., Jensen, E. and Gusset, M. 2017. Impact of a global biodiversity education campaign on zoo and aquarium visitors. *Frontiers in Ecology and the Environment* 15: 243–247.
74. Jensen, E. 2014. Evaluating children's conservation biology learning at the zoo. *Conservation Biology* 28: 1004–1011.
75. Mellish, S., Ryan, J. C., Pearson, E. L. and Tuckey, M. R. 2019. Research methods and reporting practices in zoo and aquarium conservation-education evaluation. *Conservation Biology* 33: 40–52.
76. Grajal, A., Luebke, J. F., Kelly, L.-A. D. et al. 2017. The complex relationship between personal sense of connection to animals and self-reported proenvironmental behaviors by zoo visitors. *Conservation Biology* 31: 322–330.
77. Minteer, B. A. and Rojas, C. 2018. The transformative ark. In S. Sarkar and B. Minteer (Eds.), *A Sustainable Philosophy: The Work of Bryan Norton*. New York, NY: Springer.
78. Hutchins, M., Smith, B. and Allard, R. 2003. In defense of zoos and aquariums: The ethical basis for keeping wild animals in captivity. *Journal of the American Veterinary Medical Association* 223: 958–966.
79. Xu, W., Viña, A., Kong, L. et al. 2017. Reassessing the conservation status of the giant panda using remote sensing. *Nature Ecology & Evolution* 1: 1635–1638.
80. Yiming, L., Zhongwei, G., Qisen, Y. et al. 2003. The implications of poaching for giant panda conservation. *Biological Conservation* 111: 125–136.
81. Schaller, G. B. 1993. *The Last Panda*. Chicago, IL: University of Chicago Press.
82. Wei, W., Swaisgood, R. R., Dai, Q. et al. 2018. Giant panda distributional and habitat-use shifts in a changing landscape. *Conservation Letters* 11: e12575.
83. Folch, J., Cocero, M. J., Chesne, P. et al. 2009. First birth of an animal from an extinct subspecies (*Capra pyrenaica pyrenaica*) by cloning. *Theriogenology* 71: 1026–1034.
84. Shapiro, B. 2017. Pathways to de-extinction: How close can we get to resurrection of an extinct species? *Functional Ecology* 31: 996–1002.
85. Jorgensen, D. 2013. Reintroduction and de-extinction. *BioScience* 63: 719–720.
86. Cohen, S. 2014. The ethics of de-extinction. *Nanoethics* 8: 165–178.
87. Audubon Project Puffin. 2017. Eastern Egg Rock Puffins increase to 172 pairs. projectpuffin.audubon.org/conservation/eastern-egg-rock-puffins-increase-172-pairs.
88. Kramer, R. and Jenkins, A. 2009. *Ecosystem Services, Markets, and Red Wolf Habitat: Results from a Farm Operator Survey*. Durham, NC: Nicholas Institute for Environmental Policy Solutions. Duke University.
89. Hubbs, C., Edwards, R. J. and Garrett, G. P. 2002. Threatened fishes of the world: *Gambusia heterochir* Hubbs, 1957 (Poeciliidae). *Environmental Biology of Fishes* 65: 422–422.

Chapter 17

1. Du Pisani, J. 2006. Sustainable development: Historical roots of the concept. *Environmental Sciences* 3: 83–96.
2. Meadows, D.H., D.L. Meadows, J. Randers, and W.W.I. Behrens. 1972. *The Limits to Growth: A Report for The Club of Rome's Project on The Predicament of Mankind*. New York, NY: Universe Books.
3. United Nations. 1973. Report of the United Nations Conference on the Human Environment, Stockholm, 5–16 June 1972. Vol. A/CONF.48/14/Rev.1. New York, NY: United Nations.
4. World Commission on Environment and Development. 1987. *Our Common Future*. Oxford, UK: Oxford University Press.
5. Daly, H.E. 1990. Toward some operational principles of sustainable development. *Ecological Economics* 2: 1–6.
6. Keiner, M. 2005. *History, Definition(s) and Models of Sustainable Development*. Zurich, Switzerland: Eidgenössische Technische Hochschule Zürich.
7. Goodland, R. 1995. The concept of environmental sustainability. *Annual Review of Ecology and Systematics* 26: 1–24.
8. Robinson, J. 2004. Squaring the circle? Some thoughts on the idea of sustainable development. *Ecological Economics* 48: 369–384.
9. Redclift, M. 2005. Sustainable development (1987–2005): An oxymoron comes of age. *Sustainable Development* 13: 212–227.
10. Robinson, J. and J. Tinker. 1997. Reconciling ecological, economic, and social imperatives: A new conceptual framework. In T. Schrecker (Ed.), *Surviving Globalism: Social and Environmental Dimensions*. London: Macmillan, St. Martin's Press.
11. Holden, E., K. Linnerud, and D. Banister. 2017. The imperatives of sustainable development. *Sustainable Development* 25: 213–226.
12. Costanza, R. and B.C. Patten. 1995. Defining and predicting sustainability. *Ecological Economics* 15: 193–196.
13. United Nations. 2015. *Charter of the United Nations and Statute of the International Court of Justice*. New York, NY: United Nations Publications.
14. United Nations. 1993. *Report of the United Nations Conference on Environment and Development. Vol. II Proceedings of the Conference*. New York, NY: United Nations. A/CONF.151/26/Rev.1 (Vol. II).
15. United Nations. 1993. *Agenda 21: Rio Declaration and Forest Principles*. New York, NY: United Nations Publications.
16. United Nations General Assembly. 1992. *Rio Declaration on Environment and Development*. United Nations, A/CONF.151/26 (Vol. I).
17. United Nations General Assembly. 1992. *Non-legally binding authoritative statement on principles for a global consensus on the management, conservation and sustainable development of all types of forests*. United Nations, A/CONF.151/26 (Vol. III).
18. United Nations General Assembly. 2000. *United Nations Millennium Declaration*. United Nations, A/RES/55/2.
19. United Nations. 2015. *The Millennium Development Goals Report*. New York, NY: United Nations Publications.
20. United Nations General Assembly. 2015. *Transforming Our World: the 2030 Agenda for Sustainable Development*. United Nations, A/RES/70/1.
21. United Nations. 1992. *Convention on Biological Diversity*. United Nations.
22. Whitehorn, P.R., et al. 2019. Mainstreaming biodiversity: A review of national strategies. *Biological Conservation* 235: 157–163.
23. Conference of the Parties to the Convention on Biological Diversity. 2010. *The Strategic Plan for Biodiversity 2011–2020 and the Aichi Biodiversity Targets*. United Nations Environment Programme, UNEP/CBD/COP/DEC/X/2.
24. United Nations. 1992. *United Nations Framework Convention on Climate Change*. United Nations.
25. United Nations. 1998. *Kyoto Protocol to the United Nations Framework Convention on Climate Change*. United Nations.
26. United Nations. 2015. *Paris Agreement*. United Nations.
27. United Nations. 1994. *United Nations Convention to Combat Desertification in Those Countries Experiening Serious Drought and/or Desertification, Particularly in Africa*. United Nations.

28. International Bank for Reconstruction and Development. 2017. *The World Bank Environmental and Social Framework.* Washington, DC: World Bank.

29. International Finance Corporation. 2012. *Performance Standard 6 Biodiversity Conservation and Sustainable Management of Living Natural Resources.* World Bank Group.

30. Qian, N. 2015. Making progress on foreign aid. *Annual Review of Economics* 7: 277–308.

31. Moyo, D. 2009. *Dead Aid: Why Aid Is Not Working and How There Is A Better Way for Africa.* New York, NY: Farrar, Straus and Giroux.

32. Gulrajani, N. 2017. Bilateral donors and the age of the national interest: what prospects for challenge by development agencies? *World Development* 96: 375–389.

33. Robinson, J.G. 2012. Common and conflicting interest in the engagements between conservation organizations and corporations. *Conservation Biology* 26: 967–977.

34. Honey, M. 2008. Ecotourism and Sustainable Development: Who Owns Paradise? (2nd ed.). Washington, DC: Island Press.

35. Kappelle, M. 2016. *Costa Rican Ecosystems.* Chicago, IL: University of Chicago Press.

36. Jones, G. and A. Spadafora. 2017. Creating ecotourism in Costa Rica, 1970–2000. *Enterprise & Society* 18: 146–183.

37. World Travel and Tourism Council. 2018. *Economic Impact 2018, Costa Rica.* London: World Travel and Tourism Council.

38. Convention on International Trade in Endangered Species of Wild Fauna and Flora. 2018. *Report on the Monitoring the Illegal Killing of Elephants.* Colombo, Sri Lanka: Convention on International Trade in Endangered Species of Wild Fauna and Flora, CoP18 Doc. 69.2.

39. Hauenstein, S., M. Kshatriya, J. Blanc, C.F. Dormann and C.M. Beale. 2019. African elephant poaching rates correlate with local poverty, national corruption and global ivory price. *Nature Communications* 10: 2242.

40. Heinrich, S., et al. 2017. *The global trafficking of pangolins: a comprehensive summary of seizures and trafficking routes from 2010–2015.* TRAFFIC, Southeast Asia Regional Office, Petaling Jaya, Selangor, Malaysia.

41. Kawarazuka, N. and C. Béné. 2011. The potential role of small fish species in improving micronutrient deficiencies in developing countries: building evidence. *Publich Health Nutrition* 14: 1927–1938.

42. Food and Agriculture Organization of the United Nations. 2018. *The State of the World Fisheries and Aquaculture 2018: Meeting the Sustainable Development Goals.* Rome: United Nations Publications.

43. United Nations Environment Programme. 2018. *Emissions Gap Report 2018.* Nairobi, Kenya: United Nations Publications.

44. United Nations Conference on Trade and Development. 2018. *Promoting investment in the Sustainable Development Goals.* United Nations, Investment Advisory Series, Series A, Number 8, Geneva, Switzerland.

45. Costanza, R., et al. 2016. The UN Sustainable Development Goals and the dynamics of well-being. *Solutions* 7: 20–22.

46. Nilsson, M., D. Griggs, and M. Visbeck. 2016. Map the interactions of sustainable development goals. *Nature* 534: 320–322.

47. Nilsson, M., et al. 2018. Mapping interactions between the sustainable development goals: Lessons learned and ways forward. *Sustainability Science* 13: 1489–1503.

48. Weitz, N., H. Carlsen, M. Nilsson, and K. Skanberg. 2018. Toward systemic and contextual priority setting for implementing the 2030 Agenda. *Sustainability Science* 13: 531–548.

49. Pradhan, P., L. Costa, D. Rybski, W. Lucht, and J.P. Kropp. 2017. A systematic study of Sustainable Development Goal (SDG) interactions. *Earth's Future* 5: 1169–1179.

50. Costanza, R., et al. 2014. Time to leave GDP behind. *Nature* 505: 283–285.

51. Kubiszewksi, I., et al. 2013. Beyond GDP: Measuring and achieving global genuine progress. *Ecological Economics* 93: 57–68.

52. Costanza, R., et al. 2004. Estimates of the Genuine Progress Indicator (GPI) for Vermont, Chittenden County and Burlington, from 1950 to 2000. *Ecological Economics* 51: 139–155.

53. Centre for Bhutan Studies & GNH Research. 2016. *A compass towards a just and harmonious society 2015 GNH survey report.* Centre for Bhutan Studies & GNH Research, Thimphu, Bhutan.

54. Xiao, H., J. Cheng, and X. Wang. 2018. Does the Belt and Road Initiative promote sustainable development? Evidence from countries along the Belt and Road. *Sustainability* 10: 4370.

55. Ascensão, F., et al. 2018. Environmental challenges for the Belt and Road Initiative. *Nature Sustainability* 1: 206–209.

56. Li, N. and E. Shvarts. 2017. The Belt and Road Initiative: WWF recommendations and spatial analysis. Briefing Paper, 2017), https: //go.nature.com/2v3SwoG.

57. International Bank for Reconstruction and Development. 2019. *Belt and Road Economics Opportunities and Risk of Transport Corridors.* Washington, DC: World Bank.

58. McCarthy, T., D. Mallon, R. Jackson, P. Zahler, and K. McCarthy. 2017. *Panthera uncia.* IUCN Red List of Threatened Species e.T22732A50664030:

59. Stiles, D. 2009. CITES: Approved ivory sales and elephant poaching. *Pachyderm* 45: 150–153.

60. Wasser, S., et al. 2010. Elephants, ivory, and trade. *Science* 327: 1331–1332.

61. Naidoo, R., B. Fisher, A. Manica, and A. Balmford. 2016. Estimating economic losses to tourism in Africa from the illegal killing of elephants. *Nature Communications* 7: 13379.

62. Stiglitz, J.E., A. Sen, and J.-P. Fitoussi. 2009. *Report by the Commission on the Measurement of Economic Performance and Social Progress.* Paris: Commission on the Measurement of Economic Performance and Social Progress.

63. Barrington-Leigh, C. and A. Escande. 2018. Measuring progress and well-being: A comparative review of indicators. *Social Indicators Research* 135: 893–925.

64. Stern, S., A. Wares, and T. Epner. 2018. 2018 *Social Progress Index Methodology Summary.* Washington, DC: Social Progress Imperative.

65. United Nations. 2017. Resolution adopted by the General Assembly on 6 July 2017. 71/313 Work of the Statistical Commission pertaining to the 2030 Agenda for Sustainable Development. United Nations, A/RES/71/313.

Index

The letter *b* after a page number indicates that the entry is included in a box;
f indicates that the entry is included in a figure; *t* indicates that the entry is included in a table.

B

C

F

G

H

I

J

K

L

M

Q

R

S